Table of atomic weights listed alphabetically

Scaled to relative atomic mass $^{12}C = 12$ exactly. A number in parentheses is the atomic mass number of the isotope of longest known half-life.

Element	Symbol	Atomic number	Atomic weight	Element	Symbol	Atomic number	Atomic weight
Actinium	Ac	89	(227)	Mercury	Hg	80	200.59
Aluminum	Al	13	26.9815	Molybdenum	Mo	42	95.94
Americium	Am	95	(243)	Neodymium	Nd	60	144.24
Antimony	Sb	51	121.75	Neon	Ne	10	20.179
Argon	Ar	18	39.948	Neptunium	Np	93	(237)
Arsenic	As	33	74.9216	Nickel	Ni	28	58.70
Astatine	At	85	(210)	Niobium	Nb	41	92.9064
Barium	Ba	56	137.33	Nitrogen	N	7	14.0067
Berkelium	Bk	97	(247)	Nobelium	No	102	(255)
Beryllium	Be	4	9.01218	Osmium	Os	76	190.2
Bismuth	Bi	83	208.980	Oxygen	O	8	15.9994
Boron	B	5	10.81	Palladium	Pd	46	106.4
Bromine	Br	35	79.904	Phosphorus	P	15	30.9738
Cadmium	Cd	48	112.40	Platinum	Pt	78	195.09
Calcium	Ca	20	40.08	Plutonium	Pu	94	(244)
Californium	Cf	98	(251)	Polonium	Po	84	(209)
Carbon	C	6	12.011	Potassium	K	19	39.098
Cerium	Ce	58	140.12	Praseodymium	Pr	59	140.907
Cesium	Cs	55	132.905	Promethium	Pm	61	(145)
Chlorine	Cl	17	35.453	Protactinium	Pa	91	(231)
Chromium	Cr	24	51.996	Radium	Ra	88	(226)
Cobalt	Co	27	58.9332	Radon	Rn	86	(222)
Copper	Cu	29	63.546	Rhenium	Re	75	186.207
Curium	Cm	96	(247)	Rhodium	Rh	45	102.905
Dysprosium	Dy	66	162.50	Rubidium	Rb	37	85.4678
Einsteinium	Es	99	(254)	Ruthenium	Ru	44	101.07
Erbium	Er	68	167.26	Samarium	Sm	62	150.35
Europium	Eu	63	151.96	Scandium	Sc	21	44.959
Fermium	Fm	100	(257)	Selenium	Se	34	78.96
Fluorine	F	9	18.9984	Silicon	Si	14	28.086
Francium	Fr	87	(223)	Silver	Ag	47	107.868
Gadolinium	Gd	64	157.25	Sodium	Na	11	22.9898
Gallium	Ga	31	69.72	Strontium	Sr	38	87.62
Germanium	Ge	32	72.59	Sulfur	S	16	32.06
Gold	Au	79	196.967	Tantalum	Ta	73	180.948
Hafnium	Hf	72	178.49	Technetium	Tc	43	(97)
Helium	He	2	4.0026	Tellurium	Te	52	127.60
Holmium	Ho	67	164.930	Terbium	Tb	65	158.925
Hydrogen	H	1	1.0079	Thallium	Tl	81	204.37
Indium	In	49	114.82	Thorium	Th	90	232.038
Iodine	I	53	126.904	Thulium	Tm	69	168.934
Iridium	Ir	77	192.22	Tin	Sn	50	118.69
Iron	Fe	26	55.847	Titanium	Ti	22	47.90
Krypton	Kr	36	83.80	Tungsten	W	74	183.85
Lanthanum	La	57	138.905	Uranium	U	92	238.03
Lawrencium	Lr	103	(260)	Vanadium	V	23	50.9414
Lead	Pb	82	207.19	Xenon	Xe	54	131.30
Lithium	Li	3	6.941	Ytterbium	Yb	70	173.04
Lutetium	Lu	71	174.97	Yttrium	Y	39	88.9059
Magnesium	Mg	12	24.305	Zinc	Zn	30	65.38
Manganese	Mn	25	54.938	Zirconium	Zr	40	91.22
Mendelevium	Md	101	(258)				

Chemistry
with Inorganic
Qualitative Analysis

Chemistry

with Inorganic Qualitative Analysis

Therald Moeller
Arizona State University

John C. Bailar, Jr.
University of Illinois

Jacob Kleinberg
University of Kansas

Cyrus O. Guss
University of Nevada at Reno

Mary E. Castellion
Stamford, Connecticut

Clyde Metz
Indiana University–Purdue University at Indianapolis

ACADEMIC PRESS

New York   San Francisco   London

A Subsidiary of Harcourt Brace Jovanovich, Publishers

Cover Photograph

Citric acid crystals magnified $80\times$ and photographed under polarized light.
See Section 32.8

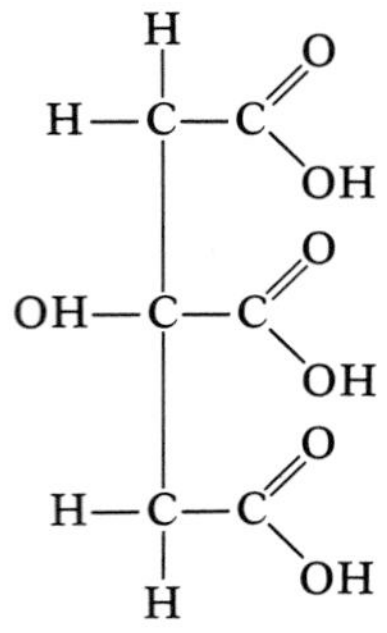

citric acid

Design and art direction by Betty Binns Graphics/Martin Lubin
Illustrations by Rino Dussi

ACADEMIC PRESS, INC.
111 Fifth Avenue, New York, New York 10003

United Kingdom Edition published by
ACADEMIC PRESS, INC. (LONDON) LTD.
24/28 Oval Road, London NW1 7DX

ISBN: 0-12-503350-8
Library of Congress Catalog Card Number: 79-89282

PRINTED IN THE UNITED STATES OF AMERICA

To everyone who helped.

CONTENTS

CONTENTS

ix

CONTENTS

x

10 PERIODIC
PERSPECTIVE: THE
REPRESENTATIVE
ELEMENTS
page 273

CONTENTS

**13   WATER AND THE
HYDROSPHERE**
page 354

**14   SOLUTIONS
AND COLLOIDS**
page 380

CONTENTS

xiv

CONTENTS

PREFACE
TO CHEMISTRY WITH
INORGANIC QUALITATIVE
ANALYSIS

A chemistry textbook is not, like a marble statue, something to be carved out and allowed to stand and be admired for all time. First, the needs of teachers and students are varied and changing. Second, those who use textbooks are not likely to stand back and admire. Instead, they become intimately involved with the details on every page and they freely comment upon them.

The authors of this textbook plan, through this and subsequent editions, to meet the varied needs of chemistry teachers and their students. In doing so, we remain committed to the philosophy expressed in the first edition of *Chemistry*. Both descriptive chemistry and the principles of chemistry are necessary to the study of the subject, and these topics should be covered with scientific honesty and in the clearest language possible.

This edition of *Chemistry*, incorporating inorganic qualitative analysis, is a logical result of our belief in the value of descriptive chemistry and practical experience. Our aim is to emphasize the ingenious application of the principles of equilibrium represented in qualitative analysis and the properties of the ions shown by the reactions of the analysis scheme. In order to focus on the most important principles and reactions, we have covered these topics in the textbook, but have left the details of the laboratory work for a separate manual.

We are convinced that the periodic table provides the best basis for learning descriptive chemistry. Therefore in this edition, as in the original edition of *Chemistry*, the properties of the elements and their compounds are taken up in the order of the periodic table position of the elements. In other words, we have chosen not to rearrange the discussion of the metals in the order of their appearance in the cation groups.

Specifically, we have made the following changes in *Chemistry* for this edition.

Chapter 2 has been simplified—chemical nomenclature is now presented in an appendix (Appendix D), rather than in Chapter 2. This change has allowed us to expand the coverage of nomenclature.

The subject can now be taught as a whole, or individual sections of the Appendix can be assigned as the need arises.

The subjects of equilibrium (Chapter 16), acids and bases (Chapter 17), and ions in aqueous solution (Chapter 18) have been expanded to provide the essential basis for "qual." Many new worked examples, some of them quite simple, have been added. In addition, we now approach equilibrium via experimental data, rather than through kinetics and rate constants.

In Chapter 22, which completes the coverage of the nonmetallic elements (except for carbon), we have added a general introduction to inorganic qualitative analysis. This is followed by a brief review of the principles applied in anion analysis and a discussion of the chemistry of an analysis scheme for eleven anions.

Two new chapters concentrate on the equilibria, ionic properties, and chemical reactions relevant to cation analysis. Chapter 30 reviews the principles of equilibrium and shows how these principles are applied in qual. The Examples and many of the Exercises in Chapter 30 are drawn from the reactions of the qual scheme. Chapter 31 discusses the chemistry of a scheme for the analysis of twenty-two anions in five groups. Flow charts are presented, the procedures are outlined in terms of what is happening chemically, and equations are written for all of the important reactions. Laboratory directions are not included in the text but are given in the separate laboratory manual.

We are grateful to the following reviewers whose suggestions have been invaluable in the conception, preparation, and completion of this project: Dr. Don L. Armstrong, Whittier College; Dr. Latha M. Barnes, West Georgia College; Dr. Patricia A. Boaz, Indiana University–Purdue University at Indianapolis; Dr. Loren S. Carter, Boise State Idaho; Dr. G. Mattney Cole, Jr., University of Georgia; Dr. Edward R. Covington, Tennessee State University; Dr. Marion C. Day, Louisiana State University; Dr. Rolla M. Dyer, Indiana State University–Evansville; Dr. Morton P. Eisner, West Los Angeles College; Dr. Gordon Ewing, New Mexico State University; Professor Michael Guttman, Miami–Dade Community College; Dr. Helmi S. Habib, Central Washington State College; Dr. Lonnie Haynes, Tennessee State University; Dr. Tom Hays, Arizona State University; Dr. David D. Holder, Tennessee State University; Dr. Ronald Johnson, Emory University; Dr. James L. Kroon, Bethel College; Dr. Robert H. Marshall, Memphis State University; Dr. David Moseley, Washington State University; Dr. Robert Nakon, West Virginia University; Mr. Paul D. Neumann, P.E., Nashville State Technical Institute; Dr. John Newey, American River College; Dr. Edward J. O'Reilly, University of North Dakota; Dr. Charles A. Reynolds, University of Kansas; Dr. Jerry D. Skelton, Oakland City College; Dr. Jimmy C. Stokes, West Georgia College; Dr. Philip K. Welty, Miami University, Ohio; Dr. David L. Wilson, Valencia Community College.

THERALD MOELLER CYRUS O. GUSS
JOHN C. BAILAR, JR. MARY E. CASTELLION
JACOB KLEINBERG CLYDE METZ

PREFACE
TO CHEMISTRY

The science of chemistry touches practically every aspect of our lives. It is also important in the successful pursuit of many careers. Both students and instructors face several challenges in a chemistry course: How can we choose among the many topics that might be included? Which information is essential? How can we present, and learn, all of the essential information and still include some subjects that simply are of daily practical value or are interesting?

This book represents the authors' answer to these challenges, done in a way that we feel will be helpful to both students and instructors. Throughout the book, we have kept two major goals in mind. Our first goal has been to provide a reasonable balance between the theory of chemistry and descriptive chemistry—the description of the appearance, structure, and behavior of the chemical elements and their compounds. Our second goal has been to present each topic as logically, simply, and directly as possible.

Descriptive chemistry and the theory of chemistry are two sides of the same coin. We feel strongly that they should not be treated as though they are separate subjects. Theory must always stand the test of experiment, and practical chemistry that does not take theory into account can often be wasted effort. Our approach to descriptive chemistry is not encyclopedic. However, we have presented enough descriptive material so that students can understand the characteristics of important chemicals, some modern methods for their industrial preparation, and the impact they have on our world. By referring often to applications of chemistry in our daily lives, we emphasize the practical value of knowing something about chemistry. A large number of tables and figures complement the text, and the physical properties of the elements and compounds are illustrated as they are introduced.

We have endeavored to be scientifically honest in our approach to theory, while not going beyond what is appropriate in a general first-year course. For example, we have developed an effective but simple approach to thermodynamics through chemistry rather than physics. As much as possible, we have included only those concepts that can be described fully enough to allow for reasonable understanding. This text does not use calculus, and problems involving arithmetic calculations are illustrated with worked-out examples. When it is necessary to introduce topics based on mathematics beyond the level of this book, we have pointed this out. We do not want students to feel that they have missed the reason for something, when in truth it is not there.

The introduction to each chapter gives a glimpse of the general nature of the subject matter to follow, or a look at a particularly intriguing or entertaining aspect of the subject. The specific topics to be discussed are briefly listed. Thus, the student begins the chapter with a focus on what lies ahead. Many new terms must be introduced in the study of chemistry; each significant new term in this book is defined in a declarative sentence. The new terms are printed in boldface type at the point where they are defined, and they are listed at the end of each chapter. Some of the more important terms and equations are also given as "flying definitions" at the upper corners of the pages, where they can be referred to quickly.

Each chapter concludes with a generous number of exercises, many of which are real-life applications of the topics covered in the chapter. Some exercises logically extend the concepts of the chapter to new areas. In each of the descriptive chapters, a number of exercises are designed to integrate principles presented earlier. The exercises are graded in difficulty; harder questions are marked with an asterisk and brain teasers are marked with two. They include all necessary data so that students will not have to search through the text, and answers to selected problems are given in the back of the book.

A great deal of thought has gone into the sequence of the chapters. The descriptive chemistry, as we have noted, is not lumped together at the end of the book. The essential aspects of chemical bonding and of chemical thermodynamics are presented in early chapters; the more abstract and difficult parts of these subjects come later, when the students are better prepared to deal with them.

The first five chapters of the book are an introduction to the subject. The basic vocabulary of chemistry and the principles of stoichiometry are fully presented in these chapters. Also, atoms, molecules, and ions, gases, thermochemistry, and types of reactions are discussed and defined. By the end of these five chapters, all students should be on an equal footing. This approach should be helpful to both students and instructors by bringing students with varying backgrounds, including those who have had no high school chemistry, to the same level early in the course. These chapters also provide the fundamentals needed for laboratory work.

Chapter 6, on the atmosphere, gives students a chance to apply some of the principles of chemistry to the world around them. The balance between conceptual and descriptive aspects continues with Chapters 7 and 8, which present a coherent and continuous development of the topics of atomic structure and nuclear chemistry, with emphasis on the applications of nuclear chemistry.

Chapter 9 presents the basics of chemical bonding—what students must get out of this introductory course—while Chapter 21 goes beyond the basics to explore the valence bond and molecular orbital theories. Just as Chapter 9 prepares the reader for Chapter 21, so Chapter 10 gives a broad view of the periodic table in preparation for the more detailed descriptive coverage of the elements in Chapters 11, 20, 22, 23, 26, 27, and 29. Chapter 12 includes both crystal structure and some of the more up-to-date aspects of solid state chemistry. Chapter 5 introduces elementary thermochemistry, while Chapter 25 explores the more difficult aspects of thermodynamics—free energy and entropy. Chapter 13, Water and the Hydrosphere, logically follows the theoretical discussion of Chapter 12, The Liquid and Solid States: Changes of State. Chapter 14, Solutions and Colloids; Chapter 15, Chemical Kinetics; and Chapter 16, Chemical Equilibrium, all expand on the basic foundations established in Chapters 1 through 5.

Many chapters of *Chemistry* are grouped so that the principles chapters are followed by chapters that discuss applications of these principles. The chapter on hydrogen and oxygen (11) shows how the periodic trends explained in Chapter 10 are applied when discussing specific elements and their compounds. Chapter 20 on the halogens gives some practical examples of the principles of oxidation–reduction discussed in Chapter 19. The chapters on nitrogen, phosphorus, and sulfur (22) and on carbon and hydrocarbons (23) illustrate some of the covalent bonding theories presented in Chapters 9 and 21.

Recognizing that each chemistry instructor has an individual approach to the subject, we have structured the book so that variations in the chapter sequence are possible. For example, Chapter 8, Nuclear Chemistry, could be taught later. The chapters on oxidation–reduction (Chapter 19) and electrochemistry (Chapter 24) could be taught consecutively, as could the chapters on thermochemistry (Chapter 5) and chemical thermodynamics (Chapter 25), or those on bonding (Chapter 9) and the covalent bond (Chapter 21). Where time is a limiting factor, a selected few of the chapters on metals and nonmetals could be presented. For example, the chapters on hydrogen and oxygen (10), nitrogen, phosphorus, and sulfur (22), and transition metals (28) would provide a good picture of what descriptive chemistry is and how it is used.

The coming of SI units is causing more than a little confusion in chemistry at present. We have chosen units for our examples and exercises that reflect what the student is likely to encounter *today* in lecture halls, journals, books, and supply catalogs. Therefore, we have used some SI units, such as the nanometer and the joule, but have not given up, for example, liters in favor of cubic decimeters. Because we feel it is logical and simple, we have adopted the mole as the sole unit for the "amount of a substance." SI units are introduced in Chapter 1, as are factor label (dimensional analysis) calculations. There are separate appendixes on mathematics, logarithms, and significant figures, as well as SI units in the back of the book.

Throughout the book, we have included special sections designed to broaden the reader's view of the topics being discussed. The Tools of Chemistry essays describe how such tools as infrared and ultraviolet spectroscopy, x rays, and chromatography can be used. Essays called Asides discuss such subjects as metals as poisons, the names of the elements, and photography—topics that are of interest but need not be studied and learned. The Thoughts on Chemistry quotations at the end of each chapter are intended to be entertaining and informative, and to show that chemistry and science have been written about with elegance and style.

The dedication of this book is our way of acknowledging the assistance and encouragement of the many individuals who have contributed to the preparation of this book. At this point, we want to mention by name the reviewers and consultants who have given us so much sound advice during the course of this book's development. Dr. Bolesh Skutnik of Fairfield University, Fairfield, Connecticut, contributed greatly during the early stages of the manuscript, particularly in developing the chapter-by-chapter outline. Dr. Mark P. Freeman of Dorr-Oliver, Inc., Stamford, Connecticut, made a significant contribution as a consultant on the first principles of chemistry. Dr. Floyd James of Miami University, Oxford, has read every word, and provided many excellent criticisms and suggestions. The following professors, who read the manuscript at various stages of its development, gave us many valuable suggestions reflecting their years of teaching experience: Professor William G. Bailey, Broward Community College, Fort Lauderdale; Professor Anthony M. Dean, University of Missouri, Columbia; Professor James A. Fries, Northern State College, Aberdeen; Professor Verl G. Garrard, University of Idaho, Moscow; Professor Morton Z. Hoffman, Boston University; Professor Ken Whitten, University of Georgia, Athens.

Finally, we want to thank all our friends from the textbook department of Academic Press; their steadfast support and their creative contributions have done much to bring this book into being.

JOHN C. BAILAR, JR.
THERALD MOELLER
JACOB KLEINBERG
CYRUS O. GUSS
MARY E. CASTELLION
CLYDE METZ

1 CHEMISTRY: THE SCIENCE OF MATTER

Look around you. That's how chemistry began—in the limitless curiosity of human beings about their surroundings.

Possibly you are sitting at your desk with some paper and a wooden pencil or a plastic pen at hand to take notes. Maybe there are some metal paper clips and a pottery coffee cup or a glass soft-drink bottle or an aluminum can on your desk. What could you do to investigate the materials in your paper, your pencil or pen, the paper clips, the cup, the bottle, or the can? Scratch them. Which is harder? Put a drop of water, or alcohol, or acid on each one. What happens? Weigh pieces of equal size. Which is heavier? Try to burn a small piece of each. Which ones burn? What is left afterwards?

You could work your way around your room cataloging how everything in it responds to these and other tests. You could go outside and do the same for the rocks and plants. Pretty soon you would be able to draw conclusions about which things are similar to each other and which differ from each other.

If you have a curious nature, your next questions should begin with "Why. . .?" and "How. . .?" Why does wood burn, but pottery not? How can I predict whether other things will or won't burn? Why are some things heavier or harder than others? Why don't these things dissolve in water? How does acid change a paper clip?

Chemistry has its roots in just this kind of speculation about the nature of simple things. In early times, people wondered about air, and water, and rocks, and fire, and looked for magical and mystical answers to questions about the physical world around them. Once the importance of systematic observation was recognized, the foundation was laid for chemistry and all the sciences.

In this chapter we first give some basic definitions—science, matter, chemistry—and then an explanation of the subdivisions of chemistry. We review the modern units of measurement and a problem-solving method—the factor-dimensional method—that will be used throughout this book. In the last section of the chapter we take a brief look at the role of chemistry in coping with the major problems facing mankind.

1

Science and matter

1.1 Science

A natural science is a classification of knowledge about things that are observable in nature, in the material world, and in the universe. Each branch of science organizes a multitude of facts and answers to "How. . .?" and "Why. . .?" questions. In the **biological sciences,** questions are asked mainly about things that are alive. In the **physical sciences,** the questions pertain mainly to things that are not alive. Botany and zoology are biological sciences, geology and meteorology are physical sciences.

Chemistry is in a central position. It applies mathematics and the laws of physics, and is thought of most often as a physical science. But the chemical elements are the building blocks for everything in the universe, living or not. Chemistry, therefore, legitimately asks questions about life itself.

The boundaries between the biological sciences and chemistry are becoming increasingly blurred. These boundaries will continue to fade as scientists solve such puzzles as the structure of the genes that carry the message of life to future generations. But despite such changes, the scientific method will remain, for all of science is built upon its principles. The scientific approach to solving a problem frequently begins with systematic observations. Once enough observations have been made a hypothesis can be formulated. At this stage a **hypothesis** is a tentative explanation for a set of observations.

Experiments are next devised to test the validity of the hypothesis in as many ways as possible. If a hypothesis survives the test of many experiments, it may be accepted as a law. A **law** is a statement of a relation between phenomena that, so far as is known, is always the same under the same conditions. As knowledge in a particular area of science grows, further experiments may lead to further laws. Eventually it may be possible to formulate a theory. Theories have larger scope than laws. A **theory** is a unifying principle or group of principles that explains a body of facts or phenomena and those laws that are based upon them.

Central to the work of scientists and to progress in science is a constant testing of current assumptions as well as a constant search for answers to new problems. However, to talk of "the scientific method" may lead you to believe that scientific inquiry always moves forward in a rigid, step-by-step progression. Seldom is this true. Sometimes law precedes theory, sometimes theory precedes law. Facts and observations may come first, or a purely theoretical idea may come first. Usually everything is happening at once.

1.2 States and properties of matter

Almost all definitions of chemistry contain the word "matter" or the word "materials." Matter is all around us. Matter is everything of which the world is made. It occupies space, it has mass, and, except for some gases, it can be seen and touched.

[1.1] CHEMISTRY: THE SCIENCE OF MATTER

There are three **states of matter**—the gaseous state, the liquid state, and the solid state. Some of the general properties that distinguish gases, liquids, and solids from each other are listed in Table 1.1.

At ordinary temperatures and pressures oxygen, nitrogen, hydrogen, carbon dioxide, chlorine, ammonia, and methane (the major component of natural gas) are all gases. Water, ethyl alcohol, mercury, and gasoline are liquids at ordinary temperatures and pressures. And solids, of course, are everywhere we look, most of them combinations of several kinds of substances. Some common simple substances are solids at ordinary temperatures and pressures, including most metals, such as iron, copper, and gold; carbon, either as diamond or graphite; sodium chloride (common salt); and sucrose (common sugar). But pottery, the substance from which your coffee cup is made, is a mixture of complex materials that contain silicon, aluminum, oxygen, and small amounts of various metals.

Some substances can exist in all three states. Water is known in the solid state as ice, in the liquid state (at room temperature), and in the gaseous state as steam. Many metals, which are usually solid, can be melted, and if heated to higher temperatures, can become gaseous. Some substances, however, cannot exist in the gaseous state; others cannot exist in the liquid state, and some cannot exist in either the gaseous or the liquid state. For example, calcium carbonate, a solid, cannot be melted or vaporized, for upon heating it decomposes into calcium oxide, which is another solid, and carbon dioxide. Upon gen-

The gaseous state is considered in detail in Chapter 3, the liquid and solid states in Chapter 12.

TABLE 1.1
General properties of the gaseous, liquid, and solid states

Property	Gaseous state	Liquid state	Solid state
Compressibility[a]	Very great	Slight	Almost nil
Expandability[b]	Infinite	Slight	Almost nil
Shape	That of container	That of container, but with flat surface and fixed volume	Fixed, with no relationship to container
Flow[c]	Rapid, viscosity very small	Slower, viscosity variable	Almost nil, except under high pressures; very high viscosity
Structure	Completely disordered	Ordered only in limited regions	Completely or nearly completely ordered throughout
Energy content (states of same substance)	Largest (removal of energy yields liquid)	Intermediate (removal of energy yields solid; addition of energy yields gas)	Smallest (addition of energy yields liquid or, in a few instances, gas)

[a] Compressibility is a decrease in volume with increase of pressure.
[b] Expandability is an increase in volume with increase of temperature.
[c] Viscosity is the resistance of a substance to flow.

tle heating, sugar melts to the liquid state. Upon heating to higher temperatures sugar does not become gaseous, but instead decomposes into a variety of products that contain carbon. However, all gases and liquids can be condensed to the solid state.

The careful examination and measurement of the properties of different substances are important in chemistry. Color, melting point, boiling point, density, and hardness are among the observable properties of matter. These are physical properties. **Physical properties** can be exhibited, measured, or observed without resulting in a change in the composition and identity of a substance. **Chemical properties,** by contrast, can only be observed in chemical reactions. A **chemical reaction** is a process in which at least one substance is changed in composition and identity. The burning of wood in air and the tarnishing of silver demonstrate chemical properties. The burning wood is combining with oxygen in the air to form carbon dioxide, water, and carbon-containing substances, which we see as smoke and ashes. The shiny silver is combining with sulfides in the air to form silver sulfide, the black substance that stains the surface. Before we can define chemistry further, however, we must distinguish between the different kinds of matter that we will encounter.

1.3 Kinds of matter

Every kind of matter can be classified as either a pure substance or a mixture. **A pure substance** is a form of matter that has identical physical and chemical properties no matter what its source. For instance, pure water is colorless and odorless, boils at 100°C and freezes at 0°C at atmospheric pressure, weighs 1 gram per milliliter at 4°C, and does not burn. It has these properties whether it is distilled from sea water, dipped from a mountain stream, or prepared in a chemical reaction by the union of hydrogen and oxygen. And it is these characteristic properties that enable us to distinguish water from other substances.

A **mixture** is any combination of two or more substances in which the substances combined retain their identity. The substances in a mixture can be combined in any proportions. The properties of the mixture are the properties of its components, and the components can be retrieved intact from the mixture without a chemical reaction. A **heterogeneous mixture** is a mixture in which the individual components of the mixture remain physically separate and can be seen as separate components, although in some cases a microscope is needed. Concrete and granite are heterogeneous mixtures; so are milk and cake batter. Powdered iron and powdered sulfur, no matter how well stirred, form a heterogeneous mixture. They can be separated by using a magnet to attract the powdered iron, as shown in Figure 1.1a.

The substances in a **homogeneous mixture** are thoroughly intermingled; the composition and appearance of the mixture are uniform throughout. Air is a homogeneous mixture of gases, and motor oil is a homogeneous mixture of liquid petroleum derivatives. Homogeneous mixtures, whether gaseous, liquid, or solid, also can be separated by physical means.

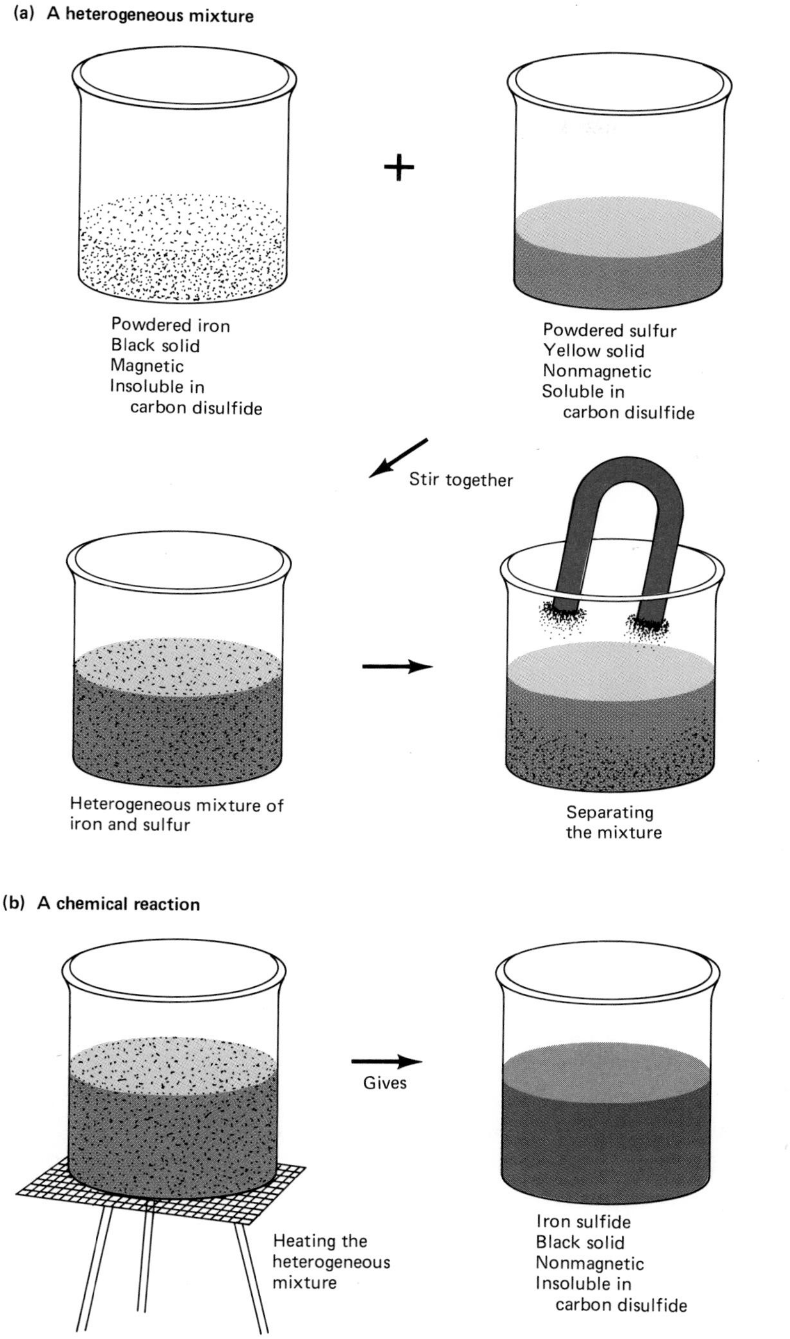

Powdered iron
Black solid
Magnetic
Insoluble in
    carbon disulfide

+

Powdered sulfur
Yellow solid
Nonmagnetic
Soluble in
    carbon disulfide

Stir together

Heterogeneous mixture of
iron and sulfur

Separating
the mixture

(b) A chemical reaction

Gives

Heating the
heterogeneous
mixture

Iron sulfide
Black solid
Nonmagnetic
Insoluble in
    carbon disulfide

FIGURE 1.1

Iron plus sulfur: a mixture or a chemical compound. *If iron and sulfur powders are thoroughly mixed but not heated, a mixture results. The iron can be completely removed from the sulfur by a magnet. Conversely, the sulfur can be completely removed from the iron by dissolving it in carbon disulfude, a liquid in which iron is not soluble. Powdered iron and powdered sulfur when heated together give a chemical compound, iron sulfide.*

Pure substances can be either elements or compounds. An *element* is a pure substance that cannot be converted into a simpler form of matter by any chemical reaction. Hydrogen and oxygen are elements; so are iron and sulfur. All of the known elements are listed on the endpapers of this book.

Under the right conditions hydrogen and oxygen combine to form water. Heating a mixture of iron and sulfur causes them to com-

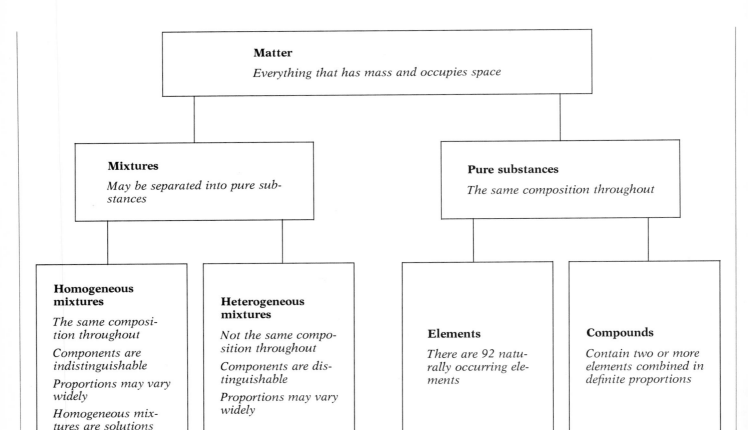

Matter

Everything that has mass and occupies space

Mixtures

May be separated into pure substances

Pure substances

The same composition throughout

Homogeneous mixtures

The same composition throughout

Components are indistinguishable

Proportions may vary widely

Homogeneous mixtures are solutions

Heterogeneous mixtures

Not the same composition throughout

Components are distinguishable

Proportions may vary widely

Elements

There are 92 naturally occurring elements

Compounds

Contain two or more elements combined in definite proportions

FIGURE 1.2
Classification of matter. *By naturally occurring elements, we mean elements that are present in nature in even the minutest amounts. Four elements— technetium, astatine, francium, and promethium—although present, are so scarce that collecting a measurable amount is exceedingly difficult or impossible.*

bine to form iron sulfide, a substance with properties clearly different from those of either iron or sulfur (Figure 1.1b). Both of these processes are chemical reactions, and the products are chemical compounds. **A compound** is a substance of definite composition in which two or more elements are chemically combined; a compound can be separated into its components by chemical reactions but not by purely physical methods.

The general classification of matter is summarized in Figure 1.2. We shall define the types of matter more accurately in later chapters.

Chemistry: The science of matter

1.4 Chemistry

Simply defined, chemistry is the science of matter. Chemistry is necessary in the study and manipulation of any material, and in our modern world we are surrounded by materials that have been studied, manipulated, and even invented by chemistry. Our automobiles are made of metals, fabrics, and plastics. Few of these materials can be obtained from natural sources in forms that are readily usable.

The metals are recovered from mineral ores; the fabrics are produced from plant, animal, or synthetic fibers; the plastics are made by combining simple materials to form new and more complex materials.

We travel on highways surfaced with asphalt and concrete. We drink water that has been purified with chlorine. We read from paper made from cellulose and printed with highly colored inks. We watch television screens that have been coated with tiny chemical spots, each of which responds with a color when energy is applied. With chemistry, we alter an almost endless variety of materials to form other materials with characteristics that we desire.

Not all chemistry has a practical product as its objective, of course. Chemists search for understanding and the ability, based on understanding, to predict what will happen when changes occur. And some chemists pursue inquiries into the nature of matter solely because they enjoy it.

With these observations in mind, we may consider a more formal definition of chemistry: **Chemistry** is the branch of science that deals with matter, with the changes that matter can undergo, and with the laws that describe these changes. As this definition implies, chemistry is both a theoretical and an applied science. The **principles of chemistry** are the explanations of the chemical facts; this is where you meet the hypotheses, the laws, and the theories. **Descriptive chemistry,** as you might expect, is the description of the elements and compounds, their physical states, and how they behave. No matter how chemistry is used, a good balance between principles and descriptive chemistry is a necessity. We will attempt to maintain such a balance throughout this textbook.

1.5 Subdivisions of chemistry

The five major subdivisions of chemistry are listed in Table 1.2. Originally, organic chemistry dealt only with substances obtained from living materials, but this distinction has long since vanished. **Organic chemistry** has become the chemistry of carbon compounds and their derivatives. Organic compounds that contain only the two elements carbon and hydrogen are called **hydrocarbons.** Almost all other

Subdivision	Subjects covered
Organic chemistry	**Compounds of carbon and hydrogen (hydrocarbons)** **All compounds derived from hydrocarbons**
Inorganic chemistry	**All elements** **Compounds of all elements** *except* **the hydrocarbons and their derivatives**
Analytical chemistry	**Measurements of amounts of substances** **Measurements of chemical composition of materials** **Separation of the components of mixtures**
Physical chemistry	**Measurements of physical properties** **Theoretical interpretation of physical and chemical properties**
Biochemistry	**Pure substances and chemical reactions in living systems**

TABLE 1.2

Major subdivisions of chemistry

organic compounds can be thought of as derivatives of hydrocarbons. **Inorganic chemistry** is the chemistry of all elements (including carbon) and their compounds, with the exception of hydrocarbons and hydrocarbon derivatives. The other major subdivisions—analytical chemistry, physical chemistry, biochemistry—are defined briefly in Table 1.2.

The boundaries between organic and inorganic chemistry, like the boundaries between the different sciences, and indeed between all areas of chemistry, are disappearing. Many compounds between organic and metallic substances—organometallic compounds—are now known, and organic compounds have been made that behave in ways traditionally attributed only to inorganic compounds.

Some specialized areas of chemistry, which provide bridges to other major fields of study and endeavor, are geochemistry; microbial and medicinal chemistry; agricultural, fertilizer, soil, and food chemistry; polymer chemistry; cellulose, paper, and textile chemistry; and industrial and environmental chemistry. As we have pointed out, whenever people explore materials, living or nonliving, they eventually become concerned with chemistry.

Units of measure; problem solving

1.6 Systems of measurement

Since 1793, when the metric system was devised by the French National Academy of Sciences to overcome the profusion of units handed down from medieval times, various international bodies have been defining and redefining units of measurement and attempting to gain widespread uniformity in their use. For many years scientists everywhere and people in most European countries have used the metric system, which is a decimal system.

In 1960, the International Bureau of Weights and Measures adopted the International System of Units, known as SI units (for *Système Internationale*). The SI system is a revision and extension of the metric system; it provides basic units for each type of measurement. Scientists and engineers throughout the world in all disciplines are now being urged to use the SI system exclusively; most major countries have adopted it.

The United States, until 1975, officially stayed with a weights and measures system based upon the English system of inches and feet, ounces and pounds, pints and quarts, and so on. The Metric Conversion Act of 1975 commits this country to a policy of voluntary conversion to the metric system by 1985. The United States Metric Board has the job of coordinating the changeover and educating the public to the use of the new units. Everyone must learn how to buy a liter of milk and what to wear when the temperature is 4°C.

The kilogram and the liter are standard units in both the original metric system and the new SI system, and scientists have dealt in kilograms and liters for a long time. However, some of the units

Quantity	Name of unit	Abbreviation
Length	meter	**m**
Mass	kilogram	**kg**
Time	second	**s** (sec)
Electric current	ampere	**A**
Temperature	kelvin	**K** (°K)
Luminous intensity	candela	**cd**
Amount of substance	mole	**mol** (mole)

TABLE 1.3
SI base units

in the SI system are supposed to replace other units that are still widely used. The SI system is finding acceptance, but because scientists are human and, like everyone else, resistant to changing their ways, it will undoubtedly be a long time before the changeover to the new system is complete.

The SI system defines a single base unit to measure each of seven quantities—length, mass, time, electric current, temperature, amount of substance, and luminous intensity (see Table 1.3). Other necessary units, such as that for volume, are derived from the base units. A standard set of prefixes indicates variations in the magnitude of these units, as shown in Table 1.4. Some equivalents between SI units and other units are given in Table 1.5. (See also Appendix B.)

The fractions and multiples represented by the prefixes in Table 1.4 and some of the unit equivalents in Table 1.5 have been written as exponential numbers. Very large and very small numbers arise often in chemistry. Such numbers are easier to handle and easier to comprehend when they are given in exponential form. If you do not know how to express numbers in exponential form, also called scientific notation, you should study Appendix A.2 at this point.

In the following sections, we discuss the units encountered most often in chemistry.

1.7 Length

In both the metric and SI systems, the meter is the basic unit of length or distance. Chemists often have to talk about very small lengths, such as the dimensions of particles of matter. In this book, we use the nanometer (nm), an SI unit, for this purpose. A nanometer is one billionth of a meter, or 1×10^{-9} m.

1.8 Volume

The milliliter and the liter are the standard units of volume in most chemical laboratory work. One milliliter is exactly equal to one cubic centimeter (cm^3), and one liter exactly equals 1000 milliliters (ml) or 1000 cubic centimeters. The SI system recommends replacing the

TABLE 1.4
Unit prefixes for SI units
The prefix is written in front of the unit name with no hyphen, e.g., micrometer, nanometer, or the symbol is written in front of the unit abbreviation, e.g., μm (micrometer), nm (nanometer); μ is the lower case Greek letter "mu."

d	deci-	10^{-1}
c	centi-	10^{-2}
m	milli-	10^{-3}
μ	micro-	10^{-6}
n	nano-	10^{-9}
k	kilo-	10^{3}
M	mega-	10^{6}

TABLE 1.5
Equivalence between units

Length
1 inch = 2.54 cm
1 ft   = 30.48 cm
1 yd   = 0.9144 m
1 mi   = 1.609 km
1 cm   = 0.3937 inch
1 m   = 39.37 inches
1 Å   = 0.1 nm

Volume (liquid)
1 qt   = 0.946 liter
1 gal = 3.785 liters

Weight, mass
1 oz   = 28.35 g
1 lb   = 453.6 g
1 g   = 2.205×10^{-3} lb
1 kg   = 2.205 lb

Heat, energy
1 cal = 4.184 J
1 erg = 1×10^{-7} J
1 erg = 2.39×10^{-8} cal

TABLE 1.6

Units in chemistry

Included here are the units most often encountered in chemistry. Those given in color are used throughout this book. Those given in black may be encountered in other chemistry books or in your study of other sciences.

Symbol	Unit	Unit equivalent
Length		
m	meter	—
cm	centimeter	1×10^{-2} m
mm	millimeter	1×10^{-3} m
nm	nanometer	1×10^{-9} m
Å	Ångstrom	1×10^{-10} m
μ	micron	1×10^{-6} m
mμ	millimicron	1×10^{-9} m
Mass		
kg	kilogram	1×10^{3} g
g	gram	—
mg	milligram	1×10^{-3} g
Volume		
l	liter	—
ml	milliliter	1×10^{-3} liter
cm³	cubic centimeter	1 ml
m³	cubic meter	—
dm³	cubic decimeter	—
μl	microliter	1×10^{-6} liter
Heat		
kcal	kilocalorie	1×10^{3} cal
cal	calorie	—
J	joule	—
kJ	kilojoule	1×10^{3} J

liter with the cubic decimeter, but in this book we have chosen to use milliliters and liters. (From 1901 to 1964, a liter was defined as the volume of 1 kilogram of water at 4°C. During this period a milliliter was very slightly larger than a cubic centimeter. In 1964, the liter was redefined as exactly equal in volume to 1000 cubic centimeters, thereby getting rid of some confusion.)

1.9 Mass vs. weight

The distinction between mass and weight should be clear to anyone who watched the astronauts bounding over the surface of the moon. The moon's gravity is smaller than that of the Earth, and so the weight of the astronauts was less there. Their bodies, however, were unchanged and had the same mass as always. **Mass** is an intrinsic property and represents the quantity of matter in a body. **Weight** is the force a body exerts because of the pull of gravity on the body's mass. Now that we have carefully made this distinction, we must admit that in talking and writing about chemistry, the distinction is often ignored, and the terms "mass" and "weight" are used imprecisely.

Both the SI and metric systems rely on the gram and its fractions and multiples for units of mass, with the kilogram designated the basic mass unit by the SI system.

Mass and volume are properties that by themselves disclose nothing about the identity of a substance. But when they are combined as mass per unit volume, called **density,** these units give one of those properties which does describe a substance. For example, the density of aluminum is 2.7 g/cm³, the density of iron is 7.8 g/cm³, and the density of mercury is 13.6 g/cm³; osmium and iridium, the densest of the elements, have densities of 22.61 and 22.65 g/cm³, respectively.

1.10 Heat

When a warm body and a cold body come into contact, heat energy (thermal energy) flows from the body with the higher temperature to that with the lower temperature. Heat is given off or taken up in almost all chemical changes and in many physical changes. The calorie (cal) is a unit of measure of heat. One calorie is close to the amount of heat needed to raise the temperature of 1 gram of water by 1°C. Because the magnitude of the numbers is more convenient, the heat in chemical changes is usually expressed in kilocalories (kcal). The "calories" counted by dieters are really kilocalories.

In the SI system, the basic unit of energy is the joule (J). In this book, you will be given the opportunity to use both kilojoules and kilocalories, because kilocalories are still in widespread use. The units of length, mass, volume, and heat used in chemistry are summarized in Table 1.6.

1.11 Temperature

You should be familiar with three temperature scales: the Kelvin scale, °K; the Celsius scale, °C; and the Fahrenheit scale, °F (see Figure 1.3). On the Fahrenheit scale, the one that has been in everyday use in the United States, the freezing point of water is 32°, the boiling point of water (at the average atmospheric pressure at sea level) is 212°, and the temperature range between these two points is divided into 180 equal degrees. The Celsius and Fahrenheit scales coincide at −40° (see Figure 1.3).

The Celsius scale has been used in most scientific work; its reference points are 0° for the freezing point of water and 100° for the boiling point of water (at the average atmospheric pressure at sea level). Since the same temperature range is represented by 180° on the Fahrenheit scale and 100° on the Celsius scale, a Fahrenheit degree is 100/180 or $\frac{5}{9}$ of a Celsius degree. Anders Celsius described a thermometer using this scale in 1742. The same scale has been called the centigrade scale. Fortunately, both are represented by the symbol, °C.

Temperatures on the Celsius and Fahrenheit scales can be interconverted by the following equations (Note the difference in placement of the parentheses.)

$$°C = \frac{5}{9}(°F - 32) \qquad °C = \left[(°F + 40)\frac{5}{9}\right] - 40$$
$$\text{or}$$
$$°F = \frac{9}{5}(°C) + 32 \qquad °F = \left[(°C + 40)\frac{9}{5}\right] - 40$$

The Kelvin temperature unit is one of the basic SI units. On the Kelvin scale, 0° represents absolute zero, the lowest temperature attainable. The freezing point of water is 273.15°K and the boiling point is 373.15°K. Therefore, the Kelvin degree and the Celsius degree are the same size, and a Celsius temperature can be converted to a Kelvin temperature by adding 273.15. For general purposes, 273° is sufficiently accurate:

$$°K = °C + 273$$

Although SI recommends the Kelvin, K, without a degree sign, we have chosen to use °K.

1.12 The factor-dimensional method of calculation

The factor-dimensional method of calculation is a powerful and useful technique for solving problems. Units are retained with all numbers in setting up the solution to a problem; then the units are canceled in the same manner as numbers are canceled. With this method, a seemingly complicated problem can be reasoned out step by step. This process leads to a much better understanding than if the

$$°C = \left[(°F + 40)\frac{5}{9}\right] - 40$$
$$°F = \left[(°C + 40)\frac{9}{5}\right] - 40$$
$$°K = °C + 273$$

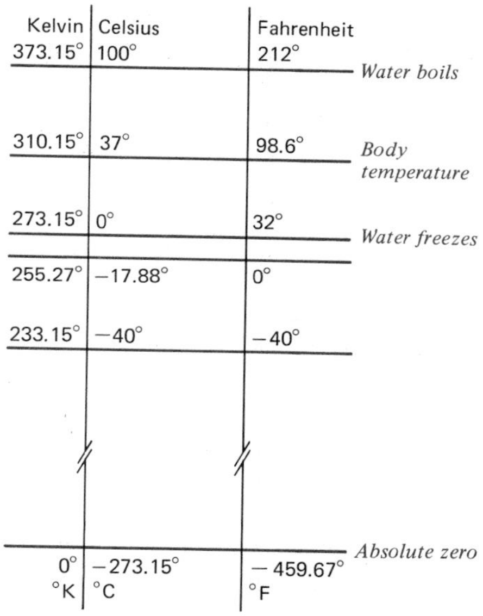

Kelvin	Celsius	Fahrenheit	
373.15°	100°	212°	Water boils
310.15°	37°	98.6°	Body temperature
273.15°	0°	32°	Water freezes
255.27°	−17.88°	0°	
233.15°	−40°	−40°	
0° °K	−273.15° °C	−459.67° °F	Absolute zero

FIGURE 1.3
Comparison of °C, °F, and °K.

problem were solved with some memorized formulas. If a problem is set up correctly, the answer will be in the correct units. A wrong method of solving the problem is instantly recognizable, for the answer will be in the wrong units.

In this section, some unit-conversion problems are solved by the factor-dimensional method. The principles remain the same when the method is used for solving many other kinds of problems in chemistry.

If a measurement in centimeters—say, 94.0 cm—must be converted to meters, the factor-dimensional equation looks like this:

$$\text{distance (in meters)} = 94.0 \text{ cm} \times \underbrace{\frac{1 \text{ m}}{100 \text{ cm}}}_{\text{conversion factor}}$$

cancel units as you would numbers

$$= 0.940 \text{ m}$$

The conversion factor is obtained from the relationship 1 m = 100 cm by noting that this expression means that there is 1 m per 100 cm, or 1 m/100 cm. To convert from meters to centimeters, the conversion factor would be 100 cm/1 m.

Here is a more complicated example:

EXAMPLE 1.1

■ Convert 16.1 km to miles using the following unit equivalents:

1 km = 1000 m	1 ft = 12 inches
1 m = 100 cm	1 mi = 5280 ft
1 inch = 2.54 cm	

The calculation is set up as follows:

$$\text{distance (in miles)} = 16.1 \text{ km} \times \underbrace{\frac{1000 \text{ m}}{1 \text{ km}}}_{km \text{ to } m} \times \underbrace{\frac{100 \text{ cm}}{1 \text{ m}}}_{m \text{ to } cm} \times \underbrace{\frac{1 \text{ inches}}{2.54 \text{ cm}}}_{cm \text{ to inches}}$$

$$\times \underbrace{\frac{1 \text{ ft}}{12 \text{ inches}}}_{inches \text{ to } ft} \times \underbrace{\frac{1 \text{ mi}}{5280 \text{ ft}}}_{ft \text{ to } mi}$$

$$= 10.0 \text{ mi}$$

In each multiplication, a unit from the preceding step is canceled and replaced by a new unit. Note that the answer to this problem, 10.0 mi, is given to three places, that is, as 10.0 rather than 10 or 10.00. The choice of how many digits to give in an answer is governed by the rules of significant figures, which reflect the precision of measurements. In these days when nearly everyone has a pocket calculator, knowing how many of the eight or ten digits that appear in an answer are significant is very important. If you are not familiar with the concept of significant figures, you should study Appendix A.1.

∎

The factor-dimensional method of calculation can be summarized in four steps:

1. Determine what is to be calculated and what its units should be.
2. Using the data in the problem and any necessary conversion factors, set up the calculation.
3. Include units in each factor and cancel units just as you would numbers.
4. Write the numerical answer with its units. If the units obtained are incorrect, recheck the conversion factors and the cancellations.

As Example 1.2 shows, it is sometimes necessary to square or raise to another power one or more of the conversion factors.

EXAMPLE 1.2

∎ Engine capacities in American cars are usually given in cubic-inch displacement; for European cars they are usually given in cubic-centimeter or liter displacement. (Displacement is the total volume swept by the pistons in one stroke.)

A typical European luxury sports car might have an engine displacement of 3.20 liters. How does this compare with the displacement of a typical American family car, which is about 400 cubic inches?

Unit equivalents:

1 inch = 2.54 cm $1.000 \text{ cm}^3 = 1.000 \text{ ml}$ 1000 ml = 1.000 liter
1 cu inch = $(2.54)^3 \text{ cm}^3$

In order to make the comparison, convert 400 cubic inches to liters:

$$\text{displacement (in liters)} = 400 \text{ cu inches} \times \underbrace{\frac{(2.54)^3 \text{ cm}^3}{1 \text{ cu inch}}}_{\substack{cu \ inches \\ to \ cm^3}} \times \underbrace{\frac{1.000 \text{ ml}}{1.000 \text{ cm}^3}}_{cm^3 \ to \ ml}$$

$$\times \underbrace{\frac{1.000 \text{ liter}}{1000 \text{ ml}}}_{ml \ to \ liters}$$

$$= 6.55 \text{ liters}$$

A typical American family car has a displacement about twice that of a European luxury sports car.

∎

With the factor-dimensional method, a unit-conversion calculation and the calculation needed to solve a problem can be set up in a single expression, as in Example 1.3.

EXAMPLE 1.3

■ What is the weight in pounds of a gold bar 12.0 inches long, 6.00 inches wide, and 3.00 inches thick? The density of gold is 19.3 g/cm³.

Volume of gold bar = 12.0 inches × 6.00 inches × 3.00 inches
  (cu inches)
    = 216 cu inches

Unit equivalents:

1 inch = 2.54 cm   1 lb = 453.6 g
1 cu inch = (2.54)³ cm³
    = 16.4 cm³

$$\text{Weight (in pounds)} = \frac{19.3 \text{ g}}{1 \text{ cm}^3} \times \frac{16.4 \text{ cm}^3}{1 \text{ cu inch}} \times \frac{1 \text{ lb}}{453.6 \text{ g}} \times 216 \text{ cu inches}$$
$$= 151 \text{ lb}$$

Note that the first two steps in this calculation convert density in g/cm³ to density in lb/cu inch. ■

Chemistry and the future

The number of people living on Earth has been growing at the rate of 2% per year. If this rate of growth continues, the world population will double every 35 years. By 2000 A.D., between six and eight billion people will be living here (Table 1.7). All of them will need food, clothing, and shelter. How much each man, woman, and child can have is going to depend on how well we learn to manage the available natural resources. Highly industrialized countries like the United States, England, and Japan place great demands on natural resources in order to provide materials to keep the machines of industry at work. In the United States in 1970, 6% of the world's population used between one-third and one-half of all the energy supplies and natural resources consumed that year.

Most products of our industrial civilization wind up as trash sooner or later. Gaseous wastes drift out of our smokestacks and exhaust pipes and linger in the air over our cities. Liquid and solid trash is dumped into our streams, rivers, and oceans, where it is distributed around the world. We have been very successful in producing new materials for our industrial products. As a result, much of our trash will be with us far into the future. Aluminum cans, pesticides, and some plastics, for example, do not decay as readily as do the natural materials they have replaced.

The demands for clean air and water, the resulting environmental protection laws, and the economics of raw-material availabil-

TABLE 1.7
World population

2000 B.C.	1.08×10^8
A.D. 1	1.38×10^8
1000	2.75×10^8
1650	5.00×10^8
1850	1×10^9
1950	2.49×10^9
1960	2.98×10^9
1973	3.84×10^9
1975	3.95×10^9
2000	8×10^9

ity are forcing manufacturers to look anew at how their products are produced and used, and where they wind up after they are used. The energy problem increasingly affects decisions on how to extract raw materials, how to process them, and what products to manufacture. A balance will often have to be struck between the desire to recycle and, thereby, avoid waste and depletion of resources, and the cost in energy of transporting and processing waste for recycling.

Chemistry has an important role to play in determining what happens. Energy is produced in chemical reactions. Chemistry can increase the yield of crops, prolong the storage life of food, and make protein from petroleum. Chemical methods of birth control can help limit the number of people waiting to eat the food. Manufacturers can use raw materials more efficiently when they better understand the properties of materials, and consumers also can benefit from such an understanding. Chemistry will be needed to keep our water drinkable and our air breathable, and an understanding of chemistry can help government officials and voters see to it that beneficial laws are passed.

We are not about to tell you that science, and chemistry in particular, is mankind's great hope. The hard questions about population, food and energy, natural resources, and pollution of our environment will not be answered by chemists, engineers, politicians, economists, sociologists, or any single group. But as you study chemistry, keep in mind that which is sometimes easy to forget—that chemistry has great relevance to the world outside the laboratory and the classroom.

THOUGHTS ON CHEMISTRY

Spaceship Earth

Now there is one outstandingly important fact regarding Spaceship Earth, and that is that no instruction book came with it. I think it very significant that there is no instruction for successfully operating our ship. In view of the infinite attention to all other details displayed by our ship, it must be taken as deliberate and purposeful that an instruction book was omitted. Lack of instructions has forced us to find that there are two kinds of berries—red berries that kill us and red berries that will nourish us. And we had to find out ways of telling which-was-which red berry before we ate it or otherwise we would die. So we were forced, because of the lack of an instruction book, to use our intellect, which is our supreme faculty, to devise scientific experimental procedures and interpret effectively the significance of the experimental findings. Thus, because the instruction manual was missing we are learning how we safely can anticipate the consequences of an increasing number of alternative ways of extending our satisfactory survival and growth—both physical and metaphysical.

. . . In organizing our grand strategy we must first discover where we are now; that is, what our present navigational position in

OPERATING MANUAL FOR SPACESHIP EARTH, *by R. Buckminster Fuller*

CHEMISTRY: THE SCIENCE OF MATTER

15

the universal scheme of evolution is. To begin our position-fixing aboard our Spaceship Earth we must first acknowledge that the abundance of immediately consumable, obviously desirable or utterly essential resources have been sufficient until now to allow us to carry on in spite of our ignorance. Being eventually exhaustable and spoilable, they have been adequate only up to this critical moment. This cushion-for-error of humanity's survival and growth up to now was apparently provided just as a bird inside of the egg is provided with liquid nutriment to develop to a certain point. But then by design the nutriment is exhausted at just the time when the chick is large enough to be able to locomote on its own legs. And so as the chick pecks at the shell seeking more nutriment it inadvertently breaks open the shell. Stepping forth from its initial sanctuary, the young bird must now forage on its own legs and wings to discover the next phase of its regenerative sustenance.

My own picture of humanity today finds us just about to step out from amongst the pieces of our just one-second-ago broken eggshell. Our innocent, trial-and-error-sustaining nutriment is exhausted. We are faced with an entirely new relationship to the universe. We are going to have to spread our wings of intellect and fly or perish; that is, we must dare immediately to fly by the generalized principles governing the universe and not by the ground rules of yesterday's superstitious and erroneously conditioned reflexes. And as we attempt competent thinking we immediately begin to reemploy our innate drive for comprehensive understanding.

OPERATING MANUAL FOR SPACESHIP EARTH, by R. Buckminster Fuller, 1970 (pp. 47 and 51)

Significant terms defined in this chapter

Note: Significant terms are listed in the order in which they appear in the chapter.

biological sciences	homogeneous mixture
physical sciences	compound
hypothesis	chemistry
law	principles of chemistry
theory	descriptive chemistry
states of matter	organic chemistry
physical properties	hydrocarbons
chemical properties	inorganic chemistry
chemical reaction	mass
pure substance	weight
mixture	density
heterogeneous mixture	

Exercises

Note: One asterisk before a problem indicates it is challenging; two asterisks indicate it is a brain teaser.

1.1 A "chemical magician" sets four beakers in front of you and announces that all four contain water. A small amount of the liquid from the fourth beaker is poured into the first beaker, producing a red liquid; into the second beaker, producing a white liquid; and into the third beaker, producing a blue liquid. You decide that this demonstration must be a trick and ask to investigate both the original liquids and the final solutions. You find that the original liquid in the first beaker has the obvious properties of water, of being colorless and clear. Foolishly, you drink it—a dangerous thing to do in chemistry. You observe that it is an excellent laxative. You test a sample of the original liquid in the second beaker and find that the liquid contains water and the compound lead nitrate. Upon careful observation, you notice that the third original liquid has a light blue color characteristic of a solution of the compound cupric nitrate in water. The fourth liquid is slippery to the touch—another foolish test in chemistry—and produces skin burns rather quickly. A deep whiff—another unwise move in chemistry—of the air above the liquid readily identifies it as ammonia in water. The red and blue liquids are quite stable, but the white liquid eventually separates into a white solid on the bottom of the beaker with a clear, colorless liquid above it. Your investigations prove that you were correct in assuming the demonstration to be a trick.

You have applied the scientific method to this problem. What was your (a) hypothesis, (b) experiment, and (c) theory? List the (d) physical and (e) chemical properties mentioned for the various liquids. Which beakers contained, either before or after mixing, (f) matter, (g) mixtures, (h) homogeneous mixtures, (i) heterogeneous mixtures, (j) only a pure substance, (k) only a pure compound, and (l) only an element?

1.2 Which of the following types of chemist—(i) analytical chemist, (ii) biochemist, (iii) inorganic chemist, (iv) organic chemist, or (v) physical chemist—would primarily be interested in (a) the preparation of a compound containing carbon, hydrogen, and nitrogen; (b) the determination of the exact formula of the compound; (c) the behavior of the compound in biological systems; (d) the preparation of a new compound containing the carbon-hydrogen-nitrogen compound combined with copper and chlorine; (e) the determination of the exact molecular structure of these compounds? This exercise illustrates the overlap and teamwork between the various disciplines of chemistry.

1.3 Which statements are true? Rewrite any false statement so that it is correct.

(a) The individual components in a homogeneous mixture remain physically separate and can be seen as separate components.
(b) A generalization about various facts or observations is known as a hypothesis.
(c) The density of a substance depends on the total amount of substance present.
(d) A theory is a single statement or a series of statements that are so well documented that no experiment is likely to yield data that will refute it.
(e) Chemistry is the branch of science that deals with matter, with the changes that matter can undergo, and with the laws that describe these changes.

1.4 Which of the following metric units are used to express (a) mass, (b) energy, (c) length, (d) volume, and (e) temperature: (i) erg, (ii) cm, (iii) cm^3, (iv) °K, (v) Å, (vi) J, (vii) km, (viii) ml, (ix) g, (x) °C, (xi) nm, (xii) liter, (xiii) cal, (xiv) m, (xv) dm^3, (xvi) kg, and (xvii) mg?

1.5 Which of the following units are correct for the property that is being measured: (a) the area of a football field in m^2, (b) the volume of an apple juice bottle in cu liters, (c) the density of wood in kg/m^3, (d) the length of an eraser in ml, (e) the radius of a basketball in kg, (f) the length of time of a TV commercial in Msec, and (g) the height of an evergreen tree in cm^3?

1.6 Which of the following conversions are incorrect: (a) 1 m = 2.54 cm, (b) 1000 m = 1 km, (c) 1 ml = 1 cm^3, (d) 454 lb = 1 g, (e) 1 Å = 0.1 nm, (f) 1 dm^3 = 1 liter, (g) 1 g/cm^3 = 1×10^3 kg/m^3, (h) 1°F = 1°K, (i) 1 cal = 4.184 J, (j) 1 J = 1×10^{-7} erg, and (k) 1 mμ = 1 nm?

1.7 Clearly distinguish between (a) physical and chemical properties, (b) mass and weight, (c) substances and mixtures, (d) elements and compounds, and (e) homogeneous and heterogeneous mixtures.

1.8 In which of these processes is a chemical reaction taking place: (a) magnetizing a nail, (b) fermenting grape juice, (c) burning of gasoline, (d) cutting meat, (e) cooking meat, (f) baking a cake, (g) "turning" of leaves in autumn, (h) setting of concrete, and (j) inflating a balloon?

1.9 Using the factor-dimensional method of calculation, make each of the following conversions: (a) 100.0 yd to cm, (b) 20.0 liter to gal, (c) 1.00 ft^3 to mm^3, (d) 1.429 g/liter to lb/ft^3, (e) 45 mi/hr to m/sec, (f) 440 yd to m, (g) 7 lb 10 oz to kg, (h) 25 gal to liters, (i) 0.0025 inch to nm, and (j) 14300 cal to J.

1.10 For each case, which is greater: (a) 1.0 cm or 1.0 inch, (b) 35°C or 70.0°F, (c) 300°K or 85°F, (d) 1.0 mg or 1.0 cg, (e) 1.0 m or 1.0 yd, (f) 32 km or 20.0 mi, (g) 325 kcal or 100. J. (h) 50 nm or 0.5 m, (i) 1.0 gal or 3.5 liters (j) 80.0 km/hr or 55 mi/hr, (k) 10 lb or 10 kg, (l) 800.0 mg or 1.0 kg, (m) 0.8 nm or 8 Å?

	(i) Matter	(ii) Heterogeneous	(iii) Homogeneous	(iv) Mixture	(v) Substance	(vi) Compound	(vii) Element
(a) Sugar	X		X		X	X	
(b) Brine							
(c) Air	X		X			X	
(d) Pencil lead	X		X				
(e) Steam							
(f) Diamond							
(g) Gasoline			X			X	
(h) Dry Ice							
(i) Oil on a rain puddle		X					

1.11 Place x's in as many boxes as are appropriate for each case. *Note:* Brine is sea water, pencil lead is a clay-graphite mixture, gasoline contains many hydrocarbons, and Dry Ice is solid carbon dioxide.

1.12 Express the following measurements for an "ideal" human body in the metric system: (a) female: 5′6″ tall, 125 lb in weight and 2.0 ft³ in volume and (b) male: 5′11″ tall, 185 lb in weight and 2.8 ft³ in volume.

1.13 The label on a package from Europe gave the dimensions as 42 cm × 84 cm × 7 cm and the weight at 3 kg. A local postal clerk wanted to find out if postal regulations had been violated by shipping such a large package. What dimensions did the clerk arrive at after converting these metric values to the English system?

1.14 An athlete runs 100.00 yd in 9.4 sec. If this velocity is maintained exactly, how long will it take the athlete to run 100.00 m?

1.15 A very important constant that we will encounter in this book is known as the ideal gas constant. It is numerically equal to 8.314 J/mole °K. Express the value of this constant in (a) erg/mole°K, (b) cal/mole°K, and (c) liter atm/mole°K given 1 J = 10⁷ erg, 1 cal = 4.184 J, and 1 liter-atm = 24.2 cal.

1.16 One hundred cubic centimeters of uranium metal weighs 1.897 kg. What is the density in lb/ft³?

1.17 Assuming the density of water to be 1.0 g/cm³, what is the mass of a gallon of water in pounds?

1.18 Convert the following temperatures which are commonplace in our daily lives to the Celsius scale: (a) normal body temperature, 98.6°F; (b) a cold, wintry day, 10°F; (c) a warm fall day, 78°F; and (d) the running temperature of a modern auto engine, 250°F.

1.19 Convert each of the following melting point temperatures to values on the Kelvin scale: (a) water, 32°F; (b) cesium, 84°F; (c) white phosphorus, 111°F; and (d) nitrogen, −346°F.

1.20 Convert each of the following boiling point temperatures to values on the Fahrenheit scale: (a) water, 100.00°C; (b) nitric oxide, −151.8°C; (c) sulfur, 444.6°C; (d) iron, 2750°C; and (e) sulfuric acid, 338°C.

1.21* A letter, weighed using a triple-beam balance, was 53.5 g. What would be the amount of postage required to mail the letter at these first-class rates: 15¢ for the first ounce and 13¢ for each additional ounce?

1.22* A dairy buys exactly 1000 qt of milk at 45¢ per quart. During the night the country changes to the metric system and the dairy sells the milk for 45¢ per liter. How much profit or loss was made?

1.23* The radius of a hydrogen atom is about 0.58 Å, and the distance between the sun and the Earth is about 93 million miles. Find the ratio of the radius of the hydrogen atom to the sun–Earth distance so that the units cancel.

1.24* The radius of a neutron (a subatomic particle) is approximately 1.5×10^{-15} m. Find the density of a neutron if its mass is 1.675×10^{-24} g. How does this compare to the density of the heavy metal uranium, which is 19 g/cm³?

1.25* A container weighs 68.31 g empty and dry, 93.34 g filled with water, and 88.42 g filled with a second liquid. If the density of water is 1.0000 g/ml, find the density of the second liquid.

1.26* Assuming that the density of water is 1.000 g/ml, compute the density of a metal sample from the following data: (a) weight of empty container = 66.734 g, (b) weight of container and sample = 87.807 g, (c) weight of container, water and sample = 105.408 g, and (d) weight of container and water to occupy the same volume as in (c) = 91.786 g.

1.27* At what temperature will a Fahrenheit thermometer give (a) the same reading as a Celsius thermometer, (b) a reading that is twice that on the Celsius thermometer, and (c) a reading that is numerically the same but opposite in sign from the Celsius scale?

1.28* Confirm the values given in Figure 1.3 for the absolute zero shown on the Fahrenheit and Celsius scales by assuming the value of 0°K and calculating the corresponding value on each scale.

1.29* The British Thermal Unit, BTU, is the common unit for rating home air conditioners and furnaces. It is defined as the quantity of heat required to raise the temperature of one pound of water by one degree Fahrenheit. A calorie is the quantity of heat required to raise the temperature of one gram of water by one degree Celsius. How many calories are equivalent to one BTU?

1.30** A buyer of aluminum for a canning company suspected that a supply of metal cubes was not solid. A typical cube weighed 42.22 g and was 2.5 cm along an edge; (a) calculate its density. The result was compared to the value determined using a cylinder of metal known to be solid which had a radius of 2.50 cm, a length of 10.00 cm and a mass of 0.5305 kg. (b) Calculate the density of the cylindrical piece and (c) compare the results of the two densities. *Note:* The volume of a cube is equal to the cube of its edge, and the volume of a cylinder is equal to π(pi), or 3.1416, times the square of the radius times the length. (d) Is density a chemical or physical property? The buyer reported the density to the company's engineering department in lb/ft^3 units. (e) What was this value?

This exercise illustrates an application of the scientific method to an everyday problem. What was the (f) hypothesis, (g) experiment, and (h) theory?

Atomic theory is presented in this chapter. Modern definitions of atoms, molecules, and ions—the three types of particles of which all substances are composed—are given. The symbols for the elements and the formulas of chemical compounds are introduced. Finally, atomic and molecular weights and the weight relationships that can be derived from chemical formulas are presented, with emphasis on Avogadro's number and the concept of the mole and molar weight.

How many times can you divide a piece of iron into smaller and smaller pieces that retain the properties of iron? Will you reach a point where no further division is possible, or can the process go on indefinitely? Such questions about the nature of matter were debated by the Greek philosophers. Democritus of Abdera, in about 400 B.C., came remarkably close to the modern answer to these questions— atomic theory. He argued that all matter is composed of tiny homogeneous particles which are hard and impenetrable, differ in size and shape, can come together in different combinations, and are constantly in motion. Democritus named the particles "atoms" from the Greek word meaning "indivisible."

Aristotle opposed these ideas. Instead of atoms he favored the concept that Earth, air, fire, and water form the basis of all matter. Aristotle's teachings were widely accepted and the theory of atoms fell into disrepute for over 1500 years. In the Middle Ages, progress in what was to become chemistry fell to the alchemists. Unlike the Greeks, the alchemists were experimentalists—they did not concern themselves with theories about atoms. The alchemists never reached their goals of transforming "baser" metals into gold, or finding the Elixir that would impart eternal life. However, from their laboratories came much practical information about metals and minerals, information that was later useful in the growth of the science of chemistry.

Chemistry: where to begin?

What might you do if you were about to take a trip to a foreign country? Probably you would first get a book or two from which to learn some practical things such as a few phrases in the language and the value of the money. If you planned to stay for quite a while, you could assume that these basics would help to get the trip started, but that more would have to be learned along the way. In this chapter we deal with a similar situation as we embark on the study of chemistry. There are certain things that we need to know in order to get started. Yet much must be left to be learned further down the road.

The starting point for this journey is the atomic theory. (Some of the history of the atomic theory is given in the Aside, Toward Atomic Theory Through History.) Atoms, and the molecules and ions derived from them, are the "currency" of chemistry. It is not possible to go very far without understanding what they are and how to "use" them. The purpose of this chapter is to introduce these species and their weight relationships, while recognizing that fuller explanations of some aspects of the subject must be presented later.

First we examine the behavior of atoms in the light of what was known when John Dalton presented his atomic theory. Then symbols used for atoms are introduced. We next define the terms "molecule" and "ion," and extend the discussion of symbolism to these species. The very important subject of the relative weights of atoms and molecules is then presented, followed by a discussion of how formulas are determined. Finally, we briefly discuss solutions and their concentrations for the benefit of those who will soon be using solutions in the laboratory.

2.1 What is an atom?

We do not use the word "atom" in quite the same way that the ancient Greeks did. (Our word "molecule" is nearer to their meaning, as we will see in the next section.) However, their perception of an atom was remarkably close to what we know today. In modern terminology, the word **atom** is reserved for the smallest particle of an element that can participate in a chemical reaction.

A one-sentence definition of an atom is not very informative. We can learn more about atoms by examining how atomic theory explains many observations about the behavior of matter. To begin with, atomic theory allows us to define an **element** as a substance composed of only one kind of atom. In Chapter 1, we defined an element as a pure substance that cannot be converted to a simpler substance by any chemical reaction. The atomic definition explains why this is so: Elements are already in the simplest form possible.

In the following section we discuss three laws describing the behavior of matter that were known before the acceptance of atomic theory. We will see how atomic theory applies to these laws and helps to explain them.

Toward the atomic theory through history

The philosophers of ancient Greece were the first to speculate about the existence of atoms—ultimate small, indivisible particles that make up all of matter. Lucretius (96–55 B.C.) said it this way, ". . . all nature as it is in itself consists of two things—for there are atoms and there is the void."

Over the centuries until the 1800s the theory of the existence of atoms came in and out of favor. The remarkable way in which atomic theory can explain the behavior of matter could not be fully appreciated, however, until chemistry became an experimental science. Robert Boyle (1627–1691) was a strong advocate for the experimental approach. He demonstrated how to learn by careful experimentation, and had only disdain for those who expounded theories about matter without testing their theories against facts. Boyle did not accept atomic theory nor did he believe in such "elements" as fire, water, salt, or sulfur, which some scientists of the time thought to be present in all matter. He did, however, show how various types of chemical substances could be distinguished from each other, and he carefully and accurately recorded his experiments so that they could be duplicated.

For more than 100 years after Boyle, knowledge was accumulated about the behavior of minerals and other pure substances, and especially about gases. John Dalton (1766–1844), following in Boyle's tradition of careful observation, began in 1787 to keep a journal about the weather, a habit that he continued for 57 years. A few years later Dalton began to experiment with water vapor in the air, the solubility of gases, and the combining weights of gases. It is thought that this work eventually led him to revive atomic theory.

Dalton's atomic theory, first published in 1808, can be summarized as follows:

1. All matter consists of tiny particles. Dalton, like the Greeks, called these particles atoms.

2. Atoms of one element can neither be subdivided nor changed into atoms of any other element.

3. Atoms cannot be created or destroyed.

4. All atoms of the same element are identical in mass, size, and other properties.

5. Atoms of one element differ in mass and other properties from atoms of other elements.

6. Chemical combination is the union of atoms of different elements; the atoms combine in simple, whole-number ratios to each other.

Chemists in Dalton's time were struggling to understand the "affinity" of some substances for others, which leads to their chemical combination, and the weight relationships among such substances. Joseph Gay-Lussac provided strong support for Dalton's atomic theory in his observation that the volumes of combining gases always had simple, whole-number ratios to each other. Avogadro clearly explained the significance of this observation with the following hypothesis, now accepted as a law: Equal volumes of gas (at the same temperature and pressure) must contain equal numbers of particles. (In Chapter 3 we will further examine properties of gases.)

Confusion was mounting during this period over the meaning of the terms "atom" and "molecule" as they were used by different groups of chemists, and also over the weights of the combining substances. Stanislao Cannizaro opened the door to a clarification of the situation when, in 1858, he brought Avogadro's ideas, which had been largely ignored, to the attention of the scientific community. Cannizaro also proved the usefulness of Avogadro's hypothesis by devising a method for determining the relative weights of atoms.

Sodium + chlorine ⟶
  39.34 g   *60.66 g*
                sodium chloride
                    100.00 g

Sodium + chlorine ⟶
  50.0 g   *50.0 g*
     sodium chloride +   sodium
          82.4 g   *17.6 g left over*

FIGURE 2.1
The law of definite proportions.
Sodium chloride NaCl *is 39.34%*
sodium by weight and 60.66% chlorine
by weight.

2.2 Atoms and mass in chemical combination

Three laws summarize what was known about atoms and mass in chemical combination up to the time of Dalton. These laws remain valid today for most chemical reactions and compounds.

The first law, the **law of conservation of mass,** may be stated as follows: In chemical reactions, matter is neither created nor destroyed. Put another way, the total mass of the materials that react is equal to the total mass of the materials produced in the reaction. In ordinary chemical reactions it is *atoms* that are neither created nor destroyed. In Section 1.2 we defined a chemical reaction as a process in which at least one substance is changed in identity. This is accomplished by the *rearrangement* of atoms, but the *atoms themselves* remain unchanged.

The **law of definite proportions** reflects the unvarying composition of compounds. In a pure compound, two or more elements are combined in definite proportions by weight. For example, consider sodium chloride (see Figure 2.1). We all know it as white, crystalline table salt. One way to prepare it in the laboratory is to pass chlorine, a greenish gas, over sodium, a silvery metal. The sodium burns in the chlorine, and crystalline sodium chloride appears in the container. Sodium chloride made in this way or in any other way, or sodium chloride obtained from natural sources, has been examined often enough for us to state with certainty that it is a compound containing 39.34% sodium and 60.66% chlorine by weight. Every 100.00 g of sodium chloride is the product of the combination of 39.34 g of sodium and 60.66 g of chlorine. Here is evidence that atoms of different elements have different masses. We can assume, based on the properties of sodium chloride, that each atom of sodium combines with one atom of chlorine. From the combining weights of sodium and chlorine we can then conclude that the weight of a sodium atom is roughly two-thirds that of a chlorine atom.

The **law of multiple proportions,** first stated by Dalton, is a direct consequence of his atomic theory: If two elements combine to form more than one compound, a fixed weight of element A will combine with two or more different weights of element B so that the different weights of B are in the ratio of small whole numbers. This is just a way of saying that one atom of A will combine only with one atom of B (forming AB), or with two atoms of B (forming AB_2), or with three atoms of B (forming AB_3), and so on. For example, carbon (symbolized by C) and oxygen (symbolized by O) form two different compounds: carbon monoxide, CO, and carbon dioxide, CO_2. Experiments have proven that 10.0 g of carbon will combine with either 13.3 g of oxygen or 26.6 g of oxygen. The ratio 13.3:26.6 reduces to 1:2. This small-whole-number ratio is just what the laws of definite proportions and multiple proportions predict—a specific amount of carbon will combine only with a specific amount of oxygen, or with twice that amount of oxygen. [Note that the percentage by weight (Section 2.10) of oxygen in CO_2 is *not* twice that in CO.]

As explained at the beginning of this chapter, we are introducing here subjects about which more will be learned later. The atomic theory as stated by Dalton and the three laws discussed in this section marked the beginning of the growth of modern chemistry. These ideas also will guide us as we proceed in the study of chemistry. Later, we will find that atoms are not indivisible under all conditions, that not all atoms of the same element are exactly the same, and that there are a few substances that have other than whole-number ratios of atoms. However, these exceptions do not diminish the importance or validity of the atomic theory.

2.3 The symbols for the elements

Elements and compounds often are represented more conveniently by symbols and formulas, respectively, than by names. The symbols for the elements are given on the endpapers of this book. Single-letter symbols are capital letters. Two-letter symbols are always written with the first letter capitalized and the second, a small letter. Table 2.1 gives the symbols for some of the more familiar elements. You will pick up others as you study chemistry. (An extended discussion of symbols is given in Appendix D.)

We describe chemical reactions by *chemical equations* that show by the use of symbols how atoms are rearranged or combined when new substances are formed. Whether we were interested in the reaction of one atom of nitrogen with one atom of oxygen, or the reaction of larger numbers of these atoms, we would write the chemical equation as follows:

arrow means "yields" or "gives"

$$N + O \longrightarrow NO \qquad (1)$$

(Chemical equations are discussed further in Section 4.1.) A compound like NO is called a **binary compound** because it contains atoms of only two elements.

2.4 Molecules and ions

Every substance is composed of atoms, molecules, ions, or some combination among them. In fact, many more substances are composed of molecules and ions than of uncombined atoms.

In modern terms, a **molecule** is the smallest particle of a pure substance that has the composition and chemical properties of that substance. This makes a molecule the ultimate particle of a pure substance that the Greeks were looking for. Suppose we had a pile of crystalline sulfur. If we could divide the pile enough times, we would eventually wind up with individual sulfur molecules, with each containing eight sulfur atoms. We write the formula of this molecule as S_8; the subscript 8 shows that eight sulfur atoms are chemically combined. The S_8 molecule is a **polyatomic molecule**—a molecule containing more than two atoms. Any further division of the S_8 molecules would give a substance with different properties from those of ordinary crystalline sulfur.

TABLE 2.1

Symbols for some of the elements
The asterisks mark all of the elements whose symbols are not based on their current names in English.

Aluminum	Al
Antimony*	Sb
Arsenic	As
Bromine	Br
Carbon	C
Chlorine	Cl
Copper*	Cu
Gold*	Au
Hydrogen	H
Iron*	Fe
Lead*	Pb
Mercury*	Hg
Nitrogen	N
Oxygen	O
Phosphorus	P
Potassium*	K
Silver*	Ag
Sodium*	Na
Sulfur	S
Tin*	Sn
Tungsten*	W
Zinc	Zn

TABLE 2.2

Chemical formulas for molecules of some elements

The subscript gives the number of atoms in a molecule of the element.

Monatomic molecules

He	Helium
Ne	Neon
Ar	Argon

Diatomic molecules

O_2	Oxygen
N_2	Nitrogen
Cl_2	Chlorine

Polyatomic molecules

P_4	Phosphorus
S_8	Sulfur

TABLE 2.3

Some common monatomic ions

Li^+	Bi^{3+}
Na^+	Cr^{3+}
K^+	Fe^{3+}
Cu^+	Co^{2+}
Ag^+	Mn^{2+}
Mg^{2+}	Sn^{2+}
Ca^{2+}	F^-
Ba^{2+}	Cl^-
Cu^{2+}	Br^-
Zn^{2+}	I^-
Cd^{2+}	O^{2-}
Hg^{2+}	S^{2-}
Fe^{2+}	N^{3-}
Al^{3+}	P^{3-}

To further clarify this distinction between atoms and molecules, let us look at the molecules of some other elements (Table 2.2). Gaseous oxygen usually exists as a **diatomic molecule**—a molecule made of two atoms. An oxygen molecule, O_2, can split in a chemical reaction into two oxygen atoms, which can then combine with other atoms or molecules. In fact, this happens very quickly because the oxygen atoms cannot exist independently for more than an instant. Hydrogen, nitrogen, and chlorine molecules are also diatomic. A few elements, notably helium, neon, argon, krypton, and xenon, ordinarily exist as single atoms. An atom of, say, helium, can therefore also be thought of as a molecule of helium; and molecular helium is **monatomic,** that is, made of uncombined atoms.

When one atom of element A combines with one atom of element B, they can form a molecule of a compound, written AB, or A—B, as in Equation (1). Molecules of compounds may contain from two atoms to hundreds of atoms. There are simple diatomic molecules such as nitric oxide, NO, which contains one nitrogen atom and one oxygen atom. There are giant molecules such as chlorophyll *a*, a natural compound that traps solar energy in green plants, which is composed of 137 atoms of the elements carbon, hydrogen, nitrogen, and oxygen. And there are polymer molecules that resemble endless chains. The weights of individual molecules range from 10^{-24} to 10^{-21} g. Molecular diameters are on the order of 10^{-8} cm or 0.1 nm.

The line drawn between A and B, A—B, represents a chemical bond. Until we give a more formal definition later (Chapter 9), you can think of the chemical bond simply as the force that holds the atoms together in compounds.

One fact about atoms that we will investigate more fully later is that they are not indivisible. For the moment this is relevant only as it applies to the existence of ions. Atoms have an internal structure of smaller particles, including equal numbers of protons, which have a positive charge, and electrons, which have a negative charge. An **ion** can be formed from a neutral atom by the loss or addition of one or more electrons. Positively charged ions are called **cations,** and negatively charged ions are called **anions.**

For example, a sodium ion, represented by Na^+, is a sodium atom that has lost one electron and thus has a positive charge. A barium ion, Ba^{2+}, is a barium atom that has lost two electrons and therefore has twice the positive charge of a sodium ion. A chlorine atom gains one electron to form Cl^-, the chloride ion, and an oxygen atom gains two electrons to form the negatively charged oxide ion, O^{2-}, which has twice the negative charge of a chloride ion. Table 2.3 lists some of the common monatomic ions.

Positive and negative ions combine in ionic compounds in sufficient numbers so that electrical neutrality is maintained—the total positive charge equals the total negative charge. This means, for example, that Na^+ combines in a 1-to-1 ratio with Cl^- ions, but Ba^{2+} combines with Cl^- ions in a 1-to-2 ratio. The resulting ionic compounds are sodium chloride, NaCl, and barium chloride, $BaCl_2$. Most ionic compounds are crystalline solids like sodium chloride, NaCl, which is table salt. An ionic compound is just a collection of ions held together by the mutual attraction of positive and negative charges. Independent molecules of NaCl and other ionic compounds do not ordinarily exist.

In summary, atoms are found joined together in independent particles called molecules, as ions formed by the gain or loss of electrons from the atoms, and as free atoms. Ions are held together in compounds by the attraction between the positive and the negative charges of cations and anions.

2.5 Formulas for chemical compounds

Chemical compounds are represented by chemical formulas. A **chemical formula** gives the symbols for the elements combined, with subscripts indicating how many atoms of each element are included. Table 2.4 gives some simple formulas. For molecular compounds such as H_2O and SO_2 the formula can represent one molecule or larger numbers of molecules. As explained in the preceding section, ionic compounds do not consist of separate molecules; they consist of very large numbers of ions mutually attracted to each other. Formulas for these substances represent only the ratios in which the ions are combined. As we saw in the previous section, for example, the formula $BaCl_2$ indicates that in this substance, barium and chloride ions are present in the ratio 1 : 2. From observations of the properties of this ionic compound, it is known that $BaCl_2$ molecules are not present—but we cannot tell this from the formula. The term **formula unit** refers to the simplest unit indicated by the formula of a nonmolecular compound. For $BaCl_2$, one formula unit is one barium ion plus two chlorine ions.

Chemical formulas are used to represent ions as well as neutral compounds. For example, one sulfur atom and four oxygen atoms form an ion with a charge of -2, the sulfate ion, SO_4^{2-}. Such ions that incorporate more than one atom are called **polyatomic ions,** or ionic radicals, or simply radicals. Table 2.5 lists some of the common polyatomic ions, most of which are negatively charged, together with the names by which these ions are known.

Formulas give only a limited amount of information about the structure of compounds, that is, about which atoms are joined to each other and what their arrangement in space is. Since the sulfur atom and four oxygen atoms frequently stay together as a group, they are written together, as in the ammonium sulfate and sulfuric acid formulas in Table 2.6. Here again, we cannot tell from the formula what we know from the properties of the compounds—that H_2SO_4 is a molecular compound but that $(NH_4)_2SO_4$ contains NH_4^+ and SO_4^{2-} ions.

The CH_3CH_2OH formula for ethyl alcohol in Table 2.6 shows that an OH group is attached to a CH_2 group, which is attached to a CH_3 group. Structural formulas like the second one for sulfuric acid or the second one for ethyl alcohol in Table 2.6 go one step further in representing the arrangement of atoms. Drawings of molecules showing the relative sizes of atoms, like the third representation of sulfuric acid in the table, show the geometry of a molecule even more clearly.

+ ions are cations
− ions are anions

TABLE 2.4

Chemical formulas for some simple compounds
The formulas and names of chemical compounds are discussed in Appendix D.

Water	H_2O
Potassium chloride	**KCl**
Sulfur dioxide	SO_2
Silver sulfide	Ag_2S
Aluminum oxide	Al_2O_3
Methyl alcohol	CH_3OH
Diethyl ether	$CH_3CH_2OCH_2CH_3$

TABLE 2.5

Polyatomic ions
These are some of the more common polyatomic ions. Such ions are also called "radicals." The PO_4^{3-} ion is often called simply the phosphate ion.

NH_4^+	**Ammonium**
CN^-	**Cyanide**
CO_3^{2-}	**Carbonate**
ClO_3^-	**Chlorate**
ClO_4^-	**Perchlorate**
CrO_4^{2-}	**Chromate**
$Cr_2O_7^{2-}$	**Dichromate**
MnO_4^-	**Permanganate**
NO_2^-	**Nitrite**
NO_3^-	**Nitrate**
O_2^{2-}	**Peroxide**
OH^-	**Hydroxide**
PO_4^{3-}	**Orthophosphate**
SO_3^{2-}	**Sulfite**
SO_4^{2-}	**Sulfate**

The names of the elements

The names of the elements provide a fascinating glimpse into the history of chemistry. Gold, silver, mercury, copper, iron, lead, tin, mercury, antimony, carbon, and sulfur, although not then understood as elements, have been known as pure substances since ancient times. Therefore, the symbols for these elements are based on their Latin names. Gold, Au, was called *aurum* by the Romans, meaning shining dawn, and mercury, Hg, was *hydrargum*, liquid silver. *Ferrum* was the Latin name for iron, Fe; *argentum*, for silver, Ag; *plumbum*, for lead, Pb; and *stibium*, for antimony, Sb. Sulfur, S, from *sulfurium;* carbon, C, from *carbo*, meaning charcoal; and copper, Cu, from *cuprum*, for the island we now know as Cyprus, have retained names as well as symbols based on the Latin.

In the eighteenth century, chemists were intrigued with studies of the atmosphere, and the gases they discovered were given names based on what was then believed about their chemistry. "Hydrogen" is from the Greek words meaning "water-former," and "nitrogen" and "oxygen" are from the Greek words meaning "soda-former" and "acid-former."

The names of metals, except for those known from ancient times, all end in *-ium*. Aluminium was the name first given to the metal called "aluminum" in the United States. It is still called "aluminium" in England and in many other parts of the world. We are blessed with the four tongue-twisters, terbium, erbium, ytterbium, and yttrium, because these metals were all isolated from ores found in Ytterby, a small town not far from Stockholm.

When faced with the challenge of naming new elements, chemists have turned to the heavens for plutonium, neptunium, uranium, and cerium (Ceres, an asteroid discovered at about the same time as the element). They looked to mythology for thorium (Thor, the Scandinavian god of war), promethium (Prometheus, the bringer of fire), and, for two elements that were tantalizingly difficult to separate, tantalum and niobium (Tantalus and his daughter, Niobe).

Names of some elements honor the places where the elements were discovered: californium, berkelium, europium, americium, francium, germanium. And in recent years a series of man-made elements have been named in honor of famous scientists: einsteinium (Albert Einstein), fermium (Enrico Fermi), mendelevium (Dimitri Mendeleev), nobelium (Alfred Nobel), and lawrencium (Ernest Lawrence).

TABLE 2.6

Some chemical formulas and structures

Sodium nitrate	$NaNO_3$	*read "N-A-N-oh-three"*

Ammonium sulfate	$(NH_4)_2SO_4$	*read "N-H-four-taken twice-S-oh-four"*

note parentheses to avoid confusion about what is taken twice

Sulfuric acid	H_2SO_4	*read "H-two-S-oh-four"*

	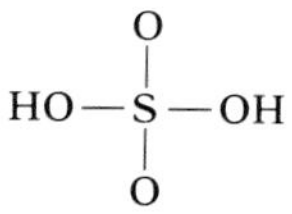	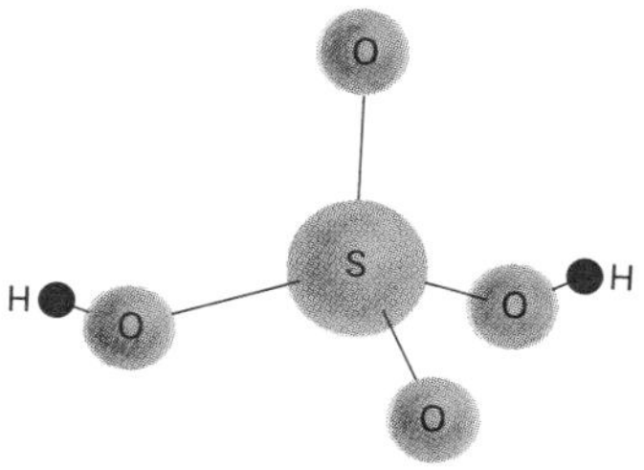	
Ethyl alcohol	C_2H_5OH	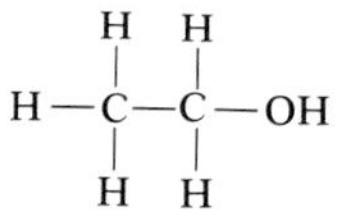	CH_3CH_2OH

Benzene	C_6H_6 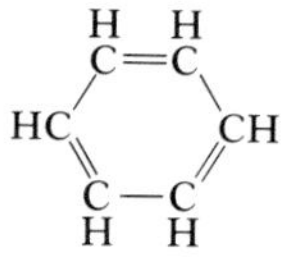 	

The structural formula for the giant chlorophyll *a* molecule is given in Figure 2.2. As you can imagine, drawing the geometry of this molecule is a formidable undertaking. In structural formulas such as the one for chlorophyll *a* in Figure 2.2, which represent carbon atoms joined in rings, the symbol for carbon is usually omitted—a carbon atom and whatever hydrogen atoms are attached to it are assumed to be present at each unlabeled corner. The last formula for benzene in Table 2.6 provides a simple example of this practice.

Ideally, every chemical compound should have a unique name. For ionic compounds and relatively small molecules this requirement is not too difficult to fulfill. But when large molecules have to be named, and particularly when there are several molecules which differ only slightly, the situation becomes complicated. **Chemical nomenclature** is the collective term for the rules and regulations that govern naming chemical compounds. Appendix D gives a thorough review of nomenclature and should be consulted whenever you have a question. Specific points of nomenclature are discussed in the text where appropriate.

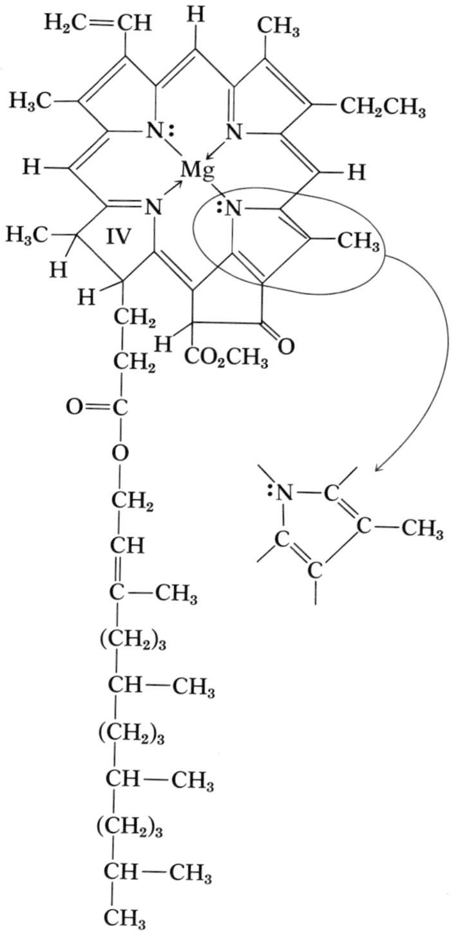

FIGURE 2.2

The structure of chlorophyll *a*, $C_{55}H_{72}MgN_4O_5$. *Carbon atoms occur at each unlabeled corner in the rings, as shown in the insert.*

29

EXAMPLE 2.1

■ The mineral malachite, $Cu_2(CO_3)(OH)_2$, is a common copper mineral. What elements make up this mineral? What is the ratio of the atoms of the different elements in this compound?

The formula tells us that in addition to copper, the mineral contains carbon, oxygen, and hydrogen. The number of atoms of each shown by the formula unit is two copper atoms, one carbon atom, five oxygen atoms (three from the carbonate ion and a total of two from the hydroxide ions), and two hydrogen atoms. Any amount of this mineral will contain copper, carbon, oxygen, and hydrogen atoms in the ratio $2:1:5:2$. ■

Atomic, molecular, and molar weight relationships

In chemistry we deal with quantities of substances that we can see or weigh, and we also deal with atoms, molecules, and ions that we cannot directly see or weigh. The *most important* thing for you to learn from this chapter is how we relate seeable, weighable amounts of chemical compounds to the masses of individual atoms and molecules that cannot be seen or weighed directly.

Avogadro's law—that equal volumes of gases (at the same temperature and pressure) contain equal numbers of molecules—provided the key to the initial determination of atomic weights. Recall that in our discussion of the formation of sodium chloride (Figure 2.1), we assumed, from our modern knowledge, that one sodium atom and one chlorine atom combined to form one NaCl unit. But early chemists did not have this luxury. They did not know the relative weights of atoms and they *also* did not know the formulas of the substances that were reacting or being formed.

However, by comparing the weights of equal volumes of gases and accepting Avogadro's law, the weights of individual molecules could be compared. If a weight was assigned to a molecule of one gas, then values could be assigned to the weights of molecules of other gases. Eventually by using this and additional increasingly sophisticated methods, a complete scale of relative atomic weights was developed.

In chemistry it is frequently of vital importance to know the number of atoms or molecules in a weighable amount of a substance or, conversely, to know the weight of a specific number of atoms or molecules. The conversion factor between these two levels of information is provided by the mole. The mole is a unit that represents the exact number of particles in a specific weight of a substance in grams. The following sections explain how we express the weights of atoms and molecules and how we relate them to larger amounts of substances via the mole concept.

2.6 Atomic weight

For the early chemists using a relative scale of atomic weights was a necessity—they had no method for weighing an atom. Today we have sophisticated techniques for determining the masses of atoms with great accuracy. We know that one atom of uranium, one of the heaviest naturally occurring elements, has a mass of 3.9527×10^{-22} g, and that one atom of hydrogen, the lightest element, has a mass of 1.67380×10^{-24} g. The **true mass of an atom** is the value of its mass in mass units, such as grams. But we still use a relative atomic weight scale because it is more convenient. It is much easier to think of a uranium atom as about 238 times heavier than a hydrogen atom than to deal with numbers with more than 20 zeros after the decimal point.

Relative atomic weights are proportional to the true masses of the atoms. One element is chosen as a standard, it is assigned an atomic weight, and all other atomic weights are expressed relative to that standard. Hydrogen, which was assigned an atomic weight of 1, and oxygen, which was assigned an atomic weight of 16, have both been used as standards. Hydrogen was chosen because it is the lightest element. Oxygen was chosen because it forms compounds with nearly all of the elements and leads to atomic weights for many elements that are close to whole numbers.

However, we now know that most elements occur naturally as mixtures of atoms with slightly different masses, called isotopes (explained in Section 7.10.) By international agreement, the standard for atomic weights is now carbon-12, the most abundant isotope of carbon, to which an atomic weight of exactly 12 units has been assigned. The atomic weights given in the tables and used for most chemical calculations are averages; they reflect the composition of the naturally occurring isotopic mixtures of each element. Therefore, **atomic weight** is now defined as the average weight of the atoms of the naturally occurring element relative to $\frac{1}{12}$ the weight of an atom of carbon-12. (Modern values for the atomic weights of the elements are given on the endpapers of this book.)

An **atomic mass unit** (amu) is defined as $\frac{1}{12}$ of the weight of one carbon-12 atom, and atomic weights are given in this unit. For example, the atomic weight of lead is 207.19 amu, and the atomic weight of uranium is 238.03 amu. One atomic mass unit is equal to 1.6606×10^{-24} g.

EXAMPLE 2.2

■ In naturally occurring neon gas, 90.92% of the atoms have a mass of 19.99244 amu, 0.257% have a mass of 20.99395 amu, and 8.82% have a mass of 21.99138 amu. What is the average atomic weight of neon?

The average atomic weight can be calculated by multiplying each mass by the fraction of the atoms having that mass and adding these masses together.

Average atomic weight
(amu) $= (0.9092)(19.99244 \text{ amu})$
$+ (0.00257)(20.99395 \text{ amu})$
$+ (0.0882)(21.99138 \text{ amu})$
$= 18.177 \text{ amu} + 0.0540 \text{ amu} + 1.940 \text{ amu}$
$= 20.171 \text{ amu}$

The average atomic weight of neon, based on the given isotopic mixture, is 20.171 amu. ■

2.7 Molecular weight

The **molecular weight** is the sum of the atomic weights, in atomic mass units, of the atoms in the formula of a chemical compound. To calculate molecular weight we must know the correct formula of the compound and the atomic weight of each different element in the compound. The "molecular" weight of an ionic compound, or other compounds that do not contain discrete molecules, is found in the same way. Sometimes the term *formula weight* is used instead of "molecular weight" in referring to such compounds.

EXAMPLE 2.3

■ Find the molecular weights of (a) phosphorus pentachloride, PCl_5; (b) calcium phosphate, $Ca_3(PO_4)_2$; (c) tetraamminecopper(II) ion, $[Cu(NH_3)_4]^{2+}$; and (d) copper(II) sulfate pentahydrate, $CuSO_4 \cdot 5H_2O$. The atomic weights can easily be found in the alphabetical table inside the front cover of this book.

The atomic weight of each element in the compound is multiplied by the number of atoms of that element. The total for all the atoms in the formula is the molecular weight.

(a) For PCl_5:

P (1 atom) $\left(30.97 \dfrac{\text{amu}}{\text{atom}}\right)$ = 30.97 amu

Cl (5 atoms) $\left(35.453 \dfrac{\text{amu}}{\text{atom}}\right)$ = 177.27 amu

molecular wt PCl_5 = 208.24 amu

(b) For $Ca_3(PO_4)_2$:

Ca (3 atoms) $\left(40.08 \dfrac{\text{amu}}{\text{atom}}\right)$ = 120.24 amu

P (2 atoms) $\left(30.97 \dfrac{\text{amu}}{\text{atom}}\right)$ = 61.94 amu

O (8 atoms) $\left(15.999 \dfrac{\text{amu}}{\text{atom}}\right)$ = 127.99 amu

molecular wt $Ca_3(PO_4)_2$ = 310.17 amu

(c) For $[Cu(NH_3)_4]^{2+}$:

Cu   (1 atom) $\left(63.55 \dfrac{amu}{atom}\right)$ = 63.55 amu

N   (4 atoms) $\left(14.01 \dfrac{amu}{atom}\right)$ = 56.04 amu

H   (12 atoms) $\left(1.008 \dfrac{amu}{atom}\right)$ = $\dfrac{12.10 \text{ amu}}{}$

molecular wt $[Cu(NH_3)_4]^{2+}$ = 131.69 amu

(d) For $CuSO_4 \cdot 5H_2O$:

Cu   (1 atom) $\left(63.55 \dfrac{amu}{atom}\right)$ = 63.55 amu

S   (1 atom) $\left(32.06 \dfrac{amu}{atom}\right)$ = 32.06 amu

O   (9 atoms) $\left(15.999 \dfrac{amu}{atom}\right)$ = 143.99 amu

H   (10 atoms) $\left(1.008 \dfrac{amu}{atom}\right)$ = $\dfrac{10.08 \text{ amu}}{}$

molecular wt $CuSO_4 \cdot 5H_2O$ = 249.68 amu ■

The respective molecular weights, calculated to the nearest 0.01 amu, are (a) 208.24 amu, (b) 310.17 amu, (c) 131.69 amu, and (d) 249.68 amu.

2.8 Avogadro's number, the mole, and molar weight

Now we can return to the problem we mentioned earlier: How do we relate visible weighable masses to numbers and masses of individual atoms and molecules? Three quantities are at the heart of the matter in answering this important question. We define these quantities here so that you can see how they are related to each other. We discuss each of them separately in the following sections. The three quantities are Avogadro's number, the mole, and the molar weight:

Avogadro's number is the number of atoms in exactly 12 g of carbon-12.

The **mole** is an Avogadro's number of anything.

The **molar weight** of a substance is the weight in grams of one mole of that substance.

Avogadro's number. Consider one atom of oxygen and one atom of sulfur, and their modern atomic weights:

O   atomic weight = 15.9994 amu
S   atomic weight = 32.06 amu

Sulfur atoms weigh about twice as much as oxygen atoms. If we could take one gram of each substance, we would have roughly one sulfur atom for every two oxygen atoms. But if we took 2 g of sulfur and 1 g of oxygen, then we would have roughly the same number of atoms of each element. And if we took *exactly* 15.9994 g of oxygen and *exactly* 32.06 g of sulfur, then we would have *exactly* the same number of atoms. By the same reasoning, 1.0079 g of hydrogen or 238.03 g of uranium contain this same number of atoms. In other words, an amount of any element with a weight in grams that is numerically equal to the atomic weight of the element contains the same number of atoms as a similar amount of any other element.

Recall that carbon-12 is the standard for the scale of atomic weights. Since Avogadro's number is defined as the number of atoms in exactly 12 g of carbon-12, it follows that each of the quantities we have been discussing—15.994 g of oxygen, 32.06 g of sulfur, or a weight of any element equal to its atomic weight—contains an Avogadro's number of atoms. By similar reasoning, if the molecular weight of a substance is y, then y grams will contain an Avogadro's number of molecules (or formula units) of that substance.

Avogadro's number has been determined by many methods and with great accuracy. The most up-to-date value is 6.022045×10^{23}. In most calculations in this book, it will be sufficient to use 6.022×10^{23}. A number this large is beyond comprehension. It may help to tell you that 6.022×10^{23} baseballs would cover the entire surface of the Earth to a depth of 60 miles. Similarly, 6.022×10^{23} grains of rice would weigh 13.2 quadrillion tons and fill a cube 146 miles long on each side.

EXAMPLE 2.4

■ How many ozone molecules and how many oxygen atoms are present in 48.00 g of ozone, O_3?

The molecular weight of ozone is (3 atoms)(15.999 amu/atom) = 48.00 amu; thus in 48.00 g of ozone there will be an Avogadro's number of molecules. Each molecule contains three oxygen atoms, so the number of oxygen atoms is

$$(6.022 \times 10^{23} \text{ molecules}) \left(3 \frac{\text{atoms}}{\text{molecule}} \right) = 1.807 \times 10^{24} \text{ atoms}$$

The 48.00 g sample of ozone contains 6.022×10^{23} molecules and 1.807×10^{24} oxygen atoms. ■

The mole. The mole is the most meaningful unit of measure for the amounts of substances that are represented by chemical formulas or that take part in chemical reactions. One mole represents a definite number of particles, or entities. We speak of a mole of ions, a mole of atoms, a mole of electrons, a mole of molecules—theoretically, even a mole of butterflies. The mole is one of the seven basic SI units (Section 1.6). Here is the complete SI definition of a mole:

The mole is the amount of substance of a system that contains as many elementary entities as there are atoms in 0.012 kg (12 g) of carbon-12.

Note: When the mole is used, the elementary entities must be specified and may be atoms, ions, electrons, other particles, or specified groups of such particles.

In other words, as we said above, the mole is an Avogadro's number of anything.

The symbol often used for the amount of substance expressed in moles is n and the substance can be specified by writing it as a subscript. For example,

One mole of hydrogen	$n_{H_2} = 1$ mole
One-half mole of calcium atoms	$n_{Ca} = 0.5$ mole
Two moles of phosphorus pentachloride	$n_{PCl_5} = 2$ moles

In calculations where only one kind of entity is expressed in moles, the subscript is often omitted.

The SI system recommends the abbreviation mol for mole. We have chosen to write mole instead.

Once we know the number of moles of a substance, it is always possible to find the number of "elementary entities" that are present. Avogadro's number provides the conversion factor. For example, for a molecular compound:

$$\text{no. of molecules} = \overline{\text{moles}} \times \left(6.022 \times 10^{23} \, \frac{\text{molecules}}{\overline{\text{mole}}} \right)$$

EXAMPLE 2.5

■ A one-liter flask of air contains 0.040 mole of N_2. How many molecules of nitrogen are present?

The number of molecules is given by

$$\begin{aligned}
\text{no. of molecules} &= \text{moles} \times (6.022 \times 10^{23} \text{ molecules/mole}) \\
&= (0.040 \text{ mole})(6.022 \times 10^{23} \text{ molecules/mole}) \\
&= 2.4 \times 10^{22} \text{ molecules}
\end{aligned}$$

The flask contains 2.4×10^{22} molecules of N_2. ■

EXAMPLE 2.6

■ During the purification of metallic copper by electrolysis, 1.2×10^{22} atoms of copper metal were transferred from one electrode to another. How many moles of copper metal were transferred?

mass
= moles × molar weight

Molar weights of some different substances

	Molar weight (g)
Atoms	
O atoms	16.00
Fe atoms	55.85
S atoms	32.06
Molecules	
O₂	32.00
H₂O	18.02
SO₂	64.06
Ions	
Na⁺	22.99
Fe³⁺	55.85
NO₃⁻	62.01
Ionic compounds	
Na⁺NO₃⁻	85.00
Fe³⁺(NO₃⁻)₃	241.88

The number of moles of atomic copper is

$$n_{Cu} = \frac{\text{no. of atoms}}{6.022 \times 10^{23} \text{ atoms/mole}} = \frac{1.2 \times 10^{22} \text{ atoms}}{6.022 \times 10^{23} \text{ atoms/mole}}$$

$$= 0.020 \text{ mole}$$

During the electrolysis, 0.020 mole of metallic copper was transferred. ■

Molar weight. The molar weight of an atomic substance is the weight in grams numerically equal to the atomic weight in atomic mass units. (This quantity is sometimes called a *gram atomic weight.*) For a compound, the molar weight is the weight in grams numerically equal to the molecular weight or formula weight in atomic mass units. (This quantity is sometimes called the *gram molecular weight* or the *gram formula weight.*) Table 2.7 gives the molar weights of a variety of substances. Think for a moment about how the molar weights are related to the atomic weights of the elements involved. Note that, although ions are formed from atoms by the loss or gain of electrons, the molar weight of an ion is taken as equal to the molar weight of the atom or group of atoms in the ion. This is possible because the weight of an electron is negligibly small compared to the weights of atoms.

From Table 2.7 it should be clear why it is always necessary to specify exactly what substance is being referred to when moles are used. For example, one mole of oxygen molecules contains two moles of oxygen atoms, and one mole of $NaNO_3$ contains one mole of Na^+ ions and one mole of NO_3^- ions, or a total of two moles of ions.

The molar weight of a substance has the units of grams per mole, and it provides the means of conversion between grams and number of moles

$$\text{mass(g)} = n(\text{moles}) \times \text{molar weight} \left(\frac{\text{grams}}{\text{mole}} \right)$$

EXAMPLE 2.7

■ What is the molar weight of chlorophyll *a*, $C_{55}H_{72}MgN_4O_5$?

	moles		$\frac{g}{mole}$		g
C	55	×	12.01	=	660.55
H	72	×	1.008	=	72.58
Mg	1	×	24.31	=	24.31
N	4	×	14.01	=	56.04
O	5	×	16.00	=	80.00
			molar wt	=	893.48 g

The molar weight of chlorophyll *a* is 893.48 g. ■

EXAMPLE 2.8

■ What is the mass of the nitrogen in the flask described in Example 2.5?

The molecular weight of N_2 is (2 atoms)(14.01 amu/atom) = 28.02 amu and the molar weight is 28.02 g/mole. Thus the mass of the 0.040 mole of nitrogen is

$$mass = n_{N_2} \times \text{(molar weight)}$$
$$= (0.040 \text{ mole})(28.02 \text{ g/mole})$$
$$= 1.1 \text{ g}$$

The 2.4×10^{22} molecules in the flask weigh 1.1 g. ■

EXAMPLE 2.9

■ What mass of copper metal was transferred from one electrode to another in the electrolysis process described in Example 2.6?

The mass of copper is

$$mass = n_{Cu} \times \text{(molar weight)}$$
$$= (0.020 \text{ mole})(63.55 \text{ g/mole})$$
$$= 1.3 \text{ g}$$

The mass of copper was 1.3 g. ■

2.9 Using Avogadro's number, moles, and molar weight

The three quantities that we have just defined—Avogadro's number, moles, and molar weight—can be used separately and in combination to answer a wide variety of questions. The following examples represent a few more of the types of problems that can be solved by using these quantities.

EXAMPLE 2.10

■ Which is the larger amount of substance in moles, 3.5 g of carbon dioxide, CO_2, or 3.5 g of sodium chloride, NaCl?

For CO_2:

	moles	$\frac{g}{mole}$		g
C	1	$\times$ 12.0	=	12.0
O	2	$\times$ 16.0	=	32.0
		molar wt CO_2 =		44.0 g

For NaCl:

	moles	$\frac{g}{mole}$		g
Na	1	$\times$ 23.0	=	23.0
Cl	1	$\times$ 35.5	=	35.5
		molar wt NaCl =		58.5 g

Setting the calculation up using the factor-dimensional method,

$$n_{CO_2} = 3.5 \text{ g} \times \frac{1 \text{ mole}}{44.0 \text{ g}} \qquad\qquad n_{NaCl} = 3.5 \text{ g} \times \frac{1 \text{ mole}}{58.5 \text{ g}}$$

$$= 0.080 \text{ mole} \qquad\qquad\qquad\qquad = 0.060 \text{ mole}$$

The amount of CO_2 in moles is larger than the amount of NaCl. ■

Equal molar amounts of substances contain equal numbers of atoms, molecules, or formula units—whatever their "elementary entity" is.

EXAMPLE 2.11

■ What weight of barium contains the same number of atoms as 10. g of calcium?

First, find the number of moles of calcium.

$$n_{Ca} = 10. \text{ g Ca} \times \frac{1 \text{ mole}}{40. \text{ g}}$$

$$= 0.25 \text{ mole}$$

Then find the weight of barium from its molar weight. Introducing the symbol m for weight (mass) in grams,

$$m_{Ba} = 0.25 \text{ mole} \times \frac{137 \text{ g}}{\text{mole}}$$

$$= 34 \text{ g}$$

To have the same number of atoms as 10. g of calcium requires over three times as much barium; that is, 34 g. ■

The molar weight and Avogadro's number can be used together in calculations dealing with the numbers and weights of atoms.

EXAMPLE 2.12

■ The nutritional recommended daily allowance (rda) of iron is 18 mg for an adult. How many iron atoms is this?

Setting the problem up using the factor-dimensional method gives the following expression:

$$\text{Fe atoms} = (18 \text{ mg}) \left(\frac{10^{-3} \text{ g}}{\text{mg}}\right)\left(\frac{1 \text{ mole}}{55.85 \text{ g}}\right)\left(6.022 \times 10^{23} \frac{\text{atoms}}{\text{mole}}\right)$$

$$= 1.9 \times 10^{20} \text{ atoms}$$

The rda of iron for an adult contains 1.9×10^{20} Fe atoms. *Note:* To solve this problem, it was necessary to multiply numbers given in exponential form, or scientific notation. At this point, you may wish to review the rules for using such numbers in arithmetic, which are given in Appendix A.2. ■

EXAMPLE 2.13

■ What is the mass of one atom of gold?

$$\text{mass of Au atom (in g)} = \left(\frac{197.0 \text{ g}}{\text{mole}}\right)\left(\frac{1 \text{ mole}}{6.022 \times 10^{23} \text{ atoms}}\right)$$
$$= 3.271 \times 10^{-22} \text{ g/atom}$$

A gold atom weighs 3.271×10^{-22} g. ■

EXAMPLE 2.14

■ A chemical reaction required 4.25 moles of potassium chlorate, $KClO_3$. What mass of $KClO_3$ was needed?

The molar weight of $KClO_3$ is

	moles	$\frac{\text{g}}{\text{mole}}$		g
K	1	× 39.10	=	39.10
Cl	1	× 35.45	=	35.45
O	3	× 16.00	=	48.00
		molar wt $KClO_3$	=	122.55 g

The mass of $KClO_3$ needed is

$$\text{mass} = n \times (\text{molar weight})$$
$$= (4.25 \text{ moles})(122.55 \text{ g/mole})$$
$$= 521 \text{ g}$$

A sample of $KClO_3$ weighing 521 g was needed. ■

2.10 Weight relationships and chemical formulas

The correct chemical formula for a compound is the source of information about the atomic, molar, and weight relationships within the substance. Table 2.8 summarizes the kinds of information that can be derived from the formula of one compound, ethyl alcohol.

Percentage composition is the percent by weight of each element in a compound. It is found by dividing the weight of each element in one mole of the compound by the molar weight of the compound and multiplying by 100.

TABLE 2.8

Information from a chemical formula

The formula is given for ethyl alcohol, molecular weight 46.07. The percentage composition was found as follows:

% C = $(24.02/46.07) \times 100$
% O = $(16.00/46.07) \times 100$
% H = $(6.05/46.07 \times 100$

C_2H_5OH

1 molecule of ethyl alcohol contains:

2 carbon atoms

1 oxygen atom

6 hydrogen atoms

1 mole of ethyl alcohol contains:

2 moles of carbon atoms

1 mole of oxygen atoms

6 moles of hydrogen atoms

46.07 g of ethyl alcohol contains:

24.02 g of carbon

16.00 g of oxygen

6.05 g of hydrogen

Ethyl alcohol is:

52.14% carbon

34.73% oxygen

13.13% hydrogen

EXAMPLE 2.15

■ Which iron oxide has the higher percentage of iron, FeO, iron(II) oxide, or Fe_2O_3, iron(III) oxide?

For FeO:

	moles	$\frac{g}{mole}$	g
Fe	1	× 55.85 =	55.85
O	1	× 16.00 =	16.00
		molar wt FeO =	71.85 g

For Fe_2O_3:

	moles	$\frac{g}{mole}$	g
Fe	2	× 55.85 =	111.70
O	3	× 16.00 =	48.00
		molar wt Fe_2O_3 =	159.70 g

$\dfrac{55.85 \text{ g}}{71.85 \text{ g}} \times 100\% = 77.73\%$ Fe

$\dfrac{16.00 \text{ g}}{71.85 \text{ g}} \times 100\% = 22.27\%$ O

100.00%

$\dfrac{111.70 \text{ g}}{159.70 \text{ g}} \times 100\% = 69.94\%$ Fe

$\dfrac{48.00 \text{ g}}{159.70 \text{ g}} \times 100\% = 30.06\%$ O

100.00%

The FeO has the higher iron content. Note that in each case the entire percentage composition was determined and checked by making sure that the percentages added up to 100%. ■

Determining the correct formula of a chemical compound newly prepared in the laboratory or newly discovered in nature is an essential part of its identification. The first step in the study of such a compound is frequently the determination of the weight or the percent by weight of each element that is present. This information is used to find the simplest, or empirical, formula of the compound.

The **simplest formula** of a compound gives the simplest whole-number ratio of atoms in the compound. It represents the chemical composition of the compound in terms of the smallest possible number of atoms of each element present. For example, the simplest formula of diborane, a compound known to contain boron and hydrogen, is BH_3.

The **true formula** of a compound represents the actual number of atoms which are combined in each molecule of the compound. In the case of diborane it has been found that each molecule contains 2 boron atoms and 6 hydrogen atoms, so that its true formula is B_2H_6. For ionic and other nonmolecular compounds, only simplest formulas can be written; the term "true formula" has no meaning for such compounds.

An experimentally determined simplest formula is usually called an **empirical formula,** implying that the true formula is not yet known. To find the true formula from the empirical formula, the molecular weight of the compound in question must be determined. This molecular weight will be equal or nearly equal to some multiple of the weight calculated from the empirical formula. If the multiple is 1, the empirical and true formulas are identical. For diborane, which would give an empirical formula of BH_3, the multiple is two.

The objective in an empirical formula problem is to find the number of moles of atoms of each element present from the data given. Then the ratio of atoms is found from the ratios of moles by dividing the number of moles of each element by the number of moles of the element that is present in the smallest amount. Some experimental error may occur in the ratios, which can then be rounded off.

EXAMPLE 2.16

■ An oxide of phosphorus contains 0.5162 g of phosphorus and 0.6667 g of oxygen. What is the empirical formula of this oxide?

	Phosphorus (P)	Oxygen (O)
Actual weights (g)	0.5162	0.6667
Moles of atoms $\left(\dfrac{g}{g/mole}\right)$	$\dfrac{0.5162}{30.97} = 0.01667$	$\dfrac{0.6667}{16.00} = 0.04167$
Mole ratios (mole/mole)	$\dfrac{0.01667}{0.01667} = 1.000$	$\dfrac{0.04167}{0.01667} = 2.500$
Relative number of atoms	1	2.5

On this basis, the empirical formula could be written as $PO_{2.5}$. However, fractional numbers of atoms are usually avoided when writing a formula. Hence, the relative numbers are doubled, and the empirical formula becomes P_2O_5. ■

EXAMPLE 2.17

■ The approximate molecular weight of the oxide described in Example 2.16 was found to be 280 amu. What are the true formula and the exact molecular weight of this oxide?

The molecular weight of the oxide based on its empirical formula is

$$\left(2 \frac{atoms}{molecule} \times 30.97 \frac{amu}{atom}\right) + \left(5 \frac{atoms}{molecule} \times 16.00 \frac{amu}{atom}\right)$$
$$= 141.94 \text{ amu/molecule}$$

A comparison of this value with the approximate value of 280 amu shows that the true molecular formula is double the empirical formula, or P_4O_{10}. Therefore, the exact molecular weight is (141.94 amu) × 2 = 283.88 amu. ■

If the experimentally determined composition of a compound is obtained as the weight percentage of each element present, it is convenient to take 100 g of the compound as the basis for calculation. Each weight percentage is then taken as the weight in grams of an element.

ATOMS, MOLECULES, AND IONS [2.10]

41

EXAMPLE 2.18

■ The hormone adrenaline is released in the human body during stress periods and increases the body's metabolic rate. Like many biochemical compounds adrenaline is composed of carbon, hydrogen, oxygen, and nitrogen. Its composition by weight is 56.8% C, 6.56% H, 28.4% O, and 8.28% N. Determine the formula of adrenaline.

In the following table, 100 g of adrenaline is taken as the basis for calculation.

	Carbon	Hydrogen	Oxygen	Nitrogen
Actual weights (g)	56.8	6.56	28.4	8.28
Moles of atoms $\left(\dfrac{g}{g/mole}\right)$	$\dfrac{56.8}{12.01}$	$\dfrac{6.56}{1.008}$	$\dfrac{28.4}{16.00}$	$\dfrac{8.28}{14.01}$
	= 4.73	= 6.51	= 1.78	= 0.591
Mole ratios (mole/mole)	$\dfrac{4.73}{0.591}$	$\dfrac{6.51}{0.591}$	$\dfrac{1.78}{0.591}$	$\dfrac{0.591}{0.591}$
	= 8.00	= 11.02	= 3.01	= 1.00
Relative number of atoms	8	11	3	1

On the basis of this analysis, the formula for adrenaline is $C_8H_{11}O_3N$. Without additional experimental information it is not possible to determine whether this is an empirical formula or the true formula. (It is the true formula.) ■

Solutions

2.11 Molecules and ions in solution

A **solution** is a homogeneous mixture of the molecules, atoms, or ions of two or more substances. We commonly use the word "solution" to refer to a liquid, usually water, with something dissolved in it. A solution of any substance in water is called an **aqueous solution.** (Solutions of all types are discussed in Chapter 14. Note that we rarely deal with atoms in aqueous solutions.)

Solutions are spoken of as having two components: the solvent and the solute (or solutes). If a handful of salt is dissolved in a bucket of water, the salt is the solute and the water is the solvent. The **solvent** is the component of a solution usually present in the larger amount. The solvent is the medium in which the **solute**—the component of a solution usually present in the smaller amount—has dissolved. However, the terms "solvent" and "solute," while convenient, are often imprecise and do not have a fixed scientific meaning.

When an ionic compound dissolves completely in water, all of the ions simply separate from each other and move about independently. The **dissociation of an ionic compound** is the transformation of a neutral ionic compound into positive and negative ions in solution.

Care must be exercised in dealing with molar quantities of ionic compounds. For example, one mole of potassium chloride, KCl, is one mole of KCl formula units (molar weight = 39.10 g + 35.45 g = 74.55 g). However, KCl dissociates in water to yield 2 moles of ions per mole of KCl.

$$K^+Cl^- \rightarrow K^+ + Cl^-$$

1 mole   *1 mole*   *1 mole*

6.022 × 10²³ formula units   *6.022 × 10²³ ions*   *6.022 × 10²³ ions*

One mole of ammonium sulfate dissolves to give 3 moles of ions in aqueous solution.

$$(NH_4^+)_2SO_4^{2-} \rightarrow 2NH_4^+ + SO_4^{2-}$$

1 mole   *2 moles*   *1 mole*

6.022 × 10²³ formula units   *2 × 6.022 × 10²³ ions*   *6.022 × 10²³ ions*

Two extremes in behavior are possible when a molecular compound dissolves in water. Either the molecules of the compound simply become mixed with the water molecules or the molecule is converted into ions in solution. For example, although hydrogen chloride, HCl, is a gaseous molecular compound, its molecules are converted to H^+ and Cl^- ions in aqueous solution. **Ionization** is the formulation of ions from a nonionic substance. We would say that hydrogen chloride is ionized in aqueous solution.

$$HCl \xrightarrow{H_2O} H^+ + Cl^-$$

Often the atoms, molecules, or ions in a solution attract solvent molecules to themselves. The combination of solute molecules with solvent molecules is called **solvation.** For aqueous solutions, the combination of solute with water molecules is called **hydration.** The association between the solvent and solute is not usually considered as the formation of a new chemical compound, for the forces involved are too weak.

2.12 Concentration of solutions

In Table 2.8 we illustrated the information that can be obtained from a chemical formula. The amounts of substances needed in a chemical reaction can be determined with the aid of formulas if the pure substances are to be combined. However, chemical reactions are frequently performed by mixing aqueous solutions of the substances involved. Because solutions are mixtures, they can contain widely varying proportions of solutes and solvent. Therefore, we need an additional kind of information—we need to know the concentration of the solution, that is, the specific amount of solute dissolved in a specific amount of solvent or solution.

Any solution of known concentration is called a **standard solu-**

$$\text{molarity} = \frac{\text{moles of solute}}{\text{volume of solution (liters)}}$$

tion. The concentration of a solute in a solvent is most often expressed in a combination of units that represents the weight or the volume of the solute per weight or volume of the solution. The usual weight unit is the gram; the usual volume unit is the liter. For use in chemistry, it is most convenient to convert the grams to moles and to express the concentration as **molarity,** which is the number of moles of solute per liter of solution.

$$\text{molarity } (M) = \frac{\text{moles of solute}}{\text{volume of solution (liters)}} \tag{2}$$

A 1-molar solution, written $1M$, of a substance is made by adding to 1 mole of the substance enough water to make exactly 1 liter of solution. For example, a $1M$ solution of NaCl would be made by weighing out 58.5 g of NaCl and in a calibrated flask adding enough water to the NaCl to give 1 liter of solution. On the outside of the bottle of this solution we would write "$1M$ NaCl."

EXAMPLE 2.19

■ If 4.0 g of NaOH is dissolved in water to give 500. ml of solution, what will be the molarity of the solution?

$$n_{\text{NaOH}} = \frac{4.0 \text{ g}}{40.0 \text{ g/mole}} = 0.10 \text{ mole}$$

$$M = \frac{\text{moles solute}}{\text{liters solution}}$$

$$= \frac{0.10 \text{ mole}}{0.500 \text{ liter}} = 0.20 \text{ mole/liter}$$

It is a $0.20M$ NaOH solution. ■

EXAMPLE 2.20

■ What is the molarity of a 32 wt % solution containing 95 g, or 1.5 moles, of nitric acid (HNO_3) in 250. ml of solution?

$$M = \frac{\text{moles } HNO_3}{\text{liters of solution}} = \frac{1.5 \text{ moles}}{0.250 \text{ liter}} = 6.0 \text{ mole/liter}$$

It is a $6.0M$ HNO_3 solution. ■

EXAMPLE 2.21

■ What is the molarity of chloride ion (Cl^-) in a solution prepared by dissolving 16.7 g of $CaCl_2$ in sufficient water to obtain 400. ml of solution? Note that 2 moles of Cl^- are released when 1 mole of $CaCl_2$ dissolves in water.

Because $CaCl_2$, not Cl^-, has been weighed out, the number of moles of $CaCl_2$ must be determined first.

$$n_{CaCl_2} = \frac{16.7 \text{ g}}{111.0 \text{ g/mole}} = 0.150 \text{ mole}$$

$$n_{Cl^-} = 0.150 \text{ mole } CaCl_2 \times \frac{2 \text{ mole } Cl^-}{1 \text{ mole } CaCl_2} = 0.300 \text{ mole } Cl^-$$

$$M = \frac{\text{moles } Cl^-}{\text{liters of solution}} = \frac{0.300 \text{ mole}}{0.400 \text{ liter}} = 0.750 \text{ mole/liter}$$

This is a $0.750M$ Cl^- solution. ■

Sometimes the concentration of solutions is expressed as a percentage. In this case it must be clear whether it is **volume percent** or **weight percent.**

$$\text{volume percent (vol \%)} = \frac{\text{volume of solute}}{\text{volume of solute} + \text{volume of solvent}} \times 100\% \quad (3)$$

$$\text{weight percent (wt \%)} = \frac{\text{weight of solute}}{\text{weight of solution}} \times 100\% \quad (4)$$

To convert from volume percent to weight percent, or to find the weight of solute in a given volume of solution of a known weight percent, the density of the solution, usually expressed in grams per milliliter of solution, must be known.

EXAMPLE 2.22

■ What is the volume percent of methanol, CH_3OH, in the solution formed when 2.0 liters of methanol are mixed with 3.0 liters of diethyl ether, $(C_2H_5)_2O$? (Assume that there is no change in the volume of either liquid when they are mixed.)

$$\text{vol \% } CH_3OH = \frac{\text{vol } CH_3OH}{[\text{vol } CH_3OH] + [\text{vol } (C_2H_5)_2O]} \times 100\%$$

$$= \frac{2.0 \text{ liters}}{(2.0 + 3.0) \text{ liters}} \times 100\%$$

$$= 40.\%$$

The solution is 40. vol % methanol. ■

EXAMPLE 2.23

■ What is the weight percent of methanol in the solution with diethyl ether described in Example 2.22? The density of methanol is 0.793 g/ml and that of diethyl ether is 0.714 g/ml.

$$m_{CH_3OH} = 2.0 \text{ liters} \times 1000 \frac{ml}{\text{liter}} \times 0.793 \frac{g}{ml}$$

$$= 1.6 \times 10^3 \text{ g}$$

$$m_{(C_2H_5)_2O} = 3.0 \text{ liters} \times 1000 \frac{ml}{\text{liter}} \times 0.714 \frac{g}{ml}$$

$$= 2.1 \times 10^3 \text{ g}$$

$$\text{wt \% } CH_3OH = \frac{\text{wt } CH_3OH}{\text{total solution wt}} \times 100\%$$

$$= \frac{1.6 \times 10^3 \text{ g}}{(1.6 \times 10^3 \text{ g}) + (2.1 \times 10^3 \text{ g})} \times 100\%$$

$$= 43\%$$

The solution is 43 wt % methanol. ∎

EXAMPLE 2.24

∎ The density of a 32% by weight nitric acid solution is 1.19 g/ml. How many grams of pure HNO_3 are present in 250. ml of solution? How many moles of pure HNO_3 are there in this volume of solution?

$$\text{total wt of solution (g)} = \text{vol} \times \text{density}$$
$$= 250. \text{ ml} \times 1.19 \text{ g/ml}$$
$$= 298 \text{ g}$$
$$m_{HNO_3} = \text{wt fraction } HNO_3 \times \text{total wt}$$
$$= 0.32 \times 298 \text{ g} = 95 \text{ g}$$
$$n_{HNO_3} = \frac{95 \text{ g}}{63.0 \text{ g/mole}} = 1.5 \text{ moles } HNO_3$$

The solution contains 95 g, or 1.5 moles, of HNO_3. ∎

THOUGHTS ON CHEMISTRY

The requisites of a good hypothesis

THE REQUISITES OF A GOOD HYPOTHESIS,
by Robert Boyle

The Requisites of a Good *Hypothesis are:*
1. *That it be Intelligible.*
2. *That it neither Assume nor suppose anything Impossible, Unintelligible, absurd, or demonstrably False.*
3. *That it be Consistent with it self.*
4. *That it be fit and sufficient to Explicate the Phaenomena, especially the chief.*
5. *That it be, at least consistent, with the rest of the Phaenomena it particularly relate to; and do not contradict any other known Phaenomena of Nature, or manifest Physical Truth.*

The Qualityes & Conditions of an Excellent *Hypothesis are*
1. *That it be not Precarious, but have sufficient Grounds in the Nature of the Thing itself, or at least be well recommended by some Auxiliary Proofs.*
2. *That it be the* simplest *of all the Good ones we are able to frame, at least containing nothing that is superfluous or Impertinent.*
3. *That it be the* only *Hypothesis that can Explicate the Phaenomena; or at least, that dos explicate them so well.*
4. *That it enable a skilful Naturalist to foretell future Phaenomena by their Congruity or Incongruity to it; and especially the events of such Experiments as are aptly devis'd to examine it, as Things that ought or ought not, to be consequent to it.*

THE REQUISITES OF A GOOD HYPOTHESIS, by Robert Boyle, Royal Society Boyle Papers Vol. XXXVII. (Quoted from p. 134, in *Robert Boyle on Natural Philosophy*, by Marie Boas Hall, Indiana University Press, 1965.)

Significant terms defined in this chapter

atom
element
law of conservation of mass
law of definite proportions
law of multiple proportions
binary compound
molecule
polyatomic molecule
diatomic molecule
monatomic
ion
cation
anion
chemical formula
formula unit
polyatomic ion
chemical nomenclature
true mass of an atom
atomic weight
atomic mass unit

molecular weight
Avogadro's number
mole
molar weight
percentage composition
simplest formula
true formula
empirical formula
solution
aqueous solution
solvent
solute
dissociation of an ionic compound
ionization
solvation
hydration
standard solution
molarity
volume percent (for solutions)
weight percent (for solutions)

Exercises

2.1 One of the most important commercial uses of lead has been in the production of tetraethyllead, an anti-knock additive for gasoline. This compound is produced by the reaction of ethyl chloride, C_2H_5Cl, with a sodium-lead alloy. An approximate analysis of the alloy showed that it was Na_4Pb, but very detailed work revised the formula to $Na_{31}Pb_8$. The ethyl chloride is produced by the reaction of ethyl alcohol, C_2H_5OH, with hydrochloric acid, a solution of HCl in H_2O. An analysis of the tetraethyllead indicated that the molecule had the formula $Pb(C_2H_5)_4$. Further investigation of the products of the reaction showed that an ionic by-product was formed that had the formula NaCl.

List the (a) empirical and (b) true formulas for the various substances. Identify the (c) cations and (d) anions for solutions of NaCl and HCl. Which substance would be considered the (e) solvent and which the (f) solute in the hydrochloric acid solution? (g) Which substances are binary compounds?

2.2 Which statements are true? Rewrite any false statement so that it is correct.
(a) Dalton's atomic theory states that chemical combination is the union of atoms of different elements such that the ratios of the weights of the combined atoms are simple whole numbers.
(b) Even though an element may be defined as a substance composed of only one kind of atom, the molecule is the smallest particle for several elements.
(c) The molar weight is the same thing as the molecular weight.
(d) The empirical formula represents the actual number of atoms that are combined in each molecule of the compound.
(e) A solution contains only one solvent and one solute.
(f) Dalton's law of definite proportion states that if two elements combine to form more than one compound, the weights of one element that combine with a fixed weight of the other will be related to each other as ratios of small whole numbers.
(g) A molecule of oxygen, O_2, has a mass of 32.0 g.

2.3 What is the numerical value of Avogadro's number? What is the relation of Avogadro's number to the mole? How many atoms are present in a molar weight of an element that is not molecular?

2.4 Distinguish clearly between (a) atoms and molecules, (b) atoms and ions, (c) molecular weight and molar weight, (d) atomic weight and actual weight, (e) molecules and polyatomic ions, and (f) anions and cations.

2.5 What is the difference between (a) 2H and H_2, (b) C_2H_2 and C_6H_6, (c) Hg_2Cl_2 and $HgCl_2$, and (d) O, O^{2-}, O_2, $O_2{}^{2-}$, and O_3?

2.6 Name the elements and state in what atomic ratios these elements make up the following substances: (a) $Ba(OH)_2 \cdot 8H_2O$, (b) $CuBr_2 \cdot 3Cu(OH)_2$, (c) Mg_3N_2, (d) $NOBF_4$, (e) KHC_2O_4, (f) $NaMnO_4$, (g) $SnO \cdot Sn(NO_3)_2$, (h) $3ZrO_2 \cdot CO_2 \cdot H_2O$.

2.7 Calculate the average atomic weight using the following compositions of the naturally occurring element: (a) carbon, which consists of 98.89% atoms having a mass of 12.00000 amu and 1.11% of 13.00335 amu; (b) magnesium, which consists of 78.70% atoms having a mass of 23.98504 amu, 10.13% of 24.98584 amu, and 11.17% of 25.98259 amu; (c) sulfur, which consists of 95.0% atoms having a mass of 31.97207 amu, 0.76% of 32.97146 amu, 4.22% of 33.96786 amu, and 0.014% of 35.96709 amu.

2.8 What is the molar weight of the following substances: (a) H_2SO_4, (b) $Ca(OH)_2$, (c) $Ca_3(PO_4)_2$, (d) Al_2O_3, (e) $(NH_4)_2SO_4$, (f) Cl_2, (g) Fe, (h) $C_{12}H_{22}O_{11}$, (i) $KClO_3$, (j) C_2H_5OH, (k) $CuSO_4 \cdot 5H_2O$ where $\cdot 5H_2O$ means that there are five molecules of water for every $CuSO_4$?

2.9 Calculate the molar weight for each of the following substances: (a) H_3PO_4, (b) $(NH_4)_3AsO_4$, (c) Fe_3O_4, (d) $UO_2(SO_4)$, (e) $HgBr_2$, (f) NO, (g) NO_2, (h) H_2O_2, (i) $Ba(OH)_2$, (j) XeF_6, (k) $C_{12}H_{22}O_{11}$, (l) Mn^{2+}, (m) $H_2PO_4{}^-$, (n) $CO_3{}^{2-}$, and (o) N.

2.10 Calculate the number of moles in (a) 10.0 g of calcium carbonate, $CaCO_3$; (b) 14 g of iron, Fe; (c) 24.5 g of formaldehyde, H_2CO; (d) 34 g of sucrose, $C_{12}H_{22}O_{11}$; and (e) 33.5 g of acetic acid, CH_3COOH.

2.11 How many atoms are present in each sample of the substances mentioned in Exercise 2.10?

2.12 The chemical formula for pyrite is FeS_2. What is the (a) molecular weight and (b) molar weight of this substance? (c) How many moles are contained in 10.0 g of FeS_2? How many (d) cations and (e) anions are present in the 10.0 g sample?

2.13 Carbon tetrachloride, CCl_4, was formerly used as a dry cleaning fluid. What is the (a) molecular weight and (b) molar weight of this substance? (c) Calculate the mass of one molecule of CCl_4. How many (d) moles and (e) molecules of CCl_4 are present in 17.93 g of the compound? How many (f) carbon atoms and (g) chlorine atoms are present in the 17.93 g sample?

2.14 What weight in grams corresponds to (a) 0.50 mole of plaster of paris, $CaSO_4 \cdot \frac{1}{2}H_2O$; (b) 1.01 mole of quartz,

SiO_2; (c) 3 moles of quicksilver, Hg; (d) 0.42 mole of saccharin, $C_6H_4(CO)(SO_2)NH$; (e) 0.25 mole of saltpeter, KNO_3; and (f) 1×10^{-10} mole of HCl?

2.15 Determine the percent composition of (a) KClO, (b) $KClO_2$, (c) $KClO_3$, and (d) $KClO_4$.

2.16 Calculate the weight percent of phosphorus in (a) $Ca_3(PO_4)_2$, (b) $Ca(H_2PO_4)_2$, (c) $Na_5P_3O_{10}$, (d) P_4O_{10}, and (e) $N_3P_3Cl_6$.

2.17 Calculate the percent composition of (a) acetone, CH_3COCH_3; (b) alum, $K_2SO_4 \cdot Al_2(SO_4)_3 \cdot 24H_2O$; (c) alumina or corundum, Al_2O_3; (d) aspirin, $CH_3COOC_6H_4COOH$; (e) beryl, $Be_3Al_2(SiO_3)_6$; (f) carborundum, SiC; (g) cellulose, $(C_6H_{10}O_5)_n$, where n is a large number; (h) DDT, $(ClC_6H_4)_2CHCCl_3$; (i) diethyl ether, $(C_2H_5)_2O$; and (j) LSD, $C_{20}H_{25}N_3O$.

2.18 Compare the weight percentages of iron in FeO, Fe_2O_3, and Fe_3O_4. Assuming that the cost of shipping 1.00 lb of each oxide is the same, which oxide would be the most economical to ship in terms of maximum amount of iron delivered?

2.19 What weight of oxygen is contained in 5.5 g of $KClO_3$?

2.20 A 3.56 g sample of iron powder was heated in a chlorine atmosphere and 10.39 g of a dark substance presumed to be iron chloride was formed. Determine the empirical formula of the iron chloride that was formed.

2.21 Barium combines with oxygen to form barium oxide, BaO, and barium peroxide, BaO_2. (a) What is the composition of these compounds expressed in weight percent? (b) Divide the weight percent of oxygen by the weight percent of barium for each compound to determine the mass ratios of oxygen to barium in each compound. (c) Do these compounds illustrate the law of multiple proportions?

2.22 Find the empirical formula of each of the following minerals: (a) talc, which is used for talcum powder, ceramics, and laundry tubs, and contains 19.23% Mg, 29.62% Si, 42.18% O, and 8.97% OH; and (b) borax, which is used for softening water and washing clothes, and contains 12.1% Na, 11.3% B, 29.4% O, and 47.3% H_2O. Note that in both instances above, oxygen appears twice. This is the manner in which geological analyses are stated.

2.23 An organic compound containing 52.18% C, 13.04% H, and 34.78% O has an observed molecular weight of 91.6 amu. Determine (a) the empirical formula, (b) the molecular formula, and (c) the exact molecular weight of the compound.

2.24 What masses of NaCl and H_2O are present in 160 g of a 12 wt % aqueous solution of NaCl?

2.25 How would you prepare 150 g of a 6 wt % solution of KOH in water?

2.26 Sodium fluoride has a solubility of 4.22 g in 100.0 g of water at 18°C. Express the concentration in weight percent.

2.27 What weight of each solute is needed to prepare 250 ml of a 1.0M solution of (a) KCl, (b) LiBr, (c) $FeSO_4 \cdot 7H_2O$, (d) $NiCl_2$, and (e) NH_3?

2.28 Solutions containing (a) 10.0 g of Na_2SO_4, (b) 56 g of $CaCl_2$, and (c) 42.6 g of $Al(NO_3)_3$ dissolved in sufficient water to make a total volume of 1.00 liter of solution were prepared. What is the concentration expressed in molarity that should be written on the respective bottles?

2.29 How many moles of HCl are contained in (a) 437 g, (b) 2.39×10^{54} molecules, and (c) 73.5 ml of a 6.0M solution?

2.30 Which sample contains the largest number of moles of substance: (a) 50 ml of 12M HCl, considering HCl only, (b) 1.39×10^{17} molecules of H_2O_2, or (c) 42.187 g of H_2O?

2.31* The average atomic weight of copper is 63.546 amu. Atoms having masses of 62.9298 amu and 64.9278 amu contribute to this average. Calculate the fraction of each type of atom making up naturally occurring copper.

2.32* A highly purified sample of carbon tetrabromide, CBr_4, contains 96.379% bromine and 3.621% carbon by weight. Using the atomic weight of carbon as 12.0111 amu, find the exact atomic weight of bromine.

2.33* One type of artificial diamond (commonly called YAG for yttrium aluminum garnet) can be represented by the formula $Y_3Al_5O_{12}$. (a) Calculate the weight percentage composition of this compound. (b) What is the weight of yttrium present in a 200-carat YAG (1 carat = 200 mg)?

2.34* Copper is obtained from ores containing the following minerals: azurite, $Cu_3(CO_3)_2(OH)_2$; chalcocite, Cu_2S; chalcopyrite, $CuFeS_2$; covellite, CuS; cuprite, Cu_2O; and malachite, $Cu_2(CO_3)(OH)_2$. Which mineral has the highest copper content on a weight percentage basis?

2.35* Test tubes containing 1.00 g samples of (a) lime, CaO; (b) slaked lime, $Ca(OH)_2$; (c) magnesium, Mg; (d) nitroglycerine, $C_3H_5(NO_3)_3$; and (e) water, H_2O, are placed in front of you. Which contains the largest number of moles? Which contains the largest number of atoms?

2.36* A sample of an organic compound weighing 2.00 g gave 4.86 g of CO_2 and 2.03 g of H_2O upon quantitative combustion. If the compound is known to contain only C, H, and O, what is its empirical formula?

2.37* Calculate the atomic weight of a metal that forms an oxide having the empirical formula M_2O_3 and contains 68.4% of the metal by weight. Identify the metal.

2.38* Tungsten forms a series of chlorides containing 46.4 wt %, 50.9 wt %, 56.5 wt %, and 72.2 wt % of tungsten, respectively. (a) Determine the empirical formulas of these compounds. (b) Using a basis of one gram of tungsten, calculate the weight of chlorine present in each compound. (c) Show that this series of chlorides illustrates the law of multiple proportions.

2.39* A solution that is 37.2 wt % HCl has a density of 1.19 g/ml. What weight of HCl is contained in 30.0 ml of this solution?

2.40* A 40.0 ml sample of diethyl ether, $(C_2H_5)_2O$, is dissolved in enough methyl alcohol, CH_3OH, to make 250. ml of solution. If the density of the ether is 0.714 g/ml, what is the molarity of this solution?

2.41** A 10.0 g sample of an orange powder was piled on an asbestos sheet and ignited by a piece of burning magnesium metal. A chemical reaction took place that produced a miniature volcano. As sparks flew, a fluffy green powder was formed and steam was generated. Find the formula of (a) the orange powder if it contains 11.1% N, 3.2% H, 41.3% Cr, and 44.4% O and (b) the green powder if it contains 68.4% Cr and 31.6% O. (c) How many moles of orange powder were used? (d) How many atoms were present in the sample of orange powder? (e) If the density of the orange powder is 2.15 g/ml, what is the volume of the powder used? If 6.03 g of the green powder and 2.86 g of water were formed, (f) identify the third product of the reaction and (g) determine the mass of this third product released.

In this chapter we first briefly review the properties of gases and discuss some practical aspects of measuring pressure. Next, the kinetic-molecular theory of gases is presented, so that subsequently we can show how the theory relates to each gas law. Boyle's law, Charles' law, and the combined gas law, all of which deal with pressure, volume, and temperature changes in a fixed quantity of gas, are given in the following sections. Through Avogadro's law, Gay-Lussac's law, and the ideal gas law, the amount of gas involved is introduced into the pressure–volume–temperature relationships. Finally, Dalton's law of partial pressures, effusion and diffusion, and deviations from the gas laws are discussed.

The word "gas" was created in 1662 to describe a specific form of matter. Johann van Helmont, a Belgian physician, derived the word from the Greek word "chaos," which referred to the original matter from which the Earth was formed. (Some have suggested that he chose "chaos" because his equipment kept blowing up when he tried to make gases.)

Forty years later, the first of the "gas laws" describing the behavior of gases was discovered by Robert Boyle. But after Boyle, more than one hundred years passed before a great period of discovery transformed the young science of chemistry. In the mid-eighteenth century, individual gases were recognized as different from air, which was discovered to contain several different gases.

In 1776, Antoine Lavoisier, basing his conclusions partly on the discoveries of others, explained combustion as the chemical combination of oxygen from the air with combustible materials. Once the role of oxygen in combustion was understood, many chemical changes could be explained, such as the rusting of iron and the burning of hydrogen to give water. Lavoisier went on to sort out the confusing nomenclature left over from the days of the alchemists, and to list and clearly name the 33 elements known at that time.

Once the identity of gases was understood, their physical behavior was subjected to careful experimentation. Four more "gas laws" were discovered between 1787 and 1829. During this period Dalton also published his atomic theory. In the mid-nineteenth century—which marked the beginning of physical chemistry—the kinetic-molecular theory of gases (the theory of moving molecules) was presented in mathematical form. All of the gas laws were explained by kinetic-molecular theory, and all of the postulates of kinetic-molecular theory were in harmony with the atomic theory. Both in terms of a model of moving particles, which can be visualized, and in terms of mathematics, the kinetic-molecular theory explained what had already been observed about gas behavior and predicted much of what has since been observed. This theory is still providing new information about the gaseous state.

The nature of gases

3.1 General properties of gases

Most familiar gases are colorless and odorless—the oxygen and nitrogen of the atmosphere, the bubbles of carbon dioxide that rise in a glass of carbonated beverage, the hydrogen or helium gas that is used to fill balloons, and the poisonous carbon monoxide that occurs, along with other gases, in the exhaust fumes of automobiles. A few gases are colored and are thus quite visible; for example, nitrogen dioxide (NO_2) is red-brown, and iodine vapor has a beautiful violet color. And all substances that have odors can exist as gases or vapors, for a material must reach the nostrils as a gas in order to be perceived as having an odor.

The names and formulas of some gases are given in Table 3.1. "Gas" refers to a substance that is entirely gaseous at ordinary temperatures and pressures. "Vapor" is reserved for a gas that evaporates from a material that is usually solid or liquid at ordinary temperatures. The term "volatile" is applied to a substance that forms a vapor very readily. For example, ether is a volatile liquid.

All gases expand uniformly to occupy whatever space is available, whether large or small (see Table 1.1). Think of how quickly the smell of frying onions can fill a large house. Because molecules in the

TABLE 3.1
Gases

This is a list of some compounds that are gases at ordinary temperatures and pressures. Pranksters often ask for a harmless gas with a very bad odor, but there is no such thing—all evil-smelling gases are poisonous. Some odorless gases, such as CO, are also poisonous, of course.

CO	Carbon monoxide	Odorless, poisonous
CO_2	Carbon dioxide	Odorless, nonpoisonous
NH_3	Ammonia	Pungent odor, poisonous
PH_3	Phosphine	Terrible odor, very poisonous
CH_4	Methane	Odorless, flammable
C_2H_2	Acetylene	Mild odor, flammable
HCl	Hydrogen chloride	Choking odor, harmful and poisonous
SO_2	Sulfur dioxide	Suffocating odor, irritating to eyes
NO_2	Nitrogen dioxide	Red-brown, irritating odor, very poisonous
H_2S	Hydrogen sulfide	Rotten egg odor, very poisonous

Priestley, Lavoisier, and the phlogiston theory

Before chemistry could move forward, it had to be purged of the phlogiston theory. Phlogiston was thought to be present in every material that would burn. The process of burning was the release into the air of phlogiston, a hypothetical substance that could not be isolated.

The phlogiston theory, originated by Johann Becher in 1669 and later popularized by George Ernest Stahl, dominated chemistry for almost a hundred years because in some ways it made sense. In fact, it was the exact opposite of the correct interpretation of combustion. But it did make a positive contribution in stating that the calcination of metals—the heating of a metal to give a powdery, nonmetallic substance, a calx—is the same process as combustion. According to phlogiston theory,

$$\text{combustible substance} \xrightarrow{\text{burning}} \text{ashes} + \text{phlogiston}$$

$$\text{metal} \xrightarrow{\text{heat}} \text{metal ash (calx)} + \text{phlogiston}$$

However, phlogistonists had to struggle to explain why metal calxes were heavier than the metals from which they were formed. Some theorists resorted to proposing that phlogiston had a negative weight.

The discovery of oxygen, for which Carl Scheele, a Swedish apothecary, and Joseph Priestley share the credit, ultimately killed the phlogiston theory. Joseph Priestley was an English clergyman, a multi-talented man who developed unorthodox religious and political ideas, with a heavy emphasis on individualism—ideas for which he was forced to flee England to the United States at the time of the French Revolution. While in England, Priestley founded a school, wrote books on English grammar, was elected to the Royal Society on the basis of his experiments on electricity, and investigated "airs," or gases.

On August 1, 1774, Priestley heated the red powder we know as mercury(II) oxide and obtained a colorless gas in which substances burned more vigorously than in air. He assumed that the gas was air which had lost all of its phlogiston, and was therefore more ready to absorb phlogiston than ordinary air. When Priestley described this experience to Antoine Lavoisier, a brilliant Frenchman whose accomplishments and interests were as diverse as Priestley's Lavoisier immediately grasped the significance of "dephlogisticated air."

Lavoisier already knew that the white powder formed when phosphorus burns is heavier than phosphorus and that the metallic lead made from lead calx weighs less than the lead calx. In 1776 he named "dephlogisticated air" oxygen and identified it as a component of the atmosphere that combines with burning substances. He explained that oxygen is the gas formed when calxes, which he called oxides, give metals upon heating. Ironically, Priestley never accepted Lavoisier's explanation of combustion and calcination; he remained a staunch defender of the phlogiston theory long after most chemists had abandoned it. Perhaps for this reason, Lavoisier did not credit Priestley for his work with oxygen. Lavoisier, often called the father of modern chemistry, ". . . discovered no new body—no new property—no natural phenomenon previously unknown. . . . His merit, his immortal glory consisted in this—that he infused into the body of science a new spirit" (Justus von Liebig in *Chemische Briefe*). Antoine Lavoisier, who had improved French gunpowder, run a model farm, helped to develop the metric system, served in the government assembly, and founded modern chemistry by explaining combustion and reforming nomenclature, was sent on May 7, 1794 to the guillotine by the leaders of the French Revolution because he had also been a tax collector. On that day a scientific colleague of Lavoisier's said, "It took but a moment to cut off that head; perhaps a hundred years will be required to produce another like it."

gaseous state are widely separated, gases are much lighter than liquids or solids. Yet gases have mass, as is evident from the force of the wind on a stormy day. A balloon filled with hydrogen or helium rises in the air because these gases are lighter than air. The densities of some common gases are listed in Table 3.2.

Gases expand or contract with changes in temperature and can be compressed by increased pressure. Therefore, the variables of temperature, pressure, and volume must always be specified when handling gases and discussing their behavior. With cooling and, in some cases, with increased pressure, all gases can ultimately be condensed to liquids. The condensation of a gas to a liquid is called **liquefaction.**

3.2 Units of pressure

Pressure is force per unit area. Uniformity in the choice of units for the measurement of gas pressures seems particularly far off. Each branch of science and engineering has its preferences.

In chemistry, millimeters of mercury is often the unit given. For example, 20 mm Hg, 760 mm Hg. However, this unit has caused some distress because it makes pressure look like a function of length, which it is not. To get around this problem the torr was defined as a pressure unit. (It was named for Evangelista Torricelli, Galileo's assistant, who invented the mercury barometer.) One **torr** is the pressure exerted by a column of mercury 1 mm high. This means that, for example, 20 mm Hg = 20 torr.

The average pressure of the atmosphere at sea level is 760 torr. Therefore, the unit of pressure called an **atmosphere** is equal to 760 torr; it is often used to express high pressure, for example, 25 atm (meaning 25 × 760 torr). The bar is another unit sometimes used for high pressures (see Table 3.3).

A common unit that gives pressure its dimensions of force per unit area is pounds per square inch. When, for example, tire pressure is given in pounds per square inch, say 30 lb/sq inch, or 30 psi, the meaning is 30 lb/sq inch more than atmospheric pressure. Pressure given in this way, that is, pressure above atmospheric, is sometimes called "gauge pressure," abbreviated psig.

The unit of pressure in the SI system is the pascal, abbreviated Pa (see Appendix B). One atmosphere is equal to 101,325 Pa.

3.3 Measuring pressure

The pressure of the atmosphere is measured by instruments called **barometers.** A simple barometer consists of a glass tube, closed at one end, filled with mercury, and inverted in a dish of mercury (Figure 3.1). The tendency of the liquid mercury to run out of the tube into the dish is opposed by the pressure exerted by the atmosphere on the surface of the mercury in the dish. The mercury levels in the tube and the dish quickly adjust so that the pressure exerted by the mercury column is balanced by the pressure of the atmosphere. At sea level,

1 atm = 760 torr

TABLE 3.2

Densities of some gases
The values given are the densities at 0°C and atmospheric pressure at sea level.

Gas	Density (g/liter)
Hydrogen (H_2)	0.08987
Helium (He)	0.179
Nitrogen (N_2)	1.256
Carbon monoxide (CO)	1.256
Air	1.297
Oxygen (O_2)	1.429
Argon (Ar)	1.783
Carbon dioxide (CO_2)	1.965
Chlorine (Cl_2)	3.165
Sulfur hexafluoride (SF_6)	6.602 (20°C)

TABLE 3.3

Units of pressure
The units given in color are used most often in chemistry and in this book.

mm Hg	millimeters of mercury
torr	torr
atm	atmospheres
bar	bars
lb/sq inch	pounds per square inch
dynes/cm²	dynes per square centimeter
Pa	pascals

The average pressure of the atmosphere at sea level is equal to

760 mm Hg

760 torr

1 atm

1.013 bar

14.7 lb/sq inch

1.013 × 10⁶ dynes/cm²

101,325 Pa

FIGURE 3.1

Mercury barometer. *The height of the column of mercury in millimeters is read as the atmospheric pressure. The space above the mercury is not a complete vacuum, for a little mercury evaporates into it. But at ordinary temperatures this vapor pressure is so small that it can be disregarded. For example, at 20°C it would depress the mercury column only 0.0012 mm. Other liquids can be used to make barometers, the column height varying with the density of the liquid. Mercury is used most often because the column is of a convenient height. Water would give a column 34 ft high—which explains why 34 ft is the maximum depth for a well from which water is drawn by a suction pump.*

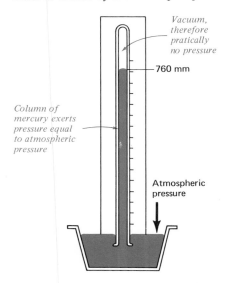

Vacuum, therefore pratically no pressure

760 mm

Column of mercury exerts pressure equal to atmospheric pressure

Atmospheric pressure

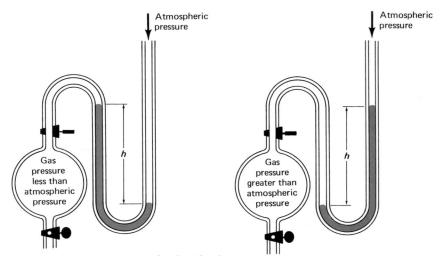

Gas pressure = atmospheric pressure (torr) − *h*(torr)   Gas pressure = *h*(torr) + atmospheric pressure(torr)

FIGURE 3.2

Manometer. *The difference in the height of the mercury columns in the two arms of the tube is measured, and atmospheric pressure is read from a standard barometer. The unknown pressure in the flask or whatever gas system is attached to the manometer is then calculated as shown.*

the difference in the height of the two levels averages 760 mm. On a high mountain, it might be only 500 mm.

Another type of barometer, an aneroid barometer, has a container with flexible walls like those of an accordion. The walls move up and down with changes in atmospheric pressure and a mechanical connection causes the movement to be recorded on a calibrated dial.

Manometers measure gas pressures other than that of the atmosphere by comparing an unknown pressure with a known pressure. A manometer that uses atmospheric pressure as the known pressure is shown in Figure 3.2. To measure very low pressures sensitive manometers of more complicated design than shown are needed.

Low pressures are used in many industrial operations, such as in the refining of crude metals and in the degassing of molten steel, and also in a wide variety of laboratory experiments. The regions of pressure are referred to as follows:

Low vacuum	760 to 25 torr
Medium vacuum	25 to 1×10^{-3} torr
High vacuum	1×10^{-3} to 1×10^{-6} torr
Very high vacuum	1×10^{-6} to 1×10^{-9} torr
Ultra-high vacuum	1×10^{-10} torr and beyond

3.4 Kinetic-molecular theory of gases

The kinetic-molecular theory is one of the outstanding achievements of science. The theory states that molecules in the gaseous state are continuously moving about, colliding with each other and with the walls of their container. The theory has been verified in many ways, both by experiment and by mathematics. (Table 1.1 presents some of the observations about the properties of gases that led to the kinetic-molecular theory.)

The following five postulates summarize the kinetic-molecular theory of gases:

1. Molecules in the gaseous state are relatively far apart and, in comparison with their size, the spaces between them are large.

A substance in the gaseous state occupies a much larger volume than the same amount of that substance occupies in the liquid or solid state. For example, when a given weight of liquid water at 100°C changes to steam at the same temperature, the volume of the steam is 1600 times greater than the volume of the water. The same number of water molecules have spread out over a much larger space. Gases can be compressed easily because the distances between the molecules decrease when pressure is applied. The size of the molecules of a gas is so small compared to the size of the spaces between them that the volume occupied by the molecules is, for many purposes, considered negligible. For the same reason, forces of attraction or repulsion between the molecules of a gas can, under many conditions, be considered negligible.

2. The molecules of a gas are in constant motion.

In a blown-up balloon, it is the impact of multitudes of molecules in the gas that pushes back the sides of the balloon and keeps them taut (Figure 3.3). According to kinetic-molecular theory, molecules in the gaseous state are constantly moving in random, straight lines. The molecules change direction only when they collide with each other or with the walls of their container. The gas pressure is the result of impacts with the walls. The pressure does not seem to fluctuate with each blow because the number of gas molecules is very large, the molecules move rapidly, and the impacts are so frequent that they cannot be detected individually. A molecule of a gas at 25°C and 1 atm undergoes about 1×10^9 collisions per second. Although the spaces between gas molecules are great, gases fill any container by constantly moving about through its entire volume.

3. The average speed of gas molecules increases with rising temperature and decreases with falling temperature.

If a gas is confined in a vessel of fixed volume, the gas pressure increases as the temperature is raised and decreases as the temperature is lowered. Alternatively, if the gas is confined under constant pressure (as in a cylinder with a movable piston), the volume varies directly with the temperature. The kinetic-molecular theory explains these phenomena as follows: As the temperature rises, the molecules move faster (Figure 3.4) and hit the fixed walls of a container harder and more often, thereby increasing the pressure. (The total pressure is directly proportional to the number of particles hitting the wall per second per unit area.) If the temperature falls, the molecules hit the fixed walls with less force and less frequently, and the pressure decreases. In a vessel that can expand, such as a balloon or a cylinder with a movable piston, the volume increases with temperature. The pressure stays the same, because even though the molecules hit the wall with greater force, there are fewer impacts per unit area per sec-

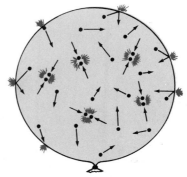
The pressure on the container walls is the sum of the force of the individual impacts.

FIGURE 3.3
Kinetic-molecular theory. *Gas molecules are far apart, constantly moving in straight lines, and colliding with each other and the walls of the container.*

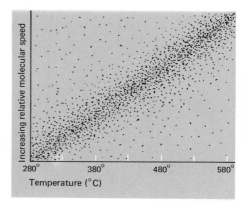

FIGURE 3.4
Temperature vs. molecular speed. *As the temperature of the gas rises, the average speed at which the molecules move also rises. Each dot represents an individual molecule.*

THE GASEOUS STATE [3.4]

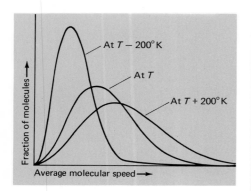

FIGURE 3.5

Distribution of molecular speeds.
Relatively few molecules have the highest or the lowest speeds—most fall in the middle. These curves are for argon gas; T = 300°K.

ond. To maintain constant pressure with decreasing temperature, the volume must decrease, thereby allowing more frequent collisions with the wall per unit area.

Consider a practical example: Why does a cake fall if the oven door is opened too soon? As the colder air rushes into the oven, the velocity of the gas molecules trapped inside the cake batter decreases, and so the pressure inside the gas bubbles decreases. Since the batter has not yet cooked long enough to be set, the pressure no longer holds it up, and the cake collapses.

Note that not all the molecules in a gas sample are flying about at the same speed (Figure 3.5). Some go much faster than others. Collisions between molecules continually cause exchanges in energy, slowing some molecules down and speeding up others. It is the *average* speed of all the molecules in a gas that increases as the temperature is raised.

4. At the same temperature, the molecules of different gases, whether they are light or heavy, have the same average kinetic energy.

The energy of a moving body—kinetic energy, E_k—is a function of the body's mass, m, and its velocity, v.

$$E_k = \tfrac{1}{2}mv^2$$

Temperature is a measure of the average kinetic energy of the molecules in a gas. Therefore, molecules of different gases at the same temperature must have the same average kinetic energy. In other words, at any given temperature, the average kinetic energy of any gas is a constant. What this means is that molecules with larger mass must, on the average, move more slowly at a given temperature than molecules with smaller mass. In short, the larger the m, the smaller the v. As a consequence, a light gas released into a container will spread throughout the container more quickly than will a heavy gas.

5. Collisions of gas molecules with each other or with the walls of a container are perfectly elastic.

When gas molecules collide with each other or with the walls of a container, they bounce off with no loss in energy—they undergo frictionless, or perfectly elastic, collisions. Energy may be exchanged between colliding molecules, but the *total* energy of the molecules that have collided is the same after the collision as it was before. If this were not so, gas molecules would gradually lose energy, slow down, and eventually come to a halt, just as a bouncing tennis ball does. We can see that this does not happen. Gas molecules do not eventually collect at the bottom of a container, the atmosphere remains suspended in the Earth's gravitational field, and the pressure in a leak-free container of gas does not decrease, no matter how long the container stands.

3.5 Ideal vs. real gases

The kinetic-molecular theory provides a model of how gases *ideally* behave—the molecules are point masses with negligible volume and move in random straight-line paths. No forces of attraction or repulsion are exerted between the molecules or between the molecules and their surroundings, and the molecules bounce off of each other or the walls of their container with no loss of kinetic energy. Experimentally and in mathematical theory, molecules with all of these properties would exactly fulfill the predictions of the gas laws, which you will be learning in later sections of this chapter. A gas that perfectly obeys the gas laws is called an **ideal gas.**

Real gas molecules always have real volume, of course, and they also have forces of attraction and repulsion toward each other. Gas molecules of small mass that are far apart, have behavior that is closest to what theory predicts for an ideal gas. As gases approach liquefaction and the molecules come closer together because of volume and temperature decreases, or pressure increases, deviations from ideality become greater. Deviations from ideality are discussed further at the end of this chapter (Section 3.17). In the examples and problems in the following sections, the gases are treated as ideal.

Volume, pressure, and temperature relationships

3.6 Volume vs. pressure: Boyle's law

The volume of a gas decreases when the pressure on the gas increases, and vice versa. If the pressure on a gas is doubled while the temperature remains unchanged, the gas volume decreases to one-half the original volume; if the pressure is tripled, the gas volume decreases to one-third the original volume; and so on. This relationship was first described by Robert Boyle in 1662. Boyle's experimental apparatus is shown in Figure 3.6.

Boyle's law is stated as follows: *At constant temperature, the volume of a given mass of gas is inversely proportional to the pressure upon the gas.* The temperature must be constant; otherwise, changing temperature would cause additional changes in volume. Stated mathematically, Boyle's law is

At constant temperature,

volume of the gas → $V = \text{constant} \times \dfrac{1}{P}$ ← *pressure of the gas* (1)

an inverse proportion

Or, stated another way

$PV = \text{constant}$ (2)

These relationships hold for any given mass of a gas at a fixed temperature, whether it is pure or a mixture.

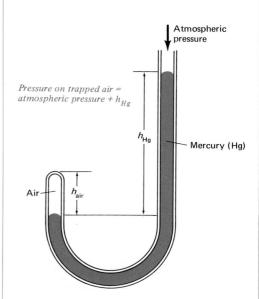

FIGURE 3.6

Boyle's apparatus for measuring volume changes with changing pressure. *The pressure on the trapped air is varied by adding mercury to the open side of the J tube. Atmospheric pressure is read from a standard mercury barometer. The height h_{air} is proportional to the volume of the trapped gas, and the total pressure on the trapped gas is atmospheric pressure plus the pressure of the mercury column. Boyle said that the particles of air behaved like coiled springs.*

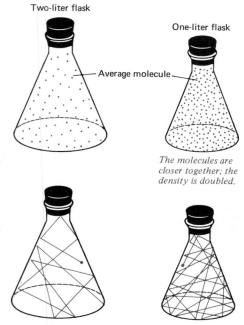

Two-liter flask

One-liter flask

Average molecule

The molecules are closer together; the density is doubled.

The average molecule hits the wall twice as often. The total number of impacts with the wall is doubled and the pressure is doubled.

FIGURE 3.7

Effect of volume change on an equal number of gas molecules at constant temperature. *For equal amounts of the same gas at the same temperature, the molecules moving at the same speed collide with the walls twice as often at one-half the volume, and so on.*

FIGURE 3.8

Boyle's law. *In (a) volume is plotted versus pressure for a gas at constant temperature. If, as in (b), volume is plotted versus the reciprocal of the pressure (1/P), a straight line is obtained with slope equal to the constant, PV = constant. Plots like these are always obtained when the two plotted quantities are inversely proportional to each other.*

From kinetic theory, we know that the space between gas molecules becomes smaller when gas volume decreases (Figure 3.7). With less space to move around in, but no change in the speed with which they are moving (constant temperature), the gas molecules are bound to collide with the walls more often, leading to greater pressure.

This relationship of pressure to volume is plotted in Figure 3.8. If the pressure or volume of a given sample of gas is changed without changing the temperature, the new pressure or volume can be calculated from the relationship

$$P_1 V_1 = P_2 V_2 \qquad (3)$$

indicate initial values — *indicate new values*

Equation (3) results because P_1V_1 and P_2V_2 are equal to the same constant.

EXAMPLE 3.1

■ A common application of Boyle's law is in finding the volume of an irregularly shaped container by (a) evacuating the container, (b) connecting to it a container holding a gas of known volume at a known pressure, (c) allowing the gas to fill both containers, (d) measuring the final pressure of the combined system, and (e) using Boyle's law to calculate the unknown volume (see Figure 3.9). A 515 ml container holding air at 735 torr was connected to an evacuated container to be calibrated, and the stopcock was opened. What is the volume of the second container if the final pressure was measured at 432 torr?

The original conditions of the gas were

$$P_1 = 735 \text{ torr} \qquad V_1 = 515 \text{ ml}$$

and the final conditions were

$$P_2 = 432 \text{ torr} \qquad V_2 = V + 515 \text{ ml}$$

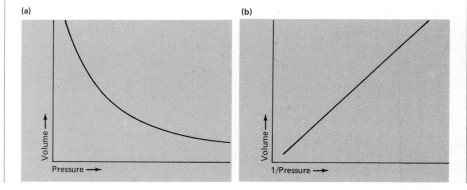

(a) Volume vs. Pressure

(b) Volume vs. 1/Pressure

where V is the unknown volume. First, solve Equation (3) for V_2:

$$P_1V_1 = P_2V_2$$

$$V_2 = \frac{P_1V_1}{P_2}$$

Then substitute the values given above and solve for V.

$$V + 515 \text{ ml} = \frac{(735 \text{ torr})(515 \text{ ml})}{432 \text{ torr}}$$

$$V = 876 \text{ ml} - 515 \text{ ml} = 361 \text{ ml}$$

The volume of the second container is 361 ml. ∎

3.7 Volume vs. temperature: Charles' law

As a gas is heated its volume becomes larger, and as a gas is cooled its volume becomes smaller. Jacques Charles, a French scientist, found in 1787 that at constant pressure, all gases expand or contract in the same way when the temperature is raised or lowered, and that the change in volume is quite uniform over large ranges of temperature. At constant pressure, for each Celsius degree of change in temperature, the volume of any gas changes by 1/273 of the volume that it has at 0°C. For example, the volume increase is the same whether the temperature change is from 20°C to 21°C or from 85°C to 86°C.

A plot of temperature versus volume at constant pressure for an ideal gas is given in Figure 3.10. The dashed portion of the plot illustrates a "logical" conclusion from Charles' observations: If a gas were cooled to −273.15°C, it would disappear entirely. This, of course, does not happen. Charles' law becomes a less and less accurate description of gas behavior as the gas molecules get closer together. With decreasing volume, the force of attraction between molecules increases and the real volume of the molecules occupies a larger percentage of the total volume—that is, the gas becomes less ideal. Be-

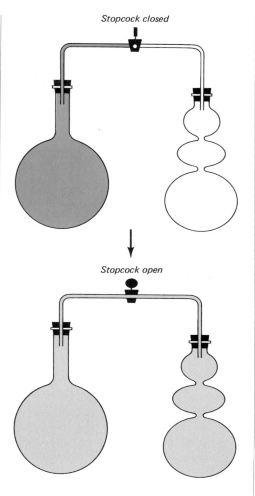

FIGURE 3.9
Example 3.1: *Irregularly shaped containers.*

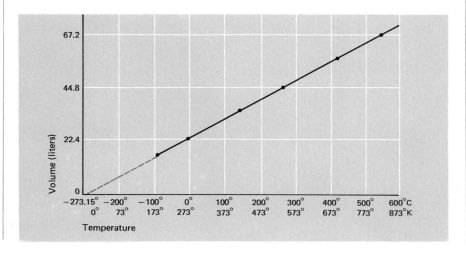

FIGURE 3.10
Charles' law. *Volume is plotted versus temperature at constant pressure. This graph shows the variation in the volume of 1 mole of an ideal gas with changing temperature. The equivalent temperatures on the Celsius and Kelvin scales are shown. Note that when the temperature doubles (2 × 273°K) the volume occupied by 1 mole doubles.*

Charles' law: At constant P, V is directly proportional to T

fore $-273.15°C$ is reached, the gas molecules get close enough together for the force of attraction between them to overcome the energy of their random motion; as a result, the gas liquefies or solidifies.

Absolute zero, $-273.15°C$, is the lowest possible temperature. It has been approached as closely as $-273.148°C$, but never attained. About 100 years after Charles formulated his law, the British physicist Lord Kelvin hit upon the idea of an **absolute temperature scale**—a scale that takes absolute zero as its zero point. The Kelvin scale (see Figure 1.3) is the absolute temperature scale based on the Celsius scale. (There are other absolute temperature scales; the only requirement is that on an absolute scale the zero point is absolute zero.) Kelvin's scale simplified the statement of **Charles' law:** *At constant pressure, the volume of a given mass of gas is directly proportional to the absolute temperature.* Mathematically, Charles' law takes the following forms:

At constant pressure,

$$\underbrace{V = \text{constant} \times \overset{\text{temperature of the gas}}{T}}_{\text{a direct proportion}} \quad \text{or} \quad \frac{V}{T} = \text{constant} \qquad (4)$$

If the volume or temperature of a given mass of gas is changed at constant pressure, the following expression relates the initial to the final conditions:

$$\frac{V_1}{T_1} = \frac{V_2}{T_2} \qquad (5)$$

EXAMPLE 3.2

■ A cylinder with a movable piston is filled at 24°C with a gas that occupies 36.2 cm^3. If the maximum capacity of the cylinder is 65.2 cm^3, what is the highest temperature to which the cylinder can be heated at constant pressure without having the piston come out?

The original conditions of the gas are

$$V_1 = 36.2\ \text{cm}^3 \qquad T_1 = 24° + 273°\text{K} = 297°\text{K}$$

and the final conditions are

$$V_2 = 65.2\ \text{cm}^3 \qquad T_2 = ?$$

Solving Charles' law for T_2 and substituting the above values gives

$$T_2 = (T_1)\left(\frac{V_2}{V_1}\right) = (297°\text{K})\left(\frac{65.2\ \cancel{\text{cm}^3}}{36.2\ \cancel{\text{cm}^3}}\right) = 535°\text{K}$$

The highest temperature to which the gas in the cylinder can be heated is 535°K or 262°C. Let's apply a quick common sense check of our solution to make sure that we have not made an error: We are increasing the volume of the gas in this problem. This, according to Charles' law, means that the temperature must increase, so our answer of 262°C seems reasonable. ■

3.8 Standard temperature and pressure (STP)

STP = 0°C (273°K),
760 torr (1 atm)

Since the volume of a gas changes with both pressure and temperature changes, stating only the volume of a gas is not sufficient to indicate the quantity of gas present. The temperature and pressure at which the volume was measured must also be given. For example, 4 liters of hydrogen at 760 torr and 30°C is about three times as much hydrogen as 4 liters of hydrogen at 380 torr and 200°C (see Example 3.3).

Combined gas law:
For a fixed amount of gas,
$$\frac{P_1V_1}{T_1} = \frac{P_2V_2}{T_2}$$

To simplify comparisons among quantities of gases, scientists have agreed always to state the volume that a sample of gas would occupy at a specified temperature and pressure. The **standard temperature and pressure** (abbreviated STP) universally used for this purpose are 0°C (273°K) and 760 torr.

The volume of a gas is often reported as, for example, 25 liters (STP). When the pressure and temperature are not given with a gas volume, you can assume that the conditions were STP. Measurements of gas volumes need not, of course, always be made at STP conditions. But the volume measured under any conditions of temperature and pressure may be converted to the value at STP by the combined expression of Boyle's and Charles' laws given in the next section.

3.9 P, V, and T changes in a fixed amount of gas

The relationships among pressure, volume, and temperature expressed in Boyle's law and Charles' law can be combined mathematically. The resulting expression is commonly called the **combined gas law:**

For a fixed amount of gas

$$\frac{P_1V_1}{T_1} = \frac{P_2V_2}{T_2} \qquad (6)$$

This equation applies to any changes in the variables P, V, and T for the same amount of a gas. Given any five of the quantities in Equation (6) the sixth may be calculated. When one variable remains unchanged ($P_1 = P_2$, or $V_1 = V_2$, or $T_1 = T_2$), that variable can be cancelled from both sides of the equation.

To determine the value of one quantity in Equation (6), first solve the equation for the quantity. For example, suppose you know the initial conditions P_1, V_1, and T_1 and the final conditions P_2 and T_2, and you wish to calculate V_2. First determine that

$$V_2 = \frac{P_1}{P_2} \times \frac{T_2}{T_1} \times V_1$$

Then insert the numerical values and solve for V_2.

EXAMPLE 3.3

◼ In Section 3.8 we stated that a sample of hydrogen having a volume of 4.0 liters at 760 torr and 30.°C contains about three times as much hydrogen as a sample having a volume of 4.0 liters at 380 torr and 200.°C. Prove this statement by changing the pressure and temperature conditions of the first sample (sample 1) to those of the second sample (sample 2) and then comparing the volumes that result.

The initial conditions for sample 1 are

$$P_1 = 760 \text{ torr} \qquad T_1 = 30.° + 273°K = 303°K \qquad V_1 = 4.0 \text{ liters}$$

and the final conditions are

$$P_2 = 380 \text{ torr} \qquad T_2 = 200.° + 273°K = 473°K \qquad V_2 = ?$$

Solving the combined gas law, Equation (6), for V_2 and substituting the above values gives

$$V_2 = (V_1) \left(\frac{P_1}{P_2}\right)\left(\frac{T_2}{T_1}\right)$$

$$= (4.0 \text{ liters}) \left(\frac{760 \text{ torr}}{380 \text{ torr}}\right)\left(\frac{473°K}{303°K}\right) = 12.5 \text{ liters}$$

The volume of sample 1 under the same conditions as sample 2 is 12.5 liters. Thus we have

$$\frac{(12.5 \text{ liters})}{(4.0 \text{ liters})} = 3.1$$

a little over three times as much hydrogen. Even in this more complicated problem, we can apply the common sense check of our answer for V_2: The factor P_1/P_2 should be greater than unity because we are decreasing the pressure on the hydrogen which, according to Boyle's law, means the volume must increase. And the factor T_2/T_1 should be greater than unity because we are increasing the temperature on the gas which, according to Charles' law, means the volume must increase. And indeed, our factors of 760/380 and 473/303, both greater than unity, do seem reasonable. ◼

Mass, molecular, and
molar relationships

3.10 Combining volumes of gases: Gay-Lussac's law

Joseph Gay-Lussac was a French contemporary of John Dalton. After carefully studying reactions involving gases, he established in 1808 the following law, known as **Gay-Lussac's law of combining volumes:** *When gases react or gaseous products are formed, the ratios of the volumes of the gases involved, measured at the same temperature and pressure, are small whole numbers.*

Consider, for example, the reaction between gaseous hydrogen and gaseous chlorine to form gaseous hydrogen chloride. When all measurements are made at the same temperature and pressure, equal volumes of hydrogen (H_2) and chlorine (Cl_2) react and the volume of the hydrogen chloride formed is twice the volume of the hydrogen or the volume of the chlorine:

Hydrogen (H_2) + chlorine (Cl_2) ⟶ hydrogen chloride (2HCl)
1 volume   *1 volume*   *2 volumes*

Similarly, if the water produced in the reaction between hydrogen and oxygen is maintained as steam, it is found that 2 volumes of steam require 2 volumes of hydrogen and 1 volume of oxygen:

Hydrogen ($2H_2$) + oxygen (O_2) ⟶ steam ($2H_2O$)
2 volumes   *1 volume*   *2 volumes*

Note that the relative gas volumes are the same as the coefficients of the gaseous compounds in the equations.

The simple whole-number ratios of gaseous volume in gas reactions, like the whole-number ratios of the law of multiple proportions, reflect the existence of atoms and molecules. But Gay-Lussac's contemporaries had not clearly grasped the idea of atoms and molecules. As a result, Gay-Lussac's law remained an unexplained puzzle to most chemists for the next 50 years, even though an Italian scientist, Amadeus Avogadro, had suggested the correct solution in 1811.

3.11 Equal volumes of gases: Avogadro's law

What is now known as **Avogadro's law** can be stated as follows: *Equal volumes of gases, measured under the same conditions of temperature and pressure, contain equal numbers of molecules.* Avogadro was the first to recognize that not only do atoms of different elements combine, but that atoms of the same element can combine to form molecules. On the basis of Avogadro's law, if equal volumes of hydrogen

> Avogadro's law:
> Equal volumes of gases, measured at the same T and P, contain equal numbers of molecules

TABLE 3.4

Gay-Lussac's law of combining volumes and Avogadro's law

The coefficients in an equation indicate the relative numbers of molecules involved in a reaction and, on the basis of Avogadro's law, the relative volumes of the reactants and products that are gaseous.

(1)	H_2	+	Cl_2	$\longrightarrow$	2HCl
	hydrogen		*chlorine*		*hydrogen chloride*
	1 volume		*1 volume*		*2 volumes*
(2)	$2H_2$	+	O_2	$\xrightarrow{100°C}$	$2H_2O$ (steam)
	hydrogen		*oxygen*		*water*
	2 volumes		*1 volume*		*2 volumes*
(3)	$3H_2$	+	N_2	$\longrightarrow$	$2NH_3$
	hydrogen		*nitrogen*		*ammonia*
	3 volumes		*1 volume*		*2 volumes*
(4)	P_4	+	$6Cl_2$	$\xrightarrow[\text{above } 280°C]{}$	$4PCl_3$
	phosphorus		*chlorine*		*phosphorus(III) chloride*
	1 volume		*6 volumes*		*4 volumes*

and chlorine (reaction 1, Table 3.4), react to form hydrogen chloride, then equal numbers of molecules of each element must be taking part in the reaction. From the relative volumes of hydrogen chloride, hydrogen, and chlorine, another point is clear: The number of molecules of hydrogen chloride formed is twice the number of hydrogen molecules or of chlorine molecules taking part in the reaction. In simplest terms, one molecule of hydrogen must react with one molecule of chlorine to produce two molecules of hydrogen chloride. This can happen only if each hydrogen molecule and each chlorine molecule divides into two (Figure 3.11).

Additional evidence is necessary to determine whether the hydrogen molecule is H_2, H_4, H_6, or contains some other multiple of 2 of hydrogen atoms. Because no case has been found in which the volume of a gaseous product formed from hydrogen is more than twice the volume of the hydrogen used, and because modern determinations of atomic and molecular weight support it, there is now no doubt that the correct formula of hydrogen gas is H_2. The same reasoning can be used to show that the formula of chlorine is Cl_2, so that the correct equation for the reaction of hydrogen and chlorine is that given in Table 3.4, reaction 1.

In a similar way, the formation of two volumes of steam from one volume of oxygen and the formation of two volumes of ammonia from one of nitrogen indicate that oxygen and nitrogen must be present as the diatomic molecules O_2 and N_2.

Early chemists found it difficult to accept the fact that a mole of a relatively heavy gas does not, under identical conditions of temperature and pressure, occupy a larger volume than a mole of lighter gas. In terms of the kinetic-molecular theory it makes sense. A gas is mostly empty space occupied by molecules moving around with an average energy that varies with the temperature. Since molecules of different gases at the same temperature have the same average kinetic energy, they can spread out to occupy the same volume.

A mole is defined as an Avogadro's number of molecules, atoms, or ions. Therefore, equal volumes of gases at the same temperature

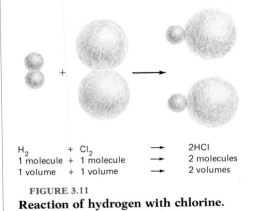

H_2 + Cl_2 $\longrightarrow$ 2HCl
1 molecule + 1 molecule $\longrightarrow$ 2 molecules
1 volume + 1 volume $\longrightarrow$ 2 volumes

FIGURE 3.11
Reaction of hydrogen with chlorine.

[3.11] THE GASEOUS STATE

and pressure, because they contain equal numbers of molecules, must also contain equal numbers of moles of gases. Mathematically, Avogadro's law may be stated as follows:

At constant temperature and pressure,

$$V = \text{constant} \times n \quad \text{or} \quad \frac{V_1}{n_1} = \frac{V_2}{n_2} \tag{7}$$

number of moles of gas *gas 1* *gas 2*

That is, the volume of a gas at constant temperature and pressure is directly proportional to the number of moles of gas present.

molar volume of a gas
= 22.4 liters (at STP)

EXAMPLE 3.4

■ Incandescent light bulbs contain inert gases such as nitrogen, in which the filament will not burn out so quickly. The approximate volume of a 100-watt bulb is 130 cm³, and the bulb contains 3×10^{-3} mole of nitrogen. How many moles of nitrogen would be inside a 150-watt bulb under the same pressure and temperature conditions if the volume of the larger bulb is 185 cm³?

For the 100-watt bulb:

$n_1 = 3 \times 10^{-3}$ mole $V_1 = 130$ cm³

For the 150-watt bulb:

$n_2 = ?$ $V_2 = 185$ cm²

Solving the mathematical statement of Avogadro's law for n_2 and substituting the above values gives

$$n_2 = (n_1)\left(\frac{V_2}{V_1}\right) = (3 \times 10^{-3}\ \text{mole})\left(\frac{185\ \text{cm}^3}{130\ \text{cm}^3}\right)$$
$$= 4 \times 10^{-3}\ \text{mole}$$

There will be 4×10^{-3} mole of dinitrogen in the 150-watt bulb. Again, we can check our answer by using the commonsense method: Because the 150-watt bulb has a larger volume, according to Avogadro's law, there should be more moles of gas in it—our answer of 4×10^{-3} seems reasonable. ■

3.12 Molar volume

Standard molar volume is defined as the volume occupied by one mole at standard temperature and pressure (STP). According to Avogadro's law, one mole of any gas should occupy the same volume as one mole of any other gas, when the volumes are measured at the same conditions of temperature and pressure.

Molar volume may be found for a gas of known molecular weight by determining the gas density at STP and dividing the molar weight by the density.

$$\text{molar volume (liters/mole)} = \frac{\text{molar weight (g/mole)}}{\text{density (g/liter)}} \tag{8}$$

On the basis of many measurements of gas densities, it has been found that for all gases at standard conditions the molar volume averages 22.4 liters.

3.13 Ideal gas law

Boyle's law, Charles' law, and Avogadro's law establish relationships for gases among four variables—the volume, the pressure, the temperature, and the amount of gas, which we express in moles, n.

$$V = \text{constant} \times \frac{1}{P}, \text{ at fixed } T \text{ and } n$$
$$V = \text{constant} \times T, \text{ at fixed } P \text{ and } n$$
$$V = \text{constant} \times n, \text{ at fixed } P \text{ and } T$$

These proportionalities combine to give a single relationship

$$V = \text{constant} \times \frac{1}{P} \times T \times n \quad \text{or} \quad PV = \text{constant} \times T \times n$$

The constant is represented by the symbol R and, written in its usual form, we have the **ideal gas law**

$$PV = nRT \tag{9}$$

Based as it is on the individual gas laws, which assume ideal behavior of gases, this equation summarizes the behavior of all ideal gases. R, called the ideal gas constant, or universal gas constant, has the same value (within experimental error) for all gases.

The ideal gas constant, R, is not simply a number—it has dimensions that depend upon the dimensions of P, V, and T. The value of R can be calculated from molar volume and the standard temperature and pressure. The molar volume of an ideal gas has been found by extrapolation of measurements of real gases at conditions as close to ideal as possible; the most accurate value is 22.41383 liters.

$$R = \frac{PV}{nT} = \frac{(1 \text{ atm})(22.414 \text{ liters})}{(1 \text{ mole})(273.15°\text{K})} = 0.082057 \text{ liter-atm/mole } °\text{K}$$

Usually, this value of R is rounded off to 0.0821 liter-atm/mole °K. If P is expressed in torr

$$R = \frac{(760 \text{ torr})(22.414 \text{ liters})}{(1 \text{ mole})(273.15°\text{K})} = 62.364 \frac{\text{liter-torr}}{\text{mole } °\text{K}}$$

As Table 3.5 shows, P, V, and T can be expressed in many different units, and for each combination of units R has a different numerical value. In working problems with the ideal gas law, it is essential that the units of P, V, and T and the units of the gas constant are in agreement. Then, if the values of three of the four terms P, V, T, and n are known, the other one can be calculated.

EXAMPLE 3.5

■ Prove the statement made about the two gas samples described in Example 3.3 by using the ideal gas law and comparing the number of moles of gases. The statement was that 4.0 liters of hydrogen at 760 torr and 30.°C contains about three times as much hydrogen as do 4.0 liters at 380 torr and 200.°C.

For sample 1:

P = (760 torr)(1 atm/760 torr) = 1.0 atm V = 4.0 liters
T = 30.° + 273°K = 303°K

which, upon substituting into the ideal gas law, gives

$$n = \frac{PV}{RT} = \frac{(1.0 \text{ atm})(4.0 \text{ liters})}{(0.0821 \text{ liter-atm}/\text{mole °K})(303°K)} = 0.16 \text{ mole}$$

For sample 2:

P = (380 torr)(1 atm/760 torr) = 0.50 atm V = 4.0 liters
T = 200.° + 273°K = 473°K n = ?

which gives

$$n = \frac{(0.50 \text{ atm})(4.0 \text{ liters})}{(0.0821 \text{ liter-atm}/\text{mole °K})(473°K)} = 0.052 \text{ mole}$$

Thus sample 1 contains

$$\frac{(0.16 \text{ mole})}{(0.052 \text{ mole})} = 3.1$$

a little over three times as much hydrogen as sample 2.   ■

3.14 Weight, density, and molecular weight relationships

Densities and molecular weights can be calculated by using Avogadro's law. We know that equal volumes of gases contain equal numbers of molecules. Therefore, the weights of equal volumes of gases at the same conditions of temperature and pressure must be proportional to the weights of the molecules. This fact allows us to

$$R = 0.0821 \text{ liter-atm}/\text{mole °K}$$

TABLE 3.5
Ideal gas constant values

0.0821	liter-atm/mole °K
62.36	liter-torr/mole °K
1.987	cal/mole °K
8.314	J/mole °K
8.314×10^7	ergs/mole °K

calculate the density of a gas of known molecular weight or the molecular weight of a gas of known density, by taking as a basis of comparison the known values of these quantities for other gases.

$$\frac{\text{density gas A}}{\text{density gas B}} = \frac{\text{molecular wt gas A}}{\text{molecular wt gas B}} = \frac{\text{molar wt gas A}}{\text{molar wt gas B}} \tag{10}$$

EXAMPLE 3.6

■ The density of oxygen, O_2, is 1.42904 g/liter at STP. What is the molecular weight of methane, a gas containing only carbon and hydrogen, if it has a density of 0.7168 g/liter at STP?

For the oxygen,

$$d_{O_2} = 1.42904 \text{ g/liter} \qquad \text{molecular wt} = 32.00 \text{ amu}$$

For the methane,

$$d_{\text{methane}} = 0.7168 \text{ g/liter} \qquad \text{molecular wt} = ?$$

The densities and molecular weights of these gases are related as follows:

$$\frac{d_{O_2}}{d_{\text{methane}}} = \frac{\text{molecular wt } O_2}{\text{molecular wt methane}}$$

First, we solve this relationship for the molecular weight of methane (by cross multiplying and dividing both sides by d_{O_2}).

$$\text{molecular wt methane} = (\text{molecular wt } O_2) \left(\frac{d_{\text{methane}}}{d_{O_2}} \right)$$

Next, we substitute the numerical values and calculate the answer.

$$\text{molecular wt methane} = (32.00 \text{ amu}) \frac{(0.7168 \text{ g/liter})}{(1.42904 \text{ g/liter})} = 16.05 \text{ amu}$$

The molecular weight of methane is 16.05 amu. Because carbon has an atomic weight of 12 amu and hydrogen of 1 amu, it is possible to deduce that the formula of methane is CH_4. ■

Molecular weight or molar weight, gas density, or sample weight can all be determined from P, V, T data by substitution in the rearrangement of the ideal gas law. To use the ideal gas law in this way it is convenient to replace n, moles, with the weight of a sample in grams m divided by the molar weight of the gas (grams per mole).

$$PV = \frac{m(\text{g})}{\text{molar wt (g/mole)}} RT$$

EXAMPLE 3.7

■ A sample of gas that occupies 270 ml at 740 torr and 98°C weighs 0.276 g. It is known to be either methyl alcohol, CH_3OH, or ethyl alcohol, C_2H_5OH. Which alcohol is present?

For our alcohol,

P = (740 torr)(1 atm/760 torr) = 0.97 atm
V = (270 ml)(10^{-3} liter/ml) = 0.27 liter
T = 98° + 273°K = 371°K
m = 0.276 g

Solving

$$PV = \frac{m}{\text{molar wt}} RT$$

for the molar weight and substituting the above values, we get

$$\text{molar wt} = \frac{m\ RT}{PV} = \frac{(0.276\ g)(0.0821\ \text{liter-atm}/\text{mole °K})(371°K)}{(0.97\ \text{atm})(0.27\ \text{liter})}$$
$$= 32\ g/\text{mole}$$

The molar weight of the unknown alcohol is 32 g/mole, which agrees with the value calculated from atomic weights for the formula of methyl alchohol. ■

EXAMPLE 3.8

■ What is the density of acetone vapor, C_3H_6O, at a temperature of 95°C and a pressure of 650 torr?

For the gas:

P = (650 torr)(1 atm/760 torr) = 0.86 atm
T = 95° + 273°K = 368°K
molar wt = 3(12) + 6(1) + 1(16) = 58 g/mole

Recognizing that $d = m/V$, we solve the ideal gas law for m/v as follows

$$d = \frac{m}{V} = \frac{P(\text{molar wt})}{RT}$$

which upon substitution of the above values gives

$$d = \frac{(0.86\ \text{atm})(58\ g/\text{mole})}{(0.0821\ \text{liter-atm}/\text{mole °K})(368°K)} = 1.7\ g/\text{liter}$$

The density of acetone under the given conditions is 1.7 g/liter. ■

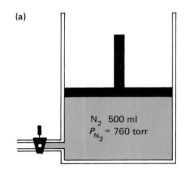

(a)

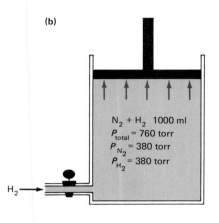

(b)

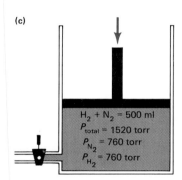

(c)

FIGURE 3.12
Dalton's law of partial pressures. *A cylinder with a movable piston contains 500 ml of nitrogen at a pressure 760 torr (a). Hydrogen is introduced. The piston rises in the cylinder until the volume is 1000 ml, when the addition of hydrogen is stopped (b). Each gas now exerts half of the total pressure, or 380 torr. The piston is next forced down until the volume in the cylinder is again 500 ml (c). This doubles the total pressure and doubles the pressure of each of the two gases.*

EXAMPLE 3.9

■ The mass of an "empty" flask is actually the sum of the masses of the flask and the air in the flask. In some experiments it is necessary to know how much air is in the flask. Calculate the weight of air in an "empty" flask if the flask has a volume of 270 ml and the temperature and pressure of the air in the flask are 22°C and 755 torr, respectively. Assume the "molar weight" of air to be 29.0 g/mole.

For the air in the flask:

V = (270 ml)(10^{-3} liter/ml) = 0.27 liter
P = (755 torr)(1 atm/760 torr) = 0.993 atm
T = 22° + 273°K = 295°K
molar wt = 29.0 g/mole

Rearranging the equation

$$PV = \frac{m}{\text{molar wt}} RT$$

solving for m, and substituting the above values gives

$$m = \frac{PV(\text{molar wt})}{RT} = \frac{(0.99 \text{ atm})(0.270 \text{ liter})(29.0 \text{ g/mole})}{(0.0821 \text{ liter-atm/mole °K})(295°K)} = 0.32 \text{ g}$$

Our "empty" flask contains 0.32 g of air. ■

3.15 Pressure in gas mixtures: Dalton's law

In a mixture of gases that do not react with each other, each molecule moves about independently just as it would in the absence of molecules of other kinds. Each gas in a mixture distributes itself uniformly throughout the entire space available as though no other gas were present. The molecules strike the walls as frequently, with the same energy, and therefore with the same pressure as they do when there is no other gas present.

Among the studies which led John Dalton to the atomic theory were experiments on gas pressures in mixtures. He summarized his conclusions in 1803 in what is known as **Dalton's law of partial pressures:** *In a mixture of gases, the total pressure exerted is the sum of the pressures that each gas would exert if it were present alone under the same conditions.*

If several different gases, 1, 2, 3 . . . are placed in the same container they form a homogeneous mixture. The pressure of a single gas in a mixture is called its **partial pressure.** The small letters p_1, p_2, p_3, . . . represent partial pressures. The total pressure of a gas mixture is given by

partial pressures of gas$_1$, gas$_2$ · · ·

$$P_{\text{total}} = p_1 + p_2 + p_3 + \cdots \tag{11}$$

Consider a volume of air under a pressure of 1000 torr. Approximately one-fifth of the molecules are oxygen molecules and approximately four-fifths are nitrogen molecules. The approximate partial pressures are, therefore, 200 torr for oxygen and 800 torr for nitrogen. It makes no difference that the mass of an oxygen molecule is greater than the mass of a nitrogen molecule. Molecules of any gas have, on the average, the same kinetic energy at the same temperature (recall the fourth postulate of the kinetic-molecular theory from Section 3.4).

In the experiment shown in Figure 3.12, hydrogen is pumped into a cylinder containing nitrogen until the gas volume is doubled. The total pressure is then doubled and the space occupied by the gas mixture returns to what it was at the beginning of the experiment. Thus, the nitrogen occupies its original volume and exerts its original pressure, completely independent of the hydrogen that also occupies the cylinder. The hydrogen occupies the same space it did before and exerts the same pressure. The old adage that two things cannot occupy the same space at the same time does not seem to apply to gases.

In a closed container partly filled with water, the air over the liquid soon becomes mixed with water vapor. Eventually, the air holds all of the water vapor that it can, and some of the water vapor condenses back into liquid. For example, in a terrarium, water is constantly evaporating from the surface of the plants and soil, and in an equal but opposite change, water is condensing on the walls and lid and running back into the soil. In the terrarium, as in any other closed vessel containing water, or any other liquid, the processes of evaporation and condensation eventually reach a state of **dynamic equilibrium**—a state of balance between exactly opposite changes occurring at the same rate.

For water or any other liquid, **vapor pressure** is the pressure exerted by the vapor over a liquid once evaporation and condensation have reached equilibrium. The total pressure over a liquid is the sum of the pressure of the gas present plus the vapor pressure of the liquid at the existing conditions of temperature and pressure.

Frequently a gas is collected in the laboratory over some liquid in which it is not very soluble. The procedure is always basically the same: The vessel in which the gas is to be confined is filled with the liquid and inverted in a container of the same liquid. The gas is led through a tube from the apparatus in which it is generated, under the liquid, and to the mouth of the vessel, where it bubbles up into the vessel through the liquid (see Figure 3.13). The pressure of the gas col-

---

Dalton's law:
$$P_{\text{total}} = p_1 + p_2 + p_3 \cdots$$

FIGURE 3.13
Collecting a gas over a liquid. *Atmospheric pressure on the surface in dish B keeps vessel A full (a) until the liquid in it is displaced by the gas (b). When all of the gas is collected it may have a greater (c) or a smaller (b) pressure than the atmosphere. By raising or lowering vessel A, the liquid levels in A and B are made equal (d). In this way, the gas is brought to atmospheric pressure (which can be read from a barometer). The gas volume is determined by marking the level of the liquid in vessel A, setting the vessel upright, and measuring the volume of water required to fill the vessel to the mark.*

(a)
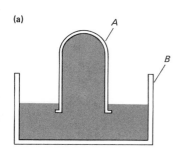

(b)
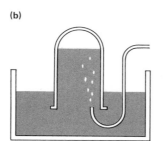

(c)
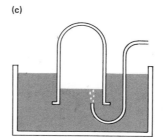

(d)
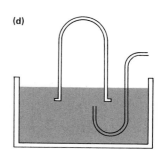

lected in this way can be found from the total pressure over the liquid in the outer container by subtracting from it the vapor pressure of the liquid at the existing temperature.

EXAMPLE 3.10

■ Potassium chlorate ($KClO_3$) was heated to give oxygen gas

$$2KClO_3 \longrightarrow 2KCl + 3O_2$$

A volume of 550 ml of oxygen was collected over water at 21°C at an atmospheric pressure of 743 torr. The vapor pressure of water at 21°C is 19 torr. How many moles of oxygen were collected?

The pressure of the oxygen will be the total pressure less the partial pressure of the water vapor. Thus

$$P = (743 \text{ torr} - 19 \text{ torr})(1 \text{ atm}/760 \text{ torr}) = 0.95 \text{ atm}$$
$$V = (550 \text{ ml})(10^{-3} \text{ liter/ml}) = 0.55 \text{ liter}$$
$$T = 21° + 273°K = 294°K$$

Substituting these values into the ideal gas law gives us

$$n = \frac{PV}{RT} = \frac{(0.95 \text{ atm})(0.55 \text{ liter})}{(0.0821 \text{ liter·atm}/\text{mole °K})(294°K)} = 0.022 \text{ mole}$$

The amount of oxygen collected was 0.022 mole or about 0.70 g. ■

Sometimes the concentrations of the components of a solution are expressed in mole fractions, and this is particularly useful with gaseous solutions. The **mole fraction** is simply the number of moles of a component in a solution divided by the total number of moles of all components in the solution.

$$X_1 = \frac{n_1}{n_1 + n_2 + \cdots} \quad \text{(12)}$$

moles of component 1

mole fraction of component 1

The partial pressure of a gas is equal to the total pressure of the gas mixture multiplied by the mole fraction of that gas

$$p_1 = X_1 P_{\text{total}}$$

and the mole fractions of two components in a mixture are related as follows:

$$\frac{p_1}{p_2} = \frac{X_1}{X_2} \quad \text{(13)}$$

EXAMPLE 3.11

■ What is the mole fraction of the water vapor in the sample described in Example 3.10?

The partial pressure of the water in the mixture is $p_{H_2O} = 19$ torr and the total pressure is

$$P_{total} = 19 \text{ torr} + 724 \text{ torr} = 743 \text{ torr}$$

Therefore, the mole fraction of water vapor is

$$X_{H_2O} = \frac{p_{H_2O}}{P_{total}} = \frac{19 \text{ torr}}{743 \text{ torr}} = 0.026$$

The mole fraction of water in the gaseous mixture is 0.026. ■

3.16 Effusion and diffusion; Graham's laws

Effusion is the escape of molecules in the gaseous state one by one, without collisions, through a hole of molecular dimensions. Effusion is of interest for three reasons: (1) It is a simple demonstration of the validity of kinetic-molecular theory; (2) it provides one method for determining molecular weights; and (3) it provides a practical method for the separation of gases.

In 1846 Graham found experimentally that the rate of effusion was dependent upon the density of a gas. **Graham's law of effusion** may be stated as follows: *The rates of effusion of two gases at the same pressure are inversely proportional to the square roots of their densities.*

$$\frac{\text{rate}_1}{\text{rate}_2} = \frac{\sqrt{d_2}}{\sqrt{d_1}} \qquad (14)$$

We now know from Avogrado's law that density is directly proportional to molecular weight, and therefore

$$\frac{\text{rate}_1}{\text{rate}_2} = \frac{\sqrt{(\text{molecular wt})_2}}{\sqrt{(\text{molecular wt})_1}} \qquad (15)$$

The determination of molecular weight from an effusion experiment is illustrated in Example 3.12.

From kinetic-molecular theory, we also know that two gases of different masses have the same average kinetic energy, $E_k = \frac{1}{2}mv^2$. Therefore, their average molecular speeds must differ and are related for gases 1 and 2 as follows:

$$m_1 v_1^2 = m_2 v_2^2$$

$$\frac{v_1^2}{v_2^2} = \frac{m_2}{m_1}$$

$$\frac{v_1}{v_2} = \frac{\sqrt{m_2}}{\sqrt{m_1}}$$

Since the masses of molecules are directly proportional to their molecular weights, Graham's law of effusion is a demonstration that the number of molecules escaping in collision-free effusion is directly proportional to their average speed, a consequence of the kinetic-molecular theory.

$$\frac{\text{rate}_1}{\text{rate}_2} = \frac{v_1}{v_2}$$

When two gases of different masses are confined in two halves of a container divided by a porous membrane, the lighter gas effuses through the membrane faster than the heavier one. As a result, a pressure difference is built up between the two chambers. The experiment described in Figure 3.14 demonstrates the difference in the effusion rates of hydrogen, and the nitrogen and oxygen in the air.

Effusion through a membrane can be used to separate gases of different weights. Separation of forms of uranium of slightly different weight by effusion played an important role in the dramatic effort by the United States to develop an atomic bomb during World War II (see Section 8.15; also An Aside: The Atomic Bomb, in Chapter 8). Naturally occurring uranium contains only 0.72% of uranium-235 in a mixture with uranium-238. This is not a high enough percentage of uranium-235 for use of the uranium in a nuclear fission bomb. The uranium is converted to a mixture of the gaseous hexafluorides, $^{236}UF_6$ and $^{238}UF_6$. By successive passage of the mixture through a series of porous membranes in what is called a uranium enrichment process, the percentage of uranium-235 can eventually be raised to a useful level. This method of uranium enrichment is commonly referred to as the "gaseous diffusion" of uranium. However, diffusion is a different physical process.

Diffusion is a more complicated phenomenon that, like effusion, is based on molecular motion. **Diffusion** is the mixing of molecules of different gases by random motion and collisions until the mixture be-

FIGURE 3.14
Apparatus for demonstrating Graham's law of effusion. *The porous cup A contains air and is surrounded by air until bell jar B, which contains hydrogen, is placed over it. The hydrogen effuses into A faster than the oxygen and nitrogen effuse out. A slight pressure thus builds up in the porous cup, and water is forced from the tip C. As hydrogen accumulates in A and air in B, the pressures equalize, and the flow of water from C stops. If the bell jar is then removed, there is an escape of hydrogen through the walls of A which is more rapid than the diffusion of air into A. This lowers the pressure in vessel A below the atmospheric pressure, and air is drawn back through the water.*

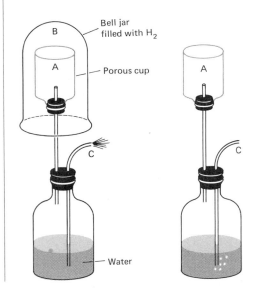

comes homogeneous. Graham also investigated diffusion, and in 1831 stated what is now known as **Graham's law of diffusion:** *The rates of diffusion of two gases are inversely proportional to the square roots of their densities (or, their molecular weights).* The mathematical outcome of this law is exactly the same as that of the law of effusion:

$$\frac{\text{rate}_1}{\text{rate}_2} = \frac{\sqrt{d_2}}{\sqrt{d_1}} = \frac{\sqrt{(\text{molecular wt})_2}}{\sqrt{(\text{molecular wt})_1}} \tag{16}$$

EXAMPLE 3.12

■ Many times it is more convenient to measure the time for a gas sample to effuse than to measure its actual rate of effusion. Because the rate is defined as the number of molecules moving through a molecular-sized hole per unit time, rate is inversely proportional to time, giving

$$\frac{\text{rate}_1}{\text{rate}_2} = \frac{\text{time}_2}{\text{time}_1} = \sqrt{\frac{\text{molar wt}_2}{\text{molar wt}_1}}$$

as long as we measure the same number of molecules moving through the hole for each gas.

Find the molar weight of a gas that takes 33.5 sec to effuse from a porous container if an identical molar amount of CO_2 takes 25.0 sec.

Solving the above equation for the molar weight of the unknown gas and substituting the respective times and molar wt = 44.0 g/mole for CO_2, gives

$$\text{molar wt}_2 = (\text{molar wt}_{CO_2})\left(\frac{\text{time}_2}{\text{time}_{CO_2}}\right)^2$$

$$= \left(44.0\ \frac{\text{g}}{\text{mole}}\right)\left(\frac{33.5\ \text{sec}}{25.0\ \text{sec}}\right)^2 = 79.0\ \text{g/mole}$$

The molar weight of the unknown gas is 79.0 g/mole. ■

Nonideal gases

3.17 Deviations from the gas laws

We have already pointed out (Section 3.4) that the gas laws are not perfectly followed by all gases. The deviations from ideal behavior are small for gases that do not liquefy easily, such as oxygen, hydrogen, and nitrogen, but are fairly large for more readily condensable gases, such as carbon dioxide and ammonia. The deviations become greater for all gases as conditions of liquefaction are approached.

The inexactness of the ideal gas laws can be understood by close examination of the assumptions that were made, or implied, in deriving the laws. First, it was assumed that the molecules in a gas are so far apart that they are independent of each other. This is not quite true, for the positive nuclear charge of the atoms in each molecule has some attraction for the negatively charged electrons of the atoms in all of the other molecules. This attraction tends to draw the molecules together and as a result decrease the volume of the gas. As the pressure is increased and the molecules come closer together, this effect is enhanced. Therefore as the pressure is increased, the volume of the gas tends to decrease more than would be anticipated. Lowering the temperature causes a similar effect.

In deriving the gas laws, it was assumed that changes in temperature and pressure affect the entire volume occupied by a gaseous material. This is not quite correct, for the molecules themselves do not expand or contract, but only the space between them does. The volume occupied by molecules is so small a fraction of the total, that the error introduced by this assumption can be disregarded in most work. However, at high pressures or low temperatures, this error becomes appreciable. As the pressure is increased or the temperature decreased, the fraction of the total volume that is compressible becomes smaller, which tends to make the volume decrease less than the gas laws predict.

These two sources of error thus tend to offset each other; but in very exact work, the gas law equation must be corrected to eliminate the inaccuracies. One of the several more exact ways of expressing the gas law is the van der Waals equation,

$$\left(P + \frac{an^2}{V^2}\right)(V - nb) = nRT$$

in which a is a constant expressing the molecular attractions and b is related to the volume actually occupied by the molecules of the substance. The corrections represented by a and b are different for different molecules. Therefore, standard tables must be consulted to find the values of a and b needed for different gases.

THOUGHTS ON
CHEMISTRY

On the constitution of bodies

A NEW SYSTEM OF CHEMICAL PHILOSOPHY, by *John Dalton*

*When any body exists in the elastic state [the gaseous state]
its ultimate particles are separated from each other to a much greater
distance than in any other state; each particle occupies the centre of a
comparatively large sphere, and supports its dignity by keeping all the
rest, which by their gravity, or otherwise are disposed to encroach upon
it, at a respectful distance. When we attempt to conceive the number
of particles in an atmosphere, it is somewhat like attempting to conceive the number of stars in the universe; we are confounded with the
thought. But if we limit the subject, by taking a given volume of any*

gas, we seem persuaded that, let the divisions be ever so minute, the number of particles must be finite; just as in a given space of the universe, the number of stars and planets cannot be infinite.

Chemical analysis and synthesis go no farther than to the separation of particles one from another, and to their reunion. No new creation or destruction of matter is within the reach of chemical agency. We might as well attempt to introduce a new planet into the solar system, or to annihilate one already in existence, as to create or destroy a particle of hydrogen. All the changes we can produce, consist in separating particles that are in a state of cohesion or combination, and joining those that were previously at a distance.

In all chemical investigations, it has justly been considered an important object to ascertain the relative weights of the simples which constitute a compound. But unfortunately the enquiry has terminated here; whereas from the relative weights in the mass, the relative weights of the ultimate particles or atoms of the bodies might have been inferred, from which their number and weight in various other compounds would appear, in order to assist and to guide future investigations, and to correct their results. Now it is one great object of this work, to shew the importance and advantage of ascertaining the relative weights of the ultimate particles, both of simple and compound bodies, the number of simple elementary particles which constitute one compound particle, and the number of less compound particles which enter into the formation of one more compound particle.

A NEW SYSTEM OF CHEMICAL PHILOSOPHY, by John Dalton, 1808 (Quoted from pp. 135–136 in *A Treasury of Scientific Prose*, H. M. Jones and I. B. Cohen, eds., Little, Brown & Co., New York, 1963)

Significant terms defined in this chapter

liquefaction	Gay-Lussac's law of combining volumes
pressure	Avogadro's law
torr	standard molar volume
atmosphere (unit)	ideal gas law
barometer	Dalton's law of partial pressures
manometer	partial pressure
ideal gas	dynamic equilibrium
Boyle's law	vapor pressure
absolute zero	mole fraction
absolute temperature scale	effusion
Charles' law	Graham's law of effusion
standard temperature and pressure (STP)	diffusion
combined gas law	Graham's law of diffusion

Exercises

3.1 About 20 years ago it was discovered that the "inert" gases really were not so inert after all. In particular, xenon reacts with fluorine under various conditions to form a whole series of compounds. Consider the preparation of XeF_4: a 1 to 5 mixture by volume of Xe to F_2 is heated in a nickel can at 400°C and 6 atm pressure for a few hours and cooled, producing colorless crystals of XeF_4. The equation for the reaction is

$$Xe + 2F_2 \longrightarrow XeF_4$$

The nickel container after the reaction contains gaseous F_2 and a little XeF_4 vapor above the crystals of XeF_4.

According to Gay-Lussac's law of combining volumes, what volume of (a) F_2 would react and (b) gaseous XeF_4 be formed for every milliliter of Xe that reacts? (c) We can deduce that the partial pressure of Xe is 1 atm and the partial pressure of F_2 is 5 atm in the original reaction mixture. This deduction is based on what law? Predict what will happen to the gases in the nickel container after the reaction if (d) the pressure is increased at constant temperature (using Boyle's law) and (e) the temperature is increased at constant pressure (using Charles' law). (f) Using Graham's law, predict the increasing order for the rate of effusion of the gases: Xe, F_2, and XeF_4. (g) Basing your argument on actual volumes of molecules, which of the three gases, Xe, F_2, or XeF_4, would you predict would deviate most from ideal gas behavior at high pressures?

3.2 Which of the following properties are characteristic of a gas: (a) negligible compressibility, (b) infinite expandability, (c) shape is that of its container but with a flat surface and fixed volume, (d) rapid flowing because of very small viscosity, (e) molecules are moving with complete disorder, and (f) removal of energy produces either the liquid or solid state?

3.3 List the five postulates of the kinetic-molecular theory for an ideal gas. Explain how these might not be valid for a liquid.

3.4 Why would you expect gases to obey simpler laws than other states of matter? Base your answer on a molecular-level interpretation of the postulates of the kinetic-molecular theory for an ideal gas. Under what conditions will real gases most resemble ideal gases?

3.5 Which statements are true? Rewrite any false statement so that it is correct.
(a) A gauge on a cylinder of compressed gas reads zero pressure. This means the tank is empty.
(b) An increase from 10°C to 20°C will approximately double the volume of an ideal gas.
(c) Pressure is a force exerted per unit volume.
(d) Avogadro's law states that equal volumes of gases, measured under the same conditions of temperature and pressure, contain equal numbers of atoms.
(e) At the same temperature, the molecules of different gases, whether they are light or heavy, have the same average kinetic energy.

3.6 Distinguish clearly between (a) a barometer and a manometer, (b) an ideal and a real gas, and (c) actual molecular speed and average molecular speed.

3.7 A local disk-jockey commented one morning: "The present temperature is 21 degrees and the outlook for today is a high of 42 degrees, just twice as hot." Comment on his statement, basing your discussion on absolute temperature.

3.8 What experimental data could be used to prove that molecular chlorine is diatomic?

3.9 What three types of experiments could you use to establish that the molecular formula for butadiene is C_4H_6 and not C_2H_3?

3.10 The manometers shown are filled with mercury. Find the pressure in torr of the gas, P_{gas}, in each case:

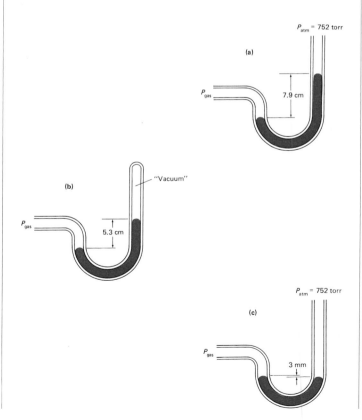

3.11 A typical laboratory atmospheric pressure reading is 745 torr. Convert this value to (a) lb/sq inch, (b) mm Hg, (c) inches Hg, (d) bar, (e) Pa, (f) atm, and (g) ft H_2O.

3.12 For an ideal gas, calculate the
(a) pressure needed to confine the gas at 75 liters if it is expanding from 25 liters and 1.00 atm at constant temperature,
(b) volume of the gas at 75 kPa if it is expanding from 10.0 dm^3 and 140 kPa at constant temperature,
(c) pressure as the gas is heated from 0°C and 1.00 atm to 135°C in a fixed volume,
(d) temperature as the gas is cooled from 14 lb/sq inch and 830°C to 1 lb/sq inch in a fixed volume,
(e) volume as the gas is heated from 1.000 liter and 0°C to 1°C under constant pressure,
(f) temperature as the gas is cooled from 10.0 liter and 950°K to 1.0 liter under constant pressure,
(g) volume of the gas at −14°C and 367 torr if it occupies 3.65 liters at 25°C and 745 torr,
(h) pressure of the gas at 78°F and 37.5 ft^3 if it occupies 375 ft^3 at 135 lb/sq inch and 85°F,
(i) temperature of the gas at 5.0 bar and 75 liters if it occupies 175 liters at 15.0 bar and 298°K.

3.13 A gas occupies a certain volume at 600. torr and 27°C. If the pressure is reduced to 500. torr, what will the temperature have to be to keep the volume constant?

3.14 A steel cylinder of fixed volume has an attached pressure gauge which reads 298 torr at 25°C. At what temperature should it register 1000. torr? Could you use this sort of device as a thermometer?

3.15 A gas occupies a volume of 600. ml at 25°C and 640. torr. If the
(a) temperature remains constant, what will be the pressure when the volume is 800. ml?
(b) pressure remains constant, at what temperature will the volume be 800. ml?
(c) volume remains constant, at what temperature will the pressure be 800. torr?

3.16 Calculate the volume of an ideal gas at Dry Ice (−78.5°C), liquid N_2 (−195.8°C), and liquid He (−268.9°C) temperatures if it occupies 10.00 liters at 25°C. Plot your results and extrapolate to zero volume. At what temperature does this occur?

3.17 A basketball was inflated to 8 psig (lb/sq inch gauge pressure) in a garage at 20°C. Outside on the driveway at a temperature of −5°C, the ball seemed "flat." Calculate the pressure of the air in the ball assuming the atmospheric pressure to be 14.4 lb/sq inch.

3.18 What volume of chlorine will react with 2 liters of

each of the following gases, H_2, CO, C_2H_4, and C_2H_2, under constant temperature and pressure conditions? The equations are

$$H_2 + Cl_2 \longrightarrow 2HCl \qquad CO + Cl_2 \longrightarrow COCl_2$$
$$C_2H_4 + Cl_2 \longrightarrow C_2H_4Cl_2 \qquad C_2H_2 + 2Cl_2 \longrightarrow C_2H_2Cl_4$$

3.19 One liter of sulfur vapor at 500°C and 1 atm is burned in pure molecular oxygen to give 8 liters of sulfur dioxide gas, SO_2, measured at the same temperature and pressure. How many atoms are there in a molecule of sulfur in the gaseous state?

3.20 One liter of PCl_3 in the gaseous state at 200°C and 1 atm reacts with an equal volume of molecular chlorine gas measured under the same conditions. The product, also a gas, occupies 1 liter when measured under the same conditions. What is the formula of this gas?

3.21 The density of ethane, C_2H_6, at STP is 1.34 g/liter. Predict the density of ethylene, C_2H_4, under these same conditions.

3.22 The density of molecular oxygen is 1.42904 g/liter at STP and that of ozone is 2.144 g/liter. Find the molecular weight of ozone. What is the formula of ozone given the fact that it contains only oxygen atoms?

3.23 For an ideal gas, calculate
(a) the number of moles in 1.00 liter at 25°C and 1.00 atm,
(b) the pressure of the gas if 5.29 moles occupies 3.45 liters at 45°C,
(c) the volume occupied by 1.00 mole at −75°C and 12.5 atm,
(d) the temperature at which 0.013 mole occupies 74 liters at 450 atm,
(e) the density at 25°C and 10.0 atm if the molecular weight is 18 amu,
(f) the molecular weight if 0.52 g occupies 610 ml at 385 torr and 45°C.

3.24 What is the density of gaseous XeF_2 at STP?

3.25 A gas has a molecular weight of 90 amu. What volume will 5.0 g of it occupy at 20°C and 776 torr?

3.26 Six moles of molecular nitrogen are confined in a 4.5-liter vessel. What is the temperature if the pressure is 3.0 atm?

3.27 A sample of gas is confined in a 3.0-liter container at a pressure of 2280 torr and a temperature of 27°C. How many moles of gas are there?

3.28 Ten moles of oxygen are confined in a vessel with a capacity of 8.0 liters. If the temperature is 0°C, what is the pressure?

3.29 A barge containing 640 tons of liquid chlorine was involved in an accident on the Ohio River. What volume would the chlorine occupy if it were converted to a gas at 740 torr and 15°C?

3.30 A liquid sample weighs 0.800 g. When converted to vapor at 100.°C and 720. torr, it occupies a volume of 100. ml. What is the molecular weight of this material?

3.31 The molar volume of an ideal gas at STP is 0.792 ft^3. Calculate the gas constant, R, in units of ft^3 atm/mole °K.

3.32 A sample of molecular oxygen, weighing 24.0 g, is confined in a vessel at 0°C and 1000. torr. If 6.00 g of molecular hydrogen is now pumped into the vessel at constant temperature, what will be the final pressure in the vessel (assuming only simple mixing)?

3.33 A mixture of 4.18 g of chloroform, $CHCl_3$, and 1.95 g of ethane, C_2H_6, at 375°C exerts how many atmospheres of pressure inside a 50.0-ml metal bomb? How much pressure is contributed by the $CHCl_3$?

3.34 A cyclopropane-oxygen mixture can be used as an anesthetic. If the partial pressures of cyclopropane and oxygen are 150 torr and 550 torr, respectively, what is the ratio of the number of moles of cyclopropane to the number of moles of oxygen?

3.35 A 5.00-liter flask containing He at 5.00 atm was connected to a 4.00-liter flask containing N_2 at 4.00 atm. Using Boyle's law for each gas, (a) find the partial pressures of the gases after they are allowed to mix, and, using Dalton's law, (b) find the total pressure of the mixture. (c) What is the mole fraction of the helium?

3.36 A sample of hydrogen was collected over water at 25°C. The vapor pressure of water at this temperature is 23.8 torr. A dehydrating agent (something that absorbs water) was added to remove the water. If the original volume of wet hydrogen was 33.3 liters and the original pressure of the wet hydrogen was 738 torr, what volume would the dry hydrogen occupy at 743 torr?

3.37 A sample of unknown gas flows through the wall of a porous cup in 39.9 min. An equal volume of molecular hydrogen, measured at the same temperature and pressure, flows through in 9.75 min. What is the molecular weight of the unknown gas?

3.38 What would be the relative rates of effusion of gaseous H_2, HD, and D_2? D is a chemical symbol used by many chemists for deuterium, an isotope of hydrogen having an atomic weight of 2.0140 amu as compared to that of 1.007825 amu for H.

3.39 A sample of a gas has a molar volume of 10.3 liters at a pressure of 745 torr and a temperature of $-138°C$. Is the gas acting ideally?

3.40* The radius of a typical molecule of gas is 0.2 nm. (a) Find the volume of a molecule assuming it to be spherical. (b) Calculate the volume actually occupied by a mole of these molecules. (c) If a mole of this gas is at STP, find the fraction of the volume of the gas actually occupied by the molecules. (d) Comment on your answer to (c) in view of the first postulate of the kinetic-molecular theory of an ideal gas.

3.41* What is the average kinetic energy of an oxygen molecule at 25°C if it has an average velocity of 4.44×10^2 m/sec? What is the molar average kinetic energy?

3.42* The average velocity of oxygen molecules at 25°C is 4.44×10^2 m/sec. What is the average velocity of nitrogen molecules at this temperature?

3.43* Show that the following data for the experimental apparatus shown in Figure 3.6 prove Boyle's law:

$P_{atm} = 745$ torr

h_{Hg} in cm	1.0	2.0	3.0	4.0	5.0
h_{air} in cm	25.00	24.68	24.34	24.05	23.75

3.44* Assume that under a constant pressure an ideal gas occupies a volume of 518.4 ml at 0.0°C and 708.3 ml at 100.0°C. Show how you would determine absolute zero on the Celsius scale from these data.

3.45* What temperature would be necessary to double the volume of an ideal gas initially at STP if the pressure decreased by 25%?

3.46* Trapped air is used as a ballast in water storage tanks in homes using water pumps so that large quantities of water come from the tank at fairly uniform pressure. In a typical system, water is pumped into the storage tank until a pressure of 60 psig is reached, water is withdrawn until a pressure of 30 psig is reached, and then the tank is refilled until a pressure of 60 psig is again reached. If the volume of trapped air at 60 psig is about 15 gal, what volume of water is delivered before the pump turns on for refilling?

3.47 A laboratory technician forgot what the color coding on some commercial cylinders of gas meant, but remembered that each of two tanks in particular contained one of the following gases: He, Ne, Ar, or Kr. Density measurements at STP were made on samples of the gases from these cylinders and were found to be 0.178 g/liter and 0.900 g/liter. Which of these gases were present?

3.48* Cyanogen is 46.2% carbon and 53.8% nitrogen by weight. At a temperature of 25°C and a pressure of 750 torr, 1.00 g of cyanogen occupied 0.476 liter. What is the empirical formula and the correct molecular formula of cyanogen?

3.49* A highly volatile liquid was allowed to vaporize completely into a 250-ml flask immersed in boiling water. From the data given, calculate the molar wt of the liquid: wt of empty flask = 65.347 g; wt of flask filled with water at room temperature = 327.4 g; wt of flask and condensed liquid = 65.739 g; atmospheric pressure = 743.3 torr; temperature of boiling water = 99.8°C; and density of water at room temperature = 0.997 g/ml.

3.50* Consider the following arrangement of interconnecting flasks. Assuming no temperature change, what would be the final pressure after all stopcocks were open?

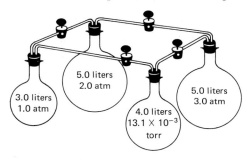

3.51* The van der Waals constants for carbon tetrachloride are $a = 20.39$ liter²-atm/mole² and $b = 0.1383$ liter/mole. Find the pressure of a sample if one mole occupies 30.0 liters at 77°C (just slightly above the boiling point), assuming CCl_4 obeys the (a) ideal gas law and (b) van der Waals gas law. (c) Calculate the percent difference and (d) comment.

3.52* Values of molar weight calculated using the ideal gas law are good only to the extent that the gas behaves as an ideal gas. However, all real gases approach ideal gas behavior at very low pressures, so a common technique for obtaining very accurate molar weights is to measure the density of a gas at various low pressures, calculate d/P, plot d/P against P, extrapolate to $P = 0$ to find the intercept, and calculate the molar wt using

molar wt = (intercept)RT

where $R = 0.0820568$ liter-atm/mole °K. Find the molar wt for SO_2 from the following data at 0°C:

P in atm	0.1	0.01	0.001	0.0001
(d/P) in g/liter-atm	2.864974	2.858800	2.858183	2.858121

3.53* The density of dry air at STP is 1.2929 g/liter, and that of molecular nitrogen is 1.25055 g/liter. (a) Find the "molecular wt" of air from these data. If air consists of 78.08% N_2, 20.95% O_2, 0.93% Ar, and 0.03% CO_2, (b) calculate the average "molecular wt" of air and (c) compare your answers from (a) and (b). (d) Why can't Graham's law be used to find the molecular wt? (e) What is the density of dry air at 745 torr and 25°C? (f) What is the density of air at 745 torr and 25°C if it contains water vapor at a partial pressure of 13 torr? (g) Explain why your answer to (f) should be less than that of (e). (h) What is the partial pressure of CO_2 in a room containing dry air at 1.00 atm? (i) What mass of CO_2 will be present at 25°C in a closet 5 × 11 × 10 ft if the air pressure is 752 torr? (j) What is the predicted pressure of a molar sample of air at −150°C in a 1.00-liter container assuming ideal gas behavior? (k) Repeat the calculation of (j) assuming air to obey the van der Waals gas law with $a = 1.38$ liter²-atm/mole² and $b = 0.037$ liter/mole. (l) Is there much difference in your answers to (j) and (k)?

4 ATOMS, MOLECULES, AND IONS IN ACTION; STOICHIOMETRY

This chapter presents definitions and mathematical tools that will be useful throughout your study of chemistry. The chemical equation is introduced and the balancing of chemical equations is described. The relationships between the amounts of substances represented by a correctly written chemical equation are explained. Next, some of the most commonly encountered dynamic situations in chemistry are introduced—equilibrium; the interactions of water, acids, and bases; and oxidation–reduction reactions.

Each of these types of reactions is the subject of more extensive discussion in later chapters.

This chapter is an introduction to the dynamics of chemistry— an introduction to what happens when atoms, ions, and molecules go into action and chemical change takes place. Chemical change was observed long before it was understood. Fires burned, grapes fermented, iron rusted. From the time of the Greek philosophers, on through the age of the alchemists, and into the beginning of the nineteenth century, bits and pieces of information were gathered about chemical change and the chemical properties of substances.

For example, a class of substances called "acids" was recognized very early in the history of chemistry. The word "acid" was derived from the Latin, acidus, meaning "sour," because sourness is a property of acids. An acid also could be identified by its participation in specific chemical changes, such as the production of a gas when the acid is mixed with limestone. Legend has it that Hannibal tried to clear trails in the Alps for his elephants by pouring hot vinegar, an acid, over the limestone rocks. Substances called "bases" were also identified by their common properties. And it was suspected long ago that acids and bases were related chemically, because the characteristic properties of an acid and a base disappear when the two are mixed together.

Two major questions arise when substances are mixed together and react chemically. These questions are: What changes do the substances undergo? What amounts of the substances are involved? The ability to predict the changes grew from observations. Once atomic theory and the laws of chemical combination were understood, it became possible to predict the amounts of substances that take part in chemical change.

Reactions and equations

4.1 Chemical equations

Any process in which the atoms, molecules, or ions of at least one substance become the atoms, molecules, or ions of a chemically different substance is a chemical reaction. The atoms, molecules, or ions may be in the gaseous, liquid, or solid states, or they may be in solution. The substances that are changed in a reaction are called the **reactants.** The new substances that are produced in a chemical reaction are called the **products.**

Chemical equations are the sentences of the language of chemistry. A **chemical equation** represents with symbols and formulas the total chemical change that occurs in a chemical reaction. In the preceding chapter, we presented the chemical equations for reactions of gases (Table 3.4) to illustrate Gay-Lussac's law of combining volumes. For example,

$$P_4 + 6\,Cl_2 \rightarrow 4\,PCl_3 \tag{1}$$
$$\underset{reactants}{} \quad \underset{product}{}$$

This equation tells us that the reactants elemental phosphorus and elemental chlorine combine to give phosphorus(III) chloride, the product. A chemical equation must be *balanced*—that is, the number of atoms of each kind must be the same on each side of the equation. For Equation (1) there are 4 phosphorus atoms and 12 chlorine atoms in the reactants and in the products. The reason that equations must be balanced is, of course, the law of conservation of mass: *atoms can be rearranged in chemical reactions, but not created or destroyed.*

The general scheme for writing any chemical equation is

$$\text{reactant}_1 + \text{reactant}_2 + \text{reactant}_3 + \ldots \longrightarrow$$
$$\text{product}_1 + \text{product}_2 + \text{product}_3 + \ldots \tag{2}$$

There may be only one reactant or only one product, or there may be several substances as both products and reactants. The order in which the reactants or products are written has no significance—the following two equations mean exactly the same thing:

$$\underset{\substack{silver\\sulfate}}{Ag_2SO_4} + \underset{\substack{hydrogen\\sulfide}}{H_2S} \longrightarrow \underset{\substack{silver\\sulfide}}{Ag_2S} + \underset{\substack{sulfuric\\acid}}{H_2SO_4}$$

$$\underset{\substack{hydrogen\\sulfide}}{H_2S} + \underset{\substack{silver\\sulfate}}{Ag_2SO_4} \longrightarrow \underset{\substack{sulfuric\\acid}}{H_2SO_4} + \underset{\substack{silver\\sulfide}}{Ag_2S}$$

To see how the process of writing a chemical equation works, we can use the reaction by which both Priestley and Scheele discovered

FIGURE 4.1

Priestley's preparation of oxygen from mercuric oxide. *When the mercuric oxide was heated by the sun's rays, as when focused by a glass lens, the gas that Priestley thought of as "eminently respirable air" was trapped under the bell jar and liquid mercury remained in the dish.*

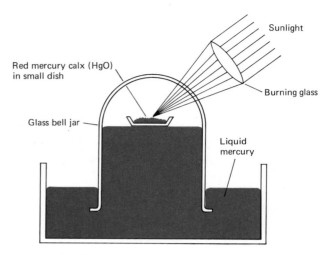

Sunlight

Red mercury calx (HgO) in small dish

Burning glass

Glass bell jar

Liquid mercury

oxygen (Figure 4.1). They did not have enough information to write a correct chemical equation for the transformation of mercury oxide to oxygen and mercury, but we do. The reaction Priestley and Scheele observed may be represented in words:

Mercury oxide, a red powder, when heated, is transformed into mercury and oxygen.

Mercury oxide is the reactant, and mercury and oxygen are the products. The first step toward writing the equation for the reaction is to determine the correct formulas for the reactants and products. Using the description of mercury oxide as a red powder, we can look in a handbook of chemistry and find that the compound used by Priestley and Scheele must have been mercury(II) oxide, HgO, because mercury(I) oxide, Hg_2O, the only other mercury oxide, is a black powder. We already know that the product, gaseous oxygen, exists as diatomic molecules, O_2. Elemental mercury is a metal and is simply represented in an equation by its symbol, Hg.

The second step in writing the equation is to put down the symbols and formulas on either side of an arrow according to the general scheme of Equation (2). In this case, we have only one reactant

$$HgO \longrightarrow Hg + O_2 \qquad (3)$$
mercury(II) oxide yields mercury + oxygen

The third step is to balance the equation. This equation is not correct because there are two oxygen atoms on the product side and only one oxygen atom on the reactant side. To remedy this we try putting the coefficient 2 in front of HgO

$$2HgO \longrightarrow Hg + O_2 \qquad (4)$$

Now there are two oxygen atoms on each side, but the mercury atoms are out of balance. The correct equation emerges when the coefficient 2 is added in front of Hg on the product side

[4.1] STOICHIOMETRY

$$2HgO \longrightarrow 2Hg + O_2 \qquad (5)$$

This equation says that two molecules of mercury(II) oxide decompose for every two atoms of mercury and one molecule of oxygen formed. Since each chemical formula can also represent one mole of a substance, the coefficients in an equation also tell how many moles of each reactant are required in a reaction and how many moles of product they can produce. Two moles of mercury(II) oxide decompose to give 2 moles of mercury and 1 mole of dioxygen.

EXAMPLE 4.1

■ Plaster of paris is produced by heating gypsum to drive off some of the water that is combined with calcium sulfate in gypsum crystals. The equation for the reaction is

$$2CaSO_4 \cdot 2H_2O \longrightarrow (CaSO_4)_2 \cdot H_2O + 3H_2O$$
gypsum   plaster of paris   water

Show how this equation is an example of the law of conservation of mass.

The equation tells us that 2 moles of gypsum are reacting. This corresponds to a mass equal to two times the molar weight of gypsum, or

(2 moles gypsum)(172.0 g/mole) = 344.0 g gypsum

The equation tells us that 1 mole of plaster of paris and 3 moles of water are being produced. Thus the total mass of products is

(1 mole plaster of paris)(290.0 g/mole) = 290.0 g plaster of paris
(3 moles water)(18.0 g/mole)   =   54.0 g water
                                       344.0 g products

The mass of the reactants is 344.0 g and the mass of the products is 344.0 g. Thus the equation is an example of the law of conservation of mass. ■

The steps in writing a correct chemical equation are summarized in Table 4.1. Additional examples of balancing equations are given in the next section.

4.2 Balancing chemical equations

Chemical equations are balanced by adding coefficients in front of the formulas until the number of atoms of each kind is the same on both sides of the arrow. The coefficients chosen are usually the smallest whole numbers that will achieve a balanced equation. A coefficient multiplies everything in a formula. For example, changing H_2O to $2H_2O$ puts 4 hydrogen and 2 oxygen atoms into the balancing act.

TABLE 4.1
Balancing chemical equations

(1) **Decide what the reactants and products are**

(2) **Write the formulas of the reactants on the left and the formulas of the products on the right, with an arrow between them**

(3) **Balance the equation**

Note that the subscripts in a chemical formula may not be changed to achieve balance. Changing the subscripts is equivalent to changing the identity of the substance represented by the formula. For example, Equation (3) in the preceding section could not have been balanced by changing HgO to HgO_2—HgO_2 would be something different from the red oxide of Priestley and Scheele, if it existed, which it does not. Changing H_2O to H_2O_2 to balance an equation would be equivalent to changing water to hydrogen peroxide, a very different substance from water.

In this section we further illustrate the method of balancing an equation used for the mercury(II) oxide reaction. The method essentially involves balancing masses. It is a commonsense method in which coefficients are added, the equality of the number of atoms on each side is checked, and the process is repeated as necessary until balance is achieved.

Suppose we know only the formulas for the reactants and products of the reaction between aluminum(III) sulfide and water

$$\text{\textit{not balanced}} \qquad Al_2S_3 + H_2O \longrightarrow Al(OH)_3 + H_2S \qquad (6)$$
$$\quad\quad\quad\underset{\text{aluminum}}{} \underset{\text{water}}{} \underset{\text{aluminum}}{} \underset{\text{hydrogen}}{}$$
$$\quad\quad\quad\underset{\text{sulfide}}{} \qquad\qquad \underset{\text{hydroxide}}{} \underset{\text{sulfide}}{}$$

Where possible, it is often easiest to begin with atoms that appear in only one formula on each side of the equation. In Equation (6), there are three, Al, S, and O. Starting with Al—we must have two Al atoms on the right

$$\text{\textit{not balanced}} \qquad Al_2S_3 + H_2O \longrightarrow 2Al(OH)_3 + H_2S \qquad (7)$$

Next choosing S, we see that we need to increase the number of S atoms on the right to 3

$$\text{\textit{not balanced}} \qquad Al_2S_3 + H_2O \longrightarrow 2Al(OH)_3 + 3H_2S \qquad (8)$$

Going on to oxygen, there are now six oxygen atoms on the right (2×3; the coefficient times the subscript), so six are needed on the left.

$$\text{\textit{balanced}} \qquad Al_2S_3 + 6H_2O \longrightarrow 2Al(OH)_3 + 3H_2S \qquad (9)$$

Now it is necessary to check the number of hydrogen atoms to see if they balance. On the left there are $6 \times 2 = 12$ hydrogen atoms and on the right there are $2 \times 3 = 6$ plus $3 \times 2 = 6$ hydrogen atoms, also giving a total of 12 hydrogen atoms. Equation (9) is completely balanced.

EXAMPLE 4.2

■ Some cigarette lighters use butane, C_4H_{10}, as fuel. When butane burns, it reacts with oxygen from the air to produce water and carbon dioxide. Write a balanced chemical equation for this reaction.

The first step in writing the correct equation is to identify the reactants and products. These were given in the statement of the problem: C_4H_{10} and O_2 for the reactants and H_2O and CO_2 for the products. We next set up the equation with the reactants on the left and the products on the right

$$C_4H_{10} + O_2 \longrightarrow CO_2 + H_2O$$

All that remains is to find the coefficients needed to balance the equation. On the left side of the equation, we have 4 C atoms, and on the right side of the equation, we have 1 C atom, so we put a 4 in front of the CO_2, giving

$$C_4H_{10} + O_2 \longrightarrow 4CO_2 + H_2O$$

On the left, we have 10 H atoms, and on the right, we have 2 H atoms, so we put a 5 in front of the H_2O, giving

$$C_4H_{10} + O_2 \longrightarrow 4CO_2 + 5\,H_2O$$

There are 13 O atoms on the right and 2 O atoms on the left. We can balance the number of oxygens by using $\frac{13}{2}$ (or $6\frac{1}{2}$) in front of the O_2, giving

$$C_4H_{10} + (\tfrac{13}{2})O_2 \longrightarrow 4CO_2 + 5H_2O$$

Ordinarily we do not use fractional coefficients, so we clear the fractions by multiplying all coefficients by 2, giving

$$2C_4H_{10} + 13\,O_2 \longrightarrow 8CO_2 + 10H_2O$$

Let's make a quick check:

no. carbon atoms	2(4) = 8 on left	8(1) = 8 on right
no. hydrogen atoms	2(10) = 20 on left	10(2) = 20 on right
no. oxygen atoms	13(2) = 26 on left	8(2) + 10(1) = 26 on right

The equation is balanced. ∎

4.3 Reaction conditions

Information about the physical state of the reactants and products or about some of the conditions under which a reaction occurs is often added to chemical equations. The states of the substances involved are indicated by placing after the formulas the symbols *g* for gas, *l* for liquid, and *s* for solid in parentheses (Table 4.2). These designations refer to the states of the pure substances under the conditions of the reaction. The symbol *l* is used for a pure liquid; the symbol *aq* is used for substances in aqueous solution.

The capital Greek letter delta, Δ, is often placed over the arrow to indicate that heat is required to make a reaction take place.

TABLE 4.2
Symbols for the states of reactants and products

(*s*)	**Solid**
(*l*)	**Liquid**
(*g*)	**Gas**
(*aq*)	**Aqueous solution**

$$2HgO(s) \xrightarrow{\Delta} 2Hg(l) + O_2(g) \tag{10}$$

solid | when | liquid | gaseous
mercury(II) | heated | mercury $+$ | oxygen
oxide | yields

The exact temperature required, as well as the pressure, might be written over the arrow, instead, as in the following equation for the industrial preparation of methyl alcohol.

$$CO(g) + 2H_2(g) \xrightarrow{350°C, \ 200-300 \ atm} CH_3OH(g) \tag{11}$$

carbon | hydrogen | methyl
monoxide | | alcohol

This reaction occurs more easily, as do many reactions, in the presence of a **catalyst**—a substance that affects the rate of a reaction, but is recovered from the reaction unchanged. For reaction (11), the catalyst is a mixture of zinc and chromium oxides, and this information can also be placed over the arrow.

$$CO(g) + 2H_2(g) \xrightarrow[ZnO-Cr_2O_3]{350°C, \ 200-300 \ atm} CH_3OH(g) \tag{12}$$

Equation (13) represents the industrial preparation of polypropylene from propylene. The reaction conditions are 70°C and 3 atm pressure, the catalyst is a triethylaluminum–titanium (IV) chloride mixture, and the solvent is heptane. In this equation, n stands for any large number. Polypropylene is a **polymer**—a large molecule made of many units of the same structure linked together.

$$n CH_3CH{=}CH_2 \xrightarrow[(C_2H_5)_3Al\text{-}TiCl_4]{70°C, \ 3 \ atm, \ heptane} \left(\!\begin{array}{c} CH_2CH \\ | \\ CH_3 \end{array}\!\right)_n \tag{13}$$

propylene   polypropylene

4.4 Ionic reactions and ionic equations

The reaction between sodium chloride and silver nitrate in aqueous solution is represented by the following equation:

$$AgNO_3(aq) + NaCl(aq) \longrightarrow AgCl(s) + NaNO_3(aq) \tag{14}$$

silver | sodium | solid | sodium
nitrate | chloride | silver | nitrate
in aqueous | in aqueous | chloride | in aqueous
solution | solution | | solution

When the two solutions are mixed, solid silver chloride immediately forms. The appearance of a solid when a reaction occurs in a solution is called **precipitation**. The solid that forms during a reaction in solution is called a **precipitate**. A precipitate is indicated in an equation by drawing a line under it, <u>AgCl</u>, by printing it in boldface type, **AgCl**, by drawing an arrow pointing down after the formula, AgCl ↓ , or, as we do, simply by showing (s) that a solid is formed.

Because silver nitrate, sodium chloride, and sodium nitrate are all soluble ionic substances, they are dissociated into ions in aqueous solution.

$$AgNO_3(s) \xrightarrow{H_2O} Ag^+(aq) + NO_3^-(aq) \qquad (15)$$

$$NaCl(s) \xrightarrow{H_2O} Na^+(aq) + Cl^-(aq) \qquad (16)$$

$$NaNO_3(s) \xrightarrow{H_2O} Na^+(aq) + NO_3^-(aq) \qquad (17)$$

Therefore, another way of writing the overall equation for the reaction of silver nitrate and sodium chloride in aqueous solution would be to show the individual ions.

$$Ag^+(aq) + NO_3^-(aq) + Na^+(aq) + Cl^-(aq) \longrightarrow$$
$$AgCl(s) + Na^+(aq) + NO_3^-(aq) \qquad (18)$$

Two ions did not undergo any chemical change in this reaction; the Na^+ and NO_3^- ions, in aqueous solution appear on both sides of the equation. Ions that are present during a reaction but are unchanged are called **spectator ions.** These ions can be canceled out of the ionic equation given above and the complete chemical change represented as

$$Ag^+(aq) + Cl^-(aq) \longrightarrow AgCl(s) \qquad (19)$$

This is a **net ionic equation**—an equation that shows only the species involved in the chemical change, not all of the ions present in the solution. Equation (19) tells us that a silver ion and a chloride ion, when brought together in solution, will give solid silver chloride. The silver and chloride ions can be from any source, not just from the compounds we started with in Equation (14).

In net ionic equations, the algebraic sum of the charges on the left and of the charges on the right must be balanced. Electrical charge, like mass, must be conserved. Equations (15) to (19) each have a net charge of zero on each side. As long as charge balance is maintained, the charge on each side of an ionic equation can have any value. In the following equation, the charge on each side is + 17. Note that in the following equation and throughout this book, we often leave out the designation (aq) after ions; you may always assume that ions are in aqueous solution unless it is explicitly stated that they are not.

$$MnO_4^- + 5Fe^{2+} + 8H^+ \longrightarrow 5Fe^{3+} + Mn^{2+} + 4H_2O(l)$$

In every balanced equation, both mass and charge must be balanced.

EXAMPLE 4.3

■ Balance the net ionic equation for the reaction in aqueous solution of $Ag(NH_3)_2^+$ with hydrochloric acid to give solid silver chloride and ammonium ion. Hydrochloric acid, HCl, is completely ionized in solution.

Balanced chemical equation:
mass on left = mass on right

$$\frac{charge}{on\ left} = \frac{charge}{on\ right}$$

The reactants are $Ag(NH_3)_2^+$ and hydrochloric acid. Hydrochloric acid in solution should be written as H^+ and Cl^-. The products are $AgCl$ and NH_4^+. Thus the unbalanced equation is

$$Ag(NH_3)_2^+ + H^+ + Cl^- \longrightarrow AgCl + NH_4^+$$

This equation is balanced with respect to Ag and Cl. On the left side we have 2 N atoms and on the right only 1 N atom, so we place a 2 before the NH_4^+ giving

$$Ag(NH_3)_2^+ + H^+ + Cl^- \longrightarrow AgCl(s) + 2NH_4^+$$

There are 8 H atoms on the right and 7 on the left, so if we place a 2 in front of H^+, we obtain our final result

$$Ag(NH_3)_2^+ + 2H^+ + Cl^- \longrightarrow AgCl(s) + 2NH_4^+$$

A check of the atoms on each side shows that the equation is balanced. Because this is an ionic equation, the charges on the species must also balance. On the left side, we have $+1$ from $Ag(NH_3)_2^+$, $2(+1) = +2$ from $H+$, and -1 from Cl^- for a total of $+2$ for the reactants. On the right side we have 0 from AgCl and $2(+1) = +2$ from NH_4^+ giving a total of $+2$ for the products. The equation is balanced with respect to both atoms and charge. The overall equation for this reaction is

$$Ag(NH_3)_2Cl(aq) + 2HCl(aq) \longrightarrow AgCl(s) + 2NH_4Cl(aq) \qquad \blacksquare$$

EXAMPLE 4.4

■ Chlorine gas can be prepared in the laboratory by the reaction of manganese dioxide with hydrochloric acid:

$$\underset{\substack{\text{manganese} \\ \text{dioxide}}}{MnO_2(s)} + \underset{\substack{\text{hydrochloric} \\ \text{acid}}}{4HCl(aq)} \longrightarrow \underset{\substack{\text{manganese(II)} \\ \text{chloride}}}{MnCl_2(aq)} + \underset{\text{water}}{2H_2O(l)} + \underset{\text{chlorine}}{Cl_2(g)}$$

Aqueous solutions of HCl and $MnCl_2$ contain only ionic species. Write the net ionic equation for this preparation.

First of all, we write the overall equation using the ionic form for all ionic species in solution

$$MnO_2(s) + 4H^+ + 4Cl^- \longrightarrow Mn^{2+} + 2Cl^- + 2H_2O(l) + Cl_2(g)$$

We can see that two chloride ions are acting as spectator ions. Therefore these ions can be canceled from each side to give the net ionic equation as

$$MnO_2(s) + 4H^+ + 2Cl^- \longrightarrow Mn^{2+} + 2H_2O(l) + Cl_2(g) \qquad \blacksquare$$

4.5 Information from chemical equations: stoichiometry

The word "stoichiometry" was coined in 1792 by Jeremias Richter as a name for the science of measuring the proportions of the chemical elements (see Thoughts on Chemistry). Richter was one of the first chemists to realize that the masses of the elements and the amounts in which they combine have a fixed relationship to each other. Today the term **stoichiometry** is used for the derivation of quantitative information from symbols, formulas, and equations.

A chemical equation is essentially a ratio—it shows the relative amounts of reactants and products involved in a reaction. The coefficients in a reaction can be multiplied or divided by any factor without changing the meaning of the equation. The following equations all give the same information:

$$2H_2(g) + O_2(g) \longrightarrow 2H_2O(l)$$
$$6H_2(g) + 3O_2(g) \longrightarrow 6H_2O(l)$$
$$H_2(g) + \tfrac{1}{2}O_2(g) \longrightarrow H_2O(l)$$

The ratios of a chemical equation can be expressed as ratios of molecules, moles, and masses, and, where gases are involved, volumes. Table 4.3 illustrates the kind of information about molecular, molar, mass, and volume relationships that can be derived from a single chemical equation.

Each time a reaction is carried out for the purpose of preparing a chemical compound, questions such as these must be asked: How much of each reactant has to be weighed out and allowed to react in order to produce the desired amount of product? If only a few grams of one reactant is available, how much of the other reactant will be needed? How much product can be prepared from a given amount of starting materials? Will any amount of the reactants be left unchanged? All of these questions and many more can be answered with

---

TABLE 4.3

Information from a chemical equation

This reaction between sulfur dioxide and oxygen accounts for the presence of sulfur trioxide in the air as a pollutant and is a step in the sequence that leads to formation of H_2SO_4 in the atmosphere.

$$2SO_2(g) + O_2(g) \longrightarrow 2SO_3(g)$$
sulfur dioxide   oxygen   sulfur trioxide

Each	*can react with*	*to yield*
2 molecules of SO_2	1 molecule of O_2	2 molecules of SO_3
2 moles of SO_2	1 mole of O_2	2 moles of SO_3
128 g of SO_2	32 g of O_2	160 g of SO_3
44.8 liters (STP) of SO_2	22.4 liters (STP) of O_2	44.8 liters (STP) of SO_3
2 volumes of SO_2[a]	1 volume of O_2[a]	2 volumes of SO_3[a]

[a] Volumes all measured at same T and P.

TABLE 4.4

Solving stoichiometry problems

(1) **Write a balanced chemical equation**

(2) **Convert the information given to moles**

(3) **Inspect the chemical equation for molar relationships**

(4) **Convert from moles to the answer desired**

information from the chemical equation for the reaction under consideration. Whatever the question, the balanced equation and the factor-dimensional method can be relied upon to yield the answers needed.

To solve stoichiometry problems requires the four steps summarized in Table 4.4. We have already discussed how to accomplish the first step, writing a correctly balanced equation. In many cases, of course, the equation is already known. Since it is easiest to use ratios of moles as a basis for calculation, the second step is to convert the known quantities of the substances involved into moles. Then, inspection of the equation provides the information about ratios of reactants and products needed to calculate the answers to the questions being asked.

Examples 4.5 to 4.7 deal with amounts of reactants and products. In Example 4.5 the required amount of one reactant is found from the known amount of the other reactant. Example 4.6 answers the general question, how much reactant will be needed to obtain a specific amount of product? And Example 4.7 illustrates a case in which there is a **limiting reactant**—a single reactant that determines the maximum amount of product because it is present in the smallest stoichiometric amount. The expression **stoichiometric amount** refers to the exact amount of a substance required according to a chemical equation.

EXAMPLE 4.5

■ What mass of oxygen is consumed in the combustion of 100.0 g of gasoline assuming the gasoline to be isooctane, C_8H_{18}? The balanced equation is

$$2C_8H_{18}(l) + 25O_2(g) \longrightarrow 16CO_2(g) + 18H_2O(g)$$
$$\text{isooctane} \quad \text{oxygen} \quad \text{carbon dioxide} \quad \text{water}$$

The number of moles of isooctane in 100.0 g is

$$\frac{100.0 \text{ g}}{114.2 \text{ g/mole}} = 0.8757 \text{ mole } C_8H_{18}$$

The chemical equation above states that 25 moles of oxygen react with 2 moles of isooctane. Thus find that the amount of oxygen reacting is

$$(0.8757 \text{ mole } C_8H_{18}) \left(\frac{25 \text{ moles } O_2}{2 \text{ moles } C_8H_{18}} \right) = 10.95 \text{ moles } O_2$$

This corresponds to

$$(10.95 \text{ moles})(32.00 \text{ g/mole}) = 350.4 \text{ g } O_2$$

[4.5] STOICHIOMETRY

Thus 100.0 g of isooctane requires 350.4 g of O_2. ■

EXAMPLE 4.6

■ What mass of hydrogen is needed to produce 100.0 g of iron by the following reaction?

$$Fe_3O_4(s) \; + \; 4H_2(g) \xrightarrow{\Delta} 3Fe(s) \; + \; 4H_2O(g)$$
iron oxide   *hydrogen*   *iron*   *water*

The 100.0 g of Fe corresponds to

$$\frac{100.0 \; g}{55.85 \; g/\text{mole}} = 1.791 \; \text{moles Fe}$$

The equation tells us that 3 moles of Fe are produced for every 4 moles of H_2, thus the amount of H_2 needed is

$$(1.791 \; \text{moles Fe}) \left(\frac{4 \; \text{moles } H_2}{3 \; \text{moles Fe}} \right) = 2.388 \; \text{moles } H_2$$

which corresponds to

$$(2.388 \; \text{moles})(2.016 \; g/\text{mole}) = 4.814 \; g \; H_2$$

For each 100.0 g of Fe produced, 4.814 g of H_2 is required. ■

EXAMPLE 4.7

■ What mass of $PbCl_2$ can be obtained from a reaction mixture containing 25.0 of PCl_3 and 41.7 g of PbF_2? The reaction is

$$3PbF_2(s) \; + \; 2PCl_3(l) \longrightarrow 2PF_3(g) \; + \; 3PbCl_2(s)$$
lead(II) fluoride   *phosphorus(III)*   *phosphorus(III)*   *lead(II) chloride*
            chloride   *fluoride*

The number of moles of each of the reactants is

$$\frac{41.7 \; g}{245.2 \; g/\text{mole}} = 0.170 \; \text{mole } PbF_2 \qquad \frac{25.0 \; g}{137.4 \; g/\text{mole}} = 0.182 \; \text{mole } PCl_3$$

If all of the PbF_2 reacts, the chemical equation tells us that

$$(0.170 \; \text{mole } PbF_2) \left(\frac{3 \; \text{moles } PbCl_2}{3 \; \text{moles } PbF_2} \right) = 0.170 \; \text{mole } PbCl_2$$

will be formed. Similarly, if all of the PCl_3 reacts,

$$(0.182 \; \text{mole } PCl_3) \left(\frac{3 \; \text{moles } PbCl_2}{2 \; \text{moles } PCl_3} \right) = 0.273 \; \text{mole } PbCl_2$$

will be formed. Obviously both answers are not correct. In this problem, PbF_2 is the limiting reactant because it is present in the smaller stoichiometric amount. The molar amount of PbF_2 determines the molar amount of $PbCl_2$ that can be formed.

$$(0.170 \text{ mole}) (278.1 \text{ g/mole}) = 47.3 \text{ g } PbCl_2$$

The amount of $PbCl_2$ produced will be 47.3 g. Only

$$(0.170 \text{ mole } PbF_2) \left(\frac{2 \text{ moles } PCl_3}{3 \text{ moles } PbF_2} \right) = 0.113 \text{ mole } PCl_3$$

will react, leaving

$$25.0 \text{ g} - (0.113 \text{ mole}) (137.4 \text{ g/mole}) = 9.5 \text{ g } PCl_3$$

in excess. ■

The maximum amount of a product that can, according to the chemical equation, be obtained from a known amount of reactants is called the **theoretical yield.** For many reasons, the amount of a product actually obtained may be less than the theoretical amount. Perhaps some product was lost or spilled in handling, or perhaps not all of the reactant was converted to product because the reaction conditions were not right, for instance, the temperature was too low. Sometimes undesired reactions also occur and use up some of the reactant or product.

The actual yield in a reaction is expressed as the **percent yield**

$$\% \text{ yield} = \frac{\text{wt of product obtained}}{\text{theoretical yield}} \times 100\% \qquad (20)$$

Example 4.8 illustrates finding theoretical yield and percent yield.

EXAMPLE 4.8

■ The final step in the industrial production of aspirin (acetylsalicylic acid) is the reaction of salicylic acid with acetic anhydride:

$$HOC_6H_4COOH(s) + (CH_3CO)_2O(l) \longrightarrow$$
salicylic   acetic
acid   anhydride

$$CH_3OCOC_6H_4COOH(s) + CH_3COOH(l)$$
acetylsalicylic   acetic
acid   acid

To test a new method of handling the materials, a chemist ran the reaction on a laboratory scale with 25.0 g of salicylic acid and excess acetic anhydride (over 20 g). He obtained 24.3 g of aspirin. What was the percent yield?

The number of moles of salicylic acid is

$$\frac{25.0 \text{ g}}{138 \text{ g/mole}} = 0.181 \text{ mole HOC}_6\text{H}_4\text{COOH}$$

The equation tells us that one mole of aspirin is produced for each mole of salicylic acid that reacts, so

$$(0.181 \text{ mole HOC}_6\text{H}_4\text{COOH}) \left(\frac{1 \text{ mole CH}_3\text{OCOC}_6\text{H}_4\text{COOH}}{1 \text{ mole HOC}_6\text{H}_4\text{COOH}}\right)$$
$$= 0.181 \text{ mole aspirin}$$

which is equivalent to

$$(0.181 \text{ mole aspirin})(180. \text{ g/mole}) = 32.6 \text{ g aspirin}$$

which would be formed if the reaction were 100% efficient (the theoretical yield). The percent yield is found as follows:

$$\% \text{ yield} = \frac{\text{wt of product obtained}}{\text{theoretical yield}} \times 100\%$$

$$= \frac{24.3 \text{ g}}{32.6 \text{ g}} \times 100\% = 74.5\%$$

The percent yield in this reaction was 74.5%. ■

When either all or some of the reactants and products in a reaction are gases, relationships between the amounts of the gaseous species can be obtained by using Gay-Lussac's law of combining volumes, Avogadro's law, and the molar volume. Examples 4.9 and 4.10 illustrate how to find the volumes of gaseous reactants and products, respectively. In Example 4.11, gas volumes are used in the analysis of a mixture.

EXAMPLE 4.9

■ Many campers use small propane stoves to cook meals. How many liters of air (assumed to be 20.% O_2 by volume) will be required to burn 10.0 liters of propane? Assume all gas volumes are measured at the same temperature and pressure. The equation is

$$\underset{propane}{C_3H_8(g)} + \underset{oxygen}{5O_2(g)} \longrightarrow \underset{carbon\ dioxide}{3CO_2(g)} + \underset{water}{4H_2O(g)}$$

The chemical equation tells us that 1 volume of C_3H_8 requires 5 volumes of O_2 for complete combustion according to Gay-Lussac's law of combining volumes. Thus

$$(10.0 \text{ liters propane}) \left(\frac{5 \text{ volumes O}_2}{1 \text{ volume propane}}\right) = 50.0 \text{ liters O}_2$$

Recognizing that 1.00 liter of air contains 0.20 liter of O_2, the amount of air needed is

$$\frac{50.0 \text{ liters } O_2}{0.20 \text{ liter } O_2/\text{liter air}} = 250 \text{ liters air}$$

The amount of air needed is 250 liters. ■

EXAMPLE 4.10

■ What volume of carbon dioxide at 735 torr and 22°C can be obtained by the reaction of 15 g of calcium carbonate with excess hydrochloric acid? The equation is

$$CaCO_3(s) + 2HCl(aq) \longrightarrow CaCl_2(aq) + H_2O(l) + CO_2(g)$$

calcium carbonate	*hydrochloric acid*	*calcium chloride*

    water   *carbon dioxide*

The number of moles of $CaCO_3$ is

$$\frac{15 \text{ g}}{100.0 \text{ g/mole}} = 0.15 \text{ mole } CaCO_3$$

The chemical equation tells us that 1 mole of CO_2 is produced for every mole of $CaCO_3$, so

$$(0.15 \text{ mole } CaCO_3) \left(\frac{1 \text{ mole } CO_2}{1 \text{ mole } CaCO_3} \right) = 0.15 \text{ mole } CO_2$$

is produced in this reaction. Using the ideal gas law, this corresponds to

$$V = \frac{nRT}{P} = \frac{(0.15 \text{ mole})(0.0821 \text{ liter-atm/mole }°K)(295°K)}{(735 \text{ torr})(1 \text{ atm}/760 \text{ torr})} = 3.8 \text{ liters}$$

This reaction will produce 3.8 liters of CO_2. ■

EXAMPLE 4.11

■ A by-product of an industrial process was a mixture of sodium sulfate, Na_2SO_4, and sodium bicarbonate, $NaHCO_3$. To determine the composition of the mixture, a sample weighing 8.00 g was heated until constant weight was achieved, indicating that the heat-induced reaction was complete. Under these conditions the sodium bicarbonate undergoes the following reaction

$$2NaHCO_3(s) \xrightarrow{\Delta} Na_2CO_3(s) + CO_2(g) + H_2O(g)$$

sodium bicarbonate   *sodium carbonate*   *carbon dioxide*   *water*

and the sodium sulfate is unchanged. The volume of CO_2 released by the sample was 775 ml measured at 1.00 atm and 22°C. What was the weight percent of $NaHCO_3$ in the original by-product?

The number of moles of CO_2 can be calculated using the ideal gas law as

$$n = \frac{PV}{RT} = \frac{(1.00 \text{ atm})(0.775 \text{ liter})}{(0.0821 \text{ liter-atm}/\text{mole °K})(295°K)} = 3.20 \times 10^{-2} \text{ mole } CO_2$$

The chemical equation tells us that 2 moles of $NaHCO_3$ react to form 1 mole of CO_2, so

$$(3.20 \times 10^{-2} \text{ mole } CO_2) \left(\frac{2 \text{ moles } NaHCO_3}{1 \text{ mole } CO_2} \right)$$
$$= 6.40 \times 10^{-2} \text{ mole } NaHCO_3$$

was originally present. This corresponds to

$$(6.40 \times 10^{-2} \text{ mole } NaHCO_3)(84.1 \text{ g}/\text{mole}) = 5.38 \text{ g } NaHCO_3$$

in the sample, or

$$\frac{5.38 \text{ g}}{8.00 \text{ g}} \times 100\% = 67.3\%$$

The by-product contained 67.3% $NaHCO_3$. ∎

There is an important point to be noted about Example 4.12, in which consecutive reactions are involved: Because the products of the first reaction are the reactants in the second reaction, the masses of substances formed in the first reaction need not be calculated. Instead, the molar relationships can be used to go directly to the information desired about the final reaction.

EXAMPLE 4.12

∎ The production of potassium permanganate requires two steps. The first reaction involves converting manganese dioxide (a naturally occurring mineral known as pyrolusite) to potassium manganate.

$$2MnO_2(s) + 4KOH(aq) + O_2(g) \longrightarrow 2K_2MnO_4(aq) + 2H_2O(l)$$

manganese dioxide   potassium hydroxide   oxygen   potassium manganate   water

and the second reaction involves changing potassium manganate to potassium permanganate

$$2K_2MnO_4(aq) + Cl_2(g) \longrightarrow 2KMnO_4(aq) + 2KCl(aq)$$

potassium manganate   chlorine   potassium permanganate   potassium chloride

What mass of $KMnO_4$ will be produced from 100.0 g of MnO_2?

The number of moles of MnO_2 is

$$\frac{100.0 \text{ g}}{86.94 \text{ g/mole}} = 1.150 \text{ moles } MnO_2$$

The first chemical equation tells us that 2 moles of K_2MnO_4 are produced for every 2 moles of MnO_2 reacting and, in turn, the second chemical equation tells us that 2 moles of $KMnO_4$ are produced for every 2 moles of K_2MnO_4, so

$$(1.150 \text{ moles } MnO_2) \left(\frac{2 \text{ moles } K_2MnO_4}{2 \text{ moles } MnO_2}\right)\left(\frac{2 \text{ moles } KMnO_4}{2 \text{ moles } K_2MnO_4}\right)$$

$$= 1.150 \text{ moles } KMnO_4$$

This corresponds to

$$(1.150 \text{ moles } KMnO_4)(158.04 \text{ g/mole}) = 181.7 \text{ g } KMnO_4$$

For every 100.0 g of MnO_2 reacting, 181.7 g of $KMnO_4$ will be produced.

The determination of the empirical formula of a substance from data about its weight composition or percentage composition was described in Section 2.10. Example 4.13 shows how an empirical formula can be determined from information revealed by the reaction used in analyzing the compound. ∎

EXAMPLE 4.13

∎ A 2.47 g sample of a compound containing only C, H, and O was burned, giving 4.73 g of CO_2 and 2.90 g of H_2O. Determine the empirical formula of the compound.

Even though we don't know the formula of the compound, we can write the following unbalanced equation representing the reaction

$$C_xH_yO_z + wO_2 \longrightarrow xCO_2 + \left(\frac{y}{2}\right) H_2O$$

The number of moles of CO_2 and water are

$$\frac{4.73 \text{ g}}{44.0 \text{ g/mole}} = 0.108 \text{ mole } CO_2 \qquad \frac{2.90 \text{ g}}{18.0 \text{ g/mole}} = 0.161 \text{ mole } H_2O$$

Our chemical equation tells us that for every mole of CO_2, there is a mole of C atoms in the compound, so we have 0.108 mole of C atoms in our sample. Similarly, for the hydrogen, we have 2(0.161) = 0.322 mole of H atoms in the compound. These two elements account for

$$(0.108 \text{ mole})(12.0 \text{ g/mole}) + (0.322 \text{ mole})(1.00 \text{ g/mole}) = 1.62 \text{ g}$$

of the sample. The remaining weight, 2.47 g − 1.62 g = 0.85 g, must be the amount of O in the sample. This is equivalent to

$$\frac{0.85 \text{ g}}{16.0 \text{ g/mole}} = 0.053 \text{ mole O atoms}$$

The formula of the compound is $C_{0.108}H_{0.322}O_{0.053}$; converted to whole-number subscripts, this is C_2H_6O. ■

Important types of reactions

4.6 Reversible reactions and chemical equilibrium

Equilibrium is a state of balance between equal and opposing forces. When you stand on your bathroom scales to weigh yourself, you and the scales are in equilibrium—your body exerts a downward force and the scales exert an equal force in the opposite direction. This is **static equilibrium,** a state of balanced forces with no motion.

Because molecules are constantly in motion, any equilibrium in a chemical system is a dynamic equilibrium, like the liquid-vapor equilibrium in a terrarium (Section 3.15). Equal and opposite changes occur at the same rate in a dynamic equilibrium.

A chemical equilibrium is set up by any chemical reaction that can proceed in either direction—a **reversible reaction.** The classic example of a reversible reaction involves iron and iron oxide (Fe_3O_4), and hydrogen and water. In 1766, Sir Henry Cavendish discovered the element hydrogen by passing steam through a red-hot iron gun barrel.

$$3Fe(s) + 4H_2O(g) \xrightarrow{\Delta} Fe_3O_4(s) + 4H_2(g) \tag{21}$$

When the hydrogen is removed from the reaction system, the reaction proceeds in the direction shown by Equation (21).

If instead, hydrogen gas is passed over hot iron oxide, the reaction that occurs is exactly the reverse of the one that Cavendish carried out.

$$Fe_3O_4(s) + 4H_2(g) \xrightarrow{\Delta} 3Fe(s) + 4H_2O(g) \tag{22}$$

Here also, the reaction proceeds in the direction shown when one of the products, in this case the steam, is removed from the reaction system.

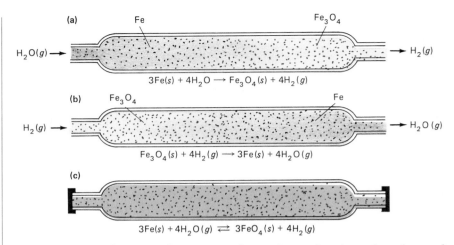

FIGURE 4.2

A reversible reaction. *The direction of the reaction in* (a) *and* (b) *is determined by the removal of a component of the reaction system. Equilibrium is reached in the closed system* (c).

TABLE 4.5

Classic properties of solutions of acids and bases

Early chemists blithely tasted their chemicals. Experience has shown this can be a fatal mistake. Never *taste any chemicals.*

  Acids
Sour taste
Change color of indicator dyes
  e.g., litmus turns from blue to red;
  phenolphthalein turns from red to
  colorless
React with active metals to give hydrogen
  e.g., Zn(s) + 2H⁺ ⟶ Zn²⁺ + H₂(g)
**Acidic properties disappear in reaction
with a base**

  Bases
Bitter taste
Slippery feeling
Change color of indicator dyes
  e.g., litmus turns from red to blue;
  **phenolphthalein turns from colorless
  to red**
**Basic properties disappear in reaction
with an acid**

However, if iron and steam are heated together in a closed vessel the situation is different (Figure 4.2). As soon as traces of iron oxide and hydrogen have formed, they begin to react with each other to form iron and steam; reactions (21) and (22) proceed simultaneously.

At first, the concentration of steam is relatively high and reaction (21) is faster. As steam is consumed, this reaction slows down. At the same time, since a supply of hydrogen is building up, the reverse reaction (22) between H_2 and Fe_3O_4 speeds up. Eventually, both reactions proceed with equal speed. Once this point is reached, the amounts of reacting materials do not change. The two reactions have reached an equilibrium. The reactions have not stopped but each exactly undoes the work of the other.

In **chemical equilibrium** the rates of the forward and reverse reactions are the same, and the amounts of the species present do not change with time. An equilibrium reaction is written with a double arrow to indicate reversibility.

$$3Fe(s) + 4H_2O(g) \rightleftharpoons Fe_3O_4(s) + 4H_2(g)$$

A common misunderstanding upon first encountering an equilibrium reaction is to think that the materials on the two sides of the equilibrium equation are present in equal amounts or in equal concentrations. This happens rarely. Most often the substances on one side of the equation are present in greater amounts than the substances on the other side. It can be either side. The temperature, the amounts of reactants initially present, the reaction conditions, and the characteristics of the particular reaction determine which side will be favored.

4.7 Water, a slightly ionized liquid

Water molecules in the liquid state are in equilibrium with hydrogen ions, H^+, and hydroxide ions, OH^-.

$$H_2O(l) \rightleftharpoons H^+ + OH^- \tag{23}$$

The concentration of the ions is ordinarily very small. The forward and reverse reactions in Equation (23) balance out with only a small number of the water molecules ionized.

However, the equilibrium concentration of H^+ and OH^- can be changed by dissolving other substances in the water. Therefore, this equilibrium is important in determining the properties of aqueous solutions, particularly solutions of acids and bases.

At equilibrium, amounts of reactants and products are constant; forward and reverse rates are equal.

4.8 Acids and bases: H^+ and OH^-

Acids and bases were first recognized as specific classes of compounds because of the distinctive properties exhibited by their aqueous solutions. The classic acid and base properties are given in Table 4.5, and a few common acidic and basic substances are listed in Table 4.6.

In the 1880s Svante Arrhenius, a Swedish chemist, developed a theory of ionization. Part of his theory explains that the classic acid properties are imparted to a solution by the hydrogen ion, H^+, and the basic properties by the hydroxide ion, OH^-. An **acidic aqueous solution** contains a greater concentration of H^+ ions than OH^- ions. And a basic aqueous solution, better called an **alkaline aqueous solution,** contains a greater concentration of OH^- ions than H^+ ions. The term "alkaline" is preferable because "basic" also has other meanings. "Alkaline" derives from "alkali," an old name for substances with the classic properties.

According to the theory of Arrhenius, a base is any compound that contains a hydroxide group and yields a hydroxide ion in water.

The acidity of vinegar is due to acetic acid ($HC_2H_3O_2$). Fruits and vegetables derive their acidity from various organic acids. Gastric juice contains hydrochloric acid (HCl). Carbonated beverages are acidic because of the reaction of CO_2 with water. Aspirin and vitamin C are organic acids, acetylsalicylic acid and ascorbic acid, respectively. Household ammonia is basic because of the reaction of ammonia gas (NH_3) with water (see Table 4.7), and milk of magnesia contains magnesium hydroxide [$Mg(OH)_2$]. Soap and detergents contain various organic and inorganic bases.

TABLE 4.6

Common acidic and alkaline substances

Some acidic substances	Some alkaline substances
Vinegar	**Household ammonia**
Tomatoes	**Baking soda solution**
Citrus fruits	**Soap**
Carbonated beverages	**Detergents**
Black coffee	**Milk of magnesia**
Gastric fluid	**Oven cleaners**
Vitamin C	**Lye**
Aspirin	**Drano**
Saniflush	

TABLE 4.7

Some H^+ acids and OH^- bases

Arrhenius acids

HCl	Hydrochloric acid
HBr	Hydrobromic acid
HI	Hydroiodic acid
HNO_3	Nitric acid
H_2SO_4	Sulfuric acid
$HClO_4$	Perchloric acid
HCN	Hydrocyanic acid
$HC_2H_3O_2$	Acetic acid
H_3PO_4	Phosphoric acid

Arrhenius bases

NaOH	Sodium hydroxide
KOH	Potassium hydroxide
$Ba(OH)_2$	Barium hydroxide
$Ca(OH)_2$	Calcium hydroxide
$Mg(OH)_2$	Magnesium hydroxide

Other bases

Na_2O + $H_2O \longrightarrow 2Na^+ + 2OH^-$
sodium
oxide

CaO + $H_2O \longrightarrow Ca^{2+} + 2OH^-$
calcium
oxide

NH_3 + $H_2O \rightleftharpoons NH_4^+ + OH^-$
ammonia

The classic bases are ionic hydroxides that dissociate in water, such as sodium hydroxide (see Table 4.7 for others).

$$NaOH(s) \xrightarrow{H_2O} Na^+ + OH^-$$

We also think of compounds such as ammonia and certain metal oxides as bases, because they react with water to give hydroxide ions.

$$\underset{\text{calcium oxide}}{CaO(s)} + H_2O(l) \longrightarrow Ca^{2+} + 2OH^-$$

$$NH_3(g) + H_2O(l) \rightleftharpoons NH_4^+ + OH^-$$

A **base** is a substance that in aqueous solution increases the hydroxide ion concentration.

The classic Arrhenius acids are nonionic compounds containing hydrogen that ionize in water to give H^+, for example, sulfuric acid. (See Table 4.7 for others.)

$$H_2SO_4(l) \xrightarrow{H_2O} 2H^+ + SO_4^{2-}$$

An **acid** is a substance that in aqueous solution increases the hydrogen ion concentration.

Compounds such as sodium hydroxide and sulfuric acid, which give high concentrations of hydroxide and hydrogen ions, respectively, are strong bases and strong acids.

Broader definitions of acids and bases have been found useful for explaining many types of chemical reactions. These definitions are discussed in Chapter 17. For now, we can work with this picture of acids and bases as substances that affect the hydrogen ion and hydroxide ion concentrations of aqueous solutions.

4.9 Neutralization

Neutralization is the reaction of an acid with a base. Here are the equations for two neutralization reactions.

$$\underset{\substack{\text{hydrochloric} \\ \text{acid}}}{HCl(aq)} + \underset{\substack{\text{sodium} \\ \text{hydroxide}}}{NaOH(aq)} \longrightarrow \underset{\substack{\text{sodium} \\ \text{chloride}}}{NaCl(aq)} + \underset{\text{water}}{H_2O(l)} \qquad (24)$$

$$\underset{\text{sulfuric acid}}{H_2SO_4(aq)} + \underset{\substack{\text{potassium} \\ \text{hydroxide}}}{2KOH(aq)} \longrightarrow \underset{\substack{\text{potassium} \\ \text{sulfate}}}{K_2SO_4(aq)} + \underset{\text{water}}{2H_2O(l)} \qquad (25)$$

The products of the reaction of an H^+ acid with an OH^- base are water and an ionic compound called a salt. **Salts** are ionic compounds composed of the cations of bases and the anions of acids. The salts formed in reactions (24) and (25) are soluble in water and remain in solution as dissociated ions. Crystalline sodium chloride and potassium sulfate would be obtained if the water were evaporated.

If chemically equivalent amounts (see next section) of a strong base and a strong acid react, both the acidic and the basic properties are "neutralized"—they disappear completely. A **neutral aqueous solution** contains equal numbers of H^+ and OH^- ions.

When reactions (24) and (25) are written as ionic equations

$$H^+ + Cl^- + Na^+ + OH^- \longrightarrow Na^+ + Cl^- + H_2O(l)$$
$$2H^+ + SO_4{}^{2-} + 2K^+ + 2OH^- \longrightarrow 2K^+ + SO_4{}^{2-} + 2H_2O(l)$$

it becomes apparent that the Na^+ and Cl^- ions, and the K^+ and $SO_4{}^{2-}$ ions are spectators, not participants. The net ionic equation for every neutralization reaction between H^+ acids and OH^- bases is the reverse of the ionization of water.

$$H^+ + OH^- \rightleftharpoons H_2O(l)$$

Hydrogen ions and hydroxide ions combine to form water molecules until the normal pure water equilibrium between the ions and molecules is reestablished.

Some salts formed in neutralization reactions are not soluble in water

$$\underset{\substack{\text{barium} \\ \text{hydroxide}}}{Ba(OH)_2(aq)} + \underset{\substack{\text{sulfuric} \\ \text{acid}}}{H_2SO_4(aq)} \longrightarrow \underset{\substack{\text{barium} \\ \text{sulfate}}}{BaSO_4(s)} + \underset{\text{water}}{2H_2O(l)}$$

In this reaction, two chemical changes are taking place at the same time: the neutralization reaction $H^+ + OH^- \rightleftharpoons H_2O$ and the formation of crystalline barium sulfate from Ba^{2+} and $SO_4{}^{2-}$ ions.

4.10 Equivalent weight and normality for acids and bases

The concept of equivalents was historically very important to chemistry. Information about the specific amounts of one element or compound that were chemically equivalent to specific amounts of other compounds eventually led to our understanding of atomic weight, mass in chemical combination, and stoichiometry. For example, Richter and Ernst Fischer in the 1800s published tables of equivalents of acids and bases. They knew that 1000 parts by weight of sulfuric acid or 1405 parts of nitric acid would both be neutralized by 672 parts of ammonia. These quantities of sulfuric acid, nitric acid, and ammonia were chemically equivalent in neutralization reactions, although Richter and Fischer did not know why.

Our present knowledge of stoichiometry and the concept of the mole has made the use of equivalents unnecessary in most circumstances. The molar weight of a substance is more useful; it is based solely on the identity of the substance and has only one value, whatever the fate of the substance in a reaction. Equivalent weight, however, is a quantity based on the behavior of a substance in a particular reaction. The definition of equivalent weight varies with the type of reaction considered, and the equivalent weight of the same substance may also be different in different reactions.

Equivalent weights are likely to be encountered today mainly in connection with acid–base reactions. The **equivalent weight of an acid** is the weight of the acid that will give one mole of H^+ ions. The **equivalent weight of a base** is the weight of a base that will give one mole of OH^- ions. One equivalent of any acid will neutralize one equivalent of any base.

The equivalent weight of an acid or a base is its molar weight divided by the number of "equivalents" the compound supplies per mole—that is, the number of moles of H^+ or OH^- ions available in the reaction under consideration.

$$\text{equiv wt} = \frac{\text{molar wt}}{\text{equiv/mole}} \tag{26}$$

In the neutralization reactions in Table 4.8, one mole of hydrochloric acid supplies one mole of hydrogen ions in each of the first three reactions. Therefore, the equivalent weight of hydrochloric acid is the same as its molar weight

$$\text{equiv wt HCl} = \frac{36.45 \text{ g/mole}}{1 \text{ equiv/mole}} = 36.45 \text{ g/equiv}$$

Sodium hydroxide and ammonia each take up 1 mole of hydrogen ion and their equivalent weights are also the same as their molar weights. However, 1 mole of calcium hydroxide supplies 2 moles of OH^- ion (reaction 2, Table 4.8) and 1 mole of sulfuric acid supplies 2 moles of hydrogen ion (reaction 4, Table 4.8). Their respective equivalent weights, therefore, are

$$\text{equiv wt Ca(OH)}_2 = \frac{74.10 \text{ g/mole}}{2 \text{ equiv/mole}} = 37.05 \text{ g/equiv}$$

$$\text{equiv wt H}_2\text{SO}_4 = \frac{98.08 \text{ g/mole}}{2 \text{ equiv/mole}} = 49.04 \text{ g/equiv}$$

Note that the number of equivalents per mole for the same compound

TABLE 4.8
Equivalents of acids and bases

All of the reactants and products in these reactions are in aqueous solution.

---

(1)   **HCl**   +   **KOH**   ⟶ **KCl + H₂O**
    1 equiv/mole   *1 equiv/mole*

(2)   **2HCl**   +   **Ca(OH)₂**   ⟶ **CaCl₂ + 2H₂O**
    1 equiv/mole   *2 equiv/mole*

(3)   **HCl**   +   **NH₃**   ⟶ **NH₄Cl**
    1 equiv/mole   *1 equiv/mole*

(4)   **H₂SO₄**   +   **2NaOH**   ⟶ **Na₂SO₄ + 2H₂O**
    2 equiv/mole   *1 equiv/mole*

(5)   **H₃PO₄**   +   **3NaOH**   ⟶ **Na₃PO₄ + 3H₂O**
    3 equiv/mole   *1 equiv/mole*

(6)   **H₃PO₄**   +   **2NaOH**   ⟶ **Na₂HPO₄ + 2H₂O**
    2 equiv/mole   *1 equiv/mole*

---

can vary—in reaction 5 (Table 4.8) H_3PO_4 gives up three hydrogen ions, but in reaction 6 it gives up only two hydrogen ions.

A solution containing one equivalent of a substance in 1 liter of solution (Figure 4.3) is called a "one-normal solution", written $1N$. **Normality** is the concentration of a solution expressed in equivalents per liter, as opposed to molarity, which is moles per liter. A solution with 2 equivalents per liter is $2N$, and 0.5 liter of a $2N$ solution contains one equivalent of solute. From the equivalent weights just discussed, we know that a $1M$ solution of hydrochloric acid is also a $1N$ solution, but that a $1M$ solution of sulfuric acid is $2N$ in reactions in which it yields two equivalents of hydrogen ion per mole.

If the normality of a solution is expressed as N and the volume of the solution in liters as V, then NV is equal to the number of equivalents of solute in the given volume of solution.

$$V \text{ (liters)} \times N \text{ (equiv/liter)} = \text{no. of equivalents}$$

Often it is convenient to express a smaller volume in milliliters. The relationship then becomes

$$V(\text{ml}) \times \frac{N \text{ equiv/liter}}{1000 \text{ ml/liter}} = \text{no. of equivalents}$$

Equal volumes of solutions of acids and bases of equal normality will exactly neutralize each other. If the normality of one solution is unknown, it can be determined from the volume of the other solution used in the reaction,

$$N_1V_1 = N_2V_2$$

where 1 and 2 indicate the two solutions.

EXAMPLE 4.14

■ What is the mass of H_2SO_4 in 100.0 ml of commercially available concentrated sulfuric acid that is $35.9N$? Assume H_2SO_4 acts as an acid that will lose two hydrogen ions, so that 1 mole = 2 equivalents.

The number of equivalents in 100.0 ml of $35.9N$ acid is

$$\text{no. of equiv} = NV = (35.9 \text{ equiv/liter})(0.1000 \text{ liter}) = 3.59 \text{ equiv}$$

The equivalent weight is one-half the molecular weight, or 49.0 g/equiv. Thus

$$(3.59 \text{ equiv})(49.0 \text{ g/equiv}) = 176 \text{ g } H_2SO_4$$

There is 176 g of H_2SO_4 in each 100.0 ml of concentrated sulfuric acid. Note that this concentration in molarity is

$$\text{molarity} = \frac{n}{V} = \frac{176 \text{ g}/(98.0 \text{ g/mole})}{0.100 \text{ liter}} = 18.0M$$

■

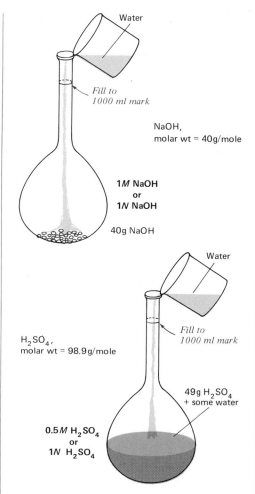

FIGURE 4.3
Preparation of standard solutions.
Equal volumes of these two solutions will completely neutralize each other.

$$2NaOH + H_2SO_4 \longrightarrow Na_2SO_4 + 2H_2O$$

4.11 Oxidation state

In Chapter 2 we described Stock nomenclature, in which Roman numerals are used to indicate the oxidation number of an element in a compound (Appendix D). For monatomic ions, the oxidation number is equal to the charge on the ion. However, not all compounds contain monatomic ions. By use of the rules given in this section, oxidation numbers can be assigned to almost every atom in every molecule or polyatomic ion. The **oxidation number** or **oxidation state**—the terms are used interchangeably—is a number that represents the positive or negative character of atoms in a compound. Oxidation state can be thought of as expressing the tendency of an atom to attract electrons, that is to become more negative, or to give up electrons, that is to become more positive.

The oxidation number is useful in writing formulas, in recognizing oxidation–reduction reactions (Section 4.12), and in balancing oxidation–reduction equations. It also helps to predict the properties of compounds. Most elements can have more than one oxidation state, and there tend to be similarities among compounds with a given element in the same oxidation state.

The rules for assigning oxidation numbers are given in the following paragraphs. Table 4.9 summarizes the rules and gives examples. Study the table while you study the rules.

1. The oxidation number of any element in the free state is zero. Whether the element is polyatomic or not, the oxidation number of a free element is zero.

2. The oxidation number of a monatomic ion is equal to the charge on the ion. For example, in $CaCl_2$ the oxidation number of the chlorine (Cl^-) is -1 and the oxidation number of the calcium (Ca^{2+}) is $+2$.

3. Oxygen has an oxidation number of -2. Oxygen is assigned an oxidation number of -2 in all ionic and nonionic compounds and in all polyatomic ions. The only exceptions are peroxides and superoxides, which contain O–O bonds, and the few compounds that contain O–F bonds.

4. Hydrogen has an oxidation number of $+1$. In virtually all of its compounds, the oxidation number of hydrogen is $+1$. The only exception is in the binary metal hydrides, such as NaH, where hydrogen is assigned an oxidation number of -1.

5. For a neutral compound, the algebraic sum of the oxidation numbers of all the atoms must equal zero. By using the preceding four rules together with this rule or rule 6, it becomes possible to calculate oxidation numbers for atoms in many compounds without any additional information.

6. For a polyatomic ion, the algebraic sum of the oxidation numbers of the atoms must equal the charge on the ion. For example, in the carbonate ion, CO_3^{2-}, oxygen contributes a total oxidation number of 3×-2, or -6; so to have a total charge of -2, the oxidation number of carbon must be $+4$.

To find the oxidation number of sulfur in H_2SO_4, use the assigned oxidation numbers of $+1$ for hydrogen and -2 for oxygen.

$$\text{total oxid. no. H} + \text{oxid. no. S} + \text{total oxid. no. O} = 0$$
$$[2(+1)] + \text{oxid. no. S} + [4(-2)] = 0$$
$$\text{oxid. no. S} = -2 + 8 = +6$$

The oxidation number of sulfur in H_2SO_4 is $+6$. Although we have written out an equation to illustrate this process, an equation is not

In (5) and (6) the oxidation numbers calculated using the rules are shown in color.

---

(1) Free elements, oxid. no. = 0

$\overset{0}{Ca}$ $\overset{0}{O_2}$ $\overset{0}{S_8}$

(2) Monatomic ions, oxid. no. = ionic charge

$\overset{+1}{Na}\overset{-1}{Cl}$ $\overset{+2}{Ca}\overset{-1}{F_2}$ $\overset{+1}{Cu}\overset{-1}{Br}$ $\overset{+2}{Cu}\overset{-1}{Br_2}$

(3) Oxygen, oxid. no. = −2 (except, e.g., $\overset{+1\ -1}{H_2O_2}$; $\overset{+2-1}{OF_2}$)

$\overset{-2}{H_2O}$ $\overset{-2}{BaO}$ $\overset{-2}{H_2SO_4}$ $\overset{-2}{CO_2}$

(4) Hydrogen, oxid. no. = +1 (except, e.g., $\overset{+1\ -1}{NaH}$)

$\overset{+1}{H_2O}$ $\overset{+1}{H_2SO_4}$ $\overset{+1}{Ba(OH)_2}$ $\overset{+1}{CH_4}$

(5) Neutral compounds, sum of oxid. nos. = 0

$\overset{2(+5)\ \ 5(-2)}{As_2\quad O_5}$ $\overset{+1\ +7\ \ 4(-2)}{K\ Cl\ O_4}$ $\overset{-3\ \ 3(+1)}{N\ H_3}$ $\overset{+4\ \ 4(-1)}{C\ Cl_4}$

(6) Polyatomic ions, sum of oxid. nos. = charge of ion

$\left(\overset{-2\ +1}{O\ H}\right)^{-}$ $\left(\overset{+6\ \ 4(-2)}{S\ O_4}\right)^{2-}$ $\left(\overset{+3\ \ 2(-2)}{Cl\ O_2}\right)^{-}$ $\left(\overset{2(+5)\ \ 7(-2)}{P_2\quad O_7}\right)^{4-}$

---

really necessary. Just by examining the formula and the known oxidation states, you can see what is needed to maintain neutrality.

$\overset{2(+1)}{H_2}\quad \overset{+6}{S}\quad \overset{4(-2)}{O_4}$

In a binary compound, the element that usually forms negative ions is assigned an oxidation number equal to the charge it would have if it were present as a negative ion (whether it is ionic or not). The other element is given the positive oxidation number that will make the molecule neutral. For example, in PCl_3 and PCl_5 chlorine is assigned an oxidation number of − 1, resulting in oxidation numbers of + 3 and + 5 for phosphorus.

It is important to understand that oxidation numbers, while useful for bookkeeping, do not have exact physical significance, as do the charges on ions. For example, in a compound like H_2S, the oxidation states of + 1 for hydrogen and − 2 for sulfur do not mean that hydrogen and sulfur are present as ions with these charges.

EXAMPLE 4.15

■ Using the rules discussed above, find the oxidation states of sulfur in the following species: (a) S_8, (b) S, (c) S^{2-}, (d) H_2S, (e) SO_2, (f) SO_3, (g) HSO_4^-, (h) H_2SO_4, and (i) SO_3^{2-}.

The first rule tells us that the oxidation number for S in (a) and (b) is zero. The oxidation number of S in (c) is − 2 according to our second rule. The fourth and fifth rules tell us that in (d) the sulfur must be − 2 so that the total charge on the molecule is zero. Our third and fifth rules tell us that in (e) the sulfur is + 4 and in (f) the sulfur is + 6 so that these molecules are neutral.

TABLE 4.9
Rules for assigning oxidation numbers

Oxidation: increase in
oxidation number
Reduction: decrease in
oxidation number

If we apply rules three through six to the remainder of the substances in a more mathematical form, we obtain

(g) HSO_4^- $1(+1) + 1(x) + 4(-2) = -1$ $x = +6$

(h) H_2SO_4 $2(+1) + 1(x) + 4(-2) = 0$ $x = +6$

(i) SO_3^{2-} $1(x) + 3(-2) = -2$ $x = +4$

The oxidation number of sulfur is $+6$ in HSO_4^- and H_2SO_4, and it is $+4$ in SO_3^{2-}. ∎

4.12 Oxidation–reduction reactions

Any process in which an oxidation number increases algebraically is **oxidation.** Any process in which an oxidation number decreases algebraically is **reduction** (see Figure 4.4). These modern definitions hold for all the many types of reactions, both simple and complicated, that have been classified as oxidations or reductions.

For example, look at potassium in the reaction of potassium with iodine to form potassium iodide, an ionic compound. Applying the rules for the assignment of an oxidation number

$$\overset{0}{2K}(s) + I_2(g) \longrightarrow \overset{+1}{2KI}(s) \tag{27}$$

and the definition given above, we find that potassium has been oxidized—it has gone from an oxidation state of 0 to an oxidation state of $+1$. In going to a more positive oxidation state, an atom becomes more positive and less negative in character. In the case of potassium in reaction (27), this is easily explained by the formation of the ionic compound K^+I^-. The neutral potassium atoms have become positive potassium ions by the loss of one electron each.

The iodine has also undergone a change in oxidation state

$$2K(s) + \overset{0}{I_2}(g) \longrightarrow 2K\overset{-1}{I}(s) \tag{28}$$

In changing from an iodine molecule to iodide ions, the iodine atoms have been reduced. Each iodine atom has undergone a decrease in oxidation number that occurred because an electron was added to give a negative iodide ion.

Equations (27) and (28) illustrate an important point: Oxidation and reduction always occur together. Every electron that is shifted in the oxidation part of a reaction is utilized in the reduction process: every electron that is used in the reduction part of a reaction is supplied by the oxidation process. **Oxidation–reduction reactions,** sometimes called *redox reactions,* are reactions in which oxidation and reduction occur together. The total increase in oxidation number equals the total decrease in oxidation number. The utility of this concept in balancing oxidation–reduction equations is illustrated in Section 19.4. All chemical reactions can be classified as either oxidation–reduction reactions or non-oxidation–reduction reactions (reactions in which no oxidation number change occurs).

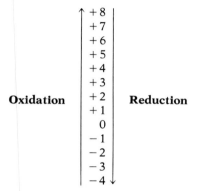

FIGURE 4.4

Direction of oxidation state change in oxidation and reduction. *The values +8 and −4 represent the maximum and minimum known oxidation states.*

An atom, molecule, or ion that can cause another substance to undergo an increase or decrease in oxidation state is called an **oxidizing agent** or a **reducing agent,** respectively. In the reaction between ammonia and oxygen to give nitrogen(II) oxide,

$$\overset{-3}{4NH_3}(g) + \overset{0}{5O_2}(g) \rightleftharpoons \overset{+2-2}{4NO}(g) + \overset{-2}{6H_2O}(l)$$

ammonia is the reducing agent and oxygen is the oxidizing agent. The oxidizing agent is itself reduced; the reducing agent is itself oxidized. Note that the entire compound—not just the atom that changes oxidation number—is usually called the oxidizing or reducing agent.

**THOUGHTS ON
CHEMISTRY**

Stoichiometry

Mathematics includes all those sciences which refer to magnitude, and consequently a science lies more or less in the province of mathematics (geometry), according as it requires the determination of magnitudes. In chemical experiments this truth has often led me to the question, whether and how far chemistry is a part of applied mathematics; and especially in considering the well-known fact, that two neutral salts, when they decompose each other, form again neutral compounds. The immediate consequence, in my opinion, could only be, that there are definite relations between the magnitude of the component parts of neutral salts. From that time I considered how these proportions could be made out, partly by exact chemical experiments, partly combining chemical with mathematical analysis. In my inaugural dissertation, published at Konigsberg in 1789, I made a slight attempt, but was not then supplied with the requisite chemical apparatus, nor was I sufficiently ready with all requisite information bearing on my present system, imperfect as it may be. The result, therefore, was very imperfect. I promised, however, not to let the matter rest with that imperfect essay, but to work out this branch with all the accuracy and profundity of which I was capable, as soon as I was supplied with the requisite conveniences. This promise, I hope in the present volume, to make good, although I am far from believing that what I am now going to say will not be in need of still more thorough and accurate elaboration, for who will venture to limit the extent and the power which is the destination of a young and budding science. . . .

. . . As the mathematical portion of chemistry deals in a great measure with bodies which are either elements or substances incapable of being decomposed and as it teaches also their relative magnitudes, I have been able to find no more fitting name for this scientific discipline than the word stoechiometry, from στοιχεῖον which, in

STUDIES ON STOICHIOMETRY,
by Jeremias B. Richter

STOICHIOMETRY [4.12]

the Greek language, means something which cannot be divided, and
μετρειν which means to find out relative magnitudes. . . .

DEFINITION 1.

Stoechiometry (stoechyometria) is the science of measuring the quantitative proportions, or the proportions of the masses in which chemical elements stand in regard to each other. The mere knowledge of these relations might be called quantitative stoechiology.

STUDIES ON STOICHIOMETRY, by Jeremias B. Richter, 1792–1794 (Quoted from pp. 205–207 in *A Source Book in Chemistry*, H. M. Leicester and H. S. Klickstein (eds.), McGraw-Hill, New York, 1952)

Significant terms defined in this chapter

reactants	acidic aqueous solution
products	alkaline aqueous solution
chemical equation	base
catalyst	acid
polymer	neutralization
precipitation	salts
precipitate	neutral aqueous solution
spectator ions	equivalent weight of an acid
net ionic equation	equivalent weight of a base
stoichiometry	normality
limiting reactant	oxidation number
stoichiometric amounts	oxidation state
theoretical yield	oxidation
percent yield	reduction
equilibrium	oxidation–reduction reactions
static equilibrium	oxidizing agent
reversible reaction	reducing agent
chemical equilibrium	

Exercises

4.1 Consider the hypothetical photosynthesis reaction described by the equation

$$CO_2(g) + H_2O(l) \xrightarrow[\text{chlorophyll}]{h\nu \text{ (light)}} C_{12}H_{22}O_{11}(aq) + O_2(g)$$

Identify the (a) reactants, (b) products, and (c) the catalyst. (d) What additional factor is indicated by the equation as necessary for the reaction to occur? (e) Balance the equation. How many molecules of (f) CO_2 and (g) H_2O are necessary to produce one molecule of $C_{12}H_{22}O_{11}$? How many moles of (h) CO_2 and (i) H_2O are necessary to produce one mole of $C_{12}H_{22}O_{11}$? How many atoms of (j) C, (k) H, and (l) O are necessary to produce one molecule of $C_{12}H_{22}O_{11}$? If the reaction were carried out in a closed container such as a biological cell, (m) does the reaction conserve mass and (n) would there be a pressure change?

4.2 Which statements are true? Rewrite any false statement so that it is correct.
(a) Reactants are substances that are changed in a chemical reaction.
(b) A catalyst is a compound that forms as a solid and comes out of solution during a reaction.
(c) The net ionic equation shows only the species involved in the chemical change and not all of the ions present.
(d) A limiting reagent is present in largest stoichiometric amount and determines the maximum amount of product that can be formed.
(e) The theoretical yield is found by dividing the mass of the product obtained by the maximum amount possible and multiplying by 100%.
(f) Under conditions of chemical equilibrium, the rates of the forward and reverse reactions are the same so that, even though this is a dynamic system, the concentrations or amounts of the species present do not change with time.
(g) Dalton's law of partial pressures allows us to find relative ratios of volumes of gases reacting and being produced in a chemical equation.
(h) Ionization is the formation of ions from an ionic material.
(i) Spectator ions have mass and charge and participate in a reaction.
(j) A reducing agent undergoes reduction.

4.3 Distinguish clearly between (a) a chemical reaction and a chemical equation, (b) a net ionic equation and an overall equation, (c) static and dynamic equilibrium, (d) limiting and excess reagents, (e) an acid and a base, (f) oxidation and reduction, (g) an oxidizing agent and a reducing agent, and (h) normality and molarity.

4.4 Balance each of the following chemical equations:
(a) $Sn(s) + NaOH(aq) \longrightarrow Na_2SnO_2(aq) + H_2(g)$
(b) $Sn(s) + NaOH(aq) + H_2O(l) \longrightarrow Na_2SnO_3(aq) + H_2(g)$

(c) $Cl_2O_7(g) + H_2O(l) \longrightarrow HClO_4(aq)$
(d) $Br_2(l) + H_2O(l) \longrightarrow HBr(aq) + HOBr(aq)$
(e) $Ca_3(PO_4)_2(s) + H_2SO_4(aq) \longrightarrow CaSO_4(s) + H_3PO_4(aq)$
(f) $Fe_3O_4(s) + H_2(g) \longrightarrow Fe(s) + H_2O(l)$
(g) $KClO_3(s) \longrightarrow KCl(s) + O_2(g)$
(h) $MnO_2(s) + HCl(aq) \longrightarrow MnCl_2(aq) + Cl_2(g) + H_2O(l)$
(i) $Fe^{2+}(aq) + H^+(aq) + Cr_2O_7^{2-}(aq) \longrightarrow$
$$Cr^{3+}(aq) + H_2O(l) + Fe^{3+}(aq)$$
(j) $Zn(s) + H^+(aq) + NO_3^-(aq) \longrightarrow$
$$NH_4^+(aq) + H_2O(l) + Zn^{2+}(aq)$$
(k) $H_2S(aq) + H^+(aq) + NO_3^-(aq) \longrightarrow$
$$NO(g) + H_2O(l) + S(s)$$
(l) steam and hot carbon reacting to form hydrogen and carbon monoxide
(m) potassium reacting with water to give potassium hydroxide and hydrogen
(n) magnesium carbonate decomposing to form magnesium oxide and carbon dioxide

4.5 A student wrote the following equation to describe the decomposition of hydrogen peroxide, H_2O_2, into oxygen and water:

$$H_4O_4 \longrightarrow 2H_2O + O_2$$

What is wrong?

4.6 How many moles of oxygen can be obtained by the decomposition of one mole of reactant in each of the following reactions?
(a) $2KClO_3(s) \longrightarrow 2KCl(s) + 3O_2(g)$
(b) $2H_2O_2(aq) \longrightarrow 2H_2O(l) + O_2(g)$
(c) $2HgO(s) \longrightarrow 2Hg(l) + O_2(g)$
(d) $2NaNO_3(s) \longrightarrow 2NaNO_2(s) + O_2(g)$

4.7 Ethylene can be produced from ethyl alcohol using phosphoric acid:

$$CH_3CH_2OH(g) \xrightarrow[210°C]{H_3PO_4} CH_2{=}CH_2(g) + H_2O(g)$$
$$\textit{ethyl alcohol} \qquad\qquad \textit{ethylene}$$

(a) How many moles of ethyl alcohol are needed to produce 0.1 mole ethylene? (b) Would 4 liters or 2 ml of water be produced when 0.1 mole of ethyl alcohol reacts? (c) How many molecules of water are produced for each mole of ethylene? (d) How many molecules of hydrogen are involved in this reaction for each mole of ethylene? (e) How many atoms of hydrogen are involved in this reaction for each mole of ethylene? (f) What is the catalyst?

4.8 Consider the reaction

$$NH_3(g) + O_2(g) \longrightarrow NO(g) + H_2O(l)$$

For every 1.50 moles of NH_3, how many moles of (a) O_2 are required, (b) NO are produced, and (c) H_2O are produced? *Hint:* Before blindly answering these questions, balance the equation.

113

4.9 Which reaction uses the most nitric acid to form one mole of water?

(a) $3Cu(s) + 8HNO_3(aq) \longrightarrow$
$$3Cu(NO_3)_2(aq) + 2NO(g) + 4H_2O(l)$$
(b) $Al_2O_3(s) + 6HNO_3(aq) \longrightarrow 2Al(NO_3)_3(aq) + 3H_2O(l)$
(c) $4Zn(s) + 10HNO_3(aq) \longrightarrow$
$$4Zn(NO_3)_2(aq) + NH_4NO_3(aq) + 3H_2O(l)$$

4.10 Consider the following unbalanced equation:

$$HBr(aq) + H_2SO_4(aq) \longrightarrow H_2O(l) + SO_2(g) + Br_2(l)$$

(a) Balance the equation. (b) Assuming that 1.00 mole of HBr is reacting with 2.00 moles of H_2SO_4, (i) how many moles of SO_2 will be formed, (ii) how many molecules of H_2O will be formed, and (iii) how many molecules of bromine will be formed?

4.11 Give an interpretation for the following equation

$$C_7H_{16}(l) + 11\ O_2(g) \longrightarrow 7CO_2(g) + 8H_2O(g)$$

based on (a) moles, (b) molecules, (c) volumes of gases, and (d) mass.

4.12 Find the oxidation number of Cr in the following substances: (a) $K_2Cr_2O_7$, (b) Na_2CrO_4, (c) Cr, (d) Cr^{3+}, (e) $CrBr_2$, (f) $[Cr(H_2O)_6]I_3 \cdot 3H_2O$, (g) CrN, (h) CrO, (i) CrO_2, (j) Cr_2O_3, (k) CrO_2Cl_2, (l) CrS, and (m) $Cr(OH)_4{}^-$.

4.13 Give the oxidation numbers of the elements in the following species: (a) KH, (b) $MnCl_2$, (c) $NH_4{}^+$, (d) $SO_3{}^{2-}$, (e) Na_2O_2, (f) $Cr_2O_7{}^{2-}$, (g) MnF_3, (h) ICl_3, (i) H_2Se, (j) P_4, (k) $Mg(NO_3)_2$, (l) $KHCO_3$, (m) $(NH_4)_2SO_4$, (n) $Fe(ClO_4)_3$, (o) Al_2O_3, and (p) P_4O_{10}.

4.14 In each of the following equations pick out the (a) oxidizing agent, (b) reducing agent, and (c) show the change in oxidation state which occurs for each.
(i) $4Al(s) + 3O_2(g) \longrightarrow 2Al_2O_3(s)$
(ii) $Cr_2O_7{}^{2-} + 3SO_3{}^{2-} + 8H^+ \longrightarrow$
$$2Cr^{3+} + 3SO_4{}^{2-} + 4H_2O(l)$$
(iii) $Sn^{2+} + 2Fe^{3+} \longrightarrow Sn^{4+} + 2Fe^{2+}$
(iv) $MnO_2(s) + 4HCl(aq) \longrightarrow$
$$MnCl_2(aq) + Cl_2(g) + 2H_2O(l)$$

4.15* Aqueous ammonia and nitric acid react to form ammonium nitrate

$$HNO_3(aq) + NH_3(aq) \longrightarrow NH_4NO_3(aq)$$

The solution can be evaporated to yield the white crystalline NH_4NO_3. (a) Is this an oxidation–reduction reaction or an acid–base reaction? If the solid NH_4NO_3 is heated, it undergoes decomposition

$$NH_4NO_3(s) \longrightarrow N_2O(g) + 2H_2O(g)$$

(b) What are the oxidation states of N in HNO_3 and NH_3? (c) What is the value of the oxidation state for N in N_2O?

4.16* Fluorescein is an orange-red powder which dissolves in alkaline solution producing a strong green fluorescence. (The alkaline solution looks yellow-green from the reflected light, but is reddish-orange by the transmitted light.) Fluorescein is produced by heating a mixture of phthalic anhydride and resorcinol

$$C_6H_4(CO)_2O + 2C_6H_4(OH)_2 \xrightarrow[\substack{190-200°C \\ 10\ hr}]{\substack{anhydrous \\ ZnCl_2}} C_{20}H_{12}O_5$$
phthalic anhydride   *resorcinol*   *fluorescein*

(a) Identify the missing product in the above equation and (b) write the complete balanced equation. (c) How many molecules of resorcinol are needed to react with 1.0 mole of phthalic anhydride? (d) What are the necessary conditions for the reaction? (e) Show that the equation is an example of the law of conservation of mass.

4.17* A simple way to determine the presence of zinc in a sample of Al is to (a) dissolve the metal sample in acid to form H_2, Zn^{2+}, and Al^{3+}; (b) treat the solution with a dilute aqueous solution of NH_3, which precipitates the Al as $Al(OH)_3$ and ties up the Zn^{2+} as $Zn(NH_3)_4{}^{2+}$; (c) separate the precipitate from the soluble $Zn(NH_3)_4{}^{2+}$; and (d) add $S^{2-}(aq)$ to the zinc ion solution (if any is present) to give $ZnS(s)$. Write the net ionic equations corresponding to each reaction in the process.

4.18 Zirconium is obtained industrially using the Kroll process

$$ZrCl_4(s) + 2Mg(s) \longrightarrow Zr(s) + 2MgCl_2(s)$$

Calculate the weight of Zr obtainable for each kilogram of Mg consumed.

4.19 Calculate the stoichiometric amount of sodium required to produce 80.0 g of sodium hydroxide by direct reaction with water

$$2Na(s) + 2H_2O(l) \longrightarrow 2NaOH(aq) + H_2(g)$$

Although this reaction is performed frequently, it is rather dangerous because the hydrogen can form an explosive mixture with the oxygen in the air.

4.20 Find the weight of chlorine that will combine with 1.38 g of hydrogen to form hydrogen chloride

$$H_2(g) + Cl_2(g) \longrightarrow 2HCl(g)$$

4.21 What mass of solid AgCl would be precipitated from a solution containing 1.50 g of $CaCl_2$, assuming an excess amount of $AgNO_3$ is added?

$$CaCl_2(aq) + 2AgNO_3(s) \longrightarrow 2AgCl(s) + Ca(NO_3)_2(aq)$$

4.22 A sample of magnetic iron oxide, Fe_3O_4, was completely reduced with hydrogen at red heat. The water vapor formed by the reaction

$$Fe_3O_4(s) + 4H_2(g) \longrightarrow 3Fe(s) + 4H_2O(g)$$

was condensed and found to weigh 7.5 g. Calculate the weight of Fe_3O_4 that had been reduced.

4.23 An impure sample of $CuSO_4$ weighing 5.52 g was dissolved in water and reacted with excess zinc

$$CuSO_4(aq) + Zn(s) \longrightarrow ZnSO_4(aq) + Cu(s)$$

What is the % $CuSO_4$ in the sample if 1.49 g of Cu were produced?

4.24 What mass of potassium can be produced from 100.0 g of Na added to 100.0 g of KCl?

$$Na(l) + KCl(l) \longrightarrow NaCl(l) + K(l)$$

4.25 What is the maximum amount of sodium chloride that can be formed from 5.00 g of sodium and 7.10 g of chlorine? Which substance is the limiting factor? Which substance is in excess?

$$2Na(s) + Cl_2(g) \longrightarrow 2NaCl(s)$$

4.26 What mass of $BaSO_4$ will be produced by reacting 33.2 g of Na_2SO_4 with 43.5 g of $Ba(NO_3)_2$?

$$Ba(NO_3)_2(aq) + Na_2SO_4(aq) \longrightarrow BaSO_4(s) + 2NaNO_3(aq)$$

4.27 Consider the reaction for making DDT

$$CCl_3CHO + C_6H_5Cl \longrightarrow (ClC_6H_4)_2CHCCl_3 + H_2O$$

(a) Balance the equation. If 100.0 g of CCl_3CHO were reacted with 100.0 g of C_6H_5Cl, (b) how much DDT would be formed, (c) how many moles of water would be formed, and (d) what would happen if the amount of CCl_3CHO was doubled?

4.28 What weight of water gas is formed by passing 1.00 kg of steam over white-hot carbon, assuming that 45.0% of the steam reacts?

$$H_2O(g) + C(s) \rightarrow \underbrace{CO(g) + H_2(g)}_{water\ gas}$$

4.29 Assuming a coal sample contains 1.2 wt % S, what mass of SO_2 is released to the atmosphere for each gram of coal burned?

$$S(s) + O_2(g) \longrightarrow SO_2(g)$$

4.30 A 0.78 g sample of an organic compound containing only carbon and hydrogen produced 2.64 g of CO_2 and 0.54 g of H_2O when burned. What is the empirical formula for the compound?

4.31 An organic compound containing C, H, and O was being tested as a possible gasoline substitute. If a 1.00 g sample of the compound gave 1.375 g of CO_2 and 1.127 g of H_2O upon combustion, find its empirical formula.

4.32 Nitrous oxide undergoes decomposition when heated

$$2N_2O(g) \longrightarrow 2N_2(g) + O_2(g)$$

What is the molar composition of the gaseous mixture produced? Compare this composition to that of air and predict whether it will support combustion.

4.33 Assuming that all gases in the reaction are measured at the same temperature and pressure conditions, calculate the volume of water vapor obtainable by the explosion of a mixture of 250. ml of hydrogen gas and 125 ml of oxygen gas.

4.34 What volume of N_2 is required to convert 15.0 liters of H_2 to NH_3? Assume all gases to be at STP.

$$N_2(g) + 3H_2(g) \longrightarrow 2NH_3(g)$$

4.35 Calculate the weight of hydrogen gas required to produce 450 liters of ammonia (see Exercise 4.34). Assume the volumes are measured at 210°C and 550 atm.

4.36 What volume of carbon dioxide, measured at STP, can be obtained by reacting 50.0 g of $CaCO_3$ with excess hydrochloric acid?

$$CaCO_3(s) + 2HCl(aq) \longrightarrow CaCl_2(aq) + CO_2(g) + H_2O(l)$$

What would the volume be, had it been measured at 25°C and 740 torr?

4.37 Ammonium nitrite decomposes to water and nitrogen when heated. What volume of N_2, at STP, will be released by the decomposition of 80.0 g of NH_4NO_2?

$$NH_4NO_2(s) \longrightarrow N_2(g) + 2H_2O(g)$$

4.38 Calculate the volume of CO_2, measured at 0°C and 2.0 atm, required to precipitate 2.00×10^2 g of calcium carbonate from excess limewater (aqueous calcium hydroxide).

$$Ca(OH)_2(aq) + CO_2(g) \longrightarrow CaCO_3(s) + H_2O(l)$$

4.39 What volume of hydrogen fluoride at 743 torr and 24°C will be released by reacting 47.2 g of xenon difluoride with a stoichiometric amount of water (that is, the exact amount required)? The unbalanced equation is

$$XeF_2(s) + H_2O(l) \longrightarrow Xe(g) + O_2(g) + HF(g)$$

What volumes of oxygen and xenon will be released?

4.40 Chlorine can be prepared in the laboratory by the reaction

$$2NaCl(aq) + MnO_2(s) + 3H_2SO_4(aq) \longrightarrow$$
$$MnSO_4(aq) + 2NaHSO_4(aq) + Cl_2(g) + 2H_2O(l)$$

What weight of NaCl is needed to prepare 500.0 ml of Cl_2 at 730.0 torr and 25°C? Will 1.50 g of MnO_2 be enough to prepare this amount of Cl_2?

4.41 Calculate the volume in liters of CO_2 at 306°C and 1.2 atm produced by burning one ton of coke.

$$C(s) + O_2(g) \longrightarrow CO_2(g)$$

4.42 What weight of potassium chlorate would be required to give the proper amount of oxygen to burn 35.0 g of methane?

$$2KClO_3(s) \longrightarrow 2KCl(s) + 3O_2(g)$$
$$CH_4(g) + 2O_2(g) \longrightarrow CO_2(g) + 2H_2O(g)$$

4.43 Hydrogen, obtained by the electrolysis of water, was combined with chlorine to produce 51.0 g of hydrogen chloride. Calculate the weight of water electrolyzed.

$$2H_2O(l) \longrightarrow 2H_2(g) + O_2(g) \qquad H_2(g) + Cl_2(g) \longrightarrow 2HCl(g)$$

4.44 An electrochemical process for making hydrogen peroxide, H_2O_2, is

$$2H_2SO_4 \longrightarrow \underset{\substack{peroxysulfuric \\ acid}}{H_2S_2O_8} + H_2$$
$$\underset{\substack{sulfuric \\ acid}}{}$$

$$H_2S_2O_8 + 2H_2O \longrightarrow 2H_2SO_4 + H_2O_2$$

(a) Combine the two equations to obtain the overall equation. (b) What mass of H_2O_2 can be produced from one mole of H_2SO_4 going through the cycle? (c) Is H_2SO_4 a catalyst?

4.45 What mass of H_2SO_4 can be produced in the process given below if 1.00 kg of FeS_2 were used? The unbalanced equations for the process are

$$FeS_2(s) + O_2(g) \longrightarrow Fe_2O_3(s) + SO_2(g)$$
$$SO_2(g) + O_2(g) \longrightarrow SO_3(g)$$
$$SO_3(g) + H_2O(l) \longrightarrow H_2SO_4(aq)$$

4.46 What is the equivalent weight of each of the following: (a) HBr, (b) $Ca(OH)_2$, (c) KOH, (d) H_2SO_4, (e) H_3PO_4, (f) $Al(OH)_3$ assuming complete reaction of all H^+ or OH^-?

4.47 If 100.00 g of $Ba(OH)_2 \cdot 8H_2O$ is dissolved in enough water to make 1.000 liter of solution, what is the normality of the solution?

4.48 How many moles of acid are present in 108 ml of 0.62M solution? If we add enough water to make 0.300

liter of acid solution, how many moles of acid will it now contain? What is the molarity of the final solution?

4.49 How many equivalents of base are present in 1.50 liters of a 0.15N solution? If enough water is added to make 5.00 liters of base solution, how many equivalents of base will the solution contain? What is the normality of the final solution?

4.50 What is the concentration of an HCl solution if 23.65 ml neutralizes 25.00 ml of a 0.1037M solution of NaOH?

$$HCl(aq) + NaOH(aq) \longrightarrow NaCl(aq) + H_2O(l)$$

4.51 What volume of 0.324M HNO_3 solution is required to completely react with 22.0 ml of 0.0612M $Ba(OH)_2$?

$$Ba(OH)_2(aq) + 2HNO_3(aq) \longrightarrow Ba(NO_3)_2(aq) + 2H_2O(l)$$

4.52 What volume of 0.50M HBr is required to react with 0.75 mole of $Ca(OH)_2$?

$$2HBr(aq) + Ca(OH)_2(aq) \longrightarrow CaBr_2(aq) + 2H_2O(l)$$

4.53 Calculate the theoretical yield of AgCl formed from the reaction of 5.23 g of $ZnCl_2$ with 35.0 ml of 0.325M $AgNO_3$.

$$ZnCl_2(aq) + 2AgNO_3(aq) \longrightarrow Zn(NO_3)_2(aq) + 2AgCl(s)$$

4.54 An excess of $AgNO_3$ reacts with 100.0 ml of an $AlCl_3$ solution to give 0.275 g of AgCl. What is the molarity of the $AlCl_3$ solution? The unbalanced equation is

$$AlCl_3(aq) + AgNO_3(aq) \longrightarrow AgCl(s) + Al(NO_3)_3(aq)$$

4.55* You are to design an experiment for the preparation of hydrogen. For the same weight of hydrogen, which metal, Zn or Al, is cheaper to use if Zn costs about half as much as Al on a weight basis?

$$Zn(s) + 2HCl(aq) \longrightarrow ZnCl_2(aq) + H_2(g)$$
$$2Al(s) + 6HCl(aq) \longrightarrow 2AlCl_3(aq) + 3H_2(g)$$

4.56* A standard qualitative analysis scheme for the separation and identification of a solution containing Ba^{2+}, Sr^{2+} and Ca^{2+} is

$$\underset{mixture}{[Ba^{2+}, Sr^{2+}, Ca^{2+}]} + CrO_4^{2-} \xrightarrow{HC_2H_3O_2, \ NH_4C_2H_3O_2}$$

$$BaCrO_4(s) + \underset{mixture}{[Sr^{2+}, Ca^{2+}]}$$

$$[Sr^{2+}, Ca^{2+}] + CrO_4^{2-} \xrightarrow{OH^-} SrCrO_4(s) + Ca^{2+}$$
$$Ca^{2+} + C_2O_4^{2-} \longrightarrow CaC_2O_4(s)$$

What is the composition of a mixture of these ions if 1.00 g of each of the precipitates was collected?

4.57* A spark has been passed through a mixture of 1.00 g H_2 and 1.00 g O_2 and water has been formed. What are the amounts of substances present?

4.58* Consider the reaction

$$3HCl(g) + 3HNF_2(g) \longrightarrow 2ClNF_2(g) + NH_4Cl(s) + 2HF(g)$$

A mixture of 8.00 g of HCl and 10.00 g of HNF_2 is allowed to react. If the reaction is only 15% efficient, what is the composition of the final mixture?

4.59* Hydrogen reacts with some of the more active metals to form crystalline ionic hydrides, e.g., lithium hydride

$$2Li(s) + H_2(g) \longrightarrow 2LiH(s)$$

(a) What mass of LiH would be produced by allowing 10.0 g Li to react with 10.0 liters of H_2 (measured at STP)? (b) If the actual yield was 6.7 g, what is the % yield?

4.60* A common laboratory preparation of oxygen is

$$2KClO_3(s) \xrightarrow[\Delta]{MnO_2} 2KCl(s) + 3O_2(g)$$

If you were designing an experiment to generate four bottles (each containing 250 ml) of O_2 at 25°C and 723 torr and allowing for 50% waste, what weight of potassium chlorate would be required? As a laboratory instructor, how would you explain the symbol Δ and MnO_2 written near the arrow?

4.61* The thickness of zinc plate on a piece of galvanized iron was determined by allowing acid to react with the zinc and collecting the resulting hydrogen.

$$Zn(s) + 2H^+ \longrightarrow Zn^{2+} + H_2(g)$$

Determine the thickness of the zinc plate from the following data: sample size = 1.50 cm × 2.00 cm; volume of dry hydrogen = 30.0 ml; temperature = 25°C; pressure = 747 torr; and density of zinc = 7.11 g/cm³. *Note:* The acid solution contained an "inhibitor" ($SbCl_3$) which prevented the iron from reacting.

4.62* Consider the following equation for photosynthesis

$$CO_2(g) + H_2O(l) \longrightarrow \underset{hexose}{C_6H_{12}O_6(aq)} + O_2(g)$$

(a) Balance the equation. (b) What volume of CO_2 at 25°C and 745 torr is necessary to react with 1.00 g of water? (c) How many grams of hexose will be formed in the reaction between sufficient CO_2 and 1.00 g water? (d) What volume of O_2 at 25°C and 745 torr is produced by the reaction described in (c)?

4.63* Consider the following equation

$$HNO_3(aq) + Cu(s) \xrightarrow{\Delta} Cu(NO_3)_2(aq) + NO_2(g) + H_2O(l)$$

(a) Balance the equation. (b) A piece of Cu metal 1.35 inch × 0.72 inch × 0.39 inch reacts with 157 ml of 1.35M nitric acid solution. The density of copper is 8.92 g/cm³; find the number of moles of each reactant. (c) How many molecules of H_2O are formed? (d) What volume of NO_2 at 1.01 atm and 297°K will be formed? (e) Describe what would happen if the amount of Cu were doubled.

4.64* Two moles of acetone undergo a reaction giving diacetone alcohol (an industrial solvent). This compound, however, can lose water to give mesityl oxide.

$$2\underset{acetone}{CH_3COCH_3} \xrightarrow[71\% \text{ yield}]{Ba(OH)_2} \underset{diacetone\ alcohol}{HOC(CH_3)_2CH_2COCH_3} \xrightarrow[65\% \text{ yield}]{I^-}$$

$$\underset{mesityl\ oxide}{(CH_3)_2CCHCOCH_3}$$

(a) Are there any products not shown for the reactions? (b) Assuming the reactions to be 100% efficient, how many moles of diacetone alcohol and mesityl oxide are produced in the reaction of 1 mole of acetone? (c) What are the amounts of diacetone alcohol and mesityl oxide produced in the reaction of 1 mole of acetone? The yields for the reactions are given in the equation. (d) Using your answers for (b) and (c), what is the % yield for the overall process?

4.65* The Ostwald process for preparing nitric acid is

$$4NH_3(g) + 5O_2(g) \xrightarrow[600°-1000°C]{Pt-Rh} 4NO(g) + 6H_2O(g)$$
$$2NO(g) + O_2(g) \longrightarrow 2NO_2(g)$$
$$3NO_2(g) + H_2O(l) \longrightarrow 2HNO_3(aq) + NO(g)$$

What is the maximum amount of HNO_3 that can be produced from 1.00 mole of O_2? What would be done with the NO produced in the third reaction?

4.66* What volume of hydrogen will be released at 745 torr and 25°C if a 19.6 g piece of Zn is placed in 25.0 ml of 0.275M H_2SO_4? The equation for the reaction is

$$Zn(s) + H_2SO_4(aq) \longrightarrow H_2(g) + ZnSO_4(aq)$$

4.67** Consider the following three-step preparation of acetylene, C_2H_2:

$$\underset{limestone}{CaCO_3(s)} \xrightarrow{\Delta} \underset{unslaked\ lime}{CaO(s)} + CO_2(g) \qquad (i)$$

$$CaO(s) + C(s) \longrightarrow \underset{\substack{calcium \\ carbide}}{CaC_2(s)} + CO(g) \qquad (ii)$$

$$CaC_2(s) + H_2O(l) \longrightarrow Ca(OH)_2(s) + C_2H_2(g) \qquad \text{(iii)}$$

and the subsequent use of acetylene as a fuel

$$C_2H_2(g) + O_2(g) \longrightarrow CO_2(g) + H_2O(g) \qquad \text{(iv)}$$

(a) Find the oxidation numbers of the elements in the reactions. (b) Identify which reactions are oxidation–reduction reactions. (c) Balance the equations. (d) What volume of acetylene at 25°C and 745 torr would be produced from a reaction mixture containing 1.00 kg each of C and CaO? (e) What is the general relationship between the volume of O_2 and the C_2H_2 needed to react in (iv) if they are measured under the same temperature and pressure conditions? (f) What is the relationship between the volume of air (20% O_2 by volume) and the C_2H_2 needed to react in (iv)?

5 THERMOCHEMISTRY

What is heat? What is temperature? Answering these questions
ought to be easy. Everyone knows what heat and temperature are.
Heat is what makes you pull back your hand from a radiator in the
winter. Temperature is what you read from a thermometer. And yet,
when we try to go beyond our intuitive understanding of heat and
temperature, the subject becomes harder. We have to use our imagination and mathematics to picture and explain events that we can observe only indirectly.

Kinetic-molecular theory tells us that heat is simply the motion of
tiny particles. Thermal energy, or heat, is the lowest form of energy—
all other types of energy eventually degrade to heat. For example, the
radiant energy from a light bulb is converted to thermal energy in the
objects on which the light falls. The electrical energy that escapes from
a frayed electrical cord can create enough heat to start a fire.

If heat is the energy in the particles of which a substance is composed, then a bigger mass of something ought to contain more heat
than a smaller mass of the same substance at the same temperature.
Also, different substances should contain different amounts of heat.
Can we tell from looking at a big block of metal and a small one, or
from looking at blocks of two different metals, which ones "contain"
more heat? No. Can we measure the amount of heat, or internal energy, in these samples? Not really.

But if we burn a block of wood, we can measure how much heat
is given off in the reaction of the wood with oxygen. If a hot block of
metal is placed in contact with a cold block of metal, the hot one will
cool down and the cool one will warm up until they both reach the
same temperature, and we can measure the heat lost by one and
gained by the other. This demonstrates a law about the behavior of
heat: Heat always flows from a hot body to a colder body. In fact, this

This chapter first discusses heat and
heat capacity, and the heat of chemical and physical changes. Then enthalpy, the primary quantity by which
we deal with heat in chemistry, is introduced. Standard enthalpies are defined and the use of enthalpies in calculating heats of formation of compounds and heats of reaction is explained.

law leads to one way to define temperature, albeit the definition is a negative one: If heat can flow from one body to another, they are at different temperatures; if no heat flows between two bodies, they are at the same temperature. Although we cannot measure the amount of heat within a single body, we can measure temperature changes and calculate the amount of heat involved in those changes. The subject of this chapter is thermochemistry—the study of the thermal energy associated with chemical and physical changes of substances.

Heat

5.1 Heat and heat capacity

As an ice cube melts, heat is flowing to the cube from the air, or a surrounding liquid, or the table top. In every process, the heat gained in one place is lost in another. This is simply a consequence of the law of conservation of energy, which may be stated this way: *The energy of the universe is constant.* This is known as the first law of thermodynamics. Neither thermal energy nor any other kind of energy can disappear. Using q as a symbol for heat, the mathematical statement is this:

$$q_{\text{gained}} + q_{\text{lost}} = 0 \tag{1}$$

When heat is absorbed, or flows toward a body, q is given a positive sign—heat has been *added* to the body under consideration. When heat is released, or flows away from a body, q is given a negative sign—heat has been *subtracted* from the body under consideration.

We observe heat flow by measuring changes in temperature. To calculate q we must know the change in temperature ΔT; the mass or amount of the substance that has changed temperature; and a factor that expresses how much heat is required to raise by a certain amount the temperature of a given amount of the substance. The name for this factor is **heat capacity**, C—the amount of heat required to raise the temperature of a given amount of material by 1 Celsius degree. When this amount of the substance is expressed in moles, we refer to the *molar* heat capacity. Each substance has its own characteristic heat capacity; for example, it takes about 2 cal to raise the temperature of one mole of carbon in the form of coal by 1°C; for coal, $C = 2$ cal/mole °C.

The relationships among q, ΔT, and C, and the amount of substance are as follows:

$$q(\text{cal}) = n(\text{moles}) \times C\,\frac{(\text{cal})}{(\text{mole °C})} \times \Delta T(\text{°C}) \tag{2}$$

EXAMPLE 5.1

■ Exactly 100 J of heat was transferred to one mole of iron and the resulting temperature increase was 3.98°C. Find the molar heat capacity of iron in both joules and calories.

The molar heat capacity is

$$C = \frac{q}{n\,\Delta T} = \frac{100.0 \text{ J}}{(1 \text{ mole})(3.98°C)} = 25.1 \text{ J/mole °C}$$

Using the exact conversion factor, 1 cal = 4.184 J, gives 25.1 J/mole °C × (1 cal/4.184 J) = 6.00 cal/mole °C. The heat capacity of iron is 25.1 J/mole °C or 6.00 cal/mole °C. ■

When heat capacity is given in terms of the mass of a substance in grams it is called **specific heat**—the amount of heat required to raise the temperature of 1 g of a substance by 1 Celsius degree. (We choose to use units with specific heat, as many do, because it is desirable in factor-dimensional calculations. Many others think of specific heat as a dimensionless quantity since it was originally defined as the ratio of the heat capacity of a substance to the heat capacity of water. In calorie/gram °C units the value of the heat capacity of water is 1, and taking the ratio cancels the units while leaving the numerical value unchanged.)

EXAMPLE 5.2

■ What is the specific heat of iron (in both joules and calories) if the heat capacity is 25.1 J/mole °C?

The relationship between specific heat and heat capacity is

$$\text{sp heat} = \frac{C}{\text{molar weight}} = \frac{25.1 \text{ J/mole °C}}{55.85 \text{ g/mole}} = 0.449 \text{ J/g °C}$$

The specific heat of iron is 0.449 J/g °C or 0.107 cal/g °C. ■

EXAMPLE 5.3

■ What is the final temperature of the system prepared by pouring 100.0 g of hot water originally at 98.6°C into 50.0 g of cold water originally at 25.0°C? The specific heat of water is 1.00 cal/g °C.

The heat loss by the hot water is

$$q_{\text{lost}} = m \text{ (sp heat) } \Delta T = (100.0 \text{ g})(1.00 \text{ cal/g °C})(T - 98.6°C)$$

1 cal = 4.184 J

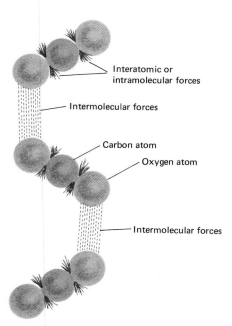

Interatomic or
intramolecular forces

Intermolecular forces

Carbon atom

Oxygen atom

Intermolecular forces

FIGURE 5.1

Intermolecular and intramolecular forces in carbon dioxide. *When the forces acting between atoms—the interatomic forces—are strong enough to hold the atoms together, we say that a chemical bond has been formed.*

and the heat gain by the cold water is

$$q_{gain} = (50.0 \text{ g})(1.00 \text{ cal/g °C})(T - 25.0°C)$$

where T represents the final temperature of the 150.0 g of water. The law of conservation of energy states that all of the heat must be accounted for, or

$$q_{gain} + q_{lost} = 0$$

Thus, substituting the above expressions for q_{gain} and q_{lost} and solving for T gives

$$(100.0)(1.00 \text{ cal/°C})(T - 98.6°C) + (50.0)(1.00 \text{ cal/°C})(T - 25.0°C) = 0$$
$$150.0 \, T = 11,110°C$$
$$T = 74.1°C$$

The final temperature of the water is 74.1°C, assuming no heat loss to the surroundings. ∎

5.2 Heat in chemical reactions and changes of state

In many familiar chemical reactions heat is given off to the surroundings. When we huddle around a campfire to keep warm, we are taking advantage of the heat released as the carbon–hydrogen–oxygen compounds in the wood are converted to carbon dioxide, water, and other products. Once the fire has been lit, the chemical reaction keeps going and heat continues to be given off until all of the wood has been consumed.

The campfire reaction—the reaction of wood with oxygen—is an **exothermic process,** a process that releases heat. Like the burning of the campfire, many exothermic reactions keep going once they get started. Some chemical reactions, however, will only proceed if heat is available from an outside source, either the atmosphere or something hotter than that. An **endothermic process** is a process that absorbs heat. Endothermic processes slow down and stop when not enough heat is available.

Physical processes involving atoms, molecules, and ions are also accompanied by thermal energy exchange. Such processes do not produce new chemical compounds, but change the form or state of a substance. For example, characteristic amounts of heat can be measured for pure substances undergoing condensation of a gas to a liquid, or the melting of a solid, or vaporization of a liquid. Such interconversions among the solid, liquid, and gaseous states are called **changes of state.** All changes of state are either exothermic or endothermic.

The amount of thermal energy evolved or absorbed in a change of state reflects the strength of **intermolecular forces**—the various forces of attraction or repulsion between individual molecules (Figure 5.1). For example, consider the conversion of water to steam at

100°C. It may be a surprise at first to realize that heat added to water at 100°C does not raise the temperature of the water. The added thermal energy is used to separate the molecules in the liquid state from each other's influence and drive them off into the gaseous state, where they are further apart. Similarly, heat removed from water vapor at 100°C does not cool the vapor, but only causes it to condense. The amount of heat added in order to vaporize a given mass of liquid water is equal to the amount of heat removed in order to condense the same mass of water vapor.

In a chemical reaction the heat exchanged is a measure of the forces that hold atoms or ions together in chemical compounds. For a nonionic compound, the forces that hold the atoms together in the molecule are called **intramolecular forces** (see Figure 5.1). A chemical reaction that is exothermic in one direction is endothermic if it is carried out in the opposite direction, and the amount of heat required in the endothermic reaction is exactly equal to that released in the exothermic direction.

Thermochemistry is the study of the thermal energy changes associated with the physical or chemical changes of pure substances.

Enthalpy

5.3 Enthalpy defined

Chemicals are often handled in the laboratory in vessels open to the atmosphere—that is, at constant pressure. And a chemical process is usually considered complete when no further changes take place and the substances have returned to the temperature at which they started, usually room temperature. The amount of heat absorbed or released in such a thermochemical change is referred to as the change in enthalpy, represented by the symbol

$$\underset{\Delta H}{\overset{\text{the change in} \quad \text{enthalpy}}{}}$$

Only *changes* in enthalpy can be measured; values of H cannot be found, and they are fortunately not needed. Under the stated conditions of constant pressure and a return to the original temperature, $\Delta H = q$. (If a pressure change occurred a correction would have to be made to the value of ΔH.)

Enthalpy is most conveniently thought of as the heat content of a substance—the amount of energy within a substance that is available for conversion to heat. The **enthalpy change** is the difference between the heat content of substances in the final state, after the change is complete, and the heat content of the substances in the initial state, before any change has occurred.

For a chemical reaction, the change in enthalpy (frequently referred to just as the enthalpy of reaction) is

$$\Delta H_r = H_{\text{products}} - H_{\text{reactants}} \qquad (3)$$

EXOTHERMIC

Heat is evolved

Products have lower heat content than reactants

Products, in general, are more stable than reactants

Reactants

$N_2H_4(l) + 2H_2O_2(l)$

$\Delta H = -153.52$ kcal
$(-642.33$ kJ$)$

Products

$N_2(g) + 4H_2O(g)$

ENDOTHERMIC

Heat is absorbed

Products have higher energy content than reactants

Products, in general, are less stable than reactants

Products

$2Hg(l) + O_2(g)$

$\Delta H = +43.42$ kcal
$(+181.7$ kJ$)$

Reactants

$2HgO(s)$

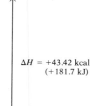

FIGURE 5.2
Exothermic and endothermic chemical reactions. *The mixture of hydrazine N_2H_4 and hydrogen peroxide H_2O_2 in the exothermic reaction is used as a rocket fuel.*

For a change of state, for example the melting of ice, the enthalpy change is

$$\Delta H = H_{H_2O(l)} - H_{H_2O(s)} \qquad (4)$$

In an exothermic change, heat is given up and ΔH is *negative;* the final state has a smaller heat content and ΔH states the amount of heat given up. In an endothermic change heat is added and ΔH is *positive;* the final state has a greater heat content and ΔH states the amount of heat added (Figure 5.2). A negative enthalpy change means that the final state is more thermally stable than the initial state. If the enthalpy change is positive, the reverse is true. Exothermic changes *often* take place spontaneously and endothermic changes *usually do not* take place spontaneously.

The total value of the change in enthalpy depends on the amounts of the substances involved in the change, as well as on the type of change. ΔH, therefore, must be expressed as the quantity of heat per quantity of the substance or substances in question. The heat units are usually kilocalories or kilojoules. ΔH is given as either the amount of heat per mole or the amount of heat for the molar quantities of the substances in a balanced chemical equation. For example, the enthalpy of vaporization of water at its normal boiling point is written $\Delta H_{vap} = 9.7$ kcal/mole, meaning that 9.7 kcal must be added to convert 1 mole of liquid water at 100°C to one mole of water vapor at 100°C. For a chemical reaction we write a thermochemical equation, an equation that includes the enthalpy for the reaction as written. For example, the following equation

$$2HgO(s) \xrightarrow{\Delta} 2Hg(l) + O_2(g) \qquad \Delta H = 43.42 \text{ kcal } (181.7 \text{ kJ)} \qquad (5)$$

tells us that 43.42 kcal of heat must be added (because ΔH is positive) to convert 2 moles of solid mercury(II) oxide to 2 moles of liquid mercury and 1 mole of dioxygen. Alternatively, we could say that the enthalpy for the decomposition of mercury(II) oxide is 21.71 kcal/mole.

5.4 Standard enthalpies

A characteristic quantity of heat is associated with every chemical reaction. The ΔH's for many reactions have been measured directly (see Tools of Chemistry: The Calorimeter) and by using these data the ΔH's for many more reactions can be calculated. Comparisons of the heats of a large number of different reactions provide information about the relative strengths of chemical bonds, the characteristics of atoms of different elements, and, as we mentioned above, the possibility of specific reactions taking place. All of these topics are of central importance in chemistry and are discussed in later chapters.

In order to make meaningful comparisons among reaction enthalpies it is necessary to know the temperature at which ΔH was measured and the physical state of the reactants. If the starting ma-

TOOLS OF CHEMISTRY

The calorimeter

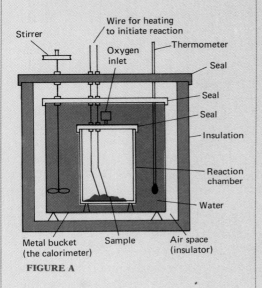

Stirrer

Wire for heating to initiate reaction

Oxygen inlet

Thermometer

Seal

Seal

Seal

Insulation

Reaction chamber

Water

Metal bucket (the calorimeter)

Sample

Air space (insulator)

FIGURE A

A device for measuring the heat given off or absorbed in a chemical reaction is called a **calorimeter.** Many different types of calorimeters have been designed for various purposes. One of the most common is a bomb calorimeter, illustrated in Figure A. The bomb calorimeter is so named because the reaction takes place in a "bomb," a chamber strong enough to withstand high temperatures and sudden changes in pressure.

Heats of combustion and the caloric value of foods are frequently measured in a bomb calorimeter. The substance to be burned is weighed and placed in the reaction chamber, which is then filled with high-pressure oxygen in order to assure complete combustion. The reaction chamber is then sealed and placed in the water bath. Once the temperature of the water has become constant, it is accurately measured. The stirrer keeps the water in motion so that heat is evenly distributed and the temperature is uniform.

The reaction is initiated and the increase in the water temperature is recorded. The total heat produced by an exothermic reaction is equal to the heat absorbed by the water plus the heat absorbed by all of the parts of the calorimeter within the insulated chamber. The heat evolved is calculated from the temperature change by using the known specific heat of water and the heat capacity of the calorimeter.

A typical bomb calorimeter might contain 2000 g of water and have a total heat capacity for the calorimeter plus the water of 2.42 kcal/°C. (The specific heat of water is 1.00 cal/g °C; therefore, 2000 g × 1.00 cal/g °C = 2000 cal/°C, or 2.00 kcal/°C of the total heat capacity is contributed by the water, and the rest by the parts of the calorimeter.) The total heat capacity of a calorimeter is called the **calorimeter constant.**

If 0.640 g of naphthalene, $C_{10}H_8$, were burned in this calorimeter and a temperature rise of 2.54°C recorded, the molar heat of combustion of $C_{10}H_8$ would be calculated as follows:

$$\begin{aligned}\text{total heat evolved} &= \text{temperature rise (°C)} \times \text{heat capacity} \\ \text{(kcal)} & \qquad\qquad\qquad\qquad\qquad\qquad\quad \text{(kcal/°C)} \\ &= 2.54°C \times \frac{2.42 \text{ kcal}}{1°C} \\ &= 6.15 \text{ kcal}\end{aligned}$$

Since 6.15 kcal of heat was liberated by 0.640 g of naphthalene, it is necessary to find the amount of heat that would be liberated by one mole of naphthalene.

$$\begin{aligned}\text{heat of combustion of 1 mole of } C_{10}H_8 &= \frac{6.15 \text{ kcal}}{0.640 \text{ g}} \times \frac{128 \text{ g}}{1 \text{ mole}} \\ \text{(kcal)} \\ &= 1.23 \times 10^3 \text{ kcal/mole}\end{aligned}$$

The heat of combustion of one mole of naphthalene at (note) *constant volume,* is 1.23×10^3 kcal/mole. To find the enthalpy of this reaction, which is the heat of the reaction at constant pressure, a small correction must be made. The two values are not greatly different.

terial in a reaction of water molecules was ice, some of the thermal energy would be used up in melting the ice. Chemists have agreed to report reaction enthalpies for reactants in what is called their standard state. The **standard state** of any substance is the physical state in which it is most stable at 1 atm pressure and a specified temperature. The usual specified temperature is 298°K. For gases, this means that the pressure of a pure gas, or its partial pressure in a mixture, is 1 atm. For substances that are solid at this temperature and pressure, and can also have several different crystalline forms, it is necessary to specify for which form the enthalpy is given.

Standard enthalpies are enthalpies expressed for changes of substances in their standard states. The symbol for a standard enthalpy change is $\Delta H°$, and the specified temperature is often given as a sub-

TABLE 5.1

Standard enthalpies of formation

The values of ΔH_f° are for the formation of one mole of the given compound by direct combination of the elements in their standard states at 298°K. A compound that is a gas or a liquid at 25°C and 1 atm has that form as its standard state. For solids that can have several different crystalline states, it is necessary to specify for which form the value of ΔH_f° is given. To convert these values to kJ/mole multiply by 4.184 kJ/kcal.

Substance	ΔH_f° (kcal/mole)	Substance	ΔH_f° (kcal/mole)
(Standard form of all elements)	0.000	**Ca (OH)$_2$** (s)	−235.80
C (graphite)	0.000	**HCl** (g)	−22.062
C (diamond)	0.4533	**PCl$_3$** (g)	−68.6
P (s, red)	−4.2	**PCl$_5$** (g)	−89.6
P (s, black)	−9.4	**PCl$_5$** (s)	−106.0
P (s, white)	0.000	**AgCl** (s)	−30.370
H$_2$O (g)	−57.796	**NaCl** (s)	−98.232
H$_2$O (l)	−68.315	**KCl** (s)	−104.175
SO$_2$ (g)	−70.944	**HBr** (g)	−8.70
SO$_3$ (g)	−94.58	**HI** (g)	6.33
NO (g)	21.57	**H$_2$S** (g)	−4.93
NO$_2$ (g)	7.93	**NH$_3$** (g)	−11.02
CO (g)	−26.416	**CS$_2$** (l)	21.44
CO$_2$ (g)	−94.051	**CH$_4$** (g)	−17.88
PbO (s, yellow)	−51.94	**C$_2$H$_2$** (g)	54.19
PbO (s, red)	−52.34	**H$_2$SO$_4$** (l)	−194.548
PbO$_2$ (s)	−66.3	**HNO$_3$** (l)	−41.61
Al$_2$O$_3$ (α − s)	−400.5	**PbSO$_4$** (s)	−219.87
Ag$_2$O (s)	−7.42	**CaCO$_3$** (calcite)	−288.45
Fe$_2$O$_3$ (s)	−197.0	**CaCO$_3$** (aragonite)	−288.49
Fe$_3$O$_4$ (s)	−267.3	**CaSO$_3$·2H$_2$O**	−421.2
CaO (s)	−151.9	**CaSO$_4$·2H$_2$O**	−483.06
CaC$_2$ (s)	−15.0	**(CaSO$_4$)$_2$·H$_2$O**	−752.94

script ΔH°_{298}. We have chosen to omit specifying the temperature, and, unless otherwise stated, all ΔH° values in this book are given for 298°K. (It is important to note that the standard state of a substance is a specific physical state and is in no way related to the "standard conditions" of temperature and pressure (STP) used most often in discussing gases.)

One of the simplest chemical reactions is the formation of a compound from its elements, called a **direct combination** reaction. The **standard enthalpy of formation** of a compound, ΔH°_f, is the heat of formation of one mole of the compound by direct combination of the elements in their standard states at the specified temperature. Some representative standard enthalpies of formation, often referred to simply as heats of formation, are given in Table 5.1. By convention, the standard enthalpy of formation of a pure element in its standard state is taken to be zero.

5.5 Using enthalpies

Suppose we want to know the chemical equation for the formation of mercury(II) oxide from its elements and the standard enthalpy of formation of this compound. Equation (5), in which the substances are in their standard states, gives all the necessary information.

$$2HgO(s) \longrightarrow 2Hg(l) + O_2(g) \qquad \Delta H^\circ = 43.42 \text{ kcal } (181.7 \text{ kJ}) \qquad (5)$$

First, the equation must be written in the reverse direction and divided by a factor of 2, so that we are dealing with the formation of only one mole of HgO from the elements

$$Hg(l) + \tfrac{1}{2}O_2(g) \longrightarrow HgO(s)$$

Writing an equation for the formation or reaction of one mole of a substance for the purpose of giving the enthalpy per mole sometimes requires using fractional coefficients.

Next, the sign of the enthalpy must be reversed. This illustrates an important principle of thermochemistry: ΔH *for a reaction in one direction is equal in magnitude and opposite in sign to* ΔH *for the reaction in the reverse direction.*

Then, the ΔH value given in Equation (5) must be divided by 2 because the number of moles reacting has been divided by 2. This illustrates the second important principle of thermochemistry: ΔH *is directly proportional to the quantities of reactants or products.*

The complete thermochemical equation is

$$Hg(l) + \tfrac{1}{2}O_2(g) \longrightarrow HgO(s) \qquad \Delta H^\circ_f = -21.71 \text{ kcal } (-90.8 \text{ kJ})$$

The enthalpy of formation, or heat of formation, ΔH°_f of HgO(s) is -21.71 kcal/mole (-90.8 kJ/mole).

It might be helpful to think of the heat added to, or produced by, a reaction as one of the products or reactants. If the amounts of materials are changed, the amount of heat changes. For example, consider the combustion of carbon disulfide in oxygen, an exothermic reaction for which $\Delta H° = -257$ kcal (-1075 kJ).

$$CS_2(l) + 3O_2(g) \longrightarrow CO_2(g) + 2SO_2(g) + 257 \text{ kcal}$$

If three times as much CS_2 reacts, then $\Delta H° = 3 \times (-257 \text{ kcal}) = -771$ kcal (-3226 kJ).

$$3CS_2(g) + 9O_2(g) \longrightarrow 3CO_2(g) + 6SO_2(g) + (3 \times 257 \text{ kcal})$$

The reverse of an exothermic reaction is always an endothermic reaction. The heat appears on the opposite side in the equation, therefore the sign of the enthalpy changes. For the exothermic formation of hydrogen chloride $\Delta H° = -44.124$ kcal (-184.61 kJ)

$$H_2(g) + Cl_2(g) \longrightarrow 2HCl(g) + 44.124 \text{ kcal}$$

For the endothermic decomposition of hydrogen chloride, $\Delta H° = +44.124$ kcal ($+184.62$ kJ).

$$2HCl(g) + 44.124 \text{ kcal} \longrightarrow H_2(g) + Cl_2(g)$$

From the standard enthalpies of formation listed in Table 5.1, equations such as the following can be written:

$$H_2(g) + \tfrac{1}{2}O_2(g) \longrightarrow H_2O(l)$$
$$\Delta H_f° = -68.315 \text{ kcal/mole} (-285.83 \text{ kJ/mole})$$
$$3 \text{ Fe}(s) + 2O_2(g) \longrightarrow Fe_3O_4(s)$$
$$\Delta H_f° = -267.3 \text{ kcal/mole} (-1118 \text{ kJ/mole})$$
$$\tfrac{1}{2}H_2(g) + \tfrac{1}{2}I_2(s) \longrightarrow HI(g)$$
$$\Delta H_f° = +6.33 \text{ kcal/mole} (+26.5 \text{ kJ/mole})$$
$$Pb(s) + \tfrac{1}{8}S_8(s) + 2O_2(g) \longrightarrow PbSO_4(s)$$
$$\Delta H_f° = -219.87 \text{ kcal/mole} (-919.94 \text{ kJ/mole})$$

5.6 Hess' law

Hess' law is the third important principle of thermochemistry: *The total enthalpy change in a chemical reaction is independent of the number and nature of intermediate steps that may be involved in that reaction.* In other words, the heat absorbed or evolved in going from the initial state to the final state is the same no matter by what route the reactants get there.

Hess' law and the other thermochemical principles are consequences of the law of conservation of energy. It is impossible to reverse a reaction and get more heat out than was put in, or to go from the same reactants to the same products by different intermediate steps and get more heat out by one route than by another.

The very useful result of Hess' law is that chemical equations and their enthalpies can be dealt with algebraically, a process that allows the calculation of reaction enthalpies that are not known or are difficult to measure. The desired enthalpy is calculated from the enthalpies of a sequence of reactions that can be combined algebraically to give the desired reaction. It doesn't matter whether the sequence is real or exists only on paper.

EXAMPLE 5.4

■ Calculate the standard enthalpy of formation of SO_3 at 25°C using the following reactions and enthalpies:

$$S_8(s) + 8O_2(g) \longrightarrow 8SO_2(g) \qquad \Delta H° = -2375 \text{ kJ}$$
$$2SO_2(g) + O_2(g) \longrightarrow 2SO_3(g) \qquad \Delta H° = -198 \text{ kJ}$$

The equation for the formation of one mole of SO_3 from the elements is

$$\tfrac{1}{8}S_8(s) + \tfrac{3}{2}O_2(g) \longrightarrow SO_3(g)$$

This equation and its enthalpy change can be obtained by dividing the first equation and $\Delta H°$ by 8, dividing the second equation and $\Delta H°$ by 2, and adding the results:

$$\tfrac{1}{8}S_8(s) + O_2(g) \longrightarrow SO_2(g) \qquad \Delta H° = -2375/8 = -296.9 \text{ kJ}$$
$$\underline{SO_2(g) + \tfrac{1}{2}O_2(g) \longrightarrow SO_3(g) \qquad \Delta H° = -198/2 = -99 \text{ kJ}}$$
$$\tfrac{1}{8}S_8(s) + \tfrac{3}{2}O_2(g) \longrightarrow SO_3(g) \qquad \Delta H° = -396 \text{ kJ}$$

The enthalpy of formation of $SO_3(g)$ at 25°C is -396 kJ/mole or -94.6 kcal/mole. ■

5.7 $\Delta H_r°$ from $\Delta H_f°$

The standard enthalpies of formation of hundreds of compounds have been tabulated. These tables have great utility because the standard enthalpy of a chemical reaction can be calculated by subtracting the total standard enthalpies of formation of all the reactants from the total standard enthalpies of formation of all the products.

$$\Delta H_r° = (\text{sum } \Delta H_f° \text{ products}) - (\text{sum } \Delta H_f° \text{ reactants}) \qquad (6)$$

Note that since $\Delta H_f°$ values are given in kilocalories per mole of each species, $\Delta H_f°$ in Equation (6) for each reactant and product must be multiplied by the number of moles of that species taking part in the reaction.

In the following example, it is shown that using Equation (6) is equivalent to adding or subtracting the thermochemical equations of formation of the reactants and products in the reaction for which the $\Delta H_r°$ is desired.

EXAMPLE 5.5

■ The Thermit reaction used for welding metals involves the reduction of Fe_2O_3 by Al

$$2Al(s) + Fe_2O_3(s) \longrightarrow Al_2O_3(s) + 2Fe(s)$$

What is $\Delta H°$ at 25°C for this reaction?

The equation given can be considered as the sum of the equation for the formation of Al_2O_3 from the elements and the equation for the decomposition of Fe_2O_3 into the elements (the reverse of the formation reaction). Using values from Table 5.1 gives

$$\begin{array}{ll} 2Al(s) + \tfrac{3}{2}O_2(g) \longrightarrow Al_2O_3(s) & \Delta H° = -400.5 \text{ kcal} \\ \underline{Fe_2O_3(s) \longrightarrow 2Fe(s) + \tfrac{3}{2}O_2(g)} & \underline{\Delta H° = 197.0 \text{ kcal}} \\ 2Al(s) + Fe_2O_3(s) \longrightarrow Al_2O_3(s) + 2Fe(s) & \Delta H° = -203.5 \text{ kcal} \end{array}$$

The heat of reaction is -203.5 kcal or -851.4 kJ. We would arrive at the same answer using

$$\begin{aligned} \Delta H° &= (\text{sum } \Delta H_f° \text{ products}) - (\text{sum } \Delta H_f° \text{ reactants}) \\ &= [(1 \text{ mole})(\Delta H_f° \text{ } Al_2O_3) + (2 \text{ mole})(\Delta H_f° \text{ Fe})] - [(2 \text{ mole})(\Delta H_f° \text{ Al})] \\ &\qquad\qquad + (1 \text{ mole})(\Delta H_f° \text{ } Fe_2O_3)] \\ &= [(1 \text{ mole})(-400.5 \text{ kcal/mole}) + (2 \text{ mole})(0)] - [(2 \text{ mole})(0) \\ &\qquad\qquad + (1 \text{ mole})(-197.0 \text{ kcal/mole})] \\ &= -203.5 \text{ kcal} \end{aligned}$$

Because this large amount of heat cannot be rapidly dissipated to the surroundings, the reaction takes place with extreme violence and the reacting mass may reach temperatures near 3000°C. ■

5.8 Other useful enthalpies

The enthalpy for the removal of one electron from an atom or an ion in the gaseous state is called the **ionization energy.** For example, for the formation of gaseous Na^+,

$$Na(g) \longrightarrow Na^+(g) + e^- \qquad \Delta H = 118 \text{ kcal } (494 \text{ kJ})$$

The ionization energy of sodium is 118 kcal/mole. The ionization energy is always positive—it is the energy required to pull an electron away. (Sometimes this quantity is called ionization potential and is given in electron volts.) The energy for removal of the first electron is called the first ionization energy, for the second electron, the second ionization energy, and so on. Just "ionization energy" means the energy of the first ionization.

The energy liberated as heat in the addition of an electron to an atom in the gaseous state is called the **electron affinity**. For example, for the formation of gaseous chloride ions

$$Cl(g) + e^- \longrightarrow Cl^-(g) \qquad \Delta H = -88 \text{ kcal}$$

(A word of caution is necessary about the sign of electron affinities. Always remember that the addition of an electron to a gaseous atom is an exothermic reaction. However, because electron affinities are determined by measuring the heat of the reverse reaction, they are cited in many tables as positive.)

Comparison of the values of ionization energies and also, to a lesser extent, the values of electron affinities, provides useful information about the types of compounds formed by different elements (Chapter 10).

The enthalpies for changes of state (Figure 5.3) are usually referred to as heats of vaporization, heats of fusion, and so on. These quantities are nothing more nor less than enthalpies of physical changes. They change sign for reverse processes, are directly proportional to the amount of material involved, and obey Hess' law, just as enthalpies of reaction do.

Thermal energy is also exchanged when a substance goes into solution. Heats of solution, or enthalpies for reactions between solvent and solute such as

$$AB(s) \xrightarrow{\text{solvent}} A^+(solv) + B^-(solv)$$
$$AB(s) \xrightarrow{\text{solvent}} AB(solv)$$

may be either positive or negative.

EXAMPLE 5.6

■ There are two crystalline forms of PbO—one is yellow and the other is red. Using the heats of formation given in Table 5.1, find ΔH_r° at 25°C for the solid–solid phase transition

$$PbO(\text{yellow}) \longrightarrow PbO(\text{red})$$

The heat of reaction is given by

$$
\begin{aligned}
\Delta H^\circ &= (\text{sum } \Delta H_f^\circ \text{ products}) - (\text{sum } \Delta H_f^\circ \text{ reactants}) \\
&= [1 \text{ mole})(\Delta H_f^\circ \text{ red})] - [(1 \text{ mole})(\Delta H_f^\circ \text{ yellow})] \\
&= [(1 \text{ mole})(-52.34 \text{ kcal/mole})] - [(1 \text{ mole})(-51.94 \text{ kcal/mole})] \\
&= -0.40 \text{ kcal}
\end{aligned}
$$

About 400 cal or 1700 J will be released during this phase change.

■

Solid $\underset{\text{Heat of crystallization}}{\overset{\text{Heat of fusion}}{\rightleftharpoons}}$ Liquid

Liquid $\underset{\text{Heat of condensation}}{\overset{\text{Heat of vaporization}}{\rightleftharpoons}}$ Gas

Solid $\underset{\text{Heat of deposition}}{\overset{\text{Heat of sublimation}}{\rightleftharpoons}}$ Gas

FIGURE 5.3
Changes of state. *The enthalpy values are usually given in kilocalories per mole of substance changed at constant temperature. The heat added to raise the substance to the point where, e.g., vaporization begins is not included. Sometimes values for changes of state are given in calories per gram. Usually the enthalpies of fusion, vaporization, and sublimation are tabulated in handbooks of chemistry. For a further discussion of changes of state, see Section 12.5.*

THERMOCHEMISTRY [5.8]

Early observations on changes of state

LECTURES ON THE ELEMENTS OF
CHEMISTRY, *by Joseph Black*

The opinion I formed from attentive observation of the facts and phenomena is as follows. When ice or any other solid substance is melted, I am of the opinion that it receives a much larger quantity of heat then what is perceptible in it immediately afterwards by the thermometer. A large quantity of heat enters into it, on this occasion, without making it apparently warmer, when tried by that instrument. This heat must be added in order to give it the form of a liquid; and I affirm that this large addition of heat is the principal and most immediate cause of the liquefaction induced.

On the other hand, when we freeze a liquid, a very large quantity of heat comes out of it, while it is assuming a solid form, the loss of which heat is not to be perceived by the common manner of using the thermometer. The apparent temperature of the body, as measured by that instrument, is not diminished, or not in proportion to the loss of heat which the body actually suffers on this occasion; and it appears, from a number of facts, that the state of solidity cannot be induced without the abstraction of this large quantity of heat. And this confirms the opinion that this quantity of heat, absorbed, and, as it were, concealed in the composition of liquids, is the most necessary and immediate cause of their liquidity. . . .

I, therefore, set seriously about making experiments, conformable to the suspicion that I entertained concerning the boiling of liquids. My conjecture, when put into form, was to this purpose. I imagined that, during the boiling, heat is absorbed by the water and enters into the composition of the steam produced from it, in the same manner as it is absorbed by ice in melting and enters into the composition of the resulting water. And, as the ostensible effect of the heat in this latter case consists, not in warming the surrounding bodies, but in converting the ice into water; so, in the case of boiling, the heat absorbed does not warm surrounding bodies, but converts the water into steam. In both cases, considered as the cause of warmth, we do not perceive its presence: it is concealed, or latent, and I gave it the name of LATENT HEAT. *. . .*

Many have been the speculations and views of ingenious men about this union of bodies with heat. But, as they are all hypothetical, and as the hypothesis is of the most complicated nature, being in fact a hypothetical application of another hypothesis, I cannot hope for much useful information by attending to it. A nice adaptation of conditions will make almost any hypothesis agree with the phenomena.

THERMOCHEMISTRY

This will please the imagination, but does not advance our knowledge. I therefore avoid such speculations, as taking up time which may be better employed in learning more of the general laws of chemical operations. . . .

LECTURES ON THE ELEMENTS OF CHEMISTRY, by Joseph Black, 1803 (Quoted from p. 120 ff., *Harvard Case Histories in Experimental Science* (J. B. Conant, ed.), Vol. 1, Case 3 (prepared by Duane Roller), Harvard University Press, Cambridge, 1957)

Significant terms defined in this chapter

heat capacity
specific heat
exothermic process
endothermic process
changes of state
intermolecular forces
intramolecular forces
thermochemistry
enthalpy

enthalpy change
calorimeter
calorimeter constant
standard state
standard enthalpies
direct combination
standard enthalpy of formation
ionization energy
electron affinity

Exercises

5.1 A sleepy student (although probably unaware of all this) performs the following series of thermochemical processes each time a breakfast of coffee and toast is prepared: (i) a pot of water is put on the stove and heated to boiling, (ii) the student burns a finger on the steam coming from the pot, (iii) hot water is poured into a cup containing some instant coffee, (iv) the solution is stirred, (v) the warm brew burns his tongue, (vi) the coffee cools to drinking temperature, (vii) frozen bread is put in a toaster, (viii) the bread catches on fire because the toaster was set wrong, (ix) butter is spread on a piece of properly cooked toast and "melts," and (x) the digestive system converts the toast to energy, fat, and waste.

In which steps (a) did an exothermic process occur, (b) did an endothermic process occur, (c) did a phase change occur, (d) was the heat capacity important, (e) did a spontaneous process occur, and (f) was the law of conservation of energy obeyed? (g) If a very precise thermometer was placed in the cup of coffee immediately before (iv) was done, what would have been observed? (h) What would have happened if hot water had been used instead of cold water in (i) assuming the stove delivers a constant amount of heat in a given time?

5.2 Which statements are true? Rewrite any false statements so that it is correct.
(a) A positive sign on an enthalpy change means that an endothermic process has occurred.
(b) The calorimeter constant is the amount of heat required to raise 1 g of a substance by one degree.
(c) The enthalpy of fusion is numerically equal, but opposite in sign to the enthalpy of crystallization.
(d) The standard state of any substance is the physical state in which it is most stable at 1 atm pressure and a specified temperature.
(e) The heat transferred under constant volume conditions is the enthalpy change.

5.3 Distinguish clearly between (a) exothermic and endothermic, (b) intermolecular and intramolecular forces, (c) ionization energy and electron affinity, and (d) heat of formation and heat of reaction.

5.4 Why is the enthalpy of neutralization essentially constant for the reaction between one equivalent of a strong acid and one equivalent of a strong base?

5.5 State Hess' law. Why is it important in thermochemistry?

5.6 Write an equation showing the reaction associated with the ionization energy of element M. Write a similar equation for the electron affinity of element X.

5.7 Which reaction will be more exothermic? Why?

(i) $A(g) + B(s) \longrightarrow C(l) + H_2O(g)$
(ii) $A(g) + B(s) \longrightarrow C(l) + H_2O(l)$

5.8 The ionization energy of oxygen is 315.517 kcal/mole. To remove a second electron requires an additional 812.19 kcal/mole (the second ionization energy). Why is this second value much larger than the first?

5.9 The enthalpy for the addition of one electron to an oxygen atom is −33.6 kcal/mole (an electron affinity of +33.6). To add a second electron, 210 kcal/mole is required. Explain the difference between these numbers.

5.10 Arrange the following amounts of energy in increasing order: (a) 3 kcal; (b) 2 MJ; (c) 14 liter-atm; (d) 7.2 erg; (e) 15 J; and (f) 4 cal. Note that 1 litter-atm = 24.2 cal.

5.11 What would be the enthalpy change for (a) heating one mole of ethyl alcohol (heat capacity = 26.64 cal/mole °C) from −5°C to 35°C, (b) evaporating one gram of water (heat of vaporization = 9.7171 kcal/mole) at 100°C, (c) cooling 15.0 g of red phosphorus (heat capacity = 5.07 cal/mole °C) from 25°C to 23°C, and (d) heating 10.0 g of water from −5°C to 25°C (specific heat = 0.5 cal/g °C for ice and 1.0 cal/g °C for water; heat of fusion = 1436.3 cal/mole).

5.12 The specific heat of aluminum is 0.215 cal/g °C and that of lead 0.031 cal/g °C. Which substance would drop more in temperature if 50. cal were removed from (a) 1-g samples and (b) 1-mole samples?

5.13 What is the final temperature of a system prepared by placing a 10.00-g block of aluminum (specific heat = 0.215 cal/g °C) originally at 75.0°C on top of a 100.00-g block of aluminum originally at 25.0°C? Assume that no heat escapes to the surroundings.

5.14 Find the final temperature of the system prepared by placing 100.0 g of hot copper metal (specific heat = 0.092 cal/g °C) originally at 98.6°C into 100.00 g of water (specific heat = 1.00 cal/g °C at 25.0°C?

5.15 A calorimeter is calibrated by adding 100.0 ml of 0.100M HCl to a like amount of 0.100M NaOH. The resulting temperature change of the 200.0 ml NaCl solution formed as a result of the reaction (sp heat = 0.97 cal/g °C, density = 1.01 g/ml) and the calorimeter was 0.627°C. Find the calorimeter constant. For a strong acid reacting with a strong base $\Delta H° = -13.345$.

5.16 A bomb calorimeter was constructed so that any contribution of the products of the chemical reaction occurring within to the heat capacity (the calorimeter constant) could be neglected. A 1.298-g sample of benzoic acid was burned and the observed temperature change was 4.32°C.

What is the calorimeter constant if the heat of combustion of benzoic acid is -6320 cal/g?

5.17 A 1.02-g sample of urea, $(NH_2)_2CO$, was burned in a bomb calorimeter and the observed temperature increase was 0.381°C. If the calorimeter constant is 28.1 kJ/°C, find the molar heat of combustion of urea.

5.18 Write the chemical equation for the formation of one mole of methane, $CH_4(g)$, from the elements. Find the heat of formation of CH_4 from the following equations and values of ΔH_r° at 20°C:

$$H_2(g) + \tfrac{1}{2}O_2(g) \longrightarrow H_2O(l) \qquad \Delta H^\circ = -68.38 \text{ kcal}$$
$$C(s) + O_2(g) \longrightarrow CO_2(g) \qquad \Delta H^\circ = -94.38 \text{ kcal}$$
$$CH_4(g) + 2O_2(g) \longrightarrow CO_2(g) + 2H_2O(l)$$
$$\Delta H^\circ = -210.8 \text{ kcal}$$

5.19 Find ΔH_r° for making chloroform from methane

$$CH_4(g) + 3Cl_2(g) \longrightarrow CHCl_3(l) + 3HCl(g)$$

using the following equations

$$\tfrac{1}{2}H_2(g) + \tfrac{1}{2}Cl_2(g) \longrightarrow HCl(g) \qquad \Delta H^\circ = -22.062 \text{ kcal}$$
$$C(s) + 2H_2(g) \longrightarrow CH_4(g) \qquad \Delta H^\circ = -17.88 \text{ kcal}$$
$$C(s) + \tfrac{1}{2}H_2(g) + \tfrac{3}{2}Cl_2(g) \longrightarrow CHCl_3(l)$$
$$\Delta H^\circ = -32.14 \text{ kcal}$$

5.20 Find the heat of formation of liquid hydrogen peroxide at 25°C from the following thermochemical equations:

$$H_2(g) + \tfrac{1}{2}O_2(g) \longrightarrow H_2O(g) \qquad \Delta H^\circ = -57.796 \text{ kcal}$$
$$2H(g) + O(g) \longrightarrow H_2O(g) \qquad \Delta H^\circ = -221.539 \text{ kcal}$$
$$2H(g) + 2O(g) \longrightarrow H_2O_2(g) \qquad \Delta H^\circ = -255.88 \text{ kcal}$$
$$2O(g) \longrightarrow O_2(g) \qquad \Delta H^\circ = -119.106 \text{ kcal}$$
$$H_2O_2(l) \longrightarrow H_2O_2(g) \qquad \Delta H^\circ = 12.30 \text{ kcal}$$

5.21 Calculate the enthalpy change at 25°C for the reaction described by the equation

$$2NO(g) + O_2(g) \longrightarrow 2NO_2(g)$$

using data from Table 5.1.

5.22 When a welder uses an acetylene torch, it is the combustion of acetylene that liberates the intense heat for welding metal together. The equation for this process is

$$2C_2H_2(g) + 5O_2(g) \longrightarrow 4CO_2(g) + 2H_2O(g)$$

The heat of combustion of acetylene is -1256 kJ/mole. What amount of heat is liberated when 0.260 kg of C_2H_2 is burned?

5.23 Compare the quantities of heat liberated per mole of iron formed when the oxides of Fe_3O_4 and Fe_2O_3 are reduced by aluminum.

$$3Fe_3O_4 + 8Al \longrightarrow 4Al_2O_3 + 9Fe \qquad \Delta H^\circ = -800.1 \text{ kcal}$$
$$Fe_2O_3 + 2Al \longrightarrow Al_2O_3 + 2Fe \qquad \Delta H^\circ = -203.5 \text{ kcal}$$

5.24 The heat of formation for $Ca(C_2H_3O_2)_2(s)$ is -355.0 kcal/mole. Using the heats of formation given in Table 5.1 for $CO_2(g)$, $H_2O(g)$ and $CaO(s)$, find the heat of reaction for

$$Ca(C_2H_3O_2)_2(s) + 4O_2(g) \longrightarrow 4CO_2(g) + 3H_2O(g) + CaO(s)$$

5.25 Aragonite, a mineral, undergoes the following reaction in a very dilute solution of carbonic acid (CO_2 in H_2O).

$$CaCO_3(\text{aragonite}) + H_2CO_3(aq) \rightleftharpoons Ca(HCO_3)_2(aq)$$

This means that the $CaCO_3$ can be dissolved in one place and transported to another in the form of $Ca(HCO_3)_2$—a very important step in the formation of stalagmites and stalactites. Calculate ΔH_r° at 25°C if the heat of formation is -1207 kJ/mole for $CaCO_3$ (aragonite), -699 kJ/mole for $H_2CO_3(aq)$, and -1925 kJ/mole for $Ca(HCO_3)_2(aq)$.

5.26 The first ionization energy for Na is 118 kcal/mole and the enthalpy of addition of the first electron to Cl is -83.3 kcal/mole ($+83.3$ electron affinity). Find ΔH° for the process

$$Na(g) \longrightarrow Na^+(g) + e^-$$
$$\underline{Cl(g) + e^- \longrightarrow Cl^-(g)}$$
$$Na(g) + Cl(g) \longrightarrow Na^+(g) + Cl^-(g)$$

The large, positive value of ΔH would indicate NaCl to be unstable, but yet we know better. What is wrong with the prediction?

5.27 Using the heats of formation of gaseous and liquid water given in Table 5.1 calculate the heat of vaporization of water at 25°C.

5.28 Find the heat of sublimation of PCl_5 using the heats of formation given in Table 5.1. Which physical state has the lower enthalpy?

5.29 Using the standard heats of formation given in Table 5.1, predict which crystalline form of the following substances is more stable at 25°C and 1 atm; (a) red, white, or black phosphorus; (b) diamond or graphite; (c) calcite or aragonite.

5.30* (a) A student heated a sample of a metal weighing 32.6 g to 99.8°C and put it into 100.0 g of water at 23.6°C in a calorimeter. The final temperature was 24.4°C. The student calculated the specific heat of the metal, neglecting to use the calorimeter constant. What was the answer? The metal was known to be either Cr, Mo, or W and by comparing the value of the specific heat to those of the metals (Cr, 0.107; Mo, 0.0597; W, 0.0322 cal/g °C), the student identified the metal. What was the choice? (b) A student at the next laboratory bench did the same experiment, got the same data, and used the calorimeter constant, which was 98 cal/°C. Was the identification of the metal different?

5.31* The neutralization of a weak acid such as hydrocyanic acid, HCN, with a strong base can be considered as two steps:

$$HCN(aq) \longrightarrow H^+(aq) + CN^-(aq) \qquad \Delta H^\circ_{\text{ionization}}$$
$$\underline{H^+(aq) + OH^-(aq) \longrightarrow H_2O(l) \qquad \Delta H^\circ_{\text{neutralization, strong}}}$$

$$HCN(aq) + OH^-(aq) \longrightarrow H_2O(l) + CN^-(aq)$$
$$\Delta H^\circ_{\text{neutralization, weak}}$$

If ΔH°_f at 25°C in kcal/mole is 25.6 for HCN(aq), 0 for H$^+$(aq), 36.0 for CN$^-$(aq), -54.970 for OH$^-$(aq), and -68.315 for H$_2$O(l), find:

(a) $\Delta H^\circ_{\text{ionization}}$
(b) $\Delta H^\circ_{\text{neutralization,strong}}$
(c) $\Delta H^\circ_{\text{neutralization, weak}}$

5.32* As a person walks along a level sidewalk, approximately 2.0 kJ of energy is dissipated for each mile traveled for each pound the person weighs. How far should a 120-lb person walk to overcome the weight-gaining effects of a 1-lb box of candy? Assume the reaction

$$C_{12}H_{22}O_{11}(s) + 12O_2(g) \longrightarrow 12CO_2(g) + 11H_2O(l)$$
$$\text{\small sucrose}$$

is 50% efficient in the body and that $\Delta H^\circ_f = -531.0$ kcal/mole for sucrose.

5.33 The heat of formation of H$_2$SO$_4$ (1M) is -212 kcal/mole and of H$_2$SO$_4$ (16M) is -207 kcal/mole. What is the heat of dilution—that is, the enthalpy change for the reaction?

$$H_2SO_4(16M) + \text{solvent} \longrightarrow H_2SO_4(1M)$$

If the solution cannot gain or lose this heat fast enough to remain at 25°C, what sensation would you feel if you touch the beaker in which this reaction was performed?

5.34* The heat of formation of NH$_4$NO$_3$(s) is -366 kJ/mole and for NH$_4$NO$_3$(1M) it is -341 kJ/mole. What is the heat of solution that corresponds to ΔH°_r for the reaction

$$NH_4NO_3(s) + \text{solvent} \longrightarrow NH_4NO_3(1M)$$

If the solution cannot gain or lose this heat fast enough to remain at 25°C, what sensation would you feel if you touch the beaker in which this reaction was performed?

5.35* A sample of coke contains 90.9% carbon by weight. Assuming that all of the heat produced by the burning of this coke comes from its combustion to CO$_2$, calculate the total quantity of heat obtainable at 25°C through the burning of exactly 1 kg of the coke. If the coke sample contained 0.1% sulfur by weight and this sulfur burned completely to sulfur dioxide, SO$_2$, what is the total quantity of heat that would result from this source when the coke was burned? The heats of formation of CO$_2$ and SO$_2$ are -94.051 kcal/mole and -70.944 kcal/mole, respectively.

5.36* The heat of formation of HCl(g) is -22.062 kcal/mole at 25°C. The value of ΔH°_f at 500°K can be found by (a) calculating the enthalpy change for cooling $\frac{1}{2}$ mole of H$_2$(g) and $\frac{1}{2}$ mole of Cl$_2$(g) from 500°K to 298°K; (b) adding the heat of reaction at 298°K to the answer of part (a); and (c) adding the enthalpy change for heating one mole of HCl(g) from 298°K to 500°K to the result of part (b). Find ΔH°_f at 500°K given the heat capacities are 8.104 cal/mole °C for Cl$_2$(g), 6.96 cal/mole °C for HCl(g) and 6.889 cal/mole °C for H$_2$(g). Is the reaction more or less exothermic at 500°K than at 298°K?

$$\frac{1}{2}H_2(g) + \frac{1}{2}Cl_2(g) \xrightarrow{\Delta H^\circ_{500} = ?} HCl(g)$$

$$\Big\downarrow \Delta H^\circ_a \qquad\qquad \Big\uparrow \Delta H^\circ_c$$

$$\frac{1}{2}H_2(g) + \frac{1}{2}Cl_2(g) \xrightarrow{\Delta H^\circ_b = \Delta H^\circ_{298}} HCl(g)$$

5.37* The heat of formation at 1000°K for Al(s) is -2.514 kcal/mole, for Al(l) it is 0.000 kcal/mole, and for Al(g) it is 74.119 kcal/mole. Find the heat of (a) fusion, (b) vaporization, and (c) sublimation at this temperature. (d) What is the relationship between these three enthalpy values?

5.38** A series of experiments was designed to determine the enthalpy of formation of KBr(s). These experiments were all carried out in the same styrofoam coffee cup calorimeter. The first experiment consisted of pouring 10.0 g of water at 37.42°C into the cup containing 10.0 g of water at 25.16°C, giving a final temperature of 29.64°C. (a) Find the calorimeter constant.

The second experiment consisted of pouring 10.00 ml of 0.100M HCl at 24.96°C into the cup containing 10.00 ml of 0.100M NaOH, also at 24.96°C, giving a final temperature of 25.45°C. (b) Using the value of $\Delta H^\circ_r = -13.35$ kcal/mole for the neutralization, confirm the value of the calorimeter constant found in the first experiment. The third experiment consisted of reacting 10.00 ml of 0.100M HBr at 25.06°C with 10.00 ml of 0.100M KOH in the cup at 25.06°C, giving a final temperature of 25.55°C. (c) Find ΔH°_r for this neutralization and, using $\Delta H^\circ_f = -28.77$ kcal/mole for HBr(0.1M), -114.86 kcal/mole for KOH(0.1M), and -68.32 kcal/mole for H$_2$O(l), calculate ΔH°_f for KBr(0.05M). The fourth experiment consisted of dissolving one millimole of KBr(s) in 20.00 ml of water at 24.33°C to give a final temperature of 24.16°C. (d) Find ΔH°_r for this solution process

$$KBr(s) + \text{solvent} \longrightarrow KBr(0.05M)$$

and using the ΔH°_f for KBr(0.05M) determined in part (c), find ΔH°_f for KBr(s). All solutions are dilute enough that the specific heats can be considered to be 1.00 cal/g °C and the densities to be 1.00 g/ml.

6 THE ATMOSPHERE

The parts of the Earth of interest to chemistry can be divided into three spherical shells. The lithosphere is the solid crust of the Earth; it is about 35-km deep and contains about 1% of the mass of the Earth. The hydrosphere includes all the water on the surface of the planet. And the atmosphere is the gaseous envelope of the Earth that begins at our feet and extends out to the point where atoms and molecules are no longer held by gravity.

These three spheres include what is collectively called the ecosphere (or sometimes the biosphere)—the portions of the air, water, and soil spheres that support life. The ecosphere is our source for all the elements and the sink to which our wastes return.

If we could impart to you only one idea about what happens to chemicals in the ecosphere it would be this: nothing stands still. The oxygen we breathe, the parts of our automobiles, the detergent in our dishwashers—each came from somewhere in the ecosphere and is on its way somewhere, passing through the atmosphere, the hydrosphere, and the lithosphere in a cycle that is dependent upon its chemical and physical properties, and upon the way it is treated during its passage.

This chapter describes in general terms the atmosphere and its chief components, both natural and man-made. The regions and composition of the atmosphere are defined. Some of the properties of the gases in the atmosphere are discussed and the chemistry of the noble gases is presented. Air pollution—a complex subject that is here to stay as an important part of chemistry—is outlined very briefly in terms of the five major pollutants: carbon monoxide, hydrocarbons, nitrogen oxides, sulfur oxides, and particulates.

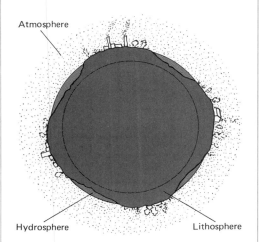

Atmosphere

Hydrosphere

Lithosphere

FIGURE 6.1

The three regions of the Earth. (*Not drawn to scale.*)

TABLE 6.1

Average atmospheric pressure and density at various altitudes

Altitude (km)	Pressure (torr)	Density (particles/cm³)
0	7.6×10^2	2.5×10^{19}
5	3.7×10^2	1.3×10^{19}
10	1.8×10^2	7.7×10^{18}
15	85	3.9×10^{18}
25	18	7.7×10^{17}
50	0.66	2.3×10^{16}
100	2.4×10^{-4}	1.1×10^{13}
160	1.7×10^{-6}	2.9×10^{10}

The atmosphere and the air

6.1 Functions of the atmosphere

The atmosphere includes all the gases that surround the Earth. All of the gases essential to life exist in the lower atmosphere (Figure 6.1). Carbon dioxide is necessary to plants for photosynthesis; oxygen is required by animals for respiration. Nitrogen is also vital as a component of living material, although it must first be transformed from gaseous N_2 to a soluble compound (such as ammonia, NH_3) before it can be utilized. Water, used by all forms of life, passes through the atmosphere as water vapor. (The cycles of oxygen, carbon, nitrogen, and water are described in later chapters.)

The atmosphere is bombarded by cosmic radiation from outer space and by sunlight. A large part of the radiation, which would be very damaging to life on Earth, is prevented from reaching us by interaction with the atoms, molecules, and ions of the atmosphere. Only two bands of radiation from the sun get through. One includes visible radiation and portions of the infrared and ultraviolet radiation; the other is in the radiowave region (see Tools of Chemistry: The Electromagnetic Spectrum, Chapter 7).

The heat balance of the Earth is controlled by the atmosphere, which uniformly redistributes the energy from the sun. Without the atmosphere, we would have 180°F temperatures by day and −220°F temperatures by night. Of the incoming solar radiation, 30% is reflected back into space by the atmosphere, clouds, and the surface of the Earth. This energy is lost to us. About 50% of the radiation reaches the ground and the oceans where it is absorbed as heat. The remaining 20% is absorbed by the components of the atmosphere and, in this way, affects the temperature of the atmosphere at various levels (Section 6.4).

6.2 Pressure of the atmosphere

The air at the surface of the Earth is compressed by the weight of the air above it, which it must support. As you go to increasingly higher altitudes, less and less of the atmosphere is above you and the pressure decreases. For each increase in altitude of about 5 km (16,500 ft), the pressure and density of the air decrease by about one-half (Table 6.1).

At 10 km (33,000 ft), the pressure is only one-fourth as much as at sea level. This is a common cruising altitude for jet aircraft. At this altitude, a normal inhaled breath takes into the lungs only one-fourth as much weight of air as at sea level, and therefore only one-fourth as much oxygen. This is not enough oxygen to meet the normal demands of the body. More oxygen for each breath is made available either by pressurizing the air in the cabin of the plane, or by providing pure oxygen.

6.3 Composition of the air

The air is in constant motion, and any substance that stays in the atmosphere a reasonable length of time becomes widely distributed. Nitrogen and oxygen make up 99% of the air by volume. Nitrogen and oxygen plus argon, the next most abundant gas in the air, make up a total percentage of more than 99.9% of the air by volume.

In a sample of air taken at any altitude up to 85 km, the amounts of the first eight gases listed in Table 6.2 will be just about constant. (The composition of the air is measured on dry samples because the amount of water vapor varies greatly from place to place and from day to day.)

The amounts of the substances below xenon in Table 6.2 can be considered variable. Note that these gases make up less than 0.1% of the total. All but ozone, which is generated in the upper atmosphere, enter the air as the products of natural processes on Earth, mainly from the decay of dead plant and animal matter. Many of these substances are also man-made pollutants, as we shall see in Sections 6.6–6.11. Although the concentration of carbon dioxide is quite stable as an average over the course of, say, a year, its concentration varies widely over cities, where it is generated by the combustion of fossil fuels and the breathing of people and animals.

TABLE 6.2
Average composition of clean, dry air [a]

Symbol or formula	Name	Percent by volume
N_2	Nitrogen	78.084
O_2	Oxygen	20.946
Ar	Argon	0.934
CO_2	Carbon dioxide	0.0325
Ne	Neon	1.818×10^{-3}
He	Helium	5.24×10^{-4}
Kr	Krypton	1.14×10^{-4}
Xe	Xenon	8.7×10^{-6}
CH_4	Methane	1.6×10^{-4}
H_2	Hydrogen	5×10^{-5}
CO	Carbon monoxide	8×10^{-6} to 5×10^{-5}
N_2O	Nitrous oxide	2 to 4×10^{-7}
SO_2	Sulfur dioxide	7×10^{-7} to $>1 \times 10^{-4}$
NO	Nitric oxide	10^{-6} to 10^{-4}
NO_2	Nitrogen dioxide	10^{-6} to 10^{-4}
HCHO	Formaldehyde	$\leq 10^{-5}$
NH_3	Ammonia	$\leq 10^{-4}$
O_3	Ozone	0 to 5×10^{-5}

[a] Adapted from "An Introduction to Air Chemistry," S. S. Butcher and R. J. Charlson, pp. 4, 5, Academic Press, 1972.

6.4 Liquefaction and distillation of air

At a low enough temperature and a high enough pressure, intermolecular forces of attraction overcome the kinetic energy of gas molecules, and any gas can be liquefied. At one time hydrogen, nitrogen, and oxygen were called the "permanent" gases because no one had been able to liquefy them. The main problem was that no way had been found to cool the gases sufficiently. By the 1880s this problem was solved, and it became possible to liquefy air and the permanent gases. By 1902, the first commercially practical methods for preparing oxygen from liquid air had been perfected.

When a gas is compressed, the molecules are forced closer to each other. The intermolecular forces increase, and unless the heat is removed, energy is released that raises the temperature of the gas. (Remember that heat is the energy of moving molecules.) Temperatures low enough for liquefaction are obtained (as shown in Figure 6.2) by compressing a gas, cooling it to remove the heat generated by compression, then allowing it to expand through a small opening, or valve—a process that cools the gas further. (Cooling a gas by expansion is called the Joule–Thomson effect.) Compressed gases cool upon free expansion because energy is required to overcome the intermolecular forces.

The cooled gas is passed back over the incoming compressed gas, lowering its temperature some more. As the gas passes through the compression–cooling–expansion–cooling cycle a number of times the temperature is gradually lowered to the point of liquefaction.

Liquid air is separated into its components by **fractional distillation**—separation of liquid mixtures based upon the different boiling points of different components of the mixture. As liquid air is allowed to warm up, the first portions, or fractions, that come off are rich in nitrogen (b.p. − 195.80°C), the lowest-boiling component of the mixture. The liquid left behind grows richer in oxygen (b.p.

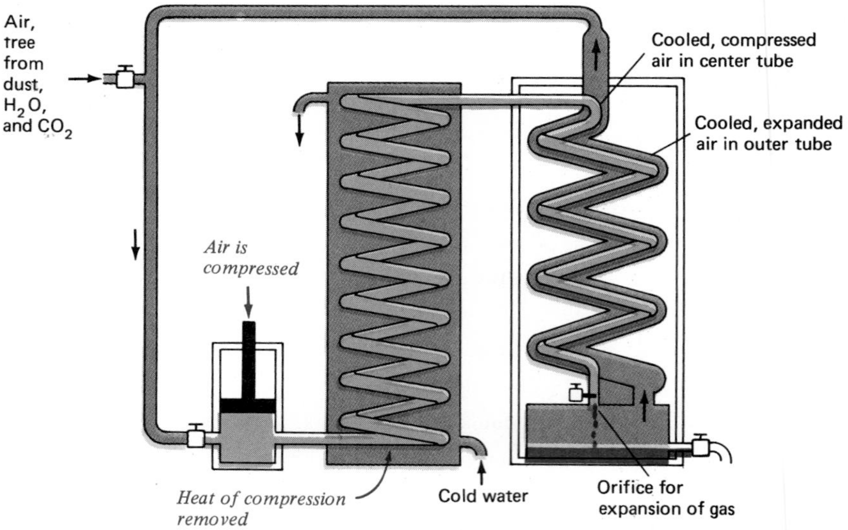

FIGURE 6.2

A simple liquid air machine. *The temperature of the gas leaving the orifice becomes lower and lower as the cycle is repeated. Once liquefaction of the gas escaping the orifice begins, the process is continuous.*

Air, tree from dust, H_2O, and CO_2

Cooled, compressed air in center tube

Cooled, expanded air in outer tube

Air is compressed

Heat of compression removed

Cold water

Orifice for expansion of gas

− 182.97°C) and argon (b.p. − 189.34°C). Modern air liquefaction plants produce pure nitrogen, oxygen, argon (and neon, krypton, and xenon; Section 6.13) by careful fractional distillation of liquid air.

There are two ways to store gases. A gas may be stored under high pressure at ordinary temperatures in a heavy steel cylinder. Or it may be cooled until it liquefies and then stored as a liquid at ordinary pressure, but at or below the boiling point of the liquid. If at the boiling point, boiling away of part of the liquid keeps the rest of it cool, and the container must be vented to the outside air to prevent a build-up of pressure. Tanks of this type must be well-insulated, but need not be of heavy construction.

6.5 Regions of the atmosphere

Four regions of the atmosphere—the troposphere, the stratosphere, the mesosphere, and the thermosphere—are defined by temperature variations with height (Figure 6.3). In addition to the decreasing density with altitude, there are variations in the chemical composition of these regions. Oxygen, nitrogen, and argon are present throughout the atmosphere; the other gaseous species vary with altitude.

The *troposphere* is the region just above the surface of the Earth. The temperature in the troposphere decreases with altitude up to an area of minimum temperature at about 11 km called the *tropopause*. The weather, the air we breathe, and 80% of the mass of the atmosphere are located in the troposphere. This is the region most greatly affected by air pollution and the region that contains most of the water vapor (above 10 km water is present as ice crystals).

The carbon dioxide and water vapor in the troposphere absorb a large part of the 20% of the incoming solar radiation that is taken up by the atmosphere. The surface of the Earth, as it cools, reemits the energy that it has absorbed. Some of this reemitted radiation is also absorbed by tropospheric water and carbon dioxide, and once again radiated back to the surface. In this way additional heat is kept within the lower atmosphere. Such warming by absorption and re-emission of radiation is called the **greenhouse effect.**

Owing to the activities of people, mainly the burning of **fossil fuels**—natural gas, coal, and petroleum—the carbon dioxide content of the atmosphere has been increasing. It has been estimated that worldwide an 18-fold increase in man-made emissions of carbon dioxide will have occurred between the years 1890 and 2000.

If the rate at which solar radiation arrives at the Earth is constant, an increase in carbon dioxide content of the atmosphere will increase the amount of heat radiated back to Earth via the greenhouse effect. A 10% increase in carbon dioxide could, it has been estimated, increase the average temperature of the Earth by 0.5°C. Based on the rising carbon dioxide content of the atmosphere, predictions of impending disaster by melting of the polar ice caps, flooding of coastal lands, and changing climate are made with regularity.

A period of increasing annual average temperature did occur

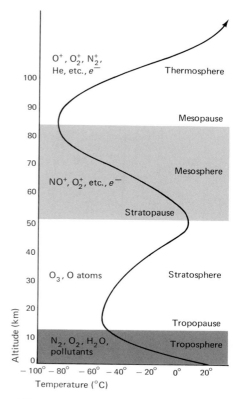

FIGURE 6.3

Regions of the atmosphere. *The four regions of the atmosphere are defined by the temperature minima of the tropopause and the mesopause, and the maximum of the stratopause. Oxygen and nitrogen are the most abundant gases in the atmosphere, and, roughly up to the mesopause, the concentrations of the first eight gases listed in Table 6.2 are constant. The substances listed in this figure are those that, in addition, are present in the regions above the troposphere.*

from 1880 to 1940. However, this was followed by a period of decreasing annual average temperature. Other things must also be happening. One suggestion is that the amount of dust in the atmosphere is increasing, and that this limits the amount of sunlight reaching the Earth and counteracts the effect of carbon dioxide.

The *stratosphere* (11–50 km) is characterized by increasing amounts of ozone, O_3. From 0.04 parts per million, in the troposphere, ozone increases to a maximum of about 10 ppm at 25–30 km. The ozone is formed in a two-step process initiated by ultraviolet radiation, which provides enough energy to split the dioxygen into two oxygen atoms. The oxygen atoms then react with more dioxygen to give ozone.

$$O_2 \xrightarrow{h\nu} O + O$$
$$O + O_2 \longrightarrow O_3$$

(The symbol $h\nu$, read "H-nu," is used to indicate radiation.) The ozone cycle in the stratosphere is completed by ozone absorbing ultraviolet radiation and breaking down into dioxygen and an oxygen atom. Heat is given off in this reaction and serves to warm the stratosphere. (Note in Figure 6.3 that the stratosphere is a zone of increasing temperature with altitude.)

$$O_3 \xrightarrow{h\nu} O_2 + O + heat$$

Reactions such as these that are initiated by radiation are called **photochemical reactions.**

Concern has arisen in recent years over pollutants that can find their way into the stratosphere and destroy ozone there. The primary reason for concern is that a decrease in stratospheric ozone would allow more ultraviolet radiation to pass through the stratosphere and reach the surface of the Earth. This type of radiation is known to be a cancer-causing agent, and the occurrence of skin cancer has been shown to increase with increased exposure to ultraviolet radiation.

Two types of compounds have received the most attention as potential ozone-depleting agents—the chlorofluorocarbons (carbon–chlorine–fluorine containing compounds), known as **Freons,** and nitrogen oxides. Nitric oxide, NO, is produced in the engines of supersonic transport planes, and their exhaust releases large quantities of this oxide directly into the stratosphere (NO is also produced in other processes near ground level; Section 6.9).

The ozone concentration is decreased by the following reaction sequence in which, it should be noted, no NO is consumed. The oxygen atoms are available in the stratosphere from the decomposition of ozone and oxygen (see equations above).

$$NO + O_3 \longrightarrow NO_2 + O_2$$
$$NO_2 + O \longrightarrow NO + O_2$$

Freons are introduced into the atmosphere from aerosol sprays, in which they function as propellants, and from refrigerating equipment, in which they act as coolants. In the stratosphere they are thought to first yield chlorine atoms in a photochemical reaction

$$CF_2Cl_2 \xrightarrow{h\nu} CF_2Cl + Cl$$
$$CFCl_3 \xrightarrow{h\nu} CFCl_2 + Cl$$

The reactive chlorine atoms then destroy ozone in the following reaction sequence, which can be repeated over and over since the chlorine atoms are regenerated in the second reaction.

$$Cl + O_3 \longrightarrow ClO + O_2$$
$$ClO + O \longrightarrow Cl + O_2$$

In the *mesosphere* and the *thermosphere* large quantities of free ions and electrons are formed in photochemical reactions. Table 6.3 lists, roughly in the order of increasing frequency with altitude, some of the reactions that occur in the ion-forming regions of the mesosphere and the thermosphere (together called the *ionosphere*). Free ions and electrons do not last long below the mesosphere, where pressure and density are greater. They immediately collide with other ions, atoms, or molecules to form neutral species. However, active species can survive for a long time in the upper atmosphere because they do not soon encounter something with which to combine.

The *exosphere*—the highest region of the atmosphere—contains mainly atomic and ionic oxygen, hydrogen, and helium. A particle here is as likely to escape from Earth's gravity as be held by it. Few reactions occur in the exosphere because there are so few atoms and ions present. It has been said that one particle could travel around the Earth in the exosphere and not collide with another.

Beyond the exosphere lies the unbounded area that we call interstellar space. A surprising number of different molecules have been detected here, although, of course, in very low concentration. Several are organic molecules, such as methane, methanol, and formaldehyde (Table 6.4), that could combine to form the larger molecules necessary to support life.

Air pollution

6.6 Pollutants and their sources

A **pollutant** is an undesirable substance added to the environment, usually by the activities of Earth's human inhabitants. Some pollutants, such as the insecticide DDT [dichlorodiphenyltrichloroethane, $CCl_3CH(C_6H_5Cl)_2$] are unknown in nature. Others, such as nitrogen(II) oxide, NO, occur naturally (from bacterial decay and in lightning flashes), but become pollutants when they are added to the environment in excessive amounts. Still other substances, called **secondary pollutants,** are harmful materials formed by chemical reactions in the atmosphere or hydrosphere.

TABLE 6.3

Atom- and ion-forming reactions in the ionosphere
These reactions occur, roughly, with increasing frequency with altitude in the order given here

$$NO \xrightarrow{h\nu} NO^+ + e^-$$
$$O_2 \xrightarrow{h\nu} O_2^+ + e^-$$
$$N_2 \xrightarrow{h\nu} N_2^+ + e^-$$
$$N_2^+ + O \longrightarrow NO^+ + N$$
$$O \xrightarrow{h\nu} O^+ + e^-$$
$$He \xrightarrow{h\nu} He^+ + e^-$$
$$O_2 \xrightarrow{h\nu} O + O$$
$$N_2 \xrightarrow{h\nu} N + N$$

TABLE 6.4

Some molecules observed in interstellar space
A total of about 30 species have been detected.

Dihydrogen	H_2
Carbon monoxide	CO
Hydrogen cyanide	HCN
Water	H_2O
Ammonia	NH_3
Methanol	CH_3OH
Methane	CH_4
Formaldehyde	HCHO
Formic acid	H_2CO_2

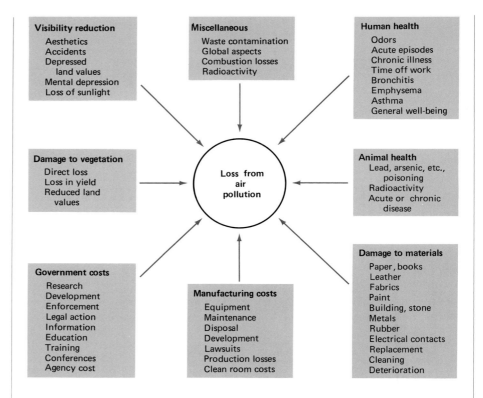

FIGURE 6.4

Effects of air pollution. [*Source: H. C. Wohlers, Air, a priceless resource, in* Environmental Health (*P. W. Purdeom, ed.*), *Academic Press, New York, 1971.*]

TABLE 6.5

The five major air pollutants

Name	Properties	Name	Properties
Carbon monoxide, **CO**	**Colorless, odorless gas; m.p. $-199°C$; b.p. $-191.5°C$**	*Sulfur oxides* **Sulfur(IV) oxide (sulfur dioxide), SO_2**	**Colorless gas; suffocating odor; m.p. $-73°C$; b.p. $-10°C$**
Hydrocarbons	**Compounds of carbon and hydrogen only, e.g., methane, CH_4; ethylene, $CH_2 = CH_2$**	**Sulfur(VI) oxide (sulfur trioxide), SO_3**	**White crystalline solid at room temperature; reacts rapidly with H_2O to give H_2SO_4; b.p. $45°C^a$**
Nitrogen oxides **Nitrogen(II) oxide (nitric oxide) NO**	**Colorless, odorless gas; m.p. $-161°C$; b.p. $-152°C$**	*Particulates*	**Solids or liquids suspended in air**
Nitrogen(IV) oxide (nitrogen dioxide) NO_2	**Brown gas; suffocating odor; m.p. $-11°C$; b.p. $21°C$**		

^{a} m.p. = 16.8°C, 32.5°C, or 62.3°C, depending upon crystalline form.

To estimate the overall damage from air pollution is difficult—there are so many ways in which such pollution can be harmful and costly (Figure 6.4). In one study of the economic effects of air pollution in 17 counties in the New York–New Jersey area, it was estimated that pollution cost $620 per family per year.

The five major air pollutants (Table 6.5) are carbon monoxide, hydrocarbons, the nitrogen oxides, the sulfur oxides, and **particulates**—airborne solid particles and liquid droplets. Each of these pollutants is discussed in the sections that follow. Pollutants enter the troposphere and, with a few exceptions, their effects are felt there.

Combustion of fossil fuels accounts for well over half of the pollutants produced each year. Such combustion takes place both in motor vehicles and in stationary sources, such as homes, where the fuel is burned for heating, and power plants. In comparing pollutants and their hazards, it should be remembered that different substances become harmful at different concentrations. For this reason, Figure 6.5 compares the major pollutants both with respect to the amount produced and the relative health effects of each.

6.7 Carbon monoxide

The largest source of man-made carbon monoxide is automobile exhaust. The fate of the carbon in the complete combustion of hydrocarbon fuels can be pictured as follows:

$$C + O_2 \longrightarrow CO_2$$

When the oxygen supply is low or combustion is incomplete for other reasons, such as poor mixing of fuel and oxygen or a low combustion temperature, carbon monoxide rather than carbon dioxide is formed (see Figure 6.6).

$$2C + O_2 \longrightarrow 2CO$$

Levels of CO as high as 100 ppm have been measured in downtown urban areas at peak traffic hours. Carbon monoxide is also produced in high-temperature industrial processes by the decomposition of carbon dioxide, and in solid waste disposal.

Carbon monoxide is a poison because it combines with hemoglobin up to 300 times more easily than does oxygen. The normal function of hemoglobin is to combine with oxygen in the lungs to form oxyhemoglobin. The oxyhemoglobin travels to the cells, where it gives up oxygen and picks up carbon dioxide for the return trip to the

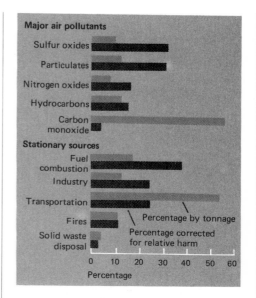

FIGURE 6.5

Amounts and sources of major pollutants. *The gray bars show the percent by a weighted index that takes into account the relative health effects of each pollutant. The colored bars for tonnage are based on Council on Environmental Quality Data (1972). "Fires" include agricultural burning to clear land, burning of coal refuse, forest fires, and structural fires. [From* Living in the Environment: Concepts, Problems, and Alternatives, *by G. Tyler Miller, Jr. © 1975 by Wadsworth Publishing Company, Inc., Belmont, California. Reprinted by permission of the publisher.]*

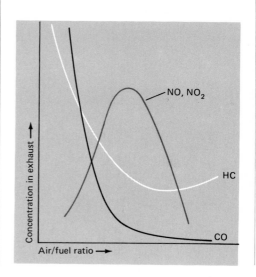

FIGURE 6.6

Effects of air–fuel ratio on automobile exhaust. *Carburetor design has been modified to avoid low air–fuel ratios, where hydrocarbon (HC) and CO emissions are at a maximum, but a complicating factor is the peak in NO$_x$ emissions. (Not drawn to scale.)*

TABLE 6.6

National primary ambient air quality standards

Standards are those recommended by the Environmental Protection Agency in 1975

Carbon monoxide
Maximum 8-hr concentration of 9 ppm, not to be exceeded more than once a year

Hydrocarbons
Maximum 3-hr concentration (6–9 am) of 0.24 ppm not to be exceeded more than once a year

Nitrogen dioxide
Annual arithmetic mean, 100 μg/m^3

Photochemical oxidants (O_3, PAN, etc.)
Maximum 1-hr concentration of 0.08 ppm not to be exceeded more than once a year

Sulfur dioxide
Annual arithmetic mean of 0.03 ppm

Particulates
Annual geometric mean, 75 μg/m^3

lungs. As carbon monoxide levels in the blood increase, more and more hemoglobin is tied up and is not free to carry oxygen. The body becomes oxygen-starved.

The amount of carbon monoxide–hemoglobin complex in the blood is proportional to the amount of carbon monoxide in the atmosphere. At low concentrations (50–100 ppm), responses are slowed down. Higher concentrations lead to headaches, dizziness, and, ultimately, at concentrations of 750 ppm or more, to coma and death. Carbon monoxide is especially dangerous because it is colorless and odorless. Many carbon monoxide–caused deaths result each year from defective exhaust systems in cars or poorly designed or defective space heaters and furnaces. An automobile engine should never be run in a closed or poorly ventilated garage, or in a closed, stationary car in order to keep the heater on.

The Environmental Protection Agency (EPA) has recommended "national primary ambient air quality standards" for major pollutants. "Ambient air" is the air outside of buildings "to which the general public has access." The standard defines levels of air quality judged "necessary, with an adequate margin of safety, to protect the public health." These standards, set in 1975, are given in Table 6.6.

Until recently it was believed that the major source of *all* atmospheric carbon monoxide was the combustion of gasoline in the automobile. It has now been found that at least ten times more carbon monoxide enters the atmosphere from natural processes than from all man-made sources combined. According to one investigation, about 3.7 billion tons of carbon monoxide are produced annually. Of this amount 80% comes from the conversion of methane (CH_4) released in the bacterial decay of living matter, 3% from the synthesis and decay of chlorophyll, and about 10% from the oceans and other as yet undetermined sources. Only 7% (270 million tons) arises from man-made sources. Since much of this is released in centers of high population density, local carbon monoxide levels 50–100 times higher than the worldwide concentration of 0.1–0.5 ppm are produced.

The concentration of carbon monoxide in the atmosphere is not increasing. Although the fate of atmospheric carbon monoxide is not fully understood, it is known that much of it is converted to carbon dioxide by soil bacteria or by reactions in the atmosphere.

6.8 Hydrocarbons

The internal combustion engine is the largest single source of man-made hydrocarbon pollutants, just as it is of carbon monoxide. Motor fuels are mixtures of hydrocarbons, and these compounds are introduced to the atmosphere when unburned or partially burned fuel is emitted in automobile exhaust. Gasoline also escapes by evaporation from the fuel tank and the engine. Little hydrocarbon pollution comes from stationary fuel-burning uses, since the combustion temperature tends to be higher and combustion tends to be more complete than in automobile engines. However, industrial processes in which hydrocarbons are lost during handling make a substantial contribution to the total of hydrocarbon pollutants.

Many different hydrocarbons are present in the air. Most of these hydrocarbons are of low molecular weight (up to roughly 10 carbon atoms) and are gases or volatile liquids at ordinary temperatures. Methane (CH_4), the simplest hydrocarbon, is the most abundant hydrocarbon pollutant. For example, it was found at an average of about 3 ppm in Los Angeles air in 1965.

Harmful effects from hydrocarbons at present concentrations in the air are unlikely. Several thousand times larger amounts would be necessary to cause irritation or physiological damage in human beings. However, because of photochemical reactions with oxygen and nitrogen oxides, hydrocarbons play a crucial role in the formation of photochemical oxidants and photochemical smog, both of which have a far greater potential for doing damage.

Steps obviously are necessary to reduce pollution from automobile engines. How to accomplish this and how quickly it can be done are questions that have brought government, the automobile industry, and consumer and environmental protection groups into continuing conflict.

The Clean Air Act of 1970 called for a 90% reduction in automobile emission of three pollutants—hydrocarbons, carbon monoxide, and nitrogen(II) oxide (next section) between 1970 and 1975. The amounts of emission of each pollutant were to be decreased in stages, with the toughest standard enforceable in 1976 model automobiles. A variety of technical problems has caused postponement of enforcement of the final set of standards.

However, some reductions have been achieved. By changing engine design to improve combustion efficiency and to better control the air-to-fuel ratio (Figure 6.6), losses of unburned gasoline have been cut back. To decrease the concentration of hydrocarbons and carbon monoxide in the exhaust gases major reliance thus far has been placed upon catalytic converters. Exhaust gases pass over a catalyst containing a metal such as platinum or rhodium, and are oxidized to harmless substances by additional air taken in at the converter.

$$\text{hydrocarbons} + O_2 \longrightarrow H_2O + CO_2$$
$$2CO + O_2 \longrightarrow 2CO_2$$

Most cars manufactured since 1975 are equipped with such oxidizing catalytic converters.

Introduction of the catalytic converters has led to another change that should be very beneficial to the quality of our air. Lead, used in gasoline in the form of tetraethyllead to prevent "knocking," renders the metallic catalyst in the converters useless. To solve this problem cars that can run on lead-free gasoline have been put on the market, and such fuel has been made available by the oil companies. Thus, a major source of lead, a toxic pollutant of the air and water, may be eliminated in a few years.

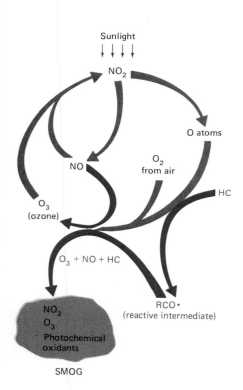

Photochemical smog requires sunlight, hydrocarbons, and nitrogen oxides

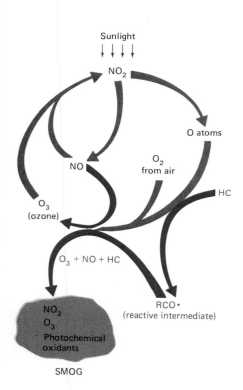

FIGURE 6.7
Interaction of hydrocarbons with natural NO_2 in photochemical reaction cycle. *The natural cycle is shown in black.*

6.9 Nitrogen oxides

Two nitrogen oxides (Table 6.5) are produced by oxidation of atmospheric nitrogen. (There are other nitrogen oxides, but only these two are pollutants.)

$$N_2 + O_2 \longrightarrow 2NO$$
nitric oxide

$$2NO + O_2 \longrightarrow 2NO_2$$
nitrogen dioxide

The first of these reactions takes place at high temperatures (about 1200–1750°C); therefore, NO is a by-product of any high-temperature combustion in the presence of air. Most fossil fuel combustion in both motor vehicles and stationary sources achieves the necessary temperatures and almost 90% of the pollutant NO is introduced in this way. Pollution by NO reaches its highest levels over highly industrialized cities with high automobile populations.

On their own, neither of these nitrogen oxides is a health hazard at present pollution levels. Nitrogen dioxide is potentially the more dangerous of the two, and in high enough concentration can damage the lungs. But the real hazard of nitrogen oxides lies in the formation of the unpleasant mixture of gases and particulates that make up photochemical smog. The essential ingredients for initiation of **photochemical smog** are sunlight, nitrogen oxides, and hydrocarbons. This is the type of smog for which Los Angeles is famous.

A normal nitrogen dioxide reaction cycle is initiated by sunlight

$$NO_2 \xrightarrow{h\nu} NO + O$$
$$O + O_2 \longrightarrow O_3$$
$$O_3 + NO \longrightarrow NO_2 + O_2$$

This cycle produces, but also uses up, NO and O_3 (ozone) and adds no new NO_2 to the atmosphere. The intervention of hydrocarbons in the cycle starts the trouble (Figure 6.7). Hydrocarbons react with the oxygen atoms to form highly reactive intermediates of a type called **free radicals**—reactive species that contain unpaired electrons. The free radicals (symbolized here by a dot on the formula, e.g., RCO·) initiate a variety of reactions, among which may be the following:

$$RCO\cdot + O_2 \longrightarrow RCO_3\cdot$$
$$RCO_3\cdot + \text{hydrocarbons} \longrightarrow RCHO, \quad R_2C = O$$
$$RCO_3\cdot + NO \longrightarrow RCO_2\cdot + NO_2$$
$$RCO_3\cdot + O_2 \longrightarrow O_3 + RCO_2\cdot$$
$$RCO_3\cdot + NO_2 \longrightarrow RCO_3NO_2$$
peroxyacylnitrates (PAN)

As a result, concentrations of ozone, the peroxyacylnitrates (PAN), and aldehydes (RCHO) build up. These substances are all irritants to the respiratory system and also can damage many materials. Ozone is particularly destructive to rubber and fabrics, and is harmful to crops and ornamental plants. Nitrogen dioxide also accumulates as a result of the free radical reactions and is responsible for the characteristic brown color of photochemical smog. Airplane pilots are familiar with the sight of a brown pall hanging over cities.

The word "smog" is a combination of "smoke" and "fog." It is something of a misnomer here, because photochemical smog includes nitrogen oxides, ozone, PAN, aldehydes, hydrocarbons, and carbon monoxide, but not smoke or fog. The term is more accurate for the London-type smog, which is described in the next section.

6.10 Sulfur oxides

Sulfur dioxide, SO_2, is produced when sulfur-containing coal and fuel oil are burned. Very little sulfur dioxide pollution can be blamed on the automobile, because in the production of gasoline most of the sulfur is removed. One-half of the annual sulfur dioxide emission comes from power plants, and the second most substantial contribution is made by industrial processes. A major polluting industrial process is metal smelting, in which sulfur is removed from sulfide ores by oxidation, for example,

$$2ZnS(s) + 3O_2(g) \longrightarrow 2ZnO(s) + 2SO_2(g)$$

Sulfur trioxide is formed in the air by oxidation of the lower oxide.

$$2SO_2(g) + O_2(g) \rightleftharpoons 2SO_3(l \text{ or } s)$$

In the presence of water vapor, sulfur trioxide immediately reacts to form sulfuric acid, which remains suspended in droplets in the air.

$$SO_3(l \text{ or } s) + H_2O(g) \longrightarrow H_2SO_4(l)$$

Sulfur dioxide and sulfur trioxide are both strongly irritating to the respiratory tract. At SO_2 concentrations of 5 ppm almost everyone suffers throat and eye irritation, and SO_3 at a concentration of 1 ppm can cause severe discomfort. Elderly persons and those with respiratory diseases are most seriously affected. Sulfuric acid is a strong acid; it is corrosive and damages building materials, especially marble, corrodes iron and steel, and is very destructive to lung tissue.

London smog, or gray smog, is initiated by a mixture of sulfur oxides, particulates, and high humidity. Many of the chemicals in particulates can catalyze the formation of SO_3 from SO_2, and high humidity assures the formation of a fog containing many sulfuric acid droplets. The sulfuric acid also coats the surface of particulates and is drawn into the lungs with them. The smog of 1952 in London, during which SO_2 reached a concentration of 1.34 ppm, is blamed for 4,000 deaths. (For comparison, "normal" concentrations of SO_2 in six major U.S. cities in one recent year was less than 0.15 ppm.)

Sulfur oxides in the atmosphere are also the major contributors to the increasing acidity of rain. Particularly over industrial areas,

London smog requires SO_2 and SO_3, particulates, and H_2O.

rain brings down with it an undesirable concentration of acid by washing sulfuric acid out of the atmosphere. Nitric acid, HNO_3, formed by oxidation of NO_2, and other acid substances are also present in acid rain. In addition to the destructive effects of the acid on materials, acid rain has been blamed for declining fish populations and crop damage.

The best hope for decreasing sulfur oxide pollution is the wider use of low-sulfur or sulfur-free fuels. Coal from west of the Mississippi River is lower in sulfur than that mined in the eastern United States. However, because of the greater distance it must be transported to larger cities, and also because this coal yields somewhat less energy per ton, it is more expensive in the East. It is possible to remove sulfur from fuel oil or to make sulfur-free liquefied gaseous fuel from coal, although neither process is yet economically attractive. Systems to remove sulfur dioxide from smokestack gases are also used. One method converts the sulfur dioxide to calcium sulfate by reaction with limestone

$$2CaCO_3(s) + 2SO_2(g) + O_2(g) \longrightarrow 2CaSO_4(s) + 2CO_2(g)$$
limestone

Although this process is successful in removing the pollutant SO_2 from stack gases, it creates the problem of disposal of large quantities of solid calcium sulfate. The solution of one problem creates another. Dilemmas of this type are not uncommon in the field of environmental protection.

6.11 Particulates

The range in size and chemical nature of particulates is great. There are 5-nm particles of soot and 500–1000-nm droplets of sulfuric acid mist. Particles produced in burning coal, called fly ash, can be up to 500,000 nm in diameter and are the oxides of various metals and nonmetals, such as Fe_2O_3, TiO_2, and P_4O_{10}.

Table 6.7 lists the annual production of particulates from both natural sources and from human activities. Man-made particulates are contributed to the atmosphere in roughly equal amounts from stationary fuel combustion, industrial processes, and fires, such as forest fires and agricultural burning of wastes.

Persistent irritation of the lungs by particulates may be a contributing factor in the rising incidence of emphysema, although the causes of the disease are not completely understood. Chronic bronchitis is also aggravated by particulates. And some particulates, notably those of certain heavy metals, may exert long-term toxic effects specific to their individual chemistry.

An untold amount of damage is done each year by particulates settling out on buildings and within homes. Costs of cleaning go up and corrosion is often accelerated. Particulates are also suspected of influencing weather patterns by serving as nuclei for cloud forma-

TABLE 6.7
Sources of particulates in the atmosphere: worldwide summary [a]

The units are 10⁶ metric tons; e.g., fly ash is emitted at the rate of 36 × 10⁶ metric tons per year

Source	Natural (10^6 metric tons/year)	Manmade (10^6 metric tons/year)
Primary particle production		
Fly ash from coal	—	36
Iron and steel industry emissions	—	9
Non-fossil fuels (wood, mill wastes)	—	8
Petroleum combustion	—	2
Incineration	—	4
Agricultural emission	—	10
Cement manufacture	—	7
Miscellaneous	—	16
Sea salt	1000	—
Soil dust	200	—
Volcanic particles	4	—
Forest fires	3	—
Subtotal	1207	92
Gas-to-particle conversion		
Sulfate from H_2S	204	—
Sulfate from SO_2	—	147
Nitrate from NO_x	432	30
Ammonium from NH_3	269	—
Organic aerosol from hydrocarbons, etc.	200	27
Subtotal	1105	204
Total	2312	296

[a] From "An Introduction to Air Chemistry," Samuel S. Butcher and Robert J. Charlson, p. 163, Academic Press, 1972.

tion, and by altering the amount of radiation reaching the Earth's surface.

To decrease pollution by particulates, the particles must be captured before they enter the atmosphere. Various devices are in use to wash out, settle out, or precipitate out pollutants before waste gases leave power plants or industrial plants. As is usually the case with pollution control measures, they add to the cost of the process or the product being manufactured, a factor that retards the introduction of such measures.

6.12 Discovery of the noble gases

Helium (He), neon (Ne), argon (Ar), krypton (Kr), xenon (Xe), and radon (Rn) make up the group of elements called the noble gases. All except radon are normally present in the atmosphere (Table 6.2), although in such small concentration that they are sometimes called the "rare gases." Radon is radioactive (Chapter 8) and decomposes soon after it is formed from other radioactive elements.

William Ramsay was a Scottish chemist who was convinced that there were other gases in the air besides oxygen, nitrogen, argon, and helium. In 1894 Ramsay had, in collaboration with Lord Rayleigh, already discovered argon. He had allowed a sample of nitrogen from the atmosphere to react with magnesium to form magnesium nitride

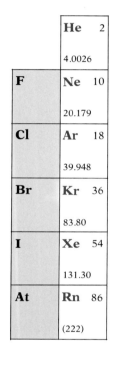

$$3Mg(s) + N_2(g) \xrightarrow{\Delta} \underset{magnesium\ nitride}{Mg_3N_2(s)}$$

A small amount of the gas did not react with magnesium. Ramsay examined this gas in a relatively new instrument, a spectroscope, in which lines of light characteristic of the different elements are visible. The pattern of lines was different than that of nitrogen; Ramsay also found the new gas to be denser than nitrogen. At the suggestion of a friend, the new gas was named "argon," meaning the lazy one, because it was so unreactive.

The existence of helium was recognized from spectra of sunlight, and for a while it was thought to be present only in the sun. Ramsay first observed it on Earth in the gases formed by a radioactive mineral; and a German chemist, H. Kayser, 1895, identified helium in the atmosphere.

The liquefaction of air was achieved at just the right time for Ramsay and his assistant, William Travers. By carefully separating the components of liquid air, they soon isolated and identified with the spectroscope three more gases in the atmosphere—krypton (in May 1898), named from the Greek meaning "hidden"; neon (in June 1898), named from the Greek meaning "new"; and xenon (in July 1898), named from the Greek meaning "stranger."

6.13 Properties of the noble gases

The most outstanding property by which the noble gases were known up until 1962 was their inability to form compounds. In fact, they were called the "noble" gases because of their aloofness from interaction with other elements. "Inert gases" was another collective term for the noble gases. However, in 1962, $XePtF_6$ and XeF_4 were prepared and found to be stable compounds. A large family of xenon compounds has now been synthesized, and a few compounds of krypton and at least one of radon are known. The noble gases are certainly not

TABLE 6.8
Properties of the noble gases

The noble gases are all colorless, odorless gases

Gas	Gas density (g/liter; 0°C, 1 atm)	Boiling point (°C)	Melting point (°C)
He	0.179	−268.93	−272.2
Ne	0.900	−246.08	−248.7
Ar	1.78	−185.88	−189.2
Kr	3.75	−153.4	−156.6
Xe	5.90	−108.12	−111.9
Rn	—	−62	−71

the most reactive of elements; however, they can no longer be called "inert." (The electronic structure and bonding in noble gases and noble gas compounds are discussed in Chapter 9.)

The physical properties of the noble gases are summarized in Table 6.8. Helium is the second lightest gas (hydrogen is lighter) and has the lowest boiling point of any substance, only about four degrees above absolute zero. This property makes helium valuable in **cryogenics**—the study of phenomena at low temperatures.

6.14 Preparation and uses of the noble gases

Natural gas from some wells contains up to 6% of helium. Helium is separated from the other components of natural gas by liquefaction followed by fractional distillation. Most of the helium used commercially is obtained from this source. The other noble gases (except radon) are all prepared by the liquefaction and distillation of air.

Helium is much less dense than air and is nonflammable, so it is used to fill balloons and dirigibles. An oxygen–helium mixture is breathed by those who work at high pressure in tunnel-building or deep-sea diving. Helium dissolves less readily than nitrogen in body fluids. Therefore, when workers return from a high-pressure area to atmospheric pressure, they are less likely to suffer from the "bends"—a painful condition caused by dissolved gases bubbling out of the blood.

When an electric current is passed through neon under low pressure in a closed tube, the neon glows bright red. All signs made from such gas-filled tubes are called "neon" signs, though some contain mixtures of neon and argon or other gases. Argon is also used to fill incandescent light bulbs, where its inertness keeps it from reacting with the filament, and its density prevents evaporation of the metal in the filament. Argon, being the most plentiful, is the least expensive of the noble gases. The use of krypton and xenon is limited by their scarcity and high cost. They find application mainly in light bulbs designed for special uses, such as extremely long life, or in flash bulbs or electronic flashes for high-speed photography.

THOUGHTS ON CHEMISTRY

The Laws of Ecology

The First Law of Ecology:
Everything is connected to everything else.
The Second Law of Ecology:
Everything must go somewhere.
The Third Law of Ecology:
Nature knows best.
The Fourth Law of Ecology:
There is no such thing as a free lunch.

THE CLOSING CIRCLE: NATURE, MAN AND TECHNOLOGY by Barry Commoner (pp. 33–48, Alfred A. Knopf, New York, 1971)

Significant terms defined in this chapter

lithosphere
hydrosphere
atmosphere
ecosphere
fractional distillation
greenhouse effect
fossil fuels
photochemical reactions
Freons
pollutant
secondary pollutant
particulates
photochemical smog
free radicals
London smog
cryogenics

Exercises

6.1 In many cases coal having a high sulfur content is burned in air by electrical power plants to generate electricity. If we consider the coal to contain only carbon, sulfur, and trace amounts of hydrocarbons, and air to contain only nitrogen, oxygen and argon, the products of the combustion found in the exhaust gas are CO_2, CO, H_2O, SO_2, N_2, NO_2, Ar, and "fly ash." Which of these products are (a) "major components" of air, that is, greater than 0.1 vol %; (b) "trace components" of air, that is, less than 0.1 vol % and (c) pollutants? (d) Which of the five major classifications of air pollutants are present? Which pollutant could be removed by passing the exhaust gas through (e) a bed of limestone, and (f) a combustion chamber containing excess oxygen?

6.2 Which of the following properties of air indicate that air is a mixture and not a true compound: (a) gaseous nature (b) variable composition, (c) components retain individual properties, (d) liquid form can be separated into various components by fractional distillation, and (e) compressible?

6.3 Write the chemical symbol and name for each of the noble gases. How is each prepared? Identify one use for each of these gases.

6.4 Name the five classes of air pollutants. Identify a source of each. What are a few of the effects of each of these pollutants? Briefly discuss one technique for controlling each of these pollutants.

6.5 Which statements are true? Rewrite any false statement so that it is correct.
(a) The lithosphere includes all of the water on the surface of the Earth as well as the gaseous envelope surrounding the Earth.
(b) During the fractional distillation of liquid air, the oxygen (b.p. $-182.97°C$) vaporizes leaving the nitrogen (b.p. $-195.80°C$).
(c) Judging from the trend shown in Table 6.8 for the boiling points of the noble gases, it appears that there is a relationship between the boiling point and atomic weight.
(d) Because neon is rather inert chemically, it is commonly used to fill incandescent light bulbs to keep the filament from burning out.
(e) The atomic and ionic species present in the exosphere are at high temperatures and high pressures.

6.6 Describe in your own words what happens as (a) air is liquefied, (b) a catalytic converter is used to reduce harmful auto emissions, (c) air leaks into an argon-filled incandescent light bulb, (d) carbon monoxide enters the lungs, (e) Freons interact with ozone, (f) photochemical smog is produced, and (g) the relative amount of CO_2 in the atmosphere increases.

6.7 Each day the sun radiates 4×10^{29} J of energy, of which only a small fraction, 5×10^{-8}, impinges on the Earth. How much solar energy is reflected, absorbed by the atmosphere, and absorbed by the lithosphere?

6.8 Relative humidity (expressed in %) is defined as

$$\text{relative humidity} = \frac{p_{H_2O}}{p_{H_2O}^0} \times 100\%$$

where p_{H_2O} is the partial pressure of water in air and $p_{H_2O}^0$ is the maximum vapor pressure of water in air at that temperature. What is the partial pressure of water in an air sample at 25°C if the relative humidity is 75% and $p_{H_2O}^0 = 23.756$ torr?

6.9 If $p_{H_2O}^0 = 9.209$ torr at 10°C, would precipitation take place if the air sample described in Exercise 6.8 were cooled to this temperature?

6.10 If dry air is considered to contain 75.53 wt % N_2, 23.14 wt % O_2, and 1.282 wt % Ar, find the average "molar weight" of dry air.

6.11 Using the ideal gas law, calculate the density of air at 744 torr and 23°C. Assume that the "molar weight" of air is 29.1 g/mole.

6.12 The lifting capacity of a balloon is found by subtracting the mass of the balloon filled with the gas to be used (hot air, He, H_2, etc.) from the mass of the balloon filled with air at the same pressure and temperature as the air on the outside of the balloon. What is the lifting capacity of a 1000.-liter balloon filled with helium at 0°C and 1.00 atm? The density of air under these conditions is 1.29 g/liter.

6.13 An atmosphere containing approximately 750 ppm of CO is considered lethal. What mass of CO corresponds to this concentration in a room 25 ft by 25 ft by 11 ft if the temperature is 25°C and the total pressure is 1.00 atm?

6.14 What mass of $CaCO_3$ would be required to react with each gram of SO_2 that is present in the exhaust gases emitted by a power plant?

6.15 The enthalpy of formation at 25°C of $CaCO_3$(calcite) is -288.45 kcal/mole, of $SO_2(g)$ is -70.944 kcal/mole, of $O_2(g)$ is 0 kcal/mole, of $CaSO_4(s)$ is -342.42 kcal/mole, and of $CO_2(g)$ is -94.051 kcal/mole. What is the heat of reaction for

$$2CaCO_3 \text{ (calcite)} + 2SO_2(g) + O_2(g) \longrightarrow$$
$$2CaSO_4(s) + 2CO_2(g)$$

155

6.16* Assuming dry air to contain 78.1 vol % N_2, 20.9 vol % O_2, and 1.0 vol % Ar, express the composition of air in terms of mass percent.

6.17* Would you expect the density of moist air (air with a high humidity) to be greater or less than the density of dry air at the same temperature and total pressure? To confirm your expectations, calculate the density of dry air at 25°C and 1.00 atm using the composition given in Exercise 6.10 and calculate the density of air having 75% relative humidity—see Exercise 6.8—under the same conditions.

6.18* Magnesium reacts with N_2 and O_2 to form Mg_3N_2 and MgO. Calculate the masses of Mg_3N_2 and MgO that would be formed by the reaction of 1.00 liter of air at 23°C and 749 torr with excess Mg. What volume of gas would remain after the reactions?

6.19* Consider the balloon described in Exercise 6.12 to be filled with H_2 instead of He, and calculate its lifting capacity. Why is H_2 seldom used in balloons even though the use of He reduces the amount of cargo that the balloon could carry? What would happen if the balloon were filled with Ar? The density of H_2 under the conditions of the problem is 0.0896 g/liter.

6.20* Write the equation for the decomposition of $XeF_6(s)$ to $XeF_4(g)$ and $F_2(g)$. What volume of gaseous products at 300.°C and 1.00 atm would be formed by the decomposition of 1.00 g of XeF_6?

6.21* What is the mass of hydrocarbons present in a cubic foot of air at 20°C and 745 torr if the hydrocarbon concentration is 3 ppm by weight?

6.22* A sample of coal was found to contain 0.3 wt % sulfur. Assuming the only other component to be carbon, what amount of $CaCO_3$ is needed to remove the SO_2 produced when 1.00 ton of the coal is burned? Also, assume that the SO_2–$CaCO_3$ reaction is only 79% efficient.

6.23* The enthalpy of formation is 59.553 kcal/mole for $O(g)$, 0 kcal/mole for $O_2(g)$ and 34.1 kcal/mole for $O_3(g)$. Find ΔH_f° for the reaction between atomic O and molecular O_2 to give ozone

$$O(g) + O_2(g) \longrightarrow O_3(g)$$

6.24* A weather balloon using He is to be designed to maintain a height of 10 km above sea level once it has risen that high. (a) What is the average temperature of the atmosphere at this altitude (see Figure 6.3)? The atmospheric pressure, P (expressed in atm), at any altitude, h (expressed in km), can be calculated to a good approximation by using

$$\log P = -6.1 \times 10^{-2} h$$

(b) Find the pressure at 10 km. (c) Calculate the density of the air under these conditions. (d) What must be the average density of the balloon under these conditions? (e) What must be the average density of the balloon if it were filled with H_2 instead of He?

The mean free path, l (expressed in meters), for a He atom is given by

$$l = 3.5 \times 10^{-9}(T/P)$$

where the pressure P is in atmospheres and the temperature T is in Kelvin degrees; l represents the average distance that a He atom will travel before colliding with another atom. Calculate l for He in the balloon at (f) 10 km and (g) 1.00 atm and 25°C. (h) How many times farther does a molecule travel (on the average) at an altitude of 10 km before striking another than at 1.00 atm and 25°C?

7 ATOMIC STRUCTURE

The discovery of the electron in 1897 unleashed a host of questions that had to be answered. What other particles might be present within an atom? What holds these particles together? What is the arrangement of these particles within an atom?

During the years from the discovery of the electron to the discovery of the neutron in 1932 (see Table 7.1), physicists and chemists struggled with the problem of atomic structure. It was an exciting time. Although there were false leads and controversy, it was clear that they were getting closer and closer to a workable model for the atom. It seemed certain that once that was achieved, many seemingly unrelated facts about elements and compounds would fall into place.

The story of the growth of atomic theory is in part the story of a series of classic experiments each designed to help "see the unseeable and know the unknowable." Much of this work was done or directed by two of the outstanding scientists and teachers of the time. One was J. J. Thomson, who headed the Cavendish Laboratory of Experimental Physics at Cambridge University, a position he assumed when he was only 28 years old. The other was Ernest Rutherford, who succeeded Thomson at the Cavendish Laboratory after many productive years at McGill University in Canada and at the University of Manchester in England.

Thomson discovered the electron and measured the ratio of its charge to its mass. Six of his students won Nobel prizes for work related to atomic structure. Rutherford, who himself had been Thomson's student, explained radioactivity, discovered the α-particle (alpha-particle), and showed that the mass of an atom is concentrated in the nucleus.

Another group of scientists became famous during this period for their contributions to the theory of matter. Max Planck, Louis de

In this chapter we present the development of atomic structure and atomic theory. First we describe the discovery of the three fundamental particles of greatest interest to chemistry: the electron, the proton, and the neutron. Next we discuss the three important atomic structure models: the Rutherford nuclear atom, the Bohr atom, and the wave-mechanical atom. There are several sections on nuclear arithmetic, and the last section relates the electronic structure of atoms to the periodic table.

Broglie, Werner Heisenberg, and Erwin Schrödinger were among those who raised the description of atoms to a high level of mathematical sophistication. This process continues; though we have an atomic model that works well, our concepts of atomic structure are not considered complete. They are being altered and refined constantly, and there is always the possibility that tomorrow someone will perform an experiment that will require great changes in our ideas.

7.1 Early periodic table

As the elements were discovered and their properties determined, similarities and differences became apparent. For example, sodium and potassium, both first prepared from hydroxides (NaOH and KOH), look alike (soft, silvery solids) and have similar properties. Fluorine and chlorine (both gases) also resemble each other. But fluorine and potassium or sodium and chlorine are quite dissimilar. It seemed that it should be possible to classify or arrange the elements in some way that would make sense of such similarities and differences.

It often happens in science that two individuals independently present equivalent ideas. In 1869, Dimitri Mendeleev, a Russian, and Lothar Meyer, a German, both found ways to arrange the elements in the order of increasing atomic weight so that those with similar properties were placed together in a table (Figure 7.1). Mendeleev stated that, *"the elements if arranged according to their atomic weights, show a distinct periodicity of their properties."* This statement came to be known as the *periodic law*. A table in which elements are arranged according to the periodic law is called a *periodic table*. (A revised and modern statement of the periodic law is given in Section 7.19.)

Where the then-known elements did not fill out his scheme, Mendeleev had the courage to leave gaps and in some cases to predict the properties of the elements that should fill the gaps. The agreement

FIGURE 7.1

Mendeleev's periodic table (*as it was published in* Zeitschrift fur Chemie *in 1869*). *The tables of Mendeleev and Meyer were very similar although Mendeleev based his primarily on chemical properties and Meyer based his primarily on physical properties. Note the spaces for unknown elements, as indicated by question marks in the symbol column. Two pairs of elements–tellurium and iodine (symbolized by J here), and gold and bismuth–are out of atomic weight order because Mendeleev recognized that their chemical properties fit better in other places.*

			Ti = 50	Zr = 90	? = 180
			V = 51	Nb = 94	Ta = 182
			Cr = 52	Mo = 96	W = 186
			Mn = 55	Rh = 104,4	Pt = 197,4
			Fe = 56	Ru = 104,4	Ir = 198
		Ni =	Co = 59	Pd = 106,6	Os = 199
H = 1			Cu = 63,4	Ag = 108	Hg = 200
	Be = 9,4	Mg = 24	Zn = 65,2	Cd = 112	
	B = 11	Al = 27,4	? = 68	Ur = 116	Au = 197?
	C = 12	Si = 28	? = 70	Sn = 118	
	N = 14	P = 31	As = 75	Sb = 122	Bi = 210?
	O = 16	S = 32	Se = 79,4	Te = 128?	
	F = 19	Cl = 35,5	Br = 80	J = 127	
Li = 7	Na = 23	K = 39	Rb = 85,4	Cs = 133	Tl = 204
		Ca = 40	Sr = 87,6	Ba = 137	Pb = 207
		? = 45	Ce = 92		
		?Er = 56	La = 94		
		?Yt = 60	¦Di = 95		
		?In = 75,6¦	Th = 118?		

between Mendeleev's predictions and the properties of the later-discovered elements that filled these places is remarkable. No doubt because of this success, Mendeleev is most often given credit as the designer of the periodic table.

The importance of the periodic table in the development of chemistry cannot be overemphasized. With its help, information about familiar elements and their compounds can be used as an aid in studying unfamiliar species. We shall return to the periodic table (Section 7.19) after studying the atom and what has been learned about its structure, and we shall see how Mendeleev's ordering of the elements is completely explained by modern atomic theory.

Particles

7.2 Electricity and matter

By the end of the nineteenth century it was becoming apparent that electricity had some sort of direct relationship with matter. In 1833 Michael Faraday had found that when electricity passes through a molten salt, the salt breaks down. The amount of a salt that undergoes this reaction is directly proportional to the amount of electricity that flows through the system.

Later in the 1800s, the passage of electricity through a gas was studied. The apparatus devised for this purpose was the **gas-discharge tube,** a glass tube that can be evacuated and into which are sealed two electrodes (Figure 7.2). An **electrode** is a conductor through which electrical current enters or leaves a conducting medium. William Crookes, an English inventor, editor, and scientist, studied the behavior of many gases in these tubes, which are now often called Crookes tubes.

A potential difference of 10,000 V (volts) between electrodes in the open air will barely cause a spark to jump a few millimeters. If the electrodes are enclosed in a glass tube and the air is gradually pumped out, even if the electrodes are some distance apart, a glow ap-

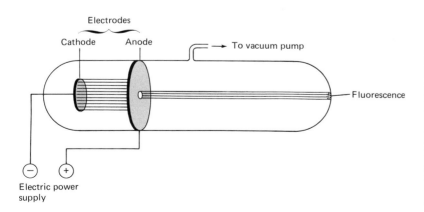

FIGURE 7.2
A gas-discharge tube showing cathode rays. *The nature of cathode rays was studied by varying the gases in the tube and placing various targets within the tube. (Reproduced by permission of the publisher from* Chemistry: Experiments and Principles, *by O'Connor et al., © 1968 by D. C. Health and Company, Lexington, Massachusetts.)*

pears in the tube and current flows when the pressure gets down to 1–20 torr. Different gases in the tube give different colors to the glow.

When more gas is pumped out of the tube, down to 10^{-2} torr and lower, the light fades out. However, current continues to flow through the tube and the glass wall at the end of the tube opposite the negative electrode gives off a greenish glow.

Crookes concluded, in 1879 (Table 7.1), that rays of particles were flowing from the negative electrode in a gas-discharge tube. He called these rays "cathode rays"; their properties are summarized in Table 7.2. Cathode rays are the same, Crookes found, no matter what

Custer's Last Stand	1876		
		1879	Cathode rays, William Crookes
		1881	"Electron," G. Johnstone Stoney
		1885	Hydrogen spectrum, Johann Balmer
		1886	Canal rays, Eugen Goldstein
Klondike gold rush	1897	1897	Electron e/m, J. J. Thomson
Spanish–American War	1898		
McKinley assassinated; Theodore Roosevelt becomes U.S. President	1901	1900	$E = h\nu$, Max Planck
Wright Brothers flight at Kitty-Hawk	1903		
		1905	Quantized radiation, Albert Einstein
		1909	Charge on electron, Robert A. Millikan
		1911	Nuclear atom model, Ernest Rutherford
Woodrow Wilson elected U.S. President	1912	1913	Hydrogen atom model, Niels Bohr Meaning of atomic number, Henry Moseley
World War I begins	1914		
Russian Bolshevik Revolution	1917		
Treaty of Versailles ends World War I	1919		
		1923	$\lambda = \dfrac{h}{mv}$, Louis de Broglie
		1926	Wave-mechanical atomic model Erwin Schrödinger Uncertainty principle, Werner Heisenberg
Transatlantic flight by Lindbergh	1927	1927	Electron diffraction, C. J. Davisson and L. H. Germer
U.S. stock market crash; beginning of Depression	1929		
		1932	Neutron, James Chadwick
Hitler becomes dictator in Germany	1933		

material is used for the negative electrode or what gas is used in the tube.

The Crookes tube was the ancestor of neon signs, fluorescent lights, and television tubes. What we call "neon" signs may contain neon, which gives a red glow. They may also contain other gases. Helium gives a white light. If a little mercury is present in an argon-filled tube, the light is blue. Argon in a yellow glass tube gives a yellow light, and mercury in a yellow tube gives green light.

Fluorescent light bulbs are produced by coating the inside walls of long cathode ray tubes with material that emits visible light when struck by cathode rays. A similar coating is placed on the inside of a television tube; a rapidly moving cathode ray continuously scans the coated tube and the television picture is created by varying the intensity of the ray.

7.3 Electrons

George Johnstone Stoney, a physicist, proposed in 1881 that electricity is carried by individual, negatively charged particles. This was not a completely new idea. Benjamin Franklin had suggested it in about 1750. However, in Stoney's time electricity was thought of by many as a fluid. Stoney estimated the size of the fundamental unit charge—he was quite close to the correct value—and named the particle the "electron" (from the Greek for amber, a material known from ancient times to acquire a charge when rubbed with silk).

For many years J. J. Thomson studied cathode rays, pondering the meaning of the properties they showed in his own experiments and those of others. In 1897 he was ready to announce his conclusion: cathode rays are streams of negative particles of electricity—exactly the same as Stoney's electrons. Since cathode rays are the same for any gas or any electrode material, electrons must be present in all matter. With this discovery, Dalton's model of the atom as hard and indivisible had to be put aside forever.

Thomson provided the experimental proof that cathode rays are particles. Obviously, if the rays are particles with mass and charge, it should be possible to measure these properties. Thomson could not determine the mass or charge separately. However, by an ingenious use of the balancing forces of electric and magnetic fields, he was able to measure the ratio of the charge of an electron (e) to its mass (m) (Figure 7.3). He found that the value of e/m was constant for different gases and different electrode materials.

The first accurate determination of the charge of the electron was made in another classic experiment—the oil drop experiment, performed in 1909 by an American, Robert A. Millikan. Tiny oil droplets, some of which were charged by friction while passing through an atomizer, were released between two plates on which the charge could be turned on and off (Figure 7.4). The rate of fall of the droplets under the influence of gravity alone, and then their rate of rise and fall when the charge was on, was closely observed. Sometimes the rate changed abruptly, showing that the droplet had captured an ion from the air.

TABLE 7.2
Properties of cathode rays

1. Travel in straight lines from the cathode to the anode
2. Cast shadows when metal objects are placed in their path
3. Produce fluorescence where they strike the glass walls of the tube
4. Heat thin metal foils to incandescence
5. Produce mechanical motion in pinwheels placed in their path
6. Cause ionization of gas molecules
7. "Expose" photographic films or plates
8. Produce highly penetrating radiation (x rays, Section 7.12) when directed against a target
9. Impart a negative charge to such a target
10. Undergo deflection parallel to an applied electrostatic field (away from the negative electrode) and perpendicular to an applied magnetic field

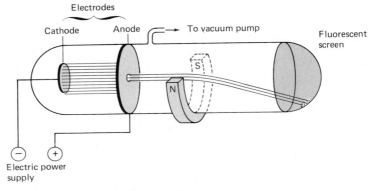

(a) Deflection of cathode ray by magnetic field

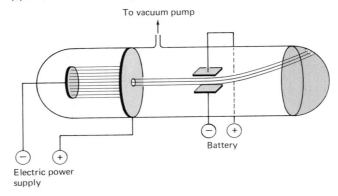

(b) Deflection of cathode ray by electric field

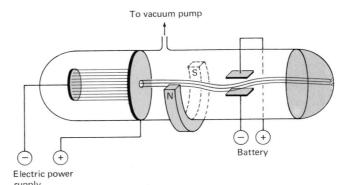

(c) Balanced deflection of cathode ray by electric and magnetic fields

FIGURE 7.3

Determination of *e/m* of the electron.
The magnetic and electric fields deflect the cathode ray in opposite directions. By varying these two opposing forces until they balance (an often-used principle in the design of experiments), Thomson was able to calculate from the field strengths, the e/m value of the electron. (Reproduced by permission of the publisher from Chemistry: Experiments and Principles, *by O'Connor et al., © 1968 by D. C. Health and Company, Lexington, Massachusetts.)*

In some experiments the air was ionized by x-rays, assuring the capture of ions. After many measurements on droplets with different charges, Millikan found that the charge was *always* a small-whole-number multiple of -1.60×10^{-19} C (coulomb). This therefore, is the smallest possible negative charge—the charge of an individual electron.

Once the charge on an electron was known, the mass could be calculated from the value of *e/m*. The mass of the electron is 1/1837 of the mass of a hydrogen atom. **A fundamental particle** is one that is present in all matter. **A subatomic particle** is a particle smaller than

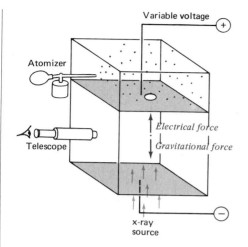

FIGURE 7.4
Millikan oil drop experiment.
The velocity of the rise and fall of charged oil drops between the + and − plates is observed through the microscope. Both the charge on the drops and the electric field can be varied. Here again (see Figure 7.3) an experiment makes use of the balancing of opposing forces.

the smallest atom. The **electron** is a fundamental, subatomic, negatively charged particle, and **cathode rays** are streams of electrons flowing from the cathode toward the anode in a gas-discharge tube. Modern values for the properties of the electron and other fundamental particles are given in Table 7.3.

7.4 Protons

Atoms are electrically neutral. Therefore, the identification of positive charges to balance the negative charge of the electrons became essential. Also, since electrons have such small mass, they could not account for the much greater mass of atoms.

The first observations of positively charged particles were made in gas-discharge tubes. In 1886, Eugen Goldstein discovered that rays with a positive charge were flowing in the opposite direction from the cathode rays in such tubes. These *canal rays*, named because they

---

Electrons and protons are both stable outside of an atom. Neutrons eventually decompose spontaneously.

TABLE 7.3

Fundamental particles of importance to chemistry

Particle	Symbol	Mass[a]	Charge In coulombs	Relative
Electron	e^-	9.109534×10^{-28} g	-1.60×10^{-19}	-1
		5.485802×10^{-4} amu		
Proton	p^+	$1.6726485 \times 10^{-24}$ g	$+1.60 \times 10^{-19}$	$+1$
		1.0072764 amu		
Neutron	n	$1.6749543 \times 10^{-24}$ g	0	0
		1.0086650 amu		

[a] Strictly speaking, we should use the term *rest mass*. Physicists have shown that moving particles have greater mass than particles that are "resting." The distinction has little bearing here, and throughout this book we use "mass" to refer to the rest mass of particles.

TABLE 7.4

Properties of canal rays

These rays are so named because they pass through a hole or "canal" in the cathode.

1. Travel in straight lines toward the cathode
2. Produce fluorescence when they strike the walls of the tube
3. Are deflected in the opposite direction from cathode rays by both electric and magnetic fields
4. Are deflected less than cathode rays by fields of equal strength
5. Expose photographic plates
6. Differ for different elements

pass through a "canal" in the cathode, have the properties listed in Table 7.4.

Evaluation of the charge and mass of canal-ray particles showed that they are much heavier than electrons; they vary in mass according to the gas present in the tube; and they have positive charges that are small-whole-number multiples of $+1.60 \times 10^{-19}$ C. The lightest and simplest positive particle in a canal ray, with a charge of $1 \times (+1.60 \times 10^{-19})$C, is formed when the gas is hydrogen. Figure 7.5 illustrates schematically what happens in a gas-discharge tube in the formation of both canal rays and cathode rays. Canal rays are streams of positive ions created by collision of electrons with atoms; electrons are knocked off the atoms to give positive ions.

Since the lightest canal-ray particles were hydrogen ions, H^+, named protons by Rutherford, it was concluded that this was the lightest positively charged particle in any atom. The proton was assumed to be the positive particle needed to balance the negative charge of the electrons in all matter. The mass of the proton is 1837/1838 of the mass of a hydrogen atom; in other words, a proton is 1837 times heavier than an electron. The **proton** is a fundamental, subatomic particle with a positive charge equal in magnitude to the negative charge of the electron (see Table 7.3).

7.5 α-Particles

During the years that Rutherford was studying atomic structure, he was also studying radioactive elements—elements that spontaneously break down because they are unstable. (The next chapter is devoted to radioactivity and nuclear chemistry.) One result of this work was Rutherford's identification of α-particles ("alpha-particles"), which are given off by some radioactive elements. An **α-particle** is a dipositive helium ion, He^{2+}. α-Particles are also formed in gas-discharge tubes that contain helium. With a positive charge of 3.20×10^{-19} C, and a mass of 4.002 amu, α-particles have twice the charge of a proton and about four times its mass.

While the α-particle is not fundamental to the model of atomic structure, it was used in some pioneering experiments. α-Particles

FIGURE 7.5

Cathode rays and canal rays. *Electrons from the atoms of the cathode material are accelerated toward the anode, forming cathode rays. These electrons collide with the gaseous atoms or molecules, knocking off other electrons, and leaving positively charged ions, a process that causes a glow. The positive ions are attracted to the cathode, forming canal rays. As the pressure becomes lower the electrons encounter fewer and fewer gas molecules and eventually the glow disappears. However, the electron beam continues to flow from the cathode.*

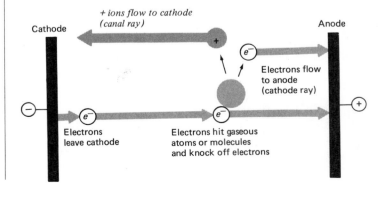

are emitted from radioactive elements with very high energy. For example, they come from radium atoms at a speed of 1.5×10^7 m/sec. Rutherford and others recognized the value of such particles. Because of their high energy, they could be fired like bullets at atoms and from what happened, information about the structure of atoms could be obtained.

One result of Rutherford's experiments was the production of H^+ ions, that is, of protons, in the bombardment of nitrogen and other elements, thereby showing that protons are present in atoms other than hydrogen atoms. But the most important result came from the fate of the α-particles themselves in the bombardment of thin foils of metals (Section 7.7).

7.6 Neutrons

The neutron is the third fundamental particle of importance to chemistry. The existence of a neutral particle of mass comparable to that of the proton was suspected as early as 1920 by Rutherford. However, it was not until 1932 that such a particle was discovered. James Chadwick, an English physicist, directed a stream of α-particles at a beryllium target. Very penetrating radiation that had no charge was produced. Chadwick identified this radiation as a beam of neutral particles with a mass almost the same as the mass of the proton. He called these particles "neutrons."

The discovery of the neutron was a significant contribution to the developing picture of atomic structure. The **neutron** is a fundamental, subatomic particle that has a mass almost the same as the mass of the proton and has no charge (Table 7.3).

The nucleus and nuclear arithmetic

7.7 Rutherford's atomic model: The nuclear atom

After his famous measurement of e/m for the electron, J. J. Thomson suggested that an atom is a sphere of positive electricity in which electrons are embedded like raisins in a plum pudding. Since the electrons are so light, the mass of this atom would have to result largely from the positively charged part of it. The electrons would be distributed at maximum distances from each other because of their like charges.

But Thomson's model failed to explain the results of experiments on α-particle scattering, done in 1911 at Manchester by Hans Geiger and Ernest Marsden. A narrow beam of α-particles from a radioactive source was aimed at a thin metal foil (e.g., Au, Pt, Cu, Ag) (Figure 7.6a). Most of the particles passed through the foil without changing direction. However, a few were deflected, some of them through rather large angles. And a small number bounced back toward the source. Rutherford was surprised by the result: "It was almost as incredible as if you fired a 15-inch shell at a piece of tissue paper and it came back and hit you."

If the mass and positive charge were uniformly distributed throughout each atom in the metal foil, as suggested by the Thomson model, no concentration large enough to deflect the charged particle could exist. Recall that the α-particles are traveling very fast and have four times the mass of a proton.

Faced with this disagreement between fact and theory, Rutherford did what a good scientist must do. He abandoned the old theory and devised a better one. He proposed that each atom has a dense central core, which he called the nucleus. The **nucleus** is a central region, very small by comparison with the total size of an atom, in which all of the mass and positive charge of the atom are concentrated.

FIGURE 7.6

α-Particle scattering. (a) *Where α-particles strike the zinc sulfide screen, a flash of light (a "scintillation") can be observed through the microscope. Most α-particles pass through the metal foil undeflected. Some are deflected through varying angles, and a few are deflected back toward the source.* (b) *This diagram shows how α-particles are scattered by a single atom with a dense positive nucleus.*

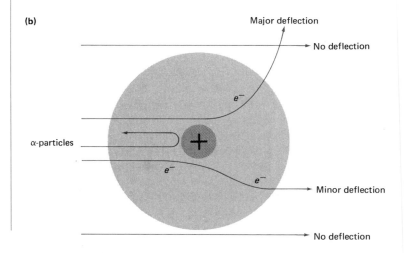

In α-particle bombardment of a foil of such nuclear atoms, the majority of the α-particles encounter only empty space and pass through undeflected (Figure 7.6b). Only those coming close to concentrations of mass and positive charge can be deflected by the great forces of repulsion between like charges. And only those few that collide directly with a center of mass and charge are returned in the direction from which they came.

Rutherford accounted for nuclear charge in terms of protons located in the nucleus. The positive charge of the protons is balanced by the negative charge on the electrons residing outside the nucleus. However, this was not completely satisfactory because the mass of protons and electrons equal in number to the nuclear charge did not account for the observed atomic masses. For example, a nucleus of two protons to balance the two electrons could not explain the helium atom, which has a mass *four* times that of a proton. Rutherford therefore assumed that nuclei contained enough additional protons to account for an atom's mass, plus enough electrons inside the nucleus to maintain electrical neutrality.

After Chadwick's discovery of the neutron in 1932, a more satisfactory model of the atom emerged. A nucleus contains protons equal in number to the number of electrons in the atom plus enough neutrons to account for the remaining atomic mass. For example, the helium atom contains two protons and two neutrons in its nucleus, and two electrons outside the nucleus.

Rutherford's concept of the nuclear atom is now universally accepted. Atomic diameters are on the order of 0.1 nm and nuclear diameters are on the order of only 10^{-6} nm. Rutherford did not know the whole story of what was inside the nucleus. We still do not know it. But he was right in picturing the atom as mostly empty space occupied by electrons moving around a very small, dense central core.

7.8 Atomic numbers and x-ray spectra

Numbering the elements in the order in which they appear in the periodic table, beginning with 1 for hydrogen, 2 for helium, and so on, gives what are called "atomic numbers." Henry G. J. Moseley, a young Englishman who studied x-ray spectra (see Tools of Chemistry: Electromagnetic Radiation and Spectra) in Rutherford's laboratory, proved that atomic numbers have a significant physical meaning.

x-Rays were first observed in 1895 by W. C. Roentgen. These rays are produced when an electron beam strikes a target placed in a cathode ray tube. Moseley's task, which he accomplished in the remarkably short time of six months in 1913, was to photograph the x-ray spectra of as many elements as possible.

Several series of lines appear in the x-ray spectrum of each element. Moseley found that as the atomic weight of the target elements increased, the position of any specific line in a specific series moved

Electromagnetic radiation and spectra

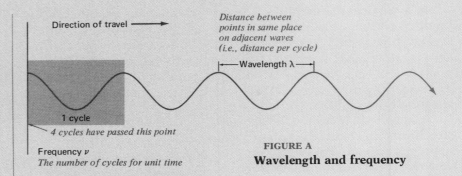

FIGURE A
Wavelength and frequency

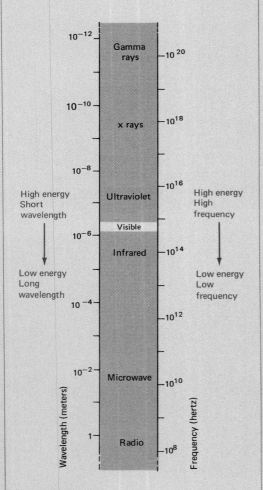

When a rainbow forms in the sky or gasoline produces colored patterns as it floats on a puddle, sunlight is being broken up into a continuous spectrum. The colors from violet, through green and yellow, to red blend into each other without separation. The light from the filament of an incandescent light bulb gives the same array of colors when it passes through a prism.

The colors visible to the human eye form only a small part of the entire spectrum of electromagnetic radiation. **Radiation** is energy traveling through space—traveling as a series of waves or as a stream of particles. The properties of electromagnetic radiation are described in terms that come from the picture of radiation as a series of waves traveling through space. The alternate picture of radiation as a stream of particles is discussed in Section 7.12.

The distance between any two points in the same relative location on adjacent waves is the **wavelength** of the radiation, symbolized by λ (Greek lambda) (Figure A). The number of wavelengths passing a given point in a unit of time is the **frequency,** symbolized by ν (Greek nu). Wavelength is given in units of length—nanometers (nm), or centimeters (cm), or angstroms (Å) (see Table 1.6)—meaning the length per cycle. Frequency is usually measured as cycles per second and is expressed as reciprocal second, $\sec^{-1}$ (Figure A), the word "cycle" being left out. (In the SI system, one cycle per second is called a "hertz"; 1 hertz (Hz) = 1 $\sec^{-1}$.)

A **spectrum** is an array of waves or particles spread out according to the increasing or decreasing magnitude of some physical property. The entire electromagnetic spectrum is sketched in Figure B. It extends from short-wavelength, high-frequency gamma rays to long-wavelength, low-frequency radio waves. An increase in frequency or a decrease in wavelength represent an increase in energy. (Think of it this way: It takes more energy to cover a distance in a lot of short jumps than in a fewer long jumps.) The regions of the electromagnetic spectrum that are useful in chemistry are listed in Table A.

The velocity of light in a vacuum is a constant, 3.00×10^{10} cm/sec. Wavelength and frequency are related by the equation

$$\underset{\text{velocity of light}}{c} = \lambda \underset{\text{wavelength}}{\overset{\text{frequency}}{\nu}}$$

Frequency can be calculated from wavelength and wavelength can be

TABLE A

Regions of the electromagnetic spectrum of chief interest to chemistry

Gamma rays	Emitted in radioactive decay
x Rays	Diffraction of x rays used in determining crystal structure
Ultraviolet	Absorption spectra used for structure determination in covalent organic and inorganic compounds
Visible	Emission and absorption spectra used for qualitative and quantitative identification of elements
Infrared	Absorption spectra used for structure determination in covalent organic and inorganic compounds

TABLE B

Properties of yellow and red light

Yellow light has higher energy than red light.

Yellow light	Red light
Wavelength (distance per cycle)	
5800 Å	7000 Å
580 nm	700 nm
5.8×10^{-5} cm	7.0×10^{-5} cm
Frequency (cycles per second)	
$\nu \ (\text{sec}^{-1}) = \dfrac{c \ (\text{cm sec}^{-1})}{\lambda \ (\text{cm})}$	
$\dfrac{3.00 \times 10^{10} \text{ cm sec}^{-1}}{5.8 \times 10^{-1} \text{ cm}}$	$\dfrac{3.00 \times 10^{10} \text{ cm sec}^{-1}}{7.0 \times 10^{-5} \text{ cm}}$
$= 5.2 \times 10^{14} \text{ sec}^{-1}$	$= 4.3 \times 10^{14} \text{ sec}^{-1}$
(or Hz)	(or Hz)

calculated from frequency by using the preceding equation (Table B). Another quantity used to characterize radiation is the wave number.

$$\bar{\nu} = \frac{1}{\lambda}$$

wave number

The **wave number** is the number of wavelengths per unit of length covered. Its unit is the reciprocal of the wavelength unit, and is usually given in cm^{-1}.

In a **continuous spectrum,** like the visible spectrum from sunlight, radiation at all wavelengths is emitted. In a **line spectrum** radiation is emitted or absorbed only at certain wavelengths (Figure C). For example, when a vaporizable substance such as sodium chloride is placed in a flame, yellow radiation is visible. This radiation yields a line spectrum by which the presence of even tiny amounts of sodium or sodium compounds can be detected. A molecule emits or absorbs radiation over certain ranges of wavelengths, rather than individual wavelengths, and therefore produces a **band spectrum.**

The spectrum of the radiation emitted by a substance is called an **emission spectrum.** The spectrum of radiation after a continuum of radiation has passed through a substance is called an **absorption spectrum.** Sodium vapor gives an emission spectrum. White light passed through sodium vapor gives an absorption spectrum consisting of a continuous spectrum interrupted by dark lines at the same places as the yellow lines in the sodium emission spectrum.

Spectra result from the emission and absorption of energy by matter. Therefore, spectra are an important source of information about the structure of matter.

FIGURE C

Continuous and line spectra. *The continuous spectrum in the visible region is what we see in a rainbow. Continuous spectra are emitted when solids, liquids, or very highly compressed gases have been heated to incandescence. Line spectra are emitted by atoms in the gaseous state that have absorbed extra energy. Molecules under similar conditions give band spectra—groups of lines spaced very close together.*

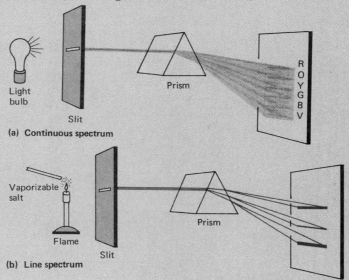

(a) Continuous spectrum

(b) Line spectrum

Atomic number
= no. of protons in nucleus

Mass number (A)
= atomic no. (Z)
    + neutron number (N)
= (protons + neutrons)

FIGURE 7.7
Atomic x-ray spectra. *A sketch of the relative positions of emission lines in some of Moseley's spectra. The element scandium, not available to Moseley, is missing between titanium and calcium.*

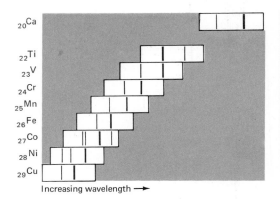

Increasing wavelength →

by regular intervals toward shorter wavelengths (Figure 7.7). There was a simple mathematical relationship between the frequency of these lines and the atomic number. Here, in his own words, are his conclusions:

1. Every element from aluminum to gold is characterized by an integer N which determines its x-ray spectrum. Every detail in the spectrum of an element can therefore be predicted from the spectra of its neighbors.

2. This integer N, the atomic number of the element, is identified with the number of positive units of electricity contained in the atomic nucleus.

The **atomic number** of an element, symbolized by Z, is equal to the number of protons in the nucleus of each atom of that element.

7.9 The nucleus: Mass number and atomic number

Now we can summarize our knowledge of the atomic nucleus in terms of protons, neutrons, and atomic number. The atomic number Z is equal to the number of protons in the nucleus of every atom of an element. The atomic number determines the identity of an atom. Every atom with an atomic number of 8 is an oxygen atom. Atoms with atomic numbers of 9 or 7 are atoms of fluorine or nitrogen, respectively. The atomic number is, of course, also equal to the number of electrons in an atom.

The **mass number,** A, is the sum of the number of protons and the number of neutrons in a nucleus, and is the whole number closest to the atomic weight of an element (next section).

$$\underset{\displaystyle \text{mass number}}{} A = Z + N \qquad (1)$$

atomic number

number of neutrons

The number of neutrons, or **neutron number**, N, is the difference between the mass number and the atomic number.

$$N = A - Z \qquad (2)$$

170

For example, helium ($Z = 2$) atoms have two protons in their nuclei. For helium, with mass number 4, the nuclei each contain $4 - 2 = 2$ neutrons. An atom of fluorine with $Z = 9$ and $A = 19$ contains 9 protons and 10 neutrons.

The numbers of neutrons and protons in the nuclei of light elements are about equal. Atoms of the heavier elements have more neutrons than protons in their nuclei. Bismuth atom nuclei, for example, each contain 83 protons and 126 neutrons.

In the shorthand notation for atoms or ions (Figure 7.8) the left superscript position is reserved for the mass number and the left subscript for the atomic number. The atoms described above are represented by

$$_2^4He \qquad _9^{19}F \qquad _{83}^{209}Bi$$

7.10 Isotopes and atomic weight

The number of protons is the same for all atoms of the same element, but the number of neutrons may vary. In other words, the atomic number of an element has only one value, but the mass number may have several values. Atoms with different mass numbers but the same atomic number are called **isotopes.**

Oxygen in the air is a mixture of atoms of mass numbers 16, 17, and 18—$_8^{16}O$, $_8^{17}O$, $_8^{18}O$. The nucleus of each of these atoms contains 8 protons, together with 8, or 9, or 10 neutrons, respectively. Each isotopic oxygen atom also contains 8 charge-balancing electrons, of course. Naturally occurring hydrogen ($Z = 1$) is almost entirely a mixture of atoms of mass numbers 1 and 2—$_1^1H$, $_1^2H$. Each hydrogen atom contains a single proton and a single electron, but the nucleus of each atom of the heavier hydrogen isotope, $_1^2H$, called deuterium, also contains a single neutron. A third and much scarcer isotope of hydrogen, called tritium, $_1^3H$, has a mass number of 3. Natural uranium ($Z = 92$) is a mixture of atoms of mass numbers 234, 235, and 238, with 92 protons in each nucleus and, respectively, 142, 143, and 146 neutrons.

Because the atomic weight is an average based on the percentage of atoms of each isotope in the naturally occurring isotopic mixture (Section 2.6, Example 2.2), atomic weights are not whole numbers even though atomic numbers and mass numbers are whole numbers. Isotopic masses also differ from whole numbers. For one reason, on the atomic mass unit scale based on $\frac{1}{12}$ of the mass of an atom of the carbon-12 isotope, the masses of the proton and the neutron differ slightly from 1 (see Table 7.3). Furthermore, isotopic mass includes not only the mass of the nucleus, but also the masses of the electrons in the atom. In addition, the sum of the masses of the neutrons, protons, and electrons in an atom does not exactly equal the mass of the atom itself. A relatively small amount of mass is converted into the energy that binds the particles together, in accordance with the Einstein relationship between mass and energy, $E = mc^2$ (Section 8.2).

The total effect of all the factors discussed in this section is quite

Isotopes:
same atomic number
different mass number

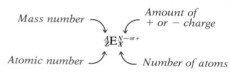

For example

$^{18}O_2$	Two atoms of oxygen of mass number 18 combined
$C^{18}O_2$	Carbon dioxide with two atoms of oxygen of mass number 18 per molecule
$^{44}Ca^{2+}$	The positive ion of calcium of mass number 44
$_{48}^{113}Cd$, $_{49}^{113}In$	Cadmium ($Z = 48$) and indium ($Z = 49$) of the same mass number

FIGURE 7.8

Notation for an atom or ion of element E. *Only the subscripts or superscripts needed for the purpose at hand are used. Often the atomic number is not included. (This is the notation as agreed upon today. Some older publications place the mass number at the right, e.g., C^{14}.)*

Mass spectrometer

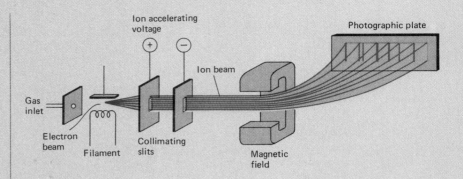

FIGURE A

Mass spectrograph. *The electron beam ionizes the gas sample; here the ions of different e/m are recorded on a photographic plate. In a mass spectrometer they are recorded on a chart or graph.*

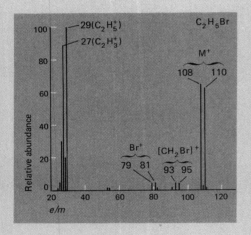

FIGURE B

The mass spectrum of ethyl bromide, C_2H_5Br (mol wt = 108.96 amu). *The principal ions formed in the mass spectrometer by ethyl bromide are $[C_2H_5Br]^+$, symbolized M^+; $[CH_2Br]^+$; Br^+; $[C_2H_5]^+$; and $[C_2H_3]^+$. The natural isotopic distribution of the bromine isotopes, ^{79}Br (50.54%) and ^{81}Br (49.46%), is reflected in the pairs of peaks for each of the bromine-containing ions. (From K. Biemann,* Mass Spectrometry, *McGraw Hill, New York, 1962, p. 127.)*

A spectrum, you will recall (Tools of Chemistry: Electromagnetic Radiation and Spectra), shows how radiation or particles spread out in an order based on the increase or decrease in magnitude of some physical property of the radiation or particles. The spectrum might be of the wavelength or frequency of radiation or of one of many properties of atoms, ions, molecules, or subatomic particles. A **spectrometer** is an instrument used to produce a spectrum. A spectrometer usually has four basic sections: (1) a source of the beam of radiation or particles; (2) an analyzer, which separates the beam according to the property being analyzed; (3) a detector, which measures quantity (e.g., the intensity of radiation or the number of particles); and (4) a display device, which makes the results visible as a graph, chart, or photograph.

A **mass spectrometer** spreads a beam of positive ions into a spectrum according to the charge-to-mass ratio of the ions (Figure A). The modern mass spectrometer operates on the same principles as the tube in which Thomson examined canal rays (see Figure 7.3). Under vacuum, a beam of positive ions is produced either by the bombardment of a gas by an electron beam or by the ionization of a solid by an electric spark. The positive ion beam is lined up ("collimated") by a slit system and accelerated by an electrical potential of variable strength. The ions then pass through a magnetic field of variable strength, which is perpendicular to their path. This force deflects the ions into curved paths in which the radius of curvature depends upon the mass and charge of the ions.

By varying both the accelerating electric potential and the strength of the magnetic field, ions are made to arrive at the ion detector in the order of their charge-to-mass ratio *e/m*. In a *mass spectrograph* the spectrum is recorded on a photographic plate. In a *mass spectrometer* the spectrum is electrically recorded as a graph (Figure B) or, with the aid of a computer coupled to the instrument, the spectrum is provided directly as a printout of the masses and abundances of the ions present.

The mass spectrometer can be used to determine the masses and relative abundances of isotopes (see Figure B) and also to identify even small amounts of compounds by their "fingerprints," that is, by the distinctive pattern of ions formed when the compound breaks down in the electron beam.

The atomic weight of each element, which is the weighted average of the isotope atomic masses, is given in color in the last column. The percentages of the isotopes in naturally occurring mixtures are given in parentheses. The true mass in grams of one atom of an isotope can be found by multiplying the atomic mass by 1.6605×10^{-24} g/amu. For example, the true mass of 1 atom of ^{16}O is 2.66×10^{-25} g.

TABLE 7.5
Isotopes and masses

Element and isotopes (%)	Protons	Neutrons	Atomic weight, or isotope atomic mass (amu)
Carbon ($Z = 6$)			12.011
^{12}C (98.892)	6	6	12.0000[a]
^{13}C (1.108)	6	7	13.0003
Oxygen ($Z = 8$)			15.9994
^{16}O (99.759)	8	8	15.9949
^{17}O (0.037)	8	9	16.9991
^{18}O (0.204)	8	10	17.9992
Bromine ($Z = 35$)			79.904
^{79}Br (50.52)	35	44	78.9183
^{80}Br (49.48)	35	45	80.9163
Uranium ($Z = 92$)			238.03
^{234}U (0.0057)	92	142	234.041
^{235}U (0.72)	92	143	235.044
^{238}U (99.27)	92	146	238.051

[a] By definition.

small, and the mass in atomic mass units of each isotope is quite close to the sum of the number of protons and the number of neutrons in the nucleus of an atom of that isotope. This makes the atomic weight close to the mass numbers. Table 7.5 illustrates the relationships discussed in this section and the two preceding sections.

Quantum theory and the atom

7.11 The role of the electron

The Rutherford atom left an inexact picture of how the electrons are arranged about the nucleus. Rutherford recognized that the electrons must be in motion or they would be pulled into the nucleus by the attraction between opposite charges. He suggested that the electrons might orbit the nucleus as planets orbit the sun. However, the laws of physics known in Rutherford's time predicted that charged particles like electrons, when moving in a field of force like that of the nucleus, would give off radiation and gradually lose energy. The orbiting electrons therefore would ultimately collapse into the nucleus.

Planck's equation:
$$E = h\nu$$

Since electrons never fall into the nucleus, new laws were obviously needed. The "old" physics was based on the behavior of large, easily observed objects. New experimental techniques were making it possible to observe more closely than ever before the behavior of electrons and other small particles, and of radiation and spectra. The old laws just do not work in this submicroscopic world. In the early 1900s several remarkable theoretical advances were made and a new physics was born. Based on what is called quantum theory, the new physics deals successfully with particles of the size of molecules, atoms, and electrons, and with the energy of radiation.

7.12 Waves as particles—particles as waves: Quantum theory

The introduction of quantum theory revolutionized our understanding of matter and energy, and led directly to modern chemistry. Central to quantum theory and the new physics is the concept of *wave–particle duality*. Here it is, stated as simply as we know how: In some ways electromagnetic radiation behaves like continuous waves and in some ways it behaves like individual particles. Then again, in some ways matter behaves like individual particles and in some ways it behaves like waves.

(Do not worry if, after reading this section, you feel that you do not totally understand wave–particle duality. No course in general chemistry or general physics can give you enough information to provide a basis for total understanding. The concept is largely mathematical and the mathematics required is too complex for consideration here. However, the concept is also too important to be ignored.)

A quantum can be thought of as a package or bundle of something that is only available in specific and separate amounts. The quantum is somewhat like the ice cream in an ice cream cone that you buy at a store. You can order one scoop, or two scoops, or three scoops, but not one and one-half or two and one-half scoops. The amount of ice cream you get is one, or two, or three times the size of the scoop (Figure 7.9).

Something that is **quantized** is restricted to amounts that are multiples of the basic unit, or quantum, for the particular system. The idea that *energy* is quantized was introduced in 1900 by Max Planck. Planck's equation, which was destined for fame, is

$$energy \quad E = h\nu \qquad (3)$$

Planck's constant

frequency

Planck assumed that the energy of radiation is proportional to its frequency ν (nu), and that energy and frequency are related by a constant h, now called Planck's constant. The value of $h\nu$ gives the size of a quantum of radiation, and Planck stated that a body could only absorb or give off radiation in units of $h\nu$. Like the scoops of ice cream,

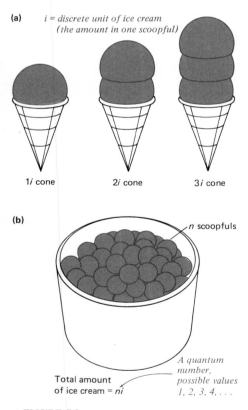

(a) i = discrete unit of ice cream
(the amount in one scoopful)

$1i$ cone $2i$ cone $3i$ cone

(b)

n scoopfuls

Total amount of ice cream = ni

A quantum number, possible values 1, 2, 3, 4, . . .

FIGURE 7.9
If ice cream were quantized . . .
(a) Quantized ice cream in cones.
(b) Quantized ice cream in a tub.

the radiation could, according to Planck, be transferred only in small-whole-number multiples of the unit, $1h\nu$, or $2h\nu$, or $3h\nu$, and so on. Whole-number multipliers that specify the amounts of energy are called **quantum numbers.**

Planck's constant h is a fundamental, universal constant. It appears in the mathematics of many systems. Modern values for h are

$$h = 6.6262 \times 10^{-34} \text{ J sec} \quad = 6.6262 \times 10^{-27} \text{ erg sec} \quad (4)$$

The units in Equation (3) would be

$$E(\text{J}) = h(\text{J sec})\nu(\text{sec}^{-1})$$

The term *quantum theory* refers to the theory and mathematics of finite quanta of energy on a molecular or atomic scale.

Planck applied quantum theory only to the body giving off the radiation. In 1905 Albert Einstein extended the quantum concept to the electromagnetic radiation itself. This was a bit like deciding that the ice cream in the tub is already frozen in scoop-sized portions (see Figure 7.9b). Here was the first suggestion that radiation could behave like a wave, yet also be divided into packets of energy like a stream of particles.

Einstein succeeded in explaining another phenomenon that had stumped classical physics. In the **photoelectric effect,** electrons (called photoelectrons) are given off by certain metals (particularly Cs and the other alkali metals, Li, Na, K, and Rb) when light shines on them. The dilemma, in part, was this: Light of any frequency should, according to classical theory, carry increasing energy as the intensity of the light increases. Electrons should be able to accumulate energy from the light until they have acquired enough to break away from the metal as photoelectrons. But, in reality, this does not happen with light of any frequency. For each metal there is a characteristic minimum frequency—the *threshold frequency*—below which the photoelectric effect does not occur.

A red light ($\nu = 4.3$–$4.6 \times 10^{14} \text{ sec}^{-1}$) of great intensity can shine on a piece of potassium for hours and no photoelectrons will be released. But as soon as even a very weak yellow light ($\nu = 5.1$–$5.2 \times 10^{14} \text{ sec}^{-1}$) shines on potassium, the photoelectric effect begins. The threshold frequency of potassium is $5 \times 10^{14} \text{ sec}^{-1}$.

Einstein showed that the energy of the incoming radiation is quantized according to $E = h\nu$. A single quantum of radiant energy ($1h\nu$) is called a **photon.** Individual photons of high-intensity light of a given frequency have no greater energy than individual photons of low-intensity light of the same frequency. High-intensity light just has *more* photons, each with the *same* energy (Figure 7.10).

Each photon can act like a particle in a collision with a single electron and deliver to that electron a maximum of $1h\nu$ of energy. The amount of energy delivered by one photon can increase only if the frequency increases. This explains the threshold frequency. When the frequency and, therefore, the energy of the incoming light are too low, not even one electron can acquire enough energy to escape. Red

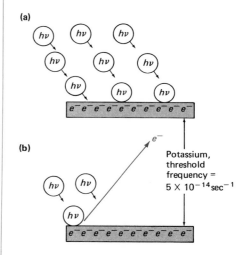

(a)

(b)

Potassium, threshold frequency = $5 \times 10^{-14} \text{ sec}^{-1}$

FIGURE 7.10

The photoelectric effect. *It takes energy equal to or greater than the $h\nu$ threshold to knock electrons out of a metal. (a) High-intensity red light, $\nu = 4.5 \times 10^{14} \text{ sec}^{-1}$; (b) low-intensity yellow light, $\nu = 5.2 \times 10^{14} \text{ sec}^{-1}$.*

light can never initiate the photoelectric effect in potassium, because each photon has too little energy.

Note that all photons travel with the speed of light. This is one difference between photons and particles such as electrons, whose speed can vary.

So far in this section we have presented the evidence for the behavior of *waves as particles*. The quantum theory of radiation was accepted because it was successful in explaining the photoelectric effect and other phenonmena.

Louis de Broglie (pronounced duh braẃ-glee) in 1923 turned up the other side of the coin—the *particles as waves* side. He reasoned that if radiation has particle-like properties, then particles in motion should have wave-like properties. de Broglie predicted that the wavelength of a moving particle, λ (lambda), could be calculated from the equation

$$\lambda = \frac{h}{mv} \qquad \text{(5)}$$

Planck's constant — h

mass of particle — m

velocity of particle — v

Two American physicists at the Bell Telephone Laboratory—C. J. Davisson and L. H. Germer—soon did an experiment that proved that de Broglie was right. In 1927 Davisson and Germer found that what they had first observed as an unexpected result in an electron-scattering experiment was the diffraction of the electron beam. When their electron beam was aimed at a nickel crystal they obtained a diffraction pattern (see Tools of Chemistry: Diffraction, in Chapter 12) similar to the pattern obtained from the diffraction of x-rays (electromagnetic radiation) by a crystal. The wavelength of the electron beam calculated from the diffraction pattern agreed within 1% with the wavelength calculated from de Broglie's equation. Electrons can indeed behave like waves. *Wave mechanics* is the part of quantum theory that deals with the wavelike properties of particles.

The wavelength of a body of any mass can be calculated with the de Broglie equation and presumably bodies of any mass have wavelengths. However, the wave properties of matter are observable only for particles of very small mass. An electron of mass 9.11×10^{-28} g moving at the speed of light has a wavelength of about 2×10^{-8} cm, a value comparable to the diameter of an atom. A 1-g mass moving at 1 cm/sec has a wavelength of the order of 10^{-26} cm, much too small to be observed.

7.13 Atomic spectra

In the early 1900s atomic spectra seemed as hard to explain as the photoelectric effect. Atoms excited by heat or by an electric discharge, that is, atoms that have absorbed energy over and above their normal energy content, radiate energy at specific wavelengths, not just anywhere in the electromagnetic spectrum. All atoms in the gaseous state give line emission spectra and there is a characteristic line spectrum for each element. This regularity in the line spectra, so puzzling at first, was the key to understanding electronic structure.

The simplest line spectrum is found for hydrogen. The spectra of heavier atoms become more and more complex. The hydrogen spectrum consists of five series of lines, named for the men who discovered them (Figure 7.11).

The wave numbers of all the lines in the hydrogen spectrum can be accurately calculated from a simple formula

$$\bar{\nu} = R \left(\frac{1}{n_1^2} - \frac{1}{n_2^2} \right) \tag{6}$$

where R is a constant called the Rydberg constant (named for spectroscopist Johannes Rydberg), and n_1 and n_2 are whole numbers. There is a characteristic single value of n_1 for each series of lines in the hydrogen emission spectrum, while for each series n_2 can be any number greater than n_1, that is, $n_2 = n_1 + 1, n_1 + 2, \ldots$. J. J. Balmer in 1885 first developed this empirical relationship for the series bearing his name, where $n_1 = 2$. (An **empirical relationship** is based solely on experimental facts and is derived without the use of any theory or explanation of the facts.) For the later-discovered series n_1 has the values 5, 4, 3, and 1, respectively. The experimentally determined value of the Rydberg constant for the hydrogen spectrum is 109,677 cm^{-1}.

7.14 Bohr model of the hydrogen atom

The great Danish physicist Niels Bohr in 1913 presented the first model of an atom based on the quantization of energy. Bohr's model explained the structure of the hydrogen atom and its spectrum. He overcame the difficulties of the Rutherford atomic model by simply assuming that classical physics was wrong. There is no reason, Bohr decided, to expect electrons in atoms to radiate energy as long as they are not *given* any extra energy. And from the mathematics of his theory he derived the equation for the wave numbers of the lines in the hydrogen spectrum by a completely nonexperimental route.

The Bohr theory of the hydrogen atom is based upon three assumptions:

1. The electron in the hydrogen atom can move about the nucleus in any one of several fixed circular orbits, but only in these orbits.

2. The momentum of the electron in an orbit is quantized; it is a whole-number multiple of $h/2\pi$, where h is Planck's constant. (**Momentum** is mass times velocity.) By this condition of quantization, the electron is assigned a total energy for each orbit, and no other energies are possible. Each orbit in the Bohr hydrogen atom is called a **stationary state.** The lowest energy orbit, the one in which the electron normally resides, is the **ground state** for the electron. The states

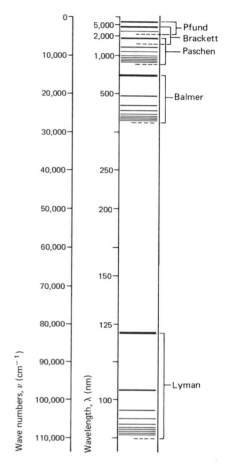

FIGURE 7.11
The hydrogen emission spectrum. *A schematic drawing showing the five series of lines. The dashed lines indicate the limits of each series. The Balmer series, which is in the visible region, was the first to be discovered.*

of energy higher than the ground state are **excited states,** reached by the electrons when an atom has absorbed extra energy.

3. The electron does not radiate energy as long as it remains in one of the orbits, or stationary energy states. When the electron drops from a higher energy state to a lower one, a definite quantity of energy is emitted as one photon of radiation. The energy change and the frequency of this radiation are proportional to each other according to Planck's relationship.

$$E_2 - E_1 = h\nu \tag{7}$$

higher lower
energy energy
state state

The same quantity of energy is *absorbed* if the electron is raised from state 1 to state 2.

Bohr pictured each allowed electron energy level as a circular orbit around the nucleus. In the normal, or ground state, the electrons in hydrogen atoms are at the lowest energy level. The quantum number n for the ground state is equal to 1. When the atoms absorb energy, for example, from the electric current in a gas-discharge tube, they become excited and the electrons jump to higher levels ($n > 1$, n greater than 1). As they fall back, the energy is emitted and spectral lines corresponding to the energy changes can be observed. How each of the five series of lines in the hydrogen spectrum originates is shown in Figure 7.12.

Two of Bohr's ideas—quantization of energy levels for electrons in atoms, and energy radiation by the electron only when changing energy levels—have stayed with us. However, the Bohr model could not explain the emission spectra of atoms containing more than one electron. Neither could it cope with the appearance of additional lines in the hydrogen spectrum when more refined spectrographs were used, or with the splitting of some emission lines in the presence of a strong magnetic field. The next atomic model that we consider here is able to handle these phenomena. The wave-mechanical model of the atom is currently the model of most importance to chemistry.

7.15 Wave-mechanical model of the atom

Just as the Bohr atom was developed from Planck's quantum theory, so the wave-mechanical model of the atom was developed from de Broglie's treatment of particles as waves. In 1926, Werner Heisenberg in Germany and Erwin Schrödinger in Austria studied the mathematics of the wave properties of the electrons in atoms. From their results has come our most successful model of atomic structure. It shows in detail how the electrons are arranged about the nucleus, successfully predicts the hydrogen emission spectrum, including the previously unexplained lines, and can be adapted to atoms other than hydrogen.

The wave-mechanical theory of the atom is a mathematical

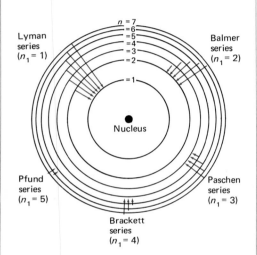

FIGURE 7.12
Bohr hydrogen atom (*not drawn to scale*). *The single electron can reside in any circular orbit without emitting energy. The electron reaches orbits with n > 1 only when the atom is excited. The series of lines in the hydrogen spectrum arise from transitions of the electron from orbit to orbit. The n_1 value for each series refers to the n_1 in Equation (6). The n_2 values are all of the n values greater than n_1 in each series.*

theory—it is best stated in mathematics beyond the level of our discussion. Our approach here is to describe the most important physical consequences of the theory, and to present with very little explanation the few mathematical parts of the theory that we need, primarily dealing with quantum numbers. One success of wave mechanics is that the quantization of the energy of the electron need not be assumed to make the mathematics work—quantization arises as a natural consequence of the mathematics.

Heisenberg is particularly remembered for what we call the **Heisenberg uncertainty principle:** *It is impossible to know simultaneously both the exact momentum and the exact position of an electron.* Despite many efforts, no one has yet been able to disprove this principle. It has not been possible to accurately measure both the position and the momentum of an electron in such a way that the conditions of the experiment do not change either the position or the momentum of the electron.

The Heisenberg principle, theoretically, holds for all objects, including those that we see around ourselves all the time. However, like the wavelength of particles, it becomes significant only at the subatomic level. Suppose you shine a flashlight into a dark closet and observe a mouse running across the floor. The light from the flashlight does not noticeably slow the mouse down or change the direction of his flight. Trying to look at an electron is a different story. The electron is so small that no matter what kind of radiation we shine on it, the speed and direction of the electron will be changed the instant the radiation hits it.

If we know exactly where, say, a rocket is at a given moment plus its speed, its mass, and the direction in which it is going, we can calculate the entire path of the rocket through space. A consequence of the Heisenberg principle is that we cannot know enough to calculate a path for an electron moving about a nucleus.

Schrödinger's fame lies in a set of equations that show what can be calculated for an electron—its energy level and the probability of finding it within a particular area surrounding the nucleus. The space in which an electron with a specific energy is most likely to be found—the energy level for a particular electron—is called an **atomic orbital.** Note the distinction between an orbit, as in the Bohr atom, and an orbital. An orbit is a clearly defined path through space; an orbital is just a region in space—we know that the electron is somewhere within that region, but according to the Heisenberg principle, we cannot know its exact location. The energies of the orbitals can be calculated from lines in spectra that correspond to transitions between the orbital energy levels. The terms "ground state" for the lowest energy level and "excited state" for higher energy levels are also applied to orbitals in atoms and molecules.

7.16 Orbitals and quantum numbers

To describe the placement of electrons in a Bohr circular orbit requires only one quantum number that indicates the principal energy level of the electron (see Figure 7.12). A complete description of the

TABLE 7.6

Angular momentum quantum number (l) values

The letters s, p, d, and f come from the designation of lines in emission spectra as belonging to the sharp, principal, diffuse, or fundamental series. The number of sublevels in each quantum level is equal to n.

For n =	$l = 0$ to $n - 1$ (subshell letters)
1	0 (s)
2	0, 1 (s, p)
3	0, 1, 2 (s, p, d)
4	0, 1, 2, 3 (s, p, d, f)

TABLE 7.7

Magnetic quantum number (m) values

For l =	$m = +l \ldots 0 \ldots -l$ (total)
0 (s)	0 (one s orbital)
1 (p)	+1, 0, −1 (three p orbitals)
2 (d)	+2, +1, 0, −1, −2 (five d orbitals)
3 (f)	+3, +2, +1, 0, −1, −2, −3 (seven f orbitals)

atomic orbital occupied by a specific electron in the wave mechanical atom requires three quantum numbers. We might say that the energy of an electron is quantized in three ways. A fourth quantum number must be added to represent the rotation of an electron about its own axis.

We need to know the quantum numbers of electrons for two main reasons: to be able to understand how electrons are distributed in the atoms of the different elements and to be able to picture the interactions of atoms in terms of the arrangement of their orbitals in space. As we have stated, the quantum numbers arise from the mathematics of Schrödinger's equation. Since any description in words of their origin and significance is inadequate, we present them here as simply as possible.

1. The *principal quantum number n* designates the main energy level with respect to the nucleus. It is roughly equivalent to the *n* of the energy levels in the Bohr atom. Electrons with higher values of *n* have higher energy and are further from the nucleus than those with lower *n* values. Values of *n* from 1 to 7 are known.

2. The *angular momentum quantum number l* is required because electrons within each principal energy level occupy slightly different energy levels called subshells, or sublevels. The number of possible subshells increases with *n*. The possible values of *l* are 0 to *n* − 1. For example, the *n* = 3 shell has three *l* levels, *l* = 0, 1, and 2 (Table 7.6). The first four subshells are designated by the letters *s*, *p*, *d*, and *f*; these levels correspond to *l* values of 0, 1, 2, and 3, respectively. Transitions between electrons in the subshells can be observed in spectra. The *s*, *p*, *d*, and *f* orbitals each represent different characteristic regions in space.

3. The *magnetic quantum number m* is required because orbitals for electrons in the same *l* subshell are oriented differently about the nucleus. The allowed values for *m* are *l* to 0 to −*l* (Table 7.7). For example, since *p* orbitals have *l* = 1, there are always three different *p* orbitals, for *m* = +1, 0, and −1 (Table 7.7). Electrons with the same *n* and *l* values, but different *m* values are at the same energy level. Their differences show up in spectra observed in the presence of a magnetic field. This appearance of additional lines in a magnetic field is called the Zeeman effect.

4. The *spin quantum number* m_s designates two possible spin orientations for the individual electrons. It represents the motion of an electron about its own axis, not with respect to the nucleus (Figure 7.13).

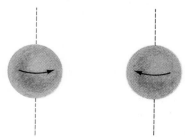

FIGURE 7.13
Electron spin. *Two possible orientations of the spin of an electron about its own axis.*

This spinning causes each electron to behave like a micromagnet. The two possible values of m_s are $+\frac{1}{2}$ and $-\frac{1}{2}$. Each atomic orbital can be occupied by two electrons, one with $m_s = \frac{1}{2}$ and one with $m_s = -\frac{1}{2}$.

7.17 Picturing orbitals

There is no good way to draw pictures of atomic orbitals. Artists' best efforts at representing the three-dimensional areas sometimes called "electron clouds" still convey the idea of a distinct "shape". These "shapes" are only regions in space within which electrons can be found with a certain probability. The shapes have no physical existence; nor are they independent of each other. The overall distribution of electron density for a given principal quantum level is probably close to a spherical form for any isolated atom.

With that said, we can look at pictures of orbitals. All s orbitals are spherically symmetrical about the nucleus. The s orbital for the $n = 1$ level is quite close to the nucleus, the s orbital for $n = 2$ is further away from the nucleus, and so on, like the layers in an onion (Figure 7.14).

There are three p orbitals, one for each value of the quantum number m. Each p orbital is an ellipsoidal region with two lobes (Figure 7.15). The lobes can be thought of as directed along the x, or y, or z axes of a set of coordinates; that is, the electrons in each p orbital stay as far from the electrons in the other p orbitals as possible.

The d orbitals—five of them for the five possible values of m—are more complex in shape (Figure 7.16). One d orbital resembles the p orbital along the z axis, but with the addition of a doughnut-shaped area in the center. The other four orbitals each have four lobes; in one case these lobes are oriented along the x and y axes and in the other three cases the lobes are oriented between the axes. The seven f orbitals are even more complex—two are shown in Figure 7.17.

The shapes of atomic orbitals, or more accurately, their orientations in space, often determine the nature of chemical bonding and the geometry of molecules.

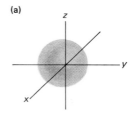

(a)

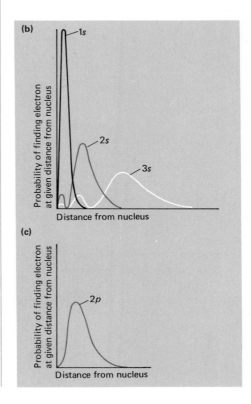

(b)

(c)

FIGURE 7.14

s **Orbitals of the hydrogen atom.** *Each s orbital is a sphere surrounding the nucleus (a). The radii of the 1s, 2s, and 3s orbitals are the distances where the probability of finding the electron is greatest. The curves in (b) and (c) are plots of the wave equations for each orbital. The 2p orbital plot (c) is included to show that 2s and 2p electrons reside at roughly the same distance from the nucleus and therefore have roughly the same energy. The orbital plots for other atoms have similar shapes, although, of course, the values of the maximum radii are different.*

181

FIGURE 7.15
p Orbitals.

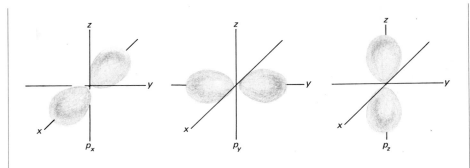

FIGURE 7.16

d Orbitals. *Note that the first two orbitals lie along the axes and the last three lie between the axes.*

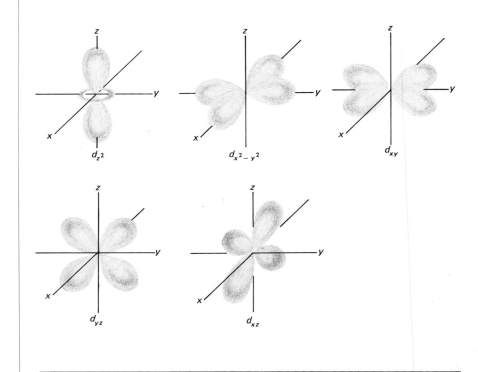

FIGURE 7.17
f Orbitals. *Three of the seven f orbitals have the shape shown in (a), with different orientations in space; the other four have the shape shown in (b), also with different orientations in space.*

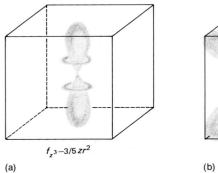

Electrons and the periodic table

7.18 Electronic configurations

The **electronic configuration** of an atom is the designation of the orbitals occupied by all of the electrons in the atom. The notation used to represent electrons in atomic orbitals is the following

$$nl^{x}$$

angular momentum quantum number

principal quantum number 1, 2, 3, . . .

number of electrons in l level

where *n* represents the principal quantum level as a number (1, 2, 3, . . .); *l* represents the *l* sublevel as a letter (*s, p, d, f,*); and *x* indicates the number of electrons in the *l* sublevel. For example,

$$3d^{8}$$

In the n = 3 level   are occupied by 8 electrons

the d orbitals

The complete electronic configuration for an atom is given by the series of symbols for each occupied orbital. For a neutral atom, the sum of all the *x* values is the atomic number of the element. For example, the complete ground-state configuration of an argon atom, Z = 18, is written

Ar $1s^2 2s^2 2p^6 3s^2 3p^6$

showing that the 1*s*, 2*s*, and 3*s* orbitals each contain two electrons and that the 2*p* and 3*p* orbitals each contain six electrons.

The electronic configurations for any atom can be written by symbolically putting electrons into orbitals one by one according to the following three principles:

1. *Aufbau principle: Electrons first occupy the lowest-energy orbitals available to them; they enter higher-energy orbitals only when the lower-energy orbitals are filled.* The sequence of the filling of the orbitals, illustrated in Figure 7.18, is as follows:

1*s* 2*s* 2*p* 3*s* 3*p* 4*s* 3*d* 4*p* 5*s* 4*d* 5*p* 6*s* 4*f* 5*d* 6*p* 7*s* 5*f* 6*d* 7*p*

This sequence is determined by the general order of orbital energy levels (Figure 7.19). It is followed exactly as far as vanadium (Z = 23) (see Table 7.10). After that there are a few exceptions to the sequence due to variations in the energy level order. Because electrons in each atom are influenced by different nuclear charges and different total

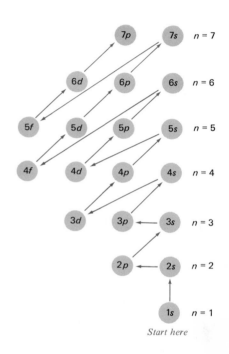

FIGURE 7.18

Orbital occupancy sequence—a diagram. *The corresponding energy levels are shown in Figure 7.19.*

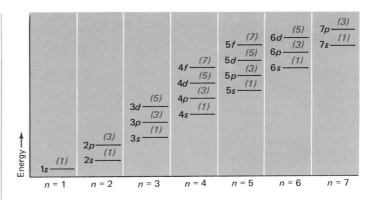

FIGURE 7.19
Orbital energy levels. *The relative energies for filling the orbitals for each principal quantum level. The colored numbers show the number of orbitals available at each energy sublevel.*

surrounding electronic charges, energy levels that are close together (e.g., $5f$ and $6d$; see Np and Pu in Table 7.10) sometimes alternate in energy and therefore in order of filling.

2. Pauli exclusion principle: No two electrons can have the same four quantum numbers, $n, l, m_l,$ and m_s. The maximum number of electrons that can reside in each orbital is set by this principle, which is a way of saying that no two electrons can have the same wave pattern. The significance of the spin quantum number here is that it allows two electrons of opposite spin to occupy each atomic orbital. For example, each s subshell has only one orbital, that is, one value of the magnetic quantum number, and therefore can accommodate a maximum of 2 electrons. Each p level, with three possible orientations (three magnetic quantum numbers), can accommodate 6 electrons; each d level 10 electrons; and each f level 14 electrons. These relation-

TABLE 7.8

Electron distribution and maximum electron population of each atomic orbital.

The number of s, p, d, and f orbitals is determined by the number of m_l values for each sublevel (see Table 7.5). Two electrons of opposite spin can reside in each orbital.

n	l	Number of l orbitals	Maximum number of electrons per l sublevel	Maximum number of electrons per n level
1	0(s)	One s orbital	2	2
2	0(s)	One s orbital	2	
	1(p)	Three p orbitals	6	8
3	0(s)	One s orbital	2	
	1(p)	Three p orbitals	6	
	2(d)	Five d orbitals	10	18
4	0(s)	One s orbital	2	
	1(p)	Three p orbitals	6	
	2(d)	Five d orbitals	10	
	3(f)	Seven f orbitals	14	32
5	0(s)	One s orbital	2	
	1(p)	Three p orbitals	6	
	2(d)	Five d orbitals	10	
	3(f)	Seven f orbitals	14	
	4("g")[a]	Nine "g" orbitals[a]	18[a]	50[a]

[a] No element has yet been found that uses an $l = 4$ ("g") energy sublevel.

ships and the resulting maximum number of electrons in each principle energy shell are summarized in Table 7.8.

3. *Hund principle: Orbitals of equal energy are each occupied by a single electron before the second electron of opposite m_s enters the orbital.* The electrons thus stay as far apart as possible. This principle is important only for p, d, and f orbitals ($l > 0$) since an s orbital can hold only two electrons.

For example, consider the nitrogen atom ($Z = 7$). It has the configuration

$$N \qquad 1s^2 2s^2 2p^3$$

The Hund principle requires that the $2p$ electrons be unpaired, if possible, so one electron goes into each of the three p orbitals

$$N \qquad 1s^2 2s^2 2p_x^1 2p_y^1 2p_z^1$$

If a small box is used to represent each orbital and an arrow to represent each electron and the direction of its spin, this configuration can be depicted as

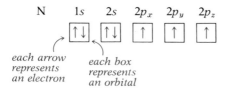

each arrow represents an electron   each box represents an orbital

As a result of the Hund principle, it is possible to have an atom or ion (or a molecule) with unpaired electrons even though the total number of electrons is even. There can be a *maximum* of one unpaired s electron, three unpaired p electrons, five unpaired d electrons, and seven unpaired f electrons. The presence or absence of unpaired electrons is found experimentally by the behavior of a substance in the presence of a magnetic field. **Paramagnetism** is the property of attraction to a magnetic field shown by substances containing unpaired electrons. **Diamagnetism** is the property of repulsion by a magnetic field and shows the absence of unpaired electrons.

Using the three principles just given, the electronic configuration of zirconium ($Z = 40$) is worked out as follows: Orbitals must be used to accommodate 40 electrons. The $n = 1$ and $n = 2$ orbitals are filled to maximum capacity first.

$$1s^2 2s^2 2p^6 \qquad \text{(10 electrons)}$$

The next orbitals in the energy order are $3s$, $3p$, $4s$, and $3d$ (it is easier for an electron to enter the $4s$ orbital than the $3d$ orbital). There are enough electrons to fill up all of these orbitals too. Now the $n = 1, 2$, and 3 levels have their maximum populations and $4s$ is also full.

TABLE 7.9

Some ground-state electronic configurations

In each case note that the superscripts add up to Z. Of these elements, atoms of B, Sc, Sb, and Pr are paramagnetic, because they contain unpaired electrons.

	Z	Configuration
B	5	$1s^2\,2s^2\,2p^1$
Ne	10	$1s^2\,2s^2\,2p^6$
Ar	18	$1s^2\,2s^2\,2p^6\,3s^2\,3p^6$
Ca	20	$1s^2\,2s^2\,2p^6\,3s^2\,3p^6\,4s^2$
Sc	21	$1s^2\,2s^2\,2p^6\,3s^2\,3p^6\,3d^1\,4s^2$
Zn	30	$1s^2\,2s^2\,2p^6\,3s^2\,3p^6\,3d^{10}\,4s^2$
Sb	51	$1s^2\,2s^2\,2p^6\,3s^2\,3p^6\,3d^{10}\,4s^2$ $4p^6\,4d^{10}\,5s^2\,5p_x^1\,5p_y^1\,5p_z^1$
Pr	59	$1s^2\,2s^2\,2p^6\,3s^2\,3p^6\,3d^{10}\,4s^2$ $4p^6\,4d^{10}\,4f^3\,5s^2\,5p^6\,6s^2$

$$1s^2 2s^2 2p^6 3s^2 3p^6 3d^{10}4s^2 \quad \text{(30 electrons)}$$

(Note that configurations are written here in the order of the principal quantum number, not the filling order.)

Ten electrons remain to be placed. The next orbital in the sequence is $4p$, which takes a maximum of six electrons; followed by $5s$, which takes two electrons. The remaining two electrons go into the $4d$ orbitals. The total electronic configuration of Zr is

$$\text{Zr} \quad 1s^2 2s^2 2p^6 3s^2 3p^6 3d^{10}4s^2 4p^6 4d^2 5s^2 \quad \text{(40 electrons)}$$

The two $4d$ electrons reside in separate orbitals. Some further examples of complete configuration are given in Table 7.9.

The energy sequence and the three principles discussed above are for symbolically filling in electrons in order to determine the configuration of any atom. They represent the relative energies needed to add an electron at the same time that a proton is being added to a nucleus. But note that in any atom, the energy required to *remove* an electron is first determined by the principal quantum level n. When an atom is ionized the electrons of greatest n value always come off first. For example, in the formation of Zr^{2+}, the $5s^2$ electrons would be lost. Within the same shell, p electrons are removed more easily than s electrons.

The ground-state configurations of all the elements are given in Table 7.10. As we have already pointed out, there are some departures from the regularity of orbital filling suggested by Figures 7.18 and 7.19. These exceptions, however, have little effect upon chemical characteristics. Some of the irregularities result from the apparent stability of half-filled and completely filled l subshells. For example, the Cr atom has an outer configuration of $3d^5 4s^1$ instead of $3d^4 4s^2$. Other irregularities arise from small differences in the energies of orbitals, particularly at large n values. For example, Ce ($Z = 58$) has the outer configuration of $4f^1 5d^1 6s^2$ instead of $4f^2 6s^2$, whereas Pr ($Z = 59$) has the arrangement $4f^3 6s^2$.

EXAMPLE 7.1

■ Compare the electronic configurations of K^+ and Cl^-.

An atom of potassium has 19 electrons, but the single positive charge tells us that one electron has been removed. So for the 18 electrons in the K^+ ion we write

$$\text{K}^+ \quad 1s^2\ 2s^2\ 2p^6\ 3s^2\ 3p^6$$

An atom of chlorine has 17 electrons, but the single negative charge tells us that one electron has been added. So for the 18 electrons in the Cl^- ion we write

$$\text{Cl}^- \quad 1s^2\ 2s^2\ 2p^6\ 3s^2\ 3p^6$$

The electronic configuration for K^+ and Cl^- are identical and are the same as for the Ar atom. ■

TABLE 7.10

*Electronic configurations of the elements**

Atomic number	Symbol	Electronic configuration
1	H	$1s^1$
2	He	$1s^2$
3	Li	$1s^2\,2s^1$
4	Be	$1s^2\,2s^2$
5	B	$1s^2\,2s^2\,2p^1$
6	C	$1s^2\,2s^2\,2p^2$
7	N	$1s^2\,2s^2\,2p^3$
8	O	$1s^2\,2s^2\,2p^4$
9	F	$1s^2\,2s^2\,2p^5$
10	Ne	$1s^2\,2s^2\,2p^6$
11	Na	$1s^2\,2s^2\,2p^6\,3s^1$
12	Mg	$1s^2\,2s^2\,2p^6\,3s^2$
13	Al	$1s^2\,2s^2\,2p^6\,3s^2\,3p^1$
14	Si	$1s^2\,2s^2\,2p^6\,3s^2\,3p^2$
15	P	$1s^2\,2s^2\,2p^6\,3s^2\,3p^3$
16	S	$1s^2\,2s^2\,2p^6\,3s^2\,3p^4$
17	Cl	$1s^2\,2s^2\,2p^6\,3s^2\,3p^5$
18	Ar	$1s^2\,2s^2\,2p^6\,3s^2\,3p^6$
19	K	$1s^2\,2s^2\,2p^6\,3s^2\,3p^6\quad 4s^1$
20	Ca	$1s^2\,2s^2\,2p^6\,3s^2\,3p^6\quad 4s^2$

Transition metals, Period 4

21	Sc	$1s^2\,2s^2\,2p^6\,3s^2\,3p^6\,3d^1\,4s^2$
22	Ti	$1s^2\,2s^2\,2p^6\,3s^2\,3p^6\,3d^2\,4s^2$
23	V	$1s^2\,2s^2\,2p^6\,3s^2\,3p^6\,3d^3\,4s^2$
24	Cr	$1s^2\,2s^2\,2p^6\,3s^2\,3p^6\,3d^5\,4s^1$
25	Mn	$1s^2\,2s^2\,2p^6\,3s^2\,3p^6\,3d^5\,4s^2$
26	Fe	$1s^2\,2s^2\,2p^6\,3s^2\,3p^6\,3d^6\,4s^2$
27	Co	$1s^2\,2s^2\,2p^6\,3s^2\,3p^6\,3d^7\,4s^2$
28	Ni	$1s^2\,2s^2\,2p^6\,3s^2\,3p^6\,3d^8\,4s^2$
29	Cu	$1s^2\,2s^2\,2p^6\,3s^2\,3p^6\,3d^{10}\,4s^1$
30	Zn	$1s^2\,2s^2\,2p^6\,3s^2\,3p^6\,3d^{10}\,4s^2$
31	Ga	$1s^2\,2s^2\,2p^6\,3s^2\,3p^6\,3d^{10}\,4s^2\quad 4p^1$
32	Ge	$1s^2\,2s^2\,2p^6\,3s^2\,3p^6\,3d^{10}\,4s^2\quad 4p^2$
33	As	$1s^2\,2s^2\,2p^6\,3s^2\,3p^6\,3d^{10}\,4s^2\quad 4p^3$
34	Se	$1s^2\,2s^2\,2p^6\,3s^2\,3p^6\,3d^{10}\,4s^2\quad 4p^4$
35	Br	$1s^2\,2s^2\,2p^6\,3s^2\,3p^6\,3d^{10}\,4s^2\quad 4p^5$
36	Kr	$1s^2\,2s^2\,2p^6\,3s^2\,3p^6\,3d^{10}\,4s^2\quad 4p^6$
37	Rb	$5s^1$
38	Sr	$5s^2$

Transition metals, Period 5 (Krypton core)

39	Y	$4d^1\quad 5s^2$
40	Zr	$4d^2\quad 5s^2$
41	Nb	$4d^4\quad 5s^1$
42	Mo	$4d^5\quad 5s^1$
43	Tc	$4d^5\quad 5s^2$
44	Ru	$4d^7\quad 5s^1$
45	Rh	$4d^8\quad 5s^1$
46	Pd	$4d^{10}$
47	Ag	$4d^{10}\quad 5s^1$
48	Cd	$4d^{10}\quad 5s^2$
49	In	$4d^{10}\quad 5s^2\,5p^1$
50	Sn	$4d^{10}\quad 5s^2\,5p^2$
51	Sb	$4d^{10}\quad 5s^2\,5p^3$
52	Te	$4d^{10}\quad 5s^2\,5p^4$
53	I	$4d^{10}\quad 5s^2\,5p^5$
54	Xe	$4d^{10}\quad 5s^2\,5p^6$

Atomic number	Symbol	Electronic configuration
55	Cs	$4d^{10}\quad 5s^2\,5p^6\quad\quad 6s^1$
56	Ba	$4d^{10}\quad 5s^2\,5p^6\quad\quad 6s^2$

Lanthanides (Krypton core)

57	La	$4d^{10}\quad 5s^2\,5p^6\,5d^1\quad 6s^2$
58	Ce	$4d^{10}\,4f^1\quad 5s^2\,5p^6\,5d^1\quad 6s^2$
59	Pr	$4d^{10}\,4f^3\quad 5s^2\,5p^6\quad 6s^2$
60	Nd	$4d^{10}\,4f^4\quad 5s^2\,5p^6\quad 6s^2$
61	Pm	$4d^{10}\,4f^5\quad 5s^2\,5p^6\quad 6s^2$
62	Sm	$4d^{10}\,4f^6\quad 5s^2\,5p^6\quad 6s^2$
63	Eu	$4d^{10}\,4f^7\quad 5s^2\,5p^6\quad 6s^2$
64	Gd	$4d^{10}\,4f^7\quad 5s^2\,5p^6\,5d^1\quad 6s^2$
65	Tb	$4d^{10}\,4f^9\quad 5s^2\,5p^6\quad 6s^2$
66	Dy	$4d^{10}\,4f^{10}\quad 5s^2\,5p^6\quad 6s^2$
67	Ho	$4d^{10}\,4f^{11}\quad 5s^2\,5p^6\quad 6s^2$
68	Er	$4d^{10}\,4f^{12}\quad 5s^2\,5p^6\quad 6s^2$
69	Tm	$4d^{10}\,4f^{13}\quad 5s^2\,5p^6\quad 6s^2$
70	Yb	$4d^{10}\,4f^{14}\quad 5s^2\,5p^6\quad 6s^2$
71	Lu	$4d^{10}\,4f^{14}\quad 5s^2\,5p^6\,5d^1\quad 6s^2$

Transition metals, Period 6 (Krypton core)

72	Hf	$4d^{10}\,4f^{14}\quad 5s^2\,5p^6\,5d^2\quad 6s^2$
73	Ta	$4d^{10}\,4f^{14}\quad 5s^2\,5p^6\,5d^3\quad 6s^2$
74	W	$4d^{10}\,4f^{14}\quad 5s^2\,5p^6\,5d^4\quad 6s^2$
75	Re	$4d^{10}\,4f^{14}\quad 5s^2\,5p^6\,5d^5\quad 6s^2$
76	Os	$4d^{10}\,4f^{14}\quad 5s^2\,5p^6\,5d^6\quad 6s^2$
77	Ir	$4d^{10}\,4f^{14}\quad 5s^2\,5p^6\,5d^7\quad 6s^2$
78	Pt	$4d^{10}\,4f^{14}\quad 5s^2\,5p^6\,5d^9\quad 6s^1$
79	Au	$4d^{10}\,4f^{14}\quad 5s^2\,5p^6\,5d^{10}\quad 6s^1$
80	Hg	$4d^{10}\,4f^{14}\quad 5s^2\,5p^6\,5d^{10}\quad 6s^2$
81	Tl	$4d^{10}\,4f^{14}\quad 5s^2\,5p^6\,5d^{10}\quad 6s^2\,6p^1$
82	Pb	$4d^{10}\,4f^{14}\quad 5s^2\,5p^6\,5d^{10}\quad 6s^2\,6p^2$
83	Bi	$4d^{10}\,4f^{14}\quad 5s^2\,5p^6\,5d^{10}\quad 6s^2\,6p^3$
84	Po	$4d^{10}\,4f^{14}\quad 5s^2\,5p^6\,5d^{10}\quad 6s^2\,6p^4$
85	At	$4d^{10}\,4f^{14}\quad 5s^2\,5p^6\,5d^{10}\quad 6s^2\,6p^5$
86	Rn	$4d^{10}\,4f^{14}\quad 5s^2\,5p^6\,5d^{10}\quad 6s^2\,6p^6$
87	Fr	$4d^{10}\,4f^{14}\quad 5s^2\,5p^6\,5d^{10}\quad 6s^2\,6p^6\quad 7s^1$
88	Ra	$4d^{10}\,4f^{14}\quad 5s^2\,5p^6\,5d^{10}\quad 6s^2\,6p^6\quad 7s^2$

Actinides (Krypton core)

89	Ac	$4d^{10}\,4f^{14}\quad 5s^2\,5p^6\,5d^{10}\quad 6s^2\,6p^6\,6d^1\,7s^2$
90	Th	$4d^{10}\,4f^{14}\quad 5s^2\,5p^6\,5d^{10}\quad 6s^2\,6p^6\,6d^2\,7s^2$
91	Pa	$4d^{10}\,4f^{14}\quad 5s^2\,5p^6\,5d^{10}\,5f^2\quad 6s^2\,6p^6\,6d^1\,7s^2$
92	U	$4d^{10}\,4f^{14}\quad 5s^2\,5p^6\,5d^{10}\,5f^3\quad 6s^2\,6p^6\,6d^1\,7s^2$
93	Np	$4d^{10}\,4f^{14}\quad 5s^2\,5p^6\,5d^{10}\,5f^4\quad 6s^2\,6p^6\,6d^1\,7s^2$
94	Pu	$4d^{10}\,4f^{14}\quad 5s^2\,5p^6\,5d^{10}\,5f^6\quad 6s^2\,6p^6\quad 7s^2$
95	Am	$4d^{10}\,4f^{14}\quad 5s^2\,5p^6\,5d^{10}\,5f^7\quad 6s^2\,6p^6\quad 7s^2$
96	Cm	$4d^{10}\,4f^{14}\quad 5s^2\,5p^6\,5d^{10}\,5f^7\quad 6s^2\,6p^6\,6d^1\,7s^2$
97	Bk	$4d^{10}\,4f^{14}\quad 5s^2\,5p^6\,5d^{10}\,5f^9\quad 6s^2\,6p^6\quad 7s^2$
98	Cf	$4d^{10}\,4f^{14}\quad 5s^2\,5p^6\,5d^{10}\,5f^{10}\,6s^2\,6p^6\quad 7s^2$
99	Es	$4d^{10}\,4f^{14}\quad 5s^2\,5p^6\,5d^{10}\,5f^{11}\,6s^2\,6p^6\quad 7s^2$
100	Fm	$4d^{10}\,4f^{14}\quad 5s^2\,5p^6\,5d^{10}\,5f^{12}\,6s^2\,6p^6\quad 7s^2$
101	Md	$4d^{10}\,4f^{14}\quad 5s^2\,5p^6\,5d^{10}\,5f^{13}\,6s^2\,6p^6\quad 7s^2$
102	No	$4d^{10}\,4f^{14}\quad 5s^2\,5p^6\,5d^{10}\,5f^{14}\,6s^2\,6p^6\quad 7s^2$
103	Lr	$4d^{10}\,4f^{14}\quad 5s^2\,5p^6\,5d^{10}\,5f^{14}\,6s^2\,6p^6\,6d^1\,7s^2$
104	—	$4d^{10}\,4f^{14}\quad 5s^2\,5p^6\,5d^{10}\,5f^{14}\,6s^2\,6p^6\,6d^2\,7s^2$
105	—	$4d^{10}\,4f^{14}\quad 5s^2\,5p^6\,5d^{10}\,5f^{14}\,6s^2\,6p^6\,6d^3\,7s^2$
106	—	$4d^{10}\,4f^{14}\quad 5s^2\,5p^6\,5d^{10}\,5f^{14}\,6s^2\,6p^6\,6d^4\,7s^2$

* Noble gases are shown against a color background. Unfilled orbitals are shown in color.

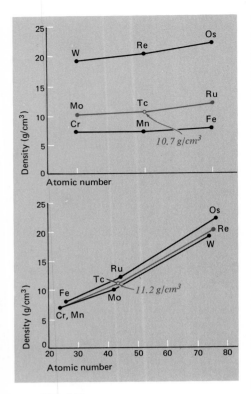

FIGURE 7.20

Example 7.3. *Plots of densities vs. atomic number.*

EXAMPLE 7.2

■ Many commercial periodic tables list oxidation states for the elements in addition to the density, color, crystal structure, melting points, and other properties. Using the electronic configuration for phosphorus

$$P \qquad 1s^2\, 2s^2\, 2p^6\, 3s^2\, 3p^3$$

show how the values of -3, $+3$, and $+5$ for the oxidation states of phosphorus can be explained.

We must keep in mind that the oxidation state of an element is only a bookkeeping method for electrons. A few compounds containing P^{3-} ions are known and the -3 oxidation state can be easily explained by the phosphorus atom gaining three electrons to finish filling the $3p$ subshell. Compounds containing P^{3+} and P^{5+} ions are not known (although $+3$ ions of some of the other members of the nitrogen family are known). However, these oxidation states can be explained by the involvement of the three most weakly held electrons in the $3p$ subshell in bonds to give the $+3$ oxidation state, followed by the involvement of the two electrons in the $3s$ subshell to give the $+5$ value. ■

7.19 The modern periodic table

The physical and chemical properties of the elements are periodic functions of their atomic numbers. That is a modern statement of the **periodic law.** Mendeleev had arranged the elements in his periodic table according to increasing atomic weight (but see Figure 7.1 caption). However, after the work of Moseley and the discovery of isotopes, the atomic number was recognized as the more fundamental property, and the property on which the periodic table should be based.

EXAMPLE 7.3

■ Find the approximate density of Tc by applying the periodic law to the densities of Cr (7.2 g/cm³), Mn (7.2), Fe (7.86), Mo (10.2), Ru (12.3), W (19.35), Re (20.53), and Os (22.48).

The periodic law states that physical and chemical properties are functions of atomic number. To see this function for density, plots are made of density versus atomic number (Figure 7.20). Assuming the same trend (or curve) for Mo–Tc–Ru as for W–Re–Os and Cr–Mn–Fe, the density of Tc is found from the graph to be about 10.7 g/cm³.

We can check our answer by making a similar plot for the periodic functions of Fe–Ru–Os and Cr–Mo–W so that we may obtain a trend for Mn–Tc–Re (Figure 7.20). From the plot the density is found to be about 11.2 g/cm³ for Tc.

The average of the two results, 10.7 and 11.2 g/cm³, is 11.0 g/cm³ and differs by only

$$\frac{11.0 \text{ g/cm}^3 - 11.50 \text{ g/cm}^3}{11.50 \text{ g/cm}^3} \times 100\% = 4\%$$

from the known value of 11.50 g/cm³ for the density of technetium.

■

A horizontal row of elements in the periodic table (Table 7.11) is called a **period.** There are seven periods in the table containing, respectively, 2, 8, 8, 18, 18, and 32 elements and a theoretical maximum of 32 elements in the seventh period, which is incomplete. Atoms of each element in a period have one more electron and one more proton than the atoms of the element to the left in the table (or, for the first element in a period, at the end of the period above).

In each successive period, the next highest principal energy level ($n = 1$, $n = 2$, etc.) is being filled by electrons in atoms of the elements of the period. The number of elements in each period is determined by the number of atomic orbitals available in the energy level that is being filled. The 1s orbital is filled at helium ($Z = 2$), limiting the first period to two elements. One electron enters the 2s orbital at lithium, and successive filling of the 2s and 2p orbitals is complete at neon ($Z = 10$). The process is then repeated with the 3s and 3p orbitals, giving the third period of elements, sodium through argon ($Z = 18$). Then the 4s orbital is occupied (potassium and calcium), but at scandium ($Z = 21$) it is easier for the next electron to enter a 3d orbital than a 4p orbital (see Figure 7.18). Continuing occupancy of the 3d orbitals gives the elements of what is called the first d transition series. The 4p orbitals then fill, and the fourth period ends at krypton ($Z = 36$). After krypton, a similar sequence is repeated as the 5s and then 4d orbitals are filled to give the second d transition series beginning at yttrium ($Z = 39$). The fifth period ends at xenon ($Z = 54$).

The third d transition series begins at lanthanum ($Z = 57$). However, the next electrons enter and fill 4f orbitals (beginning with cerium, $Z = 58$) to give the first f transition series (the lanthanide series). After lutetium ($Z = 71$), the third d transition series is completed, followed by addition of 6p electrons to complete the sixth period at radon ($Z = 86$). A similar sequence is then repeated, with the second f transition series (the actinide series) beginning at actinium ($Z = 89$).

The seventh period includes most of the man-made radioactive elements. The series limit should be $Z = 118$, where the 5f, 6d, 7s, and 7p orbitals would be filled.

To save space, and also to confine elements with closely similar properties to a single column, the two f transition series are generally placed at the bottom of the periodic table.

The elements in a single vertical column in the table are referred to together as **groups** or **families.** The atoms of elements in the same column have identical or very similar electronic configurations in the highest occupied orbitals.

Representative Elements — *d*-Transition Elements — **Representative Elements** — Noble gases

Group numbers. Period number, highest occupied electron level.

Main periodic table (columns: Group I = ns^1; Group II = ns^2; d-transition outer configurations as noted; Representative III–VII = ns^2np^1 … ns^2np^5; Group VI = ns^2np^4; Noble gases = ns^2np^6). Each cell: Symbol, atomic number; outer configuration; atomic weight.

I (ns^1)	II (ns^2)	III ($(n-1)d^1ns^2$)	IV ($(n-1)d^2ns^2$)	V ($(n-1)d^3ns^2$)	VI ($(n-1)d^5ns^1$)	VII ($(n-1)d^5ns^2$)	($(n-1)d^6ns^2$)	($(n-1)d^7ns^2$)	($(n-1)d^8ns^2$)	($(n-1)d^{10}ns^1$)	($(n-1)d^{10}ns^2$)	III (ns^2np^1)	IV (ns^2np^2)	V (ns^2np^3)	VI (ns^2np^4)	VII (ns^2np^5)	Noble (ns^2np^6)
H 1 $1s^1$ 1.0079																	**He** 2 $1s^2$ 4.0026
Li 3 $2s^1$ 6.941	**Be** 4 $2s^2$ 9.01218											**B** 5 $2s^22p^1$ 10.81	**C** 6 $2s^22p^2$ 12.011	**N** 7 $2s^22p^3$ 14.0067	**O** 8 $2s^22p^4$ 15.9994	**F** 9 $2s^22p^5$ 18.9984	**Ne** 10 $2s^22p^6$ 20.179
Na 11 $3s^1$ 22.9898	**Mg** 12 $3s^2$ 24.305											**Al** 13 $3s^23p^1$ 26.9815	**Si** 14 $3s^23p^2$ 28.086	**P** 15 $3s^23p^3$ 30.9738	**S** 16 $3s^23p^4$ 32.06	**Cl** 17 $3s^23p^5$ 35.453	**Ar** 18 $3s^23p^6$ 39.948
K 19 $4s^1$ 39.098	**Ca** 20 $4s^2$ 40.08	**Sc** 21 $3d^14s^2$ 44.959	**Ti** 22 $3d^24s^2$ 47.90	**V** 23 $3d^34s^2$ 50.9414	**Cr** 24 $3d^54s^1$ 51.996	**Mn** 25 $3d^54s^2$ 54.938	**Fe** 26 $3d^64s^2$ 55.847	**Co** 27 $3d^74s^2$ 58.9332	**Ni** 28 $3d^84s^2$ 58.70	**Cu** 29 $3d^{10}4s^1$ 63.546	**Zn** 30 $3d^{10}4s^2$ 65.38	**Ga** 31 $4s^24p^1$ 69.72	**Ge** 32 $4s^24p^2$ 72.59	**As** 33 $4s^24p^3$ 74.9216	**Se** 34 $4s^24p^4$ 78.96	**Br** 35 $4s^24p^5$ 79.904	**Kr** 36 $4s^24p^6$ 83.80
Rb 37 $5s^1$ 85.4678	**Sr** 38 $5s^2$ 87.62	**Y** 39 $4d^15s^2$ 88.9059	**Zr** 40 $4d^25s^2$ 91.22	**Nb** 41 $4d^45s^1$ 92.9064	**Mo** 42 $4d^55s^1$ 95.94	**Tc** 43 $4d^55s^2$ (97)	**Ru** 44 $4d^75s^1$ 101.07	**Rh** 45 $4d^85s^1$ 102.905	**Pd** 46 $4d^{10}$ 106.4	**Ag** 47 $4d^{10}5s^1$ 107.868	**Cd** 48 $4d^{10}5s^2$ 112.40	**In** 49 $5s^25p^1$ 114.82	**Sn** 50 $5s^25p^2$ 118.69	**Sb** 51 $5s^25p^3$ 121.75	**Te** 52 $5s^25p^4$ 127.60	**I** 53 $5s^25p^5$ 126.904	**Xe** 54 $5s^25p^6$ 131.30
Cs 55 $6s^1$ 132.905	**Ba** 56 $6s^2$ 137.33	**La*** 57 $5d^16s^2$ 138.905	**Hf** 72 $4f^{14}5d^26s^2$ 178.49	**Ta** 73 $5d^36s^2$ 180.948	**W** 74 $5d^46s^2$ 183.85	**Re** 75 $5d^56s^2$ 186.207	**Os** 76 $5d^66s^2$ 190.2	**Ir** 77 $5d^76s^2$ 192.22	**Pt** 78 $5d^96s^1$ 195.09	**Au** 79 $5d^{10}6s^1$ 196.967	**Hg** 80 $5d^{10}6s^2$ 200.59	**Tl** 81 $6s^26p^1$ 204.37	**Pb** 82 $6s^26p^2$ 207.19	**Bi** 83 $6s^26p^3$ 208.980	**Po** 84 $6s^26p^4$ (209)	**At** 85 $6s^26p^5$ (210)	**Rn** 86 $6s^26p^6$ (222)
Fr 87 $7s^1$ (223)	**Ra** 88 $7s^2$ (226)	**Ac**** 89 $6d^17s^2$ (227)	**Rf** 104 (260)	**Ha** 105 (260)													

* Lanthanides ~ $4f^n5d^{0-1}6s^2$

Ce 58 $4f^15d^16s^2$ 140.12	**Pr** 59 $4f^35d^06s^2$ 140.907	**Nd** 60 $4f^45d^06s^2$ 144.24	**Pm** 61 $4f^55d^06s^2$ (145)	**Sm** 62 $4f^65d^06s^2$ 150.35	**Eu** 63 $4f^75d^06s^2$ 151.96	**Gd** 64 $4f^75d^16s^2$ 157.25	**Tb** 65 $4f^95d^06s^2$ 158.925	**Dy** 66 $4f^{10}5d^06s^2$ 162.50	**Ho** 67 $4f^{11}5d^06s^2$ 164.930	**Er** 68 $4f^{12}5d^06s^2$ 167.26	**Tm** 69 $4f^{13}5d^06s^2$ 168.934	**Yb** 70 $4f^{14}5d^06s^2$ 173.04	**Lu** 71 $4f^{14}5d^16s^2$ 174.97

** Actinides ~ $5f^n6d^{0-1}7s^2$

Th 90 $5f^06d^27s^2$ 232.038	**Pa** 91 $5f^26d^17s^2$ (231)	**U** 92 $5f^36d^17s^2$ 238.03	**Np** 93 $5f^46d^17s^2$ (237)	**Pu** 94 $5f^66d^07s^2$ (244)	**Am** 95 $5f^76d^07s^2$ (243)	**Cm** 96 $5f^76d^17s^2$ (247)	**Bk** 97 $5f^96d^07s^2$ (247)	**Cf** 98 $5f^{10}6d^07s^2$ (251)	**Es** 99 $5f^{11}6d^07s^2$ (254)	**Fm** 100 $5f^{12}6d^07s^2$ (257)	**Md** 101 $5f^{13}6d^07s^2$ (258)	**No** 102 $5f^{14}6d^07s^2$ (255)	**Lr** 103 $5f^{14}6d^17s^2$ (260)

f-Transition Elements

Symbol / Atomic number / Outer configuration / Atomic weight:
H 1 $1s^1$ 1.0079

TABLE 7.11 *The modern periodic table*
The atomic number, atomic weight, and electronic configuration are given here for each element. The information given in a periodic table varies with the purpose of the table and the amount of space available. For three pairs of elements (Ar and K; Co and Ni; Te and I) the element with the larger atomic weight precedes the element with the smaller atomic weight. This is the result of varying isotope abundances. Values in parentheses are for longest-lived isotopes.

Elements with each of the *four basic types of electronic configuration* are grouped together in the modern periodic table. These four basic configurations are shown together in Table 7.12 and defined as follows.

a. Noble gas configuration. All of the energy levels that are occupied by electrons are completely filled. These elements, with the exception of helium, have ns^2np^6 configurations, where n is the number of the outermost shell; helium, the lightest of the noble gases has a $1s^2$ electronic configuration.

b. Representative element configuration. The outermost shell of electrons is incomplete and all occupied energy levels underlying this shell are filled to capacity with electrons. In these elements only s and p energy levels of the outermost shell are being occupied, and configurations varying from ns^1 to ns^2np^5 are observed. In a sense, the noble gases may be regarded as the terminal elements of the various periods (horizontal series) of representative elements.

The groups of representative elements are given Roman numerals, starting at the left in the periodic table with I for Li–Fr. (Hydrogen stands alone and is not a member of any group.) In this book we refer to Representative Group I, Representative Group II (Be–Ra), and so on. (In many books representative element groups are indicated by the letter A after the Roman numeral; Be–Ra would be Group IA. The letter B is then used for some of the transition element families.)

Across each period of representative elements there is a change from metallic to nonmetallic properties for the elements. This reflects the increase in the number of electrons in the outermost energy levels of the atoms. Metal atoms have a small number of easily removed outer electrons, while nonmetal atoms have more electrons in the outer shells, but these electrons are held more firmly than in metals. The semiconducting elements fall on a zig–zag diagonal line across the middle of the representative element periods, and in terms of their properties these elements are intermediate between metals and nonmetals (see Table 10.1).

TABLE 7.12

The four basic types of electronic configurations

	Inner levels	Outermost level (n)	Outer configuration
Noble gas elements e.g., **He, Ne, Kr**	**Filled**	s **or** s **and** p **filled**	ns^2np^6 (except He, $1s^2$)
Representative elements e.g., **Na, Sr, Br**	**Filled**	s **and** p **being filled**	ns^1 to ns^2np^5
d-**Transition elements** e.g., **Sc, Pd, W**	$(n-1)d$ **being filled, others filled**	s^1 **or** s^2	$(n-1)d^{1-10}ns^{1,2}$
f-**Transition elements** e.g., **Nd, Tm, Pu**	$(n-2)f$ **being filled,** $(n-1)d$ **or** d^2, **others filled**	s^2	$(n-2)f^{1-14}$ $(n-1)d^{0-2}ns^2$

c. d-Transition element configuration. An inner *d* level is being occupied by electrons. Elements with this configuration are also called simply the transition elements. Atoms of these elements generally have the outer configurations $(n - 1)d^{1-10}ns^{1-2}$, with all other energy levels that are occupied filled to capacity. The *d*-transition elements are all metals.

d. f-Transition element configuration. An *f* level in the second from the outermost shell is being occupied. Elements of this configuration are also called the inner transition elements. In atoms of such elements the outer configurations are $(n - 2)f^{1-14}(n - 1)d^{0-1}ns^2$ with all other energy levels that are occupied being completely filled. The *f*-transition elements are all metals.

Electronic configurations determine the nature of the bonds that hold atoms together in molecules and ions, the crystal and molecular structures of pure substances, and the physical and chemical properties of the elements and their compounds. Much of what remains in this book relates directly or indirectly to electronic configurations.

THOUGHTS ON CHEMISTRY

Quantum behavior

Because atomic behavior is so unlike ordinary experience, it is very difficult to get used to, and it appears peculiar and mysterious to everyone—both to the novice and to the experienced physicist. Even the experts do not understand it the way they would like to, and it is perfectly reasonable that they should not, because all of direct, human experience and of human intuition applies to large objects. We know how large objects will act, but things on a small scale just do not act that way. So we have to learn about them in a sort of abstract or imaginative fashion and not by connection with our direct experience. . . .

We would like to emphasize a very important difference between classical and quantum mechanics. We have been talking about the probability that an electron will arrive in a given circumstance. We have implied that in our experimental arrangement (or even in the best possible one) it would be impossible to predict exactly what would happen. We can only predict the odds! This would mean, if it were true, that physics has given up on the problem of trying to predict exactly what will happen in a definite circumstance. Yes! physics has given up. We do not know how to predict what would happen in a given circumstance, and we believe now that it is impossible—that the only thing that can be predicted is the probability of different events. It must be recognized that this is a retrenchment in our earlier ideal of understanding nature. It may be a backward step, but no one has seen a way to avoid it. . . . So at the

THE FEYNMAN LECTURES ON PHYSICS, *by Richard P. Feynman, Robert B. Leighton, and Matthew Sands*

[7.19] ATOMIC STRUCTURE

present time we must limit ourselves to computing probabilities. We say "at the present time," but we suspect very strongly that it is something that will be with us forever—that it is impossible to beat that puzzle—that this is the way nature really is.

THE FEYNMAN LECTURES ON PHYSICS, *by Richard P. Feynman, Robert B. Leighton, and Matthew Sands* (Vol. 3, Addison-Wesley, Reading, Mass., 1965)

Significant terms defined in this chapter

gas-discharge tube	atomic number
electrode	mass number
fundamental particle	neutron number
subatomic particle	isotopes
electron	quantized
cathode rays	quantum numbers
proton	photoelectric effect
α-particle	photon
neutron	empirical relationship
nucleus	momentum
radiation	stationary state
wavelength	ground state
frequency	excited state
spectrum	Heisenberg uncertainty principle
wave number	atomic orbital
continuous spectrum	electronic configuration
line spectrum	paramagnetism
band spectrum	diamagnetism
emission spectrum	periodic law
absorption spectrum	period
spectrometer	group
mass spectrometer	family

Exercises

7.1 Each of the following scientists made a major contribution to the understanding of atomic structure. For each, briefly describe the experiment or theory and briefly interpret the significance of the results: (a) William Crookes, (b) Johann Balmer, (c) Eugen Goldstein, (d) J. J. Thomson, (e) Max Planck, (f) Albert Einstein, (g) Robert A. Millikan, (h) Ernest Rutherford, (i) Niels Bohr, (j) Henry Moseley, (k) Louis de Broglie, (l) Erwin Schrödinger, (m) Werner Heisenberg, (n) C. J. Davisson and L. H. Germer, and (o) James Chadwick.

7.2 Distinguish clearly between (a) the properties of electrons, neutrons, protons, and α particles; (b) emission and absorption spectra; (c) continuous and line spectra; (d) atomic number and atomic weight; (e) atomic number, mass number, and neutron number; (f) particle and wave nature of electrons; (g) stationary, excited, and ground states; (h) orbit and orbital; (i) paramagnetism and diamagnetism; (j) a period and a group or family in the period table; (k) wavelength, wave number, and frequency; (l) canal rays and cathode rays; (m) fundamental and subatomic particles; and (n) quantized and nonquantized properties.

7.3 Which statements are true? Rewrite any false statement so that it is correct.
(a) The properties of elements are periodic functions of their atomic numbers.
(b) Cathode rays are streams of electrons flowing from the cathode toward the anode in a gas-discharge tube.
(c) The distance between any two points in the same place on adjacent waves is the wavelength.
(d) The atomic number gives the position of the element in the periodic table and is equal to the positive charge on the nucleus of each atom of that element.
(e) Wave mechanics is that part of quantum theory that deals with wavelength properties of particles.
(f) An electronic configuration of an atom is a description of the orbitals used by all the electrons in the atom.
(g) Paramagnetism is the attraction of a substance toward a magnetic field as a result of unpaired electrons.
(h) Quantum theory allows us to simultaneously predict the exact location and momentum of an electron.

7.4 What property of atoms did both Mendeleev and Meyer use to organize the elements in their periodic tables? How was this changed by the results obtained by Moseley? How did Mendeleev extend our chemical knowledge using his table?

7.5 Choose from the following list the symbols that (a) make up a group of isotopes of an element, (b) have the same number of neutrons, and (c) have the same mass number (4 different sets): (i) ^{12}N, (ii) ^{13}B, (iii) ^{13}N, (iv) ^{14}C, (v) ^{14}N, (vi) ^{15}N, (vii) ^{16}N, (viii) ^{16}O, (ix) ^{17}N, (x) ^{17}F, and (xi) ^{18}Ne.

7.6 Complete the following table:

Elemental symbol	Number of protons	Number of neutrons	Number of electrons	Charge on species	Name
$^{41}_{20}Ca^{2+}$					
$^{190}_{78}Pt$					
$^{223}_{87}Fr$					
$^{139}_{53}I^{-}$					
$^{3}_{2}He^{2+}$					
		7	6		Carbon
	14	15		0	
		18	18		Sulfur
		30	24		Iron
		118		+3	Gold

7.7 What are the four major sections of a spectrometer? Briefly describe the purpose of each.

7.8 State the postulates of the Bohr theory. Why is this theory *not* today's working model of the atom?

7.9 Which situations illustrate quantization; (a) the length of a rope, (b) the energy of an electron, (c) seating in an auditorium, (d) the wavelength of light, (e) seating in a classroom containing movable chairs, (f) grains of corn?

7.10 Prepare sketches for the (a) s, (b) p_x, (c) p_y, (d) p_z, (e) d_{z^2}, (f) $d_{x^2-y^2}$, (g) d_{xy}, (h) d_{yz}, and (i) d_{xz} atomic orbitals.

7.11 Complete the following table of quantum numbers for atomic orbitals:

Symbol	Name	Permitted values	Physical interpretation
	Magnetic quantum number		
			Angular momentum or "spin" of electron
l			
		1,2,3, . . .	

7.12 What values can m_l take for a (a) $4d$ orbital, (b) $1s$ orbital, and (c) a $3p$ orbital?

7.13 Using the rules given in Section 7.18, write reasonable electronic configurations for the atoms (a) K, (b) Sc, (c) Si, (d) F, (e) U, (f) Ag, (g) Ge, (h) Ca, (i) B, (j) C, (k) Cr, (l)

Zn, (m) P, (n) Ne, (o) Al, (p) Mo, (q) I, (r) Cu, (s) Mn, (t) Au, (u) Mg, (v) Fe, (w) Pr, (x) Sn, (y) Ga, and (z) W. Compare your answers to the known configurations given in Table 7.10. How many are different and why? Which of these elements are paramagnetic?

7.14 Using the electronic configurations you wrote for the atoms of the elements in Exercise 7.13, predict (a) what positively charged ions K, Sc, Ag, Ca, Zn, Al, Mg, Fe, and Sn might form by the loss of the outermost electrons and (b) what negatively charged ions F, C, P, and I might form by the gain of electrons to complete the outer shell.

7.15 Write reasonable electronic configurations for (a) K^+, (b) Fe^{2+}, (c) Fe^{3+}, (d) Pb^{2+}, (e) Cd^{2+}, (f) P^{3+}, (g) F^-, (h) O^{2-}, (i) P^{5+}, (j) Cr^{3+}, (k) Zn^{2+}, (l) S^{2-}, (m) Se^{2-}, (n) N^{3-}, (o) H^+, and (p) H^-. Which of these are diamagnetic?

7.16 In a suitable reference such as the Table of Isotopes in the *Handbook of Chemistry and Physics* (The Chemical Rubber Co.) look up the following information for Kr: (a) the number of known isotopes, (b) the average atomic weight, and (c) the percentage of natural abundance of each of the isotopes.

7.17* Write the electronic configuration for the Cu atom using the order of filling predicted by the Aufbau principle. Because a configuration of d^{10} is very favorable, one of the $4s$ electrons is used to complete the $3d$ subshell. Write the electronic configuration for this alternative arrangement of electrons. Assuming the loss of the outermost electrons, the important oxidation states of Cu can be predicted using both of these configurations: What are they? Do the different configurations predict a difference in the magnetic properties of Cu?

7.18* In what respects does our solar system represent a Bohr atom having nine electrons? How does it differ from the fluorine atom?

7.19 The boiling point of PH_3 is $-87.7°C$, GeH_4 is $-88.5°C$, H_2Se is $-41.5°C$, and SbH_3 is $-17.1°C$. Predict the boiling point of AsH_3. Compare your predicted value to the observed value of $-55°C$.

7.20 Find the average atomic weight of each of the following elements using the following percent of natural abundances and actual atomic weights of the isotopes: (a) Li, which contains 7.42% of ^6Li (6.01512 amu) and 92.58% of ^7Li (7.01600 amu); (b) Mg, which contains 78.70% of ^{24}Mg (23.98504 amu), 10.13% of ^{25}Mg (24.98584 amu), and 11.17% of ^{26}Mg (25.98259 amu); and (c) Sr, which contains 0.56% of ^{84}Sr (83.9134 amu), 9.86% of ^{86}Sr (85.9094 amu), 7.02% of ^{87}Sr (86.9089 amu), and 82.56% of ^{88}Sr (87.9056 amu).

7.21 Using the Rydberg equation for atomic hydrogen, calculate $\bar{v}$ for an atom as an electron changes from the $n = 3$ state to the $n = 1$ state. What wavelength does this value of $\bar{v}$ correspond to? Is this energy change in the visible region of the spectrum?

7.22 Find the de Broglie wavelength of (a) a 1.2-g bullet moving with a velocity of 1.5×10^4 cm/sec, (b) a 2.6 ton automobile moving with a velocity of 55 miles/hr, and (c) an electron ($m_e = 9.11 \times 10^{-31}$ kg) moving with a velocity of 3.00×10^7 m/sec.

7.23 The approximate radius of a hydrogen atom is 0.58 Å and of a proton is 1.5×10^{-15} m. Calculate the fraction of atomic volume occupied by the nucleus assuming both to be spherical.

7.24* The approximate radius of a neutron is 1.5×10^{-15} m. From the mass of a neutron, 1.67×10^{-27} kg, find the density of a neutron.

7.25* Each of the ions formed by the elements in Representative Group I of the periodic table has a spectral line in the visible region of the spectrum, which can be used to identify the element. Using the following data for the most intense lines, predict what color would be observed (violet, 400–450 nm; blue, 450–510 nm; green, 510–550 nm; yellow, 550–590 nm; orange, 590–620 nm; red, 620–700 nm): (a) Li, $\lambda = 6708$ Å for $2p \rightarrow 2s$; (b) Na, $\bar{v} = 1698$ cm^{-1} for $3p \rightarrow 3s$; (c) K, $v = 3.90 \times 10^{14}$/sec for $4p \rightarrow 4s$ and 7.41×10^{14}/sec for $5p \rightarrow 4s$; (d) Rb, $\lambda = 7.9 \times 10^{-7}$ m for $5p \rightarrow 5s$ and 4.2×10^{-7} m for $6p \rightarrow 5s$; and (e) Cs, $\bar{v} = 3.45 \times 10^{14}$ Hz for $6p \rightarrow 6s$ and 6.53×10^{14} Hz for $7p \rightarrow 6s$.

7.26* Predict the melting point, boiling point, and density of cesium and the density of cesium oxide using the following data. For each element the density of the element is the third entry in the column and the density of its oxide is the fourth entry in the column.

K	Ca	Sc
63.65°C	839°C	1539°C
774°C	1484°C	2832°C
0.86 g/ml	1.54 g/ml	2.992 g/ml
2.32 g/ml	3.38 g/ml	3.864 g/ml

Rb	Sr	Y
38.89°C	769°C	1523°C
688°C	1384°C	3337°C
1.532 g/ml	2.6 g/ml	4.34 g/ml
3.72 g/ml	4.7 g/ml	5.01 g/ml

	Cs	Ba	La
melting point		725°C	920°C
boiling point		1640°C	3454°C
density of element		3.51 g/ml	6.194 g/ml
density of oxide		5.72 g/ml	6.51 g/ml

7.27 The average atomic weight of chlorine is 35.453 amu. There are only two isotopes in naturally occurring chlorine—^{35}Cl, at. wt. = 34.96885 amu and ^{37}Cl at. wt. = 36.96712 amu. Calculate the percent composition of naturally occurring chlorine.

7.28** At any given time, there is probably less than one ounce of francium in the crust of the Earth. There is only one isotope that occurs in nature ($^{223}_{87}Fr$) although there are 20 isotopes known for this element ($^{204}_{87}Fr$ to $^{224}_{87}Fr$). How many (a) electrons, (b) protons, and (c) neutrons are in ^{223}Fr? (d) How do these numbers change for the other isotopes? (e) Write the electronic configuration for the ground state of Fr and predict (f) reasonable oxidation states for the element based on electron gain or removal, and (g) whether it is paramagnetic or diamagnetic. (h) Repeat questions (e) and (g) for the major ion of Fr.

The two major electronic transitions in Rb^+ are $5p \rightarrow 5s$ (7900 Å) and $6p \rightarrow 5s$ (4201 Å) and in Cs^+ they are $6p \rightarrow 6s$ (8700 Å) and $7p \rightarrow 6s$ (4593 Å). (i) Predict the major transitions in Fr^+ and the corresponding wavelengths. (j) Will either of these be observed by the human eye?

Look up the formulas of the (k) chlorides, (l) oxides and (m) hydrides of the Representative Group I metals and (n) predict the formulas of the corresponding compounds of Fr. (o) What is the name usually given to this group of elements?

The melting points of several elements near Fr in the table are 38.89°C for Rb, 769°C for Sr, 1523°C for Y, 28.40°C for Cs, 725°C for Ba, 920°C for La, 700°C for Ra, and 1050°C for Ac. (p) Predict the melting point of Fr from these data.

8 NUCLEAR CHEMISTRY

The story of the discovery of radioactivity and the exploitation of nuclear energy holds more high drama than any other area of science. We think of Pierre and Marie Curie, reducing several tons of ore to a few specks of previously unknown radioactive elements. And carrying out this laborious task in a shed once used by a medical school for dissecting cadavers, a shed freezing and damp in the winter, and hot and stinking in the summer. There are the honors that came to Pierre and Marie and to their daughter, Irène, and Irène's husband Frédéric Joliot-Curie who carried on the research; and there are the deaths of both Marie and Irène from leukemia, quite likely caused by radiation from their own experiments.

World War II thrust a generation of physicists, chemists, and engineers into what was for many of them a moral dilemma—rushing to build a weapon that, while certain to end the war, would unleash a force too terrible to comprehend. But there was also drama and excitement in the triumph of pure science surrounding the development of the bomb. For example, there was the day when Enrico Fermi supervised the start-up of the first nuclear chain reaction, when he ordered what he calculated to be the last control rod drawn out and then announced with a smile, "The reaction is self-sustaining."

Drama of another kind was there when the first bomb was exploded over the New Mexico desert, and J. Robert Oppenheimer, the brooding, intellectual physicist who had overseen the birth of the bomb, remarked that at the moment of the explosion there came to his mind these words from the Bhagavad Gita, the sacred book of the Hindus, "I am become Death, the Shatterer of Worlds."

Many benefits to mankind have developed from the application of radioactivity in medicine and industry, and from the utilization of nuclear energy for power. But, 30 years after the beginning of the nuclear

In this chapter we look first at the forces within the nucleus, the conditions under which a nucleus becomes unstable and radioactive, and the relationship of nuclear stability to the abundance of the elements. Next we survey the types of nuclear changes—α, β, and γ decay—of natural radioactive elements; the bombardment reactions by which many radioactive isotopes and the synthetic elements can be synthesized; and fission and fusion. Last, we describe the nuclear power plants in use at the present time and those that may become important in the future.

age, the horror of nuclear weapons seems to be fading from men's minds, nations continue to test and stockpile enough nuclear weapons to wipe each other out many times over, and two Japanese Nobel prize-winning physicists have made this plea: "The gravity of the nuclear peril is such that we must work to achieve nuclear disarmament at the earliest possible time. We appeal to the people of the world, and especially to scientists and engineers, to join us in this urgent undertaking before it is too late. We insist as a first step that all governments renounce forever and without condition the threat or use of nuclear weapons." (Sin-Itiro Tomonaga and Hideki Yukawa, in a statement presented at a symposium on "A New Design To Complete Nuclear Disarmament," Kyoto, Japan, August 28–September 1, 1975.)

Nuclear stability and instability

8.1 The nucleus

To review briefly (see Sections 7.9 and 7.10), the nucleus of an atom contains protons equal in number to the atomic number of the element (Z) plus neutrons equal in number to the difference between the mass number (A) and the atomic number. Atoms with the same number of protons but different numbers of neutrons are isotopes of the same element. **Nuclide** is a general term used to refer to any isotope of any element.

Protons and neutrons, collectively called **nucleons**, are packed tightly together in a nucleus, and there is little free space. As a result, the volume of a nucleus is directly proportional to its mass. Most nuclei are either spherical or slightly football-like in shape. Nuclear radii are generally slightly less than 10^{-12} cm—about 10,000 times smaller than atomic radii.

The force of electrical attraction or repulsion between charged particles is the **Coulomb force,** also called the Coulomb interaction. Particles of opposite charge attract each other and particles of the same charge repel each other. The Coulomb force between two particles is a function of the charges on the particles and the distance between them

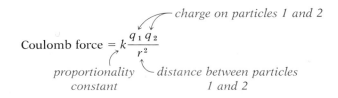

$$\text{Coulomb force} = k\frac{q_1 q_2}{r^2}$$

The **nuclear force** is the force of attraction between nucleons; it acts between protons, between neutrons, or between protons and neutrons. The force that can hold positively charged protons and uncharged neutrons so closely together in the absence of any negative charge is very different from and much stronger than the Coulomb forces between electrons and protons. In fact, the nuclear force is the strongest force yet to be discovered.

However, the exact nature of the nuclear force is not yet known. Here, it is of interest mainly that the nuclear force is very strong (30–40 times the Coulomb repulsion at 1×10^{-13} cm) and very short range, extending only to a distance of about 10^{-13} cm.

Figure 8.1 shows the potential energy changes as a proton approaches a neutron or a proton. At distances greater than 10^{-13} cm two protons repel each other. Once they reach this distance the nuclear force of attraction takes over. A neutron and a proton neither repel nor attract each other until they reach the distance at which the nuclear force can act.

8.2 Nuclear binding energy

The equivalence of mass and energy was formulated by Albert Einstein in the now-famous equation

$$E = mc^2 \tag{1}$$

where m is the *mass (kg)*, c is the *velocity of light (m/sec)*, and E is the *energy (joules = kg m²/sec²)*.

It is in dealing with nuclear forces and subatomic particles that we have to reckon most seriously with this relationship. The mass change in most chemical reactions is too small to be considered. In the formation of one mole of water from hydrogen and oxygen atoms, a mass loss of about 3×10^{-9} g occurs. And the mass loss in burning one gallon of gasoline, a process that releases 32,000 kcal, is only 1.5×10^{-6} g. For most purposes, such changes can be safely ignored, and mass is considered to be constant in chemical reactions.

In nuclear reactions, the quantities of energy released or absorbed and the mass losses or gains involved are much larger. Therefore, mass changes in nuclear reactions cannot be disregarded. However, the laws of conservation of mass and energy are not violated, as it may appear at first. Mass plus its equivalent in energy, or vice versa, energy plus its equivalent in mass, are conserved.

(a) Potential energy between pair of protons

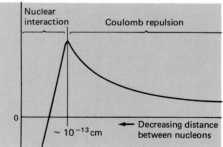

(b) Potential energy between proton and neutron

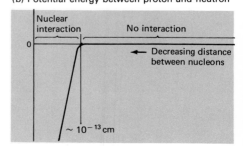

FIGURE 8.1

Potential energy barrier between nucleons. (*a*) *Coulombic repulsion acts between two protons until they reach a distance of about 10^{-13} cm. At this point the nuclear force of attraction takes over. (b) No forces act between a proton and a neutron (or between two neutrons) until a distance of 10^{-13} cm is reached.*

(Energy plus mass) is conserved

We have not found a way to combine protons and neutrons directly to form a nucleus. However, we define **nuclear binding energy** as the energy that would be released in the combination of nucleons to form a nucleus. The mass of an atom is always less than the total mass of the electrons, protons, and neutrons that form the atom (Figure 8.2). This difference in mass between the atom and its components taken separately represents the energy holding the nucleus together—the nuclear binding energy. (Recall that a system is more stable when energy is given up in its formation. Nuclear binding energy is analogous to the enthalpy of chemical reactions. However, it is customary to state nuclear binding energy as a positive number.)

The masses of neutrons and atomic hydrogen are used to calculate nuclear binding energy. This is equivalent to including electrons and calculating the energy released in forming the neutral atom. The difference that results from including the electrons is very small.

EXAMPLE 8.1

■ The actual atomic mass of $^{40}_{20}$Ca is 39.96259 amu. Find the binding energy for this nuclide, using 1.008665 amu for the mass of a neutron and 1.007825 amu for the mass of atomic hydrogen.

The change in mass in the formation of an atom of $^{40}_{20}$Ca from 20 neutrons and 20 atoms of hydrogen is

$$\Delta m = \text{(mass of nuclide)} - (20)(\text{mass of neutron})$$
$$- (20)(\text{mass of hydrogen})$$
$$= (39.96259 \text{ amu}) - (20)(1.008665 \text{ amu}) - (20)(1.007825 \text{ amu})$$
$$= -0.36721 \text{ amu}$$

To convert this mass in amu to mass in kilograms

$$(-0.36721 \text{ amu})(1.6605655 \times 10^{-27} \text{ kg/amu}) = -6.0978 \times 10^{-28} \text{ kg}$$

According to the Einstein mass–energy theory, the energy equivalent to this mass is found as follows:

$$E = mc^2$$
$$= (-6.0978 \times 10^{-28} \text{ kg})(2.9979 \times 10^8 \text{ m/sec})^2$$
$$= -5.4804 \times 10^{-11} \text{ kg m}^2/\text{sec}^2 = -5.4804 \times 10^{-11} \text{ J}$$

Usually energies are expressed in millions of electron volts (MeV) (see Appendix B) instead of joules when dealing with nuclear processes. Using the conversion factor 1 MeV = $1.6021892 \times 10^{-13}$ J gives us

$$(-5.4804 \times 10^{-11} \text{ J})(1 \text{ MeV}/1.6021892 \times 10^{-13} \text{ J}) = -342.06 \text{ MeV}$$

Because binding energy is considered positive, we have as our answer for the binding energy of $^{40}_{20}$Ca a value of 342.06 MeV. The conversion of mass changes in atomic mass units to energy in millions of electron volts is a routine calculation and the conversion factor 1

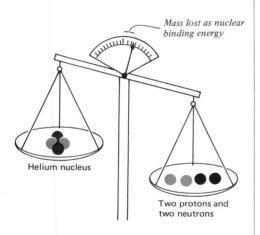

Mass lost as nuclear binding energy

Helium nucleus

Two protons and two neutrons

FIGURE 8.2
Nuclear mass vs. nucleon mass. *The mass of a nucleus is always less than the mass of the uncombined neutrons and protons. The difference in mass has been converted to nuclear binding energy. (This difference could not, of course, be weighed on a balance.)*

amu = 931.5017 MeV is commonly used. Thus for our nuclide, neglecting the negative sign,

$$(0.36721 \text{ amu})(931.5017 \text{ MeV/amu}) = 342.06 \text{ MeV} \qquad \blacksquare$$

The binding energy in one mole of the $^{40}_{20}$Ca atoms of the preceding example would be

$$5.4804 \times 10^{-11} \frac{J}{\text{atom}} \times \frac{1 \text{ kJ}}{10^3 \text{ J}} \times 6.0220 \times 10^{23} \frac{\text{atoms}}{\text{mole}}$$
$$= 3.3003 \times 10^{10} \text{ kJ/mole}$$

Compare this with the energy released in the complete combustion of one mole of acetylene, which is only 1.34×10^3 kJ. Nuclear reactions have energies on the order of one or more million times the energy of chemical reactions.

The **binding energy per nucleon** (also called average binding energy) is the nuclear binding energy of a nucleus divided by the number of nucleons in that nucleus. For example, for $^{40}_{20}$Ca from Example 8.1 the binding energy per nucleon is

$$\frac{342.06 \text{ MeV}}{40 \text{ nucleons}} = 8.5515 \text{ MeV/nucleon}$$

Binding energy per nucleon is more useful than nuclear binding energy for comparing the stability of one nucleus with that of another.

Elements beyond carbon have binding energies per nucleon that vary between about 7.5 and 8.8 MeV (Figure 8.3). The higher the binding energy, the more stable the nuclide. The highest values, and therefore the most stable nuclei, occur at intermediate mass numbers—between about 40 and 100—with the maximum for elements in the iron, cobalt, nickel region of the periodic table. Iron and nickel are among the more abundant elements in the universe, and

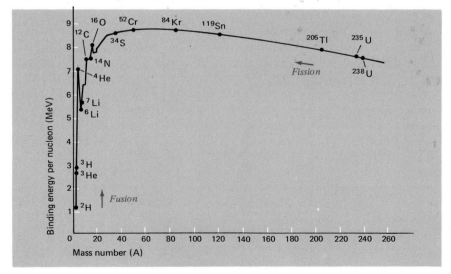

NUCLEAR CHEMISTRY [8.2]

FIGURE 8.3

The binding energy per nucleon vs. the mass number. *Nuclear fission and nuclear fusion bring nuclei into the area of greater stability by increasing the binding energy per nucleon.*

201

Nuclear fission:
splitting of a heavy nucleus
into two parts

Nuclear fusion:
combination of
two light nuclei

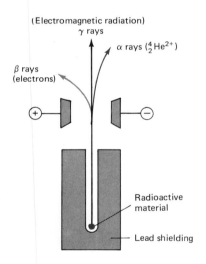

FIGURE 8.4

α, β, and γ rays in an electric field.
The α rays have a positive charge and are heavier (they are less deflected) than β rays, which have a negative charge. γ Rays are neutral and are unaffected by the field. While this sketch illustrates the behavior of all three types of rays, they are not ordinarily produced simultaneously by the same radioactive source.

their nuclear stability is a factor contributing to their abundance. For the heavier elements the binding energy per nucleon drops off slightly, to about 7.5 MeV.

Since the nuclei of elements of intermediate mass numbers are the most stable, the conversion of elements of smaller or larger mass number to elements in the middle of the curve in Figure 8.3 releases energy. **Nuclear fusion** is the combination of two light nuclei to give a heavier nucleus of intermediate mass number. **Nuclear fission** is the splitting of a heavy nucleus into two lighter nuclei of intermediate mass number; other particles may also be formed. In both fission and fusion, nuclei become more stable as they are transformed to nuclei in the middle of the curve in Figure 8.3. The total mass of the nuclei involved decreases, and energy is released in both processes.

8.3 Radioactivity

Radioactivity was discovered quite by accident in 1896 by a French physicist, Henri Becquerel. He was studying a uranium salt (potassium uranyl sulfate, $K_2SO_4 \cdot UO_2SO_4 \cdot 2H_2O$) that was **phosphorescent**—it reemitted radiation when it was exposed to light and continued to emit radiation after it was taken out of the light. (A **fluorescent** substance reemits radiation when it is exposed to light but stops reemitting when the radiation source is removed.) Becquerel found that his uranium salt left its image on a photographic plate even when the plate was wrapped in black paper to prevent the visible phosphorescent glow from reaching it. Furthermore, the image was produced even when the salt was *not* phosphorescing due to previous exposure to sunlight. What Becquerel had observed was the effect of the emission of γ rays (gamma rays).

Radioactivity is the spontaneous emission by unstable nuclei of particles or electromagnetic radiation, or both. Isotopes that spontaneously decay in this way are called **radioactive isotopes** (or radioisotopes). **Natural radioactivity** is the decay of radioactive isotopes found in nature. **Artificial radioactivity** is the decay of man-made radioactive isotopes. **Stable isotopes** do not spontaneously decay.

In 1897 Marie Curie set out to search systematically for radioactivity. Among the elements known at that time, and with the methods of measurement available, she found only uranium and thorium to be radioactive. But a uranium-containing ore, pitchblende, was more radioactive than either the uranium or the thorium contained in it. Marie Curie correctly suspected that an undiscovered radioactive element might also be present in pitchblende. By what she termed a "long, arduous, and costly process" she and her husband Pierre isolated from several tons of raw material a few tenths of a gram of each of two previously unknown radioactive elements—polonium ($Z = 84$; named for Marie Curie's native Poland) and radium ($Z = 88$; named from the Latin for "ray").

The three types of emanations from natural radioactive elements were named after the first three letters of the Greek alphabet. These emanations are characterized by their relative masses, or lack of

202

mass, and their behavior in an electric field (Figure 8.4). α Rays (alpha rays) have a positive charge and a relatively large mass compared to β rays (beta rays), which are negatively charged. γ Rays are neutral and have no mass. The α rays are streams of α particles, which were identified by Rutherford as helium atoms stripped of their two electrons ($^4_2He^{2+}$). β Rays proved to be streams of electrons, and γ rays (gamma rays) are electromagnetic radiation at the high-energy short-wavelength end of the spectrum (see Tools of Chemistry: Electromagnetic Radiation and Spectra, in Chapter 7).

It was Rutherford and Frederick Soddy who concluded in 1902 that in the process of radioactive decay, one element is transformed, or *transmuted*, into another element. To announce this conclusion to the scientific community must have taken a special form of courage. It was contrary to what had been firmly believed since the time of Dalton—that atoms of one element cannot be transformed into atoms of another element. (Rutherford is reported to have remarked, "For Mike's sake, Soddy, don't call it transmutation. They'll have our heads off as alchemists.")

8.4 Isotopes

Four-fifths of the elements occur naturally as mixtures of isotopes, in some cases of just two (e.g., nitrogen, $^{14}_7N$ and $^{15}_7N$), in some cases of many isotopes (e.g., tin, which has 10 isotopes, ranging from $^{112}_{50}Sn$ to $^{124}_{50}Sn$). **A natural isotope** is found in nature and can be either stable or radioactive. An **artificial isotope** is a stable or radioactive isotope not found in nature and known only as a man-made element.

Roughly one-third of the elements have natural radioactive isotopes. All of the isotopes of elements heavier than bismuth ($Z > 83$) are radioactive. Almost all natural radioactive isotopes of measurable abundance are very slow to decay and have been here since the Earth was formed. The exceptions are tritium (hydrogen-3), which is very rare, and carbon-14. These two nuclides are continually being formed by cosmic ray bombardment of other nuclides. Primary cosmic rays are high-energy particles from outer space—mainly protons—that continually bombard the Earth. In the atmosphere they collide with atoms to produce secondary cosmic rays—composed of photons, electrons, and other particles.

Isotopes of the same element have essentially the same chemical and physical properties. This makes the separation of one from the other very difficult. Isotopes of the same element undergo all of the same chemical reactions. However, the speed, or *rate*, of the reactions may differ slightly. For a given element and a specific reaction, the difference in rate increases as the relative difference in mass number increases. Thus rate differences are greatest with isotopes of the lighter elements and are at a maximum with those of hydrogen ($A = 1, 2, 3$).

Sometimes these differences can be used in the separation of isotopes. For example, water molecules containing deuterium are decomposed by an electric current at a slower rate than normal water molecules. More hydrogen atoms than deuterium atoms go off as gas, leaving behind water containing a higher proportion of deuterium. Some separations can be achieved by gas–liquid exchange reactions in which the concentration of one isotope in one phase is favored. For instance the following reaction concentrates $^{15}_7N$ in solution.

$$^{15}_7NH_3(g) + {}^{14}_7NH_4{}^+(aq) \longrightarrow {}^{15}_7NH_4{}^+(aq) + {}^{14}_7NH_3(g)$$

Heavier isotopes must be separated by physical methods that depend upon small absolute differences in mass. Isolating a large enough amount of uranium-235 was one of the major obstacles to be overcome in building the first atomic bomb. The uranium isotopes are separated by taking advantage of the differences in effusion rates (Section 3.16) of gaseous uranium hexafluorides.

However, new and more economical isotope separation methods are being sought. Possibilities are gas-centrifuge separation, in which the heavier isotopes are spun to the outside in a powerful centrifuge, and laser separation. A *laser* is a device that produces an intense beam of coherent radiation—radiation of a single wavelength with all of the waves in step with each other. By the use of a finely tuned laser the atoms of one isotope can be excited. Then they can be converted to ions and separated by an electric field from the nonionized atoms of a similar isotope, or else they can be taken up in a chemical reaction that does not occur with the unexcited isotope.

8.5 Neutron–proton ratio

The ratio of neutrons to protons is related to the stability of the nucleus. A plot of the number of neutrons versus the number of protons for stable natural nuclides shows that for the lighter elements the neutron-to-proton ratio is $1:1$. For heavier nuclei, as shown by Figure 8.5, the number of neutrons increases faster than the number of pro-

FIGURE 8.5

Number of neutrons vs. number of protons for stable nuclei. *Nuclides above the stable nuclei curve undergo radioactive decay mainly by positron emission or electron capture. Nuclides below this curve decay mainly by electron emission.*

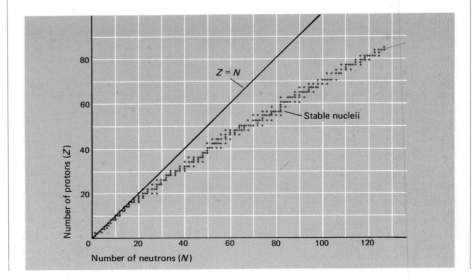

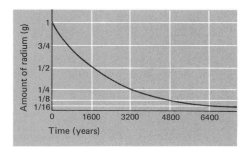

FIGURE 8.6
Radioactive decay curve for
$^{226}_{88}$**Ra.** *This isotope has a half-life of approximately 1600 yr.*

tons and the neutron–proton ratio finally reaches about 5:3. The additional neutrons apparently provide the additional nuclear force necessary to hold larger numbers of protons close together within the nucleus. Once the atomic number reaches 83, even extra neutrons are not sufficient to maintain stability and, as we have mentioned, all nuclides of $Z > 83$ are unstable and radioactive.

For each nuclear charge a neutron–proton ratio within a specific range is required for stability. Essentially, radioactivity is the transformation of unstable nuclei to nuclei with more favorable neutron–proton ratios. Nuclides with too many protons fall above the stable nuclei curve in Figure 8.5. Such nuclides decay so that the net result is conversion of a proton to a neutron. Nuclides with too many neutrons fall below the stable nuclei curve in Figure 8.5. Such nuclides decay so that the net result is the conversion of a neutron to a proton. Section 8.10 explains how these conversions take place.

8.6 Half-life

A convenient way of characterizing a radioactive isotope is by its **half-life**—the time it takes for one-half of the nuclei in a sample of a radioactive isotope to decay. For example, consider an isotope with a half-life ($t_{1/2}$) of ten years. Suppose 100 g of this isotope is put away today. Ten years from today 50 g of the original isotope will remain, along with the more stable decay products. Ten years after that, 25 g of the original isotope will be left, and so on.

A radioactive decay curve for radium-226, which has a half-life of approximately 1600 yr, is given in Figure 8.6. Half-lives of radioactive isotopes range from a few microseconds to as long as 10^{15} yr (Table 8.1).

Ingenious use has been made of carbon-14 to determine the age of plant and animal relics. The concentration of carbon-14 on Earth is kept relatively constant by a balance between its rate of formation by cosmic ray neutron (n) bombardment of nitrogen

$$^{14}_{7}\text{N} + n \longrightarrow {}^{14}_{6}\text{C} + {}^{1}_{1}\text{H}$$

and its rate of decay by emitting electrons (e^-)

$$^{14}_{6}\text{C} \longrightarrow {}^{14}_{7}\text{N} + e^-$$

TABLE 8.1

Some radioisotope half-lives
Sodium-24, iron-52, and cobalt-60 are produced in accelerators for use in medical treatment or diagnosis. Strontium-90, iodine-131, and cesium-137 have been introduced to the atmosphere as fallout from nuclear weapons testing.

Isotopes	Half-life
Radioisotopes produced on earth by cosmic rays	
$^{3}_{1}$H (tritium)	1226 yr
$^{14}_{6}$C	5770 yr
Natural radioisotopes	
$^{40}_{19}$K	1.3×10^9 yr
$^{144}_{60}$Nd	5×10^{15} yr
$^{232}_{90}$Th	1.39×10^{10} yr
$^{235}_{92}$U	7.13×10^8 yr
$^{238}_{92}$U	4.49×10^9 yr
Artificial radioisotopes	
$^{24}_{11}$Na	15 hr
$^{52}_{26}$Fe	8.3 hr
$^{60}_{27}$Co	5.26 yr
$^{90}_{38}$Sr	28 yr
$^{87}_{35}$Br	5.5 sec
$^{131}_{53}$I	8.1 days
$^{137}_{55}$Cs	30 yr
$^{239}_{94}$Pu	24,000 yr

As long as a plant or animal lives, it contains the same proportion of $^{14}_{6}C$ as its surroundings. But as soon as a plant stops utilizing CO_2 or an animal stops eating carbon-containing plants, its supply of $^{14}_{6}C$ is no longer replenished, and the ratio of radioactive $^{14}_{6}C$ to nonradioactive $^{12}_{6}C$ begins to decrease. From the half-life of carbon-14—5770 yr—and the measured ratio of $^{14}_{6}C/^{12}_{6}C$, the age of the plant or animal remains can be calculated (Section 15.11). This procedure is known as radiocarbon dating.

8.7 Cosmic abundance and nuclear stability

A plot of the cosmic abundance of the elements (Figure 8.7) reveals some interesting relationships. Except for hydrogen, elements with even atomic numbers are more abundant than their neighbors with

FIGURE 8.7

Relative cosmic abundances of the elements. *Silicon is widespread in the crust of the Earth and its abundance is used as a standard of comparison.* (*From L. H. Ahrens,* Distribution of the Elements in Our Planet, *p. 14, McGraw-Hill, New York, 1965.*)

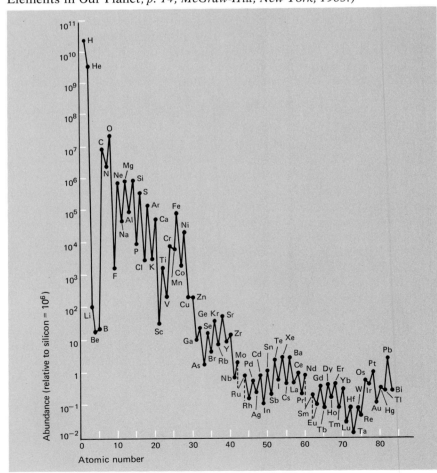

odd atomic numbers. And in general, light elements are more abundant than heavy elements.

If abundance is a reflection of stability—a reasonable assumption—then even numbers of protons and nuclear stability must be related in some way. Further examination reveals that of the ten *most* abundant elements (again with the exception of hydrogen) all have even numbers of *both* neutrons and protons. In addition, of the nuclides in the Earth's crust, 86% have even mass numbers.

Certain numbers of neutrons or protons—called **magic numbers**—impart particularly great nuclear stability. The magic numbers are 2, 8, 20, 28, 50, 82, and 126. For example, $^4_2$He, $^{16}_8$O, $^{40}_{20}$Ca, $^{88}_{38}$Sr, and $^{208}_{82}$Pb are very stable and abundant isotopes. The magic numbers were so named before enough was known about the nucleus to interpret their significance. Now, these relationships are explained by the existence of energy levels within the nucleus similar to electron energy levels outside of the nucleus. The magic numbers represent shells filled by nucleons. Nuclides with filled shells have greater nuclear stability, just as atoms with shells filled by electrons have greater chemical stability.

The elements originated, it is believed, in nuclear reactions within the stars. In stars like our sun, hydrogen is converted to helium by nuclear fusion. By successive fusion reactions, helium then yields the elements through iron and nickel. Because nuclear binding energy is at a maximum with the iron and nickel nuclides, continued fusion to produce heavier elements does not occur. The heavier elements are formed in smaller quantities by subsequent nuclear reactions of other types.

After the Earth was formed and before it cooled, natural geochemical processes concentrated iron and nickel in its core and the light elements in its crust and atmosphere. Thus, the portion of the Earth available to chemists and geologists offers an abundance and distribution pattern of the elements quite different from that in the universe.

Nuclear reactions:
(1) spontaneous decay
(2) bombardment
(3) fission
(4) fusion

Nuclear reactions

The following sections describe the various types of **nuclear reactions**—reactions that result in changes in the atomic number, mass number, or energy state of nuclei.

Nuclear reactions result from (1) the spontaneous decay of radioactive nuclides, either artificial or natural, (2) bombardment reactions, in which electromagnetic radiation or fast-moving particles are captured to form an unstable nucleus that subsequently decays, (3) the fission of unstable heavy nuclei, (4) the fusion of light nuclei, which happens spontaneously only in the sun and the stars. Although fission can and does sometimes happen spontaneously on Earth, such events are not very common (see An Aside: Nature's Nuclear Reactor). Fission is of greatest interest when it is carried out under man-made conditions that allow utilization of the energy released.

TABLE 8.2

Particles in nuclear reactions

These are the particles and the symbols most commonly encountered in writing nuclear reactions. The subscript is either the atomic number or the charge of the particle.

Name	Symbols
Alpha particle (helium nucleus)	α, $^4_2\alpha$, ^4_2He
Electron (beta minus particle)	β^-, $^0_{-1}\beta$, $^0_{-1}e$
Positron (beta plus particle)	β^+, $^0_{+1}\beta$, $^0_{+1}e$
Proton (hydrogen nucleus)	p, ^1_1p, ^1_1H
Neutron	n, ^1_0n
Neutrino	ν, $^0_0\nu$
Antineutrino	$\bar{\nu}$, $^0_0\bar{\nu}$

8.8 Writing equations for nuclear reactions

In order to write equations for nuclear reactions, we must add to our chemical notation some symbols for the particles that arise in nuclear reactions. These symbols are summarized in Table 8.2. Electrons, for example, have historically been known as beta particles in nuclear chemistry. They are often given the symbol β^- or, adding the "mass number" of an electron, which is zero, as a superscript and the charge on the electron as a subscript, the symbol $^0_{-1}\beta$. The neutrino and antineutrino, two massless chargeless particles, and the positron, the positively charged twin of the electron, are new to us here, and are encountered only in nuclear reactions. The reactions in which they are formed are discussed in following sections.

Mass number and often atomic number as well are written for all of the nuclides involved in a nuclear reaction. Including the mass number and either the atomic number or charge for the particles is not always done, but is helpful in writing and balancing nuclear equations.

Following are the equations for some nuclear reactions of historical interest:

First recognized natural transmutation of an element (Rutherford and Soddy, 1902)

$$^{226}_{88}\text{Ra} \longrightarrow {}^4_2\alpha + {}^{222}_{86}\text{Rn} \tag{2}$$

an atom of radium-226   *decays to*   *an α particle* + *an atom of radon-222*

First artificial transmutation of an element (Rutherford. 1919)

$$^{14}_7\text{N} + {}^4_2\alpha \longrightarrow {}^{17}_8\text{O} + {}^1_1p \tag{3}$$

an atom of nitrogen-14   *bombarded by α particles*   *yields*   *an atom of oxygen-17* + *a proton*

Discovery of the neutron (Chadwick, 1932)

$$^9_4\text{Be} + {}^4_2\alpha \longrightarrow {}^{12}_6\text{C} + {}^1_0n \tag{4}$$

an atom of beryllium-9   *bombarded by α particles*   *yields*   *an atom of carbon-12* + *a neutron*

Discovery of nuclear fission (Otto Hahn and Fritz Strassman, 1939)

$$^{235}_{92}\text{U} + {}^1_0n \longrightarrow {}^{141}_{56}\text{Ba} + {}^{92}_{36}\text{Kr} + 3{}^1_0n \tag{5}$$

an atom of uranium-235   *bombarded by a neutron*   *yields*   *an atom of barium-141* + *an atom of krypton-92* + *three neutrons*

Some of the species in nuclear reactions may be ions, for example, the α particle, which is $^4_2\text{He}^{2+}$. Usually, however, only the mass and atomic numbers are of concern in nuclear reactions, and the charges need not be included in the symbols. As a nuclear reaction proceeds, excited electrons within an atom undergo whatever rearrangements are necessary to return the atom to a stable state, and ions eventually pick up electrons to become neutral species.

A nuclear equation is correctly written when the following two rules are obeyed:

1. *Conservation of mass number.* The total number of protons and neutrons in the products must equal the total number of protons and neutrons in the reactants. This can be checked by comparing the sum of the mass numbers of the products with that of the reactants. For example, in Equation (4), $9 + 4 = 12 + 1$.

2. *Conservation of nuclear charge.* The total charge of the products must equal the total charge of the reactants. This can be checked by comparing the sum of the atomic numbers of the products with the sum of the atomic numbers of the reactants. For example, in Equation (4), $4 + 2 = 6 + 0$.

If the atomic numbers and mass numbers of all but one of the atoms or particles in a nuclear reaction are known, the unknown particle can be identified by using the rules given.

EXAMPLE 8.2

■ A nuclide of element 104 (for which the name rutherfordium has been suggested) with a mass number of 257, $^{257}_{104}\text{Rf}$, is formed by the nuclear reaction of $^{249}_{98}\text{Cf}$ with $^{12}_{6}\text{C}$ with the emission of four neutrons. This new nuclide has a half-life of about 5 sec and decays by emitting an α-particle. Write the equations for these nuclear reactions and identify the nuclide formed as $^{257}_{104}\text{Rf}$ undergoes decay.

The equation for the formation of $^{257}_{104}\text{Rf}$ is

$$^{249}_{98}\text{Cf} + ^{12}_{6}\text{C} \longrightarrow ^{257}_{104}\text{Rf} + 4^{1}_{0}n$$

Note that the totals of the mass numbers of the reactants and products are equal, $249 + 12 = 261 = 257 + 4(1)$, as are the totals of the atomic numbers, $98 + 6 = 104 = 104 + 4(0)$.

The equation for the radioactive decay of $^{257}_{104}\text{Rf}$ can be written as

$$^{257}_{104}\text{Rf} \longrightarrow ^{4}_{2}\text{He} + ^{A}_{Z}\text{E}$$

where E represents the unidentified element. The values of A and Z can be obtained from the equations for the totals of the mass numbers and of the atomic numbers:

$$257 = 4 + A \qquad 104 = 2 + Z$$

Solving these equations gives $A = 253$ and $Z = 102$. Thus the unknown product is $^{253}_{102}\text{No}$ and the complete equation is

$$^{257}_{104}\text{Rf} \longrightarrow ^{4}_{2}\text{He} + ^{253}_{102}\text{No}$$

■   NUCLEAR CHEMISTRY [8.8]

EXAMPLE 8.3

■ The last step in the uranium decay series (Section 8.13; Figure 8.11) is

$$^{210}_{84}\text{Po} \longrightarrow {}^{206}_{82}\text{Pb} + {}^{4}_{2}\text{He}$$

The respective masses of the nuclides are 209.9829 amu, 205.9745 amu and 4.00260 amu. Find the energy released by this reaction.

The change in mass is

$$\Delta m = (\text{mass of Pb}) + (\text{mass of He}) - (\text{mass of Po})$$
$$= 205.9745 \text{ amu} + 4.00260 \text{ amu} - 209.9829 \text{ amu}$$
$$= -0.0058 \text{ amu}$$

which gives

$$(-0.0058 \text{ amu})(931.5 \text{ MeV/amu}) = -5.4 \text{ MeV}$$

The energy released by each atom undergoing this reaction is 5.4 MeV. On a molar basis, this corresponds to

$$(-5.4 \text{ MeV/atom})(1.602 \times 10^{-13} \text{ J/MeV})(6.022 \times 10^{23} \text{ atoms/mole})$$
$$= -5.2 \times 10^{11} \text{ J/mole}$$

which is quite a bit of energy! ■

8.9 γ Decay

α Decay and β decay, described in the following two sections, are the more important routes by which radioactive nuclides decay spontaneously. Frequently, α or β decay leaves the nucleus above its ground-state nuclear energy level. To return to this most stable energy level the nucleus emits a γ ray, thereby giving up the excess energy in the form of electromagnetic radiation (see Figure 8.8). The wavelength range for x rays and γ rays is from roughly 1 to 0.001 nm. x Rays are at the longer wavelength end of this range and γ rays at the shorter wavelength end of the range. The terms refer specifically to the *origin* of the radiation—x rays from energy changes of electrons and γ rays from nuclear energy changes.

γ Rays travel with the speed of light and their energy differences are represented by the different frequencies of the radiation ($E = h\nu$). Further support for the existence of discrete nuclear energy levels is provided by γ-ray emission, in which only certain specific frequencies occur for specific radionuclides. There is no change of mass number or charge in γ decay.

indicates excited state

$$^{A}_{Z}\text{X}^{*} \longrightarrow {}^{A}_{Z}\text{X} + h\nu \qquad (6)$$

The *range* of radiation is the distance the radiation travels before losing all of its energy. γ Rays are the most penetrating type of natural radiation. Aluminum between 5- and 11-cm thick is required to stop γ rays with energies between 1 and 10 MeV, representing a penetration by γ rays that is 100 times that of β rays.

γ-Ray emission usually occurs within a nanosecond following an α or β decay—virtually at the same time. When the excited nuclide survives for a longer time it is called an "isomer" and its decay by delayed γ-ray emission is called an **isomeric transition** (see Table 8.4).

8.10 α Decay

α Decay occurs mainly in nuclides that are unstable because they are too heavy (those with $Z > 83$ and $A > 200$). **α Decay** is the emission of an α particle by a radioactive nuclide. The mass number is decreased by 4 units and the atomic number by 2, with the net result that the neutron–proton ratio increases.

$$_{Z}^{A}X \longrightarrow {}_{Z-2}^{A-4}X + {}_{2}^{4}\alpha \qquad (7)$$

Sometimes several α-particle emissions must take place before a stable nuclear mass and charge are achieved. α Particles emitted from one radionuclide have a single energy or one of a few specific energies (Figure 8.8), providing evidence for the existence of energy levels within the nucleus.

Emission of an α particle can leave a nuclide in an excited state that then emits γ radiation to reach the stable state. The total energy of the emitted α particle and the γ ray is equal to the reaction energy. For example, in Figure 8.8 the energy of α_1 plus γ_2 plus γ_3, or α_1 plus γ_1, equals $E_1 - E_2$, the total reaction energy.

The range of radiation varies with the medium. α-Particle radiation is not very penetrating. For 1–10-MeV particles, the range is several centimeters in air and on the order of 10^{-3} mm in aluminum. α Particles can be stopped by a single sheet of paper.

8.11 β Decay

There are three types of **β decay:** electron emission, positron emission, and electron capture.

a. Electron emission. The emission of an electron—a negative β particle—causes no change in the mass number of a nuclide, but *increases* the positive charge on the nucleus by 1 (Table 8.3). In electron emission the neutron–proton ratio decreases.

$$_{Z}^{A}X \longrightarrow {}_{Z+1}^{A}X + {}_{-1}^{0}\beta \qquad (8)$$

The $_{-1}^{0}\beta$ particles are emitted along a continuous spectrum of energies from zero to a maximum that is characteristic of the decaying nuclide.

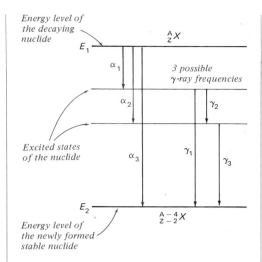

FIGURE 8.8
An α-decay scheme. *Isotope X yields α particles with three different energies and three different γ rays.*

TABLE 8.3

Net effect of β-decay reactions

An isolated neutron decays by the reaction shown here with a half-life of about 12 min. The other reactions do not occur in isolation from nuclei. The net effect of β decay is either the conversion of a neutron to a proton or the conversion of a proton to a neutron.

Electron emission
$$_{0}^{1}n \longrightarrow {}_{1}^{1}p + {}_{-1}^{0}\beta + {}_{0}^{0}\bar{\nu}$$

Positron emission
$$_{1}^{1}p \longrightarrow {}_{0}^{1}n + {}_{+1}^{0}\beta + {}_{0}^{0}\nu$$

Electron capture
$$_{1}^{1}p + {}_{-1}^{0}\beta \longrightarrow {}_{0}^{1}n + {}_{0}^{0}\nu$$

α particle = $_2^4\text{He}^{2+}$

β^- particle = electron

β^+ particle = positron

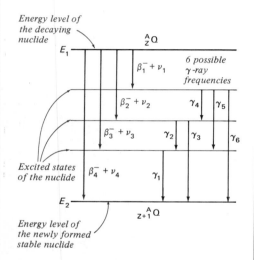

Energy level of the decaying nuclide

E_1

$_Z^A Q$

$\beta_1^- + \nu_1$

6 possible γ-ray frequencies

$\beta_2^- + \nu_2$ γ_4 | γ_5

$\beta_3^- + \nu_3$ γ_2 | γ_3 γ_6

Excited states of the nuclide $\beta_4^- + \nu_4$ γ_1

E_2

$_{Z+1}^A Q$

Energy level of the newly formed stable nuclide

FIGURE 8.9

A β-decay scheme. *Isotope Q decays by electron emission to give β⁻ rays with energies from 0 to $E_1 - E_2$ plus γ rays of six different frequencies.*

Explanation of radioactive decay by electron emission originally offered two difficulties. First, as there are no electrons in the nucleus, where does the electron come from? The answer is that the electron is created at the moment of its emission: a neutron within the nucleus decays to give a proton and an electron.

Second, how can the law of conservation of energy be obeyed if the emitted electrons carry away a continuous range of energies? Since electrons are so much lighter than nuclei, they must receive most of the reaction energy as kinetic energy. But β decay, like α decay, leaves nuclides in excited states that reach the more stable state by emission of γ rays with certain specific energies. Therefore, if one β particle can have, say, 10% of the reaction energy, and the next 11% of the reaction energy, and so on, a problem arises: The energy of the β particles, plus the specific energies of the γ rays do not equal the reaction energy. Where does the rest of the energy go? This dilemma was solved by the suggestion that a neutral massless particle that carries away the remaining energy is emitted simultaneously with an electron.

The existence of two such particles has now been proved—they are the **antineutrino** ($\bar{\nu}$), a massless chargeless particle that is emitted with an electron, and the **neutrino** (ν), a massless chargeless particle that is emitted with a positron (see below). For example, in Figure 8.9, the nuclide can be transformed to the highest excited state by the emission of β_1^- and ν_1, whose total energy equals the difference between E_1 and the first excited state. The energy of β_1^- plus $\bar{\nu}_1$, *plus* the appropriate combination of γ rays, equals the reaction energy.

β Rays are more penetrating than α rays; for example, a 3-MeV β particle can travel 10 m in air but is stopped by an aluminum sheet 0.5 mm thick. (A 3-MeV α ray is stopped by an aluminum sheet only 0.015 mm thick.)

b. Positron emission. **A positron** is a particle identical to an electron in all of its properties except for the charge, which is +1 rather than −1. Positrons were first detected during an investigation of cosmic rays. Soon afterward, in 1934, Frederic and Irène Joliot-Curie identified positrons in the radioactive decay of the artificial radionuclide phosphorus-30.

Positron, or $\beta+$, emission causes a decrease of 1 in the atomic number, but no change in the mass number, leading to an increase in the neutron–proton ratio

$$_Z^A\text{X} \longrightarrow _{Z-1}^A\text{X} + _{+1}^0\beta \tag{9}$$

Isotopes with too few neutrons (above the curve in Figure 8.5) decay in this way. Positrons, like electrons, are emitted with a continuous range of energies, and they are accompanied by neutrinos that carry away the remaining energy (Table 8.4). Only artificial radioisotopes have been observed to undergo positron emission.

When a positron and an electron interact, they annihilate each other—all of their mass is converted to energy in the form of two 0.51-MeV γ rays traveling in opposite directions.

c. Electron capture. The positive charge of an unstable nucleus

An m following the mass number superscript, as in the isomeric transition reaction, indicates an excited, or isomeric, state of an isotope. (Sometimes an asterisk is used to designate an excited state.) As is usually done, neutrinos and antineutrinos are omitted.

TABLE 8.4

Radioactive isotope decay

Reaction	Reaction symbol	Change in			Example
		A	Z	n/p	
Electron emission	β^-	0	+1	Decrease	$^{227}_{89}\text{Ac} \longrightarrow ^{227}_{90}\text{Th} + _{-1}^{0}\beta$
Positron emission	β^+	0	−1	Increase	$^{13}_{7}\text{N} \longrightarrow ^{13}_{6}\text{C} + _{+1}^{0}\beta$
Electron capture	EC	0	−1	Increase	$^{73}_{33}\text{As} + _{-1}^{0}\beta \longrightarrow ^{73}_{32}\text{Ge}$
α-Particle emission	α	−4	−2	Increase	$^{210}_{84}\text{Po} \longrightarrow ^{206}_{82}\text{Pb} + _{2}^{4}\alpha$
Isomeric transition	IT	0	0	—	$^{77\text{m}}_{34}\text{Se} \longrightarrow ^{77}_{34}\text{Se} + \gamma \text{ ray}$
Spontaneous fission	SF	—	—	—	$^{254}_{98}\text{Cf} \longrightarrow$ Intermediate mass nuclides + neutrons

can also be decreased by **electron capture**—the capture by the nucleus of one of its own inner shell electrons. The mass number is again unchanged while the neutron–proton ratio increases.

$$_{Z}^{A}\text{X} + _{-1}^{0}\beta \longrightarrow _{Z-1}^{A}\text{X} \qquad (10)$$

Here a proton captures an electron to produce a neutron and a neutrino (Table 8.3). As the electrons rearrange themselves to compensate for the electron pulled into the nucleus, x rays are emitted. β Decay and the other spontaneous nuclear decay reactions are summarized in Table 8.4.

8.12 Radiation and matter

In the decay of a radioactive nuclide, mass is lost and energy is generated. The energy is given up mainly as the kinetic energy of the particles and nuclei produced in the reaction. The energetic particles or nuclei collide with the atoms and molecules of their surroundings, gradually losing their energy. Most of this lost energy is converted to heat.

In some cases electrons in the surrounding atoms are excited in the collision and subsequently give up excess energy in the form of electromagnetic radiation—usually in the visible, ultraviolet, or x-ray regions. One type of luminescent paint takes advantage of this property. A nuclide that decays by α-particle emission is mixed with a substance that contains easily excited electrons, for example, zinc sulfide. At one time, luminescent radium-containing paint was widely used in watch and instrument dials. However, the radiation

FIGURE 8.10
Effects of radiation exposure. *The time after radiation exposure at which various effects, depending on the degree of exposure, can occur. Molecular changes in cells are caused both directly by the radiation and indirectly by free radicals formed by irradiation of water molecules. (From P. N. Tiwari,* Fundamentals of Nuclear Science, *p. 131, Wiley Eastern Limited, New Delhi, India, 1974.)*

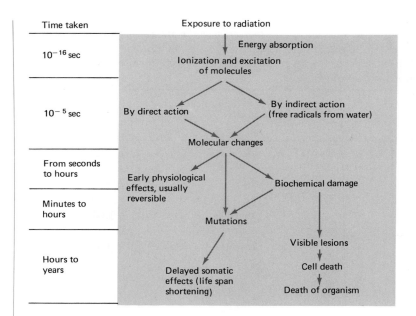

from radium is hazardous and other materials have been substituted for it. In the 1920s, many women painting watch dials developed bone cancer because they were in the habit of pointing the tips of their paintbrushes by licking them.

The energetic particles from a radioactive nuclide can also knock electrons out of the surrounding atoms, causing ionization. This is the process that is most damaging to plants and animals that are exposed to harmful radiation. α Rays, β rays, γ rays, x rays, neutrons, protons, and heavier charged particles all cause ionization, excitation, and chemical change in plant and animal cells. The irradiated organism or plant may undergo *somatic effects*—changes in its own cell structure, immediate or delayed, that may be damaging, but will not be passed on to future generations. It may also suffer *genetic effects*—changes in the genes that will produce physical changes in future generations.

How seriously an organism is damaged by any type of radiation depends on many factors, including the penetration of the radiation, the total dose received, the length of the exposure, the energy of the radiation, and the tissue irradiated. Figure 8.10 is a generalized scheme summarizing the possible effects of radiation exposure on plants and/or animals.

8.13 Natural radioactive series

Most of the heavy natural radioactive nuclides are members of one of three series. Each of these series has a long-lived parent and a stable isotope of lead as the end product. Decay from the parent to the final stable isotope proceeds through a sequence of reactions, most of which are α- or β-particle emission. The sequence for the series begin-

Uses of radioisotopes

Each distinctive property of radioisotopes has been put to use in dozens of ways in fields ranging from chemical synthesis, to medicine, to agriculture, to manufacturing and engineering.

We have already mentioned the dating of historical artifacts, a procedure based on the constancy of the rate of decay of isotopes. Many other applications arise simply because a radioisotope is continuously emitting radiation that can be detected and measured, making it possible to monitor the location of the isotope. Radioactive tracers have helped to follow the course of underground water supplies, and a large number of isotopes are used as diagnostic tools in medicine. Specific isotopes find their way in the body to specific organs, and by emitting radiation from these locations, make the organ visible to a radiation-detecting camera. For example, thallium-201 concentrates in normal heart muscle, allowing determination by the dark areas on the film of the amount of damage caused by a heart attack. Complementary information is obtained with technetium-99m, which is taken up by damaged heart cells, but not by healthy ones.

One of the first applications of radioisotopes was based on the ability of radiation to destroy cells—radiation therapy for cancer, in which the malignant cells are wiped out. In another application of this property, male insects have been irradiated to destroy their reproductive cells. For instance, millions of sterile flies were released near cattle farms, where they mated with female flies, who then produced no offspring. In this way, the screwworm fly, which does great damage to livestock, has been eliminated in several areas.

In chemistry and biochemistry, radioactive labeling of compounds has imparted much information about how reactions take place. By determining in which of the reaction products the radioactive atom winds up, the route of the atom through intermediate chemical and metabolic processes can be worked out. For example, isotopic labeling revealed that, contrary to what had been believed, protein molecules in the liver, although their number remains unchanged, are constantly being synthesized and destroyed at equal rates.

The heat generated by radioactive decay can be converted to electricity in nuclear-powered batteries for use in remote locations, such as the Moon. Plutonium-238 powers a tiny battery used in heart pacemakers implanted in the human body. The long life of the isotope allows the battery to function for almost 10 years before an operation must be performed to replace it.

The regular variation in the penetration by radiation of materials of various thicknesses has found many industrial and engineering applications. The thickness of a film of plastic passing over a set of rollers as it is formed can be monitored by radiation passing through the film. This information is then fed back electronically to control the machinery and regulate the thickness of the product.

We shall mention just one more item in this sampling of applications of radioactivity—neutron activation analysis, which is a valuable technique for measuring the concentration of small amounts of elements. The great advantage of neutron activation analysis is that the sample is not destroyed in the process. For example, irradiating an old Dutch painting with neutrons forms silver-110m from the traces of silver in white pigment. The amount of silver present can be found from the γ-ray spectrum of the isotope. A change in silver extraction methods in the mid-nineteenth century decreased the amount of silver impurity in the white pigment made after that time. Therefore, the amount of silver-110m present in the pigment reflects the date when a painting was made.

FIGURE 8.11

Uranium-238 decay series. *Some of the isotopes decay by both α- and β-particle emission. The half-lives of the nuclides are given in parentheses.*

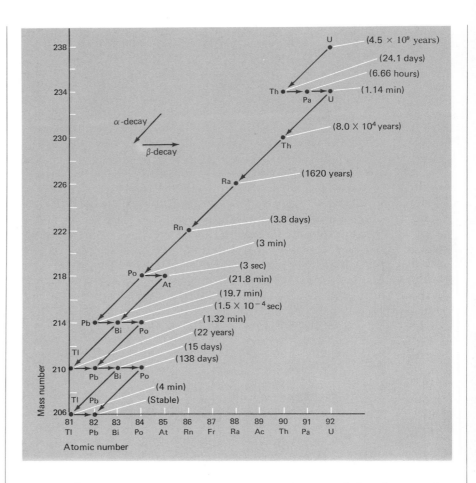

TABLE 8.5

The natural radioactive decay series

The mass number of each member of the thorium series is given by 4 times a whole number (m); of the uranium series by 4 times a whole number plus 2; of the actinium series by 4 times a whole number plus 3. ²⁴¹Pu is the parent of a fourth series, which is the 4m + 1 series that ends with ²⁰⁹Bi. This series was discovered with the advent of artificial radioisotope production and makes only a very small contribution to the quantity of natural radioisotopes

Designation	Parent	Half-life of parent (yr)
Thorium series 4m	$^{232}_{90}$**Th**	1.39×10^{10}
Uranium series 4m + 2	$^{238}_{92}$**U**	4.49×10^{9}
Actinium series 4m + 3	$^{235}_{92}$**U**	7.13×10^{8}

ning with uranium-238 is given in Figure 8.11, and the three series are summarized in Table 8.5.

The ratio of lead-206 to uranium-238 in natural minerals, together with the half-life of the uranium isotope, can be used to calculate the age of the Earth, which is currently thought to be about 4.5 billion years.

8.14 Bombardment

In **bombardment** a nucleus is struck by a moving particle, usually of about 10 MeV energy. The particle combines with the nucleus to form an unstable compound nucleus that decays either instantaneously or with a measurable half-life (Figure 8.12). If the bombarding particle has enough energy to overcome the nuclear force (in the billion electron volt range), it may smash the atom to bits. Physicists search among the debris from such reactions for information about nuclear particles and forces.

Early investigators were limited to bombardment with α particles from naturally radioactive sources. To enter a nucleus, a particle must have enough energy to overcome any repulsive forces between itself and the nucleus. The α particles have energies of less than 10 MeV and can be captured only by relatively light nuclei.

Two English physicists, Sir John Cockcroft and E.T.S. Walton in 1932 succeeded in accelerating protons in a vacuum tube. The first nuclear reaction produced by artificially accelerated particles was the splitting of lithium-7 into two α particles.

$$\,^1_1p \;+\; \,^7_3\text{Li} \longrightarrow \,^4_2\alpha \;+\; \,^4_2\alpha$$

Artificial radioactivity was discovered as the result of a bombardment reaction. Appropriately, the discovery was made by Iréne Curie, Marie Curie's daughter, in collaboration with her husband, Frédéric Joliot. In 1934 they found a curious result in the bombardment of aluminum, boron, and magnesium with α particles from a naturally radioactive source. The elements were transmuted—the result they were looking for. But when the bombardment ceased, the product nuclei continued to decay, giving off β^+ rays in the same way as the known natural radionuclides do. For aluminum, the reaction sequence is

$$\,^{27}_{13}\text{Al} \;+\; \,^4_2\alpha \longrightarrow \,^{30}_{15}\text{P} \;+\; \,^1_0n$$

$$\,^{30}_{15}\text{P} \xrightarrow{\;t_{1/2}\,=\,2.55\text{ min}\;} \,^{30}_{14}\text{Si} \;+\; \,^{\;0}_{+1}\beta$$

In shorthand notation for bombardment, the bombarding particle and the product particle are written in parentheses between the symbols for the reactant and product nuclides

$$\,^{27}_{13}\text{Al}\,(\alpha,\,n)\,^{30}_{15}\text{P}$$

Bombardment had obvious potential as a technique for producing artificial radioactive isotopes and possibly even new elements, and for studying the structure of the nucleus. In the 1930s the race began to build machines that could give higher and higher energies to such particles as protons, electrons, and deuterons (the nuclei of deuterium atoms, $\,^2_1\text{H}$). *Cyclotrons*, first built by E. O. Lawrence in California in 1932, accelerate positively charged particles in a spiral path through and between D-shaped magnets. Next came *synchrocyclotrons*, in which particles are accelerated as in cyclotrons, but using a variable electrical field with a fixed magnetic field (instead of both fields fixed). *Proton synchrotrons* accelerate protons in a fixed circle using electrical and magnetic fields that are both variable. In the new *linear accelerators* (linacs) particles pass through the center of a series of cylinders electrically charged so that the particles are continuously accelerated.

Accelerators continue to grow in size and complexity. The cost of the big accelerators is now so great that only national governments can afford to pay for them. As an example, the linear accelerator at Stanford University in California, which is 2 miles in length and can accelerate electrons to 20 BeV (billion electron volts), cost $100 million. (It was financed by the U.S. Atomic Energy Commission.)

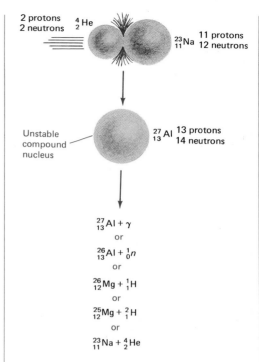

FIGURE 8.12

Bombardment of sodium-23 by α particles. *The compound nucleus may decay by one or several of the routes shown.*

The synthesis of the **transuranium elements**—elements with atomic numbers greater than that of uranium (atomic numbers > 92) was made possible by the advent of the big accelerators. The bombardment reactions by which these elements were first prepared are listed in Table 8.6. Most of the elements were first made at the famous Lawrence Radiation Laboratory of the University of California at Berkeley.

Weighable amounts of elements 93–99 have been prepared. The remaining transuranium elements have been identified literally atom by atom, an impressive achievement.

8.15 Nuclear fission

In a single nuclear fission event, one atom of a heavy fissionable isotope splits into two atoms of intermediate mass and several neutrons. Fission can be initiated by bombardment with many types of particles or by radiation, and many heavy nuclides can undergo fission. However, the neutron-induced fission of natural uranium-235, or of artificial plutonium-239, are the only fission reactions of practical importance. The history of the discovery of nuclear fission is inter-

TABLE 8.6

The transuranium elements— the reactions in which they were first observed.

The first preparation of elements 104 and 105 has been claimed by two groups— one at the University of California at Berkeley (their reactions are given here) and one in Dubna, Russia. Which group was first and therefore has the right to name these elements has not yet been resolved. The Californians suggest rutherfordium (Rf) for 104 and hahnium (Ha) for 105. The Russians suggest kurchatovium (Kv) for 104. Element 106, for which no name has yet been proposed, has also been synthesized by both Russian and American scientists.

Atomic number	Symbol	Name	Reaction
93	Np	Neptunium	$^{238}_{92}\text{U} + {}^{1}_{0}n \longrightarrow {}^{239}_{93}\text{Np} + {}^{0}_{-1}e$
94	Pu	Plutonium	$^{238}_{92}\text{U} + {}^{2}_{1}\text{H} \longrightarrow {}^{238}_{93}\text{Np} + 2{}^{1}_{0}n$
			$^{238}_{93}\text{Np} \longrightarrow {}^{238}_{94}\text{Pu} + {}^{0}_{-1}e$
95	Am	Americium	$^{239}_{94}\text{Pu} + {}^{1}_{0}n \longrightarrow {}^{240}_{95}\text{Am} + {}^{0}_{-1}e$
96	Cm	Curium	$^{239}_{94}\text{Pu} + {}^{4}_{2}\text{He} \longrightarrow {}^{242}_{96}\text{Cm} + {}^{1}_{0}n$
97	Bk	Berkelium	$^{241}_{95}\text{Am} + {}^{4}_{2}\text{He} \longrightarrow {}^{243}_{97}\text{Bk} + 2{}^{1}_{0}n$
98	Cf	Californium	$^{242}_{96}\text{Cm} + {}^{4}_{2}\text{He} \longrightarrow {}^{245}_{98}\text{Cf} + {}^{1}_{0}n$
99	Es	Einsteinium	$^{238}_{92}\text{U} + 15{}^{1}_{0}n \longrightarrow {}^{253}_{99}\text{Es} + 7{}^{0}_{-1}e$
100	Fm	Fermium	$^{238}_{92}\text{U} + 17{}^{1}_{0}n \longrightarrow {}^{255}_{100}\text{Fm} + 8{}^{0}_{-1}e$
101	Md	Mendelevium	$^{253}_{99}\text{Es} + {}^{4}_{2}\text{He} \longrightarrow {}^{256}_{101}\text{Md} + {}^{1}_{0}n$
102	No	Nobelium	$^{246}_{96}\text{Cm} + {}^{12}_{6}\text{C} \longrightarrow {}^{254}_{102}\text{No} + 4{}^{1}_{0}n$
103	Lr	Lawrencium	$^{252}_{98}\text{Cf} + {}^{10}_{5}\text{B} \longrightarrow {}^{257}_{103}\text{Lr} + 5{}^{1}_{0}n$
104	—	—	$^{249}_{98}\text{Cf} + {}^{12}_{6}\text{C} \longrightarrow {}^{257}_{104}\text{X} + 4{}^{1}_{0}n$
105	—	—	$^{249}_{98}\text{Cf} + {}^{15}_{7}\text{N} \longrightarrow {}^{260}_{105}\text{X} + 4{}^{1}_{0}n$
106	—	—	$^{249}_{98}\text{Cf} + {}^{18}_{8}\text{O} \longrightarrow {}^{263}_{106}\text{X} + 4{}^{1}_{0}n$

The atomic bomb

At the end of the 1930s, turmoil had come to Europe. Refugees were fleeing troubled countries and national energies were being turned to war and defense.

In 1938 Otto Hahn, a chemist at the Kaiser Wilhelm Institute in Berlin, identified barium in a sample of uranium that had been bombarded by slow neutrons. He dared to write of the discovery to Lise Meitner, a former colleague who had fled from Germany to Sweden. It was Christmas time and Lise Meitner's nephew Otto Frisch, a physicist working at Niels Bohr's Institute of Theoretical Physics in Copenhagen, was visiting her. They walked together in the snow near Gothenburg; while sitting on a log they figured out how fission of a heavy nucleus into smaller fragments could release a large amount of energy. Frisch went back to his laboratory, measured the speed of a barium nucleus released in uranium bombardment by neutrons, and found that it was what they had thought it would be—very fast.

Nuclear fission was announced on February 11, 1939. The news caused great excitement among physicists, and about a month later Leo Szilard made clear the implications of the discovery for a world heading for war. Szilard showed that in each fission of a uranium atom, two or three neutrons were released. This meant one thing—a terrible explosive force could be released in a nuclear fission chain reaction.

An extraordinary group of refugees in America, physicists all, drafted a letter to President Roosevelt. Szilard, Edward Teller, and Eugene Wigner, all Hungarians; Victor Weisskopf, an Austrian; and the brilliant Italian, Enrico Fermi, explained the potential of nuclear fission, and warned that Germany might already be working to build a nuclear bomb. So that their warning would carry more weight, they persuaded Albert Einstein, the world's most famous and respected scientist, to sign the letter.

A massive, secret effort was eventually mounted by the U.S. government—the Manhattan Project. Its outcome was the construction of three nuclear explosive devices. One was tested on July 16, 1945 at Alamogordo, New Mexico, in the Jornada del Muerto Valley, near the Sangre de Christo Mountains—the Journey of Death Valley and the Blood of Christ Mountains, strangely appropriate names. The second was the bomb that leveled the Japanese city of Hiroshima on August 6, 1945, six years almost to the day after delivery of the letter to President Roosevelt. The third was the bomb dropped August 9, 1945 on Nagasaki. Five days later Japan surrendered.

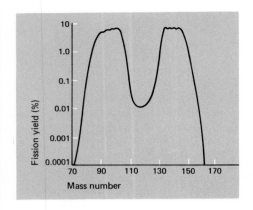

FIGURE 8.13

Distribution of isotopes in the slow-neutron-induced fission in uranium-235. *On the average, 2.5 neutrons and 200 MeV per fission are also produced.*

FIGURE 8.14

Chain reactions. *(a) In an uncontrolled chain reaction, the fission of each atom releases enough high-energy neutrons to initiate the fission of more than one new atom. Fission of all the fuel available takes place rapidly and with explosive force. (b) In a nuclear reactor the chain reaction is controlled so that each fission initiates the fission of only one more atom, and energy is released at a controlled rate.*

twined with the history of World War II (see An Aside: The Atomic Bomb). Uranium-235 and plutonium-239 are the fissionable isotopes used in the first atomic bombs.

One mode of fission of uranium-235 is shown in Equation (5) in Section 8.8; there are many others, mainly giving isotopes with mass numbers near 95 and 139 (Figure 8.13), together with varying amounts of energy. The average energy for uranium-235 fission is 200 MeV per single atom, and an average of 2.5 neutrons is given off in each fission. Most of the isotopes produced are radioactive.

The single fact of greatest significance for uses of nuclear energy is the production of more than one neutron per fission event. Enough neutrons are available to keep the fission going. A reaction or series of reaction steps that initiates repetition of itself is called a **chain reaction.** Nuclear fission can become a *self-sustaining chain reaction* when the number of neutrons emitted equals or is greater than the number of neutrons absorbed by fissioning nuclei plus those lost to the surroundings. The **critical mass** of a fissionable material is the smallest

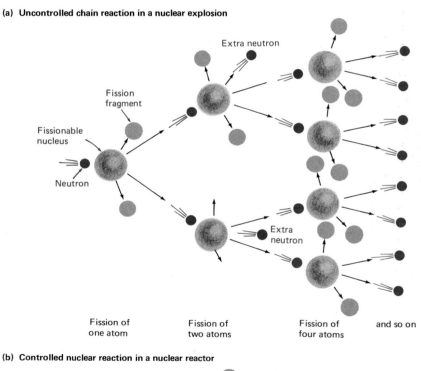

(a) Uncontrolled chain reaction in a nuclear explosion

Extra neutron

Fission fragment

Fissionable nucleus

Neutron

Extra neutron

Fission of one atom

Fission of two atoms

Fission of four atoms

and so on

(b) Controlled nuclear reaction in a nuclear reactor

Fission fragment

Fissionable nucleus

Neutron

Extra neutron

Extra neutrons

Extra neutron

Extra neutrons available for a neutron beam

mass that will support a self-sustaining chain reaction under a given set of conditions. Critical mass is affected by many factors, including the geometry of the sample of fissionable material.

In the first atomic bombs, small masses of pure plutonium-239 or uranium-235 were pushed together by conventional explosives into a critical mass, setting off a rapid chain reaction that continued until the force of the atomic explosion dispersed the fuel (Figure 8.14a). By controlling the rate of a chain reaction, energy can be produced slowly enough to allow its use for constructive rather than destructive purposes. The term **nuclear reactor** is usually applied to the equipment in which fission is carried out at a controlled rate.

8.16 Nuclear fusion

So far, *continuous nuclear fusion* has been observed only in the sun and the stars. The conversion of hydrogen to helium is the main source of solar energy.

$$4{}_{1}^{1}\text{H} \longrightarrow {}_{2}^{4}\text{He} + 2{}_{+1}^{0}\beta + 26.7 \text{ MeV} \tag{11}$$

(Reaction (11) is the sum of several intermediate reactions.)

Nuclei must approach each other with a large amount of energy if they are to overcome the electrostatic repulsion between their positive charges and be able to fuse. Particles can be given enough energy in an accelerator to fuse with target atoms, and fusion was known and understood several years before the discovery of fission. However, in an accelerator bombardment reaction, only a random few atoms collide and fuse, while a lot of energy is expended by the accelerator. Useful extraction of energy from nuclear fusion is not possible under these conditions.

The temperature in the center of the sun is about 15×10^6 °C. At such a high temperature, atoms are stripped of their electrons, forming a *plasma*—a neutral mixture of ions and electrons at a high temperature. Within a dense, very hot plasma, light nuclei are moving fast enough that large numbers of them can collide and fuse, releasing large amounts of energy in the process. Nuclear reactions at very high temperatures (roughly $>10^6$ °C) are called **thermonuclear reactions.**

To produce the temperature needed for thermonuclear fusion on Earth once seemed impossible. But, man-made nuclear fission in the form of the atomic bomb provided the match needed to light a large-scale fusion reaction. The result was the "H bomb" first tested at Bikini atoll in the Pacific Ocean in 1952.

In very general terms, a hydrogen bomb can be thought of as a fission bomb surrounded by the compound formed by deuterium (hydrogen-2) and lithium-6, ${}_{3}^{6}\text{Li}{}_{1}^{2}\text{H}$. The sequence of reactions is

$$\text{fission bomb} \longrightarrow \text{heat} + \text{neutrons} \tag{12}$$
$${}_{3}^{6}\text{Li} + {}_{0}^{1}n \longrightarrow {}_{2}^{4}\text{He} + {}_{1}^{3}\text{H} + 4.78 \text{ MeV} \tag{13}$$
$${}_{1}^{2}\text{H} + {}_{1}^{3}\text{H} \longrightarrow {}_{2}^{4}\text{He} + {}_{0}^{1}n + 17.6 \text{ MeV} \tag{14}$$

The reaction of lithium and neutrons is used to produce tritium for the fusion reaction between tritium and deuterium. The energy release per nuclear fusion is roughly one-tenth that released per nuclear fission event. (Compare 200 MeV per uranium-235 fission, with 17.6 MeV per deuterium–tritium reaction.) However, the yield from a thermonuclear explosion is much greater than in a fission explosion for several reasons: because of the smaller mass of lithium and deuterium compared to that of uranium, because new fusionable nuclei are produced in the explosion, and possibly also because a larger proportion of the fuel is consumed. Fortunately, thermonuclear weapons have so far been used only in tests.

Nuclear energy

Nuclear reactors provided 8% of the electrical energy produced in the United States in 1975. Where nuclear power generation will be in the total United States and world energy picture in the year 2000, or even 10 years from now, is difficult to predict.

All but one of the 56 commercial nuclear power plants operating in the United States in 1975 were of the type called light water reactors (see below). They produce energy primarily from the fission of uranium-235. Uranium-235 makes up 0.7% of the uranium obtained from natural uranium ores, which are mainly U_3O_8. Therefore, natural uranium is, like petroleum and coal, a fuel with a limited supply lifetime.

The highest hopes for future nuclear power ride first with breeder reactors, which produce more fuel than they use up, and eventually with fusion reactors, which draw for fuel on the virtually limitless supply of deuterium available in the oceans of the world.

8.17 Light water reactors

A nuclear fission reactor has five essential parts (Figure 8.15):

a. Fuel. A core of fissionable material; light water reactors use *enriched uranium*, U_3O_8, in which the natural 0.7% of uranium-235 has been raised to the 2%–3% uranium-235 needed for efficient operation. The fuel is formed into solid elements with a protective coating.

b. Moderator. The neutrons produced in fission must be slowed down to speeds at which they produce the fission reaction most efficiently. Neutrons must be able to lose energy by colliding with the nuclei of the moderator, without being absorbed by or reacting with the moderator.

Materials with a large proportion of hydrogen atoms, which have a similar mass to that of the neutrons, are good moderators. *Light water reactors* are so called because they use ordinary water as the moderator (D_2O is "heavy" water). Graphite is also a good moderator.

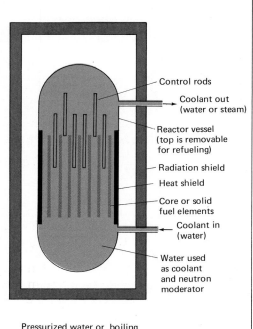

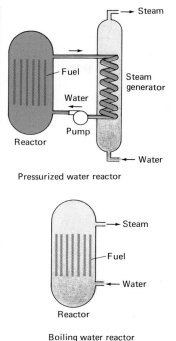

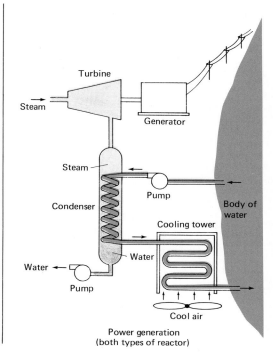

Pressurized water or boiling
water reactor (cross section)

Pressurized water reactor

Boiling water reactor

Power generation
(both types of reactor)

FIGURE 8.15

Light water nuclear reactors. *Both boiling water and pressurized water reactors generate electricity from steam in the same way as it is generated in a coal-fired power station. Note that the coolant is in a closed system and does not come in contact with the outside body of water. Both nuclear and coal-fired power stations need cooling water; the thermal pollution (temperature rise in a natural body of water used for cooling) is somewhat greater for nuclear reactors. (From* Man and Atom: Building a New World through Nuclear Technology *by Glenn T. Seaborg and William R. Corliss. Copyright © 1971 by Glenn T. Seaborg and William R. Corliss. Reprinted by permission of the publishers, E. P. Dutton.)*

c. Control system. Just enough free slow neutrons to carry on the chain reaction at a safe rate are needed (see Figure 8.14). If too many neutrons initiate fission, more heat would build up than could be carried away. This condition could lead to the most serious type of nuclear reactor accident, a meltdown, in which the fuel core and eventually its container, would melt.

In a light water reactor control rods containing boron are raised and lowered in between the fuel elements so that unneeded neutrons are removed by the reaction

$$^{10}_{5}B + ^{1}_{0}n \longrightarrow ^{7}_{3}Li + ^{4}_{2}\alpha$$

Enough control rods are available to completely stop the reaction by lowering all of them into the space between the radioactive rods.

d. Cooling system. The energy of the fission reaction must be carried away for transformation into electrical power and to keep the reactor from overheating as well. In pressurized water reactors, liquid water under pressure is the coolant; in boiling water reactors steam is the coolant (Figure 8.15).

FIGURE 8.16
The nuclear fuel cycle. (*Reprinted from the November 17, 1975 issue of* Business Week *by special permission.* © *1975 by McGraw-Hill, Inc.*)

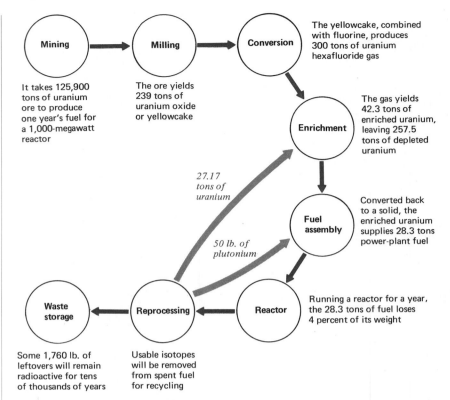

Mining

It takes 125,900 tons of uranium ore to produce one year's fuel for a 1,000-megawatt reactor

Milling

The ore yields 239 tons of uranium oxide or yellowcake

Conversion

The yellowcake, combined with fluorine, produces 300 tons of uranium hexafluoride gas

Enrichment

The gas yields 42.3 tons of enriched uranium, leaving 257.5 tons of depleted uranium

Fuel assembly

Converted back to a solid, the enriched uranium supplies 28.3 tons power-plant fuel

Reactor

Running a reactor for a year, the 28.3 tons of fuel loses 4 percent of its weight

Reprocessing

Usable isotopes will be removed from spent fuel for recycling

Waste storage

Some 1,760 lb. of leftovers will remain radioactive for tens of thousands of years

27.17 tons of uranium

50 lb. of plutonium

e. Shielding. Both the walls of the reactor, and the personnel operating the reactor must be protected from heat and radiation. In addition, the entire nuclear reactor is enclosed in a heavy steel or concrete dome that is intended to contain the radioactive materials that might be set free in a serious reactor accident.

The greatest difficulties facing the present day nuclear power industry are in the fuel cycle (Figure 8.16):

a. Isotope separation. The process is costly in both dollars and energy. All enrichment in this country is presently done in three gaseous diffusion plants operated by the U.S. government. These plants, originally built to produce fuels for weapons, cost $2.3 billion to build and use a maximum of 6000 megawatts (MW) of electricity per year. As pointed out above (Section 8.4), less costly methods are being sought.

b. Reprocessing of spent fuel. Much slower progress has been made toward getting reprocessing plants operating than in getting the reactors operating. Meanwhile, stockpiles of spent fuel are growing. As natural uranium ores are depleted, the need for reprocessing will become greater.

c. Waste storage. Permanent solutions to the storage of wastes that remain radioactive for thousands of years have not been satisfactorily worked out. Opponents of nuclear power generation argue that we can never be sufficiently protected from release of dangerously radioactive materials through technical failure, human error, or the actions of subversive groups.

Nature's nuclear reactor

A remarkable discovery in 1972 in the Gabon Republic in West Africa has shown that nature, not man, created the first nuclear fission reactor. Between 1.7 and 1.9 billion years ago, in what is now an open-pit uranium mine at a place called Oklo, a rich deposit of uranium ore began operating as a natural fission reactor.

The clue that led to discovery of the reactor was the finding in May, 1972 that the uranium-235 content of the uranium taken from the mine was significantly less than that generally found. The search for the cause of this discrepancy led to the conclusion that the uranium-235 has been depleted through nuclear fission.

Elements that are usually formed in the fission of uranium-235 were found in veins of the ore depleted in that isotope. Moreover, the quantities of these elements and their isotopic composition could *only* be accounted for by their origin in fission.

A truly extraordinary series of events had to occur for nature to have assembled the reactor. Uranium from an entire watershed had to be washed into the Oklo region to yield concentrated deposits. Operation of the nuclear chain reaction would have required a concentration of uranium in the ore of at least 10%, with a minimum of 1% of uranium-235. Since uranium-235 decays more rapidly ($t_{1/2}$ = 700 million years) than uranium-238 ($t_{1/2}$ = 4.5 billion years), uranium-235 must have been much more abundant in the past. It has been estimated that at the time of the formation of the Earth the total natural abundance of uranium-235 was about 17% and that 1.7–1.9 billion years ago when the reactor began to function, the abundance was about 3%. On the basis of the necessary 1% of uranium-235, a natural fission reactor could have operated at Oklo up until about 400 million years ago.

To sustain an appropriate flow of neutrons required a seam of uranium ore at least 2 m thick. Also required was a natural mechanism for moderating the high-energy neutrons given off by the fissioning uranium nuclei. Such moderation was apparently provided by the hydrogen atoms of water. About 6% water by weight would give the optimum moderation for an ore about 2 billion years old. Water of crystallization (water chemically bound to the minerals in the ore) would have satisfied this condition. The ore deposit was probably also saturated with ground water, which might have slowed down the neutrons too much. However, once a chain reaction had begun, some loss of water by evaporation due to the heat released could have supplied conditions favorable for moderation.

The evidence indicates that besides the fission of uranium-235, conversion of uranium-238 to plutonium-239 through the same sequence of reactions used in modern breeder reactors [see Equations (15)–(17)] also occurred at Oklo. Plutonium-239 decays by α-particle emission to uranium-235.

8.18 Breeder reactors

A **breeder reactor** produces more fissionable atoms than it consumes. The primary advantages of a breeder reactor are that it is able to convert uranium-238 or thorium-232, both naturally more plentiful than uranium-235, into fissionable fuel, and that it is thermally more efficient than the light water reactors because it can operate at higher temperatures. The excess fissionable artificial isotopes (uranium-233 or plutonium-239) produced in a breeder can also presumably be used as fuel for other reactors. The first commercial-scale breeder reactor for power generation began operation in 1972 in the Soviet Union.

One type of breeder reactor is the liquid metal fast breeder. It uses a liquid metal such as sodium as the coolant, and the neutrons need not be moderated (hence "fast breeder"). The following equations show the reaction cycle in a fast breeder reactor fueled by plutonium-239 bred from uranium-238.

$$^{238}_{92}U + ^{1}_{0}n \longrightarrow ^{239}_{92}U \tag{15}$$

$$^{239}_{92}U \xrightarrow{t_{1/2} = 24 \text{ min}} ^{239}_{93}Np + ^{0}_{-1}\beta \tag{16}$$

$$^{239}_{93}Np \xrightarrow{t_{1/2} = 2.4 \text{ days}} ^{239}_{94}Pu + ^{0}_{-1}\beta \tag{17}$$

$$^{239}_{94}Pu + ^{1}_{0}n \longrightarrow \text{fission products} + 2\text{--}3\ ^{1}_{0}n + \text{energy} \tag{18}$$

It has been predicted that breeder reactors could be available for commercial power generation in the United States by 1990. But a few major questions remain about the advisability of developing large-scale breeder reactors. First, do we need the power and fuel economy of the breeders enough to risk having to handle large amounts of plutonium-239, a material so poisonous that inhaling a tiny speck of it can be fatal? Second, how great a risk is incurred by making available in various parts of the country and in transit plutonium-239, which is much more readily used in nuclear weapons than enriched uranium? The answers to these questions lie as much in the realm of politics and sociology as in the realms of science and technology.

8.19 Fusion reactors

The problem of igniting, controlling, and maintaining a fusion reaction in a power plant has been called the greatest technological challenge taken up by man. Light atoms must be held together in a vacuum at high density, at a high enough temperature, and for a long enough time for the thermonuclear fusion to take place.

In the sun, the large gravitational field holds the plasma together. No comparable gravitational field exists on Earth and no known substance can withstand plasma temperatures. Two approaches are being taken toward controlling nuclear fusion—magnetic field containment and laser fusion.

If a large enough current is passed through a plasma parallel to the axis of a magnetic field, the plasma is pinched in and pulled away from the walls of the container, forming a so-called magnetic bottle. Plasma behavior in such a field has turned out to be very complex, but gradually it is being understood, and experiments are coming closer and closer to plasma confinement under the conditions needed for fusion.

In laser fusion, a powerful laser beam strikes a pellet of fuel with such energy that some of the surface material is vaporized. The resulting shock wave compresses the fuel at the center of the pellet, both bringing the nuclei close enough together and raising the temperature to the point where fusion can occur.

The fusion of deuterium and tritium to give helium (the reaction used in the hydrogen bomb; Section 8.15) is the prime candidate for the major reaction in thermonuclear power plants.

Deuterium is available in huge quantities in the oceans and obtaining it would be easy and quite inexpensive. Two reaction schemes have been proposed for fusion reactors, one involving lithium as a breeder for tritium, as in reactions (12) to (14) and the other involving only deuterium.

$$\begin{aligned}
{}_1^2H + {}_1^2H &\longrightarrow {}_1^3H + {}_1^1H + 4.0 \text{ MeV} \\
{}_1^2H + {}_1^2H &\longrightarrow {}_2^3He + {}_0^1n + 3.3 \text{ MeV} \\
{}_1^2H + {}_1^3H &\longrightarrow {}_2^4He + {}_0^1n + 17.6 \text{ MeV} \\
5\,{}_1^2H &\longrightarrow {}_2^4He + {}_2^3He + {}_1^1H + 2\,{}_0^1n + 24.9 \text{ MeV}
\end{aligned}$$

The heat produced in fusion reactions can be converted to electricity by conventional turbines, as in light water reactors. However, converting fusion energy more directly and efficiently to other forms of energy is being considered. One plan is to use fusion energy directly to split water into hydrogen and oxygen, and then to produce methane (CH_4) from the hydrogen (see Section 11.7). Methane can be used as a fuel, or built up into larger carbon–hydrogen compounds like those derived from petroleum.

Fusion reactors would have a number of advantages over fission reactors: (1) the lower cost and unlimited supply of the fuel; (2) the production of a far smaller and more manageable amount of radioactive waste; (3) the worldwide availability of the fuel, thereby avoiding international tensions over obtaining it; (4) the potential for production of energy at a lower cost. The most optimistic predictions see commercial-scale thermonuclear power plants in operation by the year 2000. This will require the solution of many complex physics and engineering problems.

The atomic age begins

A description of the first atomic bomb explosion—the test explosion at Alamogordo, New Mexico on July 16, 1945.

MEN AND ATOMS,
by William L. Laurence

Suddenly, at 5:29:50, as we stood huddled around our radio, we heard a voice ringing through the darkness, sounding as though it had come from above the clouds: "Zero minus ten seconds!" A green flare flashed out through the clouds, descended slowly, opened, grew dim, and vanished into the darkness.

The voice from the clouds boomed out again: "Zero minus three seconds!" Another green flare came down. Silence reigned over the desert. We kept moving in small groups in the direction of Zero. From the east came the first faint signs of dawn.

And just at that instant there rose as if from the bowels of the earth a light not of this world, the light of many suns in one. It was a sunrise such as the world had never seen, a great green supersun climbing in a fraction of a second to a height of more than eight thousand feet, rising even higher until it touched the clouds, lighting up earth and sky all around with a dazzling luminosity.

Up it went, a great ball of fire about a mile in diameter, changing colors as it kept shooting upward, from deep purple to orange, expanding, growing bigger, rising as it expanded, an elemental force freed from its bonds after being chained for billions of years. For a fleeting instant the color was unearthly green, such as one sees only in the corona of the sun during a total eclipse. It was as though the earth had opened and the skies had split.

A huge cloud rose from the ground and followed the trail of the great sun. At first it was a giant column, which soon took the shape of a supramundane mushroom. Up it went, higher and higher, quivering convulsively, a giant mountain born in a few seconds instead of millions of years. It touched the multicolored clouds, pushed its summit through them, and kept rising until it reached a height of 41,000 feet, 12,000 feet higher than the earth's highest mountain.

All through the very short but long-seeming time interval not a sound was heard. I could see the silhouettes of human forms motionless in little groups, like desert plants in the dark. The newborn mountain in the distance, a giant among the pygmies of the Sierra Oscuro range, stood leaning at an angle against the clouds, like a vibrant volcano spouting fire to the sky.

Then out of the great silence came a mighty thunder. For a brief interval the phenomena we had seen as light repeated themselves in terms of sound. It was the blast from thousands of blockbusters going off simultaneously at one spot. The thunder reverberated all through the desert, bounced back and forth from the Sierra Oscuro, echo upon echo. The ground trembled under our feet as in an earthquake. A wave of hot wind was felt by many of us just before the blast and warned us of its coming.

The big boom came about a hundred seconds after the great flash—the first cry of a newborn world.

Significant terms defined in this chapter

nuclide
nucleon
Coulomb force
nuclear force
nuclear binding energy
binding energy per nucleon
nuclear fusion
nuclear fission
phosphorescent
fluorescent
radioactivity
radioactive isotope

natural radioactivity
artificial radioactivity
stable isotope
natural isotope
artificial isotope
half-life
magic numbers
nuclear reaction
isomeric transition
α decay
β decay

antineutrino
neutrino
positron
electron capture
bombardment
transuranium elements
chain reaction
critical mass
nuclear reactor
thermonuclear reaction
breeder reactor

Exercises

8.1 The first artificially produced element was technetium. (a) Write the equation for the production of $^{97}_{43}$Tc by the reaction of $^{96}_{42}$Mo with $^{2}_{1}$H and (b) identify the other product of the reaction. How many (c) protons, (d) neutrons, and (e) nucleons are present in an atom of $^{97}_{43}$Tc? (f) What is the nuclide temporarily formed by $^{96}_{42}$Mo and $^{2}_{1}$H before it breaks up to form the products you have written in your equation? All known isotopes of Tc are radioactive and $^{97}_{43}$Tc decays by the process of electron capture. (g) Write the equation for this decay and (h) identify its product. (i) Is this decay an example of natural or artificial radiation? (j) In the process of electron capture, essentially what is happening to a proton? (k) How is the neutron–proton ratio changing in this process? The half-life of $^{97}_{43}$Tc is 2.6×10^{6} yr. (l) If you paid \$100 for a gram of $^{97}_{43}$Tc, would you worry about your economic loss as a significant portion of it undergoes decay in a week's time? All of the isotopes of Tc with mass numbers between 92 and 97 decay by electron capture or positron emission. (m) What element do all these isotopes form when they decay? All of the isotopes of Tc with mass numbers between 98 and 107 decay by electron emission. (n) What element do all these isotopes form upon decay?

8.2 Distinguish clearly between (a) α, β, and γ radiation; (b) fusion and fission; (c) binding energy and binding energy per nucleon; (d) natural and artificial radioactivity; (e) radioactive and stable isotopes; (f) positrons and electrons; (g) somatic and genetic effects.

8.3 Which statements are true? Rewrite any false statement so that it is correct?
(a) Radioactivity is the spontaneous emission of particles or electromagnetic radiation by unstable nuclei.
(b) Natural isotopes cannot be radioactive.

(c) At the end of one half-life, 50% of a radioactive substance remains and at the end of the second half-life, the remaining 50% undergoes radioactive decay leaving no original sample.
(d) Certain combinations of atomic numbers and neutron numbers (known as magic numbers) result in extremely stable nuclei.
(e) A chain reaction involves the emission of γ radiation from excited nuclei.
(f) A self-sustaining nuclear fission reaction generates a number of neutrons equal to or larger than the number of neutrons absorbed by the fissioning nuclei and lost to the surroundings.
(g) The critical mass is the neutral mixture of ions and electrons at a high temperature undergoing a thermonuclear reaction.

8.4 Why do the various isotopes of an element (such as $^{1}_{1}$H, $^{2}_{1}$H, and $^{3}_{1}$H) undergo the same chemical reactions with other elements (such as O)?

8.5 Consider a radioactive nuclide with a neutron–proton ratio that is higher than those for the stable isotopes of that element. What mode(s) of decay might be expected for this nuclide and why?

8.6 Describe what happens to the (a) atomic number, (b) mass number, (c) neutron number, and (d) neutron–proton ratio during (i) α emission, (ii) electron emission, (iii) positron emission, (iv) electron capture, and (v) isomeric transition.

8.7 Write the equations for the following nuclear processes: (a) $^{228}_{90}$Th undergoing α decay, (b) $^{110}_{49}$In undergoing positron emission, (c) $^{110}_{49}$In undergoing EC, (d) $^{127}_{53}$I$(p,7n)$X and identify X, (e) $^{10}_{5}$B(X,p)$^{10}_{4}$Be and identify X, (f) X$(n,\alpha)^{7}$Li and identify X, (g) $^{95}_{42}$Mo$(p,$ X$)^{95}_{43}$Tc and identify X, and (h) $^{96m}_{43}$Tc undergoing isomeric transition.

8.8 An alkaline earth element (Representative Group II) is radioactive. It and its daughters decay by emitting three α particles in succession. In what periodic table group is the resulting element found?

8.9 The nuclide $^{159}_{64}$Gd is a β^- emitter. The energies of the β^- are 0.59, 0.948, and 0.89 MeV and the energies of the accompanying γ radiation are 0.362 and 0.058 MeV. Prepare an energy diagram consistent with these data. Be sure to write the equation.

8.10 At one time, each isotope of each element had a unique name (until the known isotopes became too numerous to name). What nuclide corresponds to the name "radioactinium" if this "element" is produced in the actinium series from $^{235}_{92}$U by the successive emission of an α particle, a β^- particle, an α particle and a β^- particle?

8.11 List the five primary components of an atomic reactor and briefly describe their functions. Do any of these components present ecological or environmental problems?

8.12* The nuclide $^{19}_{8}$O is radioactive. (a) Predict the mode of decay for this isotope and (b) write the equation. (c) A 4.60-MeV particle is emitted followed by a 0.20-MeV γ ray and a 3.25-MeV particle is emitted followed by subsequent 1.37- and 0.20-MeV γ rays. Prepare a decay scheme for this nuclide.

8.13 Find the nuclear binding energy (in MeV) for the following nuclides: (a) $^{14}_{7}$N, (b) $^{20}_{10}$Ne, (c) $^{32}_{16}$S, (d) $^{45}_{21}$Sc, (e) $^{56}_{26}$Fe, (f) $^{59}_{27}$Co, (g) $^{64}_{30}$Zn, (h) $^{81}_{35}$Br, (i) $^{106}_{46}$Pd, and (j) $^{130}_{52}$Te. The respective atomic masses in amu are: 14.00307, 19.99244, 31.97207, 44.95592, 55.9349, 58.9332, 63.9291, 80.9163, 105.9032, 129.9067. Which of these has the largest binding energy per nucleon?

8.14 Which reaction produces the larger amount of energy from 1 g of reactants?

fission: $^{235}_{92}$U + $^{1}_{0}n \longrightarrow \, ^{94}_{40}$Zr + $^{140}_{58}$Ce + 6 $_{-1}^{0}\beta$ + 2 $^{1}_{0}n$
fusion: $2\,^{2}_{1}$H $\longrightarrow \, ^{3}_{1}$H + $^{1}_{1}$H

The atomic masses are 235.0439 amu for $^{235}_{92}$U, 1.00867 amu for $^{1}_{0}n$, 93.9061 amu for $^{94}_{40}$Zn, 139.9053 amu for $^{140}_{58}$Ce, 0.00055 amu for $_{-1}^{0}\beta$, 3.01605 amu for $^{3}_{1}$H, 1.007825 amu for $^{1}_{1}$H, and 2.0140 amu for $^{2}_{1}$H.

8.15 The reaction for the first fusion bomb was $^{7}_{3}$Li(p,α). (a) Write the complete reaction for the process and identify the other product. (b) The atomic masses are 1.007825 amu for $^{1}_{1}$H, 4.00260 amu for α, and 7.01600 amu for $^{7}_{3}$Li. Find the energy for the reaction of one atom of Li.

8.16 Rutherford's first nuclear transformation can be represented by the shorthand notation ^{14}N(α,p)^{17}O. (a) Write the corresponding nuclear equation for this process. The

respective atomic masses are 14.00307 amu for $^{14}_{7}$N, 4.00260 amu for $^{4}_{2}$He, 1.007825 amu for $^{1}_{1}$H, and 16.99913 amu for $^{17}_{8}$O. (b) Calculate the energy released by this reaction. (c) What would happen chemically if an electrical spark were passed through the mixture of products?

8.17* Calculate the binding energy per nucleon for the following isotopes: (a) $^{15}_{8}$O with a mass of 15.00300 amu; (b) $^{16}_{8}$O with a mass of 15.99491 amu; (c) $^{17}_{8}$O with a mass of 16.99913 amu; (d) $^{18}_{8}$O with a mass of 17.99915; and (e) $^{19}_{8}$O with a mass of 19.0035 amu. Which of these would you expect to be most stable?

8.18* Consider the following possible reactions when $^{27}_{13}$Al is bombarded with neutrons:

$$^{27}_{13}\text{Al} + {}^{1}_{0}n \longrightarrow {}^{28}_{13}\text{Al} \begin{cases} {}^{26}_{13}\text{Al} + 2\,^{1}_{0}n \\ {}^{28}_{13}\text{Al} + \gamma \\ {}^{27}_{12}\text{Mg} + {}^{1}_{1}\text{H} \\ {}^{24}_{11}\text{Na} + {}^{4}_{2}\text{He} \end{cases}$$

The nuclide masses are 26.98153 amu for $^{27}_{13}$Al, 1.008665 amu for $^{1}_{0}n$, 25.9858 amu for $^{26}_{13}$Al, 27.98193 amu for $^{28}_{13}$Al, 26.98437 amu for $^{27}_{12}$Mg, 1.007825 amu for $^{1}_{1}$H, 23.99102 amu for $^{24}_{11}$Na, and 4.00260 amu for $^{4}_{2}$He. Based on energy considerations, predict which reaction is most favorable.

8.19* The thickness of a shielding material for γ radiation, X, to decrease the amount of radiation from A_0 to A is given by

$$X = -\frac{2.303 \, \log(A/A_0)}{\mu}$$

where μ is a parameter that depends on the substance being used as the shielding material and on the energy of the radiation. If $\mu = 48.9$ m^{-1} for stopping 3.0 MeV γ radiation in Pb, calculate the thickness needed to decrease A_0 to $0.5\,A_0$. Repeat the calculation to find X for decreasing A_0 to $0.01\,A_0$.

8.20** The alchemist's dream of turning lead into gold is today's reality. Refer to the table on the next page for clues in answering the following questions about this process: (a) the process involves the following steps—(i) $^{206}_{82}$Pb(n,α)A, (ii) A decays to form B, (iii) B(n,α)C, (iv) C decays to form D, (v) D(n,α)E, and (vi) E decays to Au. Write the complete equations for the six reactions. (b) The overall equation for the process is

$$^{206}_{82}\text{Pb} + 3\,^{1}_{0}n \longrightarrow 3\,_{-1}^{0}\beta + 3\,^{4}_{2}\text{He} + {}^{197}_{79}\text{Au}$$

The atomic masses are 205.9745 amu for $^{206}_{82}$Pb, 1.008665 amu for $^{1}_{0}n$, 0.00055 amu for $_{-1}^{0}\beta$, 4.00260 amu for $^{4}_{2}$He, and 196.9666 amu for $^{197}_{79}$Au. What is the energy of this reaction? (c) Note that the radioactive isotopes to the right of the stable nuclides undergo β^- emission. Correlate these facts with the relationship of the neutron–proton ratio to stability.

A ⟶

	193	194	195	196	197	198	199	200	201	202	203	204	205	206	207	208	209	210	211	212
$_{82}$**Pb**		EC	EC	EC	EC	EC	EC or β^+	EC	EC or β^+	EC	EC	α	EC	23.1%	22.6%	52.3%	β^-	β^-	β^-	β^-
$_{81}$**Tl**	EC	EC	EC	EC	EC	EC	EC	EC or β^+	EC	EC	29.50%	β^- or EC	70.50%	β^-	β^-	β^-	β^-	β^-		
$_{80}$**Hg** (189–192 decay by EC or β^+)	EC	EC	EC	0.146%	EC	10.02%	16.84%	23.13%	13.22%	29.80%	β^-	6.85%	β^-	β^-						
$_{79}$**Au** (186–192 decay by EC)	EC	EC or β^+	EC	EC	100%	β^-	β^-	β^-	β^-	β^-	β^-	β^-	β^-							
$_{78}$**Pt** (184–192 decay by EC or α)	EC	32.9%	33.8%	25.3%	β^- 20.0 hr	7.21%	β^-	β^-												

Z ⟵

TABLE FOR EXERCISE 8.20

This is a portion of a typical isotope table. Where percentages appear, the isotopes are stable and occur naturally in the percentage given. Where symbols appear, the isotopes are radioactive and decay in the way symbolized (see Table 8.4).

9 THE CHEMICAL BOND

In this chapter we define the three basic types of chemical bonds that join atoms and ions together, and discuss the properties associated with each type of bond—metallic, ionic, and covalent. The weaker forces that act between molecules are also described. And we look at the properties of the bonds themselves—bond energies, bond lengths, covalent and ionic radii. In addition, we discuss a subject of great usefulness—the geometry of molecules.

Among the people that you know there are probably some that you think of as cheerful individuals, some that you think of as pessimistic or gloomy, others who are purposeful and always busy, and still others who are lazy or devoted mainly to having fun. We put people in categories in many ways–leaders and followers, winners and losers, good guys and bad guys.

Yet no collection of adjectives, no list of generalities, can give the whole picture of what a single individual is like. There isn't any substitute for getting to know people by talking with them, being in their company, and seeing how they react in different situations.

Here and in the next chapter we are going to deal with many generalities about chemical species—about bonding types in this chapter and about the properties of representative elements related to their placement in the periodic table in the next chapter. Understanding the bonding and the properties of a compound is valuable and can lead to predictions about chemical behavior. But, as with human beings, to know a single chemical element or compound requires getting to know it as an individual. In later chapters on descriptive chemistry we shall become better acquainted with the personalities of the individual elements and some of their compounds.

Some definitions

9.1 Definition of the chemical bond

Few terms in chemistry are as difficult to define precisely as "the chemical bond." We have collected at the end of this chapter (Thoughts on Chemistry: Chemical Bond Definitions) several of the ways that the concept of the chemical bond has been put into words. The best known and most often quoted is that of Linus Pauling.

Any theory of the chemical bond must explain why some atoms join together and others do not. The theory must also explain the varying degrees of stability of chemical compounds; the arrangement of bonded atoms or ions with respect to each other; and, in terms of the electrons, just what is represented by that line we have been drawing to show two or more atoms joined together.

How the electrons play their roles determines which type of chemical bond will form between two atoms. Here is our entry into the catalog of chemical bond definitions: A **chemical bond** is a force that acts strongly enough between two atoms or groups of atoms to hold them together in a different, stable species having measurable properties. A chemical bond can be described in terms of the change in energy that results when atoms, ions, or groups of atoms approach each other closely enough to interact. As we learned in Chapter 5, chemical changes tend to take place when the products of the change are energetically more stable than the reactants. We shall return to this interaction at the end of the chapter.

9.2 Valence electrons and Lewis symbols

The term "valence" has had various meanings during the development of theories of chemical bonding. It is now used in a general way to mean several things. The charge on an ion is spoken of as its valence: "Calcium has a valence of two in Ca^{2+}." Sometimes "valence" is used to refer to the number of bonds that an atom forms in a molecule: "The oxygen atom has a valence of two in H_2O."

Valence electrons are the electrons that take part in chemical bonding of any type. We say that "Carbon has four valence electrons," meaning that carbon has four electrons available for bonding. For the representative elements the valence electrons all reside in the outermost electron shell, and the term **valence shell** is used for the highest electron shell. For example, the valence electrons of carbon are the $2s^2$ and $2p^2$ electrons, and they reside in the $n = 2$ shell, which for the carbon atom is the valence shell. The number of valence electrons in each of the representative element atoms is equal to the representative group number, for example, four for carbon from Group IV and five for nitrogen from Group V.

The Lewis symbol is a convenient notation for showing both the valence electrons in atoms and the electrons involved in bonding in compounds. In this notation the symbol of the element represents the

number of valence electrons
= group number for
representative elements

nucleus plus the underlying, normally filled, electron shells of the atom. The valence shell electrons or bonding electrons are indicated by dots (or circles, or Xs, etc.) arranged around the atomic symbol so that the pairing of two electrons is represented by two dots on the same side of the symbol; for example, for the phosphorus atom

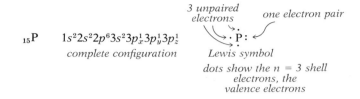

$_{15}P \qquad 1s^2 2s^2 2p^6 3s^2 3p_x^1 3p_y^1 3p_z^1$

complete configuration

Table 9.1 gives the Lewis symbols for atoms of the second period elements. How Lewis symbols are used in writing the formulas of compounds is shown in later sections.

The use of the terms "valence electrons" and "valence shell" becomes a bit ambiguous when we get to the transition elements. For these elements, electrons in the second from the outermost shell, the $(n - 1)d$ level, often take part in bonding. For example, to form the Cu^{1+} ion, a Cu atom has given up a single $4s$ electron, but to form the Cu^{2+} ion, a Cu atom has lost the $4s$ electron plus a $3d$ electron.

9.3 Noble gases and the stable octet

The noble gases come at the end of each period in the periodic table and are the least reactive of all the elements. The outermost energy level of the helium atom is fully occupied by two electrons. The outermost energy levels of atoms of all of the other noble gases are occupied by eight electrons, two s electrons and six p electrons (Table 9.2). Each orbital holds two electrons, and all electrons in the noble gases are paired.

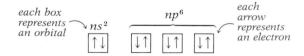

The chemical stability of the noble gas configuration is shown by the high ionization energies of the noble gases (Table 9.2)—that is, the difficulty with which an electron is removed—and by their lack of reactivity. A rule that explains the formation of many compounds, although by no means all compounds, is based upon this stability of the ns^2np^6 noble gas configuration. According to the **octet rule** non-noble-gas atoms tend to combine by gain, loss, or sharing of electrons so that the outermost energy level of each atom holds or shares four pairs of electrons in an ns^2np^6 configuration.

TABLE 9.1

Lewis symbols for the second-period elements

Element	Representative group number	Valence shell configuration	Number of valence electrons
Li·	I	$2s^1$	1
Be:	II	$2s^2$	2
Ḃ:	III	$2s^2 2p^1$	3
·Ċ:	IV	$2s^2 2p^2$	4
·Ṅ:	V	$2s^2 2p^3$	5
·Ö:	VI	$2s^2 2p^4$	6
·F̈:	VII	$2s^2 2p^5$	7
:N̈e:	VIII	$2s^2 2p^6$	8

TABLE 9.2

The noble gases

Helium and neon have the highest ionization energies of all of the elements; only fluorine has a higher ionization energy than argon.

Element	Atomic number	Valence shell configuration	Lewis symbol	Ionization energy (kcal/mole at 0°K)
He	2	$1s^2$	He:	567.0
Ne	10	$2s^2 2p^6$	:N̈e:	497.3
Ar	18	$3s^2 3p^6$	:Är:	363.4
Kr	36	$4s^2 4p^6$	:K̈r:	322.8
Xe	54	$5s^2 5p^6$	:Ẍe:	279.7
Rn	86	$6s^2 6p^6$	:R̈n:	247.9

9.4 Types of chemical bonds

The reaction between gaseous chlorine and solid sodium to form sodium chloride

$$2Na(s) + Cl_2(g) \longrightarrow 2NaCl(s) \qquad \Delta H = -196.5 \text{ kcal } (-822.0 \text{ kJ})$$

is the formation of an ionic compound (NaCl) from a metal (Na) and a nonmetal (Cl_2). The substances involved in this reaction, whose prop-

Noble gas compounds

By 1900 all of the noble gases had been discovered (see Chapter 6). For the next 60 years it was believed by almost all chemists and taught in almost all chemistry courses that helium, neon, argon, krypton, xenon, and radon were inert, that because of their stable filled outer electron shells they neither gained, lost, nor shared electrons to form compounds. Isaac Asimov has pointed out a fine semantic distinction: Chemists were thinking and writing, "The noble gases cannot form compounds under any conditions," when they should have been saying, "As far as we know, the noble gases do not form compounds."

The preparation in 1962 of xenon platinum hexafluoride ($Xe[PtF_6]$), followed shortly thereafter by the preparation of xenon tetrafluoride (XeF_4), amazed many chemists. Obviously, they had forgotten to remain open to the possibility of discovery of new facts.

The preparation of $Xe[PtF_6]$ came about when Neil Bartlett, who had just made $O_2^+[PtF_6]$ looked, "quite by chance" he said, at a chart of ionization energies plotted against atomic number (a similar chart appears in most general chemistry textbooks; see Figure 10.3). He saw that the ionization energy of xenon was almost equal to that of oxygen. Bartlett had the thought that xenon might undergo the same reaction as did O_2 with PtF_6, a very strong oxidizing agent.

The rest is history. The $Xe + PtF_6$ reaction worked. And xenon tetrafluoride could also be made quite simply in another way—xenon and fluorine were heated together at 750°C for one hour. In fact, if xenon and fluorine are mixed in a flask and the flask is allowed to stand in the sunlight, some of the compound eventually forms. Xenon tetrafluoride, XeF_4, turned out to be a quite ordinary crystalline substance which can be melted, recrystallized, and stored in a bottle on the shelf. A barrage of research was set off by the initial discoveries of xenon compounds and within a year 50 publications on noble gas compounds had appeared in scientific journals.

By now, it has been shown that the bonding in noble gas compounds is not exotic, but fits into existing theories. Most of the compounds known are those of xenon combined with fluorine or oxygen. This is reasonable since xenon has the second lowest ionization energy of the noble gases, and fluorine and oxygen are highly electronegative elements. Some of the well-characterized compounds are XeF_2, XeF_4, XeF_6, $XeOF_4$, XeO_2F_2, XeO_3, XeO_4, $XeF^+Sb_2F_6^-$, and $XeF_5^+BF_4^-$. Krypton difluoride has also been prepared, and a very small amount of a radon fluoride. (Although radon has an even lower ionization energy than xenon and should form stable compounds, its radioactivity and scarcity make it very difficult to work with.) As yet, no comparable compounds of helium, neon, or argon have been prepared.

TABLE 9.3

Properties of sodium, chlorine, and sodium chloride

Property	Sodium Na (a metal)	Chlorine Cl₂ (a covalent molecule)	Sodium chloride NaCl (an ionic compound)
Appearance	Silvery solid	Greenish yellow gas	Colorless crystals
Atomic or mol wt (amu)	22.99	70.90	58.44
Melting point (°C)	97.83	-100.98	808
Boiling point (°C)[a]	882.9	-34	1465
Density (g/cm³)[b]	0.967 (s, 25°C)	1.9 (s, -101°C)	2.16 (s, 25°C)
Heat of fusion at m.p.[c] kcal/mole (kJ/mole)	0.63 (2.6)	1.53 (6.40)	6.8 (28.4)
Heat of vaporization at b.p.[c] kcal/mole (kJ/mole)	19.4 (81.2)	4.88 (20.4)	40.8 (171)
Electrical conductivity Liquid Solid	Very large Very large	Very small Very small	Large Very small

[a] Boiling point at atmospheric pressure.
[b] The notation (s, 25°C), (l, 25°C), etc., means that the density is given for a solid (s), or a liquid (l), or a gas (g) at the temperature indicated.
[c] See Figure 5.3.

erties are given in Table 9.3, can be used to illustrate the three basic types of chemical bonds—metallic, ionic, and covalent bonds—which are the subject of this chapter.

In the molten state both elemental sodium and sodium chloride are excellent conductors of electricity, whereas chlorine is a nonconductor. By contrast, in the solid state only elemental sodium is a conductor. Since electric current is a flow of charged particles—either electrons or ions—liquid and solid sodium and liquid sodium chloride must contain either freely mobile electrons or ions, whereas solid sodium chloride and chlorine in either state can not contain mobile electrons or ions.

The following observations can be made:

1. Both solid and liquid sodium have conductivities larger than that of molten sodium chloride by a factor of about 100,000 (Figure 9.1). Indeed the conductivity of sodium is so large that the metal must offer almost no resistance to the flow of electric current.

2. Differences in conductivity between liquid and solid sodium are comparatively small, and for both states the conductivity decreases as the temperature increases.

3. Sodium undergoes no chemical change as the current flows through it.

4. The conductivity of solid and liquid sodium chloride increases with increasing temperature, and the conductivity changes dramatically from a very small value to a very large one when the solid melts.

5. As current passes through it, liquid sodium chloride decomposes to sodium metal and chlorine gas.

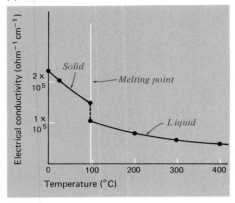

(a) Sodium

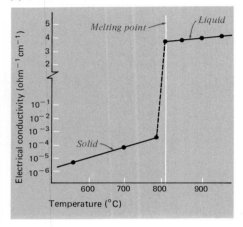

(b) Sodium chloride

FIGURE 9.1
Electrical conductivity of sodium and sodium chloride. *Note that the scales of these two graphs are quite different. At the melting point the conductivity of sodium chloride increases 10,000-fold. However, the conductivity of sodium at its lowest point in (a) is still 100,000 times greater than the conductivity of molten sodium chloride.*

All of these observations suggest that in molten or solid sodium the current is carried by mobile electrons and the sodium atoms or ions are not involved, while in molten sodium chloride the current is carried by ions, which are much less mobile than electrons.

These and other observations also indicate that the easily available outermost electrons of metal atoms are free to move and do not "belong" to a specific atom. If electrons are pushed in at one end of a piece of metal, electrons come out at the other end, but the metal undergoes no detectable change. Interference with the flow of electrons by the increased vibration of the metal atoms at higher temperatures accounts for the decrease in conductivity with increasing temperature.

The electronic configuration of the sodium atom

$$Na \qquad 1s^2 2s^2 2p^6 3s^1$$

shows the easily available outer $3s$ electron. Each sodium atom in solid or liquid sodium loses its valence electron, which joins the electrons lost by all the other sodium atoms in what has been called an "electron sea." The force of metallic bonding is the attraction between the positive metal ions and the surrounding sea of electrons. These highly mobile electrons account for the high conductivity of metals.

In ionic compounds, positive and negative ions formed by the gain and loss of electrons are held together by the force of electrostatic attraction between the oppositely charged particles. This constitutes ionic bonding. The configuration changes in the formation of sodium chloride are

$$Na\cdot \quad + \quad \cdot \overset{\cdot\cdot}{\underset{\cdot\cdot}{Cl}} : \quad \longrightarrow \quad [Na]^+ \quad \left[: \overset{\cdot\cdot}{\underset{\cdot\cdot}{Cl}} : \right]^-$$
$$(1s^2 2s^2 2p^6 (3s^1)) \quad (1s^2 2s^2 2p^6 3s^2 3p^5) \quad (1s^2 2s^2 2p^6) \quad (1s^2 2s^2 2p^6 3s^2 3p^6)$$

In a crystal of sodium chloride or any other ionic compound, the positive and negative ions occupy rigidly fixed positions and vibrate only a little about these positions. This allows practically no pathway for the conduction of electricity. When the crystal melts, the ions become mobile, leading to the sharp increase in conductivity (Figure 9.1b). The ions can then react by gaining or losing electrons at the electrodes.

When a chemical bond forms between two atoms of a nonmetal, such as in H_2 or Cl_2, the force holding the atoms together is obviously neither the electrostatic force of the ionic bond, nor the same force that occurs in a metallic bond. Nonmetals are most often nonconductors, and therefore neither ions nor free electrons can be present. Instead, the electrons of the atoms in the compound must somehow be rearranged to yield an electrically neutral molecule that has a lower energy content than the two separate atoms. It is believed that this is accomplished by the sharing of valence electrons between atoms, leading to the name "covalent" for such bonding.

In the Cl_2 molecule, the unpaired $3p$ electrons are shared by the two chlorine atoms. Pairs of valence shell electrons not involved in the bonding are called **nonbonding electron pairs,** or **lone pairs.**

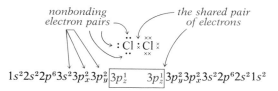

$$1s^2 2s^2 2p^6 3s^2 3p_x^2 3p_y^2 \boxed{3p_z^1 \qquad 3p_z^1} 3p_y^2 3p_x^2 3s^2 2p^6 2s^2 1s^2$$

Such sharing allows each atom to achieve a stable octet configuration, a state that is often, but not always, the result of covalent bonding. The ·'s and ×'s show the atomic origin of the electrons and thus aid in electron bookkeeping. However, the electrons in the bond are completely equivalent and indistinguishable from each other. The electrons can be thought of as occupying atomic orbitals that overlap. (A full discussion of the covalent bond in terms of orbitals is deferred to Chapter 21.)

To summarize:

Metallic bonding is based upon the presence of electrons that are free to move. Sodium and other metals, and also most alloys, display this type of interaction.

Ionic bonding is based upon the electrostatic attraction between positive and negative ions. Sodium chloride and other salts display this type of interaction.

Covalent bonding is based upon forces that do not require the presence of free electrons or ions; in covalent bonding electrons are shared. Chlorine (Cl_2), carbon dioxide (CO_2), and other molecular substances exemplify this type of interaction.

The metallic bond

9.5 Configuration and bonding in metals

Sodium, iron, copper, tin, gadolinium, hafnium, platinum, bismuth, and thorium (Table 9.4) are typical metals from various parts of the periodic table. The highest quantum level in the atoms of many metals holds one or two *s* electrons and the inner electronic shells are completely or partially filled.

In Section 9.4, we described the electronic state of metals as a sea of electrons surrounding a network of positive ions. This is in some ways an overly simple picture of metallic bonding, yet it succeeds, in a qualitative way, in explaining most of the characteristic properties of metals. (Metallic bonding is further discussed in Section 21.13.)

Metallic bonding is the attraction between positive metal ions and surrounding freely mobile electrons (Figure 9.2). The representative metals contribute their outermost *s* or their outermost *s* and *p* electrons to the sea of free electrons. In transition metals some of the *d* level electrons, close in energy to the outermost electrons, can also become free electrons.

TABLE 9.4

Electron configurations of metal atoms

The noble gas symbol in square brackets indicates that beneath its outermost energy level the atom has all the filled orbitals of that noble gas, e.g., [Ne] = $1s^2 2s^2 2p^6$; [Ar] = $1s^2 2s^2 2p^6 3s^2 3p^6$; [Kr] = $1s^2 2s^2 2p^6 3s^2 3p^6 3d^{10} 4s^2 4p^6$, and so on. Metals are not limited to only one or two electrons in the valence shell, although most have that configuration.

	Z	Configuration	
Na	11	[Ne]	$3s^1$
Fe	26	[Ar]	$3d^6 4s^2$
Cu	29	[Ar]	$3d^{10} 4s^1$
Sn	50	[Kr]	$4d^{10} 5s^2 5p^2$
Gd	64	[Xe]	$4f^7 5d^1 6s^2$
Hf	72	[Xe]	$4f^{14} 5d^2 6s^2$
Pt	78	[Xe]	$4f^{14} 5d^9 6s^1$
Bi	83	[Xe]	$4f^{14} 5d^{10} 6s^2 6p^3$
Th	90	[Rn]	$6d^2 7s^2$

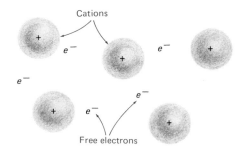

FIGURE 9.2
Metallic bonding. *The attraction between positive ions and surrounding freely mobile electrons.*

THE CHEMICAL BOND [9.5]

9.6 Properties imparted by the metallic bond

The high boiling points and heats of vaporization of metals show that it is difficult to free metal ions from the surrounding free electrons. In general, melting points and boiling points are higher for metals with a larger number of valence electrons that can become "free" metallic electrons. Such metals are also harder and more dense than those with fewer valence electrons. In Table 9.5 compare the properties of potassium, the atoms of which have one valence electron, with those of calcium, atoms of which have two valence electrons. Melting points are also influenced by the size of the atoms and the interatomic distances in the metal. Pure metals and many alloys retain the properties of high conductivity and metallic luster in the liquid state. The general properties of materials that have metallic bonding are given in Table 9.6.

Metals are usually dense, but when a mechanical force is applied to a metal, the cations can just move over, sliding along on a "cushion" of free electrons. No specific bonds need be broken, the forces between the cation and the free electrons need not be disrupted, and no additional repulsive forces are encountered (as is the case in ionic solids.) This explains the ease with which metals are hammered into shape (malleability) or drawn out into wires (ductility).

The free electrons are not limited to a few specific energy levels, but have many different energies. Therefore, they are able to absorb and reemit light of all wavelengths, leading to the luster that is characteristic of metals and alloys. When the surface of a metal is dull rather than lustrous, it is usually because a layer of metal oxide or sulfide has formed on the surface.

TABLE 9.5

Properties of some metallic elements

Property	Potassium K	Calcium Ca	Copper Cu	Zinc Zn	Tin[a] Sn
Appearance	Silvery, lustrous	Silvery, lustrous	Reddish, lustrous	Bluish white, lustrous	White, lustrous
Atomic weight (amu)	39.099	40.08	63.546	65.38	118.69
Melting point (°C)	63.4	851	1083	419.5	231.9
Boiling point (°C)	757	1482	2582	907	2270
Density (g/cm³)	0.856 (s, 18°C)	1.54 (s, 20°C)	8.94 (s, 20°C)	7.133 (s, 25°C)	7.29 (s, 25°C)
Heat of fusion at m.p. kcal/mole (kJ/mole)	0.57 (2.4)	2.2 (9.2)	3.11 (13.0)	1.59 (6.65)	1.69 (7.07)
Heat of vaporization at b.p. kcal/mole (kJ/mole)	19.4 (81.2)	36.7 (156)	72.8 (305)	27.4 (115)	61.8 (259)
Electrical conductivity Solid Liquid	Very large Very large	Very large Very large	Very large Very large	Very large Very large	Very large Very large

[a] In the form of white tin.

The high electrical conductivity of metals is provided by the presence of mobile electrons, which begin to flow when an electric potential is applied across a piece of metal. Metals also conduct heat very well. The thermal conductivity of metals is 10 to 10,000 times better than for most other substances. The electrons can move freely as the temperature rises, and they readily pass along their increased kinetic energy to other electrons. Also, the ions in a metal are freer to vibrate in place than the ions in an ionic compound, and so they contribute to high thermal conductivity. Silver is one of the best conductors of heat and electricity. If a teaspoon handle feels hot as soon as you stir your tea, you will know that you are using sterling silver (92.5% Ag) and not silver-plated flatware.

The ionic bond

9.7 Configurations of ions

The octet rule correctly predicts the configurations of the monatomic ions formed by all nonmetals and by the representative metals of Groups I and II. The alkali metals lithium through francium lose their single outermost s electron (the ns electron) to give ions with a $+1$ charge. Similarly, the alkaline earth metals beryllium through radium lose both ns electrons to give $+2$ ions.

The nonmetals with five, six, and seven electrons in their outermost shells form negative ions with noble gas configurations by gaining three electrons, two electrons, or one electron, respectively (Table 9.7). The hydrogen atom also forms a noble-gas-type anion, the hydride ion, H^-, by gaining one electron to achieve the helium configuration ($1s^2$).

Atoms of the representative metals of Groups I and II have **noble gas cores**—two or eight electrons in the $n - 1$ shell, the shell underneath the valence shell—so that by losing one or two ns electrons the noble gas configuration is achieved. The aluminum atom also has a noble gas core. In those compounds where aluminum is ionic (many of its compounds are covalent) it forms Al^{3+} by losing its $3s^2$ and $3p^1$ electrons. The remaining representative metals have **pseudo-noble-gas cores**—s, p, and d levels completely filled with a total of 18 electrons in the $n - 1$ shell. For example,

$_{49}$In	$2 - 8 - 18$	$4s^24p^64d^{10}$	$5s^25p^1$
	$n = 1$–3 levels fully occupied by 28 electrons	$n - 1$ level with 18 electrons	valence electrons
$_{83}$Bi	$2 - 8 - 18 - 32$	$5s^25p^65d^{10}$	$6s^26p^3$
	$n = 1$–4 levels fully occupied by 60 electrons	$n - 1$ level with 18 electrons	valence electrons

TABLE 9.6

General properties of metals and metal-like compounds

Lustrous appearance

Malleable and ductile

High melting point

High boiling point

High heat of vaporization

High density

Excellent conductor of heat and electricity

TABLE 9.7

Nonmetal ions

The roman numerals are the representative group numbers. The outermost configurations are shown; lower levels are filled. All of these ions have noble gas configurations. In simple binary ionic compounds, the nonmetals always have the configuration shown here. Carbon does not form C^{4-} ions, but only C_2^{2-} ions. Astatine does form an ion (At^-, $6s^26p^6$), but it is a radioactive element and is so short-lived that very little of it can be accumulated.

I	VII
H^- (hydride)	F^- (fluoride)
($1s^2$)	($2s^22p^6$)
V	Cl^- (chloride)
N^{3-} (nitride)	($3s^23p^6$)
($2s^22p^6$)	Br^- (bromide)
P^{3-} (phosphide)	($4s^24p^6$)
($3s^23p^6$)	I^- (iodide)
VI	($5s^25p^6$)
O^{2-} (oxide)	
($2s^22p^6$)	
S^{2-} (sulfide)	
($3s^23p^6$)	

TABLE 9.8

Representative group metal ions

The outermost configurations are shown; lower levels are filled. The roman numerals are the representative group numbers.

Noble gas configuration (2 or 8) ns^2 or ns^2p^6			Pseudo-noble gas configuration (18) $(n-1)s^2p^6d^{10}$	18 + 2 Configuration $(n-1)s^2p^6d^{10}ns^2$		
I	II	III	III	III	IV	V
Li^+	Be^{2+}					
Na^+	Mg^{2+}	Al^{3+}				
K^+	Ca^{2+}		Ga^{3+}	Ga^+		
Rb^+	Sr^{2+}		In^{3+}	In^+	Sn^{2+}	Sb^{3+}
Cs^+	Ba^{2+}		Tl^{3+}	Tl^+	Pb^{2+}	Bi^{3+}
Fr^+	Ra^{2+}					

TABLE 9.9

Some transition metal ions

The outermost level configurations are shown; lower levels are filled.

Sc^{3+}	$3s^23p^6$
Cu^+	$3s^23p^63d^{10}$
Cu^{2+}	$3s^23p^63d^9$
Zn^{2+}	$3s^23p^63d^{10}$
Co^{2+}	$3s^23p^63d^7$
Tb^{4+}	$4f^75s^25p^6$
U^{3+}	$5f^36s^26p^6$

TABLE 9.10

General properties of ionic compounds

The boiling point range of ionic compounds is approximately 700°–3500°C. Polar solvents are those having some charge separation (see Section 9.16).

Crystalline solids

Hard and brittle

High melting point

High boiling point

High heat of vaporization

High heat of fusion

Good conductor of electricity when molten

Poor conductor of electricity as a solid

Many are soluble in water and other polar solvents

Atoms of such representative metals give ions in which either all of the valence electrons are lost, to give a pseudo-noble-gas configuration, or only the *p* electrons are lost, to give what is called an 18 + 2 configuration.

Tables 9.7 and 9.8 show the ions formed by representative metals and the nonmetals *when they participate in ionic bonding*. This does not mean that only ionic compounds are formed by these elements. Most of them also participate in covalent bonding, and many may have additional oxidation states in covalent bonds.

The *d* and *f* transition metals, in contrast to the representative metals, rarely form ions with noble gas configurations. Only the ions of the scandium family elements have octet configurations. In the other *d* transition metals the $(n-1)d$ and the ns levels differ very little in energy and electrons in both levels are available for bonding (either ionic or covalent). In a number of ions the *d* level has its maximum of 10 electrons and the ion has a pseudo-noble-gas configuration. Table 9.9 gives some examples of transition metal ions.

The requirements for the formation of an ionic bond are an atom that will lose one or more electrons and an atom that will gain one or more electrons. Neither of these processes can require a large amount of energy. For this reason, ions with charges of +1, -1, +2, and -2 are much more common than those with charges of +3 or -3 or more. Removing an electron from a positively charged species or adding an electron to a negatively charged species are both energetically more difficult than similar processes in neutral species.

9.8 Properties imparted by the ionic bond

Ionic bonding is based on electron transfer and is the attraction between positive and negative ions (Figure 9.3). The strong electrostatic force of attraction in ionic compounds reaches out in all directions, and each ion in a solid ionic, crystalline substance is sur-

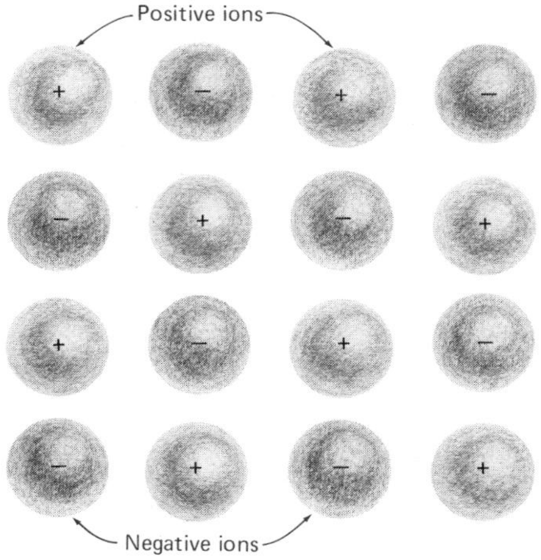

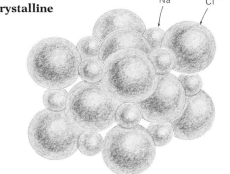

FIGURE 9.3
Ionic bonding. *Electron transfer and the attraction between positive and negative ions.*

FIGURE 9.4
Arrangement of ions in crystalline sodium chloride, NaCl.

rounded by other ions of opposite charge. For example, in a sodium chloride crystal (Figure 9.4) each Na^+ ion is surrounded by six Cl^- ions and each Cl^- ion is surrounded by six Na^+ ions. Any changes that require disrupting the arrangement of ions in a crystalline ionic compound require a large amount of energy.

As a result, ionic compounds (Tables 9.10, 9.11) have high melting points and boiling points, and high heats of vaporization and fusion. For the same reasons, ionic crystalline substances are hard—a

TABLE 9.11
Properties of some ionic compounds

Property	Lithium fluoride LiF	Potassium bromide KBr	Calcium nitrate $Ca(NO_3)_2$	Barium sulfate $BaSO_4$	Praseodymium iodide PrI_3
Appearance	White crystals	White crystals	White crystals, hygroscopic[a]	White crystals	Green crystals, hygroscopic[a]
Formula weight (amu)	25.939	119.01	164.096	233.4	521.62
Melting point (°C)	870	730	561	1580	733
Boiling point (°C)	1670	1435	dec.[b]	dec.	1377
Density (g/cm³)	2.295 (s, 21.5°C)	2.75 (s, 25°C)	2.36 (s, 20°C)	4.5 (s, 15°C)	2.31 (s)
Heat of fusion at m.p. kcal/mole (kJ/mole)	2.4 (10)	7 (30)	5.1 (21)	9.7 (41)	12.7 (53.1)
Heat of vaporization at b.p. kcal/mole (kJ/mole)	51.0 (213)	37.1 (155)	dec.	—	41 (172)
Electrical conductivity Liquid Solid	Large Very small	Large Very small	Large Very small	Large Very small	Large Very small

[a] A *hygroscopic* substance takes up water from the air to become a wet solid.
[b] dec. stands for decomposes. Means that at the melting point or boiling point a chemical change takes place, usually a breakdown to simpler substances.

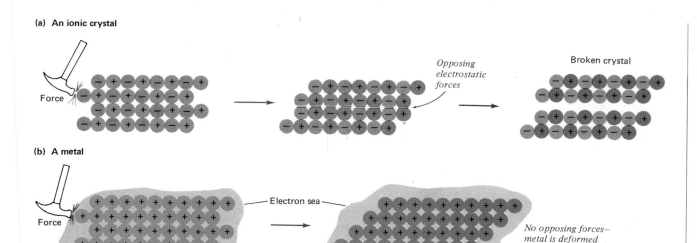

(a) An ionic crystal

Force

Opposing
electrostatic
forces

Broken crystal

(b) A metal

Force

Electron sea

No opposing forces—
metal is deformed

FIGURE 9.5

Shattering an ionic crystal; bending a metal.

strong force is needed to break up the crystal lattice. Ionic solids are, however, brittle and when struck hard enough they shatter along the planes between rows of ions (Figure 9.5).

The ratio of the different numbers of ions in a crystal is the ratio shown in the formula of the compound—the numbers of ions necessary for electroneutrality. The geometrical arrangement of the ions is determined by the size and number of ions of each kind. The density of ionic compounds varies with the spacing of ions in the crystal lattice and also, of course, with the weights of the ions.

Solid ionic compounds are poor conductors of electricity because the ions are rigidly fixed in their positions. The ions become free to move in the molten state, and then the salts become conductive. Ionic solids are not very good conductors of heat either, since the ions do not easily pass kinetic energy along to their neighbors.

Many, although by no means all, ionic compounds are soluble in water. The ions are freed from the crystal lattice by hydration, the surrounding of ions by water molecules. The energy given up in hydration is used to free the ions from their positions in the crystal.

The covalent bond

9.9 Configurations in covalent bonding

A **single covalent bond** is a bond in which two atoms are held together by sharing two electrons. The two electrons in the bond spend most of their time in the space between the two atoms and thus provide the "glue" that holds the atoms together. Each positive nucleus is attracted toward the region of high electron density between them. **Covalent bonding** is based upon electron-pair sharing and is the attraction between atoms that share electrons.

Covalent bonds tend to form between two atoms when the loss of electrons by one atom and/or the gain of electrons by the other atom is not easy; in other words, when ionic or metallic bonding is unlikely. The most obvious example is the bonding of nonmetal atoms with themselves (H_2, N_2, Cl_2, C in diamond, etc.) and with each other. In what might be called a normal single covalent bond each atom contributes one electron to the bond.

The Lewis formulas for some simple molecules are given in Table 9.12. Different symbols are used there for the valence electrons of each atom, in order to show the origin of the electrons. Of course, in a real molecule the valence electrons cannot be distinguished from each other. For covalently bonded atoms, we can see that the line we have been drawing between the symbols for atoms represents one shared pair of electrons.

In the structures given in Table 9.12 each atom in each molecule has achieved a noble gas configuration of two (for H) or eight electrons in the valence shell. For example, the nitrogen atom in a molecule of ammonia is surrounded by eight electrons, six of them in three electron-pair bonds with hydrogen atoms and two of them as a lone, or nonbonded, pair.

The elements of the first and second rows of the periodic table can accommodate a maximum of two ($1s^2$) or eight ($2s^2 2p^6$) electrons in their valence shells. Once beyond the second row, the possibility for the valence shell to accommodate more than eight electrons exists. The empty outer d energy level of the representative elements is available for occupation by electrons in bonds. In sulfur(VI) fluoride, SF_6, for example, six electron pairs are shared by the sulfur atom. Since the sulfur atom has the configuration

of necessity some of the shared electrons in the SF_6 molecule must have moved into the empty $3d$ orbitals. Another example is chlorine(III) trifluoride, ClF_3. The central chlorine atom has two unshared pairs of electrons and three shared pairs of electrons in its valence shell. Therefore, the $3s^2$, $3p^6$, and $3d^2$ levels must all be used.

sulfur(VI) fluoride, SF_6
colorless gas, m.p. $-50°C$
b.p. $64°C$

chlorine(III) fluoride, ClF_3
colorless gas,
m.p. $83°C$
b.p. $11.3°C$

TABLE 9.12

Some covalent compounds
We have given the properties for the compounds used as illustrations in this table and elsewhere in this chapter not because you should memorize them, but as a reminder that the symbols do stand for real substances that can be seen or handled (with varying degrees of ease or safety).

ammonia, NH_3
colorless gas, pungent odor; m.p. $-78°C$, b.p. $-33°C$

hydrazine, N_2H_4
colorless liquid or white crystals; m.p. $1.4°C$, b.p. $113.5°C$; explosive

ethane, C_2H_6
colorless gas; m.p. $-183°C$; b.p. $-89°C$

fluorine, F_2
greenish yellow poisonous gas; m.p. $-220°C$, b.p. $-188°C$

carbon tetrachloride, CCl_4
colorless liquid; m.p. $-23°C$, b.p. $76.8°C$; poisonous vapor

THE CHEMICAL BOND [9.9]

The valence shell of an atom in a molecule can also be occupied by *less* than an octet of electrons. Such compounds are formed by atoms with fewer than four valence electrons, for example, boron, $[He]2s^22p^1$, in boron trifluoride.

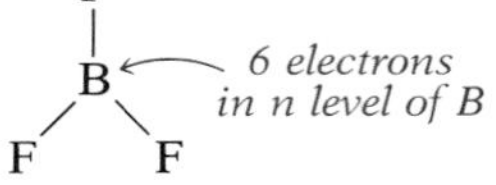

$$F$$
$$|$$
$$F \diagdown B \diagup \longleftarrow \text{6 electrons in n level of B}$$
$$F$$

boron trifluoride, BF$_3$
colorless gas, m.p. −127°C, b.p. −100°C

While there are many exceptions to the octet rule among transition elements and for compounds such as BF_3, SF_6, and ClF_3, the presence of four pairs of electrons in the valence shell is the most common condition among simple covalent compounds.

9.10 Multiple covalent bonds

More than one pair of electrons can be shared between the same two atoms, resulting in what is called a **multiple covalent bond**. In a **double covalent bond** two electron pairs are shared between the same two atoms.

double bond

$$H : C :: C : H \quad \text{or} \quad HC {=} CH$$
$$\quad\ddot{H}\ \ddot{H} \qquad\qquad\quad |\quad\ |$$
$$\qquad\qquad\qquad\qquad\quad H\ \ H$$

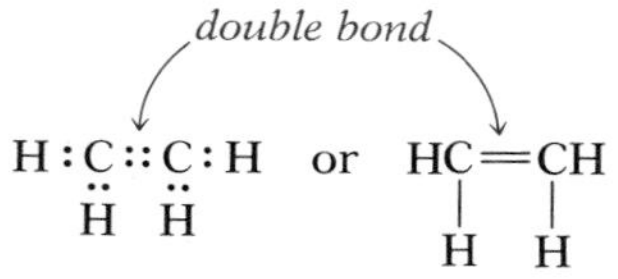

ethylene, C$_2$H$_4$
colorless gas,
m.p. −170°C, b.p. −104°C

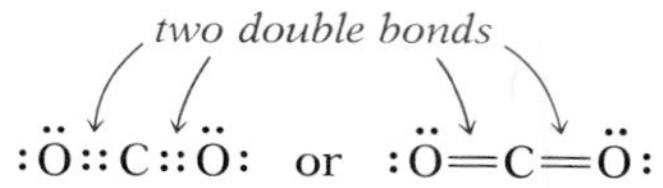

two double bonds

$$:\ddot{O} :: C :: \ddot{O}: \quad \text{or} \quad :\ddot{O} {=} C {=} \ddot{O}:$$

carbon dioxide, CO$_2$
solid goes directly to gas at
−78°C

In a **triple covalent bond** three electron pairs are shared between the same two atoms.

triple bond

$$I : C \!\vdots\vdots\! C : I \quad \text{or} \quad I {-} C {\equiv} C {-} I$$

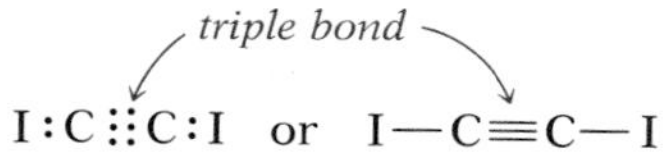

diiodoacetylene, C$_2$I$_2$
colorless crystals, m.p. 79°C,
gives explosive liquid

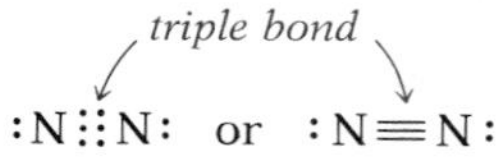

triple bond

$$:N \!\vdots\vdots\! N: \quad \text{or} \quad :N {\equiv} N:$$

nitrogen, N$_2$
colorless gas, m.p. −210°C,
b.p. −196°C

Each of the two atoms in a multiple covalent bond usually contributes one electron to each of the multiple covalent bonds and achieves an octet configuration in this way. Carbon, nitrogen, and oxygen are the most common participants in multiple covalent bonds. The **bond order** of a covalent bond between two atoms is the number of electron pairs shared between those two atoms. Thus, for nitrogen the bond order is 1 in $H_2N{-}NH_2$ (hydrazine), 2 in $HN{=}NH$ (diimine), and 3 in $N{\equiv}N$ (dinitrogen).

9.11 Coordinate covalent bonds

A single covalent bond in which <u>both electrons in the shared pair come from the same atom</u> is called a **coordinate covalent bond.** To indicate a coordinate covalent bond an arrow is sometimes drawn from the atom that donates the electron pair toward the atom with which the pair is shared.

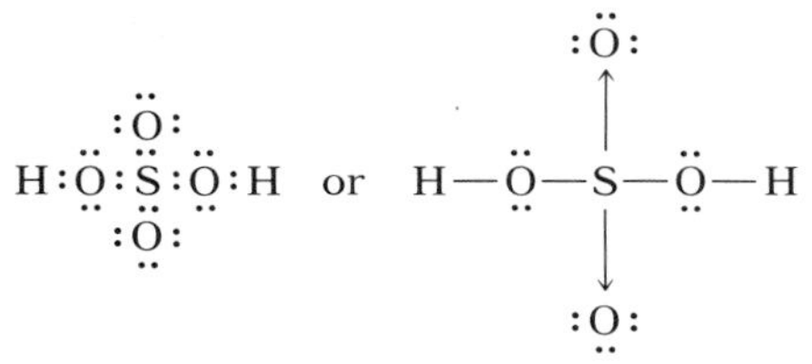

$$\begin{array}{ccc}
 & & :\ddot{O}: \\
 & & \uparrow \\
:\ddot{O}: & & \\
H:\ddot{O}:\ddot{S}:\ddot{O}:H \quad \text{or} \quad & H{-}\ddot{O}{-}S{-}\ddot{O}{-}H \\
:\ddot{O}: & & \\
 & & \downarrow \\
 & & :\underset{\cdot\cdot}{O}:
\end{array}$$

sulfuric acid (H_2SO_4)
colorless liquid, m.p. 10°C,
b.p. 338°C; burns skin

The **donor atom** provides both electrons to a coordinate covalent bond and the **acceptor atom** accepts an electron pair for sharing in a coordinate covalent bond. For coordinate covalent bonds, as for any other kind of bond, it is impossible to distinguish among the electrons once the bond has formed. For example, a hydrogen ion unites with an ammonia molecule by a coordinate covalent bond to form the ammonium ion

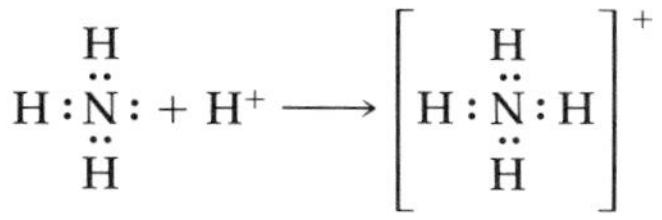

$$H:\overset{\cdot\cdot}{\underset{H}{N}}: + H^+ \longrightarrow \left[H:\overset{H}{\underset{H}{\overset{\cdot\cdot}{N}}}:H \right]^+$$

but all four hydrogens in the ammonium ion are alike.

9.12 Unpaired electrons

Stable compounds that contain one or more unpaired electrons are known. A sufficient number of unpaired electrons should appear in the Lewis formula to agree with the observed magnetic properties of the compounds. For example, the paramagnetism of chlorine(IV) oxide shows that it has one unpaired electron per molecule. The Lewis formula for ClO_2, which contains a total of 19 valence electrons (6 each from two oxygen atoms and 7 from a chlorine atom), is

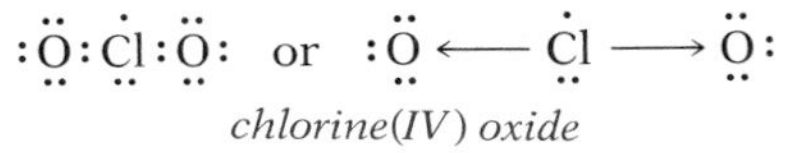

$$:\ddot{O}:\dot{C}l:\ddot{O}: \quad \text{or} \quad :\ddot{O} \longleftarrow \dot{C}l \longrightarrow \ddot{O}:$$
chlorine(IV) oxide

Most, but not all, species that contain even numbers of electrons are diamagnetic. A notable exception is the O_2 molecule, which contains two unpaired electrons. (An explanation of the paramagnetism of O_2 must be deferred to Chapter 21.)

9.13 Resonance

The representation of bonding by Lewis structures runs into difficulty when more than one formula can be written that agrees with the electronic requirements and the properties of a compound. A typical example is nitrogen(I) oxide, N_2O, a diamagnetic molecule in which the two nitrogen atoms are bonded to each other. The 16 valence electrons of N_2O can be arranged in two reasonable ways, both of which satisfy the octet rule:

$$:\ddot{N}\!=\!N\!=\!\underset{\cdot\cdot}{O}: \longleftrightarrow :N\!\equiv\!N\!-\!\underset{\cdot\cdot}{\ddot{O}}:$$

nitrogen(I) oxide

Each of these formulas accommodates the 16 valence electrons that are available, each formula gives to each atom eight valence shell electrons, and each is electrically neutral. Two structures can also be written for ozone, O_3

$$:\ddot{O}\!=\!\ddot{O}\!-\!\underset{\cdot\cdot}{\ddot{O}}: \longleftrightarrow :\underset{\cdot\cdot}{\ddot{O}}\!-\!\ddot{O}\!=\!\ddot{O}:$$

ozone

The observed properties of both nitrogen(I) oxide and ozone are those that would be expected if the two possible structures exist simultaneously, or, put another way, if the real structure is an average of two possibilities. For example, the bond distance for both oxygen-to-oxygen bonds in ozone is identical. If one *or* the other of the two structures was correct, the bond distances would differ—double bonds are usually shorter than single bonds.

The term **resonance** refers to the intermediate electronic state of a molecule for which several electronic arrangements are possible. "Resonance" does *not* mean that the molecule constantly flips from one structure to another. *Resonance is not equilibrium.* A double-headed arrow is used between resonance structures to emphasize that the structures are not in equilibrium with each other. In a sense, the concept of resonance is necessary because of limitations in the way we write structures. The actual molecular structure of a molecule for which resonance structures can be written is called a **resonance hybrid** since it has the characteristics of all possible structures. Sometimes a dashed line in a single structure is used to indicate resonance, for example, for ozone,

$$O\text{-----}O\text{-----}O$$

This shows that the oxygen–oxygen bonds are less than double bonds, but more than single bonds.

9.14 Polyatomic ions

The atoms in polyatomic ions such as hydroxide ion, ammonium ion, and nitrate ion, are held together by covalent bonds.

$$\left[\, :\!\ddot{\mathrm{O}}\!:\mathrm{H}\,\right]^{-}$$

hydroxide ion, OH⁻

$$\left[\begin{array}{c}\mathrm{H}\\[-2pt]\ddot{}\\[-6pt]\mathrm{H}\!:\!\overset{\cdot\cdot}{\underset{\cdot\cdot}{\mathrm{N}}}\!:\!\mathrm{H}\\[-2pt]\ddot{\mathrm{H}}\end{array}\right]^{+}$$

ammonium ion, NH₄⁺

$$\left[\;:\!\ddot{\mathrm{O}}\!:\;\overset{\ddot{\mathrm{O}}\!:}{\underset{:\!\ddot{\mathrm{O}}\!:}{\mathrm{N}}}\;\right]^{-}\longleftrightarrow\left[\;:\!\ddot{\mathrm{O}}\!:\;\overset{:\!\ddot{\mathrm{O}}\!:}{\underset{:\!\ddot{\mathrm{O}}\!:}{\mathrm{N}}}\;\right]^{-}\longleftrightarrow\left[\;:\!\ddot{\mathrm{O}}\;\;\ddot{\mathrm{O}}\!:\overset{}{\underset{:\!\ddot{\mathrm{O}}\!:}{\mathrm{N}}}\;\right]^{-}$$

nitrate ion, NO₃⁻
3 resonance structures

We write brackets around these structures and show the charge as a superscript to avoid putting the charge on a specific atom. The charge is distributed over the whole ion. The oxygen atom has six valence electrons and the hydrogen atom one, a total of seven valence electrons. Since the hydroxide ion has eight valence electrons, it has one extra electron and a charge of -1. One valence electron for each of four hydrogen atoms and the normal number of 5 for the nitrogen atom would give a total of 9 electrons. With 8 electrons present, the ammonium ion is short one electron and thus has a $+1$ charge. Similarly, the nitrate ion has one more valence electron (24) than the total for three oxygen atoms and one nitrogen atom, and thus a -1 charge.

Compounds like $K^{+}OH^{-}$ or $NH_4^{+}Cl^{-}$ incorporate both covalent and ionic bonds. The covalent bonds within each polyatomic ion are so strong that the ion remains a single unit in a crystal or in water solution. Therefore, compounds containing polyatomic ions are usually classified as ionic compounds.

9.15 Writing Lewis formulas for covalent species

The proper approach to writing a Lewis formula for a covalent compound can be summarized as follows:

1. Write down the correct arrangement of the atoms in space, using single bonds.

2. Calculate the total number of valence electrons available by adding the number of valence electrons for each atom, subtracting 1 for each unit of positive charge, or adding 1 for each unit of negative charge.

3. Assign two electrons to each covalent bond.

4. Distribute the remaining electrons so that each atom has the appropriate number of nonbonded electrons. For representative elements this number is often enough electrons so that each atom is surrounded by an octet.

5. If there are not enough electrons to go around, change some single bonds to multiple bonds.

EXAMPLE 9.1

■ Write suitable Lewis electron dot formulas for (a) the chlorate ion, ClO_3^{-}; (b) chlorous acid, $HClO_2$; and (c) the nitrosyl ion, NO^{+}.

(a) The arrangement of atoms in ClO_3^- is

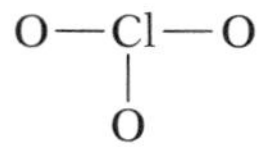

$$O—Cl—O$$
$$|$$
$$O$$

because compounds containing ClO_3^- are known not to have properties typical of peroxide bonds, that is, O—O. The number of valence electrons we have to work with is

$$\text{total valence electrons} = \underset{Cl}{7} + \underset{O}{3(6)} + \underset{charge}{1} = 26$$

The three covalent bonds we have already put in the structure account for 6 electrons, leaving 20 electrons to be distributed. If we place 6 on each of the oxygen atoms and 2 on the chlorine atom, we find that all of the octets are filled using the available electrons:

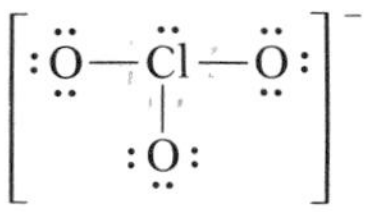

$$\left[:\ddot{O}—\ddot{C}l—\ddot{O}: \atop \qquad :\ddot{O}: \right]^-$$

Note that we put brackets and the charge on our final answer if the species is an ion.

(b) The arrangement of atoms in $HClO_2$ is

$$H—O—Cl—O$$

because $HClO_2$ does not have peroxide properties and is known to contain the O—H bond (as do all oxo acids). The number of valence electrons we have to work with is

$$\text{total valence electrons} = \underset{Cl}{7} + \underset{O}{2(6)} + \underset{H}{1} = 20$$

Of these, we have already used 6 in the above structure, leaving 14 to be distributed. If we complete the octets on the O atoms and Cl atom, it works out just right:

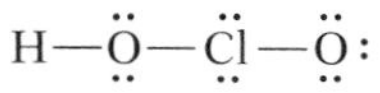

$$H—\ddot{O}—\ddot{C}l—\ddot{O}:$$

(c) The only possible arrangement of atoms in NO^+ is

$$N—O$$

In this structure we must distribute

$$\text{total valence electrons} = \underset{N}{5} + \underset{O}{6} - \underset{charge}{1} = 10$$

so that both octets are satisfied. A single bond between N and O would use 2 electrons and leave both the N and the O atoms 6 electrons short of an octet. A double bond would use 4 electrons and leave both N and O four electrons short of an octet. Since there are only 10 electrons available, neither of these structures is correct. Only a triple bond between the atoms will allow an octet on each atom, so we have

$$\left[:N\equiv O: \right]^{+}$$

■

9.16 Polar covalent bonds and electronegativity

In the H_2, Cl_2, and N_2 molecules each of the bonding electrons spends an equal amount of time in the vicinity of each atom. In a covalent bond of this type—a **nonpolar covalent bond**—the electrons are shared equally. (Figure 9.6a).

Whenever atoms of two different elements are covalently bonded, the sharing of the electrons becomes unequal, because no two different atoms have exactly the same electron-attracting ability. The electron density around one atom becomes greater than around the other. How unequally the electrons are shared depends on the relative abilities of the two atoms to attract electrons.

A covalent bond in which electrons are shared unequally is called a **polar covalent bond** (Figure 9.6b); one atom acquires a *partial* negative charge ($\delta-$; Greek delta and a minus sign) and the other acquires a *partial* positive charge ($\delta+$). These are not unit charges, but only represent a reorientation, a sort of pushing around, of the total electron density of the two bonded atoms. The entire molecule remains electrically neutral.

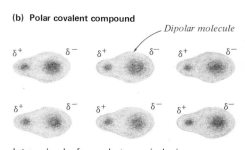

(a) **Nonpolar covalent compound**

Neutral molecule

Intramolecular force: electron pair sharing
Intermolecular force: London force

(b) **Polar covalent compound**

Dipolar molecule

Intramolecular force: electron pair sharing
Intermolecular force: dipole-dipole force
hydrogen bonding
London force

FIGURE 9.6
Covalent bonding. *The sharing of a pair of electrons of opposite spin between two atoms. (The intermolecular forces are described in Sections 9.20–9.22.)*

$$\overset{\delta+}{H}\overset{\delta-}{-F} \quad \substack{polar \\ covalent \\ bonds} \quad \overset{\delta+}{H}\overset{\overset{\delta-}{O}}{\diagup}\underset{H}{}$$

In hydrogen fluoride and water molecules, electrons are attracted away from hydrogen and toward fluorine and oxygen atoms, respectively. Polar covalent bonds also occur between atoms of the same element, if the rest of the molecule differs in electron-attracting ability. For example, the carbon–carbon bond in trifluoroethane is polar (although less so than a carbon–fluorine bond) because of the strong electron-attracting ability of fluorine atoms

$$\overset{\delta+}{H_3C}\overset{\delta-}{-CF_3}$$

trifluoroethane

The nonpolar covalent bond is at one end of the continuum of variation in the polar nature of bonding. At the other end of this continuum one atom attracts electrons so strongly that the electrons depart completely from the other atom and the bonding is ionic. Polar covalent bonds can be thought of as having partial ionic character (Section 9.23).

High electronegativity
means high electron affinity
and high ionization energy

The ability of an atom to attract electrons to itself is called **electronegativity.** An **electronegative atom** tends to have a partial negative charge in a covalent bond, or to form a negative ion by gaining electrons. Elements with high electron affinities are highly electronegative. The word "electropositive" is used in a general sense as the opposite of "electronegative." An **electropositive atom** tends to have a partial positive charge in a covalent bond, or to form a positive ion by losing electrons. Elements with low ionization energies are highly electropositive.

Various properties of molecules have been utilized for assigning numerical values to the electronegativities of atoms. Values for some representative elements from the Pauling scale of electronegativity (the most widely used scale) are given in Table 9.13. (A complete list is given in Table 10.7.) These are relative values based upon the arbitrarily assigned value of 4.0 for fluorine, the most electronegative element.

Two atoms with greatly different electronegativities (see Na and Cl values in Table 9.13) form an ionic bond. Pairs of atoms with small differences in electronegativity form polar covalent bonds with the partial negative charge on the atom with the greater electronegativity (see H and Cl in Table 9.13). The difference for identical bonded atoms, as in Cl_2, is zero and the bond is a nonpolar covalent bond. A frequently used "rule of thumb" is that an electronegativity difference of 2 or more is necessary for the formation of an ionic bond between two atoms.

TABLE 9.13

Electronegativities of some representative elements
Given here are the Pauling values, derived from bond energy relationships.

Nonmetals		Metals	
F	4.0	Be	1.5
O	3.5	Al	1.5
N	3.0	Mg	1.2
Cl	3.0	Li	1.0
Br	2.8	Ca	1.0
C	2.5	Sr	1.0
S	2.5	Ba	0.9
I	2.5	Na	0.9
P	2.1	K	0.8
H	2.1	Rb	0.8
Si	1.8	Cs	0.7

9.17 Dipole moment

A polar bond, as a result of the concentration of positive charge at one end and negative charge at the other end, is a **dipole**—a pair of opposite charges of equal magnitude at a specific distance from each other. The degree of polarity of a bond or a molecule is measured by its **dipole moment** μ (Greek mu, pronounced "mew"). The unit for dipole moments is the debye, D (pronounced "de-buy").

FIGURE 9.7

Polar molecules in an electric field. *One way of measuring dipole moments is by comparing current flow in a vacuum with current flow through the gaseous polar molecules, which turn in the field so that their charges align with the field.*

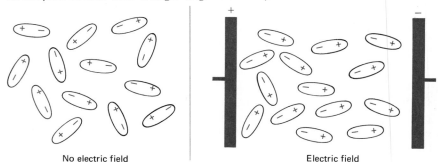

No electric field   Electric field

TABLE 9.14

Dipole moments of some gaseous polyatomic molecules

The direction of the dipole for each bond and each molecule is shown by the arrows.

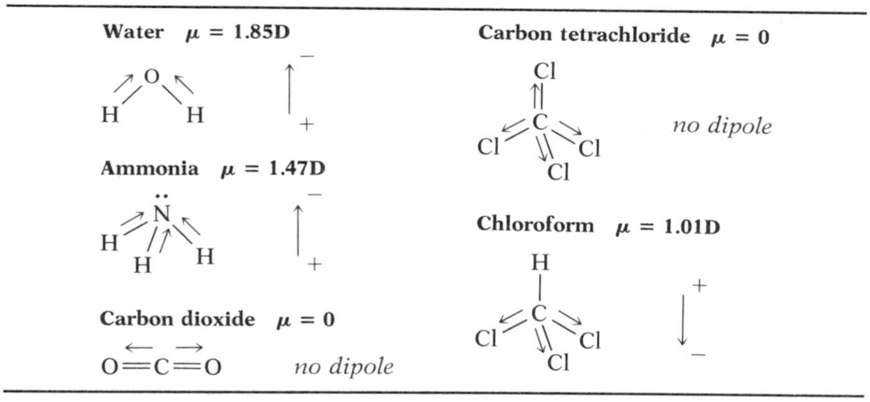

Water $\mu = 1.85D$

Ammonia $\mu = 1.47D$

Carbon dioxide $\mu = 0$

O=C=O   *no dipole*

Carbon tetrachloride $\mu = 0$

no dipole

Chloroform $\mu = 1.01D$

For a diatomic molecule in the gaseous state, the measured dipole moment (Figure 9.7) is a direct indication of the polarity of the bond. The decrease in bond polarity with decreasing electronegativity of the halogen atom is shown by the dipole moments of the hydrogen halides.

	$\delta+$ $\delta-$	$\delta+$ $\delta-$	$\delta+$ $\delta-$	$\delta+$ $\delta-$
	H—F	H—Cl	H—Br	H—I
$\mu =$	1.9 D	1.04 D	0.79 D	0.38 D
electronegativity of halogen =	4.0	3.0	2.8	2.5

The dipole moment of a polyatomic molecule is determined by both bond polarity and molecular geometry (Table 9.14). In the angular water molecule, negative charge is concentrated on the oxygen atom, but the balancing positive charge is effectively centered between the two hydrogen atoms, leading to a large dipole moment. The ammonia molecule, which has triangular pyramidal geometry (Section 9.26) also has a large dipole moment.

Internal compensation of partial charges in a molecule of suitable geometry can lead to an overall absence of polarity ($\mu = 0$) even though the individual bonds are themselves polar. For example, in the linear arrangement of atoms in carbon dioxide (Table 9.14) one strongly polar carbon–oxygen double bond cancels the other.

The symmetrical regular tetrahedral geometry of carbon tetrachloride (CCl_4) provides the same internal compensation (see Table 9.14). However, in chloroform ($CHCl_3$), the hydrogen atom is both less electronegative than the carbon atom and markedly different in electronegativity, size, and electronic atmosphere from the chlorine atoms. Thus, negative charge is concentrated among the chlorine atoms, and the molecule has a dipole moment.

9.18 Properties imparted by the covalent bond

Most covalently bonded substances form discrete molecules—individual molecules that retain their structure and identity when they are separated. The covalent bond is a highly directed bond, and

TABLE 9.15

General properties of covalent compounds

Melting and boiling points, and heats of vaporization and fusion all tend to increase with increasing molecular weight for similar covalent compounds. The boiling point range of covalent compounds (not including network covalent compounds like diamond; Section 9.19) is approximately −250° to 600°C.

Gases, liquids, or solids

Solids brittle and weak, or soft and waxy

Low melting points

Low boiling points

Low heats of vaporization

Low heats of fusion

Low densities

Poor conductors of heat and electricity

Nonpolar compounds generally soluble in nonpolar solvents

Polar compounds generally soluble in polar solvents

each covalent molecule has a distinct geometry (Section 9.26) determined by the bonding in the molecule. For example, the three atoms of CO_2 are arranged in a straight line, but the H_2O molecule is bent (see Table 9.14). A common arrangement of four atoms about a fifth is the tetrahedron (Table 9.14, carbon tetrachloride and chloroform).

Most liquid and solid covalent substances have relatively low melting and boiling points, and heats of fusion and vaporization because not much energy is required to separate the molecules from each other (Tables 9.15, 9.16). Solid covalent substances are usually either soft and waxy, or brittle and easily broken up.

The melting and boiling points, and the heats of vaporization and fusion, of covalent substances (with the exception described in the next section) are generally much lower than those of ionic or metallic substances. Among covalent substances, just how low or high these properties range does not depend on the force of the covalent bond itself, for the covalent bond remains intact during melting and vaporization. The properties listed depend instead on intermolecular forces (Sections 9.20–9.22), the weaker forces acting *between* molecules.

9.19 Network covalent substances

A small number of substances such as diamond (crystalline carbon), borazon (one form of boron nitride, BN), carborundum (silicon carbide, SiC), and quartz (silicon dioxide, SiO_2) have most of the characteristics of covalent compounds, except that they are very hard and have very high melting points (approximate range 2000°C–6000°C). In these materials no discrete molecules are present. The diamond

TABLE 9.16

Properties of some typical nonpolar covalent substances

Property	Oxygen O_2	White phosphorus P_4	Carbon tetrachloride CCl_4	Iodine I_2	Silicon tetrabromide $SiBr_4$
Appearance	Colorless gas	White waxy solid	Colorless liquid	Violet-black crystals	Colorless fuming liquid
Molecular weight (amu)	32.0	123.8952	153.823	253.8090	347.72
Melting point (°C)	−218.4	44.1	−22.96	113.6	5.4
Boiling point (°C)	−183	280.5	76.75	185	154
Density (g/cm³)	1.429 (g/liter) (g, 0°C)	1.83 (s, 25°C)	1.59 (l, 20°C)	4.93 (s, 20°C)	2.77 (l, 25°C)
Heat of fusion at m.p. kcal/mole (kJ/mole)	0.106 (0.444)	0.15 (0.63)	0.60 (2.5)	3.74 (15.6)	2.0 (8.4)
Heat of vaporization at b.p. kcal/mole (kJ/mole)	1.63 (6.82)	2.97 (12.4)	7.17 (30.0)	19.95 (83.47)	9.1 (38)
Electrical conductivity Solid Liquid	Very small Very small	Very small Very small	Very small Very small	Very small Small	Very small Very small

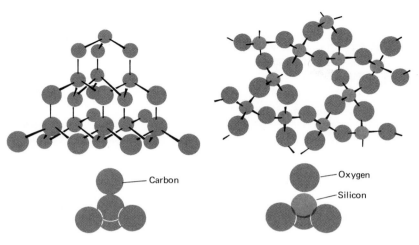

(a) Diamond, m.p. > 3550°C, b.p. 4827°C   (b) Quartz, m.p. 1610°, b.p. > 2200°C

— Carbon

— Oxygen

— Silicon

FIGURE 9.8

Network covalent substance. *Three-dimensional array of covalently bonded atoms. Diamond and quartz structures are shown as examples. On a hardness scale from 1 to 10 diamond has a hardness of 10 and quartz of 7. For comparison, rock salt has a hardness of 2; marble. of 3; and iron, of 4–5 on the same scale.*

and the other materials like it are **network covalent substances**—three-dimensional arrays of covalently bonded atoms.

In diamond (Figure 9.8a) each carbon atom is covalently bonded to four other carbon atoms, each of which is bonded to three more carbon atoms, and so on. In quartz (Figure 9.8b), a form of silica, SiO_2, each Si atom is bonded to four O atoms and each O atom is bonded to two Si atoms in an infinite three-dimensional array. Many covalent bonds must be broken to disrupt the crystals, resulting in the characteristic hardness, high melting and boiling points, and high heats of fusion and vaporization of these substances.

For network covalent substances as for ionic substances, the simplest formula represents only the stoichiometric composition. A piece of the substance is a giant molecule of molecular weight determined only by the size of the piece.

Intermolecular forces

Intermolecular forces act between molecules. They are much weaker than the metallic, ionic, or covalent bonding forces. The strength of the intermolecular forces at a particular temperature determines whether a molecular covalent substance is a gas, liquid, or solid at that temperature. There are three principal types of intermolecular forces: dipole–dipole forces, hydrogen bonding, and London forces. Collectively these are called **van der Waals forces**—the short-range intermolecular forces.

9.20 Dipole–dipole forces

In **dipole–dipole interaction,** molecules with permanent dipoles attract each other electrostatically; the positive end of one molecule attracts the negative end of another molecule, and so on, leading to an

TABLE 9.17

Properties of some nonmetal hydrides

The data for the second-period elements are given in black, those for the
third-period elements in color.

Property	Methane CH_4 (nonpolar)	Silane SiH_4 (nonpolar)	Ammonia NH_3 (polar, H bonded)	Phosphine PH_3 (polar)	Water H_2O (polar, H bonded)	Hydrogen sulfide H_2S (polar)	Hydrogen fluoride HF (polar, H bonded)	Hydrogen chloride HCl (polar)
Appearance	Colorless gas	Colorless gas	Colorless gas	Colorless gas	Colorless liquid	Colorless gas	Colorless fuming gas, or liquid	Colorless gas
Molecular weight (amu)	16.042	32.092	17.030	33.998	18.015	34.076	20.006	36.461
Melting point (°C)	−182.48	−185.0	−77.7	−133.81	0.00	−85.60	−83.04	−114.2
Boiling point (°C)	−161.49	−111.2	−33.4	−87.78	100.00	−60.75	19.54	−84.9
Density of liquid (g/ml)	0.415 (−164°C)	0.68 (−185°C)	0.683 (−33.4°C)	0.746 (−87.78°C)	0.958 (100°C)	0.950 (−61.3°C)	0.991 (19.5°C)	1.187 (−84.9°C)
Heat of fusion at m.p. kcal/mole (kJ/mole)	0.225 (0.941)	0.159 (0.665)	1.351 (5.52)	0.270 (1.13)	1.44 (6.02)	0.568 (2.38)	1.09 (4.56)	0.476 (1.99)
Heat of vaporization at b.p. kcal/mole (kJ/mole)	1.96 (8.20)	2.9 (13)	5.58 (23.3)	3.49 (14.6)	9.72 (40.7)	4.46 (18.7)	6.13 (25.6)	3.86 (16.11)
Equivalent conductance of liquid (ohm^{-1} cm^{-1})	ca. 0	ca. 0	5×10^{-9} (33.4°C)	ca. 0	6×10^{-8} (25°C)	3.7×10^{-11} (−78.3°C)	1.4×10^{-5}	—
Dipole moment (D)	0	0	1.47	0.55	1.85	1.10	1.9	1.04

alignment of the molecules (see Figures 9.6 and 9.9). Dipole–dipole forces are not very effective in the gaseous state, where molecules are far apart. As molecules approach each other (when the temperature drops or pressure increases) the dipole attraction can pull them together into a liquid or solid.

The dipole moment is a measure of the polarity of a molecule. Compare the dipole moments (Table 9.17) and the melting and boiling points of SiH_4, PH_3, and H_2S, which have similar molecular weights. SiH_4 is nonpolar while H_2S has a dipole moment twice that of PH_3. As would be expected on the basis of dipole–dipole interaction, SiH_4 has the lowest melting and boiling points and H_2S has the highest.

9.21 Hydrogen bond

Hydrogen bonding is also an electrostatic attraction, but it is stronger than ordinary dipole–dipole interactions and occurs only between a few kinds of atoms. When a hydrogen atom is covalently bonded to an electronegative atom that strongly attracts the single electron pair, the small hydrogen atom has little electron density around it. Under these circumstances, the hydrogen atom can act as a bridge to another electronegative atom. For example, hydrogen fluoride crystals contain infinitely long chains in which each hydrogen atom can be thought of as covalently bonded to one fluorine atom and hydrogen-bonded to another.

$$\text{H} \diagdown_{\text{F}} \diagup \text{H} \overset{\text{F}}{\underset{140°}{\curvearrowright}} \text{H} \diagdown_{\text{F}} \diagup \text{H}$$

The hydrogen bond in HF is strong enough to persist in the liquid state, where the chains are shorter and vary in length, and even in the gaseous state, where the chains are still shorter—in $(HF)_n$, $n = 2-6$.

A **hydrogen bond** is the attraction of a hydrogen atom covalently bonded to an electronegative atom for a second electronegative atom (Figure 9.9). The strongest hydrogen bonds form between H and F, N, or O, which are small atoms with their negative charge highly concentrated in a small volume.

The effect of hydrogen bonding on the properties of compounds is clearly illustrated by the nonmetal hydrides (see Table 9.17). Monosilane, SiH_4, is less volatile (shown by its higher boiling point) than methane, as would be expected from its larger molecular weight. However, in each of the other comparable pairs from a periodic family (NH_3–PH_3, H_2O–H_2S, HF–HCl), the compound with the smaller molecular weight is less volatile and has the higher melting point, the larger heats of fusion and vaporization, and the larger electrical conductance, because of the extra energy needed to break the intermolecular hydrogen bonds in NH_3, H_2O, and HF.

Hydrogen bonds can form between atoms in different molecules or in the same molecule. The large molecules in living systems are often held in exactly the shape they need for their specific biochemical function by hydrogen bonds. DNA (deoxyribonucleic acid) is the molecule that passes on the genetic code to the next generation. Hydrogen bonding is one of the two forces that hold the DNA double helix structure in shape. The ease and speed with which hydrogen bonds can be produced and reformed makes them ideal for this purpose. (Hydrogen bond energies are about 5–10 kcal/mole; Section 9.24.) The energy required to make and break equivalent covalent bonds in these large, biochemically important molecules would be much too great. (See Table 9.19 for some covalent bond energies.)

9.22 London force

The third type of intermolecular force, and the weakest, is the London force. This force acts on all atoms and molecules, polar or nonpolar, and is responsible for the liquefaction of even the monatomic noble gases at low enough temperatures. (The London force is one of the short-range forces represented by the constant a in van der Waals equation for nonideal gases; Section 3.16. Sometimes the phrase "van der Waals attraction" is used incorrectly for the London force alone; more correctly all of the short-range forces that help to liquefy gases are called van der Waals forces.)

The London force is the result of a nonpermanent dipole in one

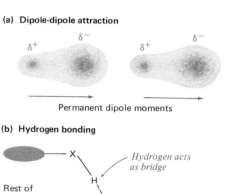
(a) Dipole-dipole attraction

δ^+ δ^- δ^+ δ^-

Permanent dipole moments

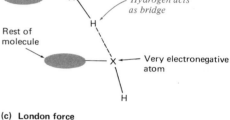
(b) Hydrogen bonding

Rest of molecule

Hydrogen acts as bridge

Very electronegative atom

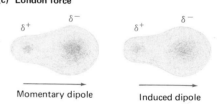
(c) London force

δ^+ δ^- δ^+ δ^-

Momentary dipole   Induced dipole

FIGURE 9.9
Intermolecular forces.

TABLE 9.18

London forces and boiling points
The decrease in boiling point in each of the series of compounds is due to decreasing London forces.

1. Same molecular weight, increasingly compact shape

$CH_3CH_2CH_2CH_2CH_3$
n-pentane, C_5H_{12}
b.p. 36°C

$CH_3CH_2CHCH_3$
      |
     CH_3
Isopentane, C_5H_{12}
b.p. 28°C

```
        CH_3
         |
CH_3 — C — CH_3
         |
        CH_3
```
Neopentane, C_5H_{12}
b.p. 9.5°C

2. Similar shape, increasing molecular weight

CH_4
Methane, CH_4
b.p. −161.5°C

$CH_3CH_2CH_3$
n-propane, C_3H_8
b.p. −44.5°C

$CH_3CH_2CH_2CH_3$
n-Butane, C_4H_{10}
b.p. −0.5°C

atom inducing a dipole in the next atom, and so on. For example, at a given instant one atom might have more of its electron density on one side than on the other (see Figure 9.9). The slight positive charge on one side would attract electrons from a neighboring atom, and, for that instant, a force of attraction would exist between these atoms. The **London force** is the attraction between fluctuating dipoles in atoms and molecules that are very close together. The fluctuating dipoles occur, of course, because electrons are constantly in motion.

The strength of London forces is influenced by the size and shape of the molecules involved. Close contact between the molecules over a larger area gives greater opportunity for dipole interaction than when only a small amount of contact is possible. In the series of pentanes in Table 9.18, the London forces get weaker, as shown by the lower boiling points, as the molecules get closer to spherical in shape (spheres can make contact at only one point). Among molecules of similar shape (Table 9.18b), London forces increase with an increasing number of electrons, that is, with increasing molecular weight. The difference in boiling point between CH_4 and SiH_4 (Table 9.17) is due to the different strengths of the London forces. Melting points are affected by crystal geometry and other factors in addition to London forces, and therefore may not show as good a correlation with molecular weight and shape as do boiling points.

Properties of bonds

9.23 The continua of bond types

Although we often talk and write about "ionic compounds," or "covalent bonds," or "metallic bonding," the bonding in most chemical species is *not* 100% ionic, or 100% covalent, or 100% metallic. Instead, the bonding is somewhere in between—anywhere along a continuum of bonding from covalent to ionic, or from ionic to metallic, or from metallic to covalent. We speak of a bond as "covalent" if the properties expected for covalently bound atoms predominate, or we speak of an alloy as "metallic" if it exhibits mainly the properties we expect of metals.

The triangular diagram in Figure 9.10 gives examples of compounds that fall along the bond type continua. At the corners are lithium, a metal that is a good conductor of electricity, has a low electronegativity, and easily gives up its valence electron to form ionic compounds; cesium fluoride, CsF, a highly ionic compound between the most electronegative nonmetal and the least electronegative metal; and fluorine, F_2, a molecule with a 100% covalent bond between two small very electronegative atoms.

Only by determining many properties of a substance can the nature of its bonding be understood. Information about the spectra, bond energies, interatomic distances, dipole moments, and structures of substances can be useful. But no single experimental method

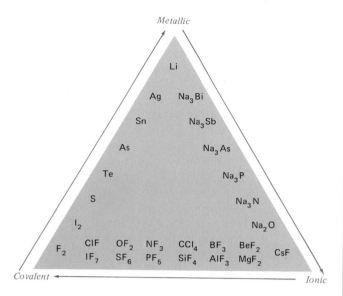

FIGURE 9.10
The metallic–covalent–ionic bonding continua. *Examples of substances that exhibit the variation of bonding from purely metallic in lithium, to purely covalent in fluorine, to highly ionic in cesium fluoride (the compound between the least electronegative metal and the most electronegative nonmetal). (From Chemical Constitution, J. A. A. Ketelaar, Elsevier, Amsterdam, 1958, p. 21.)*

or type of calculation has yet proved totally reliable in determining or predicting quantitatively what type of bond exists in a particular species.

One approach to finding out where a compound lies on the ionic–covalent continuum is the theoretical calculation of the percent contribution of ionic and covalent bonding to the total bond energy. Figure 9.11 shows the results of such calculations done by one of several possible methods. Here only Cl_2 has a 100% bond type—it is 100% covalent. Among the compounds shown, HCl, CCl_4, NCl_3, Cl_2O, and ClF, for example, are thought of as covalent compounds. LiCl, NaCl, $MgCl_2$, KCl, and CaCl are thought of as ionic compounds.

9.24 Bond strength and bond length

The changes in energy that occur when two ions, A^+ and B^-, approach each other can be expressed as a function of the distance between their nuclei. At an infinite distance no forces act between the two particles. As they approach, forces of both attraction and repulsion gradually come into play.

The attraction between the positive charge of A^+ and the negative charge of B^- is the major influence on energy in the approach of two oppositely charged ions. Additional forces arise from the attraction of electrons in one atom for the nucleus of the other, the London force, and electron–electron and nucleus–nucleus repulsions. When the ions get close enough together so that their filled inner shells overlap, an additional large repulsive force is added.

FIGURE 9.11
Chloride bond energies. *The colored areas represent the contribution of ionic bonding to the total element-to-chlorine bond energy, as calculated by the method of Sanderson (Chemical Bonds and Bond Energy, P. T. Sanderson, Academic Press, New York, 1971, p. 103).*

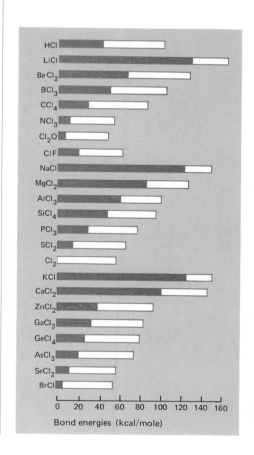

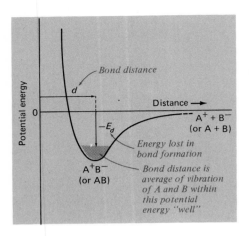

FIGURE 9.12

Energy–distance relationship in bond formation. *The values of d and E_d vary for different molecules, but if bond formation is to occur the general shape of the curve is as shown.*

The sum of all the forces acting between two ions yields a curve with a general shape like that in Figure 9.12. The potential energy is at a minimum when the ions are at distance d, which is the bond distance, or **bond length**—the average distance from nucleus to nucleus in the stable compound. (It is necessary to say *average* because the bonded atoms or ions always vibrate back and forth within the region of bond stability.) The energy $-E_d$ is the energy given up in formation of the bond, showing that the bonded ions are in a more stable state than the nonbonded ions.

In the approach of two neutral atoms about to form a covalent bond, the large positive-ion–negative-ion attraction is missing. However, all of the other forces described above act between two approaching neutral atoms A and B. Plotting the sum of these forces alone would give a very shallow curve. Not enough energy would be lost to lead to the formation of a stable bond. The additional energy needed for covalent bond formation arises when the outermost, incompletely filled orbitals overlap. Electrons are shared in pairs, they can move freely in the space surrounding both atoms, and in this condition they have lower energy than when they are not shared. The general shape of the curve for the approach of A and B to form a covalent bond is similar to that in Figure 9.12. In a covalent bond most of the energy comes from electron-pair sharing.

The enthalpies of reactions in which bonds are broken are a measure of the strength of chemical bonds. The values are always reported for reactions in the gaseous state, where the atoms formed are distinctly separate and free from attraction or repulsion by other atoms, molecules, or ions.

In some reactions it is possible to rupture a single chemical bond to give the free atoms.

$$H_2(g) \longrightarrow H(g) + H(g) \qquad \Delta H° = 104.2 \text{ kcal } (436.0 \text{ kJ})$$
$$HI(g) \longrightarrow H(g) + I(g) \qquad \Delta H° = 71.3 \text{ kcal } (298 \text{ kJ})$$

Bond dissociation energy is the enthalpy per mole for breaking exactly one bond of the same type per molecule. The bond dissociation energy of the H–H bond is 104.2 kcal/mole; the bond dissociation energy of the H–I bond is 71.3 kcal/mole.

When it is only possible to rupture several bonds in a single reaction, bond energies, also called average bond energies, must be used. For example, four carbon–hydrogen bonds are broken in the gas-phase dissociation of methane.

$$CH_4(g) \longrightarrow C(g) + 4H(g) \qquad \Delta H° = 397.6 \text{ kcal } (1664 \text{ kJ})$$

One-fourth of the enthalpy of the preceding reaction is the average bond energy of the C–H bond, 99.4 kcal/mole. **Bond energy** is the *average* enthalpy per mole for breaking one bond of the same type per molecule.

In general, the larger the bond energy, the more resistant the bond is to thermal rupture and the stronger the bond. The amount of energy released in the formation of a particular bond between two atoms is numerically the same, but with the opposite sign, as that re-

TABLE 9.19

Average bond energies in kcal/mole

To convert these values to kJ/mole, multiply by 4.184 kJ/kcal.

	F	O	Cl	N	Br	I	S	C	H
H—	136	111	103	86	88	71	88	99	104
C—	117	78	78	68	65	52	69	79	—
C=	—	192	—	123	—	—	139	141 / 121 if alternating — and =	

174 if —C(=O)—

	F	O	Cl	N	Br	I	S	C	H
C≡	—	257	—	205	—	—	—	194	
S—	82	101	65	—	50	—	59		
S=	—	125	—	—	—	—			
I—	67	—	50	—	42	36			
Br—	68	—	52	39	46				
N—	67	55	48	38					
N=	—	141	—	113					

97 if —NO$_2$
88 if —NO$_3$

	F	O	Cl	N	Br	I	S	C	H
N≡	—	—	—	226					
Cl—	61	49	58						
O—	51	34							
O=	—	119							
F—	38								

quired to break the same bond. This makes possible the use of bond energies to find reaction enthalpies.

The enthalpy of a reaction can be estimated by finding the difference between the energy needed to form all of the bonds in the products and the energy needed to form all of the bonds in the reactants. This is equivalent to the following:

$$\Delta H_r^\circ = -\left(\begin{array}{c}\text{sum of bond energies of}\\\text{all bonds in products}\end{array}\right) + \left(\begin{array}{c}\text{sum of bond energies of}\\\text{all bonds in reactants}\end{array}\right)$$

The accuracy of such calculations is limited. The strength of a chemical bond is not independent of its surroundings, that is, of the molecule in which it exists. Bonds between A and B in different compounds, or even in the same compound, may have different strengths, and the energy to break them may vary with the reaction sequence. For example,

$$\text{NH}_3(g) \longrightarrow \text{NH}_2(g) + \text{H}(g) \qquad \Delta H^\circ = 104 \text{ kcal } (435 \text{ kJ})$$
$$\text{NH}_2(g) \longrightarrow \text{NH}(g) + \text{H}(g) \qquad \Delta H^\circ = 90 \text{ kcal } (377 \text{ kJ})$$
$$\text{NH}(g) \longrightarrow \text{N}(g) + \text{H}(g) \qquad \Delta H^\circ = 85 \text{ kcal } (356 \text{ kJ})$$

For single covalent bonds, the energies range from about 40 to 140 kcal/mole (160–600 kJ/mole) (Table 9.19) and the bond lengths from roughly 0.05 to 0.2 nm (Table 9.20). In general, stronger bonds have shorter bond lengths. Multiple bonds, with much greater electron density between the atoms, are always stronger and shorter than single bonds.

TABLE 9.20

Some bond lengths

Bond lengths are given here in the SI unit, nanometers; they are also often given in angstroms (Å): 1 nm = 10 Å.

H–H	0.074
H–Cl	0.127
H–C	0.109
C–C	0.154
C=C	0.134
C≡C	0.120
C–O	0.143
C=O	0.120
C≡O	0.113
Si–O	0.166
N–N	0.145
N=N	0.125
N≡N	0.110
Cl–Cl	0.198

THE CHEMICAL BOND [9.24]

By comparison with the bond energies, the weak intermolecular force energies range from about 0.1 to 1.0 kcal/mole for the London force to about 5–10 kcal/mole for hydrogen bonds (except for the stronger hydrogen–fluorine hydrogen bond, which has an energy of about 35 kcal/mole).

EXAMPLE 9.2

■ Estimate the heat released as one mole of *n*-butane burns.

$$
\begin{array}{cccc}
\text{H} & \text{H} & \text{H} & \text{H} \\
| & | & | & | \\
\text{H}-\text{C}-\text{C}-\text{C}-\text{C}-\text{H} \\
| & | & | & | \\
\text{H} & \text{H} & \text{H} & \text{H}
\end{array}
\ (g) + (13/2)\ \text{O}\!=\!\text{O}\ (g) \longrightarrow
$$

$$
4\ \text{O}\!=\!\text{C}\!=\!\text{O}\ (g) + 5\ \text{H}-\text{O}-\text{H}\ (g)
$$

In this reaction we find that 3 C—C, 10 C—H and (13/2) O=O bonds are being broken and 8 C=O and 10 O—H bonds are being formed. Thus ΔH_r° is

$$
\Delta H_r^\circ = -\left[(8\ \text{mole})\left(192\ \frac{\text{kcal}}{\text{mole}}\right) + (10\ \text{mole})\left(111\ \frac{\text{kcal}}{\text{mole}}\right)\right]
$$

$$
+ \left[(3\ \text{mole})\left(79\ \frac{\text{kcal}}{\text{mole}}\right) + (10\ \text{mole})\left(99\ \frac{\text{kcal}}{\text{mole}}\right)\right.
$$

$$
\left. + \left(\frac{13}{2}\ \text{mole}\right)\left(119\ \frac{\text{kcal}}{\text{mole}}\right)\right]
$$

$$
= -646\ \text{kcal}
$$

We would predict $\Delta H_r^\circ = -646$ kcal. The actual value is -635 kcal (-2660 kJ) for the reaction as written. ■

9.25 Atomic and ionic radii

Since atoms are approximately spherical in shape, their sizes can be expressed in terms of their radii. Of course, a single atom cannot be isolated so that it can be measured. Radii are determined by measuring the distances between atoms in the combined state and assigning part of the distance to each atom. For atoms of the same element the atomic radius is one-half of the covalent single bond distance. For example, the bond distance in the chlorine molecule :C̈l: C̈l: is 0.1986 nm and the chlorine atom is assigned an atomic radius of one-half the bond distance, or 0.0993 nm. When the bond distance between two atoms of different elements is measured, the predetermined radius of one atom must be used to find the radius of the other atom. The term **atomic radius** is commonly used for the radius of an atom in a single covalent bond (Figure 9.13). (Some confusion exists over this term. It has also been used for *van der Waals radii*, the radii of *nonbonded* atoms held together only by weak intermolecular forces. In this book the "atomic radii" are covalent single bond radii.) The **bond angle** is the angle between two atoms that are both bonded to the same third atom.

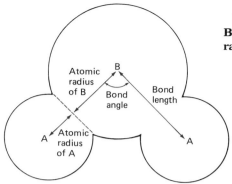

FIGURE 9.13
Bond length, bond angle, and atomic radii.

When an atom loses an electron its radius decreases and when an atom gains an electron its radius increases.

Na atomic radius = 0.157 nm   Cl atomic radius = 0.0993 nm
Na^+ ionic radius = 0.095 nm Cl^- ionic radius = 0.181 nm

The **ionic radius** is the radius of a cation or anion. Some atomic and ionic radii are given in Table 9.21. Tables of atomic and ionic radii are used in the prediction of bond distances in compounds.

Tables also sometimes give the **metallic radius**—the radius assigned to a metal ion as it exists in the pure metal.

9.26 Molecular geometry: Valence-shell electron-pair repulsion theory

The most important determining factor in the geometry of covalently bonded molecules is the number of electron pairs in the valence shell of the atoms involved. Because of electrostatic repulsion, valence-shell electron pairs arrange themselves around an atom so that they are as far apart as possible. The approach to predicting molecular geometry based on this principle is called the valence-shell electron-pair repulsion (VSEPR) theory.

Consider a molecule AB_n, in which A is the central atom and has no lone pair, or nonbonded, electrons. The ideal arrangements—the

Species	Radius (nm)	Species	Radius (nm)	Species	Radius (nm)	Species	Radius (nm)
						S	0.104
H	0.037	Be	0.0889	C	0.0771	S^{2-}	0.184
		Be^{2+}	0.035	Si	0.1173		
Li	0.1225			Sn	0.1412	F	0.072
Li^+	0.068	Mg	0.1364	Sn^{2+}	0.093	F^-	0.133
		Mg^{2+}	0.066				
Na	0.1572			Sn^{4+}	0.071	Cl	0.099
Na^+	0.095	Ba	0.1981			Cl^-	0.181
		Ba^{2+}	0.134	N	0.074		
Rb	0.216			P	0.110	Br	0.114
Rb^+	0.147	B	0.080			Br^-	0.196
Cs	0.235	Al	0.1248	O	0.074	I	0.133
Cs^+	0.167	Al^{3+}	0.052	O^{2-}	0.132	I^-	0.220

TABLE 9.21

Some typical atomic and ionic radii

Listing radii as fixed numbers in tables should not lead us to forget that atoms and ions are not solid spheres, but electron clouds, and that the radii are always influenced to some extent by the environment of a specific atom or ion. Values of radii also may vary with the method of determination.

TABLE 9.22

Geometry of covalent molecules with no unshared electron pairs on the central atom

In the molecules shown here all valence electrons on the central atom (A) participate in covalent bonds. Geometry could possibly be limited by the size of A and B. Note: The geometry summarized here and in Table 9.23 applies to most cases where A is a representative element and often when A is a transition element. When d orbitals enter into bonding, there are additional geometric possibilities (Chapter 28).

Formula type	Shared electron pairs	Arrangement of B atoms relative to A atoms (ideal BAB bond angle)	Structures
AB_2	2	Linear (180°)	 B——A——B
AB_3	3	Triangular planar (120°)	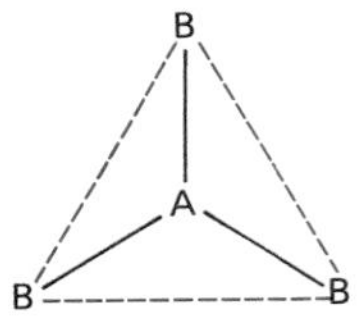
AB_4	4	Tetrahedral (109°28′)	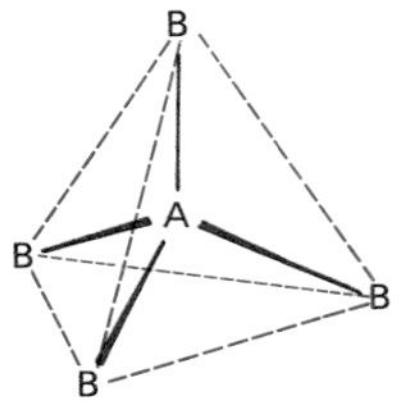
AB_5	5	Triangular bipyramidal (B^aAB^a, 180°) (B^eAB^e, 120°) (B^eAB^a, 90°)	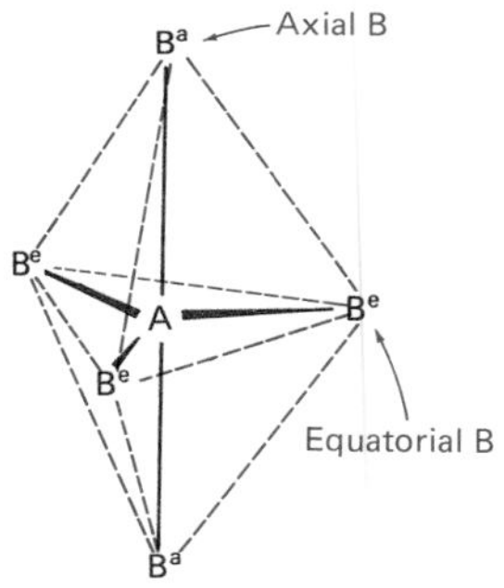
AB_6	6	Octahedral (90°)	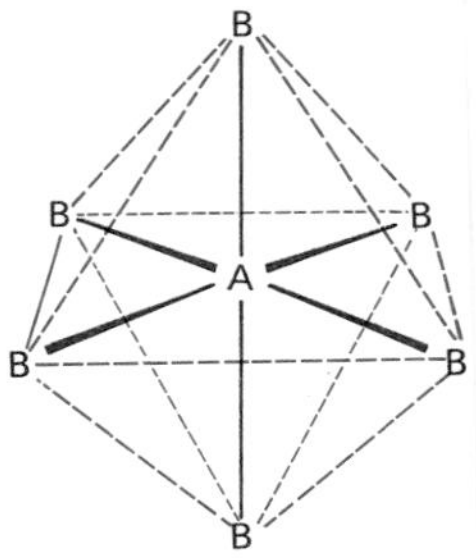

ones that minimize repulsion among electron pairs in the valence shell of atom A—are linear for $n = 2$, triangular planar for $n = 3$, tetrahedral for $n = 4$, triangular bipyramidal for $n = 5$, and octahedral for $n = 6$. Table 9.22 illustrates these shapes.

In the molecules illustrated in Table 9.22 all valence electrons on the A atoms are involved in electron-pair bonds, and the B—A—B bond angles are very close to the ideal. What happens when atom A has one or more pairs of nonbonding electrons? The *general* shape of the molecule is simply that in which the space of one electron-pair bond is filled by the nonbonding pair. For example (Figure 9.14), replacing one bond in AB_3 by a nonbonding pair (the compound is symbolized AB_2E) leaves the two B atoms attached to A at an angle, ideally 120°.

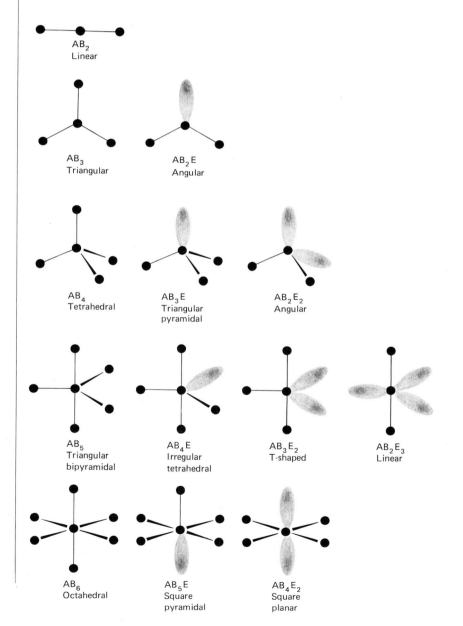

FIGURE 9.14

General shapes of covalent molecules with bonding and nonbonding electron pairs. *A nonbonding electron pair is symbolized by E. With a nonbonded electron pair on A, the BAB bond angles tend to be smaller than ideal. (From R. J. Gillespie, 1963, Journal of Chemical Education, Vol. 40, p. 295.)*

TABLE 9.23

Geometry of covalent molecules AB_n and AB_nE_m

E represents an unshared electron pair. The observed molecular shape is derived
from the geometry as shown in Figure 9.14. See Note in Table 9.22 caption.

Type formula	Shared electron pairs	Unshared electron pairs	Observed molecular shape	Idealized molecular geometry	Examples
AB_2	2	0	Linear	Linear	$CdBr_2$
AB_2E	2	1	Angular	Triangular planar	$SnCl_2$, PbI_2
AB_2E_2	2	2	Angular	Tetrahedral	OH_2, OF_2, SCl_2, TeI_2
AB_2E_3	2	3	Linear	Triangular bipyramidal	XeF_2
AB_3	3	0	Triangular planar	Triangular planar	BCl_3, BF_3, GaI_3
AB_3E	3	1	Triangular pyramidal	Tetrahedral	NH_3, NF_3, PCl_3, $AsBr_3$
AB_3E_2	3	2	T-shaped	Triangular bipyramidal	ClF_3, BrF_3
AB_4	4	0	Tetrahedral	Tetrahedral	CH_4, $SiCl_4$, $SnBr_4$, ZrI_4
AB_4E	4	1	Irregular tetrahedral	Triangular bipyramidal	SF_4, $SeCl_4$, $TeBr_4$
AB_4E_2	4	2	Square planar	Octahedral	XeF_4
AB_5	5	0	Triangular bipyramidal	Triangular bipyramidal	PF_5, $PCl_5(g)$, SbF_5
AB_5E	5	1	Square pyramidal	Octahedral	ClF_5, BrF_5, IF_5
AB_6	6	0	Octahedral	Octahedral	SF_6, SeF_6, $Te(OH)_6$, MoF_6

The nonbonding electron pairs, however, spread out around atom A more than electron pairs in bonds (because nonbonding pairs are attracted by only one nucleus, not two). The lone pair on atom A crowds both the shared electron pairs of A–B bonds and other unshared pairs on atom A. So in most cases the BAB bond angles in compounds where atom A has one or more lone pairs are compressed and are smaller than the ideal. For example, the bond angle in CH_4, with no lone pair, is the ideal tetrahedral angle 109.47°. In NH_3, with one lone pair, the extra repulsion of the lone pair decreases the angle to 106.67° and in H_2O with two lone pairs, the angle is 104.45°.

Table 9.23 summarizes the geometry of molecules with and without nonbonding electron pairs on the central atoms and gives further examples. When d and/or f orbitals enter into the molecular bonding of transition metal atoms, other types of molecular geometry are possible. For example, the ion $[PtCl_4]^-$, which falls in the AB_4 class, has square planar geometry—all of the five atoms lie in the same plane. (The bonding and geometry of such species are discussed in Chapter 28.)

By using Tables 9.22 and 9.23, good predictions can be made for the general shape of many molecules, once the Lewis formula and the number of lone pairs have been determined. An excellent demonstration of this was the correct prediction of the structures of the previously unknown compounds of xenon, shown in Figure 9.15 (see An Aside: Noble Gas Compounds, in this chapter).

EXAMPLE 9.3

■ What are the molecular geometries of (a) BrF, (b) BrF$_3$, and (c) BrF$_5$? Which of these compounds are dipoles?

The Lewis structures of these compounds are

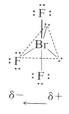

(a) Any diatomic molecule is linear. Because of the difference in the electronegativities of F and Br, the covalent bond between the atoms will be polar and the molecule will be a dipole.

$$: \overset{..}{\underset{..}{Br}} — \overset{..}{\underset{..}{F}} :$$
$$\delta+ \qquad\qquad \delta-$$

(b) Table 9.23 tells us that the geometric arrangement of the three F atoms and two unshared pairs of electrons around the Br in BrF$_3$ will be in the shape of a capital T

(c) According to Table 9.23, the geometric arrangement of five F atoms and one unshared pair of electrons around the Br in BrF$_5$ will be in the shape of a square pyramid

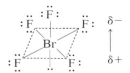

Again, each Br–F bond will be polar and the entire molecule will be a dipole. ■

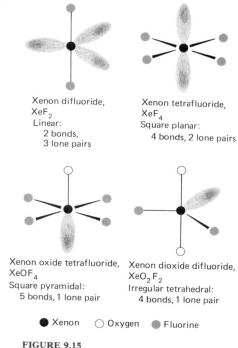

Xenon difluoride, XeF$_2$
Linear:
  2 bonds, 3 lone pairs

Xenon tetrafluoride, XeF$_4$
Square planar:
  4 bonds, 2 lone pairs

Xenon oxide tetrafluoride, XeOF$_4$
Square pyramidal:
  5 bonds, 1 lone pair

Xenon dioxide difluoride, XeO$_2$F$_2$
Irregular tetrahedral:
  4 bonds, 1 lone pair

● Xenon ○ Oxygen ● Fluorine

FIGURE 9.15
Structures of the xenon fluorides and oxyfluorides.

As indicated in the drawing, each Br–F bond will be polar and the entire molecule will be a dipole.

Chemical bond definitions

There is a chemical bond between two atoms or groups of atoms in case that the forces acting between them are such as to lead to the formation of an aggregate with sufficient stability to make it convenient for the chemist to consider it as an independent molecular species.

THE NATURE OF THE CHEMICAL BOND, by Linus Pauling (p. 6, 3rd ed., Cornell University Press, Ithaca, New York, 1960)

Particles attract one another by some force, which in immediate contact is exceeding strong, at small distances performs the chymical operations, and reaches not far from the particles with any sensible effect.

OPTICKS, by Sir Isaac Newton, 1730 (Quoted from p. 166 in *A Short History of Chemistry*, by J. R. Partington, Harper, New York, 1937)

It is clear that the intimate description of a chemical bond of which we have spoken, must be essentially electronic. It is the behaviour and distribution of electrons around the nucleus that gives the fundamental character of an atom: it must be the same for molecules. In one sense, therefore, the description of the bonds in any molecule is simply the description of the electron distribution in it.

VALENCE, by C. A. Coulson (p. 3, Clarendon Press, Oxford, England, 1952)

Chemical bonds result when the electron structure of an atom is altered sufficiently to link it with the electron structure of another atom or atoms.

CHEMICAL BONDING, in the *Encyclopedia Brittanica* (15th ed., Vol. 4, p. 84, 1974)

If the net attractive interaction between two atoms or among more than two atoms is sufficiently strong so that the unique properties of the combination can be studied experimentally before it decomposes, the atoms are said to be held together by chemical bonds.

CHEMICAL BONDS AND BOND ENERGY, by R. T. Sanderson (p. 1, Academic Press, New York, 1971)

THE CHEMICAL BOND

Significant terms defined in this chapter

chemical bond
valence electron
valence shell
octet rule
nonbonding electron pairs
lone pair
metallic bonding
noble gas core
pseudo-noble-gas core
ionic bonding
single covalent bond
covalent bonding
multiple covalent bond
double covalent bond

triple covalent bond
bond order
coordinate covalent bond
donor atom
acceptor atom
resonance
resonance hybrid
nonpolar covalent bond
polar covalent bond
electronegativity
electronegative atom
electropositive atom
dipole
dipole moment

network covalent substance
van der Waals force
dipole–dipole interaction
hydrogen bond
London force
bond length
bond dissociation energy
bond energy
atomic radius
bond angle
ionic radius
metallic radius

Exercises

9.1 A common laboratory technique to remove oxygen from a gas is to pass the gas over hot finely divided copper

$$2Cu(s) + O_2(g) \longrightarrow 2CuO(s)$$

The copper can be regenerated for further use by reducing the CuO with hydrogen

$$CuO(s) + H_2(g) \longrightarrow Cu(s) + H_2O(g)$$

What type of intramolecular bonding would be found in (a) Cu, (b) O_2, (c) CuO, (d) H_2, and (e) H_2O? Write the Lewis structures for (f) O_2, (g) CuO, (h) H_2 and (i) H_2O. What will be the geometric shapes of (j) O_2 and H_2 and (k) H_2O? (l) Are any of these substances paramagnetic? (m) Are any of the substances polar molecules? What types of intermolecular bonding would be found in (n) O_2, (o) H_2, and (p) H_2O?

9.2 Distinguish between: (a) intermolecular and intramolecular bonding, (b) covalent and ionic bonding, (c) polar and nonpolar covalent bonding, (d) covalent and coordinate covalent bonding, (e) electrical conduction in a metal and by an ionic substance, (f) valence and bond order, (g) donor and acceptor atom, (h) paramagnetic and diamagnetic molecules, (i) atomic and ionic radii, and (j) bond dissociation energy and average bond energy.

9.3 Which statements are true? Rewrite any false statement so that it is correct.
(a) The chemical bonds in HCl and DCl (deuterium chloride) are essentially the same.

(b) Once formed, a coordinate covalent bond is no different from an ordinary covalent bond.
(c) The presence of polar covalent bonds in a molecule will make it a dipole.
(d) Electronegativity is the energy released when an electron is added to a gaseous neutral atom to produce a gaseous anion.
(e) Electrons are shared equally in a nonpolar covalent bond, shared unequally in a polar covalent bond, and transferred in an ionic bond.
(f) Because of intramolecular vibrations, we might expect the bond lengths of the four C–Cl bonds in CCl_4 to be slightly different at any given instant.
(g) The covalent bonds in the so-called "network covalent substances" are considerably different from those in ordinary molecules.
(h) The higher value of the constant a in the van der Waals constant for aniline, $C_6H_5NH_2$, predicts that there are stronger intermolecular forces present than in toluene, $C_6H_5CH_3$.
(i) The Lewis electron dot structures shown for a case of resonance really represent the limiting situations and the true electronic distribution is somewhere in between these limits.

9.4 What are the three basic types of chemical bonding? Give an example of each. What are the differences among these types? What general properties are associated with each type of bonding?

9.5 Describe three types of intermolecular forces. Give an example of each. What general properties are associated with each of these types?

9.6 Name two compounds that contain network covalent bonds. What kinds of properties are associated with these substances? How do we calculate the molar weight of these substances?

9.7 Why are metallic and ionic bonds nondirectional?

9.8 What is the octet rule? In which of the following species is this rule violated?

(a) H—H

(b) $:\!\ddot{F}\!-\!\ddot{F}\!:$

(c) $:\!P\!\!<\!\!\stackrel{\ddot{P}}{\underset{\ddot{P}}{|}}\!\!>\!\!P\!:$

(d) $H\!-\!\underset{\underset{H}{|}}{\overset{\overset{H}{|}}{C}}\!-\!H$

(e) $:\!\ddot{O}\!-\!\dot{\underset{..}{Cl}}\!-\!\ddot{O}\!:$

(f) $:\!\ddot{F}\!-\!\dot{\ddot{Xe}}\!\dot{}\!-\!\ddot{F}\!:$

(g) $:\!\ddot{O}\!=\!C\!=\!\ddot{O}\!:$

(h) $\left[:\!\ddot{O}\!-\!\underset{\underset{:\ddot{O}:}{|}}{\overset{\overset{:\ddot{O}:}{|}}{S}}\!-\!\ddot{O}\!:\right]^{2-}$

9.9 Classify the chemical bonds that would form between the following pairs of atoms as mainly (i) ionic, (ii) polar covalent, or (iii) nonpolar covalent: (a) Li, O; (b) Br, I; (c) Mg, H; (d) O, O; (e) H, O; (f) Si, O; (g) N, O; and (h) Sr, F.

9.10 Draw Lewis structures for the following molecules and polyatomic ions: (a) H_2O_2, (b) OH^-, (c) H_2S, (d) IO_4^-, (e) C_2H_3Cl, (f) NO, (g) SeF_4, (h) ONCl, (i) F_2O, (j) BeH_2, (k) ClO_4^-, (l) SO_2, (m) NOF, (n) ICl_4^-, (o) NH_4^+, (p) H_2SO_4, (q) CH_3COOH, (r) NCl_3, (s) $AsCl_5$, and (t) POF_3. Which of these are diamagnetic?

9.11 Describe the molecular geometry of the species given in Exercise 9.10.

9.12 Two resonance structures for benzene, C_6H_6, can be written as

What would you predict for the carbon–carbon bond length in benzene if the carbon–carbon single bond length is 0.154 nm and the double bond length is 0.134 nm?

9.13 Identify the major intermolecular force for each of the uncharged species given in Exercise 9.10.

9.14 What intermolecular forces are present in liquid ammonia, NH_3, and methane, CH_4? Which of these compounds should have the lower freezing and boiling points? Which substance would you expect to be a liquid over the larger temperature range?

9.15 There are two different molecular structures of butane, C_4H_{10}. Which would you expect to have the higher boiling point?

$CH_3\!-\!\underset{\underset{CH_3}{|}}{\overset{\overset{H}{|}}{C}}\!-\!CH_3 \qquad CH_3CH_2CH_2CH_3$

9.16 The enthalpy of vaporization of H_2S is 4.463 kcal/mole, of H_2Se, 4.62 kcal/mole, and of H_2Te, 5.55 kcal/mole. (a) Explain this general trend. (b) Predict the heat of vaporization of H_2O using these data. (c) Why is there a large difference between your answer for part (b) and the measured value of 9.7171 kcal/mole?

9.17 Which substance has the greatest intermolecular force to overcome: (a) NaF, NaCl, NaBr, or NaI—considering electronegativity; (b) P_4, S_8, or Cl_2—considering mass; (c) CO_2 or SO_2—considering dipole moment; (d) F_2 or Ar—considering shape; (e) *n*-octane and isooctane

considering shape; and (f) CH_4 or CCl_4—considering mass?

9.18 Write the Lewis structure for molecular $AlCl_3$. Note that the octet rule is not obeyed by this molecule. In the gaseous phase two molecules of $AlCl_3$ are joined together (dimerized) to form Al_2Cl_6. Write the Lewis structure for this molecule.

9.19 The structures for three molecules having the formula $C_2H_2Cl_2$ are

Describe the intermolecular forces in each of these compounds.

9.20 Draw a likely structure for the dimer of acetic acid, which is formed as a result of two hydrogen bonds between the two individual molecules. The structure of acetic acid is

9.21 Would ΔH_r° for both of the following reactions be the same if you used average bond energies to find the enthalpy change? Give a reason for your answer.

$$H-C-C-C-H + (13/2)\,O_2(g) \rightarrow 4CO_2(g) + 5H_2O(l)$$

$$H-C-C-C-C-H + (13/2)\,O_2(g) \rightarrow 4CO_2(g) + 5H_2O(l)$$

9.22 Why are ΔH_f° values calculated using bond energies only approximate? Even though values of bond energies are given for molecules and atoms in the gaseous phase, they are often used for reactions in solid, liquid, and solution phases. What additional information is needed so that these calculations are more valid?

9.23* There are five compounds having the formula CH_xCl_y. (a) Write the Lewis structures for these substances. (b) Discuss the intermolecular bonding in each. (c) Try to arrange these substances in order of increasing boiling point.

9.24* Why does HF have a lower boiling point and lower heat of vaporization than H_2O even though the molecular weights are nearly the same and the hydrogen bonds between molecules of HF are stronger?

9.25* Correctly assign the boiling points to the substances within each group on the basis of intermolecular forces: (a) Ne, Ar, Kr, $-246.048°C$, $-185.7°C$, $-152.30°C$; (b) N_2, HCN, C_2H_6, $-195.8°C$, $-88.63°C$, $26°C$; (c) NH_3, H_2O, HF, $-33.35°C$, $19.54°C$, $100.00°C$; and (d) CaO, KF, CsI, $1280°C$, $1505°C$, $2850°C$.

9.26 One way to calculate the approximate amount of ionic character in a chemical bond between two atoms is by using the following equation:

% ionic character = (16)(electronegativity difference) + (3.5)(electronegativity difference)²

Note that we always find the electronegativity difference by subtracting the value for the less electronegative element from the value for the more electronegative element (see Table 9.13). Calculate the % ionic character in chemical bonds between: (a) Cs and F, (b) Na and Cl, (c) S and O, (d) I and Cl, and (e) N and N using this equation.

9.27 Following are some heats of formation at 25°C: $C(g)$, 171.291 kcal/mole; $CH(g)$, 142.4 kcal/mole; $CH_2(g)$, 93.7 kcal/mole; $CH_3(g)$, 33.2 kcal/mole; $CH_4(g)$, -17.88 kcal/mole; and $H(g)$, 52.095 kcal/mole. Find ΔH_r° for each of the following equations for the stepwise decomposition of CH_4 and find the C–H bond energy

$$CH_4(g) \longrightarrow CH_3(g) + H(g)$$
$$CH_3(g) \longrightarrow CH_2(g) + H(g)$$
$$CH_2(g) \longrightarrow CH(g) + H(g)$$
$$CH(g) \longrightarrow C(g) + H(g)$$

9.28 Calculate the S–F bond energy if $\Delta H_f^\circ = -289$ kcal/mole for $SF_6(g)$, 66.636 kcal/mole for $S(g)$ and 18.88 kcal/mole for $F(g)$.

9.29 Using the bond energies given in Table 9.19, predict the heats of combustion for the following fuels:

$$CH_4(g) + 2O_2(g) \longrightarrow CO_2(g) + 2H_2O(g)$$
methane

$$C_2H_4(g) + 3O_2(g) \longrightarrow 2CO_2(g) + 2H_2O(g)$$
ethylene

$$H_2(g) + \tfrac{1}{2}O_2(g) \longrightarrow H_2O(g)$$
hydrogen

$$C_2H_6(g) + \tfrac{7}{2}O_2(g) \longrightarrow 2CO_2(g) + 3H_2O(g)$$
ethane

$$C_2H_2(g) + \tfrac{15}{2}O_2(g) \longrightarrow 2CO_2(g) + H_2O(g)$$
acetylene

$$CH_3OH(g) + \tfrac{3}{2}O_2(g) \longrightarrow CO_2(g) + 2H_2O(g)$$
methyl alcohol

Which fuel delivers the most heat per gram of fuel? Which fuel delivers the most heat per gram of total reactants?

9.30* Using the equation given in Exercise 9.26, find the difference in electronegativity that is required to give a bond (a) 60% ionic and 40% covalent character and (b) 40% ionic and 60% covalent character.

9.31* The van der Waals constants for n-pentane are $a = 19.01$ liter^2 atm/mole^2 and $b = 0.1460$ liter/mole and for isopentane are $a = 18.05$ liter^2 atm/mole^2 and $b = 0.1417$ liter/mole. Basing your reasoning on intermolecular forces, why do you expect a for n-pentane to be higher? Basing your reasoning on molecular size, why do you expect b for n-pentane to be higher? Find the pressure of one mole of each of these gases at 325°K in a container of 25.0 liters. Which gas deviates more from ideality?

9.32* Using the bond energies given in Table 9.19, estimate the heat of reaction for

$$H_2(g) + \tfrac{1}{2}O_2(g) \longrightarrow H_2O(g)$$

Since the heat of reaction is released by the steam as it cools from the temperature of the flame to 25°C, what is the flame temperature if the specific heat of steam is 0.5 cal/g °C?

9.33* The entry in Table 9.19 for the O–H bond was calculated from the following reaction

$$H_2(g) + \tfrac{1}{2}O_2(g) \longrightarrow H_2O(g) \qquad \Delta H° = -57.796 \text{ kcal}$$

using the equation

$$\Delta H_r° = -(\text{sum of bond energies of all bonds in products}) + (\text{sum of bond energies of all reactants})$$

$$
\begin{aligned}
-57.796 &= -[(2 \text{ mole})(\text{O—H})] \\
\text{kcal} &\quad + [(1 \text{ mole})(\text{H—H}) + (0.5 \text{ mole})(\text{O}{=}\text{O})] \\
&= -[(2 \text{ mole})(\text{O—H})] \\
&\quad + [(1 \text{ mole})(104 \text{ kcal/mole}) \\
&\quad + (0.5 \text{ mole})(119 \text{ kcal/mole})]
\end{aligned}
$$

giving 111 kcal/mole for each bond. Using the same technique, confirm the value for a C$=$O bond in CO$_2$ using the reaction

$$CH_4(g) + 2O_2(g) \longrightarrow CO_2(g) + 2H_2O(g)$$
$$\Delta H° = -191.76 \text{ kcal}$$

assuming that the values for the O$=$O, O—H, and C—H bonds are known.

9.34** The electronic configuration of each element in the third period of the periodic table, Na–Ar, gives us much information about the compounds these elements form with each other. (a) Write the electron dot formulas for the valence electrons for these elements.

Several of these elements exist in molecular form: (i) Na as Na$_2$ in the gaseous state, (ii) P$_4$, (iii) S$_8$, and (iv) Cl$_2$. (b) Write the Lewis dot structures for these molecular forms. What type of bonding would you expect to find (c) in solid Na, Mg, and Al; (d) in solid Si; (e) between the molecules of Na$_2$, P$_4$, S$_8$, and Cl$_2$; and (f) in Ar?

Compounds between the following pairs of elements are known to exist: (v) Na, P; (vi) Na, Cl; (vii) Mg, Si; (viii) Mg, P; (ix) Mg, S; (x) Mg, Cl; (xi) Al, S; (xii) Al, Cl; (xiii) Si, S; (xiv) Si, Cl; (xv) P, S; (xvi) P, Cl; and (xvii) S, Cl. Which of these pairs form bonds that are (g) ionic, (h) polar covalent, and (i) nonpolar covalent? (j) Predict the formulas of these compounds and (k) write the Lewis structures. (l) Determine the geometry of the molecular compounds and (m) predict which will be dipoles. (n) Why are there no compounds listed containing Ar? (o) Are there any compounds that have resonance structures? (p) List the paramagnetic substances.

272

10 PERIODIC PERSPECTIVE: THE REPRESENTATIVE ELEMENTS

At the beginning of this book we talked about examining the world around us and discovering as much as possible about the materials to be found there. Such a process, applied with diligence and imagination, has led to the periodic table, to quantum theory, and beyond, into realms of mathematics.

Sometimes chemistry becomes so theoretical that it is easy to forget its relationship to real substances that can be seen and touched. The periodic table provides the ideal focal point at which invisible electrons and orbitals can be related to seeable, touchable matter—the elements. The word "aluminum" means an element whose atoms have 13 protons, 14 neutrons (in its stable isotope), and 13 electrons. Aluminum atoms have an outer shell configuration of $3s^2 3p$ and can readily form both ionic and covalent bonds. "Aluminum" also means a silvery white metal. Aluminum occurs naturally combined with oxygen and silicon in silicates. Aluminum is quite soft and light; is used in pots and pans, airplanes, and packaging materials; and its oxide, alumina, forms rubies and sapphires and is used in making glass.

In individual chapters later in this book we deal with the descriptive chemistry of the elements in detail—what the substances themselves look like, and how they behave; what kinds of compounds they form and what the compounds look like and how they behave; and in what ways we can manipulate these substances and put them to use. Here, we pause to look up and down and across the columns and rows of the periodic table.

In this chapter we examine the significant relationships that exist among the representative elements. Emphasis is given to those properties of the elements that can be correlated with electronic configurations. Among the topics considered are metallic and nonmetallic character, atomic and ionic sizes, ionization energies and electron affinities, the types of chemical bonds formed between the elements, and factors that determine bond formation. The importance of the role of the environment in which chemical combination occurs is also considered. We conclude the chapter with a broad overview of significant trends in behavior among the elements of the individual representative groups.

Nonmetals: Poor conductors
High electron affinities
High ionization energies

Types of representative elements

The representative elements (Table 10.1) include all of the nonmetals, all of the semiconducting elements, and the representative metals. Each representative period consists of eight elements, except for the first period, which is limited to hydrogen and helium, since at helium the first electronic shell is complete with its pair of s electrons. In the other periods, s and p energy levels are being filled in a regular fashion until the $ns^2 np^6$ noble gas configuration is reached. As we have seen (Section 9.3), this configuration is one of very great chemical stability, and the noble gases do not readily enter into chemical combination.

In three places the atomic numbers of successive representative elements are not consecutive. Between calcium and gallium, and between strontium and indium lie the elements of the 3d- and 4d-transition series, in which the 3d and 4d energy levels are being filled. Between barium and thallium are both the lanthanide elements cerium through lutetium, in which the 4f energy level is being occupied, and the 5d-transition series, lanthanum through gold. As a result of the interposition of the d- and f-transition series, beginning with the potassium–krypton period and continuing for the remaining periods, the elements of families III through VII have atoms with pseudo-noble-gas cores, that is, with eighteen electrons in the next to outermost shell. The atoms of all of the other representative elements have noble gas cores.

10.1 Nonmetals

Seventeen of the representative elements are nonmetals (see Table 10.1). At ordinary temperatures some are gases (e.g., hydrogen and oxygen), others are solids (e.g., sulfur and iodine) and one, bromine, is a liquid. Nonmetals are ordinarily very poor conductors of electric current. Their atoms have high electron affinities (Section 5.8) and high ionization energies, reflecting the ease with which they gain electrons and the difficulty with which they lose them. Nonmetal atoms tend to form simple monatomic anions of noble gas configuration by acquiring electrons from metal atoms. This tendency increases across a sequence of nonmetals (e.g., from carbon to fluorine) and decreases down a group (e.g., from fluorine to iodine). In line with their large electron affinities, nonmetals, with the exception of the noble gases, have large electronegativities. The most electronegative elements are at the right in the periodic table; these elements often react as oxidizing agents.

When nonmetals react with each other, covalent substances are almost always formed. Whether a compound between a nonmetal and a metal is primarily ionic or covalent is influenced by many factors (Section 10.9), two of importance being the electronegativity

Group I	Group II	Group III	Group IV	Group V	Group VI	Group VII	Noble Gases
$_1$H $1s^1$							$_2$He $1s^2$
$_3$Li $2s^1$	$_4$Be $2s^2$	$_5$B $2s^22p^1$	$_6$C $2s^22p^2$	$_7$N $2s^22p^3$	$_8$O $2s^22p^4$	$_9$F $2s^22p^5$	$_{10}$Ne $2s^22p^6$
$_{11}$Na $3s^1$	$_{12}$Mg $3s^2$	$_{13}$Al $3s^23p^1$	$_{14}$Si $3s^23p^2$	$_{15}$P $3s^23p^3$	$_{16}$S $3s^23p^4$	$_{17}$Cl $3s^23p^5$	$_{18}$Ar $3s^23p^6$
$_{19}$K $4s^1$	$_{20}$Ca $4s^2$ *	$_{31}$Ga $3d^{10}4s^24p^1$	$_{32}$Ge $3d^{10}4s^24p^2$	$_{33}$As $3d^{10}4s^24p^3$	$_{34}$Se $3d^{10}4s^24p^4$	$_{35}$Br $3d^{10}4s^24p^5$	$_{36}$Kr $3d^{10}4s^24p^6$
$_{37}$Rb $5s^1$	$_{28}$Sr $5s^2$ *	$_{49}$In $4d^{10}5s^25p^1$	$_{50}$Sn $4d^{10}5s^25p^2$	$_{51}$Sb $4d^{10}5s^25p^3$	$_{52}$Te $4d^{10}5s^25p^4$	$_{53}$I $4d^{10}5s^25p^5$	$_{54}$Xe $4d^{10}5s^25p^6$
$_{55}$Cs $6s^1$	$_{56}$Ba $6s^2$ *	$_{81}$Tl $4f^{14}5d^{10}6s^26p^1$	$_{82}$Pb $4f^{14}5d^{10}6s^26p^2$	$_{83}$Bi $4f^{14}5d^{10}6s^26p^3$	$_{84}$Po $4f^{14}5d^{10}6s^26p^4$	$_{85}$At $4f^{14}5d^{10}6s^26p^5$	$_{86}$Rn $4f^{14}5d^{10}6s^26p^6$
$_{87}$Fr $7s^1$	$_{88}$Ra $7s^2$						

semiconducting elements

metals — nonmetals

TABLE 10.1

Electronic configurations of the representative elements and noble gases

*The heavy line separates the metals from the semiconducting elements, shown in color, and the nonmetals. d and f transition series intervene at the points marked *.*

of the nonmetal and the energy needed to remove electrons from the metal atoms.

Because of the tendency of their atoms to achieve noble gas configurations by covalent bonding, pure nonmetals exist in a variety of molecular states of aggregation. Only the noble gases are monatomic. Covalent bonding between atoms of the same element is called **catenation**. For example, white phosphorus, P_4, is a catenated form of phosphorus. In the catenated forms of the nonmetallic elements adjacent atoms are linked by covalent bonds and each atom has an octet configuration. Before a catenated nonmetallic element can take part in a chemical reaction, covalent bonds must be broken. The energy required for this process is an important factor in determining their reactivity.

The noble gases, nitrogen, and oxygen occur in the atmosphere. Carbon and sulfur are found in pure elemental forms in nature. All of the other nonmetals are too reactive to remain uncombined, and occur naturally only in compounds.

Carbon, nitrogen, oxygen, and fluorine, the first members of their respective periodic groups, are significantly different in chemical behavior from the other members of their groups. These differences arise mainly because in these second-period elements *only* the $2s$ and the three $2p$ orbitals are available for bonding and no more than four covalent bonds can be formed. For the heavier members of each group, not only are s and p orbitals available, but also the d orbitals of the valence shell.

The descriptive chemistry of the nonmetals is presented in the following chapters of this book: Chapter 11, hydrogen and oxygen; Chapter 20, the halogens; Chapter 22, nitrogen, phosphorus, and sulfur; and Chapter 23, carbon and the hydrocarbons. The noble gases have been discussed in Chapter 6.

THE REPRESENTATIVE ELEMENTS [10.1]

Metals: Good conductors
Low ionization energies
Low electron affinities

10.2 Metals

Metals have low ionization energies, low electron affinities, and relatively large atomic radii. Some general properties of metals and nonmetals are contrasted in Table 10.2. The most reactive metals are at the left in the periodic table; they tend to behave as reducing agents toward other substances. Among the representative metals, only bismuth is found as a free element in nature. The other representative metals are reactive enough to be found only in the combined form, unlike the transition metals, many of which occur naturally as free metals.

Across a period of representative elements the elements become less metallic and more nonmetallic in their properties. Metallic properties persist further to the right in the periodic table for the heavier elements than for the lighter elements. In the final complete period of representative elements, even in the fifth family, bismuth ($4f^{14}$ $5d^{10}$ $6s^2$ $6p^3$) is a metal, although its conductivity is low. Note, therefore, that metallic character in not limited to atoms with one, two, or three electrons in the outermost shell. Instead, as the atoms become more complex and the outermost electrons are further from the nucleus, metallic properties become more pronounced. Even iodine, a nonmetal, has luster and can conduct electricity, although very

TABLE 10.2

General properties of metallic and nonmetallic elements

The boiling point range for metals is approximately 650°C to 6000°C; for nonmetals it is approximately −250°C to 450°C (except for carbon, which is a network covalent substance). Note that the properties of many elements do not agree with these general properties. For example, sodium melts below 100°C, mercury melts at −38.87°C, carbon as diamond melts at about 3700°C and is the hardest of all substances, tin is brittle above 161°C, lithium has a low density, carbon as graphite is a metallic conductor, and iodine can also conduct electricity, although only weakly.

Metals	Nonmetals
Physical properties	
Solids of high melting point, yielding high-boiling liquids	Gases or low-melting solids, yielding low-boiling liquids
Opaque	Often transparent or translucent
Lustrous, reflecting light of many wavelengths	Dull, reflecting light poorly or absorbing strongly
High density	Low density
Often hard	Often soft
Malleable, ductile, strong	Brittle, weak
Conductors of heat and electricity	Insulators
Chemical properties	
Binary compounds with hydrogen uncommon	Binary compounds with hydrogen common, often complex
Oxides and hydroxides are alkaline	Oxides and hydroxides are acidic
Halides often ionic	Halides often covalent
Form simple cations; complex cations and anions	Form simple and complex anions; few cations

poorly. The descriptive chemistry of the representative metals is presented in Chapter 26.

10.3 Semiconducting elements

In appearance, the seven semiconducting elements boron, silicon, germanium, arsenic, antimony, selenium, and tellurium (see Table 10.1) resemble metals. They reflect visible and infrared light much less effectively than metals, but they do have a gray, metallic luster. The electrical and optical differences between metals and semiconducting elements are related to differences in bonding (Chapter 27). In the semiconducting elements, the "free" electrons are much less mobile than those in metals.

In chemical behavior, the semiconducting elements are primarily nonmetallic in character. For example, their binary compounds with hydrogen and the halogens are covalent. The ranges of the first ionization energies and electron affinities of the semiconducting elements are intermediate between those of metals and nonmetals (Table 10.3).

The descriptive chemistry of the semiconducting elements is discussed in Chapter 27.

Properties of atoms and ions

10.4 Atomic and ionic radii

The radius of an atom or an ion is determined primarily by the arrangement of the electrons and how strongly they are pulled in toward the nucleus. If only one electron is present in an atom, the total force acting on the electron is that of the total nuclear charge, that is, of the number of protons in the nucleus. However, when many electrons are present there is a **screening effect**—a decrease in the nuclear charge acting on an electron mainly due to the effect of other electrons in inner shells.

Furthermore, the closer an electron comes to the nucleus, the stronger the nuclear attraction for the electron. This is called the **penetration effect**—the increase in the nuclear charge that acts on electrons that approach the nucleus more closely. For a given n value, the degree of penetration by an electron depends on which orbital the electron occupies (see Figures 7.14–7.17), and decreases in the order of $s > p > d > f$. The s electrons get closer to the nucleus than the p electrons in the same shell and so on.

The **effective nuclear charge** is the portion of the nuclear charge that acts on an electron; it is influenced by both the screening and penetration effects and is less than the actual nuclear charge.

The nuclear charge increases regularly across a period because protons are being added to the nucleus of the atom. Each electron added to the outermost shell produces a very small screening effect.

TABLE 10.3

Comparison of ranges of electron affinities and ionization energies
For the electron affinity ranges we have considered only experimentally measured values (see Table 10.6).

First ionization energies, in kcal/mole
(kJ/mole)

Nonmetals	239–567	(1000–2372)
Semiconducting elements	199–226	(833–946)
Metals	89.8–215	(376–900)

Electron affinities, in kcal/mole
(kJ/mole)

Nonmetals	0–83.3	(0–349)
Semiconducting elements	18.4–46.6	(77.0–195)
Metals	8.07–28.8	(33.8–120.5)

Therefore the effective nuclear charge increases regularly across the period, and the electrons are pulled progressively closer to the nucleus. As a result, atomic radii generally decrease across a period of representative elements.

Within a family of representative elements, however (see, for example lithium–cesium in Figure 10.1), there is a general increase in atomic radii as the atomic number increases. Nuclear charge increases down the family at the same time that new shells of electrons are added (see Table 10.1). These two factors work in opposite directions in their effect on radii, but the addition of new electronic shells predominates.

Because of the intervention of the transition elements, gallium, indium, and thallium have much smaller atomic radii, relative to their Group II neighbors, than do boron and aluminum. Another result of the transition element intervention and the relatively greater atomic numbers of the elements following them is the closeness in size of the second and third period elements in a family, beginning with aluminum and gallium and going to the right in the table (see Figure 10.1).

Atomic and ionic radii are related in many important ways to the

FIGURE 10.1

Atomic radii of the representative elements (in nm). *The values for the noble gases for which compounds are not known were found by extrapolation. Atomic radii of the radioactive elements are not known either because of the impossibility of obtaining measurable quantities (At, Fr) or because of experimental problems caused by intense radioactivity (Ra).*

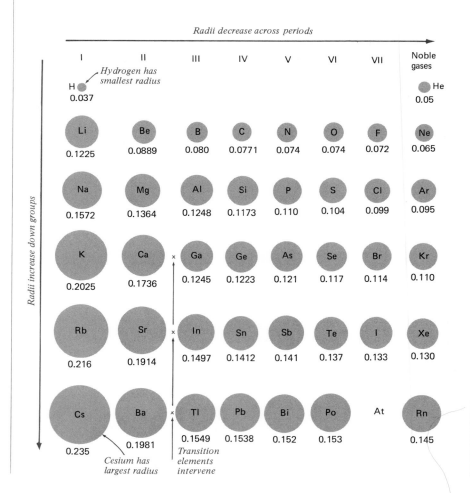

[10.4] THE REPRESENTATIVE ELEMENTS

278

chemical behavior of the elements. In general, smaller atoms give up their valence electrons less readily than larger atoms. The closer electrons are to the positive nucleus and the smaller the screening effect of electrons in inner shells, the less easily are the outer electrons lost. This effect contributes to the decrease in metallic character in going from left to right across a period.

For the representative elements, many similarities in chemical behavior are found within each family. These similarities are mainly due to the common type of electronic configuration in the valence shells. In each family, however, the first member differs in many respects from the other elements in the family, partly due to its comparatively small atomic size.

In their chemistry, lithium, beryllium, and boron bear many resemblances to the elements diagonally below them in the next families.

This diagonal relationship can be explained as follows: As the atomic size decreases from left to right across a period, there is a general decrease in the ease with which electrons are lost. The decrease in metallic character in moving from an element to the next one at the right is at least partially compensated for by moving one step down in the latter family to a larger atom; thus the elements on the diagonal are roughly similar to each other.

The ionic radii of the representative elements are plotted by periods in Figure 10.2. Species that have identical configurations are

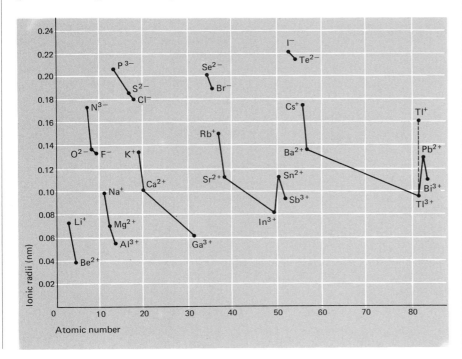

FIGURE 10.2
Variation of ionic radii of representative elements with atomic number. *The lines connect isoelectronic series.*

said to be **isoelectronic.** In an isoelectronic series such as the sodium, magnesium, and aluminum ions, all of which have the $1s^2\, 2s^2\, 2p^6$ configuration, the increasing effective nuclear charge pulls the electron shells progressively closer, and the ionic radii decrease. The radii of the isoelectronic negative ions of phosphorus, sulfur, and chlorine of configuration $1s^2\, 2s^2\, 2p^6\, 3s^2\, 3p^6$ show a similar decrease with increasing nuclear charge.

10.5 Ionization energies

Ionization energies and electron affinities (next section) are of use in two ways: (1) Their relative values show the relative tendencies of atoms to lose and gain electrons, properties of importance in compound formation; and (2) their specific values are needed for the calculation of the heats of various reactions and physical changes. [Recall that the ionization energy and electron affinity are measured for electron loss and gain in the gas phase (Section 5.8). Also recall that Hess' law allows calculations for the heat of a reaction by any route. Therefore, an *actual* gas-phase reaction need not be part of an actual chemical change for these quantities to be useful in calculating the heat of that reaction.]

The energy necessary to remove the least tightly bound electron from a gaseous atom is called the **first ionization energy.** Loss of this electron gives an ion with a single positive charge. The energy necessary to remove the most loosely held electron from the gaseous $+1$ ion is the **second ionization energy.** Third, fourth, and so on, ionization energies can be measured similarly.

Ionization energy values are influenced by four interdependent atomic characteristics and often do not vary uniformly with atomic structure.

1. *The actual charge on the nucleus.* An increase in nuclear charge without any other major change is generally responsible for an increase in ionization energy because of the greater attraction of the nucleus for electrons.

2. *The screening effect.* An increase in the number of electrons in the shells between the nucleus and the valence shell results in an increase in the extent to which the outermost electrons are shielded from the nucleus and thus in a decrease in ionization energy. Factors 1 and 2 are represented by the effective nuclear charge. *All other factors being equal,* the greater the effective nuclear charge, the larger the ionization energy.

3. *The size of the atom.* It would be predicted that an increase in atomic radius *alone* would be accompanied by a decrease in ionization energy, as the electron would be further from the nuclear charge. However, the size of an atom is dependent upon the number of electrons *and* the nuclear charge, and strict comparisons of the effect of change of size on ionization energies are therefore limited.

4. *The nature of the energy level (s, p, d, or f) from which an electron is being removed.* In any one shell, s electrons (those in the lowest energy

Group I	Group II	Group III	Group IV	Group V	Group VI	Group VII	Noble gases
H 1							**He** 2
314							567
Li 3	**Be** 4	**B** 5	**C** 6	**N** 7	**O** 8	**F** 9	**Ne** 10
124	215	191	260	335	314	402	497
Na 11	**Mg** 12	**Al** 13	**Si** 14	**P** 15	**S** 16	**Cl** 17	**Ar** 18
118	176	138	188	253	239	300	363
K 19	**Ca** 20	***Ga** 31	**Ge** 32	**As** 33	**Se** 34	**Br** 35	**Kr** 36
100	141	138	182	226	225	273	323
Rb 37	**Sr** 38	***In** 49	**Sn** 50	**Sb** 51	**Te** 52	**I** 53	**Xe** 54
96	131	133	169	199	208	241	280
Cs 55	**Ba** 56	***Tl** 81	**Pb** 82	**Bi** 83	**Po** 84	**At** 85	**Rn** 86
90	120	141	171	168	194		248
Fr 87	**Ra** 88						
	122						

ionization energies increase across period

highest first ionization energy

ionization energies decrease down group

lowest first ionization energy

TABLE 10.4

First ionization energies of the representative elements

(*) *Transition elements intervene. The values are given in kilocalories per mole at 0°K. To convert these values to kJ/mole, multiply by 4.184 kJ/kcal. Note that these values differ somewhat from the values at 25°C used for several calculations in this chapter.*

level) are expected to be most tightly bound and f electrons least tightly bound.

Among the representative elements first ionization energies generally increase across a period as the elements become less metallic (Table 10.4). The nuclear charge is regularly increasing in this direction, while the screening effect of the underlying shells is constant; therefore, the effective nuclear charge is regularly increasing. Ionization energies generally decrease down a periodic group, as the size of the atom increases, while the outer electronic configuration remains the same and the effective nuclear charge remains about the same.

Successive electrons are usually removed from an atom with increasing difficulty. This is illustrated by the ionization energies for the third period elements in Table 10.5. The effective nuclear charge of the atom increases with the loss of each electron, and consequently the remaining electrons are more tightly held. Once the electrons in the valence shell are lost and a noble gas or pseudo-noble-gas configuration is obtained, there is a very large increase in the quantity of energy required to remove the next electron. The energy required is well beyond that obtained by chemical means, and no positive ions are formed in which an atom of a representative element has lost more electrons than are found in its outermost shell.

10.6 Electron affinities

The energy released when a single electron is added to a neutral gaseous atom is the electron affinity. (Recall that electron affinity is customarily given as a positive number.)

TABLE 10.5

Ionization energies of Na–Ar period elements

Energy for the removal of successive valence electrons is given in black and for removal of inner shell electrons is given in color. Values are in kilocalories per mole at 0°K. To convert these values to kJ/mole, multiply by 4.184 kJ/kcal. Underneath the valence shell each of these elements has a neon-like ($1s^2 2s^2 2p^6$) configuration. The shielding effect across the period is approximately constant. To convert to approximate values at 298°K add 1.5 kcal per electron removed.

Ionization energy	$_{11}$Na ($3s^1$)	$_{12}$Mg ($3s^2$)	$_{13}$Al ($3s^2 3p^1$)	$_{14}$Si ($3s^2 3p^2$)	$_{15}$P ($3s^2 3p^3$)	$_{16}$S ($3s^2 3p^4$)	$_{17}$Cl ($3s^2 3p^5$)	$_{18}$Ar ($3s^2 3p^6$)
1st	118	176	138	188	253	239	300	363
2nd	1091	347	434	377	453	540	549	637
3rd	1652	1850	656	772	696	808	920	943
4th	2280	2520	2767	1041	1185	1091	1233	1379
5th	3196	3260	3547	3846	1500	1671	1564	1730
6th	3975	4300	4393	4731	5084	2030	2230	2106
7th	4807	5190	5580	5684	6074	6480	2636	2860
8th	6092	6130	6577	7009	7134	7583	8034	3309
9th	6913	7560	7615	8115	8595	8739	9244	9755

The following factors are of greatest influence on the electron affinity:

1. *Effective nuclear charge.* Electron affinity increases with increasing effective nuclear charge, other factors being equal.

2. *The size of the atom.* With essentially constant effective nuclear charge, the electron affinity generally decreases with increasing atomic radius. In other words, it is easier to add an electron to a smaller atom.

3. *Electronic configuration.* The elements with the highest electron affinities are the halogens—fluorine, chlorine, bromine, and iodine (Table 10.6)—which, with their $ns^2 np^5$ configurations, are one electron short of noble gas configurations. On the other hand, the electron affinities of the noble gases and the Representative Group II elements are negative. In atoms of these elements all of the occupied energy levels are completely filled. The outermost energy levels cannot accommodate any additional electrons and, moreover, the nucleus is extremely well shielded. The ability of such a nucleus to bind an electron in a completely unoccupied energy level is extremely small, and the negative electron affinities indicate that negative ions of these elements would be unstable.

Electron affinities would be expected to increase across a period as the atoms become smaller and the effective nuclear charge increases. Although there are large decreases in electron affinity values for the filled outer energy levels of the noble gases and the Group II elements, and smaller decreases between the Group IV and Group V elements (see below), a general increase across a period does occur

[10.6] THE REPRESENTATIVE ELEMENTS

Group I	Group II	Group III	Group IV	Group V	Group VI	Group VII	Noble gases
H 1 — 17.4							**He** 2 — (−5.1)
Li 3 — 14.3	**Be** 4 — (−57)	**B** 5 — 5.5	**C** 6 — 29.3	**N** 7 — 0.0	**O** 8 — 33.8	**F** 9 — 77.0	**Ne** 10 — (−6.9)
Na 11 — 12.6	**Mg** 12 — (−55)	**Al** 13 — 11	**Si** 14 — 28.6	**P** 15 — 18	**S** 16 — 49.9	**Cl** 17 — 83.3	**Ar** 18 — (−8.3)
K 19 — 15.6	**Ca** 20 — (−37.4)	* **Ga** 31 — (8.5)	**Ge** 32 — 27.7	**As** 33 — 18	**Se** 34 — 46.6	**Br** 35 — 77.6	**Kr** 36 — (−9.2)
Rb 37 — 11.2	**Sr** 38 — (−40.1)	* **In** 49 — 8.1	**Sn** 50 — 28.8	**Sb** 51 — 24.2	**Te** 52 — 45.4	**I** 53 — 70.6	**Xe** 54 — (−9.7)
Cs 55 — 10.9	**Ba** 56 — (−12)	* **Tl** 81 — 12	**Pb** 82 — 24.2	**Bi** 83 — 24.2	**Po** 84 — (42)	**At** 85 — (65)	**Rn** 86 — (−9.7)
Fr 87 — (10.5)	**Ra** 88						

Be has lowest electron affinity

Cl has highest electron affinity

TABLE 10.6

Electron affinities of the representative elements

(*) Transition elements intervene. The values are given in kcal/mole at 0°K; to convert to kJ/mole, multiply 4.184 kJ/kcal. Values in parentheses have been calculated from theory; others are from experimental measurements. Only eight electron affinity values, mainly for the halogens, had been experimentally measured before 1970. Since then the development of more accurate measurement techniques and better methods of producing negative ions have provided the additional values given in this table. [*Source:* E.C.M. Chen and W.E. Wentworth, *Journal of Chemical Education,* **52,** 486 (1975)]

(see Table 10.6; Figure 10.3). Variations in electron affinity upon moving down a group are not so easily explained. There is a general decrease, but notable exceptions are the electron affinities of nitrogen, oxygen, and fluorine, which are smaller than those of their neighbors in the third period.

The ionization energies and electron affinities of the third period elements are plotted versus atomic number in Figure 10.3. Similar relationships occur among the elements of the other representative element periods. Sodium, the Group I element of the period, has the lowest ionization energy. The large drop in first ionization energy from magnesium to aluminum is contrary to the general trend, but is easily accounted for. The electron first removed from magnesium lies in the *s* energy level of the third shell, whereas that removed from aluminum is located in the *p* level, a level of higher energy. The small decrease in first ionization energy from phosphorus ($3s^2 3p^3$) to sulfur ($3s^2 3p^4$) may be explained in the following manner. In phosphorus each of the three *p* orbitals is occupied by a single electron, whereas in sulfur two of the electrons are paired and two unpaired. The forces of repulsion between two electrons in the same orbital are greater than those between two electrons in different orbitals, making it easier to

FIGURE 10.3

Trends in the first ionization energies and electron affinities across the Na–Ar period. *Note the general increase in ionization energy across the period and the decrease between* Mg *and* Al *and between* P *and* S. *Note also the similar general increase across the period for electron affinity, with decreases between* Na *and* Mg, Si *and* P, *and between chlorine and argon. Similar plots would be obtained for the elements of each representative period.*

Ionization energy

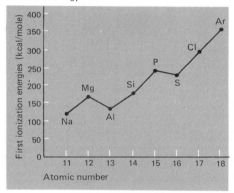

Electron affinity

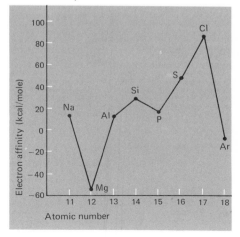

remove a p electron (one of the pair) from the sulfur atom than one from phosphorus.

The large drop in electron affinity between sodium and magnesium occurs because the ns^2 electron level of magnesium is filled, and the next electron must enter a p orbital. The smaller drop between silicon and phosphorus is, like the ionization potential drop from phosphorus to sulfur, due to spin pairing of the electrons. An electron added to phosphorus enters a p orbital already occupied by one electron. This occurs less readily than addition of an electron to the remaining unoccupied p orbital of silicon.

10.7 Electronegativity

The electronegativity (Section 9.16) values for the representative elements (Table 10.7) increase from left to right in the periodic table as the number of valence electrons increases and the size of the atoms decreases. Fluorine is the most electronegative element; its atoms are small and its electron affinity is high. As a result, it attracts electrons very strongly.

Within a family, electronegativity generally decreases with increasing atomic number and size. Cesium, the representative element with the largest size and the smallest ionization energy, is the least electronegative of the representative elements. In other words, cesium is the most metallic and most electropositive element.

10.8 Electronegativity and oxidation state

Electronegativity values are helpful in determining oxidation numbers in cases where it might otherwise be difficult to decide

TABLE 10.7

Electronegativities of the representative elements

(*) *Transition elements intervene.*

electronegativities increase across period →

electronegativities decrease down group ↓

Group I	Group II	Group III	Group IV	Group V	Group VI	Group VII
H 1 2.1						
Li 3 1.0	Be 4 1.5	B 5 2.0	C 6 2.5	N 7 3.0	O 8 3.5	F 9 4.0
Na 11 0.9	Mg 12 1.2	Al 13 1.5	Si 14 1.9	P 15 2.1	S 16 2.5	Cl 17 3.0
K 19 0.8	Ca 20 1.0	Ga 31 1.8	Ge 32 2.0	As 33 2.0	Se 34 2.4	Br 35 2.8
Rb 37 0.8	Sr 38 1.0	In 49 1.5	Sn 50 1.7	Sb 51 1.9	Te 52 2.1	I 53 2.4
Cs 55 0.7	Ba 56 0.9	Tl 81 1.4	Pb 82 1.6	Bi 83 1.9	Po 84	At 85
Fr 87	Ra 88					

which atom has the positive and which the negative oxidation number. In binary covalent compounds and polyatomic ions such as PO_4^{3-} or ClO_4^-, the more electronegative atom is arbitrarily assigned the negative oxidation number corresponding to the charge it would have if it were present as an ion. This leaves assignment of a positive oxidation number to the less electronegative element. As a result, oxidation numbers represent the relative electronegativities of atoms and the direction of the polarity of a bond. For example, in iodine trichloride, ICl_3, chlorine is assigned the -1 oxidation number because it is the more electronegative of the two atoms. To achieve the necessary neutrality, iodine is assigned an oxidation number of $+3$ in this compound.

Compound formation

10.9 Energy relationships

What determines whether chemical combination between the atoms of two elements will occur? What determines whether the compound formed will have primarily ionic or primarily covalent bonds? Answering such questions is not simple, but a great deal of light can be shed by considering the energy changes in the processes involved.

An ionic bond commonly results from the union of two atoms with widely differing abilities to attract electrons. A covalent bond results from the union of two atoms that do not differ sufficiently in their attraction for electrons to bring about the transfer of electrons.

We examine here some of the chlorides of the elements of the Na–Ar period—$NaCl$, $MgCl_2$, $AlCl_3$, $SiCl_4$, PCl_3, PCl_5, and SCl_2—to see what we can discover about bonding and energy relationships.

Chlorine has the highest electron affinity in this period (Table 10.6), and, in fact, of any element—it certainly will form ionic compounds whenever it can acquire electrons. The electron affinities of the other elements in the period are considerably smaller than that of chlorine.

The ionization energies for loss of all of the valence electrons (Table 10.5) range from a low of 118 kcal/mole (494 kJ/mole) for sodium to the impossibly high 6379 kcal/mole (26,690 kJ/mole) for sulfur. There is no question that formation of Na^+ or Mg^{2+} is relatively easy. At aluminum the value is 1228 kcal/mole (5139 kJ/mole); aluminum might or might not form an Al^{3+} ion. On the basis of ionization energies, for silicon, phosphorus, or sulfur to lose all of their valence electrons looks very difficult—this much energy is not often available in chemical reactions.

At this point, based on ionization energies alone, we might predict that $NaCl$ is definitely ionic, $MgCl_2$ is likely to be ionic, whether or not $AlCl_3$ is ionic is questionable, and from there on it becomes more likely that the compounds are covalent, with the possible exception of PCl_3 and SCl_2.

From electronegativities, we obtain support for the conclusion that NaCl and MgCl$_2$ are ionic compounds—the differences between the electronegativities of sodium and magnesium and that of chlorine are 2.1 and 1.8, respectively. For AlCl$_3$ the difference is 1.5 and it is difficult to decide whether the compound is ionic or covalent. Perhaps it will be intermediate in its behavior. As we have pointed out, the rule of thumb is that an electronegativity difference of 2 is required for an ionic compound. The electronegativity differences between chlorine atoms and silicon, phosphorus, and sulfur atoms grow smaller and smaller, ruling out ionic bonding in SiCl$_4$, PCl$_3$, PCl$_5$, and SCl$_2$.

However, using only ionization energies and electron affinities is unrealistic. We know that most chemical reactions—the ones of importance in industry or for living things—don't occur in the gaseous state.

Consider the enthalpy of formation of gaseous NaCl. This enthalpy is the sum of the ionization energy for one mole of Na atoms plus the energy released when one mole of electrons is added to one mole of chlorine atoms (equal in magnitude but opposite in sign to the electron affinity).

$$\Delta H^\circ \text{ (kcal)}$$

	ΔH° (kcal)		
Na(g) $\longrightarrow$ Na$^+$(g) + e^-	120.04	*ionization energy*	(1)
Cl(g) + e^- $\longrightarrow$ Cl$^-$(g)	−87.9	*electron affinity*	(2)
Na(g) + Cl(g) $\longrightarrow$ Na$^+$(g) + Cl$^-$(g)	32.1 (134 kJ)		(3)

The ΔH° value is positive, indicating that this reaction would not be likely to happen spontaneously—the energy released in reaction (2) is not large enough to ionize a gaseous sodium atom.

However, the spontaneous reaction of gaseous sodium and gaseous chlorine *does* occur readily—the product is solid, crystalline sodium chloride. A factor neglected in the calculation of Equations (1)–(3) is the **lattice energy**—the energy liberated when gaseous ions combine to give a crystalline ionic substance. Including the lattice energy gives a more realistic estimate of whether or not sodium and chlorine combine to yield an ionic compound.

	ΔH° (kcal)		
Na(g) + Cl(g) $\longrightarrow$ Na$^+$(g) + Cl$^-$(g)	32.1		(3)
Na$^+$(g) + Cl$^-$(g) $\longrightarrow$ NaCl(s)	−185.4	*lattice energy*	(4)
Na(g) + Cl(g) $\longrightarrow$ NaCl(s)	−153.3 (−641.4 kJ)		(5)

Obviously, the lattice energy supplies the additional energy needed to bring about transfer of electrons for the gaseous atoms. Of course, this is still a hypothetical calculation in the sense that NaCl is not ordinarily made from gaseous atoms. The reaction is usually carried out with solid sodium and gaseous chlorine—this is also energetically favorable, as shown by the following calculation:

$$\Delta H° \text{ (kcal)}$$

$Na(g) + Cl(g) \longrightarrow NaCl(s)$	-153.3	(5)
$Na(s) \longrightarrow Na(g)$	25.98	(6)
$\frac{1}{2}Cl_2 \longrightarrow Cl(g)$	29.082	(7)
$Na(s) + \frac{1}{2}Cl_2(g) \longrightarrow NaCl(s)$	$-98.2 \ (-411 \text{ kJ})$	(8)

Equation (5): $\frac{1}{2}$(bond dissociation energy). Equation (6): sublimation energy.

Students often ask questions like this one: Why does magnesium form compounds that contain Mg^{2+} rather than Mg^+, when so much less energy is required to remove only one electron instead of two? Here again, the answer lies with the lattice energy. The Mg^{2+} cations of smaller size and higher charge interact more strongly with the anions in a crystal lattice than would Mg^+. Therefore, the larger lattice energy compensates for the additional energy needed to remove two electrons rather than one.

Next, let's see what a calculation that includes the lattice energy will tell us about the possibility that solid aluminum chloride, formed by the reaction of solid aluminum with gaseous chlorine, is an ionic compound. The following is equivalent to combining the reaction steps of Equations (1), (2), (4), (6), and (7) in a single calculation:

$$\Delta H° \text{ (kcal)}$$

$Al(g) \longrightarrow Al^{3+}(g) + 3e^-$	1232.7	(9)
$3Cl(g) + 3e^- \longrightarrow 3Cl^-(g)$	-263.7	(10)
$Al^{3+}(g) + 3Cl^-(g) \longrightarrow AlCl_3(s)$	-1302.6	(11)
$Al(s) \longrightarrow Al(g)$	78.0	(12)
$\frac{3}{2}Cl_2(g) \longrightarrow 3Cl(g)$	87.246	(13)
$Al(s) + \frac{3}{2}Cl_2(g) \longrightarrow AlCl_3(s)$	$-168.4 \ (-704.6 \text{ kJ})$	(14)

A $\Delta H°$ of -168.4 kcal certainly indicates that the formation of solid ionic aluminum trichloride is possible. And, based on other observations, we can confirm that solid crystalline aluminum trichloride is an ionic compound. Interestingly, in both the gaseous and liquid states aluminum chloride forms a covalent dimeric molecule, Al_2Cl_6, suggesting that our conclusion that aluminum chloride is intermediate between ionic and covalent in its behavior was a good one.

For comparison, the following is the calculation for the formation of ionic aluminum triiodide from solid aluminum and solid iodine. Note that the electronegativity difference between aluminum and iodine is smaller than that between aluminum and chlorine—it is only 0.9.

$$\Delta H° \text{ (kcal)}$$

$Al(g) \longrightarrow Al^{3+}(g) + 3e^-$	1232.7	(9)
$3I(g) + 3e^- \longrightarrow 3I^-(g)$	-217.5	(15)
$Al^{3+}(g) + 3I^-(g) \longrightarrow AlI_3(s)$	-1100	(16)
$Al(s) \longrightarrow Al(g)$	78.0	(12)
$\frac{3}{2}I_2(s) \longrightarrow \frac{3}{2}I_2(g)$	22.385	(17)
$\frac{3}{2}I_2(g) \longrightarrow 3I(g)$	54.22	(18)
$Al(s) + \frac{3}{2}I_2(s) \longrightarrow AlI_3(s)$	$70 \ (290 \text{ kJ})$	(19)

Ionic aluminum triiodide formation does not seem likely, on the basis of this calculation. From other observations it is known that aluminum triiodide is a covalent, dimeric molecule, Al_2I_6, in the solid state as well as in the gaseous and liquid states (as is aluminum bromide).

Can it be concluded, since aluminum is often covalent in its halides, that it is not a very reactive metal—that it is reluctant to react by electron loss? Not always. In aqueous solutions, aluminum quite readily loses its valence electrons to form hydrated Al^{3+} ions. What is different? Here the environment has been changed, the reactions are taking place in a new medium—water. A new energy contribution is made by the hydration energy of the ions. By including this factor in a thermochemical calculation, we can show that aluminum and iodine will react in aqueous solution to give aluminum(III) iodide in solution as ions.

$$\Delta H° \text{ (kcal)}$$

$Al(s) \longrightarrow Al(g)$	78.0	heat of sublimation	(12)
$Al(g) \longrightarrow Al^{3+}(g) + 3e^-$	1232.7	ionization energy	(9)
$Al^{3+}(g) \xrightarrow{H_2O} Al^{3+}(aq)$	−1438	hydration energy	(20)
$Al(s) \xrightarrow{H_2O} Al^{3+}(aq) + 3e^-$	−127 (531 kJ)		(21)

$\frac{1}{2}I_2(s) \longrightarrow \frac{1}{2}I_2(g)$	7.46	$\frac{1}{2}$(heat of sublimation)	(22)
$\frac{1}{2}I_2(g) \longrightarrow I(g)$	18.074	$\frac{1}{2}$(dissociation energy)	(23)
$I(g) + e^- \longrightarrow I^-(g)$	−72.5	electron affinity	(24)
$I^-(g) \xrightarrow{H_2O} I^-(aq)$	34	hydration energy	(25)
$\frac{1}{2}I_2(s) + e^- \xrightarrow{H_2O} I^-(aq)$	−13 kcal (54 kJ)		(26)

$Al(s) \xrightarrow{H_2O} Al^{3+}(aq) + 3e^-$	−127	(21)
$3[\frac{1}{2}I_2(s) + e^-] \xrightarrow{H_2O} 3I^-(aq)$	−39	(26)
$Al(s) + \frac{3}{2}I_2(s) \xrightarrow{H_2O} Al^{3+}(aq) + 3I^-(aq)$	−166 kcal (−695 kJ)	(27)

The reaction is quite exothermic, and it is clear that without the contribution of the heat of solution this would not be the case.

Table 10.8 summarizes the formation and properties of the chlorides of the third period elements. In keeping with the trends that we have been discussing, sodium and magnesium chlorides are ionic compounds, aluminum trichloride is ionic in the solid state and covalent in the liquid and gaseous states, and the remaining compounds are covalent. (In the solid state, phosphorus(V) chloride retains covalent bonds between phosphorus and chlorine, but forms pairs of complex ions, $[PCl_4]^+$ and $[PCl_6]^-$, that give an ionic, crystalline compound.)

The electronic configurations of the atoms from which the compounds are formed are shown.

TABLE 10.8

Chlorides of third-period elements

Sodium chloride
  Colorless crystals, m.p. 808°C

$$Na\cdot + \cdot \overset{\cdot\cdot}{\underset{\cdot\cdot}{Cl}}: \longrightarrow Na^+ + :\overset{\cdot\cdot}{\underset{\cdot\cdot}{Cl}}:^-$$

$$3s^1 \qquad 3s^2 3p^5 \qquad 2s^2 2p^6 \qquad 3s^2 3p^6$$

Magnesium chloride, MgCl$_2$
  White crystals, m.p. 708°C

$$Mg: + 2\cdot \overset{\cdot\cdot}{\underset{\cdot\cdot}{Cl}}: \longrightarrow Mg^{2+} + 2:\overset{\cdot\cdot}{\underset{\cdot\cdot}{Cl}}:^-$$

$$3s^2 \qquad 3s^2 3p^5 \qquad 2s^2 2p^6 \qquad 3s^2 3p^6$$

Aluminum trichloride, AlCl$_3$ (or dimer Al$_2$Cl$_6$)
  White deliquescent crystals, m.p. 190°C

$$\overset{\cdot\cdot}{Al}\cdot + 3\cdot \overset{\cdot\cdot}{\underset{\cdot\cdot}{Cl}}: \longrightarrow Al^{3+} + 3:\overset{\cdot\cdot}{\underset{\cdot\cdot}{Cl}}:^- \overset{heat}{\longrightarrow}$$

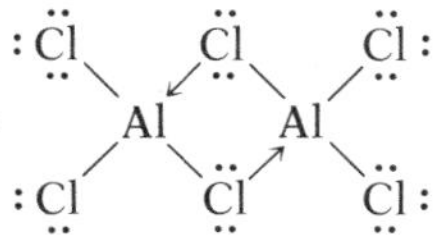

$$3s^2 3p^1 \qquad 3s^2 3p^5 \qquad 2s^2 2p^6 \qquad 3s^2 3p^6$$

Silicon tetrachloride, SiCl$_4$
  Colorless fuming liquid, m.p. −70°C

$$\overset{\cdot\cdot}{\underset{\cdot}{Si}}\cdot + 4\cdot \overset{\cdot\cdot}{\underset{\cdot\cdot}{Cl}}: \longrightarrow$$

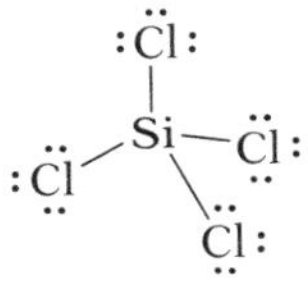

$$3s^2 3p^2 \qquad 3s^2 3p^5$$

Phosphorus trichloride, PCl$_3$
  Colorless fuming liquid, m.p. −112°C

$$\cdot \overset{\cdot\cdot}{\underset{\cdot}{P}}\cdot + 3\cdot \overset{\cdot\cdot}{\underset{\cdot\cdot}{Cl}}: \longrightarrow$$

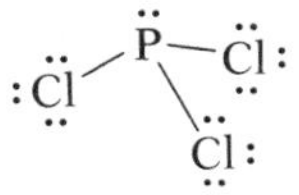

$$3s^2 3p^3 \qquad 3s^2 3p^5$$

Phosphorus pentachloride, PCl$_5$
  Yellowish white crystals, m.p. 167°C

$$\cdot \overset{\cdot\cdot}{\underset{\cdot}{P}}\cdot + 5\cdot \overset{\cdot\cdot}{\underset{\cdot\cdot}{Cl}}: \longrightarrow \longrightarrow [PCl_4]^+[PCl_6]^-$$

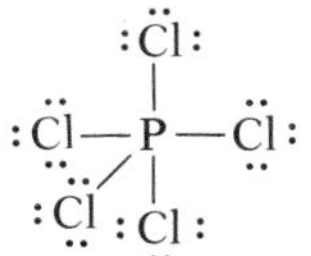

$$3s^2 3p^3 \qquad 3s^2 3p^5 \qquad \textit{(in vapor)} \qquad \textit{(in crystal)}$$

Sulfur dichloride, SCl$_2$
  Dark red, fuming liquid, m.p. −78°C

$$:\overset{\cdot\cdot}{\underset{\cdot\cdot}{S}}: + 2\cdot \overset{\cdot\cdot}{\underset{\cdot\cdot}{Cl}}: \longrightarrow$$

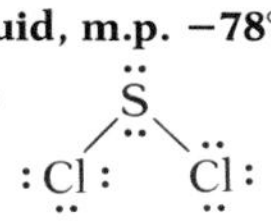

$$3s^2 3p^4 \qquad 3s^2 3p^5$$

THE REPRESENTATIVE
ELEMENTS [10.9]

289

10.10 Hydrogen

There is no truly appropriate position for hydrogen among the families of elements in the periodic system. Although, like the alkali metals (Representative Group I), hydrogen has a single *s* valence electron, it is quite different from these elements and cannot be included with them in the same periodic group. For example, the single valence electron is easily removed from any alkali metal to form a singly charged cation such as Na^+ or K^+. However, the hydrogen atom loses its electron with great difficulty (see Table 10.4), so great that the simple H^+ ion cannot be formed in ordinary chemical reactions.

Like halogen atoms (Representative Group VII), the hydrogen atom is one electron short of a noble gas configuration. However, hydrogen does not fit well with the halogen family either. Halogen atoms readily acquire an electron to form -1 anions with noble gas configurations. For example, the electron affinity of iodine, the least reactive of the halogens, is 70.6 kcal/mole, whereas that of hydrogen is only 17.4 kcal/mole. Hydrogen combines with the most reactive representative metals to form salts containing the hydride ion, H^-. Moreover, in contrast to the halide ions, the hydride ion is unstable in water, immediately reacting to give hydrogen and a hydroxide ion.

$$H^- + H_2O(l) \longrightarrow H_2(g) + OH^-$$

Group I
——
Li
Na
K
Rb
Cs
Fr

10.11 Representative Group I (the alkali metals)

The Representative Group I elements are all reactive metals. The atoms of each have noble gas cores and each has a single *s* electron in its valence shell. The most electropositive of all the metals, the alkali metals react very readily by giving up their single outer shell electron (the *ns* electron). The alkali metals (except for lithium in rare instances) always form ionic compounds and they have only one oxidation state, which is $+1$ (see Table 10.9).

Group II
——
Be
Mg
Ca
Sr
Ba
Ra

10.12 Representative Group II (the alkaline earth metals)

The atoms of the Representative Group II elements, the beryllium family, have noble gas cores and two electrons (ns^2) in the valence shells of their atoms. Like the alkali metals, they exhibit only one oxidation state (Table 10.9), $+2$, which corresponds to the loss of both valence electrons. With the exception of beryllium, these elements almost exclusively form ionic compounds. [Magnesium and beryllium form highly polar, but covalent bonds with hydrocarbon groups, for example, $(CH_3)_2$ Mg.]

The ionization energy for the removal of the two valence elec-

[10.10] THE REPRESENTATIVE
ELEMENTS

290

Except for oxygen and fluorine, the most electronegative elements, the maximum
oxidation state for each element is equal to its group number.

$_1$H
+1: HCl
−1: Na$^+$H$^-$

Group I	Group II	Group III	Group IV	Group V	Group VI	Group VII
$_3$Li	$_4$Be	$_5$B	$_6$C	$_7$N	$_8$O	$_9$F
			+2:CO	−3:(Li$^+$)$_3$N^{3-}; NH$_3$	−2:(Na$^+$)$_2$O^{2-}; H$_2$O	
+1:Li$^+$Cl$^-$	+2:BeCl$_2$	+3:BCl$_3$	+4:CO$_2$; CCl$_4$	+2:NO	−1:(Na$^+$)$_2$O$_2$$^{2-}$; H$_2O_2$	−1:Na$^+$F$^-$; CF$_4$
				+3:N$_2$O$_3$		
				+4:NO$_2$		
				+5:N$_2$O$_5$		
$_{11}$Na	$_{12}$Mg	$_{13}$Al	$_{14}$Si	$_{15}$P	$_{16}$S	$_{17}$Cl
					−2:(Na$^+$)$_2$S^{2-}; H$_2$S	−1:Na$^+$Cl$^-$; CCl$_4$
			+4:SiCl$_4$;	−3:PH$_3$; Na$_3$P	+4:SO$_2$	+1:HOCl
+1:Na$^+$Cl$^-$	+2:Mg^{2+}(Cl$^-$)$_2$	+3:Al$_2$Cl$_6$	−4:SiH$_4$	+3:PCl$_3$	+6:SO$_3$	+3:HClO$_2$
				+5:PCl$_5$		+5:HClO$_3$
						+7:HClO$_4$
$_{19}$K	$_{20}$Ca	$_{31}$Ga	$_{32}$Ge	$_{33}$As	$_{34}$Se	$_{35}$Br
			+2:GeCl$_2$	−3:AsH$_3$; Na$_3$As	−2:H$_2$Se	−1:Na$^+$Br$^-$; HBr
+1:K$^+$Cl$^-$	+2:Ca^{2+}(Cl$^-$)$_2$	+3:Ga$_2$Cl$_6$	+4:GeCl$_4$	+3:AsCl$_3$	+4:SeO$_2$	+1:HOBr
				+5:As$_4$O$_{10}$	+6:SeO$_3$	+5:HBrO$_3$
						+7:HBrO$_4$
$_{37}$Rb	$_{38}$Sr	$_{49}$In	$_{50}$Sn	$_{51}$Sb	$_{52}$Te	$_{53}$I
			+2:Sn^{2+}(F$^-$)$_2$; SnCl$_2$	−3:SbH$_3$	−2:H$_2$Te	−1:Na$^+$I$^-$; HI
+1:Rb$^+$Cl$^-$	+2:Sr^{2+}(Cl$^-$)$_2$	+3:InCl$_3$	+4:SnCl$_4$; SnO$_2$	+3:Sb^{3+}(F$^-$)$_3$; SbCl$_3$	+4:TeO$_2$	+1:HOI
				+5:SbCl$_5$	+6:TeO$_3$	+5:HIO$_3$
						+7:HIO$_4$
$_{55}$Cs	$_{56}$Ba	$_{81}$Tl	$_{82}$Pb	$_{83}$Bi	$_{84}$Po	$_{85}$At
		+1:Tl$^+$Cl$^-$	+2:Pb^{2+}(F$^-$)$_2$	−3:BiH$_3$		
+1:Cs$^+$Cl$^-$	+2:Ba^{2+}(Cl$^-$)$_2$	+3:TlF$_3$	+4:PbO$_2$	+3:Bi^{3+}(F$^-$)$_3$		
				+5:Bi$_2$O$_5$		

trons from the small beryllium atom is so large (637.8 kcal/mole)
that most of the binary anhydrous compounds of the element, like
those of aluminum in Group III, are covalent. Like aluminum, however, beryllium tends to be ionic in aqueous solution because of the
high hydration energy of the beryllium ion (-807 kcal/mole). As
mentioned in the section on ionization energies (Section 10.5), the
tendency of metals to react by electron loss regularly increases down
each group (as the atomic size increases) and decreases from an alkali
metal to the element beside it in Group II.

TABLE 10.9
*Oxidation states and typical
compounds for the representative
elements*

10.13 Representative Group III

The valence shell configuration of the Representative Group III elements is ns^2np. All of the elements in this group form compounds in
the $+3$ oxidation state. However, in this group and in Groups IV–VI,
the elements are less like each other than in Groups I, II, and VII.
Boron is a semiconducting element and its compounds are all cova-

Group III
B
Al
Ga
In
Tl

Down Representative Groups III–V: lower oxidation states become more stable

lent, both in their pure state and in aqueous solution. The energy necessary to remove the three valence electrons from boron is so large that neither lattice energies nor hydration energies can overcome it to allow the formation of B^{3+}. In its general chemical properties, boron resembles its diagonal neighbor silicon more than it resembles the other Group III elements. The remaining members of the group are metals and their compounds can be either ionic or covalent. The elements all form hydrates in aqueous solution.

Gallium, indium, and thallium can each have a +1 oxidation state. However, compounds in the +1 state are more numerous and more stable for indium and thallium than for gallium. This illustrates a general trend among the metals and semiconducting elements of Representative Groups III, IV, and V toward greater stability for lower oxidation states of the elements further down the groups.

10.14 Representative Group IV

Group IV
C
Si
Ge
Sn
Pb

In Representative Group IV carbon is a nonmetal, silicon and germanium are semiconducting elements, and tin and lead are metals. Atoms of these elements have ns^2np^2 configurations in their valence shells.

All of the members of this group exhibit an oxidation state of +4, for example, in their compounds with the halogens and oxygen (EX_4 and EO_2, where E is a Group IV element), and a state of −4 in the hydrides (EH_4). There are substantial differences in the chemistry of the compounds of these elements. For the halides and hydrides, the one common factor is the tetrahedral molecular geometry, which is predicted by the electron-pair repulsion theory (Section 9.26). The general chemical stability of the halides and hydrides of this family is greatest for carbon and decreases from silicon to lead. For example, methane, CH_4, is not decomposed into its constituent elements below about 800°C, but the corresponding compounds of the other elements of this group decompose at much lower temperatures (decomposition temperatures: SiH_4, 450°C; GeH_4, 285°C; SnH_4, 150°C; PbH_4, about 0°C).

The +2 oxidation state becomes increasingly stable down the carbon family. Carbon monoxide is the only stable compound of carbon in the +2 state. For silicon, the +2 state has been best characterized for the gases SiO and SiF_2, which can be obtained at high temperatures. The relatively few known compounds of dipositive germanium are potent reducing agents, the germanium being readily oxidized to the +4 state. The +2 oxidation state is most important in the chemistry of tin and lead, the metallic members of the family. These are the only members of the family that can form typical salts containing +2 ions. Such ions have the 18 + 2 configuration and result from the loss of the p electrons of the valence shell.

One of the outstanding properties of carbon—a property of minor significance for the other members of the family—is the ability of its atoms to bond to each other through the sharing of a pair of electrons

$$-\overset{|}{\underset{|}{C}}:\overset{|}{\underset{|}{C}}:\overset{|}{\underset{|}{C}}:\overset{|}{\underset{|}{C}}-$$

to form compounds with skeletal structures of carbon chains or rings. The relatively high bond energy of the C–C bond in comparison with the bond energies of the covalent single bonds for the other elements is undoubtedly one factor contributing to the greater ability of carbon to self-link.

	C—C	Si—Si	Ge—Ge	Sn—Sn	Pb—Pb
bond energies, kcal/mole	79	42	38	34	24
(kJ/mole) at 298 °K	(330)	(180)	(160)	(140)	(100)

Another important property possessed by the carbon atom but not by atoms of other members of the group is its ability to form multiple covalent bonds with other carbon atoms as well as with oxygen, sulfur, and nitrogen:

$$\overset{\diagdown}{\diagup}C::C\overset{\diagup}{\diagdown} \qquad -C::C- \qquad \overset{\diagdown}{\diagup}C::O \qquad \overset{\diagdown}{\diagup}C::S \qquad -C::N$$

10.15 Representative Group V

Nitrogen and phosphorus are the nonmetals of Representative Group V. Arsenic and antimony fall among the semiconducting elements, and the last element of the group, bismuth, is a metal. In the valence shell the atoms of these elements have the ns^2np^3 configuration and thus lack three electrons of a noble gas configuration.

Because of the much greater electronegativity of the nitrogen atom and its inability to accommodate more than eight electrons in its valence shell, the chemical behavior of nitrogen is substantially different from that of the other members of the group. Nitrogen is about as electronegative as chlorine; only fluorine and oxygen have larger electronegativities (Table 10.7).

Atoms of the nonmetals nitrogen and phosphorus can acquire noble gas configurations by accepting three electrons. The N^{3-} ion occurs only in nitrides formed with the most electropositive metals $[(Li^+)_3N^{3-}; (M^{2+})_3(N^{3-})_2$, where M is Be, Mg, Ca, Sr, or Ba]. Elevated temperatures are required to prepare these compounds. A few ionic phosphides, for example, $(Na^+)_3P^{3-}$, are known.

Oxidation states of $+3$, -3, and $+5$ are common for all the elements in Representative Group V, with the exception of the -3 state for bismuth. In covalent compounds with hydrogen, EH_3, the elements are regarded as in the -3 state because hydrogen is always assigned a $+1$ oxidation state in such compounds. The -3 state also occurs in binary covalent or ionic compounds with metals such as Li_3N, Na_3P, or Na_3As.

Group V
N
P
As
Sb
Bi

The +3 and +5 oxidation states are exhibited in the halides and oxides of Representative Group V elements and in the acids and salts derived from the oxides. All of the possible trihalides are formed, while not all of the pentahalides are known. Some examples of +3 oxides are N_2O_3, P_4O_6, As_4O_6, and Sb_4O_6. The +5 pentoxides, E_2O_5, are known for all of the elements in the family.

Nitrogen also differs from the other members of the family in the large number of oxidation states in which it can exist. All possible oxidation states from -3 to $+5$ are known for nitrogen.

10.16 Representative Group VI

Oxygen and sulfur are the nonmetallic members of the oxygen family. Selenium and tellurium are classed as semiconducting elements, polonium, a rare radioactive substance, has similarities to both tellurium and bismuth, but appears to be mainly metallic in its behavior.

In their valence shells atoms of the Group VI elements have the ns^2np^4 configuration and are thus two electrons short of noble gas electronic arrangements. Like the halogens (Group VII), they have a very strong tendency to attain noble gas (s^2p^6) electronic configurations when they enter into chemical combination.

By the transfer of two electrons from metallic atoms into their p orbitals, dinegative ions are formed by the Group VI elements—O^{2-}, S^{2-}, Se^{2-}, and Te^{2-}. Oxygen, next to fluorine the most electronegative of all the elements, forms ionic compounds with most of the metals; sulfur, selenium, and tellurium form dinegative anions only when they combine with extremely electropositive metals such as sodium, potassium, and calcium.

These elements can also attain the ns^2np^6 configuration by covalent union with other elements, for example, as in hydrides. Except for water, which ordinarily shows essentially no reducing power, the hydrides are reasonably good reducing agents, with the reducing ability increasing from H_2S to H_2Te. This trend parallels the decreasing bond energy of the element–hydrogen bond.

	O—H	S—H	Se—H	Te—H
bond energy, kcal/mole	111	88	66	57
(kJ/mole) at 298°K	(460)	(370)	(280)	(240)

Oxygen exhibits positive oxidation states *only* when bonded to fluorine, as in OF_2 and O_2F_2. The other elements of Group VI, which are much less electronegative than oxygen, frequently have oxidation numbers of $+4$ and $+6$ when they are covalently joined to oxygen or one of the halogens. For example, sulfur has oxidation numbers of $+4$ and $+6$ in sulfur dioxide (SO_2) and sulfur trioxide (SO_3).

Sulfur, selenium, and tellurium can accommodate *more* than eight electrons in their valence shells. As a result these elements can be bound by covalent single bonds to more then four groups, as, for example, in the hexafluorides, SF_6, SeF_6, and TeF_6.

Sulfur and selenium both form octatomic molecules, S_8 and Se_8. The atoms are joined to each other through covalent single bonds in a puckered ring (Section 22.14). Other species in which catenation occurs are the polyatomic anions of sulfur, selenium, and tellurium, for example, S_6^{2-}, Se_4^{2-}, and Te_3^{2-}. The only common similar example for oxygen is the peroxide ion, O_2^{2-}. The much greater thermodynamic stability of element–element single bonds for sulfur and selenium as compared with oxygen and tellurium explains their different abilities to self-link; the bond energies are as follows:

	O—O	S—S	Se—Se	Te—Te
bond energy, kcal/mole	34	59	44	33
(kJ/mole) at 298°K	(140)	(250)	(180)	(140)

10.17 Representative Group VII (the halogens)

With Representative Group VII we return to a group of elements which, like the metals of Representative Groups I and II, are much more similar to each other than are the elements of the intervening groups. All of the halogens are nonmetals with high electron affinities, high electronegativities, and high ionization energies. The atoms of fluorine are much smaller than those of the other halogens, and fluorine is the most electronegative of the elements. For these reasons, there are greater differences in properties and behavior between fluorine and chlorine than between the other halogens in going down the group.

The halogens have the ns^2np^5 configuration and readily react to form singly charged anions or a single covalent bond. In such compounds they have the -1 oxidation state, but in compounds among themselves (interhalogen compounds) or with bonds to oxygen they can, except for fluorine, exhibit positive oxidation states of $+1$, $+3$, $+5$, or $+7$.

Group VII
F
Cl
Br
I
At

THOUGHTS ON CHEMISTRY

The grim silence of facts

While grading a beginning graduate inorganic examination some time ago I was startled to discover that the student believed silver chloride to be a pale green gas. Now we all have our off days (does not the very first sentence of chapter one of a recent, and excellent, paperback run: "A flammable silver metal seizes electrons from a deadly green gas, and the resultant positive and negative ions line up into the tasty crystals of common salt, NaCl" ?), and I read on willing to forgive and forget, if not to allow partial credit. A little later the student launched into a long, plausible explanation as to why silver chloride is a pale green gas. I was reminded of Dr. Johnson's: "I can give you the explanation, M'am, but not the understanding of it." . . .

THE GRIM SILENCE OF FACTS, *by Derek A. Davenport, 1970*

THE REPRESENTATIVE ELEMENTS [10.17]

That we should begin by setting up a skeleton of inorganic principles is undeniable. Without it the presentation of facts becomes inefficient and their accumulation a shapeless mass of protoplasm. But these principles should not be restricted to the narrowly structural. Instability and stability, lability and inertness, oxidation and reduction, acidity and basicity, and their relationship to position in the periodic table are as much a part of modern inorganic chemistry as is ligand field theory, though admittedly they are less fashionable. . . .

But, the arguments run, teaching the facts of inorganic chemistry is dull beyond conception. Leaving aside the implicit suggestion that molecular orbital theory is intrinsically hilarious, one must demur. Anyone who can bring life to a character table should not despair of lending a little color to the oxidation states of vanadium. It takes effort—what in teaching doesn't—but the effort must be made. For as Conrad urged: "Every sort of shouting is a transitory thing, after which the grim silence of facts remains."

THE GRIM SILENCE OF FACTS, by Derek A. Davenport (Quoted from p. 271 in the *Journal of Chemical Education,* Vol. 47, 1970)

Significant terms defined in this chapter

catenation
screening effect
penetration effect
effective nuclear charge
isoelectronic
first ionization energy
second ionization energy
lattice energy

Exercises

10.1 The following "message" was found in an alien spacecraft:

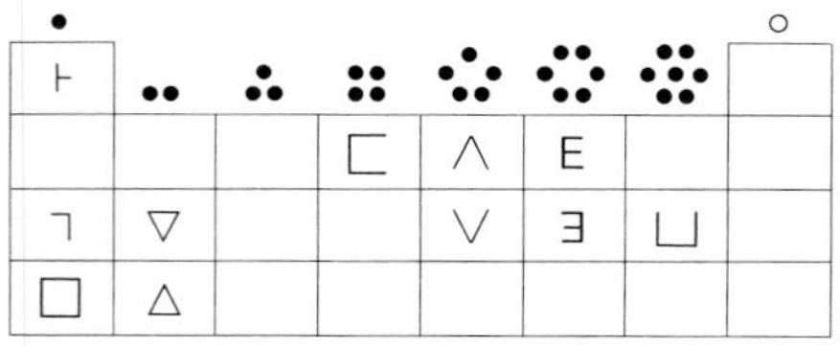

A chemist on Earth quickly recognized the "message" as a portion of the periodic table showing the elements important to their life form. Answer the following questions about the alien elements: (a) What is the symbol for the least active metal which forms a singly positive ion? (b) Which nonmetals form doubly negative ions? (c) Which metals form hydrides? (d) Which halogen is important to their life form? (e) Which elements undergo catenation? (f) Which element forms a compound with hydrogen having the formula H_2X, which is a weak acid in water? (g) Which element forms a covalently bonded compound having the formula EX_2, where E and X are nonmetals and each atom of X gains two electrons? (h) Is their life form based on carbon or silicon?

10.2 Which statements are true? Rewrite any false statement so that it is correct.
(a) Catenation is the process in which water molecules combine with gaseous positive ions to give stable hydrated cations.
(b) The second ionization energy is usually much larger than the first ionization energy because it is more difficult to remove an electron from a positively charged species than from a neutral species.
(c) The screening effect is the decrease in nuclear charge acting on an electron due mainly to the effect of other electrons in inner shells.
(d) The outer shell electronic configuration of elements in Group VI is given by ns^2nd^4.
(e) The ions O^{2-}, F^-, Na^+, Mg^{2+}, and Al^{3+} are isoelectronic with Ne.
(f) Atomic radii are always larger than ionic radii.

(g) Oxygen is the only element sufficiently electronegative that the fluorine atom will have a positive oxidation state when combined with it.

10.3 Your roommate asks you to explain what a semiconductor is and how it is different from a metal. How would you answer the question?

10.4 Identify the general trend across a period as (i) increasing, (ii) decreasing, or (iii) no significant change for the following properties: (a) atomic radius, (b) ionic radius of isoelectronic species, (c) ionization energy, (d) electron affinity, (e) electronegativity, (f) positive charge on cation, and (g) negative oxidation state. Repeat this exercise identifying the trend down a family.

10.5 Identify and give a reasonable explanation for the trend in the (a) first ionization energies of the alkaline earth metals—Group II—based on atomic sizes, (b) first ionization energies of the elements in the third period—Na to Ar—based on electronic configurations and atomic sizes, (c) electronegativities of the halogens—Group VII—based on atomic sizes, (d) amount of ionic character in chemical bonds between F and the elements in Group III based on electronegativities, and (e) atomic radii of the noble gases based on atomic structure.

10.6 Arrange the following ions in order of increasing size: F^-, Mg^{2+}, Cl^-, Be^{2+}, S^{2-}, and Na^+.

10.7 Your younger brother, who is taking high school chemistry, asks you to explain why electronegativity increases across a period from left to right with increasing atomic number, but decreases down a group while atomic number is increasing. How would you answer the question?

10.8 The second value of the electron affinity for oxygen indicates that the process of addition of an electron to O^- is highly unfavorable. Explain why the oxide ion exists in solid compounds.

10.9 Compare the following general properties for metals and nonmetals: (a) melting and boiling points, (b) density, (c) hardness, (d) color, (e) malleability and ductility, (f) conductivity of heat and electricity, (g) formation of binary compounds with hydrogen, (h) acidity or alkalinity of oxides and hydroxides in water, (i) type of bonding in halides, and (j) types of ions formed.

10.10 Choose the best answer to each of the following questions:
(a) Which compound has the most ionic character in the bonding? (i) NaCl, (ii) $MgCl_2$, (iii) HCl, (iv) Cl_2, (v) CCl_4
(b) Which element has the lowest first ionization energy? (i) Li, (ii) K, (iii) Ca, (iv) Se, (v) Kr

(c) Which element is the best oxidizing agent? (i) H_2, (ii) P_4, (iii) O_2, (iv) Cl_2, (v) F_2
(d) Which element forms a network covalent bond in the solid state? (i) Li, (ii) C, (iii) O_2, (iv) Cl_2, (v) Ne
(e) Which substance contains the least polar covalent bonding? (i) H_2O, (ii) CO_2, (iii) NO_2, (iv) O_2, (v) SO_2
(f) Which has the largest radius? (i) H, (ii) He^+, (iii) Br, (iv) Br^-, (v) Br(VII)
(g) Which has the lowest first ionization energy? (i) He, (ii) Ne, (iii) Ar, (iv) Kr, (v) Xe
(h) The elements in which group form the series of compounds having the empirical formulas E_2O_5, E_2O_3, EF_5, EF_3, and EH_3? (i) I, (ii) III, (iii) V, (iv) VII, (v) noble gases

10.11 The placement of H, Li, Be, and B in the periodic table is based mainly on the electronic configurations of the atoms of these elements. Why is H unique among the elements and rather difficult to assign to a particular group? Which groups are closest to H in general chemical properties? Why are the properties of Li, Be, and B somewhat different from those of the other elements in their respective groups?

10.12 Write the electronic configurations for atoms of elements in the nitrogen family—Group V. Specify whether the element is a metal, nonmetal, or a semiconducting metal and predict the oxidation states for each element.

10.13 What is the atomic number and electronic configuration for the following atoms: (a) the second alkali metal—Group I, (b) the third alkaline earth metal—Group II, (c) the third halogen—Group VII, and (d) the third noble gas?

10.14 Write the formulas for the oxides of the elements in the third period—Na to Ar—and specify the type of bonding in each compound. Repeat this exercise for the binary compounds containing hydrogen.

10.15* A neutral atom of element "X" has 15 electrons. Answer as many of the following questions as you can without looking at a periodic table: (a) What is the approximate atomic weight? (b) What is the atomic number? (c) What is the total number of s electrons? (d) Is the element a metal, nonmetal, or semiconducting element? (e) What is the empirical formula of the binary compound formed between sodium and this element? (f) What is the empirical formula of the binary compound formed between chlorine and this element? (g) What oxidation states can we normally expect for this element?

10.16* Once discovered, element 113 should be placed in the periodic table below thallium. Predict some of its characteristics such as electronegativity, size, oxidation numbers, and physical state.

10.17* The strength of an acid depends on how easily the bond holding the hydrogen atom to the molecule can be broken to produce $H^+(aq)$. Look carefully at each pair of acids and predict which acid is stronger:

(a) based on electronegativity of the center atom

$$H-\overset{\cdot\cdot}{\underset{\cdot\cdot}{O}}-\overset{:\overset{\cdot\cdot}{O}:}{\underset{\cdot\cdot}{Br}}-\overset{\cdot\cdot}{\underset{\cdot\cdot}{O}}: \quad or \quad H-\overset{\cdot\cdot}{\underset{\cdot\cdot}{O}}-\overset{:\overset{\cdot\cdot}{O}:}{\underset{\cdot\cdot}{Cl}}-\overset{\cdot\cdot}{\underset{\cdot\cdot}{O}}:$$

(b) based on electron distribution around the center atom

$$H-\overset{\cdot\cdot}{\underset{\cdot\cdot}{O}}-\overset{:\overset{\cdot\cdot}{O}:}{\underset{\cdot\cdot}{Cl}}-\overset{\cdot\cdot}{\underset{\cdot\cdot}{O}}: \quad or \quad H-\overset{\cdot\cdot}{\underset{\cdot\cdot}{O}}-\overset{:\overset{\cdot\cdot}{O}:}{\underset{:\overset{\cdot\cdot}{O}:}{Cl}}-\overset{\cdot\cdot}{\underset{\cdot\cdot}{O}}:$$

(c) based on the size of center atom

$$H-\overset{\cdot\cdot}{S}-H \quad or \quad H-\overset{\cdot\cdot}{\underset{\cdot\cdot}{O}}-H$$

10.18 Assume that the first ionization energy of Se is unknown, but that the values for the other elements given in Table 10.4 are known. Predict the value for Se by using both horizontal and vertical periodic trends.

10.19 The densities of gaseous chlorine, hydrogen, neon, nitrogen, and radon are 3.214, 0.08988, 0.90035, 1.25055, and 9.73 g/liter, respectively, at STP. Confirm whether these elements exist in monatomic or diatomic form.

10.20 Using the bond energies given in Table 9.19, calculate ΔH for the equations given below and predict which form of molecular oxygen would be preferred:

(a) $O(g) \longrightarrow \frac{1}{8}\begin{bmatrix} \overset{\cdot\cdot}{O}-\overset{\cdot\cdot}{O} \\ :\overset{\cdot\cdot}{O}: \qquad :\overset{\cdot\cdot}{O}: \\ :\overset{\cdot\cdot}{O}: \qquad :\overset{\cdot\cdot}{O}: \\ \overset{\cdot\cdot}{O}-\overset{\cdot\cdot}{O} \end{bmatrix}(g)$

(b) $O(g) \longrightarrow \frac{1}{2}\left[:\overset{\cdot\cdot}{O}=\overset{\cdot\cdot}{O}:\right](g)$

10.21 The enthalpy of formation of $Li(g)$ at 25°C [that is, the enthalpy for the reaction $Li(s) \rightarrow Li(g)$] is 38.410 kcal/mole and of $Li^+[Li(s) \rightarrow Li^+(g) + e^-]$ 164.236 kcal/

mole. Calculate the ionization energy of Li at this temperature.

10.22 The enthalpy of formation of $Mg^{2+}(g)$ at 25°C is 561.788 kcal/mole, $O^{2-}(g)$ is 192 kcal/mole, and $MgO(s)$ is -143.84 kcal/mole. Calculate the lattice energy for MgO. Likewise, the enthalpy of formation at 25°C of $Na^+(g)$ is 146.015 kcal/mole, $F^-(g)$ is -79.5 kcal/mole, and $NaF(s)$ is -136.0 kcal/mole. Calculate the lattice energy for NaF. Why is there such a large difference between these lattice energies even though both substances crystallize in the same pattern in the solid state and there is only about a 10% difference in the distance between the ions?

10.23* The lattice energies for NaCl and RbF are both about -184 kcal/mole. Combine these values with the values of ionization energies, see Table 10.4; electron affinities, see Table 10.6; and the following heats of formation: 25.98 kcal/mole for $Na(g)$, 20.51 kcal/mole for $Rb(g)$, 18.3 kcal/mole for $F(g)$, and 29.012 kcal/mole for $Cl(g)$ to find $\Delta H°$ for the formation reaction given by the equation

$$M(s) + \tfrac{1}{2}X_2(g) \longrightarrow MX(s)$$

Which salt is more stable based on the heat of formation values?

10.24** Let's pretend we know nothing about element 52—tellurium. From our working knowledge of the periodic law and atomic structure, we can attempt to predict many properties for this element. (a) Write a reasonable electronic configuration for the 52 electrons and (b) identify the period and family in which the element would appear in the periodic table. (c) Would we expect this element to have properties similar to those of metals, nonmetals, or semiconducting elements? (d) Would we expect the gaseous form of this element to be monatomic? Will the value of the atomic radius be larger or smaller than that of (e) I and (f) Se? (g) Are the atoms paramagnetic? (h) What is the approximate atomic weight of this element?

What would we predict for the (i) oxidation states of this element? Are the (j) positive ions and (k) negative ions larger or smaller than the atoms? What would be the empirical formula of the (l) oxides, (m) halides, and (n) hydride of this element?

Using the data given in Tables 10.4 and 10.6 for As, Se, Br, Sb, I, Bi, and Po, predict the (o) electron affinity and (p) first ionization energy for tellurium. (q) Compare your answer for the ionization energy to that calculated using the heat of formation of $Te(g)$ as 196.7 kJ/mole and $Te^+(g)$ as 1072.2 kJ/mole.

11 HYDROGEN AND OXYGEN

Chemistry, there is no doubt, is important on every level of our lives—from our personal, daily routine (e.g., to use or not to use certain products), to international politics (e.g., the worldwide petroleum supply). Therefore, in this chapter and in the other chapters devoted to the chemistry of the elements and their compounds we have two primary purposes. The first is to impart an awareness of the relationships between chemistry and, as we so fondly put it, the real world. The second is to show some of the ways that chemical theory relates to chemical fact.

A few things should be explained about the material in these chapters. Where we speak of a substance as "a gas," or "a solid," or as "reactive" or "unreactive," we are referring to the state of behavior of the substance at average room temperature and pressure. We don't bother to say "at room temperature" every time. If the conditions are different from these, it is stated. The various numerical properties of the elements and compounds are presented not so much as numbers to be memorized, but more to illustrate how measurable quantities are used and to give a feeling for the relative magnitudes of the properties.

In this chapter hydrogen and oxygen—two elements of great significance to life on Earth—are discussed. As is the case in each descriptive chapter, the properties, reactions, preparation, and uses of the elements are presented. In this chapter, compounds of each element—the hydrides and oxides—are examined, as are ozone and hydrogen peroxide. The natural cycle for oxygen is linked to that of carbon and so they are discussed together here. Another special topic considered is the potential of hydrogen as a fuel.

11.1 Origin and abundance of hydrogen and oxygen

At this time in history, hydrogen is the most abundant element in the cosmos (see Figure 8.7) and oxygen is the most abundant element in the crust of the Earth.

Oxygen makes up almost 50% by weight of the Earth's crust. In the solid crust, oxygen is mainly present as the oxides, silicates, and carbonates of the metals and semiconducting elements. Water, a compound that is 11.19% by weight hydrogen and 88.81% by weight oxygen, covers 70% of the Earth's surface as oceans, rivers, lakes, and other bodies of water. The atmosphere of the Earth contains about 20% by volume of dioxygen (O_2).

However, very little **dihydrogen**, H_2, is present in the atmosphere of the Earth today. Hydrogen is most abundant on Earth in compounds. In addition to the water of the hydrosphere, plants and animals have a large hydrogen content due to the water that they contain. A further amount of combined hydrogen is present in plants and animals in the form of the three main classes of biological substances—proteins, fats, and carbohydrates. Hydrogen is also a component of the fossil fuels. In our sun and other stars, the fusion of hydrogen nuclei to produce helium gives rise to enormous quantities of energy. This process in our sun provides all of the energy necessary for life here on Earth.

In the atmosphere of the primitive Earth, hydrogen and helium were more abundant than any of the other elements, as is still true in the cosmos. Whatever oxygen was present was probably combined with hydrogen as water, with carbon as carbon dioxide, or with sulfur as sulfur dioxide.

During the 4.5 billion years since the formation of the Earth, major changes in the atmosphere have taken place. The following processes may have occurred. The light gases—principally dihydrogen and helium—were able to achieve a velocity great enough to escape from Earth's gravity and were gradually lost to outer space. Carbon dioxide and water vapor, from the primitive atmosphere and also from hot springs and volcanoes, were removed from the atmosphere as limestone ($CaCO_3$) and the water of the oceans.

The first free oxygen to appear in the atmosphere probably came from the dissociation of water by sunlight (a photochemical reaction)

$$2H_2O(g) + h\nu \longrightarrow 2H_2(g) + O_2(g)$$

Sunlight also provided the energy needed for the formation of the organic molecules necessary to life. As the amounts of oxygen, and of ozone from the oxygen, increased, more of the radiation harmful to life as we know it was absorbed by the atmosphere (Section 6.1). Eventually, the point at which photosynthesis could begin and life could safely move out of the protective depths of the ocean was reached. Almost all of the oxygen in the atmosphere has been formed by photosynthesis, which can be summarized as

a general formula for carbohydrates

$$xCO_2 + xH_2O \xrightarrow{\text{solar energy}} (CH_2O)_x + xO_2 \tag{1}$$

(This simplified equation does not, of course, indicate the complexity of reactions and chemical compounds involved in the total process of photosynthesis.)

Although hydrogen continues to be lost from the atmosphere by escape from Earth's gravity, it is replaced by the decay of organic matter, as well as by volcanic gases and photochemical dissociation of water.

Hydrogen

Hydrogen is the only element that is not properly a member of any group in the periodic table. It has similarities to both the alkali metals and to the halogens, but is sufficiently different from them that it must stand alone. Hydrogen forms positive ions less readily than the alkali metals, and it forms negative ions less readily than the halogens. Most often, hydrogen enters into chemical combination through the formation of a single, covalent bond.

11.2 Properties of hydrogen

Hydrogen is the lightest element. Dihydrogen, H_2, is a colorless, odorless, tasteless gas. The properties of hydrogen are summarized in Table 11.1. The very low boiling and melting points (compare with those of the noble gases, Table 6.8) reflect the weakness of the forces between the molecules. The density of hydrogen is about 1/14 that of air, and it is only slightly soluble in water.

The high flammability of dihydrogen requires that it be handled with care. Hydrogen unites with oxygen to give water either at elevated temperatures or upon ignition by a spark or even by static electricity. Air containing as little as 4.1% by volume up to as much as 74.2% by volume of hydrogen is potentially explosive. The low density of the gas provides excellent buoyancy to hydrogen-filled balloons, but the disastrous explosion of the hydrogen-filled airship, the Hindenburg, in 1937, demonstrated the hazard of using it for this purpose.

The bond dissociation energy for the hydrogen–hydrogen bond is relatively high

$$H\text{—}H \rightleftharpoons H + H \qquad \Delta H° = 104.19 \text{ kcal (435.93 kJ)}$$

and the hydrogen molecule is very stable. At 1727°C molecular hydrogen is only 0.1% dissociated. Because of the large amount of energy needed to rupture the H—H bond, the reactivity of the hydrogen molecule is not great and many of its reactions are slow. Elevated temperatures and/or catalysts, which in many cases serve to split the molecule into atoms, are often required to achieve effective reactions with H_2. Hydrogen *atoms* are very reactive.

The H^+ ion is simply a proton (Figure 11.1). No other positive ion

TABLE 11.1

Properties of hydrogen

Molecular properties

Melting point (°C)	−259.2
Boiling point (°C)	−252.8
Density (g/liter)	0.0899
Bond length (nm)	0.0742
Bond dissocation energy, kcal/mole (kJ/mole)	104.19 (435.93)

Atomic properties

Ionization energy, kcal/mole (kJ/mole)	314 (1310)
Electron affinity kcal/mole (kJ/mole)	17.4 (72.8)
Electronegativity	2.1
Atomic radius (nm)	0.037
Ionic radius (H⁻) (nm)	0.21

FIGURE 11.1
Hydrogen.

Normal hydrogen, or protium, ^1_1H
Hydrogen atom, H·
  $1s^1$ configuration
  1 electron
  1 proton
Hydrogen molecule, or dihydrogen, H_2, H:H
  2 electrons
  2 protons
Hydride ion, H⁻, (H:)⁻
  $1s^2$ configuration
  2 electrons
  1 proton
Proton, H⁺
  1 proton only

Deuterium, D, ^2_1H
Deuterium atom
  1 electron
  1 proton
  1 neutron

Tritium, T, ^3_1H
Tritium atom
  1 electron
  1 proton
  2 neutrons

is so small nor has such a high charge density, for, with its single electron gone, the positive charge of the proton is completely unshielded. Although H^+ ions do exist in the exosphere (Section 6.5), we normally only find them in the solvated condition—united with one or more solvent molecules as, for example, in the hydronium ion H_3O^+, which we sometimes write to represent H^+ in aqueous solution.

The hydrogen atom has only one electron—its <u>configuration</u> is $1s$. In <u>bonding</u> it can attain the noble gas configuration of helium ($1s^2$) in two ways: (1) by addition of an electron to form the hydride ion, H^- ($1s^2$), or (2) by sharing an electron in a covalent bond. In most of its compounds hydrogen is covalently bonded. With its two protons and two electrons, the hydrogen molecule, H : H, is the lightest and simplest of all molecules. Hydrogen bonding (Section 9.21) is a characteristic property of hydrogen covalently bound to small, highly electronegative atoms. Hydride ions exist only in compounds of hydrogen with metals that are easily ionized, for example, in sodium hydride, NaH (Section 11.8a).

The hydrogen molecule often acts as a <u>reducing agent</u>, being oxidized itself from an oxidation number of 0 to an oxidation number of $+1$, as, for example, in the reduction of iron oxide to metallic iron

$$\overset{0}{H_2}(g) + \overset{+2}{FeO}(s) \overset{\Delta}{\longrightarrow} \overset{0}{Fe}(s) + \overset{+1}{H_2O}(g)$$

In all of the reactions cited in the following sections, except for the formation of metal hydrides, dihydrogen is a reducing agent. In the products it is in the $+1$ <u>oxidation state</u> and is covalently bound to a more electronegative element. Hydrogen is assigned a -1 oxidation state in metal hydrides.

11.3 Isotopes of hydrogen

The most abundant of the three isotopes of hydrogen is ordinary hydrogen (or protium), the species of mass number 1, which is characterized by a nucleus of one proton and is represented by the symbol ^1_1H (see Figure 11.1). The other isotopes of hydrogen have mass numbers of 2 and 3 (Figure 11.1). The isotope of mass number 2 has a nucleus of one proton and one neutron and is called **deuterium,** or heavy hydrogen, represented by the symbol ^2_1H or D. This isotope constitutes 1 part in 5,000 of naturally occurring hydrogen. The isotope of mass number 3, with a nucleus consisting of one proton and two neutrons, is called **tritium** and is symbolized as ^3_1H or T. Tritium is radioactive, in contrast to ordinary hydrogen and deuterium, which have stable nuclei. Tritium occurs to only an extremely small extent in nature and is commonly obtained by synthetic means—for example, by a nuclear reaction between lithium (Li) and neutrons.

$$^6_3Li + ^1_0n \longrightarrow ^3_1T + ^4_2He$$

The degree to which isotopes of any element differ is largely dependent upon their percentage difference in mass, and of all the elements this difference is greatest for hydrogen. In general, ordinary hydrogen undergoes reactions much more rapidly than deuterium, and the latter, in turn, more rapidly than tritium. It must be emphasized, however, that products formed with any particular element or compound have the same compositions, because each of the isotopic atoms has the same electronic structure. For example, union of chlorine with ordinary hydrogen, deuterium, and tritium yields HCl, DCl, and TCl, respectively. The physical properties of H_2 and D_2, and H_2O and D_2O—**heavy water**—are compared in Table 11.2.

Deuterium and tritium have proved to be extremely valuable tools in the study of chemical reactions of compounds containing hydrogen. Some of the ordinary hydrogen in a compound under investigation is replaced with deuterium or tritium. If the locations of these atoms in the reactants are known, the course of a chemical reaction can often be inferred from their location in the products. Deuterium is located by mass spectrometric analysis (Tools of Chemistry, Mass Spectrometer, in Chapter 7) and tritium by its radioactivity.

11.4 Reactions of hydrogen

a. Reactions with other elements. Dihydrogen reacts directly with many of the elements. With nonmetals at elevated temperatures, the products are covalent compounds. For example, hydrogen combines with all of the halogens to give hydrogen halides. In the laboratory hydrogen bromide that is **anhydrous**—free of water—can be prepared by passing a mixture of hydrogen and bromine over a catalyst at 350°C.

$$H_2(g) + Br_2(g) \xrightarrow[\text{catalyst}]{350°C} 2HBr(g)$$

Ammonia is commercially produced by the direct combination of nitrogen and hydrogen

$$N_2(g) + 3H_2(g) \xrightarrow[\text{pressure}]{\overset{\Delta}{\text{catalyst}}} 2NH_3(g)$$

The union of dihydrogen with oxygen to yield water is an excellent example of a reaction which, although having a favorable $\Delta H°$, proceeds at an insignificant rate at ordinary temperatures.

$$H_2(g) + \tfrac{1}{2}O_2(g) \longrightarrow H_2O(g) \quad \Delta H° = -57.79 \text{ kcal } (-241.8 \text{ kJ})$$

Even at 400°C in the presence of a catalyst, the reaction is extremely slow. At 700°C the combination of the elements occurs with explosive violence. Use is made of the vigor of the reaction to attain temperatures in the neighborhood of 2800°C by means of the oxyhydrogen torch.

At elevated temperatures dihydrogen unites with sulfur vapor to form hydrogen sulfide, a compound isoelectronic with water. The

TABLE 11.2

Comparison of properties of H_2 and D_2, and of H_2O and D_2O

D_2O *is commonly called heavy water.*

	H_2	D_2
Melting point (°C)	−259.2	−254.4
Boiling point (°C)	−252.8	−249.5
Bond length (nm)	0.0742	0.0742
Bond dissociation energy, kcal/mole (kJ/mole)	104.19 (435.93)	105.62 (441.91)
Heat of fusion, cal/mole (J/mole)	28 (120)	47 (200)
Heat of vaporization, cal/mole (J/mole)	216 (904)	293 (1230)

	H_2O	D_2O
Melting point (°C)	0.0	3.8
Boiling point (°C)	100.0	101.4
Density at 25°C (g/ml)	0.997	1.10
Heat of fusion, cal/mole (J/mole)	1436 (6008)	1500 (6276)
Heat of vaporization, cal/mole (J/mole)	9717 (40,660)	9944 (41,610)

reaction is much less vigorous than that between hydrogen and oxygen.

$$H_2(g) + S(g) \xrightarrow{\Delta} H_2S(g) \qquad \Delta H° = -4.93 \text{ kcal } (-20.6 \text{ kJ})$$

hydrogen sulfide

With <u>reactive metals</u>, hydrogen forms saltlike hydrides (Section 11.8a), for example,

$$H:H(g) + 2Li(l) \xrightarrow[\text{vacuum}]{700°C} 2Li^+ H:^-(l)$$

lithium hydride
white, crystalline
solid, m.p. 680°C

b. Reactions with metal oxides. Dihydrogen reacts with oxides of certain metals (those below Fe in the electromotive series; Section 11.5) to give the free metal and water, for example

$$CuO(s) + H_2(g) \xrightarrow{\Delta} Cu(s) + H_2O(g)$$

$$FeO(s) + H_2(g) \xrightarrow{\Delta} Fe(s) + H_2O(g)$$

$$WO_3(s) + 3H_2(g) \xrightarrow{\Delta} W(s) + 3H_2O(g)$$

This type of reaction is expensive as compared with most industrial processes, but is used to obtain costly metals, such as tungsten (W), from their oxide ores.

c. Hydrogenation. **Hydrogenation** is the addition of the two atoms of dihydrogen to a compound. Hydrogenation reactions generally require catalysts and are usually carried out at high pressures.

Most often the purpose of hydrogenation is to add hydrogen to the carbon–carbon double bond in an organic compound or a mixture of organic compounds. A simple example is the hydrogenation of ethylene.

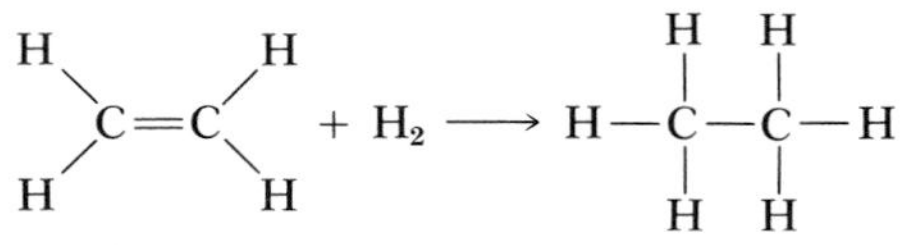

$$\Delta H° = -32.7 \text{ kcal } (-137 \text{ kJ})$$

ethylene   ethane

Vegetable oils such as soybean, cottonseed, and coconut oils are hydrogenated to produce solid fats for shortening and margarine. With the approaching shortage of petroleum, interest is growing in the liquefaction of coal, a process in which coal is hydrogenated and larger compounds are split to smaller ones of lower molecular weight (the hydrogenated carbon–carbon bond breaks). Addition of 2%–3% by weight of hydrogen to coal yields a heavy oil that can be used as a fuel in power plants. With the addition of 6% or more of hydrogen, coal is converted to a distillable mixture of light oils. The search for a method to hydrogenate coal all the way to methane gas, a substitute for natural gas as a fuel, is also under way.

Another industrially important hydrogenation is the conversion of carbon monoxide to methanol (methyl alcohol, also called wood alcohol).

$$CO(g) + 2H_2(g) \xrightarrow[\substack{200 \text{ atm} \\ \text{catalyst}}]{400°C} \underset{\substack{methanol, \ b.p. \ 65°C}}{CH_3OH(g)} \qquad \Delta H° = 25.4 \text{ kcal (106 kJ)} \quad (2)$$

Ninety percent of the methanol produced is used as an intermediate in the production of other chemicals, and the remaining 10% is used as a solvent.

11.5 Electromotive series

An **electromotive series** is a list of elements, usually metals (Table 11.3), in the order of decreasing ease of electron loss in aqueous solution. The order was first established by comparing the reactivities of metals and metal ions toward each other and toward other species. For example, a free metal will displace from aqueous solution an ion of a metal below it in the series,

$$Mg(s) + Cu^{2+} \longrightarrow Cu(s) + Mg^{2+}$$

The magnesium has been oxidized and the copper ion reduced.

Hydrogen is included in the series as a reference point. Metals above hydrogen will displace hydrogen from an acid solution [see Equations (5) and (6)]; those below it will not. The ease of formation and reduction of metal oxides also correlates with the position of a metal in the series. The least reactive metals—those at the very bottom—are more likely to be found in nature in an uncombined state, rather than as sulfides, oxides, and so on. Their oxides (not formed by direct combination; Section 11.10) are most easily reduced.

11.6 Preparation and uses of hydrogen

The main industrial sources of dihydrogen are water and hydrocarbons, both abundant and relatively inexpensive raw materials. At present, the major hydrogen-producing processes use various combinations of catalysts and elevated temperatures, with the overall result that hydrocarbons in the presence of steam and/or oxygen are converted to hydrogen plus carbon dioxide and/or carbon monoxide. Using the general formula for saturated hydrocarbons, C_nH_{2n+2}, where n is a whole number, the basic reactions are

$$C_nH_{2n+2} + nH_2O \rightleftharpoons nCO + (2n + 1)H_2$$
$$C_nH_{2n+2} + 2nH_2O \rightleftharpoons nCO_2 + (3n + 1)H_2$$

In the steam reforming of hydrocarbons, the vaporized hydrocarbons react with steam over a bed of nickel catalyst at temperatures between 600°C and 1000°C and at atmospheric pressure. To ob-

Free metal displaces ions of metals below it in electromotive series

TABLE 11.3
Electromotive series of metals
The active metals are at the top. They are easily oxidized and form ions readily. The activity of the metals decreases down the series.

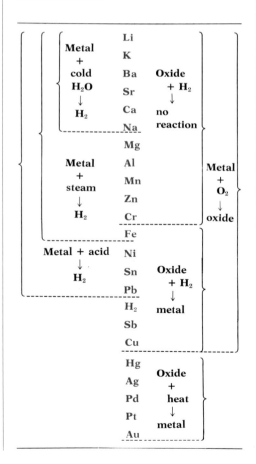

305

tain pure hydrogen the gaseous product mixture is first passed with steam over a catalyst to convert CO to CO_2 (called the "water gas shift" reaction)

$$CO(g) + H_2O(g) \xrightarrow[\text{catalyst}]{250°C} CO_2(g) + H_2(g) \qquad (3)$$

The carbon dioxide is then easily separated from hydrogen by passing the mixture under pressure into cold water or into an amine solution—the carbon dioxide dissolves, but hydrogen does not.

The industrial process for manufacturing ammonia—the largest consumer of hydrogen—requires a 1:3 mixture of nitrogen and hydrogen as the starting material, and the industrial process for manufacturing methanol requires a 1:2 mixture of carbon monoxide and hydrogen. By adjusting the conditions of steam reforming of hydrocarbons (air is added when a source of nitrogen is needed), the gaseous reaction mixtures for these two processes are produced directly. Most commercially produced hydrogen is used within the chemical industry. Table 11.4 summarizes the uses of hydrogen.

An older method for producing hydrogen was by the action of steam on coke (coke is coal that has been heated in the absence of air to leave behind mostly carbon)

$$H_2O(g) + C(s) \xrightarrow{1000°C} H_2(g) + CO(g) \qquad (4)$$

Carbon monoxide was converted to carbon dioxide by the shift reaction (Equation 3). The initial mixture of H_2 and CO is known as *water gas*. It can be used as a fuel since both gases burn readily in air $(2CO + O_2 \rightarrow 2CO_2; 2H_2 + O_2 \rightarrow 2H_2O)$, giving a very hot flame.

Dihydrogen of the greatest purity can be prepared by the electrolytic decomposition of water in the presence of a base, such as sodium hydroxide (Na^+OH^-), which forms ions that serve as the carriers of the electric current.

$$2H_2O(l) \xrightarrow[\text{base}]{\text{electrolysis}} 2H_2(g) + O_2(g)$$

The process is expensive because it consumes a large amount of electricity.

Dihydrogen is also obtained as a byproduct in the manufacture of gasoline. Hydrocarbons of high molecular weight are converted to a mixture of lower molecular weight compounds more suitable for gasoline and hydrogen by decomposition on a catalyst at elevated temperatures.

A method that can be used for preparing hydrogen in the laboratory (Figure 11.2) is by displacement from dilute mineral acids by active metals (Section 11.5), for example,

$$Zn(s) + 2HCl(aq) \longrightarrow H_2(g) + ZnCl_2(aq) \qquad (5)$$
$$Zn(s) + H_2SO_4(aq) \longrightarrow H_2(g) + ZnSO_4(aq) \qquad (6)$$

TABLE 11.4

Uses of hydrogen

Major

Ammonia manufacture

Hydrogenation
    Of petroleum products
    Of vegetable oils
    Of CO to give CH$_3$OH (methanol)

Others

Fuel (including rocket fuel)

Reduction of oxide-containing ores to give free metals

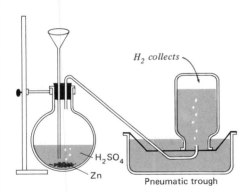

FIGURE 11.2
Laboratory preparation of hydrogen.

The zinc chloride or zinc sulfate remains in solution as ions. In terms of the ionic equation, both reactions can be represented as

$$Zn(s) + 2H^+ \longrightarrow Zn^{2+} + H_2(g)$$

Hydrogen is also produced by <u>displacement from pure water by active metals</u> (the alkali metals, and Ca, Ba, Sr), with the production of metal hydroxides. For sodium and water, the reaction is

$$2Na(s) + 2H_2O(l) \longrightarrow H_2(g) + 2NaOH(aq)$$

In general, this method is too vigorous to be suitable for the laboratory preparation of hydrogen (Section 26.4b).

A number of common metals, such as tin, aluminum, or zinc, and the semiconducting element silicon cause hydrogen <u>displacement from aqueous solutions of strongly basic hydroxides</u>, such as sodium hydroxide.

$$Sn(s) + 2NaOH(aq) + 4H_2O(l) \longrightarrow 2H_2(g) + Na_2Sn(OH)_6 \ (aq)$$
sodium stannate

$$Si(s) + 4NaOH(aq) \longrightarrow 2H_2(g) + Na_4SiO_4(aq)$$
sodium silicate

11.7 Hydrogen as a fuel

One proposed way to meet the need for new energy sources is to burn hydrogen as a fuel in industry and power plants, and possibly also in homes and motor vehicles. This proposal is referred to as the *hydrogen economy*. There are several advantages to a hydrogen economy. Hydrogen combustion yields only water as the product—no sulfur, nitrogen, or carbon oxides and no particulates, and, therefore, no pollution. And the source of the hydrogen would be water, which is in virtually unlimited supply, is regenerated in the combustion, and is uniformly distributed around the world.

Dihydrogen is envisioned not only as a fuel, but as a direct substitute for electricity, being transported, as is natural gas, through an underground pipeline network. Unlike electricity, hydrogen can be stored (usually as a liquid at low temperatures), making it possible to produce the gas continuously and call upon reserves when the demand for energy is high. The hydrogen would be burned at the site of use to provide thermal energy, or would be used in fuel cells (Section 24.7) to provide electrical energy.

Two methods of producing hydrogen for these purposes have been proposed. One is electrolysis (Section 11.6) and the other is the intriguing thermochemical reaction cycle. The first method utilizes electricity, possibly from solar energy, the second utilizes only heat, possibly from a nuclear power plant. Whether either method can produce hydrogen cheaply enough to make its use practical is an open question.

Metals above H_2 in the electromotive series displace H_2 from acid

TABLE 11.5

TABLE 11.5

Proposed thermochemical cycle for hydrogen production

1. $CaBr_2 + 2H_2O \xrightarrow{730°C} Ca(OH)_2 + 2HBr$

2. $2HBr + Hg \xrightarrow{250°C} HgBr_2 + H_2$

3. $HgBr_2 + Ca(OH)_2 \xrightarrow{100°C} CaBr_2 + H_2O + HgO$

4. $HgO \xrightarrow{600°C} Hg + \frac{1}{2}O_2$

TABLE 11.6

Binary compounds of hydrogen with the common representative elements

A thermochemical cycle for hydrogen production is a series of reactions in which the only things consumed are water and heat, and the only products are hydrogen and oxygen. All of the other species involved are recycled. One of the many proposed thermochemical cycles for hydrogen is given in Table 11.5.

11.8 Binary compounds of hydrogen

Hydrogen–metal compounds extend over the entire range from ionic to covalent. The best-characterized binary compounds of hydrogen are those it forms with the representative group elements (Table 11.6). Most of them have simple formulas that can be deduced from the electronic configurations of the elements to which hydrogen is bound. The compounds of hydrogen with beryllium and magnesium of Group II, and with all of the elements of Group III are notable exceptions. (The boron hydrides, discussed in Section 27.9a, have structures that cannot be explained in terms of our common concepts of chemical bonding.)

The Group I–III compounds, and also SbH_3, BiH_3, SeH_2, and TeH_2 are named as hydrides, e.g., lithium hydride, although SeH_2 and TeH_2 are also called hydrogen selenide and hydrogen telluride. The other compounds are named as follows (going from the top down within the groups): Group IV, methane, silane, germane, stannane, plumbane; Group V, ammonia, phosphine, arsine; Group VI, water, hydrogen sulfide; Group VII, the hydrogen halides, e.g., hydrogen fluoride.

Group	I	II	III	IV	V	VI	VII
Valence shell configuration	s^1	s^2	s^2p^1	s^2p^2	s^2p^3	s^2p^4	s^2p^5
	$Li^+ H{:}^-$	$(BeH_2)_x{}^a$	$(BH_3)_2{}^b$	$H{:}\overset{H}{\underset{H}{\overset{..}{C}}}{:}H^b$	$\overset{H}{\underset{H}{:}\overset{..}{N}}{:}H^b$	$\overset{H}{:}\overset{..}{\underset{..}{O}}{:}H^b$	$H{:}\overset{..}{\underset{..}{F}}{:}$
	$Na^+ H{:}^-$	$(MgH_2)_x{}^a$	$(AlH_3)_x{}^a$	$H{:}\overset{H}{\underset{H}{Si}}{:}H^b$	$\overset{H}{\underset{H}{:}P}{:}H^b$	$\overset{H}{:}\overset{..}{\underset{..}{S}}{:}H^b$	$H{:}\overset{..}{\underset{..}{Cl}}{:}$
	$K^+ H{:}^-$	$Ca^{2+} 2H{:}^-$	$(GaH_3)_x{}^a$	$H{:}\overset{H}{\underset{H}{Ge}}{:}H^b$	$\overset{H}{\underset{H}{:}As}{:}H$	$\overset{H}{:}\overset{..}{Se}{:}H$	$H{:}\overset{..}{\underset{..}{Br}}{:}$
	$Rb^+ H{:}^-$	$Sr^{2+} 2H{:}^-$	$(InH_3)_x{}^a$	$H{:}\overset{H}{\underset{H}{Sn}}{:}H^b$	$\overset{H}{\underset{H}{:}Sb}{:}H$	$\overset{H}{:}Te{:}H$	$H{:}\overset{..}{\underset{..}{I}}{:}$
	$Cs^+ H{:}^-$	$Ba^{2+} 2H{:}^-$		$H{:}\overset{H}{\underset{H}{Pb}}{:}H$	$\overset{H}{\underset{H}{:}Bi}{:}H$		

^{a} The subscript x indicates that the actual molecule consists of a number of recurring units of the formula shown.

^{b} Formula of most simple compound; others, e.g., C_2H_6, C_3H_8, C_4H_{10}, Si_2H_6, of a more complicated nature are known.

[11.7] HYDROGEN AND OXYGEN

a. Saltlike hydrides. With the metals of Representative Group I and with calcium, strontium, and barium of Group II, all of which form positive ions easily, hydrogen yields solid ionic compounds containing the hydride ion, H^-. These compounds are known as the *saltlike* or saline *hydrides*. When decomposed by heat, they yield the free metal and elemental hydrogen, for example,

$$2Na^+H^-(s) \longrightarrow 2Na(s) + H_2(g)$$

Another characteristic reaction of the saltlike hydrides is their instantaneous decomposition by water, also with the formation of free hydrogen.

$$H^- + H_2O(l) \longrightarrow H_2(g) + OH^-$$

Calcium hydride, CaH_2, is often used as a drying agent for organic liquids, removing H_2O by the reaction above.

b. Covalent hydrides. With the elements of Representative Groups IV through VII, most of which are nonmetals, hydrogen forms binary covalent compounds. Only the simplest of the many known hydrides for the elements of Group IV are listed in Table 11.6.

Most of the covalent hydrides shown in Table 11.6 are gaseous at room temperature. With a few exceptions, these compounds exist as simple (unassociated) molecules when cooled to the liquid and solid states. Hydrogen fluoride, water, and ammonia, because the hydrogen in them is bound to small, electronegative elements, form larger aggregates through hydrogen bonding.

In any one group, the thermal stability of the covalent hydrides decreases with increasing atomic number and size of the atom bound to hydrogen. For example, ammonia (NH_3) is decomposed by heat much less readily than is phosphine (PH_3), and the latter in turn is more stable thermally than arsine (AsH_3). This order of decreasing thermal stability parallels the order of decreasing strength of the covalent bond between hydrogen and the atom to which it is joined, and is reflected in the values for the standard enthalpies of formation of the compounds:

ΔH_f°	kcal/mole	(kJ/mole)
NH_3	−11.02	(−46.11)
PH_3	1.3	(5.4)
AsH_3	15.88	(66.44)

Ammonia functions as a base in water, water itself can behave both as an acid and a base (Section 17.2), and hydrogen fluoride acts as an acid. With an increase in group number, that is, with an increase in the electronegativity (see Table 10.7) of the element bound to hydrogen, the polarity of the covalent bond between hydrogen and the element increases. This leads to an increase in the tendency of the hydride to function as an acid. A similar trend is observed across the other sequences of hydrides of Groups V through VII (e.g., PH_3, H_2S, HCl; AsH_3, H_2Se, HBr). (The acid strength of hydrides is further discussed in Section 17.3.)

11.9 Properties of oxygen

Oxygen is the second-row element of Representative Group VI. Dioxygen is a colorless, odorless, tasteless gas; it boils 70 Celsius degrees above dihydrogen (Table 11.7). Pure gaseous oxygen is slightly more dense than air (density of dry air, about 1.2 g/liter). Liquid oxygen is pale blue.

The solubility of dioxygen in water is slight—about 30 ml dissolves in 1 liter of water at 20°C and 1 atm. However, this slight solubility makes possible both aquatic life and the destruction of waste material in bodies of water (Sections 13.6–13.8).

The electronic configuration of atomic oxygen is $1s^2 2s^2 2p^4$; there are six electrons in the valence shell. Dioxygen, despite its even number of electrons (16), has two of them unpaired and is paramagnetic (a subject taken up in Section 21.11). In bonding, the oxygen atom forms an oxide ion by gaining two electrons or it shares electrons to form single or double covalent bonds, for example,

$$Ca^{2+}(:\ddot{O}:)^{2-} \qquad H:\ddot{O}:H \qquad H:\ddot{O}:\ddot{C}l:\ddot{O}: \qquad CH_3CH_2HC::\ddot{O}$$

The oxidation number of oxygen in such compounds is -2.

Oxygen is second only to fluorine in electronegativity, and therefore has a positive oxidation number in its compounds with fluorine (e.g., OF_2, where fluorine has oxidation number -1 and oxygen therefore has $+2$). Like most nonmetals, oxygen has a high ionization energy and a high electron affinity. Only the remaining Group VI elements and the halogens have higher electron affinities than oxygen.

The oxygen–oxygen bond in O_2 is quite stable, with a bond dissociation energy similar to that of hydrogen (see Table 11.1). The reactivity of molecular oxygen, also like that of hydrogen, is relatively low, although oxygen does react spontaneously at room temperature with some strong inorganic reducing agents and with many organic compounds. The rusting of iron and the decomposition of organic matter are also reactions involving oxygen that are spontaneous, although slow, at ordinary temperatures. Elevated temperatures and sometimes also elevated pressures are required to effectively carry out most reactions of dioxygen.

11.10 Reactions of oxygen

a. Reactions with other elements. Oxygen can combine directly with all of the elements except the noble gases, the halogens, and some of the less active metals such as silver, gold, and platinum, to form **oxides**—a term generally reserved for the binary compounds of oxygen. It is often necessary to initiate the oxygen–element reaction

TABLE 11.7

Properties of oxygen

Molecular properties	
Melting point (°C)	−218.8
Boiling point (°C)	−183.0
Density (g/liter)	1.429
Bond length (nm)	0.12
Bond dissociation energy, kcal/mole (kJ/mole)	119.11 (498.36)

Atomic properties	
Ionization energy, kcal/mole (kJ/mole)	314 (1310)
Electron affinity, kcal/mole (kJ/mole)	33.8 (140)
Electronegativity	3.5
Atomic radius (nm)	0.074
Ionic radius (O^{2-}) (nm)	0.14

by heating, but in most cases, once combination begins, enough heat is evolved to maintain the reaction. In all of its reactions with nonmetals and in most of those with metals, dioxygen is reduced to the -2 state. The compounds formed with nonmetals are covalent. Those with metals may be ionic or covalent, depending upon the metal and its oxidation state (Section 11.13)

Some typical reactions of dioxygen (in excess) with metals and nonmetals are shown in Table 11.8.

b. Reactions with oxides. For elements that can exhibit more than one oxidation state, it is usually possible to convert an oxide containing the element in a low oxidation state to one in which the element is in a higher state, by reaction with dioxygen. For example, the reaction of copper(I) oxide gives copper(II) oxide, and that of tin(II) oxide gives tin(IV) oxide

$$2Cu_2O(s) + O_2(g) \xrightarrow{\Delta} 4CuO(s)$$
$$2SnO(s) + O_2(g) \xrightarrow{\Delta} 2SnO_2(s)$$

c. Other oxidation reactions. The combination of any element or compound with oxygen is an "oxidation" reaction (broader meanings of the term "oxidation" are discussed in Chapter 19). Compounds often react with oxygen to yield products in which each of the elements of the compounds have combined with oxygen.

$$4NH_3(g) + 5O_2(g) \xrightarrow{\text{catalyst}} 4NO(g) + 6H_2O(g)$$
$$CS_2(g) + 3O_2(g) \longrightarrow CO_2(g) + 2SO_2(g)$$

The slow oxidation of food in the body, catalyzed by enzymes, provides the energy necessary to sustain life.

We usually reserve the term **combustion** for any chemical change in which both heat and light are given off. All combustion reactions in air or oxygen are oxidation reactions. Combustion of fossil fuels and wood are basically oxidation reactions that release a great deal of energy. For example, methane, the major component of natural gas, yields over 200 kcal of energy for each mole that is burned.

$$CH_4(g) + 2O_2(g) \longrightarrow CO_2(g) + 2H_2O(l) \quad \Delta H° = -212.8 \text{ kcal } (-890.4 \text{ kJ})$$

Carbon dioxide and water are the sole products of the complete combustion of hydrogen–carbon, or hydrogen–oxygen–carbon compounds in the presence of sufficient oxygen. (Carbon monoxide is formed when not enough oxygen is present; Section 6.7). Many materials that will not burn, such as glass, ceramics, clays, and water, contain elements that are already fully combined with oxygen.

11.11 Preparation and uses of oxygen

On an industrial scale, most dioxygen ($>95\%$) is produced by fractional distillation of liquid air (Section 6.4). This is the method of choice because it is by far the most economical. About 85% of the

TABLE 11.8

Direct combination reactions of oxygen

Reactions with nonmetals

$$S_8(s) + 8\ O_2(g) \longrightarrow 8\ SO_2(g)$$
sulfur   sulfur dioxide

$$S_8(s) + 12\ O_2(g) \longrightarrow 8\ SO_3(s)$$
sulfur   sulfur trioxide

$$P_4(s) + 5\ O_2(g) \longrightarrow P_4O_{10}(s)$$
phosphorus   phosphorus(V) oxide

$$C(s) + O_2(g) \longrightarrow CO_2(g)$$
carbon   carbon dioxide

$$2\ C(s) + O_2(g) \longrightarrow 2\ CO(g)$$
carbon   carbon monoxide

Reactions with metals

$$2\ Mg(s) + O_2(g) \longrightarrow 2\ MgO(s)$$
magnesium   magnesium oxide

$$2\ Cu(s) + O_2(g) \longrightarrow 2\ CuO(s)$$
copper   cupric oxide (copper(II) oxide)

$$4\ Al(s) + 3\ O_2(g) \longrightarrow 2\ Al_2O_3(s)$$
aluminum   aluminum oxide

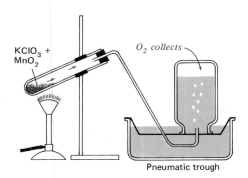

KClO₃ + MnO₂

O₂ collects

Pneumatic trough

FIGURE 11.3
Laboratory preparation of oxygen.
The test tube must be tilted to prevent the hot mixture from touching the stopper. Explosion can result from contact of this mixture with a rubber stopper.

annual output of oxygen (425 billion cubic feet in 1976) is produced on the site of its use. Liquid air plants stand beside steel mills, the single largest users of oxygen, or beside metal fabricating plants. The oxygen goes directly to the mill or plant, where it is used to speed combustion and raise the combustion temperature (Section 29.17). A relatively new use for on-site-generated oxygen is in waste water treatment, where it replaces air in the activated sludge process (Section 13.14).

Other oxygen is transported to its industrial users in cylinders, tanks, or tank cars. The greatest outlet for such oxygen is the chemical industry, where oxygen is consumed in oxidation reactions and the production of oxygen-containing compounds. Other uses are summarized in Table 11.9.

Some commercial oxygen is also derived from the electrolysis of water (Section 11.6); oxygen from this source, like electrolytic hydrogen, is very pure, but quite expensive.

Laboratory-scale preparation of oxygen usually involves heating an oxygen-containing compound, which gives up all or part of its oxygen. The thermal decomposition of potassium chlorate, the most common laboratory method for oxygen preparation, is aided by a catalyst (Figure 11.3).

$$2KClO_3(l) \xrightarrow{\Delta,\ MnO_2\ or\ Fe_2O_3} 3O_2(g) + 2KCl(s)$$
*potassium
chlorate*

Another convenient route to oxygen in the laboratory is by the reaction of water with sodium peroxide.

$$2Na_2O_2(s) + 2H_2O(l) \longrightarrow O_2(g) + 4NaOH(aq)$$
sodium peroxide

11.12 Ozone

Ozone, O_3, is oxygen in the form of gaseous, triatomic molecules. Oxygen and ozone are **allotropes**—different forms of the same element in the same state (in this case, both gases). The oxygen atoms in ozone are joined at an angle of 116°49′ (Figure 11.4).

Pure ozone is a pale blue gas, it condenses to a dark blue liquid that boils at −111.5°C. Ozone is formed when energy is supplied to gaseous oxygen in the form of radiation, electricity, or heat.

$$3O_2(g) \longrightarrow 2O_3(g) \qquad \Delta H° = 68.2\ kcal\ (285\ kJ) \qquad (7)$$

The characteristic odor of ozone is noticeable near a sparking electric motor.

Usually, the preparation of ozone is accomplished by passing gaseous oxygen through an electrical field under high voltage. The reaction is highly endothermic and the yield of ozone is low. Equation (7) shows that ozone has a much higher energy content than

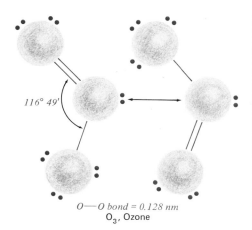

$116°\ 49'$

$O—O\ bond = 0.128\ nm$

O_3, Ozone

FIGURE 11.4
Resonance structures of ozone.
Although these structures depict both single and double oxygen–oxygen bonds, two different bond distances between these atoms are not observed. The actual bond distance is intermediate between that expected for a single bond and that for a double bond.

dioxygen. As might be expected, the stability and reactivity of ozone differ from those of oxygen. Ozone is less stable and more reactive than oxygen. Ozone tends to decompose to oxygen, in a reversal of reaction (7). The decomposition, slow at low temperatures, increases rapidly with rising temperature, is catalyzed by a variety of substances such as platinum and manganese dioxide, MnO_2, and can be explosive.

Ozone is a much more powerful oxidizing agent than ordinary oxygen. Even such relatively inactive metals as silver and mercury, which are inert to diatomic oxygen, are converted to oxides by ozone. When ozone functions as an oxidant, only part of the oxygen in the molecule is reduced and dioxygen is a product, for example, in the oxidation of lead sulfide.

$$PbS(s) + 4O_3(g) \longrightarrow PbSO_4(s) + 4O_2(g)$$

In this reaction, sulfur is oxidized from its lowest oxidation state (-2) to its highest $(+6)$. The oxidation of potassium iodide by ozone, in which diatomic oxygen is also formed, is often used as a test for ozone. The iodine liberated from KI is detected by the deep blue color it gives a suspension of starch in water.

$$2KI(aq) + O_3(g) + H_2O(l) \longrightarrow 2KOH(aq) + I_2(aq) + O_2(g)$$
$$I_2(aq) + starch \longrightarrow blue\ complex$$

(Other oxidizing agents also liberate iodine from KI.)

In Chapter 6 we discussed the important roles played by ozone in the evolution of the atmosphere and in protecting the surface of the Earth from life-damaging radiation. Commercially, the major use of ozone is as a bleaching agent. It is also a disinfectant and is used as a chemical reagent. A **reagent** is any chemical used to bring about a desired chemical reaction.

11.13 Binary compounds of oxygen

Like the hydrides, oxides span the entire range from ionic to covalent compounds, depending upon the position in the periodic table of the

TABLE 11.10

Binary compounds of oxygen with the common representative elements

Elements whose symbols are shown in color form ionic oxides. Those marked x form "polymeric" oxides to varying degrees.

Group	I	II	III	IV	V	VI	VII
Valence shell configuration	s^1	s^2	s^2p^1	s^2p^2	s^2p^3	s^2p^4	s^2p^5
General formula	$(M^+)_2O^{2-}$	$M^{2+}O^{2-}$	M_2O M_2O_3	MO MO_2	M_2O_3 M_2O_5	MO_2 MO_3	M_2O_5 M_2O_7
	Li	Be	B_x	C	N		
	Na	Mg	Al	Si_x	P_x	S	Cl
	K	Ca	Ga	Ge_x	As_x	Se_x	Br
	Rb	Sr	In	Sn_x	Sb_x	Te_x	I_x
	Cs	Ba		Pb	Bi_x		

TABLE 11.11

Melting and boiling points of some nonmetal oxides

Formula of compound	Name	m.p. (°C)	b.p. (°C)
SO_2	Sulfur dioxide	−75.46	−10.02
SO_3	Sulfur trioxide	16.85	44.8
NO	Nitric oxide (nitrogen(II) oxide)	−161	−151
NO_2	Nitrogen dioxide (nitrogen(IV) oxide)	−11.2	21.2
P_4O_6	Phosphorus(III) oxide	23.8	175.4
CO	Carbon monoxide	−207	−192
CO_2	Carbon dioxide	−78.5 (sublimes)	—
SiO_2[a]	Silicon dioxide (silica)	~1550	2230
B_2O_3[a]	Boric oxide	557	1500

[a] The formulas SiO_2 and B_2O_3 are empirical formulas and represent the smallest whole number ratios of oxygen atoms to boron and silicon atoms in the compounds. These oxides do not exist as simple molecular species, but rather as polymeric aggregates. The melting point cited for SiO_2 is for its most common crystalline form, quartz.

element with which oxygen is combined (Table 11.10). With elements to the left in the table, the oxides are ionic; with those to the right and at the top of their groups, the oxides are covalent molecules. In between, the type of bonding and also the structures of the compounds are intermediate. Often (as with the Group III hydrides) the formula represents the simplest stoichiometry, but the molecules are actually polymers—many repeating units of the same stoichiometry are linked together.

  a. Reactions with nonmetals. Most nonmetal oxides exist as simple molecules, a fact reflected in their low melting and boiling points (Table 11.11). Silicon dioxide and boric oxide form polymeric structures and the complexity of their structures is shown by their high melting points.

Most nonmetal oxides react with water to give oxygen-containing acids:

$$SO_2 + H_2O \rightleftharpoons H_2SO_3$$
sulfurous acid

$$SO_3 + H_2O \longrightarrow H_2SO_4$$
sulfuric acid

$$P_4O_6 + 6H_2O \longrightarrow 4H_3PO_3$$
phosphorous acid

$$CO_2 + H_2O \rightleftharpoons H_2CO_3$$
carbonic acid

A nonmetal oxide that combines with water to give the acid is known as an **acid anhydride** (anhydride meaning "without water"), or an acidic oxide.

b. Reactions with metals. Four classes of binary compounds of oxygen with metals are known (Table 11.12). All metals form normal oxides—binary compounds with oxygen in its most common oxidation state of -2: for example, Na_2O, sodium oxide; MgO, magnesium oxide; Al_2O_3, aluminum oxide. The alkali and alkaline earth metals also form stable compounds with anions containing two or three oxygen atoms in lower negative oxidation states (Table 11.12), two of which are fractional. The superoxides and ozonides are most readily formed with the largest of the metal ions of Representative Group I (potassium, rubidium, and cesium).

c. Properties of normal oxides. Many of the normal metal oxides are ionic. Such oxides are completely basic and, to the extent that they are soluble, react with water with essentially quantitative conversion of O^{2-} to OH^-.

$$K_2O(s) + H_2O(l) \longrightarrow 2K^+OH^-(aq)$$
$$MgO(s) + H_2O(l) \longrightarrow Mg^{2+}(OH^-)_2(s)$$
$$Na_2O(s) + H_2O(l) \longrightarrow 2Na^+OH^-(aq)$$

A metal oxide that yields a base with water is known as a **basic anhydride,** or a basic oxide.

Metals with low ionization energies tend to give ionic oxides. The greater the energy necessary to remove the valence electrons from the metal atom, the smaller the likelihood that the bonds between metal and oxygen will be ionic. This can be illustrated with the oxides Na_2O, MgO, and Al_2O_3. The valence shell configurations and ionization energies for the removal of the valence electrons for sodium, magnesium, and aluminum are

		Ionization energy, kcal/mole (kJ/mole)
Na	$3s^1$	118 (494)
Mg	$3s^2$	523 (2190)
Al	$3s^2 3p^1$	1228 (5138)

On the basis of the ionization energy values, the least ionic of the oxides should be Al_2O_3. This prediction is borne out by the chemical

---

Many nonmetal oxides are acidic

Many metal oxides are basic

TABLE 11.12

Types of metal oxides

All metals form normal oxides. Sodium and barium peroxide are the two most common peroxides. Superoxides and ozonides are seldom encountered.

Type	Ion	Average oxidation state of oxygen
Normal oxides e.g., **Na₂O, MgO**	O^{2-}	-2
Peroxides e.g., **Na₂O₂, BaO₂**	$\left(:\ddot{O}:\ddot{O}:\right)^{2-}$	-1
Superoxides e.g., **KO₂**	$\left[:\ddot{O}:\ddot{O}:\right]^{-}$ $\updownarrow$ $\left[:\ddot{O}:\ddot{O}:\right]^{-}$	$-\frac{1}{2}$
Ozonides e.g., **KO₃**	$\left[:\ddot{O}:\ddot{O}:\ddot{O}:\right]^{-}$ $\updownarrow$ $\left[:\ddot{O}:\ddot{O}:\ddot{O}:\right]^{-}$ $\updownarrow$ $\left[:\ddot{O}:\ddot{O}:\ddot{O}:\right]^{-}$	$-\frac{1}{3}$

properties of the oxides. Na_2O and MgO are basic and must therefore be essentially ionic. On the other hand, Al_2O_3 exhibits characteristics of both a weakly basic and weakly acidic oxide (Section 17.3). Although the bonds between the aluminum and oxygen atoms are largely ionic, the acidic properties of Al_2O_3 shows that the bonds are also somewhat covalent.

If a metal forms more than one oxide, the ionic character of the oxide diminishes with increasing oxidation state of the metal, since increasing amounts of energy are required to remove successive electrons from the metal atom. The decrease in ionic character is accompanied by an increase in the acidic nature of the oxide and a decrease in the basic nature of the oxide. This is exemplified by the oxides of chromium.

Oxidation state of chromium	Oxide	Basic or acidic nature
+2	CrO	basic
+3	Cr_2O_3	both acidic and basic
+6	CrO_3	acidic

Metal oxides insoluble in water often dissolve in acid solutions by reacting to form soluble salts, for example,

$$CuO(s) + 2HCl(aq) \longrightarrow H_2O(l) + CuCl_2(aq)$$

Certain acidic oxides and basic oxides can react with each other to yield salts

$$CaO(s) + SiO_2(s) \xrightarrow{\Delta} CaSiO_3(s)$$

$$6Li_2O(s) + P_4O_{10}(s) \xrightarrow{\Delta} 4Li_3PO_4(s)$$

11.14 Hydrogen peroxide

In addition to water (discussed at length in Chapter 13), hydrogen forms another oxide, hydrogen peroxide, H_2O_2. The chemistry of hydrogen peroxide is substantially different from that of water.

a. Properties and reactions of hydrogen peroxide. Hydrogen peroxide (m.p. −0.89°C, b.p. 151°C) is a polar covalent compound. The two oxygen atoms are linked by a single electron-pair bond (Figure 11.5). Hydrogen peroxide, like water, is associated through hydrogen bonding. It is completely miscible with water and is an extremely weak acid, which yields two hydrogen ions per molecule.

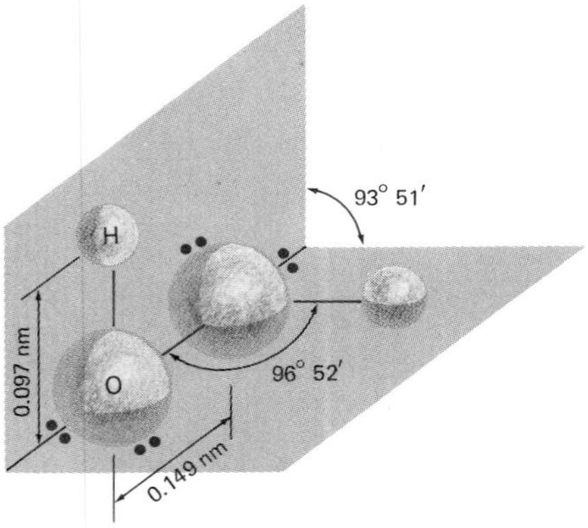

$$H_2O_2(aq) \rightleftharpoons H^+ + HO_2^-$$
hydroperoxide ion

$$HO_2^- \rightleftharpoons H^+ + O_2^{2-}$$
peroxide ion

FIGURE 11.5

The configuration of the H_2O_2 molecule.

In reactions with strong bases two types of **salts** may be formed—hydroperoxides, which contain the HO_2^- ion and peroxides, with the O_2^{2-} ion.

The decomposition of pure hydrogen peroxide and concentrated aqueous solutions of the compound occurs readily, often explosively, with the liberation of oxygen.

$$2H_2O_2(l) \xrightarrow{\text{heat or catalyst}} 2H_2O(l) + O_2(g)$$
$$\Delta H° = \text{about} -47 \text{ kcal } (-200 \text{ kJ}) \text{ at } 20°C \quad (8)$$

The decomposition is catalyzed by traces of impurities, such as certain metal ions (e.g., Fe^{2+}), finely divided metals (e.g., Pt, Au), various metal oxides (e.g., MnO_2), and also blood and saliva.

In the peroxide ion, oxygen is in an oxidation state of -1 and can be both oxidized to the zero state (in O_2) and reduced to the -2 state (in H_2O), as in reaction (8). Therefore, hydrogen peroxide is both an oxidizing and a reducing agent. Actually it is an extremely powerful oxidizing agent and a poor reducing agent, acting in the latter role only toward very strong oxidants and usually in acidic solution. The oxidations of arsenic(III) oxide and lead(II) sulfide illustrate the behavior of hydrogen peroxide as an oxidizing agent.

$$As_2O_3(s) + 2H_2O_2(l) + H_2O(l) \longrightarrow 2H_3AsO_4(aq)$$
$$PbS(s) + 4H_2O_2(l) \longrightarrow PbSO_4(s) + 4H_2O(l)$$

Paintings in which Pb^{2+} salts were used as white pigments (white lead) become dark on aging because of reaction with H_2S in the air to form PbS. They can be restored to whiteness by treatment with H_2O_2.

When hydrogen peroxide functions as a reducing agent, the oxidation product is one mole of dioxygen for each mole of hydrogen peroxide consumed. The conversion of permanganate ion, MnO_4^-, to manganous ion, Mn^{2+}, is an example of reduction by hydrogen peroxide.

$$2MnO_4^- + 5H_2O_2(aq) + 6H^+ \longrightarrow 2Mn^{2+} + 5O_2(g) + 8H_2O(l)$$

b. Preparation and uses of hydrogen peroxide. Industrially, hydrogen peroxide is prepared from concentrated sulfuric acid via the peroxodisulfate ion, $S_2O_8^{2-}$. In the two-step process, the sulfuric acid is first converted to peroxodisulfuric acid by electrolysis, and is then regenerated in the reaction with water to yield H_2O_2.

$$2H_2SO_4(aq) \xrightarrow{\text{electrolysis}} \underset{\substack{\text{peroxodisulfuric} \\ \text{acid}}}{H_2S_2O_8(aq)} + H_2(g)$$
$$H_2S_2O_8(aq) + 2H_2O(l) \longrightarrow 2H_2SO_4(aq) + H_2O_2(aq)$$

Commercial hydrogen peroxide solutions have a concentration of about 30% H_2O_2 by weight. A 3% by weight solution of H_2O_2 is used medically as an antiseptic. A stabilizer must be added to such solutions to prevent H_2O_2 decomposition. Distillation of a 30% solution under reduced pressure gives a product containing about 90% H_2O_2, which is also commercially available. The pure compound may be obtained from the 90% material by crystallization.

In the laboratory, dilute solutions of hydrogen peroxide are usually made by reaction between sodium peroxide or barium peroxide and cold dilute sulfuric acid.

$$Na_2O_2(s) + 2H^+ + SO_4^{2-} \longrightarrow H_2O_2(aq) + 2Na^+ + SO_4^{2-}$$
$$BaO_2(s) + 2H^+ + SO_4^{2-} \longrightarrow H_2O_2(aq) + BaSO_4(s)$$

Hydrogen peroxide and its solutions are used as bleaching agents for textiles, paper, and hair; as intermediates in the chemical industry; in water purification; and as disinfectants for cuts and sore throats. Pure H_2O_2 is used as an oxidizer in rocket fuel.

11.15 The oxygen and carbon cycles

The major reservoirs for oxygen as it circulates in the ecosphere are the carbon dioxide and molecular oxygen in the atmosphere, and the organic matter of plants and animals [which we represent by the highly generalized formula $(CH_2O)_x$]. One-fourth of the atoms in all living matter are oxygen atoms. Since oxygen is combined with

FIGURE 11.6
The oxygen–carbon cycle.

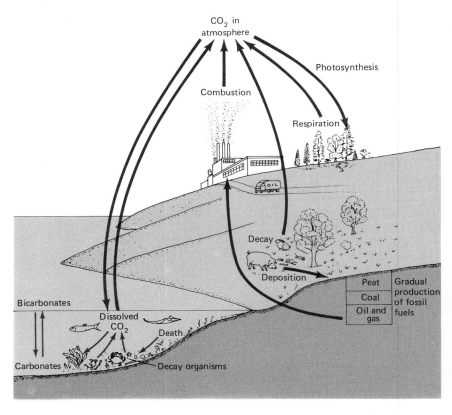

carbon through most of its cycle, the cycling of carbon and oxygen through the atmosphere, hydrosphere, and lithosphere are best discussed together.

The simplified cycle in Figure 11.6 shows the major pathways of oxygen and its compounds. Under the influence of sunlight and chlorophyll, carbon dioxide and water react in the green parts of plants to form carbohydrates in the process that we call **photosynthesis** [see Equation (1)]. These carbohydrates undergo a series of complex reactions to form the myriad of compounds of which plant matter is made. The plant matter is consumed by animals, and animals are consumed by other animals. Animals breathe in oxygen and within their cells reverse the photosynthesis reaction, breaking down organic matter to release energy, carbon dioxide, and water.

$$(CH_2O)_x + xO_2 \longrightarrow xCO_2 + xH_2O + \text{energy}$$

Some organic matter enters the land mass as fossil fuels and as limestone ($CaCO_3$). Carbon dioxide is continually being absorbed and given up by the oceans, and carbonates are dissolved by weathering and in streams as HCO_3^-. And both in the oceans and on the land, plant and animal matter decomposes to yield carbon monoxide. Oxidation in the atmosphere or by bacteria in the soil or water converts the carbon monoxide to carbon dioxide. Almost all carbon that passes through organic matter eventually returns as CO_2 to the air or water.

**THOUGHTS ON
CHEMISTRY**

The world's biggest membrane

It takes a membrane to make sense out of disorder in biology . . . When the earth came alive it began constructing its own membrane, for the general purpose of editing the sun . . . You could say that the breathing of oxygen into the atmosphere was the result of evolution, or you could turn it around and say that evolution was the result of oxygen. You can have it either way. Once the photosynthetic cells had appeared, very probably counterparts of today's blue-green algae, the future respiratory mechanism of earth was set in place. Early on, when the level of oxygen had built up to around 1 percent of today's atmospheric concentration, the anaerobic life on the earth was placed in jeopardy, and the inevitable next stage was the emergence of mutants with oxidative systems and ATP. With this, we were off to an explosive developmental stage in which great varieties of respiring life, including the multicellular forms, became feasible . . .

It is hard to feel affection for something as totally impersonal as

THE LIVES OF A CELL: NOTES OF A BIOLOGY WATCHER, *by Lewis Thomas*

HYDROGEN AND OXYGEN [11.15]

319

the atmosphere, and yet there it is, as much a part and product of life as wine and bread. Taken all in all, the sky is a miraculous achievement. It works, and for what it is designed to accomplish it is as infallible as anything in nature. I doubt whether any of us could think of a way to improve on it, beyond maybe shifting a local cloud from here to there on occasion. The word "chance" does not serve to account well for structures of such magnificence. There may have been elements of luck in the emergence of chloroplasts, but once these things were on the scene, the evolution of the sky became absolutely ordained. Chance suggests alternatives, other possibilities, different solutions. This may be true for gills and swim bladders and forebrains, matters of detail, but not for the sky. There was simply no other way to go.

THE LIVES OF A CELL: NOTES OF A BIOLOGY WATCHER, by Lewis Thomas (Copyright © 1973 by Massachusetts Medical Society. Reprinted by permission of The Viking Press.)

Significant terms defined in this chapter

dihydrogen	hydrogenation	reagent
deuterium	electromotive series	acid anhydride
tritium	oxides	basic anhydride
heavy water	combustion	photosynthesis
anhydrous	allotropes	

Exercises

11.1 A chemistry graduate student used the following series of reactions to prepare diborane, B_2H_6

$$2Na + H_2 \longrightarrow 2NaH \quad \text{(i)}$$
$$4NaH + B(OCH_3)_3 \longrightarrow NaBH_4 + 3NaOCH_3 \quad \text{(ii)}$$
$$2NaBH_4 + H_2SO_4 \longrightarrow B_2H_6 + 2H_2 + Na_2SO_4 \quad \text{(iii)}$$

However, a small amount of oxygen was present in the preparation system. The oxygen reacted with the B_2H_6 and H_2 formed in reaction (iii)

$$2H_2 + O_2 \longrightarrow 2H_2O \quad \text{(iv)}$$
$$B_2H_6 + 3O_2 \longrightarrow B_2O_3 + 3H_2O \quad \text{(v)}$$

causing a serious explosion. Answer these questions about the substances and the reactions given above. What are the oxidation states of (a) H and of (b) O in all of the above compounds? (c) Which reactions involve oxidation–reduction? In which reaction is H_2 (d) an oxidizing agent and (e) a reducing agent? Which of the above compounds are (f) ionic and (g) molecular? Into which classification of hydrides does (h) NaH, (i) B_2H_6, and (j) H_2O fall? (k) Would deuterium undergo these same reactions? Which of the above reactions are examples of (l) hydrogenation and (m) combustion reactions?

11.2 Which statements are true? Rewrite any false statement so that it is correct.
(a) The electromotive series is a list of elements in the order of decreasing ionization energies.
(b) Even though aluminum has many properties of a metal, $Al(OH)_3$ is not a particularly strong base.
(c) If we include gaseous atomic oxygen, there are three allotropic forms of oxygen.
(d) The covalent hydrides are characterized by having nonstoichiometric formulas.
(e) Oxygen is such a good oxidizing agent that it reacts directly with all metals to form oxides.
(f) A glass of D_2O and a glass of H_2O would not be significantly different in appearance.
(g) An atom of hydrogen can complete its "octet" of electrons by gaining an electron (forming H^-), sharing a pair of electrons in a covalent bond, or losing its electron (forming H^+).

11.3 Distinguish clearly among (a) protium, deuterium, and tritium; (b) saltlike and covalent hydrides; (c) acid and basic anhydrides; (d) oxidation and combustion; and (e) hydrogenation and reduction.

11.4 Write the chemical reaction for a laboratory preparation of (a) H_2, (b) O_2, (c) O_3 and (d) H_2O_2. Repeat the exercise for a commercial preparation of each substance.

11.5 A student was given two unmarked test tubes, one containing hydrogen and one containing oxygen. A glowing wooden splint was inserted into the mouth of each tube and it burst into flame in the first tube and was extinguished in the second tube. A burning splint was held near the mouth of the second tube and a minor explosion occurred which sounded similar to a sharp bark. Which gas did the student decide was in each of the test tubes?

11.6 Write the equations for the reaction of (a) $H_2(g)$ as a reducing agent with (i) $FeO(s)$, (ii) $Br_2(g)$, and (iii) $C_2H_4(g)$; (b) $H_2(g)$ as an oxidizing agent with $Li(l)$; (c) $O_2(g)$ as an oxidizing agent with (i) $S_8(s)$, (ii) $Mg(s)$, (iii) $Cu_2O(s)$, and (iv) CH_4; (d) $O_3(g)$ as an oxidizing agent with $PbS(s)$; (e) $H_2O_2(aq)$ as an oxidizing agent with $SO_3^{2-}(aq)$; and (f) $H_2O_2(aq)$ as a reducing agent with $MnO_4^-(aq)$.

11.7 Using the electromotive series given in Table 11.3, predict whether the following reactions will or will not occur as written:

(a) $Mg(s) + Cu^{2+} \longrightarrow Mg^{2+} + Cu(s)$
(b) $Pb(s) + 2H^+ \longrightarrow H_2(g) + Pb^{2+}$
(c) $2Ag^+ + Cu(s) \longrightarrow 2Ag(s) + Cu^{2+}$
(d) $2Al^{3+} + 3Zn(s) \longrightarrow 3Zn^{2+} + 2Al(s)$
(e) $Mg(s) + Ca(s) \longrightarrow Mg^{2+} + Ca^{2+}$
(f) $2Al^{3+} + 3Pb^{2+} \longrightarrow 2Al(s) + 3Pb(s)$
(g) $H_2(g) + Zn^{2+} \longrightarrow 2H^+ + Zn(s)$

11.8 Each of the following compounds is a basic or acid anhydride: (i) MgO, (ii) P_4O_6, (iii) CO_2, (iv) SO_2, and (v) Na_2O. For each substance, write the equation for the reaction between the substance and water.

11.9 The metallic element M forms a series of oxides: MO, M_2O_3, and M_2O_5. One of these oxides will dissolve only in an acidic solution while the others will dissolve in basic or neutral solutions. Which oxide is it that is different and why?

11.10* Name two allotropic forms of oxygen. Write the Lewis structures for these molecules. Can resonance structures be written? Discuss the shapes of these molecules. Is either of these allotropes a dipole? Based on intermolecular forces and mass, which would you predict to have the higher boiling point?

11.11* Write the formula of the binary compound formed between hydrogen and each of the elements in the second period (Li–F). Discuss the geometry of those that are molecular and identify the intermolecular forces that are present.

11.12* Does the electromotive series in Table 11.3 predict that the reaction given by the equation

$$Cu(s) + 2H^+ \longrightarrow Cu^{2+} + H_2(g)$$

will occur? Yet we can easily verify in the laboratory that copper will react with nitric and sulfuric acids according to the equations

$$3Cu(s) + 8HNO_3(aq) \longrightarrow 3Cu(NO_3)_2(aq) + 2NO(g) + 4H_2O(l)$$
$$Cu(s) + 2H_2SO_4(l) \xrightarrow{\Delta} CuSO_4(s) + SO_2(g) + 2H_2O(g)$$

How can you explain this behavior?

11.13 The crust of the Earth is estimated to be about 95% silicate minerals. Calculate the weight percent of oxygen in a typical (a) feldspar, $KAlSi_3O_8$, and (b) silica, SiO_2. Do these values confirm that the crust is roughly 50 wt% oxygen since the feldspars and silicas make up about 75% of the silicate minerals?

11.14 Name the three isotopes of hydrogen. How do they differ in (a) electronic and (b) nuclear structure? The average atomic weight of naturally occurring hydrogen is 1.00797 amu. Assuming only $_1^1H$ (1.007825 amu) and $_1^2H$ (2.0140 amu) to be present, (c) find the percentage of natural abundance of $_1^2H$.

11.15 What volume of H_2 at 21°C and 719 torr can be generated by allowing 3.5 g of $Zn(s)$ to react with 75 ml of $6M$ H_2SO_4?

11.16 Deuterium can be obtained by the electrolysis of heavy water. Assuming heavy water to be 95 wt% D_2O, what mass of D_2O is required to generate 25 ml of $D_2(g)$ measured at 23°C and 721 torr? The atomic weight of D is 2.0140 amu.

11.17 What volume of dry O_2 at 37°C and 765 torr can be produced from the decomposition of 25 g of $KClO_3$ by the following reaction

$$2KClO_3(s) \xrightarrow[\Delta]{MnO_2} 2KCl(s) + 3O_2(g)$$

11.18 Torches that use hydrogen and oxygen produce very high temperatures. Those that electrically decompose the hydrogen molecules into atomic hydrogen before mixing the fuel with the oxygen generate even higher temperatures. The heat of formation at 25°C is −57.796 kcal/mole for $H_2O(g)$ and 52.095 kcal/mole for $H(g)$. Calculate $\Delta H°$ for the reaction.

$$2H(g) + \tfrac{1}{2}O_2(g) \longrightarrow H_2O(g)$$

and compare the value to that for

$$H_2(g) + \tfrac{1}{2}O_2(g) \longrightarrow H_2O(g)$$

11.19 Although eight isotopes of oxygen are known, only three are nonradioactive and occur naturally: 99.759% $^{16}_8O$, 15.99491 amu; 0.037% $^{17}_8O$, 16.99912 amu; and 0.204% $^{18}_8O$, 17.99914 amu. Calculate the average atomic weight of naturally occurring oxygen.

11.20 Silver will not react directly with oxygen at room temperature to form Ag_2O even though the reaction is slightly exothermic, but will react with ozone according to the equation

$$2Ag(s) + O_3(g) \longrightarrow Ag_2O(s) + O_2(g)$$

(a) Confirm this statement by finding the enthalpy of reaction given that the $\Delta H°$ of formation at 25°C is −7.42 kcal/mole for $Ag_2O(s)$ and 34.1 kcal/mole for $O_3(g)$. (b) What mass of ozone would be required to produce 100.0 g of Ag_2O?

11.21 Following are some heats of formation at 25°C: $H_2O(g)$, −57.796 kcal/mole; $H_2O(l)$, −68.315 kcal/mole; $H_2O_2(g)$, −32.58 kcal/mole; and $H_2O_2(l)$, −44.88 kcal/mole. Calculate the heat of vaporization of both compounds at this temperature and compare the strengths of the intermolecular forces present.

11.22* Calculate the neutron/proton ratio for the stable isotopes of (a) hydrogen (H and D) and (b) oxygen (^{16}O, ^{17}O, and ^{18}O) and (c) for the radioactive isotopes of each element (T, ^{13}O, ^{14}O, ^{15}O, ^{19}O, and ^{20}O). These latter isotopes are known to undergo decay by emitting either β^+ or β^- particles. (d) Predict the respective mode of decay (see Section 8.5) and (e) write the corresponding nuclear equation. The atomic mass of ^3_1H is 3.01605 amu and of ^3_2He is 3.01603 amu. (f) Find the energy emitted as tritium undergoes nuclear decay to ^3_2He.

11.23* How many different types of water molecules can be formed using the two nonradioactive isotopes of hy-drogen (H and D) and the three nonradioactive isotopes of oxygen (^{16}O, ^{17}O, and ^{18}O)? What is the ratio of the rate of effusion of the heaviest to the lightest? Use atomic weights of 1, 2, 16, 17, and 18.

11.24* Water gas is produced by the reaction between steam and hot carbon

$$C(s) + H_2O(g) \longrightarrow CO(g) + H_2(g)$$

(a) What volume of steam at 1.53 atm and 184°C is needed to produce 100.0 liters of water gas at 820°C and 1.02 atm? The heat of combustion of $CO(g)$ is −67.635 kcal/mole and of $H_2(g)$ is −68.315 kcal/mole. (b) How much heat will be generated by the complete combustion of the 100.0 liters of water gas?

11.25** Aqueous solutions of H_2O_2 undergo the following decomposition

$$2H_2O_2(aq) \longrightarrow 2H_2O(l) + O_2(g)$$

Although some decompositions are undesirable, this is the reaction that makes H_2O_2 an important disinfectant and bleach. What are the oxidation numbers of (a) H and (b) O in each of the substances? If the reaction is an oxidation–reduction reaction, identify the (c) oxidizing and (d) reducing agents. What types of intermolecular forces are present between molecules of (e) H_2O_2, (f) H_2O_2 and H_2O, (g) H_2O and (h) O_2? If the oxygen formed from the decomposition of 1.00 g of H_2O_2 is confined to a volume of 25 ml at 25°C, (i) what will be the pressure? The enthalpy of formation at 25°C is −191.17 kJ/mole for $H_2O_2(aq)$ and −285.83 kJ/mole for $H_2O(l)$. (j) What is the heat of reaction?

The heat of formation of $H_2O_2(g)$ is −136.31 kJ/mole and of $H_2O(g)$ is −241.82 kJ/mole and the bond energy of H–H is 435.93 kJ/mole and of O=O is 498.17 kJ/mole. Calculate the (k) O–H and (l) O–O bond energies.

12 THE LIQUID AND SOLID STATES; CHANGES OF STATE

The word "solid" is another word like "heat" and "temperature," which we discussed in the introduction to Chapter 5. Everyone has an intuitive feeling for what a solid is. But the word has a subtly different significance to a mineralogist, a crystallographer, an electronics engineer, or a civil engineer.

In the popular sense, anything that is obviously not a gas and obviously not a liquid—anything that does not flow or take the shape of its container—is a solid. To many scientists, the term "solid" is appropriate only for crystalline compounds, with their orderly internal structure. The fine distinction that window glass, which certainly appears to be solid, is truly just a very viscous liquid, is of practical importance to very few people. However, an understanding of the difference will aid the study of chemistry.

Another popular use of the word "solid" was born with the revolution in the electronics industry caused by the replacement of vacuum tubes with transistors. The term "solid state" is freely applied to anything that contains a transistor or other semiconducting device. (Chapter 27 on the semiconducting elements includes an explanation of how semiconductors work.)

The growing understanding of the electronic properties of solids has stimulated investigations of other properties of solids. Since the late 1950s, several new branches of science and technology have emerged, such as solid-state physics, solid-state chemistry, and materials science and engineering. Engineers now can predict the properties of materials not just on the basis of past experience, but also on the basis of an understanding of the internal structure of the substances involved. This chapter can barely scratch the surface of what has been learned about the solid state in recent years.

In this chapter we first present the interrelationships among the gaseous, liquid, and solid phases. Gases have already been discussed in detail (Chapter 3). The middle part of this chapter deals with some specific properties of liquids. Finally, we discuss the solid state. The fundamental topic of crystal structure is explored, and the internal arrangements of several types of solids are presented. The importance of defects and the occurrence of solid-state reactions are briefly touched upon.

323

Relationships between phases

12.1 Phases

The term **phase** refers to a homogeneous part of a system in contact with but separate from other parts of the system. A glass of iced tea—we might think of the iced tea as the *system*—has a solid phase (ice, a pure substance) and a liquid phase (tea, a solution), for example. The iced tea includes substances in the solid and the liquid states. A bottle of oil and vinegar contains only substances in the liquid state, but it also has two phases, because the oil phase and the vinegar phase remain in contact with each other but do not mix. It is also possible to have a completely solid substance in which several phases that are different crystalline forms are in contact with each other.

Suppose a balloon is filled with hydrogen and oxygen in nonstoichiometric amounts—that is, in amounts that are not exactly those needed for complete reaction. The hydrogen–oxygen mixture is homogeneous and only one phase is present. What happens to the number of phases if the hydrogen and oxygen react to form water? After the reaction, two phases are present—liquid water and a gaseous phase containing water vapor and the excess hydrogen or oxygen.

12.2 Kinetic-molecular theory for liquids and solids

The atoms, molecules, and ions in liquids and solids, like those in gases (Section 3.4), are in constant motion. In a gas the molecules are far apart and relatively independent of each other. However, in liquids and solids the motion is restricted, for the atoms, molecules, or ions are packed tightly and held together either by intermolecular forces, or by metallic, ionic, or network covalent bonds.

In a liquid, a molecule cannot move even a fraction of its radius without striking another molecule. Each of the colliding molecules rebounds, but the distance which either one can move is extremely small. Because of many such collisions, the diffusion of one liquid into another in which it is soluble is much slower than the diffusion of one gas into another. In a solid the molecules or ions are even more restricted in their motions—most of them can only vibrate about their fixed positions. Diffusion, however, does also occur in solids, usually by the movement of particles into defects in the solid structure (Section 12.16).

With the exception of the first postulate (concerning the large distances between particles), the postulates of the kinetic-molecular theory for gases (Section 3.4) also pertain to particles in the solid and liquid states. Exactly as for gases (see Figure 3.5), the average kinetic energy of liquid and solid state particles is a function of their temperature. The higher the temperature, the greater the average kinetic energy of the particles, and vice versa. At low enough temperatures, even the relatively weak intermolecular forces present in nonpolar molecules can hold the molecules together in a solid. All liquids become solid if cooled to sufficiently low temperatures.

12.3 Vaporization

Molecules near the surface of a liquid may escape into the gaseous phase. Many such escaping molecules strike molecules of the gas above the liquid and bounce back into the liquid, but if the vessel is open to the air all the molecules eventually escape from the liquid. We call this process **evaporation**—the escape of molecules from a liquid in an open container to the gaseous phase. **Vaporization** is the general term for escape of molecules from the liquid or solid phase to the gas phase.

The escaping molecules are the "hottest" molecules—the ones at the high kinetic energy end of the curve. Therefore evaporation is faster when the temperature is higher, for a larger fraction of molecules have energy high enough to escape (see Figure 3.5). Escape of the highest-energy molecules leaves behind a collection of molecules with a lower average kinetic energy and, therefore, a lower temperature. When evaporation is rapid, for example, when you get out of a swimming pool on a windy day, the cooling effect of evaporation is obvious. It is less noticeable in slow evaporation from a noninsulated container because heat is gradually taken up from the surroundings as evaporation occurs.

If a liquid is in a tightly closed container, molecules escape from the liquid into the space above, but can go no further (Figure 12.1). Some of the molecules, in bouncing around the enclosed space, hit the liquid surface and enter the liquid again. This process is called **condensation,** or liquefaction—the movement of molecules from the gaseous phase to the liquid phase. (Note the *e* in liquefaction—to use *i* in its place is wrong.) Eventually, the concentration of molecules in the vapor is so great that the number of them going back into the liquid equals the number escaping from the liquid. A dynamic equilibrium (Section 3.15) has been established, and the vapor pressure reaches a constant value.

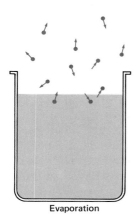

Evaporation

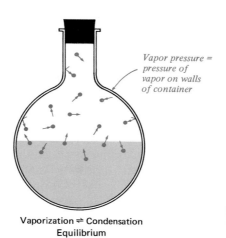

Vapor pressure = pressure of vapor on walls of container

Vaporization ⇌ Condensation
Equilibrium

FIGURE 12.1

Evaporation, vaporization, condensation, and vapor pressure.

FIGURE 12.2

Vapor pressure vs. temperature curves. *With the exception of water and carbon dioxide, which are included for comparison, each of these liquids is an anesthetic. The molecular weight and dipole moment of each substance are given on the curves.*

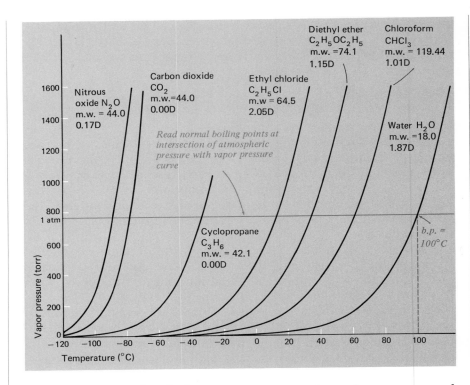

The magnitude of the vapor pressure depends upon several factors. The first is the concentration of molecules in the vapor state and their velocity, both of which are functions of temperature and increase with increasing temperature. The second factor is the structure and molecular weight of the molecules of the liquid. When intermolecular forces of attraction are high, as in a hydrogen-bonded substance or in a high molecular weight substance with relatively high London forces, molecules must achieve a high energy in order to escape from the liquid. A small fraction of them can do so, leading to a relatively lower vapor pressure at a given temperature. In contrast, when intermolecular forces are weak, the vapor pressure at any given temperature will be relatively higher. The vapor pressures of some simple compounds are given in Figure 12.2. Think for a moment about how their molecular weights and polarity might affect their respective vapor pressures. Note that the vapor pressure of a specific liquid substance is not dependent upon the size or shape of the container, or the ratio of liquid to empty space, or the pressure of any gas other than its own vapor above the liquid.

Particles with a high enough kinetic energy can also escape from the surface of a solid, and solids do have vapor pressures. They are, of course, generally lower than the vapor pressures of liquids. **Sublimation** is the vaporization of a solid followed by the condensation of the vapor to form the solid again. This process is frequently used to purify a solid. Sometimes just the transition from the solid to the gas phase is called "sublimation," and the transition from the gas phase directly to the solid phase is called "deposition."

EXAMPLE 12.1

■ Basing your argument on intermolecular forces, place the following substances in order of increasing vapor pressure above the liquids at −100°C:

$$\underset{\text{ethane}}{\text{H}-\overset{\overset{\displaystyle H}{|}}{\underset{\underset{\displaystyle H}{|}}{C}}-\overset{\overset{\displaystyle H}{|}}{\underset{\underset{\displaystyle H}{|}}{C}}-\text{H}} \qquad \underset{\text{dimethyl ether}}{\text{H}-\overset{\overset{\displaystyle H}{|}}{\underset{\underset{\displaystyle H}{|}}{C}}-\overset{..}{\underset{..}{O}}-\overset{\overset{\displaystyle H}{|}}{\underset{\underset{\displaystyle H}{|}}{C}}-\text{H}} \qquad \underset{\text{ethyl alcohol}}{\text{H}-\overset{\overset{\displaystyle H}{|}}{\underset{\underset{\displaystyle H}{|}}{C}}-\overset{\overset{\displaystyle H}{|}}{\underset{\underset{\displaystyle H}{|}}{C}}-\overset{..}{\underset{..}{O}}-\text{H}}$$

The molecules in liquid ethane and dimethyl ether are held together only by weak London forces. In liquid ethyl alcohol, there is a considerable amount of hydrogen bonding. Thus ethane and dimethyl ether should have much higher vapor pressures than ethyl alcohol at the same temperature. The strengths of the London forces are proportional to the mass, so we would choose ethane to have the higher vapor pressure of these two substances, because it has a smaller mass and smaller intermolecular forces. Our predicted order of increasing vapor pressures at −100°C is

ethyl alcohol < dimethyl ether < ethane

which is in excellent agreement with the respective vapor pressure values of 4×10^{-4} torr, 6 torr, and 360 torr at −100°C. ■

12.4 Boiling point and melting point

When the temperature of a liquid in an open vessel is increased sufficiently, a point is reached at which the escaping molecules have enough energy to sweep aside the molecules of the atmospheric gases. This temperature is the **boiling point** of the liquid—the temperature at which the vapor pressure of a liquid equals the pressure of the gases above the liquid, and bubbles of vapor form throughout the liquid. The boiling point, therefore, varies with the pressure, and it is essential to know the pressure at which a boiling point was measured. The **normal boiling point** of a liquid is its boiling point at 760 torr, the average atmospheric pressure at sea level. You can assume that a boiling point given without mention of pressure is the normal boiling point.

The vapor pressure curve of a substance (see Figure 12.2) shows its boiling point at various pressures. Below atmospheric pressure, boiling points are lower than the normal boiling point. At 9000 feet above sea level, where the pressure is about 550 torr, water boils at 91°C. At this altitude an egg must be boiled for about 5 minutes to get the same consistency as a 3-minute egg at sea level, and a hard-boiled egg requires about 18 minutes of cooking time.

It is possible for a liquid to be **superheated**—heated to a temperature above the boiling point, without the occurrence of boiling. The sudden initiation of boiling and bubble formation in a superheated liquid can be quite violent.

The addition of heat to a solid causes the particles to vibrate faster and faster in their fixed positions until they are no longer held firmly in place and are thus free enough to form a liquid. The **melting point** of a solid is the temperature at which the solid and liquid phases of a substance are at equilibrium. **Fusion** is a term also used to describe the conversion of a solid to a liquid. The freezing point is identical with the melting point but is thought of as being approached from the opposite temperature direction. In other words, freezing point is a property of a liquid, melting point a property of a solid.

12.5 Changes of state

The temperature of a substance does not change during a change of state. For example, consider vaporization of a liquid at its boiling point. Below the boiling point, the added heat increases the average kinetic energy of the molecules and the temperature rises. Once the boiling point is reached, the temperature remains constant as long as any liquid is present. All of the added heat serves to free more and more molecules from the attraction of their neighbors. Conversely, in a change of state brought about by cooling, such as crystallization (liquid to solid), molecules slow down and submit to the attraction of their neighbors. This restricts their motion. The kinetic energy which they had is transformed into thermal energy, which maintains the temperature, and the temperature remains constant until all of the liquid has solidified.

TABLE 12.1

Vaporization and fusion data for some of the substances commonly found in air

ΔH values in cal/mole can be converted to J/mole by multiplying by 4.184 J/cal.

		Fusion		Vaporization	
Substance		m.p. (°C)	ΔH(fus) (cal/mole)	b.p. (°C)	ΔH(vap) (cal/mole)
N_2	nitrogen	−210	172	−196	1333
O_2	oxygen	−219	106	−183	1630
Ar	argon	−190	281	−186	1558
CO_2	carbon dioxide	−56	1990	−78[a]	6031[a]
N_2O	nitrous oxide	−91	1563	−89	3956
I_2	iodine	114	3740	183[a]	1039[a]
NH_3	ammonia	−78	1351	−33	5581
H_2O	water	0	1436	100	9717
D_2O	heavy water, deuterium oxide	3.8	1501	101.4	9944

[a] Goes directly to vapor from solid. These are temperatures and enthalpies of sublimation.

The terms for the enthalpies, or heats, of the changes of state were introduced earlier (see Figure 5.3). The **molar heat of vaporization** is the amount of heat needed to convert one mole of a substance to its vapor. Similarly, the **molar heat of fusion** is the heat of the solid-to-liquid change for one mole of a substance. Some change-of-state enthalpies are given in Table 12.1. All such enthalpies refer to the changes of state at constant temperatures, which are the boiling points or melting points of the substances, or the temperatures at which the solid goes directly to a vapor.

12.6 Phase diagrams

The phase diagram for a pure substance shows the temperatures and vapor pressures at which the substance can exist in different phases. To construct a phase diagram for a three-phase, solid–liquid–gas system, we first combine the vapor pressure–temperature curve for the liquid phase (like those in Figure 12.2) with the vapor pressure–temperature curve for the solid phase. This second curve represents the temperatures and pressures at which molecules can escape directly from the solid into the vapor phase, in other words, where solid and vapor are in equilibrium. The general shape of such curves is shown in Figure 12.3a.

The vapor–liquid equilibrium curve in Figure 12.3a would continue upwards, if there were space, until it ended abruptly at the **critical point**—the point above which no amount of pressure is great enough to cause liquefaction. The temperature at this point is called the **critical temperature**—the temperature above which a substance cannot exist as a liquid. The **critical pressure** is the pressure that will cause liquefaction of a gas at the critical temperature.

The curve for the vapor pressure of the solid ends at its intersection with the liquid curve—at this point the solid melts. A liquid can be **supercooled**—cooled below its freezing point without the occurrence of freezing—as shown by the dashed line in Figure 12.3a. In this condition, crystallization will sometimes start if a "seed crystal" of the substance is added to provide a surface on which crystals begin to form. Supercooled liquids often crystallize very rapidly, with the noticeable evolution of heat, once crystallization is triggered in this way. To complete the simple phase diagram, the line representing the relationship between pressure and melting point is added (Figure 12.3b). In most cases, the melting point increases with pressure. Either way, the effect is usually quite small.

At any point not on a line in the phase diagram, only a single phase of the substance can exist. At any point along a line, two phases

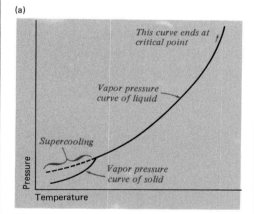

(a)

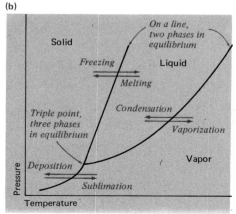

(b)

FIGURE 12.3
Phase diagram for a solid–liquid–gas system. *To the vapor pressure curves of the solid and liquid (a) are added the melting point–vapor pressure curve of the solid (b). The colored arrows show the phase changes at constant pressure that occur in crossing the lines at various points.*

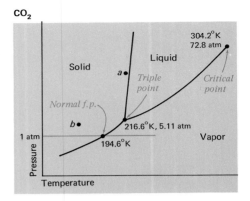

CO₂

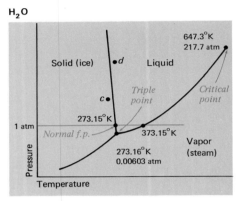

H₂O

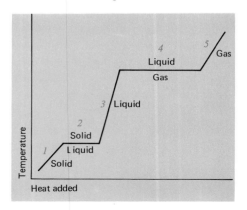

Phase diagrams for CO₂ and H₂O. *The temperature and pressure scales are not linear and the slope of the solid–liquid line
is greatly exaggerated. (The lettered points are used in various
examples and exercises in this chapter.)*

are in equilibrium. At the point where three lines intersect in a phase
diagram—called a **triple point**—three phases are in equilibrium.

The phase diagrams for carbon dioxide and water are given in
Figure 12.4. Note the location of atmospheric pressure on these diagrams. The **normal freezing point** of a liquid is the freezing point at
1-atm pressure. The temperature of the triple point between solid,
liquid, and gas usually differs slightly from the normal freezing
point, because it represents the three-phase equilibrium under only
the vapor pressure of the pure substance at that temperature. The
pressure at the triple point can be quite different from 760 torr.

For carbon dioxide, the solid–liquid–gas triple point is at
5.11 atm, well above atmospheric pressure. Therefore, as solid
carbon dioxide (known by its trade name, Dry Ice) warms up, it
by-passes the liquid phase and is converted from the solid to the
vapor phase. Liquid carbon dioxide is accessible only at pressures
greater than 5.11 atm.

EXAMPLE 12.2

■ Describe what happens to a sample of CO₂ originally at point *a* in
the phase diagram (Figure 12.4) as it undergoes a pressure decrease
at constant temperature.

For a change at constant temperature, the CO₂ will follow a vertical path on the diagram consisting of (1) a very slight expansion of
the solid; (2) the establishment of solid–liquid equilibrium until all of
the CO₂ is changed completely to the liquid state; (3) a very slight expansion of the liquid; (4) the establishment of a liquid–gas equilibrium until all of the CO₂ is completely changed to the gaseous state;
and (5) expansion of the gas. ■

EXAMPLE 12.3

■ Prepare a heating curve—a sketch of temperature vs. heat
added—for heating at constant pressure of a sample of CO₂ originally
at point *a* in the phase diagram (Figure 12.4).

For a change at constant pressure, the CO₂ will follow a horizontal path on the diagram consisting of (1) heating the solid; (2) the establishment of a solid–liquid equilibrium at constant temperature
until the CO₂ is changed completely to liquid; (3) heating the liquid;
(4) the establishment of liquid–gas equilibrium at constant temperature until the CO₂ is changed completely to gas; and (5) heating the
gas. These steps are shown on the heating curve sketch in Figure 12.5.
■

The liquid state

12.7 Structure, density, and volume

The arrangement of particles in a liquid is neither completely random as in a gas nor strictly ordered as in a crystalline solid. Liquids have what we call short-range order. Groups of particles in a liquid might be closely packed together or associated in chains or rings. These groups move past each other with equal ease in any direction, accounting for the fluidity of liquids and the way that they take the shape of their container. Because the particles in liquids are close enough to touch each other, there is little room to squeeze them together. Liquids are only slightly compressible.

Most substances that immediately come to mind as examples of liquids, such as water, alcohol, cleaning fluids, or antifreeze, are composed of rather simple molecules, containing up to about 20 atoms per molecule. Other substances that are solid at ordinary temperatures such as metals, some salts, and polymers of moderate molecular weight, become liquids at sufficiently high temperatures (Table 12.2). Despite extensive investigations, no comprehensive theory of liquid structure has yet been developed.

The difference in density between the solid and liquid states of pure substances is not great, and pressure changes have only a small effect on the volume and density of liquids and solids. The density of most substances decreases by about 10% upon melting. Water is a notable exception, for its density increases upon melting (Section 13.1). For most liquids, raising the temperature causes an increase in volume and a decrease in density, water again being the most outstanding exception.

We have already discussed many of the properties of liquids in connection with the changes of state. Some additional properties by which liquids may be characterized are discussed in the following sections.

12.8 Surface tension

In the interior of a liquid each molecule is subjected to equal forces of attraction from other molecules on all sides. Molecules at the surface feel these forces on only the liquid side, however, and as a result are pulled inward and closer together (Figure 12.6). The surface of a liquid can be thought of as behaving like a stretched membrane trying to contract to the smallest possible surface area. **Surface tension** is the property of a surface that imparts membrane-like behavior to the surface.

Surface tension causes tiny suspended liquid droplets to take the shape of a sphere—the shape with the smallest surface-to-volume ratio. And it is surface tension that supports a steel needle on the surface of a glass of water, despite the greater density of steel than water. The needle is not floating because of density differences, and if pushed under, it will sink.

TABLE 12.2
Types of liquids

Liquid helium is used in cryogenics. Molten silica and molten metals are important in glass technology and steel-making, respectively. And molten salts are used in fuel cells and as heat-transfer agents.

Type	Example
Monatomic	Noble gases at low temperatures, e.g., He, Ne, Ar
Molecular	Organic and inorganic liquids, e.g., benzene (C_6H_6), carbon tetrachloride (CCl_4), Br_2, PCl_3
Associated molecules	Hydrogen-bonded substances, e.g., H_2O, NH_3, HF
Ionic	Molten salts, e.g., Li_2CO_3, cryolite ($Na_3[AlF_6]$), UF_4
Polymeric	Organic polymers and molten oxides such as silica, SiO_2, and silicates
Metallic	Mercury and molten metals, e.g., Fe, Na

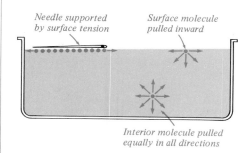

Needle supported by surface tension *Surface molecule pulled inward*

Interior molecule pulled equally in all directions

FIGURE 12.6
Surface tension.

Liquid crystals

The intermediate state of matter known as the liquid crystalline state has become familiar through the advertising for LCD—liquid crystal display—wrist watches. These watches show the time as a digital display of black numbers against a silvery background. (The watches that show red digits against a black background have LED—light-emitting diode—displays which are semiconducting devices.)

Liquid crystals flow like a liquid but have molecular order like that of a crystalline solid and exhibit some of the optical properties of crystals. All liquid crystalline substances are long and narrow, rigid organic molecules, for example,

$$CH_3O - \text{\bigcirc} - N\overset{O}{=}N - \text{\bigcirc} - OCH_3$$

p-azoxyanisole

Such substances pass through the liquid crystalline state during melting from a true crystalline solid to a true liquid. The intermolecular forces in the liquid crystals are weak and easily disrupted by temperature, mechanical stress, or an electrical field.

Three types of molecular arrangement have been identified in liquid crystals. *Smectic liquid crystals* are turbid, viscous substances with parallel molecules in planes that can slip past each other (Figure A). The curious name comes from the Greek word for soap;

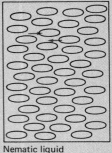

Molecules remain in their own planes, planes slide past each other — Smectic liquid crystalline substance

Parallel molecules slide past each other — Nematic liquid crystalline substance

FIGURE A
Molecular arrangement in smectic and nematic liquid crystals.

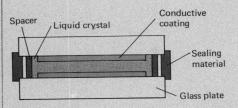

Spacer — Liquid crystal — Conductive coating — Sealing material — Glass plate

FIGURE B
A liquid crystal cell for an optical display.

soap films are smectic substances. *Nematic liquid crystals* are less turbid and more mobile than smectic liquid crystals. Frequently, a smectic substance is transformed to a nematic substance during the course of heating to a higher temperature.

In nematic liquid crystals the molecules are parallel but not ordered into layers. The word "nematic" comes from the Greek work meaning thread, and refers to a threadlike optical pattern seen in films of such substances.

Cholesteric liquid crystals, sometimes called twisted nematic crystals, have order similar to that of the nematic substances. However, the molecules in successive layers, instead of being randomly oriented, form a helical structure. Because of this structure, cholesteric crystals have the remarkable optical property of reflecting light of different colors under different conditions, for example, at different temperatures. Thin films of cholesteric liquid crystals are used to map skin temperature, as an aid to the diagnosis of diseased tissue, and to map electric circuit temperatures. In the variety stores, cholesteric liquid crystals have shown up in rings that change color with a person's "mood," that is, with one's skin temperature. The name "cholesteric" arises from the fact that these substances are all derivatives of cholesterol.

Cells like that shown in Figure B are the basic building blocks for liquid crystal displays in watches and other instruments. The cell is transparent when no current is passing through the conductive coating. When the electric field is applied, the molecular alignment changes and the cell appears opaque, or black if viewed through a polarizing film. The liquid crystals in such cells are smectic or nematic.

The term "surface tension" commonly refers to a liquid surface in contact with a gas. However, there is surface tension at the interface between any two phases. The magnitude of the surface tension at a liquid–solid interface depends upon the relative amounts of attraction between the liquid molecules themselves and between the liquid and solid molecules. A liquid is said to "wet" a solid when the attraction between the liquid and solid molecules is greater than the internal cohesive forces in the liquid. In terms of the angle θ (Greek theta), as shown in Figure 12.7, wetting occurs when $\theta < 90°$ and nonwetting occurs when $\theta > 90°$.

In a very narrow tube, called a capillary, a liquid surface takes on a curved shape—a meniscus. The curvature is caused by the surface tension of the liquid, and its shape depends upon how well the liquid wets the walls of the capillary (Figure 12.7). Because of surface tension, the pressure on the convex side of a liquid surface is greater than on the concave side. Therefore, a liquid that wets the capillary walls is pushed up in the tube, a process sometimes called *capillary action*. The liquid rises until the pressure of the liquid column equals the pressure difference between the flat surface of the liquid and the pressure under the meniscus. Water moves through fine pores in soil and also rises from the roots of a tree to its highest branches by capillary action. A liquid that does not wet capillary walls is pushed down in the capillary, until the forces are balanced in the same way.

12.9 Viscosity

Viscosity is the resistance of a liquid to flow; it is the opposite of fluidity. "Molasses in January" has a high viscosity; it will not run downhill easily or quickly. Internal forces that keep liquid particles from flowing past each other are responsible for viscosity.

Like most other properties of liquids, viscosity is dependent upon the size, shape, and chemical nature of the molecules. In general, higher intermolecular forces mean higher viscosity and, for similar types of compounds, higher molecular weight means higher viscosity. Viscosity, as illustrated by molasses, increases as the temperature decreases.

Lubricating oils are classified by their viscosity. The Society of Automotive Engineers assigns numbers to motor oils based on their relative viscosities. The SAE numbers range from 5 to 50 at a given temperature—0°F for winter (W) oils and 210°F for oils that are usable at summer temperatures. The higher the number, the greater the viscosity. Oils with a double number contain additives that moderate the change of viscosity with temperature so that they are neither too thin in summer nor too thick in winter. For example, SAE 10W/40 oil has an SAE number of 10 at 0°F and of 40 at 210°F.

12.10 Index of refraction

The three physical properties usually listed in a chemical handbook as part of the description of a liquid are the melting point, the boiling point, and the refractive index. The **refractive index** of a substance is

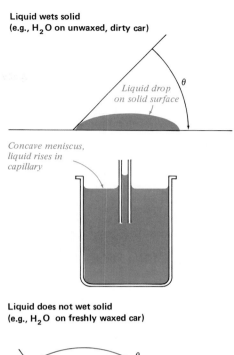

Liquid wets solid
(e.g., H$_2$O on unwaxed, dirty car)

Liquid drop on solid surface

θ

Concave meniscus, liquid rises in capillary

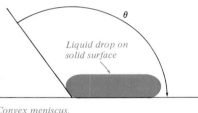

Liquid does not wet solid
(e.g., H$_2$O on freshly waxed car)

Liquid drop on solid surface

θ

Convex meniscus, liquid falls in capillary

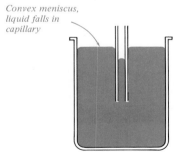

FIGURE 12.7
Wetting and capillarity. *The curved surface of a liquid in a tube is called a meniscus.*

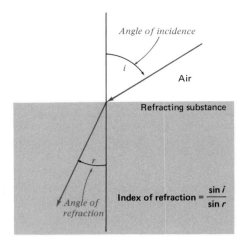

the ratio of the velocity of light in a vacuum or in air to the velocity of light in the substance. Because it is a ratio, refractive index has no units. It is usually given in reference to velocity in air and is found by comparing the angle of incidence with the angle of refraction of light of a specific wavelength (Figure 12.8).

Since the refractive index is sensitive to temperature changes (it increases with increasing temperature), and since it varies with the wavelength of the light, both of these variables must be specified in stating a refractive index. The usual symbol is as follows:

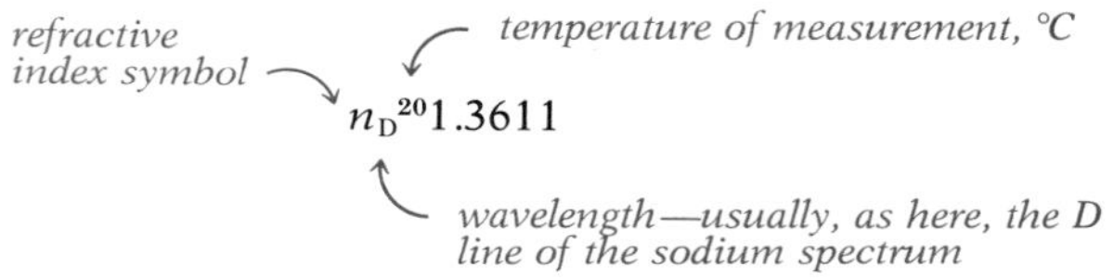

refractive index symbol ⤵   ⤷ *temperature of measurement, °C*

$$n_\text{D}^{20} 1.3611$$

⤷ *wavelength—usually, as here, the D line of the sodium spectrum*

For most liquids, the refractive index varies from 1.3 to 1.8. The accurate measurement of refractive index is simple, and therefore it is useful in many ways in the laboratory. Often, liquids with other properties that are similar can be identified by their refractive indexes. The purity of a known liquid can be monitored by variations in the refractive index, and the composition of simple liquid mixtures can be determined by refractive index measurements. For example, the amount of water in milk is determined in this way.

The solid state

12.11 Types of solids

Solids were once described as "those parts of the material world which support when sat on, which hurt when kicked, which kill when shot." In terms of their chemical and physical properties, there are

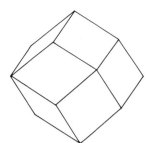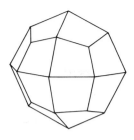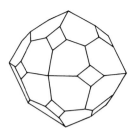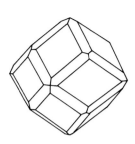

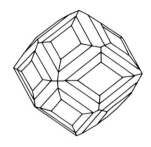

two types of solids—crystalline solids and amorphous solids. A **crystalline solid,** also called a *true solid*, is a substance in which the atoms, molecules, or ions have a characteristic, regular, and repetitive three-dimensional arrangement. Sugar and salt are crystalline solids. An **amorphous solid** is a substance in which the atoms, molecules, or ions have a random and nonrepetitive three-dimensional arrangement. Tar is an amorphous solid.

Everyone is familiar with the beautiful and varied crystalline forms of the gemstones formed in nature and collected by mineralogists (Figure 12.9). A **crystal** is a solid that has a shape bounded by plane surfaces intersecting at fixed angles. The regular geometrical shapes of crystals are the large-scale expression of the internal order of its atoms, molecules, or ions. Perfectly symmetrical single crystals of pure substances form only under conditions where the crystals have the opportunity to grow slowly in all directions. Most crystalline materials are polycrystalline—composed of small areas with plane faces and the characteristic internal order, but randomly oriented towards each other (Figure 12.10).

Crystalline substances, because of their internal order, may be *anisotropic*—they may have certain physical properties that vary with direction. For example, electrical conductivity may be greater in one direction than another. One way to identify a crystalline substance is by its cleavage into smaller pieces with planar faces when struck sharply. Amorphous substances are generally *isotropic*—their physical properties are the same in all directions. Cleavage of an amorphous material gives smaller pieces with nonplanar faces. The gemstone opal and volcanic glass are two of the very few parts of the Earth's crust that are amorphous and not crystalline.

Amorphous and crystalline materials are also quite different in their melting behavior. Crystalline solids generally have sharp melting points—when the temperature gets high enough all of the particles are released from their fixed positions and the order is destroyed all at once. Amorphous solids soften gradually as the temperature rises. Glass, tar, and many polymers are, in fact, often classified not as solids, but as liquids of such high viscosity at ordinary temperatures that their flow is not observable.

Glasses are more or less transparent supercooled liquids. Most glasses are oxides of nonmetal or semiconducting elements. In a sheet of glass taken from an old window, there is a significant (though small) difference in thickness between the bottom and the top. The bottom is thicker because the glass has flowed slightly over the years.

FIGURE 12.9
Natural crystal forms of garnet. (*From Dana's Manual of Mineralogy, C. S. Hurlburt, Jr., 18th ed., John Wiley & Sons, Publishers, New York, 1971.*)

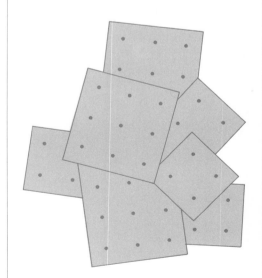

FIGURE 12.10
A polycrystalline material. *The dots represent ions or molecules. The intersections of the crystallites, or tiny crystals, are called grain boundaries.*

335

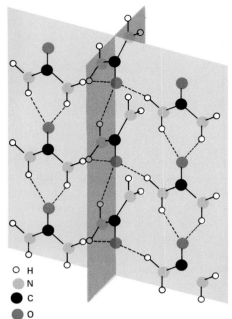

FIGURE 12.11
Crystal structure of urea, **(NH₂)₂C=O.** *This drawing represents the arrangement of urea molecules in a crystal. The dashed lines represent hydrogen bonds.*

(a) Closest packing of spheres of equal radius in a single layer. Here and in (b), all of the spheres that touch the dark central spheres are shown in light color.

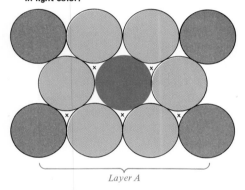

Layer A

(b) Another closest-packed layer, layer B, has been added above layer A. Layer B spheres cover the layer A spaces marked x in (a). Three spheres in layer B touch the dark central sphere of layer A.

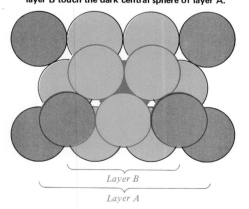

Layer B

Layer A

FIGURE 12.12
Closest packing in layers. *The light-colored spheres all touch the dark-colored sphere.*

In many scientific contexts, the word "solid" is used strictly for crystalline materials. Table 12.3 summarizes the types of crystalline solids. Network covalent substances, in which all of the atoms are continuously joined by covalent bonds, were previously described (Section 9.19). Molecular crystals include almost all organic compounds in their solid state. The intermolecular forces acting in such substances were discussed in Chapter 9 (Sections 9.20–9.22). Hydrogen bonding is particularly important in determining the arrangement in space of the molecules in crystalline organic compounds, such as in urea (Figure 12.11).

In subsequent sections we discuss the structures of pure metals (Section 12.12), ionic crystals (Section 12.14), and alloys (Section 12.15).

12.12 Crystal structure of metals

A pure metal is a crystalline solid made of identical metal atoms. Most metals have high densities, showing that in their crystal structures empty spaces are small. Think of each metal atom as a solid sphere. (We use "atom" recognizing that valence electrons from these atoms have been contributed to the "electron sea.") The simplest way to pack spheres of the same radius into a single layer as closely as possible is in a hexagonal arrangement. Each sphere touches six other spheres (Figure 12.12). This arrangement, called closest packing, can continue indefinitely in a single layer, and a crystal can be built up by superimposing one such layer over another.

Two types of superimposition maintain closest packing in three dimensions. In each, the second layer (B) is placed over the first layer (A) so that three spheres of layer B touch the same sphere of layer A

TABLE 12.3

Types of crystalline solids
Keep in mind that both covalent and ionic contributions to bonding often exist in the same substance.

Substances	Examples	Particles in crystal lattice	Major interparticle forces
Pure metals	Fe, Cu, Al	Ions	Metallic bond
Ionic crystals	All salts (e.g., NaCl, $BaSO_4$)	Monatomic or polyatomic ions	Electrostatic attraction of opposite charges
Alloys			
Simple mixtures		Ions	Metallic bond
Solid solutions		Ions	Metallic bond
Substitutional	AuAg		
Interstitial	TiN, SiC		
Intermetallic compounds	Cu_5Sn, Ag_5Al_3	Ions	Metallic bond
Molecular crystals	All covalent compounds (e.g., CO_2, ice, organic compounds)	Neutral molecules	Intermolecular forces
Network covalent materials	Silicon, diamond, quartz, germanium	Atoms	Covalent bonds

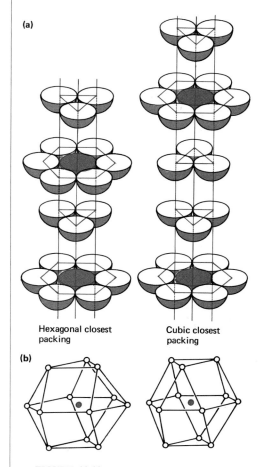

Hexagonal closest packing   Cubic closest packing

FIGURE 12.13
Hexagonal and cubic closest packing.
(a) *Exploded views of stacking of planes of spheres.* (b) *Arrangement of the 12 nearest neighbors about one sphere. In both types of packing, only the spheres in the A layers are directly above one another.*

THE LIQUID AND
SOLID STATES [12.12]

(Figure 12.12b). The third layer can then be superimposed in one of two ways. In the first, each atom in the third layer lies directly above an atom in the first layer. In this arrangement, called **hexagonal closest packing,** closest-packed layers of atoms are arranged in an ABABAB . . . sequence (Figure 12.13).

In the second way of adding the third layer (layer C), the layer is displaced so that its atoms are not directly above those of either layer A or layer B. In this arrangement, called **cubic closest packing,** closest-packed layers of atoms are arranged in an ABCABCABC . . . sequence (Figure 12.13).

In both hexagonal and cubic closest packing, each sphere touches six other spheres in its own layer, plus three in the layer above and three in the layer below. This gives each sphere twelve nearest neighbors and a crystal coordination number of 12 (Figure 12.13). The **crystal coordination number** is the number of nearest neighbors of an atom, ion, or molecule in a particular crystal structure. Seventy-four percent of the available space is occupied in metals with either of the two closest-packed structures.

About two-thirds of all metals have either hexagonal or cubic closest-packed crystal structures. Most of the remaining metals crystallize with a body-centered cubic structure (see Figure 12.18), an arrangement in which the crystal coordination number is 8, the layers are not closest-packed, and space is not quite as efficiently filled (see Example 12.4).

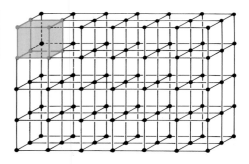

FIGURE 12.14

A simple cubic space lattice. *The unit cell is shaded in color.*

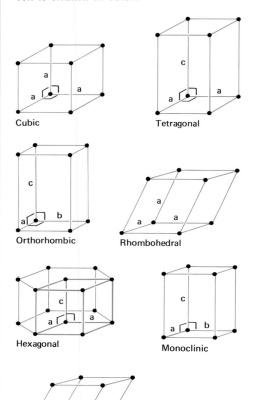

Cubic

Tetragonal

Orthorhombic

Rhombohedral

Hexagonal

Monoclinic

Triclinic

FIGURE 12.15

The primitive unit cells for the seven crystal systems. *The primitive unit cells give the basic shapes of the seven crystal systems. Where two or more of the axes are equal, the same letter is shown for each. Right angles (90°) are shown as* ⌐ *. The heavy outline shows the hexagonal unit cell; three of them form a hexagon.*

12.13 Crystal systems

The infinite repetition of an orderly arrangement of particles in a crystal means that if we start at one atom and travel in a specific direction, we must arrive at another atom with an identical environment. **A space lattice** is a system of points representing the sites with identical environments occupied by the particles in a crystal. A simple cubic space lattice is shown in Figure 12.14. The **crystal structure** of a substance is the complete geometrical arrangement of the particles that occupy the space lattice.

A **unit cell** is the part of a space lattice that, if repeated in three dimensions, will generate the entire lattice. The cube outlined in color in Figure 12.14 is the unit cell for the cubic crystal system. Each cracker box in a supermarket display is like a unit cell in the display. The overall shape of the display is governed to a certain extent by the shape of the box, with variations possible depending on how the boxes are stacked.

Crystals fall into seven different classifications or crystal systems, which are defined by starting from a common origin, proceeding in three directions until points with an identical environment are reached, and specifying the relative lengths of the three axes

TABLE 12.4

Seven crystal systems

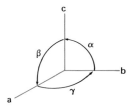

Crystal system	Relative axial length	Angles	Examples
Cubic (isometric)	$a = b = c$	$\alpha = \beta = \gamma = 90°$	Na^+Cl^-, Cs^+Cl^-, $Ca^{2+}(F^-)_2$, $Ca^{2+}O^{2-}$
Tetragonal	$a = b \neq c$	$\alpha = \beta = \gamma = 90°$	$(K^+)_2PtCl_6^{2-}$, $Pb^{2+}WO_4^{2-}$, $NH_4^+Br^-$
Orthorhombic	$a \neq b \neq c$	$\alpha = \beta = \gamma = 90°$	$(K^+)_2SO_4^{2-}$, $K^+NO_3^-$, $Ba^{2+}SO_4^{2-}$, $Ca^{2+}CO_3^{2-}$ (aragonite)
Rhombohedral (trigonal)	$a = b = c$	$\alpha = \beta = \gamma \neq 90°$	$Ca^{2+}CO_3^{2-}$ (calcite), $Na^+NO_3^-$
Hexagonal	$a = b \neq c$	$\alpha = \beta = 90°$, $\gamma = 120$	AgI, SiC, HgS
Monoclinic	$a \neq b \neq c$	$\alpha = \beta = 90°$, $\gamma \neq 90°$	$Ca^{2+}SO_4^{2-}\cdot2H_2O$, $K^+ClO_3^-$, $(K^+)_4Fe(CN)_6^{4-}$
Triclinic	$a \neq b \neq c$	$\alpha \neq \beta \neq \gamma \neq 90°$	$Cu^{2+}SO_4^{2-}\cdot5H_2O$, $(K^+)_2Cr_2O_7^{2-}$

Body-centered orthorhombic

Face-centered orthorhombic

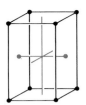
Side-centered orthorhombic

FIGURE 12.16

Orthorhombic unit cell with additional points occupied. *Because of geometrical considerations, other such unit cells are limited to the body-centered and face-centered cubic, body-centered tetragonal, and end-centered monoclinic unit cells.*

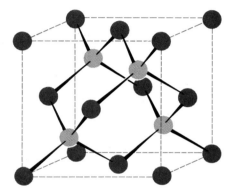

FIGURE 12.17

A diamond unit cell. *Each sphere represents a carbon atom. The unit cell belongs to the cubic crystal system, and the atoms that form the face-centered cube are shown in dark grey. The solid lines are the C—C bonds.*

and their included angles. **A primitive cell** is a unit cell in which only the corners are occupied. Sketches of the primitive cells of the seven crystal systems are given in Figure 12.15, and their defining characteristics are listed in Table 12.4.

Additional points within the unit cell can also be occupied, for example, the center of the cell, or the centers of the planar faces (Figure 12.16). A perfect crystal of diamond has the symmetry characteristics of a cube and hence belongs to the cubic crystal system. x-Ray analysis (see Tools of Chemistry, this chapter) shows that in addition to face-centered atoms, the unit cell of a diamond also contains within it four additional carbon atoms arranged in a tetrahedron (Figure 12.17).

The unit cell diagrams in Figure 12.15 are misleading because they suggest that each cell contains a lot of empty space. The packing of atoms in a crystal is more realistically shown by the sphere-based packing models for metals in Figure 12.18. Of course, when the lattice

FIGURE 12.18

Packing models and unit cells for cubic and hexagonal crystal systems.
One metal, polonium, has a simple cubic structure. All other metals have one of the remaining three structures shown. The number of atoms per unit cell for hexagonal closest packing, which is 2, is not so obvious as it is in the other cases. The unit cell for hexagonal closest packing is the hexagonal cell with one additional body-centered point.

Packing models   Unit cells

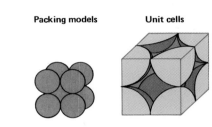
Simple cubic (e.g., Po)

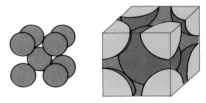
Body-centered cubic (e.g., Li, Na, Ba)

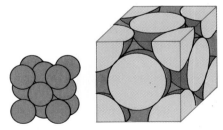
Face-centered cubic (cubic closest-packed)
(e.g., Ni, Cu, Au)

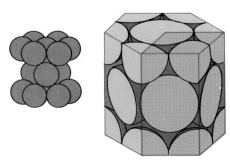
Hexagonal closest-packed (e, g., Mg, Ti, Zn)

339

points are occupied not by single atoms, but by molecules or polyatomic ions, the total structure within the unit cell is more complex.

Crystallographers talk of the number of atoms per unit cell. At first glance you might think that there are four atoms in the cubic unit cell. Not so, for the atoms must be sliced along the faces to give a single cell. For example, the simple cubic unit cell contains $8 \times \frac{1}{8}$, or 1 atom (see Figure 12.18).

EXAMPLE 12.4

■ Calculate the efficiency of packing for the body-centered lattice when it is occupied by spherical metal atoms.

The volume occupied by the atoms in the unit cell can be calculated by finding the total number of atoms in the unit cell and multiplying by the volume of one atom. For the body-centered cubic unit cell (Figure 12.18) the corners contribute $8 \times \frac{1}{8} = 1$ atom and this plus the central atom gives a total of 2 atoms for the unit cell. With each atom a sphere occupying the volume $\frac{4}{3}\pi r^3$, the total occupied volume is

$$V \text{ (occupied)} = 2 \left(\tfrac{4}{3}\pi r^3\right)$$

The volume of the unit-cell cube is a^3, where a is the unit-cell length. The body diagonal (the distance from one corner to an opposite corner) is given by $a\sqrt{3}$. This diagonal is equal to $4r$ because the corner atoms $(r + r)$ are touching the body-centered atom $(2r)$; thus $a = 4r/\sqrt{3}$ and V (unit cell) $= (4r/\sqrt{3})^3$. The fraction occupied is

$$\frac{V \text{ (occupied)}}{V \text{ (unit cell)}} = \frac{2\left(\tfrac{4}{3}\pi r^3\right)}{(4r/\sqrt{3})^3} = 0.680$$

A body-centered cubic unit cell is 68% occupied—a little less efficient packing than the closest-packed systems. Note that the percent of the volume occupied is independent of r. ■

12.14 Ionic crystal structure

In an ionic crystal, the anion and the cation are usually of different sizes. Therefore, packing is less simple than in metallic crystals. Several basic requirements govern ionic crystal structure.

1. *Electrical neutrality must be maintained.* That is, the ratio of cations to anions must be that in the formula, for example, in NaCl, the ratio is 1:1, in CaF_2, the ratio is 1:2.

2. *Anions and cations should be very close to each other, preferably just touching.* This provides the maximum stability.

3. *Each anion should have as many cation neighbors as possible.* This provides the maximum force of attraction.

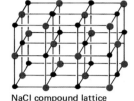
NaCl compound lattice

NaCl packing model

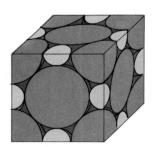
NaCl unit cell
(one Na$^+$ ion is concealed in the center of the cell)

FIGURE 12.19
Sodium chloride crystal structure.
You can see that the Na$^+$ ion in the center of the compound lattice sketch is surrounded by six Cl$^-$ ions.

4. *Ions of like charge should be relatively far from each other.* This requirement opposes and must be balanced against the preceding one.

Anions are usually larger than cations; therefore, the anions may be thought of as forming a structure into which the smaller cations must fit. Picture the Cl^- ions of NaCl as forming a cubic closest-packed array. The Na^+ ions are the right size to fit into the holes in the Cl^- lattice. The NaCl unit cell is illustrated in Figure 12.19. The crystal coordination number of both Na^+ and Cl^- is 6. Sodium chloride can equally well be pictured as a face-centered cubic lattice of Na^+ ions at sufficient distance from each other so that the Cl^- ions can fit into the spaces between Na^+ ions. Alternatively, the NaCl structure can be thought of as a face-centered cubic lattice of Na^+ cations interlocking with a face-centered cubic lattice of Cl^- ions.

The crystal structure of simple ionic compounds between monatomic ions is to a certain extent governed by the ratio of the radius of the cation to the radius of the anion—the radius ratio, r_+/r_-. The number of anions that can crowd around a cation depends upon the radius ratio. As the ratio increases, the crystal coordination number of the cations increases. For example, in two dimensions, we can see how a larger cation can be surrounded by more anions (Table 12.5).

From simple geometry, the minimum radius ratios for specific coordination numbers can be calculated (Table 12.5). The crystal

TABLE 12.5

Packing arrangements for A B ionic crystals

The minimum radius ratio for each arrangement is that in which the cation is in contact with its nearest neighbor anions. Once r_+/r_- reaches unity, the reverse of the radius ratio is used. r_-/r_+ is calculated and the choice of structure fits the corresponding geometry, except that the larger cations form the lattice and the smaller anions fit in the holes.

Minimum radius ratio, r_+/r_-	Cation coordination number	Geometry	Arrangement of nearest neighbors	Example
0.225	4	Tetrahedral	*Minimum radius ratio; cation just fits hole*	ZnS
0.414	6	Octahedral	*Larger radius ratio; anions pushed apart*	NaCl
0.732	8	Cubic	*Still larger radius ratio; room for another anion*	CsCl

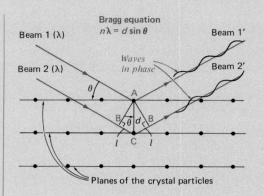

Diffraction is the bending of waves as they pass obstacles in their paths. The regularly spaced rows of atoms, ions, or molecules in a crystal diffract certain waves so that characteristic patterns of heavy and light lines result. To be diffracted, the waves must have a wavelength comparable in size to the distance between the particles in the crystal. x Rays, neutron beams, and electron beams fulfill this requirement because they have wavelengths of the order of 0.1 nm.

x-Ray diffraction is the basic tool for studying crystal structure. Only after 1912, when the German physicist Max von Laue found that x rays could be diffracted by crystals, was proof of the orderly structure of crystals possible. x Rays interact with the orbital electrons of the atoms in a crystal. x-Ray diffraction patterns are now interpreted by the simple approach developed by the English scientist, Sir Laurence Bragg. Bragg found that x rays can be thought of as rays reflected by successive planes of atoms in a crystal. The rays from the lower plane (see Figure A) travel the distance 2*l* further than the rays reflected by the upper plane. If this distance is a whole-number multiple of the wavelength, beams 1' and 2' are in phase and reinforce each other. Mathematically this condition is stated

$$n\lambda = 2l$$

whole number   *wavelength*

The angle of incidence of the beam θ is equal to the angle BAC and the triangle BAC is a right triangle, with AC equal to *d*, the distance between the planes. Therefore

$$l = d \sin \theta$$

Combining the two preceding equations gives the *Bragg equation,* or Bragg's law,

$$n\lambda = 2d \sin \theta$$

The total geometry of a crystal is worked out by rotating a crystal through many values of θ and finding *d* for rows of particles viewed in

(a) X-ray powder diffraction camera

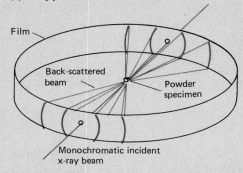

Film

Back-scattered beam

Powder specimen

Monochromatic incident x-ray beam

(b) Sketch of x-ray powder diffraction lines recorded on film — sodium chloride

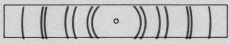

(c) Sketch of x-ray powder diffraction lines recorded by a diffractometer — sodium chloride

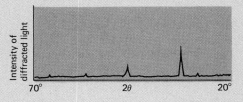

Intensity of diffracted light

70°   2θ   20°

FIGURE B

X-ray powder diffraction. *The diffraction pattern is recorded on a cylindrical film. Each line in (b) and each peak in (c) represents reflection from a different plane.*

various directions, using monochromatic x rays of known wavelength. Polycrystalline samples can also be studied, using the technique called x-ray powder diffraction (see Figure B). The intensity of the diffracted beam varies with n, usually being less intense for larger values of n.

An electron does not penetrate as far past a crystalline surface as does an x-ray beam. Therefore, low-energy electron diffraction (LEED) is useful for studying atomic arrangements and electron densities at surfaces. In electron diffraction, scattering, or reemission, of the radiation comes from both orbital electrons and nuclei. Electron diffraction studies of thin foils can yield information both about the regularity of the crystal structure and about its defects.

In neutron diffraction, the incident neutron beam is scattered solely by the nuclei, not by the orbital electrons. This makes neutron diffraction useful in two cases where x-ray diffraction will not work. (1) Neutron diffraction can locate very light atoms in the presence of heavier ones (e.g., in NaH), and (2) it can distinguish between atoms with similar numbers of electrons, such as those combined in intermetallic compounds. In x-ray diffraction, scattering is proportional to the number of orbital electrons and it fails in case (1) because the relative contribution of the light atoms is too small and in case (2) because the contributions of the two types of atoms are too similar.

The interpretation of diffraction patterns is complex. Often, particularly for x-ray diffraction, substances are identified by comparison of their diffraction patterns with those in large compilations of standard diffraction patterns.

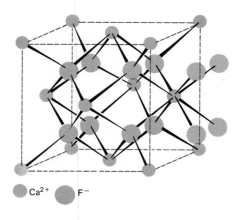

Ca²⁺   F⁻

FIGURE 12.20
Calcium fluoride (CaF₂) unit cell. *This is called the fluorite structure.*

structure of ionic compounds can in many cases be predicted from the radius ratios of the compounds. For example, it would be assumed that a radius ratio between the minima for the zinc sulfide and sodium chloride structures would give a compound with the zinc sulfide structure. However, the validity of such predictions is limited. Two complicating factors are the sometimes difficult choice of good ionic radii values and the presence in many cases of a degree of covalent bonding in the compound.

EXAMPLE 12.5

■ The ionic radii for Fe^{2+} and O^{2-} are 0.075 and 0.140 nm, respectively. On the basis of radius ratio, predict the crystal structure of iron(II) oxide, FeO.

The radius ratio is

$$\frac{r_+}{r_-} = \frac{0.075 \text{ nm}}{0.140 \text{ nm}} = 0.54$$

This ratio falls in the range for coordination number 6 (Table 12.5). Therefore, FeO should have the sodium chloride structure, with interlocking face-centered cubic Fe^{2+} and O^{2-} lattices.   ■

When the charge on the cation and the charge on the anion are numerically different, both the maximum separation of ions of like charge and the necessity for overall neutrality are all-important in determining crystal structure. For example, in a crystal of magnesium fluoride (MgF₂), the radius ratio (0.48) would allow six fluoride ions to surround each magnesium ion, but charge balance allows only three magnesium ions to surround each fluoride ion. Maximum separation of the former then places the three magnesium ions at the corners of an equilateral triangle, with a fluoride ion at the center. In a crystal of calcium fluoride (CaF₂), however, the radius ratio (0.73) allows eight fluoride ions to surround each calcium ion at the corners of a cube, but each fluoride ion can be surrounded tetrahedrally by four calcium ions (Figure 12.20). The resulting cubic unit cell is described by a calcium ion at each corner and at the center of each face of a cube and eight fluoride ions within the cube but close to the corners.

Substances that have the same crystal structure are said to be **isomorphous.** Some such substances can crystallize together to give a mixed product. For example, sodium nitrate (NaNO₃) and calcium carbonate (CaCO₃) form mixed crystals, although in terms of other physical properties and all chemical properties, they are quite different. Particularly common examples of isomorphism are found among the alums, $M^+M^{3+}(SO_4{}^{2-})_2 \cdot 12H_2O$, where M^+ may be Na^+, K^+, Rb^+, Cs^+, $NH_4{}^+$, or Tl^+; M^{3+} may be Al^{3+}, Cr^{3+}, Fe^{3+}, Co^{3+}, or Ti^{3+}; and $\cdot 12H_2O$ indicates that 12 water molecules are included in the crystal. All alums crystallize in the cubic system, often as beautiful, well-formed, octahedral crystals, and one alum can be deposited on a crystal of another alum.

Substances that can crystallize in more than one crystal system are called **polymorphous.** This phenomenon, which is less common, is exemplified by the calcite (rhombohedral) and aragonite (orthorhombic) forms of calcium carbonate.

Lattice energy reflects the stability of a crystal system. The lattice energy of a crystal of an ionic compound depends upon the electrostatic attractions and repulsions among the ions, upon the ionic radii, and upon the arrangements of the ions in the crystal.

12.15 Alloys

When two or more metals, or metals plus nonmetals, are intimately mixed to yield a substance with metallic properties, they have formed an **alloy.** Some alloys are simply heterogeneous mixtures—polycrystalline materials with small areas of different metals in contact. Other alloys are solid solutions—homogeneous mixtures in which the atoms of one metal are randomly distributed among the atoms of another metal. The requirements for solid solution formation are (1) similar chemical properties, (2) similar crystal structures, and (3) appropriate radius ratios.

A **substitutional solid solution** is an alloy in which atoms of one metal replace atoms in the crystal lattice of the other. Metals with radii that differ from each other by no more than 15% (e.g., Ni, 0.115 nm, and Cu, 0.117 nm) are completely soluble in each other in all proportions. As metal atoms become more different in size, solid solubility decreases and ultimately disappears. Silver (0.134 nm) and copper have only limited solubility in each other.

In **interstitial solid solutions** small atoms occupy holes in the crystal lattice of a metal. A series of very high melting, extremely hard, brittle, electrically conducting, and chemically inert alloys results when various d-transition metals (e.g., Ti, Zr, Hf, V, Nb, Ta, Cr, Mo, W, or Fe) combine with nitrogen, carbon, or boron at elevated temperatures. These nitrides and carbides often have or closely approach 1:1 atomic ratios. Structural analyses show that they are formed by distortions of the metallic crystal lattices by the entry of the small atoms (N, radius = 0.074 nm; C, 0.077 nm) into holes in the crystal lattice. The borides often have 1:2 ratios and structures with parallel layers of metal and boron atoms. Hydrogen combines with a number of the transition metals to give interstitial hydrides. Typical compositions for these nonstoichiometric compounds are $PdH_{0.76}$ and $TiH_{1.7}$; the compositions vary with the conditions of preparation. The hydrogen atoms occupy interstitial positions and the properties of the metal remain about the same.

Intermetallic compounds are phases of more or less fixed composition that occur in some alloys. The composition of these phases is based not on the usual concepts of valence, but upon maintaining a constant (or close to constant) ratio of the total number of valence electrons to the total number of atoms.

Some typical intermetallic compounds are AgZn, Cu_3Al, $AgCd_3$, and $FeZn_7$. Each composition is apparently the one needed to provide enough free electrons to produce stability. The intermetallic compound phases in an alloy have metallic properties, but these properties may differ substantially from those of the component metals.

The copper–zinc alloys—the brasses—illustrate the types of compositions that might be found in one type of alloy. Among the brasses, the intermetallic compounds—CuZn, Cu_5Zn_8, and $CuZn_3$— can be identified. In these compounds the ratios of atoms to valence electrons are as follows:

CuZn	2 atoms/3 electrons	or	14/21
Cu_5Zn_8	13 atoms/21 electrons	or	13/21
$CuZn_3$	4 atoms/7 electrons	or	12/21

Many alloys, particularly those of copper, silver, and gold, show the same ratios of atoms to electrons. The reasons for this are not fully understood. In addition to the intermetallic copper–zinc compounds, a solid solution of up to 32% zinc in copper in the normal face-centered cubic copper lattice, plus a solid solution of up to 5% copper in the normal zinc hexagonal closest-packed lattice, can be formed.

To summarize, the factors that influence alloy formation are atomic size, the number of valence electrons per total number of atoms, the chemical properties, and the crystal structures of the elements involved.

12.16 Defects

Except for single crystals grown under special conditions, real crystalline substances are seldom perfect. A perfect crystal is chemically pure and structurally perfect, with every lattice point occupied as specified by the unit cell.

Many physical and chemical properties of solids depend upon the presence of defects in the solid state. Perfect crystals are very strong, whereas most solids contain enough defects of various kinds to allow them to yield more easily to mechanical forces. Also, chemical reactions in the solid state (next section) require the motion of atoms or ions through a solid. In a perfect crystal there is no available pathway for such motion, whereas in real crystals, atoms or ions can move from defect to defect.

Solid-state defects can be classified as point defects, line defects, planar defects, or three-dimensional (spatial) defects.

Point defects are variations in the occupation of the lattice or interstitial sites in the crystal. Three basic types of point defects occur: (1) vacancies—unoccupied lattice sites (Figure 12.21), (2) interstitial atoms or ions—atoms or ions in the spaces in between lattice sites, and (3) impurity defects—foreign particles in regular lattice sites or interstitial sites. In ionic crystals neutrality must be maintained by the establishment of an equilibrium between positively and negatively charged defects.

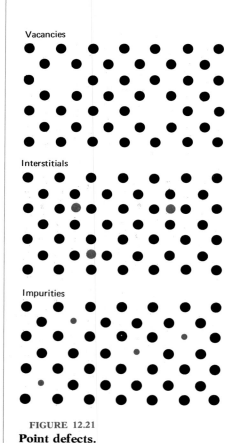

Vacancies

Interstitials

Impurities

FIGURE 12.21
Point defects.

The tendency of some substances to incorporate point defects in their lattice structures accounts for the occurrence of nonstoichiometric compounds. For example, in wustite, $Fe_{<1}O$, up to 14% of the normal cation sites can be empty. To maintain neutrality, two Fe^{2+} ions are converted to Fe^{3+} for every missing Fe^{2+}. And in stoichiometric titanium oxide, TiO, 15% of the sites of each type are vacant, leading to the occurrence of a range of nonstoichiometric titanium oxide composition, TiO_x, where x can be less than or greater than 1 depending upon the oxygen pressure during preparation of the sample. Where there are extra oxygen atoms, there are a compensating number of titanium vacancies, and vice versa.

Dislocations are line defects; they represent the displacement of rows of particles in a crystal. Edge and screw dislocations—the two types of line defects—usually occur together (Figure 12.22). Line defects account for the ease of mechanical deformation of a solid.

Planar defects include grain boundaries in polycrystalline materials, phase boundaries in multiphase solids, and stacking faults, in which the stacking sequence of crystal layers is not regular (such as the occurrence of ABCABABC . . . in a face-centered cubic crystal).

The *three-dimensional defects* include pores—open spaces within the solid—or the presence of macroscopic areas of impurities within the solid.

12.17 Solid-state reactions

The definition of a solid-state reaction is a bit difficult. Processes of chemical change in a solid are different from those in gases and liquids, where the reactants are mixed on an atomic scale. The occurrence of a solid-state reaction is dependent upon the physical state of the solid and its transport properties—those properties that govern the motion of particles through the solid. We adopt the following definition of a solid-state reaction: "A solid-state chemical reaction in the classical sense occurs when local transport of matter is observed in a crystalline phase." (This definition and the classification of solid-state reactions given here are based on *Solid State Reactions*, by H. Schmalzried, Academic Press, New York, 1974.)

The field of solid-state chemistry is expanding rapidly and includes such important areas as thermal decomposition, corrosion, reactions of glasses, phase transformations, sintering of ceramics and metals, and reactions at surfaces. **Sintering** is the transformation of a mass of small particles by controlled heating into a solid, dense, strong material. We can do little more here than classify solid-state reactions into three general categories and cite examples of each.

In *homogeneous solid-state reactions* the atoms, ions, and defects in a crystal undergo internal rearrangement among themselves, often because of a change in the temperature or pressure. For example, in a reaction important in photography, thermally excited Ag^+ ions leave their regular lattice sites in AgBr and enter interstitial positions, leaving behind cation vacancies.

Single-phase inhomogeneous solid-state reactions occur by the

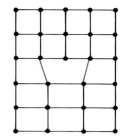
Edge dislocation

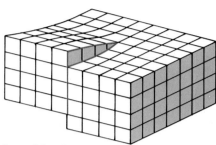
Screw dislocation

FIGURE 12.22
Line defects. *The edge dislocation is the insertion of an incomplete plane of atoms. The screw dislocation forms a spiral around the dislocation.*

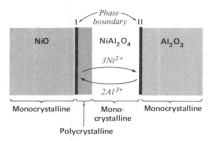

FIGURE 12.23

A solid-state reaction: NiO(s) + Al$_2$O$_3$(s) → NiAl$_2$O$_4$(s). *The reactions at the phase boundaries are*
(I) $2Al^{3+} + 4NiO \rightarrow NiAl_2O_4 + 3Ni^{2+}$
(II) $3Ni^{2+} + 4Al_2O_3 \rightarrow 3NiAl_2O_4 + 2Al^{3+}$
[*From H. Schmalzried*, Solid State Reactions, *1974, Fig. 2.1, p. 14, Verlag Chemie GMBH.*]

diffusion of one substance into another. When the substances are completely miscible and held at a temperature high enough for diffusion to occur, the diffusion continues until homogeneity is achieved. This process will occur between crystals of silver and gold, or NaBr and AgBr, placed intimately in contact with each other.

In a *heterogeneous solid-state reaction* the reactants become separated from each other by one or more product phases. For the reaction to continue, the reactants must diffuse through the product phase, so that the reaction can occur at the phase boundary. The formation of NiAl$_2$O$_4$ from NiO and Al$_2$O$_3$ (Figure 12.23) is an example of a heterogeneous solid-state reaction. Another example is the formation of a layer of tarnish on the surface of a metal. In one study a slab of silver was separated from liquid sulfur by two slabs of Ag$_2$S, the product of the silver–sulfur tarnish reaction. The Ag$_2$S slab adjacent to the sulfur gained weight exactly equal to that lost by the silver slab, showing that the silver had diffused through both Ag$_2$S slabs and reaction had occurred at the Ag$_2$S–S boundary.

The dance of the solids

by John Updike

ARGUMENT:

In stanzas associated with allegory the actual atomic structure of solids unfolds. Metals, Ceramics and Polymers. The conduction of heat, electricity and light through solids. Solidity emerges as being intricate, giddy, playful.

All things are Atoms: Earth and Water, Air
   And Fire, all, *Democritus* foretold.
   Swiss *Paracelsus*, in's alchemic lair,
Saw Sulfur, Salt, and Mercury unfold
Amid Millennial hopes of faking Gold.
Lavoisier dethroned Phlogiston; then
   Molecular Analysis made bold
   Forays into the gases: Hydrogen
Stood naked in the dazzled sight of Learned Men.

The Solid State, however, kept its grains
   Of Microstructure coarsely veiled until
   X-ray diffraction pierced the Crystal Planes
That roofed the giddy Dance, the taut Quadrille
   Where Silicon and Carbon Atoms will

Link Valencies, four-figured, hand in hand
With common Ions and Rare Earths to fill
The lattices of Matter, Glass or Sand,
With tiny Excitations, quantitively grand.

The *Metals*, lustrous Monarchs of the Cave,
   Are ductile and conductive and opaque
   Because each Atom generously gave
Its own Electrons to a mutual Stake,
A Pool that acts as Bond. The Ions take
   The stacking shape of Spheres, and slip and flow
   When pressed or dented; thusly *Metals* make
A better Paper Clip than a Window,
Are vulnerable to Shear, and, heated, brightly glow.

Ceramic, muddy Queen of human Arts,
   First served as simple Stone. Feldspar supplied
   Crude Clay; and Rubies, Porcelain, and Quartz
Came each to light. Aluminum Oxide
Is typical—a Metal is allied
   With Oxygen ionically; no free
   Electrons form a lubricating tide,
Hence, Empresslike, *Ceramics* tend to be
Resistant, porous, brittle, and refractory.

Prince *Glass, Ceramic's* son though crystal-clear,
    Is no wise crystalline. The fond Voyeur
    And Narcissist alike devoutly peer
    Into Disorder, the Disorderer
    Being Covalent Bondings that prefer
    Prolonged Viscosity and spread loose nets
    Photons slip through. The average *Polymer*
    Enjoys a Glassy state, but cools, forgets
To slump, and clouds in closely patterned Minuets.

The *Polymers*, those giant Molecules,
    Like Starch and Polyoxymethylene,
    Flesh out, as protein serfs and plastic fools,
    The Kingdom with Life's Stuff. Our time has seen
    The synthesis of Polyisoprene
    And many cross-linked Helixes unknown
    To *Robert Hooke;* but each primordial Bean
    Knew Cellulose by heart: *Nature* alone
Of Collagen and Apatite compounded Bone.

What happens in these Lattices when *Heat*
    Transports Vibrations through a solid mass?
    $C_p = 3Nk$ is much too neat;
    A rigid Crystal's not a fluid Gas.
    Debye in 1912 proposed Elas-
    Tic Waves called *phonons* which obey *Max Planck's*
    Great Quantum Law. Although amorphous Glass,
    Umklapp Switchbacks, and Isotopes play pranks
Upon his Formulae, *Debye* deserves warm Thanks.

Electroconductivity depends
    On Free Electrons: in Germanium
    A touch of Arsenic liberates; in blends
    Like Nickel Oxide, *Ohms* thwart Current. From
    Pure Copper threads to wads of Chewing Gum
    Resistance varies hugely. Cold and Light
    As well as "doping" modify the sum
    Of *Fermi* Levels, Ion scatter, site
Proximity, and other factors recondite.

Textbooks and Heaven only are Ideal;
    Solidity is an imperfect state.
    Within the cracked and dislocated Real
    Nonstoichiometric crystals dominate.
    Stray Atoms sully and precipitate;
    Strange holes, *excitons*, wander loose; because
    Of Dangling Bonds, a chemical Substrate
    Corrodes and catalyzes—surface Flaws
Help Epitaxial Growth to fix adsorptive claws.

White Sunlight, *Newton* saw, is not so pure;
    A Spectrum bared the Rainbow to his view.
    Each Element absorbs its signature:
    Go add a negative Electron to
    Potassium Chloride; it turns deep blue,
    As Chromium incarnadines Sapphire.
    Wavelengths, absorbed, are reemitted through
    Fluorescence, Phosphorescence, and the higher
Intensities that deadly *Laser Beams* require.

Magnetic Atoms, such as Iron, keep
    Unpaired Electrons in their middle shell,
    Each one a spinning Magnet that would leap
    The *Bloch* Walls whereat antiparallel
    Domains converge. Diffuse Material
    Becomes *Magnetic* when another Field
    Aligns domains like Seaweed in a swell.
    How nicely microscopic forces yield,
In Units growing Visible, the World we wield!

THE DANCE OF THE SOLIDS, by John Updike (Copyright © 1968 by John Updike. Reprinted from *Midpoint and Other Poems*, by John Updike, by permission of Alfred A. Knopf, Inc., New York.) [This poem was written after John Updike read the September 1967 issue of *Scientific American*, which was devoted to materials.]

Significant terms defined in this chapter

phase
evaporation
vaporization
condensation
sublimation
boiling point
normal boiling point
superheated
melting point
fusion
molar heat of vaporization
molar heat of fusion
critical point

critical temperature
critical pressure
supercooled
triple point
normal freezing point
surface tension
viscosity
refractive index
crystalline solid
amorphous solid
crystal
hexagonal closest packing
cubic closest packing

crystal coordination number
space lattice
crystal structure
unit cell
primitive cell
isomorphous
polymorphous
diffraction
alloy
substitutional solid solution
interstitial solid solution
intermetallic compound
sintering

Exercises

12.1 Consider the following series of steps in making a cup of instant coffee: (i) an aluminum tea kettle containing water is heated over a stove until it whistles; (ii) the hot water is poured into a cup containing the instant coffee; (iii) two lumps of sugar are added; and (iv) a chip of ice is added to cool the coffee. Excluding the kettle and cup, (a) identify the number of phases after each step. (b) What two phase changes took place for water during the process? (c) What three solid substances were used in the beverage? (d) Which of these were true crystalline solids? (e) What three liquid phases were present during the process?

12.2 Which statements are true? Rewrite any false statement so that is is correct.
(a) Gases are considered to have very little ordering, liquids to have short-range ordering, and solids to have long-range ordering.
(b) All of the postulates of the kinetic-molecular theory of gases are applicable to the liquid and solid states.
(c) The constant vaporization and condensation taking place once equilibrium has been established between a liquid and its vapor is an example of static equilibrium.
(d) A phase change is a constant pressure and constant temperature process.
(e) Covalent bonds connecting atoms are the major interparticle forces in network covalent materials.
(f) The Bragg equation describes the reflection of x rays by crystals.
(g) The space lattice, if repeated in three dimensions, will generate the entire unit cell.
(h) Intermetallic compounds are phases of more or less fixed compositions based on ratios of total number of valence electrons to total number of atoms.

12.3 Distinguish clearly between (a) the boiling point and the normal boiling point, (b) vaporization and evaporation, (c) a superheated and a supercooled liquid, (d) the triple and critical points, (e) a crystalline and an amorphous solid, (f) anisotropic and isotropic solids, (g) isomorphism and polymorphism, and (h) substitutional and interstitial solid solutions.

12.4 Name the phase transformations that are the reverse of (a) vaporization, (b) fusion, and (c) sublimation. Basing your reasoning on the law of Hess, (d) show that at a given temperature $\Delta H^\circ_{sub} = \Delta H^\circ_{fus} + \Delta H^\circ_{vap}$ for a substance.

12.5 A brown paper bag is filled with water and placed directly over the flame of a Bunsen burner. Although a small amount of surface blackening occurs, the bag does not burn while the water boils inside. Why?

12.6 The term "normal melting point" is defined as the melting point at 1 atm. Most chemists are careful to use the words "normal boiling point" but seldom use the term "normal melting point" or really worry that the atmospheric pressure does not equal 760 torr. Why?

12.7 If you pour ether (b. p. 34.5°C) on your hand, you will observe a cold sensation. Explain this effect in terms of the kinetic-molecular theory.

12.8 A half-filled "hot water bottle" is placed under a patient. If the pressure added by the weight of an arm to the outside of the bottle causes the volume of the gaseous phase to decrease by 50%, what is the new partial pressure of water in the gaseous phase? Assume that the original vapor pressure of water was 100 torr at this temperature and that the compression occurs isothermally (at constant temperature).

12.9 The equilibrium water vapor pressure at 17.2°C is 14.7 torr. Describe what happens as a warm, wet air mass having a partial pressure of water at 22.4 torr is cooled to 17.2°C.

12.10 Describe what would happen to a sample of CO_2 originally at point b in Figure 12.4 if it undergoes a sudden isothermal (constant temperature) decrease in pressure. Compare your answer to that for a sample originally at point b being heated at constant pressure.

12.11 The pressure on a sample of CO_2 at point a in Figure 12.4 and on a sample of H_2O at point c in Figure 12.4 is greatly increased under constant temperature conditions. Compare the effect on the samples.

12.12 Describe the physical state of three samples of water originally at point d in Figure 12.4. (a) Describe the changes as the pressure is decreased isothermally for the first sample. (b) Prepare a heating curve for the second sample. (c) Prepare a cooling curve for the third sample.

12.13 Name the five types of crystalline solids. For each type identify the (a) particles that are present in the crystal lattice and (b) interparticle forces between these particles.

12.14 Name the seven crystal systems. For each system discuss the restrictions on (a) unit cell axes and (b) angles between the axes and (c) prepare a sketch of the three-dimensional pattern.

12.15 Name the four classifications of crystal defects. Which of these (a) accounts for the presence of macroscopic areas of impurities within the solid, and (b) accounts for the ease of mechanical deformation of the solid?

12.16 Choose two different unit cells in the two-dimensional lattice shown. Make one as simple as possible and the other orthogonal (containing right angles). Which of these is simpler to use to calculate area, etc.?

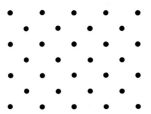

12.17 The relative humidity is defined as the actual vapor pressure of water present in air divided by the equilibrium vapor pressure at that temperature. If the relative humidity in a room 20 × 25 × 8 ft is 68% and the equilibrium vapor pressure is 19.8 torr at 22°C, find the mass of water in the air.

12.18 The apparatus shown in the sketch can be used to determine the vapor pressure of a liquid. A very small amount of liquid is injected and allowed to evaporate. More liquid is added until a small amount of liquid remains and establishes equilibrium with the vapor. The difference between the heights of the liquid in the U-shaped tube h measured in mm is the vapor pressure in torr if the tube contains mercury. At 15°C, the values read at a and b were 1.3 and 5.7 mm and at 32°C, the values were 1.7 and 47.6 mm. Find the vapor pressure at each temperature.

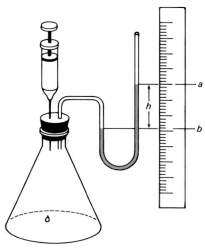

12.19 For many years drinking water has been cooled in hot climates by the evaporation of water from the surface of canvas bags. How many grams of water (heat = 1.00 cal/g °C) can be cooled from 35°C to 20°C by the evaporation of one gram of water (heat of vaporization = 540 cal/g)?

12.20 What is the amount of heat required to (a) melt one gram of ice at 0°C and (b) boil one gram of water at 100°C? See Table 12.1 for molar values of heats of transition.

12.21 A scald from steam is much more severe than a burn from hot water. To illustrate this, calculate the amount of heat released as one gram of steam at 100.0°C condenses and cools to 35°C and compare it to the heat released as one gram of water at 100.0°C undergoes the same cooling. The specific heat of water is 1.00 cal/g °C and the heat of vaporization is 540.0 cal/g.

12.22 Describe what would happen as 100.0 cal of heat was introduced under constant volume conditions to a sample of water at the triple point. Assume that the heat of fusion is 79.7 cal/g, the heat of vaporization is 540.0 cal/g, and that a negligible volume change occurs during fusion.

12.23 Calculate the packing efficiency of a primitive cubic unit cell. Assume the unit cell length to be a and the radius of the atom to be r.

12.24 Polonium crystallizes in a primitive cubic unit cell with $a = 0.336$ nm. What is the radius of a Po atom?

12.25 At 1000°C iron changes from a body-centered cubic unit cell with $a = 0.290$ nm to a face-centered unit cell with $a = 0.363$ nm. Find the radius of an iron atom in each structure.

12.26 The density of a substance depends on the mass and volume of the sample, not on the amount of material in the sample. Thus the "theoretical density" of a substance can be calculated from unit cell data by dividing the mass contained in the unit cell (number of particles multiplied by the mass of one particle) by the volume of the unit cell. Using this technique, find the theoretical density of Na that crystallizes in a body-centered cubic cell with $a = 0.424$ nm.

12.27 Predict the unit cell configurations for NH_4I and MgO, which both crystallize in the cubic system. The ionic radii are 0.065 nm for Mg^{2+}, 0.148 nm for NH_4^+, 0.140 nm for O^{2-}, and 0.216 nm for I^-.

12.28 A unit cell consists of a cube in which there are cations at each corner and anions at the center of each face. (a) Sketch the unit cell. How many (b) cations and (c) anions are present? (d) What is the simplest formula of the compound?

12.29 Three intermetallic compounds called β-, γ-, and ϵ-brass are formed between Cu and Zn. β-Brass has a weight composition of 49.3 wt % Cu, γ-brass is 37.8 wt % Cu, and ϵ-brass is 24.5 wt % Cu. Find the empirical formulas of the brasses.

12.30 A sample of a Mg–Al alloy weighs 0.100 g. Upon treatment with acid, it completely dissolves, liberating 122 ml of H_2 at 27°C and 741 torr. What is the weight percentage of Mg in the alloy?

12.31 The mineral chromite, $(Mg,Fe)Cr_2O_4$, is important for making refractories (substances resistant to high temperatures) and is the only important ore of chromium (a metal of great importance in the metallurgical industry). What is the weight fraction of Cr in pure ferrochromite, $FeCr_2O_4$, and in pure magnesiochromite, $MgCr_2O_4$? If most chromites have about equal numbers of Mg and Fe ions present, what is the average Cr content, expressed in weight percent?

12.32* Divide the heat of vaporization by the normal boiling point expressed in absolute temperature for N_2, O_2, Ar, N_2O, NH_3, H_2O, and D_2O (see Table 12.1). Compute the average of your results. You now have an empirical equation ($\Delta H_V/T$ = constant) that can be used to predict either the boiling point or the heat of vaporization if the other is

known. The heat of vaporization is 7500 cal/mole for chloroform as determined from vapor pressure measurements. Predict the normal boiling point.

12.33* The vapor pressure of a liquid can be determined by slowly bubbling an inert gas through the liquid to be analyzed (making sure equilibrium has been established) and measuring the total pressure of the gaseous mixture P_t; the number of moles of inert gas n_1; and the number of moles of liquid that vaporized n_2. Dalton's law of partial pressures (Section 3.15) gives the vapor pressure of the liquid P_2 as

$$P_2 = P_t \left(\frac{n_2}{n_1 + n_2}\right)$$

Find the vapor pressure of formic acid, HCOOH, at 25°C if the gaseous mixture collected at 745 torr, after bubbling nitrogen through the acid, contained 30.28 g of N_2 and 2.35 g of HCOOH.

12.34* A simple way to measure the surface tension of a liquid or solution is by the capillary rise method. The unknown liquid is allowed to rise in a capillary tube and the weight of its rise is compared to that of a known liquid. The surface tension γ, density d, and capillary rise height h of the two liquids are related by

$$\frac{\gamma_1}{\gamma_2} = \frac{d_1}{d_2}\frac{h_1}{h_2}$$

Ethylene glycol (permanent antifreeze) rises to a height of 2.95 cm and water to a height of 5.02 cm in a glass capillary tube. The densities of the liquids at 20°C are 1.1088 g/cm³ for the antifreeze and 0.998203 g/cm³ for water. The surface tension of water at 20°C is 73.05 dyne/cm. Find the surface tension of ethylene glycol.

12.35* A common technique for measuring the viscosity η of a liquid is to observe the time t required for a sphere to drop through it. If comparison to water is used,

$$\frac{\eta_{liq}}{\eta_{water}} = \frac{(d_{sphere} - d_{liq})t_{liq}}{(d_{sphere} - d_{water})t_{water}}$$

where d is the density and η is the viscosity in centipoises. How much longer would a pearl (density = 3.0 g/cm³) take to fall through a sample of shampoo ($d = 1.03$ g/cm³) having $\eta = 6.23$ centipoise than through water ($d = 1.00$ g/cm³) having $\eta = 0.89$ centipoise?

12.36* Crystalline silicon has the same structure as diamond, with $a = 0.54173$ nm. What is the (a) radius of a silicon atom and (b) packing efficiency of this unit cell?

12.37* Magnesium crystallizes in the hexagonal closest-packed system with $a = 0.3203$ nm and $c = 0.5196$ nm (see Figure 12.19). (a) What is the radius of a Mg atom? (b) How many atoms are in the unit cell? (c) What is the mass contained in the unit cell? The volume of the unit cell is given by $a^2c \sin 60°$. (d) Find the theoretical density of Mg. (e) What is the volume occupied by the atoms in the unit cell? (f) What is the packing efficiency?

12.38* Using quarters, derive a two-dimensional radius ratio rule as follows: (i) place three quarters in a triangle so that each is touching two others and measure the radius of the small hole; (ii) place four quarters so that the centers form a square, again each is touching two others, and measure the radius of the small hole; (iii) place five quarters in the shape of a pentagon and measure the radius of the hole; and (iv) place six coins in the shape of a hexagon and measure the radius of the hole. Divide the measured radii by the radius of a quarter and prepare a table similar to Table 12.5.

12.39* Show that the limiting radius ratio for a tetrahedral arrangement of large spherical anions about a smaller spherical cation is 0.225. Hint: Assume that the tetrahedron is fitted into a cube of edge a with anion spheres of radius r_- at alternate corners contacting each other and the cation sphere of radius r_+ at the center.

12.40** Carbon is polymorphic in that it exists both as diamond (cubic, a = 0.35597 nm) and graphite (hexago-

nal, a = 0.2455 nm and c = 0.669 nm) at room temperature. One of these states is not thermodynamically stable under room conditions although the solid–solid state phase transition takes so long that it appears stable. Using the phase diagram shown, (a) which form is stable at room conditions? (b) How would you change graphite or coal to diamond at 1 atm? The vapor pressure of solid carbon is 8.9×10^{-313} atm at 25°C. (c) What volume of air at 25°C and 1 atm would be necessary to contain one atom of $C(g)$? If the volume of the solar system is 9×10^{38} m³, (d) what are the chances of finding a gaseous carbon atom? (e) What is the radius of a carbon atom as calculated from the diamond structure? The number of atoms in the unit cell of diamond is 8 and of graphite is 4. (f) Calculate the theoretical densities of both substances.

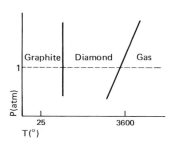

13 WATER AND THE HYDROSPHERE

The first five sections of this chapter present the descriptive chemistry of water—its properties and reactions, and the chemistry of the hydrates. The rest of the chapter is devoted to topics that should be of interest to every informed citizen of the world today: the behavior of water in nature, how man has polluted natural waters, and what can be and is being done to purify water.

From a chemical viewpoint, the most fascinating thing to be learned about water is that it is unique. In many of its properties water differs from similar compounds and has values at the outside end of the normal range. It is just these differences that often make water so ideally suited for its role in the Earth's life support system.

Water is a simple triatomic molecule, but its behavior is very complex. No chemist, or biologist, or physicist would dare to say that water has been thoroughly studied or that how it functions on a molecular scale in either living or nonliving systems is thoroughly understood.

Man is constantly withdrawing water from the hydrosphere, using the water, and then returning the water to the hydrosphere complete with pollutants. The management of water resources on a large scale is a field of growing importance. As population increases, the need for pure water grows simultaneously with the quantities of pollutants that are added to rivers, lakes, and oceans.

The chemistry of water

13.1 The water molecule and its aggregates

In the water molecule two hydrogen atoms are joined to an oxygen atom by covalent bonds, and two lone pairs of electrons remain on the oxygen atom.

$$2H\cdot + \cdot\ddot{O}\cdot \longrightarrow H:\ddot{O}:H$$

The repulsion of the lone pairs for each other and for the electrons in the O–H bonds forces the bonds together (Figure 13.1), and the H–O–H angle is 105°, which is less than the tetrahedral angle of 109.5°.

Oxygen is second in electronegativity to fluorine. Therefore, the covalent bonds in water are highly polar. The combined effects of molecular geometry and bond polarity give the water molecule a fairly large dipole moment (Figure 13.1c). The oxygen atom has a partial negative charge and the hydrogen atoms, partial positive charges, exactly the situation in which hydrogen bonding occurs. Many of the unique properties of liquid water are attributable to hydrogen bonding.

In the gas phase, water molecules are quite separate from each other. However, in liquid water, hydrogen bonding draws groups of water molecules together. Each oxygen atom can form two hydrogen bonds—one through each of the lone pairs. Although much remains to be learned about the structure of liquid water, it is safe to say that aggregates of varying numbers of water molecules held together by hydrogen bonds exist throughout liquid water. Some uncombined water molecules are probably also present (Figure 13.2a). The "flick-

(a) An isolated water molecule

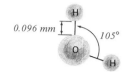

(b) Electron density

Lone pairs

(c) Water as a dipole

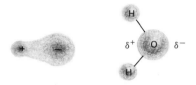

FIGURE 13.1
The water molecule.

FIGURE 13.2
Aggregates of water molecules. *The O . . . H hydrogen bonds in ice are about 0.177 nm.*

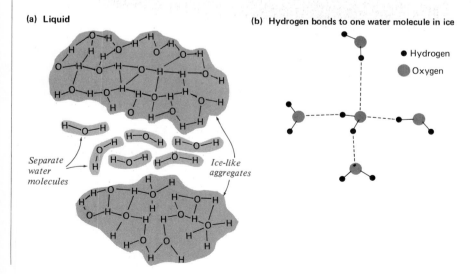

(a) Liquid

Separate water molecules

Ice-like aggregates

(b) Hydrogen bonds to one water molecule in ice

● Hydrogen
● Oxygen

(c) Ice

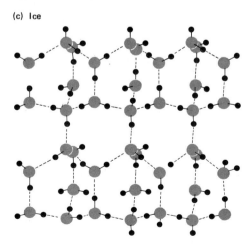

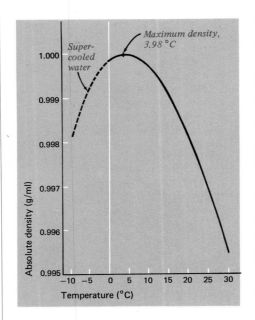

FIGURE 13.3
The effect of temperature on the density of water.

ering cluster" theory of liquid water structure pictures a dynamic equilibrium between aggregates and single molecules, with the aggregates continually forming, collapsing, and reforming, in and out of existence.

In ice, each oxygen atom is hydrogen-bonded to two hydrogen atoms, and each hydrogen atom is covalently bonded to one oxygen atom and hydrogen-bonded to another oxygen atom. The hydrogen bonds, in cooperation with the two ordinary hydrogen-to-oxygen bonds that were already present, give a structure in which each oxygen atom is connected to four other oxygen atoms through hydrogen atoms (Figure 13.2b). These units build up a big, honeycomb-like lattice with large open spaces (Figure 13.2c). When water molecules arrange themselves into this pattern they occupy more space than they did in the disorganized form of liquid water. The result is expansion on freezing, a phenomenon observed when the freezing of water in a bottle breaks the bottle. Another familiar effect of this expansion is that ice floats. Most liquids contract on freezing, so most freezing solids sink in the liquid.

When ice melts, many of the hydrogen bonds are broken, and the honeycomb structure partially collapses. This causes a decrease in volume and a corresponding increase in density. At 0°C the density of ice is 0.9168 g/cm³ and the density of water is 0.9998 g/cm³. A curious phenomenon occurs when air-free water is warmed from 0°C to 3.98°C. The water gradually becomes more dense in this temperature range, reaching a maximum density at 3.98°C (Figure 13.3). Very few liquids become more dense upon heating in any temperature range. The explanation of this phenomenon for water is that the melting of the ice does not break all the hydrogen bonds, and the honeycomb structure only partially collapses. Further heating breaks more hydrogen bonds and causes further collapse. The temperature of maximum density, 3.98°C, is merely the temperature at which contraction, as a result of collapse of the ice structure, is exactly balanced by normal thermal expansion caused by the increased energy of individual molecules. The transformation of solid ice to liquid water and then to gaseous water vapor can be summarized as passing through the following stages

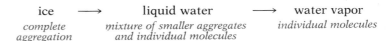

The weathering of rocks and the formation of potholes in streets are hastened by the expansion of freezing water inside cracks. More importantly, since ice stays at the top of a lake, pond, or other body of water, fish can live through the winter at the bottom. If water behaved like most liquids, lakes and streams would freeze solid, beginning at the bottom, and aquatic life would be nonexistent—or very different. The changes in temperature with the seasons in temperate climates and the resulting changes in density of water allow the water in a lake to turn over completely each spring and fall. This leads to the uniform distribution of oxygen and nutrients throughout the water (Figure 13.4).

13.2 Water as a solvent

Water is a remarkably versatile solvent. Although it may stretch our imaginations a bit to conclude, as some have, that every chemical is at least very slightly soluble in water, a great many substances certainly are water soluble in varying degrees.

Polar covalent and ionic compounds are more likely to be soluble in water than nonpolar compounds (Table 13.1). Negative centers in the dissolving substance are attracted by positive centers in the water molecules and vice versa. As a result, each dissolved particle is surrounded by water molecules held by the interparticle electrostatic attractions (Figure 13.5). The number of water molecules is usually not known and can vary with conditions.

Whether polar covalent or ionic, a substance will dissolve only if the attractions between its charge centers and the water molecules—represented by the heats of hydration—are sufficient to overcome the attractions among its own charge centers.

TABLE 13.1

Solubilities of various substances in liquid water
(*See Table 14.1 for other gas solubilities.*)

Substance			Solubility (g/100 g H_2O)	
Name	Formula	Type	20°C	50°C
Silver nitrate	$AgNO_3$	Ionic	222.0	455.0
Aluminum sulfate	$Al_2(SO_4)_3$	Ionic	36.4	52.2
Ammonium nitrate	NH_4NO_3	Ionic	192.0	344.0
Barium sulfate	$BaSO_4$	Ionic	2.5×10^{-4}	3.4×10^{-4}
Calcium acetate	$Ca(C_2H_3O_2)_2$	Ionic	34.7	33.0
Copper(II) sulfate	$CuSO_4$	Ionic	20.7	33.3
Lead(II) chloride	$PbCl_2$	Ionic	0.99	1.70
Potassium chlorate	$KClO_3$	Ionic	7.4	19.3
Sodium chloride	$NaCl$	Ionic	36.0	37.0
Zinc iodide	ZnI_2	Ionic	200.3	273.1
Carbon dioxide[a]	CO_2	Nonpolar covalent	0.169	0.076
Hydrogen[a]	H_2	Nonpolar covalent	0.00016	0.00013
Ethane[a]	C_2H_6	Nonpolar covalent	0.006	0.003
Diethyl ether	$(C_2H_5)_2O$	Polar covalent	7.5	—
Ethyl alcohol	C_2H_5OH	Polar covalent	∞	∞
Ethylene glycol	$C_2H_4(OH)_2$	Polar covalent	∞	∞
Cane sugar	$C_{12}H_{22}O_{11}$	Polar covalent	203.9	260.4

[a] Gas at 760 torr total pressure.

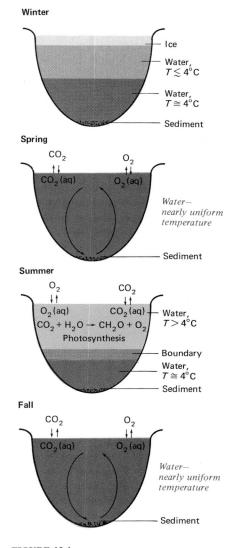

FIGURE 13.4

Lake turnover as a result of changes in water density. *During the spring thaw, surface water sinks as it is warmed and becomes more dense. During the fall freeze, warm surface water sinks as it is cooled and becomes more dense.*

Solution of ionic compound

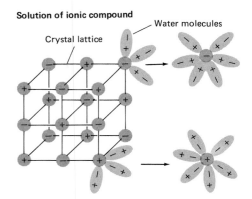

Solution of polar covalent compound

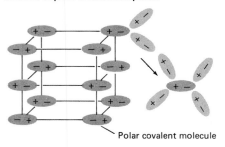

TABLE 13.2

Properties of water

358

FIGURE 13.5

Solution of ions and polar covalent molecules in water. *The polar covalent molecules and the water molecules are represented here as dipoles. Each dissolved particle is surrounded by water molecules.*

13.3 Properties of water

The polarity of water molecules is high, and in the liquid and solid states the molecules are associated by hydrogen bonding. As a result, the values for water's freezing point, boiling point, heat of vaporization, and heat of fusion (Table 13.2) are high relative to those of the hydrides of other Group VI elements (H_2S, H_2Se, H_2Te). The geometry of the water molecules and the effect of hydrogen bonding combined cause ice to have a lower density than liquid water.

Liquid water is very slightly ionized into hydroxide and hydrogen ions ($H_2O \rightleftharpoons H^+ + OH^-$), a property of great importance in acid–base chemistry (Section 4.8; Chapter 17). As is apparent from the high heat of formation of water, the union of hydrogen and oxygen to form water (Section 11.4a) is highly favored thermodynamically. The molecule is very thermally stable, and appreciable decomposition to hydrogen and oxygen occurs only at elevated temperatures (Table 13.3).

Water is ideally and uniquely suited for its role in the ecosphere by many of its properties. Water has a higher specific heat, thermal conductivity, heat of vaporization, and surface tension than almost any other liquid.

Mol wt (amu)	18.0154
Melting point (°C)	0.0
Boiling point (°C)	100.0
Density (g/cm³)	
0°C	0.9168 (ice)
	0.99984 (water)
3.98°C	0.99997
25°C	0.99704
Vapor pressure (torr; 25°C)	23.756
Triple point	0.0100°C; 4.58 torr
Dipole moment (debyes)	1.84
Surface tension (25°C) (dyne/cm)	71.97
Specific heat (25°C), cal/g °C (J/g °C)	0.99828 (4.1768)
Heat of formation (25°C), kcal/mole (kJ/mole)	−68.315 (−285.83)
Heat of fusion (0°C), kcal/mole (kJ/mole)	1.4363 (6.0095)
Heat of vaporization (100°C), kcal/mole (kJ/mole)	9.7171 (40.656)

The surface tension of water, a result of the strong attraction of water molecules for each other through hydrogen bonding, makes it easy for water to rise in the narrow vessels in plant stems and roots by capillary action (Section 12.8). The large surface tension also helps to hold water in the small spaces between soil particles.

Because water is an excellent solvent, it is the ideal medium for transporting the ions and molecules needed by plant and animal metabolism. Ions do not readily recombine in dilute aqueous solutions because the water molecules that surround them cut down the attractive force between them.

The high specific heat means that a body of water, or the body fluids, can absorb or release large quantities of heat while undergoing only small changes in temperature. The high thermal conductivity means that water readily transports heat to and from its surroundings. These properties help to moderate the temperature of the atmosphere and thereby affect the climate. In addition, warm-blooded animals maintain their necessary narrow ranges of body temperature with the aid of both the high specific heat of water and its high heat of vaporization, by which excess heat is expended in evaporating water from the skin. Also, because of the high heat of vaporization of water, less liquid evaporates with the addition of a given amount of heat than for most other liquids. This keeps to a minimum the amount of liquid that a plant or animal needs to stay alive.

13.4 Reactions of water

Water is almost as versatile in its chemical reactions as in its solvent properties. **Hydrolysis** is a general term for any reaction in which the water molecule is split and the hydrogen and oxygen atoms or OH groups from the water are added to the products. ("Hydrolysis" comes from the Greek roots *hydro*, meaning "water," and *lysis*, meaning "to break down.") Very generalized equations for hydrolysis reactions might be written as

$$AB + H_2O \longrightarrow AOH + BH$$
$$AB + H_2O \longrightarrow AO + BH_2$$

However, the actual reactions take many forms, for the reactants and products may be polyatomic molecules or ions.

To interact with water, a substance must be at least slightly polar. Hydrolysis occurs at a polar bond because the positive and negative ends of the bond are attracted to the negative and positive ends of the water dipole. Hydrolysis of inorganic polar covalent compounds is illustrated by the reactions of the nonmetal halides with water

$$PBr_3(l) + 3H_2O(l) \xrightarrow{H_2O} H_3PO_3(aq) + 3HBr(aq)$$

$$SiI_4(s) + 2H_2O(l) \xrightarrow{H_2O} SiO_2(s) + 4HI(aq)$$

Hydration: water molecule added intact
Hydrolysis: water molecule split and parts added

TABLE 13.3
Thermal dissociation of water vapor

$$2H_2O(g) \xrightarrow{\Delta} 2H_2(g) + O_2(g)$$

Temperature (°C)	Dissociation (%)
1027	0.00266
1227	0.0197
1427	0.092
1627	0.302
1927	1.21
2227	3.38
2727	11.1

WATER AND THE
HYDROSPHERE [13.4]

359

TABLE 13.4

Some hydrates

The common names are given in parentheses.

CuSO₄·5H₂O	**Copper sulfate pentahydrate (blue vitriol)**
Na₂SO₄·10H₂O	**Sodium sulfate decahydrate (Glauber's salt)**
KAl(SO₄)₂·12H₂O	**Potassium aluminum sulfate dodecahydrate (alum)**
Na₂B₄O₇·10H₂O	**Sodium tetraborate decahydrate (borax)**
FeSO₄·7H₂O	**Iron(II) sulfate heptahydrate (green vitriol)**
H₂SO₄·H₂O	**Sulfuric acid hydrate (m.p. 8.6°C)**

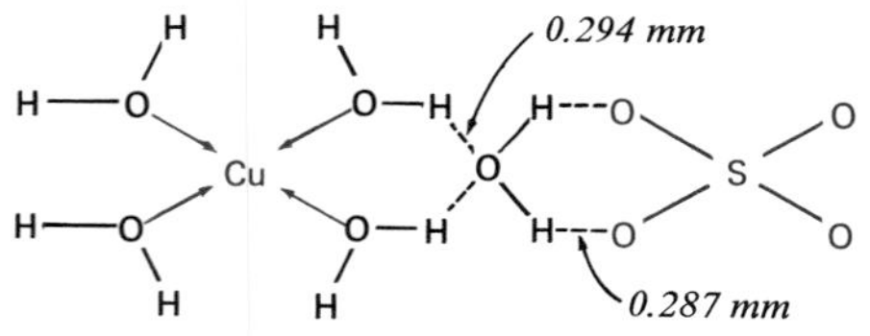

FIGURE 13.6
Copper sulfate pentahydrate, CuSO₄ · 5H₂O. *A sketch showing that four H₂O molecules are coordinated to Cu²⁺ and the fifth is held closer to SO₄²⁻ by hydrogen bonds.*

The hydrolysis of ionic compounds, for example,

$$Al_2S_3(s) + 6H_2O(l) \longrightarrow 2Al(OH)_3(s) + 3H_2S(aq)$$

is discussed in detail in Sections 18.2 and 18.3. Several processes in digestion, such as the digestion of fats and starch, involve the hydrolysis of organic compounds, with the production of smaller and simpler molecules.

Hydration is the interaction of water with other substances without splitting of the water molecule. The products might be just hydrated ions in solution, or stable hydrates, as discussed in the next section.

13.5 Hydrates

Substances that have combined with a definite proportion of water molecules are called **hydrates.** The formulas of hydrates are often written with a centered dot between the water molecules and the compound that has been hydrated, for example,

$$Ba(OH)_2 \cdot 8H_2O$$
barium hydroxide octahydrate

Such a formula does not distinguish among the various ways that water may be incorporated into the crystal.

If copper(II) sulfate is dissolved in water and the solution allowed to evaporate by standing in air, blue crystals are deposited. After careful drying at room temperature, these crystals have the chemical composition CuSO₄ · 5H₂O and remain unchanged in normally moist air. In dry air or when heated, the blue crystals lose water and crumble to a white powder of the composition CuSO₄. Treatment of this compound with water then regenerates the copper sulfate pentahydrate.

Each hydrate listed in Table 13.4 contains a fixed quantity of water and has a definite composition. Each may lose its water of hydration upon heating and may be reformed by reaction of the anhydrous substance with water.

The forces holding the water molecules in hydrates are not extremely strong, as shown by the ease with which water is lost and regained. The water molecules can (1) be held to a central atom by coordinate covalent bonds; (2) be held by hydrogen bonds; or (3) simply occupy either fixed or random positions within the crystal lattice. Not all of the water molecules in a given hydrate need be held in the same way. For example, four of the water molecules in CuSO₄ · 5H₂O are bonded to the dipositive copper ion and one is hydrogen bonded to the dinegative sulfate ion and to two of the other four water molecules (Figure 13.6). A better formulation for this compound might be [Cu(H₂O)₄]SO₄(H₂O).

Since their water molecules are held by weak forces, hydrates often have an appreciable vapor pressure. If the vapor pressure of a hydrate is greater than that of the water vapor in the air, the hydrate will undergo **efflorescence**—loss of water of hydration. It will effloresce until a state of equilibrium has been reached. For example, $Na_2SO_4 \cdot 10H_2O$ has a vapor pressure of about 14 torr at room temperature; since the partial pressure of water vapor at average room temperature and humidity is about 10 torr, $Na_2SO_4 \cdot 10H_2O$ is normally an efflorescent hydrate.

Some compounds are **hygroscopic**—they take up water from the air. Compounds that take up enough water from the air to dissolve in the water they have taken up are called **deliquescent**. For example, calcium chloride ($CaCl_2$) and sodium hydroxide (NaOH) are deliquescent. Deliquescence does not necessarily involve hydrate formation; it occurs when the saturated solution formed has a lower vapor pressure than that of water in the air. Water is often removed from gases or liquids by "drying agents" that are anhydrous salts, such as Na_2SO_4, $CaCl_2$, or $MgSO_4$. Of course, the drying agent should not react with the substance being dried.

The hydrosphere

The Earth's water supply is estimated to be about 1.4×10^9 km³. For comparison, Lake Ontario contains 1.6×10^3 km³ of water, or roughly only one-millionth of the total water on our planet. The oceans plus a few inland saltwater bodies hold 97.3% of the total water, leaving only 2.7% as fresh water. Most of the fresh water is locked up in polar ice and glaciers, lies deep in the ground, or resides in the atmosphere or the soil (Table 13.5). Furthermore, many freshwater lakes and rivers are not close enough to populated areas. This inaccessibility cuts the estimate of the portion of the world's water available for human use to only 0.003% of the total (Figure 13.7).

13.6 The water cycle

Water is constantly evaporating from the surface water of the Earth and from the soil. It is also entering the atmosphere by the process of *transpiration,* which is the release of water from the surface of plant stems and leaves. The average residence time of water in the atmosphere is about ten days before it falls as snow or rain. The residence times in bodies of water and ice are much longer—up to thousands of years. Worldwide, in a given period of time, the amount of evaporation, including transpiration, equals the amount of precipitation, so there is a constant cycling of water through the ecosphere. In the atmosphere water moves with the winds, in the hydrosphere it is moved by ocean currents and rivers, and in the lithosphere it runs

TABLE 13.5
Estimated world water supply

Source	Volume (10³ km³)	Percent of total water
Fresh water		
Polar ice and glaciers	28,200	2.04
Groundwater	8,450	0.61
Lakes	125	0.009
Soil moisture	69	0.005
Atmospheric vapor	13.5	0.001
Rivers	1.5	0.0001
Salt water		
Oceans	1,348,000	97.3
Saline lakes and inland seas	105	0.008
Total supply	**1,385,000**	**100**

If total world water supply = 4 liters

Total fresh water supply = 100 ml

Total fresh water that's not ice = 25 ml

And available fresh water = 1 drop

FIGURE 13.7
Availability of fresh water.

FIGURE 13.8
The water cycle. *In addition to the water movement shown, newly formed water molecules are added to the atmosphere from volcanoes and some hydrogen atoms are lost to outer space.*

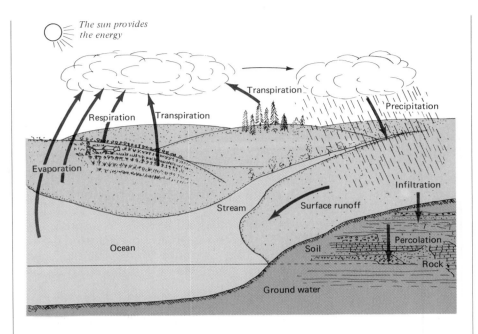

over the land to streams and percolates through the soil to the ground water (Figure 13.8).

The amount of water vapor in the atmosphere and the amount of precipitation vary widely with temperature and the weather patterns. In parts of the world where the precipitation rate is greater than the evaporation rate, water supplies are generally adequate, while in areas where the evaporation rate is greater than the precipitation rate, the residents must struggle to obtain enough water to support life and agriculture.

13.7 Natural waters

Rain, snow, hail—all forms of precipitation—make up the purest natural water. Precipitation carries dust and soluble particles down to Earth, and some gases from the atmosphere (nitrogen, oxygen, carbon dioxide, sulfur dioxide) dissolve in raindrops. Hydrogen (H^+), sulfate (SO_4^{2-}), and hydrogen carbonate (HCO_3^-) ions are formed in raindrops from sulfur dioxide (SO_2) and carbon dioxide (CO_2). When the concentrations are excessive, this process leads to acid rainfall (Section 6.10).

As the water runs over the land, it dissolves soluble ions from the minerals and soluble organic material distributed in the rocks and soil (Table 13.6) in quantities ranging from harmless and almost undetectable to objectionable and harmful. The natural processes of runoff and solution, continued over the ages, have increased the concentrations of dissolved salts in some isolated bodies of water to very high levels. Inland lakes or seas, such as the Great Salt Lake in Utah, Searles Lake in California, and the Dead Sea in Israel have become rich sources of such salts. (The salts in the oceans cannot be attributed only to runoff—many have been there since the time of volcanic activity and the original formation of the oceans.)

Natural water always contains carbon dioxide, which in solution is in equilibrium with hydrogen carbonate ion

$$CO_2(g) + H_2O(l) \rightleftharpoons HCO_3^- + H^+$$

Because this reaction is reversible, it helps to maintain a relatively constant acidity in natural waters (Section 18.9 explains how this works). This acidity is an important factor in the solubility of minerals. Limestone, a common mineral, is not very soluble in pure water, but in the presence of H^+ it dissolves by the reaction

$$\underset{\text{limestone}}{CaCO_3(s)} + H^+ \rightleftharpoons Ca^{2+} + HCO_3^-$$

Calcium ion and hydrogen carbonate ion are usually the most abundant ions in natural fresh water.

Almost all of the cations and some of the anions in fresh water are the result of the solution of minerals in runoff water. Following are a few examples of typical reactions. In these reactions only soluble ions are dissolved, leaving the composition of the mineral changed.

$$\underset{\text{anorthite}}{CaAl_2Si_2O_8(s)} + 3H_2O(l) \longrightarrow Ca^{2+} + 2OH^- + \underset{\text{kaolinite}}{Al_2Si_2O_5(OH)_4(s)}$$

$$\underset{\text{orthoclase}}{3KAlSi_3O_8(s)} + 14H_2O(l) + 2CO_2(g) \longrightarrow$$

$$2K^+ + 2HCO_3^- + 6H_4SiO_4(s) + \underset{\text{muscovite}}{KAl_3Si_3O_{10}(OH)_2(s)}$$

Nitrates and phosphates in natural waters come from the solution of organic waste and decay products, and also from fertilizers. Some sulfate ion is formed by the oxidation of sulfide (S^{2-}) minerals. Chloride presents an intriguing problem. Few minerals contain chloride ions and its source in fresh water is not known. In addition to dissolved molecules and ions, natural fresh waters contain suspended and colloidal particles (colloids are discussed in the next chapter), together with a rich variety of microorganisms (algae, bacteria, viruses, and diatoms).

The oceans are indeed "salty"—sodium and chloride ions are the most abundant ions in seawater, which contains an average of 3.5% dissolved salts. More than 70 elements have been found in seawater and there is little doubt that all of the naturally occurring elements are there. However, more than 99% of the sea salts contain only 10 elements, present as the cations Ca^{2+}, Mg^{2+}, Na^+, K^+, Sr^{2+}, and the anions CO_3^{2-}, HCO_3^-, SO_4^{2-}, Cl^-, and Br^- (Table 13.7). Evaporation of one liter of seawater would give 32.8 g of salts containing 92.4% NaCl and $MgCl_2$.

Only three substances have been recovered commercially

TABLE 13.6

Ions in natural fresh waters
These are the ions most frequently found.

Cations	Anions
Ca^{2+}	HCO_3^-, CO_3^{2-} OH^-
Na^+	SO_4^{2-}
Mg^{2+}	Cl^-
K^+	NO_3^-
Fe^{2+}, Fe^{3+}	F^-
NH_4^+	PO_4^{3-}

TABLE 13.7

Abundances of elements in seawater[a]

		Abundance (mg/liter)
>99.9%	Chlorine	19,000
	Sodium	10,600
	Magnesium	1,300
	Sulfur	900
	Calcium	400
	Potassium	380
	Bromine	65
	Carbon	28
	Oxygen	8
	Strontium	8
	Boron	4.8
	Silicon	3.0
	Fluorine	1.3
	Nitrogen	0.8
	Argon	0.6
	Lithium	0.2
	Rubidium	0.12

All other elements each less than 0.1 mg/liter.

[a] Source: *Chemical and Engineering News*, June 1, 1964.

directly from the oceans. One million gallons of seawater yield $4000 worth of magnesium, $900 worth of sodium chloride, and $200 worth of bromine. It has been estimated that 15 elements occur in seawater in concentrations high enough to make extraction possible. However, at least at present, it is not worthwhile to obtain them in this way. For example, one million gallons of seawater would yield only $3 worth of boron, $2 worth of strontium, and 2¢ worth of silver. (The prices and estimates in this paragraph are from *Water: The Web of Life,* by Cynthia A. Hunt and Robert M. Garrels, Norton, New York, 1972.)

13.8 Nature's water treatment system

As long as the quantities of impurities in water do not become excessive, water purification occurs naturally through the water cycle and the action of microorganisms. Dissolved gases and volatile impurities are flushed out by air mixed with the water as it trickles over shallow stream beds. Solids settle out in quiet pools and lakes and are filtered out as water seeps through the soil. Water that has picked up a high concentration of dissolved ions is diluted when it reaches larger bodies of water.

A host of bacteria and other microorganisms deal with the byproducts of plants and animals, living and dead. The large organic molecules, which contain primarily carbon, hydrogen, nitrogen, oxygen, sulfur, and phosphorus, are broken down by microorganisms to simple harmless molecules and ions. Decomposition of organic matter by bacteria in the presence of oxygen is called **aerobic decomposition.**

aerobic decomposition

$$\text{organic molecules containing C, H, N, O, S, P} \xrightarrow[O_2]{\text{aerobic bacteria}}$$

$$CO_2, H_2O, NO_3^-, SO_4^{2-}, HPO_4^{2-}, H_2PO_4^-$$

As long as enough oxygen is available to decompose all of the organic matter present, the microorganisms can keep a body of water sparkling and clean.

If the oxygen supply is cut down, or if the supply of organic material increases to the point where aerobic decomposition cannot keep up with it, drastic changes take place. Bacteria that depend upon oxygen die, other bacteria switch to oxygen-containing ions such as NO_3^- for their oxygen, and anaerobic bacteria, which require oxygen-free conditions, thrive. Decomposition by bacteria in the absence of oxygen is called **anaerobic decomposition.** The products are as simple as those in aerobic decomposition, but much less pleasant.

anaerobic decomposition

$$\text{organic molecules containing C, H, N, O, S, P} \xrightarrow[\text{no } O_2]{\text{anaerobic bacteria}}$$

$$CH_4, NH_3, NH_4^+, H_2S, HPO_4^{2-}, H_2PO_4^-, \text{ sometimes } PH_3$$

In a body of water that has "gone anaerobic," gas bubbles are visible and the smell of rotten eggs characteristic of hydrogen sulfide (H_2S) is in the air, sometimes combined with the smell of phosphine (PH_3), which is equally noxious. The water appears black and is filled with slime. Fish and other oxygen-consuming residents of ponds or lakes that have gone anaerobic eventually die.

Eutrophication is a natural process in which a lake grows rich in nutrients and subsequently gradually fills with organic sediment and aquatic plants. As a lake ages, the concentration of nutrients such as nitrates and phosphates—substances necessary for the growth of microorganisms—increases. The population of tiny plants and animals increases and changes in character. Huge crops of blue-green algae, called *algal blooms*, occur, during which the water resembles pea soup. Plant and animal matter, living or dead, begins to fill up the lake and the oxygen supply is no longer sufficient. The bottom of the lake may go anaerobic.

The lake becomes shallower as undecomposed sediment builds up, and plant life thrives within the lake and at its boundaries. The end result of eutrophication, a very slow process when unaffected by man's activities, is the transformation of the lake first to a marsh and then to dry land.

Water pollution

13.9 Types of pollutants

Water pollutants can be divided into three major categories: (1) substances that harm humans or animals by causing disease or physical damage; (2) substances or situations that decrease the oxygen content of water, leading to anaerobic decay and the death of aquatic life; and (3) substances that are indirectly harmful, by making water unpleasant to use or destroying the natural beauty and health of lakes, rivers, and oceans (Table 13.8).

The effects of pollutants of the third type are obvious. Nonpoisonous materials are still objectionable if they make water unattractive for drinking, bathing, or cooking, or if they make lakes unattractive for swimming. Oil spilled in the oceans often winds up as balls of black tarry material that wash up on beaches. Soil erosion, a natural process speeded up by human activities, produces most of the sediment that pollutes natural waters. Sediment can make water unfit for industrial uses in which it must pass through turbines and other machinery—the sediment is abrasive and damages the equipment. Also, water high in dissolved salt concentration is corrosive and damages water-handling pipes and equipment.

A few major pollutants of the first two types are discussed in the following sections. The U.S. Federal Water Pollution Control Administration has issued a list of criteria for public water supplies, which gives the amounts of pollutants permissible in reservoirs *before* the water is purified by removing solids and disinfecting with chlorine.

TABLE 13.8

Water pollutants
Examples are given in parentheses.

Oxygen-depleting pollutants
Organic waste (sewage)

Heat

Plant nutrients (fertilizers, detergents)
Toxic or harmful pollutants
Disease-causing agents (viruses, bacteria)

Inorganic chemicals and minerals (poisonous heavy metal ions, acids, bases)

Radioactive materials (nuclear reactor waste)

Man-made organic chemicals (pesticides)
Indirectly harmful substances
Materials that change color, odor, or taste

Sediments

Oil

High concentration of dissolved salts

TABLE 13.9

Water quality criteria for domestic water supplies [a]

Source: "Report of the Committee on Water Quality Criteria." Federal Water Pollution Control Administration, U.S. Department of the Interior, U.S. Government Printing Office, Washington, D.C., 1968.

Quality parameter	Permissible criteria (mg/liter unless otherwise indicated)
Physical	
Color (Co–Pt scale)	**75 units**
Odor	**Virtually absent**
Taste	**Virtually absent**
Turbidity	**Removable by usual water treatment**
Inorganic chemicals	
Ammonia	0.5
Arsenic	0.05
Barium	1.0
Boron	1.0
Cadmium	0.01
Chlorides	250
Chromium (hexavalent)	0.05
Copper	1.0
Dissolved oxygen	>4.0
Fluorides	0.8–1.7
Iron (filtrable)	<0.3
Lead	<0.05
Manganese (filtrable)	<0.05
Nitrates plus nitrites (as mg/liter N)	<10
Selenium	0.01
Silver	0.05
Sulfates	250
Total dissolved solids	500
Uranyl ion	5
Zinc	5

Some of these criteria are presented in Table 13.9, to give you an idea of the variety of objectionable substances that may show up in a water supply.

13.10 Oxygen-demanding pollutants

The outcome of too large a concentration of any oxygen-demanding pollutant is a switch to anaerobic decomposition, with all of its unpleasant side effects of bad smells, unattractive appearance, and dead fish. The major pollutant in this category is human sewage. Organic wastes also come from the pulp and paper industry, tanneries and slaughterhouses, and the chemical industry. Natural microorganisms can also become oxygen-demanding pollutants if they are present in excessive amounts.

TABLE 13.9 (*Continued*)

Quality parameter	Permissible criteria (mg/liter unless otherwise indicated)
Organic chemicals	
Carbon chloroform extract (CCE)	0.15
Methylene blue active substances	0.5
Oil and Grease	Virtually absent
Pesticides	
Aldrin	0.017
Chlordane	0.003
DDT	0.042
Dieldrin	0.017
Endrin	0.001
Heptachlor	0.018
Heptachlor epoxide	0.018
Lindane	0.056
Methoxychlor	0.035
Organic phosphates plus carbamates	0.1
Toxaphene	0.005
Herbicides 2,4-D plus 2,4,5-T, plus 2,4,5-TP	0.1
Radioactivity[a]	
Gross beta	1000 pCi/liter
Radium-226	3 pCi/liter
Strontium-90	10 pCi/liter

[a] Units are picocuries per liter.

In order to assess the effectiveness of wastewater treatment, a method is needed to determine the amount of oxygen-demanding wastes in water. These wastes are complex mixtures, and chemical analysis would be very difficult. However, the amount of oxygen consumed by microorganisms in decomposing organic waste is proportional to the amount of waste present. A method of assessing the organic pollution of water is based on this relationship.

The total amount of oxygen consumed by microorganisms in decomposing waste is called the **biochemical oxygen demand** (BOD) of the water. To test for BOD by waiting for total decomposition of waste would take 20–30 days—too long to be a practical test. Therefore, the standard testing procedure is to measure the BOD_5, or amount of oxygen consumed in 5 days. The procedure is as follows: A sample of the water is saturated with oxygen and incubated for 5 days at 20°C. The remaining oxygen can then be measured and subtracted from the amount of oxygen originally present to get the BOD_5, which is reported in parts per million (equivalent to milligrams per liter).

Water considered "pure" has a BOD_5 of 1 ppm, while untreated municipal sewage has a BOD_5 of 100–400 ppm. For the effluent from industrial or sewage treatment plants, a BOD_5 of less than 20 ppm is considered acceptable.

The growth of microorganisms in water is often kept under natural control by limited amounts of essential nutrients available in the water. Nitrogen and phosphorus are two nutrients that are usually in limited supply in natural ecosystems, such as lakes and streams. Nitrogen and phosphorus have come to notice as pollutants because huge quantities enter the hydrosphere in runoff from fertilized fields, municipal sewage, and industrial effluents. Phosphate-containing detergents (Section 13.12) are the principal offenders in municipal sewage, contributing up to 70% of the phosphorus present. Phosphates have been accused of speeding microorganism growth and thereby greatly accelerating eutrophication of natural waters (Section 13.8).

Heat, or thermal pollution, can contribute to oxygen depletion in three ways: (1) Relatively small increases in temperature will kill certain species of fish, leaving oxygen-demanding wastes to decay; (2) higher temperatures raise the metabolic rate of surviving fish and microorganisms, leading to increased oxygen consumption; and (3) oxygen is less soluble at higher temperatures. Most thermal pollution comes from the discharge of water that has been used for cooling in industrial plants and power plants. Nuclear power plants, in particular, have large quantities of warm water to dispose of. In locations where thermal pollution is undesirable, the obvious solution is to cool the water before releasing it; unfortunately, this is expensive. Warming natural waters need not always be harmful, especially if sufficient oxygen is available for establishment of a different, possibly desirable ecosystem, such as one in which fish can breed.

13.11 Toxic or harmful pollutants

At one time, the only dangerous water pollutants were the bacteria and viruses that cause water-borne diseases such as cholera and typhoid. In the United States, chlorination of municipal water supplies (see Section 20.5a) has largely eliminated such diseases. However, disease-causing agents always abound in sewage and must never be allowed to accumulate in the water supply.

In recent years, one substance after another has come to public attention as a toxic or harmful pollutant. Some of the substances are of relatively recent origin, most prominently the synthetic organic compounds used to kill insects, weeds, or fungi (Figure 13.9). The door to the synthesis and widespread use of such compounds was opened with the discovery of the effectiveness of DDT as an insecticide. DDT was first used during World War II for killing body lice and disease-carrying insects.

During the 1950s and 1960s DDT was sprayed everywhere to kill insects—from backyard mosquitoes to the crop-devastating cotton boll weevil. Several things have now been learned about DDT and

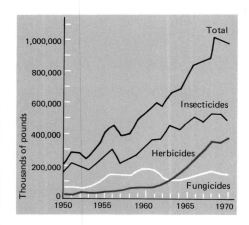

FIGURE 13.9

Production of synthetic pesticides in the United States. (*Source: "Integrated Pest Management," Council on Environmental Quality, Washington, D.C., 1972, p. 3.*)

FIGURE 13.10
Food chain concentration of DDT. *An example of how DDT concentration is increased by each species along a food chain.*

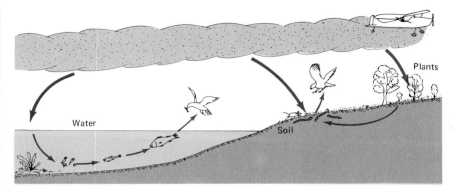

many other pesticides. They are not biodegradable. Substances that are **biodegradable** can be decomposed by natural bacterial decay. Because they are not broken down by natural processes, pesticides persist for a long time. They have become distributed throughout the environment; DDT has even been found in polar ice. Pesticides can be accumulated in high concentrations by passage along a food chain (Figure 13.10). For example, in one study the DDT concentration in lake sediments was 0.0085 ppm, in invertebrates in the area it was 0.40 ppm, in fish that feed on the invertebrates it was 5 ppm, and in the gulls that feed on these fish it was 3200 ppm. At these higher concentrations DDT can damage animal and possibly human life. One well-documented effect of DDT is the thinning of the shells of the eggs of certain birds. The use of DDT has now been severely restricted by the governments of many countries. The insecticides aldrin and dieldrin have been banned because they cause cancer in test animals. Balancing out the possible long-term harmful effects of pesticides against the need for increased food production is a difficult and continuing problem.

Other toxic substances to receive attention in recent years as water pollutants are heavy metal ions, which are not as new to the environment as are the pesticides. However, the heavy metals (see An Aside: Metals as Poisons, Chapter 26) are more hazardous and widespread than once thought. Some of the heavy metals of concern as pollutants are included in Table 13.9.

13.12 The soap and detergent opera

Hydrolysis of natural fats and oils in the presence of a base, usually sodium hydroxide, produces a soap. Long before the chemical structure of the reactants was known, soaps were made by the treatment of animal fats with lye. What we call soap, once used in bar or flake form for all household cleaning and laundry, is a mixture of the salts of straight-chain organic acids. A soap can be considered to have two parts: a long, straight hydrocarbon chain that is oil soluble and the water-soluble sodium salt of an acid. For example, a common soap is sodium stearate,

$$CH_3CH_2CH_2CH_2CH_2CH_2CH_2CH_2CH_2CH_2CH_2CH_2CH_2CH_2CH_2CH_2CH_2COO^-Na^+$$

A molecule of soap or synthetic detergent

Water-soluble anionic end — Oil-soluble end

Action of soap or synthetic detergent on oily dirt on clothing

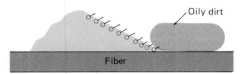
Oily dirt

Fiber

Oily dirt in wash water

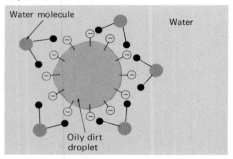
Water molecule

Water

Oily dirt droplet

FIGURE 13.11
The action of soap and synthetic detergents on oily and greasy dirt.

In the washing process, the soap molecules become oriented on the surface of a small droplet of oily or greasy dirt (Figure 13.11). The water-soluble groups on the surface of the droplets keep the droplets suspended (emulsified) in the water, and rinsing carries the droplets away.

The great disadvantage of soap is that in hard water (Section 13.13) the sodium ion is replaced by magnesium or calcium ions from the water to give insoluble compounds. They form bathtub scum and a dull film on laundry washed with soap in hard water.

Synthetic detergents, often simply called detergents, are synthetic cleaning agents (not made from animal fats) used as substitutes for soap. Most laundry detergents contain a surface-active agent that, like a soap, consists of a long hydrocarbon chain with a water-soluble group at the end of the chain, for example,

$$CH_3CH_2CH_2CH_2CH_2CH_2CH_2CH_2CH_2CH_2CH_2CH_2 - \bigcirc - SO_3^-Na^+$$

a straight-chain sodium alkylbenzenesulfonate

Most detergents also include a "builder," a compound that, in ways not completely understood, aids in softening the water and in emulsifying dirt and grease.

Detergents were introduced in 1945 and quickly became very popular. Their calcium and magnesium salts are soluble in hard water, and they are more effective cleaning agents than soap. In the first generation of detergents, the hydrocarbon side chains were not straight, as they are in soap and in the sodium alkylbenzenesulfonate written above, but were branched, for example,

$$CH_3CHCH_2CHCH_2CHCH_2CH - \bigcirc - SO_3^-Na^+$$
$$\quad | \qquad | \qquad | \qquad |$$
$$\quad CH_3 \quad CH_3 \quad CH_3 \quad CH_3$$

a branched-chain sodium alkylbenzenesulfonate

It was soon discovered that these compounds were not biodegradable. The bacteria in natural bodies of water and in sewage treatment plants rejected a branched-chain diet. The detergents remained effective, and sudsy water was found in streams and even ran out of faucets. This problem was solved by a second generation of detergents with straight hydrocarbon chain surface-active agents, a more acceptable diet to bacteria.

Another environmental problem of detergents, however, appeared in the 1970s. The builders in most detergents were polyphosphates, mainly sodium tripolyphosphate, $Na_5P_3O_{10}$, which is hydrolyzed to phosphates,

$$P_3O_{10}^{5-} + 2H_2O(l) \longrightarrow 2HPO_4^{2-} + H_2PO_4^-$$

As we have learned, phosphates are plant nutrients that have been accused of speeding up the natural process of eutrophication.

The phosphate problem has not been fully resolved. In fact, whether or not phosphates are truly harmful in natural waters is still a matter of debate. Some communities have banned phosphate detergents, but others have not. Somewhat less effective builders, for example, washing soda, $Na_2CO_3 \cdot 10H_2O$, or borax, $Na_2B_4O_7 \cdot 10H_2O$, have been substituted, but these also have drawbacks. They make the washwater strongly alkaline, which can harm the skin and eyes and which weakens the fibers of clothing. In addition, borax in sizeable concentrations is probably toxic to both plant and animal life. Another solution to the problem is to retain phosphate detergents and provide tertiary sewage treatment (Section 13.14) to remove phosphates from wastewater.

Water purification

13.13 Hard water

What we call "hard water" is recognizable by the grayish white scum it forms with soap, the dull appearance of clothes washed in it using soap, the hard scale it forms in boilers and tea kettles, and, sometimes, by an unpleasant taste. **Hard water** contains dipositive ions that form precipitates with soap or upon boiling. The principal metal ions that cause hardness are Ca^{2+}, Mg^{2+}, and Fe^{2+}. The negative ions in hard water are most often HCO_3^- and SO_4^{2-}. The reaction between the cations in hard water and the long-chain hydrocarbon salts in soaps yields the scum that forms in the washing machine or in the bathtub ring by the following reaction, where M^{2+} is Fe^{2+}, Mg^{2+}, or Ca^{2+}

$$M^{2+} + 2Na^+(C_{17}H_{35}COO^-) \longrightarrow 2Na^+ + M^{2+}(C_{17}H_{35}COO^-)_2(s)$$

Since the organic anion is largely responsible for the cleansing action of soap (Figure 13.11), its removal from solution destroys the effectiveness of the soap.

There are two types of water hardness. **Temporary**, or **carbonate, hardness,** is caused by the presence of HCO_3^- ions in hard water and can be removed by boiling the water. Boiling drives off carbon dioxide and leaves behind insoluble carbonates, called *boiler scale*, on boiler and tea kettle walls

$$M^{2+} + 2HCO_3^- \longrightarrow MCO_3(s) + CO_2(g) + H_2O(g)$$

The deposit is a poor conductor of heat and the efficiency of a boiler is decreased. Removing hardness by boiling is expensive and not practical for the large quantities of water necessary for industrial uses. **Permanent**, or **noncarbonate, hardness** is caused by the presence of sulfate ions or other ions in hard water, which do not precipitate upon boiling.

Water softening, the removal of the ions that cause hardness in water, can be accomplished in several ways for either type of hard water. Chemicals may be added to precipitate the dipositive ions. This was accomplished in home laundries for many years by adding washing soda ($Na_2CO_3 \cdot 10H_2O$) to the water before putting in the soap. Calcium ions are removed by the reaction

$$Ca^{2+} + CO_3{}^{2-} \longrightarrow CaCO_3(s)$$

The ions may also be tied up in stable ions that do not react with soap. Apparently one function of the builders in detergents is to form ions such as $[CaP_3O_{10}]^{3-}$.

Another method for softening water involves the replacement of the objectionable cations by sodium ion, which does not give a precipitate with soap or form scales. The process of replacement of one ion by another is known as **ion exchange,** and naturally occurring zeolites (Section 27.14) or similar synthetic materials are used. The zeolites, which are complex sodium aluminum silicates, possess a three-dimensional open structure, with negatively charged alumino-silicate frameworks, and sodium ions in the openings of the frameworks. When hard water is passed through a column of zeolite, the Ca^{2+}, Mg^{2+}, and Fe^{2+} ions present in the water exchange with the sodium ions and are thus removed from the liquid phase.

$$M^{2+} + 2NaZ(s) \longrightarrow 2Na^+ + M(Z)_2(s) \qquad (Z = \text{anion of zeolite})$$

When its exchange capacity is depleted, the zeolite can be regenerated by running concentrated sodium chloride solution through the column. The reverse of the reaction shown above then occurs and the column is ready for further use.

Synthetic organic materials known as ion-exchange resins have proved to be even better water softeners than the zeolites. The resins have a hydrocarbon framework of high molecular weight to which are chemically bound either negatively charged groups, for example, $SO_3{}^-$ or positively charged groups, for example, $NR_3{}^+$. The former are counterbalanced by positive ions, usually Na^+ or H^+, and the latter by anions, for example, OH^-. It is possible to remove virtually all the ions from water by passing it successively through a cation-exchange resin containing H^+ as the positive ion and then through an anion-exchange resin having OH^- as the negative ion. The H^+ replaced from one resin by cation impurities in the water combines with the OH^- replaced from the other by anion impurities. Water with almost all ions removed (except, of course, the H^+ and OH^- ions normally present from the H_2O ionization equilibrium) is called **deionized water.**

Ion-exchange resins or zeolites are used in home water-softening units and also industrially. People on restricted diets should be aware that water softened in such units has a high concentration of sodium ions.

A water-softening method sometimes used in municipal water treatment removes both carbonate and noncarbonate hardness. Called the *lime–soda process*, it relies on the addition of hydrated lime and soda, and removal of insoluble calcium, magnesium, and carbonate salts by reactions that can be summarized

$$Ca^{2+} + 2HCO_3^- + Ca(OH)_2(s) \longrightarrow 2CaCO_3(s) + 2H_2O(l)$$

hydrated lime

$$Mg^{2+} + Ca(OH)_2(s) + Na_2CO_3(s) \longrightarrow Mg(OH)_2(s) + CaCO_3(s) + 2Na^+$$

soda

The equations make this look deceptively simple. At least four separate operations are involved. At the end, the water is not completely free of hardness, for some $CaCO_3$ and $Mg(OH)_2$ remain in solution. The process is used only where extremely hard water must be purified.

13.14 Sewage treatment

Municipal sewage is what drains from our sinks, washing machines, bathtubs, and toilets, combined with industrial wastewater and, in some places, with the water that runs into storm sewers. Surprisingly, average domestic sewage is 99.94% pure water and only 0.06% dissolved and suspended solids. Taking out as much of that 0.06% as possible before returning the water to the hydrosphere is increasingly important as the population increases.

The first step in primary sewage treatment (Figure 13.12) is to filter out obvious debris such as pieces of paper and wood, and then to get rid of sand, cinders, and gravel (collectively called "grit") that might clog or damage pipes and machinery further along in the treatment plant. Next organic and inorganic solids are allowed to settle out in large sedimentation tanks. Sometimes chemicals are added to speed up settling (Section 26.15c) and sometimes provision is made for further purification of the sludge (the solids that settle out of sewage) by sludge digestion, which utilizes anaerobic decomposition. At the end of primary treatment 40%–60% of the suspended solids and 25%–35% of the oxygen-demanding wastes have been removed.

FIGURE 13.12

Primary sewage treatment. *The sludge is partially digested and the water partly removed before it is disposed of.*

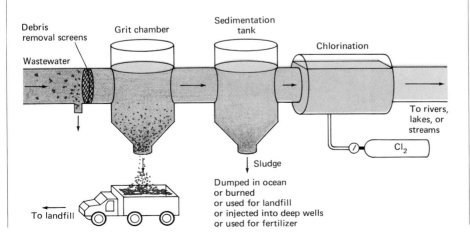

Debris removal screens
Grit chamber
Sedimentation tank
Chlorination
Wastewater
To rivers, lakes, or streams
Cl_2
To landfill
Sludge
Dumped in ocean or burned or used for landfill or injected into deep wells or used for fertilizer

FIGURE 13.13
Secondary sewage treatment.

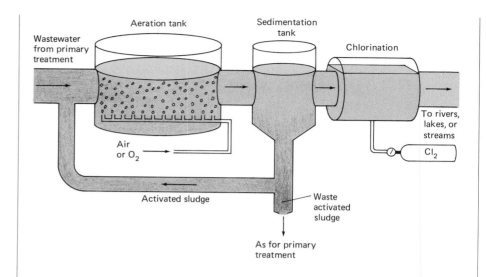

Secondary sewage treatment (Figure 13.13) goes on to remove up to 90% of the oxygen-demanding wastes. Two methods are used, both of which expose the sewage to a vigorous population of aerobic bacteria and a plentiful supply of oxygen. In the trickling filter method, which doesn't really filter anything, the water runs over a bed of stones 5–10 cm in diameter. Aerobic bacteria and other microorganisms attach themselves to the stones and take organic matter out of the water as it trickles past, while oxygen from the air mixes into the running water. The other method relies on vigorously mixing air (or oxygen) with the sewage and adding activated sludge—sludge from previous batches of sewage that has developed a high population of microorganisms. This is necessary because one batch of sewage cannot produce enough microorganisms during a short stay in the plant to purify itself sufficiently. The effluent from secondary sewage treatment is usually chlorinated to be sure that all disease-causing organisms have been destroyed.

Primary plus secondary sewage treatment does a good job of removing solids, oxygen-demanding wastes, and disease-causing organisms. Dissolved ions, including heavy metals and plant nutrients still remain, as do organic compounds, such as pesticides, that are nonbiodegradable, and also radioactive materials.

Sewage treatment designed to remove specific pollutants that remain after secondary treatment is called tertiary sewage treatment, or *advanced wastewater treatment*. There are about 21,000 sewage treatment plants in the United States. Communities that once discharged untreated sewage into bodies of water are now being forced by Federal government regulations to build treatment plants. Also, all existing treatment plants must, if necessary, upgrade their treatment to meet Federal water pollution standards. When all plants are in conformance with the government requirements, 58% of the plants will have secondary treatment and 43% will have tertiary treatment. Primary treatment alone is no longer sufficient. The major problems to be solved by tertiary treatment are the removal of excessive oxygen-demanding wastes, and of nitrogen- and phosphorus-containing materials.

13.15 Desalination

The removal of ions, especially Na$^+$ and Cl$^-$, from water is called **desalination.** Desalination is necessary to obtain very pure water from ordinary tap water, to obtain drinking water from seawater or brackish water (less salty than seawater), and to remove excessively high concentrations of ions in purifying industrial wastewater or in tertiary sewage treatment.

Desalination methods utilize one of several different principles. In distillation processes, the salty water is boiled and the pure water vapor condensed. Distillation has long been used to make small quantities of water of high purity in laboratories. It is also the basis for the majority of the large-scale seawater desalination plants. To get the maximum efficiency out of the energy input needed to vaporize the water, the vapor is passed back over the incoming seawater and the heat of condensation given up by the vapor aids in heating the seawater (Figure 13.14).

Desalination by crystallization also utilizes a phase change. Salty water is partially frozen to a slurry, then the ice crystals are separated and melted. Seawater desalination by crystallization is still in the experimental stage.

Several processes utilize membranes that in essence filter out the ions. Reverse osmosis is just what it says—when a pressure greater than the osmotic pressure is applied (Section 14.19) on the solution side of a semipermeable membrane, the water molecules diffuse through the membrane to the fresh water side. This is the opposite of the direction in which water molecules move in osmosis. Reverse os-

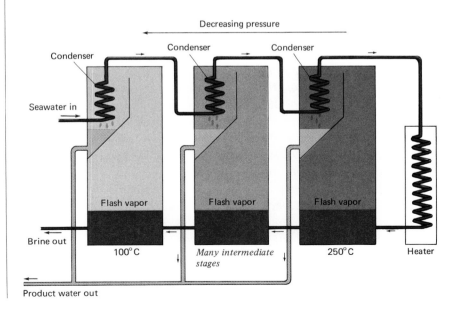

Decreasing pressure

Condenser

Condenser

Condenser

Seawater in

Flash vapor

Flash vapor

Flash vapor

Brine out

100°C

Many intermediate stages

250°C

Heater

Product water out

FIGURE 13.14

Seawater desalination by multistage flash distillation. *This is the process most widely used for seawater desalination. A plant at Key West, Florida, produces 2.6 million gallons/day by this method. Hot seawater enters a chamber where the pressure is low enough to cause some water to vaporize instantly —to "flash" into steam. The temperature of the remaining seawater drops as it moves over to the next chamber, where low pressure causes more water to flash distill. The Key West plant takes water from 250°F to 100°F in 50 stages.*

FIGURE 13.15

An electrodialysis cell. *Anions move to the right and cations to the left under the influence of a strong electric field. By alternating anion- and cation-permeable membranes, ions collect in some subcells, while all ions leave other sub-cells.*

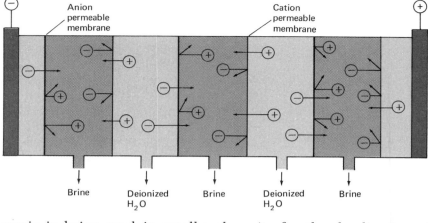

mosis is being used in small-scale units for the desalination of brackish water. The process will retain solutes of molecular weight below 500. (It does not work well with seawater; the ion concentration is too high.)

In <u>electrodialysis</u> (Figure 13.15), ion-selective membranes—membranes that let only positive or only negative ions pass through—are combined with an electrochemical cell. A few seawater desalination plants use electrodialysis. It can also be used in tertiary sewage treatment.

THOUGHTS ON CHEMISTRY

Second-hand molecules

SIZING UP SCIENCE, *by R. Houwink*

Inspector Veryfit of the Public Health Department was having a drink with his old friend Dr. Pure, who was in charge of the Waterworks. Veryfit was worried. 'I have an uncomfortable feeling,' he said, 'that sometimes I am drinking water that has been drunk before.' Dr. Pure consulted his note-book and replied that things looked very bad indeed; the total number of people ever to have lived he estimated at a hundred thousand million (10^{11}), and their average lifespan was only 25 years (—the infant mortality rate was high until very recently). If all these people excreted two litres of water every day of their lives, roughly one molecule in every million has been excreted by someone at some time. 'Assuming,' Dr. Pure added, 'that all the water on the Earth is thoroughly mixed up.'

'So the number of "dirty" molecules in your glass of beer, my dear Veryfit, is in the region of 15 million million million—far more than you can imagine. To give you some idea of the numbers involved, think of the cavity in a hollow tooth: it could contain 10^{20} molecules, which means more than 10^{14} "dirty" ones—200,000 molecules each for everyone now living.

'Of course,' he went on, 'some of these molecules have been ex-

creted more than once, so there are "double-dirty" ones; and on top of that, the "dirty" water is not distributed evenly over the Earth and it is likely to be much dirtier in densely-populated areas like this.'

The Inspector poured his beer down the sink, calculating that if there was one 'dirty' molecule per million there would be enough in his pint to provide every living person with a hundred million of them! A young scientist had been listening to this conversation from behind his newspaper, but this waste of beer prompted him to point out to Inspector Veryfit that the concentration of dirty molecules was really very low—the same, for instance, as the amount of fluoride added to the water to prevent tooth decay. 'In the air you breathe,' he continued, 'about one molecule in every 500 has been exhaled by someone else: this is 2000 times the concentration of the dirty molecules in the Inspector's beer.

'And in every breath you inhale about half a litre of air, or 10^{22} molecules, so about 300 million of these molecules have previously been exhaled by some particular person who lived to the age of 40 — Socrates, Shakespeare or Einstein, for example. Isn't it rather wonderful to think that we are breathing the same air as these people?'

'Another round!' said Inspector Veryfit, encouraged by these inspiring words, 'but nothing for Dr. Pure, who has made no attempt to size up his knowledge!'

SIZING UP SCIENCE, by R. Houwink (p. 135, John Murray (Publishers) Ltd., London, 1975)

Significant terms defined in this chapter

hydrolysis
hydration
hydrates
efflorescence
hygroscopic
deliquescent
aerobic decomposition
anaerobic decomposition
eutrophication
biochemical oxygen demand

biodegradable
synthetic detergents
hard water
temporary hardness
permanent hardness
water softening
ion exchange
deionized water
desalination

Exercises

13.1 Prepare a list of the numerous ways that you came in contact with water yesterday.

13.2 Distinguish clearly between (a) hydrated and anhydrous salts, (b) hygroscopic and deliquescent substances, (c) aerobic and anaerobic decomposition, (d) soaps and detergents, (e) hard and soft water, (f) temporary (or carbonate) hardness and permanent (or noncarbonate) hardness, and (g) primary and secondary sewage treatment.

13.3 Which statements are true? Rewrite any false statement so that it is correct.
(a) The chemical formula for a hydrate represents the average number of water molecules present.
(b) A hydrate which takes up enough water from the air to dissolve is undergoing efflorescence.
(c) Molecules of water that are chemically bonded to a hydrated salt can be held by different types of bonding.
(d) The biochemical oxygen demand is the total amount of oxygen consumed by bacteria in decaying organic waste.
(e) Transpiration is the release of water by the roots and stems of plants and from the surface of leaves.
(f) The decomposition of organic matter by bacteria in the absence of oxygen is known as aerobic decomposition.
(g) Hard water contains dipositive ions that form precipitates with soap or upon boiling.
(h) All ions can be removed from water if it is doubly distilled and then passed through a reverse osmosis chamber.
(i) Primary sewage treatment is essentially a process to eliminate specific pollutants by a series of chemical reactions.

13.4 Prepare a series of diagrams showing how ionic and polar substances dissolve in water.

13.5 Write chemical equations representing the (a) thermal dissociation of steam at high temperatures into atomic hydrogen and hydroxy radicals, (b) ionization of water at normal temperatures into hydrogen and hydroxide ions, (c) hydrolysis of $AlCl_3$, (d) reaction between a metal such as Na that is above H in the activity series and water, (e) reaction with water—if any—between a metal, such as Cu, that is below H in the activity series, (f) reaction between an acid anhydride like $P_4O_{10}(s)$ and water, (g) reaction between a basic anhydride like $CaO(s)$ and water, and (h) reaction between $CaCl_2 \cdot H_2O(s)$ and water to form $CaCl_2 \cdot 6H_2O(s)$.

13.6 What are the three ways that water molecules can be bonded to an anhydrous salt to produce a hydrate? Prepare sketches showing these ways.

13.7 Describe how "nature's water treatment system" functions for purifying water containing (a) dissolved gases and other volatile impurities, (b) solids, (c) dissolved ions, and (d) plant and animal by-products in the presence and in the absence of oxygen.

13.8 Name the three major categories of water pollutants. Give a couple of examples of each and possible ways to remove these pollutants.

13.9 Prepare a series of sketches showing the action of soap or detergent on a small droplet of greasy or oily dirt.

13.10 What are the disadvantages in using hard water for (a) cleansing and (b) producing steam? What ions are commonly found in (c) temporarily and (d) permanently hard water? (e) Write chemical equations showing the water-softening method using slaked (hydrated) lime and washing soda to remove Ca^{2+} and Mg^{2+}.

13.11 Describe the processes that occur in a municipal sewage treatment plant that performs all three treatment steps. Identify the types of materials that are removed in each step and possible uses for the recovered materials.

13.12 The heat of combustion of a typical coal is 7.5 kcal/g. Utilizing this heat, calculate the mass of steam at 125°C that can be prepared from water at 25°C for each gram of coal burned. Assume the heat of vaporization of water to be 540. cal/g, the specific heat of water to be 1.00 cal/g °C, and the specific heat of steam to be 0.50 cal/g °C.

13.13 Calculate the wt % of water in (a) gypsum, $CaSO_4 \cdot 2H_2O$; (b) epsomite, $MgSO_4 \cdot 7H_2O$; (c) turquoise, $CuAl_6(PO_4)_4(OH)_8 \cdot 4H_2O$; (d) natrolite, $Na_2Al_2Si_3O_{10} \cdot 2H_2O$; and (e) sucrose—recognizing that water molecules as such are not present in the molecule, $C_{12}H_{22}O_{11}$.

13.14 A sample of a pale green hydrated salt was dissolved in water and standard qualitative analysis tests showed it to contain Fe^{2+} and Cl^-. A 4.04-g sample of the hydrate was heated in a stream of HCl, which produced an anhydrous salt weighing 2.18 g. What is the complete formula of the hydrate?

13.15 The vapor pressure of water at 25°C over a sample of $CuSO_4 \cdot 5H_2O(s)$ is 7.8 torr. This vapor pressure is the result of the reaction

$$CuSO_4 \cdot 5H_2O(s) \longrightarrow CuSO_4 \cdot 3H_2O(s) + 2H_2O(g)$$

What is the minimum amount of $CuSO_4 \cdot 5H_2O$ needed to produce a partial pressure of 7.8 torr in a 1.00-liter container of dry air?

13.16 Table 13.7 tells us that there is 19,000 mg of chlorine in a liter of seawater. Assuming this to be in the form of Cl^-, express this concentration in molarity.

13.17 A typical analysis of water from the Mississippi River showed that it contains 34 mg Ca^{2+}/liter and 7.6 mg Mg^{2+}/liter. What amounts of $Ca(OH)_2$ and Na_2CO_3 are needed to remove these ions from one liter of river water using the soda–lime process?

13.18* The crystal structure of ordinary ice is hexagonal with a = 0.4535 nm and c = 0.741 nm. The number of molecules in the unit cell is 4. Using these data, calculate the theoretical density of ice. Hint: the volume of the unit cell is $a^2c \sin 60°$.

13.19* What will be the final temperature of the mixture prepared from 100.0 g of ice at 0°C and 100.0 g of steam at 100°C? For water, the heat of vaporization is 540. cal/g the heat of fusion is 79.7 cal/g, and the specific heat is 1.00 cal/g °C.

13.20* The scum-producing reaction between soap and Ca^{2+} represented by the equation

$$Ca^{2+} + 2C_{17}H_{35}COONa \longrightarrow (C_{17}H_{35}COO)_2Ca + 2Na^+$$

sodium stearate   calcium stearate
a typical soap   an insoluble scum

can be used to quantitatively determine water hardness by slowly adding a standardized soap solution to a water sample until a lasting sudsing effect is observed. A soap so-lution was standardized by treating it with 20.0 ml of a solution that contained 495 ppm of Ca^{2+} (1 ppm = 10^{-6}g Ca^{2+}/ml of hard water). Express the concentration of the soap solution in terms of grams of Ca^{2+}/ml soap solution if 35.2 ml of the soap solution were required for the reaction. A 25.0 ml sample of well water required 3.0 ml of the soap solution before a lasting suds was formed. What is the concentration of Ca^{2+} in the well water (expressed in ppm)?

13.21* The enthalpy of formation at 25°C of $D_2O(l)$ is -70.411 kcal/mole and of $D_2O(g)$ is -59.560 kcal/mole. (a) Calculate the heat of vaporization of D_2O. The corresponding values for H_2O are -68.315 and -57.796 kcal/mole. (b) Calculate the heat of vaporization for H_2O. (c) Do these values of the heat of vaporization imply that strong hydrogen bonding is present in D_2O? (d) Which substance has the higher value and why? (e) Divide the normal boiling point of H_2O (expressed in absolute temperature) by the enthalpy of vaporization and multiply the result by the enthalpy of vaporization for D_2O to obtain an estimate for the normal boiling point of D_2O.

The bond dissociation energy at 25°C of $H_2(g)$ is 104.190 kcal/mole, that of $D_2(g)$ is 105.962 kcal/mole, and that of $O_2(g)$ is 119.106 kcal/mole. Find the (f) D–O and (g) H–O bond energies. (h) Which bond is stronger? (i) What is the ratio of the rate of effusion of D_2O to H_2O in the gaseous state? The differences in bond energies and rates of movement allow the separation of heavy from regular water.

14 SOLUTIONS AND COLLOIDS

In this chapter we first give a qualitative description of the types of solutions—gases in gases, gases in liquids, and so on—and of the properties of ideal vs. nonideal, and electrolyte vs. nonelectrolyte solutions. Next we introduce two more ways of expressing solution concentrations (mole fraction and molality), illustrate dilution problems, and summarize the types of standard solutions. The next sections treat the vapor-pressure related properties of liquid solutions quantitatively. Finally, we discuss the properties and types of colloids.

*E*nclosed within your skin is a complex biological–physical–chemical system. Molecules are organized into nerves, muscles, bones, cartilage, and connective tissue. Maintaining all of these structures are the body fluids–blood, lymph, tissue fluids, intestinal fluids, urine–each a complex mixture of molecules and ions carried in an aqueous medium.

When substances are mixed with liquids, there are several possible results. At one extreme is a suspension—a heterogeneous mixture in which the particles are usually large enough to settle out by gravity. At the other extreme is a true solution—a homogeneous mixture in which no individual particles are detectable. Intermediate is a colloidal suspension—a heterogeneous mixture, but containing molecules or ions united in aggregates so small that they stay suspended.

Protoplasm—the living matter of all cells—is both a true solution of inorganic salts, amino acids, and simple sugars and a colloidal suspension of large protein molecules and fat globules. The behavior of protoplasm and all body fluids is governed by the basic physical and chemical principles that apply to all solutions and colloids.

Osmosis is the transport of the molecules of a solvent through a membrane. It is a process of vital importance to the distribution of nutrients and wastes in the body, and its action depends upon the concentration of solutes in the solutions on either side of the membrane.

Medicine to be administered intravenously must be dissolved in physiological saline solution, an aqueous solution of the same ionic concentration as blood, called an "isotonic" solution. If the salt concentration is not equal to that of blood, variations occur in the osmotic pressure of the solution. Blood cells bathed in such solutions *can literally explode if the salt concentration of the solution is less than that of the blood cells, or collapse if the salt concentration of the solution is greater than that of the blood cells.*

Curiously, the concentrations of ions dissolved in seawater are similar to those found in the body fluids of both marine and terrestrial animals. The Na^+ and Cl^- ions are most abundant, and K^+, Ca^{2+}, and Mg^{2+} are always present, although in varying amounts.

Types of solutions

14.1 Some definitions

If one stirs a spoonful of clean sand into a glassful of water, the sand becomes wet, but is not further changed. No matter how long the stirring is continued, the individual grains of sand retain their identity. On the other hand, if a spoonful of table salt is treated in the same way, the crystals of salt disappear into the water, and with adequate stirring the salt becomes uniformly dispersed throughout the water. We say that the salt has *dissolved* in the water, and that a *solution* has been formed. Salt is *soluble* in water, but sand is *insoluble*. The salt is the *solute* and the water is the *solvent* (Section 2.11).

As we have pointed out, the word "solution" is commonly used to refer to a solution that is liquid. However, any mixture with only a single phase is a solution. A solution may be in the liquid phase, the gas phase, or the solid phase.

If more and more salt is added to water, the point is eventually reached where no more salt will dissolve; the water can dissolve only a limited quantity of salt. **Solubility** is a measure of the amount of solute that can dissolve in a given amount of solvent. It depends upon the nature of the solute and the solvent, upon the temperature, in some cases upon the pressure, upon the presence of other solutes, and upon a variety of additional factors. There is no limit to the amount of alcohol that will dissolve in a given amount of water. Liquids that are completely soluble in each other are called **infinitely miscible.** Liquids that are mutually insoluble or very nearly so (the more usual case) are called **immiscible.** Liquids that have limited solubility in each other are **partially miscible.**

Returning to the solution of salt in water—if enough salt is added, no more seems to dissolve. In reality, it is still dissolving, but at the same time, an equal amount of salt is coming out of solution and crystallizing on the surface of the solid salt or on the sides of the container. A condition of dynamic equilibrium has been reached— the processes of dissolving and crystallizing are proceeding at the same rate, so that each exactly offsets the other (Figure 14.1). This can be shown by placing a broken crystal of salt in the solution. As ions from this broken crystal go into solution, ions from the solution precipitate on the crystal. These ions do not precipitate uniformly on the surface of the broken crystal, but on those areas which will give a more symmetrical crystal. Over a period of several days, the broken

FIGURE 14.1
Crystals in a saturated solution. *The mass of the crystals remains the same as long as the temperature is unchanged. Molecules or ions join and leave the crystal surface in a dynamic equilibrium.*

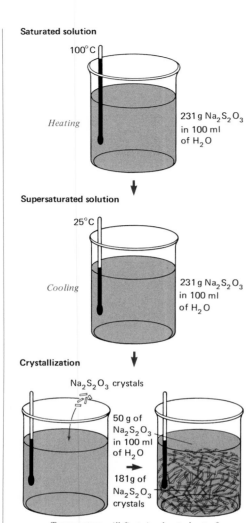

Saturated solution

100°C

Heating

231 g Na₂S₂O₃ in 100 ml of H₂O

Supersaturated solution

25°C

Cooling

231 g Na₂S₂O₃ in 100 ml of H₂O

Crystallization

Na₂S₂O₃ crystals

50 g of Na₂S₂O₃ in 100 ml of H₂O

181 g of Na₂S₂O₃ crystals

Temperature will first rise due to heat of crystallization then return to 25°C

FIGURE 14.2

Sodium thiosulfate solution. *During careful cooling of a saturated solution no solid crystallizes out, and the solution becomes supersaturated.*

corners fill in and a perfect crystal of the same mass as the broken crystal results.

A **saturated solution** is a solution in which the concentration of dissolved solute is equal to that which would be in equilibrium with undissolved solute under the given conditions. The undissolved solute need not be there. It could have been filtered off, or the solution could have been made by dissolving exactly the right amount of solute for saturation. An **unsaturated solution** is a solution that can still dissolve more solute under the given conditions.

Students often make the mistake of defining a saturated solution as one in which "the solution holds all of the solute that it can." This is not correct. Although nature always strives to reach equilibrium situations, it is quite possible to have a **supersaturated solution**—a solution that holds more dissolved solute than would be in equilibrium with undissolved solute. Many solids are more soluble at higher temperatures than at lower temperatures. Sometimes when a solution of such a material dissolved at a high temperature is then allowed to cool quietly, all of the solute remains dissolved. This happens even though it would not dissolve to that extent at the lower temperature. If the solution is shaken, or if a crystal like that of the solute is dropped into it, the solute may quickly separate from the solution as crystals. The solution will then be in equilibrium with those crystals—it has become a saturated solution. Sodium acetate ($NaC_2H_3O_2$) and sodium thiosulfate ($Na_2S_2O_3$) are notable examples of compounds that readily form supersaturated solutions. Honey is essentially a supersaturated solution of various sugars in water. It is formed in the beehive when some of the water of an unsaturated solution evaporates. Honey does not readily revert to saturation, but on long standing it may do so and become at least partly crystalline.

Supersaturated solutions of solids form because the molecules or ions of solute do not always spontaneously orient themselves into a crystal lattice. The more complex the lattice of a particular compound and the more soluble the compound, the more apt that compound is to form a supersaturated solution. If a crystal with the proper lattice is introduced into the solution, the orientation is readily achieved on the surface of the crystal (Figure 14.2). Inducing crystallization in a supersaturated solution by adding a crystal is called "seeding."

The terms "dilute" and "concentrated" are used very loosely to indicate solutions with relatively little solute, or a relatively large amount of solute. There is no definite line of demarcation between them. These terms should never be confused with "unsaturated" and "saturated," for they refer to quite different properties of the solution. A *saturated* aqueous solution of barium sulfate ($BaSO_4$) is extremely *dilute* (0.002 g of solute per liter of water at 18°C), but a *saturated* solution of sodium thiosulfate ($Na_2S_2O_3$) is very *concentrated* (500 g of solute per liter of water at 25°C).

In a liquid solution, the particles of solute are free to move. If these particles are ions, the solution will be a conductor of electricity, for the positive ions are attracted toward the negative electrode and the negative ions toward the positive electrode of an electrolytic cell (Section 19.5). Any substance forming a solution that conducts an electric current is called an **electrolyte;** any substance forming a solution that does not conduct electricity is called a **nonelectrolyte.**

14.2 Gas–gas solutions

A mixture of gases is a solution, for all gases are infinitely miscible with all others. Air is a solution of nitrogen, oxygen, carbon dioxide, argon, water vapor, and other gases. Since air is about four-fifths nitrogen by volume, one might refer to nitrogen as the solvent, but such terminology really has little meaning.

The composition of the atmosphere varies somewhat with location, but the infinite miscibility of gases ultimately allows air pollution to be shared worldwide. Radioactive gases from a test in the Pacific Ocean or hydrocarbons from heavy city traffic enter the vast atmospheric gaseous solution and eventually can be found even over Antarctica.

14.3 Gas–liquid solutions

The solubility of a gas in a liquid varies with the temperature, the pressure, and the nature of the gas (Table 14.1). The relative solubilities of gases stay about the same for different solvents. For example,

TABLE 14.1
Solubility of various gases in water

Values given are for solubility of the gas at 1 atm pressure above the solution. Equilibria are shown for those gases that ionize in aqueous solution. The length of the arrows shows the predominant reaction of the equilibrium. CO_2, e.g., is only slightly ionized and is present in solution mainly as hydrated CO_2 molecules.

	Solubility at 0°C (g/100 g H_2O)	*Solubility at higher temperatures (°C) (g/100 g H_2O)*
Ammonia, NH_3 ($NH_3 + H_2O \rightleftharpoons NH_4^+ + OH^-$)	89.5	7.4 (100°)
Argon, Ar	1.01×10^{-2}	3.96×10^{-3} (80°)
Carbon dioxide, CO_2 ($CO_2 + H_2O \rightleftharpoons H^+ + HCO_3^-$)	0.3346	5.76×10^{-2} (60°)
Chlorine, Cl_2 ($Cl_2 + H_2O \rightleftharpoons H^+ + Cl^- + HOCl$)	1.46	0.219 (80°)
Hydrogen, H_2	1.922×10^{-4}	7.9×10^{-5} (80°)
Hydrogen chloride, HCl ($HCl \rightleftharpoons H^+ + Cl^-$)	82.3	56.1 (60°)
Hydrogen sulfide, H_2S ($H_2S \rightleftharpoons H^+ + HS^-$)	0.7066	7.65×10^{-2} (80°)
Nitrogen, N_2	2.942×10^{-3}	6.60×10^{-4} (80°)
Oxygen, O_2	6.945×10^{-3}	1.381×10^{-3} (80°)
Ozone, O_3	3.9×10^{-3}	1×10^{-4} (50°)
Sulfur dioxide, SO_2 ($SO_2 + H_2O \rightleftharpoons H^+ + HSO_3^-$)	22.83	4.5 (50°)

in water, carbon dioxide is more soluble than oxygen, which is more soluble than helium (Table 14.1). This solubility order is the same in *n*-heptane (C_7H_{16}), cyclohexane (C_6H_{12}), benzene (C_6H_6), carbon tetrachloride (CCl_4), and carbon disulfide (CS_2).

A relationship between the vapor pressure of a solute gas and its solubility is expressed by **Henry's law:** *at constant temperature, the solubility of a gas is directly proportional to the pressure of that gas over the solution.* Doubling the oxygen pressure, for example, doubles the amount of oxygen that will dissolve in a given amount of solvent. Henry's law is most closely followed by dilute solutions of gases that do not react with the solvent.

Mathematically, Henry's law is expressed as

$$C = k\,p \tag{1}$$

where *gas concentration in solution* → C, *constant* → k, and *partial pressure of solute gas* → p.

where k is a constant characteristic of the specific combination of solvent and gas, p is the partial pressure of the solute gas in the gas phase over the solution, and C is the concentration of the gas in the solution.

The most important gas–liquid solution is undoubtedly that of oxygen in water. Dissolved oxygen is essential for the destruction of organic waste in sewage and for supporting aquatic life. Probably the most familiar gas–liquid solution is carbon dioxide in water. With the appropriate additional solutes that turn it into soft drinks, or beer, or champagne, this solution seems to have universal appeal.

In manufacturing soft drinks the sweetened and flavored water is saturated at a carbon dioxide pressure greater than that in the atmosphere. The carbon dioxide does not escape immediately when the pressure is relieved, for carbon dioxide readily forms supersaturated solutions in water. If the solution is stirred or shaken, however, most of the gas escapes rapidly, leaving a solution that is saturated at the newly established pressure of the carbon dioxide above the solution. The total atmospheric pressure is of no consequence in determining the solubility of the carbon dioxide in the liquid; only the partial pressure of carbon dioxide affects the solubility. This is in accordance with Henry's law and Dalton's law of partial pressures (Section 3.15). Since carbon dioxide in the atmosphere has a partial pressure of only 4×10^{-4} atm, the soft drink winds up tasting "flat" when it has come to equilibrium with the atmosphere.

EXAMPLE 14.1

■ During bottling, a carbonated beverage was made by saturating flavored water at 0°C with CO_2 at a pressure of 4.0 atm. Later, the bottle was opened and the soft drink allowed to come to equilibrium at 25°C with air containing CO_2 at a pressure of 4.0×10^{-4} atm. Find the concentration of CO_2 in the freshly bottled soda, and in the soda after it had stood open and come to equilibrium. The Henry's law constants for aqueous solutions of CO_2 are

at 0°C,

$$k = 7.7 \times 10^{-2} \frac{\text{mole/liter}}{\text{atm}}$$

at 25°C,

$$k = 3.2 \times 10^{-2} \frac{\text{mole/liter}}{\text{atm}}$$

Substituting into Henry's law gives, for the bottled soft drink,

$$C = kp$$
$$= \left(7.7 \times 10^{-2} \frac{\text{mole/liter}}{\text{atm}}\right) (4.0 \text{ atm})$$
$$= 0.31 \text{ mole/liter}$$

and for the opened soft drink, at equilibrium with atmospheric CO_2,

$$C = \left(3.2 \times 10^{-2} \frac{\text{mole/liter}}{\text{atm}}\right) (4.0 \times 10^{-4} \text{ atm})$$
$$= 1.3 \times 10^{-5} \text{ mole/liter}$$

The concentration of CO_2 decreased from 0.31 mole/liter to 1.3×10^{-5} mole/liter, a decrease of more than 99.99%, and the drink had definitely become "flat." ∎

The solubility of gases in liquids decreases with rising temperature (Figure 14.3; Table 14.1). The decrease is not at the same rate for all gases or all solvent liquids. (Note that the first bubbles that appear as a liquid is heated are not bubbles of vapor, but the dissolved gases coming out of solution.) A gas dissolved in a liquid can always be driven out of solution by boiling the solution, unless the gas reacts to a great extent with the solvent, as, for example, HCl with water. Boiling drives off a gas not only because of the decreased solubility at higher temperatures, but also because the vapors escaping from the boiling liquid carry the solute gas away and thus lower its partial pressure above the solution.

Every liquid exposed to the atmosphere contains dissolved oxygen. Often it is necessary to remove this oxygen before the liquid can be used as a solvent for a particular chemical reaction. The oxygen removal is accomplished by boiling the liquid in the presence of a nonreactive gas, nitrogen for example, or just by bubbling the nonreactive gas through the liquid for an adequate period of time so that it will carry away with it all of the oxygen. Both methods reduce the partial pressure of oxygen over the solution to practically zero, so that the oxygen readily comes out of solution.

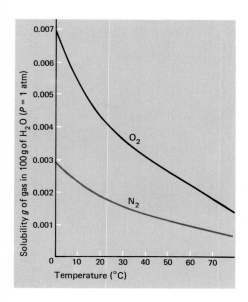

FIGURE 14.3
Solubility of nitrogen and oxygen in water at 1 atm. *Data for solubility at pure oxygen or pure nitrogen pressure.*

14.4 Gas–solid solutions

Surprisingly, there are solutions of gases in solids. The most common of these is dihydrogen in palladium metal. The volume of dihydrogen dissolved at room temperature may be as much as 900 times the volume of the palladium. One method of purifying dihydrogen consists of putting it in a thin-walled palladium vessel under slight pressure. The hydrogen dissolves in the metal and escapes on the other side of the wall where the pressure is less. Other gases are not dissolved by palladium and remain behind. Several other heavy metals and their alloys will also dissolve dihydrogen. Apparently gases such as H_2 dissociate into atoms at a metal surface. The atoms then diffuse into the metal and occupy interstitial or vacant lattice positions, forming interstitial hydrides (Section 12.15).

The equipment for the manufacture of ammonia by the Haber process (see An Aside: The Haber Process, Chapter 16) and for other high-pressure reactions involving hydrogen must be made of a special steel; ordinary steel dissolves hydrogen under pressure and allows it to escape from the synthesis chamber.

14.5 Solid–solid solutions

Many alloys are solid solutions of metals in each other (Section 12.15). For example, copper and nickel are infinitely miscible. Mixed crystals constitute another kind of solid solution. For example, the alums (Section 12.14) readily form mixed crystals since they have the same crystal structure. A solution containing both ordinary alum, $KAl(SO_4)_2 \cdot 12H_2O$, and chrome alum, $KCr(SO_4)_2 \cdot 12H_2O$, on slow evaporation, yields mixed crystals of aluminum–chrome alum, $(K(Al,Cr)(SO_4)_2 \cdot 12H_2O$. The two alums are infinitely soluble in each other, and crystals containing the two tripositive ions in any proportion are easily obtained. Many minerals are mixed crystals. For example, forsterite (Mg_2SiO_4) and fayalite (Fe_2SiO_4) are isomorphous and infinitely soluble in each other in the solid state. The olivine minerals are a series of solid solutions of forsterite and fayalite.

In other cases, however, minerals show quite limited solubility in each other. Anhydrite ($CaSO_4$) and barite ($BaSO_4$) are good examples. The radius of the Ca^{2+} ion is 0.099 nm and that of the Ba^{2+} ion is 0.135 nm. This difference is so great that when either ion substitutes for the other, the crystal is distorted. Only a certain amount of distortion can be tolerated—beyond that point, the two substances will not mix.

14.6 Liquid–liquid solutions

The miscibility of liquids depends on the similarity of their molecules. The more similar the intermolecular forces within each of two liquids, the more likely they are to be miscible.

Ethyl alcohol, C_2H_5OH, and water, HOH, are infinitely miscible.

The primary intermolecular force in each is hydrogen bonding. On mixing, the C_2H_5OH and HOH molecules readily form strong hydrogen bonds with each other. Another infinitely miscible pair is benzene and toluene.

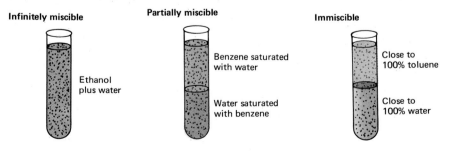

Here the weaker intermolecular London forces are the most important attractive forces between the molecules.

Benzene and *water*, however, have limited solubility in each other. They are partially miscible. Shaking equal volumes of benzene and water together in a test tube (Figure 14.4) gives two liquid phases, one a saturated solution of water in benzene, the other a saturated solution of benzene in water. A single phase results when a small amount of one of a pair of partially miscible liquids is added to a large amount of the other. However, toluene and water are immiscible, and even a small amount of one will not dissolve in the other. The molecules of water are so much more strongly attracted to each other than to those of toluene, that toluene molecules cannot come in between them.

Heat is often absorbed when two partially miscible liquids are combined (a positive heat of solution; next section). Raising the temperature increases their miscibility, often to the point where the two phases merge into one.

14.7 Solid–liquid solutions

When a crystalline solid dissolves in a liquid, the crystal lattice breaks down slowly, as individual atoms, molecules, or ions leave the surface. Therefore, a finely divided solid, like confectioner's sugar, dissolves faster than a large piece of solid, such as a crystal of rock candy. Shaking a solvent–solute mixture will hasten the dissolution of the solute because the already dissolved solute particles are carried away from the surface, decreasing their concentration in the solvent surrounding the undissolved material.

Infinitely miscible   **Partially miscible**   **Immiscible**

Ethanol plus water

Benzene saturated with water

Water saturated with benzene

Close to 100% toluene

Close to 100% water

FIGURE 14.4
Miscibility of liquids. *Very few liquids are totally immiscible.*

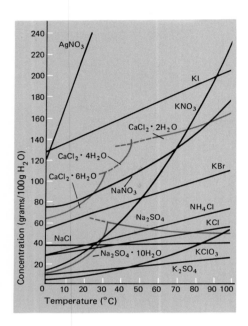

FIGURE 14.5

The effect of temperature on the solubility of several inorganic substances in water. *The solubility of the sodium sulfate hydrate increases with temperature up to the point where conversion to anhydrous sodium sulfate occurs.*

Although the *rate* of solution varies with the shaking or size of the pieces of solid, the *solubility* of the solid is unaffected. Whether it takes a few seconds with shaking or weeks without shaking, dissolution will stop when saturation for that solute–solvent combination at that temperature is reached.

The effect of temperature on the solubility of solids is quite variable. Sodium chloride (Figure 14.5) is almost as soluble at 0°C as it is at 100°C, but the solubility of potassium nitrate varies nearly 20-fold over that temperature interval (13.3 g per 100 g of water at 0°C, but 247 g per 100 g of water at 100°C).

Below 32.4°C, the solubility of sodium sulfate in water increases rapidly with rising temperature, but above 32.4°C, the solubility decreases steadily as the temperature rises. This occurs because below 32.4°C, sodium sulfate exists as a hydrate, $Na_2SO_4 \cdot 10H_2O$, but above that temperature, even in contact with water, it is anhydrous Na_2SO_4. The solubilities of the hydrate and the anhydrous salt are different, and their solubility curves intersect at the transition temperature of 32.4°C. In general, if heat is absorbed during solution, raising the temperature will increase solubility (LeChatelier's principle applies here; Section 16.7). This is the most common situation. Some substances, however, give off heat during solution, and for such substances solubility decreases with increasing temperature.

To make a thermochemical calculation of the heat of solution, the solution of a solid can be thought of as taking place in two steps. The first is the separation of the atoms, molecules, or ions from their positions in the crystal lattice and the formation of gaseous ions—the reverse of lattice formation. For example, for an ionic substance

$$AB(s) \longrightarrow A^+(g) + B^-(g) \qquad \Delta H = \text{lattice energy} \qquad (2)$$

This process requires the input of heat to overcome the ionic forces holding the ions in the crystal. The heat of reaction (2) is positive and is equal to the lattice energy. The second step is the surrounding of the atoms, molecules, or ions by solvent molecules—solvation, or in the case of water, hydration. Heat is usually evolved in the solvation of ions.

$$A^+(g) + B^-(g) \xrightarrow{\text{H}_2\text{O}} A^+(aq) + B^-(aq) \qquad \Delta H = \Delta H_{\text{hydration}} \qquad (3)$$

The heat of solution (or enthalpy of solution)—the heat released or absorbed in producing a given solution at a given temperature—is the algebraic sum of the heats of reactions (2) and (3)

$$AB(s) \xrightarrow{\text{H}_2\text{O}} A^+(aq) + B^-(aq) \qquad (4)$$

$$\Delta H_{\text{solution}} = \text{lattice energy} + \Delta H_{\text{hydration}}$$

Whether the heat of solution is positive or negative depends upon the balance between the heat required to break down the lattice and the heat given off in hydration. Heats of solution are recorded either for a specific number of moles of solvent per mole of solute, as shown in the following equations, or for a very large quantity of solvent (the heat of solution at infinite dilution).

Endothermic solution formation: *solute + solvent + heat $\longrightarrow$ solution*

$$AgNO_3(s) \xrightarrow[\text{H}_2\text{O}]{\text{50 moles}} Ag^+(aq) + NO_3^-(aq) \qquad \Delta H = 4.82 \text{ kcal } (20.17 \text{ kJ})$$

Exothermic solution formation: *solute + solvent $\longrightarrow$ solution + heat*

$$Ca(NO_3)_2(s) \xrightarrow[\text{H}_2\text{O}]{\text{10 moles}} Ca^+(aq) + 2NO_3^-(aq) \quad \Delta H = -5.48 \text{ kcal } (-22.93 \text{ kJ})$$

14.8 Is solubility a physical or a chemical phenomenon?

In many solutions, the solute and the solvent are quite independent of each other, and solution takes place only because the motion of the molecules causes them to intermingle. This is exemplified by solutions of gases in each other. It is also shown by solutions of paraffin wax or naphthalene in gasoline or benzene. The case is a little different with alloys of nickel in copper, for the two metals share electrons, but on a very mobile basis. However, we probably would not say that the metals in such alloys react with each other; solution in this case is thought of as a physical process. When sugar dissolves in water, it does so because the hydroxyl groups of the sugar molecules form hydrogen bonds with the water molecules, and this is definitely a chemical reaction. However, it is easily reversed—when the water is allowed to evaporate the hydrogen bonds are broken and crystalline sugar is formed again. In this case a chemist would probably think of solution as a physical phenomenon because the sugar is retrieved unchanged. Sodium chloride dissolves in water because the sodium ions are attracted by the negative end of the water dipole (the oxygen atoms) and the chloride ions are attracted by the positive end (the hydrogen atoms). Sodium chloride will not dissolve in liquids like benzene, because the benzene is nonpolar and does not attract the sodium and chloride ions. Salt will not dissolve very much in alcohol, which is somewhat polar, because the polar attractions are not great enough to overcome the attraction of the ions in the crystal for each other.

Although water is an excellent solvent for many substances, there are many others that it will not dissolve. It will not dissolve paraffin wax, benzene, chloroform, or other nonpolar or slightly polar covalent substances, because there are no atoms in those substances that carry strong enough positive or negative charges to be attracted by the dipolar water molecule. It will not dissolve barium sulfate ($BaSO_4$), even though that substance is completely ionic, because the attractions of the doubly charged barium and sulfate ions for each other are so great that the attraction of the water molecules cannot overcome them.

Sulfuric acid and ammonia are polar covalent molecules. They dissolve freely in water because in the presence of the polar water molecules they are ionized:

$$H_2SO_4(l) \underset{}{\overset{H_2O}{\rightleftharpoons}} H^+ + HSO_4^-$$

$$HSO_4^- \underset{}{\overset{H_2O}{\rightleftharpoons}} H^+ + SO_4^{2-}$$

$$NH_3(g) + H_2O(l) \rightleftharpoons NH_4^+ + OH^-$$

These reactions are readily reversed by boiling the solutions. In the case of sulfuric acid, the water vaporizes; in the case of ammonia, it is the ammonia that escapes as a gas. We might consider the processes of solution in these cases to be either physical or chemical.

When carbon dioxide dissolves in water, a small proportion of it reacts with the solvent according to the equation

$$CO_2(g) + H_2O(l) \rightleftharpoons HCO_3^- + H^+$$

but most of the CO_2 molecules simply intermingle with the water molecules. Most chemists would say that solution, in this case, is a physical phenomenon. However, when P_4O_{10} dissolves in water a reaction takes place, which is reversed only upon strong heating:

$$P_4O_{10}(s) + 6H_2O(l) \rightleftharpoons 4H_3PO_4(aq)$$

Most chemists probably would not say that P_4O_{10} "dissolves" in water, but that it "reacts" with the water. A similar situation arises when a piece of zinc is placed in hydrochloric acid. The zinc disappears and is uniformly dispersed throughout the liquid, but it no longer consists of zinc atoms. The metal has reacted with the acid to form zinc ions

$$Zn(s) + 2HCl(aq) \longrightarrow Zn^{2+} + 2Cl^- + H_2(g)$$

It would not be proper to say that zinc "dissolves" in hydrochloric acid; rather, zinc reacts with the acid.

Finally, insolubility may result if the molecules of a liquid solvent are so strongly attached to each other that they cannot let go to attach themselves to the molecules of a potential solute. Thus, hydrogen chloride will not dissolve in liquid hydrogen fluoride, even though the two are closely related in structure and properties. Hydrogen fluoride is strongly hydrogen bonded—much more so than water—and the molecules show little tendency to separate from each other in order to attach themselves to hydrogen chloride molecules or chloride or hydrogen ions.

There is an old adage that "like dissolves like"—that is, that covalent substances will dissolve other covalent substances, and that polar solvents will dissolve polar solutes. There is much truth in this, but the phenomenon of solubility is more complex than the rule suggests. One must take into account not only the forces of attraction between the solvent and the solute, but also the forces of attraction between ions or molecules of the solute and between molecules of the solvent, the relative sizes of the ions or molecules of the solute and the molecules of the solvent, and many other factors.

14.9 Solutions of electrolytes

The behavior of their solutions shows that there are two types of electrolytes—strong and weak. The electrical conductivity of solutions of strong electrolytes is high and the conductivity of comparable solutions of weak electrolytes is low.

Most salts are strong electrolytes. In solution they dissociate completely into ions. An aqueous NaCl solution contains hydrated Na^+ and Cl^- ions, but no NaCl molecules. Strongly basic hydroxides such as KOH and NaOH are also ionic solids and dissociate completely in solution. Many polar covalent substances, including most strong acids such as HCl, HNO_3, or H_2SO_4, are also completely ionic in solution. **Strong electrolytes** are 100% present as ions in aqueous solution. As we have pointed out (Section 2.11), the number of particles in a strong electrolyte solution is dependent upon the stoichiometry of the compound. Each mole of NaCl yields 2 moles of ions; each mole of $MgBr_2$ yields 3 moles of ions; each mole of AlF_3 yields 4 moles of ions. Note that a strong electrolyte is always entirely dissociated whether it is in a *concentrated* or a *dilute* solution.

Other polar covalent substances—the weak electrolytes—yield only a small number of ions in solutions. Their dissociation reaches equilibrium at a point where only a fraction of the molecules have ionized. **Weak electrolytes** are only partially ionized in aqueous solution. Weak electrolyte solutions contain neutral molecules in solution in addition to hydrated ions. Most organic acids are weak electrolytes. For example, only 4 of every 1000 molecules of acetic acid are ionized in a $1M$ solution at 25°C.

$$HC_2H_3O_2(aq) \rightleftharpoons H^+(aq) + C_2H_3O_2^-(aq)$$

As a solution becomes more dilute, more molecules of a weak electrolyte ionize (Table 14.2).

TABLE 14.2

Ionization of acetic acid

Concentration of $HC_2H_3O_2$	Percent ionization, $\dfrac{\text{number of ionized molecules}}{\text{original number of molecules}} \times 100\%$
$1M$	0.4
$0.1M$	1.3
$0.01M$	4.3
$0.001M$	12.4
$0.0001M$	34.0
$0.00001M$	71.1

14.10 Ideal vs. nonideal solutions

Real atoms, ions, and molecules influence each other, and their behavior rarely is exactly what theory predicts. It is useful to have a convenient way to compare real to theoretical behavior.

Ideal gas molecules are thought of as far enough apart to be independent of each other's influence. A different approach to ideality is needed for solutions, where the solvent and solute particles are inti-

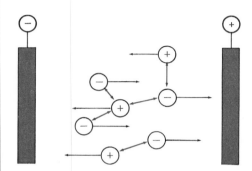

FIGURE 14.6
Ions in solution in an electrical field.
The black arrows show the force on the ions due to the electrical field. The colored arrows show the interionic forces holding the ions back.

mately mixed. An **ideal solution** is defined as one in which the forces between all particles of both solvent and solute are identical. In other words, in an ideal solution of A and B the forces between particles of A and A, or A and B, or B and B are the same.

If A and B form an ideal solution and are both liquids, intermolecular forces in pure A are so similar to those in pure B that molecules of one added to the other find themselves in a virtually identical environment. No change in total volume and no emission or absorption of heat take place when the components of ideal liquid–liquid solutions are mixed. In an ideal solution of a solid in a liquid, the forces between solvent molecules are unchanged by the presence of the solute particles.

In a nonideal solution, forces between atoms, ions, and molecules must be taken into account. Consider, for example, the electrical conductivity of a solution of a strong electrolyte, say, a salt. As we have stated, strong electrolytes are 100% present as ions in solution. If the electrolyte solution is very dilute—about $0.01M$—the conductivity is just about what we would expect on the basis of complete ionization. But the difference between the expected and observed values becomes greater and greater as the concentration of the solution is increased. For many years, this was interpreted to mean that dissociation was not complete except in dilute solutions. We know now, however, that dissociation of salts is complete, even in the solid state. Another explanation has been found, and it is quite simple—ions of opposite charge attract each other, and this mutual attraction increases as the ions come closer together in more concentrated solutions. Each ion becomes surrounded by ions of opposite charge; these tend to hold the original ion back and so retard its journey toward the electrode (Figure 14.6). The dissociation of the salt is complete, but the oppositely charged ions interfere with each other and the conductivity of the solution is decreased. The effect is larger if the solution is more concentrated or if the ions have charges greater than one, for example, for $MgSO_4$, as compared with NaCl.

To deal with nonideal solutions, whether they involve atoms, molecules, or ions, two quantities are defined. **Activity** a is the effective concentration of a component of a solution. The **activity coefficient** γ is the ratio between the activity, or effective concentration, and the actual concentration of a component in a solution. These quantities are related as follows:

$$\underset{\substack{\text{activity}\\\text{coefficient}}}{}\gamma = \overset{\text{activity, or effective concentration}}{\underset{\text{concentration}}{\frac{a}{C}}} \quad \text{or} \quad a = \gamma C \tag{5}$$

Components in an ideal solution make a fully effective contribution to the concentration, therefore $\gamma = 1$ and the activity is equal to the concentration. When interactions occur, the contributions of individual particles are affected, the activity coefficient differs from 1, and the activity differs from the concentration.

Think of a lone swimmer in a swimming pool. He can go any-where in the pool he wants to at any speed he chooses. He makes a fully effective contribution to the concentration of swimmers in the pool (1 swimmer/pool). Now put 24 other swimmers in the pool with him. The effectiveness of each is decreased by bumping into or going around the others. The total activity is less than 25 swimmers/pool and the activity coefficient is less than 1. Furthermore, the activity coefficient and the activity of the swimmers would change if 3 sea lions were thrown into the pool.

Activities and activity coefficients can be experimentally deter-mined and are collected in tables of data. Each value is specific to the system for which it was measured at the temperature of measure-ment. The activity coefficients of strong electrolytes are all less than 1, and in most cases approach 1 as the solutions became increasingly dilute. For many ions, the activity coefficient is about 0.9 in a $0.1M$ solution and about 0.7 in a $0.2M$ solution. For a $0.0992M$ solution of NaCl at 25°C, for which $\gamma = 0.782$,

$$a = \gamma C$$
$$= (0.782)(0.0992M)$$
$$= 0.0776M$$

Now that we have taken this trouble to explain, in general terms, what activities and activity coefficients are, we can say with a clear conscience that, except for extremely accurate experimental work, for the mathematics of physical chemistry, or for work with quite con-centrated solutions, activity and activity coefficients need not be used. Our main concern with them from here on will be to point out where they would fit in *if* they had to be used.

In most calculations, the following approximations can be made (Table 14.3). For gases, the activity is equal to the partial pressure of the gas in atmospheres. For pure solids and liquids, activity is equal to 1. For solutes, activity is equal to the concentration expressed in moles per liter. And for solvents, activity is equal to the mole fraction of the solvent, which in dilute solutions can be assumed to be 1. Why is the activity of pure solids and liquids, or solvents in dilute solu-tions equal to 1? Think of it this way—their effective concentrations and actual concentrations are equal, so the activity coefficient is equal to 1. And the concentration of a pure liquid or solid in, say, mole fraction or weight percent, must also be equal to 1.

TABLE 14.3

Activities for substances assumed to be close to ideal

Gases	$a = p$ **(partial pressure in atm)**
Pure liquids and solids	$a = 1$
Solutes	$a = M$ **(molarity)**
Liquids in solutions	$a = X$ **(mole fraction)**
Solvent in dilute solution	$a = 1$

Concentration of solutions

14.11 Standard solutions: Mole fraction

The mole fraction, which we introduced in our discussion of vapor pressure (Section 3.15), is the ratio of the number of moles of one component to the total number of moles of all components of the so-lution. The sum of the mole fractions of all solution components is 1.

For a two-component solution, where n represents number of moles,

$$X_A = \frac{\text{moles of A}}{\text{moles of A} + \text{moles of B}} = \frac{n_A}{n_A + n_B} \qquad (6)$$

$$X_B = \frac{n_B}{n_A + n_B} \qquad (7)$$

$$X_A + X_B = 1 \qquad (8)$$

The mole fraction expresses the ratio of the number of particles of one component to the total number of particles in a solution.

EXAMPLE 14.2

■ Find the mole fraction of table sugar, which is sucrose ($C_{12}H_{22}O_{11}$ molar wt = 342.30 g/mole) in an aqueous solution containing 30.00 g of the sugar and 70.00 g of water.

The number of moles of each component is found as follows:

$$n_{\text{sucrose}} = \frac{30.00 \text{ g}}{342.30 \text{ g/mole}} = 0.08764 \text{ mole of sucrose}$$

$$n_{\text{water}} = \frac{70.00 \text{ g}}{18.015 \text{ g/mole}} = 3.886 \text{ moles of water}$$

The mole fraction of the sugar is given by

$$X_{\text{sucrose}} = \frac{n_{\text{sucrose}}}{n_{\text{sucrose}} + n_{\text{water}}} = \frac{0.08764 \text{ mole}}{0.08764 \text{ mole} + 3.886 \text{ moles}}$$

$$= \frac{0.8764 \text{ mole}}{3.973 \text{ moles}} = 0.02206$$

The mole fraction of sucrose in the sugar solution is 0.02206.

■

14.12 Standard solutions: molality

Molality m is the number of moles of solute per kilogram of solvent. Both mole fraction and molality are concentration units that are independent of temperature and the volume changes caused by temperature. Molality is a useful concentration unit in the study of properties based on freezing point and boiling point measurements (Sections 14.17 and 14.18).

A $1m$ solution of NaOH would be prepared by dissolving 40 g (1 mole) of NaOH in 1000 g of water, or 20 g (0.5 mole) of NaOH in 500 g of water, or any combination that maintains this proportion. The molality of a solution of any number of moles of solute B, n_B, in any mass of solvent A, is found from the following relationship:

$$m(\text{moles/kg}) = \frac{n_B(\text{moles})}{\text{mass}_A(\text{g})} \times \frac{1000 \text{ g}}{1 \text{ kg}} \qquad (9)$$

EXAMPLE 14.3

■ Find the molality of the sugar solution described in Example 14.2.

Substituting 70.00 g of solvent and 0.08764 mole of sucrose into Equation (9) for molality yields

$$m = \frac{n_B}{mass_A} \times \frac{1000 \text{ g}}{1 \text{ kg}} = \frac{0.08764 \text{ mole}}{70.00 \text{ g}} \times \frac{1000 \text{ g}}{1 \text{ kg}} = 1.252 \text{ moles/kg}$$

The sugar solution is 1.252*m*.   ■

14.13 Preparation of solutions

It is often necessary in the laboratory to prepare solutions of a specific concentration from solids or from solutions of other concentrations. The facts needed to accomplish this are the weight or volume of solute and the weight or volume of solvent that must be combined.

In making dilute solutions from more concentrated solutions, the amount of solute present in a given volume of the concentrated solution must be known. Solvent is then added to change the concentration of the solution, but note that the *amount* of solute is unchanged.

We remind you here of some relationships that are useful in solution and dilution problems.

Moles of solute = V(volume of solution, liters)
$\times M$(molarity, moles/liter)

Equivs of solute = V(volume of solution, liters)
$\times N$(normality, equiv/liter)

V(volume of solution) = $\dfrac{m\text{(mass)}}{d\text{(density)}}$

The simplest way to solve solution and dilution problems is to choose a "basis" from which to work the problem using the information given. For example, the basis might be 100 g of solution, or 1 kg of solution, or 100 g of solute. If the problem has several parts, the basis is used in solving each part.

EXAMPLE 14.4

■ Describe the process for preparing 1 liter of 6.0*N* H_2SO_4 from commercially available concentrated H_2SO_4, which has a concentration of 35.9*N*.

In 1 liter of 6.0*N* H_2SO_4 (the basis) there are 6 equivalents of H_2SO_4. The volume of commercial acid corresponding to 6.0 equivalents of H_2SO_4 is

$$V = \frac{\text{equiv}}{N} = \frac{6.0 \text{ equiv}}{35.9 \text{ equiv/liter}} = 0.17 \text{ liter}$$

The dilute acid is prepared by first slowly adding 170 ml of concentrated acid to about 600 ml of water. Sulfuric acid is always added to water because a lot of heat is given off when H_2SO_4 and water are combined and the bulk of the water helps to dissipate the heat. This solution is mixed thoroughly, and then diluted with additional water until the volume is exactly 1 liter. ∎

EXAMPLE 14.5

∎ A stockroom attendant wants to prepare 3 liters of a 5 wt % hydrogen peroxide solution ($d = 1.0131$ g/ml) from the commercially available solution, which has a concentration of 30 wt % ($d = 1.1122$ g/ml). Describe the dilution process.

The mass of 3 liters of 5 wt % H_2O_2 (the basis) is

$$m = dV = \left(1.0131 \frac{g}{ml}\right)(3.00 \text{ liters})\left(1000 \frac{ml}{liter}\right) = 3040 \text{ g}$$

and the mass of H_2O_2 present is $(0.05)(3040 \text{ g}) = 150$ g. The mass of the 30 wt % H_2O_2 solution containing this amount of H_2O_2 is 150 g/0.30 = 500 g, corresponding to

$$V = \frac{m}{d} = \frac{500 \text{ g}}{1.1122 \text{ g/ml}} = 450 \text{ ml}$$

The attendant needs to take 450 ml of the 30 wt % solution and add sufficient water to obtain the 3 liters of the dilute solution. ∎

EXAMPLE 14.6

∎ Describe the steps in preparing 500 ml of 0.10 M $CuSO_4$ if the only available source of $CuSO_4$ is the hydrated form, $CuSO_4 \cdot 5H_2O$.

For 500 ml of 0.10M solution (the basis), the required number of moles of $CuSO_4$ is

$$n_{CuSO_4} = V \times M$$
$$= 0.500 \text{ liter} \times 0.10 \frac{mole}{liter}$$
$$= 0.050 \text{ mole}$$

Each mole of $CuSO_4 \cdot 5H_2O$ contains 1 mole of $CuSO_4$, so the amount of $CuSO_4 \cdot 5H_2O$ needed is 0.050 mole or $(0.050 \text{ mole})(249.68 \text{ g/mole}) = 12.5$ g. Thus the solution is prepared by adding sufficient water to 12.5 g of $CuSO_4 \cdot 5H_2O$ to make 500 ml of solution. Once the solution is formed, the water molecules from the $CuSO_4 \cdot 5H_2O$ are indistinguishable from the water molecules added in liquid form. ∎

Unit	Symbol		Calculation
Weight percent (Section 2.16)	wt %	wt % =	$\dfrac{\text{mass of B (g)}}{\text{mass of A(g) + mass of B(g)}} \times 100\%$
Volume percent (Section 2.16)	vol %	vol % =	$\dfrac{\text{volume of B (liters or ml)}}{\text{(volume of A + volume of B) (liters or ml)}} \times 100\%$
Molarity (Section 2.17)	M	M =	$\dfrac{\text{moles of B}}{\text{volume of (A + B) solution (liters)}}$
Mole fraction (Section 14.11)	X	X_B =	$\dfrac{\text{moles of B}}{\text{moles of A + moles of B}}$
Molality (Section 14.12)	m	m =	$\dfrac{\text{moles of B}}{\text{mass of A (g)}} \times \dfrac{1000\ \text{g}}{\text{kg}}$
Normality (Section 4.10)	N	N =	$\dfrac{\text{equivalents of B}}{\text{volume of (A + B) solution (liters)}}$

TABLE 14.4
Concentration of solutions
Calculation is shown for two components, solute B in solvent A.

14.14 Summary of concentration units

We have now introduced all of the major ways of expressing solution concentrations that are used in chemistry. They are summarized in Table 14.4, where cross references to the definitions are given.

EXAMPLE 14.7

■ A sample of commercial vinegar was analyzed and found to contain 4.00 wt % acetic acid ($HC_2H_3O_2$, molar wt = 60.05 g). The density of this aqueous solution is 1.0058 g/ml. Find the concentration of acetic acid expressed in mole fraction, molality, molarity, normality (assuming acetic acid to give one H^+ per molecule), and volume percent. How could an acetic acid solution of the same acid concentration be prepared from commercial acetic acid (d = 1.0492 g/ml) and water (assume d = 0.9982 g/ml).

For this example we choose the basis as 100.00 g of solution, so that the mass of acetic acid present is 4.00 g and the mass of water present is 96.00 g, giving

$$n_{HC_2H_3O_2} = \frac{4.00\ \text{g}}{60.05\ \text{g/mole}}$$
$$= 0.0666\ \text{mole}\ HC_2H_3O_2$$

$$n_{H_2O} = \frac{96.00\ \text{g}}{18.0153\ \text{g/mole}}$$
$$= 5.329\ \text{moles}\ H_2O$$

In addition, the *volume of solution V* corresponding to this basis is

$$V = \frac{m(\text{mass})}{d(\text{density})} = \frac{100.00\ \text{g}}{1.0058\ \text{g/ml}} = 99.42\ \text{ml} = 0.09942\ \text{liter}$$

The *mole fraction, X,* of the acetic acid in the vinegar is

$$X_{HC_2H_3O_2} = \frac{n_{HC_2H_3O_2}}{n_{HC_2H_3O_2} + n_{H_2O}}$$

$$= \frac{0.0666 \text{ mole}}{0.0666 \text{ mole} + 5.329 \text{ mole}}$$

$$= 0.0123$$

The *molality* of the solution is

$$m = \frac{n_{HC_2H_3O_2}}{\text{mass}_{H_2O}} \times \frac{1000 \text{ g}}{1 \text{ kg}} = \frac{0.0666 \text{ mole}}{96.0 \text{ g}} \times \frac{1000 \text{ g}}{1 \text{ kg}}$$

$$= 0.694 \text{ mole/kg}$$

The *molarity* is

$$M = \frac{n}{\text{volume of solution (liters)}} = \frac{0.0666 \text{ mole}}{0.09942 \text{ liter}}$$

$$= 0.670 \text{ mole/liter}$$

To determine *normality,* recall that the equivalent weight and molar weight are identical for an acid that gives one H^+ per molecule (see Section 4.10). Thus the number of equivalents of acid present is 0.0666, giving

$$N = \frac{\text{equiv of } HC_2H_3O_2}{\text{volume of solution (liters)}} = \frac{0.0666 \text{ equiv}}{0.09942 \text{ liter}}$$

$$= 0.670 \text{ equiv/liter}$$

A solution of the same acid concentration as this vinegar solution could be prepared by mixing pure glacial acetic acid ($d =$ 1.0492 g/ml) with water ($d =$ 0.9982 g/ml). The required volumes would be

$$V_{HC_2H_3O_2} = \frac{4.00 \text{ g}}{1.0492 \text{ g/ml}} = 3.81 \text{ ml acetic acid}$$

$$V_{H_2O} = \frac{96.00 \text{ g}}{0.9982 \text{ g/ml}} = 96.17 \text{ ml water}$$

The *volume percent* of acetic acid in this solution is

$$\text{vol \%} HC_2H_3O_2 = \frac{V_{HC_2H_3O_2}}{V_{HC_2H_3O_2} + V_{H_2O}} \times 100\% = \frac{3.81 \text{ ml}}{3.81 \text{ ml} + 96.17 \text{ ml}} \times 100\%$$

$$= 3.81 \text{ vol \% acetic acid}$$

To summarize, the concentration of acetic acid in the sample of commercial vinegar can be expressed as 4.00 wt %, $X_{HC_2H_3O_2} = 0.0123$, 0.694*m*, 0.670*M*, 0.670*N*, or 3.81 vol %. ∎

Vapor pressures of liquid solutions and related properties

14.15 Liquid–liquid solution vapor pressures: Raoult's law

The molecules in a liquid are, it has been pointed out (Section 12.2), in rapid constant motion. This motion causes molecules to escape from the surface into the vapor phase, creating a vapor pressure above the liquid. If more than one liquid is present in a solution, the total vapor pressure of the solution is the sum of the vapor pressures of each liquid. Molecules of each liquid are present at the solution surface, and for an ideal liquid–liquid solution, where the forces between all molecules are identical, the tendency of either type of molecule to escape depends only on the relative numbers of each (Figure 14.7).

The effect on vapor pressure of mixing two liquids to form an ideal solution is shown in Figure 14.8. Adding either liquid to the other (i.e., starting at either side of Figure 14.8) causes a decrease from the vapor pressure of that pure liquid because fewer molecules of the original liquid are at the surface. The partial vapor pressure of either liquid is equal to its concentration in the solution times the vapor pressure of the pure liquid. Mathematically, for an ideal solution of liquids A and B at a given temperature,

$$\underset{\substack{\text{partial} \\ \text{vapor} \\ \text{pressure} \\ \text{of A} \\ \text{over} \\ \text{solution}}}{p_A} = X_A P_A^\circ \qquad\qquad p_B = X_B P_B^\circ \qquad (10)$$

where X_A is the mole fraction of A in solution and P_A° is the vapor pressure of pure A.

The total vapor pressure of the solution is the sum of the partial pressures.

$$\begin{aligned} P_t &= p_A + p_B \\ &= X_A P_A^\circ + X_B P_B^\circ \end{aligned} \qquad (11)$$

Equations (10) and (11) express what is known as **Raoult's law,** which may be stated: *The vapor pressure of a liquid in a solution is directly proportional to the mole fraction of that liquid in the solution.*

Another way of defining an ideal solution is the following: *An ideal solution perfectly obeys Raoult's law.* There are a few miscible liquids that form solutions that come quite close to ideality. Solutions that obey Raoult's law are formed, for example, by

ethyl bromide, C_2H_5Br	n-hexane, $CH_3(CH_2)_4CH_3$
+	+
ethyl iodide, C_2H_5I	n-heptane, $CH_3(CH_2)_5CH_3$

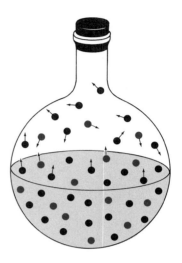

FIGURE 14.7
An ideal solution of two liquids.

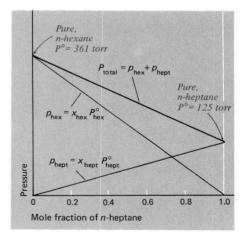

FIGURE 14.8
Vapor pressure of a solution that follows Raoult's law. *The top line is the total vapor pressure of the hexane–heptane solution. The other two lines are the partial vapor pressures of each component of the solution.*

399

Properties of real solutions of two liquids, A and B

Positive deviation is the more common situation.

Positive deviation from Raoult's law	Negative deviation from Raoult's law
A–B forces less than A–A or B–B forces	A–B forces greater than A–A or B–B forces
Solution process endothermic	Solution process exothermic
Heating increases solubility	Heating decreases solubility

It is apparent that the forces between the molecules in these pairs should be similar, for the molecules are very similar in molecular weight and structure.

The composition of the vapor over a solution of two liquids is not the same as the composition of the solution. The vapor composition depends upon the relative vapor pressures of the two liquids in an ideal liquid–liquid solution as they are related by Equation (11). The vapor over such a solution of any composition and at any temperature always holds more of the more volatile component than does the solution.

EXAMPLE 14.8

■ Find the composition of the vapor phase in equilibrium with a benzene–toluene solution of composition $X_{benz} = 0.50$, $X_{tol} = 0.50$ assuming ideal behavior at the temperature of the experiment,

$$P^{\circ}_{benz} = 73 \text{ torr} \quad \text{and} \quad P^{\circ}_{tol} = 27 \text{ torr}$$

First, find the total vapor pressure of the solution.

$$
\begin{aligned}
P_t &= X_{benz}P^{\circ}_{benz} + X_{tol}P^{\circ}_{tol} \\
&= (0.50)(73 \text{ torr}) + (0.50)(27 \text{ torr}) \\
&= 37 + 14 \\
&= 51 \text{ torr}
\end{aligned}
$$

The mole fraction of benzene in the vapor phase is

$$X_{benz} = \frac{p_{benz}}{P_t} = \frac{37}{51} = 0.73$$

Thus $X_{benz} = 0.73$ and $X_{tol} = 0.27$ in the vapor phase. There is about a 50% increase in the amount of the more volatile component in the vapor phase (benzene, b.p. 80.1°C; toluene, b.p. 110.6°C). ■

Nonideal liquid–liquid solutions show either positive or negative deviations from Raoult's law. When the different molecules of two liquids (A--B) attract each other less than they do other molecules of the same kind (A--A, B--B), the deviations are positive. Most liquid–liquid solutions behave in this way. Molecules of A or B escape from these solutions more easily than from pure liquids. Heat is absorbed when A and B are mixed (Table 14.5), and the partial vapor pressures of both A and B are greater than those predicted by Raoult's law, as shown in Figure 14.9a.

Negative deviations from Raoult's law (Figure 14.9b) occur when the different molecules attract each other more strongly than they do other molecules of the same kind. The molecules have a smaller tendency to escape from the solution surface than from their pure liquid surfaces and their partial vapor pressures are lower than predicted

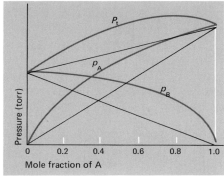

(a) **Positive deviation**

Pressure (torr) — Mole fraction of A

P_t, p_A, p_B

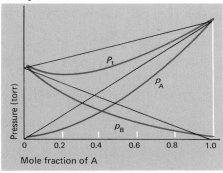

(b) **Negative deviation**

Pressure (torr) — Mole fraction of A

P_t, p_A, p_B

FIGURE 14.9

Deviations from Raoult's law. *The black lines represent ideal, Raoult's law, behavior. The colored lines show the deviations from ideal behavior. The degree of deviation, and therefore the shapes of the real solution vapor pressure curves, can vary widely. Methyl alcohol (CH₃OH) and water form a solution that deviates positively from Raoult's law. Nitric acid (HNO₃) and water solutions deviate negatively.*

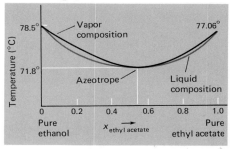

FIGURE 14.10

Boiling point diagram for a solution of two liquids, A and B (at atmospheric pressure). *The asterisks mark the beginning of distillation of a 50–50 mixture of A and B. To find the composition of vapor or liquid when the solution is boiling at T_1, read $X_{A,l}$, $X_{B,l}$, and $X_{A,v}$ $X_{B,v}$.*

by Raoult's law. The solution process is exothermic for solutions that deviate negatively from Raoult's law.

The liquids in a solution can be separated from each other by making use of the fact that the vapor is always richer in the more volatile component of the mixture than in the solution. **Distillation** involves heating a liquid or a solution to boiling, and collecting and condensing the vapors (see Tools of Chemistry, Distillation, this chapter). As a mixture of liquids is distilled, the composition of the liquid phase, the composition of the vapor phase, and the boiling point of the solution change continuously. Figure 14.10 illustrates these changes for an ideal mixture of liquids A and B. The three asterisks mark the temperature, vapor composition, and liquid composition at the beginning of the distillation of a 50–50 mixture of A and B, in which B is the more volatile component. The colored arrows mark the directions of change during distillation. The composition of the vapor at any temperature is read from the top curve and the composition of the solution is read from the bottom curve.

Solutions that deviate sufficiently from Raoult's law cannot be totally separated by fractional distillation. At some definite composition they form **azeotropes**—constant-boiling mixtures that distill without change in composition. A minimum boiling point azeotrope has a boiling point below that of either component, and a maximum boiling point azeotrope has a boiling point above that of either component (Figure 14.11).

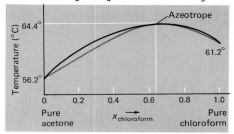

(a) **Minimum boiling point azeotrope— ethyl acetate (CH₃CO₂C₂H₅) + ethanol (C₂H₅OH)**

(b) **Maximum boiling point azeotrope— acetone (CH₃COCH₃) + chloroform (CHCl₃)**

FIGURE 14.11

Boiling point diagrams for azeotropic solutions (at 760 torr).

Distillation

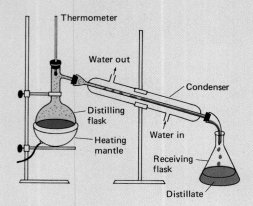

FIGURE A
Laboratory apparatus for simple distillation.

The objective of distillation is to obtain from mixtures pure liquids, or materials that are solid at ordinary temperatures but can be melted and then distilled. When the mixture is a solution of nonvolatile impurities in a liquid to be purified, simple distillation is sufficient. A laboratory scale set up for simple distillation is shown in Figure A. This type of apparatus might be used, for example, to prepare distilled water or to purify a liquid to be used as a solvent in a chemical reaction. The mixture to be distilled is heated, the vapor passes through the water-cooled condenser, and the **distillate**—the product of distillation—is collected in a receiving flask. Collection is continued as long as the thermometer shows a constant temperature, indicating that the desired liquid is distilling over.

From the boiling point diagram (see Figure 14.10) it is apparent that a single simple distillation cannot completely separate two or more volatile liquids from each other. A *series* of simple distillations would provide distillates richer and richer in the more volatile component, but the procedure would be tedious. **Fractional distillation** is a process that separates liquid mixtures into fractions that differ in boiling points. A fractionating column is designed so that a series of simple distillations is achieved over its length. The column might have an internal structure of projections or trays or perforated plates, or it may be packed with small pieces of ceramic, metal, or glass in the shape of rings, or beads, or saddles. The purpose of all these items

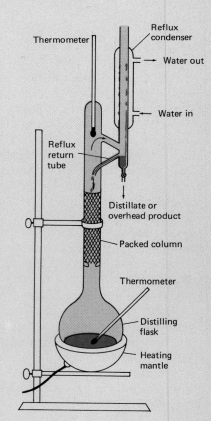

FIGURE B
Laboratory apparatus for batch fractional distillation.

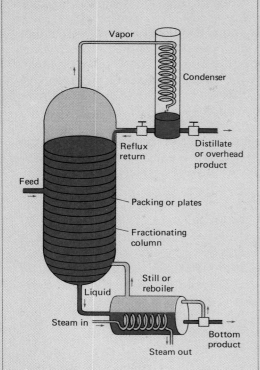

Vapor

Condenser

Reflux
return

Distillate
or overhead
product

Feed

Packing or plates

Fractionating
column

Still or
reboiler

Liquid

Steam in

Bottom
product

Steam out

FIGURE C
Industrial-scale equipment for continu-ous fractional distillation.

is to provide surfaces at which the rising vapor meets descending liq-uid and heat exchange occurs. Vapor moves up the column and some of it condenses on the packing. This condensate is richer in the less volatile components of the mixture. The condensate trickles down the column, where it meets rising hotter vapor. A heat exchange occurs—more of the more volatile component leaves the condensate and joins the rising vapor, while more of the less volatile component leaves the vapor and joins the descending liquid. The net result is that pure vapor can be collected at the top of the column and pure liquid at the bottom of the column. The process of vapor moving up the column, condensing, and trickling down the column is called **refluxing.**

A laboratory-scale fractional distillation apparatus is shown in Figure B. During distillation some of the condensing vapor is re-turned to the column and some is collected as product. Batch distillation—the distillation of one batch of the mixture at a time—is carried out in such an apparatus.

In most industrial-scale distillations, continuous operation is more economical. A simple continuous distillation column for se-parating two liquids is shown in Figure C. The material to be distilled is added continuously at the middle of the column, the more volatile product is collected continuously at the top, and the less volatile product at the bottom. To maintain the fractionation process small amounts of condensate and the higher boiling bottom product are re-turned to the column during its operation. Distillation of any type can be carried out at reduced pressure, to lower the boiling point and to protect the mixture from decomposition at higher temperatures.

When a several-component mixture is fractionally distilled, fractions that differ in boiling point are collected in the order of in-creasing boiling point, either one by one at the top of the column or continuously at successively higher points along the column. The dis-tillation of petroleum (Section 23.9) is a complex process in which many fractions that are mixtures of hydrocarbons with similar boil-ing points are collected.

Distillation of a mixture with exactly an azeotropic composition would yield no separation at all. Otherwise, one of two situations occurs. Suppose the mixture can form a minimum boiling point azeotrope. Since the azeotropic mixture boils lower than either component of the solution, the azeotropic mixture distills over first. This mixture keeps distilling over till one of the components is all gone and the composition of the azeotrope can no longer be maintained. Then the excess of the single remaining component distills at a higher temperature. The situation is the opposite if a maximum boiling point azeotrope can form. Since both components distill at a lower temperature than the mixture, whichever component is in excess distills first as the temperature is raised. This component comes over until a mixture of the azeotropic composition is left in the distillation flask. Then only the azeotropic mixture remains to be distilled at the higher temperature. Therefore, if A and B can form an azeotrope, the best separation possible from distillation of a mixture of A and B is to obtain pure A *or* pure B plus the azeotrope.

14.16 Vapor pressure lowering by nonvolatile solutes

For the moment we confine our discussion to ideal dilute solutions of *nonvolatile nonelectrolytes.* Since a nonvolatile solute has essentially zero vapor pressure, from Raoult's law the total solution vapor pressure is proportional to the mole fraction of the solvent,

$$P_t = X_A P_A^\circ \tag{12}$$

Substituting to introduce the mole fraction of solute rather than solvent, where A is again the solvent and B the solute

$$X_A + X_B = 1 \quad \text{so} \quad X_A = 1 - X_B$$
$$P_t = (1 - X_B)P_A^\circ \tag{13}$$

Equation (13) makes it clear that the total vapor pressure over a solution of a nonvolatile solute is always *less* than the vapor pressure of the pure solvent. Only the solvent contributes to the vapor pressure, and since fewer solvent molecules are present at the surface, fewer can escape into the vapor. The vapor pressure lowering, which is the difference between the vapor pressure of pure A and the total vapor pressure over the solution, can be expressed by rearranging Equation (13)

$$P_t = P_A^\circ - X_B P_A^\circ$$

vapor pressure change
$$\hookrightarrow \Delta P = P_A^\circ - P_t = X_B P_A^\circ$$

$$= \left(\frac{n_B}{n_A + n_B}\right) P_A^\circ \tag{14}$$

$\longleftarrow$ *mole fraction of B*

For dilute solutions the number of moles of solute is much smaller than the number of moles of solvent, and n_B in the denominator can be dropped to give

$$\Delta P = P_A^\circ - P_t = \left(\frac{n_B}{n_A}\right) P_A^\circ \qquad (15)$$

Therefore, *the lowering of the vapor pressure in a dilute solution of a nonvolatile nonelectrolyte is directly proportional to the amount of solute dissolved in a definite amount of solvent* (another way of stating Raoult's law). The nature of the solute particles is of no consequence—a polar molecule and a nonpolar molecule are equally effective. As we shall see (Section 14.20), in a sufficiently dilute solution, an ion, either positive or negative, exerts the same effect as a molecule.

EXAMPLE 14.9

■ What is the vapor pressure at 25°C above the aqueous sucrose solution described in Example 14.2, for which $X_{sucrose} = 0.02206$? Assume $P_{H_2O}^\circ = 23.756$ torr.

Using $X_{sucrose} = 0.02206$, the vapor pressure change is

$$\begin{aligned}
\Delta P &= P_{H_2O}^\circ - P_t \\
&= X_{sucrose} P_{H_2O}^\circ = (0.02206)(23.756 \text{ torr}) \\
&= 0.5241 \text{ torr}
\end{aligned}$$

Solving for P_t

$$\begin{aligned}
P_t = P_{H_2O}^\circ - \Delta P &= 23.756 \text{ torr} - 0.5241 \text{ torr} \\
&= 23.232 \text{ torr}
\end{aligned}$$

The vapor pressure above the sugar solution has been decreased to 23.232 torr, about a 2% decrease from that of the pure solvent, as a result of adding the solute. ■

Any property of a solvent that is affected in proportion to the concentration of solute particles, but that is independent of the nature of those particles, is called a **colligative property.** Thus, the lowering of the vapor pressure is a colligative property. *The change in any colligative property of a solution with changing concentration is directly proportional to the amount of solute dissolved in a definite amount of solvent.* This is an extremely important generalization, for it means that any colligative property can be used to measure the number of moles of solute in a definite weight of solvent. For example, if the total weight of solute in a given weight of solvent is known, and the number of particles is determined by measuring the lowering of the vapor pressure, the weight of each particle can be calculated. This furnishes a method of measuring molecular weights of substances in solution (Section 14.18).

The following are the four colligative properties—they are all related to vapor pressure: (1) vapor pressure lowering, (2) freezing point depression, (3) boiling point elevation, and (4) osmotic pressure.

14.17 Boiling point elevation and freezing point depression

Comparison of the phase diagrams for pure water and for an aqueous solution shows how boiling points are raised and freezing points lowered by the addition of a nonvolatile solute to a solvent (Figure 14.12). The solution vapor pressure curve is below the curve for pure water at all points. As a result of this lowering, the solution must be heated to a higher temperature to reach a vapor pressure equal to atmospheric pressure and therefore to boil.

The triple point also occurs at a lower temperature, and as a result the freezing point of the solution is lowered. Note that the freezing point of a solution is the temperature at which crystals of pure solvent appear. The curve representing the solid–vapor equilibrium portion of the curve is changed very little, as it still represents the equilibrium between the solid and vapor phases of the solvent. If a compound or a solid solution were formed between the solvent and solute as cooling occurred, the phase diagram would be more complex.

The definition of colligative properties in the preceding section stated that these properties are independent of the nature of the solute particles. Thus, ideally one mole of any nonvolatile nonelectrolyte solute, dissolved in a given weight of solvent, say 1000 g, will have the same effect on the colligative properties. This is true because one mole of each solute contains the same number of particles, Avogadro's number. The effects on boiling point and freezing point caused by one mole of solute per 1000 g of solvent (this is a 1 molal solution) are called the *molal elevation of the boiling point* and the *molal depression of the freezing point*, respectively. These are constants that can be experimentally determined for any solvent. Actually, it is not necessary to prepare an exactly 1 molal solution. Because the effect is directly proportional to the amount of solute dissolved in a definite weight of solvent, we can use any convenient concentration and make appropriate calculations. Molal boiling point elevation and freezing point lowering constants for several solvents are given in Table 14.6. Once the constant for a given solvent has been determined, we can calculate the boiling point elevation and the freezing point depression of a given solution of known molality by using the formulas

$$\Delta T_b = K_b\, m \qquad (16)$$

boiling point elevation

$$\Delta T_f = K_f\, m \qquad (17)$$

freezing point depression

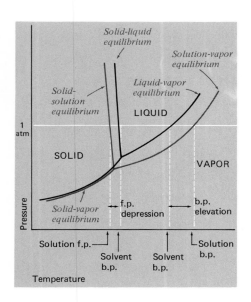

FIGURE 14.12
Comparison of phase diagrams of pure water and an aqueous solution of a nonvolatile solute.

As we have said, all particles contribute equally to a colligative property. Therefore, Equations (16) and (17) can be used for solutions that contain more than one solute; in such cases m is the total molality of all solutes.

EXAMPLE 14.10

■ What are the boiling and freezing points at 1 atm of the aqueous sucrose solution described in Examples 14.2 and 14.3 (Sections 14.11 and 14.12)?

The increase in the boiling point is found by substituting $m = 1.252$ (see Example 14.3) and $K_b = 0.512°C$ (Table 14.6) into Equation (16)

$$\Delta T_b = K_b m = (0.512°C)(1.252) = 0.641°C$$

Likewise for the freezing point lowering, Equation (17)

$$\Delta T_f = K_f m = (1.86°C)(1.252) = 2.33°C$$

Therefore, since the freezing point is lowered 2.33°C, the solution will freeze at $0.00°C–2.33°C = -2.33°C$; and since the boiling point is raised 0.641°C, the solution will boil at $100.000°C + 0.641°C = 100.641°C$.

14.18 Molecular weight determination

Any of the colligative properties can be used to determine the molecular weights of nonvolatile nonelectrolytes. Experimentally, freezing point lowering is easiest to measure, and therefore it is most often applied to the determination of molecular weights. The freezing point lowering of a solution of a known weight of solute in a known weight of solvent is found experimentally. The molality of the solution is found from $\Delta T_f = K_f m$. The molar weight of the solute is found from the calculated molality of the solution and the known amount of solute. The molar weight in grams is, of course, numerically equal to the molecular weight in atomic mass units.

EXAMPLE 14.11

■ Either camphor ($C_{10}H_{16}O$, mol wt = 152.24 amu) or naphthalene ($C_{10}H_8$, mol wt = 128.19 amu) can be used in mothballs. A 5.0-g sample of mothballs was dissolved in 100.0 g of ethyl alcohol and the resulting solution had a boiling point of 78.91°C. Determine whether the mothballs were made of camphor or naphthalene. The pure alcohol had a boiling point of 78.41°C and $K_b = 1.22°C$ for this solvent.

TABLE 14.6

Molal boiling point elevation (K_b) and molal freezing point lowering (K_f) constants for several solvents

Solvent	K_b (°C)	K_f (°C)
Water	0.512	1.86
Benzene	2.53	4.90
Nitrobenzene ($C_6H_5NO_2$)	5.24	7.00
Biphenyl	—	8.00
Ethylene dibromide	—	11.80
Naphthalene	—	6.8
Carbon disulfide	2.34	—
Carbon tetrachloride	5.03	32
Ethyl alcohol	1.22	—
Methyl alcohol	0.83	—

The boiling point elevation was

$$\Delta T_b = \text{b.p. solution} - \text{b.p. solvent}$$
$$= 78.91°C - 78.41°C$$
$$= 0.50°C$$

The molality of the solution was

$$m = \frac{\Delta T_b}{K_b} = \frac{0.50°C}{1.22°C} = 0.41$$

The number of moles of solute is found as follows:

$$n_{\text{solute}} = \frac{0.41 \text{ mole solute}}{1 \text{ kg solvent}} \times \frac{1 \text{ kg}}{1000 \text{ g}} \times 100 \text{ g solvent}$$
$$= 0.041 \text{ mole}$$

Thus the calculated molar weight is

$$\text{molar weight} = \frac{5.0 \text{ g}}{0.041 \text{ mole}} = 120 \text{ g/mole}$$

The value of 120 g/mole for the molar weight indicates that naphthalene, molecular weight 128.19 amu, was used to make the mothballs.

■

It should be pointed out that the boiling point elevation constant of a solvent is not really a constant. It varies slightly with the nature of the solute and the concentration of the solution. A great many so-called scientific constants are subject to minor variations with changing conditions, and the results of calculations involving such constants are therefore only approximate, as shown by the result in Example 14.11.

Usually molecular weights over 1000 are difficult to measure because of the large mass of material that must be dissolved in order to produce significant changes in the vapor pressure, freezing point, or melting point. For high molecular weight substances, osmotic pressure offers a better method for obtaining molecular weight.

14.19 Osmotic pressure

Liquids can diffuse through skin and other biological membranes, as well as through parchment, cellophane, and polyvinyl chloride membranes. However, most dissolved particles do not readily pass through such **semipermeable membranes**—membranes that allow diffusion of solvent and some small solute molecules, but not larger solute molecules. The passage of *solvent* molecules through a semipermeable membrane from a more dilute solution into a more concentrated solution is called **osmosis.**

If a vessel is divided into two compartments by a semipermeable membrane and the same pure liquid is placed in both halves of the vessel, molecules of the liquid diffuse through the membrane with equal freedom in both directions, and there is no apparent change. If, however, a substance is dissolved in the liquid on one side of the membrane, the effective concentration of the solvent on that side is lowered and fewer solvent molecules will diffuse through the membrane. The molecules on the other side of the membrane can still diffuse freely, however, with the result that the rates of diffusion of liquid in the two directions are unequal. The liquid levels on the two sides of the membrane soon become different. The process continues until the weight of the liquid on the solution side (Figure 14.13) of the membrane exerts a sufficiently great pressure that diffusion of solvent molecules in the two directions again proceeds at the same rate. Solutions of different concentrations would behave in the same way, with more molecules moving from the more dilute solution into the more concentrated solution until the diffusion rates and the concentrations of solvent become equal.

Osmotic pressure is defined as the external pressure exactly sufficient to oppose osmosis and stop it. Note that osmotic pressure would be a pressure on the side of the membrane with the more concentrated solution. Osmotic pressure is a colligative property and in dilute solutions is proportional to the concentration of solute particles and independent of the nature of these particles. Osmosis can be reversed by applying a pressure greater than the osmotic pressure, a principle utilized in the desalination of seawater (Section 13.15).

The mathematical expression for osmotic pressure, Π, in dilute solution resembles that for the ideal gas law.

$$\Pi = \frac{n_B RT}{V} = M_B RT \qquad (18)$$

where n_B is moles of solute, V is solution volume in liters, R is the ideal gas constant (0.0821 liter atm/mole deg for Π in atmospheres), T is the absolute temperature, and M is the molarity of the solution. Osmotic pressures have a large range of magnitudes, depending on the nature of the membrane and the solutions. In the leaves of some trees, osmotic pressures are as large as 20 atm.

EXAMPLE 14.12

■ What pressure would have to be applied to the solution side of a semipermeable membrane separating pure water from the aqueous sucrose solution described in Example 14.2 (Section 14.11) to keep solvent flow from taking place? The solution contains 0.08764 mole of sucrose in 100 g of solution. Assume the temperature to be 25°C and the density of the solution to be 1.1290 g/ml.

In order to use Equation (18) for Π, we must find the molarity of the

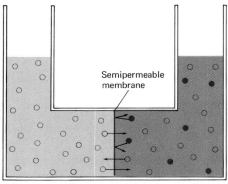

Semipermeable membrane

Pure solvent of diffusible molecules

Solution containing nondiffusible solute molecules

Equilibrium

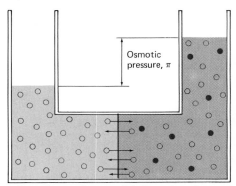

Osmotic pressure, π

Pure solvent, volume decreased because of diffusion through membrane

Solution, volume increased and solution concentration decreased because of solvent diffusion

FIGURE 14.13
Osmosis.

solution. For 100.00 g of solution (the basis), we have 0.08764 mole of sucrose and

$$V = \frac{m}{d} = \frac{100.00 \text{ g}}{1.1290 \text{ g/ml}} = 88.57 \text{ ml} = 0.08857 \text{ liter}$$

giving

$$M = \frac{n_{\text{sucrose}}}{V_{\text{solution}}(\text{liters})} = \frac{0.08764 \text{ mole}}{0.08857 \text{ liter}} = 0.9895 M$$

The osmotic pressure is

$$\Pi = MRT = \left(0.9895 \frac{\text{mole}}{\text{liter}}\right)\left(0.0821 \frac{\text{liter-atm}}{\text{mole}^\circ \text{K}}\right)(298^\circ\text{K})$$

$$= 24.2 \text{ atm}$$

In order to stop water flowing from the pure phase through the membrane to the solution phase, 24.2 atm of pressure would have to be applied to the solution side of the membrane. ∎

14.20 Colligative properties of electrolyte solutions

Both strong and weak electrolytes yield more than one particle per formula unit in solution. Since colligative properties depend solely upon the number of particles, they offer a way of experimentally exploring how many particles are present in electrolyte solutions.

We have arrived at this discussion of the colligative properties of electrolytes with an understanding of the nature of ionic bonding and ionic compounds. Historically, however, it was the study of boiling point elevation and freezing point lowering, osmotic pressure, and vapor pressure that led to the modern theory of the dissociation of ionic compounds in solution.

In the 1880s a Dutch chemist, Jacobus van't Hoff, found that it was necessary to introduce a factor i (now called the van't Hoff factor) into the equation for any colligative property in order to bring about agreement with experimental results for electrolytes. For example, the freezing point depression found from experiment is related to that calculated from molarity as follows:

$$\Delta T_f(\text{exptl}) = i \, \Delta T_f(\text{calc}) = iK_f m \tag{19}$$

It should be apparent to us that the factor i is just the number of particles formed by an electrolyte in solution. For, say, $BaCl_2$ each formula unit gives 3 ions in solution and i should equal 3. A nonelectrolyte yields one particle for each molecule and $i = 1$.

As a solution is made more dilute, the values of i for strong electrolytes approach the whole numbers expected on the basis of com-

TABLE 14.7

van't Hoff factor, i, for solutions of strong electrolytes

These values were calculated from freezing point depression measurements.

Concentration of solution	van't Hoff factor		
	NaCl	MgSO₄	K₂SO₄
0.1 m	**1.87**	**1.21**	**2.32**
0.01 m	**1.94**	**1.53**	**2.69**
0.001 m	**1.97**	**1.82**	**2.84**
	approaches 2	approaches 2	approaches 3

plete dissociation (Table 14.7). The deviation of i from whole numbers occurs because interionic attractions decrease the effectiveness of the individual ions. In other terminology, as we have seen (Section 14.10), the activity coefficients of electrolytes are less than 1.

Weak electrolytes are not, of course, expected to be completely ionized. Their degree of ionization (represented by α, alpha) can be calculated from the value of i found from colligative property measurements. The following is the simple relationship between the van't Hoff factor i; the degree of ionization, α; and ν ("nu"), the number of ions per formula unit formed by the electrolyte:

$$\alpha = \frac{i - 1}{\nu - 1} \tag{20}$$

EXAMPLE 14.13

■ Find the degree of ionization for HF, hydrofluoric acid, in an $0.100m$ aqueous solution if the freezing point of the solution is $-0.197°C$.

The van't Hoff factor is

$$i = \frac{\Delta T_f}{K_f m} = \frac{0.197°C}{(1.86°C)(0.100)} = 1.06$$

If HF acted as a strong electrolyte, it would form two ions for each molecule. Substituting $\nu = 2$ and $i = 1.06$ into the expression for α [Equation (20)] gives

$$\alpha = \frac{i - 1}{\nu - 1} = \frac{1.06 - 1}{2 - 1} = 0.06$$

At this concentration, HF is only 6% ionized—definitely a weak acid.
■

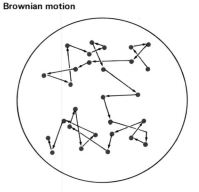

Brownian motion

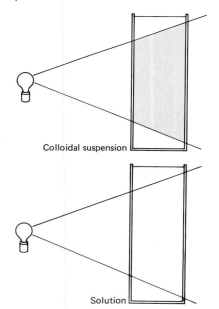

Tyndall effect

Colloidal suspension

Solution

FIGURE 14.14
Light and colloids.

14.21 Properties of colloids

If a sample of mud is stirred into water, the mud may be kept in suspension, but as soon as the stirring is stopped, the larger particles fall to the bottom of the vessel. The smaller particles remain in suspension longer, but most of them settle out before long. There are usually some, however, that stay in suspension for several days, or perhaps indefinitely. The water containing this part of the suspension may look murky, but since individual particles cannot be seen in it even under powerful magnification, they must be exceedingly small. This conclusion is supported by the fact that the suspension cannot be clarified by filtering, even through a filter with very fine pores.

A mixture in which particles remain suspended and cannot be removed by filtration is called a **colloidal suspension,** a *colloidal dispersion,* or simply a *colloid.*

Strictly speaking, a **colloid** is a substance made up of particles larger than most molecules, but too small to be seen in an optical microscope. The colloid size range is not too clearly defined, but is approximately 1 to 200 nm.

The very small particles of mud in the example cited are heavier than water, yet they do not sink. This phenomenon exists because the molecules of water are in constant motion and subject the suspended particles to frequent and continued bombardment. The bombardment is not vigorous enough to keep large particles in suspension, but it does keep the smaller ones in motion, and prevents them from settling.

When a colloidal suspension is brightly illuminated by a beam of light at right angles to the line of sight and examined under a microscope, the individual particles cannot be seen, but they are detected as tiny flashes of light dancing in the liquid. This motion, called *Brownian motion* after its discoverer Robert Brown (1773–1858), is constant but quite irregular (Figure 14.14). For many years after its discovery, the cause of the Brownian motion was not known. Several theories were advanced to explain it, Brown himself suggesting that the dancing particles were alive. Recognition that the movement of the suspended particles is due to bombardment by rapidly moving molecules (other forces are now also thought to contribute) and that the speed increases with temperature led to the development of the kinetic-molecular theory.

When a strong beam of light passes through a colloid, the reflection from the suspended particles makes the beam clearly visible. This is called the *Tyndall effect* and is frequently used to distinguish between true solutions and colloids. A familiar example of the Tyndall effect occurs in a dark, dusty room through which a ray of light passes. The dust particles are colloidally suspended in air, and even though they are not individually visible, the path of the beam is easily seen by the reflection of light from the particles.

Colloidal particles can coalesce into larger particles when they collide with each other. However, in some cases the particles do not coalesce even when the suspension stands for a long time. In spite of their incessant motion, collisions between the particles are apparently very rare. Such behavior would hardly be expected unless the particles actually repel each other. The validity of this suggestion is confirmed by the observation that under the influence of an electric potential colloidal particles migrate toward one of the electrodes. The suspended particles adsorb or collect on their surfaces hydrogen ions or negative ions such as hydroxide ions from the liquid in which they are suspended, and thus become either positively or negatively charged. In a given colloidal suspension, all of the suspended particles have the same charge. Because of their like charges, the particles repel each other and are kept from accumulating into larger, noncolloidal particles. Some colloids are positively charged and some negatively charged. The sign of the charge depends largely upon the composition of the colloid. For some substances, however, it is possible to obtain suspensions in which the colloidal particles can be either positive or negative, depending upon the mode of preparation. If a negative colloid and a positive colloid are mixed, both of them precipitate because the particles no longer repel each other. Colloidal particles may also be precipitated by neutralizing their charges in other ways, for example, by adding electrolytes or by electrolysis.

Other properties of colloidal particles depend upon the fact that they have very large surface areas in comparison with their diameters, a geometrical property characteristic of small objects. It can be understood by considering a series of cubes of different sizes. For a cube 1 cm on each edge, the ratio of surface to volume is 6 cm^{-1} (surface, 6 cm^2; volume, 1 cm^3); for a cube 0.5 cm on each edge, the ratio is 24 cm^{-1}. As the size of the cube is decreased further, the ratio of surface to volume becomes larger and larger. For a cube 10^{-7} cm on each edge, which is approximately the diameter of the smallest colloidal particles, the ratio of surface to volume is 6×10^7 cm^{-1}, or 60,000,000 cm^{-1}.

Colloidal particles are not necessarily cubic, of course, but similar geometric ratios could be calculated for objects of other shapes, and it is evident that the total surface area of the suspended particles in a colloidal suspension is tremendous. For this reason, colloidal particles are very efficient at adsorbing other substances on their surfaces.

14.22 Types of colloids

Rather than use the terms solute and solvent for colloidal suspensions, we may speak of the suspending medium and the dispersed substance. All three states of matter may serve in either capacity (Table 14.8).

The colloid described earlier (mud) consists of solid particles suspended in a liquid—a type of colloid called a **sol.** Sols are very

TABLE 14.8
Types of colloids

Suspending medium	Dispersed substance	Type of colloid	Examples
Liquid	Gas	Foam[a]	Soap suds, whipped cream, beer foam
	Liquid	Emulsion	Mayonnaise, milk, face cream
	Solid	Sols, gels[b]	Protoplasm, starch, gelatin and jelly, clay
Gas	Liquid	Liquid aerosol	Fog, mist, aerosol spray
	Solid	Solid aerosol	Smoke, airborne bacteria and viruses
Solid	Gas	Solid foam	Aerogels, polyurethane foam
	Liquid	Solid emulsions, some gels	Cheese
	Solid	Solid sol	Ruby glass, some alloys

[a] Stable foams are usually formed only when a foaming agent, such as a soap, is present in addition to the pure liquid and gas.
[b] Sols contain individual dispersed particles; in gels the particles link together in a structure of some strength.

common and examples of them are easily prepared in the laboratory. If, for example, a small amount of a solution of iron(III) chloride is poured slowly into a large volume of boiling water, and the boiling is continued for some time, a clear red sol is formed:

$$2\ FeCl_3(aq) + (x + 3)H_2O(l) \longrightarrow Fe_2O_3 \cdot xH_2O(\text{red sol}) + 6\ HCl(aq)$$

Similarly, a beautiful yellow sol of As_2S_3 may be prepared by passing hydrogen sulfide gas through a 0.5% solution of arsenous acid (H_3AsO_3) for a few minutes, and then boiling the liquid to drive out the excess hydrogen sulfide:

$$2\ H_3AsO_3(aq) + 3\ H_2S(g) \longrightarrow As_2S_3(\text{yellow sol}) + 6\ H_2O(l)$$

Under other conditions, the same reaction leads to precipitation of the arsenic sulfide instead of formation of a sol.

Sols in which the suspended particles are metallic can be prepared by reducing a salt of the metal in solution:

$$4NaAuCl_4(aq) + 3N_2H_4(aq) \longrightarrow 4Au(\text{red sol}) + 3N_2(g) +$$
$$\underset{\substack{\text{sodium} \\ \text{tetrachloroaurate(III)}}}{} \quad \underset{\text{hydrazine}}{} \qquad 4NaCl(aq) + 12HCl(aq)$$

or by passing an electric current under very high voltage between two strips of metal beneath the surface of water. Under these conditions, small particles vaporize from the surface of the metal and condense to colloidal size.

A gel is a special type of sol in which the solid particles, usually very large molecules, unite in a random and intertwined structure that gives rigidity to the mixture. For example, the carbohydrate pectin from fruit is the dispersed substance that stiffens grape jelly, or any other kind of jelly.

A colloid in which particles of a liquid are suspended in another liquid is called an **emulsion**. Colloidal suspensions of oil in water, and of water in oil, are readily prepared if an emulsifying agent is added to change the surface tensions of the two liquids appropriately. Mayonnaise is a familiar example of an emulsion. It consists chiefly of a salad oil and either lemon juice or vinegar; the emulsifying agent is uncooked egg. Milk is an emulsion in which particles of butterfat and protein are suspended in water. The particles of butterfat are a little too large to stay in suspension indefinitely, and in a few hours, they rise to the surface as cream. In homogenized milk, the droplets of butterfat have been broken into particles of colloidal dimensions, so the cream does not rise. Face creams and many other cosmetics are also emulsions.

Foam is a colloid consisting of tiny bubbles of gas suspended in a liquid. Most foams in which the liquid phase is pure water are short lived, but if the surface tension of the water is reduced by the addition of a surface-active agent, very stable foams may be generated. Common substances that can be added to water to produce this effect are soap and licorice, both of which lower the surface tension of water. Whipped cream is both an emulsion and a foam, for in it both butterfat and air bubbles are colloidally suspended.

Smoke is a colloidal suspension of solid particles in air, and fog is a colloid consisting of suspended droplets of moisture in the air. A colloidal suspension in which air (or any gas) is the suspending medium is called an **aerosol.**

The precipitation of colloidal particles is illustrated by the operation of electrostatic smoke and dust removers. In the Cottrell precipitator, for example, metal plates carrying a strong electrostatic charge are mounted in a smokestack. Because of their electrical charge, the colloidal smoke particles that pass through the stack are attracted to the plate of opposite charge and, upon contact with it, lose their charge and precipitate (Figure 14.15).

Many colloids are known in which the suspending medium is a solid, and the suspended particles are either gaseous or solid. The familiar "aerogels," which are used to adsorb odoriferous gases, consist of silica (SiO_2) containing pores of colloidal dimensions. Gases are strongly adsorbed on the surfaces of the pores. Air is adsorbed, but is readily displaced from the colloid by gases containing heavier molecules.

A very efficient device for cleaning grease spots from clothes consists of a can containing colloidal silica suspended in a mixture of a volatile oil and a pressurized gas. When the pressure is released by opening the valve, a jet of the fluid is ejected from the can and is directed on the grease spot. The liquefied gas immediately evaporates; the oil dissolves the grease in the spot and this solution is adsorbed on the silica. The oil evaporates in a few minutes, and the silica with the adsorbed grease is brushed from the garment.

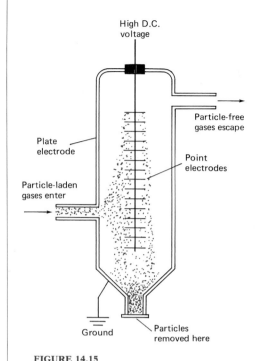

FIGURE 14.15

The Cottrell precipitator. *Installed near the top of the smokestack, the precipitator removes charged colloidal particles from the smoke.*

Except for the bones and the teeth, our own bodies are largely colloidal, the protein and fat existing as an aqueous gel. The proteins form chains that are only a few atoms in diameter, but many thousands of atoms long. Such chains present a tremendous amount of surface in relation to their weight and hence they show colloidal behavior. The fats are of lower molecular weight, but they too are colloidally dispersed in the aqueous medium. The colloidal particles, whether they be proteinaceous or fatty, are hydrogen bonded to each other and to water molecules, and so form a semirigid mass. Under some conditions, the body may retain too much water. This results in swelling (called edema) and often, a considerable amount of discomfort. The condition can frequently be relieved by including potassium salts in the diet, for it is the K^+/Na^+ ratio that determines how tightly the colloidal materials in the tissues bind water.

THOUGHTS ON CHEMISTRY
To tell a chemist

ASIMOV ON CHEMISTRY, *by Isaac Asimov*

How does one distinguish quickly and easily between a specialist and a well-primed nonspecialist? It seems to me you must find little things no one would ever think to prime the nonspecialist upon.

Since I know the chemical profession best I devised two questions, for instance, to tell a chemist from a nonchemist. Here they are:

(1) How do you pronounce UNIONIZED?

(2) What is a mole?

In response to the first question, the nonchemist is bound to say "YOON-yun-ized," which is the logical pronunciation and the dictionary pronunciation, too. The chemist, however, would never think of such a thing: he would say without a moment's hesitation: "un-EYE-on-ized."

In response to the second question, the nonchemist is bound to say, "A little furry animal that burrows underground," unless he is a civil engineer who will say, "a breakwater." A chemist, on the other hand, will clear his throat, and say "well, it's like this—" and keep talking for hours. . . .

[Now some chemists], in the nervous realization that "mole" is a little furry animal that burrows underground, try to use "mol" instead. I was forced to use "mol," in a textbook I once wrote, by the overriding vote of my two coauthors, a state of affairs which led to internal bleeding. The word is universally pronounced with a long "o" and "mol" must clearly have a short "o." Consequently, where I am my own master, I use "mole." Do you hear me world? "Mole." [We hear you. The Authors.]

ASIMOV ON CHEMISTRY, by Isaac Asimov (pp. 78–79. Copyright © 1974 by Isaac Asimov. Reprinted by permission of Doubleday & Co.)

[14.22] SOLUTIONS
AND COLLOIDS

Significant terms defined in this chapter

solubility	weak electrolytes	molality
infinitely miscible	distillation	Raoult's law
immiscible	distillate	osmosis
partially miscible	fractional distillation	osmotic pressure
saturated solution	refluxing	colloidal suspension
unsaturated solution	azeotropes	colloid
supersaturated solution	colligative property	sol
electrolyte	semipermeable membrane	gel
nonelectrolyte	ideal solution	emulsion
Henry's law	activity	foam
strong electrolytes	activity coefficient	aerosol

Exercises

14.1 A chemist wanted to remove traces of chlorine from a sample of air. In his first attempt he bubbled the gaseous mixture through a technical grade sample of carbon tetrachloride, which contained chloroform (m.p. $-63.5°C$, b.p. $61.7°C$) and naphthalene (m.p. $80.55°C$) as impurities. After smelling the "purified" gas, the chemist noted that a faint smell of chlorine was still present, so he next passed the gas through a stainless steel (18% Cr, 77% Fe) column containing "activated" charcoal. After the second process, the air was essentially chlorine-free. There are at least seven different solutions involved in the above procedures—for example, the chlorine-free air is a solution of a gas (oxygen, the major solute) dissolved in a gas (nitrogen, the solvent). Identify solutions that are examples of (a) a gas dissolved in a gas, (b) a gas dissolved in a liquid, (c) a gas dissolved in a solid, (d) a liquid dissolved in a liquid, (e) a solid dissolved in a liquid, and (f) a solid dissolved in a solid. In each case, specify which component is the solvent and which is the solute.

14.2 In each case distinguish clearly among: (a) a suspension, a solution, and a colloidal suspension; (b) solvent and solute; (c) infinitely miscible, partially miscible, and immiscible substances; (d) saturated, unsaturated, and supersaturated solutions; (e) electrolytes and nonelectrolytes; (f) strong and weak electrolytes; (g) a sol and a gel; and (h) smoke and fog.

14.3 Which statements are true? Rewrite any false statement so that it is correct.
(a) The solubility of naphthalene would be predicted to be greater in water than in benzene.
(b) As a solution freezes, (i) the solvent crystallizes, (ii) the solution becomes more concentrated, (iii) the freezing point is depressed even more, and (iv) the freezing takes place over a range of temperatures.

(c) An emulsifying agent stabilizes a colloidal suspension of a solid in a liquid.
(d) Solubility is the maximum amount of solute that can be dissolved in a solvent.
(e) The van't Hoff factor is the proportionality constant between the predicted or ideal concentration and the activity.
(f) Nonideal liquid–liquid solutions show positive and/or negative deviations from ideality.

14.4 For each of the following solutions, predict whether the relative solubility is "high" or "low" and whether the solute will be a strong electrolyte, a weak electrolyte, or a nonelectrolyte: (a) HCl in H_2O, (b) HCl in CCl_4, (c) S in CS_2, (d) C_2H_5OH in H_2O, (e) H_2O in C_2H_5OH, and (f) NH_3 in H_2O.

14.5 Describe the processes by which NaCl, HCl, and CH_3OH dissolve in water. Prepare sketches showing how these compounds exist in aqueous solution.

14.6 Why is dry HCl in the liquid state a nonelectrolyte, but the same substance in aqueous solution is a good electrical conductor?

14.7 A solution shows positive deviation from Raoult's law for both components A and B. What can be implied about the intermolecular forces between A and A, B and B, and A and B?

14.8 How does the solubility of a solution of a gas in a liquid vary with increasing (a) temperature and (b) pressure? (c) Is there an appreciable pressure dependence of the solubility of a liquid or solid solute in a liquid? (d) What determines whether the heat of solution of a liquid in a liquid is endothermic or exothermic? (e) What will be the temperature dependence of the solubility of a liquid or solid in a liquid if the heat of solution is endothermic?

14.9 Those concentration units that are based on volume of solution, solvent, solute, etc., are dependent on temperature. Which of these units depends on temperature: molality, molarity, normality, mole fraction, wt %, and vol %?

14.10 The solubility of NH_3 in water at 25°C and 1 atm pressure of NH_3 is 48 g NH_3/100 g H_2O. Would a solution containing 45 g NH_3/100 g H_2O under these conditions be considered (a) concentrated or dilute, (b) saturated or unsaturated, and (c) a strong or weak electrolyte? Repeat the exercise for a solution containing 4×10^{-3} g $AgIO_3$/100 g H_2O at 20°C (this value is the solubility).

14.11 The cleansing action of soaps and detergents (see Section 13.12) illustrates how an insoluble material can be suspended in a solvent. Prepare a sketch showing this mechanism.

14.12* If NaCl is completely dissociated in aqueous solution, why does it give less than twice the molal freezing point depression than does an equivalent number of moles of a nonelectrolyte?

14.13* The formula of the mineral chromite (the only ore mineral of chromium) is written as $(Mg,Fe)Cr_2O_4$. Why would we expect Mg^{2+} and Fe^{2+} to be able to replace each other in this mineral?

14.14* The enthalpy change for dissolving $MgBr_2(s)$ in 25 moles of H_2O is −43.0 kcal/mole and for dissolving $MgBr_2 \cdot 6H_2O(s)$ in 25 moles of H_2O is −1.2 kcal/mole. (a) Compare the relative magnitudes of the hydration and lattice energies for these compounds. (b) Will the solubilities of these substances increase or decrease with increasing temperature? (c) For which substance will the solubility vary more with temperature?

14.15* What quantity is commonly used to approximate the activity of (a) a pure solid, (b) a pure liquid, (c) a solute, (d) a gas, and (e) a liquid in a solution?

14.16 Henry's law constants at 25°C for N_2 and O_2 in water are 4.34×10^5 torr/(g/100 g H_2O) and 1.93×10^5 torr/(g/100 g H_2O), respectively. Assume the partial pressure of N_2 to be 608 torr and of oxygen to be 152 torr. Find the solubility of these gases in water. Is the ratio of the mass of O_2 to N_2 greater or less in water than in air?

14.17 Hydrogen chloride and water form an azeotrope at 1 atm containing 20.2 wt % HCl. This azeotrope has a density of 1.102 g/cm³. What is the molarity of this solution?

14.18 What is the molarity of a solution containing 100.00 g of $Ba(OH)_2 \cdot 8H_2O$ dissolved in enough water to make exactly one liter of solution?

14.19 Which aqueous solution has a higher Cl^- concentration: 0.05m HCl, 15 wt % NaCl, or a $CaCl_2$ solution in which the mole fraction of $CaCl_2$ is 0.10?

14.20 Given a sufficient quantity of 4.37M NaOH, describe how you would prepare exactly 250 ml of 0.874M NaOH.

14.21 What is the vapor pressure above a solution containing 100.00 g of water and 10.0 g of urea, $CO(NH_2)_2$, a nonvolatile solute? The vapor pressure of pure water at this temperature is 23.76 torr.

14.22 Using Raoult's law, predict the partial pressures and total pressure over a solution containing 0.300 mole acetone ($P° = 345$ torr) and 0.200 mole chloroform ($P° = 295$ torr).

14.23 An aqueous solution containing 0.502 g of an organic compound (molar wt = 60.0 g/mole) dissolved in 7.50 g of water had a boiling point of 100.57°C. What is the molality of the solution? Calculate the boiling point constant for water at 760 torr.

14.24 The molal freezing point constant for copper is 23°C. If pure copper melts at 1083°C, what will be the melting point of a brass made of 10 wt % Zn and 90 wt % Cu?

14.25 Methyl alcohol, CH_3OH, and ethylene glycol, $CH_2OH–CH_2OH$, are both used to prevent the freezing of water in automobile radiators in cold weather. Which will be more effective in a given radiator, 100 g of methyl alcohol or 100 g of ethylene glycol? Which will be more effective, a 5m solution of methyl alcohol or a 5m solution of ethylene glycol?

14.26 A solution made by dissolving 3.75 g of a nonvolatile solute in 95.0 g of acetone boiled at 56.60°C. The boiling point of pure acetone is 55.95°C and $K_f = 1.71$°C. Calculate the molar weight of the solute.

14.27 If a sugar maple tree grows to a height of 40 ft, what must be the concentration of sugar in the sap so that osmotic pressure forces the sap to the top of the tree at 0°C? The density of the sap can be considered to be 1.00 g/cm³ and of mercury to be 13.6 g/cm³.

14.28 The street departments of many cities are replacing NaCl by $CaCl_2$ for use on icy streets. The solubility of $CaCl_2$ in water at 0°C is 59.5 g/100 g H_2O and of NaCl is 35.7 g/100 g H_2O. Which solution has the lower freezing point? Assume complete dissociation of the ions.

14.29 A 0.050m aqueous solution of $K_3Fe(CN)_6$ has a freezing point of −0.2800°C. Calculate the van't Hoff factor and interpret your results.

14.30 What is the surface area of a sugar cube 1 cm along an edge? What is the surface area if the cube is divided into 1000 identical cubes each 0.1 cm along an edge? Repeat the calculation for a cube divided into 10^6 cubes each 0.01 cm along an edge.

14.31* At 1 atm the solubility of CO in water at 25°C is 0.002603 g/100 g H_2O and that of SO_2 in water is 9.41 g/100 g H_2O. What is the Henry's law constant for each gas? What is the molality of each solution? What would be the freezing point of each solution? Why is there such a large difference in the solubility of these gases?

14.32* At 1 atm pressure ethyl alcohol and water form an azeotrope at 78.7°C which contains 95.6 wt % CH_3CH_2OH. What is the "proof" of this solution if "proof" is defined as twice the volume percent measured at 60°F? The densities of pure water and alcohol at 60°F are 0.9991 and 0.793 g/cm³, respectively.

14.33* The density of a sulfuric acid solution taken from a car battery is 1.225 g/cm³. This corresponds to a $3.75M$ solution. Express the concentration of this solution in molality, mole fraction of H_2SO_4, and wt % of water.

14.34* The solubilities of $NaClO_3$ and KNO_3 in water at 50°C are 140 and 85.5 g/100 g H_2O, respectively. However, the solubilities of $KClO_3$ and $NaNO_3$ at this temperature are 19.3 and 114 g/100 g H_2O, respectively. Describe what would happen if you tried to prepare a solution that was saturated with respect to both $NaClO_3$ and KNO_3 at 50°C.

14.35* The heat of formation of $NH_4NO_3(s)$ at 25°C is -87.37 kcal/mole, of NH_4NO_3 ($1M$) is -82.0 kcal/mole, and of NH_4NO_3 ($0.1M$) is -81.2 kcal/mole. Calculate the heat of solution for preparing a $1M$ solution,

$$NH_4NO_3(s) + nH_2O \longrightarrow NH_4NO_3(1M)$$

and the heat of dilution for preparing a $0.1M$ solution from a $1M$ solution,

$$NH_4NO_3(1M) + nH_2O \longrightarrow NH_4NO_3(0.1M)$$

14.36* One gram each of NaCl, NaBr, and NaI was dissolved in 100.0 g water. What is the vapor pressure above the solution at 100°C?

14.37* The molecular weight of an organic compound was determined by measuring the freezing point depression of a benzene solution. A 0.500-g sample was dissolved in 50.0 g of benzene ($K_f = 2.53$°C) and the resulting depression was 0.42°C. What is the approximate molecular weight? The compound gave the following elemental analysis: 40.0

wt % C, 6.67 wt % H, and 53.3 wt % O. Determine the formula and exact molecular weight of the substance.

14.38* An 11.5-qt automobile cooling system containing ethylene glycol (molar wt = 62.07 g/mole) was "safe" to -20°F. A very cold night was expected and the owner decided to change the protection limit to -40°F. How was it done? The density of the original solution is 1.063 g/cm³ and of pure antifreeze 1.116 g/cm³.

14.39* A sample of heroin ($C_{21}H_{23}O_5N$, molar wt = 369 g/mole) "cut" with "milk sugar" (lactose, $C_{12}H_{22}O_{11}$, molar wt = 342 g/mole) was analyzed by osmotic pressure to determine the amount of sugar present. If 100.00 ml of solution containing 1.00 g of the cut heroin had an osmotic pressure of 539 torr at 25°C, what was the percent sugar present?

14.40* The major components of seawater are 18,980 ppm Cl^-, 10,561 ppm Na^+, 2652 ppm SO_4^{2-}, 1272 ppm Mg^{2+}, 400 ppm Ca^{2+}, 380 ppm K^+, 142 ppm HCO_3^-, and 65 ppm Br^-. A ppm (part per million) means 1 g of solute in 10^6 g of solution. Calculate the molality of seawater with respect to each ion. What would be the predicted normal freezing and boiling points? Assuming molarity and molality to be equal, what pressure would be required in order to obtain pure water by reverse osmosis?

14.41* An aqueous solution contained 1.00 g each of acetic acid (CH_3COOH, $\alpha = 0.04$); ammonium sulfate [$(NH_4)_2SO_4$, $\alpha = 1$]; and formic acid ($HCOOH$, $\alpha = 0.06$) in 100.0 g H_2O. What is the freezing point of this solution?

14.42* The freezing point of a 1.00 wt % aqueous solution of acetic acid, CH_3COOH, is -0.31°C. What is the approximate molar weight of acetic acid in water? A 1.00 wt % solution of acetic acid in benzene ($K_f = 4.90$°C) has a freezing point depression of 0.441°C. What is the molar weight of acetic acid in this solvent? Explain the difference.

14.43** The density of a commercial vinegar (5.00 wt % CH_3COOH) is 1.0055 g/cm³. (a) Express the concentration of the solute in molarity, molality, and mole fraction. (b) If the freezing point of the vinegar is -1.65°C, calculate the percent ionization of the weak acid. (c) Calculate the osmotic pressure of the solution at 25°C. (d) Assuming an ideal solution, use Raoult's law to calculate the total vapor pressure of the vinegar. The vapor pressure of water is 23.8 torr and of acetic acid is 15.5 torr at 25°C. The heat of formation of $CH_3COOH(l)$ at 25°C is -484.5 kJ/mole and for CH_3COOH(5.00 wt %) it is -485.54 kJ/mole. (e) Calculate the heat of solution.

15 CHEMICAL KINETICS

In this chapter we first look at two atoms, molecules, or ions as they collide, the conditions under which they might react with each other, and the energy changes involved. Next we examine in a general way how the concentration of the reactants, the temperature, the amount of contact between the reacting substances, and catalysis can affect reaction rates. Finally, we introduce the terminology and mathematics of chemical kinetics and briefly discuss reaction mechanisms and the interpretation of rate equations. In the next chapter we go on to discuss chemical equilibrium.

A chemical equation tells us what the reactants and products are in a particular reaction. It tells us nothing about the pathway from reactants to products, or how much of the products will actually be formed, or the speed of the reaction.

For many reactions, under the right conditions virtually all of the reactants present can be converted to products. Under other conditions, or for other reactions, only small amounts of the products are formed. A reaction proceeds until chemical equilibrium (Section 4.6) is achieved—a dynamic state in which both reactants and products are present in unchanging (but not necessarily equal) amounts. The study of chemical equilibria allows the prediction of what will happen in a reaction—the maximum amounts of products that will be formed.

However, understanding the equilibrium state of a reaction tells us nothing about how fast the reaction takes place. This is the province of chemical kinetics—the study of reaction rates and reaction mechanisms. Knowing the rate of a chemical reaction and the effect on the reaction rate of conditions such as temperature and pressure allows us to predict how fast a reaction will reach equilibrium. And reaction rate studies often give us greater understanding of the exact pathway from reactants to products—the reaction mechanism. This pathway is not always what it appears to be from the equation. For example, the overall reaction

$$A + B \longrightarrow C + D$$

might take place in two steps

$$
\begin{array}{l}
A + A \longrightarrow I + D \\
\underline{I + B \longrightarrow A + C} \\
A + B \longrightarrow C + D
\end{array}
$$

involving an intermediate I that is used up as fast as it is formed and is therefore never observed. These two steps together are the reaction mechanism of the overall reaction.

Knowledge is power, as the saying goes, and the more chemists or chemical engineers know about the equilibria, rates, and mechanisms of reactions, the better they can control events to produce the products they want in desirable quantities under economically attractive conditions.

How reactions take place

15.1 Collision theory

Chemical reactions take place by the transfer of electrons from one atom to another or by a change in the way electrons are shared. For this transfer or sharing to occur the atoms involved must be in contact. It has already been pointed out that atoms and molecules are in constant motion and that there are frequent collisions between them. This is especially apparent in the gaseous and liquid states, but it is true of the solid state, too.

The **collision theory** of chemical reaction assumes that reactions occur only when atoms, or ions, or molecules collide, but that only a very small portion of the collisions result in chemical reaction. In the great majority of collisions, the particles simply strike each other and bounce away unchanged. If a reaction is to take place during a collision, several conditions must be fulfilled.

(1) The transfer or sharing of electrons must give a structure that is capable of existence under the conditions of the collision; that is, stable bonds or new stable species must be formed.

(2) The collision must take place with enough energy that the outer electron shells of the atoms concerned penetrate each other to some extent and that bonding electrons can be rearranged.

(3) The orientation of the molecules when they collide must be such that the atoms directly involved in the transfer or sharing of electrons come into contact with each other.

Reactions between simple positively and negatively charged ions are quite straightforward and uncomplicated. When oppositely charged ions collide, the mutual attraction between them tends to hold them together. In general, reactions of ions are faster than reactions of molecules. Even in ionic reactions, however, only a small fraction of the collisions result in reaction.

Condition (3) becomes more important as the reacting species become larger. For example, molecules of hydrogen chloride and ammonia (both gases) have a much better chance to react if they collide as in (a) in Figure 15.1 than if they collide when they are in other relative positions, such as (b) or (c). The hydrogen atom of the HCl

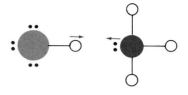

(a) Effective collision, $HCl + NH_3$

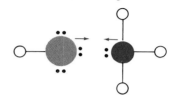

(b) Ineffective collision, $HCl + NH_3$

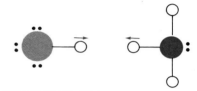

(c) Ineffective collision, $HCl + NH_3$

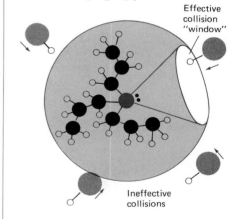

(d) $HCl + N(CH_2CH_2CH_3)_3$

Effective collision "window"

Ineffective collisions

Cl   N   C   H

FIGURE 15.1
Collisions of hydrogen chloride with ammonia and tripropylamine.

molecule must present itself to the unshared pair of electrons on the nitrogen atom. No other approach is apt to lead to reaction. This effect is more pronounced in the reaction of hydrogen chloride with tri-*n*-propylamine, $N(CH_2CH_2CH_3)_3$. Here the bulky hydrocarbon groups further hinder the approach of the HCl molecule to the nitrogen atom (Figure 15.1d). The HCl molecule must approach through a narrow "window" for an effective collision to take place.

The majority of chemical reactions are much more complex than the examples we have given, and the processes involved are often not fully understood. In fact, the number of chemical reactions whose mechanisms are completely understood is very small relative to the number of known chemical reactions.

15.2 Activation energy

Colliding molecules react only if they bring enough energy into the collision to exceed the **activation energy** ΔE_a of the reaction—the energy that the reactants must have for reaction to occur. The activation energy is like a hill that has to be climbed before getting down the other side to the destination. It represents the energy necessary to break existing bonds or to free valence electrons to move from one atom to another.

Figure 15.2 illustrates the role of the activation energy in the reaction of ozone, O_3, and nitric oxide, NO, one of the ways that ozone reacts in the upper atmosphere. The reaction liberates a large quantity of heat—an exothermic reaction (Section 5.2)

$$O_3(g) + NO(g) \longrightarrow NO_2(g) + O_2(g) \qquad \Delta H = -47.7 \text{ kcal}(-199.6 \text{ kJ})$$

Point *A* of Figure 15.2 represents the energy possessed by a mixture of ozone and nitric oxide, and point *D* represents the energy possessed by the products of the reaction, oxygen and nitrogen dioxide. The vertical distance between *B* and *C* is a measure of the energy of activation ΔE_a of this reaction. Point *C* is the energy level of "activated" molecules—molecules that have collided with enough energy to proceed to reaction products. Collisions that do not carry the molecules up to energy level *C* do not produce any reaction.

The vertical distance between *B* and *X* is a measure of the total energy change for the reaction. For a reaction like this one, a gas-phase reaction in which the number of moles of products and reactants are equal, $\Delta E° = \Delta H°$. (This is not the case for other types of reactions.) The distance from *X* to *C* represents the activation energy for the *reverse* of the reaction that we are considering,

$$NO_2(g) + O_2(g) \longrightarrow O_3(g) + NO(g)$$

This reverse reaction is endothermic—the products are at a higher energy level than the reactants.

Every reaction has a characteristic activation energy, and the values can be determined accurately by experiment. Most activation energies fall in the range of 10–50 kcal/mole (about 40–200 kJ/mole)

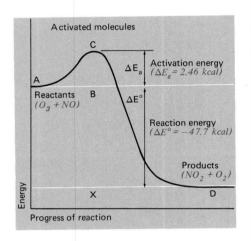

FIGURE 15.2

Energy path of the O_3 + NO reaction. *For a reaction with a smaller energy of activation, the distance BC would be smaller. For a less exothermic reaction the distance BX would be smaller. For an endothermic reaction, point D would be above point A. For this reaction* $\Delta E° = \Delta H°$ *[since* $\Delta H° = \Delta E° + RT\Delta n_g$, *where* Δn_g *is the number of moles of gaseous products minus the number of moles of gaseous reactants, and* $\Delta n_g = 0$ *for this reaction].*

with some close to zero and some as high as 150 kcal/mole (about 625 kJ/mole). The principal factor controlling the magnitude of the activation energy is the nature of the reacting species.

Not all exothermic reactions, it should be noted, proceed rapidly as soon as the reactants are mixed, even though thermodynamics indicates that the reactions are potentially spontaneous. For example, a mixture of hydrogen and chlorine can be kept indefinitely at room temperature. However, if the mixture is exposed to an electric spark or to light of short wavelength, the reaction takes place with explosive speed and evolution of heat. What happens is this: The spark or radiation provides enough energy to "activate" a few chlorine molecules—to get them to the top of the activation energy curve, where they split into chlorine atoms that react rapidly with hydrogen.

$$Cl_2(g) \longrightarrow 2Cl(g)$$
$$2Cl(g) + H_2(g) \longrightarrow 2HCl(g)$$

This reaction releases enough energy to activate more chlorine molecules. Therefore, once the first few molecules get over the activation energy hill, the reaction is self-sustaining. In the reaction of hydrogen and bromine, however, although the mechanism is similar, the amount of energy released is not enough to activate many additional bromine molecules, and the reaction is very slow.

To summarize—if at a given temperature many molecules have enough energy to climb the activation energy hill when they collide, a reaction is rapid at that temperature. In some cases substances react at room temperature as soon as they are mixed because enough energetic molecules are present at room temperature. In other cases, energy must be added to activate a few molecules, which can then react and liberate enough energy to activate the remaining molecules. In still other cases, only the continuous input of energy will push enough molecules up the activation energy hill to maintain the reaction.

15.3 Transition state theory

A different but complementary theory of how reactions take place pictures, instead of a collision between two molecules, the formation of a transition state by the two (or possibly more) reacting molecules. Approaching molecules are pictured as gradually coming under each other's influence until a transition state is formed at the peak of the energy curve (point C in Figure 15.2). The **transition state,** or **activated complex,** is a combination of reacting molecules intermediate between reactants and products. Some bonds have been weakened and new bonds have begun to form. For example, in the reaction of ammonia and methyl bromide

$$NH_3 + CH_3Br \xrightarrow{\text{polar solvent}} [CH_3NH_3]^+ + Br^-$$

the N–C bond has begun to form in the transition state, the C–Br bond has begun to weaken, and partial charges have developed on the nitrogen and bromine atoms.

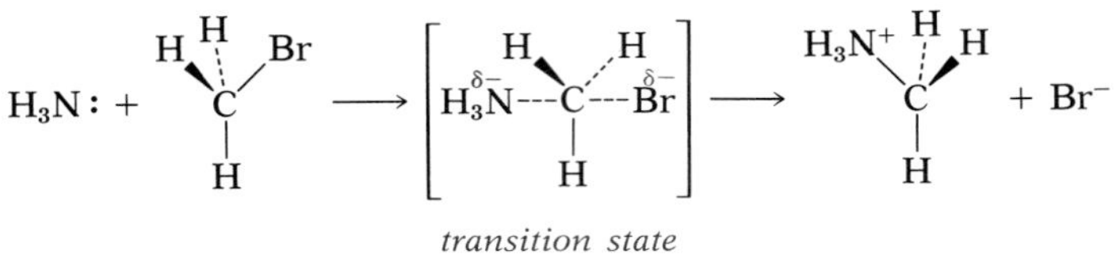

transition state

An activated complex cannot ordinarily be isolated. It breaks down to give either reactants or products, depending on the conditions of the reaction. In other words, it may go either way down the energy curve in Figure 15.2. With knowledge of the geometry of reacting molecules and good chemical intuition, it is frequently possible to determine what the products of a reaction will be by speculating upon the arrangement of atoms in the transition state and the route by which the transition state might break down.

Reaction rates and factors that influence them

15.4 Reaction rates

The **reaction rate** is the speed with which the products are produced and the reactants consumed in a specific reaction. It is usually discussed in terms of the rate at which one component of the mixture is used up or formed. The units for reaction rates are concentration per unit of time, most commonly moles/liter per time unit—seconds, minutes, days, and so on, depending upon how slow the reaction is. The units are written, for example, moles/liter sec, or moles liter^{-1} sec^{-1}, where the -1 superscript indicates the reciprocal of the unit, such as sec^{-1} = 1/sec. Here we simply write moles/liter sec.

The reaction rate is a constantly changing quantity. As the reactants are used up, fewer collisions are possible and the rate decreases. Therefore, reaction rate can be expressed only for a particular moment in time. For example, the rate at the instant of mixing the reactants might be, say, 2.5×10^{-2} mole/liter sec.

The rate of a reaction, of course, depends upon the nature of the reactants. If pieces of sodium, zinc, and tin of the same size and shape are dropped into solutions of hydrochloric acid of the same concentration, the sodium will react violently, the zinc will react at a moderate rate, and the tin will react very slowly.

For any given reaction, there are also external factors that determine the rate at which the reaction will take place. These are the concentration, the temperature, the degree of contact between the reacting substances, and, in the cases where it is occurring, catalysis. These factors are discussed in the following sections.

15.5 Concentration

Any change in the concentration of reactants causes a change in the rate of a reaction. According to collision theory, more molecules, atoms, or ions in a given volume mean more collisions per unit of time and therefore a greater chance of reaction; fewer molecules, atoms, or ions in a given volume mean fewer collisions and less chance of reaction.

To increase the concentration of a gas, the pressure need only be increased. The reaction between ethylene and hydrogen at atmospheric pressure is very slow, even under the best conditions.

$$C_2H_4(g) + H_2(g) \longrightarrow C_2H_6(g)$$
ethylene   *ethane*

At 30–40 atm pressure the number of collisions is increased and the reaction rate is increased proportionately.

The concentrations of solutions are changed, of course, by changing the ratio of solute to solvent. The concentrations of pure solids and liquid are fixed by their densities. The amount of change in a reaction rate with a change in concentration cannot be deduced from the reaction equation. It depends upon the reaction mechanism and is calculated by using the rate equation (Section 15.9), which can be found only by experiment.

15.6 Temperature

An old "rule of thumb" states that the rate of a chemical reaction is doubled for each 10° Celsius rise in temperature (Figure 15.3). However, this is only a very rough approximation; the rates of some reactions are much more than doubled by a 10° rise in temperature, but the rates of others are not nearly doubled.

The effect of changing the temperature of a reacting system is twofold. As the temperature is increased, the molecules move more rapidly, so collisions are more frequent. However, calculations show that only a few percent of the rate increase with temperature can be accounted for by the larger number of collisions. More importantly, the population of molecules in the high-energy end of the kinetic energy distribution increases with temperature. Think of E_1 in Figure 15.4 as the energy a molecule needs before a collision will lead to reaction. Obviously many more molecules are capable of effective collisions at higher temperatures. This principle is valid whether the reacting substances are in the solid, liquid, or gaseous states.

15.7 Contact between reacting substances

In a reaction between substances in the same gaseous or liquid phase—a **homogeneous reaction**—the question of contact between the reactive molecules is not an important one, for the molecules and ions are free to move and collisions are frequent. However, for reac-

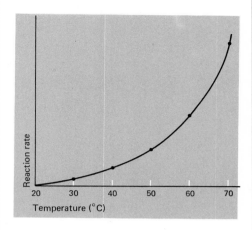

FIGURE 15.3
Reaction rate increase with rising temperature. *For a case where the rate exactly doubles for each 10°C rise in temperature.*

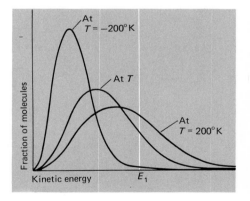

FIGURE 15.4
Average kinetic energy at three temperatures. *At E_1, compare the fraction of molecules at each temperature.*

tions between substances in different phases—**heterogeneous reactions**—bringing the reacting molecules or ions together may be difficult. For example, the reaction between steam and red hot iron

$$3Fe(s) + 4H_2O(g) \longrightarrow Fe_3O_4(s) + 4H_2(g)$$

proceeds very slowly if the iron is in one large block, but goes rapidly if the metal is powdered and spread out so as to expose a large surface to the steam. The rates at which solids react with each other are often limited by the amount of contact between them. The reaction of mercury(II) chloride and potassium iodide is a good illustration.

$$\underset{white}{HgCl_2(s)} + \underset{white}{2KI(s)} \longrightarrow \underset{red}{HgI_2(s)} + \underset{white}{2KCl(s)}$$

If large crystals of the two solids are shaken together, there is no sign of reaction; if the crystals are ground together, reaction takes place fairly rapidly, as shown by the appearance of the bright red mercury(II) iodide. If the reactants are dissolved separately in water and the solutions are mixed, the insoluble red mercury iodide precipitates almost instantly.

There are many practical examples of how the degree of contact influences the rate of reaction, such as dust explosions in flour mills and coal mines or the effect of tetraethyllead in an automobile engine. When the volatile liquid tetraethyllead is drawn into the hot cylinder, it decomposes, giving an extremely fine suspension of metallic lead. This exposes a very large surface to the burning gases and produces the antiknock effect. Knock is premature ignition of the fuel–air mixture. It hinders smooth engine performance. Antiknock agents are thought to interrupt the reactions that lead to spontaneous premature ignition. Placing a sheet of lead in the cylinder would not have a noticeable effect, because the sheet would not have enough surface in contact with the gases in the cylinder.

15.8 Catalysis

A catalyst increases the rate of a chemical reaction, but can be recovered in its original form when the reaction is finished. A catalyst cannot initiate or maintain a reaction that would not proceed in its absence. At any given temperature, a catalyst cannot change the concentrations in an equilibrium mixture; it can only change the time needed to reach equilibrium. "Catalyst" comes from the Greek meaning "to dissolve." The word was once used to mean destruction or ruin.

A catalyst works by providing an alternate and easier pathway from reactants to products. Catalysts accomplish this by a variety of mechanisms, but in each case the sole function of a catalyst is to lower the activation energy of a reaction. The role of a catalyst is highly specific to a particular reaction. A substance that catalyzes one reaction may well have no effect on another reaction, even if that reaction is very similar.

Catalysts are classified as homogeneous or heterogeneous. **Homogeneous catalysts** are present in the same phase as the reactants. They function by combining with reacting molecules or ions to form unstable intermediates. These intermediate compounds combine with other reactants to give the desired product and to regenerate the catalyst.

The energy pathway from reactants to products for the decomposition of hydrogen peroxide

$$2H_2O_2(aq) \longrightarrow 2H_2O(l) + O_2(g)$$

is shown in Figure 15.5. The upper curve is for the uncatalyzed reaction. The lower curve shows the change in the energy pathway after the addition of iodide ion, I^-, which serves as a homogeneous catalyst by the mechanism

$$\begin{array}{ll} H_2O_2 + \cancel{I^-} \longrightarrow H_2O + \cancel{IO^-} & (1) \\ \cancel{IO^-} + H_2O_2 \longrightarrow H_2O + O_2 + \cancel{I^-} & (2) \\ \hline 2H_2O_2 \longrightarrow 2H_2O + O_2 & (3) \end{array}$$

The distance BC represents the activation energy of the uncatalyzed reaction and the distance BC′, the activation energy of the catalyzed reaction. Note that a catalyst decreases by an equal amount the energies of activation of both the forward and reverse reactions. Note also that the overall energy change in the reaction (BX) remains the same.

The catalyst I^- takes part in the mechanism of the reaction, but is regenerated in its original state and can be recovered. In other words, it is not correct to say that a catalyst "changes the rate of a reaction without taking part in it."

Heterogeneous catalysts are present in a different phase than are the reactants. They are usually solids in the presence of gaseous or liquid reactants. Reactions occur at the surface of heterogeneous catalysts. For this reason the catalysts are finely divided solids or have particle shapes that provide a high surface-to-volume ratio—a property of importance in heterogeneous catalysis.

One or both reactants are chemically adsorbed at the surface of a heterogeneous catalyst (Figure 15.6). **Adsorption** is the adherence of atoms, molecules, or ions (the adsorbate) to a surface. It is called **physical adsorption** (or physisorption) when the forces between surface and adsorbate are van der Waals forces, and it is called **chemical adsorption** (or chemisorption) when the forces between adsorbate and surface are of the magnitude of chemical bond forces. The process of chemisorption on a catalyst can alter the structure and activity of a reactant in the same way as the formation of an activated complex—by weakening some bonds and allowing the formation of others.

Because heterogeneous catalysts stay in position and can be recovered readily by filtration (say, when they need purification), they are

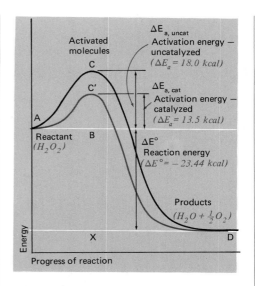

FIGURE 15.5
Energy path of the decomposition of H_2O_2, catalyzed and uncatalyzed.

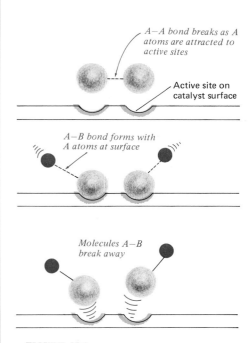

FIGURE 15.6
How the reaction A–A + 2B → 2AB might take place at a catalyst surface.

much more widely used in industry than are homogeneous catalysts. The cracking of petroleum and the reforming of hydrocarbons to give antiknock fuels (Section 23.9), and the preparation of ammonia (An Aside: The Haber Process for the Manufacture of Ammonia, Chapter 16) are carried out with the aid of heterogeneous catalysts.

For two substances that can react with each other in two different ways, a catalyst may change the rate of one reaction, but not the other. This is often of great importance in the chemical industry. For example, ammonia burns in air to give primarily nitrogen and water vapor; a competing reaction gives very small quantities of nitrogen(II) oxide (NO).

$$4NH_3(g) + 3O_2(g) \longrightarrow 2N_2(g) + 6H_2O(g) \tag{4}$$
$$4NH_3(g) + 5O_2(g) \longrightarrow 4NO(g) + 6H_2O(g) \tag{5}$$

In the presence of a large-surface-area alloy of platinum (99%) and rhodium (1%), the rate of reaction (5) is so greatly increased that it seems to proceed almost to the exclusion of the oxidation of ammonia to nitrogen, Equation (4). This conversion of ammonia to nitrogen(II) oxide is used in the manufacture of nitric acid.

Sometimes the general nature of the intermediate compound formed at a catalyst surface can be determined, even though the intermediate cannot be isolated. For example, manganese dioxide catalyzes the decomposition of potassium chlorate to form oxygen. If manganese dioxide containing an isotopic form of oxygen (^{17}O or ^{18}O) is used, the heavier oxygen isotope (O^* in the equation) appears in the gaseous product, as shown by mass spectrometry,

$$2KClO_3(l) \xrightarrow{\text{MnO}_2{}^*} 2KCl(s) + 3O_2{}^*(g)$$

This strongly suggests that the catalyst has actually reacted with the potassium chlorate. The mechanism of the reaction is not known, but one possibility is that the very unstable MnO_3 is formed. Two of the oxygen atoms in this molecule would be the original, labeled ones, but the third one would come from the potassium chlorate. The unstable MnO_3 molecule would quickly lose an atom of oxygen and revert to MnO_2. Since all of the oxygen atoms have the same chemical properties and are bound to the manganese atom by equal forces, any one of the three would be lost equally easily. Thus, in the resultant MnO_2 molecule, the manganese atom is not necessarily attached to the same oxygen atoms as it was before the reaction took place.

One difficulty in the use of heterogeneous catalysts is that most of them are readily "poisoned"—that is, impurities in the reactants coat the catalyst with unreactive material or modify its surface, so that the catalytic activity is lost. Frequently, but not always, the poisoned catalyst can be purified and used again.

Inhibitors are substances that slow down a catalyzed reaction—usually by tying up the catalyst, possibly by also tying up a reactant. For example, the development of rancidity in butter is greatly retarded by the addition of very small amounts of certain organic substances. The added organic materials apparently combine with the traces of copper ion that get into nearly all butter from the processing equipment. The copper ion catalyzes the development of rancidity, but in combination with the organic material, the copper ion does not exert a catalytic effect.

Promoters are substances that make a catalyst more effective. In the case of solid catalysts, a small amount of a promoter may encourage the formation of lattice defects, which are often the active sites on a catalyst surface.

Mechanisms and rate equations

15.9 Rate equations

Rate equations are based upon the following principle: The reaction rate is dependent upon the number of possible collisions and is therefore directly proportional to the concentrations of the reactants raised to the appropriate powers.

$\swarrow$ *"is directly proportional to"*
rate $\propto$ number of collisions $\propto$ concentration of reactants   (6)

A **rate equation** gives the relationship between the reaction rate and the concentration of the reactants. Consider the simple reaction

A + B $\longrightarrow$ products

If two molecules each of A and B are present, A–B collisions can occur in four ways (2×2), if three molecules of A and two of B are present, they can collide in six ways (2×3) (Figure 15.7), and so on. Doubling the concentration of A doubles the number of collisions and doubling the concentrations of both A and B quadruples the number of collisions.

Obviously, the number of collisions equals the product of the concentration of A and the concentration of B. Adopting the notation in which square brackets indicate concentration in moles per liter, and based upon the proportionality of Equation (6) we can write

rate $\propto$ [A] [B]   (7)

If all of the molecules are of the same type and an A–A collision is required for reaction (2A $\rightarrow$ products), the following relationship holds

rate $\propto$ [A]^2 (8)

The rate of a reaction, as we have pointed out, is different at every instant because the reactant concentration is constantly chang-

[A] = concentration of A in moles/liter

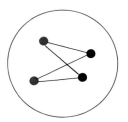
4 A–B collisions

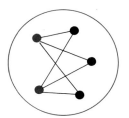
6 A–B collisions

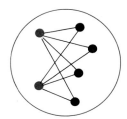
8 A–B collisions

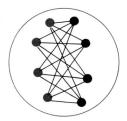
16 A–B collisions

FIGURE 15.7
Collisions of A and B.

ing. Fortunately, for the sake of studying and comparing reaction rates, the **rate constant**—the proportionality constant between the rate and the reactant concentrations—has a characteristic value for each reaction at constant temperature.

rate = k(concentration of reactants)

Because the rate constant k varies with temperature, the temperature at which a rate constant was measured must always be stated. (How k varies with temperature is discussed in Section 25.10.)

Putting together everything that we have learned for the reactions A + B → products, and A + A → products, we obtain the following two rate equations, also called rate laws.

$$\text{rate} = k\,[A]\,[B] \tag{9}$$
$$\text{rate} = k\,[A]^2 \tag{10}$$

For a reaction represented by Equation (9), doubling the concentration of A would double the rate. For a reaction represented by Equation (10), doubling the concentration of A would quadruple the rate. Relationships such as these are important in the experimental determination of rate equations.

As we shall see in the following sections, rate equations take different forms for different types of reactions. The value of k and the form of the rate equation must be found experimentally for every reaction.

15.10 Reaction order, molecularity, and elementary reactions

An **elementary reaction** is a simple reaction that occurs in a single step. **Molecularity** is the number of individual atoms, or molecules, or ions that must simultaneously react in an elementary reaction.

A + B $\longrightarrow$ products
an elementary bimolecular reaction

If this reaction happens just as it is written, it is an elementary reaction and it is bimolecular—two atoms, ions, or molecules are needed to make the reaction happen.

Most elementary reactions are either unimolecular, that is, requiring only one reactive molecule, or bimolecular. The probability of three molecules simultaneously colliding in just the right orientation to react is much smaller than the probability of two molecules colliding.

The "order of a reaction," or "reaction order" is a term that refers to the *rate equation*. **Reaction order,** also called "overall order," is the sum of the exponents of the concentration terms in the rate equation. The A + B → products reaction, for which Equation (9) is the rate equation, has an order of (1 + 1) = 2. It is called a "second-order reaction." First- and second-order reactions are the most common ones, and we shall discuss only these in the following sections. However, it is possible to have higher-order reactions, zero-order reactions, and even fractional-order reactions.

When the equation represents an elementary reaction, the molecularity and order of the reaction are both determined by the same number, for example, the A + B → products and the 2A → products reactions are each bimolecular and second order. The iodide-ion-catalyzed decomposition of hydrogen peroxide has the rate law, rate = $k[H_2O_2][I^-]$. It is therefore a second-order reaction. However, since it proceeds by two steps [see Equations (1) and (2)] *each* of which is bimolecular, it would be wrong to call the overall reaction bimolecular. Only elementary reactions are categorized by their molecularity. The need for a distinction between molecularity and reaction order will become more clear in the following sections.

15.11 First-order reactions

In the simplest possible type of reaction, the rate is determined by the concentration of a single substance. This is a first-order reaction, and if it is an elementary reaction it is unimolecular.

A $\longrightarrow$ products

Many decomposition reactions are first order, for example,

$CH_3CH_2Cl(g) \longrightarrow H_2C{=}CH_2(g) +$ $HCl(g)$
chloroethane   *ethylene*   *hydrogen chloride*

Rearrangement of the atoms of a molecule of one substance into a molecule of a different substance can also be first order

$$\underset{\text{cyclopropane}}{H_2C\overset{\overset{\displaystyle H_2}{\underset{\displaystyle |}{C}}}{\diagup\!\!\diagdown}CH_2} \longrightarrow \underset{\text{propylene}}{CH_3CH{=}CH_2}$$

And all radioactive decay reactions, such as the decay of strontium-90, are first order.

$$^{90}_{38}Sr \longrightarrow\; ^{90}_{39}Y + ^{\;\;0}_{-1}e$$

The rate of a first-order reaction depends solely upon the instability of the reacting species, not upon the collision of any two atoms, molecules, or ions. Therefore, the rate equation includes only the concentration of the reactant.

$$\text{rate} = k\,[A]$$
first-order rate equation   (11)

The rate constant for a first-order equation has reciprocal time units, for example, seconds^{-1}, minutes^{-1}, and so on, so that the *rate* is given in moles/liter per time unit. Table 15.1 lists some first-order rate constants.

Experimentally, the rate equals the decrease in the concentration of A divided by the time interval in which the decrease took place, as shown in the following equation, where Δ means "the change in"

$$\text{rate} = -\frac{\Delta[A]}{\Delta t} \tag{12}$$

By setting Equations (11) and (12) equal to each other, the following relationship is obtained:

$$k[A] = -\frac{\Delta[A]}{\Delta t} \tag{13}$$

concentration of A in moles/liter

first-order rate constant

time

In a more useful form, Equation (13) becomes

logarithm of [A] at t = 0

logarithm of [A] at t

$$\log[A] = \log[A]_0 - \frac{kt}{2.303} \tag{14}$$

log conversion factor

or

$$\log\frac{[A]_0}{[A]} = \frac{kt}{2.303} \tag{15}$$

The reactions given here and in Table 15.2 are all gas-phase reactions. They are easier to study than condensed-phase reactions, and the kinetics of many gas-phase reactions are known.

Reaction	k at 1000°C (sec^{-1})
$CO_2 \longrightarrow CO + O$	1.8×10^{-13}
$CS_2 \longrightarrow CS + S$	2.8×10^{-7}
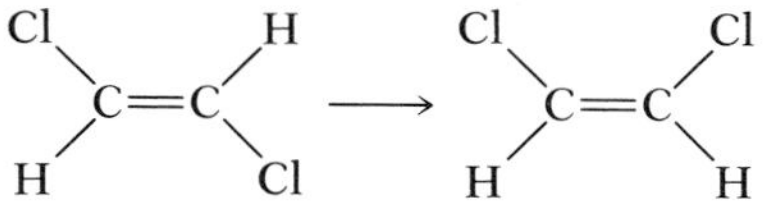	3.3×10^3
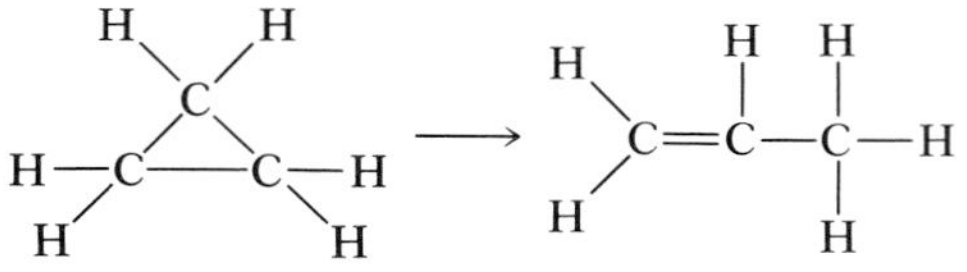	9.2
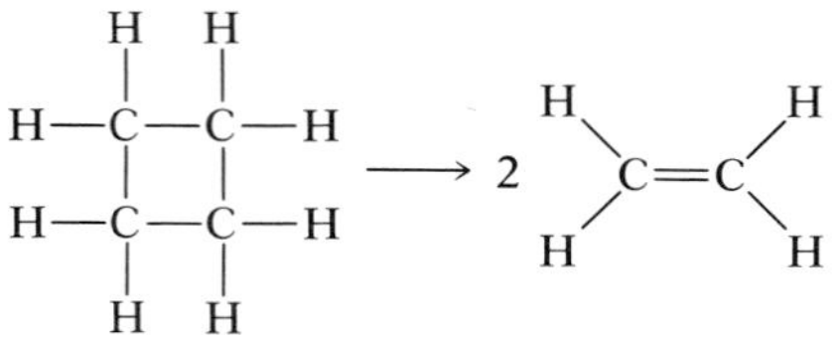	87

Equations (14) and (15) relate the concentration of A at time zero, $[A]_0$, the beginning of the reaction, to the concentration of A at any given time t, represented by $[A]$.

The transformation of Equation (13) to Equation (14) is achieved by mathematical techniques that we need not go into here. However, it will be necessary to use logarithms in problems in this and forthcoming chapters. At this point, you might want to review logarithms, which are discussed in Appendix A.

EXAMPLE 15.1

■ The rate equation describing the decomposition of N_2O_5 dissolved in CCl_4 at 45°C (Section 15.13)

$$2N_2O_5 \longrightarrow 4NO_2 + O_2$$

is

$$\text{rate} = (6.2 \times 10^{-4} \text{ min}^{-1})[N_2O_5]$$

What is the initial rate of reaction if the original concentration of N_2O_5 is 0.40M?

The reaction rate at the initial instant is

$$\text{rate} = (6.2 \times 10^{-4} \text{ min}^{-1})(0.40 \text{ mole/liter}) = 2.5 \times 10^{-4} \text{ mole/liter min}$$

The rate of disappearance of N_2O_5 at the start of the reaction is 2.5×10^{-4} mole/liter min. ■

EXAMPLE 15.2

■ What will be the concentration of N_2O_5 in the 0.40M solution described in Example 15.1 at the end of an hour?

The equation relating concentrations and time for this first-order reaction is

$$\log \frac{[N_2O_5]_0}{[N_2O_5]} = \frac{kt}{2.303} = \frac{(6.2 \times 10^{-4} \text{ min}^{-1})(1.00 \text{ hr})(60 \text{ min/hr})}{2.303}$$
$$= 1.6 \times 10^{-2}$$

Taking antilogarithms of both sides gives

$$\frac{[N_2O_5]_0}{[N_2O_5]} = 1.04$$

Solving for the concentration of N_2O_5 and using the initial N_2O_5 concentration of 0.40M,

$$[N_2O_5] = \frac{[N_2O_5]_0}{1.04} = \frac{0.40M}{1.04} = 0.38M$$

The concentration of N_2O_5 remaining is 0.38M. Only about 5% of the N_2O_5 has decomposed. ■

(a) Concentration vs. time for a first-order reaction

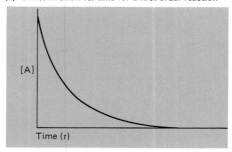

(b) Plot of the equation $\log[A] = \log[A]_0 - \dfrac{kt}{2.303}$ **for a first-order reaction**

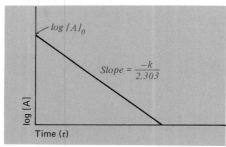

(c) Finding the slope of a line

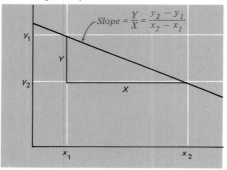

FIGURE 15.8

Plots for a first-order reaction. *Part (c) shows how the slope of a line is found. The slope is equal to y divided by x, as shown in the drawing, for any two points on the line. Slope is equivalent to the ratio of the rate of change of what is plotted on the y axis to the rate of change of what is plotted on the x axis.*

Equation (14) can be used in the experimental determination of the value of k, the rate constant. The initial concentration of A and the concentration of A at time t are measured and the values substituted into the equation for calculation of k. Since Equation (14) is the equation of a straight line, it may also be used in a graphical determination of reaction order. Obtaining a straight line by plotting experimental values of $\log[A]$ vs. t is taken as proof that a reaction is first order. The slope of this line (see Figure 15.8c for definition of slope), multiplied by -2.303, gives the rate constant of the reaction. Equation (15) is also the equation for a straight line and may be used in the same way, by plotting $\log[A]_0/[A]$ vs. t. Plots for a typical first-order reaction are shown in Figure 15.8.

In using Equations (14) and (15), [A] may be replaced by any quantity that is proportional to the concentration in moles per liter. For example, mass in grams, or, for radioactive substances, number of nuclei or even the number of counts per minute per gram measured by a Geiger counter.

An expression for finding half-life, the quantity that is used to characterize the rate of decay of radioactive isotopes, can be derived from Equation (15). This is done by replacing t by $t_{1/2}$, the half-life, and expressing the concentration of A at $t_{1/2}$ as $0.5[A]_0$. In other words, $t_{1/2}$ is the time when half of the original A is left.

$$\log \frac{[A]_0}{0.5\,[A]_0} = \frac{kt_{1/2}}{2.303}$$

$$\log 2 = \frac{kt_{1/2}}{2.303}$$

$$t_{1/2} = \frac{0.693}{k} \tag{16}$$

The interesting fact emerges that in a first-order reaction, the half-life is independent of the concentration.

EXAMPLE 15.3

■ After 2 hr, a solution originally containing 1.30×10^{-6} mole/liter of $^{240}AmCl_3$ contained only 1.27×10^{-6} mole/liter of the radioactive substance. What is the half-life of ^{240}Am?

Before we can find the half-life, we must find the first-order rate constant by solving Equation (14) for k

$$k = -2.303 \frac{\log [^{240}Am] - \log [^{240}Am]_0}{t}$$

$$= -2.303 \frac{\log (1.27 \times 10^{-6}) - \log (1.30 \times 10^{-6})}{2.00 \text{ hr}}$$

$$= 1.17 \times 10^{-2} \text{ hr}^{-1}$$

Now we can find $t_{1/2}$ from Equation (16)

$$t_{1/2} = \frac{0.693}{1.17 \times 10^{-2} \text{ hr}^{-1}} = 59 \text{ hr}$$

The half-life of americium-240 is 59 hr. ∎

EXAMPLE 15.4

∎ The half-life of carbon-14 and the rate equation for its decay provide the basis for the radiocarbon dating of historical objects discussed in Section 8.6. A sample of wood from a cyprus beam found in the tomb of Pharaoh Sneferu had a carbon-14 activity of 7.2 counts per minute for each gram of carbon. We can write this unit as min^{-1} $(\text{g C})^{-1}$. Current animal and plant tissue has a carbon-14 activity of $12.6 \text{ min}^{-1} (\text{g C})^{-1}$, and the half-life of carbon-14 is 5730 yr. How old is the Egyptian tomb?

First we find the rate constant for carbon-14 decay

$$k = \frac{0.693}{t_{1/2}} = \frac{0.693}{5730 \text{ yr}} = 1.21 \times 10^{-4} \text{ yr}^{-1}$$

Recall that we said that the concentrations in Equation (15) could be replaced by any quantity that was directly proportional to concentration. Therefore, solving the first-order rate equation for time and substituting the carbon-14 activities for concentrations

$$t = \frac{2.303 \log ([A]_0/[A])}{k}$$

$$= \frac{2.303 \log [12.6 \text{ min}^{-1}(\text{g C})^{-1}/7.2 \text{ min}^{-1}(\text{g C})^{-1}]}{1.21 \times 10^{-4} \text{ yr}^{-1}} = 4630 \text{ yr}$$

The Egyptian tomb is 4630 years old, that is, it was constructed about 2650 B.C. Historical records place Sneferu at 2625 ± 75 B.C. ∎

15.12 Second-order reactions

In Section 15.9 we gave the second-order rate equations

rate = $k[A][B]$
rate = $k[A]^2$

and showed how the reaction of A with B can be pictured as arising from collision theory. The following examples show elementary second-order reactions, that is, reactions that are bimolecular and proceed in a single step. Note that these are all reactions in the gas phase, where such simple reactions are most likely to occur.

$2NO_2(g) \longrightarrow 2NO(g) + O_2(g)$ rate = $k[NO_2]^2$
$NO(g) + O_3(g) \longrightarrow NO_2(g) + O_2(g)$ rate = $k[NO][O_3]$
$CO(g) + Cl_2(g) \longrightarrow COCl(g) + Cl(g)$ rate = $k[CO][Cl_2]$

TABLE 15.2

Rate constants for some second-order reactions

Reaction	k at 298°K (liter/mole sec)
$H_2 + I_2 \longrightarrow 2HI$	1.7×10^{-18}
$2HI \longrightarrow H_2 + I_2$	2.4×10^{-21}
$2NO_2 \longrightarrow 2NO + O_2$	1.4×10^{-10}
$O_3 + NO \longrightarrow NO_2 + O_2$	2.2×10^7
$O_3 + NO_2 \longrightarrow NO_3 + O_2$	8.4×10^4
$NO + Cl_2 \longrightarrow NOCl + Cl$	7.3×10^{-6}
$CO + Cl_2 \longrightarrow COCl + Cl$	1.3×10^{-28}
$Cl + COCl \longrightarrow Cl_2 + CO$	9.3×10^{10}

(a) Concentration vs. time for a second-order reaction

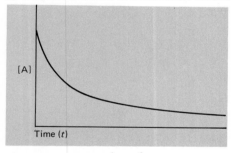

(b) Plot of the equation $\frac{1}{[A]} = \frac{1}{[A]_0} + kt$ for a second-order reaction

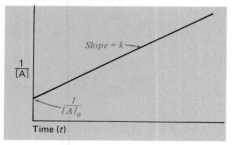

FIGURE 15.9
Plots for a second-order reaction.

There are many *overall* second-order reactions that occur by mechanisms involving more than one elementary step. The rate constants for some overall second-order reactions are given in Table 15.2. Since rate is expressed as change in concentration per unit time, the second-order rate constants must have units of reciprocal concentration times reciprocal time. For example, using seconds, k has units of $(\text{moles/liter})^{-1}(\text{sec})^{-1}$

$$\text{rate} = k(\text{moles/liter})^2$$
$$= \frac{1}{(\text{moles/liter})(\text{sec})} \times (\text{moles/liter})^2$$
$$= \frac{\text{moles/liter}}{\text{sec}}$$

The units for a second-order rate constant are also sometimes written in the following ways:

$$\text{liter mole}^{-1}\text{sec}^{-1} \quad \text{or} \quad \text{liter/mole sec}$$

Reactions in solution are more complicated than gas-phase reactions. Molecules cannot simply collide and bounce away, but may stay in each other's vicinity for a relatively long time, colliding often while they are nearby. Also, the role of the solvent molecules is not completely understood, and the nature of the solvent often affects the *rate* of the reaction. However, the general principles of kinetics that we have been discussing do also apply to reactions in solution.

By following a method similar to the one we described for the first-order rate equation, an equation can be derived that relates initial concentration to concentration at time t for A in a second-order reaction of A with A.

$$\frac{1}{[A]} = \frac{1}{[A]_0} + kt \tag{17}$$

A second-order reaction can be identified from experimental data by a linear plot of $1/[A]$ vs. time (Figure 15.9). The slope of such a plot is k.

For a second-order reaction of the type A + B → products, plotting either $1/[A]$ or $1/[B]$ versus time will give a straight line if the initial concentrations of A and B are equal. In the experimental determination of reaction rates, it is easier to follow the rate of change of one reactant than of two. One way to do this is to have one reactant, say B, present in such a large amount that the decrease in its concentration during the reaction is insignificant. The reaction then becomes a pseudo-first-order reaction in the other component, in this case A. A plot of log [A] vs. time for data taken under these conditions is a straight line. Next, the conditions would be reversed. If a straight line is also obtained for the opposite experiment, where A is in great excess and log [B] vs. time gives a straight line, it can be concluded that the reaction is first order in A, first order in B, and overall second order, with the rate = $k[A][B]$.

EXAMPLE 15.5

■ The recombination of Br atoms

$$2Br(g) \longrightarrow Br_2(g)$$

is considered to be an elementary bimolecular reaction. In one experiment the concentration of bromine atoms was $1.04 \times 10^{-5}\,M$ at 320 μsec and the original concentration was $12.26 \times 10^{-5}\,M$. Find the rate constant.

Solving the second-order rate equation for k gives

$$k = \frac{1/[Br] - 1/[Br]_0}{t}$$

$$= \frac{(1.04 \times 10^{-5}\,\text{mole/liter})^{-1} - (12.26 \times 10^{-5}\,\text{mole/liter})^{-1}}{320 \times 10^{-6}\,\text{sec}}$$

$$= 2.75 \times 10^8\ \text{liter/mole sec.}$$

The second-order rate constant is 2.75×10^8 liter/mole sec. ■

EXAMPLE 15.6

■ Show that the reaction rate for the following reaction

$$I^- + ClO^- \xrightarrow[25\,^\circ C]{OH^-} Cl^- + IO^-$$

is given by the second-order rate equation

$$\text{rate} = k[I^-][ClO^-]$$

from the following data. Determine the rate constant.

t (sec) =	0	2	4	6
$[I^-] = [ClO^-]$ (mole/liter) =	0.00200	0.00169	0.00147	0.00123

If the rate equation is correct, a plot of $1/[I^-]$ against t will be linear and the slope will be equal to k. The plot is shown in Figure 15.10, using 5.00×10^2, 5.92×10^2, 6.80×10^2, and 8.13×10^2 (mole/liter)$^{-1}$, respectively, for the values of $1/[I^-]$. The plot is linear within experimental error. The value of k is determined by the slope of the line as calculated using widely spaced points from the line.

$$k = \text{slope} = \frac{\text{change } (1/[I^-])}{\text{change } (t)}$$

$$= \frac{(7.80 - 5.00)\,10^2 (\text{mole/liter})^{-1}}{(5.50 - 0.40)\,\text{sec}} = 54.9\ \text{liter/mole sec}$$

The graphical method of finding k is preferable to the method of Example 15.5 if enough data are available. By plotting the data and

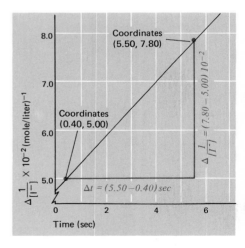

FIGURE 15.10
Plot of data for Example 15.6.

choosing widely spaced points, experimental error is smoothed out and the best possible answer obtained.

The data indicate that the reaction is first order with respect to I^-, first order with respect to ClO^-, second order overall, and the rate constant is 54.9 liter/mole sec at 25°C. ∎

15.13 Kinetics and mechanism

Studying the kinetics and mechanism of a reaction starts with experiments in which the change in the concentration of one reactant or product is measured as time passes. A recent book for chemical engineers, who must have practical information about kinetics and mechanisms in order to design chemical plants, lists 17 methods of following the changing concentration of a substance in a reaction. The methods range from total pressure change in a gas-phase reaction, to color intensity changes for colored substances, to changes in other spectroscopic properties, to changes in the radioactivity of isotopically labeled substances.

Tables of experimental data on the change in concentration of a reactant with time are assembled. The rate equation is then determined from the data by mathematical or graphical analysis. The next job is to devise a reaction mechanism that gives a rate equation in agreement with the experimentally determined rate equation.

The first question to be asked about the reaction is—Is it an elementary reaction? A simple rate equation and a simple reaction order indicate that the reaction may be elementary, but they do not prove it. For example, the rate of the reaction shown by the equation

$$\underset{\text{ethane}}{C_2H_6(g)} \longrightarrow \underset{\text{ethylene}}{C_2H_4(g)} + H_2(g)$$

depends only upon the concentration of C_2H_6. However, the first products formed are not ethylene and hydrogen, but are short-lived free radicals. The reaction is not, therefore, an elementary one. In general, we have no way to prove that a reaction is elementary; if we can find no evidence to the contrary, we assume that it is elementary, but this can be only a tentative conclusion.

Nor is it usually possible to determine the order of a reaction from the equation that represents it. For example, the equation

$$2N_2O_5(g) \longrightarrow 4NO_2(g) + O_2(g)$$

might lead one to believe that the decomposition of N_2O_5 is a second-order reaction, with rate = $k[N_2O_5]^2$. In fact, this reaction is first order, for the experimental data fit a straight line when log $[N_2O_5]$ is plotted against time, as in Figure 15.8.

The reaction does not depend upon the collisions of two molecules of N_2O_5 with each other, nor does it have a simple mechanism. One mechanism considered to be possible is the rapid establishment of the equilibrium of Equation (18a) followed by two additional steps (these are all gas-phase reactions).

$$2N_2O_5 \underset{k_2}{\overset{k_1}{\rightleftharpoons}} 2NO_2 + 2NO_3 \qquad \text{(18a)}$$

$$NO_2 + NO_3 \xrightarrow{k_3} NO + O_2 + NO_2 \qquad \text{(18b)}$$

$$NO + NO_3 \xrightarrow{k_4} 2NO_2 \qquad \text{(18c)}$$

$$\overline{\phantom{NO + NO_3 \xrightarrow{k_4} 2NO_2}}$$

$$2N_2O_5 \longrightarrow 4NO_2 + O_2$$

The second reaction [Equation (18b)] is the **rate-determining step**—the slowest step in the reaction mechanism. The overall reaction rate can be no faster than this step. Since NO_3 is an intermediate, its concentration does not appear in the rate law. The mechanism given above yields the experimentally determined rate law in such a way that k is related to the rate constants of the steps in the mechanism as follows:

$$\text{rate} = k[N_2O_5] \qquad \text{where} \qquad k = \frac{2k_1k_3}{k_2 + 2k_3}$$

Any mechanism can be assumed for a reaction if it is consistent with the rate equation and the stoichiometry of the reaction. A mechanism is usually selected by showing that no other mechanism is possible or probable. Even the periodic table is of little help in determining reaction mechanisms. For example, each of the halogens—fluorine, chlorine, bromine, and iodine—reacts in the gas phase with dihydrogen by a different mechanism (Section 20.8c).

Once the initial rate data are collected, the chemist studying kinetics spends his time trying to eliminate as many possible explanations as he can. The mechanism is often arrived at by a negative approach that first proves what the mechanism cannot be. One hopes that after the elimination process, there will be only one remaining explanation. If there are more than one, the most probable mechanism is selected. Different chemists may have different opinions as to which mechanism is most probable, a fact that sometimes leads to lively controversy in chemical publications.

**THOUGHTS ON
CHEMISTRY**

A classical view of kinetics and thermodynamics

Kinetics is a part of the science of motion. In physics the science of motion is termed dynamics and is subdivided into kinematics which treats of the motion of bodies, and kinetics which deals with the effect of forces on motion. In chemistry, no such distinction is made. Kinetics deals with the rate of chemical reaction, with all factors which influence the rate of reaction, and with the explanation of the rate in terms of the reaction mechanism. Chemical kinetics might very well be called chemical dynamics.

KINETICS AND MECHANISM, *by A. A. Frost and R. G. Pearson*

CHEMICAL KINETICS [15.13]

439

Chemical kinetics with its dynamic viewpoint may be contrasted with thermodynamics with its static viewpoint. Thermodynamics is interested only in the initial and final states of a system; the mechanism whereby the system is converted from one state to another and the time required are of no importance. Time is not one of the thermodynamic variables. The most important subject in thermodynamics is the state of equilibrium, and, consequently, thermodynamics is the more powerful tool for investigating the conditions at equilibrium. Kinetics is concerned fundamentally with the details of the process whereby a system gets from one state to another and with the time required for the transition. Equilibrium can also be treated in principle on the basis of kinetics as that situation in which the rates of the forward and reverse reactions are equal. The converse is not true; a reaction rate cannot be understood on the basis of thermodynamics alone. Therefore, chemical kinetics may be considered a more fundamental science than thermodynamics. Unfortunately, the complexities are such that the theory of chemical kinetics is difficult to apply with accuracy. As a result, we find that thermodynamics will tell with precision the extent of reaction, but only kinetics will tell (perhaps crudely) the rate of the reaction.

KINETICS AND MECHANISM, by A. A. Frost and R. G. Pearson (p. 1, Wiley, New York, 1953)

Significant terms defined in this chapter

chemical kinetics

reaction mechanism

collision theory

activation energy

transition state

activated complex

reaction rate

homogeneous reaction

heterogeneous reaction

homogeneous catalyst

heterogeneous catalyst

adsorption

physical adsorption

chemical adsorption

inhibitors

promoters

rate equation

rate constant

elementary reaction

molecularity

reaction order

rate-determining step

Exercises

15.1 The various stages in the life of a campfire are (a) striking a match, rubbing two sticks together, or striking a piece of flint with steel to generate a spark; (b) igniting tinder; (c) blowing air on the tinder to generate flames; (d) igniting kindling from the tinder; (e) igniting fuel from the kindling; (f) adding additional fuel as the fire dies down; (g) extinguishing the fire by dowsing with water and covering with dirt. Identify which of these four factors that influence the rate of a chemical reaction—(i) concentration, (ii) temperature, (iii) contact, (iv) catalyst—are involved in each of the stages.

15.2 Distinguish clearly between (a) the collision theory and the transition state theory; (b) activation energy and heat of reaction; (c) homogeneous and heterogeneous reactions; (d) physical and chemical adsorption; (e) an inhibitor, a promoter, and a catalyst; and (f) molecularity and reaction order.

15.3 Which statements are true? Rewrite any false statement so that it is correct.
(a) An elementary reaction is one that is unimolecular.
(b) The activation energy is released by the products after the reaction has taken place.
(c) A transition state complex may include the catalyst as part of its structure.
(d) Even though a catalyst is recovered in its original form after the completion of the reaction, it may have actively participated in the reaction and even undergone temporary chemical changes during the reaction.
(e) A promoter, because of its presence, speeds up the rate of a chemical reaction.
(f) Molecularity and reaction order could be the same for a given reaction, but, in general, are not the same.
(g) The rate of a reaction which occurs in a series of steps will depend on the slowest step in the reaction mechanism.

15.4 What are the three conditions for effective collisions between reacting species? Which condition is influenced most by (a) temperature, (b) molecular geometry, and (c) relative bond energies?

15.5 The rate of a reaction depends on the number of collisions between molecules which, in turn, depends on the velocity at which the molecules travel. Recalling that at a given temperature, molecular velocities are inversely proportional to mass, predict which reaction will have the higher value of k. (D is the symbol for deuterium.)

$$2HI(g) \longrightarrow H_2(g) + I_2(g) \qquad 2DI(g) \longrightarrow D_2(g) + I_2(g)$$

15.6 What are the five factors that affect the reaction rate? Illustrate each of these factors by giving an example.

15.7 What is a catalyst? How does a catalyst do its job? What is the difference between heterogeneous and homogeneous catalysis? Will a catalyst change the relative amounts of reactants and products once equilibrium has been attained? Can a catalyst change the thermodynamic spontaneity of a reaction?

15.8 You have just returned from laboratory and want to write up the results of your kinetics experiment while everything is fresh in your mind. What kind of graphs will you prepare for your concentration–time data to determine if the reaction was (a) first order or (b) second order? Suppose your results showed the reaction to be first order. (c) How would you determine the rate constant? A part of the experiment was designed to check the catalytic effect of $Cu^{2+}(aq)$ on the reaction. (d) How would the value of k differ for this trial than for the "uncatalyzed" trial?

15.9* Given the following steps in the mechanism for a chemical reaction:

$$A + B \xrightarrow{\text{fast}} C$$

$$B + C \xrightarrow{\text{slow}} D + E$$

$$D + F \xrightarrow{\text{fast}} A + E$$

(a) What is the stoichiometric equation for the reaction? (b) Which species, if any, are catalysts in this reaction? (c) Which species, if any, are intermediates in this reaction? (d) Write the rate law for the rate-determining step. (e) At any time [C] is directly proportional to [A]. Write the rate law for this reaction so that it includes only reactants and products, not intermediates. (f) What is the overall order of the reaction?

15.10 For the hypothetical equation

$$A + 2B \xrightarrow{c} D + E$$

the rate equation is

$$\text{rate} = \frac{k[A][B][C]}{[D]}$$

What will happen to the reaction rate if (a) [A] is doubled, (b) [B] is doubled, (c) more catalyst is added, (d) D is removed as it is formed to keep [D] at a small value, (e) [A] and [B] are both doubled, and (f) the temperature is increased?

15.11 Ozone reacts with both NO_2 and NO in the atmosphere according to the equations

$$O_3 + NO \longrightarrow NO_2 + O_2 \qquad O_3 + NO_2 \longrightarrow NO_3 + O_2$$

Both reactions have been found to be first order in ozone and first order in the respective oxide of nitrogen with

$k = 2.2 \times 10^7$ liter/mole sec. and 8.4×10^4 liter/mole sec., respectively. What will be the respective rates of reaction if $[O_3] = 3 \times 10^{-5}$ mole/liter in each case and both $[NO]$ and $[NO_2]$ are 0.4 mole/liter?

15.12 Consider the reaction

A + B → products

The following table gives the initial rates of reaction at various concentrations of A and B:

initial rate (mole/liter sec)	0.0090	0.036	0.018	0.027
[A] (mole/liter)	0.10	0.20	0.10	0.10
[B] (mole/liter)	0.10	0.10	0.20	0.30

Find the order of the reaction with respect to (a) A, with respect to (b) B, and (c) the overall order.

15.13 For the reaction

A + B → products

the following data were obtained:

initial rate (mole/liter sec)	0.030	0.059	0.060	0.090	0.089
[A] (mole/liter)	0.10	0.20	0.20	0.30	0.30
[B] (mole/liter)	0.20	0.20	0.30	0.30	0.50

Write the rate equation for this reaction. Be sure to evaluate k.

15.14 For the reaction

2A + B → products

the following data were obtained in a solution containing a large excess of B:

t (sec)	0	10	20	30	40	50	60	70
[A] (mole/liter)	10.00	9.00	8.35	7.70	7.15	6.60	6.25	5.95

By graphical means, determine the order of reaction for component A.

15.15 The half-life of $^{193}_{81}Tl$ is 23.0 min. Calculate the (a) rate constant k, and (b) mass of radioactive material left if a 1.00-g sample is allowed to decay for 1.00 hr.

15.16 A 3.50-mg sample of a new isotope was analyzed and found to contain only 2.73 mg of the isotope after a period of 6.3 hr. Calculate $t_{1/2}$.

15.17 A sample of bone had an activity of 2.93 min^{-1} (g C)$^{-1}$ (see Example 15.4). If the activity found in living plants and animals is 12.6 min^{-1} (g C)$^{-1}$ and $t_{1/2} = 5.73 \times 10^3$ yr for $^{14}_6C$, what is the age of the bone specimen?

15.18 If the half-life of $^{19}_8O$ is 29 sec, what fraction of the amount present of the isotope originally present would be left after 5.0 sec, the period of a typical breath?

15.19 The activation energies for the forward and reverse second-order reactions given by the equation

$$H_2(g) + I_2(g) \rightleftharpoons 2HI(g)$$

are 38.9 kcal and 42.5 kcal, respectively. (a) Using these energies prepare a plot of energy against progress of reactions. (b) What is $\Delta E°$ as calculated by the difference between these energies for this reaction? (c) Assuming $\Delta E° = \Delta H°$, how does this value of $\Delta E°$ compare to the observed value of $\Delta H° = -2.26$ kcal?

15.20 The activation energy for the reaction between O_3 and NO is 2.3 kcal/mole.

$$O_3(g) + NO(g) \rightarrow NO_2(g) + O_2(g)$$

The heat of formation at 25°C of $NO_2(g)$ is 7.93 kcal/mole, $NO(g)$ is 21.57 kcal/mole, and of $O_3(g)$ is 34.1 kcal/mole. Calculate the heat of reaction and sketch the energy–progress of reaction plot (similar to Figure 15.2).

15.21* The following concentration data were obtained for the decomposition of N_2O_5 at 45°C:

t (sec)	0	200	400	600	800	1000	1200
$[N_2O_5]$ (mole/liter)	0.250	0.222	0.196	0.173	0.153	0.136	0.120

(a) Prepare a plot of concentration against time. (b) Estimate the concentration of N_2O_5 at 1500 sec from your plot. (c) Determine the rate of reaction at 600 sec by calculating the slope of a line drawn tangent to the curve at 600 sec. (d) Repeat the calculation of (c) at 200 and 1000 sec. Does the rate seem to depend on concentration? (e) Divide your answers for (c) and (d) by the respective values of $[N_2O_5]$ and $[N_2O_5]^2$. Is the reaction first or second order?

15.22* The following data were obtained for the decomposition of diazomethane, CH_2N_2, at 600°C:

t (min)	0	5	10	15	20	25
$[CH_2N_2]$ (mole/liter)	0.100	0.076	0.058	0.044	0.033	0.025

(a) Prepare a plot of $[CH_2N_2]$ against time. (b) From the plot, determine the rate of decomposition at $t = 10$ min by calculating the slope of a line drawn tangent to the curve at time t. (c) Show graphically that the reaction is first order and determine the rate constant.

15.23* A current nuclear theory suggests that the ratio of $^{235}_{92}U/^{238}_{92}U$ was nearly unity at the time of the formation of the elements. If the current ratio is 7.25×10^{-3}, calculate the age of the elements if $t_{1/2} = 7.1 \times 10^8$ yr for $^{235}_{92}U$ and

4.51 × 10⁹ yr for $^{238}_{92}U$. Hint: (i) use Equation (14) twice, leaving log $[N]_0$, where N_0 = number of nuclei and the t's as unknowns, (ii) solve one for log$[N]_0$, (iii) substitute into the second equation, and (iv) solve for t.

15.24* For the equation

2A + B → products

the following data were obtained for a solution containing a large excess of B:

t (sec)	0	10	20	30	40	50	60	70
[A] (mole/liter)	10.00	7.95	6.31	5.00	3.92	3.16	2.50	1.90

(a) Calculate log[A] and $[A]^{-1}$ for the data. (b) Plot your results from (a) against t and determine the order of reaction with respect to A. (c) What is the value of the rate constant? (d) Which are possible mechanisms?

(1) 2A + B $\longrightarrow$ products

(2) A + B $\xrightarrow{\text{slow}}$ C   C + A $\longrightarrow$ products

(3) A $\longrightarrow$ C   C + A + B $\xrightarrow{\text{slow}}$ products

(4) A $\xrightarrow{\text{slow}}$ C   C + A + B $\longrightarrow$ products

15.25* A proposed mechanism for the aqueous reaction given by the equation

I⁻ + ClO⁻ $\longrightarrow$ IO⁻ + Cl⁻

consists of the following steps:

ClO⁻ + H₂O $\rightleftharpoons$ HClO + OH⁻   (rapid)
I⁻ + HClO $\longrightarrow$ HIO + Cl⁻   (slow)
OH⁻ + HIO $\rightleftharpoons$ H₂O + IO⁻   (rapid)

(a) Show that the sum of the three steps gives the stoichiometric equation. (b) Write the rate equation for the rate-determining step. The concentration of HClO in excess water is proportional to [ClO⁻]/[OH⁻]. (c) Rewrite the rate equation in terms of [ClO⁻], [OH⁻] and [I⁻]. (d) What is the order of reaction? (e) What would happen to the reaction rate if the concentration of OH⁻ were doubled?

15.26** An important gas-phase reaction which occurs in the atmosphere is given by the equation

$2O_3(g) \longrightarrow 3O_2(g)$ $\Delta H^\circ_{298} = -68.2$ kcal

(a) A set of experiments gave the following data

p_{O_3} torr	0.20	0.20	0.40
p_{O_2} torr	0.50	1.0	1.0
initial rate (torr sec⁻¹)	6.0	3.0	12

What are the values of x and y in the rate equation

rate $= k p^x_{O_3} p^y_{O_2}$

Under these conditions, O₂ is acting as an _____.
(b) For each of the following proposed mechanisms, write the rate law in terms of p_{O_3} and p_{O_2}. Which are acceptable based on your results for (a)?
(1) $2O_3 \longrightarrow 3O_2$
(2) $O_3 \longrightarrow O_2 + O$ (slow) $O + O_3 \longrightarrow 2O_2$ (fast)
(3) $O_3 \rightleftharpoons O_2 + O$ (fast) $O + O_3 \longrightarrow 2O_2$ (slow) where p_O is proportional to p_{O_3}/p_{O_2}
(4) $2O_3 \longrightarrow O_2 + O_4$ (slow) $O_4 \longrightarrow 2O_2$ (fast)
(5) $2O_3 \rightleftharpoons O_2 + O_4$ (fast) $O_4 \longrightarrow 2O_2$ (slow) where p_{O_4} is proportional to $p^2_{O_3}/p_{O_2}$
(c) The reaction was investigated in the presence of CO₂ to determine any catalytic effect. The results of two such experiments at 50°C are

$p_{CO_2} = 100$ torr

t (sec)	0	1800	3600	7200
p_{O_3} (torr)	200	140	100	50

$p_{CO_2} = 180$ torr

t (sec)	0	1800	3600
p_{O_3} (torr)	220	120	70

Find the order of reaction with respect to p_{O_3} for each experiment and the rate constant in each case. Does the catalyzed reaction have a different mechanism? (d) The complete rate equation for the catalyzed reaction is

rate $= k' p^x_{O_3} p^y_{CO_2}$

where the rate constants k found in part (c) are equal to $k' p_{CO_2}$. Find k' and y.

16 CHEMICAL EQUILIBRIUM

In this chapter we introduce equilibrium constant expressions and show how to formulate them. The influence on equilibrium of temperature, pressure, and concentration are examined. The last section presents a general method for solving equilibrium constant problems. Equilibrium reactions are important in most areas of chemistry and calculations based on them will reappear in the chapter on acids and bases (Chapter 17), in the chapter on ions in aqueous solutions (Chapter 18), in the chapter on complex ions (Chapter 28), and in the chapters on qualitative analysis (Chapters 22, 30, 31).

Picture a large chicken house divided into two parts by a chicken-wire wall. On one side live 200 white chickens. On the other side live 200 brown chickens. What would happen if someone left the connecting door open? Gradually the brown and white chickens would mix together as their random paths took them through the door. How fast this happens—the "reaction rate"—would depend upon how fast the chickens are moving. The process might be quicker in the morning than just before sundown when the chickens are drowsy.

Eventually, we can assume, a state of dynamic equilibrium would be reached. The brown and white chickens would be pretty well mixed together on each side, and at any given moment some brown and some white chickens would be going through the door in each direction. There would be a roughly unchanging number of chickens on each side. If we knew how long it would take to reach this state, we would still know nothing about how many chickens were on each side at equilibrium. It might be an equal number, or it might be any combination adding up to 400, depending upon the "conditions" of the "reaction." Suppose the water buckets on one side were empty. Then at equilibrium there might be only 50 chickens on that side and 350 on the other.

The purpose of this analogy, about which you are probably wondering by now, is to point out that the "rate" of a reaction and the equilibrium state of a reaction are different concepts, and that for a complete description of a reaction, information about both is necessary.

The law of chemical equilibrium

16.1 Reversibility and equilibrium

Earlier (Section 4.6) we defined equilibrium as a state of balance between equal and opposing forces. Any reversible chemical reaction can achieve equilibrium, which for such a reaction is a dynamic state. It is best for us to approach an understanding of equilibrium reactions by looking at some experiments—this is how the early chemists did it. A theoretical approach that completely verifies what was first found experimentally is available in the study of thermodynamics. However, we are not quite ready for that. (We discuss the relationship between equilibrium and thermodynamics in Section 25.6.)

Reactions in the gas phase are particularly easy to study as examples of equilibrium. Suppose some nitrogen dioxide, NO_2, is confined in a closed vessel of fixed volume at a constant temperature. Immediately, the formation of dinitrogen tetroxide, N_2O_4, begins. We choose to call this the "forward" reaction:

$$2NO_2(g) \longrightarrow N_2O_4(g)$$

As soon as some N_2O_4 has formed the "reverse" reaction begins—the breakdown of N_2O_4 molecules to give NO_2 molecules. The rates of the forward and reverse reactions and the amounts of the two gases continue to change until equilibrium is reached. At this point the "opposing forces"—the forward and reverse reactions—are balanced.

$$2NO_2(g) \rightleftharpoons N_2O_4(g)$$

The forward and reverse reactions proceed at equal rates, with the net result that the amounts of the two gases are constant. It is a dynamic state. Each time an N_2O_4 molecule is formed, two molecules of NO_2 disappear; each time an N_2O_4 molecule decomposes, two new NO_2 molecules appear. As long as the temperature and pressure remain constant and nothing is added to or taken from the mixture, the equilibrium state remains unchanged.

What happens if we pump more NO_2 molecules into the vessel? With more NO_2 available, whatever collisions precede the formation of N_2O_4 occur more often. If the formation of N_2O_4 proceeds faster than its decomposition, the amount of N_2O_4 increases. As the concentration of N_2O_4 increases, the rate of its decomposition also increases. Changes in the concentrations and in the forward and reverse reaction rates continue until equilibrium has been reached once more (Figure 16.1).

The conditions for this reaction can easily be changed. The temperature and the pressure can be varied; we can start with pure NO_2, pure N_2O_4, or any mixture of the two. Experience has shown that no matter what the conditions, equilibrium will be established.

Let's look at some data obtained by measuring the concentrations of these two gases in equilibrium mixtures (at the same constant temperature after equilibrium was reached in a closed vessel). In each of

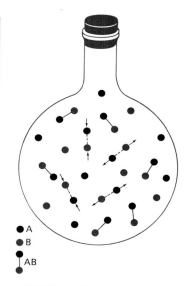

FIGURE 16.1
State of dynamic equilibrium for the reaction A + B $\rightleftharpoons$ AB. *The mixture includes A, B, and AB in amounts determined by their initial concentrations and the equilibrium constant. The forward and reverse reactions occur at equal rates.*

the four experiments the initial concentrations of the two gases were different.

Experiment No.	At equilibrium	
	$[NO_2]$ (moles/liter)	$[N_2O_4]$ (moles/liter)
1	0.052	0.595
2	0.024	0.127
3	0.068	1.02
4	0.101	2.24

It would be satisfying to find a mathematical relationship that brings some meaning to these numbers. The table shows that the concentrations of the reactant and product are *not* equal to each other at equilibrium. A little arithmetic shows that these concentrations do not maintain a constant ratio, either.

Experiment No.	$[N_2O_4]/[NO_2]$
1	11
2	5.3
3	15
4	22

We might try several dozen manipulations with such data before coming up with a meaningful relationship. Let's move ahead by skipping directly to the right one. Dividing the concentration of N_2O_4 by the square of the concentration of NO_2 gives a constant.

Experiment No.	$[N_2O_4]/[NO_2]^2$
1	220
2	220
3	220
4	220

Remember that the initial concentrations of the gases were different in each experiment. The existence of this mathematically constant relationship shows that whatever the initial conditions, the forward and reverse reactions adjust in reaching equilibrium so that $[N_2O_4]/[NO_2]^2$ remains the same.

Experiments with many different equilibrium reactions over the years have led to the formulation of an expression for the relationships among equilibrium concentrations for any reaction. For the general reaction

$$a\text{A} + b\text{B} \cdots \rightleftharpoons r\text{R} + s\text{S} \cdots \qquad (1)$$

the "equilibrium constant" is

$$K = \frac{[R]^r[S]^s \cdots}{[A]^a[B]^b \cdots} \qquad (2)$$

It is more difficult to define the equilibrium constant in words than with an equation: The **equilibrium constant** is the product of the concentrations of the reaction products, each raised to the power equal to its stoichiometric coefficient, divided by the product of the concentrations of the reactants, each raised to the power equal to its stoichiometric coefficient. Equation (2) is sometimes called the *law of chemical equilibrium.*

The NO_2–N_2O_4 reaction fits into the general scheme of Equations (1) and (2) as

$$a\mathrm{A} \rightleftharpoons r\mathrm{R} \qquad 2NO_2 \rightleftharpoons N_2O_4$$

and

$$K = \frac{[R]^r}{[A]^a} = \frac{[N_2O_4]}{[NO_2]^2}$$

Some carefully gathered experimental data on the reaction between hydrogen and iodine

$$H_2(g) + I_2(g) \rightleftharpoons 2HI(g)$$

illustrate several additional important points about chemical equilibrium. In one set of experiments, the amount of hydrogen in the initial mixture was held relatively constant (at about 0.001 mole/liter), while the amount of iodine was varied. The concentrations of H_2, I_2, and HI at equilibrium were measured (at 425.4°C). These concentrations and the values of K calculated from

$$K = \frac{[HI]^2}{[H_2][I_2]}$$

were as follows (data from A. H. Taylor and R. H. Crist, *J. Amer. Chem. Soc.* **63**, 1377, 1941):

At equilibrium

$[H_2]$	$[I_2]$	$[HI]$	K
0.001831	0.003129	0.01767	54.5
0.003560	0.001250	0.01559	54.6
0.004565	0.0007378	0.01354	54.4

The small variations in the value of K are due to small variations in experimental conditions, referred to as "experimental error," and are not significant. Therefore, we see that changing the concentration of one of two reactants does not alter the constancy of K. Additional experiments show that K remains constant with simultaneous variations in the concentrations of both reactants. Essentially, an infinite number of combinations of the concentrations of reactants and products is possible, but at equilibrium at a specific temperature and pressure, there is only one value for K.

Similar experiments at a higher temperature (490.65°C instead of 425.4°C) yielded a constant value for K of about 45.6, which illustrates the variation of K with temperature.

In further experiments with this same system, hydrogen iodide was placed in a sealed vessel and allowed to reach equilibrium with its decomposition products hydrogen and iodine—the same reaction but approached from the reverse direction. Similar data and calculations (for 425.4°C also) show by the constancy of K that the equilibrium state is the same, no matter from which direction it is approached.

At equilibrium

$[H_2]$	$[I_2]$	$[HI]$	K
0.0004789	0.0004789	0.003531	54.3
0.001141	0.001141	0.008410	54.3

To summarize, experimental data have shown clearly that

1. The equilibrium concentrations of reactants and products in a reversible reaction are related by the law of chemical equilibrium

$$K = \frac{[R]^r[S]^s \cdots}{[A]^a[B]^b \cdots} \qquad \text{where } aA + bB \cdots \rightleftharpoons rR + sS \cdots$$

2. An initial mixture of reactants of any concentrations will reach equilibrium (at a specific temperature and pressure) so that the concentrations of reactants and products are related by the law of chemical equilibrium.

3. The value of K is dependent upon the temperature.

4. Approaching equilibrium from either direction leads to the same equilibrium point (at the same temperature).

16.2 Units and equilibrium constant values

In our calculation of K for the hydrogen–iodine reaction, we bypassed the subject of units. The units are determined by the form of the K expression. For the hydrogen–iodine reaction,

$$K = \frac{[HI]^2}{[H_2][I_2]} = \frac{(\text{mole/liter})^2}{(\text{mole/liter})(\text{mole/liter})}$$

we see that K is dimensionless. This is true for any reaction in which the number of reactant molecules equals the number of product molecules.

Table 16.1 gives some additional examples of K expressions for gas-phase reactions. For Reaction (a), the units would be (liter/mole)^2; for Reactions (b) and (c), they would be liter/mole; and for Reaction (d), they would be mole/liter. The confusion that might be caused by this diversity is handled very simply—it is customary to omit the units with K values, no matter what the form of the equilibrium expression, and we do so in this book.

When the reactants and products of a reaction are all gases, partial pressures can be used in the law of chemical equilibrium instead of concentrations in moles per liter, giving K_p for the general reaction of Equation (1) as

$$K_p = \frac{p_R{}^r p_S{}^s \cdots}{p_A{}^a p_B{}^b \cdots} \longleftarrow \text{partial pressures in atm}$$

$\nearrow$ *indicates K for pressure*

For example, for the reaction

$$PCl_3(g) + Cl_2(g) \rightleftharpoons PCl_5(g)$$

we have

$$K_p = \frac{p_{PCl_5}}{p_{PCl_3} p_{Cl_2}}$$

Note that pressures in equilibrium expressions are always given in atmospheres. Frequently, the symbol K_c is used to distinguish an equilibrium constant found by using concentrations from an equilibrium constant found by using partial pressures. The value of K_p is related to the value of K_c by the ideal gas law. For substance A, because n/V is moles per liter,

$$p_A V_A = n_A RT$$

$$p_A = \frac{n_A RT}{V_A}$$

$$= [A]RT \tag{3}$$

Substituting relationships such as the ones in Equation (3) into the expression for K_p gives

$$K_p = \frac{([R]RT)^r([S]RT)^s \cdots}{([A]RT)^a([B]RT)^b \cdots}$$

$$= \frac{[R]^r[S]^s \cdots}{[A]^a[B]^b \cdots} \times (RT)^{(r+s\cdots)-(a+b\cdots)}$$

which yields the general expression

$$K_p = K_c(RT)^{\Delta n}$$

TABLE 16.1

Equilibrium expressions for homogeneous gas-phase reactions
(Brackets indicate concentration in moles per liter.)

$N_2(g) + 3H_2(g) \rightleftharpoons 2NH_3(g)$ (a)
$$K_c = \frac{[NH_3]^2}{[N_2][H_2]^3}$$

$2NO_2(g) \rightleftharpoons N_2O_4(g)$ (b)
$$K_c = \frac{[N_2O_4]}{[NO_2]^2}$$

$2CO(g) + O_2(g) \rightleftharpoons 2CO_2(g)$ (c)
$$K_c = \frac{[CO_2]^2}{[CO]^2[O_2]}$$

$COCl_2(g) \rightleftharpoons CO(g) + Cl_2(g)$ (d)
$$K_c = \frac{[CO][Cl_2]}{[COCl_2]}$$

CHEMICAL
EQUILIBRIUM [16.2]

where

$$\Delta n = \text{moles of gaseous products } (r + s \cdots)$$
$$- \text{moles of gaseous reactants } (a + b \cdots)$$

EXAMPLE 16.1

■ Methyl alcohol can be prepared commercially by the reaction of hydrogen with carbon monoxide

$$CO(g) + 2H_2(g) \rightleftharpoons CH_3OH(g)$$

Under equilibrium conditions at 700°K, $[H_2] = 0.072$ mole/liter, $[CO] = 0.020$ mole/liter, and $[CH_3OH] = 0.030$ mole/liter. What are the values of K_c and K_p at this temperature?

Writing the equilibrium expression for K_c and then substituting the concentrations into it, we get

$$K_c = \frac{[CH_3OH]}{[CO][H_2]^2} = \frac{(0.030)}{(0.020)(0.072)^2} = 290$$

We must determine the value of Δn, the change in the number of moles of gaseous reactants and products, in order to convert the value of K_c to K_p. For this reaction

$$\Delta n = \text{(moles of gaseous products)} - \text{(moles of gaseous reactants)}$$
$$= (1) - (1 + 2) = -2$$

Thus

$$K_p = K_c(RT)^{\Delta n} = (290)[(0.0821)(700)]^{-2} = 0.088$$

The values of the equilibrium constants are $K_c = 290$ and $K_p = 0.088$.

■

TABLE 16.2
Some equilibrium constants

The last two types of equilibria are discussed further in Chapters 17 and 18.

	K_c at 25°C
Gas-phase reactions	
$H_2(g) + \frac{1}{2}O_2(g) \rightleftharpoons H_2O(g)$	5.89×10^{40}
$H_2(g) + I_2(g) \rightleftharpoons 2HI(g)$	8.7×10^2
$CO_2(g) \rightleftharpoons CO(g) + \frac{1}{2}O_2(g)$	1.72×10^{-46}
Ionization of weak acids in water	
$CH_3COOH(aq) \rightleftharpoons CH_3COO^- + H^+$	1.76×10^{-5}
$HCN(aq) \rightleftharpoons H^+ + CN^-$	4.93×10^{-10}
Solubility of precipitates	
$CaSO_4(s) \rightleftharpoons Ca^{2+} + SO_4^{2-}$	2.5×10^{-5}
$AgCl(s) \rightleftharpoons Ag^+ + Cl^-$	1.56×10^{-10}
$Ag_2S(s) \rightleftharpoons 2Ag^+ + S^{2-}$	7.1×10^{-50}

Equilibrium constants are usually written as exponential numbers and they range greatly in magnitude (Table 16.2). Large values of K indicate that large amounts of products are present at equilibrium. Small values of K indicate the opposite—at equilibrium large amounts of reactants are present. For $K > 1$, the amounts of products are greater than the amounts of reactants; for $K < 1$, the amounts of reactants are greater than the amounts of products. Arrows of different length are often used to indicate the predominant direction in equilibrium reactions, as shown in Table 16.2.

$K > 1$, products $>$ reactants
$K < 1$, products $<$ reactants

16.3 Homogeneous, heterogeneous, and solution equilibria

Reversible reactions can take place and equilibrium is established no matter what the physical state of the materials involved. In the iron–steam reaction

$$3Fe(s) + 4H_2O(g) \rightleftharpoons Fe_3O_4(s) + 4H_2(g)$$

a solid and a gas are in equilibrium with another solid and gas. In the reaction of phosphorus(III) chloride with chlorine at elevated temperature,

$$PCl_3(g) + Cl_2(g) \rightleftharpoons PCl_5(g)$$

the reactants and product are all gases. The solid substances, lead(II) oxide and tin, when both are finely powdered and heated to just below the melting point, slowly reach equilibrium with lead and tin(II) oxide according to the equation

$$PbO(s) + Sn(s) \rightleftharpoons Pb(s) + SnO(s)$$

Some important equilibria of different types are shown in Table 16.3.

When all of the components of an equilibrium mixture are in the

TABLE 16.3
Some equilibrium reactions important to the chemical industry

The quantities given are the amounts of these chemicals produced in 1978.

Haber process for ammonia production
$$N_2(g) + 3H_2(g) \rightleftharpoons 2NH_3(g)$$
33.9 billion lb   **(a)**

Lime (CaO) from limestone
$$CaCO_3(s) \rightleftharpoons CaO(s) + CO_2(g)$$
38.7 billion lb   **(b)**

Preparation of ammonium sulfate (a fertilizer)
$$2NH_3(g) + H_2SO_4(l) \rightleftharpoons (NH_4)_2SO_4(s)$$
3.9 billion lb   **(c)**

Preparation of ethyl acetate (a solvent for organic chemicals)
$$CH_3COOH(l) + C_2H_5OH(l) \rightleftharpoons CH_3COOC_2H_5(l) + H_2O(l)$$
acetic acid   ethyl alcohol   *ethyl acetate*   **(d)**
227 million lb

same phase, we have a homogeneous equilibrium, as in the gas-phase reactions given in Table 16.1. In heterogeneous equilibria the reactants and products are *not* all in the same phase, as shown in Reactions (b) and (c) in Table 16.3.

Strictly speaking, equilibrium constants are valid only for ideal gases or ideal dilute solutions. The law of chemical equilibrium in its theoretically correct form is expressed in terms of the activities of the reactants and products. For

$$aA + bB \cdots \rightleftharpoons rR + sS \cdots$$

$$K = \frac{a_R{}^r a_S{}^s \cdots}{a_A{}^a a_B{}^b \cdots} \qquad \text{where } a = \text{activity}$$

For our purposes, the approximations to ideal behavior and to activities discussed in Section 14.10 (see Table 14.3) can be used in equilibrium expressions. To review: the activities of pure solids or pure liquids, or the solvent in dilute solutions, can be taken as equal to 1. For gases, activities are approximated by partial pressures in atmospheres. For solutes in dilute solutions, activities can be taken as approximately equal to molarities.

We made use of these approximations when we calculated equilibrium constants for homogeneous gas-phase reactions by using partial pressures to find K_p and when we calculated equilibrium constants for homogeneous solution reactions by using concentrations in moles per liter to find K_c. Now we want to consider how to write equilibrium constant expressions for heterogeneous equilibria.

The law of chemical equilibrium in its ideally correct form for the iron–steam reaction, a heterogeneous equilibrium, gives

$$K = \frac{a_{Fe_3O_4} a_{H_2}^4}{a_{Fe}^3 a_{H_2O}^4}$$

But because we have agreed that concentrations or pressures are approximately equal to activity, and that the activities of solids can be taken as 1, then $a_{Fe_3O_4} = a_{Fe}^3 = 1$, and

$$K_c = \frac{[H_2]^4}{[H_2O]^4}$$

or

$$K_p = \frac{p_{H_2}^4}{p_{H_2O}^4}$$

To give two further examples, the equilibrium constant for the reaction

$$NH_4Cl(s) \rightleftharpoons NH_3(g) + HCl(g)$$

is simply

$$K_c = [NH_3][HCl]$$

and for the reaction

$$CaCO_3(s) \rightleftharpoons CaO(s) + CO_2(g)$$

the equilibrium constant is

$$K_c = [CO_2] \quad \text{or} \quad K_p = p_{CO_2}$$

These last two simple equations indicate that as long as both $CaCO_3$ and CaO are present, the pressure of carbon dioxide cannot vary (at a given temperature). If pressure is applied to the system, CO_2 combines with CaO, so that the pressure of the CO_2 gas above the solid remains constant.

Many reactions with which we are concerned occur in dilute aqueous solutions. Frequently, as we will see in the next chapter, water is a "participant" in the reaction in the sense that ions in solution are hydrated, or associated with water molecules, or formed by reaction with water. Water is present in such solutions, however, in great excess compared to the other "reactants." Therefore, the solvent water is considered to be virtually a pure liquid; and for reactions in dilute aqueous solution, the concentration of water need not appear in the equilibrium constant expression. For example, for the dissolution of ammonia

$$NH_3(aq) + H_2O(l) \rightleftharpoons NH_4^+ + OH^-$$

we write

$$K_c = \frac{[NH_4^+][OH^-]}{[NH_3]}$$

It is possible to have both pure substances and solutions as reactants and products in the same heterogeneous reaction. Consider, for example, the reaction of hydrochloric acid with calcium carbonate:

$$CaCO_3(s) + 2HCl(aq) \rightleftharpoons CaCl_2(aq) + CO_2(g) + H_2O(l)$$

In terms of activities, K for this reversible reaction (Section 16.6) is

$$K = \frac{a_{CaCl_2} a_{CO_2} a_{H_2O}}{a_{CaCO_3} a_{HCl}^2}$$

We can approximate the activities of both the solid calcium carbonate and the solvent water as 1. Using pressure for the concentration of gaseous carbon dioxide and molarity for the aqueous solution of calcium chloride, we arrive at

$$K = \frac{p_{CO_2}[CaCl_2]}{[HCl]^2}$$

Each of the quantities in this K expression is an approximation of an activity. Therefore, such an expression in which molarities and par-

Concentrations of pure solids and liquids, and of solvents in dilute solutions, are not included in the equation for K.

tial pressures are used together to find K, although it may at first appear a bit strange, is correct.

EXAMPLE 16.2

■ A student attempted to prepare and isolate thiosulfuric acid, $H_2S_2O_3$, using the following sequence of reactions:

$$S(s) + O_2(g) \rightleftharpoons SO_2(g) \tag{a}$$
$$SO_2(g) + H_2O(l) \rightleftharpoons H_2SO_3(aq) \tag{b}$$
$$H_2SO_3(aq) + 2OH^- \rightleftharpoons SO_3^{2-} + 2H_2O(l) \tag{c}$$
$$SO_3^{2-} + S(s) \overset{\Delta}{\rightleftharpoons} S_2O_3^{2-} \tag{d}$$
$$S_2O_3^{2-} + 2H^+ \rightleftharpoons H_2S_2O_3(aq) \tag{e}$$

but found that the desired acid immediately decomposed as

$$H_2S_2O_3(aq) \rightleftharpoons H_2SO_3(aq) + S(s) \tag{f}$$

Write the equilibrium constant expressions for reactions (a) through (f).

For the pure solids and liquids, $S(s)$ and $H_2O(l)$, we recognize that the activities are unity and therefore the concentrations for these substances will not appear in the equilibrium expressions. For the aqueous solutions of ions and various weak acids, we use the molar concentrations to represent the activities. For Reaction (a) involving the combustion of $S(s)$, we can write either an expression for K_p or for K_c. The equilibrium constant expressions for Reactions (a) through (f) are

$$K_{(a)} = \frac{p_{SO_2}}{p_{O_2}} \qquad K_{(d)} = \frac{[S_2O_3^{2-}]}{[SO_3^{2-}]}$$

$$K_{(b)} = \frac{[H_2SO_3]}{p_{SO_2}} \qquad K_{(e)} = \frac{[H_2S_2O_3]}{[S_2O_3^{2-}][H^+]^2}$$

$$K_{(c)} = \frac{[SO_3^{2-}]}{[H_2SO_3][OH^-]^2} \qquad K_{(f)} = \frac{[H_2SO_3]}{[H_2S_2O_3]}$$

For Reaction (a), since the number of moles of reactants and products is equal, K will have the same value whether calculated from partial pressures or from concentrations. For Reaction (b), a heterogeneous equilibrium, the molarity of the H_2SO_3 solution and the partial pressure of SO_2 appear together as the approximations for activity. ■

EXAMPLE 16.3

■ Anhydrous calcium sulfate is often used as a desiccant—a substance that removes water vapor from an enclosed volume of air.

$K_p = 1.55 \times 10^3$ (for pressures in atmospheres) for the reaction at room temperature,

$$CaSO_4(s) + 2H_2O(g) \rightleftharpoons CaSO_4 \cdot 2H_2O(s)$$

What is the equilibrium vapor pressure of water in the container?

The equilibrium expression is

$$K_p = \frac{1}{p^2_{H_2O^2}}$$

Rearranging gives

$$p_{H_2O} = \sqrt{1/K} = \sqrt{1/1.55 \times 10^3} = 2.5 \times 10^{-2} \text{ atm}$$

The partial pressure of water will be reduced until it reaches 2.5×10^{-2} atm, or 19 torr. ■

$Q > K$, reverse reaction occurs
$Q < K$, forward reaction occurs

16.4 The reaction quotient

The equilibrium constant expression and the numerical values of K provide a means of predicting what will happen when substances with the potential for reacting with each other are mixed. The nonequilibrium concentrations are used to calculate the **reaction quotient,** Q, which takes the same form as the equilibrium constant but is used for reactions not at equilibrium.

$$Q = \frac{[R]^r[S]^s \cdots}{[A]^a[B]^b \cdots} \quad \text{where } aA + bB \cdots \rightleftharpoons rR + sS \cdots \qquad (4)$$

If Q is greater than K, the reaction will approach equilibrium from the direction of the reverse reaction. If Q is less than K, the reaction will proceed toward equilibrium in the forward direction.

EXAMPLE 16.4

■ Cyclohexane undergoes a molecular rearrangement in the presence of $AlCl_3$ to form methylcyclopentane

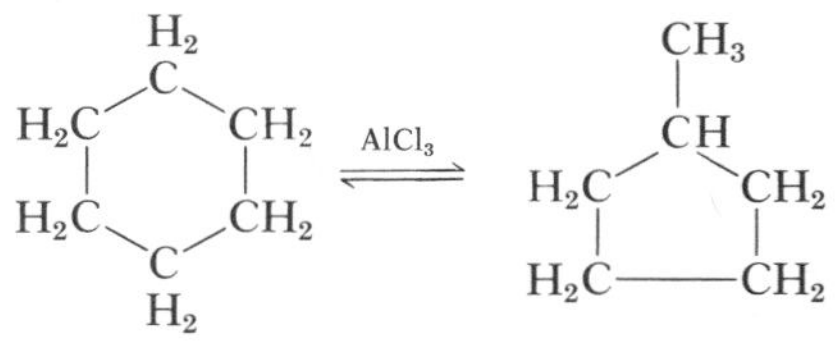

cyclohexane *methylcyclopentane*

The equilibrium constant for the reaction is 0.143 at 25°C. Describe what will happen if a solution is prepared so that it is $0.200M$ in cyclohexane and $0.100M$ in methylcyclopentane.

The reaction quotient for the solution is

$$Q = \frac{[\text{methylcyclopentane}]}{[\text{cyclohexane}]} = \frac{0.100}{0.200} = 0.500$$

Because $Q > K$, some of the methylcyclopentane will react to form additional cyclohexane. The respective concentrations will change until they are in a ratio of 0.143, at which point dynamic equilibrium is reached (see Example 16.8). Note that the reaction may or may not be instantaneous—this is the reason we also study rates of reaction.

■

16.5 Equilibrium constants and reaction equations

The first step in writing an equilibrium expression is to write the balanced equation for the reaction. How the reaction equation is written determines the form of the expression for K. For example, if Equation (c) in Table 16.1 were written instead as

$$CO(g) + \tfrac{1}{2}O_2(g) \rightleftharpoons CO_2(g)$$

its equilibrium constant would be

$$K_c = \frac{[CO_2]}{[CO]\,[O_2]^{1/2}} \equiv \overset{\textit{equivalent to}}{\left(\frac{[CO_2]^2}{[CO]^2\,[O_2]}\right)^{1/2}}$$

That is, if the coefficients in a reaction are divided by 2, the square root of the original equilibrium constant becomes the new equilibrium constant (Table 16.4).

If an equation is multiplied by a factor n, for example, $n = 3$,

$$3PCl_3(g) + 3Cl_2(g) \rightleftharpoons 3PCl_5(g)$$

its equilibrium constant becomes the original equilibrium constant raised to the nth power.

$$K_c = \frac{[PCl_5]^3}{[Cl_2]^3[PCl_3]^3} \equiv \left(\frac{[PCl_5]}{[Cl_2][PCl_3]}\right)^3$$

Writing a reaction in reverse [see Equation (b), Table 16.1]

$$N_2O_4(g) \rightleftharpoons 2NO_2(g)$$

has the effect of taking the reciprocal of its equilibrium constant.

$$K_c = \frac{[NO_2]^2}{[N_2O_4]} \equiv \frac{1}{[N_2O_4]/[NO_2]^2}$$

TABLE 16.4

Variation of equilibrium constant with how the reaction equation is written

K_0 is the original equilibrium constant.

When the equation is	the new equilibrium constant is
Reversed	K_0^{-1}
Multiplied by n	K_0^n
Divided by 2	$K_0^{1/2}$
Divided into 2 steps (a and b)	$K_0 = K_a K_b$

When the equations for two reactions are added together, the equilibrium constant for the total reaction is the *product* of the equilibrium constants of the original reactions. Consider the reaction

$$2SO_2(g) + O_2(g) + 2H_2O(g) \rightleftharpoons 2H_2SO_4(g) \tag{5}$$

The equilibrium constant for this reaction, according to the law of chemical equilibrium, is

$$K_c = \frac{[H_2SO_4]^2}{[SO_2]^2[O_2][H_2O]^2} \tag{6}$$

However, it is highly unlikely that Reaction (5) proceeds in a single step. A possible two-step mechanism is

$$2SO_2 + O_2 \rightleftharpoons 2SO_3 \tag{7}$$
$$2SO_3 + 2H_2O \rightleftharpoons 2H_2SO_4 \tag{8}$$
$$\overline{2SO_2 + O_2 + 2H_2O \rightleftharpoons 2H_2SO_4}$$

The equilibrium constants for Reactions (7) and (8) are

$$K_{(7)} = \frac{[SO_3]^2}{[O_2][SO_2]^2} \qquad K_{(8)} = \frac{[H_2SO_4]^2}{[H_2O]^2[SO_3]^2}$$

Multiplying K for Equation (7) by K for Equation (8) and canceling common terms gives the equilibrium constant expression for the reaction given in Equation (5)

$$K_{(7)}K_{(8)} = \frac{[SO_3]^2}{[O_2][SO_2]^2} \times \frac{[H_2SO_4]^2}{[H_2O]^2[SO_3]^2}$$
$$K_c = \frac{[H_2SO_4]^2}{[SO_2]^2[O_2][H_2O]^2}$$

This is the same as Equation (6), which was derived without assuming two steps in the overall reaction, and it shows that the overall equilibrium constant is equal to the product of the equilibrium constants of the steps in the mechanism.

This brings us to a very important point: *The equilibrium constant is independent of the number of steps in the reaction mechanism.* Whether Reaction (5) proceeds by 2 steps or by 50 steps, its equilibrium expression remains Equation (6).

EXAMPLE 16.5

■ Express the equilibrium constant $K_{(i)}$ for the reaction

$$2NH_3(aq) + CO_2(g) + H_2O(l) \longrightarrow 2NH_4^+ + CO_3^{2-} \tag{i}$$

in terms of the equilibrium constants for the following reactions:

k = rate constant
K = equilibrium constant

$NH_3(aq) + H_2O(l) \rightleftharpoons NH_4^+ + OH^-$ (ii)
$CO_2(g) + H_2O(l) \rightleftharpoons H_2CO_3(aq)$ (iii)
$H_2CO_3(aq) \rightleftharpoons 2H^+ + CO_3^{2-}$ (iv)
$H_2O(l) \rightleftharpoons H^+ + OH^-$ (v)

Equation (i) and the desired equilibrium constant can be obtained from the given reactions by combining equations and K's as follows:

$$\begin{array}{rlr} 2[NH_3(aq) + H_2O(l) \rightleftharpoons NH_4^+ + OH^-] & K_a = K_{(ii)}^2 & \text{(a)} \\ CO_2(g) + H_2O(l) \rightleftharpoons H_2CO_3(aq) & K_b = K_{(iii)} & \text{(b)} \\ H_2CO_3(aq) \rightleftharpoons 2H^+ + CO_3^{2-} & K_c = K_{(iv)} & \text{(c)} \\ 2[H^+ + OH^- \rightleftharpoons H_2O(l)] & K_d = 1/K_{(v)}^2 & \text{(d)} \\ \hline 2NH_3(aq) + CO_2(g) + H_2O(l) \rightleftharpoons 2NH_4^+ + CO_3^{2-} & & \end{array}$$

with $K_{(i)} = K_{(a)}K_{(b)}K_{(c)}K_{(d)} = K_{(ii)}^2K_{(iii)}K_{(iv)}/K_{(v)}^2$. As a check, let us substitute the complete expressions for $K_{(ii)}$, $K_{(iii)}$, $K_{(iv)}$, and $K_{(v)}$, and see that the correct expression for $K_{(i)}$ results:

$$K_{(i)} = \frac{K_{(ii)}^2K_{(iii)}K_{(iv)}}{K_{(v)}^2} = \frac{\left(\frac{[NH_4^+][OH^-]}{[NH_3]}\right)^2\left(\frac{[H_2CO_3]}{[CO_2]}\right)\left(\frac{[H^+]^2[CO_3^{2-}]}{[H_2CO_3]}\right)}{([H^+][OH^-])^2}$$

$$= \frac{[NH_4^+]^2[CO_3^{2-}]}{[NH_3]^2[CO_2]}$$

Comparison with Equation (i) shows that this is the correct K expression for that equation. ■

Confusion sometimes arises between the powers of the concentrations in equilibrium expressions and in rate equations. There is a fundamental difference: In *equilibrium expressions* the powers of all concentrations are always equal to the coefficients in the overall equation; in *rate equations* these powers depend upon the mechanism and may or may not equal the coefficients in the reaction equation. The values of the exponents in rate equations *must* be determined experimentally.

16.6 Are equilibria established in all reactions?

In many reactions reversibility is not readily observed. We often say that such reactions "go to completion;" that is, that they continue until one of the reactants is completely used up. Virtually all of these reactions do reach equilibrium, but one set of reactants is present in so much smaller concentration than the other set that upon superficial examination it looks as though the reaction has gone all the way. Reactions that appear to go to completion are of three main types, based on the types of products that are formed.

1. *A slightly soluble gas is formed,* for example, in the reaction between an acid and a carbonate.

$$CaCO_3(s) + 2HCl(aq) \longrightarrow CaCl_2(aq) + H_2O(l) + CO_2(g)$$

gas escapes from reaction mixture

Most of the carbon dioxide escapes from the solution and is thus effectively prevented from reacting with the materials that are still dissolved. However, a small amount of it remains in solution and is free to take part in the reverse reaction. If additional CO_2 gas is forced into the system, the reverse reaction may become the dominant one. For example, the decomposition of calcium bicarbonate, $Ca(HCO_3)_2$, in solution

$$Ca(HCO_3)_2(aq) \rightleftharpoons CaCO_3(s) + CO_2(g) + H_2O(l)$$

deposits the slightly soluble calcium carbonate. The reaction is reversed and the carbonate is dissolved if the mixture is saturated with carbon dioxide gas.

2. *A slightly soluble precipitate is formed*, for example, in the formation of mercury(II) sulfide from mercury(II) chloride and hydrogen sulfide.

solid does not go back into solution

$$HgCl_2(aq) + H_2S(g) \longrightarrow HgS(s) + 2H^+(aq) + 2Cl^-(aq)$$

Mercury(II) sulfide is one of the least soluble substances known, and it is commonly said that the reaction by which it is formed is irreversible. For all practical purposes, this is true, but finite concentrations of Hg^{2+} and S^{2-} do exist in the solution, and these ions are free to combine, respectively, with Cl^- and H^+, and a few of them undoubtedly do so.

3. *A slightly ionized material is formed*. The classic example of such a reaction is the neutralization of a strong acid by a strong base, for example, the neutralization of hydrochloric acid by sodium hydroxide,

$$H^+ + Cl^- + Na^+ + OH^- \longrightarrow Na^+ + Cl^- + H_2O(l)$$

covalent compound forms few ions

The hydrogen and hydroxide ions combine to form a covalent molecule, water, which is only very slightly ionized.

The theoretically correct answer to the question posed in the title of this section is, "yes, equilibria are established in all reactions—if enough time is available." In any system, physical or chemical forces balance each other in a state of equilibrium. However, we think of chemical reactions in which very few particles of reactants remain in the reaction mixture as "complete," and no harm is done by such thinking.

Catalysts have no effect on the position of equilibrium

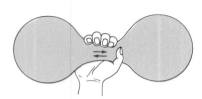

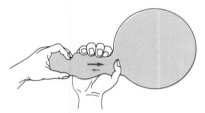

FIGURE 16.2
LeChatelier's principle demonstrated.

Factors that influence equilibria

16.7 LeChatelier's principle

The general principle that underlies all changes in equilibria is **Le-Chatelier's principle,** which states: *If a system in equilibrium is subjected to a stress, a change that will offset the stress will occur in the system.*

To visualize how LeChatelier's principle works you might think of a balloon only partly full of air. You could grasp the balloon loosely in the middle with one hand so that the amount of air is equal in both ends (Figure 16.2). This is an equilibrium situation—molecules are moving in both directions through the narrow middle section. With your other hand you could squeeze one end of the balloon—that is, apply a stress. To relieve the stress air would go into the other end, which would bulge out.

In the chemical sense, the stresses that can be applied to a system in equilibrium are changes in temperature, concentration, and pressure.

Catalysts can increase the speed with which equilibrium is reached, but they cause no change in the equilibrium constant or in the concentrations at equilibrium. The rates of both the forward and reverse reactions are increased equally by a catalyst (by lowering ΔE_a; see Figure 15.5). The effects of changes in reaction conditions on equilibrium constants, rate constants, and rates are summarized in Table 16.5. We suggest you consult this table as you read Sections 16.8 to 16.10.

16.8 Temperature

As we have stated before, the rates of chemical reactions are increased by an increase in temperature. The total effect of a tempera-

TABLE 16.5
Comparison of variations in equilibrium constants, rate constants, and rates

The effects summarized here occur in most cases, although exceptions can be found.

When	Equilibrium constant (K)	Rate constant (k)	Rate
Temperature			
Increases	Changes	Increases	Increases
Decreases	Changes	Decreases	Decreases
Catalyst is added	Does not change	Increases	Increases
Concentration of reactants			
Increases	Does not change	Does not change	Increases
Decreases	Does not change	Does not change	Decreases

[16.7] CHEMICAL EQUILIBRIUM

ture change on a reversible reaction depends upon the relative changes of the forward and reverse reaction rates. Every equilibrium involves one exothermic and one endothermic reaction. A temperature increase causes a greater increase in the rate of the endothermic reaction, and the value of the equilibrium constant changes with a change in temperature.

If we think of heat as a reactant or a product, it is easy to visualize, based on LeChatelier's principle, which direction is favored by a temperature increase. For example, the reaction between hydrogen and iodine to give hydrogen iodide is exothermic; that is, heat is a product ($\Delta H° = -12.6$ kcal). In Section 16.1 we saw that K varied with temperature as follows:

$$\overset{\xleftarrow{\hspace{3cm}} heat}{H_2(g) + I_2(g) \rightleftharpoons 2HI + 12.6 \text{ kcal}} \qquad \begin{array}{l} K \text{ at } 425.4°C = 54 \\ K \text{ at } 490.65°C = 45.6 \end{array}$$

An increase in temperature causes the decomposition of additional hydrogen iodide. The concentration of hydrogen iodide therefore becomes smaller, the concentrations of hydrogen and iodine become larger, and the value of K

$$K = \frac{[HI]^2}{[H_2][I_2]}$$

becomes smaller, as shown by the decrease in K from about 54 at 425.4°C to about 45.6 at 490.65°C.

In an endothermic reaction, we may think of heat as a reactant. The endothermic reaction of oxygen to form ozone ($\Delta H° = 68.2$ kcal) is at equilibrium at room temperature with very little ozone present. Increasing the temperature increases ozone formation and the equilibrium constant becomes larger.

$$\overset{heat \xrightarrow{\hspace{3cm}}}{68.2 \text{ kcal} + 3O_2(g) \rightleftharpoons 2O_3(g)} \qquad \begin{array}{l} K = 6.2 \times 10^{-58} \text{ at } 25°C \\ = 2.6 \times 10^{-56} \text{ at } 35°C \end{array}$$

(How K varies with temperature is discussed further in Section 25.7.)

16.9 Concentration

When the amount of a reactant or product is changed, the rates of the forward and reverse reactions adjust so that the concentrations of the other reactants and products change, while the value of K remains the same. The separation of mixtures by chromatography is based on equilibria between solutions of changing concentration (see Tools of Chemistry, Chromatography).

It is possible to visibly demonstrate the changing direction of equilibrium by changing the conditions in the reaction of bismuth chloride with water to give white insoluble bismuth oxochloride and hydrochloric acid.

Chromatography

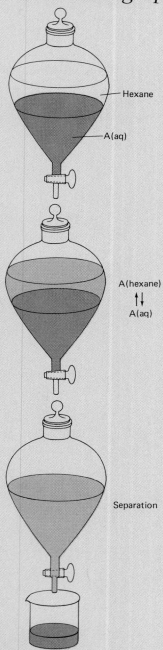

Hexane

A(aq)

A(hexane)
↑↓
A(aq)

Separation

FIGURE A
Extraction in a separatory funnel. *In this method of separation a solute is distributed between two immiscible solvents.*

Biochemical fluids, inorganic salts, organic compounds, polymers—mixtures of all kinds of substances can be separated by the various types of chromatography. The purpose of the separation might simply be to find out how many substances are in a mixture. Or it might be **qualitative analysis,** the identification of the substances in the mixture, or **quantitative analysis,** which is the determination of the amounts of substances in the mixture. Chromatography is also used to isolate the components of a mixture, although the amounts of materials that can be obtained in this way are limited. In fact, one of the great advantages of chromatographic methods is that they can be used on very small samples.

Like many separation techniques, chromatography is based on dividing substances in a mixture between phases that can then be separated. A simple illustration of the equilibrium process involved in all chromatography is the distribution of a solute between two immiscible solvents. Suppose we have an aqueous solution of A, a fairly nonpolar substance. Most likely, A is also soluble in a nonpolar solvent such as hexane. When the aqueous solution of A and hexane are placed in contact with each other (Figure A), an equilibrium is established. Some A is transferred from the aqueous solution to the hexane solution.

$$A(aq) \rightleftharpoons A(\text{hexane})$$

The equilibrium constant for the distribution of a solute between two immiscible solvents is called the **distribution coefficient.** For our system

$$K = \frac{[\text{A}]_{\text{hexane}}}{[\text{A}]_{aq}}$$

In chromatography a *series* of equilibrium distributions like the one just described is established. This occurs because one of the phases is moving. **Chromatography** is the distribution of a solute between a stationary phase and a mobile phase. The stationary phase may be a solid or a liquid supported as a thin film on the surface of an inert solid—the support. The mobile phase flowing over the surface of the stationary phase may be a gas or a liquid.

Chromatography based on solubility properties is called partition chromatography. In partition chromatography of any type the stationary phase is a liquid, and the substances being separated are distributed throughout both the stationary and the mobile phases (Figure B). Chromatographic methods in which the stationary phase is a solid are classified as adsorption chromatography. A substance leaves the mobile phase to become adsorbed on the surface of the solid phase, called the adsorbent. The equilibria in adsorption chromatography are governed by the weak intermolecular forces of physical adsorption such as London forces and hydrogen bonding. The adsorbent must have a large surface area. Alumina (Al_2O_3) and silica (SiO_2) in various forms are common adsorbents.

Partition chromatography

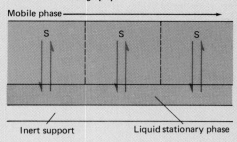

Adsorption chromatography

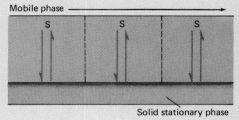

FIGURE B
Equilibria in partition and adsorption chromatography. *S represents the solute.*

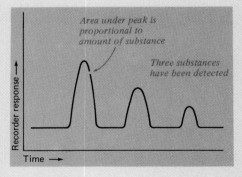

FIGURE C
A chromatogram. *This type of record of the results of chromatography is obtained in any type of chromatography when the mobile phase is monitored by instrumentation.*

The most versatile and widely used chromatographic technique is gas-liquid chromatography (GLC), which involves a stationary liquid and a mobile gas phase. Gas chromatography is a general term for chromatography in which the mobile phase is a gas. GLC is useful for studying complex mixtures (up to 50 components have been separated), for detecting very small amounts of impurities, and because it is a relatively rapid method.

Any mixture of substances that can be vaporized below about 400°C without decomposition can be examined by GLC. The sample is vaporized into a stream of inert gas that carries it past the stationary phase, which is packed in a tube, or column. The different components in the mixture are retained on the column for different characteristic time periods and then appear at the column outlet. Changes in the composition of the gas flowing from the column are monitored in many different ways. However, each involves the same principle: the change in some property of the gaseous mixture such as thermal conductivity, density, or some type of absorption spectrum is monitored by means of instruments that produce a graphic record of what happens beginning from the time the sample is injected. A chromatogram is a plot of the response of the detector to whatever property is being measured versus time (Figure C). The area under the peak in a chromatogram is proportional to the amount of the substance in a mixture.

Other terms frequently encountered in describing types of chromatography are column chromatography, paper chromatography, and thin-layer chromatography (TLC). In column chromatography, which includes gas chromatography, the stationary phase is held in a glass tube. In paper chromatography a stationary liquid phase is supported in the pores of a piece of paper. The end result of paper chromatography is a dried piece of paper with spots of the components of the mixture spread out across the paper. Thin-layer chromatography is an adsorption method in which a thin layer of the adsorbent is spread on an inert support such as a piece of glass. The sample is placed on the adsorbent and solvent is then allowed to flow past it, moving components of the mixture down the film at different rates.

$$\overset{\displaystyle \xleftarrow{\hspace{4cm} HCl}}{\underset{\displaystyle \xrightarrow[H_2O]{\hspace{4cm}}}{BiCl_3(aq) + H_2O(l) \rightleftharpoons BiOCl(s) + 2HCl(aq)}}$$

If HCl is added slowly, the BiOCl precipitate disappears as the equilibrium is displaced to the left—toward soluble $BiCl_3$. If more water is then added, the precipitate reappears as the forward reaction is once again favored.

EXAMPLE 16.6

■ A mixture of $K_2CrO_4(aq)$ and $HCl(aq)$ was allowed to come to equilibrium

$$\underset{\substack{\text{chromate ion} \\ \text{yellow}}}{2CrO_4{}^{2-}} + 2H^+ \rightleftharpoons \underset{\substack{\text{dichromate ion} \\ \text{orange}}}{Cr_2O_7{}^{2-}} + H_2O(l)$$

Describe the changes that will occur in the equilibrium system as (a) additional acid is added; (b) additional $K_2CrO_4(s)$ is added; (c) $K_2Cr_2O_7(s)$ is added; (d) Zn^{2+} is added—note that $ZnCrO_4$ is highly insoluble and $ZnCr_2O_7$ is very soluble; (e) NaOH is added.

(a) Some of the additional H^+ will react with some of the remaining chromate ion to produce more dichromate ion and water; that is, the ratio of $[Cr_2O_7{}^{2-}]$ to $[CrO_4{}^{2-}]$ will increase.

(b) Some of the additional chromate ion will react with some of the H^+ to produce more dichromate ion and water; that is, the concentration of H^+ will decrease.

(c) Some of the additional dichromate ion will react to form chromate ion; that is, the concentration of $CrO_4{}^{2-}$ will increase.

(d) Addition of Zn^{2+} will remove some of the chromate ion as $ZnCrO_4$, lowering the concentration of chromate ion. Thus some of the dichromate ion will react to form additional chromate ion and H^+.

(e) Addition of OH^- will remove some of the H^+ as a result of the acid-base neutralization. Thus some of the dichromate ion will react to form chromate ion and the additional H^+ needed to reestablish equilibrium conditions, that is, the ratio $[Cr_2O_7{}^{2-}]$ to $[CrO_4{}^{2-}]$ will decrease.

■

16.10 Pressure

Changes in pressure displace equilibria only in gas-phase reactions in which the number of moles of reactant is different from the number of moles of product. An increase in pressure shifts the equilibrium toward the reaction that produces the smaller number of molecules

The Haber process for the manufacture of ammonia

The Haber process for the manufacture of ammonia is the classic example of the role of kinetics and equilibrium in an industrial process. The reaction of nitrogen and hydrogen to give ammonia is exothermic, occurs in the gas phase, and involves a decrease in pressure.

$$N_2(g) + 3H_2(g) \rightleftharpoons 2NH_3(g)$$
$$\text{4 moles} \qquad\qquad \text{2 moles}$$

$\Delta H_{25°C} = -22.04 \text{ kcal } (-92.22 \text{ kJ})$
$\Delta H_{400°C} = -25.00 \text{ kcal } (-108 \text{ kJ})$
$K_{p,25°C} = 5.5 \times 10^5, K_{p,400°C} = 1.8 \times 10^{-4}$

LeChatelier's principle indicates that higher pressure will favor the formation of ammonia. Since the overall equilibrium is exothermic, higher temperatures will favor the decomposition of ammonia, an undesirable result, and lower temperatures will favor ammonia production.

However, a low temperature and a high pressure are not the easy answer to industrial production of ammonia by this reaction. The rate of the reaction at room temperature is very, very small. In commercial practice, both yield and rate must be considered, for a process is practical only if the plant produces a reasonable amount of the product per day.

The industrial process relies on a suitable catalyst, which increases the rate of attainment of equilibrium and permits the use of moderate temperatures. In modern industry the compromise among temperature, pressure, and yield of ammonia is struck by carrying out the reaction at about 250 atm (a pressure obtained economically with efficient centrifugal compressors) and 400°C, the lowest temperature at which the catalyst is sufficiently effective. Under these conditions about 20% ammonia is present in the gases that come off the catalytic reactor on each pass. Unconverted reactants are recycled, as shown in Figure A, which is a simplified schematic diagram of NH_3 production from N_2 and H_2.

The catalyst is iron oxide, "doubly promoted" (that is, with the activity increased by two additives) by the addition of about 0.4% potassium oxide and 0.8% aluminum oxide. As is so often the case, the mechanism of the catalytic reaction is not known exactly. However, ammonia is formed at the catalyst surface, for the rate of ammonia formation is approximately proportional to the rate of nitrogen adsorption by the catalyst.

Fritz Haber, a German chemist, developed the industrial ammonia process named for him just before World War I. Had he not done so the war might have ended much sooner. Germany's trade was cut off by a blockade, and sodium nitrate from Chile was not available. Traditionally, nitric acid for explosives had been made from sodium nitrate, but ammonia from the Haber process provided another route to nitric acid. The success of the Haber process allowed the German chemical industry to produce the explosives and fertilizers needed for the war effort despite the blockade. (The importance of the Haber process in fertilizer manufacture is discussed in An Aside: Chemical Fertilizers, Chapter 22.)

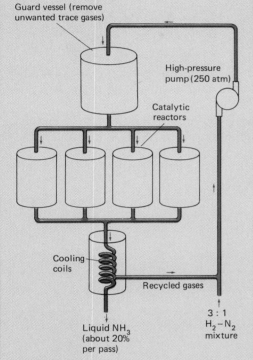

Guard vessel (remove unwanted trace gases)

High-pressure pump (250 atm)

Catalytic reactors

Cooling coils

Recycled gases

Liquid NH₃ (about 20% per pass)

3 : 1 H₂–N₂ mixture

FIGURE A

in the gas phase. For example, in the phosphorus(III) chloride–chlorine reaction

$$\text{PCl}_3(g) + \text{Cl}_2(g) \xrightleftharpoons[\text{2 moles}]{\text{pressure} \longrightarrow} \text{PCl}_5(g)$$

$$\underset{\text{2 moles}}{} \qquad \underset{\text{1 mole}}{}$$

The stress of greater pressure can be relieved by an increase in the rate of the forward reaction, which decreases the gas volume by forming one mole of gas from two. On the other hand, for the decomposition of iodopropane to give propene plus hydrogen iodide we have the opposite situation

$$\underset{\substack{\text{iodopropane} \\ \text{1 mole}}}{\text{CH}_3\text{CH}_2\text{CH}_2\text{I}(g)} \rightleftharpoons \underset{\substack{\text{propene} \\ \text{2 moles}}}{\text{CH}_3\text{CH}=\text{CH}_2(g)} + \text{HI}(g)$$

$$\xleftarrow{\hspace{3cm} \text{pressure}}$$

The reverse reaction is favored by increasing pressure. A change in pressure is essentially a change in concentration and it does not change the value of K.

The displacement of an equilibrium in the gas phase can be demonstrated visibly with the reaction of the oxides of nitrogen discussed in Section 16.1. Nitrogen dioxide, NO_2, is a brown gas that is always in equilibrium with dinitrogen tetroxide, N_2O_4, a colorless gas.

$$\underset{\text{brown}}{2\text{NO}_2(g)} \rightleftharpoons \underset{\text{colorless}}{\text{N}_2\text{O}_4(g)}$$

Increasing pressure favors the formation of N_2O_4, and the change in the concentration at equilibrium is observable by the gradual decrease in intensity of the brown color of NO_2.

Equilibrium problems

16.11 How to solve equilibrium problems

The relationship between K and the equilibrium concentrations allows us to answer many questions about equilibrium reactions, such as: What are the concentrations of all species present at equilibrium? Given one concentration at equilibrium, what are the others? How must one concentration be changed to obtain a desired change in another concentration? Whatever the unknown quantity in an equilibrium problem, it is most successfully obtained by careful and accurate bookkeeping. An excellent way to organize the solution is to follow the steps outlined below. The steps are illustrated for the following typical problem:

For the reaction $A \rightleftharpoons B + C$, what is the concentration of C at equilibrium when $K = 5 \times 10^{-6}$ and $[A] = 5$ moles/liter at the beginning of the reaction?

1. Write the balanced equation

$$A \rightleftharpoons B + C$$

2. Write the corresponding equilibrium expression

$$K = \frac{[B][C]}{[A]}$$

3. Identify the unknown. Often this will be a concentration. Write the unknown down—you may forget what you solved for.

Let $x = [B] = [C]$

4. Prepare a table consisting of
   (a) the chemical equation on the top line. Underneath the reactants and products put down their
   (b) initial concentrations and
   (c) changes in concentration—use your knowledge of stoichiometry here—and
   (d) equilibrium concentrations—the sum of steps (b) and (c)

	A	$\rightleftharpoons$ B	+ C
initial	5		
change	$-x$	$+x$	$+x$
equilibrium	$5 - x$	x	x

5. Substitute the equilibrium concentrations from step 4(d) into the equilibrium expression

$$K = \frac{x^2}{5 - x} = 5 \times 10^{-6}$$

6. Solve for the unknown, using approximations wherever the error introduced by doing so is negligible. This is the case wherever the value of K is quite small in comparison to the concentrations. Here 5×10^{-6} is small compared to 5 moles/liter. Therefore

$$x^2 = 5(5 \times 10^{-6})$$
$$x = \sqrt{25 \times 10^{-6}} = 5 \times 10^{-3}$$

7. Answer the questions in the problem

At equilibrium $[C] = 5 \times 10^{-3}$ mole/liter

This procedure is summarized in Table 16.6.

Solving equilibrium expressions for unknowns requires the use of algebra. In real laboratory and industrial situations the algebra can become very complex. In this book we restrict ourselves to problems that can be solved by using approximations and simple algebra, or to problems involving quadratic equations that can be

Quadratic equation:
$$ax^2 + bx + c = 0$$
and its solution:
$$x = \frac{-b \pm \sqrt{b^2 - 4ac}}{2a}$$

TABLE 16.6
Solution of equilibrium constant problems

1. **Write balanced chemical equation**

2. **Write equilibrium expression**

3. **Identify the unknown**

4. **Make a table:**
   line 1   **Balanced chemical equation**
   line 2   **Initial concentrations**
   line 3   **Changes in concentrations**
   line 4   **Equilibrium concentrations (line 2 + line 3)**

5. **Substitute results of step 4, line 4 into equilibrium expression**

6. **Solve for unknown**

7. **Answer questions in problem**

CHEMICAL
EQUILIBRIUM [16.11]

solved completely. For such problems, once the equilibrium expression is rearranged into the quadratic form

$$ax^2 + bx + c = 0 \qquad (9)$$

the quadratic formula may be used to solve for x

$$x = \frac{-b \pm \sqrt{b^2 - 4ac}}{2a} \qquad (10)$$

Even though an equilibrium expression may look complicated, for example,

$$\frac{(2x)^2 (0.50 + x)}{(1.32 - 2x)} = 5.46 \times 10^{-8} \qquad (11)$$

we often make approximations that greatly simplify the calculations. In solving Equation (11) we try an approximation first because the value for $K(5.46 \times 10^{-8})$ is quite small compared to the usual magnitudes of concentrations. We assume that $0.50 + x$ will essentially be 0.50 according to our rules of significant figures (Appendix A.1) and likewise $1.32 - 2x \approx 1.32$. Thus we write

$$\frac{(2x)^2 (0.50 + x)}{(1.32 - 2x)} \approx \frac{(2x)^2 (0.50)}{(1.32)} = 5.46 \times 10^{-8}$$

and we can solve for x as

$$x = \frac{1}{2} \sqrt{\frac{(5.46 \times 10^{-8})(1.32)}{(0.50)}} = 1.9 \times 10^{-4}$$

This certainly is easier than solving the original equation for x. Before we finish the problem, we should check our approximations to see if they were valid. To do this, substitute the value found for x back into the approximate equilibrium expression. Yes, $0.50 + 1.9 \times 10^{-4}$ is approximately 0.50 and $1.32 - 2(1.9 \times 10^{-4})$ is approximately 1.32. Therefore, our approximations are valid and the approximate answer is correct.

Suppose our approximate answer for x had been 0.01. Is $0.50 + 0.01 = 0.51$ still close enough to 0.50 and is $1.32 - 2(0.01) = 1.30$ still close enough to 1.32 for our approximation to be valid? When, as in this case, it is not easy to tell by inspection whether or not the approximation is a good one, the 5% error test is applied. As long as there is less than a 5% change in all of the numbers involved in the approximations, the approximation is accepted as valid. In this case, the errors were $(0.01/0.50)(100\%) = 2\%$ and $(0.02/1.32) = 1.5\%$. These errors are well within the 5% limit and therefore the answer of $x = 0.01$ is acceptable.

We can summarize the procedure for solving the equilibrium expression as follows:

1. Write the complete mathematical expression.

2. Decide what approximations can be made.

3. Write the simplified mathematical expression.

4. Solve the simplified mathematical expression.

5. Check all approximations for validity. Usually up to 5% error is acceptable.

6. Either use the approximate solution answer, if it passes the 5% error test, or solve the complete mathematical expression for the exact value of the unknown.

16.12 Some equilibrium problem solutions

To save time and energy, always try an approximate solution before attempting the more exact solution. The approximate answer is satisfactory in such a large percentage of problems that you will not be wasting your time.

EXAMPLE 16.7

■ A solution of $0.100M$ HIO_4 at 25°C was found to contain $0.038M$ H^+. Using this information, find K for the reaction

$$HIO_4(aq) \rightleftharpoons H^+ + IO_4^-$$

Using the procedure outlined in Table 16.6

$$HIO_4(aq) \rightleftharpoons H^+ + IO_4^-$$

$$K = \frac{[H^+][IO_4^-]}{[HIO_4]} \qquad K \text{ is unknown}$$

	$HIO_4(aq) \rightleftharpoons$	H^+	$+ IO_4^-$
initial	0.100	0	0
change	−0.038	+0.038	+0.038
equilibrium	0.062	0.038	0.038

$$K = \frac{(0.038)(0.038)}{(0.062)} = 2.3 \times 10^{-2}$$

The equilibrium constant is 2.3×10^{-2}. Because this is our first illustration of these calculations, let's discuss some of the entries in the table. Water is the solvent and its concentration does not enter into the equilibrium expression. The initial $0.100M$ concentration refers to how the solution was prepared, not the concentrations at equilibrium. Note that the stoichiometry of the reaction is such that $[H^+] = [IO_4^-]$ and that the concentration of HIO_4 reacting is equal to the concentration of H^+ formed. ■

EXAMPLE 16.8

■ What are the concentrations of cyclohexane and methylcyclopentane once equilibrium has been established in the solution described in Example 16.4, which is 0.200M in cyclohexane and 0.100M in methylcyclopentane?

Using our standard procedure we first write the equation and the equilibrium expression, and then choose the unknown.

$$\text{cyclohexane} \rightleftharpoons \text{methylcyclopentane}$$

$$K = \frac{[\text{methylcyclopentane}]}{[\text{cyclohexane}]} \qquad \text{Let } x = \text{change in concentration}$$

From the calculation of Q in Example 16.4 we know that to reach equilibrium some of the methylcyclopentane will react to form additional cyclohexane. Therefore, the table is written as follows:

	cyclohexane $\rightleftharpoons$ methylcyclopentane	
initial	0.200	0.100
change	$+x$	$-x$
equilibrium	$(0.200 + x)$	$(0.100 - x)$

$$K = \frac{[\text{methylcyclopentane}]}{[\text{cyclohexane}]} = \frac{0.100 - x}{0.200 + x} = 0.143$$

$$1.143x - 0.071 = 0$$
$$x = 0.062 \text{ mole/liter}$$
$$[\text{cyclohexane}] = 0.200 + 0.062 = 0.262 \text{ mole/liter}$$
$$[\text{methylcyclopentane}] = 0.100 - 0.062 = 0.038 \text{ mole/liter}$$

The concentrations will change until the solution is 0.038M in methylcyclopentane and 0.262M in cyclohexane. ■

EXAMPLE 16.9

■ The concentration equilibrium constant for the reaction

$$H_2(g) + I_2(g) \rightleftharpoons 2HI(g)$$

is 29.1 at 1000°K. What is the concentration of $I_2(g)$ under equilibrium conditions if the system originally contained $[HI] = 10$ mole/liter?

$$H_2 + I_2 \rightleftharpoons 2HI$$

$$K_c = \frac{[HI]^2}{[H_2][I_2]} \qquad \text{Let } x = [I_2]$$

$$H_2 + I_2 \rightleftharpoons 2\ HI$$

initial			10
change	$+x$	$+x$	$-2x$
equilibrium	x	x	$10 - 2x$

$$K = \frac{(10 - 2x)^2}{(x)(x)} = 29.1$$

Assuming that $10 - 2x \approx 10$, and solving for x, we get $x = 1.85$. The error in this answer is $(3.70/10)(100\%) = 37\%$, which violates our 5% rule. Therefore the complete equilibrium expression

$$\frac{100 - 40x + 4x^2}{x^2} = 29.1$$

must be solved. Rearrangement into the form of a quadratic equation, Equation (9), gives

$$25.1x^2 + 40x - 100 = 0$$

and solving using Equation (10) gives

$$x = \frac{-40 \pm \sqrt{(40)^2 - 4(25.1)(-100)}}{2(25.1)} = 1.4$$

The concentration of I_2 at equilibrium is 1.4 moles/liter. ∎

EXAMPLE 16.10

∎ What is the concentration of CO in equilibrium at 25°C in a sample of gas originally containing 1.00 M CO_2?

Using the information from Table 16.2 we write

$$CO_2(g) \rightleftharpoons CO(g) + \tfrac{1}{2}O_2(g)$$
$$K = \frac{[CO][O_2]^{\frac{1}{2}}}{[CO_2]} = 1.72 \times 10^{-46} \qquad \text{Let } x = [CO]$$

$$CO_2(g) \rightleftharpoons CO(g) + \tfrac{1}{2}O_2(g)$$

initial	1.00		
change	$-x$	$+x$	$+0.500x$
equilibrium	$1.00 - x$	x	$0.500x$

$$K = \frac{(x)(0.500x)^{\frac{1}{2}}}{(1.00 - x)} = 1.72 \times 10^{-46}$$

Assume $1.00 - x \approx 1.00$

$$\frac{(x)(0.500x)^{\frac{1}{2}}}{(1.00)} = 1.72 \times 10^{-46}$$

$$x^{\frac{3}{2}} = \frac{(1.00)(1.72 \times 10^{-46})}{(0.500)^{\frac{1}{2}}} = 2.43 \times 10^{-46}$$

$$x = 3.90 \times 10^{-31}$$

It is true that $1.00 - (3.90 \times 10^{-31}) \approx 1.00$; therefore the approximation is valid. Under equilibrium conditions, $[CO] = 3.90 \times 10^{-31}$ mole/liter. ∎

EXAMPLE 16.11

∎ A sample of air was passed through an electrical discharge, causing the following reaction

$$N_2(g) + O_2(g) \underset{1.00\ \text{atm}}{\overset{4200°K}{\rightleftharpoons}} 2NO(g)$$

Assuming that one mole of air was originally present at 1 atm pressure, calculate the partial pressure of NO at equilibrium. $K_p = 0.0123$ at 4200°K.

Assuming air to consist of 0.78 mole N_2 and 0.21 mole O_2 (see Table 6.2), we can set up the following table to describe the number of moles of NO produced:

	N_2 +	O_2	⟶ 2NO
initial	0.78	0.21	
change	$-0.5n$	$-0.5n$	$+n$
equilibrium	$0.78 - 0.5n$	$0.21 - 0.5n$	n

where n is the number of moles of NO. The total number of moles of gas (n_t) present at equilibrium is

$$n_t = (0.78 - 0.5n) + (0.21 - 0.5n) + (n) = 0.99$$

and the respective mole fractions are

$$X_{N_2} = \frac{0.78 - 0.5n}{0.99} \qquad X_{O_2} = \frac{0.21 - 0.5n}{0.99} \qquad X_{NO} = \frac{n}{0.99}$$

The corresponding partial pressures, according to Dalton's law, are

$$p_{N_2} = \left(\frac{0.78 - 0.5n}{0.99}\right)(1.00\ \text{atm})$$

$$p_{O_2} = \left(\frac{0.21 - 0.5n}{0.99}\right)(1.00\ \text{atm})$$

$$p_{NO} = \left(\frac{n}{0.99}\right)(1.00\ \text{atm})$$

Substituting these expressions for partial pressures into the expression for K_p and then solving for n gives

$$K = \frac{p_{NO}^2}{p_{N_2}p_{O_2}} = \frac{\left[\left(\dfrac{n}{0.99}\right)(1.00)\right]^2}{\left[\left(\dfrac{0.78 - 0.5n}{0.99}\right)(1.00)\right]\left[\left(\dfrac{0.21 - 0.5n}{0.99}\right)(1.00)\right]} = 0.0123$$

$$n = 0.042$$

Thus the partial pressure of NO at equilibrium is

$$p_{NO} = \left(\frac{n}{0.99}\right)(1.00 \text{ atm}) = \left(\frac{0.042}{0.99}\right)(1.00 \text{ atm}) = 0.042 \text{ atm}$$

$\blacksquare$

EXAMPLE 16.12

$\blacksquare$ Bromine and chlorine dissolve in carbon tetrachloride (CCl_4) and react to form BrCl.

$$Br_2(CCl_4) + Cl_2(CCl_4) \rightleftharpoons 2BrCl(CCl_4)$$

Under equilibrium conditions $[Br_2] = [Cl_2] = 0.0043$ and $[BrCl] = 0.0114$. Calculate K_c for this reaction. The addition of 0.0100 mole of bromine to one liter of this mixture disturbs the equilibrium. Assuming a negligible volume change, calculate the concentrations of the three species present once equilibrium has been reestablished.

The value of K_c is

$$K_c = \frac{[BrCl]^2}{[Br_2][Cl_2]}$$

$$= \frac{(0.0114)^2}{(0.0043)(0.0043)}$$

$$= 7.0$$

According to Le Chatelier's principle, addition of bromine will cause further reaction with chlorine to form BrCl. Letting x equal the additional amount of chlorine that reacts, we prepare our usual table.

	Br₂	+	Cl₂	⇌	2BrCl
initial	0.0043 + 0.0100		0.0043		0.0114
change	−x		−x		+2x
equilibrium	0.0143 − x		0.0043 − x		0.0114 + 2x

$$K_c = \frac{[BrCl]^2}{[Br_2][Cl_2]}$$

$$= \frac{(0.0114 + 2x)^2}{(0.0143 - x)(0.0043 - x)} = 7.0$$

Because we see that the value of K_c is not quite small compared to the concentrations, we solve this expression by using the quadratic equation

$$3x^2 - 0.176x + 0.000300 = 0$$

$$x = \frac{-(-0.176) \pm \sqrt{(-0.176)^2 - (4)(3)(0.000300)}}{(2)(3)}$$

$$= 0.057 \quad \text{or} \quad 0.0018$$

We find that the value 0.057 is meaningless because it would give

$$[Cl_2] = 0.0043 - 0.057 = -0.053$$

Therefore, choosing $x = 0.0018$ gives

$$
\begin{aligned}
[Cl_2] &= 0.0043 - 0.0018 \\
&= 0.0025 \\
[Br_2] &= 0.0143 - 0.0018 \\
&= 0.0125 \\
[BrCl] &= 0.0114 + 2(0.0018) \\
&= 0.0150
\end{aligned}
$$

Under the new equilibrium conditions, the concentration of Cl_2 is 0.0025 mole/liter, of Br_2 is 0.0125 mole/liter, and of BrCl is 0.0150 mole/liter. As we would predict using LeChatelier's principle, [BrCl] has increased and $[Cl_2]$ has decreased. ◼

THOUGHTS ON
CHEMISTRY

Professor Thomas Cooper's view of chemistry in 1811

INTRODUCTORY LECTURE OF THOMAS COOPER, *by Thomas Cooper*

Every person is apt to overvalue the importance of the pursuit in which he feels himself particularly interested. Yet I think it can be shewn, without much difficulty, that chemistry is of more immediate and useful application to the every day concerns of life—that it operates more upon our hourly comforts, than any other branch of knowledge whatever.

Of the manufactures which contribute to the comfort or the ornament of society, I do not know one that does not for the most part depend upon processes purely chemical: The analysis and smelting of ores; the manufactures of iron and steel; of copper and brass; and silver and pewter; the manufactures of leather, glass and pottery; of soap and candles; of drugs and medicines; the bleaching, the dying, and the printing of silks, cottons, linens and woollens; the printing of furniture paper; the composition of printer's ink as well as of common

[16.12] CHEMICAL
EQUILIBRIUM

*ink; the manufacture of parchment and of paper itself; the arts of sil-
vering and gilding; of colour making; the manufactures of wine and
vinegar; baking, brewing and distillery; even that most useful art, the
art of Cookery, is very greatly indebted to chemistry: for independent
of the late improvements of coating copper vessels with zinc, silver
and platina, instead of tin; and iron vessels with white enamel; the
whole art of cookery has undergone an economical revolution by the
experiments of Proust, Rumford, and the scientific gentlemen who
superintend the benevolent soup establishments of England . . . .*

*In every sense of the word therefore as a practical, as well as a
philosophical maxim, KNOWLEDGE IS POWER. Not only that
knowledge of human affairs and of the human character, which displays itself as well in the prudence of common life, as in the arrangement and combination to the best effect of the qualities and forces of
political communities—but that knowledge also, which subjects in the
best possible way to the use and the dominion, of man, all the powers
and properties of inferior animals and the vast range of inanimate nature. That knowledge, which multiplies a thousand fold, the physical
force of a human being—which renders every hour of existence more
desirable, by compelling every object around us, to contribute in some
way or other, to our pleasure, to our profit, to our comfort, or to our
convenience—which brings the mutual wants, and the mutual supplies of the inhabited world into immediate contact—and which multiplies not only human enjoyment, and alleviates human suffering, but
multiplies also the human species; by providing more extensively, the
means of constant employment, and comfortable subsistence.*

*It is knowledge then, that must render us respectable and re-
spected; for those only who possess it are so, whether as nations or
individuals. To acquire it, we must cherish and extend the means of
acquiring it. If not, we may vainly boast of the advantages we fancy
we enjoy, but they will ultimately fall to those who best know how to
use them. May that day not arrive, when we shall be weighed in the
balance and found wanting.*

INTRODUCTORY LECTURE OF THOMAS COOPER, by Thomas Cooper, 1811 (Archibald
Loudon, Carlisle, Pennsylvania, 1812)

Significant terms defined in this chapter

equilibrium constant	**quantitative analysis**
reaction quotient	**distribution coefficient**
LeChatelier's principle	**chromatography**
qualitative analysis	

Exercises

16.1 Predict whether the equilibrium for the photosynthesis reaction described by the equation

$$6CO_2(g) + 6H_2O(l) \rightleftharpoons C_6H_{12}O_6(s) + 6O_2(g)$$
$$\Delta H° = 669.62 \text{ kcal}$$

would (i) shift to the right—the products side, (ii) shift to the left—the reactants side, or (iii) remain unchanged if (a) $[CO_2]$ was increased, (b) p_{O_2} was increased, (c) one-half of the $C_6H_{12}O_6$ was removed, (d) the total pressure was increased, (e) the temperature was increased, and (f) a catalyst was added.

16.2 Which statements are true? Rewrite any false statement so that it is correct.
(a) The reaction quotient becomes the equilibrium constant at equilibrium.
(b) The law of chemical equilibrium states that if a system in equilibrium is subjected to a stress, the system will react in such a way as to tend to offset the stress.
(c) Once established, equilibrium conditions can be changed by changes in concentration or partial pressures, temperature, and/or addition of a catalyst.
(d) The formation of a slightly soluble gas, a slightly soluble precipitate, and/or a slightly ionizable substance will tend to force a reaction to "completion."
(e) A system after reaching chemical equilibrium can be considered to be in a static state.
(f) Because units are usually dropped from values of equilibrium constants, K_c and K_p values are interchangeable.

16.3 Write the expressions for K_c or K_p for the following reactions:

(a) $Ag^+(aq) + Cl^-(aq) \rightleftharpoons AgCl(s)$
(b) $SO_3(g) + H_2(g) \rightleftharpoons SO_2(g) + H_2O(g)$
(c) $Ag_2S(s) \rightleftharpoons 2Ag^+(aq) + S^{2-}(aq)$
(d) $CaCO_3(s) \rightleftharpoons CaO(s) + CO_2(g)$
(e) $BaCO_3(s) + C(s) \rightleftharpoons BaO(s) + 2CO(g)$
(f) $4NH_3(g) + 5O_2(g) \rightleftharpoons 4NO(g) + 6H_2O(g)$
(g) $CO(g) + 2H_2(g) \rightleftharpoons CH_3OH(l)$
(h) $MnO_4^- + 5Fe^{2+} + 8H^+ \rightleftharpoons 5Fe^{3+} + Mn^{2+} + 4H_2O(l)$
(i) $C_3H_8(g) + 5O_2(g) \rightleftharpoons 3CO_2(g) + 4H_2O(g)$
(j) $CH_3COOH(aq) \rightleftharpoons H^+ + CH_3COO^-$
(k) $Cl_2(g) + H_2O(l) \rightleftharpoons HCl(aq) + HOCl(aq)$

16.4 Silver chloride is a very slightly soluble substance, but because of the equilibrium

$$AgCl(s) \rightleftharpoons Ag^+(aq) + Cl^-(aq)$$

equal amounts of Ag^+ and Cl^- are present in the aqueous phase in contact with pure AgCl. Describe what happens as solid $AgNO_3$ is added.

16.5 Write the expression for K_c for the equation

$$2NO(g) + O_2(g) \rightleftharpoons 2NO_2(g) \qquad \Delta H°_{1000} = -27.93 \text{ kcal}$$

Will the concentration of NO_2 at equilibrium (i) increase, (ii) decrease, or (iii) remain the same if (a) additional O_2 is introduced? (b) additional NO is introduced? (c) the total pressure is increased? (d) the temperature is decreased?

16.6 Consider the reaction

$$CaCO_3(s) \rightleftharpoons CaO(s) + CO_2(g)$$

Will the mass of $CaCO_3$ at equilibrium (i) increase, (ii) decrease, or (iii) remain the same if (a) CO_2 is removed from the equilibrium system? (b) the pressure is increased? (c) solid CaO is added?

16.7 A weather indicator can be made by coating an object with $CoCl_2$ and observing the color of the object. The color change is the result of the following reaction:

$$\underset{pink}{[Co(H_2O)_6]Cl_2(s)} \rightleftharpoons \underset{blue}{[Co(H_2O)_4]Cl_2(s)} + 2H_2O(g)$$

If you were to observe a pink color, does that mean "moist" or "dry" air?

16.8 Consider an equilibrium system containing SO_2, O_2, and SO_3 where

$$SO_2(g) + \tfrac{1}{2}O_2(g) \rightleftharpoons SO_3(l)$$

Describe what happens as a small amount of radioactive oxygen is added in the form of O_2.

16.9 The reaction

$$PCl_3(g) + Cl_2(g) \rightleftharpoons PCl_5(g)$$

has come to equilibrium at a temperature at which the concentrations of PCl_3, Cl_2, and PCl_5 are 10, 9, and 12 mole/liter, respectively. Calculate the value of the equilibrium constant, K_c, for this reaction.

16.10 Nitrogen reacts with hydrogen according to the equation

$$N_2(g) + 3H_2(g) \rightleftharpoons 2NH_3(g)$$

An equilibrium mixture at a given temperature was found to contain 0.31 mole/liter N_2, 0.50 mole/liter H_2, and 0.14 mole/liter NH_3. Calculate the equilibrium constant K_c at the given temperature.

16.11 Ethyl acetate, a solvent used in lacquers, reacts with water to form acetic acid and ethyl alcohol

$$CH_3COOC_2H_5(aq) + H_2O(l) \rightleftharpoons$$
$$CH_3COOH(aq) + C_2H_5OH(aq)$$

Under certain conditions it was observed that $\tfrac{1}{3}$ mole of ethyl acetate remained at equilibrium after 1 mole of ethyl acetate was mixed with enough water to make 1 liter of solution. Calculate K_c.

16.12 Using the following equilibrium data

$[Fe^{2+}]$	$[SCN^-]$	$[Fe(SCN)_x^{2-x}]$
0.012M	0.012M	0.0012M
0.027	0.056	0.012
0.039	0.016	0.0050

determine the value of x in the equation

$$Fe^{2+} + xSCN^- \rightleftharpoons [Fe(SCN)_x]^{2-x}$$

Hint: Look for the value of x which gives a constant value for K_c.

16.13* Kinetics studies have shown that $k_f = 1.38 \times 10^{-28}$ and $k_r = 9.3 \times 10^{10}$ at 25°C for the reaction

$$CO(g) + Cl_2(g) \underset{k_r}{\overset{k_f}{\rightleftharpoons}} COCl(g) + Cl(g)$$

For this reaction, $K = k_f/k_r$.
What is the value of the equilibrium constant?

16.14 The equilibrium constant for the reaction

$$CO(g) + H_2O(g) \rightleftharpoons CO_2(g) + H_2(g)$$

is $K_c = 4.0$ at a given temperature. An equilibrium mixture contains 0.80 mole of CO, 0.35 mole of water, and 2.2 moles of CO_2. How many moles of H_2 are present? Note that the volume is not important in this calculation.

16.15 What must be the pressure of hydrogen so that the reaction

$$WCl_6(g) + 3H_2(g) \rightleftharpoons W(s) + 6HCl(g)$$

will occur if $p_{WCl_6} = 0.012$ atm and $p_{HCl} = 0.10$ atm? $K_p = 1.37 \times 10^{21}$ at 900°K.

16.16 What pressure of NH_3 is in equilibrium with H_2SO_4 and $(NH_4)_2SO_4$ in the production of ammonium sulfate, which is used in fertilizer

$$2NH_3(g) + H_2SO_4(l) \rightleftharpoons (NH_4)_2SO_4(s)$$

if $K_p = 2.23 \times 10^{31}$ for the reaction at 25°C? In order to produce ammonium sulfate, would you change the pressure of NH_3 so that it was larger, equal to, or smaller than this value?

16.17 A solution that is 1.0×10^{-3} M in Ba^{2+} is mixed with an equal amount of a solution that is $2.0 \times 10^{-6}M$ in CO_3^{2-}.
(a) Calculate the reaction quotient for the equation

$$Ba^{2+} + CO_3^{2-} \rightleftharpoons BaCO_3(s)$$

(b) If $K = 5 \times 10^8$ at 25°C for this reaction, will a precipitate form?

16.18 The value of K_p at 25°C for

$$2CO(g) \rightleftharpoons C(graph) + CO_2(g)$$

is 1.11×10^{21}. What is the value of K_c? Describe what will happen if 2.00 moles of CO and 1.00 mole of CO_2 are mixed in a 1 liter container with a suitable catalyst to make the reaction "go" at this temperature.

16.19 Citric acid reacts with water to lose H^+ in three steps

$$(C_3H_4OH)(COOH)_3(aq) \rightleftharpoons (C_3H_4OH)(COOH)_2(COO)^- + H^+$$
$$K_1 = 7.10 \times 10^{-4}$$
$$(C_3H_4OH)(COOH)_2(COO)^- \rightleftharpoons$$
$$(C_3H_4OH)(COOH)(COO)_2^{2-} + H^+ \quad K_2 = 1.68 \times 10^{-5}$$
$$(C_3H_4OH)(COOH)(COO)_2^{2-} \rightleftharpoons (C_3H_4OH)(COO)_3^{3-} + H^+$$
$$K_3 = 4.11 \times 10^{-7}$$

What is K_c for the overall reaction

$$(C_3H_4OH)(COOH)_3 \rightleftharpoons (C_3H_4OH)(COO)_3^{3-} + 3H^+$$

16.20 The reversible reaction

$$2SO_2(g) + O_2(g) \rightleftharpoons 2SO_3(g)$$

has come to equilibrium in a vessel of specific volume and at a given temperature. Before the reaction the concentrations of the reactants were 0.06 mole/liter of SO_2 and 0.05 mole/liter of O_2. After equilibrium was reached, the concentration of SO_3 was 0.04 mole/liter. What was the concentration of O_2 at equilibrium? What is the value of K_c?

16.21 What are the concentrations of cyclohexane and methylcyclopentane once equilibrium has been established in a solution containing 0.100M cyclohexane? See Example 16.4 for data.

16.22 Adenosine triphosphate (ATP) reacts with water to form adenosine diphosphate (ADP) and produces an inorganic phosphate (P_i) such as HPO_4^{2-}.

$$ATP(aq) + H_2O(l) \longrightarrow ADP(aq) + P_i \quad K_c = 4 \times 10^6$$

Both ATP and ADP are found in living cells and are important in the storage of energy for many chemical reactions in the body, for example, muscular activity. What is the concentration of ATP at equilibrium if the original solution contained $1 \times 10^{-7}M$ ATP? The calculations may be simplified by assuming that practically all of the ATP reacted and only a small portion of ATP is left, so that $x = [ATP]$ and $[ADP] = [P_i] = [(1 \times 10^{-7}) - x]$.

16.23 Chloromethane, CH_3Cl, is produced by the reaction of Cl_2 with methane, CH_4,

$$CH_4(g) + Cl_2(g) \rightleftharpoons CH_3Cl(g) + HCl(g)$$

Assuming that equal amounts of CH_4 and Cl_2 were placed in a reaction vessel so that their initial partial pressures were identical and equilibrium was allowed to be established, what was p_{CH_3Cl}/p_{CH_4} if $K_p = 7.4 \times 10^{17}$? It is not necessary to find the individual pressures of the gases at equilibrium to calculate this ratio.

16.24 What is the percent dissociation of NOCl into NO and Cl_2 if $K_p = 2.55 \times 10^{-4}$ for

$$NOCl(g) \rightleftharpoons NO(g) + \tfrac{1}{2}Cl_2(g)$$

at (a) 1.00 atm total pressure and (b) 10.0 atm total pressure? Do these values illustrate Le Chatelier's principle?

16.25 At 230°C, $K_p = 1.19$ for the equation

$$PCl_3(g) + Cl_2(g) \rightleftharpoons PCl_5(g)$$

What will be the final pressures of the gases if 1.00 mole of PCl_3 reacts with 2.00 moles of Cl_2 at a total pressure of 1.00 atm?

16.26* For the reaction

$$NO(g) + O_3(g) \underset{k_r}{\overset{k_f}{\rightleftharpoons}} NO_2(g) + O_2(g)$$

$K_p = 1.32 \times 10^{10}$ at 1000°K. For $k_f = 6.26 \times 10^8$ liter/mole sec, find k_r at this temperature, assuming that $K = k_f/k_r$.

16.27* An equilibrium exists between NO_2 and its dimer N_2O_4

$$2NO_2(g) \rightleftharpoons N_2O_4(g)$$

(a) Show that $K_p = X_{N_2O_4}/(X_{NO_2}^2 P_t)$, where X is the mole fraction and P_t is the total pressure of the system. (b) A sample of N_2O_4 undergoes 53% dissociation at 60°C and a total pressure of 1.00 atm; find K_p.

16.28* The industrial process known as solvent extraction is based on the equilibrium established for a solute between two mutually insoluble solvents. For example, when an aqueous HCl solution of $FeCl_3$ is shaken with twice its volume of ether (also containing HCl), the following equilibrium is established

$$FeCl_3(aq) \rightleftharpoons FeCl_3(ether)$$

and 99% of the $FeCl_3$ is transferred to the ether phase. Find the equilibrium constant (commonly called the *distribution ratio*) for the extraction.

16.29* There are additional equilibrium constants used in gaseous calculations. One of these is defined in terms of mole fractions, K_X (Section 3.15). Write the expression for K_X and find its value for the reaction

$$2SO_2(g) + O_2(g) \rightleftharpoons 2SO_3(g)$$

when an equilibrium mixture contains $X_{SO_2} = 0.565$, $X_{O_2} = 0.102$, and $X_{SO_3} = 0.333$.

16.30* The hydroxide ion concentration in a sample of river water is $5 \times 10^{-8} M$. What is the maximum concentration of iron (both Fe^{2+} and Fe^{3+}) that can exist in the water if $K_c = 1.64 \times 10^{-14}$ for the reaction

$$Fe(OH)_2(s) \rightleftharpoons Fe^{2+}(aq) + 2OH^-(aq)$$

and $K_c = 1.1 \times 10^{-36}$ for the reaction

$$Fe(OH)_3(s) \rightleftharpoons Fe^{3+}(aq) + 3OH^-(aq)$$

16.31* What is the value of the reaction quotient for the reaction

$$PbSO_4(s) \rightleftharpoons Pb^{2+}(aq) + SO_4^{2-}(aq)$$

if 25.0 ml of $2.0 \times 10^{-4} M$ $Pb(NO_3)_2$ is mixed with 75.0 ml of $1.0 \times 10^{-4} M$ Na_2SO_4? If $K_c = 1.06 \times 10^{-8}$ for this reaction, will a precipitate form?

16.32* At elevated temperatures, BrF_5 undergoes rapid decomposition

$$2BrF_5(g) \rightleftharpoons Br_2(g) + 5F_2(g)$$

A 0.100-mole sample of BrF_5 was placed in a 10.0-liter container and the system was allowed to equilibrate at 1500°K. If the total pressure of gases at equilibrium was 2.12 atm, (a) find the total number of moles of gas present, (b) find the number of moles of each gas present, (c) find the partial pressure of each gas present, and (d) calculate K_c and K_p.

16.33* The equilibrium constant for the reaction

$$Fe^{3+} + 3C_2O_4^{2-} \rightleftharpoons Fe(C_2O_4)_3^{3-}$$

is 1.67×10^{20}. What is the concentration of Fe^{3+} left in solution if equal volumes of $0.0010M$ $Fe(NO_3)_3$ and $1.000M$ $K_2C_2O_4$ are mixed? The calculations are simplified if you assume that all of the Fe^{3+} has reacted and write the "changes" on the basis of $Fe(C_2O_4)_3^{3-}$ undergoing decomposition.

16.34** Consider the Haber process for making ammonia

$$\tfrac{1}{2}N_2(g) + \tfrac{3}{2}H_2(g) \rightleftharpoons NH_3(g) \qquad \Delta H_{298}^\circ = -11.02 \text{ kcal}$$

(a) Complete the table at the top of the next page by indicating the effect that the changes will make on the system at equilibrium:

Change	Amount of N_2	Amount of H_2	Amount of NH_3
Increase temperature			
Increase pressure			
Increase N_2			
Add water			
Add catalyst			

(b) If $K_p = 776$ at 25°C, what is the equilibrium pressure of (i) NH_3 if $p_{N_2} = 300$ atm and $p_{H_2} = 100$ atm? (ii) H_2 if a mixture containing N_2 originally at 300 atm and H_2 originally at 100 atm reacts at constant volume? (iii) H_2 if NH_3 originally at 75 atm undergoes decomposition at constant volume? (c) What is K_c for this reaction? (d) Will K_p increase or decrease if the temperature is changed to 500°K? (e) What is K_p at 25°C for the reaction

$$N_2(g) + 3H_2(g) \rightleftharpoons 2NH_3(g)$$

In this chapter two major methods of defining acids and bases are introduced—the Brønsted–Lowry system, based on proton donation; and the Lewis system, based on electron-pair donation. The behavior of acids and bases in aqueous systems is discussed in detail. Structural factors that influence the strengths of these acids and bases are examined, as are equilibria in aqueous solutions of acids and bases.

You should learn a few practical things about acids and bases, if you have not already experienced them. (How Ira Remsen, founder of the chemistry department at Johns Hopkins University, discovered the properties of nitric acid in the 1800s is described in the Thoughts on Chemistry at the end of this chapter.) Acids and bases can hurt you— from the stinging sensation when you accidentally squirt lemon juice (citric acid) in your eye to very severe and persistent burns if you spill battery acid (sulfuric acid) or lye (sodium hydroxide) on your skin and do not immediately flush it with water.

One of the classical properties by which acids were first identified is a "sour taste." This has been well established—so we caution you not to test it by tasting any acids (or bases) in the laboratory. Think of vinegar, tomatoes, or lemon juice—that should convince you that acids are sour.

Acids and bases can also eat holes in clothing. You might first realize that an acidic or basic solution has splashed on you when your shirt or blouse comes out of the washing machine with little holes where the fabric was weakened. If you spill concentrated acid or base directly on your clothing, the effect will be immediately noticeable and the garment probably ruined. The conclusion is obvious— acids and bases must be handled with care.

Ways of defining acids and bases

Arrhenius (Section 4.8) introduced a concept of acids and bases in 1887. He proposed that the properties of acids are related to the H^+ ion and that the properties of bases are related to the OH^- ion. Over the years acids and bases have been redefined in several other ways. The result of these new definitions has been the inclusion of many more molecules and ions in the categories of acids or bases. Most chemical reactions can now be considered to be either acid-base reactions or oxidation–reduction reactions (Section 4.12 and Chapter 19).

Reactions in aqueous solution are very important in both the study of general chemistry and in practical applications of chemistry (Table 17.1). In this chapter we will stress the behavior of acids and bases in aqueous solutions. However, it is useful to understand that there are acid-base reactions comparable to that between H^+ and OH^- that take place in solvents other than water, and even in the absence of solvents. The desire to treat these reactions as acid-base reactions was one of the motivations in developing definitions other than those of Arrhenius.

Two types of acid-base definitions are of interest to us here: the Brønsted-Lowry definitions and the Lewis definitions. A Brønsted-Lowry acid-base reaction is any reaction in which a proton (H^+) is exchanged between molecules or ions. This may occur in aqueous solution

$$\overset{H^+}{\overset{\curvearrowright}{HCl}} + OH^- \xrightarrow{H_2O} Cl^- + H_2O(l)$$
$$\underset{acid}{} \quad \underset{base}{}$$

The equations here are for the behavior of these substances as acids or bases in aqueous solution. Obviously, their industrial uses take advantage of other properties as well. The remaining chemicals in the top ten are 3, oxygen; 5, nitrogen; 6, ethylene; 7, chlorine.

TABLE 17.1

Acids and bases among the top ten industrial chemicals

Rank (1978)		Billions of pounds produced in 1978
1	**Sulfuric acid** $H_2SO_4(aq) \rightleftharpoons HSO_4^- + H^+$	79.19
2	**Lime** $CaO(s) + H_2O(l) \rightleftharpoons Ca(OH)_2(s)$ $Ca(OH)_2(s) \xrightarrow{H_2O} Ca^{2+} + 2OH^-$	38.78
4	**Ammonia** $NH_3(g) + H_2O(l) \rightleftharpoons NH_4^+ + OH^-$	33.89
8	**Sodium hydroxide** $NaOH(s) \xrightarrow{H_2O} Na^+ + OH^-$	21.42
9	**Phosphoric acid** $H_3PO_4(aq) \rightleftharpoons H_2PO_4^- + H^+$	19.13
10	**Nitric acid** $HNO_3(aq) \longrightarrow H^+ + NO_3^-$	16.10

in nonaqueous solution

$$\overset{\overset{\displaystyle H^+}{\frown}}{NH_4^+} + NH_2^- \underset{}{\overset{\text{liq.}}{\underset{NH_3}{\rightleftharpoons}}} 2NH_3(l)$$
$$\quad acid \qquad base$$

or in the absence of a solvent

$$2NH_4Cl(s) + \overset{\overset{\displaystyle H^+}{\frown}}{CaO}(s) \xrightarrow{\Delta} CaCl_2(s) + 2NH_3(g) + H_2O(g)$$
$$\quad acid \qquad\quad base$$

The Lewis definitions emphasize the sharing of electron pairs and also include reactions that occur in both aqueous solutions

$$\overset{\overset{\text{electron pair}}{\frown}}{H^+ + {:}OH^-} \overset{H_2O}{\rightleftharpoons} H{:}OH$$
$$acid \quad base$$

and nonaqueous solutions

$$\overset{\overset{\text{electron pairs}}{\frown}}{Ag^+ + 2{:}NH_3} \overset{\text{liq.}}{\underset{NH_3}{\rightleftharpoons}} [Ag({:}NH_3)_2]^+$$
$$acid \quad base$$

or in the absence of a solvent

$$\overset{\overset{\text{electron pairs}}{\frown}}{Ni(s) + 4{:}CO(g)} \xrightarrow{\Delta} Ni({:}CO)_4(g)$$
$$acid \quad base$$

Before discussing the Brønsted-Lowry and Lewis acid-base definitions in detail, we mention briefly one other system, one that was developed specifically to deal with nonaqueous reactions. Called the *solvent system*, it applies both to solvents that contain protons and to those that do not.

A **solvent system acid** is any substance that gives the cation of the solvent; a **solvent system base** is any substance that gives the anion of the solvent. An **acid-base reaction** is the reaction of the solvent cation with the solvent anion. For example, in liquid anhydrous ammonia, NH_4^+ is the cation of the solvent, NH_2^- is the anion of the solvent, and a typical acid-base reaction is

$$NH_4^+ + NO_3^- + K^+NH_2^- \overset{\text{liq.}}{\underset{NH_3}{\longrightarrow}} K^+ + NO_3^- + 2NH_3(l)$$

The neutralization reaction is

$$NH_4^+ + NH_2^- \overset{\text{liq.}}{\underset{NH_3}{\longrightarrow}} 2NH_3(l)$$
$$acid \quad base$$

Some of the other solvents to which this approach applies are anhydrous acetic acid, liquid sulfur dioxide (SO_2), and liquid bromine trifluoride (BrF_3).

17.1 The unique properties of the proton

Why has the behavior of the proton so often been the criterion for acid-base behavior? Is the proton indeed a unique species, one that differs in properties from all other ionic species? After all, like any cation, the proton attracts and is attracted by anions and the negative ends of polar covalent bonds. Its association with anions and polar molecules is a consequence of this attraction. The proton, however, has a property that distinguishes it from other cations. It is the smallest cation and therefore has a larger concentration of positive charge than that of any other cation. As a result, its associations with negative species are sufficiently stronger than those of other cations to make its behavior in solution unique.

The proton's small size also increases its mobility in solution. This mobility is particularly evident in aqueous solutions in which one proton can displace another proton from its position in the water structure without altering that structure. Successive displacements of this type provide a pathway of essentially no resistance for the proton to follow through the solution (Figure 17.1).

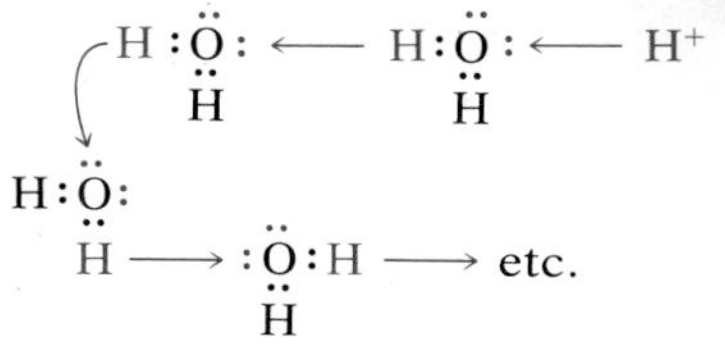

Because of both their mobility and their strong attraction to negatively charged species, a reaction involving protons in aqueous solution is likely to be extremely rapid. Reactions of the proton in highly polar nonaqueous systems (for example, liquid hydrogen fluoride or liquid ammonia) are comparably rapid; those in less polar media are understandably slower because the effect of attraction between opposite charges is decreased. However, in any system the proton generally reacts more rapidly than any other cation.

The hydroxide ion is related to water and is also particularly mobile in aqueous systems, but because of its larger mass, it is less mobile than the proton.

FIGURE 17.1
Proton pathway in aqueous solution.

17.2 Brønsted-Lowry acids and bases

The Brønsted-Lowry acid-base concept emphasizes the role of the proton. Remembering a few simple things will help you to understand how Brønsted-Lowry acids and bases function:

1. Any acid-base reaction in the Brønsted-Lowry system is essentially a competition between two bases for the proton.

2. A strong Brønsted-Lowry acid gives up its proton easily.

3. A strong Brønsted-Lowry base attracts a proton strongly.

Any molecule or ion that can lose a proton, that is, that can act as a proton donor, is a **Brønsted-Lowry acid.** Any molecule or ion that can add a proton, that is, that can act as a proton acceptor, is a **Brønsted-Lowry base.** In the reaction of hydrogen chloride with

Brønsted–Lowry:
acid—proton donor
base—proton acceptor

water to form a solution of hydrochloric acid

$$HCl(aq) + H_2O(l) \longrightarrow H_3O^+ + Cl^- \tag{1}$$

HCl is a Brønsted-Lowry acid—a strong one. All the HCl molecules give up their protons to water molecules. In this case the water molecule is the base and attracts H^+.

In Equation (1) we have shown H_3O^+, the hydronium ion. Let's look more closely at the reaction of H^+ and H_2O. No doubt exists that the water molecule can act as a base and add a proton. However, the structure of the product of this reaction is not completely understood. The proton may well be associated with more than one water molecule, for example, as $H_5O_2^+$ or $H_7O_3^+$. The number of water molecules that are part of the ion is unimportant for writing equations or for understanding acid-base reactions. With our natural desire to make things more simple, it is common to write H^+, as we will sometimes do. But in Brønsted-Lowry terms, it is important to show that H_2O is a proton acceptor by writing H_3O^+. In addition, writing H_3O^+ reminds us that H^+ is not present alone, but is always associated with water molecules.

As we have said, a Brønsted-Lowry acid-base reaction is a competition between two bases for a proton. In Reaction (1) H_2O is one base and Cl^- is the other. Chloride ion, like water, has the capacity to add a proton (to form HCl). In Reaction (1) water is the stronger of the two bases, which is to say, it has a greater attraction for the proton than do the Cl^- ions, and so the reaction goes predominantly in the direction of forming H_3O^+ and not HCl.

A Brønsted-Lowry base adds a proton to form what is called the conjugate acid of that base, for example,

base		proton		conjugate acid	
OH^-	+	H^+	$\longrightarrow$	H_2O	(2)
CN^-	+	H^+	$\longrightarrow$	HCN	(3)
CO_3^{2-}	+	H^+	$\longrightarrow$	HCO_3^-	(4)
HCO_3^-	+	H^+	$\longrightarrow$	H_2CO_3	(5)
H_2O	+	H^+	$\longrightarrow$	H_3O^+	(6)
NH_3	+	H^+	$\longrightarrow$	NH_4^+	(7)

Note that water can be either an acid or a base.

A Brønsted-Lowry acid loses a proton to form what is called the *conjugate base* of that acid, for example,

acid		proton		conjugate base	
HCl	$\longrightarrow$	H^+	+	Cl^-	(8)
H_2SO_4	$\longrightarrow$	H^+	+	HSO_4^-	(9)
HSO_4^-	$\longrightarrow$	H^+	+	SO_4^{2-}	(10)
HNO_3	$\longrightarrow$	H^+	+	NO_3^-	(11)
H_2O	$\longrightarrow$	H^+	+	OH^-	(12)

We can now see that in Reaction (1) we have two acids and two bases

$$HCl(aq) + H_2O(l) \longrightarrow H_3O^+ + Cl^-$$
$$\text{acid} \qquad \text{base} \qquad \text{acid} \qquad \text{base}$$

The point is this: In the Brønsted-Lowry system all acid-base reactions involve two conjugate acid-base pairs in equilibrium. Equations (2) to (7) or (8) to (12) represent only half of what takes place. An acid does not form its conjugate base unless another base is present, and a base does not form its conjugate acid unless another acid is present. In the following reactions, $acid_1$ gives up a proton to yield $base_1$, and $base_2$ reacts with the proton to form $acid_2$.

$acid_1$	+	$base_2$		$base_1$	+	$acid_2$	
HCl	+	H_2O	$\rightleftharpoons$	Cl^-	+	H_3O^+	(13)
HNO_3	+	H_2O	$\rightleftharpoons$	NO_3^-	+	H_3O^+	(14)
NH_4^+	+	CO_3^{2-}	$\rightleftharpoons$	NH_3	+	HCO_3^-	(15)
H_3O^+	+	OH^-	$\rightleftharpoons$	H_2O	+	H_2O	(16)
H_2O	+	NH_3	$\rightleftharpoons$	OH^-	+	NH_4^+	(17)

In each reaction, $base_1$ and $base_2$ are competing for the proton.

Water is our reference for reactions in aqueous solution. We tend to think of a base as stronger the more easily it can take a proton from a water molecule. Similarly, the more readily an acid yields a proton to a water molecule, the stronger an acid it is.

Two important conclusions can be drawn:

1. The conjugate bases of strong acids are weak bases.

2. The conjugate acids of strong bases are weak acids.

A moment's thought should convince you that the first statement is true—if the base held onto the proton strongly, the acid would not release the proton so readily. The second statement represents the opposite case—the acid is weak because its conjugate base does hold onto the proton tightly.

Table 17.2 illustrates the interrelationships among conjugate acid-base pairs. All substances that fit the Arrhenius definition of acids as substances that yield H^+ in solution are included in the Brønsted-Lowry definition. Therefore, the concept of acids is extended to include ions such as HSO_4^- and NH_4^+ that can yield a hydrogen ion in solution. You are probably used to thinking of, say, sodium hydroxide, NaOH, as "a base." Arrhenius bases, like NaOH, yield OH^- in solution. To think in Brønsted-Lowry terms you must get used to the idea that in NaOH, only the OH^- ion is the base. In addition to the OH^- ion, all anions are Brønsted-Lowry bases because they can accept H^+.

Examine the following reactions, all of which are most favorable in the directions written, in the light of Table 17.2 and the competition between bases for the proton.

$acid_1$		$base_2$		$acid_2$		$base_1$
NH_4^+	+	OH^-	$\rightleftharpoons$	H_2O	+	NH_3
CH_3COOH	+	NH_3	$\rightleftharpoons$	NH_4^+	+	CH_3COO^-
HI	+	CN^-	$\rightleftharpoons$	HCN	+	I^-

EXAMPLE 17.1

■ Consider the reactions of the hydrogen sulfate ion

$$HSO_4^- + H_3O^+ \rightleftharpoons H_2SO_4(aq) + H_2O(l)$$

and

$$HSO_4^- + H_2O(l) \rightleftharpoons SO_4^{2-} + H_3O^+$$

In which reaction is HSO_4^- an acid and in which reaction is it a base?

In the first reaction HSO_4^- acts as a proton acceptor and is therefore a base. In the second reaction HSO_4^- acts as a proton donor and is therefore an acid. ∎

EXAMPLE 17.2

■ Write the equation for the reaction between HCl and NH_3 in aqueous solution. Identify the species acting as acids. From experi-

TABLE 17.2

Conjugate acid–base pairs

The acids above hydronium ion are strong acids; the bases below hydroxide ion are strong bases.

	Acid		Base		
Increasing acid strength	Perchloric acid	$HClO_4$	ClO_4^-	Perchlorate ion	Increasing base strength
	Sulfuric acid	H_2SO_4	HSO_4^-	Hydrogen sulfate ion (bisulfate)	
	Hydroiodic acid	HI	I^-	Iodide ion	
	Hydrobromic acid	HBr	Br^-	Bromide ion	
	Hydrochloric acid	HCl	Cl^-	Chloride ion	
	Nitric acid	HNO_3	NO_3^-	Nitrate ion	
	Hydronium ion	H_3O^+	H_2O	Water	
	Trichloroacetic acid	Cl_3CCOOH	Cl_3CCOO^-	Trichloroacetate ion	
	Hydrogen sulfate ion	HSO_4^-	SO_4^{2-}	Sulfate ion	
	Phosphoric acid	H_3PO_4	$H_2PO_4^-$	Dihydrogen phosphate ion	
	Nitrous acid	HNO_2	NO_2^-	Nitrite ion	
	Hydrofluoric acid	HF	F^-	Fluoride ion	
	Formic acid	HCOOH	$HCOO^-$	Formate ion	
	Acetic acid	CH_3COOH	CH_3COO^-	Acetate ion	
	Carbonic acid[a]	$CO_2 + H_2O$	HCO_3^-	Hydrogen carbonate ion (bicarbonate)	
	Hydrosulfuric acid	H_2S	HS^-	Hydrogen sulfide ion	
	Ammonium ion	NH_4^+	NH_3	Ammonia	
	Hydrocyanic acid	HCN	CN^-	Cyanide ion	
	Hydrogen sulfide ion	HS^-	S^{2-}	Sulfide ion	
	Water	H_2O	OH^-	Hydroxide ion	
	Ammonia	NH_3	NH_2^-	Amide ion	
	Hydrogen	H_2	H^-	Hydride ion	
	Methane	CH_4	CH_3^-	Methanide ion	

[a] "Carbonic acid", H_2CO_3, has never been isolated. A solution of carbon dioxide in water is acidic by virtue of the equilibrium: $CO_2 + 2H_2O \rightleftharpoons HCO_3^- + H_3O^+$.

ence we know that the reaction is favorable in the direction written. Which of the two bases present is weaker?

$$HCl(aq) + NH_3(aq) \rightleftharpoons NH_4^+(aq) + Cl^-(aq)$$

The species that lose protons are HCl and NH_4^+; therefore these are the acids. Since the forward reaction is favored, NH_3 must be the stronger base and Cl^- must be the weaker base. ■

EXAMPLE 17.3

■ Based on the relative strengths of acid-base pairs given in Table 17.2, predict what will happen if an excess of HCl(aq) is added to a solution containing S^{2-}.

Hydrochloric acid is listed as a stronger acid than either H_2S or HS^-, which means that the following reactions will occur:

$$HCl(aq) + S^{2-} \longrightarrow HS^- + Cl^-$$
$$HCl(aq) + HS^- \longrightarrow H_2S(g) + Cl^-$$

Thus addition of HCl(aq) to a solution of S^{2-} results in the formation of $H_2S(g)$. ■

As shown by Equations (14) and (17), water can be either an acid or a base. In the presence of a molecule or an ion that can add a proton more readily than the water molecule (a stronger base than H_2O), water is an acid. In the presence of a molecule or an ion that can lose a proton more readily than the water molecule (a stronger acid than H_2O), water is a base. Ammonia (see Table 17.2) also can act as either an acid or a base. Water, ammonia, and other such substances are *amphoteric*. **Amphoterism** is the ability to act as either an acid or a base. When acid-base behavior is restricted to proton transfer, the term *amphiprotic* is sometimes used. **Amphiprotism** is the ability of a substance to either gain or lose a proton. Water and ammonia are amphiprotic.

Any substance that can give up a proton is called a **protonic acid.** Protonic acids can be *monoprotic*—capable of losing one proton,

$$HCl + H_2O \rightleftharpoons H_3O^+ + Cl^-$$

or *diprotic*—capable of losing two protons,

$$H_2SO_4 + H_2O \rightleftharpoons HSO_4^- + H_3O^+$$
$$HSO_4^- + H_2O \rightleftharpoons SO_4^{2-} + H_3O^+$$

or *triprotic*—capable of losing three protons,

$$H_3PO_4 + H_2O \rightleftharpoons H_2PO_4^- + H_3O^+$$
$$H_2PO_4^- + H_2O \rightleftharpoons HPO_4^{2-} + H_3O^+$$
$$HPO_4^{2-} + H_2O \rightleftharpoons PO_4^{3-} + H_3O^+$$

(Confusingly, the terms monobasic, dibasic, and tribasic are sometimes used in the same sense as the terms monoprotic, diprotic, and triprotic.)

The intermediate ions of diprotic and triprotic acids are amphiprotic, as illustrated by the hydrogen sulfide ion, HS^-.

$$
\begin{array}{cccccc}
acid_1 & base_2 & & base_1 & & acid_2 \\
HS^- & + \; OH^- & \rightleftharpoons & S^{2-} & + & H_2O \\
HCl & + \; HS^- & \rightleftharpoons & Cl^- & + & H_2S
\end{array}
$$

17.3 Factors affecting the strength of acids and bases

Considering the relative strengths of acids in aqueous solution comes down to considering how easily the molecules give up their ionizable protons. The strength of a base depends upon how easily a compound will give up an OH^- ion or how strongly it will attract a proton. To examine acid and base strengths we must look at the relative effects of electronegativity, atomic radii, and bond energies.

A certain ambiguity exists in using the words "strong" and "weak" with respect to acids and bases. In the absolute sense the only strong acids and bases are those that are *strong electrolytes* (Section 14.9), which are compounds that are 100% ionized in solution. For example, in aqueous solutions of the strong acid HCl

$$HCl(aq) \longrightarrow H^+ + Cl^-$$

virtually no HCl molecules are present. Thus all acids and bases that are *not* 100% ionized are on this basis "weak." However, because the extent of ionization of "weak" acids and bases varies greatly, we also speak of acids and bases as relatively "stronger" or "weaker," depending on how much H^+ or OH^- they yield in solution. In the Brønsted-Lowry system, as we have seen, the relative strengths depend upon how easily acids and bases donate or accept protons.

Nonmetal hydrides. First, let's consider the relative strengths of acids that are simply hydrides of nonmetals. Essentially, the acid strength of such hydrides depends upon the strength of the bond between the hydrogen atom and the nonmetal atom. The two most important factors influencing this bond strength are the radius of the nonmetal atom and its electronegativity.

Electronegativity is the ability of an atom in a bond to attract electrons. The ionic character of a bond increases with increasing electronegativity of the nonmetal atom. In the bond H—E, where E is a nonmetal atom, the greater the electronegativity of E, the more strongly it will attract electrons from H, thereby allowing H to separate more readily. We therefore expect acid strength to increase with the increasing electronegativity of E in H—E bonds.

The strength of the H—E bond will *decrease* as the atomic radius of E becomes larger. This decreasing bond strength will allow the proton to separate more easily. In this case, we expect acid strength to increase with the increasing size of E in an H—E bond.

In any series of metal hydrides based on the position of the nonmetals in the periodic table, both electronegativity and atomic radii must be considered in explaining or predicting acid strengths. In going down Representative Group VI, for example, the electronegativity decreases and the atomic radii increase; these trends would have opposite effects on the acid strengths of these nonmetal hydrides. The increasing acid strength of these compounds shows that the influence of the increasing radii predominates.

decreasing nonmetal electronegativity $\longrightarrow$
increasing nonmetal radii $\longrightarrow$
H_2O H_2S H_2Se H_2Te
increasing acid strength $\longrightarrow$

In general, in going *down* a periodic table group, the acidity of the hydrides increases because of the increasing atomic radii and resultant decreasing strength of the bond between H and the nonmetal atom. On the other hand, in going *across* a periodic table row, the variation in atomic radii is smaller and electronegativity becomes the predominating factor in determining acid strength. For example, the hydrides of the second-row elements nitrogen, oxygen, and fluorine display increasing acid strength:

increasing nonmetal electronegativity $\longrightarrow$
NH_3 H_2O HF
increasing acid strength $\longrightarrow$

On the basis of increasing nonmetal radii and decreasing bond energies, we would expect the acid strengths of the three strong halogen acids to increase in the order

$$HCl < HBr < HI$$

as we go down the periodic table group. However, in aqueous solution these acids and the other acids that we have labeled "strong" appear to be of equal strength. The Brønsted-Lowry acid-base system is best for visualizing how this happens. In the equilibrium

$$HA + H_2O \rightleftharpoons A^- + H_3O^+$$
strong   base$_2$ conjugate   acid$_2$
acid   base

the conjugate bases of strong acids (Cl^-, Br^-, I^-, SO_4^{2-}, ClO_4^-, and so on) have a much smaller attraction for protons than H_2O. The attraction is so small that the acids are virtually completely ionized to give H_3O^+. This effect, which causes the seemingly identical strengths of strong acids and bases in aqueous solution, is called the *leveling effect*. Bases stronger than OH^- are leveled in aqueous solution to the same strength as OH^-, just as "strong" acids are all leveled to the same strength as H_3O^+. In a less basic solvent than water, for example, in anhydrous acetic acid, the acidity of the halogen acids is found to increase in the order given above, showing that the acid strength does vary as predicted.

Oxygen-containing acids and bases. Compounds containing oxygen or hydroxide groups can be acidic, basic, or amphoteric, depending upon the position in the periodic table and the electronegativity of the element to which the oxygen atoms are bonded. An ionic or substantially ionic bond between the OH group and a metal is essential if the group is to behave as a base and give OH^- in solution. The hydroxides of Representative Groups I and II metals, such as NaOH, KOH, and $Mg(OH)_2$, are all strong bases. Covalent bonds form in the compounds between OH and the more electronegative elements. When the oxygen-to-hydrogen bonds in such compounds are polar, the compounds are acids because the hydrogen of the OH group is released in water as an ion; for example,

$$HOCl(aq) \rightleftharpoons H^+ + OCl^-$$

or

$$HOCl(aq) + H_2O(l) \rightleftharpoons H_3O^+ + OCl^-$$

For a series of oxygen-containing acids of a given element, E, acid strength increases with increasing oxygen content (for example, the oxochlorine acids, Table 17.3). The bonding of each additional strongly electronegative oxygen atom to the central atom E increases the displacement of electrons away from the hydrogen atom, thus making the H—O bond more polar and enhancing the acid strength. As a result, an oxo acid can be expected to be stronger, the larger are the values of n in the general formula $O_nE(OH)_m$. [Although we are not accustomed to thinking of acid formulas in this way, they do fit the general formula given; for example, $H_2SO_4 = O_2S(OH)_2$; $HNO_3 = O_2NOH$; $H_3PO_4 = OP(OH)_3$.] In the oxochlorine acid series of Table 17.3, $n = 0$ for the weakest acid (HOCl) and $n = 3$ for the strongest acid.

Acid strength also increases with the electronegativity of atom E, as shown by the following halogen oxo acids.

increasing nonmetal electronegativity $\longrightarrow$
HOI   HOBr   HOCl
increasing acid strength $\longrightarrow$

When the atom E in a compound containing the EOH group has an intermediate attraction for electron density, the compound can behave as either an acid or a base (Figure 17.2). This type of amphoterism is exhibited by aluminum hydroxide, a gelatinous white solid that is soluble in either acidic or alkaline solutions.

base $Al(OH)_3(s) + 3H^+ \longrightarrow Al^{3+} + 3H_2O(l)$
acid $Al(OH)_3(s) + OH^- \longrightarrow [Al(OH)_4]^-$

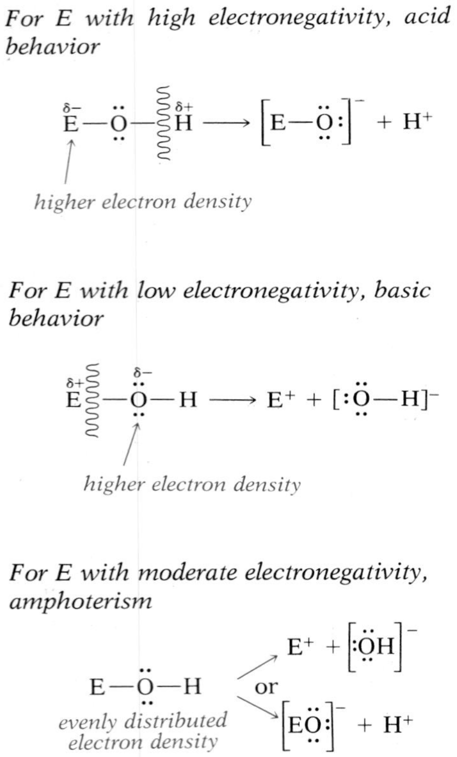

For E with high electronegativity, acid behavior

higher electron density

For E with low electronegativity, basic behavior

higher electron density

For E with moderate electronegativity, amphoterism

evenly distributed electron density

FIGURE 17.2
Acidity, basicity, and electronegativity.

Amphoterism is to be expected in the oxides or hydroxides of elements of intermediate covalent radii in intermediate oxidation states. The following trends emphasize these points:

Periodic table horizontal series—increasing acidity across the row

Group:	I	II	III	IV
	NaOH	Mg(OH)$_2$	Al(OH)$_3$	Si(OH)$_4$
	strongly basic	*basic*	*amphoteric*	*weakly acidic*

	V	VI	VII
	OP(OH)$_3$	O$_2$S(OH)$_2$	O$_3$Cl(OH)
	acidic	*strongly acidic*	*very strongly acidic*

Periodic table family—decreasing acidity down the family (same oxidation state of atom E, +3)

B(OH)$_3$	Al(OH)$_3$	Ga(OH)$_3$	In(OH)$_3$	Tl(OH)$_3$
acidic	*amphoteric*	*amphoteric*	*basic*	*basic*

Single element (different oxidation states)—increasing acidity with increasing oxidation state

Oxidation state of E:	+2	+3	+4	+5
	Mn(OH)$_2$	Mn(OH)$_3$	MnO$_2$	H$_3$MnO$_4$
	basic	*weakly basic*	*amphoteric*	*acidic*

	+6	+7
	H$_2$MnO$_4$	HMnO$_4$
	strongly acidic	*very strongly acidic*

Amphoterism in aqueous systems is not restricted to oxides or hydroxides. There are amphoteric sulfides

$$\underset{base}{As_2S_3(s)} + \underset{acid}{6H^+} \xrightarrow{H_2O} 2As^{3+} + 3H_2S(aq)$$

$$\underset{acid}{As_2S_3(s)} + \underset{base}{2SH^-} \xrightarrow{H_2O} \underset{[thioarsenate(III)\ ion]}{2AsS_2^-} + H_2S(aq)$$

and amphoteric cyanides

$$\underset{base}{Fe(CN)_2(s)} + \underset{acid}{2H^+} \xrightarrow{H_2O} Fe^{2+} + 2HCN(aq)$$

$$\underset{acid}{Fe(CN)_2(s)} + \underset{base}{4CN^-} \xrightarrow{H_2O} \underset{[hexacyanoferrate(II)\ ion]}{[Fe(CN)_6]^{4-}}$$

Variations in the strengths of anionic bases are more difficult to explain. Small and/or highly charged anions such as F$^-$, O^{2-}, N^{3-}, and CO$_3{}^{2-}$ are stronger bases than large and/or less highly charged anions such as I$^-$, S^{2-}, or MnO$_4{}^-$, because of larger attractions for protons or Lewis acids. The best foundation for predicting the base strength of

TABLE 17.3

Oxochlorine acids

Oxygen content, the oxidation state of chlorine, and the acidity all increase in this series.

Hypochlorous, HClO   very weak acid

HO : C̈l :

Chlorous, HClO$_2$ weak acid

: O :
HO : C̈l :

Chloric, HClO$_3$ strong acid

: O :
HO : C̈l : Ö :

Perchloric, HClO$_4$ very strong acid

: O :
HO : C̈l : Ö :
: O :

Lewis:
acid—electron-pair acceptor
base—electron-pair donor

anions is that anions that are conjugate bases of weak acids are strong bases; and anions that are conjugate bases of strong acids are weak bases.

17.4 Lewis acids and bases

An even broader view of acids and bases than that of Brønsted and Lowry emphasizes interactions at the valence electron level. Lewis acids and bases are defined in terms of electron pairs (this theory was devised by the American chemist G. N. Lewis). A **Lewis acid** is an atom, molecule, or ion that can act as an electron-pair acceptor. A **Lewis base** is an atom, molecule, or ion that can act as an electron-pair donor.

The electron-pair donation in the reaction of a protonic acid with water is apparent when the electron dot structures of the reactants are written

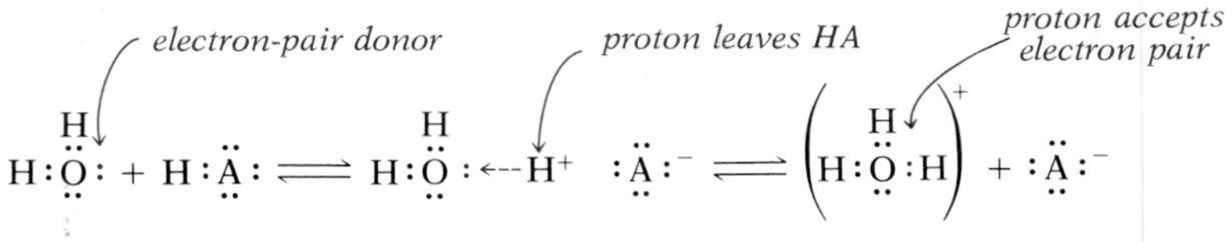

$$H:\overset{..}{\underset{..}{O}}: + H:\overset{..}{\underset{..}{A}}: \rightleftharpoons H:\overset{..}{\underset{..}{O}}:\leftarrow H^+ \quad :\overset{..}{\underset{..}{A}}:^- \rightleftharpoons \left(H:\overset{..}{\underset{..}{O}}:H\right)^+ + :\overset{..}{\underset{..}{A}}:^-$$

electron-pair donor — *proton leaves HA* — *proton accepts electron pair*

The water molecule is the base, the electron-pair donor, and the hydrogen ion is the acid, the electron-pair acceptor. An **acid–base reaction** in the Lewis system is electron-pair donation, or, essentially, the formation of a coordinate covalent bond (Section 9.11).

To see how the Lewis acid–base picture is derived, consider the formation of the salt barium sulfate from barium oxide (an ionic compound) by two different reactions, one with sulfuric acid and one with sulfur trioxide.

In aqueous solution

$$BaO(s) + H_2SO_4(aq) \longrightarrow BaSO_4(s) + H_2O(l) \tag{18}$$

In the absence of water

$$BaO(s) + SO_3(g) \longrightarrow BaSO_4(s) \tag{19}$$

If we write Reaction (18) with the reactants given as ions (barium oxide is moderately soluble; barium sulfate is only slightly soluble)

$$Ba^{2+} + O^{2-} + 2H_3O^+ + SO_4^{2-} \longrightarrow BaSO_4(s) + 3H_2O \tag{20}$$

it is possible to see that this is an acid–base reaction in Brønsted–Lowry proton-transfer terms

$$\underset{base_2}{O^{2-}} + \underset{acid_1}{2H_3O^+} \rightleftharpoons \underset{acid_2}{H_2O} + \underset{base_1}{2H_2O} \tag{21}$$

The reaction can also be written so that it is clear that electron-pair donation is involved

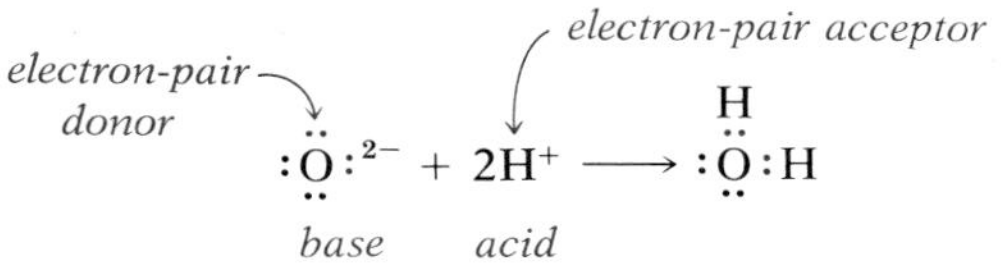

$$\underset{\substack{\text{electron-pair} \\ \text{donor}}}{\ddot{\underset{..}{O}}:^{2-}} + \underset{\substack{\text{acid} \\ \text{electron-pair acceptor}}}{2H^+} \longrightarrow :\overset{H}{\underset{..}{\ddot{O}}}:H \qquad (22)$$

base   acid

The similarity of Reaction (19), the reaction of barium oxide with sulfur trioxide, to the electron-pair sharing of Reaction (18) shows up

TABLE 17.4
Lewis acid–base reactions[a]

The shared electron pair is shown in color.

Acid	+	Base	⇌	Reaction product

Octet completion

(a)

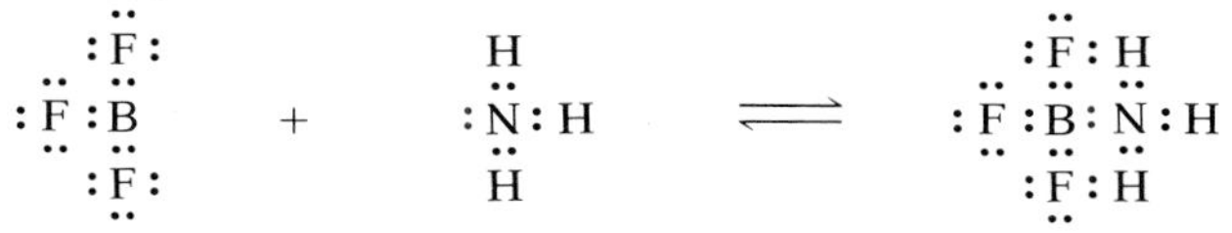

(b)

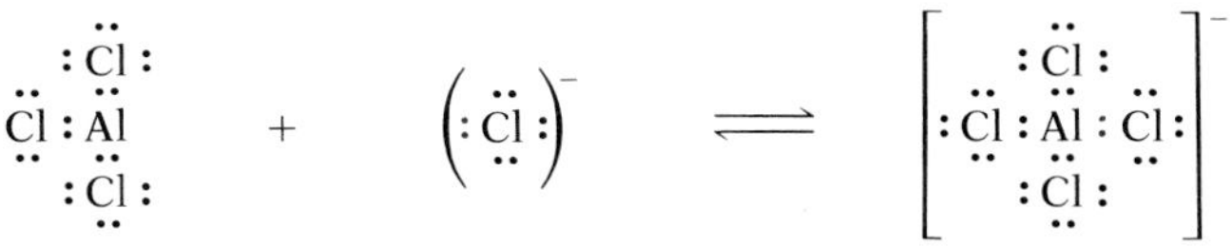

Reactions of cations

(c) 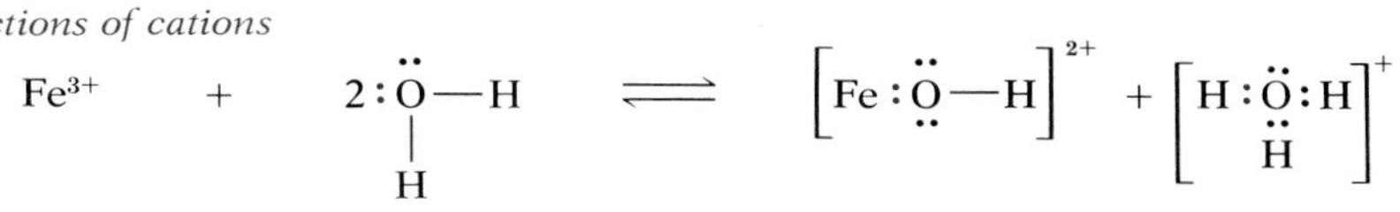 Fe^{3+} + $2:\overset{..}{\underset{|}{O}}-H$ ⇌ $\left[Fe:\overset{..}{\underset{..}{O}}-H\right]^{2+} + \left[H:\overset{..}{\underset{H}{O}}:H\right]^{+}$

(d) 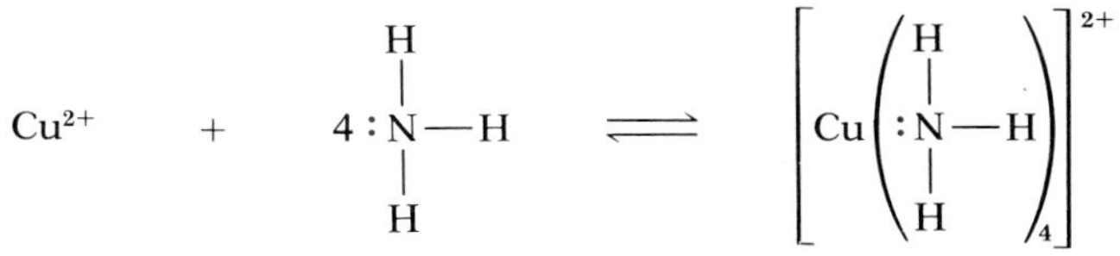 Cu^{2+} + $4:\overset{H}{\underset{H}{N}}-H$ ⇌ $\left[Cu\left(:\overset{H}{\underset{H}{N}}-H\right)_4\right]^{2+}$

Attraction of electron pair by atom in multiple bond

(e) 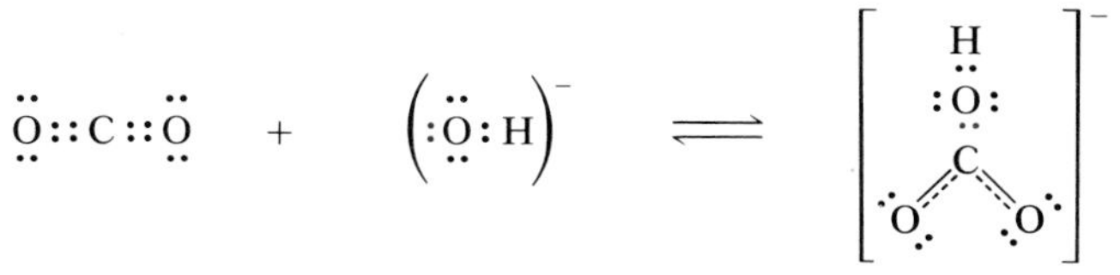 $\overset{..}{\underset{..}{O}}::C::\overset{..}{\underset{..}{O}}$ + $\left(:\overset{..}{\underset{..}{O}}:H\right)^{-}$ ⇌ $\left[\begin{array}{c} H \\ :O: \\ C \\ \overset{..}{O} \quad \overset{..}{O}: \end{array}\right]^{-}$

bicarbonate ion
HCO_3^-

Occupation of empty d orbitals

(f) 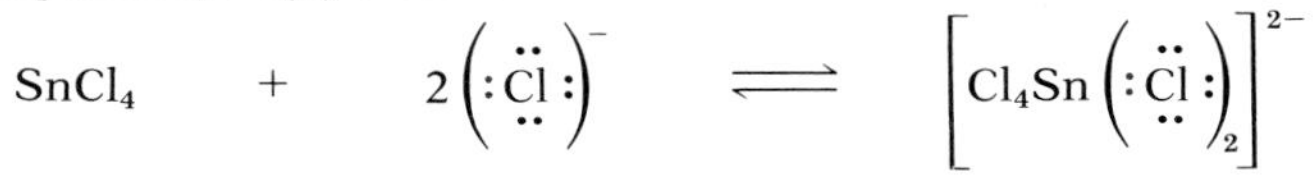 $SnCl_4$ + $2\left(:\overset{..}{\underset{..}{Cl}}:\right)^{-}$ ⇌ $\left[Cl_4Sn\left(:\overset{..}{\underset{..}{Cl}}:\right)_2\right]^{2-}$

[a] This classification is based on that of *Fundamentals of Chemistry*, 3rd ed., by F. Brescia, J. Arents, H. Meislich, and A. Turk, Academic Press, New York, 1975, p. 254.

ACIDS AND BASES [17.4]

when the significant part of Reaction (19) is written as follows:

$$(23)$$

The equations in Table 17.4 illustrate some of the many types of Lewis acid–base reactions. The Lewis definition of acids and bases is so broad that nearly all reactions except those involving the actual transfer of electrons (oxidation-reduction reactions) are of the acid-base type. All cations are acids [Reactions (c) and (d) in Table 17.4]. All anions are bases, for example,

All precipitation and dissolution reactions that depend solely upon the relative attractions between anions and cations are acid-base reactions

And all reactions that yield complex ions (Chapter 28), all of which contain coordinate covalent bonds, are also acid-base reactions. [The products of Reactions (b), (c), (d), and (f) in Table 17.4 are complex ions.]

Many Lewis acid-base reactions take place at elevated temperatures in the absence of a solvent. Following are a few examples:

base	+	acid	$\xrightarrow{\Delta}$	neutralization product
$CaO(s)$ *calcium oxide*	+	$SO_3(g)$ *sulfur(VI) oxide*	$\xrightarrow{\Delta}$	$CaSO_4(s)$ *calcium sulfate*
$2Na_2CO_3(l)$ *sodium carbonate*	+	$SiO_2(s)$ *silicon dioxide*	$\xrightarrow{\Delta}$	$Na_4SiO_4(l) + 2CO_2(g)$ *sodium silicate*
$3NaF(s)$ *sodium fluoride*	+	$AlF_3(s)$ *aluminum fluoride*	$\xrightarrow{\Delta}$	$Na_3AlF_6(s)$ *sodium hexafluoroaluminate*
$x\,CoO(s)$ *cobalt(II) oxide*	+	$(NaPO_3)_x(l)$ *polymeric sodium metaphosphate*	$\xrightarrow{\Delta}$	$x\,NaCoPO_4(l)$ *sodium cobalt(II) orthophosphate*

Expressing the strength of acids, bases, and their aqueous solutions

$$K_w = 1 \times 10^{-14} \text{ (at 25°C)}$$

17.5 Ionization equilibrium of water

Pure water, as we have learned, is a slightly ionized substance

$$H_2O(l) \rightleftharpoons H^+ + OH^- \tag{24}$$

or

$$H_2O(l) + H_2O(l) \rightleftharpoons H_3O^+ + OH^-$$

One liter of water contains 55.5 moles of H_2O molecules. The ionization of water is so slight that in each liter of water there is roughly one H^+ and one OH^- for each 555 million H_2O molecules. The equilibrium expression for Reaction (24), called the **ion product constant for water,** is

$$K_w = [H^+][OH^-] \tag{25}$$

or

$$K_w = [H_3O^+][OH^-]$$

which is the same as Equation (25) because $[H^+] = [H_3O^+]$. Water is ionized to such a small extent that the nonionized water molecules can be considered as part of a pure liquid. Therefore their concentration need not appear in the equilibrium expression (see Section 16.2). Note that we have written $[H^+]$ in Equation (25). For simplicity, we have chosen to write $[H^+]$ rather than $[H_3O^+]$ in equilibrium expressions.

Measurements show that at 25°C the value of K_w is 1.008×10^{-14}. The value increases with temperature; for example, at 60°C it is 9.6×10^{-14}. With little introduction of error, the value of K_w at 25°C (and frequently at other temperatures) is usually taken as

$$K_w = 1.00 \times 10^{-14} \tag{26}$$

From Equation (24) we know that the concentrations of hydrogen and hydroxide ions in pure water must be equal. Letting $x = [H^+] = [OH^-]$ we can see that at 25°C the concentrations of these ions are equal to 1.00×10^{-7} mole/liter.

$$[H^+][OH^-] = 1.00 \times 10^{-14}$$
$$(x)(x) = 1.00 \times 10^{-14}$$
$$x^2 = 1.00 \times 10^{-14}$$
$$x = 1.00 \times 10^{-7} \text{ mole/liter}$$

Pure water is neutral—neither acidic nor alkaline—because the concentrations of hydrogen and hydroxide ions are equal. If a substance that increases [H⁺] is added to water, then [OH⁻] must decrease to maintain equilibrium. Conversely, if a substance that increases [OH⁻] is added, [H⁺] must decrease. By using K_w, an unknown [OH⁻] can be found from a known [H⁺] and vice versa.

EXAMPLE 17.4

■ What is the concentration of OH^- in a $0.01M$ HCl solution? (HCl is 100% ionized.)

$$[H^+][OH^-] = 1.00 \times 10^{-14}$$
$$[1 \times 10^{-2}][OH^-] = 1.00 \times 10^{-14}$$
$$[OH^-] = \frac{1.00 \times 10^{-14}}{1 \times 10^{-2}}$$
$$= 1 \times 10^{-12} \text{ mole/liter}$$

The OH^- concentration in a $0.01M$ HCl solution is 1×10^{-12} mole/liter. ■

EXAMPLE 17.5

■ What is the concentration of H^+ in a $5M$ solution of KOH? Assume that $[OH^-] = 5M$ because KOH is 100% dissociated.

Solving the K_w expression for [H⁺] and substituting $[OH^-] = 5M$ gives

$$[H^+] = \frac{K_w}{[OH^-]} = \frac{1.00 \times 10^{-14}}{5} = 2 \times 10^{-15}M$$

The hydrogen ion concentration is 2×10^{-15} mole/liter. ■

Note in Example 17.4 that, as predicted, when [H⁺] is increased, equilibrium (24) shifts so that [OH⁻] decreases to less than that in pure water. Similarly, Example 17.5 shows that when [OH⁻] is increased, the equilibrium shifts so that [H⁺] decreases to less than that in pure water.

In an acidic solution (at 25°C), $[H^+] > 1 \times 10^{-7}$ mole/liter and in an alkaline solution, $[OH^-] > 1 \times 10^{-7}$ mole/liter. The larger the concentration of the hydrogen ion, the more strongly acidic is the solution. The larger the concentration of the hydroxide ion, the more strongly alkaline is the solution.

17.6 pH and pK

The concentrations of hydrogen ion and hydroxide ion in an aqueous solution of a substance are a quantitative measure of the acid or base strength of that substance and of the acidity and alkalinity of the solution. The strength of an acid or a base and of its solutions is a subject of frequent interest in chemical systems. Often these H^+ and OH^- concentrations are such small numbers that they are best expressed as exponential numbers (that is, in scientific notation). However, handling exponential numbers can be awkward. Because acidity and alkalinity are in such common use, a more convenient way for expressing these properties numerically has been devised.

Two new quantities, **pH** and **pOH,** are defined as follows:

$$pH = -\log [H^+] = \log \frac{1}{[H^+]}$$

$$pOH = -\log [OH^-] = \log \frac{1}{[OH^-]}$$

(Logarithmic notation is discussed in Appendix A.3.) The values of pH or pOH usually fall between 1 and 14 and are therefore easier to cope with than the small H^+ or OH^- concentrations.

EXAMPLE 17.6

■ What is the pH of a $0.01M$ solution of HCl? (HCl is 100% ionized.)

$$\begin{aligned} pH &= -\log [H^+] \\ &= -\log (1 \times 10^{-2}) \\ &= -(-2.0) \\ &= 2.0 \end{aligned}$$

The pH of $0.01M$ HCl is 2.0, written pH 2.0. ■

EXAMPLE 17.7

■ What is the hydrogen ion concentration in a solution of pH 1.5?

We need to rearrange the definition of pH:

$$\log [H^+] = -pH = -1.5 = 0.5 - 2$$

Taking antilogarithms gives

$$[H^+] = 3 \times 10^{-2} \text{ mole/liter}$$

The hydrogen ion concentration is 0.03 mole/liter. ■

The same notation can be used for the negative logarithm of any quantity. The expression for the ion product of water can be rewritten as follows:

pH < 7, acidic solution
pH > 7, alkaline solution

$$\begin{aligned} K_w &= [H^+][OH^-] \\ \log K_w &= \log ([H^+][OH^-]) \\ &= \log [H^+] + \log [OH^-] \\ -\log K_w &= -\log [H^+] - \log [OH^-] \\ pK_w &= pH + pOH \end{aligned} \tag{27}$$

Numerically,

$$pK_w = -\log (1.00 \times 10^{-14}) = 14.00$$

so that

$$pH + pOH = 14.00$$
$$pH = 14.00 - pOH \tag{28}$$
$$pOH = 14.00 - pH \tag{29}$$

and for a neutral solution

$$pH = pOH = 7.00$$

If the pH is smaller than 7.00, the solution is an **acidic solution;** if the pH is larger than 7.00, the solution is an **alkaline solution** (remembering that this is *strictly* accurate only at 25°C). Correspondingly, acidic solutions are described by pOH values larger than 7.00 and alkaline solutions by pOH values smaller than 7.00. The larger the departure from 7.00, the more acidic or the more alkaline the solution is (Table 17.5). In general, a solution is acidic when pH > pOH and alkaline when pOH > pH.

We stress that the pH and pOH scales are logarithmic, not linear, scales. For example, a solution of pH 1 has ten times the concentration of hydrogen ion that a solution of pH 2 has, not twice the concentration. A solution of pH 12 has 100 times the concentration of hydroxide ion that a solution of pH 10 has. Although both pH and pOH are useful, it is common practice to use pH to indicate both the acidity and the alkalinity of an aqueous solution. The pH values of some well-known substances are given in Table 17.6.

EXAMPLE 17.8

■ The pH of blood serum is 7.4. What is the hydrogen ion concentration of blood serum?

From the definition of pH

$$\begin{aligned} pH = -\log [H^+] &= 7.4 \\ [H^+] &= \text{antilog } (-7.4) \\ &= \text{antilog } (0.6 - 8) \\ &= 4 \times 10^{-8} M \end{aligned}$$

The concentration of hydrogen ion in blood serum is 4×10^{-8} mole/liter. ■

TABLE 17.5
Acidity, neutrality, and alkalinity in aqueous solutions

$[H^+]$ (mole/liter)	pH	$[OH^-]$ (mole/liter)	pOH	Nature	
10^1	-1	10^{-15}	15		
10^0	0	10^{-14}	14		
10^{-1}	1	10^{-13}	13		
10^{-2}	2	10^{-12}	12	Acidic	Acidity Increases
10^{-3}	3	10^{-11}	11		
10^{-4}	4	10^{-10}	10		
10^{-5}	5	10^{-9}	9		
10^{-6}	6	10^{-8}	8		
10^{-7}	7	10^{-7}	7	Neutral	Neutral
10^{-8}	8	10^{-6}	6		
10^{-9}	9	10^{-5}	5		
10^{-10}	10	10^{-4}	4		
10^{-11}	11	10^{-3}	3	Alkaline	Alkalinity Increases
10^{-12}	12	10^{-2}	2		
10^{-13}	13	10^{-1}	1		
10^{-14}	14	10^0	0		
10^{-15}	15	10^1	-1		

EXAMPLE 17.9

■ Express the pOH of a 0.01M solution of KOH.

Because KOH is a strong base, $[OH^-] = 0.01M$, giving

$$pOH = -\log [OH^-] = -\log (0.01)$$
$$= -(-2.0) = 2.0$$

The pOH is 2.0. ■

EXAMPLE 17.10

■ What is the $[OH^-]$ in a solution of pOH 4.18?

Substituting pOH 4.18 into the expression for pOH and solving for $[OH^-]$ gives

$$4.18 = -\log [OH^-]$$
$$[OH^-] = \text{antilog} (-4.18)$$
$$= \text{antilog} (0.82 - 5)$$
$$= 6.6 \times 10^{-5}M$$

The concentration of the hydroxide ion is 6.6×10^{-5} mole/liter. ■

TABLE 17.6
Approximate pH values of some well-known substances

0.1M HCl	1
Gastric juice	1.4
Lemon juice	2.3
Vinegar	2.9
Orange juice	3.5
Tomatoes	4.2
Coffee	5.0
Rainwater	6.2
Pure water	7.0
Blood	7.4
Seawater	8.5
Bar soap	11.0
Household ammonia	11.5
0.1M NaOH	13.0

ACIDS AND BASES [17.6]

EXAMPLE 17.11

■ What are the pH and [H$^+$] of a solution containing [OH$^-$] = 3.5 × 10^{-5}M?

The pOH of the solution is

$$pOH = -\log [OH^-] = -\log (3.5 \times 10^{-5}) = 4.46$$

giving

$$pH = 14.00 - pOH = 14.00 - 4.46 = 9.54$$

which corresponds to

$$\log [H^+] = -pH = -(9.54) = 0.46 - 10$$
$$[H^+] = 2.9 \times 10^{-10}M$$

The pH of this solution is 9.54 and the concentration of H$^+$ is 2.9 × 10^{-10} mole/liter.   ■

EXAMPLE 17.12

■ Which solution contains the highest [H$^+$]?

(a) [H$^+$] = 0.3M; (b) [OH$^-$] = 5M; (c) pH − 1.2; (d) pOH 5.9.

Solution (a) contains [H$^+$] = 0.3M. The [H$^+$] concentration for solution (b) can be found by using the expression for K_w

$$K_w = [H^+][OH^-]$$
$$[H^+] = \frac{K_w}{[OH^-]} = \frac{1.00 \times 10^{-14}}{5} = 2 \times 10^{-15}M$$

Finding [H$^+$] from the pH for solution (c) gives

$$\log [H^+] = -pH = -(-1.2) = 1.2$$
$$[H^+] = 16M$$

(Note that pH values are *not* restricted to the range of 0 to 14, but can be negative for very high concentrations of H$^+$ and greater than 14 for solutions containing very high concentrations of OH$^-$.) For solution (d) we first find pH

$$pH = 14.000 - pOH = 14.0 - 5.9 = 8.1$$

and then use this value to find [H$^+$]

$$\log [H^+] = -pH = -(8.1) = 0.9 - 9$$
$$[H^+] = 8 \times 10^{-9}$$

We see that solution (c), with pH − 1.2, has the highest [H$^+$].   ■

EXAMPLE 17.13

■ The concentration of hydrogen ion in a cup of black coffee is $1.3 \times 10^{-5}M$. Find the pH of the coffee. Is this coffee acidic or alkaline?

Substituting $[H^+] = 1.3 \times 10^{-5}M$ into the definition of pH gives

$$pH = -\log (1.3 \times 10^{-5}) = 4.89$$

This black coffee is an acidic solution—its pH is less than 7.　　■

17.7 Determination of pH: indicators

The pH of a solution is usually measured by one of two methods. The first is the observation of the color change of an indicator. The second is the use of a pH meter—an instrument that employs electrodes to measure the potential difference between solutions of known and unknown concentration (see Tools of Chemistry: The pH Meter, Chapter 24). Because of their accuracy and speed, pH meters have superseded the older indicator method in many applications. However, the indicator method remains in use because it is simple and convenient.

An **indicator** is an organic acid or base that has in its structure a group that reacts with hydrogen or hydroxide ion so that the color of the compound changes. Letting HInd represent an indicator acid, its ionization, like that of any acid, is

$$HInd(aq) \rightleftharpoons Ind^- + H^+$$

The HInd and Ind$^-$ forms must have different colors. The equilibrium constant for this reaction is

$$K_{Ind} = \frac{[H^+][Ind^-]}{[HInd]} \tag{30}$$

For the color change of an indicator to be visible, the concentration of one colored form must be about ten times greater than the concentration of the other. As a result, most indicators have a useful color change over a 2 pH unit range related to the pK_{Ind} as follows:

To see the acid color,

$$\frac{[Ind^-]}{[HInd]} = \frac{1}{10}$$

To see the alkaline color,

$$\frac{[Ind^-]}{HInd} = \frac{10}{1}$$

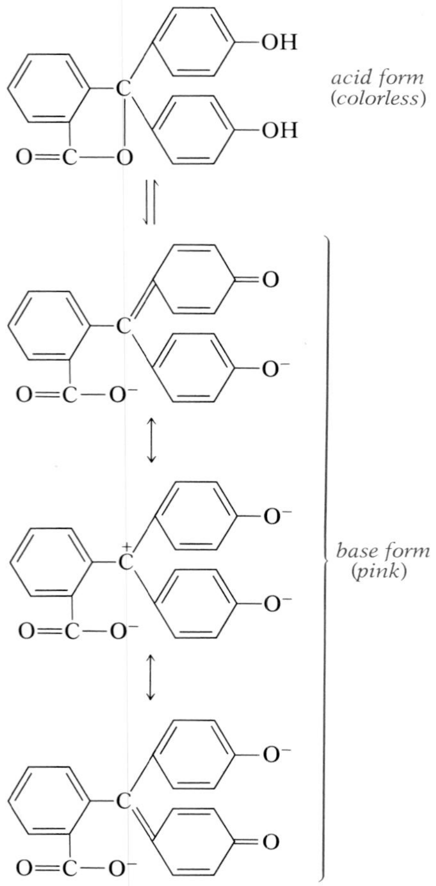

FIGURE 17.3
Phenolphthalein, an acid–base indicator. *The pink color arises from an anion with three resonance structures.*

acid form (colorless)

base form (pink)

Rearranging the expression for K_{Ind} gives

$$[H^+] = \frac{[HInd]}{[Ind^-]} K_{Ind}$$

When the acid color is visible,

$$[H^+] = 10K_{Ind}$$

or

$$
\begin{aligned}
pH &= -\log (10K_{Ind}) \\
&= -\log 10 - \log K_{Ind} \\
&= pK_{Ind} - 1
\end{aligned}
$$

It can be shown in the same way that for the alkaline color $pH = pK_{Ind} + 1$, giving a pH range of 2 for the color change of an indicator ($pH = pK_{Ind} \pm 1$). The commonly encountered indicator phenolphthalein (Figure 17.3), for example, changes from colorless to pink in the pH range from 8.3 to 10.0. Indicators are available for the entire pH range (Table 17.7).

One way to improve the accuracy of using indicators is by comparing the color of the indicator in a solution of unknown pH with the color of the indicator in a solution of known pH. Paper impregnated with an organic pH indicator is very convenient to use. Litmus paper and its pink and blue colors may be familiar to many of you. A "universal" pH paper combines several different indicators in the same paper. This paper will show a characteristic color for any point in the pH range.

The exact concentration of an acid or base in solution can be determined experimentally by titration. Acid-base titration and the use of indicators are further discussed in the next chapter (Section 18.10).

17.8 Strong and weak acids

The strength of a protonic acid in aqueous solution is measured by the extent to which the following equilibrium proceeds in the direction of ion formation:

$$HA(aq) \rightleftharpoons H^+ + A^- \tag{31}$$

or

$$HA(aq) + H_2O \rightleftharpoons H_3O^+ + A^-$$

The equilibrium constant for this reaction is called the acid ionization constant, K_a.

$$K_a = \frac{[H^+][A^-]}{[HA]} \tag{32}$$

The larger the value of K_a, the stronger is the acid.

TABLE 17.7

Indicator	Acidic color	pH range	Alkaline color
Methyl violet	Yellow	0–2	Violet
Malachite green (acidic)	Yellow	0–1.8	Blue-green
Thymol blue (acidic)	Red	1.2–2.8	Yellow
Bromphenol blue	Yellow	3.0–4.6	Purple
Methyl orange	Red	3.1–4.4	Yellow-orange
Bromcresol green	Yellow	3.8–5.4	Blue
Methyl red	Red	4.4–6.2	Yellow
Litmus	Red	4.5–8.3	Blue
Bromcresol purple	Yellow	5.2–6.8	Purple
Bromthymol blue	Yellow	6.0–7.6	Blue
Phenol red	Yellow	6.4–8.2	Red
m-Cresol purple	Yellow	7.6–9.2	Purple
Thymol blue (alkaline)	Yellow	8.0–9.6	Blue
Phenolphthalein	Colorless	8.3–10.0	Red
Thymolphthalein	Colorless	9.3–10.5	Blue
Alizarin yellow	Yellow	10.1–11.1	Lilac
Malachite green (alkaline)	Green	11.4–13.0	Colorless
Trinitrobenzene	Colorless	12.0–14.0	Orange

TABLE 17.7
Common pH indicators and their color changes

A **strong acid** is virtually 100% ionized in a dilute aqueous solution. In such a solution, Reaction (31) produces very large concentrations of H^+ and A^- and the concentration of HA is very small. As a result, the K_a values for strong acids are all greater than 1. The most common strong acids are listed in Table 17.8.

Weak acids, on the other hand, are less than 100% ionized, most of them to quite a small extent. For weak acids, the predominant direction of Reaction (31) is toward the formation of HA, and the K_a values are considerably smaller than 1.

A convenient way to compare the "strengths" of weak acids is by comparing their relative pK_a values. Since K_a values are, like hydrogen ion concentrations, frequently very small numbers, they can also be very conveniently treated as negative logarithms in the same way as defined above for K_w. The pK_a is defined as follows:

$$pK_a = -\log K_a$$

Table 17.9 gives the K_a and pK_a values for some weak acids. (A more extensive list of K_a's is given in Appendix C.) *The larger the K_a value, the stronger is the acid.* Note that acidic metal hydroxides that give H^+ by reactions such as

$$Cr(OH)_3(s) \rightleftharpoons H^+ + CrO_2^- + H_2O(l)$$

are the weakest acids.

TABLE 17.8
Common strong acids
The molarity of the most concentrated commercially available solution is given for each acid.

HClO₄
perchloric acid
hygroscopic, colorless liquid
11.7M concentrated solution

H₂SO₄
sulfuric acid
hygroscopic, colorless oily liquid
18.0M concentrated solution

HI
hydroiodic acid
aqueous solution of gaseous HI
7.6M concentrated solution

HBr
hydrobromic acid
aqueous solution of gaseous HBr
8.9M concentrated solution

HCl
hydrochloric acid
aqueous solution of gaseous HCl
11.6M concentrated solution

HNO₃
nitric acid
colorless liquid
16.0M concentrated solution

TABLE 17.9

Ionization constants of weak acids at 25°C for the reaction
$HA(aq) \rightleftharpoons A^- + H^+$

	K_a	pK_a
Hydrogen sulfate ion HSO_4^-	1.0×10^{-2}	2.00
Phosphoric acid H_3PO_4	7.5×10^{-3}	2.12
Nitrous acid HNO_2	7.2×10^{-4}	3.14
Hydrofluoric acid HF	6.5×10^{-4}	3.19
Acetic acid CH_3COOH	1.75×10^{-5}	4.757
Carbonic acid $CO_2 + H_2O(H_2CO_3)$	4.5×10^{-7}	6.35
Hydrosulfuric acid H_2S	1.0×10^{-7}	7.00
Ammonium ion NH_4^+	6.3×10^{-10}	9.20
Hydrogen sulfide ion HS^-	3×10^{-13}	12.5
Chromium(III) hydroxide $Cr(OH)_3$	9×10^{-17}	16.1
Copper(II) hydroxide $Cu(OH)_2$	1×10^{-19}	19.0
Zinc hydroxide $Zn(OH)_2$	1×10^{-29}	29.0

For pK_a values, however, *the larger the pK_a, the weaker is the acid.* For example, acetic acid, a commonly used weak acid, has a pK_a of 4.76, while zinc hydroxide, one of the weakest acids, has a pK_a of 29.0. The strong acids—those with K_a values greater than 1—have *negative* values of pK_a.

EXAMPLE 17.14

■ Hydrogen peroxide can act as a weak acid

$$H_2O_2(aq) + H_2O \rightleftharpoons H_3O^+ + HO_2^-$$

or simply

$$H_2O_2(aq) \rightleftharpoons H^+ + HO_2^- \qquad K_a = 2.2 \times 10^{-12}$$

Find the concentration of HO_2^- that is in equilibrium with $[H^+] = 0.036M$ and $[H_2O_2] = 0.100M$.

Substituting the concentrations into the equilibrium constant expression gives

$$K_a = \frac{[H^+][HO_2^-]}{[H_2O_2]} = \frac{(0.036)[HO_2^-]}{(0.100)} = 2.2 \times 10^{-12}$$
$$[HO_2^-] = 6.1 \times 10^{-12}M$$

The concentration of the hydroperoxide ion is 6.1×10^{-12} mole/liter.
■

EXAMPLE 17.15

■ Find the concentrations of the various ions present in a 0.10M solution of nitrous acid, HNO_2. $K_a = 7.2 \times 10^{-4}$ for HNO_2.

Nitrous acid establishes the equilibrium

$$HNO_2(aq) + H_2O(l) \rightleftharpoons H_3O + NO_2^-$$

or simply

$$HNO_2(aq) \rightleftharpoons H^+ + NO_2^- \qquad K_a = \frac{[H^+][NO_2^-]}{[HNO_2]}$$

Our usual approach for solving equilibrium problems applies to this and other acid or base equilibrium problems. We can see that if $x = [H^+]$, then, because $[H^+] = [NO_2^-]$, x likewise equals $[NO_2^-]$. The $[HNO_2]$ that reacts will be equivalent to the $[H^+]$ formed. Thus we write

$$HNO_2(aq) \rightleftharpoons H^+ + NO_2^-$$

initial	0.10		
change	$-x$	$+x$	$+x$
equilibrium	$0.10 - x$	x	x

Substituting into the expression for K_a gives

$$K_a = \frac{[H^+][NO_2^-]}{[HNO_2]} = \frac{(x)(x)}{(0.10 - x)} = 7.2 \times 10^{-4}$$

Assuming $(0.10 - x) \approx 0.10$ and solving gives $x = 0.0085$, which exceeds our permitted 5% error [relative error = $(0.0085/0.10) \times 100 = 8.5\%$]. Therefore we must use the quadratic formula to find the solution:

$$x^2 = (7.2 \times 10^{-4})(0.10 - x)$$
$$= 7.2 \times 10^{-5} - 7.2 \times 10^{-4}x$$
$$x^2 + 7.2 \times 10^{-4} \times (-7.2 \times 10^{-5}) = 0$$
$$x = \frac{-(7.2 - 10^{-4}) \pm \sqrt{(7.2 \times 10^{-4})^2 - (4)(1)(-7.2 \times 10^{-5})}}{2}$$
$$= 0.0081$$

Thus at equilibrium $[H^+] = [NO_2^-] = x = 0.0081M$ and $[HNO_2] = (0.10 - x) = (0.10 - 0.0081) = 0.09M$. ∎

EXAMPLE 17.16

∎ Large amounts of lactic acid ($CH_3CHOHCOOH$, present in sour milk) are used for dyeing in the wool industry and for neutralizing lime in hides in the tanning industry; $K_a = 1.37 \times 10^{-4}$ at 25°C. Find the pH of a $0.100M$ solution of lactic acid.

$$CH_3CHOHCOOH(aq) \rightleftharpoons CH_3CHOHCOO^- + H^+$$

$$K_a = \frac{[CH_3CHOHCOO^-][H^+]}{[CH_3CHOHCOOH]} \qquad \text{Let } x = [H^+]$$

	$CH_3CHOHCOOH(aq) \rightleftharpoons$	$CH_3CHOHCOO^-$	$+$	H^+
initial	0.100			
change	$-x$	$+x$		$+x$
equilibrium	$0.100 - x$	x		x

$$K_a = \frac{(x)(x)}{(0.100 - x)} = 1.37 \times 10^{-4}$$

Assuming $0.100 - x = 0.100$, we get

$$\frac{x^2}{0.100} = 1.37 \times 10^{-4}$$

$$x = 3.7 \times 10^{-3} \text{ mole/liter}$$

The approximation $0.100 - x = 0.100$ is valid to within the 5% error limit [relative error $= (3.7 \times 10^{-3}/0.1) \times 100 = 3.7\%$]. From the definition of pH

$$\text{pH} = -\log [\text{H}^+] = -\log (3.7 \times 10^{-3}) = 2.43$$

The pH of the lactic acid solution is 2.43. ∎

EXAMPLE 17.17

∎ Which solution will have the lowest pH (that is, be the strongest acid): (a) $0.10M$ $HClO_2$ with $pK_a = 1.96$; (b) $0.1M$ HCN with $pK_a = 9.21$; (c) $0.10M$ HF with $pK_a = 3.19$; or (d) $0.10M$ HClO with $pK_a = 7.538$.

We first convert the given values of pK_a to K_a values.

(a) For $HClO_2$ $\log K_a = -pK_a = -(1.96) = 0.04 - 2$
                                     $K_a = 1.1 \times 10^{-2}$

(b) For HCN $\log K_a = -(9.21) = 0.79 - 10$
                                       $K_a = 6.2 \times 10^{-10}$

(c) For HF $\log K_a = -(3.19) = 0.81 - 4$
                                       $K_a = 6.5 \times 10^{-4}$

(d) For HClO $\log K_a = -(7.538) = 0.462 - 8$
                                       $K_a = 2.90 \times 10^{-8}$

Note that the acids having larger values of K_a are those with lower values of pK_a (as is also shown in Table 17.9).

All of the acids follow the same mathematical interpretation of equilibrium:

$$\text{HX}(aq) \rightleftharpoons \text{H}^+ + \text{X}^-$$

	HX(aq)	H⁺	X⁻
initial	0.10		
change	$-x$	$+x$	$+x$
equilibrium	$0.10 - x$	x	x

(a) For $HClO_2$

$$\frac{(x)(x)}{(0.10 - x)} = 1.1 \times 10^{-2}$$

Because the value of K_a and the acid concentration are not too different in magnitude, we use the quadratic formula, which gives $x = [H^+] = 2.8 \times 10^{-2}M$ and pH = 1.55.

(b) For HCN

$$\frac{(x)(x)}{(0.10 - x)} = 6.2 \times 10^{-10}$$

$$x = [H^+] = 7.9 \times 10^{-6}M$$
$$pH = -\log (7.9 \times 10^{-6}) = 5.10$$

(c) For HF

$$\frac{(x)(x)}{(0.10 - x)} = 6.5 \times 10^{-4}$$

Here again, we assume we must use the quadratic formula, which gives $x = [H^+] = 7.7 \times 10^{-3}M$ and pH = 2.11.

(d) For HClO

$$\frac{(x)(x)}{(0.10 - x)} = 2.90 \times 10^{-8}$$

$$x = [H^+] = 5.4 \times 10^{-5}M$$
$$pH = -\log (5.4 \times 10^{-5}) = 4.27$$

The solution of $HClO_2$ has the lowest pH. Note that for these solutions of equal concentration, the lower the value of pK_a, the lower is the value of pH. ■

The percentages of ionization for some weak acids are given in Table 17.10. Although it may not be immediately obvious, there is an important difference between K_a values and percent ionization values. For a given acid (at the same temperature), K_a is always the same. However, the percentage of ionization of a weak acid or base varies with the initial concentration of the solution. The more dilute the solution, the larger is the percent ionization. For example, acetic acid (CH_3COOH; see below) in a 0.1M solution is 1.3% ionized, but in

$H_3PO_4(aq) \rightleftharpoons H^+ + H_2PO_4^-$ phosphoric acid (0.03M)	24%
$HF(aq) \rightleftharpoons H^+ + F^-$ hydrofluoric acid (0.1M)	7.7%
$CH_3COOH(aq) \rightleftharpoons H^+ + CH_3COO^-$ acetic acid (0.1M)	1.32%
$H_2O + CO_2 \rightleftharpoons H^+ + HCO_3^-$ "carbonic acid" (H_2CO_3) (0.05M)	0.21%
$H_2S(aq) \rightleftharpoons H^+ + HS^-$ hydrosulfuric acid	0.1%
$HCO_3^- \rightleftharpoons H^- + CO_3^{2-}$ hydrogen carbonate ion (0.1M)	2.2×10^{-3}%

TABLE 17.10

Percent ionization of some weak acids in aqueous solution at 25°C

ACIDS AND BASES [17.8]

a 0.1M solution it is 4.3% ionized. This increase in ionization is a consequence of LeChatelier's principle (Section 16.6). The increasing dilution is equivalent to the addition of water to the following equilibrium system:

$$HA(aq) \rightleftharpoons H^+ + A^-$$

or, written as a Brønsted-Lowry acid-base reaction,

$$HA(aq) + H_2O(l) \rightleftharpoons H_3O^+ + A^-$$

The increase in the amount of water causes the equilibrium to shift in the direction of further ionization of HA. However, that is not the whole story. The greater dilution also decreases the concentration of the hydrogen, or hydronium, ions. Therefore, the solution becomes less acidic as it is made more dilute. For acetic acid solutions:

concentration	% ionization	[H$^+$]	pH
0.1M	1.3%	0.0013M	2.88
0.01M	4.3%	0.00043M	3.88

Let's take a moment to say a few words about organic compounds that are acids. We have mentioned that organic compounds are thought of as being derived from hydrocarbons. Hydrocarbons are relatively unreactive, and organic compounds usually owe their chemical reactivity to "functional groups" (Section 32.1) attached to the hydrocarbon structures. The most common group that imparts acid properties to organic compounds is the carboxyl group, —COOH,

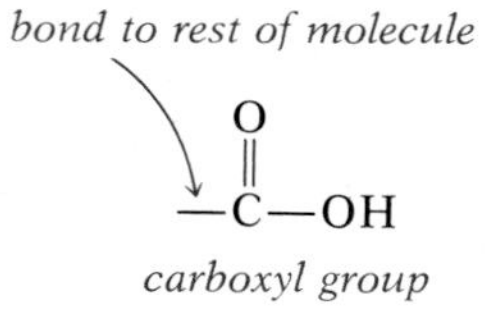

bond to rest of molecule

$$\begin{matrix} & O \\ & \| \\ -C & -OH \end{matrix}$$

carboxyl group

The frequently encountered weak acid, acetic acid, ionizes as shown:

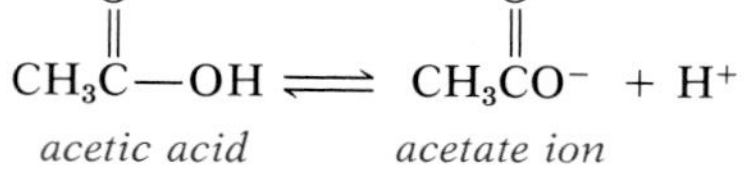

$$\underset{\text{acetic acid}}{CH_3\overset{O}{\overset{\|}{C}}-OH} \rightleftharpoons \underset{\text{acetate ion}}{CH_3\overset{O}{\overset{\|}{C}}O^-} + H^+$$

It is often written as

$$CH_3COOH \rightleftharpoons CH_3COO^- + H^+$$

(Other ways of writing the formula for acetic acid are $HC_2H_3O_2$ or HOAc. Simple monocarboxylic acids like acetic acid are all weak acids with pK_a values in the range of 4 to 5.

The variation in the degree of ionization of strong and weak acids or bases is reflected in reaction rates. The rate of any reaction in solution that depends upon hydrogen or hydroxide ion concentration is greater with a strong acid or base present than with a weak one at the same concentration. This can be demonstrated convincingly with an experiment using the reaction between magnesium metal and an acid to generate hydrogen

$$Mg(s) + 2H^+ \longrightarrow Mg^{2+} + H_2(g)$$

(Note that hydrogen is flammable and can explode; we recommend that you do not try this experiment casually.) Equal volumes of $2M$ hydrochloric acid, a strong acid, and $2M$ acetic acid, a weak acid, are placed in identical flasks. Identical balloons, each containing 1 g of magnesium, preferably as pieces of the same size and shape, are attached to the necks of the flasks. The pieces of metal are dropped simultaneously into the two solutions (Figure 17.4). Evolution of hydrogen, as measured by the rates at which the balloons inflate, is much faster with the stronger acid, although, of course, the total quantity of hydrogen released is the same in both cases.

As we have seen (Example 17.6), the pH of strong acid solutions can be found directly from the concentration of the solution. Because weak acids are less than 100% ionized, it is also necessary to know one of the three quantities, K_a, pH, or percent ionization, in order to find either of the others. The following examples illustrate the relationships among these quantities.

EXAMPLE 17.18

■ Electrical conductivity measurements show that $0.050M$ acetic acid is 1.9% ionized at 25°C. Calculate K_a.

$$CH_3COOH(aq) + H_2O \rightleftharpoons CH_3COO^- + H_3O^+$$

or simply

$$CH_3COOH(aq) \rightleftharpoons CH_3COO^- + H^+$$

$$K_a = \frac{[CH_3COO^-][H^+]}{[CH_3COOH]}$$

	$CH_3COOH(aq)$ $\rightleftharpoons$	CH_3COO^- +	H^+
initial	0.050		
change	−(0.050)(0.019)	+(0.050)(0.019)	+(0.050)(0.019)
equilibrium	(0.050)(1 − 0.019)	(0.050)(0.019)	(0.050)(0.019)

$$K_a = \frac{[(0.050)(0.019)][(0.050)(0.019)]}{[(0.050)(1 - 0.019)]} = 1.8 \times 10^{-5}$$

The ionization constant of acetic acid is 1.8×10^{-5}. ■

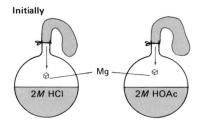

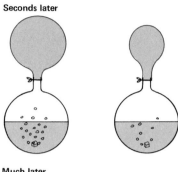

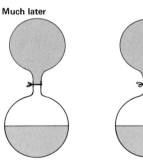

FIGURE 17.4

Variation in reaction rate with strong and weak acids. *The reaction is* $Mg(s) + 2H^+ \longrightarrow Mg^{2+} + H_2(g).$

ACIDS AND BASES [17.8]

EXAMPLE 17.19

■ What is the percent ionization in an 0.0500M solution of formic acid of $K_a = 1.772 \times 10^{-4}$?

$$HCOOH(aq) + H_2O \rightleftharpoons HCOO^- + H_3O^+$$

or simply

$$HCOOH(aq) \rightleftharpoons HCOO^- + H^+$$

$$K_a = \frac{[HCOO^-][H^+]}{[HCOOH]} \qquad \alpha = \frac{number\ of\ moles\ ionized}{number\ of\ moles\ available}$$

α, the degree of ionization, is unknown

	HCOOH(aq) $\rightleftharpoons$	HCOO$^-$ +	H$^+$
initial	0.0500		
change	-0.0500α	$+0.0500\alpha$	$+0.0500\alpha$
equilibrium	$(0.0500)(1-\alpha)$	0.0500α	0.0500α

$$K_a = \frac{(0.0500\alpha)(0.0500\alpha)}{(0.0500)(1-\alpha)} = 1.772 \times 10^{-4}$$

If we assume $1 - \alpha \approx 1$ and solve

$$\frac{(0.0500\alpha)^2}{(0.0500)(1)} = 1.772 \times 10^{-4}$$

$$\alpha = 0.0595$$

Checking our approximation, we find that it is not valid because $1 - \alpha$ is not equal to 1 within a 5% error ($\alpha = 0.0595$ is greater than 5% of 1). Therefore, this problem must be solved by putting the equilibrium constant expression into the quadratic form.

$$0.0500\alpha^2 + 1.772 \times 10^{-4}\alpha - 1.772 \times 10^{-4} = 0$$

$$\alpha = \frac{-(1.772 \times 10^{-4}) \pm \sqrt{(1.772 \times 10^{-4})^2 - (4)(0.0500)(-1.772 \times 10^{-4})}}{2(0.0500)}$$

$$= 0.0578$$

The formic acid is 5.78% ionized. ■

EXAMPLE 17.20

■ Find α, the percent ionization, and the pH of a 0.025M solution of periodic acid, HIO$_4$. $K_a = 5.6 \times 10^{-9}$.

For the equilibrium

	HIO$_4$(aq)	$\rightleftharpoons$	H$^+$	+	IO$_4^-$
initial	0.025				
change	-0.025α		$+0.025\alpha$		$+0.025\alpha$
equilibrium	$0.025(1 - \alpha)$		0.025α		0.025α

$$K_a = \frac{[\text{H}^+][\text{IO}_4^-]}{[\text{HIO}_4]} = \frac{(0.025\alpha)(0.025\alpha)}{(0.025)(1 - \alpha)} = 5.6 \times 10^{-9}$$
$$\alpha = 4.7 \times 10^{-4}$$

For this solution $\alpha = 4.7 \times 10^{-4}$ and the percent ionization is $4.7 \times 10^{-2}\%$. The concentration of H$^+$ is

$$[\text{H}^+] = (0.025)\alpha = (0.025)(4.7 \times 10^{-4}) = 1.2 \times 10^{-5}M$$
$$\text{pH} = -\log (1.2 \times 10^{-5}) = 4.92$$

The pH of the solution is 4.92. ■

EXAMPLE 17.21

■ Calculate α, the percent of ionization, and the pH for the following solutions of nitrous acid, HNO$_2$: (a) 1.0M, (b) 0.10M, (c) 0.010M, and (d) 0.0010M. $K_a = 7.2 \times 10^{-4}$ for HNO$_2$.

All of the solutions are described by the following stoichiometry, where C is the initial concentration of the HNO$_2$.

	HNO$_2$(aq)	$\rightleftharpoons$	H$^+$	+	NO$_2^-$
initial	C				
change	$-\alpha$C	+	αC	+	αC
equilibrium	C(1 $-\alpha$)		αC		αC

$$K_a = \frac{[\text{H}^+][\text{NO}_2^-]}{[\text{HNO}_2]} = \frac{(\alpha\text{C})(\alpha\text{C})}{\text{C}(1 - \alpha)} = 7.2 \times 10^{-4}$$
$$\text{pH} = -\log [\text{H}^+] = -\log (\alpha\text{C})$$

Substituting for C and solving for α and pH give the tabulated results. Note that α is close enough to K_a that the quadratic formula must be used.

		α	% ionization	pH
(a)	1.0M	0.026	2.6%	1.59
(b)	0.10M	0.081	8.1%	2.09
(c)	0.010M	0.23	23%	2.64
(d)	0.0010M	0.56	56%	3.25

These calculations show that α increases as concentration decreases, but also that the pH values increase (the solution becomes less acidic). ■

17.9 Strong and weak bases

Fewer strong bases than strong acids are encountered in aqueous solution chemistry. The most common strong base, hydroxide ion, OH^-, is derived either from the dissociation of hydroxides, for example,

$$NaOH(s) \xrightarrow{\text{H}_2\text{O}} Na^+(aq) + OH^-(aq)$$

or from the reaction of proton acceptors with water, for example,

$$NH_3(aq) + H_2O(l) \rightleftharpoons NH_4^+ + OH^- \tag{33}$$

or

$$F^- + H_2O(l) \rightleftharpoons HF(aq) + OH^- \tag{34}$$

All of the alkali metals (Li, Na, K, Rb, Cs) and alkaline earth metals except beryllium (Mg, Ca, Sr, Ba) form ionic hydroxides. Among them, sodium hydroxide, NaOH; potassium hydroxide, KOH; and calcium hydroxide, $Ca(OH)_2$ are the strong bases used most often—these **strong bases** are completely dissociated in water. For the reaction

$$MOH(aq) \rightleftharpoons M^+ + OH^-$$

the equilibrium constant, or base ionization constant, is

$$K_b = \frac{[M^+][OH^-]}{[MOH]}$$

The base strengths of the alkaline earth hydroxides are limited by their moderate solubility in water (Table 17.11).

The equilibrium between a proton-accepting base and water can be represented by

$$B(aq) + H_2O(l) \rightleftharpoons BH^+ + OH^-$$

and the base ionization constant for this equilibrium is

$$K_b = \frac{[BH^+][OH^-]}{[B]} \tag{35}$$

The proton acceptor B may be a neutral molecule as in Equation (33) for which

TABLE 17.11
Solubility of alkaline earth hydroxides (at 25°C)

	Solubility (*mole/liter*)
Be(OH)$_2$	4.6×10^{-5}
Mg(OH)$_2$	1.2×10^{-4}
Ca(OH)$_2$	2.1×10^{-2}
Sr(OH)$_2$	1.2×10^{-1}
Ba(OH)$_2$	1.5×10^{-1}

$$K_b = \frac{[NH_4^+][OH^-]}{[NH_3]} \qquad (36)$$

or an anion as in Equation (34) for which

$$K_b = \frac{[HF][OH^-]}{[F^-]} \qquad (37)$$

As with K_a, the larger the value of K_b, the stronger is the base; also, the larger the pK_b value, the weaker is the base. Table 17.10 gives the K_b and pK_b values of some common weak bases. You can see from the pK_b values that the anions of the weak acids H_2S, HCN, and H_2CO_3 are all relatively strong bases. As we would predict, the anions of the strong acids such as HNO_3 and HCl are relatively weak bases.

From Table 17.2 (Section 17.2) it can be seen that both NH_3 and F^- are weaker bases than OH^-. In the competition for a proton, OH^- is more able to attract the proton than NH_3, F^-, or any of the bases above OH^- on the right in the table. Therefore, these are all "weak bases," and at equilibrium, formation of H_2O rather than OH^- is favored. The "strong bases" are OH^- and those below it in Table 17.2. Each reacts immediately and virtually completely with water.

$$OH^- + H_2O(l) \longrightarrow H_2O + OH^-$$
$$NH_2^- + H_2O(l) \longrightarrow NH_3(g) + OH^-$$
$$H^- + H_2O(l) \longrightarrow H_2(g) + OH^-$$
$$CH_3^- + H_2O(l) \longrightarrow CH_4(g) + OH^-$$

To summarize, acids that are stronger proton donors than H_3O^+, and the OH^- ion and all bases that are stronger proton acceptors than OH^- are classified as "strong" in aqueous solution.

The amines—compounds in which the hydrogen atoms of the ammonia molecule are replaced by hydrocarbon groups—are the most common organic bases. Simple amines such as the ethylamines

$CH_3CH_2NH_2$	$(CH_3CH_2)_2NH$	$(CH_3CH_2)_3N$
ethylamine	*diethylamine*	*triethylamine*
pK_b 3.33	pK_b 3.02	pK_b 3.28

are comparable in base strength to ammonia (Table 17.12). Amines in which the nitrogen atom is bonded to a benzene ring are weaker bases because resonance (Section 9.13) withdraws electron density from the nitrogen atom and stabilizes the free amine

aniline, pK_b 9.38

Most amines can be neutralized to give salts in reactions with strong acids such as hydrochloric acid, for example,

TABLE 17.12

Ionization constants for some weak bases at 25°C for the reaction

B, *or* B$^-$ (*aq*) + H$_2$O(*l*) $\rightleftharpoons$ BH$^+$, *or* BH + OH$^-$

	K_b	pK_b
Hydroxide ion OH^-	1	0
Sulfide ion S^{2-}	7.7×10^{-2}	1.1
Triethylamine $(C_2H_5)_3N$	5.2×10^{-4}	3.28
Carbonate ion CO_3^{2-}	2.4×10^{-4}	3.62
Trimethylamine $(CH_3)_3N$	6.3×10^{-5}	4.20
Cyanide ion CN^-	1.6×10^{-5}	4.80
Ammonia NH_3	1.6×10^{-5}	4.80
Aniline $C_6H_5NH_2$	4.2×10^{-10}	9.38
Acetate ion CH_3COO^-	5.70×10^{-10}	9.24
Fluoride ion F^-	1.5×10^{-11}	10.82
Nitrate ion NO_3^-	4.0×10^{-16}	15.4
Chloride ion Cl^-	3×10^{-23}	22.5

$$CH_3NH_2 + HCl(aq) \longrightarrow CH_3NH_3{}^+Cl^-$$

methylamine   *methylammonium chloride*

The value of K_b, pH, and degree of ionization are used in calculations that involve solutions of bases, just as K_a, pH, and degree of ionization are used in problems dealing with acids in solution.

EXAMPLE 17.22

■ What is the value of K_b for CN^- given the following equilibrium concentrations: $[CN^-] = 3.7M$, $[HCN] = 1.3 \times 10^{-3}M$, and $[OH^-] = 4.6 \times 10^{-2}M$?

The equilibrium constant expression for the equation

$$CN^- + H_2O(l) \rightleftharpoons HCN(aq) + OH^-$$

is given by

$$K_b = \frac{[HCN][OH^-]}{[CN^-]} = \frac{(1.3 \times 10^{-3})(4.6 \times 10^{-2})}{3.7} = 1.6 \times 10^{-5}$$

The value of K_b for CN^- is 1.6×10^{-5}. ■

EXAMPLE 17.23

■ What are the equilibrium concentrations of the various species present in a $0.100M$ solution of the weak organic base aniline, $C_6H_5NH_2$? $K_b = 4.2 \times 10^{-10}$.

We set up our usual table with $x = [OH^-]$.

	$C_6H_5NH_2(aq) + H_2O(l) \rightleftharpoons C_6H_5NH_3{}^+ + OH^-$		
initial	0.100		
change	$-x$	$+x$	$+x$
equilibrium	$0.100 - x$	x	x

And

$$K_b = \frac{[C_6H_5NH_3{}^+][OH^-]}{[C_6H_5NH_2]} = \frac{(x)(x)}{(0.100 - x)} = 4.2 \times 10^{-10}$$

$$x = 6.5 \times 10^{-6}$$

The equilibrium concentrations are $[C_6H_5NH_3{}^+] = [OH^-] = 6.5 \times 10^{-6}M$ and $[C_6H_5NH_2] = 0.100 - 6.5 \times 10^{-6} = 0.100M$. ■

EXAMPLE 17.24

■ What is the pH of a $1.5M$ solution of NH_3? $K_b = 1.6 \times 10^{-5}$ for $NH_3(aq)$.

The concentration of OH^- can be found in the usual way.

	$NH_3(aq) + H_2O(l) \rightleftharpoons NH_4^+ + OH^-$		
initial	1.5		
change	$-x$	$+x$	$+x$
equilibrium	$1.5 - x$	x	x

$$K_b = \frac{[NH_4^+][OH^-]}{[NH_3]} = \frac{(x)(x)}{(1.5 - x)} = 1.6 \times 10^{-5}$$
$$x = [OH^-] = 4.9 \times 10^{-3}M$$

Now, solving for pOH and pH,

$$pOH = -\log[OH^-] = -\log(4.9 \times 10^{-3}) = 2.31$$
$$pH = 14.00 - pOH = 14.00 - 2.31 = 11.69$$

The pH of the ammonia solution is 11.69. ■

EXAMPLE 17.25

■ What is the percentage of ammonia molecules that react in the solution described in Example 17.24?

Assuming a liter of solution, of the 1.5 moles of NH_3 originally present, we calculated that 4.9×10^{-3} moles reacted. Thus

$$\alpha = \frac{\text{moles reacted}}{\text{moles originally present}} = \frac{4.9 \times 10^{-3}}{1.5} = 3.3 \times 10^{-3}$$

A mole represents an Avogadro's number of molecules. Therefore, we can state that only 0.33% of the ammonia molecules react with the water. ■

17.10 Relationship of K_a and K_b to K_w

The product of the ionization constant for an acid and the ionization constant of its conjugate base is the ion product constant of water

$$K_w = K_a K_b \tag{38}$$

For example, for hydrofluoric acid

$$K_w = K_a K_b$$

$$HF(aq) + H_2O \rightleftharpoons F^- + H_3O^+$$

or simply

$$HF(aq) \rightleftharpoons F^- + H^+$$

$$K_a = \frac{[F^-][H^+]}{[HF]}$$

and for fluoride ion, its conjugate base

$$F^- + H_2O(l) \rightleftharpoons HF(aq) + OH^-$$

$$K_b = \frac{[HF][OH^-]}{[F^-]}$$

The product of K_a and K_b is shown to be K_w as follows:

$$K_a \times K_b = \frac{[\cancel{F^-}][H^+]}{[\cancel{HF}]} \times \frac{[\cancel{HF}][OH^-]}{[\cancel{F^-}]}$$

$$= [H^+][OH^-] = K_w$$

EXAMPLE 17.26

■ Hydrocyanic acid, HCN, has a value of $K_a = 6.2 \times 10^{-10}$. What is the value of K_b for the conjugate base?

From the equation

$$HCN(aq) + H_2O(l) \rightleftharpoons H_3O^+ + CN^-$$

we can see that the conjugate base is the cyanide ion, CN^-. The value of K_b is given by

$$K_b = \frac{K_w}{K_a} = \frac{1.00 \times 10^{-14}}{6.2 \times 10^{-10}} = 1.6 \times 10^{-5}$$

$K_b = 1.6 \times 10^{-5}$ for CN^-.   ■

THOUGHTS ON
CHEMISTRY

Ira Remsen investigates
nitric acid

THE LIFE OF IRA REMSEN, *by Frederick H. Getman*

"While reading a textbook of chemistry," said he, "I came upon the statement, 'nitric acid acts upon copper.' I was getting tired of reading such absurd stuff and I determined to see what this meant. Copper was more or less familiar to me, for copper cents were then in use. I had seen a bottle marked 'nitric acid' on a table in the doctor's office

where I was then 'doing time!' I did not know its peculiarities, but I was getting on and likely to learn. The spirit of adventure was upon me. Having nitric acid and copper, I had only to learn what the words 'act upon' meant. Then the statement, 'nitric acid acts upon copper,' would be something more than mere words. All was still. In the interest of knowledge I was even willing to sacrifice one of the fewer copper cents then in my possession. I put one of them on the table; opened the bottle marked 'nitric acid'; poured some of the liquid on the copper; and prepared to make an observation. But what was this wonderful thing which I beheld? The cent was already changed, and it was no small change either. A greenish blue liquid foamed and fumed over the cent and over the table. The air in the neighborhood of the performance became colored dark red. A great colored cloud arose. This was disagreeable and suffocating—how should I stop this? I tried to get rid of the objectionable mess by picking it up and throwing it out of the window, which I had meanwhile opened. I learned another fact—nitric acid not only acts upon copper but it acts upon fingers. The pain led to another unpremeditated experiment. I drew my fingers across my trousers and another fact was discovered. Nitric acid acts upon trousers. Taking everything into consideration, that was the most impressive experiment, and, relatively, probably the most costly experiment I have ever performed. I tell of it even now with interest. It was a revelation to me. It resulted in a desire on my part to learn more about that remarkable kind of action. Plainly the only way to learn about it was to see its results, to experiment, to work in a laboratory."

THE LIFE OF IRA REMSEN, by Frederick H. Getman (from pp. 9, 10, published by the Journal of Chemical Education, Easton, Pennsylvania, 1940)

Significant terms defined in this chapter

solvent system acid	Lewis acid-base reaction
solvent system base	ion product constant for water
solvent system acid-base reaction	pH
Brønsted-Lowry acid	pOH
Brønsted-Lowry base	acidic solution
Brønsted-Lowry acid-base reaction	alkaline solution
amphoterism	indicator
amphiprotism	strong acid
Lewis acid	strong base
Lewis base	

Exercises

17.1 In many laboratories are large bottles labeled "FIRST AID FOR ACID BURNS—Rinse area of burn thoroughly with water. Rinse with this 5% $NaHCO_3$ solution. Rinse again with water. Seek medical attention." "FIRST AID FOR BASE BURNS—Rinse area of burn thoroughly with water. Rinse with this 5% CH_3COOH solution. Rinse again with water. Seek medical attention." (a) What is the purpose of the thorough water rinse? (b) What is the purpose of the $NaHCO_3$ rinse? Write a chemical equation to illustrate this purpose by assuming a typical laboratory acid such as HCl was spilled. (c) What is the purpose of the CH_3COOH rinse? Illustrate this purpose by writing a reaction between the CH_3COOH and a typical laboratory base such as NaOH. (d) What is the purpose of the second rinse with water? (e) What is the main reason that $NaHCO_3$ and CH_3COOH were chosen over NaOH and HCl to be put in these bottles? (f) Why should the accident be reported and additional medical attention sought?

17.2 Each of the following statements is a definition of (α) an acid, (β) a base, or (γ) an acid–base reaction in one of the theories. Match each definition to one of these categories and identify which theory is being used—(i) Arrhenius or classical theory, (ii) Brønsted–Lowry theory, (iii) Lewis theory, or (iv) solvent system theory.
(a) a substance which gives the cation of the solvent
(b) molecule or ion capable of adding a proton
(c) acid plus base yields salt and water
(d) electron–pair donation—the formation of a coordinate covalent bond
(e) produces OH^-
(f) acts as an electron-pair acceptor
(g) molecule or ion capable of losing a proton
(h) reaction of solvent anion and solvent cation to give solvent
(i) acts as an electron-pair donor
(j) reaction between an acid and base in which proton transfer takes place
(k) a substance which gives the anion of the solvent

17.3 Which statements are true? Rewrite any false statement so that it is correct.
(a) Strong acids and bases are virtually 100% ionized or dissociated in dilute aqueous solutions.
(b) The leveling effect is the seemingly identical strengths of all acids and bases in aqueous solutions.
(c) An acidic solution is characterized by pH > 7, a high value of $[H^+]$.
(d) A conjugate acid is a molecule or ion formed by the addition of a proton to a base.
(e) Amphoterism and amphiprotism are the same in aqueous solution.

17.4 Which statements are valid in describing the acidity of (a) nonmetal hydrides and (b) oxo acids? (i) Increases as atomic number increases within a periodic table family, (ii) increases as atomic number increases across a row in the periodic table, (iii) increases as atomic number decreases within a periodic table family for the same oxidation state of the elements, (iv) increases as oxidation number or oxygen content increases for the same element.

17.5 Sort the following list of chemicals into (i) acidic, (ii) basic, or (iii) amphoteric species. Assume all oxides are dissolved in water. Do not be intimidated by the way the formula of the compound is written. (a) Cs_2O, (b) N_2O_5, (c) HCl, (d) $SO_2(OH)_2$, (e) HNO_2, (f) Al_2O_3, (g) BaO, (h) H_2O, (i) CO_2.

17.6 Write the chemical equation and identify the conjugate acid for each of the following Brønsted-Lowry bases: (a) NH_3, (b) H_2O, (c) HCO_3^-, (d) CO_3^{2-}, (e) CN^-, (f) OH^-.

17.7 Write the chemical equation and identify the conjugate base for each of the following Brønsted-Lowry acids: (a) HNO_3, (b) HSO_4^-, (c) H_2SO_4, (d) HCl, (e) H_2O.

17.8 Write the chemical equation describing the following acid–base reactions in terms of the Brønsted-Lowry system: (a) H_2O and NH_3; (b) H_3O^+ and OH^-; (c) NH_4^+ and CO_3^{2-}; (d) HNO_3 and H_2O. Identify the respective conjugate acid-base pairs in each reaction.

17.9 Indicate which of the following substances, (a) HCl, (b) $H_2PO_3^-$, (c) H_2CaO_2, (d) $ClO_3(OH)$, (e) $Sb(OH)_3$ can act as (i) an acid, (ii) a base, or (iii) both according to the (α) Arrhenius or classical theory and/or (β) Brønsted–Lowry theory. Do not be confused by the way the formulas are written.

17.10 Choose (a) all Lewis acids, (b) all Brønsted–Lowry bases, and (c) all Brønsted–Lowry acids from (i) SO_3^{2-}, (ii) $AlCl_3$, (iii) Cl^-, (iv) NH_4^+, (v) NH_3, (vi) H_2O, and (vii) HBr.

17.11 Give a reaction showing (a) HSO_4^- acting as a base, (b) H_2O acting as an acid, (c) $Al(OH)_3(H_2O)_3$ acting as an acid, and (d) NH_4^+ acting as an acid.

17.12 Le Chatelier's principle can be demonstrated easily by pouring 50 ml of $0.1M$ K_2CrO_4 into a 250-ml beaker and alternately adding a little dilute HCl and a little dilute NaOH:

$$2CrO_4^{2-} + 2H_3O^+ \rightleftharpoons Cr_2O_7^{2-} + 3H_2O(l)$$
yellow   orange

(a) Identify the stronger acid in the above equation and its conjugate base. (b) What is the purpose of the NaOH? (c)

According to which theories is this considered an acid–base reaction?

17.13 For each of the reactions described by the following equations, indicate the acids and bases:

$$NH_4I + KNH_2 \xrightleftharpoons{NH_3(l)} KI + 2NH_3$$
$$La_2O_3(s) + 6NONO_3(l) \rightleftharpoons 2La(NO_3)_3 + 3N_2O_3(g)$$
$$OCN^-(aq) + H_2O(l) \rightleftharpoons HOCN(aq) + OH^-(aq)$$
$$Zn(H_2O)_4{}^{2+}(aq) + H_2O(l) \rightleftharpoons$$
$$Zn(H_2O)_3(OH)^+(aq) + H_3O^+(aq)$$
$$K_2O(s) + CO_2(g) \rightleftharpoons K_2CO_3(s)$$
$$Al^{3+}(aq) + 6F^-(aq) \rightleftharpoons AlF_6{}^{3-}(aq)$$

17.14 Based on the relative strengths of acid-base pairs given in Table 17.2, predict what will happen if (a) HI is added to NO_2^-, (b) H_2S is added to CN^-, and (c) HCOOH is added to NO_3^-. Write the appropriate equations.

17.15 Predict which acid is stronger: (a) NH_3 or CH_4; (b) HI or HCl; (c) HIO_2 or HIO_3; (d) $As(OH)_3$ or $Sb(OH)_3$; (e) H_2S or HNO_3.

17.16 List the following acids in order of increasing strength: (a) sulfuric, phosphoric, and perchloric; (b) HIO_3, HIO_2, HIO, and HIO_4; (c) selenous, sulfurous, and tellurous; (d) hydrosulfuric, hydroselenic, and hydrotelluric; (e) H_2CrO_4, H_2CrO_2, $HCrO_3$, and H_3CrO_3; (f) $H_4P_2O_7$, $HP_2O_7{}^{3-}$, $H_3P_2O_7{}^-$, and $H_2P_2O_7{}^{2-}$.

17.17 NaOH behaves as a base in water while ClOH behaves as an acid. Explain clearly this behavior and the general principles involved for any E–O–H compound.

17.18 The bicarbonate ion, HCO_3^-, is amphiprotic. (a) Illustrate this property by writing typical Brønsted–Lowry acid–base reactions. (b) Identify the conjugate acid–base pairs.

17.19 Illustrate the leveling effect of water by writing reactions for HCl and HNO_3.

17.20 Give the formula for an example chosen from the representative elements for (a) an acid oxide, (b) an amphoteric oxide, and (c) a basic oxide.

17.21 Which indicator would you choose to "signal" reaching a pH of (a) 2.4, (b) 7, (c) 10.3, (d) 5.1?

17.22 A "universal" indicator was prepared by mixing methyl red with thymolphthalein. What color would be observed at a pH value of (a) 3, (b) 7, (c) 11?

17.23* The value of K_b for CH_3NH_2 is less than that for $C_2H_5NH_2$ and for $(CH_3)_2NH$ it is less than that for

$(C_2H_5)_2NH$. Predict the relative order for the values for $(CH_3)_3N$ and $(C_2H_5)_3N$.

17.24 Convert the following values of $[H^+]$ into pH: (a) $9.5 \times 10^{-6}M$, (b) $0.0001472M$, (c) $0.00203M$, (d) $12 \times 10^{-12}M$, (e) $1.0 \times 10^1 M$. Which solutions are alkaline?

17.25 Convert the following values of $[OH^-]$ into pH: (a) $0.00035M$, (b) $11 \times 10^{-13}M$, (c) $4.5 \times 10^{-7}M$, (d) $10M$, (e) $0.00203M$. Which solutions are alkaline?

17.26 Convert the following pH values into $[H^+]$: (a) 4.67, (b) 9.05, (c) 0.0, (d) 15.97, (e) 0.153. Which solutions are acidic?

17.27 Convert the following pOH values into $[H^+]$: (a) 13.97, (b) 0.12, (c) 5.04, (d) 3.57, (e) 11.021. Which solutions are acidic?

17.28 Calculate the pH of the following solutions: (a) $0.0010M$ HNO_3, (b) $0.103M$ NaOH, (c) $0.5M$ KOH, (d) $0.104M$ HCl. Assume all solutes are completely ionized or dissociated.

17.29 The reagent bottles labeled "Dilute NaOH" and "Dilute HCl" that are commonly found in the laboratory are usually $6M$ solutions of the substances. Calculate the $[H^+]$, $[OH^-]$, pH, and pOH for each reagent, assuming that the solutes are completely ionized or dissociated.

17.30 What is the concentration of IO^- in equilibrium with $[H^+] = 0.035M$ and $[HIO] = 0.427M$? $K_a = 2.3 \times 10^{-11}$ for HIO.

17.31 Find the concentrations of the various species present in a $0.025M$ solution of hydrofluoric acid, HF. What is the pH of the solution? $K_a = 6.5 \times 10^{-4}$ for HF.

17.32 The ionization constant for acetic acid

$$CH_3COOH(aq) + H_2O(l) \rightleftharpoons CH_3COO^- + H_3O^+$$

is 1.754×10^{-5} at 25°C. What is the extent of ionization of the molecules in (a) a $0.100M$ solution and (b) a $0.0100M$ solution?

17.33 Calculate the ionization constant K_a for an acid that is 1.25% ionized in a $0.250M$ solution of the acid.

17.34 Calculate the concentrations of the various species present in a $0.0123M$ solution of diethylamine, $(C_2H_5)_2NH$. What is the pH of the solution? $K_b = 9.5 \times 10^{-4}$ for $(C_2H_5)_2NH$.

17.35 A weak base, B, reacts only to the extent of 3.2% with water in a $0.10M$ solution. Calculate K_b for this base.

17.36 Calculate the values of K_b for the conjugate base of (a) HF, $K_a = 6.5 \times 10^{-4}$ for HF; (b) HNCS, $K_a = 69$ for the acid; and (c) HIO_4, $K_a = 5.6 \times 10^{-9}$ for HIO_4.

17.37 K_w is 9.614×10^{-14} at 60°C. Find $[H^+]$ in water at this temperature and calculate the pH. Is pure water an acid at 60°C?

17.38 What volume of $0.1872M$ H_2SO_4 would be required to neutralize 25.00 ml of $0.4729M$ NaOH?

17.39 A common way to standardize an acid solution (find its exact concentration) is by titrating it against a solution containing a known weight of Na_2CO_3. For a solution of HCl, (a) write the equation for the chemical reaction (a gas is evolved) and (b) if 23.69 ml of the HCl solution neutralized 2.1734 g Na_2CO_3, calculate the number of moles of Na_2CO_3 and the molar concentration of the HCl.

17.40* Electrical conductance measurements at 18°C show that the weak base $NH_3(aq)$ is 1.4% ionized in a $0.1000M$ solution, 4.0% ionized in a $0.0100M$ solution, and 12% ionized in a $0.0010M$ solution. (a) Find the pH of each solution. (b) Calculate K_b for each solution. (The values will differ slightly because the activities of the ions are slightly different in each solution.)

17.41* A solution is prepared containing 1.20×10^{-3} mole of a weak acid, HA, in 25.0 ml of water. The freezing point is determined to be -0.100°C. Calculate the apparent molality and the van't Hoff factor. What is the percent ionization of the acid? $K_f = 1.86$°C for water.

17.42* A solution containing 3.602 g of oxalic acid, $H_2C_2O_4$, per 100.0 g of water is 27.0% ionized into H^+ and $HC_2O_4^-$ ions. Neglect the second ionization step of $HC_2O_4^-$ into H^+ and $C_2O_4^{2-}$. (a) What is the pH of the solution? (b) Calculate K_1 from the above data. (c) What is the freezing point of this solution? $K_f = 1.86$°C for water.

17.43* Consider a solution of a weak acid HA such that $K_a = 5 \times 10^{-10}$. (a) Write the equation for HA undergoing ionization in water. (b) Identify (i) the acid, (ii) the base, (iii) the conjugate acid, and (iv) the conjugate base. (c) Find the pH of a solution in which 0.100 mole of HA is dissolved in 1 liter of solution.

17.44* Sodium hydroxide is prepared commercially by the lime–soda process

$$Na_2CO_3(aq) + Ca(OH)_2(aq) \rightleftharpoons CaCO_3(s) + 2NaOH(aq)$$

and by electrolysis of brine

$$2NaCl(aq) + 2H_2O(l) \rightleftharpoons 2NaOH(aq) + H_2(g) + Cl_2(g)$$

(a) In order to produce 1 mole of NaOH, which reaction requires a smaller total mass of reactants? (b) What is a major advantage of the electrolysis process? (c) What could be done with the $CaCO_3$ formed in the lime–soda process to reduce costs?

17.45* A student prepared a solution of HCl having an approximate concentration of $0.25M$. To find the exact concentration, he collected the following standardization data: mass of $Na_2CO_3 = 0.1870$ g and 22.37 ml HCl needed for neutralization. (a) Calculate the exact concentration of the acid. He then used his acid to determine the percent of Na_2CO_3 in an unknown sample. From the data given: 0.3940 g of the unknown required 13.72 ml of HCl solution, (b) calculate the amount of Na_2CO_3 present.

17.46* The molar weight of a carbonate, $MeCO_3$, was determined by adding excess acid to the carbonate and then titrating the excess acid with a base (called "back titrating"). (a) Write the chemical equations for the reaction of HCl with $MeCO_3$ and HCl with NaOH. (b) Exactly 50.00 ml of $0.800M$ HCl was poured into a 0.844-g sample of $MeCO_3$. The excess acid was determined by back titrating with 14.30 ml of $1.400M$ NaOH. Calculate the molar weight of the carbonate.

17.47* If the flue gases from a large coal-burning power plant emit a total of 100 tons of SO_2 each day, calculate the weight of soda ash, Na_2CO_3, required to neutralize the sulfuric acid that would be recovered each day assuming a 90% conversion of SO_2 to H_2SO_4.

17.48** At 25°C, the value of pK for the reaction

$$\overset{\text{pure } D_2O}{2D_2O \rightleftharpoons D_3O^+ + OD^-}$$

involving deuterium oxide (heavy water) is 14.869. (a) What is the value of K? (b) Under equilibrium conditions, what are the values of $[D_3O^+]$ and $[OD^-]$? (c) What is the pD of pure D_2O? (d) What is the relation between pD and pOD? (e) Give a suitable definition in terms of pD for an acidic solution in this solvent system.

Consider the following acid–base reactions taking place in pure D_2O:

$$CH_3COOH + D_2O \rightleftharpoons HD_2O^+ + CH_3COO^-$$
$$NH_3 + D_2O \rightleftharpoons NH_3D^+ + OD^-$$

Which substances are acting as (f) Arrhenius or classical acids, (g) Arrhenius or classical bases, (h) Brønsted–Lowry acids, (i) Brønsted–Lowry bases, (j) solvent system acids, and (k) solvent system bases?

The equilibrium constant for the reaction

$$D_2O(l) + H_2O(l) \rightleftharpoons 2HDO(l)$$

is 3.56 at 25°C. (1) If we prepare an equimolar solution of D_2O and H_2O, which is essentially $27.7M$ in each, what will be the concentration of HDO once equilibrium has been established?

18 IONS AND IONIC EQUILIBRIA

$\mathbf{O}$ *ver the years, many people surely have been drawn to the study of chemistry by the sights and smells of a chemistry laboratory. It seems likely that this will always be the case. There is a fascination to the changing colors when solutions are mixed, the sudden appearance of solids where there were no solids, the rapid bubbling up of a gas. Such events are all part of one of the oldest areas of chemistry—the subject of this chapter—ions in solution.*

"Wet chemistry," as it is sometimes called, once provided the only way to identify the metals in an alloy or the ions in a mineral. The substance to be analyzed is dissolved, and then the ions present are separated and identified by the ingenious application of the principles of equilibrium that are discussed in this chapter. Many of you will be doing such qualitative analysis for ions in solution in the laboratory.

Today such analysis is often performed by electronic instruments. However, wet methods are quick and reliable, and remain in use in many applications. To mention just a few, methods based on the fundamentals presented in this chapter are used to determine the concentration of CO_2 in the atmosphere, the acetic acid content of vinegar, the HCO_3^- concentration in antacid tablets such as Alka Seltzer, the strength of paint removal and rust removal solutions, the ions present in hard water, and the concentration of metal ions in minerals, blood, and urine.

This chapter shows how to apply what we have learned about equilibria and acids and bases specifically to ions in aqueous solution. The major types of équilibria discussed are acid–base equilibria and equilibria between solids and their saturated solutions (heterogeneous equilibria). Included are several topics of practical importance—the titration of solutions of acids and bases, maintenance of constant pH by buffer solutions, and control of solubility by application of the principles of equilibria.

Acid-base equilibria

18.1 Ions in aqueous solution

The subject of this chapter is "Ions and Ionic Equilibria," so let's begin by reviewing the sources of ions.

Obviously, ions are present in crystalline ionic compounds. The classical examples of such compounds are the salts formed between the alkali metals (periodic table Group I) and the halogens (periodic table Group VII), compounds such as sodium chloride (NaCl), potassium bromide (KBr), lithium iodide (LiI), potassium iodide (KI), or cesium fluoride (CsF). Each of these compounds is highly soluble in water and dissociates completely as its ions are brought into solution by interaction with the water. For example,

a soluble salt

$$CsF(s) \xrightarrow{H_2O} Cs^+ + F^-$$

In other words, these salts are strong electrolytes.

Not all salts are highly soluble in water like the alkali metal halides. However, the variations in their solubility are not directly related to the degree of ionic bonding in the compounds. As we have seen (Section 14.7), the strength of the forces in the crystal lattice and the hydration energies of the ions released into solution influence just how soluble a specific compound will be. To the extent that a slightly soluble salt does dissolve in water, it also yields ions in solution. A heterogeneous equilibrium is set up in a saturated solution of a slightly soluble salt in water. Silver iodide, for example, is soluble to the extent of only about 2×10^{-6} g/liter, but all of the salt that dissolves is in the form of ions.

a slightly soluble salt

$$AgI(s) \xrightleftharpoons{H_2O} Ag^+ + I^-$$

In Chapter 17 we discussed at length the properties of acids and bases. The common strong acids (Table 17.8) are covalent compounds, but are strong electrolytes because they react with water to become essentially 100% ionized, as in the case of perchloric acid.

ionization of a strong acid

$$HClO_4(l) + H_2O(l) \longrightarrow H_3O^+ + ClO_4^-$$

The soluble metal hydroxides most often used in the laboratory to give strongly alkaline solutions, such as sodium hydroxide or potassium hydroxide, are ionic compounds and strong electrolytes.

dissociation of a soluble hydroxide

$$Na^+OH^-(s) \xrightarrow{H_2O} Na^+ + OH^-$$

Weak acids are generally covalent compounds that yield ions to the degree that they ionize in aqueous solution. Hydrosulfuric acid, for example, is only 0.1% ionized.

ionization of a weak acid
$$H_2S(g) + H_2O(l) \rightleftharpoons H_3O^+ + HS^-$$

A few covalent compounds, notably the common weak base ammonia, yield hydroxide ions by reaction with water.

ionization of a weak base
$$NH_3(g) + H_2O(l) \rightleftharpoons NH_4^+ + OH^-$$

In considering ions in aqueous solution we will be concerned with their interactions in acid-base equilibria, the extent to which various species yield ions in solution, the combination of ions to form slightly soluble salts, and situations in which ions participate simultaneously in more than one type of equilibrium.

The Brønsted-Lowry acid-base concept, introduced in Chapter 17, emphasizes the interaction of acidic and basic substances with water. We feel that this concept allows a simple and uniform approach to acid-base equilibria in aqueous solution. Throughout this chapter we will use the terms "acid" and "base" as they are defined in the Brønsted-Lowry system: acids are proton donors and bases are proton acceptors.

18.2 Reactions of ions with water

Let's think for a moment of what might happen when ions are added to pure water. The ions enter an environment in which the species H_2O, H_3O^+ (or simply H^+), and OH^- are already present. The very small concentrations of H^+ and OH^- can usually be ignored, and the outcome depends on the interaction of the ions with H_2O.

All ions in aqueous solution are hydrated—surrounded by water molecules—as a result of attraction between the ions and the polar water molecules (Sections 13.2 and 13.4).

$$M^{y+} + nH_2O \rightleftharpoons [M(H_2O)_n]^{y+} \tag{1}$$
$$A^{x-} + nH_2O \rightleftharpoons [A(H_2O)_n]^{x-} \tag{2}$$

The number of water molecules (represented by n) varies and is usually not known. The strength of the forces of attraction between the ion and the water molecules also varies.

Some hydrated ions remain unchanged in aqueous solutions. Other hydrated ions behave as either Brønsted-Lowry acids and yield acidic solutions or as Brønsted-Lowry bases and yield alkaline solutions. Let's examine these possibilities further.

a. Anions or cations simply become hydrated. Some anions and cations undergo no reaction at all with water molecules. These ions simply become hydrated and cause no change in the pH of the solution.

TABLE 18.1

Some common anions that are weaker bases than water
Compare this table with Table 17.2. Note that amphoteric anions such as HSO_4^- may yield acidic solutions and are not included here.

$$A^{x-} + nH_2O(l) \rightleftharpoons [A(H_2O)_n]^{x-}$$

ClO_4^-	I^-	Br^-	Cl^-	ClO_3^-	NO_3^-

TABLE 18.2

Some common cations that are weaker acids than water

$$M^{y+} + nH_2O(l) \rightleftharpoons [M(H_2O)_n]^{y+}$$

Li^+	Na^+	K^+	Mg^{2+}	Ca^{2+}	Ba^{2+}

TABLE 18.3

Some common anions that are stronger bases than water

$$A^- + H_2O(l) \rightleftharpoons HA(aq) + OH^-$$

NO_2^-	F^-	$HCOO^-$	CH_3COO^-	CN^-
S^{2-}	OH^-			

Anions that do not react with water molecules are all weaker bases than water. In a competition with water molecules for the proton, they are the losers. Recall from our discussion of the Brønsted-Lowry concept that the conjugate bases of strong acids are weak bases (Table 18.1; see also Table 17.2). Perchlorate ion, the conjugate base of perchloric acid, is an example of an anion that is hydrated.

anion that is a weaker base than water
$$ClO_4^- + nH_2O \rightleftharpoons [ClO_4(H_2O)_n]^-$$

The cations that do not react with water molecules are all weaker acids than water. There is no tendency for the hydrated form of such a cation, for example, Na^+,

cation that is a weaker acid than water
$$Na^+ + nH_2O \rightleftharpoons [Na(H_2O)_n]^+$$

to release a proton from one of its associated water molecules. The cations that are hydrated only in aqueous solution (Table 18.2) are from elements that fall on the left in the periodic table, coming from Representative Groups I and II, and from the lanthanides and actinides. They do not attract the polar water molecules very strongly. (These cations generally form ionic hydroxides that are strong electrolytes and yield strongly alkaline solutions.)

b. Anions react with water to give alkaline solutions. Anions of this type are stronger bases than water. They attract protons more strongly than the water molecules. These anions are the winners in the competition between bases for the protons. By taking protons from the water molecules they leave behind additional OH^- ions that make the solution alkaline.

$$A^- + H_2O \rightleftharpoons HA + OH^- \tag{3}$$
$$\text{base}_1 \quad \text{acid}_2 \quad \text{acid}_1 \quad \text{base}_2$$

Some specific examples are the following three anions:

anions that are stronger bases than water
$F^- + H_2O(l) \rightleftharpoons$	$HF(aq) + OH^-$	pK_b	10.82
$CN^- + H_2O(l) \rightleftharpoons$	$HCN(aq) + OH^-$	pK_b	4.80
$S^{2-} + H_2O(l) \rightleftharpoons$	$HS^- + OH^-$	pK_b	1.5

As you can see from the pK_b values for F^-, CN^-, and S^{2-}, there is a wide variation in the strengths of these anionic bases.

You should recognize the anions that are stronger bases than water (Table 18.3) as the conjugate bases of weak acids (see Table 17.2). This is not surprising. In a weak acid the anion holds onto the proton strongly—that is why ionization occurs to only a small extent. Because the acid yields only a limited number of ions in aqueous solution, it is a weak acid. When such an anion is set free in solution, the strength of its attraction for a proton is great enough to split a water molecule. The anion is a Brønsted-Lowry, proton-accepting base, and water is the acid. We have listed in Table 18.3 anions that are stronger bases than water in the order of increasing base strength (see Table 17.2).

c. Cations react with water to give acidic solutions. The most common example of a cation that gives up a proton to a water molecule to yield an acidic solution is the ammonium ion. The ammonium ion and all other cations that give acidic solutions are stronger acids than water.

cation that is a stronger acid than water
$$NH_4^+ + H_2O \rightleftharpoons NH_3 + H_3O^+ \qquad pK_a \ 9.20 \qquad (4)$$
$$acid_1 \qquad base_2 \qquad base_1 \qquad acid_2$$

Several hydrated metal cations give acidic aqueous solutions (Table 18.4).

hydrated cations that are stronger acids than water
$$[M(H_2O)_n]^{x+} + H_2O \rightleftharpoons [M(OH)(H_2O)_{n-1}]^{(x-1)+} + H_3O^+ \qquad (5)$$
$$acid_1 \qquad base_2 \qquad base_1 \qquad acid_2$$

In hydrated metal cations, the electrons on the negative, oxygen ends

of the water molecules are attracted to the positively charged metal ion. (The arrows indicate donation of electron pairs.) Such attraction results in general displacement of electron density toward the metal ion and enhances the polarity of the oxygen—hydrogen bonds in the water molecule (Section 13.1) exactly as for an EOH acid (Section 17.3).

Any property of the metal ion that increases its attraction for electrons favors this displacement and increases the polarity of the O—H bond. The most important of these properties are small radius and large positive charge—properties that are increasingly evident in metal ions further to the right in the periodic table. As a consequence, cations such as $[Cu(H_2O)_4]^{2+}$, $[Al(H_2O)_6]^{3+}$, and $[Fe(H_2O)_6]^{3+}$ behave as acids toward the water molecule (pK_a 8, 5, and 2.4, respectively). For example, the aluminum ion reacts with water in the following way:

$$[Al(H_2O)_6]^{3+} + H_2O(l) \rightleftharpoons [Al(H_2O)_5(OH)]^{2+} + H_3O^+$$
$$acid_1 \qquad base_2 \qquad base_1 \qquad acid_2$$

For simplicity, the bound water is often omitted in writing equations.

$$Al^{3+} + H_2O(l) \rightleftharpoons [AlOH]^{2+} + H^+$$

Cations such as $[Na(H_2O)_n]^+$ and $[Ca(H_2O)_n]^{2+}$ (pK_a 13) are too weakly acidic to react significantly with water (see Table 18.2).

TABLE 18.4

Some common cations that are stronger acids than water

$$NH_4^+ + H_2O(l) \rightleftharpoons NH_3 + H_3O^+$$
$$[M(H_2O)_n]^{x+} + H_2O(l) \rightleftharpoons$$
$$[M(OH)(H_2O)_{n-1}]^{(x-1)+} + H_3O^+$$

NH_4^+
The following cations in their hydrated forms:
$$Be^{2+} \quad Zn^{2+} \quad Al^{3+} \quad Fe^{3+} \quad Cr^{3+}$$

TABLE 18.5

Reactions of ions with H_2O

Examples of the three possibilities. Water of hydration is omitted.

Hydration only; give neutral solutions	Give alkaline solutions	Give acidic solutions
Anions that are weaker bases than H_2O, such as ClO_4^- I^- Br^- Cl^- ClO_3^- NO_3^- Cations that are weaker acids than H_2O, such as **Representative Group I cations** (Li^+, Na^+, . . .) **Representative Group II cations** (Mg^{2+}, Ca^{2+}, . . .) **Lanthanide and actinide cations** (La^{3+}, Gd^{3+}, U^{4+}, . . .)	Anions that are stronger bases than H_2O, such as NO_2^- F^- $HCOO^-$ CH_3COO^- CN^- S^{2-} OH^-	Small, highly charged cations that are stronger acids than H_2O, such as NH_4^+ Be^{2+} Zn^{2+} Al^{3+} Fe^{3+} Cr^{3+}

TABLE 18.6

Classification of salts with some examples

Salts of cations that are weaker acids than water and anions that are weaker bases than water.

NaCl   CaCl$_2$ MgBr$_2$ Ba(NO$_3$)$_2$

Salts of cations that are weaker acids than water and anions that are stronger bases than water.

Ba(CH$_3$COO)$_2$ K$_2$(CO$_3$)   NaF   KCN

Salts of cations that are stronger acids than water and anions that are weaker bases than water.

NH$_4$Cl   NH$_4$NO$_3$ Fe(NO$_3$)$_3$ AlCl$_3$

Salts of cations that are stronger acids than water and anions that are stronger bases than water.

NH$_4$(CH$_3$COO)   NH$_4$NO$_2$

Table 18.5 summarizes the reactions of various types of ions with water. The reactions of ions or ionic compounds with water to give acidic or alkaline solutions have traditionally been called *hydrolysis reactions.* "Hydrolysis" is a general term for reactions in which the water molecule is split. The **hydrolysis of an ion** is the reaction of an ion with water to give either H_3O^+ or OH^-, plus whatever is the reaction product formed by the ion.

18.3   The behavior of salts toward water

Keeping in mind the various possibilities for the behavior of individual ions toward water, we can now examine what happens when salts are dissolved in water. We classify salts according to how their ions are related to water as Brønsted-Lowry acids and bases. This classification allows us to predict whether the solution formed will be acidic, alkaline, or neutral. Table 18.6 lists the four categories of salts with some examples.

(An alternative approach to the classification of salts is based on the Arrhenius definition of acids and bases. For example, sodium hydroxide in this system is a strong base, hydrochloric acid is a strong acid, and therefore sodium chloride, NaCl, is classified as a salt of a strong base and a strong acid. Sodium acetate, Na(CH$_3$COO), is thus viewed as a salt of a strong base and a weak acid; ammonium chloride, NH$_4$Cl, is viewed as a salt of a weak base and a strong acid; and so on.)

(a) *Salts of cations that are weaker acids than water and anions that are weaker bases than water.* Since neither ion reacts significantly with water, only hydration of ions occurs and the solution is neutral.

Salts exhibiting such behavior are the perchlorates, chlorates, nitrates, chlorides, bromides, and iodides of the ions of Representative Groups I and II (except beryllium) and of the lanthanide and actinide series of elements (see Table 18.5).

(b) *Salts of cations that are weaker acids than water and anions that are stronger bases than water.* Only the anions react with water to a significant extent. Therefore, the resulting solutions are alkaline, with the anion reacting according to Equation (3). Some salts of this type are the carbonates, sulfides, and cyanides of the cations listed in paragraph (a). (See also the first and second columns of Table 18.5.)

(c) *Salts of cations that are stronger acids than water and anions that are weaker bases than water.* Only the cations react with water to a significant extent, giving up protons to form H_3O^+. Therefore, the solutions are acidic. For example, the ammonium ion from ammonium nitrate reacts according to Equation (4), and the Fe^{3+} ion in iron(III) nitrate reacts according to Equation (5). Other salts that behave similarly are the perchlorates, chlorates, nitrates, chlorides, bromides, and iodides of most metal ions *other* than those mentioned in paragraph (a).

(d) *Salts of cations that are stronger acids than water and anions that are stronger bases than water.* Except for certain ammonium salts and salts that are quite insoluble, the reactions of salts of this type with water go to completion. For example, aluminum sulfide forms only two products, aluminum hydroxide and hydrogen sulfide, leaving few ions in solution.

$$Al_2S_3(s) + 6H_2O(l) \longrightarrow 2Al(OH)_3(s) + 3H_2S(g)$$

Chromium(III) carbonate behaves similarly, leaving a solution with very few ions present.

$$Cr_2(CO_3)_3(s) + 3H_2O(l) \longrightarrow 2Cr(OH)_3(s) + 3CO_2(g)$$

A limited number of ammonium salts containing anions more basic than water will dissolve in water to give essentially neutral solutions. The cations and anions of these salts react to about the same extent with water, which is an indication that the ions are about equal in strength, one as an acid and the other as a base. For example, ammonium acetate, $NH_4(CH_3COO)$, with ions of about equal pK values,

$$NH_4^+ + H_2O(l) \rightleftharpoons NH_3(aq) + H_3O^+ \qquad pK_a\ 9.20 \qquad (6)$$
$$CH_3COO^- + H_2O(l) \rightleftharpoons CH_3COOH(aq) + OH^- \qquad pK_b\ 9.24 \qquad (7)$$

gives just about neutral solutions.

EXAMPLE 18.1

■ Predict whether aqueous solutions of the following salts would be acidic, alkaline, or approximately neutral: (a) KCl, (b) NH_4NO_2, (c) $AlBr_3$, (d) Na(HCOO).

(a) KCl is the salt of a cation, K^+, that is a weaker acid than water and an anion, Cl^-, that is a weaker base than water. Because neither ion reacts with water to any appreciable extent, the KCl solution is neutral.

(b) NH_4NO_2 is the salt of a cation and an anion both of which react with water. The pH of a solution of this salt will depend upon the relative extent of reaction of the two ions with water. For such a salt, the solution will be slightly acidic or slightly alkaline, depending upon the relative values of K_a and K_b. (In this case it will be slightly acidic because $K_a > K_b$.)

(c) $AlBr_3$ is the salt of a cation that will react with water because it is a stronger acid than water and an anion that will not react because it is a weaker base than water. Therefore, the solution will be acidic.

(d) Na(HCOO) is the salt of a cation that is a weaker acid than water and an anion that is a stronger base than water, the formate ion. The solution produced will therefore have an excess of OH^- ion and will be alkaline. ■

18.4 Equilibrium constants for the reactions of ions with water

The equilibrium constants for the reactions of ions with water are simply K_a's and K_b's for acids and bases that happen to be ions. For example, for the reaction of an anion that is a stronger base than water

$$A^- + H_2O(l) \rightleftharpoons HA(aq) + OH^-$$
$$K_b = \frac{[HA][OH^-]}{[A^-]}$$

Specifically, for acetate ion as an example,

$$CH_3COO^- + H_2O(l) \rightleftharpoons CH_3COOH(aq) + OH^-$$

Since for a base, CH_3COO^- in this case, and its conjugate acid, CH_3COOH, $K_w = K_a K_b$ (Section 17.9), the K_b for acetate ion is

$$K_b = \frac{K_w}{K_a} = \frac{[CH_3COOH][OH^-]}{[CH_3COO^-]} \tag{8}$$

This type of relationship is particularly useful because K_a values are more likely to be found tabulated than are K_b values.

EXAMPLE 18.2

■ The K_a for acetic acid is 1.754×10^{-5}. What is the equilibrium constant for the reaction of acetate ion with water?

$$K_b = \frac{K_w}{K_a} = \frac{1.00 \times 10^{-14}}{1.754 \times 10^{-5}} = 5.70 \times 10^{-10}$$

For acetate ion, the equilibrium constant for its reaction with water as a base is $K_b = 5.70 \times 10^{-10}$. ∎

Similarly, for an acidic cation like NH_4^+ [Equation (4)]

$$K_a = \frac{K_w}{K_b} = \frac{[H^+][NH_3]}{[NH_4^+]} \tag{9}$$

where K_b is the ionization constant of the base formed by the reaction of the cation with water [e.g., NH_3 in Equation (4)].

18.5 pH of salt solutions

The relationships of Equations (8) and (9) can be used to find the pH of salt solutions in which only one ion reacts with water.

EXAMPLE 18.3

∎ What are the OH^- ion concentration and the pH of an 0.10M KCN solution? For HCN, $K_a = 6.2 \times 10^{-10}$.

For this salt of a cation that is a weaker acid than water and an anion that is a stronger base than water, the only reaction of importance involves the cyanide ion. Consequently, we would expect the solution to be alkaline.

$$CN^- + H_2O(l) \rightleftharpoons HCN(aq) + OH^-$$
$$K_b = \frac{[HCN][OH^-]}{[CN^-]} = \frac{K_w}{K_a} = \frac{1.00 \times 10^{-14}}{6.2 \times 10^{-10}} = 1.6 \times 10^{-5}$$

Letting $x = [OH^-]$, we set up our usual equilibrium problem table.

	$CN^- +$	$H_2O \rightleftharpoons$	HCN +	OH^-
initial	0.10			
change	$-x$		$+x$	$+x$
equilibrium	$0.10 - x$		x	x

Substitution into the equilibrium expression gives

$$K_b = \frac{(x)(x)}{0.10 - x} \approx \frac{x^2}{0.10} = 1.6 \times 10^{-5}$$
$$x = 1.3 \times 10^{-3} \text{ mole/liter} = [OH^-]$$

The approximation $0.10 - (1.3 \times 10^{-3}) \approx 0.10$ is valid, and the OH^- ion concentration is 1.3×10^{-3} mole/liter. Next we find the pOH and the pH

$$pOH = -\log [OH^-] = -\log (1.3 \times 10^{-3}) = 2.89$$
$$pH = 14.00 - pOH = 11.11$$

The pH of the solution is 11.11. As we predicted, the solution is alkaline. ■

EXAMPLE 18.4

■ What are the H^+ ion concentration and the pH of a solution $0.10M$ in NH_4Cl? K_b for NH_3 is 1.6×10^{-5}.

For this salt, only the ammonium ion will react with water. Therefore we would expect the solution to be acidic.

$$NH_4^+ + H_2O(l) \rightleftharpoons NH_3(aq) + H_3O^+$$

or simply

$$NH_4^+ \rightleftharpoons NH_3(aq) + H^+$$
$$K_a = \frac{[NH_3][H^+]}{[NH_4^+]} = \frac{K_w}{K_b} = \frac{1.00 \times 10^{-14}}{1.6 \times 10^{-5}} = 6.3 \times 10^{-10}$$

Letting $x = [H^+]$, we set up our usual table.

	NH_4^+	$\rightleftharpoons$ NH_3	$+$ H^+
initial	0.10		
change	$-x$	$+x$	$+x$
equilibrium	$0.10 - x$	x	x

Substitution into the equilibrium expression gives

$$K_a = \frac{x^2}{0.10 - x} \approx \frac{x^2}{0.10} = 6.3 \times 10^{-10}$$
$$x = 7.9 \times 10^{-6} \text{ mole/liter} = [H^+]$$

Checking, we find that the approximation is valid since $0.10 - (7.9 \times 10^{-6}) \approx 0.10$. Therefore

$$pH = -\log (7.9 \times 10^{-6}) = 5.10$$

The pH of the solution is 5.10. The solution is acidic, as we predicted. ■

When both ions in a salt react with water, as in a salt of a cation that is a stronger acid than water and an anion that is a stronger base than water, the situation is a bit more complicated. In a solution of ammonium acetate, $NH_4(CH_3COO)$, both the ammonium and the acetate ions react with water [Equations (6) and (7)]. In addition, H^+ from the NH_4^+/H_2O reaction combines with OH^- from the CH_3COO^-/H_2O reaction. The overall reaction is therefore the sum of three reactions

$$NH_4^+ \rightleftharpoons NH_3 + \cancel{H^+} \tag{10}$$
$$CH_3COO^- + \cancel{H_2O} \rightleftharpoons CH_3COOH + \cancel{OH^-} \tag{11}$$
$$\cancel{H^+} + \cancel{OH^-} \rightleftharpoons \cancel{H_2O} \tag{12}$$
$$\overline{NH_4^+ + CH_3COO^- \rightleftharpoons NH_3(aq) + CH_3COOH(aq)} \tag{13}$$

and the equilibrium constant for Reaction (13) is equal to the product of the equilibrium constants of Reactions (10), (11), and (12).

We have shown that for an acidic cation like ammonium ion [Equation (9)]

$$K_{a,NH_4^+} = \frac{K_w}{K_{b,NH_3}}$$

and for a basic anion like acetate ion [Equation (8)]

$$K_{b,CH_3COO-} = \frac{K_w}{K_{a,CH_3COOH}}$$

We also know that for Equation (12), which is the reverse of the ionization of water, the equilibrium constant is $1/K_w$. Therefore, for the reaction of ammonium acetate with water, [Equation (13)],

$$K_{13} = \frac{[NH_3][CH_3COOH]}{[NH_4^+][CH_3COO^-]} = K_{a,NH_4+} \times K_{b,CH_3COO-} \times \frac{1}{K_w}$$
$$= \frac{K_w}{K_{b,NH_3}} \times \frac{K_w}{K_{a,CH_3COOH}} \times \frac{1}{K_w} = \frac{K_w}{K_{b,NH_3}K_{a,CH_3COOH}} \tag{14}$$

You can verify Equation (14) by writing out the expression in full in terms of the concentrations of the ions. Equation (14) is the overall equilibrium expression for the reaction of ammonium acetate with water. Similar expressions apply to the hydrolysis of any salt of a weak acid and a weak base.

EXAMPLE 18.5

■ What is the pH of an $0.010M$ solution of NH_4CN? The equilibria involved are the following:

$$NH_4^+ \rightleftharpoons NH_3 + \cancel{H^+}$$
$$CN^- + \cancel{H_2O} \rightleftharpoons HCN + \cancel{OH^-}$$
$$\cancel{H^+} + \cancel{OH^-} \rightleftharpoons \cancel{H_2O}$$
$$\overline{NH_4^+ + CN^- \rightleftharpoons NH_3(aq) + HCN(aq)}$$

To find the pH, we must (a) find the value of K for the reaction of NH_4CN with water; (b) find $[NH_3]$ (or $[HCN]$) from this overall equilibrium constant; and (c) use $[NH_3]$ (or $[HCN]$) in the K_b for NH_3 (or in the K_a for HCN) to find $[H^+]$.

(a) $K = \dfrac{[NH_3][HCN]}{[NH_4^+][CN^-]} = \dfrac{K_w}{K_{a,HCN} K_{b,NH_3}}$

$\qquad = \dfrac{1.00 \times 10^{-14}}{(6.2 \times 10^{-10})(1.6 \times 10^{-5})}$

$\qquad = 1.0$

(b) We can assume that the differences in the extent of reaction of NH_4^+ and CN^- with water are slight, and that the numbers of NH_4^+ and CN^- ions hydrolyzed are nearly equal. Therefore, let $x = [NH_3] = [HCN]$ and $(0.010 - x) = [NH_4^+] = [CN^-]$. Solving

$$K = \frac{x^2}{(0.010 - x)(0.010 - x)} = 1.0$$

$$\frac{x}{0.010 - x} = \sqrt{1.0}$$

$$x = (0.010 - x) = \sqrt{1}$$

$$x = 0.005 \ \text{mole/liter} = [NH_3] = [HCN]$$

(c) Choosing to find $[H^+]$ from $K_{a,HCN}$, and knowing that $[HCN] = 0.005$ and $[CN^-] = 0.010 - x = 0.010 - 0.005 = 0.005$,

$$K_{a,HCN} = \frac{[H^+][CN^-]}{[HCN]}$$

$$= \frac{[H^+](0.005)}{(0.005)} = 6.2 \times 10^{-10}$$

$$[H^+] = 6 \times 10^{-10} \ \text{mole/liter}$$

$$pH = -\log (6 \times 10^{-10}) = 9.2$$

A $0.01M$ solution of NH_4CN has pH 9.2. We might have predicted that this solution is slightly alkaline, because the acid strength of HCN ($K_a = 6.2 \times 10^{-10}$) is somewhat less than the base strength of NH_3 ($K_b = 1.6 \times 10^{-5}$). ∎

18.6 Reactions of acids with bases

When an acid reacts with a base, the product is a salt. The pH of the resulting solution depends upon the relative acid and base strengths of the ions present in the aqueous solution of that salt. The situation is simple for the reaction of a strong base with a strong acid. Consider the exact neutralization of 0.5 mole of KOH in solution by 0.5 mole of HCl in solution. There is only one reaction to worry about here, that is, the acid-base neutralization to form the salt KCl

$$KOH(aq) + HCl(aq) \longrightarrow KCl(aq) + H_2O(l) \tag{15}$$
$$\text{0.5 mole} \quad \text{0.5 mole} \quad \text{0.5 mole}$$

The net ionic equation for this reaction is

$$OH^- + H_3O^+ \rightleftharpoons 2H_2O(l)$$

or simply

$$OH^- + H^+ \rightleftharpoons H_2O(l) \qquad K = \frac{1}{K_w} = \frac{1}{1.00 \times 10^{-14}} = 1.00 \times 10^{14}$$

The large value of K shows that the reaction is essentially complete as written. Note that K has this same value for any strong acid–strong base reaction.

When Reaction (15) is complete we have an aqueous solution of 0.5 mole of KCl, or a solution containing 0.5 mole of K^+ ions and 0.5 mole of Cl^- ions. These ions undergo no appreciable reaction with water and the pH of the solution is just about equal to 7. When a strong acid and a strong base react in quantities that are not stoichiometrically equivalent, the pH of the resulting solution is, of course, determined by the reactant that is present in the greater amount.

Next, let's look at some reactions of weak acids and strong bases:

$$CH_3COOH(aq) + NaOH(aq) \longrightarrow Na(CH_3COO)(aq) + H_2O(l)$$
$$HCN(aq) + NaOH(aq) \longrightarrow NaCN(aq) + H_2O(l)$$
$$H_2S(aq) + 2LiOH(aq) \longrightarrow Li_2S(aq) + 2H_2O(l)$$

A general ionic equation for such reactions is

$$\underset{acid}{HA(aq)} + \underset{base}{OH^-} \longrightarrow \underset{base}{A^-} + \underset{acid}{H_2O(l)}$$

As the large number of OH^- ions available from the soluble ionic hydroxide react with H^+ from the ionization of the weak acid, this ionization is driven to completion by the formation of water, a slightly ionized substance. We know that the conjugate bases of these weak acids are stronger bases than water. Because cations such as Na^+ and Li^+ are weaker acids than water, they undergo no reaction in aqueous solution. Therefore, from what was presented in the preceding section, we can expect these solutions to be alkaline due to the reaction of the anions with water.

The situation for the neutralization of a strong acid by a weak base is similar to that for a weak acid and a strong base. Using our most common weak base, ammonia, in solution,

$$NH_3(aq) + HCl(aq) \longrightarrow NH_4Cl(aq)$$
$$2NH_3(aq) + H_2SO_4(aq) \longrightarrow (NH_4)_2SO_4(aq)$$

we see that solutions of salts of NH_4^+, which is more acidic than water, and the weakly basic conjugate bases of the strong acids are produced. Because NH_4^+ reacts with water to yield H^+ and the anions will not react appreciably, these solutions will be acidic. A general equation for the reaction of ammonia with a strong acid is

$$NH_3(aq) + H_3O^+ \longrightarrow NH_4^+ + H_2O(l) \tag{16}$$

This reaction is the reverse of the reaction of ammonium ion with water, for which we found $K_a = 6.3 \times 10^{-10}$ in Example 18.4. Therefore, the equilibrium constant for Reaction (16) is $1/K_{a,\mathrm{NH_4^+}} = 1.6 \times 10^9$. Here again the reaction goes just about to completion. The pH of the solution resulting from a strong acid–weak base reaction is calculated as illustrated in Example 18.8.

EXAMPLE 18.6

■ What is the pH of the mixture prepared by adding 50.0 ml of $0.250M$ HI to 50.0 ml of $0.100M$ KOH?

Combining the acid and base results in neutralization

$$HI(aq) + KOH(aq) \rightleftharpoons KI(aq) + H_2O(l)$$

The number of moles of acid added is

$$n = \left(0.250\ \frac{\mathrm{mole}}{\mathrm{liter}}\right)(50.0\ \mathrm{ml})\left(\frac{1\ \mathrm{liter}}{1000\ \mathrm{ml}}\right) = 0.0125\ \mathrm{mole\ HI}$$

and the number of moles of base is

$$n = \left(0.100\ \frac{\mathrm{mole}}{\mathrm{liter}}\right)(50.0\ \mathrm{ml})\left(\frac{1\ \mathrm{liter}}{1000\ \mathrm{ml}}\right) = 0.00500\ \mathrm{mole\ KOH}$$

The chemical equation for the neutralization process shows that the acid and base react in a 1-to-1 mole ratio, so there will be an excess of HI

$$(0.0125\ \mathrm{mole\ HI}) - (0.00500\ \mathrm{mole\ KOH})\left(\frac{1\ \mathrm{mole\ HI}}{1\ \mathrm{mole\ KOH}}\right)$$
$$= 0.0075\ \mathrm{mole\ HI}$$

Assuming the volume of the mixture to be 100.0 ml, the concentration of HI is

$$[HI] = \frac{(0.0075\ \mathrm{mole})}{(100.0\ \mathrm{ml})(1\ \mathrm{liter}/1000\ \mathrm{ml})} = 0.075\ \frac{\mathrm{mole}}{\mathrm{liter}}$$

Because HI is a strong acid,

$$[H^+] = [HI] = 0.075M$$

and

$$pH = -\log [H^+] = -\log (0.075) = 1.12$$

The pH of the mixture is 1.12. ■

EXAMPLE 18.7

■ A solution of $0.1M$ formic acid, HCOOH, is mixed with an equal volume of $0.1M$ NaOH solution. Find the equilibrium constant for the neutralization reaction and describe the resulting solution. $K_b = 5.643 \times 10^{-11}$ for $HCOO^-$.

The equation and equilibrium expression for the neutralization reaction are

$$HCOOH(aq) + OH^- \rightleftharpoons HCOO^- + H_2O(l)$$

$$K = \frac{[HCOO^-]}{[HCOOH][OH^-]}$$

Note that this reaction is simply the reverse of the reaction of the basic formate ion with water

$$HCOO^- + H_2O(l) \rightleftharpoons HCOOH(aq) + OH^-$$

and so K for the neutralization process is

$$K = \frac{1}{K_b} = \frac{1}{5.643 \times 10^{-11}} = 1.772 \times 10^{10}$$

Because the equilibrium constant K is so large, the neutralization reaction is essentially complete, meaning that the concentration of Na^+ is $0.05M$ (remember that we have doubled the volume of the solution) and the concentration of $HCOO^-$ is $0.05M$. Because Na(HCOO) is the salt of a cation that does not react with water plus an anion that reacts with water as a base, the solution formed in this reaction will be somewhat alkaline. ■

EXAMPLE 18.8

■ What is the pH of the resulting solution after 25.0 ml of $0.010M$ aniline, $C_6H_5NH_2$, has reacted with 25.0 ml of $0.010M$ HNO_3? $K_b = 4.2 \times 10^{-10}$ for $C_6H_5NH_2$.

The equation and equilibrium expression for the neutralization reaction are

$$H_3O^+ + C_6H_5NH_2(aq) \rightleftharpoons C_6H_5NH_3^+ + H_2O(l)$$

or simply

$$H^+ + C_6H_5NH_2(aq) \rightleftharpoons C_6H_5NH_3^+$$

and

$$K = \frac{[C_6H_5NH_3^+]}{[H^+][C_6H_5NH_2]}$$

Note that this reaction is the reverse of the reaction in which $C_6H_5NH_3^+$ acts as an acid. Therefore, $K = 1/K_a$. The value of K_a is obtained from K_w and K_b by

$$K_a = \frac{K_w}{K_b} = \frac{1.00 \times 10^{-14}}{4.2 \times 10^{-10}} = 2.4 \times 10^{-5}$$

Thus for the neutralization reaction

$$K = \frac{1}{K_a} = \frac{1}{2.4 \times 10^{-5}} = 4.2 \times 10^4$$

Because K is large, the neutralization reaction is essentially complete. Consequently, we have 50.0 ml of a solution containing

$$(25.0 \text{ ml}) \left(\frac{1 \text{ liter}}{1000 \text{ ml}}\right)\left(\frac{0.010 \text{ mole}}{\text{liter}}\right) = 2.5 \times 10^{-4} \text{ mole of } C_6H_5NH_3^+$$

or

$$(2.5 \times 10^{-4} \text{ mole}) \left(\frac{1}{50.0 \text{ ml}}\right)\left(\frac{1000 \text{ ml}}{\text{liter}}\right) = 0.0050 \frac{\text{mole}}{\text{liter}} \text{ of } C_6H_5NH_3^+$$

and also the same amount of NO_3^-. The nitrate ion does not react with water, but the cation does. To find the pH we first find H^+ by our usual method.

$$C_6H_5NH_3^+ \rightleftharpoons C_6H_5NH_2(aq) + H^+$$

	$C_6H_5NH_3^+$	$C_6H_5NH_2(aq)$	H^+
initial	0.0050		
change	$-x$	$+x$	$+x$
equilibrium	$0.0050 - x$	x	x

$$K_a = \frac{[C_6H_5NH_2][H^+]}{[C_6H_5NH_3^+]} = \frac{(x)(x)}{(0.0050 - x)} \approx \frac{x^2}{0.0050} = 2.4 \times 10^{-5}$$

$x = [H^+] = 3.3 \times 10^{-4}$ or pH = 3.48. The solution is acidic. ∎

18.7 Polyprotic acids

A polyprotic acid contains more than one ionizable hydrogen. Such an acid ionizes or undergoes neutralization in separate steps for each proton, with different equilibrium constants for each step. Sulfuric acid, for example, ionizes in two steps:

$$H_2SO_4(aq) + H_2O(l) \longrightarrow H_3O^+ + HSO_4^- \qquad \text{complete in dilute solution}$$
$$HSO_4^- + H_2O(l) \rightleftharpoons H_3O^+ + SO_4^{2-} \qquad K_{a_2} = 1.0 \times 10^{-2} \quad (pK_{a_2} \ 2.00)$$

In its first ionization, H_2SO_4 is a strong acid and readily releases its proton. The second proton, from the ion HSO_4^-, is released with much greater difficulty. For all polyprotic acids successive protons are released less readily because each proton is held more strongly by the increasingly negative anion. This effect is clearly illustrated by the pK_a's in Table 18.7.

In this section we will examine the equilibrium relationships for sulfuric acid; phosphoric acid, a triprotic acid; and hydrosulfuric acid, which is the aqueous solution of hydrogen sulfide gas (H_2S).

Sulfuric acid. The first stage of ionization of sulfuric acid, the most common diprotic acid, is essentially complete in dilute solution. The hydrogen sulfate, or bisulfate ion (HSO_4^-), however, is a weak acid. The effect of the hydrogen sulfate ion alone on pH can be seen in its salts with weakly acidic cations. Salts such as $NaHSO_4$, $NaHCO_3$, KH_2PO_4 that retain one or more of the hydrogen atoms from the anions of polyprotic acids are called **acid salts.**

(This seems a good place to point out that in the Examples and Exercises in this book we deal mainly with ions in solution at the concentrations normally encountered in laboratory work—roughly between $0.01M$ and $1M$. To be accurate about calculations for either very concentrated or very dilute solutions additional factors must be taken into account. For example, in a very dilute solution of a weak acid, the contribution of the ionization of water cannot be ignored. However, this creates the problem of solving four equations in four unknowns.)

TABLE 18.7

pK_a *values for polyprotic acids*

Hydrosulfuric	
H_2S	7.00
HS^-	12.5
Sulfuric	
H_2SO_4	Negative
HSO_4^-	2.00
Carbonic	
H_2CO_3	6.35
HCO_3^-	10.32
Arsenic	
H_3AsO_4	2.19
$H_2AsO_4^-$	6.96
$HAsO_4^{2-}$	11.5
Pyrophosphoric	
$H_4P_2O_7$	0.91
$H_3P_2O_7^-$	1.10
$H_2P_2O_7^{2-}$	6.70
$HP_2O_7^{3-}$	9.32
Phosphoric	
H_3PO_4	2.12
$H_2PO_4^-$	7.18
HPO_4^{2-}	12.0

EXAMPLE 18.9

■ Calculate the pH of a $0.10M$ solution of $NaHSO_4$. K_a for $HSO_4^- = 1.0 \times 10^{-2}$.

We are concerned with the equilibrium

$$HSO_4^- + H_2O(l) \rightleftharpoons H_3O^+ + SO_4^{2-}$$

or simply

$$HSO_4^- \rightleftharpoons H^+ + SO_4^{2-} \qquad K_a = \frac{[H^+][SO_4^{2-}]}{[HSO_4^-]}$$

Let $x = [H^+] = [SO_4^{2-}]$.

	$HSO_4^- \rightleftharpoons$	$H^+ +$	SO_4^{2-}
initial	0.10		
change	$-x$	$+x$	$+x$
equilibrium	$0.10 - x$	x	x

$$\frac{(x)(x)}{(0.10 - x)} = 1.0 \times 10^{-2}$$
$$x^2 = (1.0 \times 10^{-2})(0.10 - x)$$

Putting the result in quadratic form because the concentration and K values are not greatly different in magnitude,

$$x^2 + (1.0 \times 10^{-2}x) - (1.0 \times 10^{-3}) = 0$$

Solving the quadratic equation, we find that

$$x = 0.027 \text{ mole/liter}$$

Therefore, $[H^+] = 0.027M$, $[SO_4^{2-}] = 0.027M$, and

$$pH = -(\log 2.7 \times 10^{-2}) = 1.57$$

The solution is quite acidic. In fact, $NaHSO_4$ is so acidic that it is used to clean toilet bowls. ∎

In a $0.1M$ solution of H_2SO_4 the concentration of SO_4^{2-} ion is substantially less than the $0.027M$ concentration found in the $0.1M$ $NaHSO_4$ solution. The ionization of the HSO_4^- formed from H_2SO_4 is repressed by the H^+ ion also formed in the first stage ionization of the acid (an illustration of Le Chatelier's principle, Section 16.7; see also Section 18.8).

EXAMPLE 18.10

∎ Find $[H^+]$, $[HSO_4^-]$, $[H_2SO_4]$, and $[SO_4^{2-}]$ in equilibrium in a $0.100M$ solution of H_2SO_4. $K_{a_2} = 1.0 \times 10^{-2}$.

Because $H_2SO_4(aq)$ is a strong acid in the first ionization step,

$$H_2SO_4(aq) + H_2O(l) \longrightarrow H_3O^+ + HSO_4^-$$

we can safely state that $[H_2SO_4] \approx 0$ and that $[H_3O^+]$, or simply, $[H^+] = [HSO_4^-] = 0.100M$ before the second ionization step occurs. For the second step, letting $x = [SO_4^{2-}]$

	HSO_4^-	$\rightleftharpoons$	H^+	+	SO_4^{2-}
initial	0.100		0.100		
change	$-x$		$+x$		$+x$
equilibrium	$0.100 - x$		$0.100 + x$		x

$$K_{a_2} = \frac{[H^+][SO_4^{2-}]}{[HSO_4^-]} = \frac{(0.100 + x)(x)}{(0.100 - x)}$$
$$= 1.0 \times 10^{-2}$$

This equation must be solved by the exact method, using the quadratic equation, because the concentrations and K_{a_2} are of similar magnitude. Solving for x gives $[SO_4^{2-}] = 8.4 \times 10^{-3}M$. Thus

$$[HSO_4^-] = 0.100 - x = 0.100 - 0.0084 = 0.092M$$

and

$$[H^+] = 0.100 + x = 0.100 + 0.0084 = 0.108M \qquad \blacksquare$$

Hydrosulfuric acid. Hydrosulfuric acid, H_2S, is a weak diprotic acid, the ionization of which yields H^+ ion and HS^- (hydrogen sulfide ion) in the first step and H^+ ion and S^{2-} (sulfide ion) in the second step.

$$H_2S(aq) + H_2O(l) \rightleftharpoons H_3O^+ + HS^- \qquad K_{a_1} = 1.0 \times 10^{-7} \qquad (pK_{a_1}\ 7.00)$$
$$HS^- + H_2O(l) \rightleftharpoons H_3O^+ + S^{2-} \qquad K_{a_2} = 3 \times 10^{-13} \qquad (pK_{a_2}\ 12.5)$$

The second ionization constant of the acid is so much smaller than the first that in calculating the hydrogen ion concentration in a solution of the acid alone, the second step ionization may be neglected, and $[H^+]$ and $[HS^-]$ equated to each other.

EXAMPLE 18.11

■ A saturated aqueous hydrosulfuric acid solution at 1 atm pressure and room temperature has a concentration of approximately $0.1M$ H_2S. Find the concentration of all of the ions present in this solution. For H_2S, $K_{a_1} = 1.0 \times 10^{-7}$; $K_{a_2} = 3 \times 10^{-13}$.

For the first ionization, letting $x = [H^+] = [HS^-]$,

$$H_2S \rightleftharpoons H^+ + HS^-$$

initial	0.1		
change	$-x$	$+x$	$+x$
equilibrium	$0.1 - x$	x	x

$$K_{a_1} = \frac{[H^+][HS^-]}{[H_2S]} = \frac{(x)(x)}{(0.1 - x)} \approx \frac{x^2}{0.1} = 1.0 \times 10^{-7}$$

Making the approximation that $(0.1 - x) \approx 0.1$, since K_{a_1} is very small,

$$x^2 = 1.0 \times 10^{-8}$$
$$x = 1.0 \times 10^{-4}M = [H^+] = [HS^-]$$

The value found for the concentrations of H^+ and HS^- ions must satisfy the second step equilibrium $HS^- \rightleftharpoons H^+ + S^{2-}$. Because K_{a_2} is much smaller than K_{a_1}, we assume that the contribution to $[H^+]$ from this equilibrium is negligible. Therefore, the S^{2-} ion concentration may be calculated as follows:

$$K_{a_2} = \frac{[H^+][S^{2-}]}{[HS^-]}$$

$$= \frac{(1.0 \times 10^{-4})[S^{2-}]}{(1.0 \times 10^{-4})} = 3 \times 10^{-13}$$

$$[S^{2-}] = 3 \times 10^{-13} M$$

In a $0.1M$ aqueous H_2S solution the concentrations of the ions present are

$$[H^+] = [HS^-] = 1.0 \times 10^{-4} M$$
$$[S^{2-}] = 3 \times 10^{-13} M$$ ∎

From the preceding Example we can conclude that in a solution of H_2S alone the molar concentration of S^{2-} ion is equal to the value of K_{a_2}. Indeed *for any weak diprotic acid, the molar concentration of the ion formed by the loss of both protons is equal to the second ionization constant.*

Overall K_a values for polyprotic acids can be found from the K_a values for individual steps. For example, multiplying the equilibrium constant expressions for the two ionization steps of aqueous H_2S gives

$$K_a(\text{overall}) = \frac{[H^+][HS^-]}{[H_2S]} \times \frac{[H^+][S^{2-}]}{[HS^-]} = (1.0 \times 10^{-7})(3 \times 10^{-13})$$

$$= \frac{[H^+]^2[S^{2-}]}{[H_2S]} = 3 \times 10^{-20}$$

which represents the reaction

$$H_2S(aq) + 2H_2O(l) \rightleftharpoons 2H_3O^+ + S^{2-}$$

For a $0.1M$ saturated solution

$$\frac{[H^+]^2[S^{2-}]}{(0.1)} = 3 \times 10^{-20}$$

$$[H^+]^2[S^{2-}] = 3 \times 10^{-21}$$

This relationship *cannot be used for solutions containing only* H_2S. In such solutions, each mole of H_2S that ionizes does *not* give two moles of H^+ ions and one mole of S^{2-} ions, because only a very small fraction of the HS^- ions formed in the first step of ionization are themselves ionized. The equation is, however, very useful when the H^+ ion concentration of a saturated H_2S solution is fixed by the addition of a strong acid such as HCl or of a base such as aqueous NH_3. *Then the* S^{2-} ion concentration is inversely proportional to the square of the H^+ ion concentration. Solutions of controlled sulfide ion concentration are important in the precipitation and separation of metal sulfides in qualitative analysis (Sections 18.12, 22.16, and 30.10).

EXAMPLE 18.12

■ To what pH must a $0.01M$ H_2S solution be adjusted so that $[S^{2-}] = 1 \times 10^{-9}M$? K_a(overall) $= 3 \times 10^{-20}$ for H_2S.

The equation describing the ionization of H_2S can be written as

$$H_2S(aq) + 2H_2O(l) \rightleftharpoons 2H_3O^+ + S^{2-}$$

or simply

$$H_2S(aq) \rightleftharpoons 2H^+ + S^{2-}$$

and the equilibrium expression is

$$K_a = \frac{[H^+]^2[S^{2-}]}{[H_2S]}$$

Solving for $[H^+]$ and substituting $[S^{2-}] = 1 \times 10^{-9}M$ and $[H_2S] = 0.01M$ gives

$$[H^+] = \sqrt{\frac{K_a[H_2S]}{[S^{2-}]}} = \sqrt{\frac{(3 \times 10^{-20})(0.01)}{1 \times 10^{-9}}} = 5 \times 10^{-7}M$$

which corresponds to a pH value of

$$\begin{aligned} pH &= -\log[H^+] = -\log(5 \times 10^{-7}) \\ &= 6.3 \end{aligned}$$ ■

Phosphoric acid. Phosphoric acid is a triprotic acid that releases protons in three steps.

$$\begin{array}{lll} H_3PO_4(aq) \rightleftharpoons H^+ + H_2PO_4^- & K_{a_1} = 7.5 \times 10^{-3} & (pK_{a_1}\ 2.12) \\ H_2PO_4^- \rightleftharpoons H^+ + HPO_4^{2-} & K_{a_2} = 6.6 \times 10^{-8} & (pK_{a_2}\ 7.18) \\ HPO_4^{2-} \rightleftharpoons H^+ + PO_4^{3-} & K_{a_3} = 1 \times 10^{-12} & (pK_{a_3}\ 12.0) \end{array}$$

EXAMPLE 18.13

■ Calculate the concentration of all species (except H_2O) present in a $0.010M$ solution of H_3PO_4.

The species present are H_3PO_4, H^+, $H_2PO_4^-$, HPO_4^{2-}, and PO_4^{3-}. The concentrations of H_3PO_4, $H_2PO_4^-$ and the hydrogen ion formed from H_3PO_4 are determined from the first ionization reaction

	H_3PO_4	$\rightleftharpoons H^+$	$+ H_2PO_4^-$
initial	0.010		
change	$-x$	$+x$	$+x$
equilibrium	$0.010 - x$	x	x

Using the K_{a_1} value given just before this Example

$$K_{a_1} = \frac{[H^+][H_2PO_4^-]}{[H_3PO_4]} = \frac{(x)(x)}{(0.010 - x)} = 7.5 \times 10^{-3}$$

The first-step ionization constant is too large to allow us to disregard the breakdown of H_3PO_4. Thus, the quadratic equation must be used.

$$x^2 + (7.5 \times 10^{-3}x) - (7.5 \times 10^{-5}) = 0$$
$$x = 5.7 \times 10^{-3}M = [H^+] = [H_2PO_4^-]$$
$$[H_3PO_4] = 0.010 - x = 0.010 - 5.7 \times 10^{-3} = 4.3 \times 10^{-3}$$

The concentration of the HPO_4^{2-} is now calculated from the second-step ionization equilibrium, which depends upon $[H^+]$ and $[H_2PO_4^-]$ from the first step.

	$H_2PO_4^{2-}$	$\rightleftharpoons$	H^+	+	HPO_4^{2-}
initial	5.7×10^{-3}		5.7×10^{-3}		
change	$-x$		$+x$		$+x$
equilibrium	$(5.7 \times 10^{-3}) - x$		$(5.7 \times 10^{-3}) + x$		x

$$K_{a_2} = \frac{[H^+][HPO_4^{2-}]}{[H_2PO_4^-]} = \frac{[(5.7 \times 10^{-3}) + x](x)}{[(5.7 \times 10^{-3}) - x]} = 6.6 \times 10^{-8}$$

The value of x added to or subtracted from 5.7×10^{-3} can be disregarded. The very low value of K_{a_2} indicates that the extent of dissociation is so small that the initial concentrations of H^+ and $H_2PO_4^-$ ions are not changed significantly. Therefore,

$$\frac{(5.7 \times 10^{-3})x}{5.7 \times 10^{-3}} = 6.6 \times 10^{-8}$$

$$x = 6.6 \times 10^{-8}M = [HPO_4^{2-}]$$

The result—that the concentration of HPO_4^{2-} ion is equal to that of the ionization constant—is not surprising. (Compare with the S^{2-} ion concentration from the second-step ionization of H_2S.)

Finally, we find the concentration of PO_4^{3-} from the third step of the ionization of phosphoric acid, which is determined by $[H^+]$ from the first step and $[HPO_4^{2-}]$ from the second step. Letting $x = [PO_4^{3-}]$

	HPO_4^{2-}	$\rightleftharpoons$	H^+	+	PO_4^{3-}
initial	6.6×10^{-8}		5.7×10^{-3}		
change	$-x$		$+x$		$+x$
equilibrium	$(6.6 \times 10^{-8}) - x$		$(5.7 \times 10^{-3}) + x$		x

Here again we can neglect the effect of x on the initial concentrations.

$$K_{a_3} = \frac{[H^+][PO_4{}^{3-}]}{[HPO_4{}^{2-}]} = 1 \times 10^{-12}$$

$$\frac{(5.7 \times 10^{-3})(x)}{6.6 \times 10^{-8}} = 1 \times 10^{-12}$$

$$x = 1 \times 10^{-17} \text{ mole/liter} = [PO_4{}^{3-}]$$

To summarize,

$[H_3PO_4] = 4.3 \times 10^{-3}M$ $[HPO_4{}^{2-}] = 6.6 \times 10^{-8}$

$[H^+] = [H_2PO_4{}^-] = 5.7 \times 10^{-3}M$ $[PO_4{}^{3-}] = 1 \times 10^{-17}$ ∎

Not only do the $H_2PO_4{}^-$ and $HPO_4{}^{2-}$ ions function as acids in aqueous solution, but they are amphiprotic and also behave as bases in the following reactions with water:

$$H_2PO_4{}^- + H_2O(l) \rightleftharpoons H_3PO_4(aq) + OH^-$$
$$HPO_4{}^{2-} + H_2O(l) \rightleftharpoons H_2PO_4{}^- + OH^-$$

The basic strengths of these ionic species can be calculated from the relationship $K_w = K_a K_b$ (Section 17.10).

$$K_{b,H_2PO_4{}^-} = \frac{[H_3PO_4][OH^-]}{[H_2PO_4{}^-]} = \frac{K_w}{K_{a_1}} = \frac{1.00 \times 10^{-14}}{7.5 \times 10^{-3}} = 1.3 \times 10^{-12}$$

$$K_{b,HPO_4{}^{2-}} = \frac{[H_2PO_4{}^-][OH^-]}{[HPO_4{}^{2-}]} = \frac{K_w}{K_{a_2}} = \frac{1.00 \times 10^{-14}}{6.6 \times 10^{-8}} = 1.5 \times 10^{-7}$$

The $HPO_4{}^{2-}$ ion, although weakly basic, is a considerably stronger base than $H_2PO_4{}^-$. Comparison of K_b and K_a values shows that the $HPO_4{}^{2-}$ ion is a stronger base than it is an acid and that the reverse is true for the $H_2PO_4{}^-$ ion.

The phosphate ion, $PO_4{}^{3-}$, is by far the strongest base of the three anionic species.

$$PO_4{}^{3-} + H_2O(l) \rightleftharpoons HPO_4{}^{2-} + OH^-$$

$$K_{b,PO_4{}^{3-}} = \frac{[HPO_4{}^{2-}][OH^-]}{[PO_4{}^{3-}]} = \frac{K_w}{K_{a_3}} = \frac{1.00 \times 10^{-14}}{1.00 \times 10^{-12}} = 1 \times 10^{-2}$$

Salts such as Na_3PO_4, formed by the anion from the last stage of ionization of a polyprotic acid, are called **normal salts.** Aqueous solutions of phosphates, for example, Na_3PO_4, are good cleaning agents because of their alkalinity.

18.8 The common ion effect

In the previous section we pointed out that although sulfuric acid, H_2SO_4, is completely ionized to H^+ and $HSO_4{}^-$ ions in dilute aqueous solution, the ionization of $HSO_4{}^-$ is repressed by the H^+ formed in the first step of the ionization of H_2SO_4.

$$\overset{\longleftarrow{\scriptstyle H^+ \text{ from first step}}}{\underset{\text{in } H_2SO_4 \text{ ionization}}{}}$$
$$HSO_4^- + H_2O(l) \rightleftharpoons H_3O^+ + SO_4^{2-}$$

As a result, the concentration of SO_4^{2-} in a solution of H_2SO_4 is considerably less than that obtained from a solution of a salt containing HSO_4^- in the same molar concentration as the acid (see Examples 18.9 and 18.10). This type of repression of ionization is an illustration of the **common ion effect**—a displacement of an ionic equilibrium by an excess of one or more of the ions involved. This effect is another example of a situation in which LeChatelier's principle applies (Section 16.7).

The common ion effect is frequently used to control the pH of solutions of weak acids or bases. Consider the addition of ammonium chloride to a solution of ammonia

$$\overset{\longleftarrow{\scriptstyle NH_4^+}}{}$$
$$NH_3(aq) + H_2O(l) \rightleftharpoons NH_4^+ + OH^-$$

The NH_4^+ from the salt increases the total concentration of NH_4^+ and drives the equilibrium toward the formation of NH_3. The chloride ion is too weak a base to increase the OH^- concentration noticeably, and it does not affect the equilibrium. Therefore, the solution becomes less alkaline because of the reaction of OH^- with NH_4^+.

The addition of sodium acetate, $Na(CH_3COO)$, to an acetic acid solution similarly drives the equilibrium toward the formation of acetic acid

$$\overset{\longleftarrow{\scriptstyle CH_3COO^-}}{}$$
$$CH_3COOH(aq) + H_2O(l) \rightleftharpoons CH_3COO^- + H_3O^+$$

The Na^+ ion, a weakly acidic cation, does not yield any H^+ and therefore does not affect the equilibrium. When a salt of the conjugate base of a weak acid [for example, $Na(CH_3COO)$] is added to a solution of that acid, the result is a change in the pH of the solution. Similarly, the pH is changed when a salt of the conjugate acid of a weak base is added to a solution of that base. The concentration of the acid or base increases and the concentration of H^+ or OH^- decreases.

By rearranging the K_a expression for a weak acid

$$K_a = \frac{[H^+][A^-]}{[HA]}$$

$$[H^+] = \frac{[HA]}{[A^-]} K_a \tag{17}$$

it can be seen that the ratio of the concentration of a free acid to the concentration of the ion that is its conjugate base determines the pH of the solution. The addition of a conjugate base will always decrease the hydrogen ion concentration.

EXAMPLE 18.14

■ What are the [H⁺] and pH of $0.1M$ acetic acid that is $0.5M$ in sodium acetate?

First, we set up the usual equilibrium problem table. Let $x = [H^+]$. The final acetate ion concentration is the amount formed by acetic acid ionization plus the amount added as sodium acetate, since sodium acetate is soluble and is a strong electrolyte.

$$CH_3COOH \rightleftharpoons CH_3COO^- + H^+$$

initial	0.1	0.5	
change	$-x$	$+x$	$+x$
equilibrium	$0.1 - x$	$0.5 + x$	x

These values can be used in Equation (17)

$$[H^+] = \frac{[CH_3COOH]}{[CH_3COO^-]} \times 1.75 \times 10^{-5}$$

$$x = \frac{0.1 - x}{0.5 + x} (1.75 \times 10^{-5})$$

Since the ionization of acetic acid is decreased by the added acetate ion, we can assume that $0.1 - x$ is approximately 0.1 and therefore that x must also be negligibly small relative to 0.5. Thus

$$x = \frac{0.1}{0.5} (1.75 \times 10^{-5})$$

$$= 4 \times 10^{-6} \text{ mole/liter} = [H^+]$$

The approximation is valid, since $0.1 - (4 \times 10^{-6}) \approx 0.1$ and $0.5 + (4 \times 10^{-6}) \approx 0.5$. Next we find the pH of the solution.

$$pH = -\log (4 \times 10^{-6}) = 5.4$$

The pH of $0.1M$ acetic acid–$0.5M$ sodium acetate solution is 5.4. Compare this to a pH of 2.9 for an $0.1M$ acetic acid solution. The common ion effect has decreased the acidity of the solution. ■

EXAMPLE 18.15

■ What is the pH of a solution which is $0.10M$ in NH_3 and $0.10M$ in NH_4NO_3? $K_b = 1.6 \times 10^{-5}$ for NH_3.

Letting $x = [OH^-]$, we can write

$$K_b = \frac{[NH_4^+][OH^-]}{[NH_3]}$$

IONS AND IONIC
EQUILIBRIA [18.8]

$$NH_3(aq) + H_2O(l) \rightleftharpoons NH_4^+ + OH^-$$

initial	0.10	0.10	
change	$-x$	$+x$	$+x$
equilibrium	$0.10 - x$	$0.10 + x$	x

$$K_b = \frac{(0.10 + x)(x)}{(0.10 - x)} = 1.6 \times 10^{-5}$$

$$x = 1.6 \times 10^{-5} = [OH^-]$$

The pOH of this solution is

$$pOH = -\log [OH^-] = -\log (1.6 \times 10^{-5})$$
$$= 4.80$$

and the pH is

$$pH = 14.00 - pOH$$
$$= 14.00 - 4.80 = 9.20$$ ■

When several equilibria are established simultaneously in the same solution, each must incorporate the same concentrations of whatever ions they have in common. We have already used this principle in the preceding section for the simultaneous equilibria in the ionization of polyprotic acids, where, for example, the ion concentrations from the first ionization equilibrium determine the second equilibrium concentrations.

EXAMPLE 18.16

■ When chlorine gas is dissolved in water to make "chlorine water," the strong acid HCl and the weak acid HClO are produced in equal amounts

$$Cl_2(g) + H_2O(l) \rightleftharpoons HCl(aq) + HOCl(aq)$$

If 0.01 mole of each acid is formed in 1 liter of solution, what is the concentration of ClO^-? $K_a = 2.90 \times 10^{-8}$ for HClO.

The common ion in this case is H^+, and ClO^- is unknown. The HCl is a strong acid, so $[H^+]$ is equal to the initial [HCl].

$$HClO(aq) \rightleftharpoons H^+ + ClO^-$$

$$K_a = \frac{[H^+][ClO^-]}{[HClO]} \qquad \text{Let } x = [ClO^-]$$

$$HClO(aq) \rightleftharpoons H^+ + ClO^-$$

initial	0.01	0.01	
change	$-x$	$+x$	$+x$
equilibrium	$0.01 - x$	$0.01 + x$	x

$$K_a = \frac{(0.01 + x)(x)}{(0.01 - x)} = 2.90 \times 10^{-8}$$

Assuming that $(0.01 + x)$ and $(0.01 - x)$ are both approximately equal to 0.01, we have

$$x = [ClO^-] = 2.90 \times 10^{-8} \text{ mole/liter}$$

The approximations are valid, so the concentration of ClO^- is 2.90×10^{-8} mole/liter. ∎

Later (Section 18.15) we will examine the application of the common ion effect to reactions in which salts are precipitated or dissolved.

18.9 Buffer solutions

A **buffer solution** is a solution that resists changes in pH when small amounts of acid or base or water are added to it. Any solution of a weak acid or base plus a salt of its conjugate base or conjugate acid is a buffer solution. For example, in an $HA + M^+A^-$ buffer, added acid can react with the relatively large amount of basic anion (A^-) present

$$H^+ + A^- \longrightarrow HA$$

and added base can react with the relatively large amount of undissociated acid (HA) present

$$OH^- + HA \longrightarrow A^- + H_2O$$

The combination of a weak acid plus a salt of its conjugate base (e.g., acetic acid/sodium acetate) or of a weak base plus a salt of its conjugate acid (such as ammonia/ammonium chloride) is known as a *buffer pair*.

To demonstrate how a buffer works, we first calculate the pH of a buffer solution of 0.100M acetic acid solution that is also 0.100M in potassium acetate. Using Equation (17) and the approximations introduced in Example 18.14,

$$[H^+] = \frac{[CH_3COOH]}{[CH_3COO^-]} \times 1.75 \times 10^{-5} \qquad (18)$$

$$= \frac{0.100}{0.100} \times 1.75 \times 10^{-5}$$

$$= 1.75 \times 10^{-5} \text{ mole/liter}$$

$$pH = -\log (1.75 \times 10^{-5}) = 4.757$$

If 0.005 mole of potassium hydroxide (assume the volume does not change) is added to 1 liter of this solution, what happens? The added base neutralizes 0.005 mole of acetic acid to give

$$[CH_3COOH] = (0.100 - 0.005) \text{ mole/liter} = 0.095 \text{ mole/liter}$$

and 0.005 mole of acetate ion is formed, to give

$$[CH_3COO^-] = (0.100 + 0.005) \text{ mole/liter} = 0.105 \text{ mole/liter}$$

The resulting [H+], from Equation (18), is

$$[H^+] = \frac{0.095}{0.105} \times (1.75 \times 10^{-5})$$
$$= 1.58 \times 10^{-5} \text{ mole/liter}$$
$$pH = -\log(1.58 \times 10^{-5}) = 4.801$$

The addition of 0.005 mole of KOH to the buffer solution has changed the pH by only 0.044 unit. The pH of pure water would have changed from 7.00 to 9.30 if the same amount of KOH were added. A similar calculation for the addition of 0.005 mole of hydrochloric acid to 1 liter of this buffer solution shows a decrease in pH from 4.757 to 4.713. A series of similar calculations for the addition of acid and base to this system (Table 18.8) further illustrates its resistance to pH change.

EXAMPLE 18.17

■ What ratio of [HCOO−]/[HCOOH] is needed to make a sodium formate–formic acid buffer solution of pH 3.80? $K_a = 1.77 \times 10^{-4}$ for formic acid, HCOOH.

The equilibrium is

$$HCOOH(aq) + H_2O(l) \rightleftharpoons H_3O^+ + HCOO^-$$

A pH of 3.80M corresponds to [H₃O+], or simply, [H+] = 1.6×10^{-4} mole/liter. Using Equation (17)

$$[H^+] = \frac{[HCOOH]}{[HCOO^-]}(1.77 \times 10^{-4})$$
$$\frac{[HCOO^-]}{[HCOOH]} = \frac{1.77 \times 10^{-4}}{1.6 \times 10^{-4}} = 1.1$$

The [HCOO−]/[HCOOH] ratio must be 1.1 to attain a pH of 3.80 with this buffer system. For example, if 1M HCOOH is used, the solution must be 1.1M in sodium formate. ■

Anion/weak acid buffer pairs give solutions buffered in the acid pH range. A common ion/weak base buffer pair is used for alkaline

TABLE 18.8
Effects of added potassium hydroxide and hydrochloric acid on the pH of 0.100M CH₃COOH–0.100 M K(CH₃COO) solution

KOH added (mole/liter)	[H+] (mole/liter)	pH
0	1.75×10^{-5}	4.757
0.0001	1.75×10^{-5}	4.758
0.001	1.72×10^{-5}	4.766
0.005	1.58×10^{-5}	4.801
0.010	1.43×10^{-5}	4.844
0.050	0.583×10^{-5}	5.234

HCl added (mole/liter)	[H+] (mole/liter)	pH
0	1.75×10^{-5}	4.757
0.0001	1.75×10^{-5}	4.756
0.001	1.79×10^{-5}	4.748
0.005	1.93×10^{-5}	4.713
0.010	2.14×10^{-5}	4.670
0.050	5.25×10^{-5}	4.280

buffers. Both the pH range in which a buffer is effective and the capacity of a buffer to resist changes in pH when hydrogen ion or hydroxide ion is added are limited. The pH range can be calculated from the K_a or K_b value of the acid or base present in the buffer. Effective acid buffer solutions have values of the ratio [HA]/[A$^-$] between 1/10 and 10/1.

A relationship between pK_a and pH for a buffer can be derived from Equation (17).

$$[H^+] = \frac{[HA]}{[A^-]} K_a$$

$$-\log [H^+] = -\log K_a - \log \frac{[HA]}{[A^-]}$$

$$pH = pK_a - \log \frac{[HA]}{[A^-]} \tag{19}$$

Equation (19) is known as the Henderson-Hasselbach equation. The effective pH range of an acid buffer lies between

$$pH = pK_a - \log\left(\frac{10}{1}\right) = pK_a - 1$$

and

$$pH = pK_a - \log\frac{1}{10} = pK_a + 1$$

A buffer is best able to maintain a constant pH in the range of $pH = pK_a \pm 1$. In order to maintain a specified pH, an acid with a pK_a in this range is selected. Note that for buffers in which the acid and anion concentrations are equal, $pH = pK_a$. For a base of the type

$$B + H_2O \rightleftharpoons BH^+ + OH^-$$

the comparable relationship is

$$pOH = pK_b - \log \frac{[B]}{[BH^+]} \tag{20}$$

EXAMPLE 18.18

■ What is the pH of the $0.10M$ NH_3/$0.10M$ NH_4NO_3 solution described in Example 18.15 after (a) 0.10 g of solid NaOH or (b) 0.0010 mole of HCl has been added to 100.0 ml of the buffer?

(a) The additional NaOH

$$\frac{(0.10 \text{ g NaOH})}{(40.0 \text{ g/mole})} = 2.5 \times 10^{-3} \text{ mole NaOH}$$

$$pH = pKa - \log \frac{[HA]}{[A^-]}$$

reacts with the $(0.10 \text{ mole/liter})(0.1000 \text{ liter}) = 0.010$ mole of NH_4^+ according to the following stoichiometry:

$$NH_4^+ \quad + \quad OH^- \longrightarrow NH_3(aq) + H_2O(l)$$
$$\text{0.010 } mole \qquad \text{0.0025 } mole$$

leaving

$$(0.010 \text{ mole } NH_4^+) - (0.0025 \text{ mole } OH^-)\left(\frac{1 \text{ mole } NH_4^+}{1 \text{ mole } OH^-}\right)$$
$$= 0.008 \text{ mole } NH_4^+$$

unreacted in 100.0 ml of the buffer or $[NH_4^+] = 0.08M$, and

$$(0.010 \text{ mole } NH_3) + (0.0025 \text{ mole } OH^-)\left(\frac{1 \text{ mole } NH_3}{1 \text{ mole } OH^-}\right)$$
$$= 0.013 \text{ mole } NH_3$$

in 100.0 ml of the buffer, or $[NH_3] = 0.13M$. Thus at equilibrium, letting $x = [OH^-]$

	$NH_3(aq) + H_2O(l) \rightleftharpoons$	NH_4^+	$+ OH^-$
initial	0.13	0.08	
change	$-x$	$+x$	$+x$
equilibrium	$0.13 - x$	$0.08 + x$	x

$$K_b = \frac{[NH_4^+][OH^-]}{[NH_3]} = \frac{(0.08 + x)(x)}{(0.13 - x)} = 1.6 \times 10^{-5}$$
$$x = 2.6 \times 10^{-5}M = [OH^-]$$
$$pOH = -\log{[OH^-]} = -\log{(2.6 \times 10^{-5})} = 4.59$$
$$pH = 14.00 - pOH = 14.00 - 4.59 = 9.41$$

The pH has changed from 9.20 to 9.41.

(b) The additional 0.0010 mole of HCl reacts with the 0.010 mole of NH_3 according to the following stoichiometry:

$$NH_3(aq) \quad + \quad H_3O^+ \longrightarrow NH_4^+ + H_2O(l)$$
$$\text{0.010 } mole \qquad \text{0.0010 } mole$$

leaving

$$(0.010 \text{ mole } NH_3) - (0.0010 \text{ mole } H^+)\left(\frac{1 \text{ mole } NH_3}{1 \text{ mole } H^+}\right) = 0.009 \text{ mole } NH_3$$

unreacted in 100.0 ml of the buffer, or $[NH_3] = 0.09M$, and

$$(0.010 \text{ mole } NH_4^+) + (0.0010 \text{ mole } H^+)\left(\frac{1 \text{ mole } NH_4^+}{1 \text{ mole } H^+}\right)$$
$$= 0.011 \text{ mole } NH_4^+$$

in 100.0 ml of the buffer, or $[NH_4^+] = 0.11M$. Thus at equilibrium, letting $x = [OH^-]$

	$NH_3(aq) + H_2O(l) \rightleftharpoons$	NH_4^+	$+ OH^-$
initial	0.09	0.11	
change	$-x$	$+x$	$+x$
equilibrium	$0.09 - x$	$0.11 + x$	x

$$K_b = \frac{[NH_4^+][OH^-]}{[NH_3]} = \frac{(0.11 + x)(x)}{(0.09 - x)} = 1.6 \times 10^{-5}$$

$$x = 1.3 \times 10^{-5}M = [OH^-]$$
$$pOH = -\log [OH^-] = -\log (1.3 \times 10^{-5}) = 4.89$$
$$pH = 14.00 - pOH = 14.00 - 4.89 = 9.11$$

The pH has changed from 9.20 to 9.11.　　　　　　　　■

EXAMPLE 18.19

■ What must be the concentration of fluoride ion in a NaF/HF buffer to give pH 4.00? Assume that solid NaF is added to a $0.10M$ solution of HF without volume change. $K_a = 6.5 \times 10^{-4}$ for HF.

Although this problem can be solved in our usual manner, we instead make use of the Henderson-Hasselbach equation, Equation (19)

$$pH = pK_a - \log \frac{[HF]}{[F^-]}$$

Substituting $[HF] = 0.10M$ and

$$pK_a = -\log K_a = -\log (6.5 \times 10^{-4}) = 3.19$$

gives

$$4.00 = 3.19 - \log \frac{(0.10)}{[F^-]}$$

Solving gives

$$\log \frac{(0.10)}{[F^-]} = 3.19 - 4.00 = -0.81$$

$$\frac{(0.10)}{[F^-]} = 0.15$$

$$[F^-] = \frac{(0.10)}{(0.15)} = 0.67M$$

The concentration of F^- must be adjusted to 0.67 mole/liter.　　　■

The pH of normal human blood is about 7.4, maintained chiefly by the buffer pair H_2CO_3/HCO_3^-. The consequences of even small changes in blood pH are extremely serious. The metabolic disturbance called acidosis sets in when the blood pH drops below about 7.3. Respiratory problems, kidney failure, or poisoning with acid compounds such as aspirin (acetysalicylic acid) can induce acidosis. Coma usually results if the blood pH falls to 7.0. Alkalosis is the condition caused by a rise in the pH of blood to about 7.5 or greater. Rapid breathing decreases the CO_2 concentration in the blood and leads to alkalosis, as does the ingestion of excessive amounts of alkaline substances such as sodium bicarbonate or antacids. A blood pH of about 7.8 usually causes convulsions, cramps, and muscle twitching.

The H_2CO_3/HCO_3^- buffer system operates through the equilibria

$$CO_2 + H_2O \rightleftharpoons H_2CO_3(aq)$$
$$H_2CO_3(aq) + H_2O(l) \rightleftharpoons H_3O^+ + HCO_3^-$$

As small amounts of hydrogen ion come into the bloodstream they combine with HCO_3^- to form the weak acid H_2CO_3. Small amounts of hydroxide ion combine with H_2CO_3 to form water and HCO_3^-. Unreacted CO_2 molecules serve as a source for regeneration of H_2CO_3.

In pure water a saturated CO_2 solution at 1 atm pressure and 25°C contains about 0.034 mole/liter of CO_2 gas, *most* of which is simply dissolved in the water. Only approximately one out of every 400 CO_2 molecules in solution reacts with a water molecule to give H_2CO_3. This acid, as we have pointed out, cannot be isolated, but does exist in small concentrations in aqueous CO_2 solutions. In writing the first ionization constant expression for H_2CO_3, we take into account both the dissolved CO_2 and the small amount of CO_2 converted to H_2CO_3 by using the sum of their concentrations as the reactant concentration. That is,

$$K_1 = \frac{[H^+][HCO_3^-]}{[CO_2] + [H_2CO_3]} = 4.5 \times 10^{-7}$$

(Sometimes $[H_2CO_3]$ alone is written in the denominator; in such cases it represents $[CO_2] + [H_2CO_3]$ as described here.) The hydrogen carbonate ion has, as expected, an even smaller ionization constant,

$$K_{HCO_3^-} = \frac{[H^+][CO_3^{2-}]}{[HCO_3^-]} = 4.8 \times 10^{-11}$$

(The buffering action of the HCO_3^-/CO_3^{2-} pair is unimportant in blood chemistry.)

When CO_2 dissolves in the blood, *carbonic anhydrase*, a zinc-containing enzyme (a biological catalyst) present in red blood cells, catalyzes the reaction of the gas with water to form H_2CO_3. Carbonic anhydrase is among the fastest acting enzymes, converting tens of millions of moles of CO_2 to H_2CO_3 each minute. The equilibrium between H_2CO_3 and HCO_3^- is established very rapidly. In normal blood (pH ~ 7.4), the concentration of HCO_3^- ion is much greater than the total concentration of CO_2 and H_2CO_3.

EXAMPLE 18.20

■ Given

$$K_1 = \frac{[H^+][HCO_3^-]}{[CO_2] + [H_2CO_3]} = 4.5 \times 10^{-7}$$

calculate the ratio of the HCO_3^- concentration to the sum of the CO_2 and H_2CO_3 concentrations in normal blood (pH 7.4).

Solving the above equation for $[H^+]$ gives

$$[H^+] = K_1 \times \frac{[CO_2] + [H_2CO_3]}{[HCO_3^-]}$$

and taking negative logarithms gives

$$pH = pK_1 - \log \frac{[CO_2] + [H_2CO_3]}{[HCO_3^-]}$$

The value of pK_1 is

$$pK_1 = -\log 4.5 \times 10^{-7} = 6.35$$

Therefore

$$\log \frac{[HCO_3^-]}{[CO_2] + [H_2CO_3]} = pH - pK_1 = 7.4 - 6.35 = 1.0$$

$$\frac{[HCO_3^-]}{[CO_2] + [H_2CO_3]} = 10$$

The $[HCO_3^-]$ to $[CO_2] + [H_2CO_3]$ ratio in blood is about 10.   ■

18.10 Titration of acids and bases

Titration of an acid and a base is a technique used to determine accurately the concentration of an acidic or an alkaline solution. The practical need for such information arises frequently in medical laboratories and in industrial laboratories.

Suppose we have 50 ml of a $0.10M$ solution of HCl. The number of moles of HCl present is

$$n_{HCl} = M_{HCl} \times V_{HCl} = \left(0.1\ \frac{mole}{liter}\right)\left(\frac{1\ liter}{1000\ ml}\right)(50\ ml) = 0.005\ mole \quad (21)$$

From the stoichiometry of the reaction between HCl and a base, say NaOH,

$$HCl(aq) + NaOH(aq) \longrightarrow NaCl(aq) + H_2O(l)$$

we know that when exactly 0.005 mole of NaOH has been added to the original solution containing 0.005 mole of HCl, the acid will be exactly neutralized. The point of exact neutralization in an acid-base reaction is called the equivalence point. For any type of reaction the **equivalence point** is the point at which chemically equivalent amounts of the reactants have been mixed. For the HCl solution we started with, the equivalence point could be reached by the addition of 50 ml of $0.1M$ NaOH, or 25 ml of $0.2M$ NaOH, or 100 ml of $0.05M$ NaOH, or any other combination of concentration and volume that equals 0.005 mole of NaOH.

Titration is the measurement of the volume of a solution of one reactant that is required to react completely with a measured amount of another reactant. Frequently both reactants are in solution, and the titration is the measurement of the volume of one solution that must be added to a known volume of the other solution. Usually the concentration of one solution—a standard solution—is known.

For example, if the HCl solution described above was of unknown concentration, we could take exactly 50 ml of that solution and gradually add a standard $0.10M$ NaOH solution until the equivalence point was reached. This would occur when 50 ml of the base was added, which would indicate that the original solution contained an equal molar amount of the acid and was therefore $0.10M$.

In Chapter 4 we pointed out that equivalents in chemical reactions and in solutions whose concentrations are expressed in normality are used mainly in connection with neutralization reactions. In the titration of an acid with a base, the volumes of the solutions that are needed for complete reaction and the concentrations of the solutions are related as follows:

at the equivalence point
equivalents of acid = equivalents of base

$$\left(\text{acid concentration, } \frac{\text{equiv}}{\text{liter}}\right) (\text{acid volume, liters})$$

$$= \left(\text{base concentration, } \frac{\text{equiv}}{\text{liter}}\right) (\text{base volume, liters}) \quad (22)$$

For any titration the general relationship is

$$\underset{\substack{\text{normality} \\ \text{of reactant 1 solution}}}{\searrow} N_1 V_1 = N_2 V_2 \underset{\substack{\text{normality} \\ \text{of reactant 2} \\ \text{solution}}}{\swarrow} \quad (23)$$

volume

In a neutralization reaction the pH changes very little until all of the H^+ or OH^- present has reacted with OH^-, or H^+. At the equivalence point a small additional amount of acid, or base, causes a large rapid change in pH. To detect the equivalence point, a way must be found to observe this large change in pH. In Section 17.7 we discussed indicators, compounds that change color at specific pH's. Indicators are available for the entire range of pH (see Table 17.7). An indicator is added to the solution being titrated and the color change carefully watched for. From the volume of solution of known concentration added up to that point, the concentration of the unknown solution is calculated.

The point at which an indicator changes color in a titration is usually referred to as the **end point** of the titration. *Sometimes* the end point and the equivalence point are the same. But more often there is an end point error—a slight discrepancy between the end point and the exact equivalence point. In most titrations this error is considered negligible and is ignored, but you should be aware of the distinction between end point and equivalence point.

Students often assume that the pH at the end point of an acid-base titration must be 7. However, remember that what we have at the end point is an aqueous solution of a salt (unless, of course, the salt is insoluble). If a strong acid has been titrated with a strong base, the equivalence point does indeed occur at pH 7. When a weak acid or a weak base is involved in the titration, however, the pH at the equivalence point is determined by the reaction with water of the ions of the salt formed in the neutralization. Therefore, an indicator that changes color as close as possible to the pH of the equivalence point must be selected.

EXAMPLE 18.21

■ Exactly 23.6 ml of a 0.131N HCl solution was required for complete neutralization of 25.0 ml of an NaOH solution. What was the normality of the NaOH solution?

From the relationship between volume and normality

$$N_1V_1 = N_2V_2$$
$$\left(0.131 \ \frac{equiv}{liter}\right)(23.6 \ ml)\left(\frac{1 \ liter}{1000 \ ml}\right) = N_2(25.0 \ ml)\left(\frac{1 \ liter}{1000 \ ml}\right)$$

Note that both volumes may be expressed in liters or milliliters, because the conversion factor 1 liter/1000 ml appears on both sides of the equation and the 1000 can be canceled out.

$$N_2 = \frac{(0.131)(23.6)}{25.0} = 0.124 \ equiv/liter$$

The NaOH solution is 0.124N. ■

EXAMPLE 18.22

■ An approximately $0.1M$ HCl solution was **standardized**—its exact concentration found—by titrating it against a solution containing 0.1223 g of 99.95% pure Na_2CO_3

$$Na_2CO_3(aq) + 2HCl(aq) \longrightarrow 2NaCl(aq) + H_2O(l) + CO_2(g)$$

If 22.65 ml of the HCl solution was used to reach the equivalence point, what was the exact concentration of the acid?

The number of moles of Na_2CO_3 reacting was

$$moles = \frac{mass}{molar\ wt} = \frac{(0.1223)(0.9995)g}{(105.99\ g/mole)} = 0.001153\ mole\ Na_2CO_3$$

The chemical equation tells us that 2 moles of HCl react for every mole of Na_2CO_3, so the amount of HCl titrated is

$$(0.001153\ \cancel{mole\ Na_2CO_3}) \left(\frac{2\ moles\ HCl}{\cancel{mole\ Na_2CO_3}} \right) = 0.002306\ mole\ HCl$$

The molarity of the solution is

$$molarity = \frac{moles}{liters\ of\ solution} = \frac{0.002306\ mole}{0.02265\ liter} = 0.1018M$$

The concentration of the acid is $0.1018M$. Note that this problem could be solved using equivalents—see Example 18.23 for a typical solution in terms of equivalents. ■

EXAMPLE 18.23

■ A $0.1N$ NaOH solution was standardized by titrating it against pure potassium acid phthalate (commonly abbreviated KHP), a monoprotic acid,

If 19.61 ml of the base neutralized 0.4963 g of the KHP, molar wt = 204.23 g/mole, what is the exact concentration of the base?

The number of equivalents of KHP is found by dividing the mass used by the equivalent weight. Because KHP is monoprotic, the equivalent and molar weights are the same.

$$number\ of\ equiv = \frac{mass}{equiv\ wt.} = \frac{0.4963\ g}{204.23\ g/equiv} = 0.002430\ equiv\ KHP$$

The number of equivalents of NaOH is likewise 0.002430; thus the normality is

$$\text{normality} = \frac{\text{equiv}}{\text{liters of solution}} = \frac{0.002430 \text{ equiv}}{0.01961 \text{ liter}} = 0.1239N$$

The concentration of the base is $0.1239N$. Note that this problem could be solved using moles as was done in Example 18.22. ■

EXAMPLE 18.24

■ As a check on the standardizations described in Examples 18.22 and 18.23, 25.00 ml of HCl was transferred to a flask and titrated with the NaOH. What is the amount of base needed to reach the equivalence point if the standardizations are correct?

As with the previous examples, we may choose to solve this problem using equivalents or moles. Because we are dealing with two sets of concentrations and volumes, the easiest approach is to use equivalents. Solving Equation (22) for the volume of the base and substituting gives us

$$
\begin{aligned}
\text{base volume} &= \frac{(\text{acid concentration, } N)(\text{acid vol, liters})}{(\text{base concentration, } N)} \\
&= \frac{(0.1018 \text{ equiv/liter})(0.02500 \text{ liter})}{(0.1239 \text{ equiv/liter})} \\
&= 0.02054 \text{ liter}
\end{aligned}
$$

Thus we would expect that 20.54 ml of the base would be required. ■

EXAMPLE 18.25

■ A technician in a crime laboratory wanted to determine the barbital content in a sample by titration with NaOH. (Barbital is a sedative drug that can be addictive, and an overdose may lead to coma or death.)

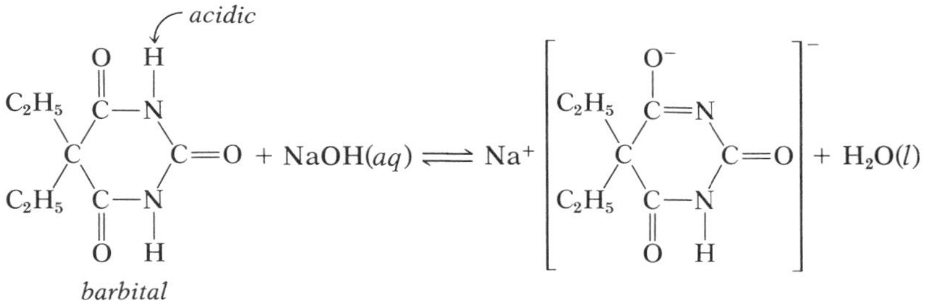

From past experience the technician knew that the concentration of the $NaC_8H_{11}N_2O_3$ at the end point would be approximately $0.001M$. If

$K_a = 3.7 \times 10^{-8}$ for barbital, at what pH would the end point be observed? What would be a good indicator for this titration?

The pH at the end point is determined by the reaction of the anion of the barbital with water

$$C_8H_{11}N_2O_3^- + H_2O(l) \rightleftharpoons C_8H_{12}N_2O_3(aq) + OH^-$$

$$K_b = \frac{[C_8H_{12}N_2O_3][OH^-]}{[C_8H_{11}N_2O_3^-]} = \frac{K_w}{K_a} = \frac{1.00 \times 10^{-14}}{3.7 \times 10^{-8}} = 2.7 \times 10^{-7}$$

Let $x = [OH^-]$.

	$C_8H_{11}N_2O_3^- + H_2O \rightleftharpoons C_8H_{12}N_2O_3 + OH^-$		
initial	0.001		
change	$-x$	$+x$	$+x$
equilibrium	$0.001 - x$	x	x

$$\frac{(x)(x)}{(0.001 - x)} \approx \frac{x^2}{0.001} = 2.7 \times 10^{-7}$$

$$x = [OH^-] = 2 \times 10^{-5} \text{ mole/liter}$$

The approximation $0.001 - x \approx 0.001$ is valid, so

$$pOH = -\log [OH^-] = 4.7$$
$$pH = 14.00 - pOH = 9.3$$

The end point should be near pH 9.3. Table 17.7 shows that a suitable indicator would be either thymol blue or phenolphthalein. ∎

18.11 Titration curves

The change of the pH throughout the course of an acid-base titration can be represented by a titration curve—a plot of pH versus the volume of acid or base added. The curves have different characteristic shapes that depend on whether strong or weak acids and bases are involved.

Figure 18.1 illustrates the titration of strong acid with a strong base. The pH here is simply the result of the amount of unreacted acid present up to the equivalence point and the amount of excess base present after the equivalence point. The pH is 7 at the equivalence point, which is at the center of the vertical part of the curve where the pH change is most rapid. This rapid change of about 8 pH units (the length of the vertical part of the curve) is what must be detected as the end point.

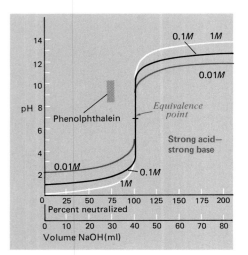

FIGURE 18.1

Strong acid–strong base titration curve. *Titration curve calculated for the titration of 40.0 ml of 0.100M HCl with 0.100M NaOH. The effect of varying dilution is shown by the colored and white curves. Calculated titration curves are useful in finding the range in which an indicator must change color to indicate the end point of a particular titration. In this case, phenolphthalein would serve fairly well as an indicator because the curve is quite vertical through its color change pH range. (Figures 18.1 to 18.3 are based upon Figures 13.1 to 13.3 in D. J. Pietrzyk and C. W. Frank,* Analytical Chemistry, *Academic Press, New York, 1974.)*

EXAMPLE 18.26

■ A 40.0 ml sample of $0.100M$ HCl is titrated with $0.100M$ NaOH. Calculate the pH of the solution (a) before the addition of NaOH and after the addition of (b) 10.0 ml, (c) 20.0 ml, (d) 30.0 ml, (e) 40.0 ml (the equivalence point), and (f) 45.0 ml of NaOH.

(a) The $[H^+]$ of the solution is equal to the initial concentration of the strong acid, or $[H^+] = [HCl] = 0.100M$. Therefore, for the pH

$$pH = -\log [H^+] = -\log (0.100) = 1.000$$

The pH of the solution before the equivalence point has been reached simply depends on how much acid has been neutralized by the base and how much H^+ is in excess in the new volume. Using $n = MV$ [Equation (21)], the original amount of H^+ present was

$$n = MV = \left(0.100 \frac{\text{mole}}{\text{liter}}\right) (40.0 \text{ ml}) \left(\frac{1 \text{ liter}}{1000 \text{ ml}}\right) = 4.00 \times 10^{-3} \text{ mole HCl}$$

The amount of HCl that reacts is based on the stoichiometry

$$H_3O^+ + OH^- \longrightarrow 2H_2O(l)$$

or simply

$$H^+ + OH^- \longrightarrow H_2O(l)$$

giving the excess as

$$n_{H^+} = (4.00 \times 10^{-3} \text{ mole HCl}) - (n_{OH^-}) \left(\frac{1 \text{ mole HCl}}{1 \text{ mole OH}^-}\right)$$

and the concentration of H^+ remaining as

$$[H^+] = \frac{n}{V} = \frac{n_{H^+}}{(40.0 \text{ ml} + V_{OH^-})(10^{-3} \text{ liter/ml})}$$

Note that to save space we have written the conversion factor 1 liter/1000 ml as 10^{-3} liter/ml. In converting milliliters to liters, say 10 ml, we will write 10×10^{-3} liter.

(b) For the addition of 10.0 ml of base

$$n_{OH^-} = MV = (0.100 \text{ mole/liter})(10.0 \times 10^{-3} \text{ liter})$$
$$= 1.00 \times 10^{-3} \text{ mole NaOH}$$

$$n_{H^+} = (4.00 \times 10^{-3} \text{ mole HCl})$$

$$- (1.00 \times 10^{-3} \text{ mole NaOH}) \left(\frac{1 \text{ mole HCl}}{1 \text{ mole OH}^-}\right)$$

$$= 3.00 \times 10^{-3} \text{ mole HCl}$$

$$[H^+] = \frac{3.00 \times 10^{-3} \text{ mole}}{(40.0 \text{ ml} + 10.0 \text{ ml})(10^{-3} \text{ liter/ml})} = 0.0600M$$
$$pH = -\log[H^+] = -\log(0.0600) = 1.222$$

Similar calculations give for the addition of (c) 20.0 ml of base, $n_{OH^-} = 2.00 \times 10^{-3}$ mole, $n_{H^+} = 2.00 \times 10^{-3}$ mole, $[H^+] = 0.0333M$, and pH = 1.477; and (d) for the addition of 30.0 ml of base, $n_{OH^-} = 3.00 \times 10^{-3}$ mole, $n_{H^+} = 1.00 \times 10^{-3}$ mole, $[H^+] = 0.0143$ and pH = 1.845.

(e) At the equivalence point, there is no excess of $[H^+]$ or $[OH^-]$ and the solution will be neutral because NaCl is the salt of ions that do not react with water. Thus the pH is 7.000. The proper indicator should change color somewhat near this pH. Suitable indicators include litmus (4.5 to 8.3), bromthymol blue (6.0 to 7.6), and phenol red (6.4 to 8.2).

(f) The presence of excess OH^- will make the solution alkaline. The amount of NaOH added is

$$n_{OH^-} = (0.100 \text{ mole/liter})(45.0 \times 10^{-3} \text{ liter}) = 4.50 \times 10^{-3} \text{ mole NaOH}$$

and the amount in excess is

$$(4.50 \times 10^{-3} \text{ mole NaOH}) - (4.00 \times 10^{-3} \text{ mole HCl})\left(\frac{1 \text{ mole NaOH}}{1 \text{ mole HCl}}\right)$$
$$= 5.0 \times 10^{-4} \text{ mole NaOH}$$

Thus $[OH^-]$ is

$$\frac{n}{V} = \frac{5.0 \times 10^{-4} \text{ mole}}{(40.0 \text{ ml} + 45.0 \text{ ml})(10^{-3} \text{ liter/ml})} = 0.0053M$$

giving

$$pOH = -\log[OH^-] = -\log(0.0053) = 2.28$$
$$pH = 14.00 - pOH = 14.00 - 2.28 = 11.72$$

Compare these results to the plot shown in Figure 18.1. ∎

The titration of a weak acid with a strong base gives a curve like the one in Figure 18.2. From the addition of the first drops of base and up to the equivalence point, the solution contains a buffer pair—in this case, acetic acid plus the acetate ion formed by the neutralization reaction. In the preceding section we found that a buffer is most effective in the range where $pH = pK_a \pm 1$. The pK_a of acetic acid is 4.76. In Figure 18.2 we can see that the flattest part of the curve before the equivalence point falls roughly in the 4.76 ± 1 range.

In fact, the pK_a of a weak acid can be determined experimentally from pH measurements by using the relationship of Equation (19)

$$pH = pK_a - \log\frac{[HA]}{[A^-]}$$

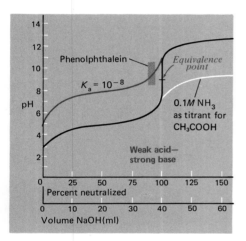

FIGURE 18.2
Weak acid–strong base titration curve. *Titration curve calculated for the titration of 0.100M CH_3COOH with 0.100M NaOH. The equivalence point occurs at an alkaline pH because a solution of $NaCH_3COO$ is present. The colored curve shows that $K_a = 10^{-8}$ is the limit for the weak acids that can be successfully titrated. The white line indicates the shape of the curve for a weak acid–weak base titration; such titrations cannot be followed with indicators.*

At the point where one-half of the acid has been neutralized by the reaction

$$HA + OH^- \longrightarrow H_2O + A^-$$

we have

$$[HA] = [A^-]$$

and

$$pH = pK_a - \log 1$$
$$= pK_a - 0 = pK_a$$

Therefore, at the half-equivalence point $pH = pK_a$. (Note this point in Figure 18.2 where at 50% neutralization $pH = 4.76 = pK_a$.) Beyond the equivalence point, the pH in a weak acid-strong base titration is determined solely by the added strong base.

EXAMPLE 18.27

■ A 40.0 ml sample of 0.100M acetic acid solution was titrated with 0.100M NaOH. Calculate the pH of the solution after the addition of (a) 10.0 ml of NaOH and (b) 40.0 ml of NaOH (the equivalence point). $K_a = 1.754 \times 10^{-5}$ for CH_3COOH.

The pH of the solution before the equivalence point can be calculated to suitable accuracy using the Henderson-Hasselbach equation [Equation (19)] for the various buffer solutions that have been formed. Originally the amount of acetic acid present was

$$n = MV = (0.100 \text{ mole/liter})(40.0 \times 10^{-3} \text{ liter})$$
$$= 4.00 \times 10^{-3} \text{ mole } CH_3COOH$$

The amount of CH_3COOH that reacts (equivalent to the amount of CH_3COO^- formed) is based on the stoichiometry

$$CH_3COOH(aq) + OH^- \longrightarrow CH_3COO^- + H_2O(l)$$

giving the amount and concentration of unreacted acid as

$$n_{CH_3COOH} = (4.00 \times 10^{-3} \text{ mole } CH_3COOH) - (n_{OH^-})\left(\frac{1 \text{ mole } CH_3COOH}{1 \text{ mole } OH^-}\right)$$

$$[CH_3COOH] = \frac{n}{V} = \frac{n_{CH_3COOH}}{(40.0 \text{ ml} + V_{OH^-})(10^{-3} \text{ liter/ml})}$$

and the amount and concentration of CH_3COO^-

$$n_{CH_3COO^-} = (n_{OH^-})\left(\frac{1 \text{ mole } CH_3COO^-}{1 \text{ mole } OH^-}\right)$$

$$[CH_3COO^-] = \frac{n}{V} = \frac{n_{CH_3COO^-}}{(40.0 \text{ ml} + V_{OH^-})(10^{-3} \text{ liter/ml})}$$

For acetic acid, the Henderson-Hasselbach equation is

$$pH = pK_a - \log \frac{[CH_3COOH]}{[CH_3COO^-]}$$

$$= -\log (1.754 \times 10^{-5}) - \log \frac{[CH_3COOH]}{[CH_3COO^-]}$$

$$= 4.7560 - \log \frac{[CH_3COOH]}{[CH_3COO^-]}$$

(a) For the addition of 10.0 ml of base,

$$n_{OH^-} = nV = (0.100 \text{ mole/liter})(10.0 \times 10^{-3} \text{ liter})$$
$$= 1.00 \times 10^{-3} \text{ mole NaOH}$$

$$n_{CH_3COOH} = (4.00 \times 10^{-3} \text{ mole CH}_3COOH)$$
$$- (1.00 \times 10^{-3} \text{ mole NaOH}) \left(\frac{1 \text{ mole CH}_3COOH}{1 \text{ mole OH}^-} \right)$$

$$= 3.00 \times 10^{-3} \text{ mole CH}_3COOH$$

$$[CH_3COOH] = \frac{3.00 \times 10^{-3} \text{ mole}}{(40.0 \text{ ml} + 10.0 \text{ ml})(10^{-3} \text{ liter/ml})} = 0.0600M$$

$$n_{CH_3COO^-} = (1.00 \times 10^{-3} \text{ mole NaOH}) \left(\frac{1 \text{ mole CH}_3COO^-}{1 \text{ mole OH}^-} \right)$$

$$= 1.00 \times 10^{-3} \text{ mole CH}_3COO^-$$

$$[CH_3COO^-] = \frac{1.00 \times 10^{-3} \text{ mole}}{(40.0 \text{ ml} + 10.0 \text{ ml})(10^{-3} \text{ liter/ml})} = 0.0200M$$

and therefore

$$pH = 4.7560 - \log \frac{(0.0600)}{(0.0200)} = 4.279$$

(b) At the equivalence point the acetic acid has been neutralized by the sodium hydroxide and we essentially have 4.00×10^{-3} mole of Na(CH$_3$COO) in 80.0 ml of solution

$$[Na^+] = [CH_3COO^-] = \frac{n}{V} = \frac{4.00 \times 10^{-3} \text{ mole}}{(80.0 \text{ ml})(10^{-3} \text{ liter/ml})} = 0.0500M$$

The acetate ion is a stronger base than water and therefore adds a proton.

$$K_b = \frac{K_w}{K_a} = \frac{1.00 \times 10^{-14}}{1.754 \times 10^{-5}} = 5.701 \times 10^{-10}$$

	CH_3COO^- + H_2O $\rightleftharpoons$ CH_3COOH + OH^-		
initial	0.0500		
change	$-x$	$+x$	$+x$
equilibrium	$0.0500 - x$	x	x

$$K_b = \frac{[CH_3COOH][OH^-]}{[CH_3COO^-]} = \frac{(x)(x)}{(0.0500 - x)} = 5.701 \times 10^{-10}$$

$$x = [OH^-] = 5.34 \times 10^{-6}M$$

$$pOH = -\log[OH^-] = -\log(5.34 \times 10^{-6}) = 5.272$$

$$pH = 14.00 - 5.272 = 8.728$$

Because the equivalence point is in the alkaline range, an indicator such as m-cresol purple (7.6 to 9.2), thymol blue (8.0 to 9.6), or phenolphthalein (8.3 to 10.0) should be used for the titration.

Compare the above answers to the plot for CH_3COOH in Figure 18.2. ∎

For the titration of an alkaline solution by an acid solution, the direction of the pH change is reversed, but the shapes of the curves are similar to those just described. Figure 18.3 shows both weak base–strong acid and strong base–strong acid titration curves. For the titration of the weak base, ammonia in this case, all the generalizations made above for a weak acid hold true: pK_b for NH_3 is 4.80, it gives a pH at the half-equivalence point of 9.2 (remember that for a base $pK_b = pOH$), the solution is buffered most effectively at pH 9.2 ± 1, and the pH after the equivalence point is passed is determined solely by the added acid. The equivalence point occurs at an acid pH because of the reaction with water of the NH_4Cl that is present.

The curves in Figure 18.2 illustrate the limits of the titration method for acids—the end points cannot be detected when the K_a is smaller than 1×10^{-8} or in the titration of a weak acid with a weak base. The K_a limit is also illustrated in the titration curve for phosphoric acid, a triprotic acid (Figure 18.4). Three stepwise neutralizations occur when base is added to this acid

$$H_3PO_4(aq) + OH^- \rightleftharpoons H_2PO_4^- + H_2O(l)$$
$$H_2PO_4^- + OH^- \rightleftharpoons HPO_4^{2-} + H_2O(l)$$
$$HPO_4^{2-} + OH^- \rightleftharpoons PO_4^{3-} + H_2O(l)$$

Only the first two equivalence points appear in the titration curve. The third, for the ionization step

$$HPO_4^{2-} \rightleftharpoons H^+ + PO_4^{3-}$$

which has $K_a = 1 \times 10^{-12}$, is not detectable.

Note that this triprotic acid releases three equivalents of hydrogen ion per mole. The 0.100M solution is therefore 0.300N and complete neutralization of 25 ml of this acid solution requires 75 ml of 0.100M (also 0.100N) NaOH.

EXAMPLE 18.28

∎ A 25.0 ml sample of 0.100M H_3PO_4 was titrated using 0.100M NaOH. Discuss the titration curve for this reaction. $pK_{a_1} = 2.12$, $pK_{a_2} = 7.18$, and $pK_{a_3} = 12.0$.

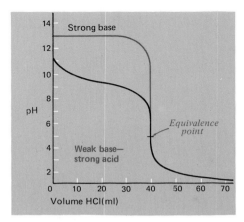

FIGURE 18.3
Weak base–strong acid titration curve.
Titration curve calculated for titration of 40.0 ml of 0.100N NH_3 with 0.100N HCl. The equivalence point occurs at an acid pH because a solution of NH_4Cl is present. The curve for titration of a strong acid by a strong base is shown for comparison.

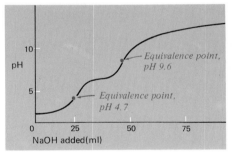

FIGURE 18.4
Titration of 25 ml of 0.100M H_3PO_4 with 0.100M NaOH. *The third equivalence point does not show up because HPO_4^{2-} is too weak an acid.*

Because of the large differences in the ionization constants, the titration of each proton can be considered separately with negligible error introduced by not considering subsequent ionizations, that is, only the species H_3PO_4, $H_2PO_4^-$ and H^+ need to be considered up to the first equivalence point; only $H_2PO_4^-$, HPO_4^{2-} and H^+ between the first and second equivalence points; and only HPO_4^{2-}, PO_4^{3-} and H^+ after the second equivalence point.

The pH of the H_3PO_4 solution before the addition of NaOH is found as follows:

$$H_3PO_4(aq) \rightleftharpoons H^+ + H_2PO_4^-$$

initial	0.10		
change	$-x$	$+x$	$+x$
equilibrium	$0.10 - x$	x	x

$$K_{a_1} = \frac{[H^+][H_2PO_4^-]}{[H_3PO_4]} = \frac{(x)(x)}{(0.10 - x)} = 7.5 \times 10^{-3}$$

Solving by using the quadratic formula gives

$$x = 2.4 \times 10^{-2} M = [H^+]$$
$$pH = -\log [H^+] = -\log (2.4 \times 10^{-2}) = 1.62$$

The pH up to the first equivalence point (0 to 25.0 ml NaOH) is given by

$$pH = 2.12 - \log \frac{[H_3PO_4]}{[H_2PO_4^-]}$$

where $[H_3PO_4]$ is the concentration of unreacted H_3PO_4 and $[H_2PO_4^-]$ is the concentration of $[H_2PO_4^-]$ formed. Both of these concentrations are calculated using the normal stoichiometric relationships described in the previous examples. The pH at the half-equivalence point (12.5 ml NaOH) is equal to $pK_{a_1} = 2.12$.

The pH of the first equivalence point, which is halfway between the first and second half-equivalence points, is

$$pH = \frac{(pK_{a_1} + pK_{a_2})}{2} = \frac{(2.12 + 7.18)}{2} = 4.65$$

The pH between the first and second equivalence points (25.0 to 50.0 ml NaOH) is given by

$$pH = 7.18 - \log \frac{[H_2PO_4^-]}{[HPO_4^{2-}]}$$

where the concentrations can be found using the usual stoichiometric methods. The pH at the point at which 37.5 ml NaOH (the second half-equivalence point) has been added is equal to $pK_{a_2} = 7.18$.

The pH of the second equivalence point is

$$pH = \frac{(pK_{a_2} + pK_{a_3})}{2}$$

$$= \frac{(7.18 + 12.0)}{2} = 9.6$$

The pH between the second and third equivalence points is theoretically given by

$$pH = 12.0 - \log \frac{[HPO_4^{2-}]}{[PO_4^{3-}]}$$

with the pH at the point at which 62.5 ml NaOH has been added equal to $pK_{a_3} = 12.0$. But the third ionization is so slight that the third proton cannot be directly titrated unless the PO_4^{3-} is removed (e.g., by precipitation with Ca^{2+}) to prevent its extensive reaction with water.

Compare these results with the curve shown in Figure 18.4. ∎

The pH change during a titration can be monitored with a pH meter, which relies on the development of a potential between electrodes dipping into the solution (see Tools of Chemistry, pH Meter, Chapter 24). The progress of a neutralization reaction can also be followed by measuring not the pH but the change in conductance of the solution during the reaction. Figure 18.5 shows the conductance during the neutralization of a dilute solution of hydrochloric acid by a dilute solution of sodium hydroxide. As NaOH is added, H^+, a highly conducting ion, is removed as H_2O. The total conductance decreases to a minimum at the equivalence point or the end point. If the salt formed in the neutralization is soluble in water, the minimum conductance is that of the salt (NaCl in Figure 18.5a). As NaOH is added beyond the equivalence point, the increase in OH^- ions which, like H^+ ions, are highly conducting, causes a rapid rise in the conductivity. If the salt formed is insoluble (e.g., barium sulfate from the neutralization of barium hydroxide by sulfuric acid, Figure 18.5b), the solution contains essentially no ions at the equivalence point and is nonconducting.

Heterogeneous equilibria

18.12 Solubility product, K_{sp}

The solubility of a salt is the amount of that salt that will dissolve in a given amount of solvent. Solubility is usually expressed as grams per 100 ml of solvent, or as moles per liter. The solubilities of salts vary widely and do not follow any predictable trends, because too many factors enter the picture, such as the lattice energies of the salts and the heats of hydration of the ions. To predict solubility from composi-

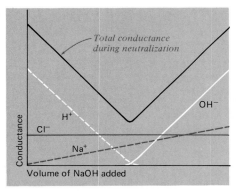

(a) HCl + NaOH

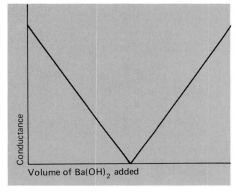

(b) H_2SO_4 + $Ba(OH)_2$

FIGURE 18.5

Change in conductance during neutralization of a strong acid by a strong base. (*a*) *Hydrochloric acid plus sodium hydroxide (dilute solutions) to give a soluble salt. The contributions of the individual ions to the conductance are shown in white and color.* (*b*) *Sulfuric acid plus barium hydroxide to give an insoluble salt. For weak acids and/or weak bases, the shapes of the curves are different.* (*From* Principles of Instrumental Analysis *by Douglas A. Skoog and Donald M. West. Copyright © 1971 by Holt, Rinehart and Winston. Reprinted by permission of Holt, Rinehart and Winston.*)

TABLE 18.9

Solubility of salts and hydroxides

These generalities apply to salts in water at room temperature; mod. sol. is moderately soluble; sl. sol. is slightly soluble.

Chlorides (Cl⁻)
soluble except
Hg_2Cl_2, $AgCl$, $PbCl_2$, $CuCl$; $PbCl_2$ sol. in hot water

Sulfates ($SO_4{}^{2-}$)
soluble except
$SrSO_4$, $BaSO_4$, $PbSO_4$; $CaSO_4$, Ag_2SO_4, mod. sol.

Nitrites ($NO_2{}^-$), nitrates ($NO_3{}^-$)
soluble except
$AgNO_2$, mod. sol.

Bromides (Br⁻), iodides (I⁻)
soluble except
those of Ag^+, $Hg_2{}^{2+}$, Pb^{2+}, and HgI_2, BiI_3, SnI_4; $HgBr_2$, mod. sol.

Sulfides (S^{2-})
insoluble except
$(NH_4)_2S$ and alkali metal sulfides

Acetates (CH_3COO^-)
soluble except
$AgCH_3COO^-$, $Cr(CH_3COO^-)_2$, $Hg_2(CH_3COO^-)_2$
sl. sol.; $Bi(CH_3COO^-)_3 \rightarrow$ insol. $BiO(CH_3COO^-)$

Chlorates ($ClO_3{}^-$), perchlorates ($ClO_4{}^-$), permanganates ($MnO_4{}^-$)
soluble except
$KClO_4$, NH_4ClO_4, mod. sol.

Carbonates ($CO_3{}^{2-}$), phosphates ($PO_4{}^{3-}$), oxalates ($C_2O_4{}^{2-}$), cyanides (CN^-)[4]
insoluble except
those of $NH_4{}^+$ and alkali metals

Hydroxides (OH^-)
insoluble except
those of alkali metals, $NH_4{}^+$, Ba^{2+}, Sr^{2+}; $Ca(OH)_2$, sl. sol.

TABLE 18.10

Some solubility products

$CaSO_4(s) \rightleftharpoons Ca^{2+} + SO_4{}^{2-}$
$K_{sp} = [Ca^{2+}][SO_4{}^{2-}] = 2.5 \times 10^{-5}$

$AgCl(s) \rightleftharpoons Ag^+ + Cl^-$
$K_{sp} = [Ag^+][Cl^-] = 1.8 \times 10^{-10}$

$Mg(OH)_2(s) \rightleftharpoons Mg^{2+} + 2OH^-$
$K_{sp} = [Mg^{2+}][OH^-]^2 = 7.1 \times 10^{-12}$

$Fe_2S_3(s) \rightleftharpoons 2Fe^{3+} + 3S^{2-}$
$K_{sp} = [Fe^{3+}]^2[S^{2-}]^3 = 1 \times 10^{-88}$

tion of a salt or hydroxide, the best we can do is to assemble a list of generalities based on observations (Table 18.9).

Many salts do not dissolve in water to a great extent. However, no salt is completely insoluble—at least a few ions always manage to escape into solution. Therefore, any salt can form a saturated solution—a solution in which the solid is in equilibrium with ions in solution. An example is the equilibrium of calcium sulfate with its ions

$$CaSO_4(s) \xrightleftharpoons{H_2O} Ca^{2+} + SO_4{}^{2-}$$

The equilibrium constant for a solid electrolyte in equilibrium with its ions in solution is called the **solubility product,** or solubility product constant, K_{sp}. For the general reaction

$$M_xA_y(s) \rightleftharpoons xM^{m+} + yA^{n-}$$

the solubility product is

$$K_{sp} = [M^{m+}]^x[A^{n-}]^y \qquad (24)$$

where m^+ and n^- are the charges of the ions. Because the activity of a solid is equal to 1 (Section 14.10), the term for concentration of the solid does not appear in the solubility product constant.

Some representative K_{sp} values are given in Table 18.10. There is a wide variation in the magnitude of K_{sp} values. Substances that have $K_{sp} < 10^{-4}$ are considered "insoluble." (Table 30.6 and Appendix C give extensive lists of K_{sp} values.)

A few things can be a bit confusing about K_{sp} values and solubility. K_{sp} reflects the solubility of substances in saturated solutions (at the given temperature), but is not numerically equal to solubility. Solubility is the actual amount of a substance present in a saturated solution. However, the K_{sp} for a salt of the type AB is the solubility squared (see Example 18.29) and for salts of other types, such as AB_2 or A_2B, it is the product of the solubility multiplied by various factors and raised to various powers (see Example 18.30). Therefore, only comparison of K_{sp} values for salts of the same general type can give information about the relative solubilities of the salts.

A further caution about solubility is the following: do not confuse the extent of dissociation of a salt with its solubility—a salt may have a very small K_{sp} and be considered essentially insoluble; yet since it is a strong electrolyte, the portion that is soluble is 100% dissociated (no "molecules" are present).

K_{sp} values vary with temperature and are usually tabulated for room temperature (20 to 25°C). We use only such K_{sp} values and omit mentioning temperature for each. The simple solubility product expression turns out to be very useful. The solubility of a salt by itself or in the presence of common ions or acids and bases can be calculated from it. K_{sp} can be used to predict the effect of pH on solubility. And for solutions containing several cations or anions, K_{sp} can be used to answer the question,"Which salt will precipitate first in reaction with other cations or anions?" The K_{sp} is also used in predicting conditions under which salts will dissolve. Examples of these applications are given in the following sections.

18.13 K_{sp} and solubility

It is possible to find K_{sp} from solubility data or to find solubilities from known K_{sp} values.

EXAMPLE 18.29

■ The solubility of silver acetate at room temperature is 11.1 g in one liter of solution. What is the solubility product for this salt?

In a saturated solution equal amounts of Ag^+ and CH_3COO^- are present, as indicated by

$$Ag(CH_3COO)(s) \rightleftharpoons Ag^+ + CH_3COO^-$$

Thus $[Ag^+] = [CH_3COO^-] = s$, where s = solubility in moles/liter.

$$K_{sp} = [Ag^+][CH_3COO^-] = (s)(s) = s^2$$

The value of s is

$$s = \frac{(11.1 \text{ g/liter})}{(166.92 \text{ g/mole})} = 0.066 \text{ mole/liter}$$

and thus

$$K_{sp} = s^2 = (0.066)^2 = 4.4 \times 10^{-3}$$

The solubility product of silver acetate is 4.4×10^{-3}. ■

EXAMPLE 18.30

■ The solubility product of Cu_2S is 3×10^{-48}. What is the solubility of this salt?

From the equation

$$Cu_2S(s) \rightleftharpoons 2Cu^+ + S^{2-}$$

we note that the concentration of Cu^+ is twice that of S^{2-} or $[Cu^+] = 2[S^{2-}]$ and that the solubility s is equal to the concentration of S^{2-} or $s = [S^{2-}]$. Thus

$$K_{sp} = [Cu^+]^2[S^{2-}] = (2s)^2(s) = 4s^3$$
$$s = \sqrt[3]{K_{sp}/4} = \sqrt[3]{(3 \times 10^{-48})/4} = 9.1 \times 10^{-17} \text{ mole/liter}$$

The solubility of Cu_2S is 9.1×10^{-17} mole/liter or roughly 1×10^{-14} g/liter. ∎

The solubility of a salt, in accordance with Le Chatelier's principle, is decreased by a common ion. K_{sp} values can be used to calculate solubilities in the presence of common ions.

EXAMPLE 18.31

∎ Which is more soluble in a $0.10M$ CrO_4^{2-} solution: $BaCrO_4$, which has $K_{sp} = 1.2 \times 10^{-10}$, or Ag_2CrO_4, which has $K_{sp} = 2.5 \times 10^{-12}$?

For the reactions

$$BaCrO_4(s) \rightleftharpoons Ba^{2+} + CrO_4^{2-}$$
$$Ag_2CrO_4(s) \rightleftharpoons 2Ag^+ + CrO_4^{2-}$$

we recognize that $[Ba^{2+}]$ is equal to the solubility of $BaCrO_4$, s_1, and that $[Ag^+]$ is equal to twice the solubility of Ag_2CrO_4, s_2. Thus

$$1.2 \times 10^{-10} = [Ba^{2+}][CrO_4^{2-}] = (s_1)(0.10)$$
$$s_1 = 1.2 \times 10^{-9} \text{ mole/liter}$$
$$2.5 \times 10^{-12} = [Ag^+]^2[CrO_4^{2-}] = (2s_2)^2(0.10)$$
$$s_2 = 2.5 \times 10^{-6} \text{ mole/liter}$$

Even though the solubility product of Ag_2CrO_4 is less than that of $BaCrO_4$, Ag_2CrO_4 is roughly 2000 times more soluble under these conditions. ∎

The total concentration of ions in solution, common or not, has a small effect on solubility because of the influence of the ions on each other. For example, $AgCl$ is slightly more soluble in KNO_3 solution than in water, because the activity coefficients are decreased. This influence of other ions is called the *salt effect*. In accurate analytical work and in biochemistry it must be taken into account.

EXAMPLE 18.32

∎ Calculate the solubility of PbS in water (a) neglecting the reaction of the ions with water and (b) considering the reaction of the sulfide ion with water. $K_{sp} = 1 \times 10^{-28}$ for PbS, $K_{b_2} = 3 \times 10^{-2}$ for S^{2-}, and $K_{b_1} = 1.0 \times 10^{-7}$ for HS^-.

(a) Letting s represent the solubility, from the equation

$$PbS(s) \rightleftharpoons Pb^{2+} + S^{2-} \qquad K_{sp} = [Pb^{2+}][S^{2-}] = 1 \times 10^{-28}$$

we can see that $s = [Pb^{2+}] = [S^{2-}]$, giving

$$K_{sp} = [Pb^{2+}][S^{2-}] = (s)(s) = 1 \times 10^{-28}$$
$$s = 1 \times 10^{-14} \text{ mole/liter}$$

The solubility—neglecting the ion reactions with water—is 1×10^{-14} mole/liter.

(b) Some of the S^{2-}, because it is a stronger base than water, reacts with water. Therefore, we would expect the solubility of PbS to be somewhat greater than the value determined in part (a). The solubility, s, will be equal to $[Pb^{2+}]$, which is also equal to the sum of the concentrations of the species containing the sulfide ion, that is,

$$s = [S^{2-}] + [HS^-] + [H_2S]$$

For the reactions with water we write

$$S^{2-} + H_2O(l) \rightleftharpoons HS^- + OH^- \qquad K_{b_2} = \frac{[HS^-][OH^-]}{[S^{2-}]} = 3 \times 10^{-2}$$

$$HS^- + H_2O(l) \rightleftharpoons H_2S(aq) + OH^- \qquad K_{b_1} = \frac{[H_2S][OH^-]}{[HS^-]} = 1.0 \times 10^{-7}$$

If we assume that the solubility of PbS is on the order of 10^{-10} to 10^{-14} mole/liter [see part (a)], the total $[OH^-]$ produced by both reactions would also be of the same order of magnitude, and would be considerably less than the $[OH^-]$ produced by the ionization of water. Thus we can assume that $[OH^-] = 1 \times 10^{-7}M$. At equilibrium the ratio of $[HS^-]$ to $[S^{2-}]$ is

$$K_{b_2} = \frac{[HS^-][OH^-]}{[S^{2-}]} = \frac{[HS^-](1 \times 10^{-7})}{[S^{2-}]} = 3 \times 10^{-2}$$

$$\frac{[HS^-]}{[S^{2-}]} = \frac{3 \times 10^{-2}}{1 \times 10^{-7}} = 3 \times 10^5$$

and the ratio of $[H_2S]$ to $[HS^-]$ is

$$K_{b_1} = \frac{[H_2S][OH^-]}{[HS^-]} = \frac{[H_2S](1 \times 10^{-7})}{[HS^-]} = 1.0 \times 10^{-7}$$

$$\frac{[H_2S]}{[HS^-]} = \frac{1.0 \times 10^{-7}}{1 \times 10^{-7}} = 1$$

These ratios tell us that the S^{2-} is tied up as H_2S and HS^- in about equal amounts and only a small portion remains as S^{2-}. Thus

$$s = [S^{2-}] + [HS^-] + [H_2S] \approx 0 + [HS^-] + [HS^-] = 2[HS^-]$$

Solving for $[HS^-]$ gives

$$[HS^-] = \frac{s}{2}$$

Ion product $> K_{sp}$:
precipitation
Ion product $< K_{sp}$:
no precipitation

and substituting into the expression for the ratio of $[HS^-]$ to $[S^{2-}]$ gives

$$\frac{[HS^-]}{[S^{2-}]} = \frac{(s/2)}{[S^{2-}]} = 3 \times 10^5$$

$$[S^{2-}] = \frac{s}{(2)(3 \times 10^5)} = (2 \times 10^{-6})s$$

Substituting $[S^{2-}] = (2 \times 10^{-6})s$ and $[Pb^{2+}] = s$ into the solubility product expression gives

$$K_{sp} = [Pb^{2+}][S^{2-}] = (s)[(2 \times 10^{-6})s] = 1 \times 10^{-28}$$
$$s = 7 \times 10^{-12} \text{ mole/liter}$$

The solubility is 7×10^{-12} mole/liter. Actually some reaction of the cation with water occurs and the solubility would be even larger than this value. ∎

18.14 K_{sp} and precipitation

The product of the concentration of ions in a nonsaturated or non-equilibrium solution is called the **ion product.** (The ion product is analogous to the reaction quotient; Section 16.4.) For example, $[Ca^{2+}][SO_4^{2-}]$ is the ion product for any solution containing calcium and sulfate ions. The ion product equals the K_{sp} when the solution is saturated. If the ion product is less than the K_{sp}, all the Ca^{2+} and SO_4^{2-} ions present are in solution. If the ion product is greater than the K_{sp}, solid calcium sulfate precipitates out until equilibrium between the solid and a solution saturated with ions is established—in other words, precipitation continues until the ion product of the ions in solution equals the K_{sp}.

EXAMPLE 18.33

■ Five grams of $Pb(NO_3)_2$ was added to one liter of 0.010M NaCl solution; $K_{sp} = 2 \times 10^{-5}$ for $PbCl_2$. Did a precipitate form?

The concentration of the lead ion, which is equal to the concentration of $Pb(NO_3)_2$, is found as follows:

$$[Pb^{2+}] = \frac{(5.0 \text{ g})/(331.20 \text{ g/mole})}{1.0 \text{ liter}} = 1.5 \times 10^{-2} \text{ mole/liter}$$

and the concentration of the chloride ion is $[Cl^-] = 0.010$ mole/liter. The ion product for the reaction

$$PbCl_2(s) \xrightarrow{H_2O} Pb^{2+} + 2Cl^-$$

is given by

$$[Pb^{2+}][Cl^-]^2 = (1.5 \times 10^{-2})(0.010)^2 = 1.5 \times 10^{-6}$$

Because the ion product is less than K_{sp}, no $PbCl_2$ was formed and we have an unsaturated solution. ■

EXAMPLE 18.34

■ A solution is $0.10M$ in Ba^{2+} and in Sr^{2+}. $K_{sp} = 2.0 \times 10^{-9}$ for $BaCO_3$ and 5.2×10^{-10} for $SrCO_3$. Describe what happens as solid Na_2CO_3 is added to the solution.

To answer this question we first write the chemical equations and the related equilibrium expressions:

$$BaCO_3(s) \rightleftharpoons Ba^{2+} + CO_3{}^{2-} \qquad SrCO_3(s) \rightleftharpoons Sr^{2+} + CO_3{}^{2-}$$
$$K_{sp} = 2.0 \times 10^{-9} = (0.10)[CO_3{}^{2-}] \qquad K_{sp} = 5.2 \times 10^{-10} = (0.10)[CO_3{}^{2-}]$$

Nothing happens as the Na_2CO_3 is added until $[CO_3{}^{2-}]$ reaches 5.2×10^{-9} mole/liter, at which point the ion product for $SrCO_3$ is equal to K_{sp} and $SrCO_3$ begins to precipitate. The $SrCO_3$ continues to precipitate as $[CO_3{}^{2-}]$ increases. When $[CO_3{}^{2-}]$ reaches 2.0×10^{-8} mole/liter, the ion product for $BaCO_3$ is equal to the K_{sp}, and as more Na_2CO_3 is added, both $SrCO_3$ and $BaCO_3$ precipitate. ■

EXAMPLE 18.35

■ What percent of the Sr^{2+} remains in solution at the point where the coprecipitation of $BaCO_3$ begins in the previous example?

At $[CO_3{}^{2-}] = 2.0 \times 10^{-8}$ mole/liter, the amount of Sr^{2+} remaining in solution is found from the expression for the solubility product

$$5.2 \times 10^{-10} = [Sr^{2+}](2.0 \times 10^{-8})$$
$$[Sr^{2+}] = 0.026 \text{ mole/liter}$$

Thus

$$\frac{0.026 \text{ mole/liter}}{0.10 \text{ mole/liter}} \times 100\% = 26\%$$

Of the original Sr^{2+}, 26% remains. ■

EXAMPLE 18.36

■ A solution that is $0.001M$ in $MgCl_2$ and $0.01M$ in NH_4Cl is made $0.01M$ in NH_3. Will $Mg(OH)_2$ precipitate? K_b for NH_3 is 1.6×10^{-5} and K_{sp} for $Mg(OH)_2$ is 7.1×10^{-12}.

The equations for the equilibria under consideration are

$$NH_3(aq) + H_2O(l) \rightleftharpoons NH_4^+ + OH^-$$

$$K_b = \frac{[NH_4^+][OH^-]}{[NH_3]} = 1.6 \times 10^{-5}$$

$$Mg(OH)_2(s) \rightleftharpoons Mg^{2+} + 2OH^-$$

$$K_{sp} = [Mg^{2+}][OH^-]^2 = 7.1 \times 10^{-12}$$

The ionization of the aqueous ammonia is repressed by the NH_4^+ ion from NH_4Cl (common ion effect). In view of the small value for the ionization constant, we can assume that all of the NH_4^+ ion in solution comes from NH_4Cl and that the equilibrium concentration of NH_3 does not differ significantly from $0.01M$. Substituting the appropriate values into the expression for K_b gives

$$K_b = \frac{(0.01)[OH^-]}{0.01} = 1.6 \times 10^{-5}$$

$$[OH^-] = 1.6 \times 10^{-5}M$$

This value is used to find the ion product $[Mg^{2+}][OH^-]^2$

$$(0.001)(1.6 \times 10^{-5})^2 = 3 \times 10^{-13}$$

The ion product is smaller than the value of K_{sp} for $Mg(OH)_2$ and therefore $Mg(OH)_2$ will not precipitate. A similar calculation will show that without the NH_4Cl present the hydroxide would precipitate. ∎

EXAMPLE 18.37

∎ Calculate the pH below which FeS will not be precipitated when one liter of solution containing 11 g of Fe^{2+} as $FeCl_2$ is saturated with H_2S. $K_{sp} = 4.2 \times 10^{-17}$; $[H^+]^2[S^{2-}] = 3 \times 10^{-21}$ (for a solution saturated with gaseous H_2S; see Section 18.7).

A solution containing 11 g of Fe^{2+} per liter is $0.20M$ in Fe^{2+}. The S^{2-} ion concentration below which FeS cannot be precipitated is calculated from the solubility product.

$$FeS(s) \rightleftharpoons Fe^{2+} + S^{2-}$$

$$K_{sp} = [Fe^{2+}][S^{2-}] = 4.2 \times 10^{-17}$$

$$(0.20)[S^{2-}] = 4.2 \times 10^{-17}$$

$$[S^{2-}] = 2.1 \times 10^{-16}M$$

The H^+ concentration in equilibrium with this concentration of sulfide ion in a solution saturated with H_2S is found from

$$[H^+]^2[S^{2-}] = 3 \times 10^{-21}$$

$$[H^+]^2 = \frac{3 \times 10^{-21}}{2.1 \times 10^{-16}} = 1 \times 10^{-5}$$

$$[H^+] = 3 \times 10^{-3}M$$

$$pH = -(\log 3 \times 10^{-3}) = 2.5$$

Below this pH, that is, in solutions containing more than $3 \times 10^{-3}M$ H^+ ion, FeS will not be precipitated by H_2S. ∎

18.15 K_{sp} and the dissolution of ionic precipitates

Ionic substances that are slightly soluble in water at room temperature can be brought into solution in several ways. In most cases, the concentration of one or both of the ions is diminished by a chemical reaction to such a degree that the ion product of the compound in solution is less than that in a saturated solution, that is, less than the solubility product. Occasionally, use can be made of a marked dependence of solubility on temperature. Concepts involved in the dissolving of ionic precipitates are described in this section, with specific illustrations pertinent to the separation and detection of cations in qualitative analysis (Chapters 30 and 31).

Cations are frequently precipitated as hydroxides, carbonates, or sulfides during qualitative analysis. The chemical reactions by which they can be brought into solution illustrate useful methods for satisfying the relationship

ion product in solution $< K_{sp}$

The hydroxide, carbonate, and sulfide anions—OH^-, CO_3^{2-}, and S^{2-}—are strong Brønsted-Lowry bases and combine with hydrogen ions to give weakly ionized conjugate acids—H_2O; HCO_3^- and H_2CO_3; HS^- and H_2S. It would thus be anticipated that hydroxides, carbonates, and sulfides could be dissolved by treatment with strong acids such as hydrochloric acid.

Practically all water-insoluble metal hydroxides that are basic or amphoteric (Section 17.9) dissolve readily in solutions of strong acid, for example,

$Fe(OH)_3(s) + 3H_3O^+ \longrightarrow Fe^{3+} + 6H_2O(l)$
iron(III) hydroxide
$Al(OH)_3(s) + 3H_3O^+ \longrightarrow Al^{3+} + 6H_2O(l)$
aluminum hydroxide

Dissolution occurs because water is a very weak electrolyte and the hydroxide ion concentration in solution is regulated by the ion product constant for water, that is, by the hydrogen ion concentration.

$K_w = [H^+][OH^-] = 1.00 \times 10^{-14}$

Thus, for example, in a solution $1M$ each in Fe^{3+} and H^+ the ion product of $Fe(OH)_3$ is

$[Fe^{3+}][OH^-]^3 = (1)(1 \times 10^{-14})^3 = 1 \times 10^{-42}$

a value much smaller than the solubility product, 3×10^{-39}.

The dissolution of slightly soluble metal carbonates is governed by the transformations

$$CO_3{}^{2-} \xrightarrow{\ H^+\ } HCO_3{}^- \xrightarrow{\ H^+\ } H_2CO_3 \longrightarrow CO_2(aq \text{ or } g) + H_2O(l)$$

The $HCO_3{}^-$ ion and H_2CO_3 are weak acids (Table 18.7 and Section 18.9). Moreover, H_2CO_3 is unstable; decomposition to CO_2 and H_2O is favored by high hydrogen ion concentration. All insoluble metal carbonates are solubilized by means of acids, even such weak acids as acetic acid.

$$MCO_3(s) + 2H^+ \longrightarrow M^{2+} + CO_2(g) + H_2O(l)$$

where M = dipositive metal ion.

The slightly soluble metal sulfides with *relatively* large solubility products (e.g., MnS, 2.3×10^{-13}; FeS, 4.2×10^{-17}; ZnS, 2×10^{-24}) are readily soluble in hydrochloric acid because of the formation of weakly ionized and volatile H_2S (K_{a_1}, 1.0×10^{-7}; K_{a_2}, 3×10^{-13}).

$$FeS(s) + 2H^+ \longrightarrow Fe^{2+} + H_2S \ (aq \text{ or } g)$$

The presence of H^+ ion from the strong acid represses the ionization of H_2S and keeps the sulfide ion concentration in solution very low (Section 18.7). Cations that form less soluble sulfides can be separated from those that give the more soluble sulfides by precipitation with H_2S from solutions about $0.3M$ in H^+ ion.

In some instances, less soluble metal sulfides are brought into solution by hydrochloric acid, because both cations and sulfide ions are converted to weakly ionized species. Examples are found with Sb_2S_3 and SnS_2:

$$Sb_2S_3(s) + 6H^+ + 8Cl^- \longrightarrow 2[SbCl_4]^- + 3H_2S(aq)$$
$$SnS_2(s) + 4H^+ + 6Cl^- \longrightarrow [SnCl_6]^{2-} + 2H_2S(aq)$$

The ions $[SbCl_4]^-$ and $[SnCl_6]^{2-}$ are soluble *complex ions* (Section 28.1) that yield concentrations of Sb^{3+} and Sn^{4+} in solution low enough to keep Sb_2S_3 and SnS_2 from forming.

Numerous other examples of the dissolution of ionic precipitates by complex ion formation are encountered in the qualitative analysis for cations.

Some very insoluble metal sulfides (e.g., PbS, $K_{sp} = 1 \times 10^{-28}$; CuS, $K_{sp} = 6 \times 10^{-36}$; HgS, $K_{sp} = 4 \times 10^{-53}$) furnish so little S^{2-} in saturated solutions that they cannot be dissolved by aqueous HCl. (HgS dissolves very slowly in concentrated HCl, S^{2-} being converted to H_2S and Hg^{2+} to the weakly dissociated $[HgCl_4]^{2-}$ complex.) In these cases S^{2-} is effectively removed by oxidation (Sections 4.12 and 19.1) to elemental sulfur. For PbS and CuS, nitric acid (Section 22.8) serves as the oxidizing agent, for example,

$$3PbS(s) + 8H^+ + 2NO_3{}^- \longrightarrow 3Pb^{2+} + 3S(s) + 2NO(g) + 4H_2O(l)$$

The much less soluble HgS does not furnish sufficient S^{2-} in saturated solution for oxidation with nitric acid to occur, and *aqua regia* (3:1, concentrated HCl:concentrated HNO_3 by volume; Section 22.8) is used. This reagent converts S^{2-} to the free element and at the same time ties up Hg^{2+} as the $[HgCl_4]^{2-}$ complex. Other examples in which oxidation-reduction (redox) reactions are useful in the separation and detection of cations are discussed in Chapters 30 and 31 and the principles of redox equilibria are also reviewed in Chapter 30.

EXAMPLE 18.38

■ A 0.010 mole sample of $Fe(OH)_3(s)$ was added to 1.0 liter of water and a strong acid was added until the precipitate dissolved. At what pH was the dissolution process complete? $K_{sp} = 3 \times 10^{-39}$ for $Fe(OH)_3$.

The concentration of OH^- at the point where dissolution is complete is found from the solubility product expression

$$Fe(OH)_3(s) \rightleftharpoons Fe^{3+} + 3OH^- \qquad K_{sp} = [Fe^{3+}][OH^-]^3 = 3 \times 10^{-39}$$
$$[OH^-] = \sqrt[3]{\frac{K_{sp}}{[Fe^{3+}]}} = \sqrt[3]{\frac{3 \times 10^{-39}}{0.010}} = 7 \times 10^{-13} M$$

Then the concentration of H^+ is found from K_w.

$$[H^+] = \frac{K_w}{[OH^-]} = \frac{1.00 \times 10^{-14}}{7 \times 10^{-13}} = 1 \times 10^{-2}$$
$$pH = -\log [H^+] = -\log (1 \times 10^{-2}) = 2.0$$

Acid would have to be added until the pH reached 2.0.   ■

EXAMPLE 18.39

■ We want to know whether $ZnS(s)$ dissolves in HNO_3 as a result of the following oxidation-reduction process

$$3ZnS(s) + 2NO_3^- + 8H^+ \rightleftharpoons 3Zn^{2+} + 3S(s) + 2NO(g) + 4H_2O(l)$$

given

$$ZnS(s) \rightleftharpoons Zn^{2+} + S^{2-} \qquad K_{sp} = [Zn^{2+}][S^{2-}] = 2 \times 10^{-24}$$
$$H_2S(aq) \rightleftharpoons 2H^+ + S^{2-} \qquad K_a = \frac{[H^+]^2[S^{2-}]}{[H_2S]} = 3 \times 10^{-20}$$
$$3H_2S(aq) + 2NO_3^- + 2H^+ \rightleftharpoons 2NO(g) + 4H_2O(l) + 3S(s)$$
$$K = \frac{[NO]^2}{[H_2S]^3[NO_3^-]^2[H^+]^2} = 1 \times 10^{83}$$

The equation and the K expression for the reaction in question must be obtained by combining the three given equations and K's as follows:

$$3ZnS(s) \rightleftharpoons 3Zn^{2+} + 3S^{2-} \qquad K = K_{sp}^3$$
$$6H^+ + 3S^2 \rightleftharpoons 3H_2S(aq) \qquad K = 1/K_a^3$$
$$3H_2S(aq) + 2NO_3^- + 2H^+ \rightleftharpoons$$
$$\underline{2NO(g) + 4H_2O(l) + 3S(s) \qquad K = 1 \times 10^{83}}$$
$$3ZnS(s) + 2NO_3^- + 8H^+ \rightleftharpoons$$
$$3Zn^{2+} + 2NO(g) + 4H_2O(l) + 3S(s) \qquad K = (K_{sp}^3)(1 \times 10^{83})/(K_a^3)$$

$$K = \frac{(2 \times 10^{-24})^3(1 \times 10^{83})}{(3 \times 10^{-20})^3} = 3 \times 10^{70}$$

Because the equilibrium constant is very large, we predict that the ZnS will dissolve. ∎

ZEN AND THE ART OF MOTORCYCLE MAIN-
TENANCE, *by Robert Pirsig*

THOUGHTS ON
CHEMISTRY

Scientific method

When I think of formal scientific method an image sometimes comes to mind of an enormous juggernaut, a huge bulldozer—slow, tedious, lumbering, laborious, but invincible. It takes twice as long, five times as long, maybe a dozen times as long as informal mechanic's techniques, but you know in the end you're going to get it. There's no fault isolation problem in motorcycle maintenance that can stand up to it. When you've hit a really tough one, tried everything, racked your brain and nothing works, and you know that this time Nature has really decided to be difficult, you say, "Okay, Nature, that's the end of the nice guy," and you crank up the formal scientific method.

For this you keep a lab notebook. Everything gets written down, formally, so that you know at all times where you are, where you've been, where you're going and where you want to get. In scientific work and electronics technology this is necessary because otherwise the problems get so complex you get lost in them and confused and forget what you know and what you don't know and have to give up. In cycle maintenance things are not that involved, but when confusion starts it's a good idea to hold it down by making everything formal and exact. Sometimes just the act of writing down the problems straightens out your head as to what they really are. . . .

The real purpose of scientific method is to make sure Nature hasn't misled you into thinking you know something you don't actually know. There's not a mechanic or scientist or technician alive who hasn't suffered from that one so much that he's not instinctively on guard. That's the main reason why so much scientific and mechanical information sounds so dull and so cautious. If you get careless or go romanticizing scientific information, giving it a flourish here and there, Nature will soon make a complete fool out of you. It does it often enough anyway even when you don't give it opportunities. One must be extremely careful and rigidly logical when dealing with Nature: one logical slip and an entire scientific edifice comes tumbling down. One false deduction about the machine and you can get hung up indefinitely.

ZEN AND THE ART OF MOTORCYCLE MAINTENANCE, by Robert Pirsig (p. 100, William Morrow & Co., New York, 1975)

Significant terms defined in this chapter

hydrolysis of an ion
acid salt
normal salt
common ion effect
buffer solution
end point

equivalence point
titration
standardized solution
solubility product
ion product

Exercises

18.1 For the sequence of events described below, answer the following questions: (a) List all the salts present in solution and as solids. (b) Identify an acid salt. (c) Which step involves the knowledge of a K_{sp}? (d) In which step was the common ion effect used? (e) Which solution would be standardized before the experiment?

A sample of water was boiled for a period of time and a scale was produced on the container. (i) The scale was analyzed and found to contain Ca^{2+} and CO_3^{2-}. (ii) With this information, the original water was analyzed and found to contain Ca^{2+} and HCO_3^{2-}. To one sample of the water, (iii) excess $C_2O_4^{2-}$ was added to assure nearly complete precipitation of CaC_2O_4, (iv) the precipitate was separated and weighed using normal quantitative techniques, and (v) the amount of Ca^{2+} present was calculated. To check this result, a second sample of the water was (vi) passed through an ion exchange column, which substituted two H^+ ions for every Ca^{2+} (vii) the resulting acid solution was titrated using NaOH, and (viii) the amount of Ca^{2+} present was calculated.

18.2 Distinguish clearly among (a) solubility and solubility product; (b) ion product and solubility product; and (c) hydrolysis and hydration.

18.3 Which of these solutions or liquids would be (a) acidic, (b) alkaline, or (c) nearly neutral: (i) pure water; (ii) aqueous solution of a salt containing a cation that is a weaker acid than water and an anion that is a weaker base than water; (iii) aqueous solution of a salt containing a cation that is a weaker acid than water and an anion that is a stronger base than water; (iv) aqueous solution of a salt containing a cation that is a stronger acid than water and an anion that is a stronger base than water; and (v) aqueous solution of a salt containing a cation that is a stronger acid than water and an anion that is a weaker base than water?

18.4 Which statements are true? Rewrite any false statement so that it is correct.
(a) Precipitation of an ionic compound will occur in a solution if the ion product exceeds the solubility product.
(b) A solution of a salt containing a cation that is a stronger acid than water and an anion that is a stronger base than water will always be neutral.
(c) The product of a titration could be a precipitate.
(d) A buffer solution is one that resists changes in pH when small amounts of acid, base, or water are added and illustrates the common ion effect.
(e) A polyprotic acid will have more than one equivalence point.

18.5 Which solutions would be (a) acidic, (b) alkaline, or (c) nearly neutral? (i) $(NH_4)_2SO_4$; (ii) KCl; (iii) LiCN; (iv) $Al(NO_3)_3$; (v) $NaClO_4$; (vi) $K_2C_2O_4$; (vii) $(NH_4)_2S$, note that $K_a < K_b$; (viii) NaH_2PO_4; (ix) $(NH_4)CN$, note that $K_a < K_b$.

18.6 Based on the general solubility rules given in Table 18.9, classify the following compounds as (a) soluble, (b) slightly soluble, or (c) insoluble: (i) silver acetate, (ii) barium sulfate, (iii) silver chloride, (iv) calcium hydroxide, (v) ammonium chlorate, (vi) zinc nitrate, (vii) potassium hydroxide, (viii) aluminum sulfate, (ix) sodium acetate, (x) copper(II) carbonate.

18.7 Write the equilibrium expression for the reaction of each of the following salts with water and calculate the numerical value using the data provided: (a) KF, $K_a = 6.5 \times 10^{-4}$ for HF; (b) NH_4NO_3, $K_b = 1.6 \times 10^{-5}$ for NH_3; (c) NH_4F.

18.8 Calculate the pH of 0.10M solutions for the compounds listed in Exercise 18.7.

18.9 Repeat the calculations of Exercise 18.8 for 0.050M and 0.010M solutions. Is there a trend of the pH with concentration?

18.10 As a result of their reaction with H_2O, which would be more acidic: a 0.10M solution of aniline hydrochloride, $C_6H_5NH_3Cl$ ($K_b = 4.2 \times 10^{-10}$ for aniline, $C_6H_5NH_2$), or a 0.10M solution of methylamine hydrochloride, CH_3NH_3Cl ($K_b = 3.9 \times 10^{-4}$ for methylamine, CH_3NH_2)?

18.11 What is the pH of a solution prepared by adding 10.0 ml of 0.100M HCl to (a) 10.0 ml of 0.100M NaOH, (b) 5.0 ml of 0.100M NaOH, (c) 15.0 ml of 0.100M NaOH, (d) 10.0 ml of 0.100M HNO_3, (e) 10.0 ml of 0.100M NH_3? $K_b = 1.6 \times 10^{-5}$ for NH_3.

18.12 A weak acid, HA, is too unstable to be made in pure form. However, the sodium salt of the acid, NaA, is stable and dissociates completely in aqueous solution. Calculate the K_a for the acid if a 0.1M solution of NaA has a pH of 8.64.

18.13 Find $[H^+]$, $[HSO_3^-]$, $[H_2SO_3]$, and $[SO_3^{2-}]$ at equilibrium in a 0.100M solution of H_2SO_3. $K_{a_1} = 1.43 \times 10^{-2}$ and $K_{a_2} = 5.0 \times 10^{-8}$ for H_2SO_3.

18.14 Find the pH of a solution of 0.250M $H_2C_2O_4$. $K_{a_1} = 5.60 \times 10^{-2}$ and $K_{a_2} = 6.2 \times 10^{-5}$ for $H_2C_2O_4$.

18.15 What is the concentration of S^{2-} present in a saturated hydrosulfuric acid solution such that the pH is 5.3? $[H^+]^2[S^{2-}] = 3 \times 10^{-21}$ for a saturated solution of H_2S.

18.16 Calculate the concentration of all species (except water) present in a 0.010M solution of H_3AsO_4. $K_{a_1} = 6.5 \times 10^{-3}$, $K_{a_2} = 1.1 \times 10^{-7}$, and $K_{a_3} = 3 \times 10^{-12}$ for arsenic acid.

18.17 What is the pH of a solution containing (a) 0.50M HCN and 0.25M KCN, (b) 0.10M $HClO_4$ and 0.10M $KClO_4$, (c) 0.013M CH_3NH_2 and 0.007M CH_3NH_3Cl, (d) 0.10M HCl and 0.10M CH_3COOH? Why is the solution described in (a) alkaline? $K_a = 6.2 \times 10^{-10}$ for HCN and 1.754×10^{-5} for CH_3COOH and $K_b = 3.9 \times 10^{-4}$ for CH_3NH_2.

18.18 Sulfamic acid, HSO_3NH_2, is a strong monoprotic acid that can be used to standardize a strong base:

$$HSO_3NH_2(aq) + KOH(aq) \rightleftharpoons KSO_3NH_2(aq) + H_2O(l)$$

If 19.35 ml of a solution of KOH neutralized a 0.179 g sample of HSO_3NH_2, what was the molar concentration of the solution of the base?

18.19 The percent of Na_2CO_3 in a sample of impure soda ash was determined by titration with 0.107N HCl.

$$Na_2CO_3(aq) + 2HCl(aq) \rightleftharpoons 2NaCl(aq) + H_2O(l) + CO_2(g)$$

If a 0.253 g sample of soda ash required 13.72 ml of HCl to reach the end point, what was the percentage of Na_2CO_3 present? Assume that the impurities do not react with the acid.

18.20 The percentage acetic acid in a vinegar sample was determined by titration with 0.103N NaOH:

$$CH_3COOH(aq) + NaOH(aq) \rightleftharpoons$$
$$Na(CH_3COO)(aq) + H_2O(l)$$

If a 10.13 g sample of vinegar required 67.43 ml of NaOH to reach the end point, did the vinegar meet the minimum specifications of 4 wt % acetic acid?

18.21 What would be the pH at the equivalence point for titrating 0.1N NaOH with (a) 0.1N HNO_3; (b) 0.1N HNO_2, $K_a = 7.2 \times 10^{-4}$; (c) 0.1N HIO, $K_a = 2.3 \times 10^{-11}$? (*Hint:* Remember the effect of dilution. In a titration, if equal volumes of each solution are present at the equivalence point, the concentrations of ions are one-half their original concentrations.) What is the trend in pH with decreasing K_a?

18.22 A buffer solution was made by mixing exactly 500 ml of 1.00M acetic acid and 500 ml of 0.500M calcium acetate solutions. $K_a = 1.754 \times 10^{-5}$ for CH_3COOH. (a) What is the total volume of the mixed solutions? What is the concentration of each of the following in the buffer solution: (b) CH_3COOH, (c) Ca^{2+}, and (d) CH_3COO^-? (e) Calculate the $[H^+]$. (f) What is the pH?

18.23 A buffer is made of 0.200M NH_4Cl and 0.100M NH_3. (a) What is the pH of this solution if $K_b = 1.6 \times 10^{-5}$ for NH_3? (b) Suppose 0.001 mole KOH were added to 1 liter of the buffer. What is the pH after this addition?

18.24 What is the pH of the buffer described in Exercise 18.22 after the addition of (a) 0.001 mole of HCl, (b) 0.001 mole of NaOH, (c) 100 ml of water?

18.25 One liter of a 0.10M solution of $NaNO_2$ was acidified by adding 0.05 mole of HCl. What is the pH of this solution? If the solution is diluted twofold with water, what is the pH? $K_a = 7.2 \times 10^{-4}$ for HNO_2.

18.26 α-Tartaric acid is a diprotic acid with $K_{a_1} = 1.04 \times 10^{-3}$ and $K_{a_2} = 4.55 \times 10^{-5}$ at 25°C. Find the pH at the equivalence points if a 0.10M solution is titrated with 0.10M NaOH. (*Hint:* Don't forget the effect of dilution.)

18.27 A 25.0 ml sample of 0.250M HNO_3 is titrated with 0.100M NaOH. Calculate the pH of the solution (a) before the addition of NaOH and after the addition of (b) 10.0 ml, (c) 25.0 ml, (d) 50.0 ml, (e) 62.5 ml, and (f) 75.0 ml of NaOH.

18.28 Calculate the K_{sp} for the following substances from the solubilities given: (a) Ag_2CrO_4, 2.8×10^{-2} g/liter; (b) $PbSO_4$, 4.5×10^{-2} g/liter; (c) CuS, 2.3×10^{-16} g/liter; (d) PbI_2, 5.6×10^{-1} g/liter; (e) $Ca_3(PO_4)_2$, 8×10^{-4} g/liter.

18.29 Calculate the solubility in grams per liter for each of the following substances from the given values of K_{sp}: (a) $Fe(OH)_2$, $K_{sp} = 8 \times 10^{-16}$; (b) $Fe(OH)_3$, $K_{sp} = 3 \times 10^{-39}$; (c) $BaSO_4$, $K_{sp} = 1.7 \times 10^{-10}$.

18.30 The solubility products at 25°C for $FeCO_3$ and CaF_2 are 2.11×10^{-11} and 2.7×10^{-11}, respectively. Calculate the solubility, expressed in molarity, for each salt. How does it happen that the solubilities are quite different, although the K_{sp}'s are almost the same?

18.31 The K_{sp} for AgBr is 4.9×10^{-13} at 25°C. What is the concentration of Ag^+ in a solution in which the concentration of Br^- is 0.010M? Express this concentration in terms of molarity and grams per liter.

18.32 What is the concentration of Cu^{2+} in a solution which was made by mixing equal volumes of 0.010M Cu^{2+} and 0.50M S^{2-}? The K_{sp} for CuS is 6×10^{-36}. How many Cu^{2+} ions are present in a liter of solution prepared as described?

18.33 If 1.0 g of $AgNO_3$ is added to 50 ml of 0.050M $Na(CH_3COO)$, will a precipitate form? (*Hint:* Calculate the molarity of Ag^+ in the solution assuming no precipitation, calculate the ion product for $Ag(CH_3COO)$, and compare the ion product to the K_{sp}, which is 4.4×10^{-3}.)

18.34 In a student-designed qualitative analysis scheme, Pb^{2+} is to be separated as PbC_2O_4 from Cu^{2+} by the addition of $Na_2C_2O_4$. If $K_{sp} = 4.8 \times 10^{-12}$ for PbC_2O_4 and 3×10^{-8} for CuC_2O_4, to what value must the concentration of $C_2O_4^{2-}$ be adjusted so that only PbC_2O_4 will form in a solution that is roughly 0.01M in each of these cations?

18.35 A solution contains 0.015M CH_3COOH and 0.0030M Ag^+. Will a precipitate form? $K_a = 1.754 \times 10^{-5}$ for acetic acid and $K_{sp} = 4.4 \times 10^{-3}$ for $Ag(CH_3COO)$.

18.36 In many qualitative analysis schemes, Pb^{2+} appears twice because sufficient Pb^{2+} remains in solution after the precipitation of the cations which form insoluble chlorides to be precipitated as the sulfide during subsequent tests. If an unknown solution contained 0.01M Pb^{2+}, what would be the concentration of Pb^{2+} in solution after precipitation by 6M HCl? $K_{sp} = 2 \times 10^{-5}$ for $PbCl_2$. If $[S^{2-}] = 0.01M$

during the second precipitation, would PbS form? $K_{sp} = 1 \times 10^{-28}$ for PbS.

18.37 What pH values could be used to separate Mn^{2+} from Zn^{2+} by selective precipitation of 0.010 mole of ZnS in 1 liter of saturated solution of H_2S? $K_{sp} = 2.3 \times 10^{-13}$ for leaving 0.10 mole of Mn^{2+} in solution MnS and 2×10^{-24} for ZnS and $[H^+]^2[S^{2-}] = 3 \times 10^{-21}$ for a saturated solution of H_2S.

18.38 A strong acid was added to a solid mixture of 0.010 mole samples of $Fe(OH)_2$ and $Cu(OH)_2$ placed in 1 liter of water. Neglecting the volume of the added acid, at what pH was the dissolution process of each hydroxide complete? $K_{sp} = 8 \times 10^{-16}$ for $Fe(OH)_2$ and 1.3×10^{-20} for $Cu(OH)_2$.

18.39 Will HgS(s) or PbS(s) dissolve in HNO_3 as a result of the oxidation-reduction process described in Example 18.39? $K_{sp} = 4 \times 10^{-53}$ for HgS and 1×10^{-28} for PbS.

18.40* What is the pH of a solution that is a mixture containing 0.10M HBrO, $K_a = 2.2 \times 10^{-9}$; 0.10M HClO, $K_a = 2.90 \times 10^{-8}$; and 0.10$M$ HIO, $K_a = 2.3 \times 10^{-11}$?

18.41* Repeat the calculations in Exercise 18.27 assuming the acid was 0.250M HCOOH. $K_a = 1.772 \times 10^{-4}$ for formic acid.

18.42* Magnesium hydroxide is a slightly soluble substance. (a) If the pH of a saturated solution of $Mg(OH)_2$ is 10.38, find the K_{sp}. (b) Calculate the solubility of $Mg(OH)_2$ in units of g/100 g of water.

18.43* A fluoridated water supply contains 1 mg/liter of F^-. What is the maximum amount of Ca^{2+}, expressed in g/liter, that can exist in this water supply? $K_{sp} = 2.7 \times 10^{-11}$ for CaF_2.

18.44* The solubility product of iron(III) hydroxide is 3×10^{-39}. (a) At what pH will a 0.001M $Fe(NO_3)_3$ solution just begin to form a precipitate? (b) If the source of the OH^- is a weak base having $K_b = 3.0 \times 10^{-6}$, what is the maximum concentration of that base which is permitted before $Fe(OH)_3$ forms?

18.45* If $K_{sp} = 2.5 \times 10^{-12}$ for silver chromate, calculate the solubility of Ag_2CrO_4 in (a) pure water, (b) a 0.20M $AgNO_3$ solution, and (c) a 0.20M K_2CrO_4 solution.

18.46* Find the concentration of Ca^{2+} and CO_3^{2-} in air-saturated water (the partial pressure of CO_2 is 3×10^{-4} atm) assuming $K_{sp} = 3.84 \times 10^{-9}$ for $CaCO_3$ and $K = 1.4 \times 10^{-6}$ for the reaction

$$CO_2(g) + H_2O(l) + CaCO_3(s) \longrightarrow Ca^{2+}(aq) + 2HCO_3^-(aq)$$

18.47** Consider the following experiment to determine the solubility product for $AgBrO_3$: (i) Ag^+ is mixed with excess BrO_3^- to form the precipitate; (ii) the solution and precipitate are allowed to establish equilibrium; (iii) the solution and precipitate are separated by filtration; (iv) the solution is titrated with SCN^- to determine the amount and concentration of unreacted Ag^+ by forming a much less soluble precipitate, $AgSCN$; (v) the amount and concentration of unreacted BrO_3^- is calculated using normal stoichiometric analysis; and (vi) the results of the fourth and fifth steps are multiplied to obtain the K_{sp}.

The experiment calls for mixing 50.0 ml of $0.010M$ $AgNO_3$ with 50.0 ml of $0.10M$ $KBrO_3$ in the first step. (a) Calculate the ion product for the solution before precipitation occurs. If the titration described in the fourth step required 12.5 ml of $0.010M$ KSCN, calculate the number of moles of (b) Ag^+ originally present, (c) BrO_3^- originally present, (d) SCN^- added, (e) unreacted Ag^+ in the solution, (f) Ag^+ that reacted, (g) BrO_3^- that reacted, and (h) unreacted BrO_3^- in the solution. From these answers, calculate the concentration of (i) unreacted Ag^+ and (j) unreacted BrO_3^-. (k) Calculate the K_{sp} for $AgBrO_3$ and (l) compare the value to the ion product determined in (a).

19 OXIDATION AND REDUCTION

The English language, like most others, is constantly changing. Words go out of fashion and disappear, new words are coined as new ideas and materials develop, and, most importantly, well-established words gradually assume new meanings. Consider, for example, the word "tremendous." It was adapted into our language from a Latin word which meant "to cause to tremble"; that is, "awesome," "dreadful," "fearful." Because we imagine awful and dreadful things to be very large, the word has come to mean "large."

"Oxidation" and "reduction" are other words that have undergone changes in meaning. Earlier (Section 4.12) we defined "oxidation" as an increase and "reduction" as a decrease in oxidation number. These definitions are broad and a great many reactions fit these definitions. Originally, the terms "oxidation" and "reduction" were applied only to the addition and removal of oxygen. How the meanings of these words have expanded illustrates how chemistry moves forward by relating what is well known to what is less well known.

In this chapter the meaning of the terms "oxidation" and "reduction" is explored, and just enough electrochemistry is presented to make clear the relationships between redox reactions and what happens in electrochemical cells. (The electrochemical cells themselves are discussed in Chapter 24.) The main additional topics covered are balancing redox equations, understanding and using standard reduction potentials, and the Nernst equation and the determination of equilibrium constants from standard reduction potentials.

Gain of electrons: reduction

Loss of electrons: oxidation

19.1 What happens in redox reactions?

Originally, "oxidation" meant "combination with oxygen" as illustrated by the equations

$$2Ca(s) + O_2(g) \longrightarrow 2CaO(s) \tag{1}$$
$$2PCl_3(l) + O_2(g) \longrightarrow 2POCl_3(l) \tag{2}$$

That meaning is still useful, but it is not always broad enough for our purposes.

In terms of the theory of ionization, we can rewrite Equation (1) as follows:

$$2Ca(s) + O_2(g) \longrightarrow 2Ca^{2+}(s) + 2O^{2-}(s) \tag{3}$$

for we know that solid calcium oxide is composed of calcium ions and oxide ions. If calcium metal is treated with chlorine, a vigorous reaction takes place, with the formation of the ionic solid, calcium chloride

$$Ca(g) + Cl_2(g) \longrightarrow Ca^{2+}(s) + 2Cl^-(s) \tag{4}$$

Equations (3) and (4) indicate that atoms of calcium are affected in the same way by oxygen and by chlorine—they are converted to calcium ions. If the first equation represents "oxidation" of the calcium atom, surely the second must also represent oxidation. The effect on the calcium atoms would, we know, be the same if the attacking agent were fluorine, bromine, iodine, or sulfur. Reactions with any of these substances convert calcium atoms to calcium ions just as oxygen does. Therefore, these are all called "oxidation reactions," though they do not involve oxygen at all. In every case, each calcium atom loses two electrons.

$$Ca \longrightarrow Ca^{2+} + 2e^-$$

These examples illustrate a broadened meaning of the term "oxidation"—a loss of electrons. Some other examples of oxidation in this sense are

$$2I^- \longrightarrow I_2 + 2e^- \tag{5}$$
$$Fe^{2+} \longrightarrow Fe^{3+} + e^- \tag{6}$$

Any substance that loses electrons is said to be oxidized. It is perhaps unfortunate that the word "oxidation" implies the use of oxygen. Some years ago, the word "de-electronation" was suggested as a replacement. This was a good word scientifically, but it was not generally adopted, perhaps because it is too long and cumbersome for common use.

Now, consider Equation (2). It clearly represents an oxidation reaction under our first definition—"combination with oxygen." Does it fit our second definition, "loss of electrons"? It would not seem to do so, for in each compound the phosphorus atom has eight electrons in its outer shell.

$$:\ddot{C}l:\qquad\qquad :\ddot{C}l:$$
$$:\overset{..}{P}:\ddot{C}l:\qquad :\ddot{O}:\overset{..}{P}:\ddot{C}l:$$
$$:\ddot{C}l:\qquad\qquad :\ddot{C}l:$$

However, by another expansion of meaning, the definition is made to fit. The pair of electrons on the phosphorus atom that is unshared in the PCl_3 molecule is shared with an oxygen atom in the $POCl_3$ molecule; in fact, the oxygen atom attracts electrons more strongly than the phosphorus atom does, so the shared pair spends more time close to the oxygen atom than to the phosphorus atom. Thus, the phosphorus atom has lost some "control" over those electrons.

An atom is said to be oxidized when it undergoes a change that makes it electronically more positive or less negative. Here is where our earlier definitions of oxidation number and of oxidation as an increase in oxidation number fit in. The rules for assigning oxidation numbers are set up so that an increase or decrease in the positive character of an atom results in an increase or decrease in oxidation number.

Reduction is the opposite of oxidation—it is the removal of oxygen from a compound, or the gain of electrons, or a change to a less positive character or a more negative character. For reasons similar to those for oxidation, reduction corresponds to a decrease in oxidation number.

We often speak of the whole molecule or ion containing the oxidized (or reduced) atom as having been oxidized (or reduced). About Equation (2) we would say that phosphorus trichloride has been oxidized.

If an atom loses an electron, where does that electron go? It is not destroyed or lost in space. The electron can do only one thing—it must go to some other atom. The "losing" atom is oxidized and the "gaining" atom is reduced. For this reason, oxidation and reduction must always proceed simultaneously, as we already know (Section 4.12). In redox reactions

electrons gained = electrons lost

and/or

oxidation number gain = oxidation number loss

The meanings of the terms used to describe oxidation and reduc-

TABLE 19.1
Vocabulary of oxidation and reduction

The quotation marks in the last column are there to remind you that electrons need not be fully lost or gained, but may only become closer to one atom or another. Note that an oxidizing agent is reduced and reducing agent is oxidized.

In:	oxidation number:	electrons are:
oxidation	increases	"lost"
reduction	decreases	"gained"
a reducing agent (a reductant)	increases	"lost"
an oxidizing agent (an oxidant)	decreases	"gained"
the oxidized substance	increases	"lost"
the reduced substance	decreases	"gained"

TABLE 19.2

Some simple oxidation–reduction reactions

The oxidation numbers are given in color for atoms that have been oxidized and in black for atoms that have been reduced.

Direct union of the elements

$$\overset{0}{2K}(s) + \overset{0}{Br_2}(l) \longrightarrow \overset{+1\ -1}{2KBr}(s)$$

$$\overset{0}{Ba}(s) + \overset{0}{H_2}(g) \longrightarrow \overset{+2\ -1}{BaH_2}(s)$$

Decomposition

$$\overset{+1\ -2}{2Ag_2O}(s) \overset{\Delta}{\longrightarrow} \overset{0}{4Ag}(s) + \overset{0}{O_2}(g)$$

$$\overset{+2\ -1}{2CuCl_2}(s) \overset{\Delta}{\longrightarrow} \overset{+1}{2CuCl}(s) + \overset{0}{Cl_2}(g)$$

Displacement

$$\overset{0}{Zn}(s) + \overset{+1}{H_2SO_4}(aq) \longrightarrow \overset{+2}{ZnSO_4}(aq) + \overset{0}{H_2}(g)$$

$$\overset{+4}{3Sn^{4+}} + \overset{0}{4Al}(s) \longrightarrow \overset{0}{3Sn}(s) + \overset{+3}{4Al^{3+}}$$

Disproportionation
(auto-oxidation-reduction)

$$\overset{+1}{3NaClO}(aq) \overset{\Delta}{\longrightarrow} \overset{+5}{NaClO_3}(aq) + \overset{-1}{2NaCl}(aq)$$

tion are summarized in Table 19.1. A species that is oxidized is called the reducing *agent* because it brings about reduction of something else. Some simple types of oxidation–reduction reactions are illustrated in Table 19.2. In the first reaction, potassium is oxidized and it is the reducing agent. Bromine is reduced and it is the oxidizing agent. Other terms sometimes used that have the same meanings as oxidizing and reducing agent are "oxidant" and "reductant."

At this point we want to say a few words about redox reactions in organic chemistry. Organic chemists tend to think of oxidation and reduction in terms of atom rearrangements. However, these reactions fulfill the definitions that we have been discussing.

"Oxidation" in organic chemistry frequently means the loss of hydrogen combined with the formation of new bonds between carbon and oxygen or another electronegative element. For example, the oxidation of an alcohol is an organic oxidation reaction.

$$\underset{2\text{-}butanol}{CH_3CH_2\underset{\displaystyle |}{\underset{\displaystyle CH_3}{C}}HOH} \xrightarrow[\overset{\displaystyle \frown}{\text{oxidizing agent}}]{K_2Cr_2O_7,\ H^+} \underset{2\text{-}butanone}{CH_3CH_2\underset{\displaystyle |}{\underset{\displaystyle CH_3}{C}}=O} + H_2O \qquad (7)$$

The fate of the oxidizing or reducing agent is not usually of interest in organic reactions and the reagent is often just written over the arrow, as we have done in Equation (7).

Sometimes the organic compound that is being oxidized is converted into more than one product by the breaking of a carbon–carbon bond

$$\underset{1,2\text{-}propanediol}{\overset{\displaystyle CH_3}{\underset{\displaystyle |}{\underset{\displaystyle H_2COH}{\overset{\displaystyle |}{HCOH}}}}} \xrightarrow{KMnO_4} \underset{acetaldehyde}{\overset{\displaystyle CH_3}{\underset{\displaystyle |}{HC=O}}} + \underset{formaldehyde}{H_2C=O}$$

Reduction in organic chemistry usually means the addition of hydrogen, the removal of oxygen or another electronegative element, or these two changes combined, for example,

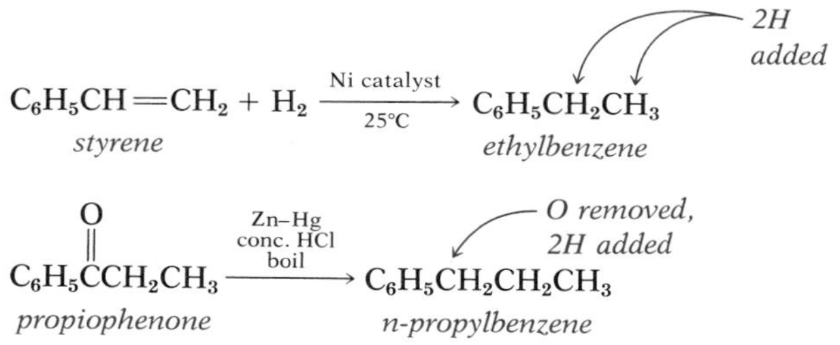

$$C_6H_5CH=CH_2 + H_2 \xrightarrow[25°C]{Ni\ catalyst} C_6H_5CH_2CH_3$$

styrene   *ethylbenzene*

2H added

$$\underset{propiophenone}{C_6H_5\overset{O}{\overset{\|}{C}}CH_2CH_3} \xrightarrow[boil]{\substack{Zn-Hg \\ conc.\ HCl}} \underset{n\text{-}propylbenzene}{C_6H_5CH_2CH_2CH_3}$$

O removed, 2H added

19.2 Relationship between electrochemistry and redox reactions

Every chemical reaction is accompanied by a characteristic energy change. Most of the time, thermal energy is exchanged—a reaction absorbs or gives off heat. Occasionally light is also given off, for example, in combustion. If an oxidizing agent and a reducing agent are mixed in the same reaction vessel and a spontaneous redox reaction occurs, it will be accompanied by the liberation of heat in most cases. For instance, spontaneous combustion of paint-soaked rags is a spontaneous redox reaction. However, if the oxidation and reduction portions of a redox reaction are physically separated from each other, but provided with external connections through which electrons and ions can flow, electrical energy rather than thermal energy is often liberated.

Electrochemistry deals with oxidation–reduction reactions that either produce or utilize electrical energy. Any device in which an electrochemical reaction occurs is called an **electrochemical cell,** or just a "cell." The reactions in electrochemical cells are all reactions in which actual electron transfer from one atom, molecule, or ion to another takes place.

The uses of electrochemistry and some of the many types of cells that are of practical value are discussed in Chapter 24. In later sections of this chapter we are mainly interested in electrochemistry as a tool for the understanding of redox reactions.

The stoichiometry of redox reactions

19.3 Half-reactions

An oxidation–reduction reaction can be thought of as the sum of two **half reactions**—reactions representing either oxidation only or reduction only. The first step in choosing the half-reactions for a given

Ion–electron equations:
both number of atoms and
charge must balance

redox reaction is to recognize, by changes in oxidation number, which species are oxidized and which are reduced. (At this point you might want to review the rules for assigning oxidation numbers, given in Section 4.11.)

In the direct combination of sodium and chlorine

$$\overset{0}{2Na}(s) + \overset{0}{Cl_2}(g) \longrightarrow \overset{+1\ -1}{2NaCl}(s)$$

sodium is oxidized (oxidation number change $0 \rightarrow 1$) and chlorine is reduced ($0 \rightarrow -1$). The half-reactions are represented by the following equations, one involving electron loss and one involving electron gain.

oxidation	$2Na \longrightarrow 2Na^+ + 2e^-$	(8)
reduction	$Cl_2 + 2e^- \longrightarrow 2Cl^-$	(9)

Such equations are called **ion–electron equations**—they include only the species directly involved in either the oxidation or reduction of a single type of atom, molecule, or ion. To simplify matters the designations (*g*), (*l*), and so on, are left out of most equations in this chapter.

Like net ionic equations (Section 4.4) each ion–electron equation must be balanced both as to number of atoms and charge. An ion–electron equation includes the oxidized and reduced forms of a single atom, ion, or molecule, plus enough electrons to convert one into the other, plus any other species needed to balance the equation. The interchangeable oxidized and reduced forms of the same species are called a **redox couple**, or just a "couple." For example, for Equations (8) and (9) the redox couples are Na(0)/Na(1) and Cl(0)/Cl(−1), where the numbers in parentheses are the oxidation numbers. Half-reactions can take place in either direction, depending upon the reaction conditions and the other species present, and the equations can be written as either oxidation or reduction. It is possible to select half-reactions and write ion–electron equations for any redox reaction, whether or not it is a reaction in which actual electron gain or loss occurs.

Here are a few more examples of half-reactions and ion–electron equations. Take a few moments to be sure that you know which ion–electron equation represents oxidation and which represents reduction.

thermal decomposition of silver(I) oxide

$$2Ag_2O \overset{\Delta}{\longrightarrow} 4Ag + O_2$$
$$2O^{2-} \longrightarrow O_2 + 4e^-$$
$$Ag^+ + e^- \longrightarrow Ag$$

displacement of copper from copper(II) sulfate by elemental iron

$$Cu^{2+} + Fe \longrightarrow Cu + Fe^{2+}$$
$$Fe \longrightarrow Fe^{2+} + 2e^-$$
$$Cu^{2+} + 2e^- \longrightarrow Cu$$

reduction of iron(III) chloride to iron(II) chloride by tin(II) chloride

$$Sn^{2+} + 2Fe^{3+} \longrightarrow 2Fe^{2+} + Sn^{4+}$$
$$Sn^{2+} \longrightarrow Sn^{4+} + 2e^-$$
$$Fe^{3+} + e^- \longrightarrow Fe^{2+}$$

Like any other chemical equation, an ion–electron equation must always describe an observable chemical change. The element undergoing a change in oxidation number must appear as part of the formula of the actual substance being oxidized or reduced. In other words, ions should not appear unless they exist as such in the reaction, either in ionic solids or as ions in solution. For example, the reduction in acidic solution of nitrate ion to nitrogen(II) oxide by hydrogen sulfide,

unbalanced $NO_3^- + H_2S + H^+ \longrightarrow NO + S + H_2O$
equation

is represented by the ion–electron equations

$$H_2S \longrightarrow S + 2H^+ + 2e^- \tag{10}$$
$$NO_3^- + 4H^+ + 3e^- \longrightarrow NO + 2H_2O \tag{11}$$

Not by $S^{2+} \longrightarrow S^0 + 2e^-$ and $N^{5+} + 3e^- \longrightarrow N^{2+}$.

You may be wondering why H^+ and H_2O appear in Equations (10) and (11). Many redox reactions occur in aqueous systems—in fact, most of the redox reactions in which we are interested do so. Whatever hydrogen or oxygen atoms are required to balance the equations must come from water and its ions. The type of product obtained often depends upon the pH of the system. Hydrogen ions and hydroxide ions never show up in the same ion–electron equation, because, obviously, a solution must be either alkaline or acidic, and only one or the other type of ion can be present in appreciable quantities.

The number of electrons gained and lost in a redox reaction must be equal. Therefore, Equation (10) must be multiplied by 3 and Equation (11) by 2 before these two ion–electron equations can be added together

$$3H_2S \longrightarrow 3S + \cancel{6H^+} + \cancel{6e^-}$$
$$2H^+$$
$$2NO_3^- + \cancel{8H^+} + \cancel{6e^-} \longrightarrow 2NO + 4H_2O$$
$$\overline{3H_2S + 2NO_3^- + 2H^+ \longrightarrow 3S + 2NO + 4H_2O}$$

to give the overall balanced equation

$$3H_2S(g) + 2NO_3^- + 2H^+ \longrightarrow 3S(s) + 2NO(g) + 4H_2O$$

Ion–electron equations do not in any way represent the mechanism of a reaction. However, knowing how to write ion–electron equations is necessary in the bookkeeping of balancing redox equations (next section), in comparing the strengths of oxidizing agents (Sections 19.5–19.8), and in the study of electrochemical cells (Chapter 24).

19.4 Balancing oxidation–reduction equations

Balancing redox equations that include the formulas of many species appears at first glance to be a formidable task. Such equations are often too complex to be balanced by inspection. However, by carefully applying a set procedure, as with equilibrium problems, finding the correct answer is greatly simplified.

The two most widely used methods for doing this are the oxidation number method and the ion–electron or half-reaction method. Both of these methods depend upon the fundamental fact that in any redox reaction the total amount of oxidation must equal the total amount of reduction.

The half-reaction method is based upon the equal gain and loss of electrons by the reduced and oxidized substances. The rules governing the half-reaction method for balancing redox equations are as follows (these rules are summarized in Table 19.3):

(1) *Write the overall equation, unbalanced, including all species that actually undergo a change in the redox reaction.* Do not include spectator ions.

(2) *Identify the oxidized and reduced substances and write the two unbalanced ion–electron equations.* If it is not immediately obvious what is oxidized or reduced, find the oxidation numbers for all species and see which ones change. Use ions only if they are present as such, usually as ions in solution. Substances that are present as solid reactants or as products that precipitate out during reaction in solution are represented by the complete compound formula in the ion–electron equation. Gases that are reactants or products in solution reactions are represented by their molecular formulas. For reactions not in solution (e.g., see Example 19.1), ions appear where they are present in ionic solids.

(3) *Balance each ion–electron equation for atoms.* Begin with the elements undergoing the oxidation number change. For neutral aqueous systems, H_2O can be added; for acidic systems, H_2O and H^+ can be added; for alkaline systems, H_2O and OH^- can be added. For non-aqueous systems, H^+ or OH^-, and H_2O can be added if they cancel in the overall equation. (These ideas will become clear in Examples 19.2–19.5.)

(4) *Balance each ion–electron equation for charge.* This is accomplished by adding electrons on one side or the other.

(5) *If necessary, multiply the half-reactions by appropriate factors so that electrons gained equal electrons lost.*

(6) *Add the ion–electron equations to get the overall equation.* The electrons should automatically cancel out. Be sure to cancel any other species that appear on both sides of the equation and to check that atoms and charges are both balanced. If you wish you may then add back any spectator ions and write the complete equation for the substances involved.

TABLE 19.3

Balancing redox equations by the half-reaction method

1. **Write overall unbalanced equation.**

2. **Identify oxidized and reduced substances and write their unbalanced ion–electron equations.**

3. **Balance atoms in each ion–electron equation.**

4. **Balance charge in each ion–electron equation.**

5. **Multiply ion–electron equations by appropriate factors so that electrons gained equals electrons lost.**

6. **Add the ion–electron equations.**

EXAMPLE 19.1

■ A student working with chlorine gas noticed a discoloration of his gold ring as a result of the formation of gold(III) chloride, $AuCl_3$, on the surface. Write the half-reactions and the balanced overall equation for this reaction.

Step 1. Identify the substances involved in the reaction and write the unbalanced equation.

$$Au(s) + Cl_2(g) \longrightarrow AuCl_3(s)$$

Step 2. Write the half-reactions.

$$\overset{0}{Au} \longrightarrow \overset{+3}{Au^{3+}} \qquad \overset{0}{Cl_2} \longrightarrow \overset{-1}{Cl^-}$$

oxidation — Au → Au³⁺   reduction — Cl₂ → Cl⁻

Step 3. Balance each ion–electron equation with respect to the number of each type of atoms that appear:

$$Au \longrightarrow Au^{3+} \qquad Cl_2 \longrightarrow 2Cl^-$$

Step 4. Balance each ion–electron equation with respect to the total electrical charge that appears on each side of the arrow. Because oxidation–reduction involves electron transfer, we write

$$Au \longrightarrow Au^{3+} + 3e^- \qquad Cl_2 + 2e^- \longrightarrow 2Cl^-$$

as the two half-reactions describing the oxidation of Au to Au^{3+} and the reduction of Cl_2 to Cl^-.

Step 5. To combine the half-reactions, the number of electrons produced by the oxidation half-reaction must equal the number of electrons supplied to the reduction half-reaction, so we write

$$2(Au \longrightarrow Au^{3+} + 3e^-) \qquad 3(Cl_2 + 2e^- \longrightarrow 2Cl^-)$$
$$2Au \longrightarrow 2Au^{3+} + 6e^- \qquad 3Cl_2 + 6e^- \longrightarrow 6Cl^-$$

Step 6. Add the two half-reactions together to get the overall equation, canceling where appropriate

$$2Au \longrightarrow 2Au^{3+} + \cancel{6e^-}$$
$$\underline{3Cl_2 + \cancel{6e^-} \longrightarrow 6Cl^-}$$
$$2Au + 3Cl_2 \longrightarrow 2Au^{3+} + 6Cl^-$$

A quick check shows that the overall equation is balanced with respect to Au, Cl, and the charge on both sides [on the left, $2(0) + 3(0) = 0$; on the right, $2(+3) + 6(-1) = 0$]. Because the $AuCl_3$ produced is in the solid form on the surface of the ring, we might choose to write the overall equation as

$$2Au(s) + 3Cl_2(g) \longrightarrow 2AuCl_3(s)$$

■

Two additional principles help in balancing ion–electron equations involving oxygen-containing anions under acidic or alkaline conditions: (1) For reactions in acidic solution, the ion–electron equation must include enough hydrogen ions to convert the oxygen in the anion to water (excluding any oxygen atoms used in other products). (2) In alkaline solutions, metal ions other than alkali metals (Representative Group I) form insoluble hydroxides. The ion–electron equation must include enough hydroxide ions to form oxo anions or insoluble hydroxides, whichever are appropriate.

Table 19.4 compares a number of redox couples under acidic and alkaline conditions. For the K^+ ions, pH has no effect since potassium hydroxide is soluble. The same situation would be true for the other alkali metal ions.

EXAMPLE 19.2

■ In each cell of the lead storage battery commonly used in automobiles, $Pb(s)$ reacts with $PbO_2(s)$ in the presence of aqueous H_2SO_4 to form $PbSO_4(s)$. Write the ion–electron equations and the overall balanced equation for this reaction.

Step 1.

$$Pb(s) + PbO_2(s) + H^+ + SO_4^{2-} \longrightarrow PbSO_4(s)$$

Step 2.

$$\overset{0}{Pb} \longrightarrow \overset{+2}{PbSO_4} \qquad \overset{+4}{PbO_2} \longrightarrow \overset{+2}{PbSO_4}$$

Step 3. Note that the only S-containing species that can be used to balance the equations is SO_4^{2-}, since it is the only S-containing species involved in the reaction; H^+ and H_2O are used in balancing since the solution is acidic.

$$Pb + SO_4^{2-} \longrightarrow PbSO_4 \qquad PbO_2 + 4H^+ + SO_4^{2-} \longrightarrow PbSO_4 + 2H_2O$$

Step 4.

$$Pb + SO_4^{2-} \longrightarrow PbSO_4 + 2e^- \qquad PbO_2 + 4H^+ + SO_4^{2-} + 2e^- \longrightarrow$$
$$PbSO_4 + 2H_2O$$

Step 5. Not needed.

Step 6.

$$Pb + SO_4^{2-} \longrightarrow PbSO_4 + 2e^-$$
$$\underline{PbO_2 + 4H^+ + SO_4^{2-} + 2e^- \longrightarrow PbSO_4 + 2H_2O}$$
$$Pb + PbO_2 + 4H^+ + 2SO_4^{2-} \longrightarrow 2PbSO_4 + 2H_2O$$

This equation is balanced with respect to Pb, S, O, H, and charge. We could rewrite it as

$$Pb(s) + PbO_2(s) + 2H_2SO_4(aq) \longrightarrow 2PbSO_4(s) + 2H_2O(l)$$

■

Couple	
K(1)/K(0)	$K^+ + e^- \longrightarrow K$
	$K^+ + e^- \longrightarrow K$
Cd(2)/Cd(0)	$Cd^{2+} + 2e^- \longrightarrow Cd$
	$Cd(OH)_2 + 2e^- \longrightarrow Cd + 2OH^-$
Fe(3)/Fe(2)	$Fe^{3+} + e^- \longrightarrow Fe^{2+}$
	$Fe(OH)_3 + e^- \longrightarrow Fe(OH)_2 + OH^-$
N(5)/N(3)	$NO_3^- + 3H^+ + 2e^- \longrightarrow HNO_2 + H_2O$
	$NO_3^- + H_2O + 2e^- \longrightarrow NO_2^- + 2OH^-$
Cr(6)/Cr(3)	$Cr_2O_7^{2-} + 14H^+ + 6e^- \longrightarrow 2Cr^{3+} + 7H_2O$
	$CrO_4^{2-} + 4H_2O + 3e^- \longrightarrow Cr(OH)_3 + 5OH^-$
Cl(5)/Cl(−1)	$ClO_3^- + 6H^+ + 6e^- \longrightarrow Cl^- + 3H_2O$
	$ClO_3^- + 3H_2O + 6e^- \longrightarrow Cl^- + 6OH^-$

EXAMPLE 19.3

■ In a Ni–Cd alkaline cell (a "ni-cad" battery) Cd(*s*) forms $Cd(OH)_2(s)$, and $Ni_2O_3(s)$ forms $Ni(OH)_2(s)$ in the presence of aqueous OH^-. Write the ion–electron equations and overall balanced equation for this reaction.

Step 1.

$$Cd(s) + Ni_2O_3(s) \longrightarrow Cd(OH)_2(s) + Ni(OH)_2(s)$$

Step 2.

$$\overset{0}{Cd} \longrightarrow \overset{+2}{Cd(OH)_2} \qquad \overset{+3}{Ni_2O_3} \longrightarrow \overset{+2}{2Ni(OH)_2}$$

Step 3. Note that OH^- and H_2O can be added because the reaction takes place in an alkaline solution.

$$Cd + 2OH^- \longrightarrow Cd(OH)_2 \qquad Ni_2O_3 + 3H_2O \longrightarrow 2Ni(OH)_2 + 2OH^-$$

Step 4.

$$Cd + 2OH^- \longrightarrow Cd(OH)_2 + 2e^- \qquad Ni_2O_3 + 3H_2O + 2e^- \longrightarrow 2Ni(OH)_2 + 2OH^-$$

Step 5. Not needed.

Step 6.

$$Cd + \cancel{2OH^-} \longrightarrow Cd(OH)_2 + \cancel{2e^-}$$
$$\underline{Ni_2O_3 + 3H_2O + \cancel{2e^-} \longrightarrow 2Ni(OH)_2 + \cancel{2OH^-}}$$
$$Cd + Ni_2O_3 + 3H_2O \longrightarrow Cd(OH)_2 + 2Ni(OH)_2$$

This equation is completely balanced and may be rewritten as:

$$Cd(s) + Ni_2O_3(s) + 3H_2O(l) \longrightarrow Cd(OH)_2(s) + 2Ni(OH)_2(s)$$

■

OXIDATION AND
REDUCTION [19.4]

EXAMPLE 19.4

■ The nitrate ion is reduced to ammonia by elemental aluminum under alkaline conditions with the formation of $Al(OH)_4^-$. Using the ion–electron method, find the balanced overall equation.

Step 1.

$$Al(s) + NO_3^- + OH^- \longrightarrow NH_3 + Al(OH)_4^-$$

Step 2.

$$\overset{0}{Al} \longrightarrow \overset{+3}{Al(OH)_4^-} \qquad \overset{+5}{NO_3^-} \longrightarrow \overset{+3}{NH_3}$$

Step 3. Balancing the oxidation equation is easy—only OH^- need be added.

$$Al + 4OH^- \longrightarrow Al(OH)_4^-$$

Here is an easy way to figure out how the OH^- ions and the H_2O molecules balance for the NO_3^- reduction (or for any half-reaction in alkaline solution that needs added H and O). First, balance the equation as though it were in acidic solution

$$NO_3^- + 9H^+ \longrightarrow NH_3 + 3H_2O$$

Then add enough OH^- ions to each side to convert the H^+ to H_2O. Do not forget that this is necessary to return the equation to alkaline conditions. Here we need $9OH^-$ ions to form $9H_2O$ molecules on the left

$$NO_3^- + 9H_2O \longrightarrow NH_3 + 3H_2O + 9OH^-$$

Then cancel any extra water molecules, to give the balanced equation

$$NO_3^- + 6H_2O \longrightarrow NH_3 + 9OH^-$$

Step 4.

$$Al + 4OH^- \longrightarrow Al(OH)_4^- + 3e^- \qquad NO_3^- + 6H_2O + 8e^- \longrightarrow NH_3 + 9OH^-$$

Step 5. To equalize the number of electrons in the ion-electron equations, there must be 24 electrons on each side

$$8(Al + 4OH^- \longrightarrow Al(OH)_4 + 3e^-)$$
$$3(NO_3^- + 6H_2O + 8e^- \longrightarrow NH_3 + 9OH^-)$$

Step 6.

$$\begin{array}{c} 5OH^- \\ 8Al + \cancel{32}OH^- \longrightarrow 8Al(OH)_4^- + \cancel{24e^-} \\ \underline{3NO_3^- + 18H_2O + \cancel{24e^-} \longrightarrow 3NH_3 + \cancel{27}OH^-} \\ 8Al + 3NO_3^- + 5OH^- + 18H_2O \longrightarrow 8Al(OH)_4^- + 3NH_3 \end{array}$$

■

EXAMPLE 19.5

■ When ignited, $(NH_4)_2Cr_2O_7(s)$ undergoes decomposition, forming $Cr_2O_3(s)$, $H_2O(g)$, and $N_2(g)$. Using the ion–electron method, write the balanced overall equation.

Step 1.

$$(NH_4)_2Cr_2O_7(s) \longrightarrow Cr_2O_3(s) + H_2O(g) + N_2(g)$$

Steps 2–6 can be combined into one set of equations:

Steps 2–6.

$$\begin{array}{l} 2NH_4^+ \longrightarrow N_2 + 8H^+ + 6e^- \\ \underline{Cr_2O_7^{2-} + 8H^+ + 6e^- \longrightarrow Cr_2O_3 + 4H_2O} \\ 2NH_4^+ + Cr_2O_7^{2-} \longrightarrow N_2 + Cr_2O_3 + 4H_2O \end{array}$$

Note that we used H^+ and H_2O to carry out the balance of H and O. As long as these terms cancel in the overall equation—except for what should be there—this is a convenient method for balancing reactions that are not even occurring in solution. The overall balanced equation may be written as

$$(NH_4)_2Cr_2O_7(s) \longrightarrow N_2(g) + Cr_2O_3(s) + 4H_2O(g) \qquad ■$$

The oxidation number method for balancing redox equations requires the following steps (summarized in Table 19.5):

Step A. Write the unbalanced overall equation. Determine from the oxidation numbers which elements are oxidized and reduced, and what the oxidation number change is for each.

Step B. Add to the equation appropriate coefficients for each of the reactants and products that include the atoms which are oxidized and reduced so that

total increase in oxidation number

= total decrease in oxidation number

This is done by finding the necessary ratio of atoms of oxidized element to atoms of reduced element that must appear on each side of the equation.

Step C. Balance the number of atoms of the remaining elements by inspection. It is best to start with elements other than O or H and to leave H_2O until last, getting the coefficient for water from the number of H atoms.

An advantage of the oxidation number method is that, since ion–electron equations do not have to be written, it is not necessary to know whether the reactants and products are covalent or ionic compounds.

TABLE 19.5

Balancing redox equations by the oxidation number method

A. **Write overall unbalanced equation. Identify oxidized and reduced substances and find the oxidation number change for each.**

B. **Add coefficients for reactants and products so that total increase in oxidation number equals total decrease in oxidation number.**

C. **Balance the atoms of the remaining elements by inspection.**

EXAMPLE 19.6

■ Ammonia is oxidized by oxygen to give nitrogen(II) oxide and water. Using the oxidation number method, find the balanced overall equation.

Step A.

$$\overset{-3}{N}H_3 + \overset{0}{O_2} \longrightarrow \overset{+2-2}{NO} + \overset{-2}{H_2O}$$

oxidation

$$\overset{-3}{N} \longrightarrow \overset{+2}{N} \qquad \text{an increase of 5 in oxidation number}$$

reduction

$$\overset{0}{O} \longrightarrow \overset{-2}{O} \qquad \text{a decrease of 2 in oxidation number}$$

Step B. Gain and loss of oxidation number can be equalized by including on each side of the equation 2 N atoms (total oxidation number increase = $2 \times 5 = 10$) for each 5 O atoms (total oxidation number decrease = $5 \times -2 = -10$). Since there are 2 atoms of oxygen in the reactant O_2 molecules, we must use 4 N atoms to 10 O atoms on this side. Therefore, equalization requires 4 NH_3 molecules to 5 O_2 molecules to the left of the arrow and 4 NO molecules to 6 H_2O molecules to the right of the arrow.

$$4(+5) = +20$$
$$4NH_3 + 5O_2 \longrightarrow 4NO + 6H_2O$$
$$10(-2) = -20$$

Step C. In this case equalization of oxidation number change balances the equation completely. The balanced equation is

$$4NH_3(g) + 5O_2(g) \longrightarrow 4NO(g) + 6H_2O(l) \qquad ■$$

EXAMPLE 19.7

■ The gas-phase reaction of HNO_3 and H_2S gives the same products as in the preceding example, plus solid sulfur. Using the oxidation number method, find the balanced overall equation.

Step A.

$$\overset{-2}{H_2S} + \overset{+5}{H}\overset{}{N}O_3 \longrightarrow \overset{+2}{N}O + H_2O + \overset{0}{S}$$

$$\overset{-2}{S} \longrightarrow \overset{0}{S} \qquad \text{an increase of 2 in oxidation number}$$

$$\overset{+5}{N} \longrightarrow \overset{+2}{N} \qquad \text{a decrease of 3 in oxidation number}$$

Step B. To equalize the oxidation number change, 3 S atoms to 2 N atoms are required on each side of the arrow

$$3H_2S + 2HNO_3 \longrightarrow 2NO + H_2O + 3S$$

Step C. Since 8 H atoms now appear to the left of the arrow, the inclusion of 4 H_2O molecules to the right of the arrow

$$\begin{array}{c} \overbrace{}^{3(+2) = +6} \\ 3H_2S + 2HNO_3 \longrightarrow 2NO + 4H_2O + 3S \\ \underbrace{}_{2(-3) = -6} \end{array}$$

both equalizes the H atoms and provides the 4 additional O atoms needed to complete balancing without altering the necessary S to N ratio.

The balanced equation is

$$3H_2S(g) + 2HNO_3(g) \longrightarrow 2NO(g) + 4H_2O(l) + 3S(s) \qquad \blacksquare$$

EXAMPLE 19.8

■ Manganese(IV) oxide and lead(IV) oxide react in the presence of aqueous nitric acid to give permanganic acid ($HMnO_4$), lead(II) nitrate, and water. Write a balanced equation for this reaction by using the oxidation number method for balancing oxidation–reduction equations.

Step A.

$$\overset{+4}{Mn}O_2 + \overset{+4}{Pb}O_2 + HNO_3 \longrightarrow \overset{+7}{H}MnO_4 + \overset{+2}{Pb}(NO_3)_2 + H_2O$$

$$\overset{+4}{Mn} \longrightarrow \overset{+7}{Mn} \qquad \text{an increase of 3 in oxidation number}$$

$$\overset{+4}{Pb} \longrightarrow \overset{+2}{Pb} \qquad \text{a decrease of 2 in oxidation number}$$

Step B. To equalize the changes in oxidation number, a ratio of 2 Mn atoms to 3 Pb atoms is needed on each side of the equation

$$2MnO_2 + 3PbO_2 + HNO_3 \longrightarrow 2HMnO_4 + 3Pb(NO_3)_2 + H_2O$$

Step C. Next we chose to balance the N atoms. This is accomplished by adding the coefficient 6 so that there are 6 NO_3 groups on the left. The next step is to determine from the 6 H atoms on the left that there should be 2 H_2O molecules on the right to balance the hydrogen atoms.

$$2MnO_2 + 3PbO_2 + 6HNO_3 \longrightarrow 2HMnO_4 + 3Pb(NO_3)_2 + 2H_2O$$

Checking the O atoms (28 on each side) shows that the equation is now completely balanced. The final, balanced equation is

$$2MnO_2(s) + 3PbO_2(s) + 6HNO_3(aq) \longrightarrow$$
$$2HMnO_4(aq) + 3Pb(NO_3)_2(aq) + 2H_2O(l)$$

The advantage of the oxidation number method shows up clearly in this example. We did not need to know whether MnO_2, PbO_2, and $Pb(NO_3)_2$ are ionic or covalent compounds. ∎

Strengths of oxidizing and reducing agents

Each ion–electron equation includes both the oxidized form and the reduced form of the same species. Therefore, a complete redox equation contains at least two oxidants—the electron acceptors—and two reductants—the electron donors. [An analogy can be made with the two acids (proton donors) and two bases (proton acceptors), in a Brønsted–Lowry acid–base reaction.]

A redox reaction can occur only if the two oxidants differ in oxidizing strengths and the two reductants differ in reducing strengths. The substances present in the greatest amount when equilibrium is reached are determined by the relative oxidizing and reducing strengths of the substances involved and the initial amounts of each.

The best way to compare the strengths of oxidants and reductants is to put them in order according to some measurable quantity that reflects their strengths. (For acids and bases, K_a values serve this purpose.) Such a list can be used qualitatively—to find, say, which of two oxidants is stronger and in which direction a reaction will go. And it can be used quantitatively, to find the magnitudes of differences in strengths.

For each redox couple, the standard reduction potential, which is mathematically related to the equilibrium constant, furnishes a measure of the direction and extent of the interconversion of reductant and oxidant. To help understand the standard reduction potential, we must first take a look at how a simple electrochemical cell works.

19.5 An electrochemical cell

When a piece of zinc is dropped into a copper sulfate solution, a spontaneous oxidation–reduction reaction occurs.

$$Cu^{2+} + Zn(s) \rightleftharpoons Zn^{2+} + Cu(s)$$

The zinc dissolves, the solution loses the blue color characteristic of the copper(II) ions, and free metallic copper appears. Figure 19.1 shows an electrochemical cell in which the same reaction takes place. On one side a zinc electrode is immersed in a Zn^{2+} solution. On the

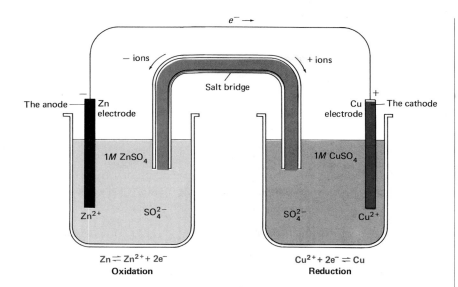

FIGURE 19.1
A zinc–copper cell. *Current flows spontaneously in this cell, for which, at 25°C, E° = +1.10 V. A typical salt bridge contains a saturated solution of potassium chloride. Positive ions from the salt bridge are attracted to the right side of this cell to replace the Cu^{2+} ions being used up there. Negative ions diffuse from the salt bridge into the left-hand compartment, attracted by the Zn^{2+} ions generated there. (See Section 24.1 for a further discussion of the salt bridge.)*

other side a copper electrode is immersed in a Cu^{2+} solution. An external conductor provides a pathway for electrons. When the circuit is complete the zinc electrode erodes away, fresh copper is deposited on the copper electrode, the copper(II) solution loses its blue color, and electrons flow from the zinc electrode to the copper electrode.

The current leaves or enters the conducting medium in an electrochemical cell through the electrodes. The medium may be a solution of an electrolyte, a molten salt, or a solid. The chemical oxidation or reduction takes place at the surface of the electrode. In some cases the electrode material takes part in the reaction (as in the cell of Figure 19.1); in other cases the electrode does not react, but only carries the current.

One electrode and its surrounding solution (for example, the copper electrode dipping into the copper sulfate solution on the right in Figure 19.1) make up a **half-cell.**

To summarize, three things happen in an electrochemical cell: (1) an oxidation reaction occurs at the surface of one electrode and a reduction reaction at the surface of the other electrode, (2) electrons flow through an external circuit, and (3) ions flow in the electrolyte solutions. All three of these processes must be able to take place simultaneously, or current will not flow through the completed circuit. Note that although the two half-reactions are physically separated, they too can only occur simultaneously.

The spontaneous flow of electrons in the external conductor (Figure 19.1) shows that electrons are attracted more strongly by one half-cell than by the other. There is a difference in electrical potential between the two half-cells. When no current is flowing in the circuit (an open circuit), the potential difference between the two half-cells is equal to what is called the "electromotive force" of the cell.

A formal definition of electromotive force, or emf, is the work per unit charge as the charge moves between two points in an electric field. In what is possibly an easier way of thinking of it, emf is the electrical force produced when any other form of energy is converted into electrical energy. For an electrochemical cell, **emf** is often said to represent the "chemical driving force" of the cell reaction. Therefore, emf, or open-circuit potential, is just what we need—a measurable quantity that reflects the relative oxidizing and reducing strengths of the species involved in a redox reaction.

The electrical potential of a cell is measured as follows: A source of variable and known voltage and a galvanometer are introduced into the external circuit. The known voltage is varied until the galvanometer shows that no current is flowing. At this point the known voltage exactly opposes the tendency of electrons to flow from one half-cell to the other and is a measure of the electromotive force, or potential, of the cell—the relative amounts of energy available in each half-cell for the production of an electric current. The unit of emf is the volt (V).

The potential measured for the zinc–copper cell of Figure 19.1 is 1.10 V when the concentration of ions is $1M$ in each half-cell. We know from observations that the reaction between Zn and Cu^{2+} proceeds spontaneously. Therefore, 1.10 V reflects the total effect of the tendency of Zn atoms to lose electrons and of Cu^{2+} ions to gain electrons under the conditions in this cell. The value of the emf would be different for different concentrations of the solutions in the cell or for different temperatures. In order to compare half-reactions, a set of standard conditions is chosen, and the voltage of each half-cell is measured under these conditions relative to a standard, or reference, electrode.

19.6 Standard reduction potentials

Each redox couple can be described by an ion–electron equation written as a reduction

$$\text{oxidant} + ne^- \longrightarrow \text{reductant}$$

with the electrons, n in number, to the left of the arrow. The emf of the half-cell in which such a reaction takes place is called an **electrode potential.** The conditions chosen for reporting electrode potentials are the **standard state conditions** of thermodynamics:

(1) All substances in solution are at $1M$ concentration
(2) All gases are at 1 atm pressure
(3) All solids and liquids are pure and in their most stable or most common state at 1 atm pressure

The symbol for an electrochemical potential measured under these conditions is $E°$. In reporting values for $E°$, the temperature must be specified, as it must always be for standard state conditions. Unless indicated differently, all $E°$ values in this book are given for 298°K; for simplicity, we omit the subscript on $E°$, which could be written as $E°_{298}$.

Concentration at the standard state would be correctly expressed thermodynamically as activity, in order to allow for the effects of interionic attraction. Here again assume that in dilute solutions, activity coefficients approach unity, and that molarity and activity are identical. Molar concentrations are used with the understanding that although the numerical results are not exact, they are of the correct orders of magnitude and thus do not alter the conclusions we draw.

The hydrogen gas–proton redox couple is chosen as the reference point for the relative scale of electrode potentials. This electrode is assigned a standard potential of 0.0000 V at 298°K.

$$2H^+(1M) + 2e^- \longrightarrow H_2(g, 1 \text{ atm}) \qquad E° = 0.0000 \text{ V}$$

The entire emf of a cell that includes an H_2/H^+ electrode plus a second electrode is assigned to the second electrode. The potential of 0.0000 V for the H_2/H^+ couple and all other potentials established relative to it under the standard state conditions are called **standard electrode potentials,** or **standard reduction potentials.**

Table 19.6 gives some standard reduction potentials. The reduction potential for a given couple has a *positive* value if the oxidant in the couple has a greater tendency to *gain* electrons than does the hy-

TABLE 19.6

Standard reduction potentials

Couple	Half-reaction	$E°(V)$
F(0)/F(−1)	$F_2 + 2e^- \longrightarrow 2F^-$	2.87
O(−1)/O(−2)	$H_2O_2 + 2H^+ + 2e^- \longrightarrow 2H_2O$	1.776
Mn(7)/Mn(2)	$MnO_4^- + 8H^+ + 5e^- \longrightarrow Mn^{2+} + 4H_2O$	1.491
Cl(0)/Cl(−1)	$Cl_2 + 2e^- \longrightarrow 2Cl^-$	1.3583
Cr(6)/Cr(3)	$Cr_2O_7^{2-} + 14H^+ + 6e^- \longrightarrow 2Cr^{3+} + 7H_2O$	1.33
O(0)/O(−2)	$O_2 + 4H^+ + 4e^- \longrightarrow 2H_2O$	1.229
Br(0)/Br(−1)	$Br_2(aq) + 2e^- \longrightarrow 2Br^-$	1.087
Fe(3)/Fe(2)	$Fe^{3+} + e^- \longrightarrow Fe^{2+}$	0.770
Cu(2)/Cu(0)	$Cu^{2+} + 2e^- \longrightarrow Cu$	0.3402
H(1)/H(0) (acidic)	$2H^+ + 2e^- \longrightarrow H_2$	0.0000
Pb(2)/Pb(0)	$Pb^{2+} + 2e^- \longrightarrow Pb$	−0.1263
Fe(2)/Fe(0)	$Fe^{2+} + 2e^- \longrightarrow Fe$	−0.409
C(4)/C(3)	$2CO_2 + 2H^+ + 2e^- \longrightarrow H_2C_2O_4$	−0.49
Zn(2)/Zn(0)	$Zn^{2+} + 2e^- \longrightarrow Zn$	−0.7628
H(1)/H(0) (alkaline)	$2H_2O + 2e^- \longrightarrow H_2 + 2OH^-$	−0.8277
Al(3)/Al(0)	$Al^{3+} + 3e^- \longrightarrow Al$	−1.67
Mg(2)/Mg(0)	$Mg^{2+} + 2e^- \longrightarrow Mg$	−2.375
Na(1)/Na(0)	$Na^+ + e^- \longrightarrow Na$	−2.7109
K(1)/K(0)	$K^+ + e^- \longrightarrow K$	−2.924
Li(1)/Li(0)	$Li^+ + e^- \longrightarrow Li$	−3.045

strongest oxidizing agent

strongest reducing agent

Increasing oxidizing strength

Increasing reducing strength

drogen ion, with both couples at standard state conditions. For example, oxygen, a strong oxidizing agent, has a fairly large positive standard reduction potential.

$$O_2 + 4H^+ + 4e^- \longrightarrow 2H_2O \qquad E° = 1.229 \text{ V}$$

A couple has a *negative* standard reduction potential if the reductant in the couple has a greater tendency to lose electrons than does hydrogen gas at standard conditions. The good reducing properties of elemental sodium are shown by its large negative standard reduction potential.

$$Na^+ + e^- \longrightarrow Na \qquad E° = -2.7109 \text{ V}$$

In general, the larger the magnitude of the electrode potential in the positive direction, the stronger the oxidant in a couple; and the larger its magnitude in the negative direction, the stronger the reductant in the couple. Couples near the top of Table 19.6 are more likely to react in the direction of being reduced and couples near the bottom are more likely to react in the direction of being oxidized. This is true either when the reaction occurs in an electrochemical cell *or* when it occurs under ordinary conditions with the reactants in direct contact. Reactions involving couples near the middle can go in either direction, depending upon the other couple present and the conditions of the reaction, particularly the pH (see Section 19.8).

Comparison of common entries in Table 19.6 and in Table 11.3, the electromotive series, will show that the elements appear in the electromotive series in the order of increasing oxidizing strength, as shown by successively less negative $E°$ values.

19.7 Using standard reduction potentials

The standard potential of a redox reaction can be calculated from the standard electrode potentials of the two half-cell reactions. Since electrode potentials are all given for half-reactions going in the reduction direction, one of the half-reactions must be reversed in direction and the sign of its $E°$ reversed before the electrode reactions and their potentials can be added. For example, to calculate the potential for our zinc–copper cell (Figure 19.1), the ion–electron equations and the potentials are combined as follows (see values in Table 19.6).

$$
\begin{array}{ll}
Cu^{2+} + 2e^- \longrightarrow Cu & E° = +0.3402 \text{ V} \\
Zn \longrightarrow Zn^{2+} + 2e^- & E° = +0.7628 \text{ V} \\
\hline
Cu^{2+} + Zn \longrightarrow Cu + Zn^{2+} & E° = \quad 1.1030 \text{ V}
\end{array}
$$

A positive value of the potential for a redox reaction means that the reaction will be spontaneous in the direction written, *either* as a reaction in solution or as a cell reaction in which the two specific half-cells are combined. A negative value means that the reaction will not go in the direction written, but will be spontaneous in the opposite direction. From the calculated value $E° = 1.1030$ V, it could be predicted that Cu^{2+} would be reduced and $Zn(s)$ oxidized, a prediction we know to be correct.

In some chemistry texts, particularly older American ones, you may find *standard oxidation potentials*. Such potentials are for the same standard state conditions relative to 0.0000 V for the H_2/H^+ couple. The differences are that the ion–electron equations are written as oxidations (electrons on the other side) and the *signs* of the potentials are opposite. In 1953 the International Union of Pure and Applied Chemistry chose to recommend the uniform, worldwide use of standard reduction potentials, which had previously been more common in Europe. The numerical values for both sets of potentials are the same; only the sign is different.

EXAMPLE 19.9

■ Find $E°$ for the reaction that occurs in an automobile lead storage battery. The standard reduction potentials that apply to the half-cell reactions (step 4 of Example 19.2) are −0.356 V for the reduction of $PbSO_4$ to Pb and 1.685 V for the reduction of PbO_2 to $PbSO_4$.

In the cell PbO_2 is reduced to $PbSO_4$, so the standard reduction potential can be used as given above for this conversion. However, in the cell Pb is oxidized to $PbSO_4$, and so the sign of the standard reduction potential must be reversed in finding the value of $E°$ for the cell reaction.

$$
\begin{aligned}
Pb + SO_4^{2-} &\longrightarrow PbSO_4 + 2e^- & E° &= 0.356\ V \\
PbO_2 + 4H^+ + SO_4^{2-} + 2e^- &\longrightarrow PbSO_4 + 2H_2O & E° &= 1.685\ V \\
\hline
Pb + PbO_2 + 4H^+ + 2SO_4^{2-} &\longrightarrow 2PbSO_4 + 2H_2O & E° &= 2.041\ V
\end{aligned}
$$

The value of $E°$ is 2.041 V and the reaction is favorable under standard state conditions. (Note that this is the voltage for one cell; a battery consists of several cells, usually 3 or 6; Section 24.6.) ■

You will recall from Section 19.4 that to get a balanced redox equation it is sometimes necessary to multiply the half-reactions by whole numbers before they can be added together. In obtaining the overall redox potential of a reaction, the values of $E°$ do not have to be multiplied by the factor used to multiply the half-reaction. The standard state potential is, by definition, the potential difference between $1M$ solutions of pure substances at 1 atm. The amount of each $1M$ solution or each pure substance involved does not change the voltage any more than does the size of the electrode. As discussed in the next section, however, the voltage does vary with concentration.

EXAMPLE 19.10

■ Suppose the half-reactions described in Example 19.1 occurred in an aqueous solution. The standard reduction potentials are 1.42 V for Au^{3+} to Au and 1.3583 V for Cl_2 to Cl^-. Find $E°$ for the reaction under these conditions.

In combining half-reactions $E°$ values are not multiplied by the same factors as equations

The voltage of a half-cell need not be multiplied by any factors because the standard state voltage does not depend on the actual quantities of chemicals involved. So we write

$$2(Au \longrightarrow Au^{3+} + 3e^-) \qquad E° = -1.42 \text{ V}$$
$$\underline{3(Cl_2 + 2e^- \longrightarrow 2Cl^-) \qquad E° = 1.3583 \text{ V}}$$
$$2Au + 3Cl_2 \longrightarrow 2Au^{3+} + 6Cl^- \qquad E° = -0.06 \text{ V}$$

The value of $E°$ is -0.06 V and the reaction is not favorable, as written, under standard state conditions in aqueous solution. However, the solid-state reaction described in Example 19.1 *does* occur spontaneously. ∎

Essentially, any redox half-reaction will proceed in the forward direction as written in Table 19.5 (i.e., reduction) in the presence of a redox couple with a less positive $E°$. And any half-reaction will proceed in the reverse of the direction written (i.e., oxidation) in the presence of a redox couple with a more positive $E°$. The tendency for such reactions to occur will be greater, the greater the difference between the potentials of the two couples.

As a further example, what can we learn from the following set of standard reduction potentials?

$$Fe^{3+} + e^- \longrightarrow Fe^{2+} \qquad E° = 0.770 \text{ V}$$
$$Sn^{4+} + 2e^- \longrightarrow Sn^{2+} \qquad E° = 0.15 \text{ V}$$
$$Cd^{2+} + 2e^- \longrightarrow Cd \qquad E° = -0.4026 \text{ V}$$

Remembering that we are comparing reactions of $1M$ solutions—since the first reaction has the most positive $E°$, it would proceed as written in the presence of either of the other two couples, while the reactions of these couples would be reversed. In other words, elemental cadmium and the tin(II) ion would both act as reducing agents toward the iron(III) ion.

$$Cd + 2Fe^{3+} \longrightarrow Cd^{2+} + 2Fe^{2+} \qquad E° = 1.173 \text{ V} \qquad (12)$$
$$Sn^{2+} + 2Fe^{3+} \longrightarrow Sn^{4+} + 2Fe^{2+} \qquad E° = 0.62 \text{ V} \qquad (13)$$

Neither the tin(IV) ion nor the cadmium(II) ion would be capable of oxidizing the iron(II) ion. These reactions would be the reverse of Equations (12) and (13) and would have $E°$ values with the same numerical values, but with negative signs, showing that they would not be spontaneous under these conditions. Elemental cadmium could be used as a reducing agent for tin(IV) ion

$$Cd + Sn^{4+} \rightleftharpoons Cd^{2+} + Sn^{2+} \qquad E° = 0.55 \text{ V}$$

The standard potential values only allow predictions of whether or not a reaction is thermodynamically possible. As for all thermodynamic quantities (e.g., ΔH or K values), the potentials have no bearing on how fast a possible reaction will proceed. For example, from the following electrode potentials

$$2H^+ + 2e^- \longrightarrow H_2 \qquad E° = 0.00 \text{ V}$$
$$Cu^{2+} + 2e^- \longrightarrow Cu \qquad E° = +0.337 \text{ V}$$

we predict that gaseous hydrogen should react with copper ions in a copper(II) solution to give elemental copper

$$Cu^{2+} + H_2(g) \longrightarrow Cu(s) + 2H^+$$

Experimentally, this reaction proves to be so slow that, except under pressure and in the presence of a catalyst, it is not observed at all.

19.8 Variation of potential with concentration

As we have pointed out, the use of standard electrode potentials to predict whether or not a given redox reaction will take place spontaneously is restricted to the unit concentration and pressure conditions required by the definition of standard potentials. This condition seldom exists in actual practice.

The standard potential $E°$ is related to the potential at other concentrations E by the **Nernst equation**

$$\tag{14}$$

Taking T as the average room temperature, 298°K, the value of the factor $2.303RT/F$ is 0.0592, and the equation becomes

$$E = E° - \frac{0.0592}{n} \log Q \tag{15}$$

The reaction quotient Q (defined in Section 16.4) is expressed in the same fashion as the equilibrium constant, but does not represent equilibrium constant concentrations. Equations (16)–(18) illustrate Q for some half-cell reactions.

$$Fe^{3+} + e^- \longrightarrow Fe^{2+} \qquad\qquad Q = \frac{[Fe^{2+}]}{[Fe^{3+}]} \tag{16}$$

$$NO_3^- + 4H^+ + 3e^- \longrightarrow NO + 2H_2O \qquad Q = \frac{p_{NO}}{[NO_3^-][H^+]^4} \tag{17}$$

$$SO_4^{2-} + H_2O + 2e^- \longrightarrow SO_3^{2-} + 2OH^- \qquad Q = \frac{[SO_3^{2-}][OH^-]^2}{[SO_4^{2-}]} \tag{18}$$

In calculating E from Equation (15), n equals the number of electrons in the balanced ion–electron equation. For Equation (16), $n = 1$, for Equation (17), $n = 3$, and for Equation (18), $n = 2$.

The Nernst equation can also be used to find the potential of a complete redox reaction in other than $1M$ solutions. In this case $E°$ is

Nernst equation:
$$E = E° - \frac{0.0592}{n} \log Q$$

the standard potential of the overall reaction and n is the number of electrons canceled in adding the two ion–electron equations.

EXAMPLE 19.11

■ The cells of the automobile battery described in Examples 19.2 and 19.9 are not constructed to operate under standard-state conditions, but rather at a concentration of about $4.0M$ for the H_2SO_4. What is the cell voltage under these conditions?

For the reaction

$$Pb(s) + PbO_2(s) + 4H^+ + 2SO_4{}^{2-} + 2e^- \longrightarrow 2PbSO_4(s) + 2H_2O(l) + 2e^-$$

the reaction quotient is

$$Q = \frac{1}{[H^+]^4[SO_4{}^{2-}]^2} = \frac{1}{[8.0]^4[4.0]^2}$$

and the number of equivalents is 2. Thus, using the value of $E°$ found in Example 19.9, which was 2.041 V, the Nernst equation gives

$$
\begin{aligned}
E &= E° - \frac{0.0592}{n} \log Q \\
&= 2.041 - \frac{0.0592}{2} \log \frac{1}{[8.0]^4[4.0]^2} \\
&= 2.041 - (-0.143) \\
&= 2.184 \text{ V}
\end{aligned}
$$

The calculated cell voltage under these conditions is 2.184 V, slightly more favorable than at standard-state conditions. ■

EXAMPLE 19.12

■ If the concentration of an $AuCl_3$ solution in contact with pure gold metal is $1.0 \times 10^{-3} M$, and chlorine is at 1 atm pressure, will the reaction described in Example 19.10 take place?

For the reaction,

$$2Au(s) + 3Cl_2(g) + 6e^- \longrightarrow 2Au^{3+} + 6Cl^- + 6e^-$$

$n = 6$ and the reaction quotient is

$$Q = \frac{[Au^{3+}]^2[Cl^-]^6}{p^3{}_{Cl_2}}$$

Note that the pressure in atmospheres and the concentrations in moles/liter are used together in the expressions for Q; they are both approximations for activity, which is a dimensionless quantity.

Using the values of $E° = -0.06$ V from Example 19.10 the Nernst equation gives

$$
\begin{aligned}
E &= E° - \frac{0.0592}{n} \log Q \\
&= -0.06 - \frac{0.0592}{6} \log \left[\frac{(1.0 \times 10^{-3})^2(3.0 \times 10^{-3})^6}{1^3}\right] \\
&= -0.06 - (-0.21) = 0.15 \text{ V}
\end{aligned}
$$

(Note that $E°$ retains its minus sign.) The positive value of 0.14 V implies that under these conditions the reaction will occur. ■

A potential difference also exists between two half-cells consisting of the same type of electrode, but with different concentrations of the same substance.

EXAMPLE 19.13

■ An electrochemical cell is designed so that at one electrode $Cl_2(g)$ is present at 1.00 atm and at the other electrode $Cl_2(g)$ is present at 0.10 atm. The two electrodes are connected by a 1.0*M* NaCl solution. What is the voltage of this cell at 25°C?

For the reactions

$$
\begin{array}{ll}
Cl_2(g,\ 1.00\ \text{atm}) + \cancel{2e^-} \longrightarrow \cancel{2Cl^-} & E° = 1.3583 \text{ V} \\
\cancel{2Cl^-} \longrightarrow Cl_2(g,\ 0.10\ \text{atm}) + \cancel{2e^-} & E° = -1.3583 \text{ V} \\
\hline
Cl_2(g,\ 1.00\ \text{atm}) \rightleftharpoons Cl_2(g,\ 0.10\ \text{atm}) & E° = 0.0000 \text{ V}
\end{array}
$$

the reaction quotient is

$$Q = \frac{p \text{ of } Cl_2 \text{ produced}}{p \text{ of } Cl_2 \text{ reacting}}$$

and the Nernst equation gives

$$
\begin{aligned}
E &= E° - \frac{0.0592}{n} \log Q \\
&= 0.000 - \frac{0.0592}{2} \log \frac{0.10}{1.00} \\
&= 0.000 - (-0.030) = 0.030 \text{ V}
\end{aligned}
$$

This cell produces 0.030 V. ■

A change in the pH has a profound effect upon the value of the reduction potential for any couple that contains the H^+ or OH^- ion. For example, for the reference couple

$$2H^+ + 2e^- \longrightarrow H_2$$

where $[H^+] = 1M$, $E°$ is 0.0000 V. A change to pure water, where $[H^+] = 1.00 \times 10^{-7}M$, while keeping the pressure of H_2 at 1.00 atm, decreases the potential for the reaction

$$E = E° - \frac{0.0592}{n} \log \frac{p_{H_2}}{[H^+]^2}$$

$$= 0.0000 - \frac{0.0592}{2} \log \frac{1.00}{[1.00 \times 10^{-7}]^2}$$

$$= 0.0000 - \frac{0.0592}{2} (14.00)$$

$$= -0.414 \text{ V}$$

The negative E value shows that the reverse of the reaction as written is favored. For the same couple in a 1.00M NaOH solution, where $[OH^-] = 1.00M$ and $[H^+] = 1.00 \times 10^{-14}M$, we have

$$E = 0.0000 - \frac{0.0592}{2} \log \frac{1.00}{[1.00 \times 10^{-14}]^2}$$

$$= -0.829 \text{ V}$$

19.9 Redox equilibria: Finding K from $E°$

When an electrochemical cell has reached equilibrium, electron flow ceases. If the cell was part of a battery, we would say the battery was dead. Because at equilibrium, $E = 0$ and $Q = K$, the Nernst equation provides a simple relationship between the equilibrium constant for a redox reaction and the standard electrode potential.

$$E = E° - \frac{0.0592}{n} \log K = 0 \qquad (19)$$

$$\log K = \frac{n}{0.0592} E° \qquad (20)$$

The magnitude of K for a half-reaction or a redox reaction is thus directly proportional to the number of electrons involved and to the standard potential for the reaction. Values of K can be calculated knowing only the standard reduction potential for a half-reaction or for an overall redox reaction. Once K is known, it may be used in any of the usual ways to determine concentrations in either redox reactions or electrochemical cells at equilibrium.

EXAMPLE 19.14

■ What will be the concentration of $AuCl_3$ at equilibrium for a gold and chlorine gas cell in which the reactions described in Example 19.10 occur?

Using $n = 6$ and $E° = -0.06$ V, as found in Example 19.10, the equilibrium constant is calculated as follows.

$$\log K = \frac{nE°}{0.0592} = \frac{6(-0.06)}{0.0592}$$
$$= -6$$
$$K = 10^{-6}$$

The expression for K is

$$K = [Au^{3+}]^2[Cl^-]^6$$

which, upon substituting $[Au^{3+}] = [AuCl_3]$ and $[Cl^-] = 3[AuCl_3]$, gives

$$K = [AuCl_3]^2(3^6[AuCl_3]^6) = 3^6[AuCl_3]^8$$

Solving for $[AuCl_3]$ and substituting the value of K gives

$$[AuCl_3] = \sqrt[8]{K/3^6} = \sqrt[8]{10^{-6}/3^6} = 0.08M$$

The reaction will proceed until the concentration of $AuCl_3$ reaches $0.08M$. ■

EXAMPLE 19.15

■ What is the equilibrium constant for the reaction that takes place in a cell of the automobile lead storage battery of Example 19.11?

The equilibrium constant depends on the value of $E°$, which was determined in Example 19.11 to be 2.041 V.

$$\log K = \frac{nE°}{0.0592} = \frac{(2)(2.041)}{(0.0592)} = 69.0$$
$$K = 1 \times 10^{69}$$

The equilibrium constant is 1×10^{69}. This reaction goes virtually to completion. ■

THOUGHTS ON CHEMISTRY

The chemical history of a candle

There is another experiment which I must give you before you are fully aquainted with the general nature of carbonic acid [CO_2]. Being a compound body, consisting of carbon and oxygen, carbonic acid is a body that we ought to be able to take asunder. And so we can. As we did with water, so we can with carbonic acid—take the two parts

THE CHEMICAL HISTORY OF A CANDLE, *by Michael Faraday*

607

asunder. The simplest and quickest way is to act upon the carbonic acid by a substance that can attract the oxygen from it, and leave the carbon behind. You recollect that I took potassium and put it upon water or ice, and you saw that it could take the oxygen from the hydrogen. Now, suppose we do something of the same kind here with this carbonic acid. . . . let me take a piece of potassium, a substance which, even at common temperatures, can act upon carbonic acid, though not sufficiently for our present purpose, because it soon gets covered with a protecting coat; but if we warm it up to the burning point in air, as we have a fair right to do, you will see that it can burn in carbonic acid; and if it burns, it will burn by taking oxygen, so that you will see what is left behind. I am going, then, to burn this potassium in the carbonic acid, as a proof of the existence of oxygen in the carbonic acid . . . you perceive that it burns in the carbonic acid—not so well as in the air, because the carbonic acid contains the oxygen combined; but it does burn, and takes away the oxygen. If I now put this potassium into water, I find that, besides the potash formed (which you need not trouble about), there is a quantity of carbon produced. I have here made the experiment in a very rough way; but I assure you that if I were to make it carefully, devoting a day to it, instead of five minutes, we should get all the proper amount of charcoal left in the spoon, or in the place where the potassium was burned, so that there could be no doubt as to the result. Here, then, is the carbon obtained from the carbonic acid, as a common black substance; so that you have the entire proof of the nature of carbonic acid as consisting of carbon and oxygen. And now, I may tell you, that whenever carbon burns under common circumstances, it produces carbonic acid.

THE CHEMICAL HISTORY OF A CANDLE: A SERIES OF LECTURES GIVEN IN 1860, by Michael Faraday, 1860 (p. 132ff, as reprinted by Crowell, New York, 1957)

Significant terms defined in this chapter

electrochemistry
electrochemical cell
half-reaction
ion–electron equation
redox couple
half-cell

electromotive force (emf)
electrode potential
standard state conditions
standard electrode potential
standard reduction potential
Nernst equation

Exercises

19.1 More sulfuric acid is produced each year than any other synthetic chemical. One process for the production of H_2SO_4 can be represented by the following equations:

$$S(s) + O_2(g) \longrightarrow SO_2(g) \qquad \text{(i)}$$
$$SO_2(g) + \tfrac{1}{2}O_2(g) \longrightarrow SO_3(g) \qquad \text{(ii)}$$
$$SO_3(g) + H_2SO_4(l) \longrightarrow H_2S_2O_7(l) \qquad \text{(iii)}$$
$$H_2S_2O_7(l) + H_2O(l) \longrightarrow 2H_2SO_4(l) \qquad \text{(iv)}$$

Answer these questions about the chemicals and reactions involved in the above process. What is the oxidation state of S in (a) S, (b) SO_2, (c) SO_3, (d) H_2SO_4, and (e) $H_2S_2O_7$ and of O in (f) O_2, (g) SO_2, (h) SO_3, (i) H_2SO_4, and (j) $H_2S_2O_7$? (k) Which reactions involve oxidation–reduction? (l) Identify the oxidizing agents in the reactions involving oxidation–reduction. (m) If the process is 100% efficient, how many moles of H_2SO_4 are produced—net—for each mole of S? [Note that some H_2SO_4 is recycled to equation (iii).]

19.2 Find the oxidation state of the specified element in each of the following compounds or ions: (a) Sb in $Sb(OH)_6^-$, (b) C in $H_2C_2O_4$, (c) P in PCl_3, (d) P in $POCl_3$, (e) V in VO^{3+}, (f) Fe in $Fe(CN)_6^{4-}$, (g) Ag in $AgNO_3$, (h) N in NH_3, (i) Br in BrF, and (j) C in HCN.

19.3 Which of the following equations describe oxidation–reduction reactions? Give your reasoning in each instance.

(a) $LiAlH_4(s) + 4H^+ \longrightarrow Li^+ + Al^{3+} + 4H_2(g)$
(b) $Cr_2O_7^{2-} + 2OH^- \longrightarrow 2CrO_4^{2-} + H_2O(l)$
(c) $3KClO(s) \longrightarrow 2KCl(s) + KClO_3(s)$
(d) $24Cu_2S(s) + 128H^+ + 32NO_3^- \longrightarrow 48Cu^{2+}$
$$+ 32NO(g) + 3S_8(s) + 64H_2O(l)$$

19.4 Hydrogen can conveniently be produced in the laboratory by the reaction of CaH_2 with water; a by-product of the reaction is $Ca(OH)_2$. What is the oxidation number of Ca in (a) CaH_2 and (b) $Ca(OH)_2$; of H in (c) CaH_2, (d) $Ca(OH)_2$, (e) H_2O, and (f) H_2; and of O in (g) $Ca(OH)_2$ and (h) H_2O? (i) Write the balanced overall equation for the reaction. Identify the (j) oxidizing and (k) reducing agents and the substances (l) oxidized and (m) reduced.

19.5 A student unsuccessfully tried to balance the following equation:

$$Mn(s) + OH^- \longrightarrow Mn(OH)_2 + O_2(g) + H_2O(l)$$

What is the oxidation state of Mn (a) in Mn and (b) in $Mn(OH)_2$ and of O (c) in OH^-, (d) in H_2O, and (e) in O_2? (f) Why couldn't the student balance this equation?

19.6 Potassium permanganate, $KMnO_4$, can be prepared using the two-step oxidation–reduction process

$$MnO_2(s) + KOH(aq) + O_2(g) \longrightarrow K_2MnO_4(aq) + H_2O(l)$$
$$K_2MnO_4(aq) + H_2O(l) \longrightarrow$$
$$KMnO_4(aq) + MnO_2(s) + KOH(aq)$$

(a) Balance these equations. (b) What is the oxidizing agent in the second reaction? (c) Is MnO_2 a catalyst? (d) Combine both balanced equations into an overall equation.

19.7 Which statements are true? Rewrite any false statement so that it is correct.
(a) A reduction is characterized by a gain of electrons, a decrease in oxidation number, and possibly the removal of oxygen from the substance undergoing reduction.
(b) The Nernst equation relates $E°$ to nonstandard state conditions.
(c) The electromotive force is the emf of a half-cell.
(d) A spontaneous oxidation–reduction reaction requires electrons from an external source.
(e) The standard reduction potential for a given couple has a positive value if the oxidant in the couple has a greater tendency to gain electrons than does the hydrogen ion.

19.8 Balance each of the following equations using the change in oxidation number method:

(a) $SiO_2(s) + Al(s) \longrightarrow Si(s) + Al_2O_3(s)$
(b) $I_2(s) + H_2S(aq) \longrightarrow I^- + S(s) + H^+$
(c) $H_2O_2(aq) + I^- + H^+ \longrightarrow I_2(s) + H_2O(l)$
(d) $H_2S(g) + O_2(g) \longrightarrow SO_2(g) + H_2O(g)$
(e) $NH_3(g) + O_2(g) \longrightarrow NO_2(g) + H_2O(g)$
(f) $SO_2(g) + H_2S(g) \longrightarrow S_8(s) + H_2O(g)$
(g) $HNO_3(aq) + Cu(s) \longrightarrow$
$$Cu(NO_3)_2(aq) + NO(g) + H_2O(l)$$
(h) $Ca_3(PO_4)_2(s) + C(s) + SiO_2(s) \longrightarrow$
$$CaSiO_3(l) + P_4(g) + CO(g)$$

19.9 Balance each of the following equations using the half-reaction method:

(a) $I^- + H^+ + NO_2^- \longrightarrow NO(g) + H_2O(l) + I_2(s)$
(b) $Al(s) + H^+ + SO_4^{2-} \longrightarrow Al^{3+} + H_2O(l) + SO_2(g)$
(c) $Zn(s) + OH^- + NO_3^- + H_2O(l) \longrightarrow$
$$NH_3(aq) + Zn(OH)_4^{2-}$$
(d) $I_2(l) + OH^- \longrightarrow I^- + IO_3^- + H_2O(l)$
(e) $H_2S(aq) + Cr_2O_7^{2-} + H^+ \longrightarrow$
$$Cr^{3+} + H_2O(l) + S_8(s)$$

19.10 For each reaction given in Exercise 19.9, identify the oxidizing agent, the reducing agent, the substance oxidized, and the substance reduced.

19.11 For each reaction in Exercise 19.9, write the expression for Q that would be used in the Nernst equation. What are the respective values of n?

19.12 Distinguish clearly between oxidation and reduction potentials. What is common and what is different about the numerical values?

19.13 When using $E°$ values to predict the course of a given oxidation–reduction reaction, what restrictions on reaction conditions must a chemist keep in mind?

19.14 Answer questions [(a)–(c)] using the following reduction potentials: 2.9 V for F_2/F^-, 0.8 V for Ag^+/Ag, 0.5 V for Cu^+/Cu, 0.3 V for Cu^{2+}/Cu, −0.4 V for Fe^{2+}/Fe, −2.7 V for Na^+/Na, and −2.9 V for K^+/K. (a) Arrange these oxidizing agents in order of increasing strength: F_2, Na^+, K^+, Cu^+, Cu^{2+}. (b) Write the half-reaction for Cu^+ acting as a reducing agent. (c) Which oxidizing agents from this list will oxidize Cu^+?

19.15 On the basis of standard reduction potentials, the reaction

$$Cu + 2Na^+ \longrightarrow Cu^{2+} + 2Na$$

is predicted to be unfavorable–see Table 19.6 for emfs. To make this nonspontaneous reaction "go," a chemist supplied electrical energy from an external source, but to his amazement the cell burst into flame. What had the chemist forgotten to consider?

19.16 The equilibrium constants of the following reactions

$$A + B^+ \longrightarrow A^+ + B \tag{i}$$
$$A + B^{2+} \longrightarrow A^{2+} + B \tag{ii}$$
$$A + B^{3+} \longrightarrow A^{3+} + B \tag{iii}$$

all have the same numerical values. Which of the following statements is correct concerning $E°$ for these reactions? (a) Reaction (i) has the largest $E°$ and reaction (iii) has the smallest value. (b) Reaction (iii) has the largest value of $E°$. (c) The $E°$ values cannot be determined unless the identities of A and B are known. (d) All three reactions have the same value of $E°$.

19.17* Based only on values of ionization potentials, we would predict the standard reduction potentials of the alkali metals to be $Li^+/Li > Na^+/Na > K^+/K >$, etc. Yet in Table 19.5 we see that this order is not correct. What additional effect have we neglected?

19.18* Balance the following equations using the half-reaction method. Note that in most cases the H^+ or OH^- and H_2O are not shown. You should be able to decide whether the system is neutral, acidic, or alkaline with a little detective work and add H^+ or OH^- and H_2O as needed to balance the equation.

(a) $Fe(CN)_6^{3-} + Cr_2O_3(s) \longrightarrow Fe(CN)_6^{4-} + CrO_4^{2-}$
(b) $ClO^- + I_2(s) \longrightarrow Cl^- + IO_3^-$
(c) $Mn(OH)_2(s) + H_2O_2(aq) \longrightarrow MnO_2(s)$

(d) $CN^- + Fe(CN)_6^{3-} \longrightarrow CNO^- + Fe(CN)_6^{4-}$
(e) $N_2H_4(aq) + Cu(OH)_2(s) \longrightarrow N_2(g) + Cu(s)$
(f) $Zn(s) + NO_3^- \longrightarrow Zn^{2+} + NH_4^+$
(g) $Cu(s) + NO_3^- \longrightarrow NO(g) + Cu^{2+}$
(h) $P_4(s) + NO_3^- \longrightarrow H_3PO_4(aq) + NO(g)$
(i) $H_2S(aq) + NO_3^- \longrightarrow S(s) + NO(g)$
(j) $Al(s) + NO_3^- \longrightarrow Al^{3+} + N_2(g)$

19.19* Write the complete half-reaction for each of the redox couples given. You may need to do a little detective work to determine the actual species that are involved in each case, e.g., Cl(3) exists as the weak acid $HClO_2$ in neutral and acidic solutions but as ClO_2^- in alkaline solutions. (a) Br(−1)/Br(5) in acidic solution, (b) Co(2)/Co(3) in alkaline solution, (c) Cl(−1)/Cl(5) in alkaline solution, (d) Pb(2)/Pb(4) in acidic solution, (e) Pb(2)/Pb(4) in alkaline solution.

19.20 Calculate the potential associated with the half-reaction

$$Co \longrightarrow Co^{2+} + 2e^-$$

if the concentration of the cobalt(II) ion is $1 \times 10^{-4}M$. The standard reduction potential for the Co^{2+}/Co couple is −0.28 V.

19.21 Calculate the reduction potential for hydrogen in a system having a perchloric acid concentration of $1 \times 10^{-4}M$ and a hydrogen pressure of 2 atm. Recall that $HClO_4$ is a strong acid in aqueous solution.

19.22 The emf of the cell for the reaction

$$M + 2H^+(1.0M) \longrightarrow H_2(1.0 \text{ atm}) + M^{2+}(0.10M)$$

is 0.500 V. What is the standard reduction potential for the M^{2+}/M couple?

19.23 Using the standard reduction potentials for the appropriate couples, indicate whether or not each of the reactions described below will occur as the equation is written at standard state conditions:

(a) $E° = -0.761$ V for $Cd(OH)_2/Cd$ and $E° = 0.0984$ V for HgO/Hg

$$Cd(s) + HgO(s) + H_2O(l) \longrightarrow Hg(l) + Cd(OH)_2(s)$$

(b) $E° = -0.036$ V for Fe^{3+}/Fe and $E° = 0.0000$ V for H^+/H_2

$$2Fe(s) + 6H^+ \longrightarrow 2Fe^{3+} + 3H_2(g)$$

(c) $E° = 1.3583$ V for Cl_2/Cl^- and $E° = 1.33$ V for $Cr_2O_7^{2-}/Cr^{3+}$

$$6Cl^- + 14H^+ + Cr_2O_7^{2-} \longrightarrow 2Cr^{3+} + 3Cl_2(g) + 7H_2O(l)$$

(d) $E° = 0.20$ V for SO_4^{2-}/H_2SO_3 and $E° = 0.96$ V for NO_3^-/NO

$$3H_2SO_3(aq) + 2NO_3^- \longrightarrow 2NO(g) + 4H^+ + H_2O(l) + 3SO_4^{2-}$$

19.24 For those reactions in Exercise 19.23 that involve the hydrogen ion, find the voltage of the electrochemical cell constructed so that the concentration of H^+ is $10M$ and everything else is under standard state conditions. Do any of the reactions undergo a change in spontaneity under these conditions?

19.25 The standard reduction potentials for the H^+/H_2 and $O_2,H^+/H_2O$ couples are 0.0000 V and 1.229 V, respectively. Write (a) the half-reactions, (b) the overall reaction, and (c) calculate $E°$ for the reaction

$$H_2(g) + \tfrac{1}{2}O_2(g) \longrightarrow H_2O(l)$$

(d) Calculate E for the cell when the pressure of H_2 is 5.0 atm and of O_2 is 9.0 atm.

19.26 Calculate the standard state voltage and the corresponding equilibrium constant for the reactions described by the following equations:

(a) $Cu(s) + Zn^{2+} \longrightarrow Zn(s) + Cu^{2+}$
(b) $2K(s) + 2H_2O(l) \longrightarrow 2K^+ + 2OH^- + H_2(g)$
(c) $2Br^- + Cl_2(g) \longrightarrow Br_2(l) + 2Cl^-$

19.27 What volume of H_2 will be generated at 745 torr and 25°C from the reaction of 75.0 g of CaH_2 with 25.0 g of water? See Exercise 19.4 for the reaction.

19.28 How many grams of $K_2Cr_2O_7$ are required to oxidize 50.0 g of HCHO to HCOOH? Assume Cr^{3+} is the preferred oxidation state for Cr in acidic medium.

19.29 A 0.500-g sample of a powder containing $SnCl_2$ was oxidized using $0.103M$ Ce^{4+}

$$Sn^{2+} + 2Ce^{4+} \longrightarrow Sn^{4+} + 2Ce^{3+}$$

Find the wt% $SnCl_2$ in the powder if 35.72 ml of the Ce^{4+} solution was used.

19.30 Assuming that when acting as oxidizing agents in acidic media $Cr_2O_7^{2-}$ forms Cr^{3+} and MnO_4^- forms Mn^{2+}, which solution can oxidize more of a given reductant: 1 liter of $1M$ $K_2Cr_2O_7$ or 1 liter of $1M$ $KMnO_4$?

19.31* What is the concentration of Ag^+ in a half-cell if the potential of the Ag^+/Ag is changed from $E° = 0.7996$ to 0.35 V?

19.32* A particular cell consists of one half-cell in which a silver wire coated with $AgCl(s)$ dips into a $1M$ KCl solution and other half-cell in which a piece of platinum dips into a solution containing $0.1M$ $CrCl_3$, $0.001M$ $K_2Cr_2O_7$, and $1M$ HCl. In the cell described the following reaction takes place

$$Ag(s) + Cr_2O_7^{2-} + Cl^- + H^+ \longrightarrow AgCl(s) + Cr^{3+} + H_2O(l)$$

The standard reduction potentials are 1.33 and 0.2223 V for the $Cr_2O_7^{2-}/Cr^{3+}$ and $AgCl/Ag,Cl^-$ couples, respectively. Write (a) the half-reactions for this cell and (b) the overall cell reaction. Determine (c) the standard cell potential and (d) the potential of the cell. (e) Calculate the equilibrium constant for the reaction.

19.33* The following reaction takes place in an electrochemical cell:

$$Sn(s) + 2AgCl(s) \longrightarrow SnCl_2(aq) + 2Ag(s)$$

For AgCl/Ag,Cl$^-$, $E° = 0.2223$ V and for Sn^{2+}/Sn, $E° = -0.1364$ V. (a) Determine $E°$ for this cell at 25°C. (b) What is the voltage if the concentration of the $SnCl_2$ is $0.10M$? (c) What would be the resulting voltage for the cell if the $SnCl_2$ were diluted tenfold? (d) At equilibrium, the cell voltage assumes a unique value. What is this value? (e) From the result in (d), calculate the equilibrium constant. (f) What is the equilibrium concentration of Sn^{2+}?

19.34* An electrochemical cell was needed in which hydrogen and oxygen would react to form water. Using the following standard reduction potentials for the couples given, determine which set of reactions gives the maximum output voltage: $E° = -0.8277$ V for H_2O/H_2, OH$^-$, $E° = 0.0000$ V for H^+/H_2, $E° = 1.229$ V for $O_2,H^+/H_2O$, and $E° = 0.401$ V for $O_2,H_2O/OH^-$.

19.35** An ore sample containing hematite, Fe_2O_3, was analyzed using the following analytical procedure: (i) the iron in the ore was dissolved in hydrochloric acid, (ii) the iron was reduced by adding Sn^{2+}, (iii) excess Sn^{2+} was removed by adding a small amount of $HgCl_2$, and (iv) the Fe^{2+} was titrated using MnO_4^-. (a) Balance the following equations for this procedure:

$$Fe_2O_3(s) + HCl(aq) \longrightarrow Fe^{3+} + Cl^- + H_2O(l)$$

$$Sn^{2+} + Fe^{3+} \overset{HCl}{\longrightarrow} Sn^{4+} + Fe^{2+}$$

$$HgCl_2(aq) + Sn^{2+} \longrightarrow Hg_2Cl_2(s) + Sn^{4+} + Cl^-$$

$$MnO_4^- + Fe^{2+} + H^+ \longrightarrow Mn^{2+} + Fe^{3+} + H_2O$$

(b) Identify which equations do not represent oxidation–reduction reactions. (c) Identify the oxidizing agent in those reactions involving redox. Using $E° = 1.491$ V for MnO_4^-, H^+/Mn^{2+}; 0.770 V for Fe^{3+}/Fe^{2+}; and 0.15 V for Sn^{4+}/Sn^{2+}, calculate (d) $E°$ for the reaction between Fe^{2+} and MnO_4^-, (e) K for the reaction between Sn^{2+} and Fe^{3+}, and (f) E for the reaction near the equivalence point, given that $[MnO_4^-] = 0.001M$, $[Fe^{2+}] = 0.0001M$, $[H^+] = 5M$, $[Mn^{2+}] = 0.02M$, and $[Fe^{3+}] = 0.01M$. Calculate the wt % of (g) Fe and (h) Fe_2O_3 in a 0.25-g sample of ore that required 12.2 ml of $0.023M$ $KMnO_4$ for the oxidation. (i) Why was the excess Sn^{2+} removed before the MnO_4^- oxidation?

20 NONMETALS: THE HALOGENS

This chapter presents the descriptive chemistry of the halogens—the Representative Group VII elements. The properties of the halogens themselves and their preparation are described. Compounds of the halogens cover the entire range from ionic to covalent. The principal classes of halogen compounds are discussed—the metal halides, the interhalogens, the hydrogen halides and their acidic aqueous solutions, and the oxo acids and their salts.

The uses of the halogens and their compounds are so varied and extensive that it is difficult to pick out the most interesting ones to mention in this chapter. Because everyone has heard of the chlorination and fluoridation of water, these are included. The widespread use of chlorine and its compounds as bleaching agents is another practical application of the chemical properties of these substances.

Colored impurities are always carried along in the process of making textiles and paper from natural fibers. Since whiteness and brightness in these materials are pleasing and desirable, ways to remove the color—to bleach the fabric or the paper—have been sought since the times of ancient Rome. Many colored compounds have covalent double bonds that are susceptible to attack by oxidizing agents. The free halogens are all good oxidizing agents, as are many of their compounds. In addition, chlorine and its derivatives are readily available and easy to handle, particularly in aqueous solutions, where they are good oxidants.

For these reasons, the chemical industry provides large quantities of chlorine and such chlorine compounds as sodium hypochlorite ($NaOCl$) to the pulp and paper industry, the textile industry, and to both home and commercial laundries to keep their products "whiter than white." Different specific reactions are involved in different types of bleaching, but in most cases the end result is the breakdown by oxidation of large colored molecules into smaller colorless ones.

The halogens

20.1 Properties of the halogens

The members of Representative Group VII—fluorine (F), chlorine (Cl), bromine (Br), iodine (I), and astatine (At)—are collectively called the halogens. They are all nonmetals. The word "halogen" comes from the Greek words meaning "salt-former." The halogens are very reactive elements and are quite similar to each other in their chemical properties. The first-row element of the group—fluorine—differs most from the others, as is true for the first element in most other groups. Because astatine is radioactive and its most stable isotope has a half-life of only 8.3 hr, little is known about its chemistry.

Fluorine and chlorine are gases, bromine is a liquid (the only liquid nonmetallic element), and iodine is a low-melting solid. The intensity of color increases from a very pale yellow for fluorine to violet-black for iodine.

Bromine and iodine have high vapor pressures, and even at room temperature red vapor can be seen above liquid bromine and violet vapor above solid iodine. The halogens all have pungent and irritating odors and attack the skin and flesh. Bromine is particularly nasty and causes burns that are very slow to heal.

The ionic and covalent radii increase down the group as new shells of electrons are added, and the melting points and boiling points also increase with increasing atomic number and atomic size (Table 20.1).

Fluorine, chlorine, bromine, and iodine all react with water (Section 20.2b), but in varying degrees. Fluorine reacts completely, leaving no F_2 in solution. The aqueous solutions of chlorine and bromine, known as chlorine water and bromine water, are acidic due to the formation of HX and HOX, where X is the usual symbol for any halogen atom. These solutions also contain dissolved Cl_2 and Br_2 in equilibrium with the acids. The general term for the binary acid, HX, is **hydrohalic acid.** The other acid, HOX, is a **hypohalous acid.** Chlorine water and bromine water are both good oxidizing agents.

	He
O	**F** 9 **Ne**
	18.9984
S	**Cl** 17 **Ar**
	35.453
Se	**Br** 35 **Kr**
	79.904
Te	**I** 53 **Xe**
	126.904
Po	**At** 85 **Rn**
	(210)

TABLE 20.1
Properties of the halogens

	Fluorine, F_2	Chlorine, Cl_2	Bromine, Br_2	Iodine, I_2
Color and state	Pale yellow gas	Yellow-green gas	Red-brown liquid	Violet-black solid
Melting point (°C)	−220	−101	−73	113
Boiling point (°C)	−188	− 34	59	184
Covalent radius (nm)	0.072	0.099	0.114	0.133
Ionic radius, X⁻ (nm)	0.131	0.181	0.196	0.222
Bond dissociation energy, kcal/mole (kJ/mole) $X_2(g) \rightarrow 2X(g)$	37.76 (157.9)	58.16 (243.4)	46.10 (192.9)	36.15 (151.2)

TABLE 20.2

**Configurations
of the halogens**

Fluorine, F
[He] $2s^2 2p^5$

Chlorine, Cl
[Ne] $3s^2 3p^5$

Bromine, Br
[Ar] $3d^{10} 4s^2 4p^5$

Iodine, I
[Kr] $4d^{10} 5s^2 5p^5$

Astatine, At
[Xe] $4f^{14} 5d^{10} 6s^2 6p^5$

Iodine is not very water soluble, and reacts with water to only a small extent. However, in nonpolar or slightly polar solvents such as carbon disulfide (CS_2), chloroform ($CHCl_3$), or carbon tetrachloride (CCl_4), it gives violet solutions, and in a polar solvent such as ethanol (C_2H_5OH) it gives brown solutions. Apparently in the violet solutions the iodine molecules are intact, while in the brown solutions there is a chemical association between the iodine and the solvent molecules.

In the free elemental state the halogens exist as covalently bonded diatomic molecules

$$:\!\ddot{X}\!:\!\ddot{X}\!:$$

The outermost electron shell of each halogen atom contains seven electrons in an $s^2 p^5$ configuration, one electron short of the noble gas configuration (Table 20.2).

The free halogen molecules are quite stable, as shown by their high bond dissociation energies (Table 20.1). Breaking the X–X bond becomes easier in going down the family from Cl_2, to Br_2, and I_2, reflecting the larger size of the molecules. The surprisingly low bond dissociation energy for fluorine is thought to be due mainly to the large repulsive forces between the small, highly electronegative fluorine atoms. These make the bond easier to break, which contributes to the extraordinary reactivity of fluorine.

The outstanding chemical characteristic of the halogens is their high reactivity. Halogen atoms readily accept a single electron to form singly charged negative ions, X^-, and they also readily share their single unpaired electron with other atoms to form covalent bonds. The ease with which halogen atoms add electrons is shown by the high values of their electron affinities (Table 20.3), higher than those of any other elements (see Table 10.6).

As would be expected from their strong tendencies to react by adding electrons, the free halogens are all strong oxidizing agents. Their standard reduction potentials (Table 20.3) demonstrate the ease with which halogen molecules react in aqueous solution to gain electrons. Fluorine is the strongest oxidizing agent among all the elements. The oxidizing ability of the free halogens decreases regularly

TABLE 20.3

**Electronic properties
of the halogens**

	Fluorine	Chlorine	Bromine	Iodine
Ionization energy (0°K) $X(g) \rightarrow X^+(g) + e^-$, kcal/mole (kJ/mole)	**402** (1680)	**300** (1260)	**273** (1140)	**241** (1010)
Electronegativity	4.0	3.0	2.8	2.5
Electron affinity (0°K) $X(g) + e^- \rightarrow X^-(g)$, kcal/mole (kJ/mole)	77.0 (322)	83.3 (349)	77.6 (325)	70.6 (295)
Standard reduction potential, $E°$ $X_2 + 2e^- \rightarrow 2X^-$ (V)	2.87	1.36	1.09	0.535

from fluorine to iodine. Free chlorine will oxidize bromide ion, and bromine will, in turn, oxidize iodide ion. For the first of these reactions, the standard potential is

$$Cl_2(g) + 2e^- \longrightarrow 2Cl^- \qquad E° = 1.358 \text{ V}$$
$$\underline{2Br^- \longrightarrow Br_2(aq) + 2e^- \qquad E° = -1.087 \text{ V}}$$
$$Cl_2(g) + 2Br^- \longrightarrow 2Cl^- + Br_2(aq) \qquad E° = 0.271 \text{ V}$$

A similar calculation for the second reaction gives

$$Br_2(l) + 2I^- \longrightarrow 2Br^- + I_2(s) \qquad E° = 0.553 \text{ V}$$

The difficulty of removing an electron from a halogen atom results in high values for the ionization energies, which decrease down the group, with the largest change between fluorine and chlorine. Only the noble gases have larger ionization energies than the halogens.

The high electronegativities of the halogens show their strong ability to attract electrons toward themselves in compounds. The bonds in covalent halogen compounds are often quite polar, with a partial negative charge residing on the halogen atom. Fluorine is the most electronegative element and can, therefore, only have a negative oxidation state (-1). The other halogens, although they do not form positive ions because of their high ionization energies, do have positive oxidation states (most often $+1$, $+3$, $+5$, or $+7$) in covalent compounds with more electronegative elements.

The electron affinity of fluorine is slightly smaller than that of chlorine. How can this be reconciled with the facts that fluorine is a better oxidizing agent than chlorine and that it is more electronegative than chlorine? In Section 10.9 we pointed out that, although electron affinities and ionization energies are useful for recognizing general trends, they do not always represent the actual behavior of atoms, molecules, and ions. Electron affinities and ionization energies are measured on isolated gaseous atoms, and not in solutions, where most reactions occur. Since electronegativities are based on compound formation, and standard reduction potentials can be measured on species in solution, these properties are much more useful and consistent with what happens in most chemical reactions than are ionization energies and electron affinities. A thermochemical calculation would show that the enthalpy of hydration of X^- ions makes a large contribution to the enthalpy of the following reaction

$$X_2(g) + 2nH_2O + 2e^- \longrightarrow 2X(H_2O)_n^-$$

20.2 Reactions of the halogens

a. Reactions with other elements. The free halogens combine directly with most elements to form **halides**—binary compounds of halogens with other elements. To give a few random examples

$$S(s) + 3F_2(g) \xrightarrow[\substack{\text{in } F_2 \\ \text{stream}}]{\substack{\text{S burns} \\ \text{with a} \\ \text{bluish flame}}} SF_6(g)$$

sulfur hexafluoride

$$2As(s) + 3Cl_2(g) \xrightarrow{\substack{\text{As burns in} \\ Cl_2 \text{ stream}}} 2AsCl_3(s)$$

arsenic trichloride

$$Mg(s) + Br_2(l) \xrightarrow{\substack{\text{cooling to} \\ \text{remove heat} \\ \text{of reaction}}} MgBr_2(s)$$

magnesium bromide

$$2Al(s) + 3I_2 \text{ (in } CS_2) \xrightarrow{\substack{45°C, \text{ in } CS_2 \\ \text{solution}}} 2AlI_3(s)$$

aluminum iodide

In general, the direct combination reactions of fluorine and chlorine are much more vigorous than those of bromine and iodine. Fluorine and chlorine are so reactive that many metals and nonmetals burn in these gases. However, with a few metals, a layer of metal fluoride rapidly forms at the surface and protects the bulk of the metal from further reaction. This property allows fluorine gas to be stored safely in containers of, for example, copper or nickel.

Each of the halogens combines directly with most metals and most semiconducting elements (Table 20.4). Binary compounds of fluorine and chlorine with oxygen and nitrogen, and of chlorine with carbon, do exist, but they are formed in reactions other than direct combination of the elements.

b. Reactions with water. Fluorine reacts vigorously and instantaneously with water to yield hydrofluoric acid, HF, and a variety of oxidation products, among which oxygen and ozone are predominant. The other halogens react with water to yield hydrohalic, HX, and hypohalous acids, HOX, in an oxidation–reduction equilibrium

$$\overset{0}{X_2}(g) + H_2O(l) \rightleftharpoons H^+(aq) + \overset{-1}{X^-}(aq) + \overset{+1}{HOX}(aq)$$

hypohalous acid

The equilibrium constants for this reaction decrease substantially from chlorine to iodine.

TABLE 20.4

Direct combination of the halogens

	With metals	With semiconducting elements	With nonmetals
F_2	**Most**	**All**	**All except O_2, N_2, and some noble gases**
Cl_2	**Most**	**All**	**All except O_2, N_2, C, and noble gases**
Br_2	**Most**	**Most**	**Only halogens, H_2, P, S**
I_2	**Most**	**Most**	**Only halogens, H_2, P**

$$K = \frac{[H^+][X^-][HOX]}{p_{X_2}}$$

At 25°C for Cl_2, $K = 4.66 \times 10^{-4}$

Br_2, $K = 5.8 \times 10^{-9}$

I_2, $K = 3 \times 10^{-13}$

Aqueous solutions of chlorine and bromine release oxygen when exposed to sunlight, due to photodecomposition of HOX

$$2HOX(aq) \xrightarrow{h\nu} O_2(g) + 2X^- + 2H^+$$

c. Reactions with covalent compounds. A **halogenation** reaction is the introduction of a halogen atom or atoms into a covalent compound. The free halogens react with many compounds in this way, demonstrating the strong oxidizing power of the elements. In each of the following examples the free halogen acts as an oxidizing agent.

$$\overset{+4}{SO_2} + \overset{0}{X_2} \longrightarrow \overset{+6\ -1}{SO_2X_2} \qquad X = F, Cl$$

sulfur dioxide — sulfuryl halide

$$\overset{+3}{PX_3} + \overset{0}{X_2} \longrightarrow \overset{+5\ -1}{PX_5} \qquad X = F, Cl, Br$$

phosphorus trihalide — phosphorus pentahalide

$$\overset{+2}{CO} + \overset{0}{X_2} \longrightarrow \overset{+4\ -1}{COX_2} \qquad X = Cl, Br$$

carbon monoxide — carbonyl halide

$$\overset{-2}{C_2H_4} + \overset{0}{X_2} \longrightarrow \overset{-1\ -1}{C_2H_4X_2} \qquad X = Cl, Br$$

ethylene — ethylene dihalide

Not every halogen takes part in every one of the many possible halogenation reactions. Fluorine is so reactive that it often behaves differently and gives other products than those shown in the preceding equations.

Halogenation reactions are of great importance in the chemical industry in the production of halogen-containing organic compounds, especially halogen-substituted hydrocarbons (Figure 20.1).

20.3 Sources of the halogens

Because of their great reactivity, the halogens are never found in the free state in nature, but commonly occur in deposits of various salts or in solutions of these salts in the oceans or in natural brine wells. The sources of the halogens are summarized in Table 20.5.

20.4 Preparation and uses of the halogens

a. Fluorine. Since fluorine is the strongest chemical oxidizing agent available, it can be obtained only by electrolytic methods. Fluorine is produced commercially by the electrolysis of hydrogen

Methane — CH_4

$\downarrow Cl_2$

Methyl chloride — $CH_3Cl + HCl$

Manufacture of silicones

$\downarrow Cl_2$

Methylene dichloride — $CH_2Cl_2 + HCl$

Paint remover

$\downarrow Cl_2$

Chloroform — $CHCl_3 + HCl$

Manufacture of fluorocarbons

$\downarrow Cl_2$

Carbon tetrachloride — $CCl_4 + HCl$

Manufacture of fluorocarbons

FIGURE 20.1

Chlorination of methane, CH_4. *An example of an industrial halogenation reaction. The products are separated by distillation. Some of their uses are listed. Ninety percent of the hydrochloric acid produced in the United States in 1970 was obtained as a by-product of such hydrocarbon chlorinations.*

TABLE 20.5
Natural occurrence of the halogens

	Percent in Earth's crust	Sources
Fluorine	0.07	**Fluorspar, CaF_2** **Cryolite, $Na_3(AlF_6)$** **Fluorapatite, $Ca_5(PO_4)_3F$**
Chlorine	0.14	**NaCl** **Salt beds, brine wells, seawater (2.8%)** **Carnallite, $KCl \cdot MgCl_2 \cdot 6H_2O$**
Bromine	2.5×10^{-4}	**NaBr, KBr, MgBr$_2$** **Brine wells** **Seawater**
Iodine	3×10^{-5}	**NaIO$_3$, NaIO$_4$** **NaNO$_3$ deposits in Chile** **Iodides** **Brine wells**

617

Industrial vs. laboratory preparations

Producing a chemical in laboratory glassware—although it sometimes seems difficult enough—is simple compared to producing several tons or several thousand tons for industrial use. In this section we take a brief look at some of the differences between laboratory preparations and industrial preparations.

To prepare a chemical compound in the laboratory, you would first take off the shelf bottles of the chemicals you need for starting materials. Next you would arrange glassware so that heat or cooling can be applied to the reaction mixture as needed, and so that condensers can keep volatile materials from escaping. When the reaction is finished you might be happy to get even a small amount of the product you wanted, and any by-products and waste solvents would go down the drain or into a waste container.

In putting up an industrial chemical plant, just assuring a large enough supply of cooling water or electricity may be a problem. And cost must be considered in all parts of the operation. It is important that as little as possible is spent on starting materials that wind up going down the drain. Waste is not just bad economics; it can require additional expenses for waste treatment to reduce pollution hazards. Therefore, the more of the starting materials converted to useful products or reclaimed to be used again, the better.

The price of starting materials is of major importance in industry because they will be purchased in large quantities. A difference of a few cents per pound may mean the difference between profit and loss on the whole operation. The cost of the equipment needed in an industrial operation must also be carefully considered. Expensive setups that will have only specialized uses are avoided if possible. On the other hand, large investments in complex equipment are not objectionable if they increase efficiency and rate of production, because their cost will be spread over millions of pounds of product.

Handling chemicals in large quantities is much more complicated than taking bottles off the shelf. In the laboratory, if 2 or 3 g of something spills or catches fire, it can certainly be a hazard, but is often just an inconvenience. However, an accident with a tank car full of a dangerous chemical is another story. Safety is, therefore, a constant concern.

It is always easiest and cheapest to deal with liquids in a chemical plant, because they can be transported within the plant in pipelines. Large plants may incorporate hundreds of miles of pipe. The pipes, reactors, and other equipment in a chemical plant must be made of sturdy corrosion-resistant materials. A reaction mixture may reside in glassware in the laboratory for a few hours, but a continuous operation in a chemical plant should be able to go on for months or even years. The chemicals should be reacting with each other as planned, not with the pipes in which they are flowing.

Chemical engineers are assigned the job of planning and designing chemical plants so that the methods developed in a small-scale laboratory preparation can be adapted to a large-scale industrial operation. Engineers must keep the principles of economics, technology, and chemistry in balance, while designing and putting into operation a plant that can deal safely and profitably with large quantities of reactants and products.

fluoride in a molten mixture of potassium fluoride and hydrogen fluoride

$$2HF \text{ (in KF)} \xrightarrow{\text{electrical energy}} H_2(g) + F_2(g)$$

The electrolytic vessel is usually constructed of one of the metals, such as nickel, copper, or the nickel–copper–iron alloy known as Monel metal, which quickly becomes protected by a metal fluoride coating. The K^+ and HF_2^- ions are formed in the HF–KF solution and conduct the current. One electrode is graphite and the other may be nickel, mild steel, or copper.

Before World War II there was no commercial production of free fluorine, and fluorine-containing chemicals were just beginning to find uses as refrigerants. The great reactivity of fluorine had made chemists reluctant to handle it. Fluorine technology developed when, in the drive to produce the atomic bomb, it was found that the uranium isotopes ^{238}U and ^{235}U could be separated by effusion of their hexafluorides (Section 3.16).

Some of the products that incorporate fluorine compounds are listed in Table 20.6. "Fluorochlorocarbon" is a general term for hydrocarbons in which some or all of the hydrogen atoms have been replaced by fluorine and chlorine atoms. The low molecular weight fluorochlorocarbons are nontoxic, stable at high temperatures, inert, odorless, and nonflammable. They are ideal for use as refrigerants and aerosol propellants because they are easily liquefied by pressure alone and have large enthalpies of vaporization.

The discovery of the useful properties of the Freons is a perfect example of problem-solving scientific research. Each of the early refrigerants was unsatisfactory. Ethylene was flammable, sulfur dioxide and ammonia were corrosive and poisonous; carbon dioxide, although a harmless gas, required cumbersome, high-pressure equipment. At General Motors (they also make refrigerators) in the 1920s a group set out to devise a better refrigerant by plotting periodic table trends of toxicity, flammability, and boiling point. The first compound they picked to try was dichlorodifluoromethane, the first Freon. However, as previously discussed (Section 6.5), fluorochlorocarbons are now suspected of being harmful to the ozone layer of the atmosphere. The end of the Freon story seems to be a government ban, at least until more information is obtained about the ozone layer depletion problem.

Polymers made from fluorocarbons, such as Teflon, are also inert and resistant to high temperatures, and they can be used in contact with very reactive chemicals. The discovery of these properties of the fluorocarbon polymers was also made during the development of the gaseous diffusion plants for uranium isotope separation.

b. Chlorine. Chlorine is prepared in industrial quantities by the electrolysis of concentrated sodium chloride solutions. Sodium hydroxide is an equally important coproduct in this process.

$$2NaCl(aq) + 2H_2O(l) \xrightarrow{\text{electrical energy}} Cl_2(g) + H_2(g) + 2NaOH(aq)$$

In one recent year, 10.7 million tons of sodium hydroxide and 10.3 million tons of chlorine gas were produced in the United States, al-

TABLE 20.6

Some fluorine-containing end products

Refrigerants

Aerosol propellants

Polyfluorocarbon resins (e.g., Teflon)

Textile finishes (e.g., Scotchgard)

Toothpaste

most all by electrolysis. (The commercial electrolytic cells are described in Section 24.8.) Eighty percent of the chlorine is used as a raw material for the manufacture of other chemicals. Some of the types of products that incorporate chlorine are listed in Table 20.7. Another important use of free chlorine is as a bleaching agent for textiles and paper. Sewage and water treatment take up about 4% of the total chlorine produced each year.

For the preparation of chlorine in the laboratory, hydrochloric acid can be oxidized by manganese dioxide

$$MnO_2(s) + 4H^+ + 4Cl^- \longrightarrow Mn^{2+} + 2Cl^- + Cl_2(g) + 2H_2O(l)$$
manganese dioxide

or sodium chloride in sulfuric acid solution can be oxidized by potassium permanganate

$$2MnO_4^- + 16H^+ + 10Cl^- \longrightarrow 2Mn^{2+} + 5Cl_2(g) + 8H_2O(l)$$
permanganate ion

c. Bromine. Almost all bromine is now prepared by oxidation of the bromide ions in water from natural brine wells. Seawater (about 75 ppm bromine), once a source of bromine, is no longer used commercially. The natural brines in Arkansas, for example, contain a much greater concentration of bromine—about 4000 ppm. The best oxidizing agent for bromide ion is chlorine.

$$2Br^- + Cl_2(g) \longrightarrow Br_2(aq) + 2Cl^-$$

Crude bromine is collected in one of two ways: With concentrated solutions, steam is blown into the bottom of the reactor along with chlorine. Then the mixture of bromine, steam, and unreacted chlorine is collected at the top of the reactor and cooled. The bromine can be separated by pouring off the water layer. With less concentrated solutions, bromine is swept out of the reaction mixture by blowing air through the solution. One method of separating bromine from the resulting mixture of air and unreacted chlorine is to treat the mixture with sulfur dioxide and steam to give a mixture of hydrobromic and sulfuric acids

$$SO_2(g) + Br_2(g) + 2H_2O(g) \xrightarrow{\Delta} 2HBr(g) + H_2SO_4(aq)$$

Reaction of the mixture of acids with chlorine regenerates the bromine. The sulfuric acid is recycled for use in the necessary adjustment of the pH of the brine that serves as starting material for the process.

The major industrial use of bromine, more than 80% of the total produced each year, has been in ethylene dibromide, $C_2H_4Br_2$, which is added to leaded gasoline. The ethylene dibromide reacts with lead oxide and lead sulfate, combustion products that would otherwise be deposited in the engine. The lead bromide formed is volatile and is

[20.4] THE HALOGENS

620

carried away with the exhaust. As antipollution laws which forbid the use of lead in gasoline go into effect, there will obviously have to be some adjustments in the industrial production of bromine. Bromine is also used in the production of agricultural chemicals such as fungicides, and in very pure chemicals for photography.

Both iodine and bromine can be prepared in the laboratory by oxidation of iodide or bromide ions by chlorine or by reagents such as MnO_2 or $KMnO_4$, that are also used in the laboratory preparation of chlorine.

d. Iodine. One source of iodine is the liquid containing sodium iodate left after removal of sodium nitrate from Chilean ores. Reduction with bisulfite ion, HSO_3^-, yields free iodine.

$$2IO_3^- + 5HSO_3^- \longrightarrow I_2(s) + 5SO_4^{2-} + 3H^+ + H_2O(l)$$
$\quad\;$ *iodate ion* $\quad\;$ *bisulfite ion*

Iodine is also commercially obtained from seaweed, which concentrates iodide ion from seawater. The seaweed is burned to yield an ash that may contain from 0.5% to 1.5% of iodide ion. The ion is oxidized to form iodine by chlorine

$$2I^- + Cl_2(g) \longrightarrow I_2(s) + 2Cl^-$$

or by other oxidizing agents such as MnO_2 or $KMnO_4$. Oil well brines are another source of iodide ion that can be oxidized to yield free iodine.

The production of iodine is expensive, limiting its use to specialized applications, mainly in photography and the pharmaceutical industry. Iodide ion is necessary for the production of thyroxine in the thyroid gland. Insufficient iodide ion in the diet leads to a condition known as a goiter, which is an enlargement of the thyroid gland. To assure the presence of iodide ion in the diet sodium or potassium iodide is added to table salt, which is sold as "iodized" salt.

20.5 Halogens and the water supply

More than half the people in the United States are drinking water to which fluoride ion has been added, and more than 90% are drinking water to which chlorine has been added. Chlorination is an essential public health measure: it prevents the spread of diseases via the water supply. Fluoridation is of greatest benefit to children: it dramatically decreases the formation of cavities in the teeth.

a. Chlorination. We have come to expect, and even take for granted, that pure and aesthetically pleasing water will run from the faucet. Chlorination of municipal water supplies is an essential step in making this possible.

Chlorine acts in two ways to improve water quality—as a disinfectant and as an oxidant. In its vitally important role as a disinfectant, chlorine kills or attenuates the activity of disease-causing bacteria or viruses. Epidemics of such water-borne diseases as dysentery, hepatitis, and typhoid have grown rare in the United States and Canada, where most cities and towns chlorinate their water.

The job of chlorine as an oxidant in water treatment is to modify the chemical nature of contaminants in the water. Substances that cause bad odors and tastes, or give undesirable color to the water, are transformed by the oxidation reactions to odorless, tasteless, and colorless materials. Many of these contaminants are organic compounds.

The disinfecting action of chlorine was once also thought to be the result of oxidation. However, it is not that simple, for it has been found that similar oxidizing agents are not as effective as chlorine in attacking disease-causing organisms. The explanation may be that the hypochlorous acid specifically attacks enzymes essential to the organisms, and in this way causes the organisms to lose their activity.

Calcium and sodium hypochlorites, $Ca(OCl)_2$, $NaOCl$, are also used in water purification, although chlorine is cheaper and easier to control, particularly in large systems. Swimming pool chlorination is based on the same principles as the chlorination of drinking water. Organic compounds that are safer to handle than free chlorine and that release chlorine in water solution have been developed for use in home swimming pools.

b. Fluoridation. Thanks to the advertising industry, it seems impossible for anyone in the United States not to know that fluorides fight cavities. Discovery of the cavity-fighting property of fluorides was made when the cause of the spotted and darkened teeth of residents of several towns in the western United States was investigated. Fluoride ion was found to be dissolving from natural mineral deposits into the town water supplies. Persistent US Public Health Service dentists soon found that along with the discolored teeth went an unusually low rate of tooth decay, and this too was attributed to fluoride ion.

At a concentration of 1 part per million in the drinking water, decay prevention occurs without discoloration, which begins at a concentration of more than 1.5 ppm. When this was learned, an obvious suggestion was made—add 1 ppm of fluoride ion to the water wherever it doesn't occur naturally and protect everyone's teeth. In 1945, Grand Rapids, Michigan, became the first city to do so.

Teeth contain calcium in the form of the mineral apatite (Figure 20.2), $Ca_5(PO_4)_3OH$. Fluoride ions replace the hydroxide ions in the apatite structure and make it resistant to attack by the acids which cause tooth decay. Fluoride toothpastes reach the surface of teeth and can build a protective coating there. But fluoride provides the greatest benefits when it is incorporated throughout the entire tooth as it forms. This is achieved when fluoride is ingested by children and gets carried to the growing teeth by the bloodstream.

The compounds used to maintain 1 ppm of fluoride ion in the water supply are sodium fluoride, NaF; sodium hexafluorosilicate, Na_2SiF_6; hexafluorosilicic acid, H_2SiF_6, which is formed only in solution; or Na_2PO_3F, sodium monofluorophosphate.

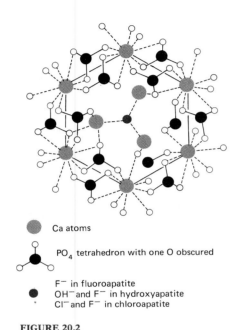

Ca atoms

PO_4 tetrahedron with one O obscured

F^- in fluoroapatite
OH^- and F^- in hydroxyapatite
Cl^- and F^- in chloroapatite

FIGURE 20.2

The crystal structure of apatite. *Apatite* $[Ca_5(PO_4)_3OH]$ *is a mineral and also occurs in teeth. Fluoride replaces* OH^- *in the structure of teeth when it is available in the diet. In the structure the* F^- *or* OH^- *is surrounded in the same plane by three* Ca^{2+} *ions. The six* Ca^{2+} *ions that delineate the hexagon are in two different planes.*

After more than 25 years of water fluoridation, it was possible to make this statement: "Fluoridation has been shown to reduce dental decay among children by two-thirds. No known deleterious effects have been noted in anyone drinking fluoridated water." (F. J. Maier, in *Water Quality and Treatment: A Handbook of Public Water Supplies*, 3rd ed., American Water Works Association, Inc., McGraw-Hill, 1971.)

Compounds of the halogens

20.6 Metal halides

Halides are known for all of the metals, and preparation of most of them is possible by the direct combination of the elements. Other preparative methods include precipitation of the insoluble salts from solutions containing the appropriate ions, for example,

$$Pb^{2+} + 2Br^- \longrightarrow PbBr_2(s)$$

or reaction of the halogen acids with active metals, for example,

$$2HCl(aq) + Mg(s) \longrightarrow MgCl_2(aq) + H_2(g)$$

Since the metals vary considerably in electropositive character and the halogens vary in electronegative character, bonding in metal halides varies with the nature of both the metal and the halogen. The metal halides, therefore, have a wide range of physical and chemical properties.

In general, metals that have low ionization energies (for example, the alkali metals, Representative Group I) form ionic halides. But, if the ionization energy for removal of all the valence electrons is high, the bonds between metal and halogen, particularly for chlorine, bromine, or iodine, will be covalent. Fluorine, the smallest and most electronegative of the halogens, frequently forms predominantly ionic compounds even with metals that have high ionization energies. For example, AlF_3 is an ionic compound. Two of the basic crystal structures for metal halides are the NaCl (Figure 12.19) and fluorite or CaF_2 structure (Figure 12.20). Another, the CdI_2 structure, is shown in Figure 20.3.

In most cases, thermal stability decreases regularly and markedly from the fluoride to the iodide salts of the same cation. This order, illustrated in Table 20.8 with some thermochemical values for the sodium halides, is a direct reflection of changes in ionic size. Interaction between the cation and the halide ion in the crystal is strongest with fluoride ion, the smallest of the halide ions, and weakest with iodide, the largest.

The greater the covalent character of the metal–halogen bond, the lower the melting and boiling points of the halides—a result of

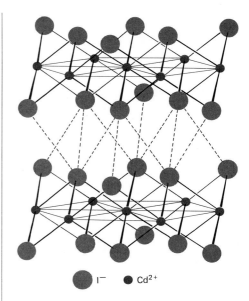

I⁻   Cd²⁺

FIGURE 20.3

The crystal structure of cadmium iodide. *This is what is called a "layer" structure.*

TABLE 20.8

Some thermochemical values for the sodium halides

	Lattice energy, ΔH_l, 25°K, kcal/mole (kJ/mole)	Standard enthalpy of formation, ΔH_f° kcal/mole (kJ/mole)
NaF	−218 (−912)	−136.0 (−569.0)
NaCl	−185 (−774)	−98.23 (−411.0)
NaBr	−176 (−736)	−86.03 (−359.9)
NaI	−168 (−703)	−68.84 (−288.0)

the weaker forces between covalent molecules than between ions in a crystal. In Table 20.9 the melting and boiling points of a number of halides are compared with the ionization energy values for the removal of the valence electrons. For compounds of a particular halide, the larger the ionization energy of the metal atom—and, therefore, the more pronounced the covalent character of the metal–halogen bond—the lower the melting and boiling points of the compound. Note also that the melting and boiling points for the metal fluorides (AlF_3, SnF_4, SbF_3) are considerably higher than those for the corresponding chlorides, reflecting the much greater electronegativity of fluorine than chlorine, which leads to more ionic character in the bonds. One further observation is pertinent. If a metal has two or more oxidation states, the halide with the metal in the lower state has the higher melting and boiling points. (Compare $SnCl_4$ with $SnCl_2$, and $SbCl_5$ with $SbCl_3$.) This results from the greater degree of ionic character in the lower-oxidation-state compound. For instance, compare the ionization energy for the removal of two electrons from tin with that for the removal of four electrons.

Frequently the solubility in water of a fluoride differs considerably from that of the corresponding chloride, bromide, and iodide. For example, whereas the fluorides of lithium, magnesium, calcium, strontium, barium, and the lanthanides (elements 58–71) are relatively insoluble in water, the other halides of these elements are

TABLE 20.9

Melting and boiling points of some metal halides

Halide	Valence shell configuration of metal atom	Ionization energy (°K), kcal/mole (kJ/mole) for removal of valence electrons of metal	Melting point; boiling point (°C)
NaCl	$3s^1$	118.5 (495.8)	801; 1413
BeCl₂	$2s^2$	634.8 (2656)	440; 520
MgCl₂	$3s^2$	522.9 (2188)	708; 1412
Al₂Cl₆	$3s^23p^1$	1228.0 (5138)	192.6 (1700 torr); 178 (sublimes)
AlF₃	$3s^23p^1$		1290; 1291 (sublimes)
SnCl₄	$5s^25p^2$	2149 (8591)	−30.2; 114.1
SnF₄	$5s^25p^2$	2149 (8591)	705 (sublimes)
SnCl₂	$5s^25p^2$	506.7[a] (2120)	246.8; 623
SbCl₅	$5s^25p^3$	3501 (14650)	2.8; 92 (30 torr)
SbCl₃	$5s^25p^3$	1162[b] (4862)	72.9; 221
SbF₃	$5s^25p^3$		290; 319 (approximate)

[a] Energy required for the removal of the two 5p electrons only.
[b] Energy required for the removal of the three 5p electrons only.

quite soluble. Most metal chlorides, bromides, and iodides are soluble, but those of silver (Ag^+), mercury(I) (Hg_2^{2+}), lead(II) (Pb^{2+}), and thallium(I) (Tl^+) are almost insoluble. The fluorides of silver and thallium(I), however, are very soluble in water.

20.7 Interhalogens

Covalent compounds formed between two different halogens are known as **interhalogens.** The interhalogens either are diatomic molecules (e.g., ICl) or molecules of the type AB_n, where A is the larger halogen atom and is surrounded by three, five, or seven atoms of the smaller halogen (Table 20.10). The smaller, more electronegative atom is assigned an oxidation state of -1, giving the central atom a positive oxidation state.

The geometry of the interhalogen molecules is what would be expected from our previous study of molecular geometry and electron pair repulsion (see Tables 9.22 and 9.23). For example, ClF_3 and IF_5 are of the types AB_3E_2 and AB_5E (where E is an unshared electron pair) and their geometries are as shown

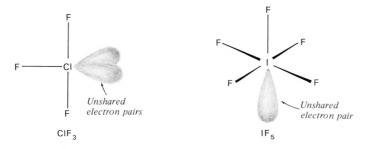

The interhalogens are all volatile, reactive compounds. Several of them are very corrosive and actively attack the skin. They are also all strong oxidizing agents and most of them have properties which lie between those of the component halogens. The interhalogens react with most other elements to yield mixtures of halides. They are readily attacked by water, with conversion of the more electronegative element to halide ion, and the less electronegative element to hypohalous acid, for example

$$ICl(s) + H_2O(l) \longrightarrow H^+ + Cl^- + HOI(aq)$$

iodide
chloride

hypoiodous
acid

20.8 Hydrogen halides and their aqueous solutions

a. Properties of the hydrogen halides. At room temperature all of the hydrogen halides (Figure 20.4) are colorless gases. They have penetrating, disagreeable odors and are strong irritants to the mucous

TABLE 20.10

Interhalogen compounds

ClF	BrF	IF_3
ClF_3	BrF_3	IF_5
ClF_5	BrF_5	IF_7
	BrCl	ICl
		ICl_3
		IBr

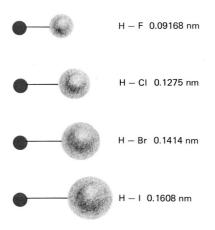

FIGURE 20.4

Internuclear distances in hydrogen halide molecules.

membranes. There is a regular increase in the melting points and boiling points, and heats of fusion and vaporization, from the chloride to the iodide (Table 20.11). Hydrogen fluoride has unexpectedly high values for these properties because the small, highly electronegative fluorine atom forms strong hydrogen bonds. In the solid and liquid states, and even in the gaseous state if the temperature is not too high, HF molecules are held together in groups by hydrogen bonding. The other hydrogen halides are simple covalent molecules in all three states.

The data of Table 20.11 for percent dissociation show that the thermal stability of the hydrogen halides decreases from hydrogen fluoride to hydrogen iodide. The fluoride is stable to dissociation into its elements at 1000°C, while hydrogen iodide is 33% dissociated. The trend in thermal stabilities is also reflected in the bond dissociation energies.

Hydrogen fluoride has no reducing properties, but the other hydrogen halides can all act as reducing agents. Hydrogen iodide is an especially strong reducing agent. The hydrogen halides are generally oxidized to halogen molecules in redox reactions. The reducing ability of the hydrogen halides decreases with increasing thermal stability.

b. Hydrohalic acids. The hydrogen halides are most familiar as their aqueous solutions, the hydrohalic acids hydrofluoric acid, hydrochloric acid, hydrobromic acid, and hydroiodic acid. The hydrogen halides are extremely soluble in water, and the solutions are acidic because the halide ions are all weaker bases than the water molecule, and therefore each gives up its proton

$$H : \ddot{X} : + H : \ddot{O} : \longrightarrow \left[H : \ddot{O} : H \right]^+ + : \ddot{X} : ^-$$

In dilute solution this reaction goes essentially to completion for all of the hydrogen halides except hydrogen fluoride. Therefore, HCl, HBr, and HI solutions are all typical strong acids, and hydrofluoric acid is a weak acid, with a K_a of 6.5×10^{-4} (comparable to the K_a of 1.75×10^{-5} for acetic acid, a typical weak acid).

Why is hydrofluoric acid a weak acid? Certainly a major reason is that the great strength of the hydrogen–fluorine bond (Table 20.11) inhibits ionization. It is interesting that concentrated solutions of hydrogen fluoride are more strongly acidic than dilute solutions. In concentrated solutions a greater degree of hydrogen bonding between the HF molecules is possible, and the hydrogen–bonded species reacts with water to give H_3O^+ and HF_2^- ions.

c. Preparation of hydrogen halides. The hydrogen halides can be obtained by the direct combination of the elements

$$H : H + : \ddot{X} : \ddot{X} : \longrightarrow 2H : \ddot{X} :$$

As would be expected, the reaction occurs with decreasing vigor from fluorine to iodine. The combination of hydrogen and fluorine is violent, even at room temperature. Chlorine and hydrogen combine very

TABLE 20.11
Some properties of the hydrogen halides

Property	HF	HCl	HBr	HI
Melting point (°C)	−83.1	−114.8	−86.9	−50.7
Boiling point (°C)	19.5	− 84.9	−66.8	−35.4
Heat of fusion at melting point, kcal/mole (kJ/mole)	1.094 (4.577)	0.476 (1.991)	0.575 (2.41)	0.686 (2.87)
Heat of vaporization at boiling point, kcal/mole (kJ/mole)	7.24 (30.3)	3.86 (16.2)	4.21 (17.6)	4.72 (19.7)
Percent dissociation into elements at 1000°C	0	0.014	0.5	33
Bond dissociation energies $HX(g) \rightarrow H(g) + X(g)$ kcal/mole (kJ/mole)	135.8 (568.2)	103.2 (431.8)	87.54 (356.3)	71.30 (298.3)

slowly in the dark, but explosively in the presence of short wave-length light, such as the ultraviolet range of sunlight. The reaction of bromine or iodine with hydrogen is slow at room temperature, but can be made to go at practical rates at elevated temperatures in the presence of a catalyst such as platinum gauze.

The first step in the reaction of hydrogen with chlorine is thought to be the photodecomposition of chlorine molecules into atoms.

$$:\ddot{C}l:\ddot{C}l: \xrightarrow{\text{sunlight}} 2 \cdot \ddot{C}l:$$

Halogen atoms contain an unpaired electron and are therefore free radicals and very reactive. This reaction initiates a several-step mechanism of the type called a chain reaction—a reaction mechanism in which reaction products initiate the repetition of one or more steps in the overall mechanism. The steps in a chain reaction may be classified into three types. The production of the first reactive inter-mediate, usually a free radical, is the **initiation** step. The **propagation** steps produce **chain carriers,** species that initiate further reactions. The chain carriers are also usually free radicals. And finally, one or several **termination** steps end the chain by the recombination of the active intermediates. The mechanism for the photochemical reaction of hydrogen with chlorine

$$H_2(g) + Cl_2(g) \xrightarrow{\text{sunlight}} 2HCl(g)$$

can be written

initiation
$Cl_2 + h\nu \longrightarrow 2Cl\cdot$

propagation
$Cl\cdot + H_2 \longrightarrow HCl + H\cdot$
$H\cdot + Cl_2 \longrightarrow HCl + Cl\cdot$

termination
$H\cdot + Cl\cdot \longrightarrow HCl$

ClO_4^- perchlorate ion

ClO_3^- chlorate ion

ClO_2^- chlorite ion

OCl^- hypochlorite ion

Other possible chain-terminating steps are

$$H\cdot + H\cdot \longrightarrow H_2$$
$$Cl\cdot + Cl\cdot \longrightarrow Cl_2$$

The direct combination of hydrogen with each of the halogens is a chain reaction, as are many other reactions of halogens. However, each hydrogen–halogen reaction has a different mechanism and a different rate equation—the hydrogen–iodine reaction mechanism fits a simple second-order rate equation (Section 15.12), rate = $k[H_2][I_2]$, while the hydrogen–bromine reaction mechanism is so complex that no simple reaction order can be assigned.

The action of warm concentrated sulfuric acid on calcium fluoride (the mineral fluorite) or sodium chloride provides a more convenient route to hydrogen fluoride or hydrogen chloride, respectively, than does direct combination of the elements.

$$CaF_2(s) + H_2SO_4(aq) \longrightarrow 2HF(g) + CaSO_4(s)$$
$$NaCl(s) + H_2SO_4(aq) \longrightarrow HCl(g) + NaHSO_4(aq)$$

These reactions are used commercially. The hydrogen halides are much more volatile (HF, b.p. 19.5°C; HCl, b.p. −84.9°C) than sulfuric acid (b.p. about 290°C with decomposition) and are easily removed from the reaction mixture.

The reaction shown for hydrogen chloride is the first step of a two-stage process and takes place even at room temperature. Further reaction occurs at about 500°C and additional hydrogen chloride is produced.

$$NaHSO_4(s) + NaCl(s) \longrightarrow HCl(g) + Na_2SO_4(s)$$

Hydrogen bromide and hydrogen iodide cannot be obtained by the reaction between concentrated sulfuric acid and a halide. As the hydrogen halides are formed, these rather strong reducing agents are oxidized by sulfuric acid, for example,

$$H_2SO_4(aq) + 2HBr(g) \longrightarrow Br_2(g) + SO_2(g) + 2H_2O(l)$$

Hydrogen bromide and hydrogen iodide can be prepared by reaction of sodium halides and orthophosphoric acid (H_3PO_4), a non-volatile acid that is a poor oxidant.

$$NaBr(s) + H_3PO_4(l) \longrightarrow HBr(g) + NaH_2PO_4(s)$$
$$NaI(s) + H_3PO_4(l) \longrightarrow HI(g) + NaH_2PO_4(s)$$

phosphoric acid   *sodium dihydrogen phosphate*

Aqueous solutions of hydrogen iodide are generally made by the reduction of an elemental iodine suspension with hydrogen sulfide

$$H_2S(g) + I_2(s) \xrightarrow{H_2O} 2HI(aq) + S(s)$$

A solution of hydrogen bromide or iodide is obtained by addition of a phosphorus trihalide to water.

$$PX_3(s) + 3H_2O(l) \longrightarrow 3HX(aq) + H_3PO_3(aq)$$
$$\text{\textit{phosphorous acid}}$$

This method is used commercially.

20.9 Oxo acids of the halogens and their salts

Oxo acids are known for all of the halogens except fluorine, which is such a strong oxidizing agent that oxofluoro species do not exist in aqueous solution. As in the interhalogens, the halogen atoms in the oxohalogen acids and their salts have positive oxidation numbers (Table 20.12). The replaceable, or acidic, hydrogen atom in each of these acids is bound to an oxygen atom, not a halogen atom. The formulas of the acids in Table 20.12 are written to emphasize this fact. However, it is more common to write the oxygen atoms together, for example, $HClO_3$ for chloric acid. The formulas of salts in Table 20.12 are written in this more customary way.

   a. Oxochloro acids and salts. Perchloric acid, $HClO_4$, is the only oxygen acid of chlorine that has been obtained as a pure compound. The others, although their salts are well characterized, are known only in solution, where they are formed in reactions of the elemental halogens with water. In a formal sense, the structures of the hypochlorite, OCl^-; chlorite, ClO_2^-; and chlorate, ClO_3^-, ions (Figure 20.5) may be regarded as derived from the tetrahedral perchlorate ion, ClO_4^-, by removal of the appropriate number of oxygen atoms. The geometry of these ions (and the equivalent ions of other halogens) is what we would predict from our previous study of geometry and electron pair repulsion (Tables 9.22 and 9.23).

Hypochlorite ion, OCl^-

Chlorite ion, ClO_2^-

Chlorate ion, ClO_3^-

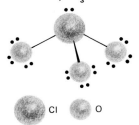
Cl   O

FIGURE 20.5

Structures of oxochloro anions. (*From Structural Inorganic Chemistry, 3rd edition, by A. F. Wells, published by Oxford University Press.*)

TABLE 20.12

Oxo acids of the halogens

Oxidation number of the halogen	Chlorine	Bromine	Iodine	Name of acid	Name of salts
+1	HOCl	HOBr	HOI	Hypohalous acid	Hypohalites, MOX
+3	HOClO	—[a]	—[a]	Halous acid	Halites, $MClO_2$
+5	$HOClO_2$	$HOBrO_2$	$HOIO_2$	Halic acid	Halates, $MClO_3$
+7	$HOClO_3$	$HOBrO_3$	$HOIO_3$	Perhalic acid	Perhalates, $MClO_4$
			$H_4I_2O_9$	Mesodiperiodic acid	Mesodiperiodates
			H_5IO_6	Paraperiodic acid	Paraperiodates

[a] The compound is unknown.

TABLE 20.13

Standard reduction potentials for oxochloro species

Alkaline reactions are shown in color.

Reaction	E°_{298} (V)
$HOCl + H^+ + 2e^- \rightleftharpoons Cl^- + H_2O$	1.49
$OCl^- + H_2O + 2e^- \rightleftharpoons Cl^- + 2OH^-$	0.90
$HOClO + 3H^+ + 4e^- \rightleftharpoons Cl^- + 2H_2O$	1.56
$ClO_2^- + 2H_2O + 4e^- \rightleftharpoons Cl^- + 4OH^-$	0.76
$ClO_3^- + 6H^+ + 6e^- \rightleftharpoons Cl^- + 3H_2O$	1.45
$ClO_3^- + 3H_2O + 6e^- \rightleftharpoons Cl^- + 6OH^-$	0.62
$ClO_4^- + 8H^+ + 8e^- \rightleftharpoons Cl^- + 4H_2O$	1.37
$ClO_4^- + 4H_2O + 8e^- \rightleftharpoons Cl^- + 8OH^-$	0.56

All of the oxochloro acids and their salts are reasonably strong oxidizing agents. The oxochloro ions function as oxidants in both acidic and alkaline solutions, as shown by the standard reduction potentials in Table 20.13. Hypochlorite ion is the most powerful oxidant in the group, and perchlorate ion is the weakest.

The strong oxidizing ability of the lower-oxidation-state oxochloro compounds makes them excellent bleaching agents (see the introduction to this chapter). The principal types of bleaching agents used for textiles, paper, and laundry bleaches are aqueous chlorine solutions (which contain hypochlorite ion, OCl^-); hypochlorite salts such as sodium hypochlorite, $NaOCl$; chlorine dioxide, ClO_2; and sodium chlorate, $NaClO_3$, which under the proper conditions releases ClO_2 in aqueous solution.

The acid strength of the oxochloro acids, as determined by their degree of ionization in solution, increases with oxygen content. Hypochlorous acid is a very weak acid, $K_a = 2.90 \times 10^{-8}$ at 25°C. Chlorous acid, although not a strong acid, is considerably stronger than hypochlorous acid, and has a K_a of about 10^{-2}. Both chloric and perchloric acids are strong acids. Perchloric acid is the strongest inorganic acid.

At room temperature about 25% of the chlorine in an aqueous solution is present in the form of HOCl. Larger concentrations of the acid can be produced by the addition of solid mercury(II) oxide, HgO, to the equilibrium mixture. The oxide combines with the hydrochloric acid to give mercury(II) chloride, a compound which is only very slightly ionized.

$$2Cl_2(g) + H_2O(l) + HgO(s) \longrightarrow HgCl_2(aq) + 2HOCl(aq)$$

On a commercial scale, solutions of sodium hypochlorite, $NaOCl$, are obtained by the electrolysis of stirred, cold, concentrated aqueous sodium chloride solutions. The chlorine produced at one electrode and the hydroxide ion produced at the other electrode react under the proper conditions as shown in the following equation:

$$Cl_2(g) + 2NaOH(aq) \longrightarrow NaOCl(aq) + NaCl(aq) + H_2O(l)$$

(In the electrolytic production of chlorine the products formed at the electrodes are kept separated so that this reaction does not occur; Section 24.8). Formation of sodium hypochlorite in this electrolytic process can also be formulated as

$$Na^+Cl^-(aq) + H_2O(l) \xrightarrow{\text{electrolysis}} Na^+OCl^-(aq) + H_2(g)$$

Chlorous acid, $HClO_2$, is an extremely unstable substance, which is formed along with chloric acid, $HClO_3$, when chlorine dioxide, ClO_2, a yellow gas that can explode unexpectedly, is passed into water.

$$2ClO_2(g) + H_2O(l) \longrightarrow HClO_2(aq) + H^+ + ClO_3^-$$
$$\text{chlorine} \qquad\qquad\qquad \text{chlorous}$$
$$\text{dioxide} \qquad\qquad\qquad\quad \text{acid}$$

Sodium chlorite, $NaClO_2$, a salt of chlorous acid used as a bleaching agent, is prepared by the reaction between chlorine dioxide and sodium peroxide, Na_2O_2.

$$2ClO_2(g) + Na_2O_2(aq) \longrightarrow 2NaClO_2(aq) + O_2(g)$$
sodium
chlorite

Solutions of chloric acid are usually made by reaction of a solution of barium chlorate with sulfuric acid; the by-product, barium sulfate, is insoluble and precipitates.

$$Ba(ClO_3)_2(aq) + 2H^+ + SO_4^{2-} \longrightarrow 2H^+ + 2ClO_3^- + BaSO_4(s)$$

Oxidation reactions in chloric acid solution can be quite violent.

Chlorates, the salts of chloric acid, are obtained by the reaction between chlorine and hot concentrated solutions of the appropriate hydroxides

$$3Cl_2(g) + 6OH^- \longrightarrow ClO_3^- + 5Cl^- + 3H_2O(l)$$
chlorate
ion

It is probable that the hypochlorite ion is an intermediate in the reaction and undergoes thermal disproportionation.

$$3\overset{+1}{OCl^-} + heat \longrightarrow \overset{+5}{ClO_3^-} + 2\overset{-1}{Cl^-}$$

Perchloric acid, $HClO_4$, can be prepared pure by distillation at low pressure from a mixture of the potassium salt and concentrated sulfuric acid. Pure perchloric acid is a colorless liquid that freezes at $-112°C$. It is a highly unstable and dangerous substance. When heated it decomposes explosively. With powerful reducing agents it also reacts violently and often explosively. If treated with strong dehydrating agents such as phosphorus(V) oxide, P_4O_{10}, perchloric acid loses water to form the anhydride, Cl_2O_7, which is even more unstable and dangerous than the acid.

Perchlorates, the salts of perchloric acid, are readily made by carefully heating chlorates in the absence of a catalyst. If the heating is too strong,

$$4ClO_3^- \overset{\Delta}{\longrightarrow} 3ClO_4^- + Cl^-$$
chlorate   perchlorate
ion   ion

the perchlorate formed decomposes with the liberation of oxygen.

$$ClO_4^- \longrightarrow Cl^- + 2O_2(g)$$

Commercially, perchlorates are made by electrolysis of hot, concentrated Cl^- solutions.

$$Cl^- + 4H_2O \xrightarrow{electrolysis} ClO_4^- + 4H_2$$

b. Oxobromo and oxoiodo compounds. Like the corresponding chloro compounds, the oxobromo and oxoiodo compounds are all strong oxidants, particularly in acidic solution. Hypobromous acid, HOBr, and hypoiodous acid, HOI, are even weaker acids than the chlorine compounds, the ionization constants at 25°C being 2.2×10^{-9} and 2.3×10^{-11}, respectively. Thus the acid strength decreases with decreasing electronegativity of the halogen, that is, with decreasing polarity of the H–O bond.

THOUGHTS ON
CHEMISTRY

Salt and civilization

Of all the basic necessities of life, salt is unique in that it is confined to certain limited locations. Men can make a living by hunting, fishing, raising animals and cultivating land over vast areas of the earth, but salt can come only from the comparatively few places where it is readily obtainable from salt lakes, mines, springs, or the ocean shores and means are at hand for extracting it. This fact, from the very beginning of civilization, has operated to make transport one of the cardinal necessities of civilized life. . . .

In Palestine agriculture spread up the Jordan River from the Dead Sea. . . . In Egypt early farming depended on boats bringing salt up the Nile from the salt swamps at the river mouth and on caravans bringing it in from the salt lakes in the desert. In France the hinterland was nourished by salt carried up the Rhone. . . .

The salt trade, particularly the overland traffic, inevitably had military, social and political consequences. The caravans and ships, and the depots to which they delivered their salt, had to be protected against bandits and marauders. It became necessary to provide them with convoys and to fortify way stations, shipping ports and trading posts. In short, the system of "protection" came into existence. As the most valuable single commodity in commerce, salt required the services of powerful protectors.

A certain political pattern seems to emerge: where salt was plentiful, the society tended to be free, independent and democratic; where it was scarce, he who controlled the salt controlled the people.

THE SOCIAL INFLUENCE OF SALT, by M. R. Bloch (pp. 89ff. in *Scientific American*, Vol. 209, No. 1, 1963. Copyright © 1963 by Scientific American, Inc. All rights reserved.)

THE SOCIAL INFLUENCE OF SALT, *by M. R. Bloch*

Significant terms defined in this chapter

hydrohalic acid
hypohalous acid
halides
halogenation
interhalogens

initiation
propagation
chain carrier
termination

THE HALOGENS

Exercises

20.1 Commercial bleaching powder is a mixture of $Ca(OCl)Cl$, $CaCl_2$, and $Ca(OCl)_2$ in approximately equimolar quantities. (a) Name these substances.

The powder is prepared by passing chlorine over dry calcium oxide

$$CaO(s) + Cl_2(g) \longrightarrow Ca(OCl)Cl(s)$$
$$2CaO(s) + 2Cl_2(g) \longrightarrow Ca(OCl)_2(s) + CaCl_2(s)$$

leaving a small amount of CaO as an impurity. (b) Find the oxidation state of chlorine in the compounds given in the equations and identify the (c) oxidizing agent and (d) reducing agent in each reaction.

The chlorine needed for bleaching is produced in neutral aqueous solution by the reaction between water, OCl^-, and Cl^-, represented by the equation

$$OCl^- + Cl^- + H_2O(l) \longrightarrow Cl_2(g) + 2OH^-$$

What element in the above equation is being (e) oxidized and (f) reduced? (g) What happens to the pH of the solution as the chlorine is being generated?

20.2 Write the atomic electronic configuration for each of the halogens. How do these change by the formation of the negative ions?

20.3 Which statements are true? Rewrite any false statement so that it is correct.
(a) A halide is a binary compound of halogen atoms formed by direct union of two different halogens.
(b) All of the hydrogen halides form strong acids when dissolved in water.
(c) Fluorine is the strongest oxidizing agent among all the elements.
(d) The shape of molecular ClF_3 will be an isosceles triangle.
(e) The acid strengths of the oxochloro acids increase with oxygen content.

20.4 Write equations representing the commercial preparation of F_2, Cl_2, Br_2, and I_2. Likewise write the equation representing the laboratory preparation of Cl_2, Br_2, and I_2 using MnO_2 and an acidic solution of NaX.

20.5 Write balanced equations for any reactions that will occur in aqueous mixtures of (a) NaI and Cl_2, (b) NaCl and Br_2, (c) NaI and Br_2, (d) NaBr and Cl_2, and (e) NaF and I_2.

20.6 Predict the type of bonding between atoms of (a) Al and I, (b) Na and F, (c) Ca and F, (d) I and Cl, and (e) Br and Br.

20.7 Why does fluorine form the anion HF_2^-, but the other halogens will not form similar species?

20.8 Arrange the oxo anions of chlorine in order of increasing thermal stability and write a balanced equation for the reaction that occurs as a salt of each is heated.

20.9 Complete and balance each of the following equations:

(a) $OCl^- + I^- \longrightarrow Cl^- + I_2$ (basic solution)
(b) $ClO_3^- + Sn^{2+} \longrightarrow Cl^- + Sn^{4+}$ (acidic solution)
(c) $I_2 + S_2O_3^{2-} \longrightarrow I^- + S_4O_6^{2-}$
(d) $IO_3^- + I^- \longrightarrow I_2$ (acidic solution)
(e) $Cl_2 + H_2O \longrightarrow$

20.10 Briefly explain why acid strength decreases in the order (a) $HI > HBr > HCl > HF$, (b) $HClO > HBrO > HIO$, and (c) $HClO_4 > HClO_3 > HClO_2 > HClO$.

20.11* An aqueous solution contains either NaBr or a mixture of NaBr and NaI. Using only aqueous solutions of I_2, Br_2, and Cl_2 and a small amount of CCl_4, describe how you might determine what is in the unknown solution.

20.12* The rate of formation of HBr(g) from $H_2(g)$ and $Br_2(g)$ between 200°C and 300°C is given by the equation

$$\text{rate} = \frac{k[H_2][Br_2]^{1/2}}{1 + k'\dfrac{[HBr]}{[Br_2]}}$$

How would you adjust the concentrations in order to make the reaction proceed as fast as possible?

20.13* What types of intermolecular forces act between molecules of (a) Cl_2, (b) HI, (c) HF, (d) $HClO_4$, (e) ICl_3, (f) HBrO, and (g) Cl_2O_7?

20.14 An excess of concentrated hydrochloric acid is added to 0.100 mole of MnO_2. Calculate the (a) volume and (b) mass of chlorine collected at 25°C at 745 torr.

20.15 The vapor pressure of $Br_2(l)$ at 25°C is 204 torr. What mass of bromine vapor would be found in a liter of air saturated with bromine?

20.16 The equilibrium constant at 25°C for the reaction

$$Cl_2(g) + H_2O(l) \rightleftharpoons H^+ + Cl^- + HClO(aq)$$

is 4.66×10^{-4}. Assuming that the HClO does not ionize appreciably, what will be the pH of a solution of "chlorine water" having a chlorine pressure of 0.5 atm?

20.17 The solubility of $CaCl_2$ in water at 0°C is 59.5 g/100 g H_2O and of NaCl is 35.7 g/100 g H_2O. Which salt is more effective in lowering the freezing point of water? $K = 1.86°C/m$ for water.

20.18 The enthalpy of formation at 25°C of HF(aq) is -76.50 kcal/mole, of OH$^-$(aq) is -54.970 kcal/mole, of F$^-$(aq) is -79.50 kcal/mole, and of H$_2$O(l) is -68.315 kcal/mole. Find the enthalpy of neutralization of HF(aq)

$$HF(aq) + OH^- \longrightarrow F^- + H_2O(l) \qquad \Delta H_{\text{neutralization}} = ?$$

Using the value of -13.345 kcal/mole as the enthalpy change for the reaction

$$H^+ + OH^- \longrightarrow H_2O(l)$$

find the enthalpy change for the reaction

$$HF(aq) \longrightarrow H^+ + F^-$$

20.19* Only one stable isotope of iodine is found in nature, $^{127}_{53}$I. Calculate the neutron–proton ratio for this stable isotope. The artificial radioisotope $^{131}_{53}$I, in the form of I$^-$, is commonly used in the treatment of the thyroid gland. What mode of decay would be predicted for this isotope? Write the nuclear equation for the decay.

20.20* In acidic solution, the dichromate ion, Cr$_2$O$_7^{2-}$, oxidizes iodide ion to elemental iodine and is itself reduced to chromic ion, Cr^{3+}. Write the equations for the half-reactions and calculate $E°$ for the reaction using the half-cell potentials given in Tables 19.5 and 20.3. Would the spontaneity of the reaction increase or decrease if the concentrations of all ions were 0.1M instead of 1M?

20.21* Powdered antimony is sprinkled into a bottle containing chlorine and tiny sparkles are generated. The white powder formed contains 47 wt % of chlorine. Find the empirical formula of the chloride and write an equation for the reaction.

20.22** The element astatine was first definitely prepared by the nuclear reaction represented by $^{209}_{83}$Bi(α,2n)$^{211}_{85}$At. (a) Write the nuclear equation for this reaction. The half-life of this isotope is 7.21 hr with 40.9% decaying by α emission and 59.1% by electron capture. (b) Write the nuclear equations for the decay process. (c) What amount of $^{211}_{85}$At would remain at the end of a week if a researcher started with 0.05 μg?

Although astatine has been detected in various minerals, most of the properties of this element have been observed from artificially produced samples or predicted from periodic relationships. Using Table 20.1, what would you predict for the (d) physical state, (e) melting point, (f) ionic radius, and (g) bond dissociation energy?

(h) What would you predict to be the common oxidation states of At? (i) Do these occur in AtI, AtBr, AtCl, HAt, AgAt, AtO$^-$, and AtO$_3^-$? Which of these compounds or ions (j) are interhalogens, (k) are hydrogen halides, (l) are metal halides, (m) will form a strong hydrohalic acid, and (n) are the anions of oxohalic acids? (o) Name these substances. Write an equation for the reaction between (p) HAt(aq) and AgNO$_3$(aq), (q) At$_2$ and KBr(aq), and (r) F$_2$ and At$^-$(aq).

21 THE COVALENT BOND REEXAMINED

We have been talking about covalent bonds since Chapter 9. By now we have established that the covalent bond forms as the result of electron pair sharing to give a lower energy state than exists in the isolated atoms. We have also established that covalent bonds have properties of bond length, bond angle, and bond energy that are quite consistent for similar bonds in different molecules.

However, there are many questions about covalent bonds that cannot be answered without a closer look at bonding theories, questions such as: What relationship exists between the sharing of electron pairs and the orbitals of the atoms that provide these electrons? How can the simultaneous sharing of the two or three electron pairs in double or triple bonds be accounted for? Can an electronic explanation of covalent bonding account for molecular geometry?

The theories that are used to answer these questions are, like quantum theory, rooted in mathematics beyond the scope of this book. Therefore, do not be concerned if you feel that you cannot completely grasp the theories presented in this chapter. The most important point is that the theories are frequently in good agreement with the experimentally determined facts about the geometry and other properties of molecules.

Two theories of covalent bonding are discussed in this chapter—valence bond theory and molecular orbital theory. Both arise from quantum mechanics and deal with discrete electron energy levels and orbitals. The chapter describes how each theory explains single and multiple covalent bonds. The concepts of hybridization of atomic orbitals and delocalized electrons are introduced. Molecular orbital energy diagrams are presented for some simple diatomic molecules and for metals, and we discuss how some of the properties of these species are accounted for by the molecular orbital theory.

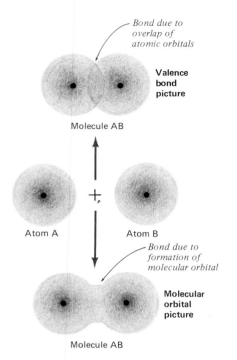

Bond due to overlap of atomic orbitals

Valence bond picture

Molecule AB

+

Atom A   Atom B

Bond due to formation of molecular orbital

Molecular orbital picture

Molecule AB

FIGURE 21.1

Comparison of valence bond and molecular orbital approaches. *In the valence bond picture the atomic orbitals maintain their identity and overlap to give an area of greater electron density between A and B. In the molecular orbital picture, the overlapping orbitals have combined and rearranged to give a bonding molecular orbital with greater electron density between A and B. (An antibonding orbital, not pictured here, is also formed; see Section 21.7.)*

21.1 Bond formation

In both the valence bond and molecular orbital theories nuclei are pictured as mutually attracted to an area of high electron density located along the line between the two nuclei—the **internuclear axis.** At the same time, the bonding electrons are mutually attracted by both nuclei.

Valence bond theory uses what we have already discussed in Chapter 6—atomic orbitals. The orbitals are, as we explained, regions in space surrounding a nucleus. Electrons are most likely to be found within the orbital. While the sum of all the orbitals on a particular atom is roughly spherical, the orbitals for individual electrons of different energy levels have distinctive, different "shapes." The shapes of the orbitals and the possible energy levels for the electrons are known through the mathematics of quantum theory. In bonding we are mainly interested in the outer shell, or valence, electrons. We know from the electronic configurations of the atoms what types of orbitals are occupied by the valence electrons.

Valence bond theory explains bond formation as the overlap of atomic orbitals (Figure 21.1). The greater the amount of overlap, the stronger the bond. To predict when and how bonding will occur according to valence bond theory, we must look at the existing configurations of the combining atoms. The possibility for bond formation exists when two atomic orbitals of similar energy level can overlap. These orbitals must be occupied by the two electrons needed for the covalent bond. The overlapping orbitals can each contain one electron, or both electrons can come from an orbital on one atom that overlaps an empty orbital on the other atom.

The molecular orbital approach, which also arises in quantum mechanics and mathematics, is different, but it also begins with atomic orbitals. Here again, only the outermost electrons need be considered. The difference is that the atomic orbitals lose their individual identities and combine to form new orbitals called molecular orbitals. **Molecular orbital theory** explains bond formation as the occupation by electrons of orbitals characteristic of the whole molecule. The difference between bonds in atomic orbital and molecular orbital theory is a little bit like the difference between two fried eggs that have run together at the edges in the frying pan and scrambled eggs. In forming molecular orbitals, the atomic orbitals, like the scrambled eggs, lose their individuality. However, atomic orbitals, unlike the scrambled eggs, are not mixed in random fashion, but form new orbitals with different specific shapes and energy levels. The requirements for successful bond formation are similar in both valence bond and molecular orbital theory:

1. The orbitals occupied by the electrons forming the bonds must have similar energies.

2. These orbitals must overlap each other enough to allow interaction.

3. The overlapping orbitals must have the same symmetry with respect to the internuclear axis.

Another bond theory—the crystal field theory—is discussed in Chapter 28 in connection with the chemistry of complexes.

Atomic orbital overlap: valence bond theory

21.2 Single bonds in diatomic molecules

The simplest example of bond formation by atomic orbital overlap is given by the H_2 molecule. The single electron associated with each hydrogen atom occupies a spherical $1s$ orbital (configuration $1s^1$). Overlap of the two $1s$ orbitals allows pairing of the electron spins.

H($1s^1$) + H($1s^1$)

End view

All bonds such as this, in which the area of highest electron density surrounds the internuclear axis, are called **σ bonds**—*sigma bonds*.

An s orbital and a p orbital each contain a single electron and can also overlap to yield a σ bond. For example, a hydrogen atom ($1s^1$) can combine with a chlorine atom ([Ne]$3s^23p^23p^23p^1$) to form an HCl molecule by overlap of the $1s$ orbital of the hydrogen atom and the $3p$ orbital of the chlorine atom that holds a single electron.

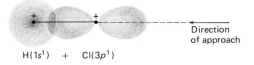

H($1s^1$) + Cl($3p^1$)

Similarly, two p orbitals can overlap to produce a σ bond, providing that the overlapping orbitals lie along the same axis. For example, the chlorine–chlorine bond in the Cl_2 molecule can be described by the overlap of two $3p$ orbitals each containing one electron.

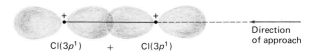

Cl($3p^1$) + Cl($3p^1$)

The p orbitals are oriented in space at 90° angles to each other; therefore, geometry forbids certain overlaps to give σ bonds between s and p orbitals on the same two atoms. Consider atoms approaching each other along the x axis. In this situation, an s orbital on one atom

FIGURE 21.2
Possibilities for σ-bond formation by s and p orbitals on the same two atoms. *The direction of approach is chosen as along the x axis.*

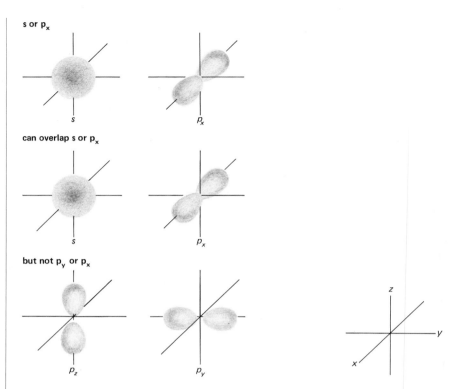

cannot overlap a p_z or a p_y orbital on the other atom sufficiently to allow bond formation. Nor, in the same situation, can a p_x orbital overlap the p orbitals extended in the y or z directions (Figure 21.2).

21.3 π Bonds

Another type of bond can be formed by the valence shell p electrons on two atoms if approach is along, say, the x axis. A p_y and a p_y, or a p_z and a p_z orbital can overlap (Figure 21.3). In such bonds the area of overlap does not completely surround the internuclear axis. Instead, there are two separate regions of high electron density, one on each side of the axis. For example, for two p_z orbitals,

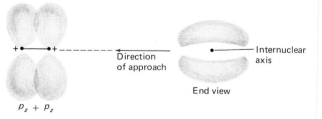

Bonds that do not completely surround the internuclear axis are called **π bonds**—*pi bonds.* The p–p π bonds leave an area of zero electron density along the internuclear axis. π Bonds are not as strong as σ bonds and almost always occur together with σ bonds, which provide the greater share (50–75%) of the force holding the atoms together. π Bonds can also result from the overlap of p and d orbitals.

21.4 Single bonds in polyatomic molecules; hybridization

Overlap of *s* and *p* orbitals as we have pictured it thus far cannot explain the bond lengths and bond angles in most molecules with more than two atoms. For example, consider the O—H and N—H bonds in H_2O and NH_3. If each bond resulted from overlap of the hydrogen $1s^1$ orbital with either a *p* orbital of O ($1s^2 2s^2 2p^4$) or a *p* orbital of N ($1s^2 2s^2 2p^3$), the HOH or HNH bond angles should be the same as the angles between the *p* orbitals, which are all 90°. But in both compounds the angles are closer to the tetrahedral angle (Figure 21.4).

Furthermore, there are difficulties in explaining the geometry of a molecule that contains bonds derived from the overlap of different types of orbitals. In methane, CH_4, two *s* electrons and two *p* electrons from C ($1s^2 2s^2 2p^2$) must each form a bond with a $1s$ electron from H. Differences in bond length and bond angle might be expected between C—H bonds from *s–p* overlap and those from *s–s* overlap. However, the four C—H bonds in CH_4 are identical in length and form equal tetrahedral angles with each other (Figure 21.4).

The geometry of CH_4, H_2O, NH_3, and other molecules was explained earlier by the valence-shell electron pair repulsion theory (Section 9.26). The theory simply states that electrons stay as far apart as possible, and that, since unshared electron pairs take up more space than electron pairs in bonds, the bond angles in molecules with lone pairs are decreased from the ideal angle.

The concept of hybridization was introduced to allow explanation of molecular geometry in terms of atomic orbitals and valence bond theory. **Hybridization** is the mixing of the atomic orbitals on a single atom to give a new set of orbitals, called hybrid orbitals, on that atom. Note that hybridization happens to the orbitals of a single atom. The hybrid orbitals are still atomic orbitals and remain oriented around a single nucleus. The difference is that several electrons that were in atomic orbitals at slightly different energy levels have settled into orbitals at the same energy level—a sort of compromise level intermediate between where the individual electrons were before. Bond formation occurs when the hybrid orbitals overlap

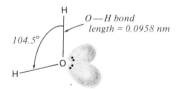

O—H bond length = 0.0958 nm

104.5°

Water, H_2O

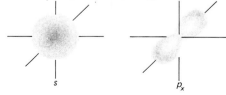

N—H bond length = 0.1012 nm

106.7°

Ammonia, NH_3

C—H bond length = 0.1091 nm

109.5°

Methane, CH_4

FIGURE 21.4

Geometry of water, ammonia, and methane molecules. (*See Figure 9.14, which shows how the same shapes arise in valence bond electron pair repulsion theory.*)

Number of hybrid orbitals = number of combined atomic orbitals

---

FIGURE 21.3

Possibilities for π-bond formation by *p* orbitals on the same two atoms. *The direction of approach is along the x axis. To form π bonds, the overlapping p orbitals must be parallel.*

p_y

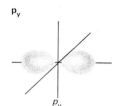

p_y

can overlap p_y but not p_z

p_y p_z

and p_z

p_z

can overlap p_z but not p_y

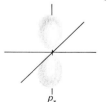

p_z

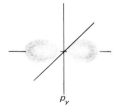

p_y

Neither p_y or p_z can overlap s or p_x

s p_x

atomic orbitals, hybrid or not, on other atoms. Conclusions about molecular geometry based on both electron pair repulsion and hybridization are in excellent agreement with each other and with the observed geometry of molecules.

By considering the approach of atoms in the gas phase toward each other, we can visualize the hybridization of atomic orbitals. We are not saying that it happens this way (or any other way), because no experimental proof of hybridization has been obtained.

The atomic orbitals of an isolated gaseous atom take the most stable arrangement in space, as calculated mathematically and as shown by the shapes we have pictured previously (Figures 7.14–7.17). The sum of the orbitals surrounding each atom is a sphere. However, as soon as other atoms approach, distortion and rearrangement of the orbitals occur. Because electrons repel each other, the areas of high electron density—the bonds—stay as far apart as possible.

Suppose two chlorine atoms approach a beryllium atom ($2s^2$ configuration) one at a time to produce $BeCl_2$ in the gas phase. First, a Be—Cl bond would be formed. What would be the best way for the second chlorine atom to approach in order to share the electron on beryllium that is still available for bonding? Obviously, from the side opposite the existing Be—Cl bond, thereby avoiding repulsion by the area of high electron density around the bond.

$$Cl\text{-----}\!\!\rightarrow\!Be\!-\!Cl \longrightarrow Cl\!-\!Be\!-\!Cl \swarrow \begin{smallmatrix}\textit{linear}\\\textit{molecule}\end{smallmatrix}$$

The same hypothetical process of building up a molecule bond by bond for boron ($2s^2 2p^1$ configuration) and chlorine would lead first to a B—Cl bond, and then to Cl—B—Cl. Boron has a still another electron available for bonding. The third chlorine atom can best approach in the same plane as the Cl—B—Cl.

The three C—B—Cl bond angles equalize in order to maintain the maximum distance between shared electron pairs.

A carbon atom ($2s^2 2p^2$ configuration) can form four covalent bonds to chlorine. Assuming that the first three hypothetical C—Cl bonds would have planar geometry similar to that of BCl_3, the obvious direction of approach to carbon for the fourth chlorine atom is from above or below the plane in which the other four atoms lie. As the fourth C—Cl bond forms, the others would move out of the plane, equalizing the angles and leading to the tetrahedral CCl_4 molecule.

These orbitals form σ bonds by overlap with other hybrid orbitals and with s and p orbitals.

TABLE 21.1
Formation of s p hybrid orbitals

Type	Constituent orbitals	Ideal bond angle	Hybrid orbitals	Geometry
sp	**One** s + **one** p **orbital**	**180°**		**Linear**
sp^2	**One** s + **two** p **orbitals**	**120°**		**Triangular planar**
sp^3	**One** s + **three** p **orbitals**	**109.5°**		**Tetrahedral**

The $BeCl_2$, BCl_3, and CCl_4 molecules illustrate the three possible types of hybridization between s and p orbitals, sp, sp^2, and sp^3 hybridization. In these symbols the superscript indicates the number of orbitals that have combined; for example

$$sp^2$$

one s orbital hybridized with two p orbitals

Note that the superscript here does not indicate the number of electrons in the orbital.

In sp **hybridization** one s and one p orbital combine to give two sp hybrid orbitals that lie at 180°. In sp^2 **hybridization** one s and two p orbitals combine to give three sp^2 hybrid orbitals that lie in a plane with 120° angles between them. In sp^3 **hybridization** one s and three p orbitals combine to give four sp^3 hybrid orbitals arranged in a tetrahedral formation. These relationships are summarized in Table 21.1. Note that the number of hybrid orbitals formed always equals the number of atomic orbitals that have combined. The hybrid orbitals have a large lobe on one side of the nucleus and a smaller lobe on the other.

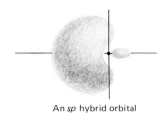

An sp hybrid orbital

Multiple covalent bonds
= σ plus π bonds

---

cis: groups on same side
trans: groups on opposite
sides

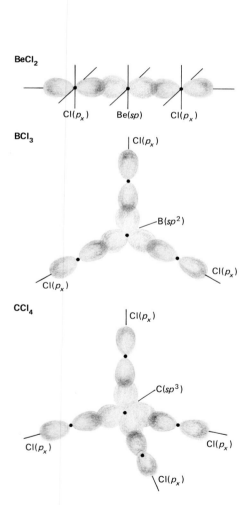

BeCl₂

Cl(p_x)   Be(sp)   Cl(p_x)

BCl₃

Cl(p_x)

B(sp^2)

Cl(p_x)

Cl(p_x)

CCl₄

Cl(p_x)

C(sp^3)

Cl(p_x)   Cl(p_x)

Cl(p_x)

FIGURE 21.5
BeCl₂, BCl₃, and CCl₄. *Valence bond pictures of bonding. The hybrid orbitals are shown in color.*

All *s–p* hybrid orbitals have this shape. For simplification, the smaller lobe is often omitted in drawings of hybrid orbitals.

Beryllium in its ground-state configuration, $1s^2 2s^2$, has no unpaired electrons. In compound formation, one $2s$ electron is thought to be promoted to an empty $2p$ orbital. Then sp hybridization can occur. Each of the resulting two sp hybrid orbitals contains a single electron and can form a single covalent bond (Figure 21.5).

In boron trichloride and carbon tetrachloride, sp^2 and sp^3 hybridization have taken place on boron and carbon, respectively (Figure 21.5). In water, ammonia, and methane (Figure 21.4), the oxygen, nitrogen, and carbon orbitals are all sp^3 hybridized, although only on the carbon atom do all four overlap to form bonds with other atoms.

Linear, triangular planar, and tetrahedral geometries are thus accounted for by hybridization of s and p orbitals. The more complex geometries involving more than four covalent bonds arise from hybridization of s, p, and d orbitals (and f orbitals in heavy elements). Triangular bipyramidal geometry and octahedral geometry (see Table 9.22) result from dsp^3 and d^2sp^3 or sp^3d^2 hybridization, respectively. In addition, dsp^2 hybridization gives square planar geometry; and dsp^3 hybridization can also give square pyramidal geometry.

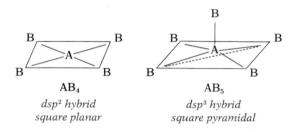

AB₄
*dsp² hybrid
square planar*

AB₅
*dsp³ hybrid
square pyramidal*

21.5 Multiple covalent bonds

One σ bond plus one π bond form a double covalent bond, and one σ bond plus two π bonds form a triple covalent bond. π Bonds, as we have discussed, are formed by the overlap of parallel p orbitals. A multiple bond always involves π-bond formation, since not more than one σ bond can occur between the same two atoms and π bonds almost always occur together with a σ bond.

Carbon–carbon covalent multiple bonds are explained by the apparent ability of carbon atoms to form either sp or sp^2 hybrid orbitals as well as tetrahedral sp^3 orbitals. In ethylene, $CH_2=CH_2$, and other doubly bonded carbon compounds, the two carbon atoms in each double bond are sp^2 hybridized. This provides a planar skeleton of σ bonds for the molecule, with each carbon atom at the center of three σ bonds arranged in a planar triangle (Figure 21.6a.). Three electrons from each carbon atom occupy the sp^2 orbitals, leaving one electron on each carbon atom in an unhybridized p orbital, which we picture as the p_z orbital. With the carbon–hydrogen skeleton in the $x–y$ plane, the two valence electrons in the p_z orbitals can be shared by sidewise overlap, leading to a π bond (Figure 21.6b).

Acetylene, HC≡CH, like ethylene, has a linear carbon skeleton.

(a) Ethylene σ–bonded skeleton

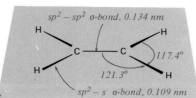

(b) Ethylene π bonds

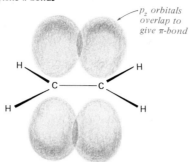

In this case, each of the carbon atoms forms an *sp* hybrid, and each forms σ bonds with the other carbon atom and with one hydrogen atom (Figure 21.7). Sidewise overlap of the p_z and p_y orbitals leads to sharing of the remaining two electrons from each carbon atom to form a pair of π bonds. The triple bond in acetylene, and all other carbon–carbon triple bonds, are, therefore, a combination of one σ bond and two π bonds.

For the π bonds in a covalent multiple bond to remain intact, the orientation of the atoms must not change—the *p* orbitals have to be parallel to each other. Turning the atoms would break the π bond. Therefore, (1) the atoms attached to the multiply bonded atoms must lie in the same plane, and (2) the atoms in multiple bonds are not free to rotate about the internuclear axis.

Because rotation about a double bond is restricted, how other atoms or groups are attached to the atoms in a double bond makes a difference. Compounds that differ in structure but have the same molecular formula are called **isomers.** The replacement of two hydrogen atoms by identical groups either on the same side or on opposite sides of the ethylene molecule yields two different isomers of the same compound. The **cis isomer** has the identical groups on the same side, and the **trans isomer** has the identical groups on opposite sides. In this type of isomerism, called cis–trans isomerism, or geometrical isomerism, the placement of groups either together or opposite each other yields different compounds. For example,

$$
\begin{array}{cc}
\underset{\mathrm{Cl}}{\overset{\mathrm{H}}{}}\!\!\diagdown\!\mathrm{C}\!=\!\mathrm{C}\!\diagup\!\!\underset{\mathrm{Cl}}{\overset{\mathrm{H}}{}} & \underset{\mathrm{H}}{\overset{\mathrm{Cl}}{}}\!\!\diagdown\!\mathrm{C}\!=\!\mathrm{C}\!\diagup\!\!\underset{\mathrm{Cl}}{\overset{\mathrm{H}}{}}
\end{array}
$$

cis-Dichloroethylene   *trans*-Dichloroethylene
b.p. 60°C   b.p. 48°C

If rotation about the double bond were possible, these two compounds would be identical.

21.6 Valence bond vs. molecular orbital theory

Molecular structure and geometry are best explained in terms of the valence bond theory. The atomic orbital overlap diagrams clearly indicate how geometry and bonding are interrelated. With such dia-

(a) Acetylene σ-bonded skeleton

(b) p orbitals, approaching to form π orbitals

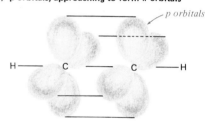

(c) π orbitals in acetylene

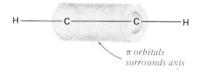

FIGURE 21.7
Acetylene molecule. (*a*) *Acetylene σ-bonded skeleton; (b) p orbitals, approaching to form π orbitals; (c) π orbitals in acetylene.*

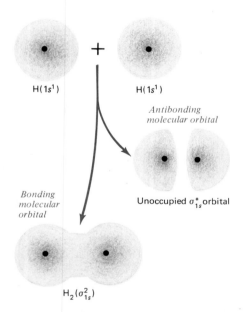

H(1s¹)   H(1s¹)

Antibonding molecular orbital

Bonding molecular orbital

Unoccupied σ_{1s}^* orbital

$H_2(\sigma_{1s}^2)$

FIGURE 21.8
The hydrogen molecular orbitals.

grams it is apparent which orbitals of the atoms are involved in bonding. Unfortunately, the relative energy levels of the orbitals are not apparent.

On the other hand, it is difficult to draw pictures of how molecular orbitals are arranged in space. But molecular orbital theory is quite useful in predicting whether or not a particular bond will form—something that is not too easy in valence bond theory. For example, molecular orbital theory explains the nonexistence of Be_2 and is the only theory that can account for the magnetic properties of the oxygen molecule (Section 21.11). The relative energy levels of the orbitals in molecules are also very successfully predicted by molecular orbital theory, and it helps theoretical chemists in interpreting spectra. It also permits a clearer understanding of multiple bonds and the prediction of the order of a bond.

Molecular orbital theory

21.7 Molecular orbitals in H_2 and "He_2"

A **molecular orbital** is the space in which an electron with a specific energy is likely to be found in the vicinity of two or more nuclei that are bonded together. The electron can be close to one nucleus at one instant and close to another nucleus at another instant. When an electron is given a larger space in which to move around, its energy decreases and under these conditions a bond can form. The spreading of an electron cloud into a larger space is called **delocalization.**

The orbitals of two atoms that are about to form a bond can be thought of as rearranging themselves as the electrons and nucleus of one atom come under the influence of the electrons and nucleus of the other atom. The number of molecular orbitals formed must equal the number of atomic orbitals that have combined. For example, when two atoms that each have three atomic orbitals combine, there will be six molecular orbitals possible in the area of the bonded atoms. This does not mean that all of these molecular orbitals will be filled with electrons. To use molecular orbital theory we first figure out the number of possible molecular orbitals. Then we go back to the atoms that combined to find out how many electrons are available. Finally, we assign the electrons to molecular orbitals, writing an electronic configuration for the molecule, somewhat as we wrote *spdf* configurations for atoms.

The process is a little like assigning dormitory rooms to freshmen in a college. The person in charge of housing first has to figure out how many rooms are available and on what floors in a particular building. But the Office of Admissions has to decide how many freshmen there will be to occupy the rooms. Once this is known, the students are assigned to the empty rooms. Maybe there will be fewer students than rooms, but the empty rooms—like the vacant molecular orbitals—are still there to be occupied if someone else comes along.

Once again, we look first at the simplest neutral molecule, the hydrogen molecule. When two hydrogen atoms approach each other, the two 1s orbitals combine to produce two molecular orbitals that differ in energy level and in the areas in space that they occupy. One is a bonding orbital—electrons in a bonding orbital draw the nuclei together. The other is an antibonding orbital—electrons in an antibonding orbital provide no attraction between the nuclei and allow the nuclei to repel each other. An analogy can be made to the attraction and repulsion of opposite poles of a magnet. Mathematically, the formation of bonding and antibonding orbitals is the result of the addition and subtraction of the wave functions of the individual atoms.

The combination of any two atomic orbitals always yields one bonding and one antibonding orbital. Note that there are four places that can be occupied by electrons in the dihydrogen molecular orbitals, since each of the two orbitals can hold two electrons. However, we have only one electron from each hydrogen atom and therefore only two places are filled.

In a **bonding molecular orbital,** most of the electron density is located between the nuclei of the bonded atoms. The bonding molecular orbital from the two 1s atomic orbitals of the two hydrogen atoms is a σ orbital—the electron cloud surrounds the internuclear axis (Figure 21.8). It is symbolized as follows, and read as "sigma one ess":

a sigma bonding
molecular orbital ⟶ σ_{1s} ⟵ formed from a 1s atomic orbital

In an **antibonding molecular orbital,** most of the electron density is located away from the internuclear axis, and there is a point, or node, of zero electron density between the nuclei. The nuclei are mutually repelled by each other's positive charges, rather than held together. Electrons in such an orbital do not form a bond; instead, they can weaken a bond. Antibonding orbitals are symbolized by adding a star to the bonding molecular orbital symbol, which is read, for example, as "sigma star one ess."

a sigma antibonding
molecular orbital ⟶ σ_{1s}^* ⟵ formed from a 1s atomic orbital

We know that a bond forms when a lower energy state is achieved by the electrons in the combining atoms. Therefore, it seems likely that a bonding molecular orbital would be at a lower energy level than the atomic orbitals from which it was formed. Similarly, an antibonding orbital ought to be at a higher energy level than the atomic orbitals that have combined.

The relative energy levels of the orbitals in hydrogen atoms and the hydrogen molecule are shown in Figure 21.9. As we thought, the bonding molecular orbitals are below the hydrogen atom 1s level and the antibonding orbitals are above. In assigning electrons to molecular orbitals, the lowest energy levels are filled first, just as for atomic orbitals. Therefore, the two electrons from the two hydrogen atoms both go into the σ orbital and the σ^* orbital remains empty. Two

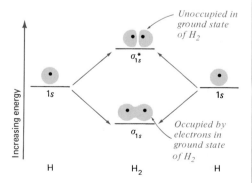

FIGURE 21.9

Relative energy levels of orbitals in H + H → H₂ reaction.

electrons in a σ orbital formed from atomic $1s$ orbitals are symbolized as

$$\sigma_{1s}^2 \qquad \textit{two electrons in a } \sigma_{1s} \textit{ orbital}$$

Now consider the approach of two helium atoms, of configuration $1s^2$. As for the hydrogen atoms, two atomic $1s$ orbitals will yield a σ_{1s} and a $\sigma_{1s}{}^*$ orbital. But here there are two more electrons to be accommodated. Both the σ and σ^* orbitals are filled by the four available electrons, and the molecular electronic configuration is $\sigma_{1s}{}^2\sigma_{1s}{}^{*2}$. The net effect of an equal number of electrons in bonding and antibonding orbitals is neutralization of the bonding orbital by the antibonding orbital. No covalent bond can form under such circumstances. This molecular orbital theory prediction agrees with what we know—gaseous helium is monatomic and He_2 molecules have never been observed.

21.8 Rules for filling molecular orbitals

In the molecular orbital treatment of bond formation we visualize electrons filling the molecular orbitals one by one in the same way as atomic orbitals are filled (Section 7.18). The rules are summarized here in terms of molecular orbitals.

1. Electrons first occupy the molecular orbitals of lowest energy; they enter higher-energy molecular orbitals only when the lower-energy orbitals are filled.

2. Each molecular orbital can accommodate a maximum of two electrons (Pauli exclusion principle).

3. Molecular orbitals of equal energy are occupied by single electrons before electron pairing begins (Hund principle).

21.9 Bond order

In a previous section, bond order was defined simply as the number of covalent bonds between two atoms and was determined by looking at the Lewis structures. Molecular orbital theory provides a way of calculating bond order:

Bond order = $\frac{1}{2}$[(the number of bonding electrons in outer shells)
$\qquad\qquad$ −(the number of antibonding electrons in outer shells)]

For the hydrogen molecule, H_2:

Bond order = $\frac{1}{2}(2 - 0) = 1$

For the helium molecule, "He_2":

Bond order = $\frac{1}{2}(2 - 2) = 0$

As we shall see, it is possible to have fractional bond orders.

Electron density diagrams

The elegant electron density diagrams presented here were plotted by a computer from quantum-mechanical calculations. The electron density in a cross-sectional plane passing through a molecule is plotted at right angles to that plane. These plots show the general shapes for some molecular orbitals of a homonuclear diatomic molecule. To save space, the peaks of some of the plots have been cut off. The chemists who produced these diagrams (John R. Van Wazer and Ilyas Absar, "Electron Densities in Molecules and Molecular Orbitals," Academic Press, 1975) cite the Confucian maxim that "one picture is worth ten thousand words."

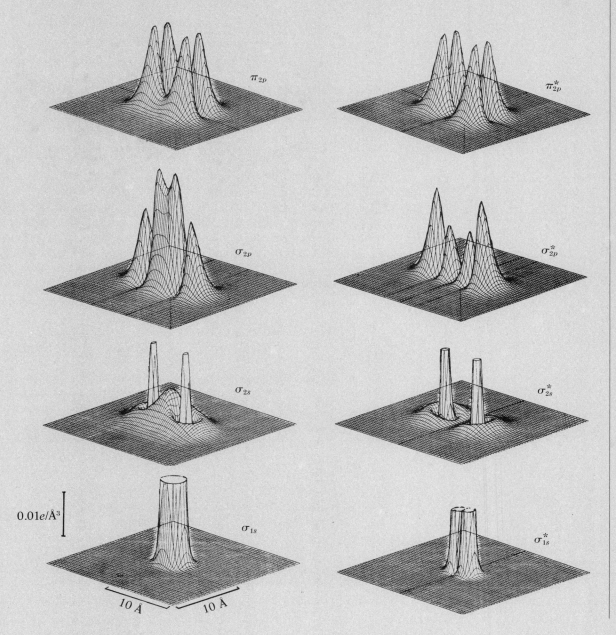

π_{2p}

π_{2p}^*

σ_{2p}

σ_{2p}^*

σ_{2s}

σ_{2s}^*

$0.01e/\text{Å}^3$

σ_{1s}

σ_{1s}^*

10 Å 10 Å

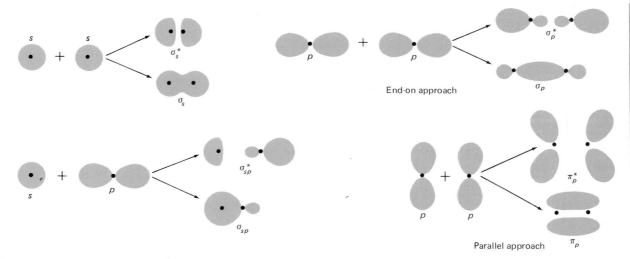

FIGURE 21.10
Molecular orbitals from *s* and *p* atomic orbitals. *The direction of approach is chosen as along the x axis. The antibonding molecular orbitals are higher in energy and the bonding molecular orbitals are lower in energy than the atomic orbitals from which they arise.*

21.10 π Molecular orbitals

In valence bond theory, we talk of π bonds as formed by parallel overlap of *p* orbitals. The molecular orbital picture is similar—the parallel *p* atomic orbitals combine to form π molecular orbitals. When the π molecular orbitals are occupied, π bonding is present in the molecule.

The *p* orbitals that approach each other end-to-end rather than from parallel positions form σ molecular orbitals. Therefore, the three *p* orbitals from each of two atoms yield six molecular orbitals —one bonding and one antibonding σ orbital, and two bonding and two antibonding π orbitals. The formation of σ and π molecular orbitals from *s* and *p* atomic orbitals is summarized in Figure 21.10.

21.11 Molecular orbitals for diatomic molecules of second-period elements, Li–Ne

The maximum number of atomic orbitals for atoms of second-period elements is five—one 1*s* orbital, one 2*s* orbital, and three 2*p* orbitals. In compound formation by these atoms, the 1*s* electrons remain essentially atomic in character and molecular orbitals need not be considered for them. Such inner-shell electrons, which are held close to the atoms and do not influence bonding, are sometimes called *nonbonding electrons*.

The four *n* = 2 atomic orbitals of the second-period elements combine to produce eight molecular orbitals. The relative energy levels of these orbitals are illustrated in Figure 21.11. Experimentally, information about orbital energy levels is usually derived from studies of spectra and wavelengths. The two π_{2p} orbitals for second-period elements lie close in energy level to the σ_{2p} orbital. It would be expected that a σ orbital would be of lower energy than a comparable π orbital, because σ bonds are generally stronger. Apparently, in some molecules, the closeness of the π_{2p} and σ_p orbitals allows their order to be reversed. The best evidence so far suggests that π_{2p} orbitals are of lower energy in the lighter second-period diatomic molecules Li_2 through N_2, while the σ_{2p} orbitals are of lower energy in O_2 and F_2.

In the following paragraphs, we consider how the molecular orbitals are filled for homonuclear gaseous diatomic molecules of the elements lithium through neon. **Homonuclear** means literally "the same nucleus," in other words, referring to atoms of the same element. The value of molecular orbital theory in predicting whether or not a bond will form and in explaining the magnetic properties, bond energies, and bond lengths of molecules is well illustrated by these molecules.

Lithium, $1s^2 2s^1$. A Li_2 molecule will contain six electrons. The first four electrons, as for all of the second-period elements, are the nonbonding $1s$ electrons. The remaining two electrons come from the $2s$ orbitals of the two lithium atoms. Just as in H_2, the s orbitals combine to give one bonding and one antibonding σ molecular orbital.

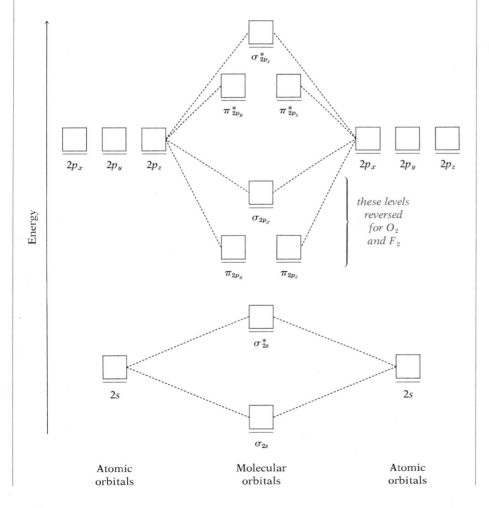

FIGURE 21.11
Relative energy levels for atomic and molecular orbitals of second-period elements. *The n = 1 electrons are essentially nonbonding, or "core" electrons and are not included in consideration of molecular orbitals. Each square represents an orbital that can accommodate two electrons. (This and other energy level diagrams are not drawn to scale.)*

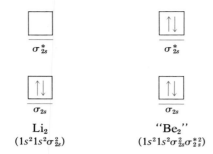

σ_{2s}^* σ_{2s}^*

σ_{2s} σ_{2s}

Li$_2$ "Be$_2$"

$(1s^2 1s^2 \sigma_{2s}^2)$ $(1s^2 1s^2 \sigma_{2s}^2 \sigma_{2s}^{*2})$

FIGURE 21.12

Molecular orbital diagrams for Li$_2$ and "Be$_2$." *In writing electronic configurations for the molecules, we have used the atomic orbital notation for the nonbonding electrons.*

TABLE 21.2

Diatomic molecules of first- and second-period elements

The lower-energy σ orbital is occupied by the two electrons (Figure 21.12), giving Li$_2$ a bond order of $\frac{1}{2}(2 - 0) = 1$.

Li$_2$ molecules can be detected in the gaseous state. They are much less stable than H$_2$ molecules because of the larger size of the lithium atom and the greater nuclear repulsions between lithium atoms than between hydrogen atoms. Compare the bond dissociation energies of H$_2$ and Li$_2$ in Table 21.2.

Beryllium, $1s^2 2s^2$. Of the eight electrons available for a Be$_2$ molecule, the first four are the nonbonding inner electrons. The next four fill the σ_{2s} and σ_{2s}^* orbitals. This gives a bond order of $\frac{1}{2}(2 - 2) = 0$; in other words, no bond at all. As predicted, a stable Be$_2$ molecule has never been observed.

Boron, $1s^2 2s^2 2p^1$. The B$_2$ molecule has six outer electrons to be distributed in molecular orbitals. Four electrons fill the σ_{2s} and σ_{2s}^* orbitals, leaving two electrons for the π orbitals. Following the Hund principle, the two electrons enter the π_{p_y} and π_{p_z} orbitals singly. This gives B$_2$ a bond order of 1.

The conclusions that may be drawn from the molecular orbital diagram for B$_2$ (Figure 21.13) are that it will exist and that it will exhibit a paramagnetism showing the presence of two unpaired electrons. Both of these predictions are correct. Note that the two-electron paramagnetism of boron is evidence for the lower energy of the π_{2p} orbitals. If the σ_{2p}^* orbital had lower energy, the last two electrons would be paired and B$_2$ would be diamagnetic.

Carbon, $1s^2 2s^2 2p^2$. Eight electrons must be placed into the molecular orbitals for C$_2$. The C$_2$ molecule exists at high temperatures or in

In general, bond length decreases and bond energy increases with bond order. Note these trends in the series B$_2$, C$_2$, N$_2$, and O$_2{}^+$, O$_2$, O$_2{}^-$.

Species	Total number of electrons	Bond order	Bond length (nm)	Bond dissociation energy, kcal/mole (kJ/mole)
H$_2$	2	1	0.074	104 (436)
"He$_2$"	4	0	—	—
Li$_2$	6	1	0.267	26.4 (111)
"Be$_2$"	8	0	—	—
B$_2$	10	1	0.159	70.6 (295)
C$_2$	12	2	0.124	142 (593)
N$_2$	14	3	0.109	226 (946)
O$_2{}^+$	15	$2\frac{1}{2}$	0.112	153 (641)
O$_2$	16	2	0.121	119 (498)
O$_2{}^-$	17	$1\frac{1}{2}$	0.130	95.2 (398)
F$_2$	18	1	0.141	37.7 (158)
"Ne$_2$"	20	0	—	—
CO	14	3	0.113	257 (1075)
NO	15	$2\frac{1}{2}$	0.115	151 (631)

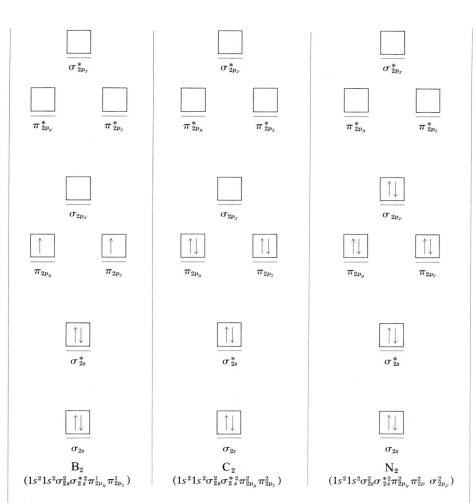

FIGURE 21.13
Molecular orbital diagrams for B$_2$, C$_2$, and N$_2$.

$\sigma^*_{2p_x}$

$\pi^*_{2p_y}$ $\pi^*_{2p_z}$

σ_{2p_x}

π_{2p_y} π_{2p_z}

σ^*_{2s}

σ_{2s}

B$_2$
$(1s^2 1s^2 \sigma^2_{2s} \sigma^{*2}_{2s} \pi^1_{2p_y} \pi^1_{2p_z})$

C$_2$
$(1s^2 1s^2 \sigma^2_{2s} \sigma^{*2}_{2s} \pi^2_{2p_y} \pi^2_{2p_z})$

N$_2$
$(1s^2 1s^2 \sigma^2_{2s} \sigma^{*2}_{2s} \pi^2_{2p_y} \pi^2_{2p}, \sigma^2_{2p_x})$

the presence of an electrical discharge. It has a bond order of $\frac{1}{2}(6 - 2) = 2$, and the higher bond order is reflected in a high bond dissociation energy (see Table 21.2). All electrons are paired, and C$_2$ is diamagnetic.

Nitrogen, 1s²2s²2p³. The N$_2$ molecule is diamagnetic and thermodynamically very stable. With eight bonding electrons and two antibonding electrons, the bond order is 3, in agreement with the Lewis formula that we have been writing for N$_2$, $:N::N:$.

Oxygen, 1s²2s²2p⁴. One of the greatest successes of molecular orbital theory is in explaining the paramagnetism of dioxygen. The theory also nicely accounts for the relative bond dissociation energies of O$_2$ and O$_2^+$. The first ten outer shell electrons from the oxygen atoms fill the molecular orbitals up to the π_{2p} orbitals. The next highest energy level has two equals π^*_{2p} antibonding orbitals available. The final two electrons of O$_2$ should enter these π^*_{2p} orbitals singly. The paramagnetism of O$_2$ shows that this does indeed happen—two unpaired electrons are present on O$_2$ (Figure 21.14). Only the molecular orbital theory was able to explain both the paramagnetism and the bond order of O$_2$, which is 2.

Removal of one electron from O$_2$ to give O$_2^+$ increases the O—O

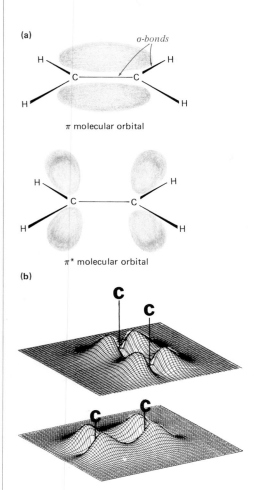

(a)

σ-bonds

π molecular orbital

(b)

π* molecular orbital

FIGURE 21.15
Molecular orbital views of ethylene.
(*a*) *Conventional drawing of ethylene* π *and* π* *molecular orbitals;* (*b*) π-*bonding orbital electron density distribution in ethylene. Views from two different angles of electron density above a plane perpendicular to the plane of the C—H skeleton. (From* Electron Densities in Molecules and Molecular Orbitals, *J. R. Van Wazer and I. Absar, Academic Press, 1975, p. 33. See An Aside: Electron Density Diagrams, this chapter.*)

FIGURE 21.14
Molecular orbital diagrams for O_2 and F_2.

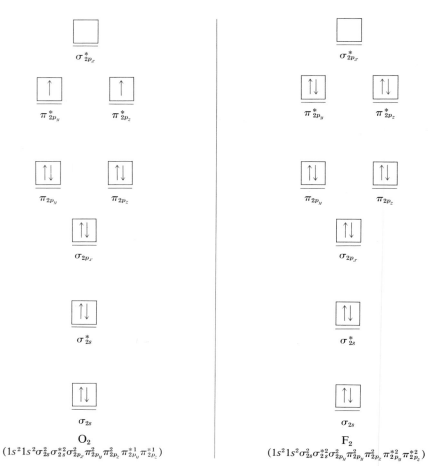

$$O_2$$
$$(1s^2 1s^2 \sigma_{2s}^2 \sigma_{2s}^{*2} \sigma_{2p_x}^2 \pi_{2p_y}^2 \pi_{2p_z}^2 \pi_{2p_y}^{*1} \pi_{2p_z}^{*1})$$

$$F_2$$
$$(1s^2 1s^2 \sigma_{2s}^2 \sigma_{2s}^{*2} \sigma_{2p_x}^2 \pi_{2p_y}^2 \pi_{2p_z}^2 \pi_{2p_y}^{*2} \pi_{2p_z}^{*2})$$

bond energy (Table 21.2). Taking away an antibonding electron has strengthened the bond, which is exactly what molecular orbital theory predicts. Note that O_2^+ has a fractional bond order of $2\frac{1}{2}$.

Fluorine, $1s^2 2s^2 2p^5$. In the F_2 molecule, all electrons are paired and the bond order, in agreement with the Lewis formula, $:\ddot{F}:\ddot{F}:$, is 1.

Neon, $1s^2 2s^2 2p^6$. The Ne_2 molecule has never been found. All electrons would be paired, and with all of the molecular orbitals in Figure 21.11 filled, the bond order would be zero.

Heteronuclear diatomic molecules. Within limits, the energy level diagram of Figure 21.11 can be used to determine the molecular orbital configuration of gaseous diatomic molecules formed between atoms of different elements—heteronuclear diatomic molecules—in the second period. **Heteronuclear** means "of different nuclei"; that is, it refers to atoms of different elements.

Consider carbon monoxide, CO, with fourteen electrons (six from C, eight from O). Placing these electrons in the energy levels in order according to the usual rules gives a molecule with the same configuration as the nitrogen molecule (Figure 21.13), and therefore a bond order of 3, $:C::O:$

The nitric oxide molecule, NO, has a total of fifteen electrons. Its configuration should be like that of nitrogen (Figure 21.13), but with one more electron added in a π_{2p}^* orbital. Nitric oxide does indeed have the paramagnetism of a molecule with one unpaired electron. *Like the $O–O^+$ case* described above, removal of an antibonding electron (the π_{2p}^* electron) from NO produces a positive ion, NO^+, more stable than its neutral parent.

The elements C and O, and N and O are not very different in atomic number, and the energies with which these atoms hold electrons are similar. Molecular orbitals can therefore form between these atoms by overlap of comparable atomic orbitals (as illustrated by Figure 21.11). When the atoms in a heteronuclear diatomic molecule have large differences in atomic number and in ionization energies, overlap occurs between orbitals of similar energy level but not necessarily of similar description (i.e., not necessarily $1s$ with $1s$, $2p$ with $2p$).

21.12 Multiple bonds; delocalized bonds

The molecular orbital picture of ethylene (Figure 21.15) is not too different from the valence bond picture (see Figure 21.6). σ Bonds determine the geometry of the carbon–hydrogen skeleton, and π bonds lie above and below the carbon–carbon internuclear axis. Three valence electrons from each carbon atom enter the σ orbitals, leaving one valence electron from each carbon atom for the π orbitals. Two p orbitals, one from each carbon atom, form two π molecular orbitals, one bonding and occupied by the two electrons, and one antibonding and unoccupied in the ground state of ethylene.

When a molecule or ion has more than one multiple bond, it is often possible to write several Lewis structures that are in agreement with the known properties of the molecule or ion. Previously we have dealt with such species—for example, NO_3^- (Section 9.13)—by the concept of resonance. In molecular orbital theory, an alternative is to picture the π orbitals as delocalized (Figure 21.16). **Delocalized orbitals** are not confined to an area near two nuclei but are spread over the entire molecule.

Benzene, C_6H_6, is the parent compound of thousands of chemical compounds. Many are of industrial importance, and many are essential to plants and animals. Delocalization of π orbitals makes an important contribution to the structure, reactivity, and stability of benzene and its derivatives. Resonance structures for benzene can be written as follows:

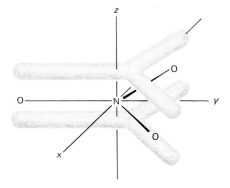

FIGURE 21.16
Delocalized π molecular orbital in nitrate ion.

THE COVALENT BOND
REEXAMINED [21.12]

653

(a) σ-bonded C—H skeleton

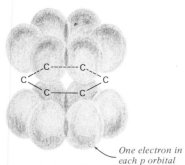

(b) p atomic orbitals available for molecular orbitals

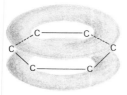

One electron in each p orbital

(c) Sum of delocalized π orbitals

(d) Symbol for benzene that emphasizes delocalized bonding

FIGURE 21.17
The benzene molecule.

To build up a molecular orbital picture of benzene, a total of 42 electrons from the six carbon atoms and six hydrogen atoms must be placed in orbitals. The first twelve electrons remain in 1s atomic orbitals on carbon. Twenty-four electrons occupy the orbitals that form the carbon–hydrogen skeleton of benzene—twelve in carbon-hydrogen bonds and twelve in carbon–carbon bonds. This leaves six electrons for the next highest energy level molecular orbitals, which must be formed from the single p orbital on each carbon atom not involved in σ bonds. Three π orbitals that add up to a region of uniform electron density above and below the ring of carbon atoms (Figure 21.17) are occupied by the six electrons in pairs; three other π* orbitals remain unoccupied.

This bonding picture accounts for the equivalence of all the carbon–carbon bonds in benzene and for the thermal stability of the molecule. As we have pointed out, electron energy is lowered when an electron is given a larger space to move about in. The six electrons in the delocalized π orbitals can be anywhere within the electron cloud shown in Figure 21.17.

21.13 Molecular orbitals in metals

When many atoms are brought together in a metal, many molecular orbitals are possible. Instead of electrons only in orbitals that encompass two, or three, or a dozen atoms in a molecule, some electrons reside in orbitals that encompass the entire piece of metal.

We have seen that two lithium atoms (Li, $1s^2 2s^1$) yield one bonding and one antibonding σ molecular orbital from combination of their individual 2s orbitals. Remember that the number of molecular orbitals formed always equals the number of atomic orbitals that have combined. Therefore, three lithium atoms produce three σ orbitals in the 2s level, ten lithium atoms produce ten molecular orbitals in the σ_{2s} level, and N lithium atoms produce $N\sigma_{2s}$ molecular orbitals.

With the energy level divided into many molecular orbitals (Figure 21.18a), the energy differences between the bonding orbitals get very small, and the energy difference between the antibonding and bonding orbitals also gets small. The point is reached where the molecular orbitals in an energy level are so close together that they are essentially continuous. A continuous group of closely spaced molecular orbitals is called a *band*.

The band formed by the lithium 2s atomic orbitals is delocalized over the whole piece of metal, much as the six atomic orbitals are delocalized over a whole benzene molecule. Such delocalized orbitals in metals hold the electrons of the "electron sea" on which our earlier, simpler explanation of metallic bonding was based. Because each lithium atom contributes one electron, this delocalized band is one-half full. The electrons enter the lowest energy orbitals within the band in pairs, leaving the higher-energy half of the band empty.

Electrons in the highest energy levels within this half-full band are close to empty levels that are not much higher in energy. Therefore, they can easily move from their own energy level to another that

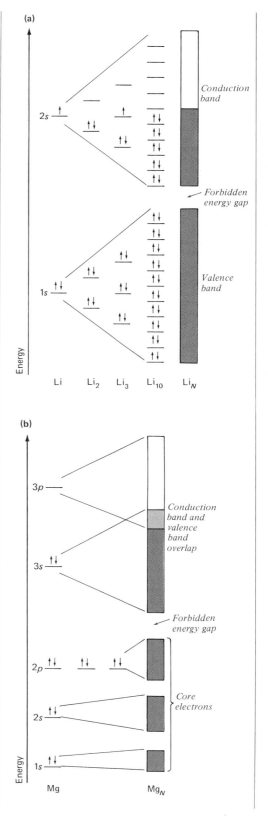

FIGURE 21.18

Formation of electron bands by lithium and magnesium.
The band structures of all alkali metals resemble that of lithium, and the band structures of all alkaline earth metals resemble that of magnesium. Metallic conduction requires the presence of either (a) a partially filled conduction bond (as in lithium) or (b) a valence band that overlaps an empty conduction band (as in magnesium).

is only slightly higher. An energy band in which electrons are free to move is a **conduction band.** The electrons in a conduction band that have enough energy so that they are not held back by attraction to the positive ions are the conduction electrons. The electrons at lower levels in a conduction band need a larger amount of energy to reach empty orbitals and in general do not participate in conduction. When an electric field is applied, the conduction electrons are accelerated in the direction of the field, and the net result is a flow of electrons.

Electrons contribute to conduction only if they are in a partially filled band. In a full band with no adjacent empty orbitals, all the electrons can do, if they have enough energy to move, is to change places with each other. The net result in the presence of an electric field is no conduction, since equal numbers of electrons are going in both directions.

At first glance, an element such as magnesium (Figure 21.18b) looks as if it should be a nonconductor, since the highest energy band is filled by two $3s$ electrons contributed by each atom. However, magnesium (and beryllium, cesium, zinc, and other metals that form 2+ ions) are metallic and do conduct electricity. What happens is that the empty $3p$ orbitals form an energy band that, because of the closeness of the atoms, overlaps the $3s$ band (Figure 21.18b). Therefore, electrons can move up out of the full $3s$ band and travel in the empty $3p$ band. The highest completely filled band in a metal is called the **valence band.** In divalent metals, the valence and conduction bands overlap. Electrons below the valence band, such as the $1s$, $2s$, and $2p$ electrons of magnesium, are sometimes called *core electrons.* They are held tightly by the nuclei.

A **forbidden energy gap**—an area of forbidden electron energies—can lie between energy bands. The energy gap is a consequence of the quantum mechanical nature of electrons. The size of the forbidden energy gap between the valence and conduction band plays an important role in the differences among conductors, semiconductors, and insulators (a subject discussed in Chapter 27).

THOUGHTS ON CHEMISTRY

The centrality of chemistry

The importance of chemistry and the apparent neglect of that importance by the public at large is sometimes seen as paradoxical or, if not paradoxical, a matter for worried comment. From time to time, groups of eminent chemists, or leaders from chemical industry, gather together to deplore the public's lack of interest and to discuss how the "image" of chemistry may be improved. Frequently their answer to the problem seems to be to try to persuade the appropriate organizations to produce more television programmes on the subject. Not only are they usually fighting a losing battle in taking this approach; they are also fighting the wrong battle. For chemistry is now of such importance to our lives that it is never likely to appear wondrous and exciting. It is like breathing—something you rarely think about, but which it is nevertheless difficult to do without.

If one judges from the coverage of science in newspapers . . . the science which sparks the public imagination is that which leaves the most questions unanswered. That, at least, is a polite way of putting it. A less polite way is to suggest that excitement and stimulation are produced by those areas of science which are sufficiently hazy around the spongy regions of speculative philosophy. The whiff of a new particle in physics causes everyone's eyes to shine. But why? Is it because quarks and "charm" open up vistas that take us far, far away from the balance of payments, or whether the roof will leak again next time it rains? . . .

As the science of chemistry has developed it has moved more from giant leaps into the unknown to small steps forward in fairly clear directions. On the other hand, the small steps nowadays often take longer to achieve than the great leaps did. This is because the great leaps were only great in the context of their own time. Today's small steps are many times more sophisticated and represent a level of scientific achievement that, even 50 years ago, would have seemed almost beyond belief. Nevertheless, as humanity's ability to manipulate molecules has increased, much of chemical research has moved towards clearly defined targets which we know will be achieved. As far as the public is concerned, if someone announces that he intends to travel to a certain destination, and then goes there in a straightforward manner, he is rather a dull fellow. If, as a result of what he has done, we are able to clothe ourselves more comfortably, or easy a pain more satisfactorily, we shall merely take that as our due. We shall still consider Perkin's accidental discovery of mauvine [a synthetic dye] in 1856, while trying to do something entirely different, as in some way more worthy than the successful synthesis of an immensely complicated molecule planned with the aid of a computer. . . .

Because chemistry is now mature enough for its own development to take place in small steps, while its use as a tool in other areas helps to produce "exciting" results, it is no longer reasonable to talk about the image of chemistry. Something so diverse, so pervasive, so central to nearly all aspects of modern life cannot expect to have any single image.

THE CENTRALITY OF CHEMISTRY, by Martin Sherwood [Quoted from pp. 782, 783, *New Scientist,* Volume 73 (No. 1045), March 31, 1977.]

THE COVALENT
BOND REEXAMINED

Significant terms defined in this chapter

internuclear axis
valence bond theory
molecular orbital theory
σ bonds
π bonds
hybridization
sp hybridization
sp^2 hybridization

sp^3 hybridization
isomers
cis isomer
trans isomer
molecular orbital
delocalization
bonding molecular orbital
antibonding molecular orbital

homonuclear
heteronuclear
delocalized orbitals
conduction band
valence band
forbidden energy gap

Exercises

21.1 The Lewis concept of atoms completing "octets" when forming chemical bonds predicts that Xe should form no compounds, and yet we know that XeF_2, XeF_4, and XeF_6 (as well as others) exist. (a) Prepare Lewis diagrams for these substances and decide what type of hybridization of the Xe atomic orbitals has taken place. (b) Draw all stereoisomers of XeF_2 and discuss your choice of molecular shape. (c) What shape do you predict for XeF_4?

21.2 Distinguish clearly between (a) a bonding and an antibonding molecular orbital, (b) valence bond theory and molecular orbital theory, (c) σ and π bonds, (d) double and triple covalent bonds, (e) cis and trans isomers, and (f) localized and delocalized π bonding.

21.3 Which statements are true? Rewrite any false statement so that it is correct.
(a) The valence bond theory considers a chemical bond to be formed by electrons occupying orbitals that are characteristic of the whole molecule.
(b) As hybridization of atomic orbitals occurs, the newly formed orbitals are identical in shape and energy and are equal in number to the original atomic orbitals that underwent hybridization.
(c) Compared to multiple bonds, the atoms in a single bond can rotate rather freely about the internuclear axis.
(d) Structural isomers are compounds that have the same formula and the same atoms attached to each other but in a different spatial arrangement.
(e) Delocalization energy in a molecule usually makes a molecule less stable than expected.
(f) A continuous group of closely-spaced molecular orbitals in a metal is known as a forbidden energy gap.

21.4 Prepare sketches of the bonding molecular orbitals formed by (a) two s atomic orbitals, (b) two p atomic orbitals interacting so that they lie parallel to each other, (c) two p atomic orbitals interacting so that they lie head to

head, and (d) an s atomic orbital and a p atomic orbital interacting so that they approach along the axis of the p orbital (e.g., p_z along the z axis). Repeat the exercise for the antibonding molecular orbitals that are also formed.

21.5 Prepare sketches of the orbitals around an atom which has undergone (a) sp, (b) sp^2, (c) sp^3, (d) sp^3d, and (e) sp^3d^2 hybridization. Show in the sketches any unhybridized p orbitals that might participate in multiple bonding.

21.6 Prepare a sketch of the cross-section taken between two atoms that have formed (a) a single σ bond, (b) a double bond consisting of a σ and a π bond, and (c) a triple bond consisting of a σ and two π bonds.

21.7 Iodine and fluorine form a series of interhalogen molecules and ions: IF (minute quantities observed spectroscopically), IF_3, IF_4^-, IF_5, IF_6^-, and IF_7. Identify the type of hybridization that occurs in the iodine atom in each substance.

21.8 Prepare a sketch of the molecule $CH_3CH\!\!=\!\!CHCH_3$ showing orbital overlaps. Identify the type of hybridization of atomic orbitals for each carbon atom.

21.9 Discuss the bonding, if any, in the following gaseous homonuclear diatomic molecules: Na_2, Mg_2, Al_2, Si_2, P_2, S_2, Cl_2, and Ar_2. Which of these molecules are paramagnetic? What is the bond order in each molecule? *Hint:* the molecular orbitals formed by the $3s$ and $3p$ atomic orbitals have the same pattern as those formed by the $2s$ and $2p$ orbitals.

21.10 List the following species in order of decreasing bond strength: O_2, O_2^+, O_2^-, O_2^{2+}, and O_2^{2-}.

21.11 Arrange these species in order of increasing bond length: CN, CN^+, and CN^-. What is the bond order in each?

21.12 Assuming the molecular orbital diagram given in Figure 21.11 is valid for NO, write the molecular orbital description for this molecule. Would NO be paramagnetic? What is the bond order? Would you predict that NO^+ would be more stable?

21.13 Relate what we learned in Chapter 9 about resonance to the concept of delocalized bonding that we learned in this chapter. Use the carbonate ion, CO_3^{2-}, to illustrate your discussion.

21.14 What type of hybridization will the carbon atom have in (a) CO, (b) CO_2, and (c) CO_3^{2-}? In which species will we find localized π bonding? In which species will there be delocalized π bonding?

21.15 Draw the Lewis diagrams for molecular oxygen and ozone. What types of hybridization do the oxygen atoms undergo in each compound? Prepare sketches of the molecules. What is the bond order of each bond in the molecules?

21.16 Explain in terms of the band theory why an alkali metal like Na or K will be a good conductor of electricity.

21.17 Even though the conduction band formed by the s orbitals in the alkaline earth metals such as Mg and Ca are filled, these metals are good conductors of electricity. Explain how conduction takes place in these metals.

21.18* What type of hybridization will the nitrogen atom(s) have in the following molecules and ions: (a) NO; (b) N_2O_2, the dimer of NO; (c) N_2O; (d) N_2O_5; (e) NO_2; (f) N_2O_4, the dimer of NO_2; (g) N_2O_3; (h) NO^+; (i) NO_2^-; (j) NO_3^-; and (k) $N_2O_2^{2-}$?

21.19* Two stereoisomers of $Pt(NH_3)_2Cl_2$ are known. Prepare sketches of this molecule using sp^3 (tetrahedral) and sp^2d (square planar) hybridization for the Pt. For which type of hybridization can you draw a second stereoisomer?

21.20 Sulfur reacts with fluorine to form two gaseous compounds: SF_4 and SF_6. The respective heats of formation at 25°C are −774 kJ/mole and −1205 kJ/mole. The heat of formation of S(g) is 279 kJ/mole and 79 kJ/mole for F(g). Calculate the S—F bond energy in each compound. What type of hybridization is present around the sulfur atom? Do the bond energies confirm the rule of thumb that stronger bonds are formed as the number of atomic orbitals involved in hybridization increases?

21.21 Calculate the C—O, C=O, and C≡O bond energies from the following heats of formation at 25°C: 171.291 kcal/mole for C(g), −26.416 kcal/mole for CO(g), −94.05 kcal/mole for $CO_2(g)$, −47.96 kcal/mole for $CH_3OH(g)$, −17.88 kcal/mole for $CH_4(g)$, 52.095 kcal/mole for H(g), −57.796 kcal/mole for $H_2O(g)$, and 59.553 kcal/mole for O(g). Do these values confirm the rule of thumb that stronger bonds are formed as the bond order increases?

21.22* Calculate the O—O and O=O bond energies from the following heats of formation at 25°C: 52.095 kcal/mole for H(g), 59.553 kcal/mole for O(g), 0 for $O_2(g)$, −57.796 kcal/mole for $H_2O(g)$, and −32.58 kcal/mole for $H_2O_2(g)$. Predict the heat of formation of $O_3(g)$ assuming that it contains one single bond and one double bond. Using this prediction and the actual value of 34.1 kcal/mole, calculate the delocalization energy in this molecule.

21.23** Write the molecular orbital description for C_2 using the order of filling the molecular orbitals as (a) σ_{2p} before π_{2p} and (b) π_{2p} before σ_{2p}. (c) Which description correctly predicts C_2 to be diamagnetic? Although the σ_{2p} orbital is higher in energy than the π_{2p} orbitals, there is little energy difference between them in C_2, and the arrangement of electrons given by the description $1s^2 1s^2 \sigma_{2s}^2 \sigma_2^{*2} \pi_{2p}^3 \sigma_{2p}$ can be observed. (d) Is this excited state of C_2 as stable as the ground state? (e) How does this excited state differ in magnetic properties from the ground state? The heats of formation at 25°C are as follows: $C_2(g)$, 200.224 kcal/mole; C(g), 171.291 kcal/mole; $H_2C=CH_2(g)$, 12.49 kcal/mole; H(g), 52.095 kcal/mole; and $CH_4(g)$, −17.88 kcal/mole. Calculate the C=C bond energy in (f) $H_2C=CH_2$ formed by a π and a σ bond and (g) C_2 formed by two π bonds. (h) Do these calculations confirm that π_{2p}^4 is more favorable than $\sigma_{2p}^2 \pi_{2p}^2$?

22 NONMETALS: NITROGEN, PHOSPHORUS, AND SULFUR

& INORGANIC QUALITATIVE ANALYSIS: ANIONS

The story of the discovery of phosphorus is worth retelling. In the seventeenth century in Hamburg, Germany, Hennig Brand, a physician and alchemist, had what seems a very curious idea. He would try to obtain from urine a liquid with which he could convert silver into gold. Brand evaporated fresh urine and let the residue stand until it had putrefied. Then he heated this potent substance vigorously and collected in water the vapors that were given off. We have no way of knowing what he expected to see, but the white, flammable, waxy solid which glowed in the dark must surely have been a surprise. Brand was the first to discover an element not among those known since antiquity—he had distilled pure white phosphorus.

Sulfur is an element that occurs in nature in elemental form and was undoubtedly known before recorded history. "Brimstone" and "sulfur" are mentioned in the Bible and in Homer's Odyssey. The alchemists sometimes used the term "sulfur" to describe anything that was combustible.

To complete the historical background of the nonmetals discussed in this chapter, nitrogen was discovered in 1772 by Daniel Rutherford. He described it as the "noxious air" left after first a mouse had died in a confined volume of air and then the "fixed air" (carbon dioxide) had been absorbed by caustic potash. Scheele, Priestley, and Cavendish all did experiments with nitrogen at about the same time.

Nitrogen, phosphorus, and sulfur—three elements essential to both plant and animal life—are discussed in this chapter. These elements are nonmetals and their covalent compounds are more numerous than their ionic compounds. Each element exhibits several different oxidation states, and the discussion of their compounds is organized by the oxidation states of the elements. Several very important industrial chemicals—ammonia, nitric acid, phosphoric acid, and sulfuric acid—are covered here.

The second part of this chapter introduces inorganic qualitative analysis. The properties of anions important to their detection by the methods of anion analysis are reviewed. We cover anions that incorporate the nonmetals nitrogen, phosphorus, sulfur, the halogens, and carbon, and include chromate ion, which contains the metallic element, chromium. The chemistry of anion analysis is discussed.

22.1 Some properties of nitrogen, phosphorus, and sulfur

Nitrogen and phosphorus are the nonmetallic members of Representative Group V, and oxygen and sulfur are the nonmetallic members of Representative Group VI. (The chemistry of oxygen was discussed in Chapter 11.)

The first element in Representative Group V—nitrogen—differs considerably from phosphorus and the other members of the group. Some of these differences are accounted for by the smaller size and greater electronegativity of the nitrogen atom (Table 22.1). Others arise because nitrogen atoms are limited to eight electrons in their valence shells, while the other elements in the group can accommodate more than eight valence shell electrons.

Nitrogen, phosphorus, and sulfur all exhibit negative <u>oxidation states</u> in their compounds with hydrogen (NH_3, PH_3, and H_2S) and in compounds in which they complete their noble gas octets by forming ions—nitride ion, N^{3-}; phosphide ion, P^{3-}; and sulfide ion, S^{2-}. Nitrogen forms compounds in all oxidation states from -3 to $+5$. In addition to the negative oxidation states, the only other important states for phosphorus are $+3$ and $+5$, and for sulfur, $+4$ and $+6$.

Nitrogen and phosphorus atoms, with valence shell configurations of s^2p^3, readily form three <u>covalent bonds</u> by sharing their p electrons, as in PCl_3 or NH_3. These atoms can take part in a fourth covalent bond by coordinate covalent bonding through their lone pair of electrons. Phosphorus and sulfur atoms also yield compounds with five and six covalent bonds, respectively, by utilizing vacant valence shell d orbitals as, for example, in PCl_5 or SF_6.

Each element forms a variety of compounds in which multiple bonding and resonance play important roles. The oxides and oxo acids are particularly interesting in this respect. In compounds containing nitrogen–oxygen bonds, π bonding is contributed by overlap of p orbitals on oxygen with p orbitals on nitrogen. In compounds containing bonds between sulfur or phosphorus and oxygen, multiple bond, or delocalized π bond, character is contributed by interaction

C	N 7	O	F
	14.0067		
Si	P 15	S 16	Cl
	30.9738	32.06	
	As	Se	

TABLE 22.1

Properties of nitrogen, phosphorus, and sulfur

Property	Nitrogen	Phosphorus	Sulfur
Configuration	$[He]2s^22p^3$	$[Ne]3s^23p^3$	$[Ne]3s^23p^4$
Formula of molecule	N_2	P_4	S_8
Melting point (°C)	-210.0	44.1 (white)	112.8 (rhombic) 119.0 (monoclinic)
Boiling point (°C)	-195.8	280 (white)	444.60
Covalent radius (nm)	0.074	0.110	0.104
Bond dissociation energy, kcal/mole (kJ/mole)	226 (946)	116 (485)	103 (427)
Ionization energy (0°K) (kcal/mole)	335	253	239
Electronegativity	3.0	2.1	2.5

of the filled oxygen *p* orbitals with empty *d* orbitals on sulfur or phosphorus atoms. Often the bonds between nitrogen, sulfur, or phosphorus and oxygen have properties intermediate between those of single and double covalent bonds.

Nitrogen is chemically very unreactive at ordinary temperatures. This lack of reactivity can be explained by the great stability of the nitrogen molecule, which has one of the highest bond dissociation energies (Table 22.1) for a diatomic molecule. Because of this great stability, many nitrogen-containing compounds decompose to produce nitrogen, often with the liberation of a considerable amount of energy. The energy release and the accompanying volume expansion of the gaseous nitrogen ideally suit nitrogen compounds to be explosives. The explosion-causing reactions are redox reactions, and many explosives contain built-in combinations of oxidant (for example, NO_3^- or NO_2^-) with reductants such as carbon–hydrogen groups.

At high temperatures nitrogen, presumably because it is then partly dissociated into atoms, undergoes <u>direct combination</u> with reactive metals and with certain nonmetals, to form nonvolatile binary nitrides.

Elemental sulfur is relatively reactive and will combine directly with most elements. Phosphorus, in its white form (Section 22.9), combines very vigorously with atmospheric oxygen, bursting into flame spontaneously. The direct combination of phosphorus also proceeds with most elements, although somewhat more vigorous conditions are required for phosphorus than for sulfur. A few examples of the direct combination of these elements are given in Table 22.2.

Nitrogen

22.2 Additional properties of nitrogen

Nitrogen is a colorless, odorless, and nontoxic gas. Like oxygen it is difficult to liquefy. It is less soluble in water than is oxygen, dissolving to the extent of 23.2 ml/liter of water at 0°C and 1 atm pressure.

The most common oxidation states for nitrogen are -3, $+3$, and $+5$. It can, however, be found in *all* of the oxidation states from -3 to $+5$. A summary of these states is given in Table 22.3 along with the names and formulas of compounds corresponding to each state.

22.3 Sources, preparation, and uses of nitrogen

Gaseous molecular nitrogen constitutes about 78% of the atmosphere by volume, and this is the major source of nitrogen, which is prepared industrially by the fractional distillation of liquid air (Section

Common oxidation states of
N: -3, $+3$, $+5$

TABLE 22.2

Some direct combination reactions of nitrogen, phosphorus, and sulfur

$$3Mg(s) + 2P(g) \xrightarrow{600°C} Mg_3P_2(s)$$
magnesium phosphide

$$3Ca(s) + N_2(g) \xrightarrow[450°C]{3–4 hr,} Ca_3N_2(s)$$
calcium nitride

$$2Al(s) + 3S(s) \xrightarrow[burning Mg]{ignite with} Al_2S_3(s)$$
aluminum sulfide

$$Al(s) + P(s)(red) \xrightarrow[burning Mg]{ignite with} AlP(s)$$
aluminum phosphide

$$2Al(s) + N_2(g) \xrightarrow{820°C} 2AlN(s)$$
aluminum nitride

TABLE 22.3

Oxidation states of nitrogen

The common names of the nitrogen oxides are given in parentheses.

Oxidation state	Compounds
-3	Nitrides, e.g., Li_3N Ammonia, NH_3 Ammonium salts, NH_4X
-2	Hydrazine, N_2H_4
-1	Hydroxylamine, NH_2OH
$+1$	Nitrogen(I) oxide, N_2O (nitrous oxide)
$+2$	Nitrogen(II) oxide, NO (nitric oxide)
$+3$	Nitrogen(III) oxide, N_2O_3 Nitrous acid, HNO_2 Nitrites, e.g., $NaNO_2$
$+4$	Nitrogen(IV) oxides, NO_2 (nitrogen dioxide), N_2O_4 (nitrogen tetroxide)
$+5$	Nitrogen(V) oxide, N_2O_5 (nitrogen pentoxide) Nitric acid, HNO_3 Nitrates, e.g., $NaNO_3$

TABLE 22.4

Uses of elemental nitrogen

Major use
Manufacture of ammonia

Other uses
Manufacture of nitrogen compounds, e.g.,
    Calcium cyanamide (a fertilizer)
    Nitrides (used in cutting and grinding tools)
    Hydrazine (a rocket fuel)

"Inerting" applications, e.g.,
    For reactive chemicals
    For foods
    For electrical equipment
    For blowing bubbles into foamed polymers

Low-temperature applications, e.g.,
    Condensing gases
    Preserving biological materials
    Cryosurgery (selective destruction of tissue)
    Freezing food

6.4). Unless purified, nitrogen from this source is contaminated with small amounts of oxygen and some of the noble gases.

In combined form, nitrogen occurs to a small extent in the Earth's crust, mainly as sodium nitrate, $NaNO_3$, and potassium nitrate, KNO_3. The major deposits of sodium nitrate are in Chile in South America, and sodium nitrate is known as Chile saltpeter. Potassium nitrate is known as saltpeter. The word "saltpeter" is thought to derive from the Latin *petrus*, "rock," referring to the compound as the "salt found in rocks."

In the laboratory, pure dinitrogen can be prepared by carefully heating an aqueous solution saturated both with ammonium chloride (NH_4Cl) and sodium nitrite ($NaNO_2$).

$$NH_4^+ + NO_2^- \xrightarrow{\Delta} N_2(g) + 2H_2O(l)$$

The industrial uses of dinitrogen fall into three major categories, based upon the utilization of the properties of the element: (1) in the manufacture of ammonia and other industrial chemicals, (2) in applications that require the presence of an inert gas ("inerting applications," as they are sometimes called), and (3) in applications that require a low temperature (Table 22.4).

22.4 Oxidation state -3: nitrides and ammonia

a. Nitrides. Ionic compounds containing the nitride ion, N^{3-}, are formed when lithium and most Representative Group II elements are heated in dinitrogen. Nitrides are white, high-melting solids that react vigorously with water to form ammonia and metal hydroxides as products, for example,

$$Li_3N + 3H_2O(l) \longrightarrow NH_3(g) + 3LiOH(aq)$$

Nitrogen also forms covalent binary compounds such as BN or P_3N_5 and interstitial nitrides, which are analogous to the interstitial hydrides (Section 12.15) and carbides (Section 23.3e). Boron nitride, BN, is a covalent compound nearly identical in structure to graphite (see Figure 23.2), with boron and nitrogen atoms alternating throughout. Just as graphite can be converted to diamond (Section 23.2), boron nitride is converted by heat and pressure to a very hard, diamondlike material called borazon.

b. Properties of ammonia. Ammonia is a colorless gas with a characteristic pungent odor. It is a powerful heart stimulant, and excessive inhalation may produce serious effects—even death. In the gaseous state, ammonia exists as discrete polar molecules of pyramidal structure (Figure 22.1). In the liquid and solid states, ammonia is extensively associated through hydrogen bonding, as would be expected from the high electronegativity of nitrogen.

Many of the chemical properties of ammonia arise from the presence in the valence shell of the nitrogen atom of an unshared pair of electrons. This permits the ammonia to act as a <u>Lewis base</u>, that is, an electron-pair donor.

The basic nature of ammonia is shown by its reaction with water. Ammonia is extraordinarily soluble in water, dissolving to the extent of about 700 volumes in 1 volume of solvent at 20°C and 1 atm pressure. The dissolving process is accompanied by the reaction

$$\text{NH}_3(g) \; + \text{H}_2\text{O}(l) \rightleftharpoons \quad \text{NH}_4^+ \; + \text{OH}^- \qquad K_b = 1.6 \times 10^{-5} \text{ at } 25°C$$
$$\qquad\quad \textit{ammonia} \qquad\qquad\qquad \textit{ammonium} \\ \textit{ion}$$

Ammonia is a weak base in water, and the solutions contain only small amounts of ammonium and hydroxide ions. Although such solutions are commonly referred to as ammonium hydroxide, the compound NH_4OH has never been isolated.

The basic nature of ammonia is also displayed in its reaction with protonic acids to form ammonium salts; for example, gaseous NH_3 and HCl yield the salt ammonium chloride.

$$\begin{array}{c} \text{H} \\ \text{H}:\overset{\cdot\cdot}{\text{N}}: + \text{H}:\overset{\cdot\cdot}{\underset{\cdot\cdot}{\text{Cl}}}: \longrightarrow \left[\text{H}:\overset{\text{H}}{\underset{\text{H}}{\text{N}}}:\text{H}\right]^+ + :\overset{\cdot\cdot}{\underset{\cdot\cdot}{\text{Cl}}}:^- \\ \text{H} \end{array}$$

Ammonia combines by electron-pair donation with most d transition metal ions and with zinc, cadmium, and mercury(II) ions to form positively charged <u>complex ions</u>. The NH_3 group in complexes is called an **ammine group.** Some examples of ammine complexes are $[\text{Ag}(\text{NH}_3)_2]^+$, $[\text{Co}(\text{NH}_3)_6]^{3+}$, and $[\text{Cu}(\text{NH}_3)_4]^{2+}$. In each case, each ammonia molecule is joined to the metal through the unshared pair of electrons on the nitrogen atom. (The nature of complexes is discussed at length in Chapter 28.)

Other important chemical reactions of ammonia depend on its ability to act as a <u>reducing agent</u>, particularly at elevated temperatures; the nitrogen-containing oxidation products vary with the reaction conditions. Although ammonia is the most stable of the hydrides of Group V with respect to decomposition to the elements, it is appreciably decomposed at high temperatures. The hydrogen formed on decomposition is responsible for the reducing action of ammonia at such temperatures.

c. Preparation and uses of ammonia. The industrial chemistry of nitrogen, ammonia, and other nitrogen-containing compounds is based to a great extent upon the Haber process for the manufacture of ammonia by the direct combination at elevated temperatures of dihydrogen and dinitrogen (see An Aside: The Haber Process, Chapter 16). Natural gas and naphtha (from petroleum) are the sources of hydrogen for the Haber process, so that countries rich in oil and natural gas are potentially big producers of fertilizers.

Second in importance to the Haber process as a commercial method for the production of ammonia is the destructive distillation of bituminous, or soft, coal. Bituminous coal is mainly a mixture of free carbon, hydrocarbons, and moderate amounts of chemically bound oxygen and nitrogen. When coal is heated in the absence of air,

FIGURE 22.1

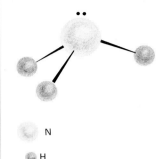

N

H

Ammonia, NH_3
m.p. $-77.7°C$
b.p. $-33.4°C$

TABLE 22.5

Uses of ammonia

Major uses
Preparation of nitric acid
Direct use as fertilizer
Synthesis of other fertilizers

Other uses
Synthesis of dyes, polymers, drugs, explosives,
   and synthetic fibers
Refrigerant
Household cleaning agent
In paper making

gases are driven off to leave a residue of coke. In the gases liberated, nitrogen is present as the free element and as ammonia.

The single greatest outlet for ammonia (Table 22.5) and for the compounds made from it is as fertilizer (see An Aside: Chemical Fertilizers, this chapter).

d. Ammonium salts. Many simple ammonium salts are very soluble in water, exceptions being ammonium perchlorate (NH_4ClO_4) and ammonium hydrogen carbonate (NH_4HCO_3). Aqueous solutions of ammonium salts of strong acids—NH_4Cl, NH_4NO_3, $(NH_4)_2SO_4$, for example—are acidic due to hydrolysis of the ammonium ion.

$$NH_4^+(aq) \rightleftharpoons NH_3(g) + H^+ \qquad K_a = 6.3 \times 10^{-10} \text{ at } 25°C$$

All ammonium salts decompose on heating, the manner of decomposition depending on the nature of the anion of the salt. Salts containing anions that do not act as oxidizing agents decompose into ammonia and the parent acids; for example,

$$\underset{\text{ammonium chloride}}{NH_4Cl(s)} \xrightarrow{\Delta} NH_3(g) + HCl(g)$$

$$\underset{\substack{\text{ammonium hydrogen} \\ \text{carbonate}}}{NH_4HCO_3(s)} \xrightarrow{\Delta} NH_3(g) + CO_2(g) + H_2O(g)$$

Decomposition of NH_4HCO_3 occurs to some extent even at room temperature, and for this reason the compound is used in smelling salts, once popular for reviving ladies prone to fainting spells. In general, ammonium salts of weak acids are much less stable thermally than those of strong acids.

Ammonium salts containing anions which are oxidizing agents undergo internal oxidation-reduction when heated, for example,

$$\underset{\text{ammonium nitrate}}{\overset{-3 \quad +5}{NH_4NO_3}(s)} \xrightarrow{170-260°C} \overset{+1}{N_2O}(g) + 2H_2O(g)$$

In this reaction, the nitrogen in the ammonium ion is oxidized, whereas that in the nitrate ion is reduced. Ammonium nitrate is a dangerous solid and should be handled with great care. Decomposition, particularly in a limited space or at elevated temperatures, may occur with explosive violence according to the equation

$$2NH_4NO_3(s) \longrightarrow 2N_2(g) + 4H_2O(g) + O_2(g)$$

Mixtures of ammonium nitrate and oxidizable materials are potentially dangerous. Accidental exposure to elevated temperatures may initiate oxidation, and this in turn may increase the temperature to the point of detonation. Mixed with organic matter, ammonium nitrate is used as a substitute for dynamite in open-pit mining. A mixture of ammonium nitrate and TNT (trinitrotoluene) is almost as powerful an explosive as TNT alone and it is much cheaper. Such mixtures are used in military bombs and shells.

[22.4] NITROGEN,
PHOSPHORUS, AND SULFUR

In April 1947, an explosion due to the decomposition of ammonium nitrate took the lives of 576 persons at Texas City, Texas. The material, intended for use as a fertilizer, was being loaded onto a ship in the harbor. Apparently, explosion occurred because the sacks of the material were piled so high and so tightly that heat built up within the pile. With proper precautions ammonium nitrate can be shipped safely and is widely used as a fertilizer.

22.5 Oxidation state −2: hydrazine

Anhydrous hydrazine, H_2NNH_2 (Figure 22.2), is a fuming colorless liquid and a potent reducing agent. It is converted mainly to nitrogen by strong oxidizing agents such as hydrogen peroxide, liquid oxygen, or nitrogen tetroxide (N_2O_4). These redox reactions are accompanied by the release of large amounts of energy,

$$H_2NNH_2(l) + O_2(g) \longrightarrow N_2(g) + 2H_2O(g)$$
$$\Delta H° = -127.70 \text{ kcal } (-534.30 \text{ kJ})$$

and, therefore, hydrazine and some of its derivatives are useful as rocket fuels.

22.6 Oxidation states +1 to +5: oxides of nitrogen

There is an oxide of nitrogen for each of the five oxidation states from +1 to +5. All but one are gases, two are paramagnetic, and two—NO and NO_2—play a significant role in air pollution (Section 6.10). In nitrous oxide and nitric oxide, nitrogen has oxidation states of +1 and +2, respectively.

Nitrogen(I) oxide, N_2O, commonly known as nitrous oxide, is obtained by heating ammonium nitrate (Section 22.4d). The N_2O molecule (Figure 22.3) is linear, with the two nitrogen atoms attached to each other, and has two resonance forms. (Molecular orbital theory would envision not two resonance forms, but sixteen electrons in molecular orbitals belonging to the whole N_2O molecule.)

Heating N_2O to high temperatures decomposes the molecule to its constituent elements, an example of the greater stability of dinitrogen. The reaction can occur with explosive force.

$$2N_2O(g) \xrightarrow{\Delta} 2N_2(g) + O_2(g) \quad \Delta H° = -39.22 \text{ kcal } (-164.1 \text{ kJ})$$
nitrous oxide

Nitrous oxide was the first synthetic anesthetic to be discovered and is still in use as a light anesthetic, especially in dentistry. When inhaled in low concentrations, the gas produces mild euphoria—the basis for its common name, laughing gas. Nitrous oxide was also the first aerosol propellant. It was introduced before World War II in whipped cream dispensers.

FIGURE 22.2

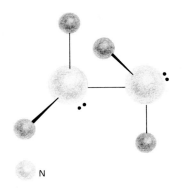

N
H

Hydrazine, H_2NNH_2
m.p. 2°C
b.p. 113.5°C

FIGURE 22.3

N
O

Nitrous oxide, N_2O
m.p. −90.8°C
b.p. −88.8°C

$$:N::N:\ddot{O}: \longleftrightarrow :\ddot{N}::N::\ddot{O}:$$

Nitrous oxide, N$_2$O
Nitric oxide, NO
Nitrogen dioxide, NO$_2$

FIGURE 22.4

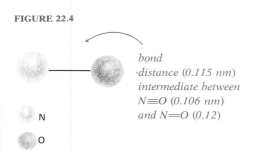

bond
·distance (0.115 nm)
intermediate between
N≡O (0.106 nm)
and N=O (0.12)

N

O

Nitric oxide, NO
m.p. −163.6°C
b.p. −151.8°C
one unpaired electron

FIGURE 22.5

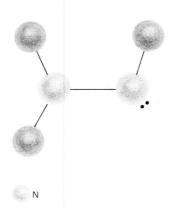

N

O

Nitrogen(III) oxide, N$_2$O$_3$
m.p.−102°C
b.p. 3.5°C

Nitrogen(II) oxide, or nitric oxide, NO, is prepared industrially by the catalytic oxidation of ammonia (Section 22.8c). In the laboratory, this oxide is prepared by the action of dilute nitric acid on elemental copper

$$3Cu(s) + 8H^+ + 2NO_3^- \longrightarrow 3Cu^{2+} + 2NO(g) + 4H_2O(l)$$

Oxygen must be excluded to prevent the oxidation of the nitric oxide to nitrogen dioxide, NO$_2$,

$$2NO(g) + O_2(g) \longrightarrow 2NO_2(g) \qquad \Delta H° = -27.9 \text{ kcal } (-117 \text{ kJ})$$
colorless   brown nitrogen
nitric oxide   dioxide

This reaction takes place readily at ordinary temperatures, and is a step in the synthesis of nitric acid.

Nitrogen(II) oxide is a fairly reactive substance. It combines directly with all of the halogens except iodine to give compounds of the general formula NOX, which are known as *nitrosyl halides*. At high temperatures it acts as an oxidizing agent toward substances that tend to react readily with oxygen. For example, ignited phosphorus and carbon continue to burn in nitric oxide.

Gaseous NO is made up of discrete molecules that contain one unpaired electron (Figure 22.4). In terms of molecular orbital theory, the unpaired electron occupies a π antibonding orbital.

Nitrogen(III) oxide, N$_2$O$_3$ (Figure 22.5), or dinitrogen trioxide, the anhydride of nitrous acid, HNO$_2$, is obtained as a blue liquid by the union of equimolar quantities of nitric oxide and nitrogen dioxide at −20°C.

$$NO(g) + NO_2(g) \rightleftharpoons N_2O_3(l)$$
nitric oxide   nitrogen dioxide   dinitrogen trioxide

The compound is never pure, for even in the liquid state it is in equilibrium with NO and NO$_2$. At 25°C and 1 atm pressure, the equilibrium mixture of gases contains only about 10% N$_2$O$_3$, which is the least stable of the nitrogen oxides.

When the trioxide, or, more precisely, an equimolar mixture of NO and NO$_2$, is passed into cold water, nitrous acid (HNO$_2$) (Section 22.7) is formed.

The nitrogen(IV) oxides are nitrogen dioxide, NO$_2$, and dinitrogen tetroxide, N$_2$O$_4$. They exist in equilibrium with each other. The formation of N$_2$O$_4$ is favored at low temperatures, as shown by the negative ΔH value.

$$2NO_2(g) \rightleftharpoons N_2O_4(g) \qquad \Delta H° = -13.67 \text{ kcal } (-57.20 \text{ kJ})$$
brown nitrogen   colorless
dioxide   dinitrogen
tetroxide

NO$_2$ is readily formed by the union of NO with atmospheric oxygen. A red-brown gas, NO$_2$ is highly toxic. The NO$_2$ molecule (Figure 22.6), like that of nitric oxide, possesses an unpaired electron. In this

FIGURE 22.6

N
O

0.1197 nm

Nitrogen dioxide, NO$_2$
m.p. $-11.20°C$
b.p. $21.2°C$
one unpaired electron
brown gas

case, the unpaired electron spends most of its time near the nitrogen atom. The N$_2$O$_4$ molecule appears to exist in a number of forms, the most stable of which is planar with the two NO$_2$ units joined through the nitrogen atoms. There are no unpaired electrons in dinitrogen tetroxide.

The nitrogen(IV) oxides are strong oxidants. The dioxide supports the combustion of ignited sulfur, phosphorus, and carbon, converting these elements to oxides. Probably the most significant reaction of nitrogen dioxide is the one with warm water to yield nitric acid and nitric oxide.

$$3NO_2(g) + H_2O(l) \longrightarrow 2HNO_3(aq) + NO(g)$$

Nitrogen(V) oxide, or dinitrogen pentoxide, N$_2$O$_5$, is the anhydride of nitric acid and the only nitrogen oxide that is a solid. It is prepared by the dehydration of pure nitric acid by phosphorus(V) oxide; metaphosphoric acid, HPO$_3$, is also formed.

$$4HNO_3(l) + P_4O_{10}(s) \longrightarrow \underset{\substack{nitrogen(V) \\ oxide}}{2N_2O_5(s)} + \underset{\substack{metaphosphoric \\ acid}}{4HPO_3(s)}$$

In the vapor state an oxygen atom serves as a bridge between two NO$_2$ groups in N$_2$O$_5$ (Figure 22.7). As might be expected, N$_2$O$_5$ reacts exothermically with water to regenerate nitric acid.

22.7 Oxidation state +3: nitrous acid and nitrites

Nitrous acid, HNO$_2$ (Figure 22.8), is unstable and is known only in solution and in the gas phase. It is a weak acid

$$HNO_2(aq) \rightleftharpoons H^+ + NO_2{}^- \qquad K_a = 7.2 \times 10^{-4} \text{ at } 25°C$$

and even in cold solutions, some decomposition occurs to give nitric acid and nitric oxide

$$3HNO_2(aq) \longrightarrow H^+ + NO_3{}^- + 2NO(g) + H_2O(l)$$

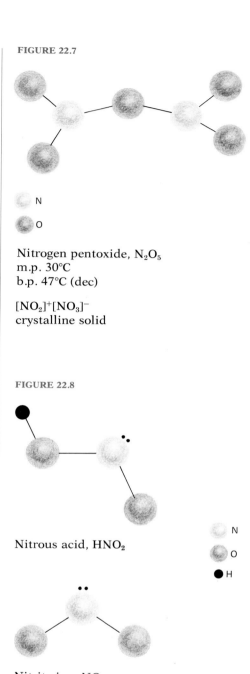

FIGURE 22.7

N
O

Nitrogen pentoxide, N$_2$O$_5$
m.p. $30°C$
b.p. $47°C$ (dec)

[NO$_2$]$^+$[NO$_3$]$^-$
crystalline solid

FIGURE 22.8

Nitrous acid, HNO$_2$

N
O
H

Nitrite ion, NO$_2{}^-$

$$\left[\ddot{\underset{..}{O}} \; \overset{..}{N} \; \overset{..}{\underset{..}{O}} .. \right]^- \longleftrightarrow \left[.. \underset{..}{O} \; \overset{..}{N} \; \ddot{\underset{..}{O}} .. \right]^-$$

Nitrites, which are the salts of nitrous acid, are colorless or pale yellow solids that are stable to heat. The best known nitrites are those of the alkali metals, which are very soluble in water, and those of calcium, strontium, and barium, which are moderately soluble in water. The nitrites are made either by passing an equimolar mixture of nitric oxide and nitrogen dioxide into a solution of metal hydroxide

$$NO(g) + NO_2(g) + 2OH^- \longrightarrow 2NO_2^- + H_2O(l)$$

or by reduction at elevated temperatures of a metal nitrate by a reducing agent such as lead or iron powder

$$NaNO_3(s) + Pb(s) \longrightarrow NaNO_2(s) + PbO(s)$$

Because of the intermediate oxidation state of the nitrogen, nitrous acid may act either as an oxidizing or as a reducing agent. In acidic solution it is a strong oxidizing agent.

$$NO_3^- + 3H^+ + 2e^- \longrightarrow HNO_2 + H_2O \qquad E° = 0.94 \text{ V}$$

$$HNO_2 + H^+ + e^- \longrightarrow NO + H_2O \qquad E° = 0.99 \text{ V}$$

The reduction product depends upon the strength of the reducing agent. For example, iodide ion reduces nitrous acid to nitric oxide, but tin(II) ion, a stronger reducing agent, reduces the acid to nitrous oxide.

$$2\overset{+3}{H}NO_2(aq) + 2H^+ + 2I^- \longrightarrow 2\overset{+2}{N}O(g) + I_2(s) + 2H_2O(l)$$

$$2\overset{+3}{H}NO_2(aq) + 4H^+ + 2Sn^{2+} \longrightarrow \overset{+1}{N_2}O(g) + 2Sn^{4+} + 3H_2O(l)$$

Nitrous acid is a relatively weak reducing agent in acidic solution and is oxidized (to nitric acid) only by such strong oxidizing agents as chlorine and permanganate ions. However, in alkaline medium it is a stronger reducing agent and can be oxidized even by such a weak oxidizing agent as oxygen.

Sodium nitrite and sodium nitrate, which simply serves as a source of additional nitrite, have a unique twofold usefulness as food additives in hot dogs, hams, bacon, cold cuts, and other cured meats. In the first place, these compounds preserve a fresh and appetizing red color in the meats. Nitric oxide from the nitrite ion apparently reacts with heme, which would give the meat a brown-gray-green appearance, and forms instead stable, bright red nitrosohemoglobin. In addition, nitrite prevents the growth of *Clostridium botulinum* spores, which cause the serious food poisoning known as botulism. Recently, nitrites and nitrates have come under suspicion because they can be converted in the stomach to nitrosamines (R_2NN=O, where R is an organic group), compounds thought to be **carcinogens,** agents that cause cancer, and **mutagens,** agents that cause genetic malformations.

22.8 Oxidation state +5: nitric acid and nitrates

a. Properties of nitric acid. Pure nitric acid can be represented by two resonance structures (Figure 22.9). In the vapor state, where the HNO_3 molecule is planar, two of the N—O bond distances are the same and different from that in the NOH group, giving evidence for the resonance structures as written, and the contribution of *p–p* π bonding. The nitrate ion is planar, with equivalent N—O bond distances (see Figure 21.16).

Nitric acid is a colorless, fuming liquid with a choking odor. Both pure nitric acid and its aqueous solutions decompose in sunlight to give NO_2.

$$4HNO_3(l) \longrightarrow 4NO_2(g) + O_2(g) + 2H_2O(l)$$

Old samples of nitric acid are often yellow or brown because of the presence of dissolved NO_2.

Miscible in all proportions with water, HNO_3 is both a strong acid and a strong oxidizing agent in aqueous solutions. The usual "concentrated nitric acid" reagent is about 70% HNO_3 by weight.

b. Nitric acid as an oxidizing agent. The product to which nitric acid is reduced depends on the concentration of the acid, the temperature of the reaction, and the nature of the reducing agent. The most common reduction products are nitrogen(IV) oxide and nitrogen(II) oxide.

$$\overset{+5}{N}O_3^- + 2H^+ + e^- \longrightarrow \overset{+4}{N}O_2 + H_2O \qquad E° = 0.81 \text{ V}$$

$$\overset{+5}{N}O_3^- + 4H^+ + 3e^- \longrightarrow \overset{+2}{N}O + 2H_2O \qquad E° = 0.96 \text{ V}$$

All metals except the least reactive ones, such as gold and platinum, are oxidized by nitric acid. In some cases, the reaction must be initiated by heat, but once started, it proceeds vigorously. In general, the stronger the reducing agent and the more dilute the acid, the lower will be the oxidation state of the nitrogen in the reduction product. For example, copper, which lies below hydrogen in the electromotive series, reduces nitric acid to nitric oxide in warm dilute nitric acid

$$3Cu(s) + 8\overset{+5}{H}NO_3(dil) \longrightarrow 3Cu^{2+} + 6NO_3^- + 2\overset{+2}{N}O(g) + 4H_2O(l)$$

while zinc, which stands above hydrogen in the electromotive series and is a much more powerful reducing agent than copper, reduces very dilute nitric acid all the way to ammonium ion.

$$4Zn(s) + 10\overset{+5}{H}NO_3(dil) \longrightarrow 4Zn^{2+} + 9NO_3^- + \overset{-3}{N}H_4^+ + 3H_2O(l)$$

With concentrated nitric acid and metals below hydrogen in the electromotive series, nitrogen dioxide is the main reduction product.

FIGURE 22.9

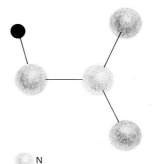

- N
- O
- H

Nitric acid, HNO_3
m.p. −42°C
b.p. 83°C

$$H:\overset{..}{O}:N\overset{:\overset{..}{O}.}{\underset{:\overset{..}{O}:}{}} \longleftrightarrow H:\overset{..}{O}:N\overset{:\overset{..}{O}:}{\underset{:\overset{..}{O}.}{}}$$

$$\overset{+5}{Cu(s)} + 4H\overset{}{NO_3}(conc) \longrightarrow Cu^{2+} + 2NO_3^- + 2\overset{+4}{NO_2}(g) + 2H_2O(l)$$

A number of metals that are above hydrogen in the electromotive series and that are attacked by dilute nitric acid—for example, iron and chromium—are inert to the very concentrated acid. These metals are said to be *passive* toward the acid.

A 3-to-1 by volume mixture of concentrated HCl and HNO₃ is known as **aqua regia** ("royal water") because of its ability to dissolve such noble metals as gold and platinum, which are inert to either acid alone. The redox reaction between concentrated nitric and hydrochloric acids gives water, elemental chlorine, and nitrosyl chloride.

$$3HCl(aq) + HNO_3(aq) \longrightarrow Cl_2(g) + \quad NOCl(g) \quad + 2H_2O(l)$$
nitrosyl chloride

The solvent action of aqua regia probably results mainly from the action of the nitrosyl chloride and chlorine, which convert the metals initially to chlorides. The metal chlorides are then transformed to stable complex anions by reaction with chloride ion. The overall reaction for gold is

$$Au(s) + 4H^+ + 4Cl^- + NO_3^- \longrightarrow [AuCl_4]^- + NO(g) + 2H_2O(l)$$

Nitric acid also oxidizes nonmetals and their compounds. Warm, concentrated acid converts iodine to iodic acid (HIO_3), carbon to carbon dioxide, sulfur to a mixture of sulfurous and sulfuric acid ($H_2SO_3 + H_2SO_4$), and phosphorus to phosphoric acid (H_3PO_4).

c. Preparation and uses of nitric acid. The major industrial method for preparation of nitric acid, known as the Ostwald process, is a three-step process requiring only anhydrous ammonia and air as the raw materials.

$$4NH_3(g) + 5O_2(g) \xrightarrow[\text{500–1000°C}]{\text{Pt–Rh catalyst}} 4NO(g) + 6H_2O(l)$$

$$2NO(g) + O_2(g) \longrightarrow 2NO_2(g)$$

$$3NO_2(g) + H_2O(l) \longrightarrow 2HNO_3(aq) + NO(g)$$

The first step, in which ammonia is the reducing agent, gives an over 90% yield of nitric oxide by passing a 90% air–10% ammonia mixture over a red-hot platinum–rhodium catalyst. The hot gases are cooled, more air is added to produce nitrogen dioxide, and then the nitrogen dioxide is passed into water to give a roughly 60% solution of nitric acid. The nitric oxide from the last step is recycled.

The acid has also been made commercially by heating sodium nitrate with concentrated sulfuric acid.

$$NaNO_3(s) + H_2SO_4(aq) \xrightarrow{\Delta} HNO_3(g) + NaHSO_4(s)$$

The success of this reaction depends upon the fact that nitric acid is much more volatile than sulfuric acid and can be distilled from the reaction mixture.

The major use of nitric acid (Table 22.6) is in the synthesis of ammonium nitrate for use as a fertilizer (see An Aside: Chemical Fertilizers).

d. Nitrates. Nitrates, the salts of nitric acid, are easily prepared by the reaction between nitric acid and metal oxides, hydroxides, or carbonates. With oxides and hydroxides, the reaction is simply neutralization and goes to completion because of the formation of water. With carbonates, the reaction is complete because of the evolution of carbon dioxide gas; for example,

$$CaCO_3(s) + 2H^+ + 2NO_3^- \longrightarrow Ca^{2+} + 2NO_3^- + CO_2(g) + H_2O(l)$$

Metal nitrates are very soluble in water. The few that appear to be insoluble, such as bismuth nitrate, $Bi(NO_3)_3$, and mercury(II) nitrate, $Hg(NO_3)_2$, are actually hydrolyzed by water with the formation of insoluble basic salts that contain the O or OH group.

$$\underset{\textit{bismuth nitrate}}{Bi(NO_3)_3(s)} + H_2O(l) \longrightarrow \underset{\substack{\textit{basic bismuth nitrate} \\ \textit{(bismuth oxynitrate)}}}{BiO(NO_3)(s)} + 2H^+ + 2NO_3^-$$

Metal nitrates are unstable toward heat and decompose in one of two ways. Nitrates of the more reactive metals evolve oxygen when heated and are converted to nitrites.

$$\underset{\textit{calcium nitrate}}{Ca(NO_3)_2(s)} \overset{\Delta}{\longrightarrow} \underset{\textit{calcium nitrite}}{Ca(NO_2)_2(s)} + O_2(g)$$

Nitrates of less active metals, such as lead and copper, decompose to give oxides, nitrogen dioxide, and oxygen.

Phosphorus

22.9 Additional properties of phosphorus

Elemental phosphorus exists in several allotropic forms. White phosphorus is a soft, low-melting, waxy, white to yellow solid, depending upon its purity. Phosphorus in this form has high reactivity—it ignites spontaneously in air to give P_4O_{10} and must be stored under water, in which it is insoluble. It is also highly toxic. Small amounts can cause severe irritation of the lungs and gastrointestinal tract and about 50 mg inhaled or ingested can be fatal. White phosphorus should *never* be touched. Its ignition temperature is about the same as the temperature of the skin, and it can cause painful, slow-healing burns.

White phosphorus is moderately soluble in diethyl ether and in benzene. In the solid, in solution, and in the vapor state below 800°C, it exists in the form of covalently bonded P_4 molecules (Figure 22.10) with a phosphorus atom at each corner of a regular tetrahedron.

FIGURE 22.10

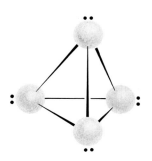

The P_4 molecule

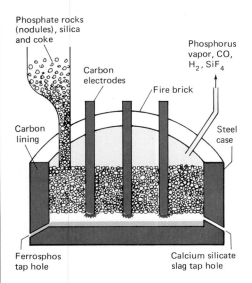

Phosphate rocks
(nodules), silica
and coke

Phosphorus
vapor, CO,
H_2, SiF_4

Carbon
electrodes

Fire brick

Carbon
lining

Steel
case

Ferrosphos
tap hole

Calcium silicate
slag tap hole

FIGURE 22.11

An electric phosphorus furnace. *A high voltage electric arc produces temperatures of 1200°–1450°C in the furnace. Molten ferrophos is heavy and sinks to the bottom, where it can be drawn off. Molten calcium silicate is less dense than the ferrophos and forms a second layer that can be drawn off separately. (From A. D. F. Toy, Phosphorus Chemistry in Everyday Living, American Chemical Society, Washington, D.C., 1976.)*

When the supply of oxygen is extremely limited and moisture is present, oxidation of white phosphorus is slow and is accompanied by phosphorescence. White phosphorus vigorously undergoes direct combination with halogens to give either trihalides (PX_3) or pentahalides (PX_5) as the main products. Reaction of white phosphorus with reactive metals (e.g., Na, Ca) at high temperatures gives phosphides (Na_3P and Ca_3P_2), which contain the P^{3-} ion.

A second allotropic form of phosphorus—red phosphorus—is obtained when the white variety is heated to about 250°C in the absence of air or when the white form is exposed to light. White phosphorus often appears yellow and is often called "yellow phosphorus" because of the presence of small amounts of this second allotrope, which forms slowly on its surfaces. Red phosphorus differs radically from the white allotrope. It is relatively nonpoisonous, is insoluble in those solvents that dissolve the white form, and melts at about 600°C. Red phosphorus is stable toward atmospheric oxidation at ordinary temperatures, but it ignites when heated to about 400°C. It undergoes the same reactions toward elemental substances as the white form; but only at considerably higher temperatures. It is less reactive because it is polymeric and the phosphorus atoms are held close together.

Phosphorus is best known in −3, +3, and +5 oxidation states. In terms of the valence-bond theory of covalent bonding, formation of five covalent bonds by a phosphorus atom requires that the two s electrons in the valence shell become unpaired and that one of them be promoted to an orbital of higher energy—a d orbital. The electronic rearrangement is from $3s^2 3p^1 3p^1 3p^1$ to $3s^1 3p^1 3p^1 3p^1 3d^1$. These five orbitals then become hybridized to give five equivalent sp^3d orbitals, and the phosphorus atom, now with five unpaired electrons in the valence shell, can combine with, for example, five chlorine or five bromine atoms.

22.10 Sources, preparation, and uses of phosphorus

Phosphorus is the twelfth most abundant element, making up 0.10% of the Earth's crust (by weight). It occurs only in combined form, mainly as phosphates in various minerals. The most common phosphate-bearing minerals are fluoroapatite, $Ca_5(PO_4)_3F$ and phosphorite, which is hydroxyapatite, $Ca_5(PO_4)_3OH$ (see Figure 20.2).

Most elemental phosphorus is produced from **phosphate rock,** as the mineral deposits containing calcium phosphate, the apatites, and silica are called, by treatment of the rock in electric furnaces (Figure 22.11). In the overall reaction, phosphate is reduced to elemental phosphorus, and coke is oxidized to carbon monoxide, with calcium being removed as $CaSiO_3$ slag.

$$2Ca_3(PO_4)_2(s) + 6SiO_2(s) + 10C(s) \xrightarrow{1200–1450°C}$$
$$6CaSiO_3(l) + 10CO(g) + P_4(g)$$

At the same time, the fluoride ion usually present in phosphate rock is converted by reaction with SiO_2 to volatile SiF_4, a toxic gas that must be scrubbed from the vapor stream. Iron oxide, usually also present in phosphate rock to a small extent, reacts with phosphorus to give iron phosphide, known as "ferrophos."

Most phosphate rock is used directly in the manufacture of phosphate fertilizers rather than in the production of the element (Table 22.7). Only about 5% of all phosphorus used in a year is in elemental form, the major use being burning of phosphorus to give phosphorus(V) oxide, which is converted to pure phosphoric acid (Section 22.13). Elemental phosphorus is directly used in fireworks and matches, as a rodent poison, and as an alloying agent.

22.11 Oxidation state −3: phosphides, phosphine, and phosphonium salts

a. Phosphides and phosphine. Nitrogen and phosphorus form similar compounds in the −3 oxidation state. Ionic phosphides are known for most of the active metals of Representative Groups I and II. These phosphides are attacked vigorously by water with the liberation of phosphine, PH_3, a volatile covalent substance that also contains phosphorus in the −3 oxidation state.

$$\underset{\text{phosphide ion}}{P^{3-}} + 3H_2O(l) \longrightarrow \underset{\text{phosphine}}{PH_3(g)} + 3OH^-$$

Phosphine can also be obtained by reaction of white phosphorus with warm aqueous solutions of strongly basic hydroxides (e.g., NaOH). (To simplify the equations, we write P, rather than P_4.)

$$4P(s) + 3OH^- + 3H_2O(l) \longrightarrow PH_3(g) + \underset{\text{hypophosphite ion}}{3H_2PO_2^-}$$

Phosphine is a colorless gas that has an offensive odor and is extremely poisonous. The pure compound ignites when heated in air at about 150°C. The product obtained by the methods described above is usually spontaneously flammable because it contains small amounts of diphosphine (P_2H_4), which is extremely reactive. Phosphine is much less soluble in water than ammonia, and is unstable in aqueous solution, decomposing to give a variety of products, including hydrogen and elemental phosphorus.

b. Phosphonium halides. Since phosphine is structurally similar to ammonia (see Figure 22.1), it should also function as an electron-pair donor toward suitable acceptors. Phosphine is indeed basic in the Lewis sense, but much less strongly so than ammonia. Like ammonia, phosphine takes up a proton from the hydrogen halides.

$$\underset{H}{\overset{H}{H:\ddot{P}:}}(g) + H:\ddot{\ddot{X}}:(g) \rightleftharpoons \left[\underset{H}{\overset{H}{H:\ddot{P}:H}}\right]^+ + :\ddot{\ddot{X}}:^-$$

TABLE 22.7
Uses of elemental phosphorus

Major uses
Production of P_4O_{10}

Other uses
Synthesis of other chemicals
Matches, bombs, and fireworks
Rodent poisons

Phosphoric acid, H_3PO_4
Phosphorous acid, H_3PO_3

FIGURE 22.12

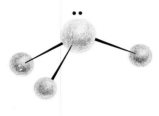

 P

X

Phosphorus trihalides, PX_3

FIGURE 22.13

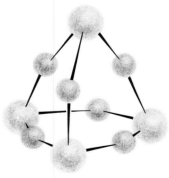

 P

O

Phosphorus(III) oxide, P_4O_6
m.p. 23.8°C
b.p. 175.4°C

The white, crystalline products of reaction, $[PH_4]X$, are known as *phosphonium halides.* These salts are much less thermally stable than the corresponding ammonium compounds. The chloride and the bromide are almost entirely dissociated into the parent compounds at room temperature. The iodide is more stable and can be sublimed at 80°C. In contrast to ammonium salts of strong acids, which are hydrolyzed only to a small extent, phosphonium iodide undergoes complete decomposition in water.

$$PH_4^+(aq) \longrightarrow PH_3(g) + H^+$$

22.12 Oxidation state +3: trihalides, oxide, and phosphorous acid

a. Phosphorus trihalides. The phosphorus trihalides can be prepared by direct union of the elements, with the phosphorus present in excess of the stoichiometric amount

$$2P(s)(\text{excess}) + 3X_2(g) \longrightarrow 2PX_3 \quad (X = F, Cl, Br, I)$$

The trihalides are simple covalent substances and have the same pyramidal molecular geometry as ammonia and phosphine (Figure 22.12). At room temperature, the fluoride is a colorless gas, the chloride and bromide are volatile, fuming liquids (PCl_3, b.p. 74.2°C; PBr_3, b.p. 175.3°C), and the iodide, a red, low-melting (m.p. 61.0°C) solid that is not stable.

The trihalides are unstable in water, the chloride, bromide, and iodide being completely and irreversibly hydrolyzed to give phosphorous acid and the corresponding hydrohalic acid.

$$PX_3 + 3H_2O(l) \longrightarrow \underset{\textit{phosphorous acid}}{H_3PO_3(aq)} + 3H^+ + 3X^-$$

b. Phosphorus(III) oxide and phosphorous acid. The oxidation of white phosphorus at about 100°C in a limited supply of air gives phosphorus(III) oxide, P_4O_6, a white, crystalline material, as the main product. The tetrahedral structure found in the P_4 molecule is maintained in P_4O_6 (Figure 22.13).

Phosphorus(III) oxide is the anhydride of phosphorous acid, H_3PO_3, and is converted to this acid when added to *cold* water.

$$P_4O_6(s) + 6H_2O(l) \longrightarrow \underset{\textit{phosphorous acid}}{4H_3PO_3(aq)}$$

(Note one of the great pitfalls in the study of phosphorus chemistry—the extra **o** appears *only* in the spelling of phosphorous acid.) Solutions of phosphorous acid are more commonly made by the hydrolysis of phosphorus trichloride (see above). From such solutions, the pure acid may be isolated as a colorless, low-melting, deliquescent solid.

The H_3PO_3 molecule is tetrahedral, with one hydrogen atom bound directly to the phosphorus atom (Figure 22.14). Since only those hydrogen atoms bound to oxygen are sufficiently polar to be neutralized by reaction with aqueous bases, phosphorous acid is a diprotic acid. It is moderately strong with respect to its first dissociation ($K_{a_1} = 1.6 \times 10^{-2}$; $K_{a_2} = 7 \times 10^{-7}$). Phosphites are salts of phosphorous acid containing either the dihydrogen phosphite ion, $H_2PO_3^-$, or the monohydrogen phosphite ion, HPO_3^{2-}.

22.13 Oxidation state +5: halides, oxide, and phosphoric acids

a. Phosphorus pentahalides. All of the pentahalides of phosphorus but the iodide are known. It is apparently impossible for five large iodine atoms to fit around the relatively small phosphorus atom. The pentahalide can be obtained as the major product when elemental phosphorus is treated with an excess of halogen (compare with preparation of phosphorus trihalides, Section 22.12a).

$$2P(s) + 5X_2(g)(\text{excess}) \longrightarrow 2PX_5 \qquad (X = F, Cl, Br)$$

However, the compounds are usually prepared by direct combination of trihalide and halogen.

$$PX_3 + X_2(g) \longrightarrow PX_5$$

In the vapor and the liquid states, the pentahalides are covalent, the PF_5 and PCl_5 molecules in the vapor having a triangular bipyramidal structure (Figure 22.15), with hybridized sp^3d orbitals on the phosphorus atom. There is excellent evidence that in the solid state the pentachloride is a saltlike substance consisting of tetrahedral $[PCl_4]^+$ and octahedral $[PCl_6]^-$ ions. The solid pentabromide apparently is also ionic and is made up of the ions $[PBr_4]^+$ and Br^-.

Like the trihalides, the pentahalides react rapidly with water. With a limited quantity of water, two of the halogen atoms in the molecule are removed, and oxohalides of the formula POX_3—phosphoryl halides or phosphorus(V) oxohalides—are formed, along with hydrohalic acids.

$$PX_5 + H_2O(l) \longrightarrow POX_3 + 2H^+ + 2X^-$$

With an excess of water, all of the halogen atoms are removed and the pentahalides are converted to phosphoric acid.

$$PX_5 + 4H_2O(l) \longrightarrow H_3PO_4(aq) + 5H^+ + 5X^-$$

b. Phosphorus(V) oxide. Phosphorus(V) oxide, P_4O_{10}, is the chief product when phosphorus is burned in an excess of air.

$$4P(s) + 5O_2(g) \longrightarrow P_4O_{10}(s)$$

FIGURE 22.14

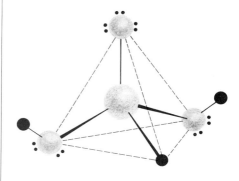

P
O
H

Phosphorous acid, H_3PO_3
m.p. 73.6°C
b.p. 200°C (dec)

$$\begin{array}{c} :\ddot{O}: \\ H:\ddot{O}:\overset{..}{P}:\ddot{O}:H \\ H \end{array}$$

FIGURE 22.15

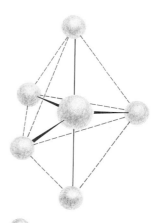

P
X

Phosphorus pentahalides, PX_5

FIGURE 22.16

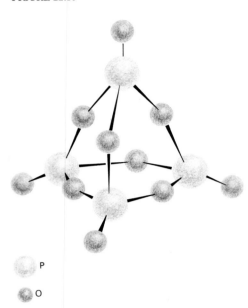

Phosphorus(V) oxide, P_4O_{10}
sublimes at 358°C

FIGURE 22.17

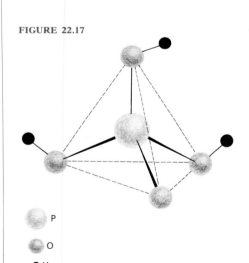

Phosphoric acid, H_3PO_4
m.p. 42.35°C

$$\begin{array}{c} :\ddot{O}: \\ H:\ddot{O}:P:\ddot{O}:H \\ :\ddot{O}: \\ \ddot{H} \end{array}$$

Phosphorus(V) oxide is a white solid that exists in a number of crystalline modifications. When heated, it sublimes as discrete P_4O_{10} molecules. In these molecules, as in P_4 and P_4O_6, the phosphorus atoms are tetrahedrally arranged (Figure 22.16). The most important chemical characteristic of phosphorus(V) oxide is its tremendous affinity for water. It reacts with water to give phosphoric acid (next section) and is therefore one of the most useful drying agents. Phosphorus(V) oxide is sometimes called phosphorus pentoxide and is written P_2O_5.

c. The phosphorus(V) acids and their salts. The formulas and names of the phosphorus(V) acids are summarized in Table 22.8. In all of these acids and their salts, each phosphorus atom is surrounded by four oxygen atoms and in none is a phosphorus atom bonded directly to another phosphorus atom. The simplest phosphorus(V) acid is phosphoric acid, H_3PO_4, formally known as orthophosphoric acid (Figure 22.17). The octet rule is satisfied by the four single, or σ, bonds around phosphorus in both phosphorous and the phosphoric acids. However, greater bond strengths and shorter bond lengths than those expected for single bonds show that there is π bond, or multiple bond, character to the phosphorus–oxygen bonds in these compounds. The π bonding is contributed by overlap of oxygen p orbitals with phosphorus d orbitals.

The higher phosphoric acids are the products of the elimination of water between successive molecules of phosphoric acid. The elimination of one molecule of water from two molecules of phosphoric acid yields pyrophosphoric, or diphosphoric acid

elimination of H_2O

$$\underset{\text{phosphoric acid}}{\text{HOP(OH} \quad \text{H)OPOH}} \longrightarrow \underset{\text{pyrophosphoric acid}}{\text{HOP—O—POH}} + H_2O$$

Triphosphoric acid, also called tripolyphosphoric acid, is formed similarly from three molecules of phosphoric acid. Theoretically this process could go on indefinitely.

Phosphoric acid is among the top ten industrial chemicals; in a recent year, fourteen billion pounds was produced. It is made by two methods. Thermal process phosphoric acid is a product of high purity, and any desired concentration may be obtained by the union of phosphorus(V) oxide with water.

$$P_4O_{10}(s) + 6H_2O \longrightarrow 4H_3PO_4(aq)$$

A less pure product, wet process phosphoric acid, is made by reaction between the calcium phosphate in phosphate rock and sulfuric acid.

$$Ca_3(PO_4)_2(s) + 3H_2SO_4(aq) \longrightarrow 2H_3PO_4(aq) + 3CaSO_4(s)$$

The acid from this process is generally concentrated by boiling off water to give an 85% aqueous solution, which is known as "syrupy" phosphoric acid because of its viscous nature.

Pure phosphoric acid is a white solid that melts at 42.35°C to a viscous liquid with a strong tendency to supercool. It is very soluble in water and is a triprotic acid. Phosphoric acid is a much weaker acid than nitric acid; the phosphoric acids and their ions, unlike nitric acid and nitrate ion, are poor oxidizing agents.

<u>Salts</u> corresponding to the three anions of phosphoric acid can be obtained by appropriate neutralization (Section 18.7) of the protons in the acid; for example,

<table>
<tr><td align="center">NaH$_2$PO$_4$
sodium dihydrogen
phosphate
(or monosodium phosphate)</td><td align="center">Na$_2$HPO$_4$
disodium monohydrogen
phosphate
(or disodium phosphate)</td><td align="center">Na$_3$PO$_4$
trisodium
phosphate
(or normal sodium
phosphate)</td></tr>
</table>

Sometimes these salts are referred to as primary, secondary, and tertiary sodium phosphate, respectively.

Aqueous solutions of normal phosphate salts are strongly alkaline as a result of hydrolysis.

$$PO_4^{3-} + H_2O(l) \rightleftharpoons HPO_4^{2-} + OH^- \qquad K_{b_1} = 5.1 \times 10^{-2}$$

Such solutions are used as cleaning agents. Dihydrogen phosphates give acidic aqueous solutions, because the ionization of the H$_2$PO$_4^-$ is of greater importance than its hydrolysis.

<u>Pyrophosphoric acid</u>, H$_4$P$_2$O$_7$ (Figure 22.18), a colorless, glassy solid, is extremely soluble in water, in which it slowly reverts to the ortho acid. Of the four types of salts that pyrophosphoric acid can form, only two are obtained conveniently. These are, for sodium,

<table>
<tr><td align="center">Na$_2$H$_2$P$_2$O$_7$
sodium acid pyrophosphate
(a "secondary pyrophosphate")</td><td align="center">Na$_4$P$_2$O$_7$
tetrasodium pyrophosphate
(a "normal pyrophosphate")</td></tr>
</table>

A mixture of polymeric acids of the empirical formula HPO$_3$ is obtained when phosphoric acid is heated.

$$nH_3PO_4 \xrightarrow{\text{325–350°C}} (HPO_3)_n + nH_2O(g)$$

$n = a$ variety of whole numbers

These acids, which are known as <u>metaphosphoric acids</u>, are difficult to separate. Metaphosphoric acids containing either rings or long chains of PO$_4$ tetrahedra have been identified. Metaphosphate salts are useful water-softening agents.

The possibilities for making <u>phosphates</u> with desirable properties are extensive—any or all of the hydrogen atoms in any of the phosphoric acids can be replaced by metal ions. Sodium, calcium, and potassium phosphates, as well as others such as magnesium phosphates, appear in an incredible variety of applications (Table 22.9). Sodium tripolyphosphate

sodium tripolyphosphate

TABLE 22.8

Phosphorus(V), or phosphoric, acids

H$_3$PO$_4$	**Orthophosphoric acid (phosphoric acid)**
H$_4$P$_2$O$_7$	**Pyrophosphoric acid (diphosphoric acid)**
H$_5$P$_3$O$_{10}$	**Triphosphoric acid (tripolyphosphoric acid)**
(HPO$_3$)$_n$	**Metaphosphoric acid**

TABLE 22.9

Uses of phosphates

In food applications, the calcium phosphates are most common. In cleaning applications, the sodium phosphates are most common.

Food uses
    Food supplement for man and animals
    Carbonated beverages (for tartness)
    Leavening agents
    Quick-cooking cereals

Cleaning agents
    Builders in detergents
    Strongly alkaline cleaners
    Abrasives in toothpastes

Water softeners

Flame retardants

In plating plastics with metals

FIGURE 22.18

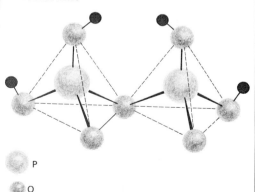

P
O
H

Pyrophosphoric acid, H$_4$P$_2$O$_7$
m.p. 61°C

Chemical fertilizers

The major end uses of nitrogen, phosphorus, and sulfur, and of their most important industrial products, ammonia, phosphoric acid, and sulfuric acid, are in the production of chemical fertilizers.

Nitrogen and phosphorus are two of the three major nutrients (Table A) supplied to plants through the soil. Sulfur plays a double role in the fertilizer industry—it is necessary to plants, although in smaller amounts than nitrogen and phosphorus, and it is vitally important in the manufacture of other chemical fertilizers (Table B), particularly the superphosphates.

Nitrogen atoms are needed for amino acids, the building blocks of proteins, and for chlorophyll, as well as for many other biochemically important molecules. The rate of growth and yield from crops is directly influenced by the nitrogen available, and it has been shown that the amount of protein in, for example, corn is increased by nitrogen fertilizers.

Plants cannot use atmospheric nitrogen, presumably because of the large amount of energy needed to break the nitrogen–nitrogen bond. Molecular nitrogen must undergo **nitrogen fixation**—the combination of molecular nitrogen with other atoms—before it can be taken up by plants from the soil, usually in the form of nitrate ion. In nature, bacteria and algae that live free in soil, ponds, and lakes, or that live in a mutually dependent relationship with specific plants, fix nitrogen by converting it to ammonia. Nitrogen in the atmosphere is fixed by the formation of NO in lightning flashes. The Haber process serves as one of our primary nitrogen fixation processes.

About 50 years ago there was, in some places, a thriving business in supplying bacteria that farmers plowed into the soil where clover or other legumes were growing. The business died out when chemical fertilizers became available at lower prices. But now interest is reviving in bacterial nitrogen fixation, and it has become an active area of research. With greater understanding of the natural fixation process, it is hoped that bacteria can again be put directly to work in the fields, cutting down the need for chemical fertilizers, which require large amounts of energy in their manufacture. Possibly better plant–microorganism combinations might be designed, or possibly a nonbiological system for nitrogen fixation that resembles the natural process might be developed.

The primary role of phosphorus in biological systems is in energy transfer—plants and animals store energy in the phosphorus–oxygen bonds. Good suppliers of phosphorus in the soil hasten the growth of young plants. Most of the phosphorus used by plants is in the form of phosphate ions, particularly the dihydrogen phosphate ion, $H_2PO_4^-$.

The raw material for phosphate fertilizers is phosphate rock. The rock is converted directly to single superphosphate—a mixture of monocalcium phosphate and calcium sulfate (gypsum),

$$2Ca_5(PO_4)_3F(s) + 7H_2SO_4(aq) + 10H_2O(l) \longrightarrow$$
$$3Ca(H_2PO_4)_2 \cdot H_2O(s) + 7CaSO_4 \cdot H_2O + 2HF(g)$$

or it is converted directly to triple superphosphate, which contains no calcium sulfate and therefore delivers a higher percentage of phosphorus,

$$Ca_5(PO_4)_3F(s) + 7H_3PO_4(aq) + 5H_2O(l) \longrightarrow$$
$$5Ca(H_2PO_4)_2 \cdot H_2O(s) + HF(g)$$

Finely ground phosphate rock can also be used as a fertilizer, or phosphoric acid can be produced and soluble salts made from the acid (Table B).

Just a word about how the other essential nutrients are delivered to crops. Most of them, including the important nutrient potassium, are available in mineral deposits. The mineral need only be mined and added in suitable amounts to nitrogen- and phosphorus-containing chemical fertilizers.

Plant nutrients

Carbon, hydrogen, and oxygen are obtained by plants from the air and the soil. All other nutrients are obtained only from the soil. Primary nutrients are needed in largest amounts (up to 200 pounds per acre), secondary nutrients are needed in smaller amounts (up to 50 pounds per acre), and others are needed only in small amounts (to < 1 pound acre).

Needed in large amounts
Carbon
Hydrogen
Oxygen
Nitrogen
Phosphorus
Potassium

Needed in moderate amounts
Calcium
Magnesium
Sulfur

Needed in trace amounts
Boron
Copper
Iron
Manganese
Zinc
Molybdenum
Chlorine

Manufacture of major chemical fertilizers[a]

Natural raw materials appear in black and manufactured raw materials in color in the left-hand column.

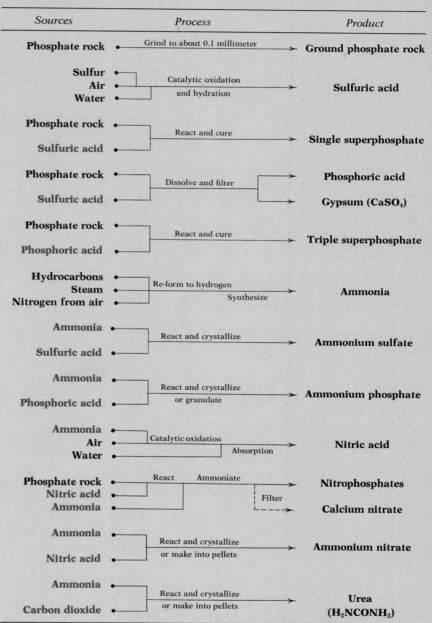

Sources	Process	Product
Phosphate rock	Grind to about 0.1 millimeter	Ground phosphate rock
Sulfur / Air / Water	Catalytic oxidation and hydration	Sulfuric acid
Phosphate rock / Sulfuric acid	React and cure	Single superphosphate
Phosphate rock / Sulfuric acid	Dissolve and filter	Phosphoric acid / Gypsum (CaSO$_4$)
Phosphate rock / Phosphoric acid	React and cure	Triple superphosphate
Hydrocarbons / Steam / Nitrogen from air	Re-form to hydrogen / Synthesize	Ammonia
Ammonia / Sulfuric acid	React and crystallize	Ammonium sulfate
Ammonia / Phosphoric acid	React and crystallize or granulate	Ammonium phosphate
Ammonia / Air / Water	Catalytic oxidation / Absorption	Nitric acid
Phosphate rock / Nitric acid / Ammonia	React Ammoniate / Filter	Nitrophosphates / Calcium nitrate
Ammonia / Nitric acid	React and crystallize or make into pellets	Ammonium nitrate
Ammonia / Carbon dioxide	React and crystallize or make into pellets	Urea (H$_2$NCONH$_2$)

[a] Adapted from "Chemical Fertilizers," Christopher J. Pratt, Copyright © 1965 by Scientific American, Inc. All rights reserved. Vol. 212, No. 6, June 1965.

is a compound used in large quantities (roughly 2200 million pounds per year) as a builder in synthetic detergents, and it is the compound held mainly responsible for the phosphate pollution of natural waters (Sections 13.8, 13.12), although run-off from tilled fields that have been treated with soluble phosphate fertilizers also adds phosphate to natural waters.

Sulfur

22.14 Additional properties of sulfur

Like phosphorus, sulfur is a solid at room temperature and exists in various allotropic forms. Although elemental sulfur is reactive and combines directly with most elements, in none of its forms is it as difficult to handle or as toxic as white phosphorus. Sulfur has been used medically as a laxative in doses up to 4 g and also in various preparations applied to the skin. Exposed to moist air, elemental sulfur slowly oxidizes to give sulfuric acid.

$$2S(s) + 2H_2O(g) + 3O_2(g) \longrightarrow 4H^+ + 2SO_4^{2-}$$

The form of sulfur stable at ordinary temperatures is rhombic sulfur (Figure 22.19), which is bright yellow in color, odorless, and tasteless. It is practically insoluble in water, but dissolves freely in carbon disulfide, CS_2 (about 50.0 g per 100 g of solvent at 25°C). When heated to 95.6°C rhombic sulfur undergoes a slow transition to monoclinic sulfur. The change is so slow that if rhombic sulfur is heated rapidly to its melting point of 112.8°C, little conversion to the monoclinic modification occurs. Like the rhombic form, monoclinic sulfur is bright yellow, odorless, tasteless, and water insoluble. Its crystals look like long, yellow needles. In both forms sulfur exists as S_8 molecules (Figure 22.20). (We have mentioned here only the more important sulfur allotropes—over thirty are known.)

The changes on melting and heating sulfur are complicated and interesting. The true melting point of rhombic sulfur is 112.8°C and that of the monoclinic form 119.0°C. When solid sulfur is heated, it melts at 115°C to give a yellow mobile liquid. The melting process is accompanied to a very small extent by the breaking of S_8 rings and the formation of chains of the composition S_6 and S_4. Between the melting point and about 160°C there is little change in the mobility of the liquid sulfur, but between 160°C and 187°C, there is a *10,000-fold increase* in viscosity and the material turns brown. The increased viscosity is due to the formation of long chains of sulfur atoms that entangle each other. Above 200°C, the viscosity decreases, probably as a result of the breaking of the long chains, and at 444.6°C, the boiling point, the sulfur is again a mobile liquid. If the liquid material at about 200°C is cooled very rapidly by pouring it into cold water, a soft rubbery substance known as plastic sulfur forms. In plastic sulfur, the long chains of sulfur atoms are in the form of coils, and the rubbery character is due to the ability of these chains to uncoil and coil again under stress. On standing, plastic sulfur slowly reverts to the rhombic allotrope.

Rhombic sulfur

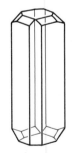

Monoclinic sulfur

FIGURE 22.19

Rhombic and monoclinic sulfur crystals.

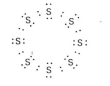

(a) Electronic structure

(b) S_8 ring structure

(c) Packing of rings in orthorhombic sulfur

FIGURE 22.20

The S_8 molecule. (a) *Electronic structure;* (b) S_8 *ring structure;* (c) *packing of rings in orthorhombic sulfur.* (*From J. Donohue,* The Structure of the Elements, *p. 335, Wiley, New York, 1974.*)

Sulfur vapor contains S_8, S_6, S_4, and S_2 molecules, with the relative amounts varying with temperature and pressure. The proportion of S_2 in the vapor increases, as expected, with increasing temperature. The diatomic molecule, like O_2, is paramagnetic.

In bonding in compounds with active metals sulfur gains two electrons to complete a noble gas octet and give sulfide ion, S^{2-}. It can also form two covalent bonds by sharing the two unpaired electrons in its p orbitals, and the sulfides cover the entire continuum from ionic to covalent bonding. In addition to the -2 oxidation state, sulfur exhibits oxidation states of $+4$ and $+6$.

Sulfur–oxygen bonds often have a considerable amount of double-bond character due to the participation of the d orbitals in π bonding. Evidence for this type of π bonding appears when sulfur–oxygen covalent bonds, as for example in H_2SO_4, are shorter and stronger than expected for the single covalent bond.

To accommodate more than eight electrons in the valence shell, as for phosphorus, valence bond theory pictures the unpairing of the electrons in the $3s$ and $3p$ orbitals of sulfur atoms, their promotion to unoccupied $3d$ orbitals, and the formation, in this case, of six equivalent sp^3d^2 hybridized orbitals. Sulfur hexafluoride has the octahedral geometry characteristic of molecules with six covalent bonds. The corresponding compounds with other halogens are unknown. Apparently, chlorine, bromine, and iodine atoms are so large that six of them cannot be accommodated in a stable octahedral arrangement around a sulfur atom.

22.15 Sources, preparation, and uses of sulfur

Sulfur is much less abundant than oxygen and one-half as abundant as phosphorus, making up 0.05% (by weight) of the Earth's crust. It is found mainly as the free element, heavy metal sulfides (e.g., PbS,

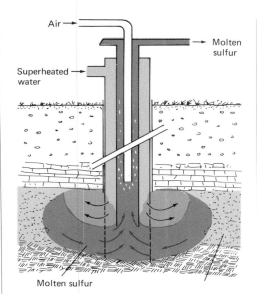

FIGURE 22.21

The Frasch process for sulfur extraction. *Three concentric pipes are sunk from the surface into the bed of sulfur. Water at a temperature of about 170°C and a pressure of 100 lb/sq inch is forced down the outer pipe to melt the sulfur. Hot compressed air is pumped down the innermost pipe and mixes with the molten sulfur, forming a froth of water, air, and sulfur. This froth rises to the surface through the middle pipe and is collected in large bins. Removal of the water leaves sulfur with a purity of about 99.5%.*

TABLE 22.10
Uses of elemental sulfur

Major use
Manufacture of sulfuric acid—roughly 85% of all elemental sulfur goes into sulfuric acid

Other uses
Pulp and paper industry
Manufacture of carbon disulfide
Rubber
Fungicides, insecticides, and medicine

ZnS), and light metal sulfates (e.g., $MgSO_4$, $CaSO_4$). Much of the sulfur currently utilized in this country is obtained from the vast deposits of the free element found in Texas and Louisana. The sulfur occurs in veins and pockets in beds of limestone and is covered with several hundred feet of quicksand and rock. It is brought to the surface by a process developed at the turn of this century and called the *Frasch process* after its inventor, the engineer Herman Frasch (the process is described in Figure 22.21) (see also Thoughts on Chemistry).

Growing concern with pollution by sulfur dioxide (Section 6.11) has led to the removal of sulfur from natural gas and petroleum. In addition, sulfur-containing compounds are being removed from the effluent and stack gases from metal-smelting plants that process sulfide ores. Elemental sulfur from these operations is now coming into the market in increasing quantities. The major use of sulfur is in the manufacture of sulfuric acid (Table 22.10).

22.16 Oxidation state −2: sulfides

a. Hydrogen sulfide. Hydrogen sulfide (Figure 22.22), H_2S, is an extremely poisonous colorless gas with an obnoxious odor resembling that of rotten eggs. The compound is readily obtained by the action of a dilute, nonoxidizing acid (for example, HCl) on metal sulfides such as iron(II) sulfide.

$$FeS(s) + 2H^+ \longrightarrow H_2S(g) + Fe^{2+}$$

Another laboratory method for preparing hydrogen sulfide consists of heating elemental sulfur with paraffin, which is a mixture of solid compounds of the general formula C_nH_{2n+2}. Sulfur combines with some of the hydrogen atoms from these compounds.

The precipitation of metal sulfides (next section) plays an important role in many qualitative analysis schemes for the identification of the elements present in a sample of unknown composition. At one time, qualitative analysis laboratories were recognizable from great distances by the strong smell of H_2S in the air as it leaked from generators that produced the gas directly. However, concentrations of H_2S greater than 13 ppm are harmful, and generation of the gas directly in solution is much wiser. This is accomplished by heating an aqueous solution of thioacetamide.

$$\underset{\text{thioacetamide}}{CH_3CSNH_2(aq)} + 2H_2O(l) \xrightarrow{\Delta} \underset{\text{acetate ion}}{CH_3COO^-} + \underset{\substack{\text{ammonium} \\ \text{ion}}}{NH_4^+} + \underset{\text{hydrogen sulfide}}{H_2S(g)}$$

Hydrogen sulfide is a good reducing agent both in the pure state and in aqueous solution. In an excess of oxygen (or air), hydrogen sulfide burns, when ignited, to give sulfur dioxide and water.

$$2H_2S(g) + 3O_2(g) \longrightarrow 2SO_2(g) + 2H_2O(g)$$

If the amount of air is limited, the oxidation product is elemental sulfur.

$$2H_2S(g) + O_2(g) \longrightarrow 2S(s) + 2H_2O(g)$$

Free sulfur is found on the slopes of volcanoes. Hydrogen sulfide issues from the volcano as one component of volcanic gases. When the hot hydrogen sulfide comes into contact with the air, some of it is oxidized to sulfur dioxide. The sulfur dioxide reacts with hydrogen sulfide to produce free sulfur, which drops to the ground.

$$2H_2S(g) + SO_2(g) \longrightarrow 2H_2O(g) + 3S(s)$$

Hydrogen sulfide dissolves in water to the extent of about 0.1 mole per liter and is a weak diprotic acid (called hydrosulfuric acid).

$$H_2S(aq) \rightleftharpoons H^+ + \underset{\substack{hydrosulfide \\ ion}}{HS^-} \qquad K_{a_1} = 1.0 \times 10^{-7}$$

$$HS^- \rightleftharpoons H^+ + \underset{\substack{sulfide \\ ion}}{S^{2-}} \qquad K_{a_2} = 3 \times 10^{-13}$$

b. Metal sulfides. The two types of salts formed by hydrogen sulfide on reaction with aqueous bases are the acid sulfides, or hydrogen sulfides, and sulfides; for example,

$$NaOH(aq) + H_2S(aq) \longrightarrow \underset{\substack{sodium \; hydrogen \; sulfide}}{NaHS(aq)} + H_2O(l)$$

$$2NaOH(aq) + H_2S(aq) \longrightarrow \underset{\substack{sodium \; sulfide}}{Na_2S(aq)} + 2H_2O(l)$$

Many sulfides, which are known for most metals, are only slightly soluble in water (Table 22.11). The only appreciably soluble sulfides are those of the alkali metals and of calcium, strontium, barium, and ammonium ion. They give solutions that are distinctly alkaline as a result of the hydrolysis of the sulfide ion:

$$S^{2-} + H_2O(l) \rightleftharpoons HS^- + OH^- \qquad K_{b_1} = 3 \times 10^{-2}$$

Soluble acid sulfides, the best known of which is NaHS, also hydrolyze, but to a much smaller extent than the normal salts, to give alkaline solutions.

$$HS^- + H_2O(l) \rightleftharpoons H_2S(aq) + OH^- \qquad K_{b_2} = 1.0 \times 10^{-7}$$

The smaller degree of hydrolysis is not unexpected—H_2S is a much stronger acid than the HS^- ion (see the K_a values above).

The addition of strong acids such as hydrochloric acid to saturated solutions of slightly soluble metal sulfides increases the solubility of the sulfides because of reaction between the sulfide ion and the hydrogen ion to give the weakly acidic ion HS^-. This reaction effectively removes sulfide ions from solution, and equilibrium is reestablished to replenish these ions; for example

$$ZnS(s) \rightleftharpoons Zn^{2+} + S^{2-}$$
$$S^{2-} + H^+ \rightleftharpoons HS^-$$

FIGURE 22.22

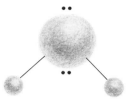

Hydrogen sulfide, H_2S
m.p. $-85.5°C$
b.p. $-60.7°C$

TABLE 22.11
Solubility product constants (K_{sp}) of some metal sulfides

Metal sulfide	K_{sp}
MnS *manganese(II) sulfide*	2.3×10^{-13}
FeS *iron(II) sulfide*	4.2×10^{-17}
ZnS *zinc sulfide*	2×10^{-24}
SnS *tin(II) sulfide*	3×10^{-27}
CdS *cadmium sulfide*	2×10^{-28}
PbS *lead sulfide*	1×10^{-28}
CuS *copper(II) sulfide*	6×10^{-36}
HgS *mercury(II) sulfide*	4×10^{-53}

NITROGEN, PHOSPHORUS,
AND SULFUR [22.16]

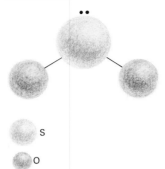

FIGURE 22.23

S

O

Sulfur dioxide, SO_2
m.p. $-75.46°C$
b.p. $-10.02°C$

In practice, treatment of insoluble sulfides with relatively small amounts of hydrogen ion dissolves all but the least soluble (e.g., CuS, HgS).

Solutions of controlled sulfide ion concentration can be obtained by regulating the hydrogen ion concentration of saturated solutions of H_2S. An increase in hydrogen ion concentration (e.g., by addition of HCl) results in a decrease in sulfide ion concentration and, conversely, a decrease in hydrogen ion concentration is accompanied by an increase in sulfide ion concentration. By careful control of the sulfide ion concentration, the precipitation of insoluble metal sulfides can be controlled, a method used in the separation of cations in qualitative analysis.

22.17 Oxidation state +4: sulfur dioxide and related compounds

a. Sulfur dioxide. Sulfur dioxide, SO_2, is a colorless gas which is liquefied at room temperature by only 5 atm pressure. The gas has a characteristic suffocating odor, and it can be very irritating to the eyes and respiratory tract. It is formed in the combustion of sulfur-containing fuels and in the smelting of sulfide ores, and is a primary air pollutant (Section 6.11).

Sulfur dioxide is extremely stable toward heat, practically no decomposition being observed below 2000°C. Sulfur dioxide exists as discrete angular molecules in the gaseous state (Figure 22.23). Only one sulfur–oxygen bond distance (0.1432 nm) is observed experimentally, and this distance is intermediate between that expected for a covalent single bond and a double bond.

Commercially, the preparation of SO_2 is accomplished by burning elemental sulfur and also by heating metal sulfides (e.g., PbS, FeS_2, ZnS) in air.

$$S(s) + O_2(g) \xrightarrow{\Delta} SO_2(g) \qquad \Delta H_f° = -70.94 \text{ kcal } (-296.8 \text{ kJ})$$

$$2ZnS(s) + 3O_2(g) \xrightarrow{\Delta} 2ZnO(s) + 2SO_2(g)$$
$$\Delta H° = -216.3 \text{ kcal } (-905.0 \text{ kJ})$$

In the laboratory, sulfur dioxide may be prepared by reaction between a strong acid (e.g., H_2SO_4) and an alkali metal sulfite (e.g., Na_2SO_3) or a hydrogen sulfite (e.g., $NaHSO_3$).

$$2H^+ + SO_3^{2-} \longrightarrow SO_2(g) + H_2O$$
$$H^+ + HSO_3^- \longrightarrow SO_2(g) + H_2O$$

Sulfur dioxide is a rather strong reducing agent. The most important reaction in which it functions as a reducing agent is its conversion to sulfur trioxide, SO_3, by molecular oxygen, a process that is slow unless suitably catalyzed. This reaction is the first step in the synthesis of sulfuric acid (Section 22.18b).

b. Sulfurous acid and sulfites. Sulfur dioxide dissolves in water to give solutions that contain sulfurous acid, H_2SO_3. The following equilibria are established by sulfurous acid, which is a weak diprotic acid known only in solution.

$$SO_2(g) + H_2O(l) \rightleftharpoons H_2SO_3(aq)$$
$$H_2SO_3(aq) \rightleftharpoons HSO_3^- + H^+ \qquad K_{a_1} = 1.43 \times 10^{-2}$$
$$HSO_3^- \rightleftharpoons SO_3^{2-} + H^+ \qquad K_{a_2} = 5.0 \times 10^{-8}$$

A large amount of the sulfur dioxide does not react with the water, but is simply physically in solution.

Sulfurous acid forms both acid sulfites (also called bisulfites), such as $NaHSO_3$, sodium hydrogen sulfite, and normal sulfites, such as Na_2SO_3, sodium sulfite. The best known and most soluble salts of sulfurous acid are those of the alkali metals, which are formed by reaction between sulfur dioxide and a solution of the alkali metal hydroxide. In essence, the reactions involve neutralization of sulfurous acid.

$$OH^- + H_2SO_3(aq) \longrightarrow \underset{\textit{hydrogen sulfite}}{HSO_3^-} + H_2O(l)$$

$$2OH^- + H_2SO_3(aq) \longrightarrow \underset{\textit{sulfite ion}}{SO_3^{2-}} + 2H_2O(l)$$

An excess of hydroxide ion is required for complete neutralization of the acid and the formation of the normal sulfite ion.

22.18 Oxidation state +6: sulfuric acid and related compounds

a. Properties of sulfuric acid. Pure sulfuric acid is a viscous liquid (density 1.85 g/ml), which begins to boil in the neighborhood of 290°C, with decomposition into sulfur trioxide and water. The molecules have a tetrahedral arrangement of oxygen atoms around the sulfur atom (Figure 22.24).

The concentrated acid ordinarily employed in the laboratory contains approximately 98% H_2SO_4 and is 18M. Acid of this concentration dissolves in water with the evolution of a large quantity of heat and great care must be taken in mixing the two substances. The danger of boiling and splattering is minimized by pouring the acid slowly into the water with constant stirring.

Sulfuric acid is a strong acid. In aqueous solution, it ionizes in two stages:

$$H_2SO_4(aq) \rightleftharpoons H^+ + \underset{\substack{\textit{hydrogen sulfate ion} \\ \textit{(or bisulfate ion)}}}{HSO_4^-}$$

$$HSO_4^- \rightleftharpoons H^+ + \underset{\textit{sulfate ion}}{SO_4^{2-}}$$

In dilute solution, the first step of the ionization is practically complete. The ionization of the hydrogen sulfate or bisulfate ion has an

FIGURE 22.24

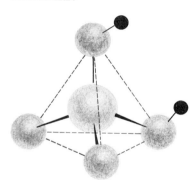

Sulfuric acid, H_2SO_4
m.p. 10.36°C
b.p. ~290°C (dec)

equilibrium constant of 1.0×10^{-2} (pK_a 2.00) at 25°C. Therefore, a liter of a $1M$ solution of sulfuric acid contains 0.99 mole of HSO_4^- but only about 0.01 mole of SO_4^{2-} ion.

The chemical properties of aqueous solutions of sulfuric acid vary considerably with the concentration of the acid. Concentrated solutions are characterized by (1) their great dehydrating power and (2) their tendency to function as moderately strong oxidizing agents, stronger than phosphoric acid, but not as strong as nitric acid.

The dehydrating ability of concentrated sulfuric acid makes it an extremely effective drying agent for wet gases and liquids, provided the gas or liquid does not react with the acid. Sulfuric acid can also remove the elements of water from organic compounds that themselves contain no free water. Table sugar, or sucrose, has the formula $C_{12}H_{22}O_{11}$. Although no water is present in the compound, the hydrogen-to-oxygen ratio is the same as that in water. When concentrated sulfuric acid is added to sugar, these elements are removed in the form of water, and a residue of carbon is left.

$$C_{12}H_{22}O_{11}(s) + 11H_2SO_4(conc) \longrightarrow 12C(s) + 11(H_2SO_4 \cdot H_2O)$$

The same type of reaction accounts for the charring of wood, paper, cotton, and wool and the destruction of skin by the concentrated acid.

The activity of concentrated sulfuric acid as an oxidizing agent is illustrated by its reactions with both metals and nonmetals. The sulfur atom is usually, but not always, reduced from its oxidation state of +6 to +4. Many metals, including those below hydrogen in the electromotive series, are attacked by a solution of hot concentrated sulfuric acid. For example, copper is oxidized to copper(II) sulfate, part of the sulfuric acid being reduced to sulfur dioxide.

$$Cu(s) + 2H_2SO_4(conc) \xrightarrow{\Delta} CuSO_4(aq) + SO_2(g) + 2H_2O(l)$$

Zinc, which stands above copper in the electromotive series and is a stronger reducing agent, reduces concentrated sulfuric acid to elemental sulfur or hydrogen sulfide. Carbon is oxidized by hot concentrated sulfuric acid to carbon dioxide.

$$C(s) + 2H_2SO_4(conc) \longrightarrow CO_2(g) + 2SO_2(g) + 2H_2O(l)$$

In dilute solutions of sulfuric acid it is the hydrogen ion from the ionization of the acid that is the oxidizing agent, not sulfate ion. For example, dilute H_2SO_4 solutions attack metals above hydrogen in the electromotive series with the liberation of hydrogen

$$Zn(s) + 2H^+ \longrightarrow Zn^{2+} + H_2(g)$$

and the metal is converted to its sulfate, in this case $ZnSO_4$. The sulfate ion is just a spectator.

b. Industrial preparation and uses of sulfuric acid; sulfur trioxide. Two processes for the manufacture of sulfuric acid are of interest. The first, by which 95% of the sulfuric acid is made today, is the contact process. The second, the lead chamber process, was discovered in 1746 and was greatly improved during the 1800s, when it was of central importance to the entire chemical industry. In both processes the major step is the catalytic oxidation of sulfur dioxide to gaseous sulfur trioxide (Figure 22.25) by atmospheric oxygen.

$$2SO_2(g) + O_2(g) \rightleftharpoons 2SO_3(g) \qquad \Delta H° = -47.27 \text{ kcal } (-197.8 \text{ kJ})$$

In the contact process the usual catalyst is vanadium(V) oxide, V_2O_5. Impure sulfur dioxide tends to poison the catalyst and decrease its effectiveness, and therefore most of the oxide used in the process is obtained by the combustion of sulfur, which provides a pure product. Since the union of sulfur dioxide and oxygen is exothermic and the reaction reaches a state of equilibrium, it would be predicted on the basis of Le Chatelier's principle (Section 16.7) that low temperatures would favor the formation of sulfur trioxide in high yield. This is true, but the rate at which combination occurs at low temperatures, even in the presence of a catalyst, is too low for the reaction to be economically practical. Therefore, the temperature employed must be high enough to effect a rapid union between sulfur dioxide and oxygen, but not so high that the equilibrium is driven toward decomposition of sulfur trioxide.

Experimental data show that at temperatures between 400° and 450°C, approximately 97% conversion to sulfur trioxide occurs; at 900°C essentially none of the compound is formed. The maximum yield can be obtained in the shortest time by carrying out the reaction in two parts. First SO_2 and O_2 are passed over the catalyst at a temperature at which the rate is high (575°C) to convert about 80% of the SO_2 to SO_3. In another pass over a second portion of catalyst at 450°C, the conversion is raised to 97%.

The sulfur trioxide is not converted to sulfuric acid by direct addition to water. Gaseous sulfur trioxide reacts with water to form a mist of sulfuric acid, which is absorbed only slowly in the liquid water. The sulfur trioxide is actually passed into 98% sulfuric acid to form pyrosulfuric acid, $H_2S_2O_7$, also called disulfuric acid. By addition of water, sulfuric acid of the desired concentration is obtained.

$$\underset{\textit{sulfur trioxide}}{SO_3(g)} + \underset{\textit{sulfuric acid}}{H_2SO_4(l)} \rightleftharpoons \underset{\textit{pyrosulfuric acid}}{H_2S_2O_7(l)}$$

$$H_2S_2O_7(l) + H_2O(l) \longrightarrow 2H_2SO_4(aq)$$

In the older process for sulfuric acid production, sulfur dioxide, atmospheric oxygen, and steam react in the presence of nitric oxide, NO, and nitrogen dioxide, NO_2, to give an aqueous solution containing 62–77% sulfuric acid. The reaction is carried out in lead-lined chambers, and hence its name as the lead chamber process. The overall reactions can be written

$$2NO(g) + O_2(g) \longrightarrow 2NO_2(g)$$
$$SO_2(g) + NO_2(g) \longrightarrow SO_3(g) + NO(g)$$
$$SO_3(g) + H_2O(g) \longrightarrow H_2SO_4(aq)$$

However, the process is much more complex than indicated by these simple reactions. The nitric oxide is a homogeneous catalyst—it

FIGURE 22.25

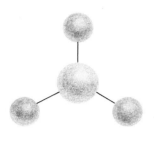

 S

O

Sulfur trioxide, SO_3
m.p. 17°C
b.p. 43°C

FIGURE 22.26

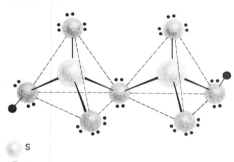

  S

  O

  H

Pyrosulfuric acid, $H_2S_2O_7$
m.p. 35°C

takes part in the reactions but is recovered. One advantage of this process is that it does not require sulfur dioxide of high purity. However, the acid obtained is contaminated with lead nitrate and lead sulfate from corrosion of the chamber walls.

As we have pointed out, sulfuric acid is the number one industrial chemical. It is necessary for the preparation of hundreds of other chemicals and in many, many industrial processes. A random and important few are noted in Table 22.12. The volume of sulfuric acid used by a country is a measure of its state of industrial advancement.

c. Sulfates. Sulfuric acid forms both acid salts, such as $NaHSO_4$, sodium hydrogen sulfate, and normal salts, such as Na_2SO_4, sodium sulfate. Most normal sulfates are easily soluble in water, the exceptions being the barium, strontium, and lead sulfates. The precipitation of barium sulfate when a solution containing Ba^{2+} is added to an acidic aqueous solution is taken as evidence for the presence of sulfate ion in the second solution. The most common hydrogen sulfates are $NaHSO_4$ and $KHSO_4$, and both are very soluble in water. Their water solutions are acidic as a result of ionization of the HSO_4^- ion.

d. Other sulfur(VI) acids and ions. Pyrosulfuric acid, $H_2S_2O_7$, is also known as "oleum" or "fuming sulfuric acid," because it evolves white fumes on contact with moist air. The fumes are caused by the immediate reaction of the escaping SO_3 with water to form ordinary sulfuric acid as a mist. Pyrosulfuric acid is a more powerful oxidizing agent than ordinary sulfuric acid. It is formed by dissolving SO_3 in H_2SO_4. As was true for phosphorus in its +5 oxo acids and salts, each sulfur atom in the +6 oxygen acids and salts is surrounded tetrahedrally by four oxygen atoms (Figure 22.26).

Peroxosulfuric acids contain the peroxo group $-\ddot{O}\!:\!\ddot{O}-$ (as in hydrogen peroxide). The two such acids that have been characterized—peroxomonosulfuric acid, H_2SO_5, and peroxodisulfuric acid, $H_2S_2O_8$—are shown in Figure 22.27.

When an aqueous solution of a metal sulfite is boiled with elemental sulfur, the thiosulfate ion is produced.

$$SO_3^{2-} + S(s) \longrightarrow S_2O_3^{2-}$$
$$\text{\textit{thiosulfate}}$$
$$\text{\textit{ion}}$$

The two sulfur atoms in the thiosulfate ion are not chemically equivalent, and the ion has a tetrahedral configuration that is, in a formal sense, derived from the tetrahedral sulfate ion by replacement of an oxygen atom with a sulfur atom (see Figure 22.24).

The most common thiosulfate is the sodium salt, which is obtained from solution as the pentahydrate, $Na_2S_2O_3 \cdot 5H_2O$. This substance is known as "hypo" and is used in photography (see **An Aside: The Photographic Process**, Chapter 29).

The thiosulfate ion is unstable in acidic solution, decomposing to give free sulfur and sulfurous acid.

$$S_2O_3^{2-} + 2H^+ \longrightarrow S(s) + H_2SO_3(aq)$$

The ion is a fairly strong reducing agent. Strong oxidants, such as chlorine, oxidize thiosulfate to sulfate ion.

$$S_2O_3^{2-} + 4Cl_2(g) + 5H_2O(l) \longrightarrow 2SO_4^{2-} + 10H^+ + 8Cl^-$$

[22.18] NITROGEN,
PHOSPHORUS, AND SULFUR

Moderately strong oxidizing agents, such as iodine, convert $S_2O_3^{2-}$ to $S_2O_6^{2-}$ (tetrathionate ion).

$$2S_2O_3^{2-} + I_2(s) \longrightarrow S_4O_6^{2-} + 2I^-$$

This reaction occurs quantitatively and rapidly, and is useful for the analytical determination of iodine.

The nitrogen, phosphorus, and sulfur cycles

Each element circulates in a cycle governed by the properties of the element and its natural sources and compounds. The primary forces in these natural cycles are the uptake of nutrients by plants and animals and the decay of dead plants and animals, the solution by runoff water of soluble materials, and the return of solids to dry land by the upwelling of mountains from the ocean floor.

The atmosphere is a primary reservoir for carbon; hydrogen and oxygen in the form of water; the carbon oxides; and methane (water cycle, Section 13.6; carbon and oxygen cycles, Section 11.15). The atmosphere is also a major reservoir for nitrogen. Sulfur spends part of its time as gaseous compounds in the atmosphere (hydrogen sulfide and sulfur oxides), but the major sulfur reservoir is as soluble sulfate in the oceans. No gaseous phosphorus compound plays a significant role in the phosphorus cycle, the primary source of phosphorus being phosphate rock.

Oxidation and reduction are important in the nitrogen cycle (Fig-

FIGURE 22.27

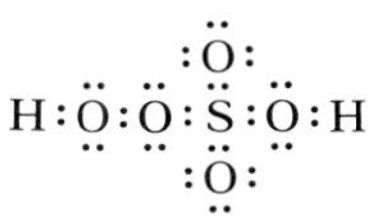

Peroxomonosulfuric acid, H_2SO_5

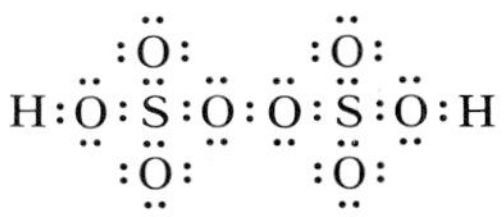

Peroxodisulfuric acid, $H_2S_2O_8$

FIGURE 22.28
The nitrogen cycle.

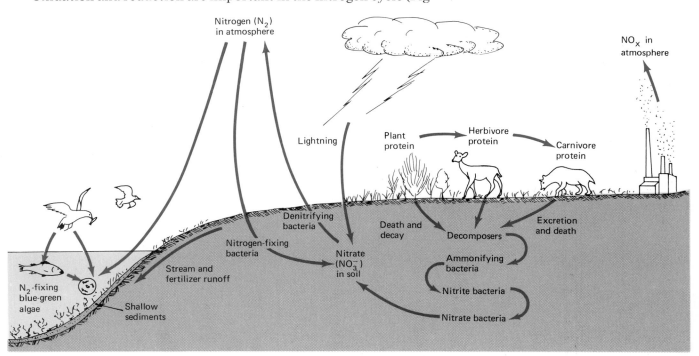

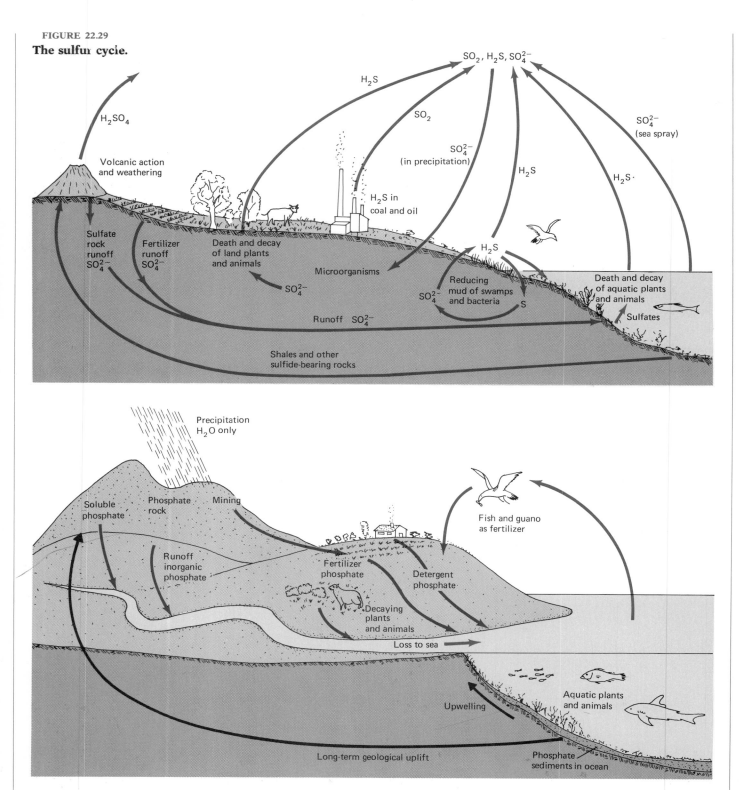

FIGURE 22.29
The sulfur cycle.

FIGURE 22.30
The phosphorus cycle.

690

ure 22.28). In plants and animals, nitrogen is present in the reduced form primarily as ammonium ion or as —NH_2 groups. Dinitrogen is fixed during electrical storms by conversion to NO and NO_2. In the soil oxidized nitrogen is present as nitrate ion. Dinitrogen from the atmosphere is converted by nitrogen-fixing microorganisms to ammonia. Other microorganisms can convert ammonia to nitrites and nitrates—a process called **nitrification.** The conversion of nitrites and nitrates to dinitrogen is called **denitrification.** Denitrification occurs during the decay of organic matter. Mankind's role in the nitrogen cycle is through the industrial fixation of nitrogen and its return to the land and water as fertilizer, and through the formation of nitrogen oxides in the combustion of fossil fuels.

Sulfur is taken up by plants as sulfate ion and from organic compounds during decay to hydrogen sulfide. Sulfur also enters the atmosphere as sulfur dioxide from fossil fuel combustion and as finely divided sulfates from sea spray. Some sulfate is dissolved from minerals and carried along by runoff water. The sulfur cycle is represented in simplified form in Figure 22.29.

The phosphorus cycle comprises only movement from land to the hydrosphere, to plants and animals, and back to the land and hydrosphere. The primary source of phosphorus is phosphate rock, which is not very soluble. Man plays an important role in the phosphorus cycle by mining large quantities of phosphate rock and returning it to the land and water as fertilizer and detergents. Phosphates are deposited in ocean sediment and after many, many years may return to dry land in the upwelling of mountains. The organic phosphate of plants and animals returns to the soil and the ocean as inorganic phosphate. The phosphorus cycle is represented in Figure 22.30.

INORGANIC QUALITATIVE ANALYSIS: ANIONS

In 1840 Karl Remegius Fresenius, a 22-year-old German, went to Bonn to study chemistry. His early apprenticeship as a pharmacist had aroused an intense interest in chemical analysis—the determination of the composition of substances both qualitatively and quantitatively.

While in Bonn, Fresenius studied the chemical literature on analysis, which was extensive but filled with unrelated procedures. Just for practice, or so he thought, he wrote an *Introduction to Qualitative Analysis* in which he presented a systematic method for identifying qualitatively the anions and the cations present in a solid or a solution.

On the insistence of his professor, Fresenius published his work. The qualitative analysis scheme that he had devised turned out to be very successful in practice and it filled a great need. By the time Fresenius died in 1897 his book had seen 17 editions in German and numerous editions in other languages, including 8 in English. Most of the qualitative analysis schemes in use today are based substantially on Fresenius' work.

With the publication of his later textbook on quantitative analysis and his founding and editing of a journal for analytical chemistry, Fresenius earned the unofficial title of Father of Analytical Chemistry.

22.19 An overview of inorganic qualitative analysis

The goal of inorganic qualitative analysis is to identify the cations and the anions present in a substance of unknown composition. Solid substances are put into solution for most parts of the qualitative analysis scheme. As such, qualitative analysis is based largely on characteristic reactions of ions in solution. The analyses for anions and for cations are carried out separately and in fundamentally different ways.

Cation analysis is highly organized. Cations are first separated into groups by precipitating the ions in one group while leaving the ions of other groups in solution. The ions in each group are further separated by carefully chosen reactions under carefully controlled conditions. Finally, each ion is positively identified by a reaction that is characteristic of that ion.

Anion analysis, on the other hand, does not require separation of the ions. First, preliminary tests that indicate which anions or groups of anions might be present and which might be absent are carried out. Distinctive reactions for each individual anion then allow verification of the presence or absence of specific anions. These distinctive reactions are not done on separated anions or groups, but on the solution (or in some cases the solid) containing all of the anions.

In this chapter and earlier chapters of this book (Sections 6.12 to 6.14; Chapter 11; Chapter 20) we have systematically examined the chemistry of all of the nonmetals except for carbon, which is discussed in the next chapter (Section 18.9 covered the H_2CO_3/HCO_3^- buffer system and Section 18.13 covered the slightly soluble metal carbonates). Where possible, our approach to the nonmetals has been to consider the properties of the elements and their compounds within the framework of the periodic table. In this way, similarities and differences in properties can frequently be explained on the basis of electronic configurations. In later chapters (Chapters 26, 27, and 29) we bring this same approach to the metals and the semiconducting elements.

However, chemical reactions often show similarities and differences that cannot be directly related to electronic configurations. Ions of elements in the same periodic table group may undergo different reactions under similar conditions. For example, Sn^{2+} is oxidized to Sn^{4+} by an acidic hydrogen peroxide solution, but Pb^{2+}, an ion of the element just below tin in the periodic table, is not oxidized under these conditions. Also, zinc sulfide (ZnS) dissolves in an acidic solution, but the sulfide of mercury (HgS), an element in the same group, does not dissolve in acidic solution.

It turns out that the equilibrium properties of ions—strengths as Brønsted-Lowry acids and bases, oxidizing and reducing ability, solubilities of salts, and tendencies to form complex ions—are sometimes more important than electronic configurations in determining the properties of ions in solution. The reactions of inorganic qualitative analysis provide excellent examples of how the principles of equilibria can be put to work to obtain desired results from chemical reactions.

Today we have an impressive array of instruments available for identifying the components of just about any type of mixture. However, "qual," as it is called, still remains of practical use in many circumstances. If the reagents are available, solution reactions often give the answer more quickly than do instrumental methods.

In addition, qualitative analysis maintains an important position in the study of general and inorganic chemistry. The laboratory work provides practice in observing chemical reactions and interpreting what you see in terms of theory. You will have to exercise judgment and make decisions when what you see does not correspond to what you expected, when you make a mistake, or when the experimental evidence is contradictory. The practical experience with equilibria and the behavior of ions in solution is a valuable addition to what we learn from our study of the elements and their compounds based on the periodic table.

22.20 Anion analysis

The purpose of qualitative analysis in the study of general chemistry is, as we have said, to impart valuable knowledge about the properties of ions in solution and to give practical experience in observing chemical reactions and interpreting what you see. Qualitative analysis *can* be carried out in a cookbook fashion, merely by carefully following the directions given in the book. Although that approach may often lead to the correct answer, it is dull and time consuming. However, if you use every clue, use your imagination to draw conclusions, and view qualitative analysis as a challenge, you will save much time and will have more fun than someone who blindly follows directions. You will also learn more.

Analysis for anions and for cations, the two parts of inorganic qualitative analysis, are carried out separately. Either part may be attacked first, but there is an advantage to analyzing for the anions first. Anion analysis is simpler, since there are fewer commonly encountered anions than cations. In anion analysis, the value of careful observation and the need to intelligently interpret what you see become obvious.

In studying an unknown, a visual examination should come first. If the unknown is a solution, what color is it? Color is important, because some inorganic ions reveal their identities through color alone. You must be careful in drawing a conclusion too quickly, however, for the color of mixtures is sometimes deceiving. For example, a solution containing pink Co^{2+} and green Ni^{2+} may be almost colorless.

If the unknown is a solid, the color is also important, but it does not necessarily indicate the color of the individual ions. Thus, Pb^{2+} and I^- are both colorless, but they combine to give the bright yellow PbI_2. It is a good idea to examine a solid unknown carefully, perhaps even under a magnifying glass. If the solid is a mixture, you may be able to see individual particles of the different substances. All such clues are helpful.

Anion analysis begins with several preliminary tests that indicate the presence or absence of individual anions or groups of anions that have similar properties. With this information at hand, the analysis proceeds to specific tests that confirm the presence of individual anions. Physical separation of the anions is generally not necessary, because many of the specific tests allow detection of one anion in the presence of the others. However, when interference by one anion in detecting another is a possibility, the interfering ion must be removed before testing for the other.

Some of the preliminary tests and some of the specific tests can be carried out on solids. Eventually, however, the unknown must be put into solution. Table 18.4 gives some general rules about the solubilities of ionic compounds. You should be familiar with these rules because they can be helpful in deciding on the presence or absence of certain ions. Insolubility of a solid unknown in water will not eliminate any ions, but it will rule out certain combinations of anions and cations. For example, if a water-soluble unknown is found to contain carbonate ion (CO_3^{2-}), the only common inorganic cations that can be present are Na^+, K^+, and NH_4^+, because the carbonates of all the others are insoluble in water.

If the unknown is insoluble in water, it must be treated with a chemical reagent to convert it to soluble substances. Some anions are not stable in acidic solution, while other ions react with each other in the presence of acid. Consequently, anion analysis is usually performed on alkaline solutions. To prepare an insoluble solid for anion analysis, the solid is boiled with a saturated sodium carbonate solution. This treatment converts the anions to soluble sodium salts and leaves the cations as insoluble carbonates or their hydrolysis products. For example,

$$MA(s) + CO_3{}^{2-} \rightleftharpoons MCO_3(s) + A^{2-}$$

or

$$MA(s) + CO_3{}^{2-} + H_2O(l) \rightleftharpoons M(OH)_2(s) + CO_2(g) + A^{2-}$$

The sodium carbonate treatment is also desirable for mixtures which contain certain heavy metal cations that interfere with some of the anion tests.

Our discussion will include analysis for 11 of the most common anions: sulfide (S^{2-}), sulfite ($SO_3{}^{2-}$), carbonate ($CO_3{}^{2-}$), nitrite ($NO_2{}^-$), iodide (I^-), bromide (Br^-), chloride (Cl^-), phosphate ($PO_4{}^{3-}$), chromate ($CrO_4{}^{2-}$), nitrate ($NO_3{}^-$), and sulfate ($SO_4{}^{2-}$). You have already encountered all of these anions, except for the chromate ion (discussed in Section 29.9), in the discussion of nonmetals (sulfides, Section 22.16; sulfites, Section 22.17; carbonates, Sections 18.9, 23.3; nitrites, Section 22.7; iodide, chloride, and bromide, Section 20.6; phosphate, Section 22.13; nitrate, Section 22.8; and sulfate, Section 22.18).

22.21 Properties of the anions

Qualitative analysis relies heavily on differences in equilibria to separate and identify ions that are similar in other ways. Acid-base equilibria, heterogeneous equilibria, redox equilibria, and complex ion equilibria are all ingeniously employed in the qualitative analysis scheme. An extensive review of these equilibria with emphasis on how they are used in cation analysis is given in Chapter 30. Here we briefly go over some of the pertinent properties of the anions included in our anion analysis scheme. (These anions are listed in Table 22.13.)

Acid-base behavior. Water-soluble salts that contain strongly basic anions in combination with weakly acidic cations give alkaline solutions. Of the anions under consideration, three—S^{2-}, $PO_4{}^{3-}$, and $CO_3{}^{2-}$—are strong Bronsted-Lowry bases. Salts of these anions with cations such as Na^+ react with water as follows to give quite alkaline solutions:

$$S^{2-} + H_2O(l) \rightleftharpoons HS^- + OH^-$$
$$CO_3{}^{2-} + H_2O(l) \rightleftharpoons HCO_3{}^- + OH^-$$
$$PO_4{}^{3-} + H_2O(l) \rightleftharpoons HPO_4{}^{2-} + OH^-$$

Four of our anions—$SO_3{}^{2-}$, $CrO_4{}^{2-}$, $NO_2{}^-$, and $SO_4{}^{2-}$—are relatively weaker bases, as shown by the K_b's listed in Table 22.14. Their salts with weakly acidic cations give less alkaline solutions than do the S^{2-}, $CO_3{}^{2-}$, and $PO_4{}^{3-}$ ions.

TABLE 22.13
Anions for analysis

Sulfide	S^{2-}
Sulfite	$SO_3{}^{2-}$
Carbonate	$CO_3{}^{2-}$
ᵛNitrite	$NO_2{}^-$
Iodide	I^-
Bromide	Br^-
Chloride	Cl^-
Phosphate	$PO_4{}^{3-}$
ᵛChromate	$CrO_4{}^{2-}$
Nitrate	$NO_3{}^-$
Sulfate	$SO_4{}^{2-}$

TABLE 22.14
K_b *values of the anions*

S^{2-}	3×10^{-2}
$PO_4{}^{3-}$	1×10^{-2}
$CO_3{}^{2-}$	2.1×10^{-4}
$SO_3{}^{2-}$	2.0×10^{-7}
$CrO_4{}^{2-}$	3.1×10^{-8}
$NO_2{}^-$	1.4×10^{-11}
$SO_4{}^{2-}$	1.0×10^{-12}
$NO_3{}^-$	5×10^{-17}
Br^-	1×10^{-23}
Cl^-	3×10^{-23}
I^-	3×10^{-24}

The remaining anions in our scheme are very weak bases (Table 22.14). Essentially these anions do not react with water at all, and thus have no effect on the pH of the solution.

Three of the anions yield thermally unstable protonic acids that decompose to evolve gases when their solutions are acidified with a dilute, strong, nonoxidizing acid such as HCl or $HClO_4$. In some cases warming these acidified solutions is necessary to ensure evolution of the gas. Writing the equations in simplified form, with H^+ instead of H_3O^+, as we shall do throughout this section on anion analysis,

$$SO_3^{2-} + 2H^+ \longrightarrow \underset{\substack{sulfurous \\ acid}}{H_2SO_3} \overset{\Delta}{\longrightarrow} \underset{colorless}{SO_2(g)} + H_2O(l)$$

$$CO_3^{2-} + 2H^+ \longrightarrow \underset{\substack{carbonic \\ acid}}{H_2CO_3} \overset{\Delta}{\longrightarrow} \underset{colorless}{CO_2(g)} + H_2O(l)$$

$$2NO_2^- + 2H^+ \longrightarrow \underset{\substack{nitrous \\ acid}}{2HNO_2} \overset{\Delta}{\longrightarrow} \underset{colorless}{NO(g)} + \underset{brown}{NO_2(g)} + H_2O(l)$$

The sulfide ion also yields a gaseous product under the same acidic conditions:

$$S^{2-} + 2H^+ \longrightarrow \underset{colorless}{H_2S(g)}$$

The CrO_4^{2-} ion, the only colored anion in our list, is yellow in alkaline solution, but changes to the orange $Cr_2O_7^{2-}$ in acidic solution (Section 29.9).

$$\underset{\substack{chromate\ ion \\ (yellow)}}{CrO_4^{2-}} \underset{OH^-}{\overset{H^+}{\rightleftharpoons}} HCrO_4^- \underset{OH^-}{\overset{H^+}{\rightleftharpoons}} \underset{\substack{dichromate\ ion \\ (orange)}}{Cr_2O_7^{2-}} + H_2O(l)$$

Redox properties. The group of anions under consideration includes oxidizing agents, reducing agents, and two ions whose redox behavior depends upon the conditions.

The NO_3^- and CrO_4^{2-} ions are quite strong oxidizing agents in acidic solution (Sections 22.8 and 29.10). The I^-, S^{2-}, and SO_3^{2-} ions are reducing agents in acidic solution.

$$I_2(s) + 2e^- \rightleftharpoons 2I^- \qquad\qquad E° = 0.535 \text{ V}$$
$$S(s) + 2H^+ + 2e^- \rightleftharpoons H_2S(aq) \qquad\qquad E° = 0.141 \text{ V}$$
$$SO_4^{2-} + 4H^+ + 2e^- \rightleftharpoons H_2SO_3(aq) + H_2O(l) \qquad E° = 0.20 \text{ V}$$

The redox properties of the SO_4^{2-} ion vary with conditions. In dilute solution this ion shows virtually no oxidizing ability. However, in highly acidic concentrated solutions, the SO_4^{2-} ion is a moderately strong oxidizing agent. (The oxidizing ability of concentrated sulfuric acid is made use of in one of the preliminary tests in anion analysis; Section 22.22.)

The nitrite ion, NO_2^-, can be either a strong oxidizing agent or a weak reducing agent in acidic solution. Only such strong oxidizing agents as permanganate ion (MnO_4^-) or chlorine (Cl_2) (see Table 19.6) are capable of oxidizing nitrite ion in acidic solution.

Heterogeneous equilibria. Precipitation reactions provide valuable information in the analysis for anions. In our scheme such reactions of several of the anions with barium ion, Ba^{2+}, are used in specific tests. The reactions of the anions with silver ion, Ag^+, are part of one of the preliminary tests.

Sulfite, carbonate, chromate, phosphate, and sulfate ions give barium salts that are only slightly soluble in water, as shown by their small K_{sp} values.

Salt	K_{sp}
$BaSO_3$	1.0×10^{-8}
$BaCO_3$	2.0×10^{-9}
$BaSO_4$	1.7×10^{-10}
$BaCrO_4$	8.5×10^{-11}
$Ba_3(PO_4)_2$	3×10^{-23}

Of these salts, only barium sulfate, $BaSO_4$, can be precipitated from solutions that have been made acidic with dilute strong acid. In such solutions the other anions form weak conjugate acids, with ionization constants as follows:

Conjugate acid	K_a
HSO_3^-	5.0×10^{-8}
HCO_3^-	4.8×10^{-11}
$HCrO_4^-$	3.2×10^{-7}
HPO_4^{2-}	1×10^{-12}

Precipitation of an ionic compound from solution begins (as explained in Section 18.12) when the ion product exceeds the K_{sp} value. In the presence of the only slightly ionized conjugate acids, the concentration of the anion is not sufficiently high for precipitation to occur. For example, the ionization of HPO_4^{2-}

$$HPO_4^{2-} \rightleftharpoons H^+ + PO_4^{3-} \qquad K_a = 1 \times 10^{-12}$$

does not provide enough PO_4^{3-} ion for the following precipitation to occur:

$$3Ba^{2+} + 2PO_4^{3-} \rightleftharpoons Ba_3(PO_4)_2(s) \qquad K_{sp} = 3.2 \times 10^{-23}$$

As a further obstacle to precipitation, the carbonate and sulfite ions decompose to give CO_2 and SO_2 when the solution has a sufficiently high H^+ ion concentration.

In neutral solution all of the anions except NO_3^- give precipitates with silver ion. Two anions, NO_2^- and SO_4^{2-}, form precipitates with relatively large values for K_{sp}.

$$AgNO_2(s) \rightleftharpoons Ag^+ + NO_2^- \qquad K_{sp} = 6.0 \times 10^{-4}$$
$$Ag_2SO_4(s) \rightleftharpoons 2Ag^+ + SO_4^{2-} \qquad K_{sp} = 1.5 \times 10^{-5}$$

Ion concentrations must be quite high before the ion products can exceed these large K_{sp} values, so that silver sulfate and silver nitrite will precipitate.

One of the silver salts, Ag_2SO_3, changes from white to black when heated, as the SO_3^{2-} ion reduces the silver ion to elemental silver. Several silver salts are colored: Ag_2CO_3, pale yellow; $AgNO_2$, pale yellow; Ag_2S, black; $AgBr$, cream; AgI, pale yellow; Ag_2CrO_4, brown-red; Ag_3PO_4, yellow.

All the silver salts except Ag_2S, AgI, $AgBr$, and $AgCl$ dissolve in strongly acidic solution (dilute $HClO_4$ is used in the analysis). The anions of these acid-soluble silver salts are either destroyed ($NO_2^- \rightarrow NO + NO_2$; $CO_3^{2-} \rightarrow CO_2$; $SO_3^{2-} \rightarrow SO_2$) or tied up in the form of their weak conjugate acids. Whichever occurs, not enough anions remain in solution to allow the ion product to exceed the K_{sp} and lead to precipitation.

We might expect that Ag_2S would dissolve in acidic solution, since the conjugate acids of the sulfide ion, HS^- and H_2S, are exceedingly weak acids. However, Ag_2S is so slightly soluble ($K_{sp} = 7.1 \times 10^{-50}$) that even these weakly acidic species supply enough S^{2-} ion to prevent the dissolution process. Silver sulfide can be dissolved only if S^{2-} is oxidized and thus effectively removed.

$$3Ag_2S(s) + 8H^+ + 2NO_3^- \longrightarrow 3S(s) + 2NO(g) + 6Ag^+ + 4H_2O(l)$$

Of the silver halides, one, $AgCl$, is soluble in aqueous ammonia as a result of the reaction

$$AgCl(s) + 2NH_3(aq) \rightleftharpoons [Ag(NH_3)_2]^+ + Cl^-$$

The complex ion, $[Ag(NH_3)_2]^+$, is only slightly dissociated into Ag^+ and NH_3 (K_d, 6.2×10^{-8}; see Table 28.5), and the equilibrium concentration of silver ion is reduced to the point at which $[Ag^+][Cl^-] < K_{sp}$. The bromide and the iodide are much less soluble than the chloride (K_{sp} values: $AgCl$, 1.8×10^{-10}; $AgBr$, 4.9×10^{-13}; AgI, 8.3×10^{-17}), and complexation with dilute aqueous ammonia ($6M$) does not remove sufficient Ag^+ for dissolution to occur. ($AgBr$ will dissolve in a concentrated ammonia solution.)

22.22 The preliminary tests

Separate preliminary tests detect the presence of anions with oxidizing or reducing properties, and of anions that fall into four different groups based on their reactions with *dilute* perchloric acid ($HClO_4$) and silver ion (Ag^+). Additional preliminary information is obtained from the action of concentrated sulfuric acid on the sample of unknown composition. In most cases, a positive result for a test consists of clearly observable phenomena such as a color change, the evo-

$$S^{2-}, SO_3^{2-}, CO_3^{2-}, NO_2^-, I^-, Br^-, Cl^-, PO_4^{3-}, CrO_4^{2-} \text{ or } Cr_2O_7^{2-}, NO_3^-, SO_4^{2-}$$

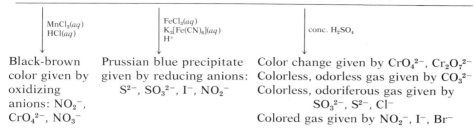

$MnCl_2(aq)$ $HCl(aq)$	$FeCl_3(aq)$ $K_3[Fe(CN)_6](aq)$ H^+	conc. H_2SO_4
Black-brown color given by oxidizing anions: NO_2^-, CrO_4^{2-}, NO_3^-	Prussian blue precipitate given by reducing anions: $S^{2-}, SO_3^{2-}, I^-, NO_2^-$	Color change given by CrO_4^{2-}, $Cr_2O_7^{2-}$ Colorless, odorless gas given by CO_3^{2-} Colorless, odoriferous gas given by SO_3^{2-}, S^{2-}, Cl^- Colored gas given by NO_2^-, I^-, Br^-

FIGURE 22.31
Flow chart of preliminary tests for anions.

lution of a gas, or the appearance of a precipitate. A flow chart for the preliminary tests for anions is given in Figure 22.31. The flow chart shows the separation of the ions, the reagents, and the observed outcome of the test. Cation flow charts are discussed in Chapter 31.

Detection of the presence of oxidizing anions. The development of a brown to black color when a few drops of an unknown solution are added to a solution of manganese(II) chloride ($MnCl_2$) in concentrated hydrochloric acid indicates the presence of oxidizing anions. Possibilities from our list of anions are NO_2^-, CrO_4^{2-}, and NO_3^- (Table 22.15). The manganese(II) is oxidized to the +3 state and the color change is due to the formation of complex ions such as $[MnCl_5]^{2-}$.

If the solution tested contains an oxidizing anion, reducing anions are probably absent, especially if the initial unknown solution is acidic. Oxidizing agents and reducing agents coexist in neutral or alkaline solutions, but not in strongly acidic solutions.

Detection of the presence of reducing anions. The appearance of a dark blue suspension or precipitate when the unknown solution is added to a solution containing $FeCl_3$, $K_3[Fe(CN)_6]$, and dilute HCl indicates the presence of reducing anions, which might be S^{2-}, SO_3^{2-}, I^-, or NO_2^- (Table 22.15). The blue product is Prussian blue, $KFe[Fe(CN)_6]$, a material containing both iron(II) and iron(III). The reaction with iodide ion, for example, is

$$2K^+ + 2Fe^{3+} + 2[Fe(CN)_6]^{3-} + 2I^- \longrightarrow 2KFe[Fe(CN)_6](s) + I_2(aq)$$
Prussian blue

Detection of anion groups. The characteristic acid-base and heterogeneous equilibria reactions of the anions allow classification of the ions into groups based on their behaviors with perchloric acid and silver ion. The reactions observed and the ions in each group are listed in Table 22.16, and a flow chart is given in Figure 22.32.

The classification of anions into groups is designed only to give preliminary information about the presence or absence of individual ions. It is *not* designed as a scheme of separation. The group tests should be done in the following order:

1. Add 6*M* perchloric acid to the sample, warm, and note any gases that are released.

2. Cool the acidic solution from step 1, add silver nitrate solution, and note the color of any precipitate that forms.

TABLE 22.15

Oxidizing and reducing anions

Oxidizing anions
NO_2^-
CrO_4^{2-}
NO_3^-

Reducing anions
S^{2-}
SO_3^{2-}
I^-
NO_2^-

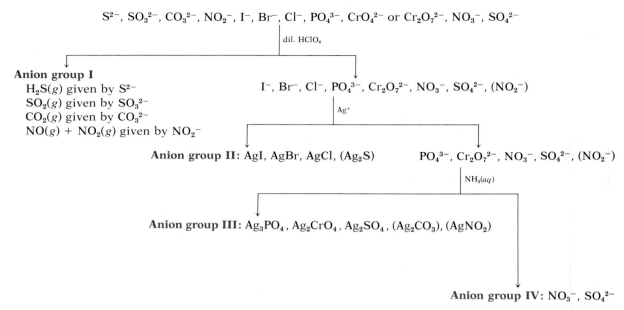

S^{2-}, SO_3^{2-}, CO_3^{2-}, NO_2^-, I^-, Br^-, Cl^-, PO_4^{3-}, CrO_4^{2-} or $Cr_2O_7^{2-}$, NO_3^-, SO_4^{2-}

dil. $HClO_4$

Anion group I
$H_2S(g)$ given by S^{2-}
$SO_2(g)$ given by SO_3^{2-}
$CO_2(g)$ given by CO_3^{2-}
$NO(g) + NO_2(g)$ given by NO_2^-

I^-, Br^-, Cl^-, PO_4^{3-}, $Cr_2O_7^{2-}$, NO_3^-, SO_4^{2-}, (NO_2^-)

Ag^+

Anion group II: AgI, AgBr, AgCl, (Ag$_2$S)

PO_4^{3-}, $Cr_2O_7^{2-}$, NO_3^-, SO_4^{2-}, (NO_2^-)

$NH_3(aq)$

Anion group III: Ag_3PO_4, Ag_2CrO_4, Ag_2SO_4, (Ag_2CO_3), $(AgNO_2)$

Anion group IV: NO_3^-, SO_4^{2-}

FIGURE 22.32
Flow chart for classifying the anions into groups. *The precipitates shown in parentheses will not appear if dilute HClO$_4$ has been added to the solution first. If the NO$_2^-$ concentration is relatively high, NO$_2^-$ will appear in Group III as well as in Group I.*

3. Remove any precipitate formed in step 2, make the solution just alkaline with 6M ammonia, add more silver nitrate if necessary, and note the color of any precipitate that forms.

This sequence is necessary so that ions from one group do not interfere with ions from another group. For example, carbonate ion will precipitate in Group III as silver carbonate if the first test carried out is the addition of silver nitrate to a neutral solution of the sample. Anions that might interfere in this way are shown in brackets in Table 22.16.

Behavior toward concentrated sulfuric acid. The use of concentrated sulfuric acid (18M) in anion analysis depends partly upon its ability to function as a strong oxidizing agent and partly upon its acid behavior. Table 22.17 shows the observations that might be made for the addition of H$_2$SO$_4$ to an unknown, and their corresponding interpretations. The test with concentrated sulfuric acid must be performed on a solid—either a solid unknown or the solid remaining after an unknown solution is evaporated to dryness.

If the unknown is a mixture of salts, the results of this test may not be easy to interpret, because the gases that form may mask each other. Also, insoluble salts (e.g., the silver halides) and combinations of metal and nonmetal that have considerable covalent character (e.g., CdI$_2$, HgCl$_2$) may react only slightly with the acid.

EXAMPLE 22.1

■ What conclusion regarding the nature of the anion or anions present in a water-soluble substance may be drawn from each of the following observations?
(a) The substance is a white solid.

(b) Treatment of the solid with cold concentrated sulfuric acid gives no apparent change.

(c) Addition of dilute perchloric acid and silver nitrate to a solution of the substance gives no precipitate.

(d) When a few drops of a solution of the solid are added to a solution of $MnCl_2$ in concentrated hydrochloric acid, no color change occurs.

Observation (a) shows that the only colored ion among those being considered, namely chromate, CrO_4^{2-}, is absent. The fact that no change occurs when cold concentrated sulfuric acid is added (b) eliminates those ions that give gases with that reagent: CO_3^{2-}, S^{2-}, SO_3^{2-}, Cl^-, NO_2^-, I^-, and Br^-. This leaves the anions PO_4^{3-}, NO_3^-, or SO_4^{2-} as possibly present. Observation (c) does not eliminate PO_4^{3-}, NO_3^-, or SO_4^{2-}, for none of these ions gives an insoluble silver salt in acidic solution. Observation (d) indicates that the substance does not contain an oxidizing anion, and so NO_3^- is eliminated. We now know that the substance may contain either PO_4^{3-} or SO_4^{2-}. We cannot decide which of these two anions is present without specific tests. ∎

22.23 The specific tests

Specific tests are described here for each of the 11 anions on our list. The preliminary tests may eliminate some anions from consider-

TABLE 22.17

Behavior of anions with concentrated sulfuric acid[a]

Observation	Interpretation		
	Cold		Hot
No apparent change	PO_4^{3-}, NO_3^-, SO_4^{2-}		PO_4^{3-}, SO_4^{2-}
Color change	CrO_4^{2-} (yellow) ⟶ $Cr_2O_7^{2-}$ (orange) $Cr_2O_7^{2-}$ ⟶ CrO_3 (red)[b]		Same
Colorless, odorless gas evolved	CO_3^{2-} ⟶ CO_2		Same
Colorless, odoriferous gas evolved	S^{2-} ⟶ H_2S SO_3^{2-} ⟶ SO_2 Cl^- ⟶ HCl		Same
Colored gas evolved	NO_2^- ⟶ NO_2 (brown) I^- ⟶ I_2 (violet) Br^- ⟶ Br_2 (red-brown)		Same Also, NO_3^- ⟶ NO_2 (if the vapors are heated)

[a] This table is adapted from T. Moeller and R. O'Connor, *Ions in Aqueous Systems*, McGraw-Hill Book Company, New York, 1972, p. 215.
[b] The red CrO_3 is seldom formed.

TABLE 22.16

The anion groups

Brackets indicate additional species that would appear if a separate sample of the solution were used for each group test. Note that S^{2-}, SO_3^{2-}, CO_3^{2-}, and NO_2^- are not stable in $HClO_4$ solution when the solution is warmed.

Group I

Anions decomposed in strongly acidic solution (dilute $HClO_4$) with the evolution, when the solution is warmed, of gases having characteristic properties.

S^{2-} ⟶ $H_2S(g)$ (*colorless; odor of decayed eggs*)

SO_3^{2-} ⟶ $SO_2(g)$ (*colorless; odor of burning sulfur*)

CO_3^{2-} ⟶ $CO_2(g)$ (*colorless; odorless*)

NO_2^- ⟶ $NO(g) + NO_2(g)$ (*brown; sharp odor*)

Group II

Anions stable in dilute $HClO_4$ and precipitated from acid solution as silver salts.

[S^{2-} ⟶ Ag_2S (*black*)]

I^- ⟶ AgI (*pale yellow*)

Br^- ⟶ $AgBr$ (*cream*)

Cl^- ⟶ $AgCl$ (*white*)

Group III

Anions stable in dilute $HClO_4$, but precipitated as silver salts only when the solution is neutralized.

[CO_3^{2-} ⟶ Ag_2CO_3 (*pale yellow*)]

[NO_2^- ⟶ $AgNO_2$ (*pale yellow*; NO_2^- *concentration must be relatively high*)]

PO_4^{3-} ⟶ Ag_3PO_4 (*yellow*)

CrO_4^{2-} ⟶ Ag_2CrO_4 (*brownish red*)

SO_4^{2-} ⟶ Ag_2SO_4 (*white*; SO_4^{2-} *concentration must be relatively high*)

Group IV

Anions stable in dilute $HClO_4$, but give soluble silver salts in both acidic and neutral media.

NO_3^-

SO_4^{2-}

ation. The specific tests for those anions suspected of being present are then performed.

Sulfide ion. The test for sulfide ion is made directly on solid samples provided they are water soluble. If the sample is in solution, the solution is evaporated to dryness before testing. The addition of dilute hydrochloric acid results in the evolution of hydrogen sulfide if the sample contains S^{2-}.

$$MS(s) + 2H^+ \longrightarrow M^{2+} + H_2S(g)$$

The H_2S is detected by its reaction with lead acetate, $Pb(CH_3COO)_2$. This test is usually carried out by exposing a paper impregnated with lead acetate to the evolving gas.

$$Pb^{2+} + H_2S(g) \longrightarrow \underset{\substack{black\ or\\silvery}}{PbS(s)} + 2H^+$$

Sulfite ion. Acidification of the test solution, or a solid sample, with dilute sulfuric acid gives sulfur dioxide if sulfite ion is present.

$$SO_3{}^{2-} + 2H^+ \longrightarrow SO_2(g) + H_2O(l)$$

Passage of the sulfur dioxide gas into a solution containing dilute nitric acid (HNO_3), barium chloride ($BaCl_2$), and a small amount of potassium permanganate ($KMnO_4$) gives a white precipitate of barium sulfate ($BaSO_4$). The permanganate, a strong oxidizing agent, converts the +4 sulfur in SO_2 to the +6 state in $SO_4{}^{2-}$. The purple $KMnO_4$ is reduced to Mn^{2+}, which is pale pink or practically colorless in dilute solution. The sulfate formed is then converted to $BaSO_4$.

$$5SO_2(g) + 2\underset{purple}{MnO_4{}^-} + 2H_2O(l) \longrightarrow 5SO_4{}^{2-} + 2\underset{\substack{pale\\pink}}{Mn^{2+}} + 4H^+$$

$$SO_4{}^{2-} + Ba^{2+} \longrightarrow \underset{white}{BaSO_4(s)}$$

Carbonate ion. If the sample to be tested is in solution, the solution is heated until it evaporates *just* to dryness. (Some metal carbonates decompose on strong heating.) The residue, or the original solid unknown, is then treated with a small amount of zinc, dilute hydrogen peroxide (H_2O_2), and dilute sulfuric acid. The resulting mixture is warmed and any gas evolved is passed into a solution of barium hydroxide ($Ba(OH)_2$). The formation of a white precipitate of barium carbonate ($BaCO_3$) shows that the sample contains $CO_3{}^{2-}$.

$$CO_3{}^{2-} + 2H^+ \longrightarrow CO_2(g) + H_2O(l)$$

$$CO_2(g) + Ba^{2+} + 2OH^- \longrightarrow \underset{white}{BaCO_3(s)} + H_2O(l)$$

It is obvious that a solution prepared by sodium carbonate treatment of a sample cannot be used in this test to determine the presence of carbonate in the original unknown.

The hydrogen peroxide is added to oxidize any SO_3^{2-} that may be present to SO_4^{2-}, which is unaffected by the sulfuric acid.

$$SO_3^{2-} + H_2O_2(aq) \longrightarrow SO_4^{2-} + H_2O(l)$$

Without the H_2O_2 addition, the SO_2 liberated upon acidification of SO_3^{2-} would also give a white precipitate, $BaSO_3$, with $Ba(OH)_2$. The zinc reacts with H_2SO_4 to generate hydrogen, and this gas helps sweep the CO_2 into the $Ba(OH)_2$ solution.

Nitrite ion. The test for nitrite ion makes use of this ion's ability to function as an oxidizing agent. The test solution is made very slightly acidic by addition of either dilute sulfuric acid or acetic acid. A few drops of freshly prepared iron(II) sulfate ($FeSO_4$) solution are then added. Part of the iron(II) is oxidized to iron(III) and the NO_2^- is reduced to NO. A reaction between this NO and some of the excess iron(II) in the solution yields a complex ion, $Fe(NO)^{2+}$, recognizable by its brown color.

$$\underset{oxidant}{NO_2^-} + \underset{reductant}{Fe^{2+}} + 2H^+ \longrightarrow Fe^{3+} + NO(aq) + H_2O(l)$$

$$Fe^{2+} + NO(aq) \longrightarrow \underset{brown}{[Fe(NO)]^{2+}}$$

The NO_3^- ion behaves similarly, but only at high hydrogen ion concentrations.

Iodide, bromide, and chloride ions. If two or more of these anions are present, the identification of each requires considerable care. The methods used for the detection of each in the presence of the others are based on the relative redox properties of the halogens (Table 20.3). The sequence of reactions is made possible because iodide ion is more easily oxidized than bromide ion, and both are more easily oxidized than chloride ion.

The solution to be tested is first made slightly acidic with dilute hydrochloric acid; then a small amount of carbon tetrachloride (CCl_4) is added. This is followed by the addition, with shaking, of a few drops of chlorine water or sodium hypochlorite (NaOCl) solution. The appearance of a violet color in the CCl_4 (bottom) layer is proof of the presence of iodide ion.

$$2I^- + Cl_2(aq) \longrightarrow \underset{violet}{I_2(in\ CCl_4)} + 2Cl^-$$

To test this solution for bromide ion, the iodine must first be oxidized to a colorless species (IO_3^-) by further addition of chlorine water.

$$I_2(in\ CCl_4) + 5Cl_2(aq) + 6H_2O(l) \longrightarrow 2IO_3^- + 10Cl^- + 12H^+$$

With still further addition of chlorine water, a yellow to brown color is formed in the CCl_4 solution if the original test solution contained bromide ion.

$$2Br^- + Cl_2(aq) \longrightarrow \underset{\textit{yellow to brown}}{Br_2(\textit{in } CCl_4)} + 2Cl^-$$

To test for chloride ion in the presence of bromide and/or iodide ions, the bromide and iodide ions must be removed. This is done by preferential oxidation of these ions in acidic (H_2SO_4) medium at elevated temperature. Peroxodisulfate ion ($S_2O_8^{2-}$) oxidizes iodide and bromide to the free elements, but is not a strong enough oxidizing agent to produce chlorine from chloride ion.

$$\underset{(X = Br,I)}{2X^-} + S_2O_8^{2-} \xrightarrow{H_2SO_4} X_2 + 2SO_4^{2-}$$

The free halogens are extracted into CCl_4. Silver nitrate is added to the acidified (HNO_3) water layer and $AgCl$ is precipitated.

$$Cl^- + Ag^+ \longrightarrow \underset{\textit{white}}{AgCl(s)}$$

Dissolution of the precipitate in aqueous ammonia unequivocally shows the presence of chloride ion.

$$AgCl(s) + 2NH_3(aq) \longrightarrow [Ag(NH_3)_2]^+ + Cl^-$$

Of course, if I^- and Br^- have been found to be absent, the test for Cl^- ion is greatly simplified. All that is necessary is to see whether a white precipitate forms when silver nitrate solution is added to a nitric acid solution of the sample and whether this precipitate dissolves when aqueous ammonia is added.

Phosphate ion. The presence of phosphate ion, PO_4^{3-}, is indicated if a bright yellow precipitate of ammonium molybdophosphate, $(NH_4)_3P(Mo_3O_{10})_4$, forms upon treatment of the test solution with ammonium molybdate–nitric acid reagent.

$$PO_4^{3-} + 12MoO_4^{2-} + 3NH_4^+ + 24H^+ \longrightarrow \underset{\textit{bright yellow}}{(NH_4)_3P(Mo_3O_{10})_4(s)} + 12H_2O(l)$$

Any reducing agents present would interfere with this test by forming blue precipitates or colors with the reagent. Consequently, before the test is carried out, reducing agents are converted to other species by oxidation with hot concentrated sulfuric acid.

Chromate ion. The presence of chromate ion, CrO_4^{2-}, is indicated by the yellow color of an alkaline sample solution. Addition of acid (e.g., dilute H_2SO_4) converts the CrO_4^{2-} to orange dichromate ion, $Cr_2O_7^{2-}$. The presence of CrO_4^{2-} ion is confirmed if a yellow precipitate of barium chromate ($BaCrO_4$) forms when the test solution is made slightly acidic with acetic acid and then treated with barium acetate, $Ba(CH_3COO)_2$, solution.

$$CrO_4^{2-} + Ba^{2+} \longrightarrow \underset{\textit{yellow}}{BaCrO_4(s)}$$

Nitrate ion. The ions NO_2^-, I^-, Br^-, and CrO_4^{2-} interfere with the detection of nitrate ion. If none of these ions is present, a dilute $FeSO_4$ solution is added to the acidified (with dilute H_2SO_4) test solution. Then *concentrated* H_2SO_4 is carefully introduced under the cooled solution so that it forms a separate layer. If after a few minutes a *brown ring* appears at the interface of the layers, nitrate ion is shown to be present.

$$NO_3^- + 3Fe^{2+} + 4H^+ \longrightarrow NO(aq) + 3Fe^{3+} + 2H_2O(l)$$

$$Fe^{2+} + NO(aq) \longrightarrow [Fe(NO)]^{2+}$$
$$\text{brown}$$

The brown ring forms at the interface between the solution and the sulfuric acid because the H^+ ion concentration there is highest.

Nitrite ion gives the same brown color, but does so throughout the entire solution because it does not require such a high H^+ ion concentration. The nitrite can be destroyed prior to testing for nitrate by reaction with solid sulfamic acid, H_2NSO_3H, in a test solution made slightly acidic with dilute H_2SO_4. Sulfamic acid is a strong acid in water. Undoubtedly there is initial proton transfer from it to NO_2^- ion, followed by oxidation of $H_2NSO_3^-$ to N_2.

$$H_2NSO_3H(aq) + NO_2^- \longrightarrow H_2NSO_3^- + HNO_2(aq)$$

$$H_2NSO_3^- + HNO_2(aq) \xrightarrow{\text{heat}} N_2(g) + H^+ + H_2O(l) + SO_4^{2-}$$

Iodide and bromide ions are oxidized by concentrated H_2SO_4 to the free halogens and these elements give brownish colors at the interface of the layers. The halide ions can be precipitated as silver salts by the addition of silver acetate, $Ag(CH_3COO)$, to a test solution acidified with acetic acid.

If CrO_4^{2-} is present in the solution it is reduced by the Fe^{2+} to dark green Cr^{3+}, thus obscuring the color of the brown ring. Chromate ion can be removed as insoluble $BaCrO_4$ by treatment of a weakly acidic (acetic acid) test solution with barium acetate.

Sulfate ion. Addition of $BaCl_2$ to a test solution made acidic with dilute HCl gives a $BaSO_4$ precipitate if SO_4^{2-} ion is present.

$$SO_4^{2-} + Ba^{2+} \longrightarrow BaSO_4(s)$$
$$\text{white}$$

EXAMPLE 22.2

■ A colorless solid dissolves readily in water to give a solution that is neutral to litmus. Addition of cold concentrated sulfuric acid to the solution results in the liberation of a brownish gas. When dilute perchloric acid is added to a solution of the original solid and the mixture is warmed, a brown gas with a sharp odor is evolved. Addition of a few drops of a solution of the solid to a solution containing iron(III) chloride, potassium hexacyanoferrate, $K_3[Fe(CN)_6]$, and dilute hydrochloric acid gives a dark blue suspension. Treatment of the original solid with dilute sulfuric acid and passage of the gas evolved into a solution containing barium chloride and potassium permanganate produce a white precipitate. Addition of barium chloride to an acidified (HCl) solution of the solid yields a white precipitate. Treatment of a solution of the solid with dilute nitric acid and silver nitrate gives a white precipitate that is completely soluble in aqueous ammonia.

What anion or anions are definitely present in the original solid? Justify your answer.

The anions in the solid are NO_2^-, SO_3^{2-}, SO_4^{2-}, and Cl^-. The absence of a yellow or orange color in the solid proves the absence of CrO_4^{2-} or $Cr_2O_7^{2-}$. The fact that an aqueous solution of the solid is neutral to litmus eliminates the strong Brønsted-Lowry bases: S^{2-}, CO_3^{2-}, and PO_4^{3-}. The brown gas liberated by the action of cold concentrated sulfuric acid is either NO_2 or Br_2, or both, indicating the presence of NO_2^- or Br^-, or both, and eliminating I^-, which gives I_2 as a violet vapor. However, those anions which form colorless gases when treated with concentrated sulfuric acid, SO_3^{2-} and Cl^-, have not been eliminated. (The CO_3^{2-} and S^{2-} ions also give colorless gases with sulfuric acid, but they have already been shown to be absent.) The evolution of a brown gas from the dilute perchloric acid shows the presence of NO_2^- ion, but does not eliminate SO_3^{2-}, which also falls in Group I in the anion group classification scheme.

The blue suspension (Prussian blue) formed in the next step shows the presence of a reducing agent and thus continues the possibility that SO_3^{2-} may be in the solid. (Remember S^{2-} and I^- have been eliminated.) The fact that a white precipitate ($BaSO_4$) is produced in the next step, which is

$$\text{unknown solution} + \text{dilute sulfuric acid} \longrightarrow \text{gas}$$
$$\text{gas} + KMnO_4(\text{oxidizing agent}) \longrightarrow SO_4^{2-}$$
$$SO_4^{2-} + Ba^{2+} \longrightarrow BaSO_4(s),$$

proves that SO_3^{2-} is present in the original solid. Sulfate ion, SO_4^{2-}, is also present, as evidenced by the formation of the white precipitate ($BaSO_4$) when a solution of the unknown is treated with an acidic solution of barium chloride. Finally, the white precipitate formed in the last test is silver chloride, $AgCl$; the complete solubility of the precipitate in dilute aqueous ammonia shows that Br^- ion is absent. No conclusion can be drawn from these tests regarding the presence or absence of NO_3^- ion. ∎

Frasch describes his
first success

After permitting the melting fluid to go into the ground for twenty-four hours, I decided that sufficient material must have been melted to produce some sulphur. The pumping engine was started on the sulphur line, and the increasing strain against the engine showed that work was being done. More and more slowly went the engine, more steam was supplied, until the man at the throttle sang out at the top of his voice, "She's pumping." A liquid appeared on the polished rod, and when I wiped it off I found my finger covered with sulphur. Within five minutes the receptacles under pressure were opened and a beautiful stream of the golden fluid shot into the barrels we had ready to receive the product. After pumping for about fifteen minutes, the forty barrels we had supplied were seen to be inadequate. Quickly we threw up embankments and lined them with boards to receive the sulphur that was gushing forth; and since that day no further attempt has been made to provide a vessel or a mold into which to put the sulfur.

When the sun went down we stopped the pump to hold the liquid sulphur below until we could prepare to receive more in the morning. The material on the ground had to be removed, and willing hands helped to make a clean slate for the next day. When everything had been finished, the sulphur all piled up in one heap, and the men had departed, I enjoyed all by myself this demonstration of success. I mounted the sulphur pile and seated myself on the very top. It pleased me to hear the slight noise caused by the contraction of the warm sulphur, which was like a greeting from below—proof that my object had been accomplished. Many days and many years intervened before financial success was assured, but the first step towards the ultimate goal had been achieved. We had melted the mineral in the ground and brought it to the surface as a liquid. We had demonstrated that it could be done.

HERMAN FRASCH, *The Journal of Industrial and Engineering Chemistry*, Feb. 1912 [with thanks to *Chemtech*, Feb. 1976]

Significant terms defined in this chapter

ammine group	phosphate rock
carcinogens	nitrogen fixation
mutagens	nitrification
aqua regia	denitrification

FRASCH DESCRIBES HIS FIRST SUCCESS, *by Herman Frasch*

QUALITATIVE ANALYSIS:
ANIONS

Exercises

22.1 Write the electronic configuration for a nitrogen atom. Assuming that the atom will gain enough electrons to attain a noble gas configuration, what would you predict for the most stable negative oxidation number of N? If the p electrons can be removed or shared more easily than the s electrons, what would you predict for the most stable positive oxidation numbers of N? Repeat this exercise for P and S.

22.2 List some of the physical properties of N_2, P_4, and S_8. What is meant by the term "allotropy"? Describe the allotropic forms of phosphorus and sulfur.

22.3 Write chemical equations for the direct combination of (i) N_2, (ii) P, and (iii) S with (a) Mg, (b) Al, and (c) O_2.

22.4 Describe commercial and laboratory preparations of nitrogen. Be sure to give any pertinent equations. Repeat the exercise for phosphorus and sulfur.

22.5 List the major use(s) of (a) N_2, (b) NH_3, (c) HNO_3, (d) P, (e) PO_4^{3-}, (f) S, and (g) H_2SO_4.

22.6 Which statements are true? Rewrite any false statement so that it is correct.
(a) Phosphorus differs in many respects from nitrogen, as does sulfur from oxygen because of the respective sizes, electronegativities, and maximum number of orbitals available in the respective valence shells.
(b) Many nitrogen-containing compounds decompose to form N_2 because of the great stability of the nitrogen–nitrogen triple bond.
(c) Most ammonium salts are water soluble and produce highly alkaline solutions.
(d) Copper will not react with hydrochloric acid but will react with nitric acid.
(e) Most metal sulfides are more soluble in an acid solution than in water.
(f) Concentrated sulfuric acid is a strong dehydrating agent.

22.7 Draw Lewis diagram(s) for (a) HNO_2, (b) HNO_3, (c) NH_3, (d) PH_3, (e) H_3PO_3, (f) H_3PO_4, (g) $H_4P_2O_7$, (h) H_2S, (i) H_2SO_3, (j) H_2SO_4, and (k) H_2SO_5.

22.8 Complete and balance the following oxidation–reduction equations:
(a) $NH_4NO_3(s) \longrightarrow$
(b) $N_2H_4(l) + O_2(g) \longrightarrow$
(c) $N_2O(g) \longrightarrow$
(d) $Cu(s) + H^+ + NO_3^-(dil) \longrightarrow$
(e) $NO(g) + NO_2(g) \longrightarrow$

(f) $NO_2(g) + H_2O(l) \longrightarrow$
(g) $H_2S(g) + H_2SO_3(aq) \longrightarrow$
(h) $SO_3^{2-} + S(s) \longrightarrow$
Identify the oxidizing agent in each case.

22.9 List the various oxidation numbers of nitrogen and give the formula of one compound or ion corresponding to each oxidation number. Name each species.

22.10 Write the electronic formulas for: (a) N_2O, (b) NO, (c) HNO_2, (d) HNO_3, (e) NO_2^-, and (f) NO_3^-. Name each species and identify the hybridization of the valence shell orbitals of the nitrogen atom in each species.

22.11 Excessive amounts of the following compounds containing nitrogen produce undesirable effects on the environment: NO, NO_2, NH_3, and NH_4NO_3. Describe the effect produced by each compound, how each enters the environment, and a possible way to reduce the amount of each in order to reduce its effect.

22.12 What structural feature do molecules of P_4, P_4O_6, and P_4O_{10} have in common? Is there a common structural feature for all of the acids containing phosphorus(V)?

22.13 Prepare Lewis structures for (i) PCl_3, (ii) PCl_4^+, (iii) PCl_5, (iv) PCl_6^-. What atomic orbitals on the phosphorus atom undergo hybridization as those species are formed? Prepare three-dimensional sketches of these species and identify the shapes.

22.14 Write the electronic formulas for the following species: (a) sulfur dioxide, (b) sulfuric acid, (c) sulfur trioxide, (d) pyrosulfuric acid, (e) peroxodisulfuric acid, (f) sulfite ion, (g) sulfate ion, (h) thiosulfate ion, (i) sulfide ion, and (j) molecular sulfur. What is the oxidation number of sulfur in each of the species?

22.15 A sample of powdered sulfur was divided into two parts—one about twice the size of the other. The smaller sample was heated until the sulfur melted and began to burn. The gaseous product (i) of this reaction was collected and mixed with water to form (ii). The larger sulfur sample was mixed with iron filings and heated in a crucible producing a dark solid (iii). This solid was placed in a container, and HCl was added producing a gas (iv), which was allowed to mix with (ii), giving a yellow to white finely divided precipitate (v). (a) Identify the five substances mentioned above and (b) write chemical equations describing the reactions.

22.16* The phase diagram of sulfur is shown in the accompanying figure. Describe what happens to a sample of sulfur (a) as it is heated very slowly at 1 atm from 25°C to 500°C; (b) that is heated to 153°C at 1420 atm; (c) as it is melted and poured slowly into boiling water at 1 atm; (d) that is in the gaseous phase between 95.31°C and 115.18°C and that has the pressure upon it increased; and (e) in the monoclinic form between 5.1×10^{-6} atm and 3.2×10^{-5} atm when it is heated.

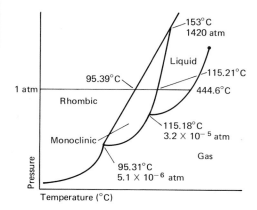

22.17 Why will anion unknowns which give strong positive tests for NO_2^- often give faint tests for NO_3^- and those containing S^{2-} or SO_3^{2-} give faint tests for SO_4^{2-}?

22.18 Certain combinations of anions are incompatible. As an example, write the equation and describe what will happen in an acidic solution of SO_3^{2-} and S^{2-} using the following half reactions and standard reduction potentials:

$$S + 2e^- \longrightarrow S^{2-} \qquad\qquad E° = -0.508 \text{ V}$$
$$SO_3^{2-} + 6H^+ + 4e^- \longrightarrow S + 3H_2O \qquad E° = 0.45 \text{ V}$$

22.19 Use the following half reactions and standard reduction potentials to show that chlorine water will oxidize I^- to I_2, I_2 to IO_3^- in $6M$ HCl, and Br^- to Br_2:

$$Cl_2(aq) + 2e^- \longrightarrow 2Cl^- \qquad\qquad E° = 1.3583 \text{ V}$$
$$Br_2(aq) + 2e^- \longrightarrow 2Br^- \qquad\qquad E° = 1.087 \text{ V}$$
$$I_2(aq) + 2e^- \longrightarrow 2I^- \qquad\qquad E° = 0.535 \text{ V}$$
$$2IO_3^- + 12H^+ + 10e^- \longrightarrow \qquad E° = 1.19 \text{ V}$$
$$I_2(aq) + 6H_2O(l)$$

Note that reaction conditions during the qualitative analysis testing are adjusted so that the oxidation of I_2 will occur before the oxidation of Br^-.

22.20 The presence of the oxidizing anions (NO_3^-, NO_2^-, and CrO_4^{2-}) can be detected using aqueous KI instead of $MnCl_2$. Write the equations representing reactions of these ions with I^- in aqueous acidic solution.

22.21 Which anions cannot be present in a water-soluble sample known to contain Cu^{2+} and Zn^{2+}? (*Hint:* Look up the solubilities of the various compounds containing the anions considered in the scheme.)

22.22 A white substance is a mixture of salts containing either Na^+ or K^+ as the cation. Among the anions possibly present are CO_3^{2-}, PO_4^{3-}, NO_3^-, S^{2-}, and SO_3^{2-}. Describe how you would determine which of these anions are actually present in the salt mixture and write equations for the reactions that confirm the presence of the ions.

22.23 A mixture is known to contain one or more of the halide ions I^-, Br^-, and Cl^-. Describe how you could test for each of these ions and write equations for the various reactions used.

22.24 A student wishes to determine whether a solution known to contain Br^- also contains I^-. A friend tells the student that this can be done simply by acidifying the solution with dilute sulfuric acid, adding $NaNO_2$ and CCl_4, and shaking the mixture. From what you can find out in the text about the redox chemistry of I^-, Br^-, and NO_2^-, do you think the friend's suggestion is a reasonable one? Defend your answer.

22.25 In the pairs of ions shown below the second interferes with the detection of the first:
(a) NO_3^- and NO_2^-
(b) $X^-(Cl^-, Br^-, or I^-)$ and S^{2-}
In each case, give the reason for the interference and tell how the offending ion may be removed. Include pertinant equations.

22.26 Identify the anion(s) present in an unknown from the following observations made on a Na_2CO_3 solution:
(a) A dark color did not form when the solution was acidified and tested with $MnCl_2$.
(b) A dark blue precipitate formed when the solution was acidified and tested with a mixture of $FeCl_3$ and $K_3Fe(CN)_6$.
(c) Bubbles of an odorless gas were observed when $HClO_4$ was added to the original solid sample.
(d) A light-colored precipitate was produced when $AgNO_3$ was added to an acidic solution.
(e) After separation of the silver precipitate in (d), negative results were obtained for the anions in Groups III and IV.
(f) A dark odoriferous gas, in addition to the gas described in (c), was generated by the addition of H_2SO_4 to the original solid sample.
(g) The sample gave a white precipitate when tested with Zn, H_2O_2, H_2SO_4, and $Ba(OH)_2$.
(h) The solution was acidified and a violet color was observed in a CCl_4 layer after the addition of chlorine water.
(i) The solution was acidified and a red-brown color was observed in the CCl_4 layer after the addition of a considerable amount of chlorine water.
(j) The tests for Cl^- were negative.

22.27 Identify the anion(s) present in an unknown from the following observations made on a colorless, water-soluble sample:

(a) A dark color was formed when the solution was acidified and tested with $MnCl_2$.

(b) A blue precipitate did not form when the solution was acidified and tested with a mixture of $FeCl_3$ and $K_3Fe(CN)_6$.

(c) No noticeable gases were generated upon addition of $HClO_4$ to the original solid sample.

(d) A white precipitate formed when the acidic solution was treated with $AgNO_3$.

(e) After separation of the silver precipitate, a second silver precipitate formed after the solution was made slightly alkaline with aqueous NH_3.

(f) A colorless odoriferous gas was generated upon addition of H_2SO_4 to the original solid sample.

(g) Addition of chlorine water to an acidified solution of the sample did not produce a violet or red-brown color in a CCl_4 layer.

(h) The white precipitate produced in (d) dissolved upon addition of NH_3.

(i) A precipitate did not form when the unknown was tested with ammonium molybdophosphate.

(j) A brown ring test was positive.

(k) A white precipitate formed upon addition of $BaCl_2$ to the acidic solution.

22.28 Tell what conclusions can be drawn from the result of parts (a) through (f) regarding *anions not present, anions possibly present,* and *anions definitely present* in the original solid.

(a) A white solid was partially soluble in water and the part that dissolved was neutral to litmus paper. The original solid gave no gaseous product when treated with dilute $HClO_4$.

A small amount of the solid was boiled with saturated Na_2CO_3 solution and the resulting mixture was filtered. Portions of the filtrate were treated with various reagents with the following results:

(b) Addition to a dilute solution of $MnCl_2$ in concentrated HCl gave no color change.

(c) Addition to an aqueous mixture of $FeCl_3$, $K_3[Fe(CN)_6]$, and HCl resulted in the formation of a deep blue precipitate.

(d) Acidification with dilute $HClO_4$, followed by warming of the solution and the addition of $AgNO_3$, yielded a pale yellow precipitate.

(e) Acidification with dilute HCl, addition of CCl_4, and dropwise addition of Cl_2 water accompanied by vigorous shaking produced a violet color in the CCl_4 (lower) layer.

(f) Acidification with a slight excess of dilute HCl followed by addition of dilute $BaCl_2$ solution gave a white precipitate.

22.29 The average atomic weight of N is 14.0067 amu. There are two isotopes which contribute to this average: $^{14}_{7}N$ (14.00307 amu) and $^{15}_{7}N$ (15.00011 amu). Calculate the percentage of $^{15}_{7}N$ atoms in a sample of naturally occurring nitrogen.

22.30 Commercial concentrated HNO_3 contains 69.5 wt % HNO_3 and has a density of 1.42 g/ml. What is the molarity of this solution? What volume of the concentrated acid should a stockroom attendant use to prepare 10.0 liters of dilute HNO_3 at a concentration of $6M$?

22.31 Calculate the pH of each of the following solutions: (a) $0.05M$ HNO_3; (b) $1M$ NH_3, $K_b = 1.6 \times 10^{-5}$; (c) $0.1M$ HNO_2, $K_a = 7.2 \times 10^{-4}$.

22.32 What would be the pH of a buffer solution prepared using equal volumes of $0.20M$ HNO_2 and $0.20M$ KNO_2? $K_a = 7.2 \times 10^{-4}$ for HNO_2.

22.33 A laundry additive contains 9.5 wt % phosphorus in the form of $Na_5P_3O_{10}$. How many grams of $Na_5P_3O_{10}$ are in a one pound box of the additive?

22.34 A sample of phosphorus was burned in air and produced a product containing 56 wt % P and 44 wt % O. Find the simplest formula of the oxide formed. The density of this oxide at 200°C and 1 atm pressure is 5.7 g/liter. What is the molecular formula of this oxide?

22.35 The enthalpy change for dissolving rhombic sulfur in 6 moles of CS_2 at 25°C is 405 cal/mole

$$S(s) + 6CS_2(l) \longrightarrow S(\text{in } 6CS_2)$$

and for monoclinic sulfur is 325 cal/mole. Which form of sulfur is more stable at 25°C?

22.36 Each day a power plant burns 2.0×10^3 tons of coal that contains 2.0 wt % of sulfur. Assuming that the sulfur is oxidized to SO_2, what mass of SO_2 is emitted each day? If it were possible to trap all of the SO_2 and convert it to 75 wt % H_2SO_4, how many tons of the acid could be obtained from the plant each day?

22.37 A gaseous mixture at 300°C in a 1.00 liter vessel originally contained 1.0 mole of SO_2 and 5.0 moles of O_2. Once equilibrium conditions were attained, 81% of the SO_2 had been converted to SO_3. What is the value of the concentration equilibrium constant for this reaction at this temperature?

22.38 Which solution has a higher mole fraction of solute: concentrated H_2SO_4, which is 96.0 wt % H_2SO_4, or concentrated "ammonium hydroxide," which is 28.6 wt % NH_3?

22.39 What is the concentration of Fe^{2+} in an aqueous solution (a) saturated with FeS, $K_{sp} = 4.2 \times 10^{-17}$; (b) which is $0.1M$ in S^{2-}; and (c) which has pH 3.5, $K_a = [H^+]^2[S^{2-}]/[H_2S] = 3 \times 10^{-20}$ where $[H_2S] = 0.1M$?

22.40 The anions of water-insoluble salts are brought into solution by boiling the solid with Na_2CO_3 solution, for example,

$$BaSO_4(s) + CO_3^{2-} \longrightarrow BaCO_3(s) + SO_4^{2-}$$

The SO_4^{2-} and CO_3^{2-} compete for Ba^{2+} as indicated by

$$[Ba^{2+}][SO_4^{2-}] = 1.7 \times 10^{-10}$$
$$[Ba^{2+}][CO_3^{2-}] = 2.0 \times 10^{-9}$$

Find the ratio of $[SO_4^{2-}]$ to $[CO_3^{2-}]$ at the point where the $BaSO_4$ just completely dissolves by eliminating $[Ba^{2+}]$ from the above equations. What is the maximum concentration of SO_4^{2-} that can be present in a saturated solution of Na_2CO_3 ($[CO_3^{2-}] = 1.5M$)?

22.41 Repeat the calculations of the previous exercise for dissolving CuS in Na_2CO_3. $K_{sp} = 2.3 \times 10^{-10}$ for $CuCO_3$ and 6×10^{-36} for CuS. If the limit of concentration for detection of the sulfide ions is $10^{-3}M$, will S^{2-} be detected using this technique?

22.42 What is the concentration of NO_2^- remaining in a solution that has been acidified with excess $6M$ HCl? Assume that the original concentration of NO_2^- was $0.1M$ and that $[H^+] = 1M$. $K_a = 7.2 \times 10^{-4}$ for HNO_2.

22.43 What mass of solid $K_2S_2O_8$ is needed to react with the I^- and Br^- in a 2 ml sample in which $[I^-] = 1 \times 10^{-2}M$ and $[Br^-] = 3 \times 10^{-2}M$?

22.44* Using molecular orbital notation, (a) write the electronic configuration for a molecule of dinitrogen. (b) What type of bonds and how many of each are predicted to be in a molecule? (c) Prepare a Lewis diagram and a three-dimensional sketch of a molecule of N_2. (d) Will the molecule by paramagnetic or diamagnetic? The $N{\equiv}N$ bond energy is 945 kJ/mole and the N—N bond energy is about 160 kJ/mole. (e) Predict whether four gaseous nitrogen atoms would form two dinitrogen molecules or a tetrahedral molecule similar to P_4, basing your prediction on the amount of energy released as the molecules are formed. (f) Repeat part (e) for phosphorus using 485 kJ/mole for $P{\equiv}P$ and 243 kJ/mole for P—P.

22.45* A careless stockroom attendant prepared a liter each of $0.1M$ Na_3PO_4, $0.1M$ Na_2HPO_4, and $0.1M$ NaH_2PO_4 but forgot to label the containers. Quickly the attendant realized that each solution had a unique value of pH and that all that was needed was to calculate the theoretical values from the values of K_a ($K_{a_1} = 7.5 \times 10^{-3}$, $K_{a_2} = 6.6 \times 10^{-8}$, $K_{a_3} = 1 \times 10^{-12}$), measure the values of the solutions using a pH meter, and properly label each container. What were the values of pH that the attendant calculated?

22.46* Assuming a molecule of SO_3 to contain one double and two single bonds between the sulfur and oxygen atoms, calculate the heat of formation of $SO_3(g)$ from the following data:

$$S(s) \longrightarrow S(g) \qquad \Delta H° = 66.636 \text{ kcal}$$
$$\tfrac{1}{2}O_2(g) \longrightarrow O(g) \qquad \Delta H° = 59.553 \text{ kcal}$$
$$\text{S—O bond energy} = 101 \text{ kcal/mole}$$
$$\text{S=O bond energy} = 125 \text{ kcal/mole}$$

Because of the delocalized π bonding in the molecule, SO_3 is more stable than predicted by this calculation. Calculate the resonance energy, given that the heat of formation of $SO_3(g)$ is -94.58 kcal/mole.

22.47** The compound hydrazoic acid, HN_3, is very explosive in the pure state, but it can be studied in aqueous solution. The acid is prepared by reacting hydrazine with nitrous acid.

$$N_2H_4(l) + HNO_2(aq) \longrightarrow 2H_2O(l) + HN_3(aq)$$

(a) What are the oxidation states of nitrogen in the compounds above? (b) What is the oxidizing agent in the reaction? The hydrazoic acid is a rather weak acid with $K_a = 2.4 \times 10^{-5}$ at 25°C. (c) Calculate the pH of a $0.01M$ solution of HN_3.

Two acceptable resonance forms can be written for the molecule

$$H{-}\ddot{N}{=}N{=}\ddot{N}{\cdot} \longleftrightarrow H{-}\ddot{N}{-}N{\equiv}N{:}$$

in which one bond between the nitrogen atoms has a bond order of 1.5 and the other, 2.5. Using the following thermodynamic data, (d) calculate the heat of formation of $HN_3(g)$ for each resonance form: $\Delta H_f° = 52.095$ kcal/mole for $H(g)$, $\Delta H_f° = 112.979$ kcal/mole for $N(g)$, average N—H bond energy = 86 kcal/mole, average N—N bond energy = 38 kcal/mole, average N=N bond energy = 113 kcal/mole, and N≡N bond energy = 226 kcal/mole. (e) Calculate the resonance energy of the molecule given that $\Delta H_f° = 70.3$ kcal/mole. The lengths of nitrogen–nitrogen bonds are 0.1449 nm for N—N, 0.123 nm for N=N, and 0.108758 nm for N≡N. (f) Estimate the bond lengths of each of the N—N bonds in HN_3. The actual values are 0.124 nm and 0.113 nm. (g) Prepare a three-dimensional sketch of the molecule showing the delocalized bonding.

23 NONMETALS: CARBON AND HYDROCARBONS

Carbon is an element with many facets to its personality. As diamond it sparkles brilliantly. As graphite it lubricates bearings, is slippery and black, and makes a mess if you get it on your clothes. As soot, it is deposited on cold surfaces from smoky flames. Carbon is one of the elements known from antiquity, and the manufacture of charcoal by heating wood in the absence of air was described in the first century A.D.

The tetrahedral bonding of carbon atoms, combined with their ability to link with each other in endless arrays, is responsible for the entire field of organic chemistry. And, of course, organic compounds, with their carbon atom-to-carbon atom skeletons, are the basis for living matter, at least as it is known on this planet.

In still another facet of its personality, carbon forms compounds with many elements, and these are the proper subject of the inorganic chemistry of carbon.

This chapter is divided into two major parts. The first considers the descriptive chemistry of carbon and some important inorganic compounds of carbon. The second part introduces the simplest of organic compounds, the hydrocarbons—the compounds composed only of hydrogen and carbon atoms. The nomenclature systems for the saturated, unsaturated, and aromatic hydrocarbons are presented. Fossil fuels, the major sources of hydrocarbons, are also discussed.

Carbon and its inorganic compounds

B	C	6	N
	12.011		
Al	Si		P

TABLE 23.1

Properties of carbon

Melting point (°C)	
Diamond	>3550
Graphite	3652–3697
	(sublimes)
Boiling point (°C)	
Diamond	4827°
Graphite	4200°
Density	
Diamond	3.56
Graphite	2.25
C—C single bond length (nm)	0.154
Covalent radius (nm)	0.0771
Average bond energy [kcal/mole (kJ/mole)]	79 (330)
Ionization energy (kcal/mole, 0°K)	260
Electron affinity (kcal/mole, 0°K)	29.3
Electronegativity	2.5

23.1 Properties of carbon

Carbon is the first member of Representative Group IV and is the only nonmetal in the group. The electronic configuration of carbon is $1s^2 2s^2 2p^2$. In compounds, carbon atoms frequently form four covalent single bonds via sp^3 hybridization (Section 21.4). Carbon can also participate in double covalent bonds through sp^2 hybridization and formation of a σ bond plus a π bond, and in triple covalent bonds through sp hybridization and formation of a σ bond plus two π bonds. Atoms of carbon form multiple bonds with themselves ($\diagdown$C$=$C$\diagdown$, $-$C$\equiv$C$-$), and with oxygen ($\diagdown$C$=$O), sulfur ($\diagdown$C$=$S), nitrogen ($\diagdown$C$=$N$-$, $-$C$\equiv$N), and phosphorus ($\diagdown$C$=$P$-$). In the binary saltlike carbides (Section 23.3e) carbon forms C^{4-} and C_2^{2-} ions. The chemistry of carbon is very different from that of the other Representative Group IV elements.

The reactivity of carbon is rather low. The carbon–carbon single bond (Table 23.1) is stronger than similar bonds for the other Group IV elements (Section 10.14). The ionization energy and electronegativity of carbon (Table 23.1) are roughly in the middle of the ranges of these values for all elements (see Tables 10.4 and 10.7). With respect to the other members of its group, carbon has the highest ionization energy and is the most electronegative element.

Naturally occurring carbon is about 99% carbon-12 and 1% carbon-13. There are also five known radioactive isotopes of carbon. Carbon-14 is present in the atmosphere in minute amounts and is used in radiocarbon dating (Sections 8.6, 15.11).

The two allotropic forms of elemental carbon, diamond and graphite, differ widely in their properties (see Table 23.1) because of differences in crystal structure.

23.2 Diamond, graphite, and other forms of carbon; their uses

Diamond is clear and colorless and is the hardest substance known. It has a very high melting point, is extremely brittle, and when struck, breaks into many pieces. It does not conduct electricity. These properties reflect the strength of network covalent bonding (Section 9.19)—each carbon atom in the crystal shares its four valence electrons with four other carbon atoms, which surround it tetrahedrally. The pattern is continued indefinitely and is terminated only at the surface of the crystal (Figure 23.1).

The refractive index of diamond is high, and when properly cut

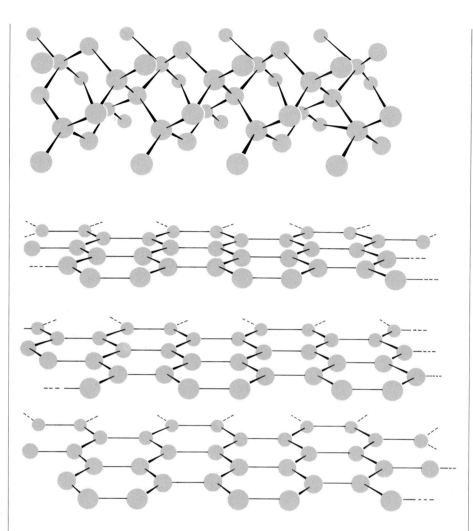

FIGURE 23.1
Arrangement of carbon atoms in diamond.

FIGURE 23.2
Arrangement of carbon atoms in graphite.

and polished, a diamond reflects light in an array of many colors. On this account, and because of its durability, diamond is highly prized as a gem stone. Impure samples are often black; these have no value as gems, but when crushed, they are used in abrasives. In rare cases a diamond has an attractive color because of traces of impurities; such a diamond, because it is unique, is especially valuable as a gem. The famous Hope diamond, for example, is blue.

Diamonds have been found in both North and South America, but nearly all of the world's supply comes from South Africa, where the diamonds exist in pipes of "blue clay" from which they are separated by washing the clay away in a stream of water.

Graphite is black and soft, and is much less dense than diamond (Table 23.1). Like diamond, graphite melts only at an extremely high temperature. It feels smooth and slippery to the touch and is an excellent lubricant. The carbon atoms are arranged in planar layers in the graphite crystal (Figure 23.2). Within each layer, each carbon atom is bonded to three other carbon atoms by covalent single bonds. The

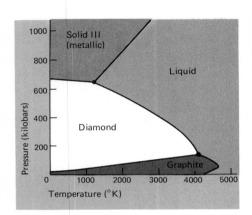

FIGURE 23.3
Phase diagram for carbon. *One kilobar is approximately equal to 1000 atm.*

fourth valence electron of each carbon atom is delocalized over the three covalent bonds and is mobile. The atoms in each layer are tightly bonded together, but the binding force between layers is weak, allowing the layers to slip over each other. The delocalized electrons give graphite metallic properties. It has a dull luster and conducts electricity moderately well. The lubricating properties of graphite depend on a film of moisture or of gas molecules adsorbed on the surface of the graphite layers. The adsorbed substance decreases friction as the layers slide past each other.

In addition to its use as a lubricant, graphite is employed on a large scale in electrodes and in molds and crucibles to be used with hot metal. Graphite is the black ingredient in pencils, a pencil "lead" being a baked rod of graphite mixed with clay.

Comparison of Figures 23.1 and 23.2 shows that the crystal structures of graphite and diamond are closely related. If alternate carbon atoms in each layer of graphite could be raised above the plane of the layer and the others depressed below it, and the layers could then be pushed a little closer together, the diamond structure would result. According to the phase diagram for carbon (Figure 23.3), this should be possible at a high enough temperature and pressure.

The difficulty of compressing graphite, however, was recognized by the famous physicist Percy Bridgeman, who said, "Graphite is nature's strongest spring," and by an unidentified scientist who put it this way, "It is easy to squeeze carbon atoms together, but very difficult to keep them squz."

After much effort, synthetic diamonds were first made in 1955 at the General Electric Company laboratory. The process requires chambers able to maintain 1.5 million lb/sq. inch pressure and 1500°K for long periods of time. A molten metal catalyst enhances the reaction rate. Synthetic diamonds find their outlet in a variety of industrial applications that mainly utilize the properties of hardness and abrasiveness in various cutting and polishing tools. For example, concrete can be cut like wood with a blade coated with tiny diamond crystals. Gem-quality diamonds have not yet been made synthetically. The diamondlike synthetic gems sold at less-than-diamond prices are either titanium dioxide (TiO_2), strontium titanate ($SrTiO_3$), or yttrium aluminum garnet ($Y_3Al_5O_{12}$). The first two give stones of fiery brilliance, but they are softer than diamond. The last, called "YAG," is less brilliant than the others, but is harder (see Figure 23.10).

In addition to diamond and graphite, several amorphous forms of carbon are familiar—charcoal, coke, carbon black (also known as lampblack and thermal black), and activated charcoal. These materials have less regular structures and in some cases are less pure than diamond or graphite.

The charcoal we burn in stoves and fireplaces is made from wood by heating it in the absence of air. Finely powdered forms of charcoal are made similarly from such materials as animal bones, coconut shells, or sugar. Fine charcoal "activated" by heating with steam to clean its surface is useful as an adsorbent. Such charcoal has a very high surface area—ranging from 600 to 2000 m²/g. Activated charcoal is used to remove bad-smelling or dangerous vapors from the air and to remove colored or bad-tasting impurities from water or other liquids. For example, many municipal water treatment plants pass water through beds of activated charcoal.

Carbon black, usually made by the thermal decomposition of hydrocarbons in an open flame, is a very finely divided, very pure form of carbon. The human eye can distinguish 260 shades of black, and 10-nm particles of carbon black are the blackest of substances. Carbon black is used as a paint pigment, and large quantities are used to reinforce and color rubber—an automobile tire is one-fourth carbon black.

23.3 Inorganic compounds of carbon

a. Carbon monoxide. When carbon is burned in a limited supply of air at 1000°C, it is almost completely converted to carbon monoxide.

$$2C(s) + O_2(g) \longrightarrow 2CO(g) \qquad \Delta H° = -52.83 \text{ kcal } (-221.0 \text{ kJ})$$

At temperatures of about 500°C, carbon monoxide is produced even if an excess of oxygen is present. The monoxide is also produced by the combustion of hydrocarbons and other carbonaceous materials in a small supply of air or by the reduction of carbon dioxide at high temperatures.

$$CO_2(g) + C(s) \longrightarrow 2CO(g) \qquad \Delta H° = 41.22 \text{ kcal } (172.4 \text{ kJ})$$
$$Fe(s) + CO_2(g) \longrightarrow FeO(s) + CO(g) \qquad \Delta H° = 2.6 \text{ kcal } (11 \text{ kJ})$$

Carbon monoxide (Figure 23.4) is a colorless, odorless gas, which is insoluble in water and in most other liquids. It is toxic because, when inhaled, it combines with the hemoglobin of the blood, displacing oxygen needed by the cells (Section 6.8). In many of its physical properties carbon monoxide closely resembles nitrogen, with which it is isoelectronic.

$$:C::O: \quad \text{and} \quad :N::N:$$

Carbon monoxide is relatively unreactive. When heated in air, it burns with the liberation of heat, forming carbon dioxide.

$$2CO(g) + O_2(g) \longrightarrow 2CO_2(g) \qquad \Delta H° = -135.3 \text{ kcal } (-566.1 \text{ kJ})$$

In sunlight, or in the presence of an appropriate catalyst, carbon monoxide combines with chlorine, yielding the highly toxic gas carbonyl chloride, better known as phosgene, $COCl_2$.

$$CO(g) + Cl_2(g) \longrightarrow COCl_2(g)$$
$$\text{\textit{phosgene}}$$

Many transition metals and some of their salts react with carbon monoxide to give metal carbonyls; for example, $[Ni(CO)]_4$ and $[Pt_2Cl_4(CO)_2]$ (these are complexes, discussed in Chapter 28).

FIGURE 23.4

c

o

Carbon monoxide, CO
m.p. − 199°C
b.p. − 192°C

$$CO_2 + H_2O \rightarrow \text{acidic solution}$$
$$CO_3{}^{2-} \text{ or } HCO_3{}^- + H_2O \rightarrow$$
$$\text{basic solution}$$

FIGURE 23.5

$$\ddot{O}::C::\ddot{O} \longleftrightarrow :\ddot{O}::C:\ddot{O}: \longleftrightarrow :\ddot{O}:C::O:$$

Carbon dioxide, CO_2
sublimes at 78.5°C

CS_2 is isoelectronic with CO_2

S

C

O

Carbon disulfide, CS_2
m.p. −110°C
b.p. 46°C

At elevated temperatures, carbon monoxide is a good reducing agent and is used in metallurgical processes, as is carbon, to free metals from their oxide ores; for example,

$$Fe_2O_3(s) + 3CO(g) \xrightarrow{\Delta} 2Fe(s) + 3CO_2(g)$$
$$ZnO(s) + C(s) \xrightarrow{\Delta} Zn(g) + CO(g)$$

b. Carbon dioxide, carbonic acid, and carbonates. Carbon dioxide, CO_2, is formed by the combustion of carbon-containing substances in an excess of oxygen. One example, the oxidation of carbon monoxide, has been noted above. Others include

$$C(s) + O_2(g) \longrightarrow CO_2(g) \qquad \Delta H° = -94.05 \text{ kcal } (-393.5 \text{ kJ})$$
$$CH_4(g) + 2O_2(g) \longrightarrow CO_2(g) + 2H_2O(g) \quad \Delta H° = -191.8 \text{ kcal } (-802.5 \text{ kJ})$$
$$CS_2(l) + 3O_2(g) \longrightarrow CO_2(g) + 2SO_2(g) \quad \Delta H° = -257.4 \text{ kcal } (-1077 \text{ kJ})$$

Carbon dioxide may also be prepared by the thermal decomposition of metal carbonates. An important industrial example is the calcination of limestone to produce quicklime, which is calcium oxide, with carbon dioxide as a by-product.

$$\underset{\text{limestone}}{CaCO_3(s)} \xrightarrow{\Delta} \underset{\text{quicklime}}{CaO(s)} + CO_2(g)$$

In the laboratory, carbon dioxide is usually produced by the action of acids on carbonates

$$CaCO_3(s) + 2H^+ \longrightarrow Ca^{2+} + CO_2(g) + H_2O(l)$$

and it is formed naturally in most fermentation reactions and by the oxidation of organic matter in air.

Carbon dioxide (Figure 23.5) is an odorless, colorless, nontoxic gas. The carbon dioxide molecule is linear and in terms of the valence bond theory of covalent bonding has three resonance structures. The gas is readily condensed to the liquid state by cooling and compression, and upon further cooling the liquid freezes to a white solid. This substance does not melt upon warming, but evaporates directly into the gaseous state (sublimes) at −78.5°C. Solid CO_2 leaves no trace when it evaporates, making it very convenient for use in cooling ice cream, chemical reactions, and a thousand other things. The common name for solid CO_2 is dry ice.

The atmosphere contains only 0.0325% carbon dioxide by volume, but atmospheric carbon dioxide plays an important role in photosynthesis and the carbon and oxygen cycles (Section 11.15). On the other hand, life is impossible in an atmosphere that contains too much carbon dioxide, because animals cannot breathe in enough oxygen to sustain themselves. A CO_2 concentration of 1% by volume in air causes headaches, 10% causes severe distress, and over 30% causes unconsciousness and death.

Since carbon dioxide is thermally stable and does not support combustion, it is frequently used to extinguish fires. Several types of fire-fighting equipment are employed to generate the gas and blanket the fire with it to exclude oxygen. In a type of fire extinguisher still found in some buildings, an acid and a solution of a bicarbonate are stored separately inside a tank (Figure 23.6). When this device is turned upside down, the acid and the bicarbonate mix and react to form carbon dioxide.

$$2NaHCO_3(aq) + H_2SO_4(aq) \longrightarrow Na_2SO_4(aq) + 2CO_2(g) + 2H_2O(l)$$

The resulting pressure drives the water and carbon dioxide from the tank through a nozzle, from which it can be directed onto the fire. In this case, both the water and carbon dioxide extinguish the fire. Fire extinguishers of this type are no longer manufactured, partly because they build up great pressures that can lead to serious accidents. Gradually, they are being replaced by tanks containing water under pressure.

Another type of fire extinguisher contains liquid carbon dioxide under pressure. When the pressure is released by opening a valve, the liquid carbon dioxide escapes and immediately evaporates. The heat that is absorbed during evaporation causes most of the escaping material to freeze, and it forms a blanket of carbon dioxide "snow."

Carbon dioxide is only slightly soluble in water (0.145 g in 100 g of water at 25°C and 1 atm pressure). Ninety-nine percent of the CO_2 stays in solution as hydrated CO_2 molecules. About 1% of the carbon dioxide reacts with water to form **carbonic acid,** H_2CO_3, a weak diprotic acid, causing an aqueous solution of CO_2 to be slightly acidic.

$$H_2CO_3 \rightleftharpoons H^+ + HCO_3^- \qquad K_{a_1} = 4.5 \times 10^{-7}$$
$$HCO_3^- \rightleftharpoons H^+ + CO_3^{2-} \qquad K_{a_2} = 4.8 \times 10^{-11}$$

Because of these equilibria, carbonated beverages have a sharp, slightly acid taste. Carbonic acid is unstable and cannot be isolated.

Two types of salts are derived from carbonic acid—the hydrogen carbonates (bicarbonates), illustrated by $NaHCO_3$ and $Ca(HCO_3)_2$, and the normal carbonates such as Na_2CO_3 and $CaCO_3$. Aqueous solutions of carbonates are mildly alkaline as a result of hydrolysis of the carbonate ion, and those of hydrogen carbonates are very weakly alkaline.

$$CO_3^{2-} + H_2O \rightleftharpoons HCO_3^- + OH^- \qquad K_{b_1} = 2.1 \times 10^{-4}$$
$$HCO_3^- + H_2O \rightleftharpoons H_2CO_3 + OH^- \qquad K_{b_2} = 2.2 \times 10^{-8}$$

The carbonate ion, CO_3^{2-}, has a planar equilateral triangle structure with the carbon atom in the center and the oxygen atoms at the corners. The ion is represented by three resonance configurations (Figure 23.7).

Minerals containing carbonate ions are plentiful in the Earth's crust, and more than half of the rock mined each year contains carbonates. The principal carbonate minerals are calcite ($CaCO_3$), which we know as limestone and marble, magnesite ($MgCO_3$), siderite

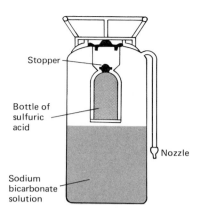

FIGURE 23.6

A CO_2 fire extinguisher. *When the tank is turned upside down, the stopper falls out of the bottle and the acid and the bicarbonate react to give CO_2, which is forced out the nozzle.*

FIGURE 23.7

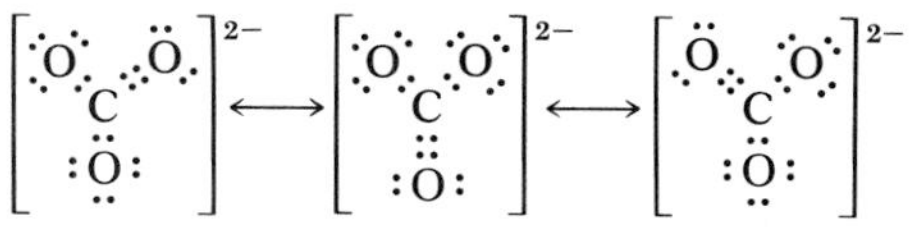

Carbonate ion, CO_3^-

FIGURE 23.8

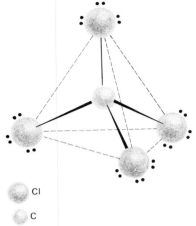

Carbon tetrachloride, CCl_4
m.p. $-23°C$
b.p. $76°C$

$$:\overset{..}{\underset{..}{Cl}}:$$
$$:\overset{..}{\underset{..}{Cl}}:\overset{..}{\underset{..}{C}}:\overset{..}{\underset{..}{Cl}}:$$
$$:\overset{..}{\underset{..}{Cl}}:$$

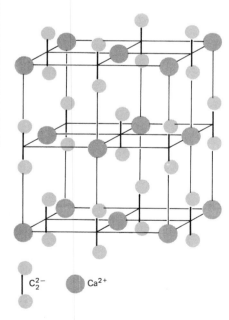

C_2^{2-} Ca^{2+}

FIGURE 23.9
The crystal structure of CaC_2. (*From A. F. Wells*, Structural Inorganic Chemistry, *3rd ed., Oxford University Press, 1962.*)

($FeCO_3$), and dolomite, which is calcite with about half of the calcium ions replaced by magnesium ions [$(Ca,Mg)CO_3$]. Calcium carbonate is also widely distributed in animal by-products such as pearls, eggshells, seashells, and coral.

c. Carbon disulfide. Carbon disulfide, CS_2, is a volatile, highly refractive liquid. The CS_2 molecule is isoelectronic with that of CO_2 (Figure 23.5). Carbon disulfide is made by the direct combination of carbon and sulfur at high temperatures. It is industrially important chiefly because of its great solvent power for waxes, greases, hydrocarbons, celluloses, and other nonpolar substances. Major disadvantages to its use are its toxicity and high flammability. Carbon disulfide vapor has been known to ignite upon contact with a hot steam pipe.

d. Carbon tetrachloride. Carbon disulfide reacts with chlorine to give carbon tetrachloride (Figure 23.8) and disulfur(I) dichloride (a compound that is sometimes erroneously called sulfur monochloride).

$$CS_2(l) + 3Cl_2(g) \longrightarrow \underset{\substack{carbon \\ tetrachloride}}{CCl_4(l)} + \underset{\substack{disulfur(I) \\ dichloride}}{S_2Cl_2(l)}$$

The products of reaction are readily separated by distillation.

At one time, carbon tetrachloride was widely used as a solvent for greases, a dry-cleaning agent, and in fire extinguishers. However, increasing awareness of its toxicity led eventually, in 1970, to a government ban on carbon tetrachloride in consumer products. The major use of carbon tetrachloride is now as a reactant and solvent within the chemical industry.

e. Carbides. The binary compounds of carbon, for example, with metals and with nonmetals such as boron, are called **carbides.** The saltlike carbides are ionic compounds between the C^{4-} or C_2^{2-} ions and metal ions. The C_2^{2-} carbides are most common. Two carbon atoms are covalently bonded to each other by three pairs of electrons.

$$[:C::C:]^{2-}$$

Carbides containing the C_2^{2-} ion readily react with water to produce acetylene and are referred to as *acetylides.*

Calcium carbide, CaC_2 (Figure 23.9), which is a typical saltlike carbide, is a colorless crystalline material prepared industrially by the reduction of calcium oxide (quicklime) with carbon in the form of coke at a very high temperature.

$$\underset{\substack{calcium\ oxide \\ (quicklime)}}{CaO(s)} + \underset{coke}{3C(s)} \overset{\Delta}{\longrightarrow} \underset{calcium\ carbide}{CaC_2(s)} + CO(g)$$

Calcium carbide produced by this reaction is the starting material in a commercial method for the production of acetylene.

$$CaC_2(s) + 2H_2O \longrightarrow Ca(OH)_2 + C_2H_2(g)$$
acetylene

Calcium cyanamide, which is an intermediate in synthesizing many organic chemicals and also a fertilizer, is made commercially from calcium carbide.

$$CaC_2(s) + N_2(g) \longrightarrow CaNCN(s) + C(s)$$
calcium cyanamide

Interstitial carbides result when carbon atoms fill open spaces in the cubic or hexagonal close-packed structures of transition metals (these compounds are analogous to the interstitial nitrides, Section 12.15). Such interstitial carbides as those of titanium, tungsten, tantalum, and niobium are very hard heat-resistant materials and are used in cutting tools. Interstitial carbides are unreactive, and the ability of the metal to conduct electricity is retained.

Boron and silicon form nonionic, diamondlike carbides. These materials come close to diamond in hardness (Figure 23.10) and are, like industrial diamonds, used as abrasives and in cutting tools. Silicon carbide is also known as Carborundum (Figure 23.11).

f. Cyanides. The major source of cyanides is hydrogen cyanide, HCN, an extremely poisonous, highly volatile liquid (b.p. 26°C), which has an odor of bitter almonds. Hydrogen cyanide is made by passing a mixture of methane, ammonia, and air over a catalyst at 800°C.

$$2CH_4(g) + 2NH_3(g) + 3O_2(g) \xrightarrow[\text{catalyst}]{800°} 2HCN(g) + 6H_2O(g)$$
methane   hydrogen cyanide

Aqueous solutions of HCN are known as hydrocyanic acid. Neutralization with a base (for example, NaOH) gives cyanide salts (for example, NaCN). The cyanide ion, like the C_2^{2-} ion, contains triply bonded carbon.

$$:C\ :N:\ ^-$$

The cyanide ion forms very stable complexes with most of the *d*-transition elements, and sodium cyanide is used in the extraction of silver and gold from their ores by complex formation as, for example, for gold.

$$4Au(s) + 8CN^- + O_2(g) + 2H_2O(l) \longrightarrow 4[Au(CN)_2]^- + 4OH^-$$

Cyanides are highly toxic and act very rapidly. As a poison, the cyanide ion functions very much like carbon monoxide, with which it is isoelectronic. Hydrogen cyanide is used as a fumigant for killing insects and rodents.

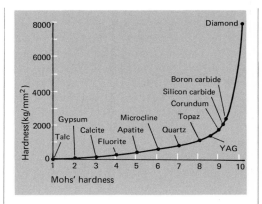

FIGURE 23.10
Hardness of diamond and other materials. *The Mohs scale for hardness assigns a value of 10 to diamond (hardest) and 1 to talc (softest).*

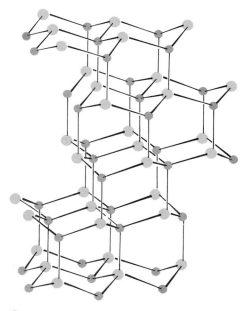

○ C
● Si

FIGURE 23.11
The crystal structure of silicon carbide. *This is one of several variations on the Carborundum crystal structure. Alternate atoms are Si and C. (From A. F. Wells,* Structural Inorganic Chemistry, *3rd ed., Oxford University Press, 1962.)*

Hydrocarbons

During the early part of the nineteenth century the compounds found in living organisms were thought to be formed only through a subtle "vital force" present in such organisms. "Organic" chemistry was named on this basis and distinguished from "inorganic" chemistry, which covered all compounds not found in living matter. It was noted that carbon atoms were present in all of the organic compounds then known.

However, in 1828 a German chemist, Friedrich Wöhler, converted ammonium cyanate, a compound that had never been found in any living organism, into urea, a substance known to be produced by animals.

$$NH_4OCN \xrightarrow{\Delta} H_2N-\overset{\overset{\displaystyle O}{\|}}{C}-NH_2$$

ammonium   *urea*
cyanate

This crucial experiment freed organic chemistry from its link to living organisms. The vital force concept gradually faded away, and organic chemistry became the chemistry of the compounds of carbon and the carbon–carbon bond. The number of organic compounds now known is in the millions. Many organic compounds do, of course, occur in nature, but even more have been made synthetically.

All organic compounds contain carbon atoms and, with a few exceptions, hydrogen atoms. Oxygen, nitrogen, the halogens, sulfur, phosphorus, silicon, and a few other elements may be present also, their frequency of occurrence in organic compounds decreasing roughly in the order given. All organic compounds can be thought of as based upon the structures of the hydrocarbons, which are introduced in the following sections. Organic compounds containing other elements are discussed later (Chapter 32).

In writing structural formulas for organic compounds, we use a dash to indicate each covalent bond between atoms. Each dash represents two electrons. No attempt is normally made to distinguish between σ bonds and π bonds. However, it is useful always to be aware that a covalent double bond is considered to be a combination of a σ bond and a π bond, and that a covalent triple bond is considered to be one σ bond and two π bonds.

23.4 Saturated hydrocarbons

Saturated hydrocarbons contain only covalent single bonds; all of the carbon atoms are sp^3 hybridized. Each tetrahedral carbon atom in a saturated hydrocarbon is joined to four other atoms, and each hydrogen atom is joined by one bond to a carbon atom. With this arrangement, it is easy to write down collections of atoms that represent actual compounds. An uninhibited approach to such an activity, limited only by the number of bonds to carbon and to hydrogen atoms, allows some chemical doodling that can be quite instructive.

For example, it is possible to arrange the atoms of C_4H_{10} in more than one way.

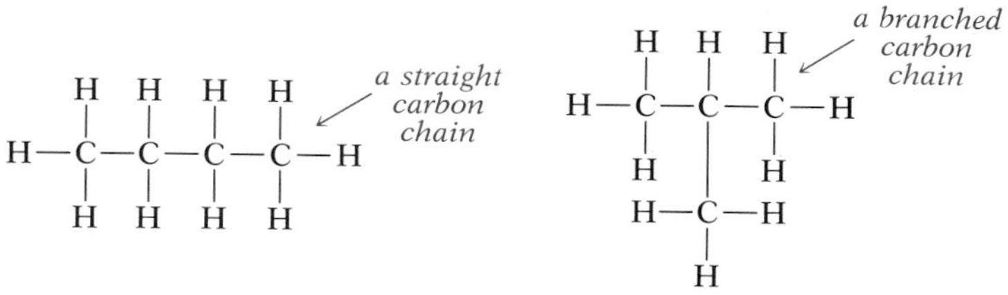

butane
(b.p. −0.5°C)

isobutane
(b.p. −12°C)

Butane and isobutane are isomers—they have the same molecular formula but are different compounds that are not readily converted into each other under ordinary conditions. Each can be isolated essentially free of the other. Saturated hydrocarbons with more than three carbon atoms all exhibit **structural isomerism**—the existence of compounds with the same molecular formula but with the atoms joined in a different order. The number of possible structural isomers increases rapidly as the number of carbon atoms in a hydrocarbon increases. One could write 75 different structures for $C_{10}H_{22}$ and (if time permitted) over 62 trillion for $C_{40}H_{82}$.

Because the saturated hydrocarbons are composed of only tetrahedral carbon atoms combined with hydrogen atoms, the bond angles are all close to 109.5° (the tetrahedral angle), the C—C bond lengths are all close to 0.154 nm, and the C—H bond lengths are all close to 0.109 nm. As a result, a continuous chain of carbon atoms with attached hydrogen atoms has a staggered, or zigzag, configuration.

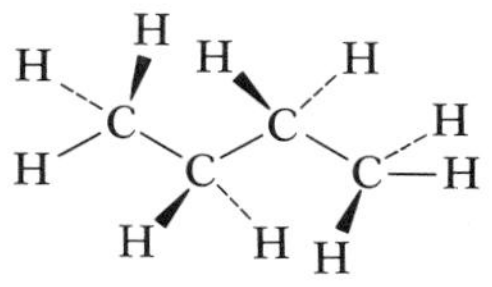

butane

In the structural formula above, the dotted bonds point toward the back of the plane of the paper and the wedge-shaped bonds point toward the viewer. Since each carbon atom can rotate about its bonds, it is possible for a long continuous chain to assume many shapes. The carbon atoms are usually written in straight lines, with the understanding that such a structural formula merely shows which atoms are bonded together.

$CH_3CH_2CH_2CH_3$
butane

Saturated hydrocarbons can have straight-chain carbon skeletons (as in butane), branched-chain carbon skeletons (as in isobutane), or cyclic, or ring-shaped, carbon chains (as in compounds 7–9, Table 23.2). The straight and branched saturated hydrocarbons are called alkanes. **Alkanes** are saturated hydrocarbons with the general molecular formula C_nH_{2n+2}. Alkanes are also sometimes referred to as **paraffin hydrocarbons** (named in 1830, when they were discovered, from the Latin meaning "barely any affinity," or not very reactive).

As n increases, a sequence of compounds is generated in which each member differs from its immediate neighbor by a CH_2 group. Such a series of compounds that can be represented by a general formula is called a **homologous series.** The first four members of the alkane series are methane, ethane, propane, and butane (1–4, Table 23.2). Note that each name ends in *ane*, indicating an alk*ane*.

It is often convenient to have a name for the group that results

TABLE 23.2

Simple saturated hydrocarbons

For the first four compounds, the common names are accepted by IUPAC.

Structure	Boiling point (°C)	Common name (IUPAC name)	Structure and common name of R groups
(1) CH_4	−162	**Methane**	CH_3—, methyl
(2) CH_3CH_3	−89	**Ethane**	CH_3CH_2—, ethyl
(3) $CH_3CH_2CH_3$	−42	**Propane**	$CH_3CH_2CH_2$—, *n*-propyl
			CH_3CHCH_3, isopropyl
			$CH_3CH_2CH_2CH_2$—, *n*-butyl
(4) $CH_3(CH_2)_2CH_3$	−0.5	**Butane**	$CH_3CH_2CHCH_3$, *sec*-butyl

(5) CH_3CHCH_3
         CH_3
	−12	**Isobutane**
		 (2-methylpropane)
			CH_3CHCH_2—, isobutyl
			       CH_3
			
			 CH_3CCH_3, *tert*-butyl
			      CH_3

(6) $CH_3(CH_2)_3CH_3$	36	**Pentane**	**(There are three isomeric R groups for this compound)**[a]

(7) H_2C——CH_2
      C
      H_2
	−33	**Cyclopropane**	H_2C——CH—, cyclopropyl
			      C
			      H_2

(8) H_2C—CH_2 H_2C—CH_2	13	**Cyclobutane**	H_2C—CH—, cyclobutyl H_2C—CH_2

(9) H_2
     C
 H_2C CH_2
 H_2C CH_2
     C
    H_2
	81	**Cyclohexane**	H_2
			     C
			 H_2C CH—, cyclohexyl
			 H_2C CH_2
			     C
			    H_2

[a] $CH_3(CH_2)_4$ is a *pentyl* group. The other R groups are not usually named as pentane-derived groups.

when one of the hydrogen atoms in a hydrocarbon molecule, say, a methane molecule, is replaced by another atom or a group of atoms. Thus, CH_3Cl could be called a chloride if the CH_3 group had a name. This would be analogous to the name, sodium chloride, for NaCl. Such names are obtained by dropping the *ane* of the hydrocarbon name and adding *yl*. Methane becomes *methyl*, ethane becomes *ethyl*, and so on. These groups containing one less hydrogen than an alkane are called **alkyl groups.** Collectively, they are very often represented by the letter R, so that R—H stands for any member of the homologous series of alkanes. Therefore, CH_3Cl is called *methyl chloride*, and RCl represents any unspecified alkyl chloride. Starting with the propyl ($n = 3$) group, isomeric R groups are possible (Table 23.2).

A systematic nomenclature for organic compounds has been formulated by the International Union of Pure and Applied Chemistry. This is known as IUPAC ("you-pack") nomenclature. Two simple rules will introduce you to the IUPAC procedure for naming alkanes with more than four carbon atoms.

1. A prefix (penta, C_5; hexa, C_6; hepta, C_7; octa, C_8; nona, C_9; deca, C_{10}) indicates the longest continuous chain of carbon atoms; the suffix *ane* is added to this prefix (an "a" is dropped).

$$CH_3CH_2CH_2CH_2CH_3 \quad or \quad CH_3(CH_2)_3CH_3$$
pentane

$$CH_3CH_2CH_2CH_2CH_2CH_3 \quad or \quad CH_3(CH_2)_4CH_3$$
hexane

2. The position and name of branches from the main chain, or of atoms other than hydrogen, are added as prefixes to the name of the longest hydrocarbon chain. The position of attachment to the longest continuous chain is given by a number obtained by numbering the longest chain from the end nearest the branch.

Some examples of this alkane nomenclature are given in Table 23.3. If there is a choice, the number at the beginning of the name should be the lowest number possible, for example

$$
\begin{array}{cccccc}
 & H & Cl & H & H & H & Cl \\
 & | & | & | & | & | & | \\
H- & C- & C- & C- & C- & C- & C-H \\
 & | & | & | & | & | & | \\
 & H & Cl & H & H & H & H
\end{array}
$$

is named 1,5,5-trichlorohexane, not 2,2,6-trichlorohexane, because 1 is smaller than 2. Common, or trivial, names—names not based on the currently recommended system of nomenclature—persist for many compounds, such as isooctane in Table 23.3. We often give both systematic and common names.

Compounds in which the carbon atoms of a saturated hydrocarbon are joined together in a ring are called cycloalkanes (also alicyclic hydrocarbons or cycloparaffins). **Cycloalkanes** are cyclic saturated hydrocarbons that have the general molecular formula C_2H_{2n}.

TABLE 23.3

Examples of systematic alkane nomenclature

$$
\overset{5}{C}H_3\overset{4}{C}H_2\overset{3}{C}H_2\overset{2}{C}HC\overset{1}{H}_3
$$

numbering begins at C nearest branch

Cl

2-chloropentane

$$
\overset{1}{C}H_3\overset{2}{C}HCH_3
$$

don't be deceived by how compound is written—look for longest chain

$$\overset{3}{C}H_2$$

$$\overset{4}{C}H_3$$

2-methylbutane

$$
\begin{array}{ccc}
CH_3 & & CH_3 \\
| & & | \\
CH_3-C-CH_2-CH \\
| & & | \\
CH_3 & & CH_3
\end{array}
$$

2,2,4-trimethylpentane

common names persist for many compounds

common name—isooctane

$$CH_2ClCH_2C(CH_3)_2CH_2CH_3$$

1-chloro-3,3-dimethylpentane

The nomenclature follows the same pattern used for the noncyclic alkanes.

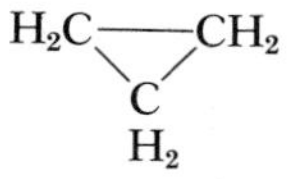

$$H_2C\text{———}CH_2$$
$$\underset{H_2}{C}$$

cyclopropane

abbreviated as ▽

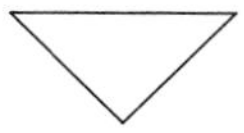

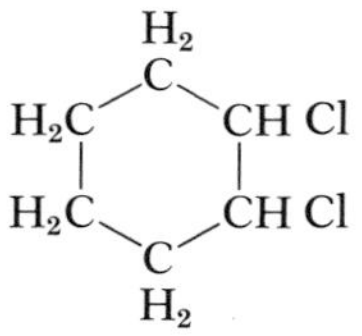

$$\underset{\underset{H_2}{|}{C}}{H_2C}\overset{\overset{H_2}{|}{C}}{\diagdown}CH\ Cl$$
$$H_2C\diagdown\underset{H_2}{C}\diagup CH\ Cl$$

1,2-dichlorocyclohexane

abbreviated as

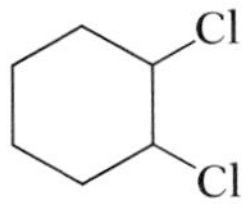

⬡ Cl
Cl

In the abbreviated expressions, we understand without writing them that there is a carbon atom at each corner and that sufficient hydrogen atoms are present to complete four bonds to each carbon atom.

At this point, close examination of Tables 23.2 and 23.3 will be very helpful. The principles of nomenclature illustrated there are used repeatedly. Note that a carbon atom joined to only one other carbon atom is called a **primary carbon atom;** one joined to two other carbon atoms, a **secondary carbon atom** (as in the *sec*-butyl group); and one joined to three carbon atoms, a **tertiary carbon atom** (as in the *tert*-butyl group). A carbon atom joined to four other carbon atoms is a **quaternary carbon atom.**

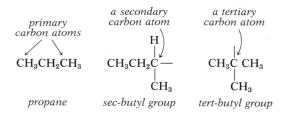

primary carbon atoms

a secondary carbon atom

a tertiary carbon atom

$CH_3CH_2CH_3$ $CH_3CH_2C\underset{|}{\overset{H}{C}}-$ $CH_3\underset{|}{C}\,CH_3$
                         $\qquad\qquad CH_3$ $\qquad CH_3$

propane   *sec-butyl group*   *tert-butyl group*

Unbranched chains are often referred to as **normal,** as in "normal hexane," written *n*-hexane. When the letter *n* is omitted, it is assumed that the normal structure is meant.

$CH_3CH_2CH_2CH_2CH_2CH_3$
n-hexane or hexane

(Nuclear magnetic resonance spectroscopy is a technique that yields information about the structure of hydrocarbons and other organic compounds; see Tools of Chemistry.)

23.5 Unsaturated hydrocarbons

Hydrocarbons that possess covalent double bonds or covalent triple bonds between carbon atoms are said to be **unsaturated hydro-**

carbons. This term arises from the ability of such bonds to accept additional atoms, or groups of atoms, with a change in the hybridization of the carbon atoms from sp^2 or sp to the sp^3 state.

Hydrocarbons with covalent double bonds are called **olefins** or **alkenes** (IUPAC). Hydrocarbons with covalent triple bonds are called **acetylenes** or **alkynes** (IUPAC). To derive the systematic names of the individual alkenes, the *ane* of the corresponding saturated hydrocarbon name is dropped and *ene* is added if one double bond is present, *diene* is added if two double bonds are present, and so on. For example,

$$CH_3CH_3 \qquad CH_2{=}CH_2 \qquad CH_2{=}CHCH{=}CH_2$$

ethane *ethene* *1,3-butadiene*

Similarly, the individual alkynes are named by dropping the *ane* and adding *yne, diyne, triyne,* and so on. In either case, the position of the multiple bond is indicated by numbering from the end of the chain, starting at the end that will assign the lower number to the first carbon atom of the multiple bond. For example,

$$CH_3CH_2CH_2CH_2CH_3 \qquad \overset{5}{C}H_3\overset{4}{C}{\equiv}\overset{3}{C}{-}\overset{2}{C}{\equiv}\overset{1}{C}H$$

pentane *1,3-pentadiyne*

In the common system of nomenclature, the *ane* ending of the saturated hydrocarbon name is replaced by *ylene* for the olefins. Compounds containing triple bonds are often named as substituted acetylenes. Table 23.4 illustrates these naming systems.

Cycloalkenes are common, but cycloalkynes exist only for C_8 or larger rings. The triple bond is not flexible enough to fit into smaller rings.

TABLE 23.4
Simple unsaturated hydrocarbons

Structure	Boiling point (°C)	Common name (IUPAC name)	Structure and common name of R group
$CH_2{=}CH_2$	−102	**Ethylene** (**ethene**)	$CH_2{=}CH{-}$, vinyl
$CH_3CH{=}CH_2$	−48	**Propylene** (**propene**)	$CH_2{=}CHCH_2{-}$, allyl
$CH_3CH_2CH{=}CH_2$	−7	**α-Butylene** (**1-butene**)	—
$(CH_3)_2C{=}CH_2$	−7	**Isobutylene** (**2-methylpropene**)	—
$CH_3CH{=}CHCH_3$	−4 (cis)	**β-Butylene** (**2-butene**)	$CH_3CH{=}CHCH_2{-}$, crotyl
$HC{\equiv}CH$	−83	**Acetylene** (**ethyne**)	$HC{\equiv}C{-}$, ethynyl
$CH_3C{\equiv}CH$	−23	**Methylacetylene** (**propyne**)	$HC{\equiv}CCH_2{-}$, propargyl
$CH_2{=}CH{-}C{\equiv}CH$	3	**Vinylacetylene** (**1-buten-3-yne**)	—

Nuclear magnetic resonance

A proton spins about its axis in the same way that an electron does, and the spinning produces a small magnetic moment. Because of this magnetic moment, the proton behaves like a tiny magnet: In an external magnetic field the proton aligns itself either parallel to the field or opposite, called antiparallel, to the field. Being parallel to the field is a somewhat lower energy condition than being antiparallel to the field. By absorbing energy in the radio frequency range of the electromagnetic spectrum, protons in a magnetic field can be made to change their alignment. This absorption is the basis for a spectroscopic technique. **Nuclear magnetic resonance spectroscopy** is the study of the absorption of radio frequency radiation by nuclei. When the nucleus being studied is the proton, it is sometimes called proton magnetic resonance. Some of the other nuclei that have spin and can be studied are deuterium (^2H), boron (^{11}B), and nitrogen (^{14}N). Here we are interested only in the proton and what can be learned from its magnetic resonance.

Nuclear magnetic resonance (NMR) is measured by placing the sample, in solution, in a constant radio frequency field (most often 60 MHz, which is 60×10^6 cycles/sec) and varying the strength of an applied magnetic field. At the appropriate combination of radio frequency and magnetic field, absorption of the radio frequency occurs and is detected and recorded. A nuclear magnetic resonance spectrum is a plot of magnetic field strength versus radio frequency absorption.

The primary use of proton magnetic resonance spectroscopy is in the determination of the structure of organic compounds. The frequency at which a proton absorbs varies with the chemical environment of that proton. Consider the methane and ethane molecules

methane

ethane

The four protons in methane all have the same chemical environment, and so do the six protons in ethane. In propane, however, the protons have two different types of surroundings

propane

Six protons have environment *a* and two protons have environment *b*. In a nuclear magnetic resonance spectrum, methane and ethane each give only one peak—absorption at only one frequency—but propane gives two peaks. And, making NMR an even more useful technique, the areas under the two peaks in the propane spectrum are proportional to the number of protons of each type—they have the ratio 6/2.

The variation in the nuclear magnetic resonance absorption by protons in different environments is called the **chemical shift.** Since it is not possible to measure the absorption of a proton that is not in some type of chemical environment, chemical shift must be measured relative to a specific standard. Most often the standard is tetramethylsilane, TMS

a compound with only one type of H environment and a peak that appears conveniently at one end of the spectrum. The absorption of TMS is set equal to zero. The chemical shift of a particular peak is the difference between its absorption and that of TMS, reported in parts per million, and called δ (delta):

$$\delta \ (\text{in ppm}) = \frac{\text{observed shift (Hz)} \times 10^6}{\text{constant spectrometer frequency (Hz)}}$$

As an example of the kinds of information that can be obtained from an NMR spectrum, we examine the spectrum of 3-methyl-1-butene, given in Figure A. Ignoring for a moment the splitting, or small peaks in each group—the four areas of absorption

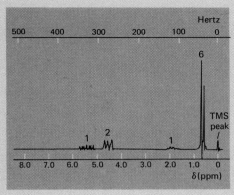

FIGURE A

Nuclear magnetic resonance spectrum of 3-methyl-1-butene.

show that there are four proton environments in the molecule, which is as it should be

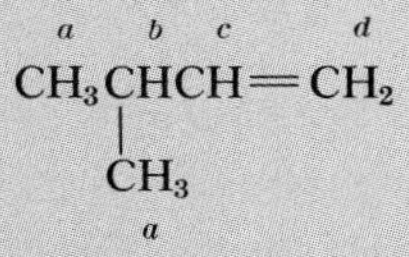

$$\overset{a}{CH_3}\overset{b}{CH}\overset{c}{CH}=\overset{d}{CH_2}$$
$$|$$
$$\underset{a}{CH_3}$$

3-methyl-1-butene

The relative areas under the peaks of $1:2:1:6$ show, as we can see in the structure, that there are six H's of one type (a), two of a second type (d), and one each of two other types (b,c). From the study of many spectra, the chemical shift ranges for different types of protons are known. Compare the following values with the spectrum: CH_3—C, 0.95–0.85; CH—, 1.6–1.4; HC=C, 4.8–6.2. With this information the complete structure of the compound could be worked out. The splitting of the absorption peak for protons in each type of chemical environment is caused by the influence of the magnetic moment of the protons on *adjacent* atoms. With experience, additional information about structure can be obtained from the splitting.

The chemical shift values for hydrogen bonded to atoms other than carbon and in different types of chemical environments have been extensively tabulated. Since its discovery in 1946, NMR spectroscopy has rapidly become an important tool in organic structure determination and also for many types of inorganic compounds.

Beginning with propylene, alkenes of the homologous series of molecular formula C_nH_{2n} are isomeric with cycloalkanes.

$$CH_3CH{=}CH_2 \qquad \overset{\displaystyle H_2C{-}\!\!-\!\!-CH_2}{\underset{\underset{\textstyle H_2}{C}}{\diagdown\ \diagup}}$$

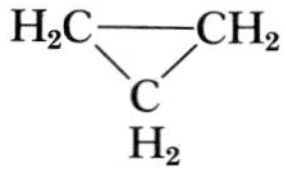

1-propene, C_3H_6   *cyclopropane, C_3H_6*

Acetylenes of the series of molecular formula C_2H_{2n-2} with $n > 2$ are isomeric with cycloalkenes.

$$CH_3C{\equiv}CH \qquad \overset{\displaystyle HC{=\!=}CH}{\underset{\underset{\textstyle H_2}{C}}{\diagdown\,\diagup}}$$

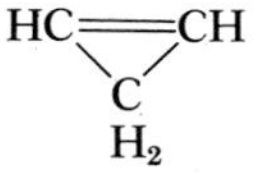

1-propyne   *cyclopropene*

The rotation of two carbon atoms joined by a double bond is inhibited by the π bond, which must be broken if rotation is to take place, because overlap of the $2p$ orbitals will be lost. As pointed out in our discussion of bonding (Section 21.5), this restricted rotation implies that cis-trans isomerism can occur. For example, 2-butene (Figure 23.12) exists in two forms that are not interconvertible without supplying the considerable amount of energy needed to break the π bond. The form with the two methyl groups on the same side of the double bond is the cis form, that with the two methyl groups on opposite sides, the trans form. When the groups are large enough to get in each other's way—called **steric hindrance**—the trans form is normally more stable than the cis form.

In cycloalkanes the carbon atoms cannot rotate, and cis–trans isomerism exists here also, as in *cis*-1, 2-dichlorocyclopropane and *trans*-1,2-dichlorocyclopropane (Figure 23.12).

The covalent double bond, but not the covalent triple bond, is very common in naturally occurring organic compounds. A notable example is a group of compounds known as *terpenes*, which have structures that are multiples of a unit of five carbon atoms, called an *isoprene unit*. Isoprene is the common name for 2-methyl-1,3-butadiene. Figure 23.13 shows the structures of two terpenes that are hydrocarbons. α-Pinene is a representative of a group of volatile sub-

FIGURE 23.12
Cis–trans isomerism.

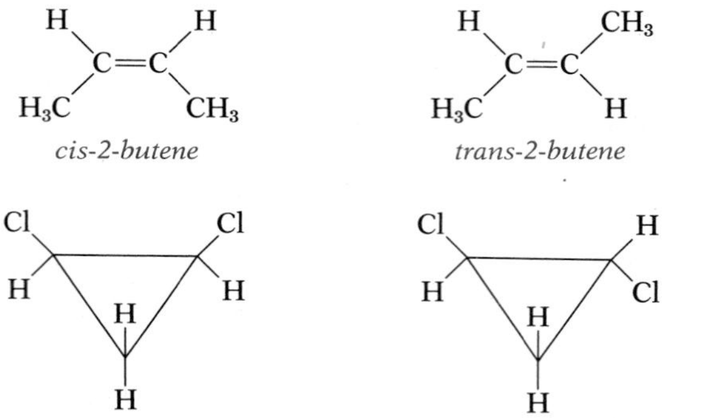

cis-2-butene   *trans-2-butene*

cis-1,2-dichlorocyclopropane   *trans-1,2-dichlorocyclopropane*

stances that contribute to the characteristic fragrance of plants and flowers. β-Carotene is the coloring material in carrots and is typical of several plant hydrocarbon pigments. The red coloring material in tomatoes is a hydrocarbon closely related to β-carotene.

Covalent double bonds that alternate with covalent single bonds are said to be **conjugated double bonds.** Compounds that contain a long system of conjugated double bonds are always *colored*.

23.6 Aromatic hydrocarbons

Benzene, C_6H_6, is the parent of a large family of compounds known as the aromatic hydrocarbons. We have described the bonding in benzene by two resonance structures (known as the Kekulé structures; see Thoughts on Chemistry, Chapter 32)

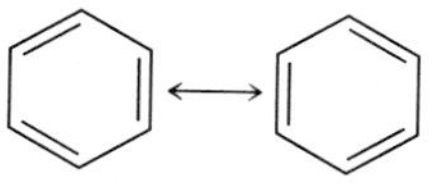

and by delocalization of electrons in molecular orbitals. (Section 21.12). We shall write the Kekulé structures to represent the benzene ring. However, remember that the benzene molecule does not actually have either Kekulé structure, and that all of the carbon–carbon bonds in a benzene ring are equivalent.

With that said, **aromatic hydrocarbons** can be defined as unsaturated hydrocarbons containing resonance-stabilized ring systems. The pleasant odor of compounds containing these ring systems originally suggested the name "aromatic," but many not-so-sweet-smelling "aromatic" compounds are now known. Instead, their similar chemical reactions are the basis for treating aromatic compounds as a group. Aromatic hydrocarbons are so distinct and different from other hydrocarbons that we have a general term for the others too. **Aliphatic hydrocarbons** are hydrocarbons that contain no aromatic rings. Saturated, unsaturated, and cyclic hydrocarbons with no aromatic rings are all aliphatic compounds.

Benzene is a colorless liquid, b.p. 80°C, that burns with a very sooty flame, a property characteristic of aromatic compounds. Atoms or groups of atoms that replace hydrogen in an organic compound are called *substituents*. Figure 23.14 lists some aromatic compounds

FIGURE 23.14
Some simple substituted benzene hydrocarbons.

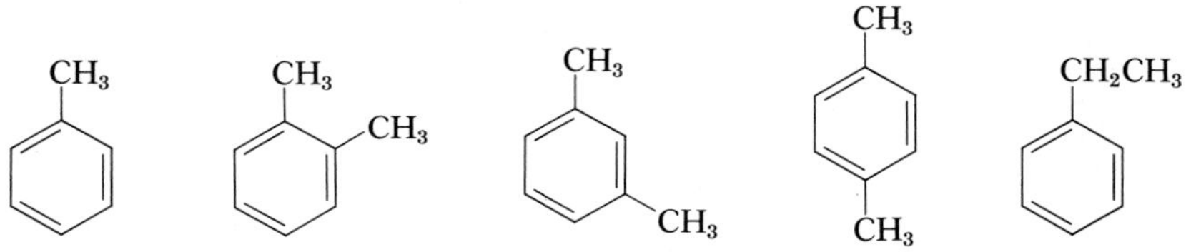

| toluene | *o*-xylene | *m*-xylene | *p*-xylene | ethylbenzene |
| b.p. 111°C | b.p. 144°C | b.p. 139°C | b.p. 138°C | b.p. 136°C |

FIGURE 23.13
Terpene hydrocarbons. *Terpene structures are multiples of the isoprene unit. The two isoprene units in pinene are shown in different colors.*

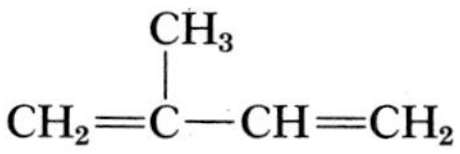

isoprene
(2-methyl-1,3-butadiene)

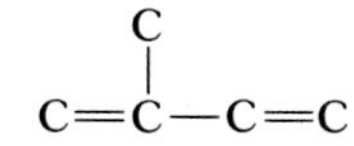

an isoprene unit

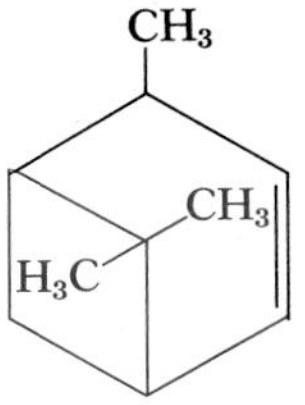

α-pinene
(occurs in turpentine)

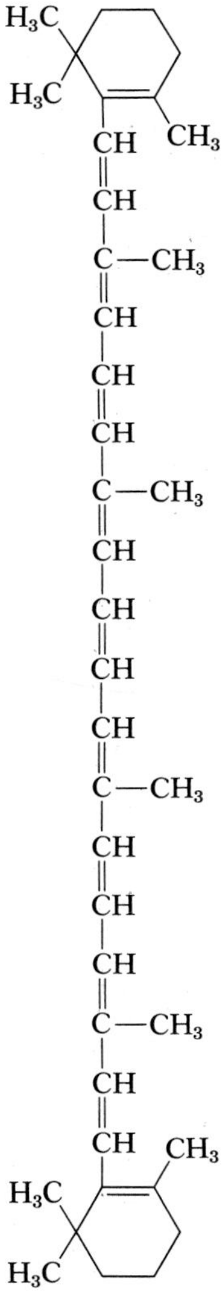

β-carotene
(pigment in carrots)

731

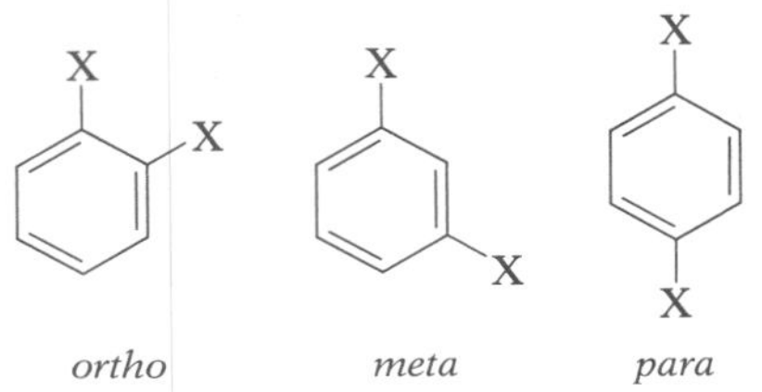

ortho   meta   para

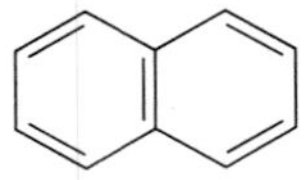

naphthalene
m.p. 80°C

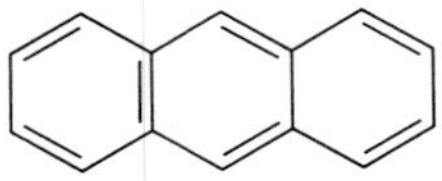

anthracene
m.p. 218°C

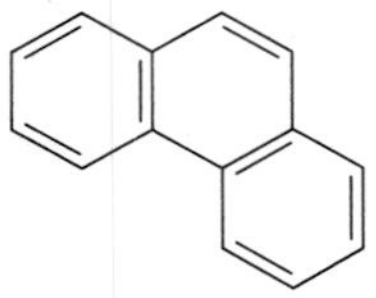

phenanthrene
m.p. 100°C

FIGURE 23.15
Some simple polycyclic aromatic hydrocarbons.

derived from benzene that contain simple substituents.

A disubstituted benzene can take one of three structurally isomeric forms, whether the substituents are alike or different. The relative positions of the two substituents are illustrated by the xylenes in Figure 23.14 and at the top of this page. The three possibilities are designated ortho (abbreviated o-), meta (m-), and para (p-).

Numbers may also be used to show the positions of substituents in aromatic compounds. An alternate name for o-xylene is 1,2-dimethylbenzene. Unless there is no question of what the structure is, as in hexachlorobenzene, numbers are always used to locate the substituents when three or more are present in the ring, for example,

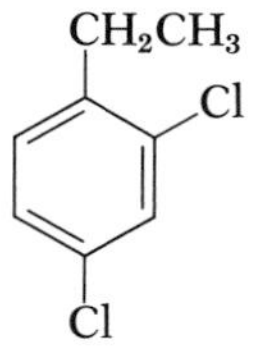

1-ethyl-2,4-dichlorobenzene

The benzene molecule less one hydrogen atom is known as the **phenyl group,** C_6H_5. Diphenylmethane, for example, is $(C_6H_5)_2CH_2$.

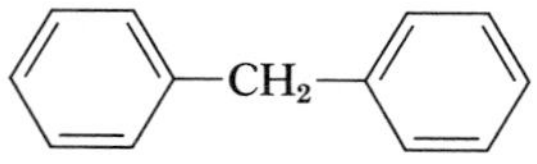

diphenylmethane

Polycyclic aromatic hydrocarbons contain two or more aromatic rings fused together, as shown by the compounds in Figure 23.15. Several resonance forms can be written for each of these compounds.

23.7 Isomerism

To review for a moment, we have thus far encountered two types of isomers, that is, compounds that have the same molecular formula but differ in the arrangement of the atoms. In structural isomerism the atoms are arranged in a different order—the same atoms are not attached to each other. Examples of structural isomers are methylcyclobutane and cyclopentane, or the three possible isomers of the diethylbenzenes (Figure 23.16).

In **cis-trans,** or geometric, **isomerism** the groups are arranged in different ways on either side of a covalent double bond, or some other rigid bond, as in a ring, for example, as in *cis-* and *trans*-2-butene (Figure 23.16). The physical properties of structural isomers and cis-trans isomers usually are different.

There is a third kind of isomerism in which the isomers are almost identical in their properties and as a result are difficult to separate. This type of isomerism is detected by polarized light and is called optical isomerism. All classes of organic compounds and many inorganic compounds can exhibit this type of isomerism.

Ordinary light rays vibrate in all planes perpendicular to the direction of travel of the rays. When such light is passed through a Nicol prism, the part of the light that is transmitted vibrates in one plane only and is called **plane-polarized light.** When a beam of plane-polarized light passes through certain substances or their solutions, the plane of vibration of the light is rotated. The amount of rotation varies from compound to compound and is measured by an instrument called a polarimeter.

What is the nature of organic substances that rotate the plane of vibration of plane-polarized light? In order to talk simply about what is a rather involved subject, we shall describe such a substance as one in which a single atom is bonded to four different kinds of atoms or groups. For example, in 3-methylhexane,

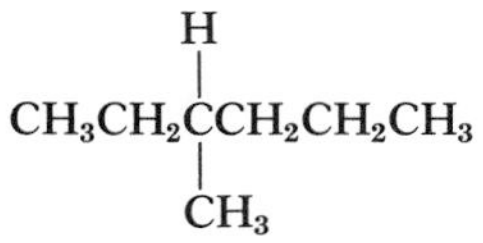

$$CH_3CH_2\overset{\overset{\displaystyle H}{|}}{\underset{\underset{\displaystyle CH_3}{|}}{C}}CH_2CH_2CH_3$$

3-methylhexane

the number 3 carbon atom in the hexane chain is bonded to an ethyl group, a methyl group, a hydrogen atom, and a propyl group. Such an atom bonded to four different kinds of atoms or groups is said to be an **asymmetric atom.** One cannot cut a molecule containing such an atom into two halves that are identical as one might, say, cut an apple into halves. An asymmetric carbon atom in a molecule causes the molecule to rotate the plane of vibration of plane-polarized light; the molecule is said to be **optically active.** The carbon atom itself is not asymmetric, of course; it is the structure containing a tetrahedral carbon atom with four different kinds of attached groups that is asymmetric. The carbon atom is asymmetrically substituted.

Optically active substances exist in pairs. One isomer rotates plane-polarized light counterclockwise, or to the left, and is called *levorotatory.* The other isomer rotates plane-polarized light an exactly equal amount clockwise, or to the right, and is called *dextrorotatory.* The members of such pairs are called **enantiomers** or **optical isomers. Optical isomerism** is the occurrence of pairs of molecules of the same molecular formula that rotate plane-polarized light equally in opposite directions. The individual molecules are referred to as the L-isomer and the D-isomer. In most properties, such as melting point, boiling point and solubility, two optical isomers behave identically and appear to have the same molecular structure. However, the dextro and levo forms can differ greatly in their reactivity toward other optically active substances. For example, levorotatory nicotine is sixty times more poisonous than dextrorotatory nicotine.

Predictably, a mixture of equal parts of the levorotatory isomer and the dextrorotatory isomer of the same substance—a **racemic mixture**—shows no net rotation of polarized light as a result of the cancellation of the equal and opposite rotation of the isomers. Such a mixture is optically inactive.

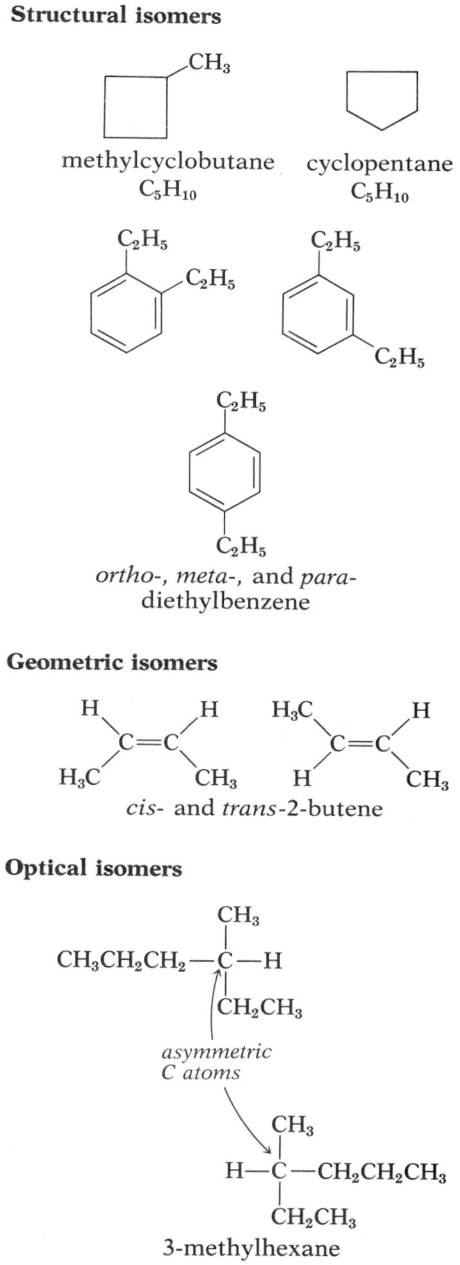

Structural isomers

methylcyclobutane C_5H_{10} cyclopentane C_5H_{10}

ortho-, meta-, and *para-* diethylbenzene

Geometric isomers

cis- and *trans-*2-butene

Optical isomers

asymmetric C atoms

3-methylhexane

FIGURE 23.16

Isomers. *Examples of the three types of isomerism introduced in this chapter.*

FIGURE 23.17

Enantiomeric forms of 3-methylhexane. (a) *Three-dimensional representation;* (b) *simulated three-dimensional models;* (c) *Fischer projections of the models in* (b) *in which horizontal lines point toward the reader, vertical lines away from the reader.*

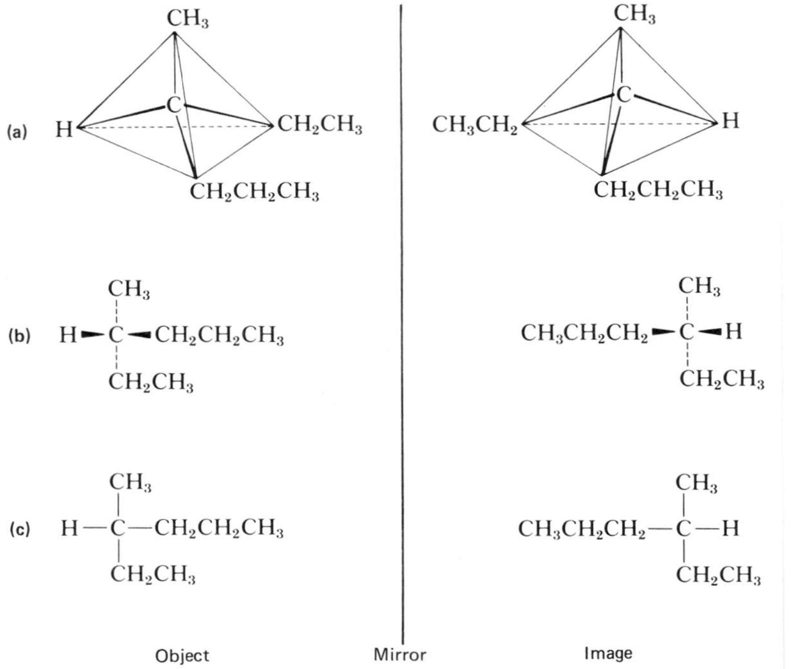

(a)

CH$_3$

H — C — CH$_2$CH$_3$

CH$_2$CH$_2$CH$_3$

CH$_3$

CH$_3$CH$_2$ — C — H

CH$_2$CH$_2$CH$_3$

(b)

CH$_3$

H ◄ C ◄ CH$_2$CH$_2$CH$_3$

CH$_2$CH$_3$

CH$_3$

CH$_3$CH$_2$CH$_2$ ► C ◄ H

CH$_2$CH$_3$

(c)

CH$_3$

H — C — CH$_2$CH$_2$CH$_3$

CH$_2$CH$_3$

CH$_3$

CH$_3$CH$_2$CH$_2$ — C — H

CH$_2$CH$_3$

Object   Mirror   Image

The members of optically active pairs have the same relationship to each other as one's left hand has to one's right hand, that is, they are mirror images of each other. Just as one's left hand is not identical with one's right hand, so molecules of optical isomers are not superimposable. An asymmetric molecule cannot be placed over its mirror image and have all like parts coincide. The two forms have different *spatial configurations*. Figure 23.17 represents the mirror image configuration of 3-methylhexane in two dimensions. (Models help in seeing the difference.) **Chirality** is the property of having nonsuperimposable mirror images.

A laboratory synthesis in which an asymmetric center is formed always gives a racemic mixture. The separation of the optical isomers in a racemic mixture, a process called **resolution,** can often be accomplished by converting the isomers temporarily into derivatives that are no longer mirror images, and hence have different solubilities.

Many naturally occurring substances exist as one or the other member of an enantiomeric pair, implying that the plant or animal that produced a given substance is discerning in this respect. Furthermore, living systems are often able to distinguish between optical isomers that are administered to them. For example, *only* the dextrorotatory form of an optically active drug might be effective in a specific case, and the levorotatory isomer might have little or no effect or even be toxic. Or just the reverse might prevail. It is this biological specificity that often requires considerable ingenuity in the synthesis of compounds that duplicate complex natural products, especially when several asymmetric atoms are present in a single molecule and the number of possible optically active isomers is quite large.

[23.7] CARBON
AND HYDROCARBONS

23.8 Properties and reactions of hydrocarbons

The properties of hydrocarbons are to a large extent those imparted by the covalent bond (Section 9.18) in combination with those imparted by intermolecular forces (Section 9.20–9.22). In solubility, the behavior of hydrocarbons reflects the fact that they are nonpolar. Hydrocarbons are virtually insoluble in water but are soluble in each other. Most hydrocarbons are less dense than water and float on the surface. Within any series of similar alkanes, alkenes, or aromatic hydrocarbons, melting points and boiling points generally increase with increasing molecular weight (as shown by the boiling points given in Tables 23.2 and 23.4). Alkanes containing eighteen or more carbon atoms are waxy solids at room temperature (the paraffins).

The reactivity of hydrocarbons varies with the presence or absence of multiple covalent bonds and with whether or not the compound is aromatic. To make an analogy between the business world and the chemical world, electrons are the "currency" in chemical reactions. An atom that is somewhat deficient in electrons and is therefore slightly positive is receptive to interaction with an atom or group of atoms that possess a pair of loosely held electrons. In other words, electron-rich, or **nucleophilic,** centers and electron-poor, or **electrophilic** centers are potential reaction sites.

A polar bond in a molecule presents a likely site for a reaction to occur. The bond-breaking process may already be in its incipient stages. The bond polarity of the C—H bond is relatively small. Therefore, saturated hydrocarbons do not show a high degree of chemical reactivity at the C—H bonds. The C—C bond is also unreactive, as might be predicted. The alkanes are not attacked at room temperature by acids, bases, or oxidizing agents such as potassium permanganate or potassium dichromate.

One reaction of alkanes that does occur readily is the replacement of a hydrogen atom by reaction with a halogen atom, a halogenation reaction. This type of reaction is typical of halogen atoms generated by the absorption of energy, for example, from light (see Section 20.8c).

$$Cl_2 \xrightleftharpoons{\text{light}} 2Cl\cdot$$
$$Cl\cdot + CH_4 \longrightarrow HCl + CH_3\cdot$$
$$CH_3\cdot + Cl_2 \longrightarrow \underset{\substack{\textit{methyl} \\ \textit{chloride}}}{CH_3Cl} + Cl\cdot$$

All hydrocarbons burn readily in air. We have already discussed in several places the large amount of energy derived from the combustion of hydrocarbons. One representative reaction is the combustion of pentane.

$$C_5H_{12}(l) + 8O_2(g) \longrightarrow 5CO_2(g) + 6H_2O(g)$$
$$\Delta H^\circ = -770.3 \text{ kcal}(-3223 \text{ kJ})$$

From the equation it is apparent that the hot products occupy a larger volume than the reactants and that combustion of hydrocarbons can provide the thrust to drive the piston in an engine.

The reactivity of double and triple bonds between carbon atoms in nonaromatic compounds is greater than that of similar single bonds. The π bond is weaker than a σ bond because of less favorable orbital overlap. Therefore, atoms in the *sp* or *sp²* hybridized state are readily transformed to the *sp³* state in aliphatic hydrocarbons. **Addition reactions**—the addition of atoms or groups to the carbon atoms in a covalent double bond or a covalent triple bond—occur readily for unsaturated hydrocarbons (Table 23.5). In such reactions, electrophilic atoms or groups readily acquire the π electrons and form σ bonds to the carbon atoms, which change, in the case of alkenes, from the *sp²* state to the *sp³* state.

Alkenes and alkynes are also quite readily oxidized, eventually leading to cleavage of the double or triple covalent bond (Table 23.5). Both the reaction with bromine and that with potassium

TABLE 23.5

Reactions of unsaturated hydrocarbons

Addition to an alkene

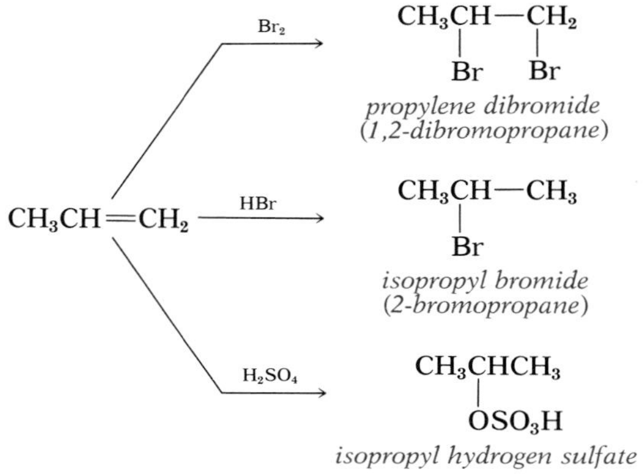

Addition to an alkyne

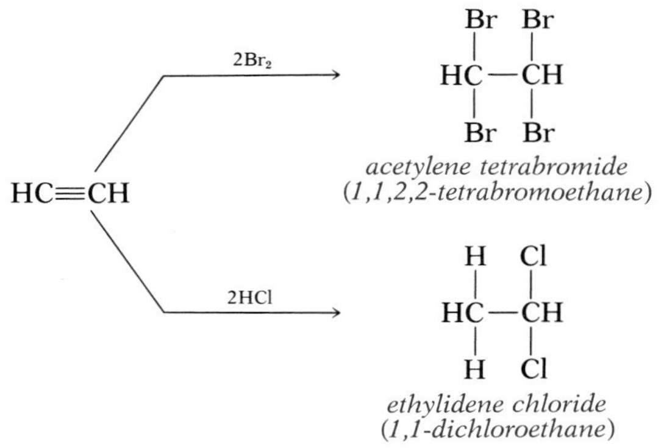

Oxidation of an alkene

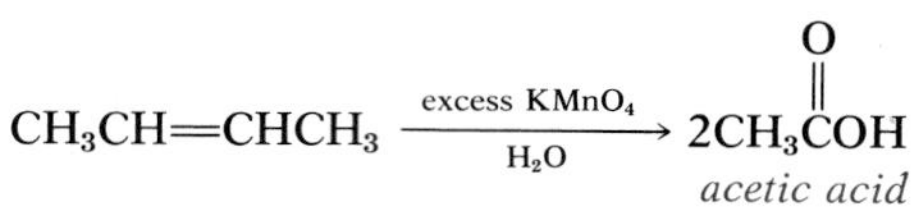

permanganate are useful to test for the presence of double and triple bonds in a molecule. The reddish color of the bromine and the purple color of aqueous permanganate are observed to disappear when in the presence of unsaturated compounds.

The reactivity of aromatic compounds differs from that of non-aromatic unsaturated compounds. In contrast to the addition reactions of olefins and acetylenes, aromatic compounds undergo **substitution reactions**—replacement of one or more hydrogen atoms of the aromatic ring with reagents such as bromine or sulfuric acid (Table 23.6). Resonance energy stabilizes the aromatic ring so strongly that the ring is preserved. The ring is also quite stable to oxidation reactions such as that with potassium permanganate (Table 23.6). The stability to oxidation and the reaction by substitution instead of addition together comprise what is called **aromatic character.**

TABLE 23.6

Reactions of aromatic compounds

The reactions shown here are typical of all aromatic compounds.

Substitution reactions

benzenesulfonic acid

bromobenzene

nitrobenzene

ethylbenzene

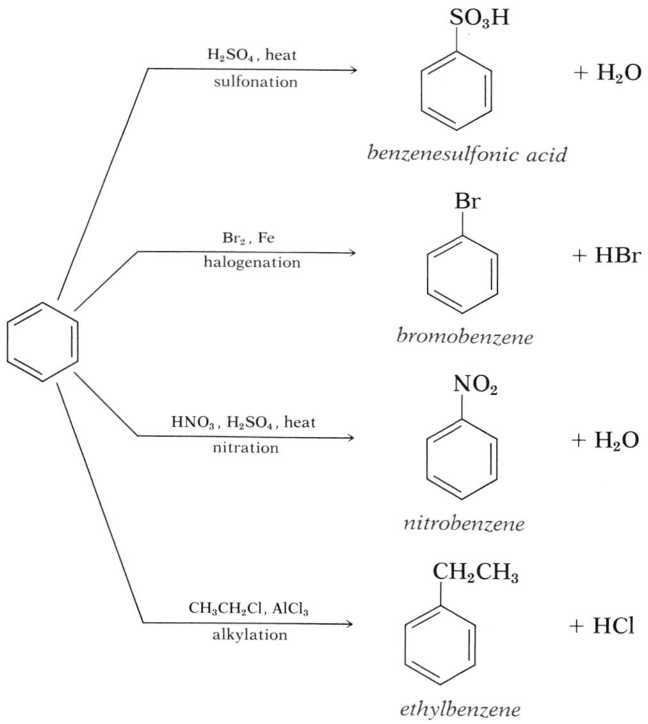

Oxidation

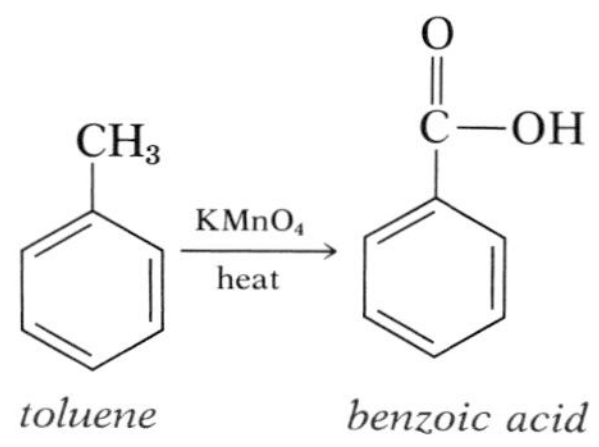

toluene   benzoic acid

23.9 Sources and uses of carbon and hydrocarbons

Carbon, the nineteenth most abundant element, makes up only 0.027% of the Earth's crust. Its importance far outweighs its abundance.

In nature, elemental carbon is found as both diamond and graphite. Naturally occurring inorganic carbon compounds include carbon monoxide and carbon dioxide cycling through the atmosphere and hydrosphere, hydrogen carbonate ion dissolved in natural waters, and limestone, dolomite, and other carbonate-bearing minerals.

The major sources of hydrocarbons and also carbon are the fossil fuels—natural gas, petroleum, and coal—formed by the decay of plants and animals over a period of millions of years. The major use of hydrocarbons and carbon is as fuel. Approximately 75% of the energy consumed in the United States is derived from natural gas and petroleum. The energy available to a society determines what that society can do and may determine what that society will do. The economics and the availability of hydrocarbons are changing. How long the current dominant role of hydrocarbons as an energy source can, or will, continue is a question that is open to debate.

Natural gas, often associated with petroleum, consists mainly of methane together with decreasing amounts of ethane, propane, butane, and isobutane. Pipelines for carrying natural gas lead from large producing fields to the major industrial areas, where it is used directly as a fuel. A major recent development has been the transportation of liquefied natural gas in refrigerated tankers from large sources of supply to those nations, such as Japan, that have a very limited domestic supply.

Petroleum, as it is pumped from underground reservoirs, is a dark, thick, smelly liquid. Up to 95% of crude petroleum (Table 23.7) is a mixture of hydrocarbons, including alkanes, cycloalkanes, and aromatics, but no alkenes. A use has been found for virtually every component of petroleum. The first step toward manufacture of petroleum products is separation of the crude petroleum by distillation and extraction into fractions possessing various boiling points and molecular weights. These fractions can be used as indicated in Table 23.8 or treated further by a variety of processes to convert the initial fractions into more desirable mixtures or pure compounds.

For example, at least two-thirds of the molecules in the gasoline that you obtain at a service station were not present as such when the petroleum was pumped from the oil well. Isomerization, cracking, alkylation, and reforming are processes used to increase the yield of gasoline from the crude oil and to improve its quality. **Petroleum isomerization,** accomplished by heat and catalysts, converts straight-chain alkanes into branched alkanes. The latter perform better as fuels. **Petroleum cracking,** also via heat and catalysts, breaks large molecules above the gasoline range into smaller molecules (alkanes and also alkenes) that are in the gasoline range. **Petroleum alkylation** combines lower-molecular-weight alkanes and alkenes to form molecules in the gasoline range. **Petroleum reforming** employs catalysts

TABLE 23.7

Approximate composition of crude petroleum

Hydrocarbons Alkanes, cycloalkanes, and aromatics	**50–95%**
Oxygen, nitrogen, and sulfur- containing compounds	**0.5–8%**
Resins and asphalts Polymers and polycyclic aromatics, among others	**5–15%**

The fractions may be divided in various ways, depending upon the temperatures in the distillation column and the needs.

TABLE 23.8
Petroleum fractions

Fraction	Boiling range (°C)	Composition	Uses
Petroleum ether	20–60	Pentanes and hexanes	Solvent
Ligroin (naphtha)	60–100	Hexanes and heptanes	Solvent; raw material
Gasoline	40–220	Butanes to tridecanes (C_4–C_{13}, mainly C_6–C_8)	Fuel for internal combustion engines
Kerosene	175–325	Octanes to tetradecanes (C_8–C_{14})	Heating fuel
Gas oil	Above 275	Dodecanes to octa-decanes (C_{12}–C_{18})	Diesel and heating fuel
Paraffin	m.p. 50–60	Tricosanes to nona-cosanes (C_{23}–C_{29})	Wax products
Lubricating oils and greases	—	Above C_{18}	Lubrication
Asphalt or petroleum coke	—	Residue	Roofing and paving

in the presence of hydrogen to convert noncyclic hydrocarbons to aromatic compounds. These processes also are used to supply higher yields of hydrocarbons needed for other purposes than gasoline.

Petroleum is the source of compounds that serve as starting materials for the syntheses of a major portion of the industrial organic chemicals, particularly those used in plastics, coatings, and synthetic rubber. However, only 10% of the petroleum processed each year goes to the chemical industry in the form of raw materials. The remaining 90% is eventually burned.

Coal, like natural gas, can be burned directly as a fuel. Coal contains carbon, hydrogen, nitrogen, oxygen, and sulfur in varying amounts. The carbon content of coal ranges from 45% to 95%. The second most important use of coal is in production of coke for the iron and steel industry (Sections 29.16, 29.17). Coke is made by heating coal in the absence of air—to drive off volatile materials—a process called carbonization. Ammonia and coal tar, from which organic chemicals, particularly aromatics, can be obtained are important by-products of this operation.

$$\text{Coal} \xrightarrow{\text{heat, no air}} \text{NH}_3 + \text{other gases (H}_2\text{, CH}_4\text{, CO, CO}_2\text{)} +$$
$$\text{coke} + \text{coal tar (source of aromatic compounds)}$$

Another source of hydrocarbons that could be developed if economics permitted (petroleum from oil wells has been cheaper in the past) is the petroleum trapped in mineral deposits—the tar sands and oil shales. Also, to obtain it in a form that would be more convenient and possibly more widely used, coal might be converted to a liquid or gaseous fuel. And it is also possible, if the price were right, to isolate carbon and organic compounds from wood and other plant sources.

The chemical history
of a candle (continued)

THE CHEMICAL HISTORY OF A CANDLE, by
Michael Faraday

*Before we leave the subject of carbon, let us make a few experiments
and remarks upon its wonderful condition as respects ordinary combustion. I have shown you that the carbon, in burning, burns only as
a solid body, and yet you perceive that, after it is burned, it ceases to
be a solid. There are very few fuels that act like this. It is, in fact, only
that great source of fuel, the carbonaceous series, the coals, charcoals,
and woods, that can do it. I do not know that there is any other elementary substance besides carbon that burns with these conditions;
and if it had not been so, what would happen to us? Suppose all fuel
had been like iron, which, when it burns, burns into a solid substance. We could not then have such a combustion as you have in
this fireplace. . . .*

*If, when the carbon burned, the product went off as a solid body,
you would have had the room filled with an opaque substance, as in
the case of the phosphorus; but when carbon burns, everything passes
up into the atmosphere. It is in a fixed, almost unchangeable condition before the combustion; but afterwards it is in the form of gas,
which it is very difficult (though we have succeeded) to produce in a
solid or a liquid state. . . .*

*It is a striking thing to see that the matter which is appointed to
serve the purpose of fuel waits in its action: it does not start off
burning, like the lead and many other things that I could show you;
but it waits for action. This waiting is a curious and wonderful thing.
Candles do not start into action at once, like the lead or iron (for iron
finely divided does the same thing as lead), but there they wait for
years, perhaps for ages, without undergoing any alteration.*

THE CHEMICAL HISTORY OF A CANDLE: A SERIES OF LECTURES GIVEN IN 1860,
by Michael Faraday, 1860 (pp. 148ff., as reprinted by Crowell, New York,
1957).

Significant terms defined in this chapter

carbonic acid
carbides
saturated hydrocarbons
structural isomerism
alkanes
paraffin hydrocarbons
homologous series
alkyl groups
cycloalkanes
primary carbon atom
secondary carbon atom
tertiary carbon atom
quaternary carbon atom
normal hydrocarbon chain
unsaturated hydrocarbon

alkenes; olefins
acetylenes; alkynes
nuclear magnetic resonance
  spectroscopy
chemical shift
steric hindrance
conjugated double bond
aromatic hydrocarbons
aliphatic hydrocarbons
phenyl group
cis–trans isomerism
plane-polarized light
asymmetric atom
optically active

optical isomerism
optical isomers; enantiomers
racemic mixture
chirality
resolution
nucleophilic
electrophilic
addition reactions
substitution reactions
aromatic character
petroleum isomerization
petroleum cracking
petroleum alkylation
petroleum reforming

Exercises

23.1 Gasoline is given an "octane number" on the basis of matching its performance to a synthetic gasoline made by mixing isooctane (octane rating = 100) with normal heptane (octane rating = 0). For example, a rating of 85 for a gasoline sample indicates that it produces the same amount of "knocking" in a test engine as a mixture containing 85% isooctane and 15% n-heptane. Assuming that a gasoline sample contains only a mixture of the various structural isomers of octane and heptane, write the structural formulas for as many of the isomers that you can and name each compound.

23.2 Write the electronic configuration of C and Si. What would you predict for the formula of the compounds formed between these elements and fluorine? Although silicon can form the SiF_6^{2-} ion, the analogous ion for carbon, CF_6^{2-}, is unknown. Why?

23.3 Prepare Lewis structures for N_2, CN^-, CO, and C_2^{2-}. What is the bond order in each? Assuming that the available molecular orbitals on CN^- and CO are similar to those on N_2 and C_2^{2-}, write the electronic configuration for all four species. What type of hybridization is present?

23.4 Complete and balance the following equations:

(a) $CaCO_3(s) \xrightarrow{\Delta}$

(b) $NaHCO_3(s) + H^+ \longrightarrow$

(c) $CS_2(g) + Cl_2(g) \longrightarrow$

(d) $Fe_2O_3(s) + CO(g) \xrightarrow{\Delta}$

(e) $CO_2(g) + C(s) \xrightarrow{\Delta}$

(f) $CaO(s) + C(s) \xrightarrow{\Delta}$

23.5 Which statements are true? Rewrite any false statement so that it is correct.
(a) The cycloalkanes are cyclic saturated hydrocarbons with the general formula C_nH_{2n+2}.
(b) An alkyl group contains one less hydrogen atom than the alkane from which it was formed.
(c) Acetylides readily react with water to produce acetylene.
(d) Optical isomers have the same molecular formula, but the atoms are joined in a different order.
(e) Resolution is the preparation of an optically inactive mixture.
(f) The stability of an unsaturated carbon ring to oxidation and to the preference of substitution reactions instead of addition reactions is known as aromatic character.

23.6 Distinguish clearly among (a) saturated and unsaturated hydrocarbons; (b) alkanes, alkenes (olefins), and alkynes (acetylenes); (c) aromatic and aliphatic hydrocarbons; (d) nucleophilic and electrophilic centers; (e) isomerization, cracking, alkylation, and reforming of petroleum.

23.7 Draw the molecular structures for each of the following compounds:

(a) n-pentane
(b) methylcyclobutane
(c) 2,2-dimethylpropane
(d) 2,4-hexadiene
(e) 2,3-dibromobutane
(f) m-dinitrobenzene
(g) phenylacetylene
(h) 1,2,4,5-tetraethylbenzene

23.8 Both α-pinene and β-carotene (Figure 23.13) have molecular structures that can be viewed as constructed from two or more isoprene units. Draw a dotted line across each carbon–carbon bond that must be broken to form the maximum number of isoprene units from these compounds.

23.9 Name each of the following groups:

(a) CH_3CH_2-

(b) CH_3CHCH_3 (with bond below)

(c)

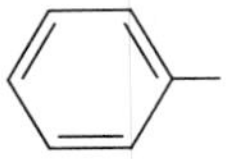

(d) $CH_3-\underset{\underset{CH_3}{|}}{\overset{\overset{CH_3}{|}}{C}}-$

(e)

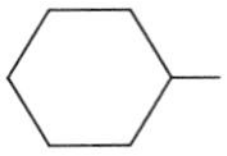

(f) CH_3CHCH_2- (with CH_3 below)

(g) $CH_3CH_2CHCH_3$ (with bond below)

(h)

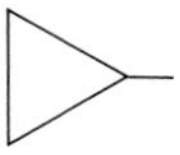

23.10 Redraw each of the following structures to show all of the atoms and bonds:

(a)

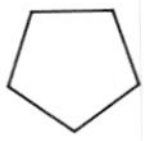

(b)

(c)

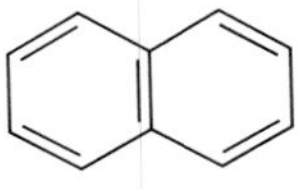

(d)

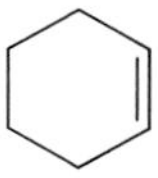

(e)

(f)

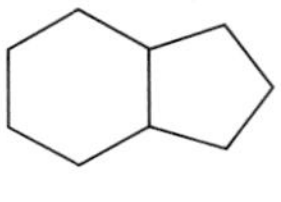

23.11 Draw all of the isomeric structures that could result from the substitution of one chlorine atom for one hydrogen atom in each of the following compounds. Name all of the compounds, including those shown.

(a) $CH_3CH_2CH_3$

(b) $CH_3-\underset{\underset{CH_3}{|}}{\overset{\overset{CH_3}{|}}{C}}-CH_3$

(c) $CH_3CH_2CH_2CH_2CH_3$

(d) $CH_3-\overset{\overset{CH_3}{|}}{CH}-CH_3$

(e) $\underset{\underset{H_2}{C}}{H_2C\overline{\qquad}CH_2}$

(f) $CH_3CH=CH_2$

(g) $CH_3\underset{\underset{CH_3}{|}}{CH}-\underset{\underset{CH_3}{|}}{CH}CH_3$

(h) $CH_3\underset{\underset{CH_3}{|}}{CH}CH_2CH_3$

23.12 Draw all of the possible molecular structures corresponding to each of the following molecular formulas. For all saturated carbon atoms, indicate whether they are primary, secondary, tertiary, or quaternary.

(a) C_5H_{12} (3 structures)
(b) C_2H_6O (2 structures)
(c) C_3H_6 (2 structures)
(d) $C_3H_6Cl_2$ (4 structures)
(e) C_6H_{14} (5 structures)
(f) C_3H_4 (3 structures)
(g) C_3H_8O (3 structures)
(h) C_4H_6 (8 structures)

23.13 Indicate the type of hybridization on each carbon atom in the following compounds:

(a) $CH_3CH=CH_2$
(b) $CH_2=CH-C\equiv CH$
(c) $CH_3-\underset{\underset{H}{|}}{C}=O$

(d) 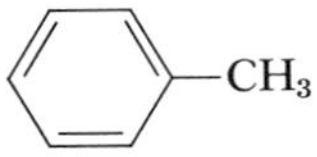 $-CH_3$

(e) $H-C\equiv N$

(f)

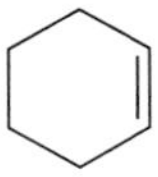

(g) $CH_3-\underset{\underset{Br}{|}}{\overset{\overset{H}{|}}{C}}-\underset{\underset{Br}{|}}{\overset{\overset{H}{|}}{C}}-CH_3$

(h) 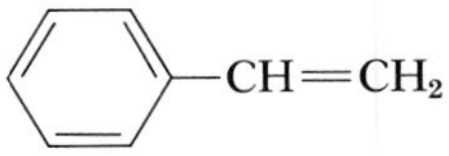 $-CH=CH_2$

(i) $CH_2=C=CH_2$
(j) $O=C=O$

23.14 Which of the following compounds can have geometric isomers? Prepare sketches of the cis and trans isomers of each.

(a) $CH_3CH=CHCH_3$
(b) $CH_3CH_2CH=CH_2$
(c) $CH_3CH_2C\equiv CH$
(d) $CH_3HC\underset{\underset{H_2}{C}}{\overline{\qquad}}CHCH_3$
(e) $CH_3CH=CHCH=CH_2$

(f) $CH_2=C\underset{\underset{CH_3}{|}}{C}CH=CH_2$

(g) $CH_3CH=CHCl$

(h) 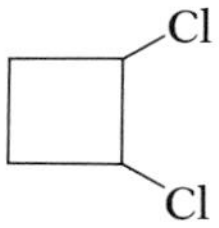 (Cl, Cl)

23.15 Complete the following equations:

(a) $CH_3CH_3 + Cl_2 \xrightarrow{h\nu}$

(b) 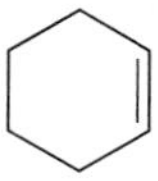 $+ Br_2 \longrightarrow$

(c) $CH_3-\underset{\underset{CH_3}{|}}{C}=CH_2 + H_2SO_4 \longrightarrow$

(d) $CH_3C\equiv CH + 2HBr \longrightarrow$

(e) 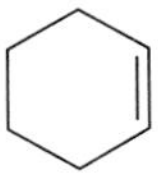 $\xrightarrow[H_2O]{KMnO_4 \text{ (excess)}}$

(f)  $\xrightarrow[H_2O, \Delta]{KMnO_4}$

(g) [toluene structure with CH$_3$] + Cl$_2$ $\xrightarrow{h\nu}$

(h) [toluene structure with CH$_3$] + Cl$_2$ $\xrightarrow{\text{Fe}}$ (show all isomers)

(i) [benzene structure] + CH$_3$CHCH$_3$ (with Cl substituent) $\xrightarrow{\text{AlCl}_3}$

(j) [naphthalene structure] + H$_2$SO$_4$ $\xrightarrow{\Delta}$ (show all isomers)

23.16 The heat of combustion at 25°C of diamond is -94.504 kcal/mole and of graphite is -94.051 kcal/mole. What is $\Delta H°$ for the reaction in which diamond forms graphite? Which crystalline state is more stable at room temperature?

23.17 Pieces of dry ice totaling 2.2 g were added to 100 ml of a 5.0M NaOH solution. What carbon-containing compound is formed and in what amount?

23.18 The production of "water gas" is represented by the equation

$$C(s) + H_2O(g) \rightleftharpoons CO(g) + H_2(g)$$

at 1000°K, $K_p = p_{CO}p_{H_2}/p_{H_2O}$ is 3.2. What are the partial pressures of CO and H$_2$ at this temperature if $p_{H_2O} = 15.6$ atm?

23.19 The heat of formation at 25°C of ZnO(s) is -348 kJ/mole, and of CO(g) is -111 kJ/mole. What is the enthalpy change for the reduction of zinc oxide to zinc metal by carbon?

23.20 The heat of formation at 25°C is -17.88 kcal/mole for CH$_4$(g), 171.291 kcal/mole for C(g), and 52.095 kcal/mole for H(g). Calculate the average C—H bond energy. The heat of formation at 25°C is -20.24 kcal/mole for C$_2$H$_6$(g), 12.49 kcal/mole for C$_2$H$_4$(g), and 54.19 kcal/mole for C$_2$H$_2$(g). Calculate the C—C, C=C, and C≡C bond energies.

23.21* The heat of combustion at 20°C is -3.487 MJ/mole for n-C$_5$H$_{12}$, -4.141 MJ/mole for n-C$_6$H$_{14}$, -4.811 MJ/mole for n-C$_7$H$_{16}$, and -5.450 MJ/mole for n-C$_8$H$_{18}$. Using -286 kJ/mole for the heat of formation of H$_2$O(l) and -395 kJ/mole for CO$_2$(g), calculate the heat of formation of each of the alkanes. Prepare a plot of $\Delta H°_{f,293}$ against the number of carbon atoms in these molecules and predict the heat of formation of n-C$_9$H$_{20}$.

$$C_nH_{2n+2}(l) + \left(\frac{3n+1}{2}\right)O_2(g) \longrightarrow nCO_2(g) + (n+1)H_2O(l)$$

23.22** You are given 1.00 kg of C(s) and asked to produce a simple fuel that yields a large amount of heat upon combustion. The first fuel you decide to try is "water gas"

$$C(s) + H_2O(g) \longrightarrow CO(g) + H_2(g)$$

which burns by

$$CO(g) + \tfrac{1}{2}O_2(g) \longrightarrow CO_2(g)$$
$$H_2(g) + \tfrac{1}{2}O_2(g) \longrightarrow H_2O(l)$$

The second fuel you decide to try is methane, made by the following process:

$$4Al(s) + 3C(s) \longrightarrow Al_4C_3(s)$$
$$Al_4C_3(s) + 6H_2O(l) \longrightarrow 3CH_4(g) + 2Al_2O_3(s)$$

which burns by

$$CH_4(g) + 2O_2(g) \longrightarrow CO_2(g) + 2H_2O(l)$$

The third fuel you decide to try is acetylene, made by the following process:

$$Ca(s) + 2C(s) \longrightarrow CaC_2(s)$$
$$CaC_2(s) + 2H_2O(l) \longrightarrow Ca(OH)_2(aq) + C_2H_2(g)$$

which burns by

$$C_2H_2(g) + \tfrac{5}{2}O_2(g) \longrightarrow 2CO_2(g) + H_2O(l)$$

and finally you decide to try benzene, made from the acetylene by

$$3C_2H_2(g) \longrightarrow C_6H_6(l)$$

which burns by

$$C_6H_6(l) + \tfrac{15}{2}O_2(g) \longrightarrow 6CO_2(g) + 3H_2O(l)$$

(a) What is the oxidation state of C in each of the above compounds? (b) Which fuel generates the most heat for the 1.00 kg of carbon used, given that the heats of combustion are: CO(g), -67.6 kcal/mole; H$_2$(g), -63.38 kcal/mole; CH$_4$(g), -210.8 kcal/mole; C$_2$H$_2$(g), -312.0 kcal/mole; and C$_6$H$_6$(l), -782.8 kcal/mole? (c) Which fuel has the best heat-to-weight ratio?

24 ELECTROCHEMISTRY

This chapter first introduces the terminology of electrodes and electrochemical cells. Faraday's laws and the quantitative relationships in electrochemistry are presented. We then go on to describe the chemistry of some of the more common cells and batteries and to discuss fuel cells, which are still at the development stage but may be important in the overall future energy picture.

Electrochemistry is of great practical value. Think of the number of batteries we have to buy to satisfy the needs of our flashlights, portable shavers and hedge clippers, radios and tape recorders, toys, and watches. Pure metals are produced from ores, inorganic and organic compounds are synthesized, metal surfaces are coated with other metals or even with paint, and items such as the flared ends of brass musical instruments are manufactured by electrochemical processes.

Such processes are attractive because they are clean and produce very pure materials. One imaginative suggestion for applying electrochemistry is in a pollution-free method for getting rid of junked cars. Reclamation of the metals in a car requires that the metals be separated from other materials. A car can't just be cut up and fed back into a steel mill. It has been estimated that a whole car could be dissolved in an electrochemical bath in about one week. Once all the metals were in solution as ions, the pure metals could be recovered one by one by deposition on the electrodes at varying electrical potentials.

You may know the story of Luigi Galvani and his accidental discovery about electricity. One day in 1791 Galvani casually placed a dissected frog on a table near an "electric machine." An assistant observed that the frog's muscles twitched when the machine discharged electricity. Curiously, what we might call "bioelectrochemistry" is still a frontier of electrochemical science. We are only beginning to understand the electrical potentials of cell membranes and the function of electrical messages in biochemistry.

Fundamentals of electrochemistry

24.1 Anodes, cathodes, and cells

To review for a moment (Sections 19.2 and 19.5), electrochemistry deals with oxidation–reduction reactions that either produce or utilize electrical energy, and electrochemical reactions take place in cells. Each cell has two electrodes—conductors through which electrons enter or leave the cell. The surfaces of the electrodes are the sites of the action in an electrochemical cell (Figure 24.1). Altogether, three things must happen in a functioning cell: (1) either oxidation or reduction occurs at each electrode, (2) electrons flow through an external conductor, (3) ions flow in the electrolyte solution.

There are two types of electrochemical cells. The first is called a "voltaic" cell after Alessandro Volta, an Italian professor of natural philosophy and a friend of Galvani. Volta constructed the first battery in about 1800, the "voltaic pile" (see Thoughts on Chemistry). Voltaic cells are also sometimes called "galvanic" cells. A **voltaic cell** generates electrical energy from a spontaneous redox reaction. The standard cell potential of a voltaic cell, $E°$, has a positive value (Sections 19.6, and 19.7). The word **battery** is generally used for two or more voltaic cells combined to provide electrical energy for a practical purpose.

In the other type of cell, a nonspontaneous redox reaction, one with a negative $E°$, is caused to occur by the addition of electrical energy from a direct current source such as a generator or a battery. An **electrolytic cell** uses electrical energy from outside the cell to cause a redox reaction to occur. We also speak of the process of "electrolysis" or of "electrolytic reactions."

By definition, in any cell the **anode** is the electrode at which oxidation occurs, and the **cathode** is the electrode at which reduction occurs. A good memory trick is this—**o**xidation and **a**node both begin with vowels, and **r**eduction and **c**athode with consonants. Also by definition, for the choice is arbitrary, the electrode that is the source of electrons is always considered to be the negative electrode. The designation of positive and negative electrodes is not vital to our study of electrochemistry, but you should know how it is done. The somewhat confusing result of designating the electron source as the negative electrode is that in a voltaic cell the anode is the negative electrode and in an electrolytic cell the cathode is the negative electrode.

Obviously, these definitions take a bit of explaining. Let's go back to the zinc–copper cell of Section 19.5. Oxidation—the loss of electrons—occurs spontaneously at the zinc electrode. Electrons are released and leave the cell through the conductor (Figure 24.2). This makes the zinc electrode the anode and the negative electrode, for it is the source of electrons. Reduction occurs at the copper electrode, where electrons reenter the cell through the external conductor and

Voltaic cell: generates electrical energy
Electrolytic cell: uses electrical energy

Anode: oxidation
Cathode: reduction

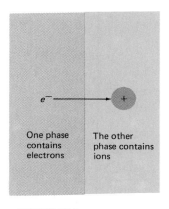

One phase contains electrons

The other phase contains ions

FIGURE 24.1
"The Fundamental Act in Electrochemistry." (*Source: J. O. M. Brockris and A. K. N. Reddy*, Modern Electrochemistry, *p. 4. Plenum, New York, 1970.*)

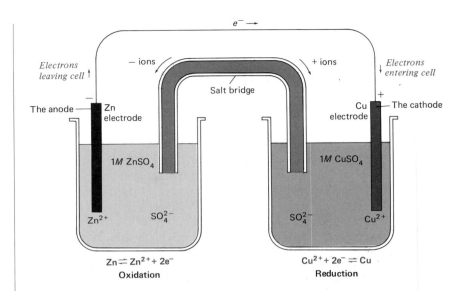

FIGURE 24.2

The zinc–copper voltaic cell. *The cell reaction is* $Zn + Cu^{2+} \rightleftharpoons Cu + Zn^{2+}$.

Salt bridge

Electrons leaving cell

Electrons entering cell

− ions

+ ions

The anode — Zn electrode

Cu electrode — The cathode

1M ZnSO₄

1M CuSO₄

Zn^{2+}

SO_4^{2-}

SO_4^{2-}

Cu^{2+}

$Zn \rightleftharpoons Zn^{2+} + 2e^-$
Oxidation

$Cu^{2+} + 2e^- \rightleftharpoons Cu$
Reduction

are utilized by the reaction with copper ions. Therefore, the copper electrode is the cathode and the positive electrode.

Figure 24.3 shows a simple cell for the electrolysis of molten sodium chloride. In this cell, the electrodes are inert—they do not take part in the reaction. Electrons are pushed into the cell from the outside source at the cathode, where they cause the reduction of sodium ions. Within an electrolytic cell, it is the cathode that is the source of electrons, and therefore, the cathode is the negative electrode. Meanwhile, chloride ions move toward the anode, where they are oxidized to give chlorine molecules and the electrons lost leave the cell.

Electrical current is carried in the external circuit of a cell by electrons and within a cell by ions. Consider the electrolysis of molten sodium chloride. At the anode, Cl^- ions are converted to chlorine molecules. The disappearance of the negative ions leaves an excess of positive ions, Na^+, around the anode. Under these conditions, the reaction would soon stop, for neutrality must be maintained. However, more Cl^- ions are attracted to the vicinity of the anode by the excess of positively charged ions and the reaction can continue. At the cathode, the story is just the opposite—positive sodium ions are consumed, leaving an excess of negative chloride ions, which attract cations to the vicinity of the cathode.

The ions that move to maintain neutrality need not be the ions that are reacting at the electrodes. In the zinc–copper cell (Figure 24.2), the salt bridge supplies the mobile ions. A salt bridge is commonly an inverted U-tube filled with a gelatin gel or aqueous solution that contains a salt. The bridge dips into the two compartments of the cell. It permits transfer of ions but limits mixing of the electrolyte solutions. At the zinc electrode, Zn^{2+} ions build up in the solution as electrons leave the cell. As a result, negative ions from the salt bridge, which might contain a saturated potassium chloride solution, are attracted into the zinc half-cell solution. Similarly, positive ions from

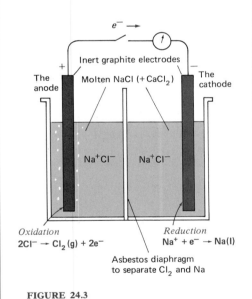

Inert graphite electrodes

The anode

Molten NaCl (+ CaCl₂)

The cathode

Na^+Cl^-

Na^+Cl^-

Oxidation
$2Cl^- \rightarrow Cl_2 (g) + 2e^-$

Reduction
$Na^+ + e^- \rightarrow Na(l)$

Asbestos diaphragm to separate Cl₂ and Na

FIGURE 24.3

The electrolysis of molten sodium chloride. *Addition of CaCl₂ to NaCl lowers the melting point enough to make the cell workable. Pure NaCl melts at about the boiling point of metallic sodium. The cell reaction is* $2NaCl(l) \rightarrow 2Na(l) + Cl_2(g)$.

[24.1] ELECTROCHEMISTRY

the salt bridge diffuse toward the copper electrode, where Cu^{2+} ions are consumed, leaving behind a solution with an oversupply of negative ions.

Note that in both the voltaic cell of Figure 24.2 and the electrolytic cell of Figure 24.3 *anions* move toward the *anode* and *cations* move toward the *cathode*. This is true for *all* electrochemical cells. Table 24.1 summarizes what happens at the electrodes in both types of cells.

24.2 Electrode reactions

Within the limits that they are valid only for standard state conditions, standard electrode potentials can be used to predict what the electrode reactions and the cell potential will be for any specific electrochemical cell.

Some typical anode and cathode reactions are listed in Table 24.2. In any given cell more than one reaction may be possible at either electrode. For reactions in aqueous solution, the oxidation or reduction of water (Reactions 4 and 8, Table 24.2) are always possi-

Anions move toward anode
Cations move toward cathode

TABLE 24.1
Electrodes

Anode	Cathode
Oxidation	**Reduction**
Attracts anions	**Attracts cations**
Electrons leave cell	**Electrons enter cell**
Negative in voltaic cell	**Negative in electrolytic cell**

TABLE 24.2
Some typical electrode reactions

Anode reactions

1. **Oxidation of an anion to a free element**

    (a) $2Cl^- \longrightarrow Cl_2(g) + 2e^-$
    (b) $4OH^- \longrightarrow O_2(g) + 2H_2O + 4e^-$

2. **Oxidation of an anion or a cation to another species in solution**

    (a) $ClO^- + 2H_2O \longrightarrow ClO_3^- + 4H^+ + 4e^-$
    (b) $Sn^{2+} \longrightarrow Sn^{4+} + 2e^-$
    (c) $2Cr^{3+} + 7H_2O \longrightarrow Cr_2O_7^{2-} + 14H^+ + 6e^-$

3. **Oxidation of a metal anode**

    (a) $Cu(s) \longrightarrow Cu^{2+} + 2e^-$
    (b) $Au(s) + 4Cl^- \longrightarrow AuCl_4^- + 3e^-$

4. **Oxidation of water**

    $2H_2O \longrightarrow O_2(g) + 4H^+ + 4e^-$ (solution becomes acidic)

Cathode reactions

5. **Reduction of a cation or a complex anion to a free metal**

    (a) $Zn^{2+} + 2e^- \longrightarrow Zn(s)$
    (b) $Ag(CN)_2^- + e^- \longrightarrow Ag(s) + 2CN^-$

6. **Reduction of an anion or cation**

    (a) $NO_3^- + 3H^+ + 2e^- \longrightarrow HNO_2 + H_2O$
    (b) $Ce^{4+} + e^- \longrightarrow Ce^{3+}$
    (c) $2H^+ + 2e^- \longrightarrow H_2$

7. **Reduction of an elemental nonmetal to an anion**

    $I_2 + 2e^- \longrightarrow 2I^-$

8. **Reduction of water**

    $2H_2O + 2e^- \longrightarrow H_2(g) + 2OH^-$ (solution becomes alkaline)

bilities, as are the reactions of hydrogen ion in acidic solution (Reaction 6c) or of hydroxide ion in alkaline solution (Reaction 1b).

The oxidation half-reaction observed at the anode should be that with the least positive standard reduction potential—in other words, the half-reaction nearest the *bottom* in a table such as Table 19.6. For example, in an aqueous solution of potassium bromide (KBr), species that might be oxidized are water or bromide ion (K^+ can be oxidized no further). The table of reduction potentials (written, note, in the opposite direction of anode reactions) provides the following:

$$O_2 + 4H^+ + 4e^- \longrightarrow 2H_2O \qquad E° = 1.229 \text{ V}$$
$$Br_2 + 2e^- \longrightarrow 2Br^- \qquad E° = 1.087 \text{ V}$$

We would predict, and correctly, that the electrolysis of a $1M$ aqueous solution of potassium bromide should yield elemental bromine, and not oxygen, as the anode product.

On the other hand, the reduction half-reaction observed at the cathode should be that with the most positive standard reduction potential; that is, the half-reaction nearest the top in a table such as Table 19.6. In the electrolysis of an acidic iron(III) chloride solution, either Fe^{3+} or H^+, or H_2O might be reduced. From the standard reduction potentials

$$Fe^{3+} + e^- \longrightarrow Fe^{2+} \qquad E° = 0.770 \text{ V}$$
$$2H^+ + 2e^- \longrightarrow H_2 \qquad E° = 0.0000 \text{ V}$$
$$2H_2O + 2e^- \longrightarrow H_2 + 2OH^- \qquad E° = -0.828 \text{ V}$$

we would predict, again correctly, that Fe^{2+}, rather than elemental hydrogen, would be the cathode product.

In addition to the restriction imposed by standard state conditions, other effects complicate predictions of the type just given. Sometimes a voltage larger than the standard voltage is required to cause an electrode reaction to occur. The voltage in addition to the calculated voltage used by an electrode reaction in a cell is called an **overvoltage.** For example, the electrolysis of acidified zinc sulfate solution results in deposition of elemental zinc at the cathode ($Zn^{2+} + 2e^- \rightarrow Zn$, $E° = -0.7628$ V) rather than the liberation of hydrogen gas ($2H^+ + 2e^- \rightarrow H_2$, $E° = 0.0000$ V), even though the reduction potential data predict the opposite. Predictions are further complicated when several possible electrode reactions have the same or nearly the same potential values. Under these conditions, two or more reactions may occur simultaneously. This may be desirable (e.g., in the simultaneous electrodeposition of several metals) or undesirable (e.g., in the simultaneous liberation of hydrogen during deposition of a metal).

24.3 Simple voltaic cells; standard cell voltage

The simplest voltaic cells are based upon spontaneous displacement or combination reactions. A cell might contain two metals and their cations, with one metal displacing the cation of the other in solution, as happens in the zinc–copper cell of Figure 24.2:

$$Zn(s) + Cu^{2+} \longrightarrow Zn^{2+} + Cu(s)$$

During the operation of this cell, both elemental zinc and copper(II) ions are consumed. Cells in which the reactants are used up irreversibly are called **primary cells.** The life of a cell utilizing this reaction will be longest when the species being consumed are present in the largest quantities. In this case, a large zinc electrode and a concentrated Cu^{2+} solution with a dilute Zn^{2+} solution would be used.

In another simple type of cell, hydrogen can be displaced from an acid by a reactive metal, for example, magnesium. If magnesium is added to an acidic solution, a strongly exothermic reaction occurs.

$$Mg(s) + 2H^+ \longrightarrow Mg^{2+} + H_2(g) \qquad \Delta H = -110.41 \text{ kcal/mole}$$
$$(-461.96 \text{ kJ/mole})$$

When magnesium and copper electrodes are immersed in dilute (about $1M$) sulfuric acid solution, the reaction given above is the cell reaction, and it develops enough electrical energy to fire a flash bulb wired into the external circuit.

The standard voltage of a cell can be found from standard reduction potentials in the usual way (Section 19.7). For example, for the cell just described, the voltage produced is calculated as follows:

cathode	$2H^+ + 2e^- \longrightarrow H_2$	$E^\circ = 0.0000$ V	*reaction direction and sign reversed*
anode	$Mg \longrightarrow Mg^{2+} + 2e^-$	$E^\circ = 2.375$ V	
cell			
voltage	$Mg + 2H^+ \longrightarrow Mg^{2+} + H_2$	$E^\circ = 2.375$ V	

This procedure is equivalent to subtracting (algebraically) the standard reduction potential of the anode from that of the cathode.

$$E^\circ_{cell} = E^\circ_{cathode} - E^\circ_{anode}$$
$$= 0.000 \text{ V} - (-2.375 \text{ V})$$
$$= +2.375 \text{ V}$$

For a cell to operate spontaneously under standard state or any other conditions, the cell potential must be positive. Note, however, that the experimentally observed potential of a specific cell seldom agrees with the calculated value. The actual potential differs from the calculated one because of the internal resistance of the cell, which varies with how it is constructed; because of differences in the concentrations of the reaction species, which affect the potential according to the Nernst equation (Section 19.9); and because of overvoltage.

To give another example, a simple combination reaction such as that between metallic zinc and iodine

$$Zn(s) + I_2(aq) \longrightarrow Zn^{2+} + 2I^-$$

will also occur in a cell. (Such a cell might be constructed by placing a zinc electrode into a dilute zinc chloride solution contained in a

$$E^\circ_{cell} = E^\circ_{cathode} - E^\circ_{anode}$$

porous cup and placing the porous cup in a concentrated aqueous solution of iodine and potassium iodide into which is inserted an inert graphite electrode.) When the external circuit is completed, electrons are released at the *zinc anode*

$$Zn \longrightarrow Zn^{2+} + 2e^-$$

and removed from the external circuit at the *carbon cathode*.

$$I_2 + 2e^- \longrightarrow 2I^-$$

This cell is described in shorthand notation as

$$Zn \mid Zn^{2+} \parallel I^-, I_2(aq) \quad (C)$$

The single vertical line indicates physical contact between species in different phases. The electrode materials are written at the extreme right and left, usually with the anode at the left. The species in solution are written in the center. The double line indicates a salt bridge, porous divider, or a similar means of permitting ion flow while inhibiting mixing of the electrolyte solutions. An inert electrode material is indicated in parentheses. For a cell in which there is no salt bridge or other boundary, such as the cell described above in which elemental magnesium reacts with hydrogen ion, a comma separates the species in the same phase.

$$Mg \mid Mg^{2+}, H^+ \mid H_2(Cu)$$

This notation can be used to describe any type of cell.

24.4 Chemical and electrical equivalence

In oxidation–reduction reactions, chemical equivalence is determined by the number of electrons available for gain, loss, or sharing. The oxidation number is, you will recall, our bookkeeping device for redox reactions. An **equivalent weight** in an oxidation–reduction reaction is equal to the molar weight of the compound oxidized or reduced divided by the algebraic change in oxidation number of the atom that is oxidized or reduced.

For example, in the reaction

$$2\overset{+2}{Fe}Cl_2(aq) + Cl_2(g) \longrightarrow 2\overset{+3}{Fe}Cl_3(aq) \tag{1}$$

iron(II) chloride is oxidized to iron(III) chloride, the total oxidation number change is $+1$, and therefore an equivalent of $FeCl_2$ as a reducing agent is equal to 1 mole, or 126.75 g. However, in a reaction where Fe^{2+} is an oxidizing agent and is itself reduced to free iron

$$Mn(s) + \overset{+2}{Fe}SO_4(sq) \longrightarrow \overset{0}{Fe}(s) + MnSO_4(aq) \tag{2}$$

the equivalent weight of the iron compound, $FeSO_4$ in this case, is one-half of the molar weight, or 75.96 g.

TOOLS OF
CHEMISTRY

pH meter

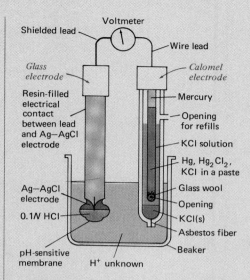

FIGURE A
Electrodes used for measuring pH.

The determination of the pH of solutions is a routine requirement in virtually all scientific research, industrial, and medical laboratories. The pH meter is the basic instrument used to accomplish this task. While the design and sensitivity of pH meters vary, the essential parts are (1) a glass electrode, (2) a reference electrode, and (3) a voltmeter calibrated to read in pH units rather than in volts.

The glass electrode is based upon a unique property of a thin membrane (about 0.5 mm) made of special glass. An electric potential develops across the membrane when its two sides are in contact with solutions of different hydrogen ion concentrations. The potential arises because of ion exchange between Na^+ ions in the glass and H^+ ions in the solutions, aided by a thin layer of hydrated glass on the membrane surface.

The glass membrane is formed into a bulb that is filled with a solution of known and constant pH (Figure A, left electrode). Sealed into the glass electrode and dipping into the solution with which the bulb is filled is an electrode, usually a silver wire coated with silver chloride. The entire glass electrode is placed in the solution of unknown pH during use, thereby putting the membrane in contact with the solutions of known and unknown pH.

The reference electrode most often used is a calomel electrode, which contains a mixture of Hg, Hg_2Cl_2 (known as calomel), and KCl. The electrode reaction for the calomel electrode is

$$Hg_2Cl_2 + 2e^- \rightleftharpoons 2Hg + 2Cl^-$$

The glass electrode and the calomel electrode together give an electrochemical cell represented by

$$Ag \left| AgCl, HCl\ (0.1N) \left| \begin{array}{l} glass \\ membrane \\ solution\ of \\ unknown\ pH \end{array} \right\| Hg_2Cl_2, KCl \right| Hg$$

The arrangement is such that the potential difference measured by the voltmeter is due only to the difference in the hydrogen ion concentration of the two solutions. Found by the Nernst equation, this potential is

$$E = a\ constant - 0.0592 \log \frac{[H^+]_{unknown}}{[H^+]_{known}}$$

or, taking $\log 1/[H^+]_{known}$ into the constant

$$E = a\ constant + 0.0592\ pH$$

In practice, a pH meter must be calibrated before each use by placing the electrodes in a solution of known pH and setting the meter to read correctly. This is necessary because the value of the constant in the above equations varies with conditions.

Electrodes based, like the glass electrode, on the movement of ions across a membrane, have recently been developed for substances other than H^+. Such electrodes, called ion-selective electrodes, can directly determine the concentrations of, for example, Ca^{2+}, Pb^{2+}, Br^-, F^-, Cl^-, S^{2-}, and SO_2. Among the many practical uses of such electrodes are the monitoring of F^- concentrations in water supplies; monitoring of NH_3, NO_x, SO_2, H_2S, HF, HCl, and HCN in stack gases; the measurement of Cl^- in sweat as a diagnostic test for cystic fibrosis; the measurement of Ca^{2+} in beer, wine, and milk; and the measurement of CN^- in metal plating baths and wastewater.

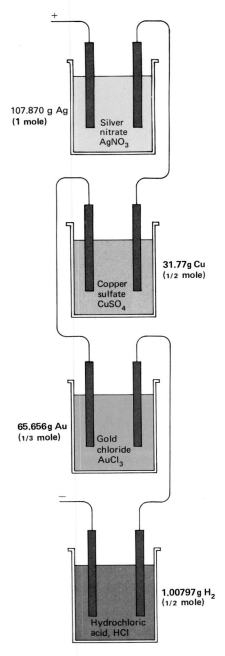

107.870 g Ag
(1 mole)

Silver
nitrate
AgNO₃

31.77g Cu
(1/2 mole)

Copper
sulfate
CuSO₄

65.656g Au
(1/3 mole)

Gold
chloride
AuCl₃

1.00797g H₂
(1/2 mole)

Hydrochloric
acid, HCl

FIGURE 24.4
Simultaneous electrolysis experiment.
This experiment illustrates chemical equivalence as the basis for Faraday's laws.

EXAMPLE 24.1

■ Determine the equivalent weight of zinc in the following reaction:

$$Zn(s) + 2HCl(aq) \longrightarrow ZnCl_2(aq) + H_2(g)$$

Zinc goes from oxidation state 0 to oxidation state $+2$. Therefore, 1 mole of Zn in this reaction corresponds to 2 equivalents.

$$\begin{aligned}\frac{\text{equiv wt Zn}}{\text{(in g/equiv)}} &= \frac{65.4 \text{ g}}{1 \text{ mole}} \times \frac{1 \text{ mole}}{2 \text{ equiv}} \\ &= 32.7 \text{ g/equiv}\end{aligned}$$

The equivalent weight of zinc in this reaction is 32.7 g. ■

Writing the ion–electron equations for the oxidation and reduction of iron(II) in Equations (1) and (2) emphasizes the role in these reactions of the number of electrons available

Fe²⁺	$\longrightarrow$	Fe³⁺	+	1e⁻

1 ion   *1 ion*   *1 electron*
6.022 × 10²³ ions   *6.022 × 10²³ ions*   *6.022 × 10²³ electrons*
1 mole of ions   *1 mole of ions*   *1 mole of electrons*

Fe²⁺	+	2e⁻	$\longrightarrow$	Fe

1 ion   *2 electrons*   *1 atom*
6.022 × 10²³ ions   *2 × (6.022 × 10²³) electrons*   *6.022 × 10²³ atoms*
1 mole of ions   *2 moles of electrons*   *1 mole of atoms*

In an electrochemical reaction, there must obviously be a direct relationship between the quantity of current passing through a cell and the amount of chemical change that takes place. This relationship can be demonstrated experimentally by the simultaneous electrolysis of solutions of silver nitrate, copper(II) sulfate, gold(III) chloride, and hydrochloric acid. When enough direct current is passed through the system in Figure 24.4 to deposit 107.868 g, or 1 mole, of silver, then $\frac{1}{2}$ mole of copper, $\frac{1}{3}$ mole of gold, and $\frac{1}{2}$ mole of molecular hydrogen are liberated simultaneously. These are the amounts of these substances equivalent to 1 mole of silver, or equivalent to the 1 mole of electrons needed to liberate 1 mole of silver. In redox reactions equivalents can be thought of as the amount of a substance liberated, dissolved, or converted to another substance per mole of electrons.

The quantitative relationship between electrical energy and chemical energy was first pointed out by Michael Faraday, an English chemist. From his experiments in 1832 and 1833, Faraday drew the following conclusions, known as *Faraday's laws* of electrolysis: (1) *The weight of a substance formed at an electrode is proportional to the amount of current passed through the cell.* (2) *The weights of different substances produced at an electrode by the same amount of current are proportional to their equivalent weights.*

The quantity of electric charge is measured in coulombs. Using the most accurate values, the charge on one electron is $1.6021892 \times 10^{-19}$ coulomb, so the total charge of 1 mole of electrons is

$$1.6021892 \times 10^{-19} \, \frac{\text{coulomb}}{\text{electron}} \times 6.022045 \times 10^{23} \, \frac{\text{electrons}}{\text{mole}}$$

$$= 96{,}484.56 \, \frac{\text{coulombs}}{\text{mole}}$$

Because of its usefulness to chemistry, this quantity of electricity is defined as a unit—the **faraday** which is equivalent to 96,485 coulombs, or the amount of electricity represented by 1 mole of electrons. For most calculations, the value of the faraday, also called the Faraday constant, can be rounded to 96,500 coulombs.

Current, measured in amperes, has the units of electric charge flowing per unit time.

1 ampere = 1 coulomb/second

By application of Faraday's laws (and the useful conversion factors given in Table 24.3), either the quantity of chemical change produced by a known quantity of electrical energy or the quantity of electrical energy required for reaction of a known quantity of a pure substance can be calculated.

EXAMPLE 24.2

■ How much elemental copper can be produced in the electrolysis of a copper(II) sulfate solution during 10.0 hours at a steady current of 100. amperes?

We suggest solving such problems in two steps. First, find the amount of electricity in faradays.

Amount of electricity (in faradays)
= current (amperes) × time (seconds)

$$\times \frac{1 \text{ coulomb}}{1 \text{ ampere sec}} \times \frac{1 \text{ faraday}}{96{,}500 \text{ coulombs}}$$

$$= 100. \text{ amperes} \times 10.0 \text{ hr} \times \frac{60 \text{ min}}{\text{hr}} \times \frac{60 \text{ sec}}{\text{min}}$$

$$\times \frac{1 \text{ coulomb}}{1 \text{ ampere sec}} \times \frac{1 \text{ faraday}}{96{,}500 \text{ coulombs}}$$

$$= 37.3 \text{ faradays}$$

Second, find the amount of product equivalent to the amount of electricity. The reaction

$$Cu^{2+} + 2e^- \longrightarrow Cu(s)$$

shows that 2 faradays of electricity will produce 1 mole of Cu. Therefore,

$$\text{wt. Cu (in g)} = 37.3 \text{ faradays} \times \frac{1 \text{ mole Cu}}{2 \text{ faradays}} \times \frac{63.55 \text{ g Cu}}{1 \text{ mole Cu}}$$

$$= 1190 \text{ g}$$

TABLE 24.3
Conversion factors

$$\frac{1 \text{ coulomb}}{1 \text{ ampere second}}$$

$$\frac{96{,}500 \text{ coulombs}}{1 \text{ faraday}}$$

$$\frac{1 \text{ faraday}}{1 \text{ equivalent}}$$

$$\frac{96{,}500 \text{ coulombs}}{\text{equivalent}}$$

The total amount of copper formed in this electrolysis is 1190 g. This assumes, of course, that the reduction reaction is 100% efficient. In practice, as for example in an industrial electrolytic copper refinery, the reaction is not quite 100% efficient because of losses of current through competing electrochemical reactions and heating. ■

EXAMPLE 24.3

■ How much electrical energy is required to liberate 5.6 liters of elemental chlorine (measured at standard temperature and pressure) from a solution containing chloride ion?

First, find the number of equivalents of chlorine produced. The reaction is

$$2Cl^- \longrightarrow Cl_2(g) + 2e^-$$

Equiv. Cl_2

$$= 5.6 \text{ liters } Cl_2 \times \frac{1 \text{ mole } Cl_2}{22.4 \text{ liters } Cl_2} \times \frac{2 \text{ moles } e^-}{1 \text{ mole } Cl_2} \times \frac{1 \text{ equiv } Cl_2}{1 \text{ mole } e^-}$$

$$= 0.50 \text{ equiv}$$

Then find the quantity of electricity required

Quantity of electricity (coulombs)

$$= 0.50 \text{ equiv } Cl_2 \times \frac{96,500 \text{ coulombs}}{1 \text{ equiv } Cl_2}$$

$$= 48,000 \text{ coulombs}$$

To produce the chlorine requires 48,000 coulombs of electricity, regardless of the time involved. As the rate of flow of the current increases, the time required decreases. ■

EXAMPLE 24.4

■ The cells in an automobile battery were charged at a steady current of 5.0 amperes for exactly 5 hr. How much Pb and PbO_2 were formed and how much $PbSO_4$ was changed in each cell? The reactions (Section 24.6) are:

$$\begin{array}{l} PbSO_4(s) + 2e^- \longrightarrow Pb(s) + SO_4^{2-} \\ PbSO_4(s) + 2H_2O \longrightarrow PbO_2 + 4H^+ + SO_4^{2-} + 2e^- \\ \hline 2PbSO_4(s) + 2H_2O \longrightarrow Pb(s) + PbO_2(s) + 2H_2SO_4(aq) \end{array}$$

First, find the amount of electricity that passed through the battery.

Amount of electricity (in faradays)

$$= 5.0 \text{ amperes} \times 5 \text{ hr} \times \frac{60 \text{ min}}{\text{hr}} \times \frac{60 \text{ sec}}{\text{min}}$$

$$\times \frac{1 \text{ coulomb}}{1 \text{ ampere sec}} \times \frac{1 \text{ faraday}}{96,500 \text{ coulombs}}$$

$$= 0.93 \text{ faraday}$$

The reactions for charging a lead storage battery show that 2 faradays of electricity produce 1 mole each of Pb and PbO_2 and consume 2 moles of $PbSO_4$. Therefore,

$$\text{wt Pb (in g)} = 0.93 \text{ faraday} \times \frac{1 \text{ mole Pb}}{2 \text{ faradays}} \times \frac{207 \text{ g Pb}}{1 \text{ mole Pb}}$$
$$= 96 \text{ g}$$

$$\text{wt } PbO_2 \text{ (in g)} = 0.93 \text{ faraday} \times \frac{1 \text{ mole } PbO_2}{2 \text{ faradays}} \times \frac{239 \text{ g } PbO_2}{1 \text{ mole } PbO_2}$$
$$= 110 \text{ g}$$

$$\text{wt } PbSO_4 \text{ (in g)} = 0.93 \text{ faraday} \times \frac{2 \text{ mole } PbSO_4}{2 \text{ faradays}} \times \frac{303 \text{ g } PbSO_4}{1 \text{ mole } PbSO_4}$$
$$= 280 \text{ g}$$

Thus, 280 g of $PbSO_4$ was converted to 96 g of Pb and 110 g of PbO_2. ∎

Practical electrochemistry

24.5 Voltaic dry cells

Although the zinc–copper cell was used extensively as a source of direct current for many years, sealed units that use no free liquid have proved to be more compact, portable, and practical. The best known and most popular of these *dry* cells is the <u>Leclanché cell</u> (Figure 24.5). This cell is an ingenious device that both allows electron transfer and consumes the gaseous products of the reactions that take place. The space between the electrodes is filled with a moist paste of NH_4Cl, $ZnCl_2$, and MnO_2. When the electrodes are connected, the zinc anode is oxidized

$$Zn(s) \longrightarrow Zn^{2+} + 2e^- \quad \text{(anode)}$$

and the electrons that result reduce the ammonium ion at the inert carbon cathode.

$$2NH_4^+ + 2e^- \longrightarrow 2NH_3(g) + H_2(g) \quad \text{(cathode)}$$

The liberated hydrogen is oxidized by the manganese(IV) oxide

$$2MnO_2(s) + H_2(g) \longrightarrow Mn_2O_3(s) + H_2O(l)$$

and ammonia is taken up as a complex salt.

$$Zn^{2+} + 2NH_3(g) + 2Cl^- \longrightarrow \underset{\textit{dichlorodiamminezinc(II)}}{Zn(NH_3)_2Cl_2}$$

The overall reaction of this primary cell is

$$2MnO_2(s) + 2NH_4Cl(s) + Zn(s) \longrightarrow$$
$$Zn(NH_3)_2Cl_2(s) + H_2O(l) + Mn_2O_3(s)$$

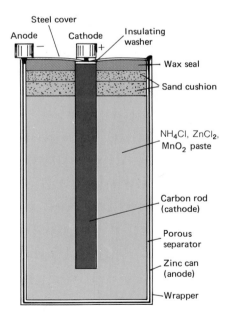

Steel cover
Anode — Cathode + Insulating washer
Wax seal
Sand cushion
NH_4Cl, $ZnCl_2$, MnO_2 paste
Carbon rod (cathode)
Porous separator
Zinc can (anode)
Wrapper

FIGURE 24.5
The Leclanché dry cell.

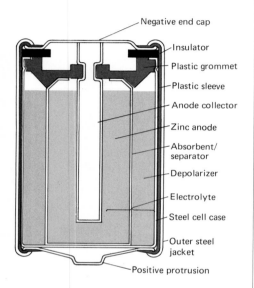

Negative end cap
Insulator
Plastic grommet
Plastic sleeve
Anode collector
Zinc anode
Absorbent/separator
Depolarizer
Electrolyte
Steel cell case
Outer steel jacket
Positive protrusion

FIGURE 24.6
An alkaline–MnO₂ battery.

Both rupture of the container by accumulating gases and a decrease in cell potential as a consequence of coating of the cathode by hydrogen are thus avoided.

The Leclanché cell delivers electrical energy at a potential of 1.5 V. Combining more than one of these cells in series (anode to cathode, cathode to anode) gives batteries with voltages that are multiples of 1.5 V. If energy is withdrawn too rapidly from a Leclanché cell, the removal of cathode gases cannot take place rapidly enough to maintain the voltage. A cell in this condition regains its voltage on standing. On the other hand, when the reactive chemicals are depleted, the cell can no longer produce energy. Unused Leclanché cells deteriorate because of a slow reaction between zinc and the ammonium ion. Common flashlight batteries are Leclanché cells.

The alkaline manganese(IV) oxide–zinc cell, which is illustrated in Figure 24.6, has replaced the Leclanché cell in many applications because it has several advantages: it yields a sustained operating voltage at large current drains, operates at low temperatures, has a longer shelf-life, and can, within limits, be recharged. Since its standard voltage is 1.54 V, it can be used in place of the Leclanché cell without problems. When the electrodes are connected, the zinc anode is oxidized

$$Zn(s) + 2OH^- \longrightarrow ZnO(s) + H_2O(l) + 2e^-$$

and the electrons that are released reduce the manganese (IV) oxide at the cathode.

$$2MnO_2(s) + H_2O(l) + 2e^- \longrightarrow Mn_2O_3 + 2OH^-$$

No gaseous products result. These cells, labeled "alkaline," are sold in retail stores as a somewhat more expensive alternative to the common Leclanché-type batteries.

The widespread use of electrically operated hearing aids and other small devices has stimulated the development of primary cells that permit the use of smaller volumes of reactants and deliver larger currents at different potentials. A particularly successful cell of this type is the mercury cell, or Ruben–Mallory cell. This cell, also an alkaline cell, contains a zinc anode and a mercury(II) oxide cathode. The anode and cathode materials are both compacted powders. The space between these electrodes is filled with a moist mercury(II) oxide paste containing some sodium or potassium hydroxide electrolyte, which is separated from the anode by a porous paper liner. The cell delivers electrical energy by the following reactions

$$Zn(s) + 2OH^- \longrightarrow ZnO(s) + H_2O + 2e^- \quad \text{(anode)}$$
$$HgO(s) + H_2O + 2e^- \longrightarrow Hg(l) + 2OH^- \quad \text{(cathode)}$$

at a potential of 1.35 V. No gaseous products result. Spent cells should be reprocessed for mercury recovery or treated to prevent mercury or mercury compounds from reentering the environment and causing contamination.

24.6 Storage cells

Cells that can be recharged by using electrical energy from an external source to reverse the initial oxidation–reduction reactions are called **storage cells** (or accumulators, or secondary cells). They are particularly useful in applications that require energy at one time but generate energy at another time, notably in the automobile. A storage battery provides electrical energy for starting the automobile and is then recharged by energy from a generator or alternator while the engine is running.

The most widely used storage cell is the lead storage cell. It is constructed of electrodes that are lead alloy grids packed with either finely divided spongy elemental lead or finely divided lead(IV) oxide. These electrodes are arranged alternately, separated by thin wooden or fiberglass sheets, and suspended in dilute sulfuric acid solution (specific gravity 1.3).

When the external circuit is closed, the spongy lead is oxidized and forms lead(II) sulfate that adheres to the electrode.

$$Pb(s) + SO_4{}^{2-} \longrightarrow PbSO_4(s) + 2e^- \quad \text{(anode)}$$

The released electrons reduce lead(IV) oxide preferentially, and in the presence of hydrogen ions and sulfate ion, adherent lead(II) sulfate forms here also,

$$PbO_2(s) + 4H^+ + SO_4{}^{2-} + 2e^- \longrightarrow PbSO_4(s) + 2H_2O(l) \quad \text{(cathode)}$$

this time on the lead(IV) oxide cathode. The net results of discharge are the conversion of the active chemicals on both electrodes to lead(II) sulfate and decrease in the concentration of the sulfuric acid solution, as indicated by a decrease in the specific gravity of the solution. (When the service station attendant checks the charge of a battery, he measures the specific gravity of the sulfuric acid solution.)

If the negative electrode of a discharged or partially discharged lead storage cell is connected to an external source of electrons (Figure 24.7), the initial oxidation process is reversed

$$PbSO_4(s) + 2e^- \longrightarrow Pb(s) + SO_4{}^{2-}$$

and the other electrode releases electrons by reversal of the initial reduction process

$$PbSO_4(s) + 2H_2O \longrightarrow PbO_2(s) + 4H^+ + SO_4{}^{2-} + 2e^-$$

Sulfuric acid is regenerated, and the specific gravity of the solution increases. During the charging operation, the cell becomes an electrolytic cell. The overall reaction in the lead storage battery is

$$Pb(s) + PbO_2(s) + 2H_2SO_4(aq) \xrightleftharpoons[\text{charge}]{\text{discharge}} 2PbSO_4(s) + 2H_2O(l)$$

(See Examples 19.2, 19.8, 19.10, 19.14, and 24.4.)

Discharging cell

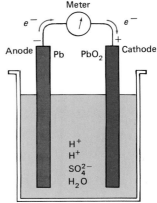

$Pb^0 + SO_4{}^{2-} \rightarrow PbSO_4 + 2e^-$
(oxidation at − electrode)
$PbO_2 + 4H^+ + SO_4{}^{2-} + 2e^- \rightarrow PbSO_4 + 2H_2O$
(reduction at + electrode)

Charging cell

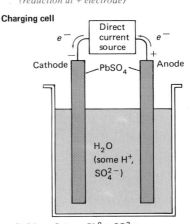

$PbSO_4 + 2e^- \rightarrow Pb^0 + SO_4{}^{2-}$
(reduction at − electrode)
$PbSO_4^2 + 2H_2O \rightarrow PbO_2 + 4H^+ + SO_4{}^{2-} + 2e^-$
(oxidation at + electrode)

FIGURE 24.7
The principles and operation of the lead storage cell during discharge and charge. *The equations under the cell at the top represent the reactions during discharge; those under the cell at the bottom represent those during recharge.*

Each lead storage cell develops a potential of 2 V. In practice, a number of these cells are connected in *parallel* (anode to anode, cathode to cathode) to increase the *quantity* of current that can be generated, and three or six such blocks of cells are connected in series to give 6-V or 12-V batteries. The resulting lead storage batteries give excellent service for long periods of time, providing their liquid levels are maintained, they are never permitted to discharge completely, they are not allowed to freeze, and they are not subjected to frequent "quick charge" procedures. (A quick charge reverses the reactions so rapidly that the regenerated Pb and PbO_2 may not completely adhere to the electrodes and may fall off. The Pb and PbO_2 can build up to form a short-circuiting sludge if the quick charge procedure is repeated frequently.)

The nickel–cadmium alkaline storage cell (used in "ni-cad" batteries) has a longer life and delivers a more nearly constant potential than the lead storage cell. In this cell, a cadmium electrode and a metal grid containing nickel(IV) oxide (or a mixture of nickel oxides with nickel in other oxidation states) are immersed in a potassium hydroxide solution. The chemical reactions during discharge and charge can be summarized by the equations

$$Cd(s) + 2OH^- \underset{\text{charge}}{\overset{\text{discharge}}{\rightleftharpoons}} Cd(OH)_2(s) + 2e^-$$

$$NiO_2(s) + 2H_2O(l) + 2e^- \underset{\text{charge}}{\overset{\text{discharge}}{\rightleftharpoons}} Ni(OH)_2(s) + 2OH^-$$

No change in concentration of the electrolyte occurs in either operation. This cell delivers current at a potential of about 1.4 V. Nickel–cadmium batteries are light and are often used in cordless appliances.

Much effort has been expended recently to develop compact storage cells that will deliver exceptionally large quantities of energy. This type of cell would be particularly useful for an electrically powered automobile that would avoid the pollution problems of the internal combustion gasoline engine. One such cell is the Weber–Kummer cell of the Ford Motor Company, a liquid sodium–sulfur cell. This cell is based upon molten sodium, separated by a sodium-ion-conducting porous ceramic barrier from molten sulfur contained in a porous electrode. The reactions occurring during discharge and charge are as follows:

$$2Na(l) \underset{\text{charge}}{\overset{\text{discharge}}{\rightleftharpoons}} 2Na^+ + 2e^- \quad \text{(negative electrode)}$$

$$S(l) + 2e^- \underset{\text{charge}}{\overset{\text{discharge}}{\rightleftharpoons}} S^{2-} \quad \text{(positive electrode)}$$

Actually, the sulfide ion and sulfur atoms probably combine to form polysulfide ions, $S_x{}^{2-}$. To produce energy, this cell must be heated initially to 250–300°C to melt the reactants; it then operates at this temperature.

The lithium–chlorine cell also delivers a high level of energy and is based upon oxidation of a strongly electropositive element (Li) and the reduction of a strongly electronegative element (Cl_2) at high temperatures. The materials in this cell are molten lithium, molten lithium chloride, and chlorine gas. As you might guess, problems arise in containing these reactive materials. Neither the sodium–sulfur cell nor the lithium–chlorine cell has yet been put to everyday use. A similar cell based upon more tractable materials—a lithium alloy and a metal sulfide—is currently being developed.

24.7 Fuel cells

Burning fuel to obtain thermal energy that is then converted into electrical energy is a wasteful process. Heat is inevitably lost to the surroundings, and there are also thermodynamic limits to the amount of useful energy that can be produced. The best conventional power plant can operate only at about 40% efficiency. Theoretically, the electrochemical conversion of traditional fuels such as hydrocarbons to combustion products could be 100% efficient. While perfect efficiency seems unattainable, 60% has been called possible.

Much research is being devoted to the development of **fuel cells,** cells that produce electrical energy directly from the oxidation of a fuel continuously supplied to the cell. The technical, economic, and practical problems to be overcome are great. These problems include difficulties in providing contact between three phases needed in a fuel cell (the gaseous fuel, the liquid electrolyte, and the solid conductor), the corrosiveness of acid electrolyte systems or the buildup of carbonates in alkaline systems, the high cost of catalysts needed for the electrode reactions (metals such as platinum, palladium, and silver), and the problems of handling gaseous fuels such as hydrogen or methane (CH_4) at either low temperatures or high pressures.

Development has proceeded furthest with cells in which hydrogen is the fuel and oxygen is the oxidant. In an alkaline system, the electrode reactions are

$$2H_2 + 4OH^- \longrightarrow 4H_2O + 4e^- \quad \text{(anode)}$$
$$O_2 + 2H_2O + 4e^- \longrightarrow 4OH^- \quad \text{(cathode)}$$

The cell reaction

$$2H_2(g) + O_2(g) \longrightarrow 2H_2O(l)$$

is simply the same reaction as the combustion of hydrogen, which yields 68.3 kcal/mole of thermal energy at 25°C under standard conditions.

Hydrogen is fed to the anode and oxygen is fed to the cathode in the hydrogen–oxygen fuel cell. Catalysts are necessary to ease breaking hydrogen–hydrogen bonds at the anode and oxygen–oxygen bonds at the cathode. In one fuel cell design, porous carbon electrodes are "activated" by the presence of metallic catalysts (Figure 24.8). The electrolyte is a concentrated aqueous potassium hydroxide solution, and the product water is removed as vapor in an excess of hydrogen. The cell operating temperature is 70–140°C and it delivers about 0.9 V.

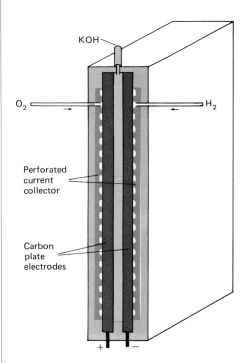

FIGURE 24.8
A hydrogen–oxygen fuel cell.

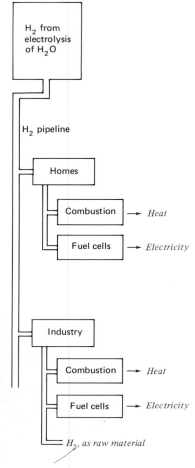

FIGURE 24.9
Scheme for a hydrogen economy.

The closest fuel cells have come to practical application is in the hydrogen–oxygen cells used to supply electricity on board Apollo and Gemini spacecraft, where the product water was used for drinking, and in an experimental fuel-cell-powered automobile. Suggestions for fuel cell use range from heart pacemaker batteries to submarine power plants. On-site generation of electricity in homes or industry (Figure 24.9) by fuel cells would play an important role in the development of a hydrogen economy (Section 11.7).

24.8 Electrolysis

The chemical reactions that occur in an electrolytic cell are often the reverse of those in a voltaic cell. Electrons must be supplied to the cell at a voltage greater than the voltage of the cell reaction (taking into account concentration effects) plus the overvoltage of the electrodes plus whatever voltage is lost through the resistance of the parts of the cell. We have mentioned (Section 11.6) that the electrolysis of water is a commercial source of very pure hydrogen. Electrolytic cells for water decomposition require 2.0 to 2.5 V, which includes 1.23 V for the standard cell voltage plus the rest for the effects described.

Electrons enter an electrolytic cell at the cathode, and the cathode itself is seldom reduced. Graphite and the noble metals (e.g., gold or platinum) are common inert electrode materials. Where electrolytic oxidation of a species in solution is the desired anode reaction, the anode is also made of an inert material.

The electrolysis of a concentrated aqueous sodium chloride solution by the reaction

$$2NaCl(aq) + 2H_2O(l) \longrightarrow Cl_2(g) + H_2(g) + 2NaOH(aq)$$

is a major industrial process for the manufacture of chlorine and sodium hydroxide in what is sometimes called the chlor–alkali industry. Two types of electrolytic cells are used commercially, the objective in both being to keep the chlorine formed at the anode from mixing with and reacting with the hydrogen and sodium hydroxide by the following reactions:

$$2NaOH(aq) + Cl_2(g) \longrightarrow NaOCl(aq) + NaCl(aq) + H_2O(l)$$
$$H_2(g) + Cl_2(g) \longrightarrow 2HCl(g)$$

In the diaphragm cell (Figure 24.10), both electrodes are inert and the electrode reactions are (see also Section 20.4b)

$$2H_2O + 2e^- \longrightarrow H_2 + 2OH^- \quad \text{(cathode)}$$
$$2Cl^- \longrightarrow Cl_2 + 2e^- \quad \text{(anode)}$$

Flowing mercury is used as the cathode in the other type of cell (Figure 24.11). Sodium ion is reduced at the mercury cathode and forms sodium amalgam

$$Na^+ + e^- + xHg(l) \longrightarrow NaHg_x(l)$$

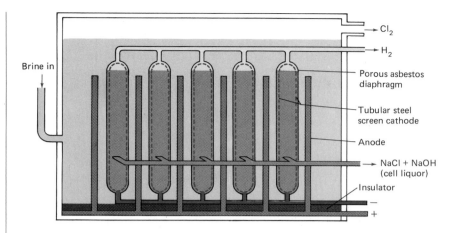

FIGURE 24.10

A diaphragm cell for electrolysis of sodium chloride. *This cell utilizes alternating rows of anodes and cathodes. The cathodes are coated with a permeable asbestos diaphragm. Ions in solution can pass through the wet asbestos diaphragm, but it stops the bubbles of hydrogen and chlorine gas. The sodium hydroxide solution produced is contaminated with unreacted salt, which precipitates as the solution is concentrated.*

which reacts with water in a separate vessel to give sodium hydroxide that is free from NaCl.

$$2NaHg_x(l) + 2H_2O(l) \longrightarrow H_2(g) + 2NaOH(aq) + 2x\,Hg(l)$$

The mercury is recycled. However, some of the mercury inevitably escapes into the surrounding air and bodies of water, and such cells have contributed to the mercury pollution problem.

At one time, graphite was used for the inert anode in both types of chlorine cells. However, a big improvement has been made by the introduction of titanium anodes coated with ruthenium oxide. The Cl_2 is released more easily and is free from organic impurities derived from the graphite electrodes. Also, the titanium anodes are dimensionally stable, whereas graphite electrodes are not.

Electrolysis is also of considerable technical importance in the following areas:

1. Preparing useful substances, including various metals (e.g., Na, Mg, Be, Ca, Al, Sn, Cu, Mn) and compounds (e.g., NaOH, Na_2CO_3, NaClO, $KClO_3$, $KClO_4$, $KMnO_4$, H_2O_2).

2. Refining metals (e.g., Zn, Al, Cu, Ni).

3. Electroplating with metals that are wear- or corrosion-resistant or yield an attractive surface (e.g., with Ag, Au, Cr, Cd, Zn, Sn).

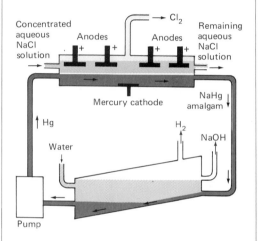

FIGURE 24.11

The mercury cell for the electrolysis of sodium chloride. *This cell is more expensive to build and operate than the diaphragm cell (Figure 24.10), but it produces a more concentrated and purer sodium hydroxide solution.*

THOUGHTS ON CHEMISTRY

Volta describes his discoveries

After a long silence, for which I shall offer no apology, I have the pleasure of communicating to you . . . some striking results I have obtained in pursuing my experiments on electricity excited by the mere mutual contact of different kinds of metal. . . .

I provide a few dozens of small round plates or disks of copper, brass, or rather silver, an inch in diameter more or less (pieces of

ON THE ELECTRICITY EXCITED BY THE MERE CONTACT OF CONDUCTING SUBSTANCES OF DIFFERENT KINDS, *by Alexander Volta coin, for example), and an equal number of plates of tin, or, what is better, of zinc, nearly of the same size and figure. . . I prepare also a pretty large number of circular pieces of pasteboard, or any other spongy matter capable of imbibing and retaining a great deal of water or moisture, with which they must be well impregnated in order to ensure success to the experiments. . . .*

Having all these pieces ready in a good state . . . I have nothing to do but to arrange them, a matter exceedingly simple and easy. . . . I place then horizontally, on a table or any other stand, one of the metallic pieces, for example one of silver, and over the first I adapt one of zinc; on the second I place one of the moistened disks, then another plate of silver followed immediately by another of zinc, over which I place another of the moistened disks. In this matter I continue coupling a plate of silver with one of zinc and always in the same order, that is to say, the silver below and the zinc above it, or vice versa, according as I have begun, and interpose between each of these couples a moistened disk. I continue to form, of several of these stories, a column as high as possible without any danger of its falling. . . .

If, by means of an ample contact of the hand (well moistened) I establish on one side a good communication with one of the extremities of my electro-motive apparatus (*we must give new names to instruments that are new not only in their form, but in their effects on the principle on which they depend); and on the other I apply the forehead, eye-lid, tip of the nose, also well moistened, or any other part of the body where the skin is very delicate: if I apply, I say, with a little pressure, any one of these delicate parts, well moistened, to the point of a metallic wire, communicating properly with the other extremity of the said apparatus, I experience, at the moment that the conducting circle is completed, at the place of the skin touched, and a little beyond it, a blow and a prick, which suddenly passes, and is repeated as many times as the circle is interrupted and restored. . . .*

What proof more evident of the continuation of the electric current as long as the communication of the conductors forming the circle is continued?—and that such a current is only suspended by interrupting that communication? This endless circulation of the electric fluid (this perpetual motion) may appear paradoxical and even inexplicable, but it is no less true and real, and you feel it, as I may say, with your hands.

ON THE ELECTRICITY EXCITED BY THE MERE CONTACT OF CONDUCTING SUBSTANCES OF DIFFERENT KINDS, by Alexander Volta. Alexander Volta's letter to the Royal Society of London, March 20, 1800 (Quoted from an English translation, *Philosphical Magazine*, September 1800, as reproduced in pp. 42ff, "Galvani-Volta," by Bern Dibner, Burndy Library, Norwalk, Connecticut, 1952)

Significant terms defined in this chapter

voltaic cell
battery
electrolytic cell
anode

cathode
overvoltage
primary cell
equivalent weight, redox

faraday
storage cell
fuel cell

Exercises

24.1 Many electronic calculators use rechargeable nickel–cadmium batteries. The overall equation for the spontaneous reaction in these cells is

$$Cd(s) + NiO_2(s) + 2H_2O \xrightarrow{KOH} Cd(OH)_2(s) + Ni(OH)_2(s)$$

What are the oxidation states of Cd in (a) Cd and (b) $Cd(OH)_2$ and of Ni in (c) NiO_2 and (d) $Ni(OH)_2$? What is the (e) oxidizing agent, (f) reducing agent, (g) substance oxidized, and (h) substance reduced? Write the shorthand notation for the (i) reduction couples and (j) the overall cell. During the middle of an examination, a student's calculator failed. (k) What happened chemically? (l) Write the overall cell equation for the recharge cycle of this cell. (m) Is the KOH present in this cell a catalyst?

24.2 Which statements are true? Rewrite any false statement so that it is correct.
(a) A voltaic cell generates electrical energy from a spontaneous oxidation–reduction reaction.
(b) The anode is the electrode at which reduction occurs.
(c) The electrode that is the source of electrons is the negative electrode.
(d) Cations are always positively charged and move toward the cathode.
(e) The cells developed early in the history of electrochemistry are known as primary cells.
(f) The same number of coulombs can be delivered during a fast charge as during a slow charge of a lead storage battery.

24.3 Distinguish clearly between the terms *anode* and *cathode*. In what type of cell can the anode be negative with respect to the cathode? Can the anode ever be positive with respect to the cathode?

24.4 Prepare a simple sketch of a voltaic cell showing the anode, cathode, the signs of the electrodes, and the direction of ion flow for the cell represented by the notation $Ag(s)|AgCl(s)|HCl(aq)|Cl_2(g)|(graphite)$. Is a salt bridge necessary for this reaction under these conditions?

24.5 Prepare a simple sketch of an electrolytic cell showing the anode, cathode, the signs of the electrodes,

and the direction of ion flow for the reaction given by the equation

$$MgF_2(l) \longrightarrow Mg(s) + F_2(g)$$

24.6 Distinguish between primary and secondary (storage) cells. Name a cell of each type that plays an important part in our everyday lives. Write the half-cell reactions and the overall cell reaction for each example.

24.7* The Edison cell, represented by $Fe|Fe(OH)_2|$ $LiOH,KOH|Ni(OH)_2|NiO \cdot OH$, is sometimes used in place of lead storage batteries when weight considerations are important. Write the half-reactions describing the oxidation and reduction processes and write the overall cell reaction.

24.8 Determine $E°$ and E for the reaction

$$Fe + 2Fe^{3+} \longrightarrow 3Fe^{2+}$$

given that $E° = -0.409$ V for Fe^{2+}/Fe and 0.770 V for Fe^{3+}/Fe^{2+} at 25°C. Assume the concentrations of iron(II) ion and iron(III) ion to be $1.0 \times 10^{-3} M$ and $1.5M$, respectively.

24.9 Consider the cell represented by the notation $Zn|ZnCl_2(aq)|Cl_2(1 \text{ atm})|(graphite)$. (a) Sketch the cell showing the anode, cathode, direction of electron flow, and ion flow, etc. (b) The standard reduction potentials are -0.7628 V for Zn^{2+}/Zn and 1.3583 V for Cl_2/Cl^- at 25°C. Calculate the emf for the cell at standard state conditions. (c) Find E for the cell when the concentration of the $ZnCl_2$ is $0.1M$.

24.10 Sketch the experimental arrangement for the cell given by $(Pt)|H_2(g)|HCl(aq)|Fe^{3+}(aq), Fe^{2+}(aq)|(Pt)$. Be sure to include the signs of the electrodes, the names of the electrodes, etc. At 25°C, $E° = 0.770$ V for Fe^{3+}/Fe^{2+}. Write the overall reaction and calculate E if $[H^+] = 0.1M$, $[Fe^{3+}] = 0.1M$, $[Fe^{2+}] = 0.01M$, and $P_{H_2} = 1.33$ atm.

24.11 Write the balanced equation for the half-reaction that takes place at each electrode as an electrical current is passed through a $1M$ aqueous solution of the following substances using inert electrodes: (a) $AgNO_3$, (b) $CuBr_2$, (c) H_2SO_4, and (d) NaOH. You may want to refer to Appendix C to decide which half-reaction is most favorable.

763

24.12 What mass of molten sodium and mass of bromine at standard conditions would be produced by electrolyzing molten sodium bromide using a current of 15 amperes for 3 hr?

24.13 What time would be required to plate an iron platter with 5.0 g of silver using a solution containing the $Ag(CN)_2^-$ ion and a current of 1.5 amperes?

24.14 How many amperes of electrical current must be passed through a $CuSO_4$ solution in order to plate out 1.0 kg of copper in 8.0 hr?

24.15 Write the chemical equation for the electrolysis of a fairly concentrated brine solution. If 1.5 amperes were passed for 5.0 hr, what volume of gaseous chlorine would be generated if it is measured at 745 torr and 85°C, assuming the process to be 80% efficient?

24.16* Consider the voltaic cell represented by $Zn|Zn^{2+}||Fe^{3+}|Fe$. (a) Write the half-reactions and overall cell reaction. (b) The standard reduction potentials for Zn^{2+}/Zn and Fe^{3+}/Fe are -0.7628 V and -0.036 V, respectively, at 25°C. Determine the standard voltage for the reaction. (c) Determine E for the cell when the concentration of Fe^{3+} is $10M$ and of Zn^{2+} is $1 \times 10^{-3} M$. (d) If 150 milliamperes is to be drawn from this cell for a period of 15 min, what is the minimum mass for the zinc electrode?

24.17* Consider the following unbalanced equation:

$$Hg(l) + Fe^{3+}(aq) \longrightarrow Hg_2^{2+}(aq) + Fe^{2+}(aq)$$

(a) Write the half reactions and the overall cell reaction. (b) Prepare a simple sketch for an electrochemical cell designed to obtain work from this reaction. Write the shorthand notation for this cell. (c) The standard reduction potentials at 25°C are 0.7961 V for Hg_2^{2+}/Hg and 0.770 V for Fe^{3+}/Fe^{2+}. Find $E°$ for the reaction. Is the reaction spontaneous under standard state conditions? (d) When $[Hg_2^{2+}] = 0.001M$, $[Fe^{2+}] = 0.1M$, and $[Fe^{3+}] = 1.00M$, what is E for the reaction? Is the reaction more, less, or identically spontaneous under these conditions than under standard state conditions?

24.18* A current passing successively through $0.5M$ aqueous solutions of $Ag(CN)_2^-$, $In_2(SO_4)_3$, and $NiSO_4$ liber-

ates 112 ml of hydrogen gas measured at standard conditions from an aqueous solution of KCl. Calculate the weight of Ag, In, and Ni deposited, assuming 100% efficiency, in each case.

24.19* Calculate the current required to deposit (a) 0.50 equivalent, (b) 0.50 mole, and (c) 0.50 g of elemental platinum from a solution containing the $PtCl_6^{2-}$ ion within a period of 5.0 hr.

24.20* A current of 250 milliamperes was passed through a slightly acidic solution of water for 5.0 min. (a) Write the equations for the reactions occurring at the anode and cathode and for the overall reaction. (b) What volumes of gases would be collected at 25°C and 1.00 atm over the water? The vapor pressure of water at this temperature is 23.756 torr.

24.21* A sample of Al_2O_3 (dissolved in cryolite) is electrolyzed using a current of 1.00 ampere. (a) What is the rate of production of Al in grams per hour? (b) The oxygen liberated at the positive carbon electrode reacts with the carbon to form CO_2. What mass of CO_2 is produced per hour?

24.22* The same quantity of electricity that deposited 0.583 g of silver was passed through a solution of a gold salt and 0.355 g of gold was formed. (a) Calculate the equivalent weight of the gold. (b) What is the oxidation state of gold in this salt? (c) If a current of 1.0 ampere was used, how long did this electrolysis last?

24.23** The production of U from purified UO_2 ore consists of the following steps:

$$UO_2 + 4HF \longrightarrow UF_4 + 2H_2O$$
$$UF_4 + 2Mg \longrightarrow U + 2MgF_2$$

What is the oxidation number of U in (a) UO_2, (b) UF_4, and (c) U? Identify the (d) oxidizing agent and (e) the substance reduced. (f) If the second reaction were performed electrochemically, predict $E°$ for the reaction given $E° = -1.50$ V for U^{4+}/U and $-2.375V$ for Mg^{2+}/Mg. (g) What current could the second reaction generate if 1.00 g of UF_4 reacted each minute? (h) What volume of HF at 25°C and 10.0 atm would be required to produce 1.00 lb of U? (i) Would 1.00 lb of Mg be enough to produce 1.00 lb of U?

25 CHEMICAL THERMODYNAMICS

Thermodynamics (*from the Greek "therme" meaning heat or energy, and "dynamis" meaning power) is the study of the transformation of energy from any one form to another. A complete study of thermodynamics involves such topics as heat, work, changes in state, and chemical energy.*

When applied to chemical systems, thermodynamics serves as a valuable tool in predicting such things as the spontaneity of a chemical reaction, the relationship between the amounts of products and reactants once equilibrium has been established, and the amount of energy absorbed or released during a reaction. However, thermodynamics cannot predict the reaction mechanism nor the reaction rate—these are in the realm of chemical kinetics, which has been explored in Chapter 15.

In Chapter 5 we discussed thermochemistry, the branch of thermodynamics that deals with heat in chemical reactions and with ΔH, the change in enthalpy, or heat content, of substances. At this point you may be wondering what more there is to be learned about thermodynamics. From Chapter 5 you know that exothermic reactions can be spontaneous and endothermic reactions are not likely to be spontaneous. Isn't this enough? As you can probably guess, the answer is no. If you go back to Section 5.3, you will see that there we emphasized the words "often" and "usually" in talking about reaction spontaneity. In this chapter we investigate all of the thermodynamic forces and explain why some changes with positive ΔH are spontaneous, while some changes with negative ΔH are not spontaneous.

The thermodynamic concepts of entropy and free energy are introduced in this chapter. We first discuss entropy, S, and entropy changes, ΔS, and their influence on the spontaneity of a reaction. Then we define the free energy change, ΔG, a parameter based on ΔH and ΔS. Free energy can be used to predict unequivocally whether or not a reaction will be spontaneous. The relationship of ΔG to equilibrium and electrochemistry is then discussed. Thermodynamic calculations of heats of reactions, equilibrium conditions, phase changes, and rates of reaction at temperatures other than 25°C are demonstrated. The last section applies thermodynamics to a biochemical reaction.

Positive ΔS: increase
in disorder

Negative ΔS: decrease
in disorder

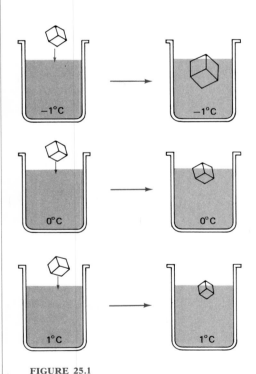

FIGURE 25.1

Changes in an ice cube at different temperatures. *At −1°C the ice cube grows larger, at 0°C it is unchanged, and at 1°C it melts a bit.*

25.1 Entropy: A quantitative measurement of randomness

In our discussion of thermochemistry in Chapter 5, we stated that exothermic changes—processes for which ΔH is negative and heat is given off—are *frequently* spontaneous. Now we are ready to discuss why some such processes are not spontaneous.

To illustrate a case in which exothermic changes are not favorable, consider the three beakers of water shown in Figure 25.1. Each beaker holds a large reservoir of water maintained at a constant temperature. Ice cubes of equal size are dropped into each beaker. After a few minutes, the ice cube in the beaker containing supercooled water at −1°C will increase in size because some of the water has frozen. No visible change will occur in the second beaker, and the ice cube in the third beaker will decrease in size because some of the ice has melted.

As we have discussed, a standard enthalpy change, $\Delta H°$, for a pure substance involves the substance in its normal state at 1 atm pressure and a specific temperature. When the temperature is not given, we assume it is 25°C, or 298°K. For a phase change such as the freezing of water, the temperature must obviously be a temperature at which the phase change will occur, rather than just 25°C, the average room temperature.

For the process

$$H_2O(l, 1 \text{ atm}) \longrightarrow H_2O(s, 1 \text{ atm})$$

the enthalpy changes at the temperatures of the three beakers of water are $\Delta H°_{272} = -1427.4$ cal/mole (-5972.2 J/mole), $\Delta H°_{273} = -1436.3$ cal/mole (-6009.5 J/mole), and $\Delta H°_{274} = -1445.2$ cal/mole (-6046.7 J/mole).

Based on the negative ΔH values alone, the water should freeze in all three beakers, not just one. Thus, we must conclude that there is a second driving force for this process that opposes the favorable enthalpy change. At −1°C, this second driving force is not strong enough to counteract the enthalpy change. At 1°C, it is strong enough to overcome the enthalpy change, and at 0°C, it must be equal in magnitude to the enthalpy change, but opposite in effect, so that an equilibrium condition exists. This new driving force that affects the spontaneity is known as the entropy change, ΔS.

Entropy, S, is a measure of the randomness or disorder of a system. The **entropy change,** ΔS, is the change in disorder accompanying any change in a system. A positive value for ΔS indicates an increase in entropy, or disorder, and a negative value for ΔS indicates a decrease in entropy, or disorder.

As for enthalpy changes, a superscript circle, $\Delta S°$, refers to an entropy change at standard state conditions and at 25°C if no other temperature is given as a subscript. Unlike enthalpy, for which only changes can be measured or calculated, values of the entropy of a substance itself—the absolute entropy—can be determined. Therefore, we can speak of the entropies of substances, S or $S°$, as well as the entropy changes for processes, ΔS or $\Delta S°$.

Consider a sample of argon for which equilibrium has been established between the liquid state and the gaseous state at the normal boiling point of argon, $-185.7°C$. Each atom in the liquid can move around freely in a space about 0.32 nm in diameter. However, the atom has only a small chance of getting out of one space and into another one in a different place (Figure 25.2). An atom in the gaseous phase moves much further—about 20 nm—before hitting another atom. The difference between these distances illustrates the significantly greater degree of randomness or disorder in the gaseous than in the liquid state.

Therefore, the transition of an argon atom from the liquid to the gaseous state should have a positive ΔS. On a molar basis, *the entropy change for this or any other phase transition* is given by

$$\overset{\text{enthalpy change for phase transition}}{\Delta S = \frac{\Delta H}{T}} \tag{1}$$

absolute temperature

Typical units for ΔS are calories per mole °K or joules per mole °K. Note that calories rather than kilocalories are commonly used for ΔS (its magnitude is smaller than that of ΔH). When $\Delta H°$ is used in Equation (1), $\Delta S°$ is obtained.

EXAMPLE 25.1

■ Calculate the standard entropy change for 1 mole of argon as it evaporates at $-185.7°C$. $\Delta H°_{87.5}$(vaporization) = 1558 cal/mole.

The entropy change at $(273.15° - 185.7°) = 87.5°K$ is given by

$$\Delta S°_{87.5} = \frac{H°_{87.5}(\text{vaporization})}{T} = \frac{1558 \text{ cal/mole}}{87.5°K} = 17.8 \text{ cal/mole °K}$$

The entropy change is 17.8 cal/mole °K or 74.5 J/mole °K.   ■

Let's return to our sample of argon to illustrate a second source of disorder—a temperature change. If we increase the temperature from $-185.7°C$ to $0°$, keeping the pressure at 1 atm, we find that an atom in the gas phase now moves about 63 nm on the average before striking another atom. Thus there has been an increase in the entropy of the sample as a result of this temperature change.

On a molar basis, the *entropy change for a change in temperature at constant pressure* is given by

$$\Delta S = 2.303 C_p \log \left(\frac{T_{\text{final}}}{T_{\text{initial}}} \right) \tag{2}$$

molar heat capacity, constant P

where C_p is the molar heat capacity for the substance at constant pressure and 2.303 is the conversion factor necessary because we

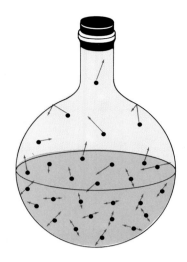

FIGURE 25.2

Randomness difference between the liquid and gaseous states of argon. *Only a few molecules in the liquid state can jump out of the space in which they move around.*

have used logarithms to the base 10 rather than natural logarithms. When C_p°, standard state molar heat capacity or molar heat capacity at 1 atm, is used in Equation (2), ΔS° is obtained. [Values of ΔS calculated using Equation (2) will be in error if C_p changes significantly with temperature.]

EXAMPLE 25.2

■ Calculate the standard state entropy change for 1 mole of argon gas as it is heated from $-185.7°C$ to $0°C$. $C_p^\circ = 5.0$ cal/mole $°K$ for argon over this temperature range.

The resulting change in entropy is given by

$$\Delta S^\circ = 2.303\, C_p^\circ \log \left(\frac{T_{final}}{T_{initial}} \right)$$

$$= (2.303)(5.0 \text{ cal/mole } °K) \log \left(\frac{273.15°K}{87.5°K} \right)$$

$$= 5.7 \text{ cal/mole } °K$$

The standard state entropy change is 5.7 cal/mole $°K$ or 23.8 J/mole $°K$. ■

If our sample of gaseous argon at $0°C$ is expanded to twice the volume by decreasing the pressure, the average distance that an atom moves before undergoing a collision is increased to about 130 nm, and more randomness is present. On a molar basis we can calculate the corresponding *entropy change for a change in volume at constant temperature* for any substance using

$$\Delta S = 2.303R \log \left(\frac{V_{final}}{V_{initial}} \right) \qquad (3)$$

where R is the gas constant (1.987 cal/mole $°K$ or 8.314 J/mole $°K$). If the substance is an ideal gas, we can substitute Boyle's law into Equation (3) to get

$$\Delta S = 2.303R \log \left(\frac{P_{initial}}{P_{final}} \right) \qquad (4)$$

EXAMPLE 25.3

■ Calculate the entropy change for doubling the volume of a sample of gaseous argon at $0°C$.

The entropy change is given by

$$\Delta S = 2.303R \log \left(\frac{V_{final}}{V_{initial}} \right) = (2.303)(1.987 \text{ cal/mole } °K) \log \left(\frac{2V_{initial}}{V_{initial}} \right)$$

$$= 1.4 \text{ cal/mole } °K$$

The entropy change is 1.4 cal/mole $°K$ or 5.8 J/mole $°K$. ■

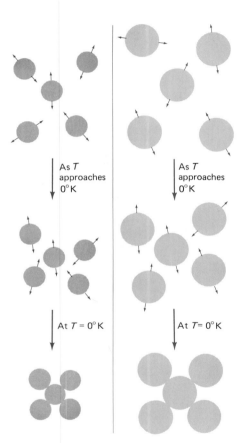

FIGURE 25.3
Decrease in randomness as absolute zero is approached. *The randomness of all perfect crystalline substances is the same at absolute zero and is $S_0^\circ = 0$. Note that this does not mean that the crystal structures of all substances are the same at absolute zero, just that all crystals are perfectly ordered.*

As T approaches $0°K$

As T approaches $0°K$

At $T = 0°K$

At $T = 0°K$

Suppose our container of gaseous argon is connected to an identical system containing neon, and the gases are allowed to mix at constant temperature and pressure. Because of the increase in volume available to a given atom after mixing, an increase in entropy occurs. The total entropy change upon mixing is the sum of the ΔS values for the change of each gas from the original pressure P of the pure gas to a partial pressure $X_i P$ in the mixture, where X_i is the mole fraction of the gas i. The expression for the entropy of mixing of gases is simply derived from Equation (4) as follows, where n_1 is the number of moles of gas 1, and n_2 is the number of moles of gas 2, and so on.

$$\Delta S = n_1\, 2.303R\, \log \frac{P}{X_1 P} + n_2\, 2.303R\, \log \frac{P}{X_2 P} + \cdots$$
$$= -2.303R(n_1 \log X_1 + n_2 \log X_2 + \cdots) \tag{5}$$

EXAMPLE 25.4

■ What is the increase in entropy when 1 mole samples of argon and neon are mixed at 0°C?

Once the gases have mixed, $X = 0.5$ for both argon and neon. Thus, the entropy change is

$$\begin{aligned}
\Delta S &= -2.303R(n_1 \log X_1 + n_2 \log X_2) \\
&= -(2.303)(1.987 \text{ cal/mole °K})[(1 \text{ mole})(\log 0.5) + (1 \text{ mole})(\log 0.5)] \\
&= -2.76 \text{ cal/°K}
\end{aligned}$$

The total entropy change is 2.76 cal/°K or 11.5 J/°K. ■

25.2 Absolute entropy of a substance

When 1 mole samples of neon and argon are cooled at 1 atm pressure to the point where they liquefy and then to the point where they solidify, forming perfect crystals, the amount of randomness decreases in both samples as the temperature decreases. What happens as the temperature approaches absolute zero? The vibrational kinetic energy of the atoms in both samples is decreasing (Figure 25.3), and the atoms are becoming more and more confined to their lattice positions in the crystal. At absolute zero, the atoms no longer move around in their crystal positions, and both substances have the same amount of disorder. By convention we choose the standard state entropy at absolute zero, S_0°, as zero for perfect crystalline substances. (The third law of thermodynamics may be stated as follows: The entropy of a perfect crystal at absolute zero is zero.)

What about a more complicated substance composed of diatomic molecules AB? If the substance forms a perfect crystal with all molecules completely ordered (Figure 25.4a), again $S_0^\circ = 0$. On the other hand, if the substance forms a crystal with the molecules oriented in a haphazard way (Figure 25.4b), the entropy is not zero, but larger

(a)

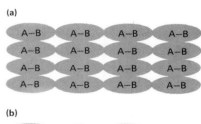

(b)

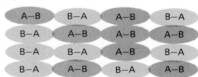

(c)

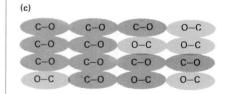

FIGURE 25.4
Disorder present at absolute zero in a nonperfect crystal.

(a) $S_0^\circ = 0$

(b) $S_0^\circ = 2.303R \log 2$

(c) $0 < S_0^\circ < 2.303R \log 2$

than zero. The amount of entropy at absolute zero in a crystal containing randomly oriented molecules is given by

$$S_0^\circ = 2.303R \log \Omega \tag{6}$$

where Ω (omega) is the number of unique ways that the molecule can orient itself in the lattice. For our sample of A—B, $\Omega = 2$ because no matter what orientation A—B has in a crystal, there is only one other unique orientation for the same molecule in the same location

A—B   B—A
orientation 1   *orientation 2*

From Equation (6), we can find that for A—B, $S_0^\circ = 1.38$ cal/mole °K or 5.76 J/mole °K.

Polar molecules like CO and NO are not completely randomly oriented in the crystalline state, but are somewhat ordered (see Figure 25.4c) as a result of dipole–dipole interactions. The value of S_0° calculated using Equation (6) for such substances will be somewhat high. The absolute entropy of such substances at absolute zero will therefore be somewhere between zero and the value given by Equation (6). There are other reasons why S_0° is not zero for a given sample of a substance, such as crystal imperfections or impurities.

The absolute entropy of a substance at any temperature above absolute zero is found as follows: Find the entropy of the substance at absolute zero given by Equation (6). Add to this the sum of the entropy changes given by Equation (1) for each phase change that the substance undergoes between absolute zero and the desired temperature. Next, add to this the sum of the entropy changes between absolute zero and the desired temperature given by Equation (2) for heating each phase over the temperature range at which it is stable. Calculation of the absolute entropy in this way is illustrated in Example 25.5.

EXAMPLE 25.5

■ Find the absolute standard state entropy of HCl(g) at 25°C. At absolute zero, 0°K, HCl exists in a solid phase called solid II. During heating, it passes through another solid phase called solid I. The necessary information to find S_{298}° for HCl is as follows:

(a) At 0°K, HCl (solid II) is a random crystalline substance, so $\Omega = 2$.

(b) ΔS° for heating HCl (solid II) from 0°K to 98.38°K is 5 cal/mole °K.

(c) The necessary enthalpy changes for HCl phase changes are
$\Delta H_{98.38}^\circ$(solid II → solid I) = 284 cal/mole,
$\Delta H_{158.94}^\circ$(solid I → liquid) = 476 cal/mole,
$\Delta H_{188.11}^\circ$ (liquid → gas) = 3860 cal/mole.

(d) The necessary standard molar heat capacities for HCl are
C_p°(solid I) = 11.9 cal/mole °K;
C_p°(liquid) = 14.0 cal/mole °K;
C_p°(gas) = 6.9 cal/mole °K.

The value of S_0° is given by Equation (6) as

$$S_0^\circ = 2.303R \log \Omega = 2.303(1.987 \text{ cal/mole } {}^\circ\text{K}) \log 2$$
$$= 1.4 \text{ cal/mole } {}^\circ\text{K}$$

Using Equation (1) gives the entropy changes for the phase transitions.

$$\Delta S^\circ = \frac{\Delta H^\circ}{T} = \frac{284 \text{ cal/mole}}{98.38^\circ\text{K}} = 2.9 \text{ cal/mole } {}^\circ\text{K}$$

$$\Delta S^\circ = \frac{476 \text{ cal/mole}}{158.94^\circ\text{K}} = 3.0 \text{ cal/mole } {}^\circ\text{K}$$

$$\Delta S^\circ = \frac{3860 \text{ cal/mole}}{188.11^\circ\text{K}} = 20.5 \text{ cal/mole } {}^\circ\text{K}$$

For heating the various phases between the respective temperatures, we use Equation (2) to find the entropy changes.

$$\Delta S^\circ = 2.303 \, C_p^\circ \log \left(\frac{T_{final}}{T_{initial}} \right)$$

$$= (2.303)(11.9 \text{ cal/mole } {}^\circ\text{K}) \log \left(\frac{158.94^\circ\text{K}}{98.38^\circ\text{K}} \right) = 5.7 \text{ cal/mole } {}^\circ\text{K}$$

$$\Delta S^\circ = (2.303)(14.0) \log \left(\frac{188.11}{158.94} \right) = 2.4 \text{ cal/mole } {}^\circ\text{K}$$

$$\Delta S^\circ = (2.303)(6.9) \log \left(\frac{298.15}{188.11} \right) = 3.2 \text{ cal/mole } {}^\circ\text{K}$$

Adding the calculated values of S_0°, ΔS° for heating solid II from 0°K to 98.38°K, and all of the calculated values of ΔS° for heating and phase transitions gives

$$S_{298}^\circ = 1.4 + 5 + 2.9 + 5.7 + 3.0 + 2.4 + 20.5 + 3.2$$
$$= 44 \text{ cal/mole } {}^\circ\text{K}$$

The value of the absolute standard state entropy of $HCl(g)$ at 25°C is 44 cal/mole $^\circ\text{K}$. More exact calculations give 44.645 cal/mole $^\circ\text{K}$. ∎

25.3 Entropy as a driving force

When a container holding a sample of benzene is placed in a large ice-water bath held at 0°C, a spontaneous process takes place—the benzene freezes at 5.53°C. For the phase change, the enthalpy change is -2379 cal/mole.

$$C_6H_6(l) \longrightarrow C_6H_6(s) \qquad \Delta H_{278.68}^\circ = -2379 \text{ cal}$$

Equation (1) gives the entropy change for the benzene

$$\Delta S^\circ = \frac{\Delta H^\circ}{T} = \frac{-2379 \text{ cal/mole}}{278.68^\circ\text{K}} = -8.537 \text{ cal/mole } {}^\circ\text{K}$$

The crystalline benzene is more ordered than the liquid, and entropy has decreased. However, this is not the only entropy change taking place. The 2379 cal given up by the benzene is absorbed by the ice-water bath at its temperature, 0°C, giving for the water

$$\Delta S° = \frac{2379 \text{ cal/mole}}{273.15°K} = 8.710 \text{ cal/mole °K}$$

The total entropy change in freezing this benzene was

$$\Delta S° = 8.710 - 8.537 = 0.173 \text{ cal/mole °K}$$

Thus, for the spontaneous freezing of benzene, the total entropy change was positive. For a spontaneous process, the total ΔS for *all* changes taking place is *always* positive. Similarly, the total entropy change for a nonspontaneous process is always negative. (The second law of thermodynamics can be stated, "All systems tend to approach a state of equilibrium," or "The entropy of the universe strives toward a maximum.")

Think of it this way. Suppose you are carrying a box of checkers in which the black pieces and the red pieces are stacked in separate piles. If you spill the checkers—causing a spontaneous change—the checkers will not land on the floor in stacks, nor will they be separated by color. In this spontaneous process the entropy, or disorder, has increased. There is no way that you can get the checkers to spontaneously return to neat, color-separated stacks. Such a return to order, a decrease in entropy, is only going to happen when you pile up the checkers one by one—a nonspontaneous process.

The entropy change for a chemical reaction may be either positive or negative, depending upon whether disorder increases or decreases in the reaction. The entropy changes for chemical reactions are calculated in one of two ways. In the first (Example 25.6), tabulated values for the absolute entropies of the substances involved in the reaction are used—the total of the entropies of the reactants (the initial entropy) is subtracted from the total entropies of the products (the final entropy)

$$\Delta S = [\text{total } S \text{ (products)}] - [\text{total } S \text{ (reactants)}] \tag{7}$$

In the second method (Example 25.7), tabulated standard entropies of formation (analogous to heats of formation; Section 5.4) and/or known entropies of reactions are combined algebraically, just as enthalpies of formation and reaction equations are combined according to Hess' law (Section 5.6).

EXAMPLE 25.6

■ Find the standard state entropy of formation of HCl (g) at 25°C. Is $\Delta S°$ favorable for this reaction? The absolute standard state entropies at 25°C are 44.646 cal/mole °K for HCl(g), 31.208 cal/mole °K for $H_2(g)$, and 53.288 cal/mole °K for $Cl_2(g)$.

The equation for the formation of HCl is

$$\tfrac{1}{2}H_2(g) + \tfrac{1}{2}Cl_2(g) \longrightarrow HCl(g)$$

The entropy change is found by subtracting the entropies of the reactants from the entropies of the products, remembering to multiply by the number of moles in the reaction

$$
\begin{aligned}
\Delta S^\circ &= [(1)S^\circ_{HCl}] - [(\tfrac{1}{2})S^\circ_{H_2} + (\tfrac{1}{2})S^\circ_{Cl_2}] \\
&= 44.646 - [(0.5)(31.208) + (0.5)(53.288)] \\
&= 2.398 \text{ cal/}^\circ K
\end{aligned}
$$

The standard entropy of formation of 1 mole of HCl(g) is 2.398 cal/°K. Because ΔS° for this reaction is positive, the entropy change in this reaction is favorable. ■

EXAMPLE 25.7

■ Using the ΔS° values at 25°C given for the following reactions

$$
\begin{array}{ll}
C(graphite) + 2Cl_2(g) \longrightarrow CCl_4(l) & \Delta S^\circ = -56.23 \text{ cal/}^\circ K \\
C(graphite) + \tfrac{3}{2}Cl_2(g) + \tfrac{1}{2}H_2(g) \longrightarrow CHCl_3(l) & \Delta S^\circ = -48.71 \text{ cal/}^\circ K \\
\tfrac{1}{2}H_2(g) + \tfrac{1}{2}Cl_2(g) \longrightarrow HCl(g) & \Delta S^\circ = 2.398 \text{ cal/}^\circ K
\end{array}
$$

find ΔS° for the reaction

$$CCl_4(l) + H_2(g) \longrightarrow HCl(g) + CHCl_3(l)$$

The desired reaction can be obtained by adding the reverse of the first equation to the second and third equations as written:

$$
\begin{array}{ll}
CCl_4(l) \longrightarrow C(graphite) + 2Cl_2(g) & \Delta S^\circ = 56.23 \text{ cal/}^\circ K \\
C(graphite) + \tfrac{3}{2}Cl_2(g) + \tfrac{1}{2}H_2(g) \longrightarrow CHCl_3(l) & \Delta S^\circ = -48.71 \text{ cal/}^\circ K \\
\tfrac{1}{2}H_2(g) + \tfrac{1}{2}Cl_2(g) \longrightarrow HCl(g) & \underline{\Delta S^\circ = 2.398 \text{ cal/}^\circ K} \\
CCl_4(l) + H_2(g) \longrightarrow HCl(g) + CHCl_3(l) & \Delta S^\circ = 9.92 \text{ cal/}^\circ K
\end{array}
$$

The entropy change for the reaction is 9.92 cal/°K under standard state conditions. ■

Free energy, G

25.4 Free energy as the criterion for spontaneity

We have now encountered two forces that will drive a change to be spontaneous. The first is the drive toward formation of more stable substances—products with a lower heat content than the reactants. This force makes exothermic processes (negative ΔH) favorable and endothermic processes (positive ΔH) not favorable. The second is the drive toward formation of products that are more disordered than the reactants. This force favors changes in which ΔS is positive and does not favor changes in which ΔS is negative. We can correctly predict that a reaction that has a negative ΔH and a positive ΔS will be spontaneous. And we can correctly predict that a reaction that has a positive ΔH and a negative ΔS will be nonspontaneous. However, when these driving forces oppose each other, it must be determined which force is more important.

At this point we call upon the thermodynamic quantity that interrelates enthalpy and entropy, the **free energy**, G, which is defined as

$$G = H - TS$$

For processes like chemical reactions that are carried out at constant temperature, the free energy change, ΔG, is

$$\Delta G = \Delta H - T \Delta S \tag{8}$$

Free energy is a quantity like enthalpy—only *changes* in free energy can be measured or calculated. For standard state conditions

$$\Delta G° = \Delta H° - T \Delta S°$$

For a spontaneous change, ΔG is always negative, and for a nonspontaneous change, ΔG is always positive. The **free energy change** is the overall chemical driving force of a reaction. It is the quantity we have been looking for—an unequivocal criterion for the spontaneity of any process.

With Equation (8) available to us, we can now return to an explanation of the ice and water systems shown in Figure 25.1. Recall that $\Delta H°$ was favorable for the process

$$H_2O(l) \longrightarrow H_2O(s)$$

at all three temperatures: $\Delta H° = -1445.2$ cal/mole at 1°C, -1436.3 cal/mole at 0°C, and -1427.4 cal/mole at -1°C. However, the values of $\Delta S°$ are unfavorable for the process: -5.291 cal/mole °K at 1°C, -5.2583 cal/mole °K at 0°K, and -5.226 cal J/mole °K at -1°C. If we use these numbers to calculate $\Delta G°$ at these temperatures, we get

$$\Delta G°_{272.15} = -(1427.4 \text{ cal/mole}) - (272.15°K)(-5.226 \text{ cal/mole °K})$$
$$= -5.1 \text{ cal/mole } (-21.9 \text{ J/mole})$$
$$\Delta G°_{273.15} = -(1436.3) - (273.15)(-5.2583) = 0.0 \text{ cal/mole}$$
$$\Delta G°_{274.15} = -(1445.2) - (274.15)(-5.291) = 5.3 \text{ cal/mole } (22.2 \text{ J/mole})$$

At -1°C, $\Delta H°$ is the more significant driving force, and we get a negative value of $\Delta G°$ for the spontaneous process. At 0°C, the $\Delta H°$ and $\Delta S°$ effects are equally important, and we get a value of zero for $\Delta G°$ for the equilibrium condition. At 1°C, $\Delta S°$ is the more important driving force, and we get a positive $\Delta G°$ for the nonspontaneous process.

The value of ΔG or $\Delta G°$ for a chemical reaction can be calculated in three different ways, depending on the type of data available: (1) from Equation (8), if ΔH and ΔS are known (Example 25.8); (2) from ΔG values if they are known for a series of reactions whose equations and ΔG values can be added or subtracted algebraically according to Hess' law to give the equation and ΔG for the desired reaction; (3) from tabulated ΔG_f° values—standard free energies of formation—for all of the reactants and all of the products, by the following relationship (Example 25.9):

$$\Delta G^\circ = (\text{sum of } \Delta G_f^\circ \text{ of products}) - (\text{sum of } \Delta G_f^\circ \text{ of reactants})$$

Note that, as for ΔH_f°, the value of ΔG_f° for an element is taken as equal to zero.

EXAMPLE 25.8

■ Determine whether the reaction

$$CCl_4(l) + H_2(g) \longrightarrow HCl(g) + CHCl_3(l)$$

is favorable at 25°C under standard state conditions. At 25°C, $\Delta H^\circ = -21.83$ kcal and $\Delta S^\circ = 9.92$ cal/°K for this reaction.

The value of ΔG° is

$$\begin{aligned}
\Delta G^\circ &= \Delta H^\circ - T\,\Delta S^\circ \\
&= (-21.83 \text{ kcal}) - (298°K)(9.92 \text{ cal/°K})(10^{-3} \text{ kcal/cal}) \\
&= -24.79 \text{ kcal}
\end{aligned}$$

The negative value of ΔG° indicates that the reaction will be spontaneous at this temperature and under standard state conditions. A very common error is to omit units on the two factors and, consequently, to omit the factor changing calories to kilocalories, giving a wrong answer. ■

EXAMPLE 25.9

■ Calculate ΔG° for the reaction in Example 25.8, given that $\Delta G_f^\circ = -15.60$ kcal/mole for $CCl_4(l)$, 0.00 kcal/mole for $H_2(g)$, -22.777 kcal/mole for $HCl(g)$, and -17.62 kcal/mole for $CHCl_3(l)$.

For the reaction

$$\begin{aligned}
\Delta G^\circ &= [(1)(\Delta G_{HCl}^\circ) + (1)(\Delta G_{CHCl_3}^\circ)] - [(1)(\Delta G_{CCl_4}^\circ) + (1)(\Delta G_{H_2}^\circ)] \\
&= [(1 \text{ mole})(-22.777 \text{ kcal/mole}) + (1 \text{ mole})(-17.62 \text{ kcal/mole})] \\
&\quad - [(1 \text{ mole})(-15.60 \text{ kcal/mole}) + (1 \text{ mole})(0 \text{ kcal/mole})] \\
&= -24.80 \text{ kcal}
\end{aligned}$$

The standard free energy change is found to be -24.80 kcal, in good agreement with the value calculated in the preceding example. ■

25.5 Conditions other than standard state

Many chemical reactions are carried out under conditions other than standard state conditions. The non-standard-state free energy, ΔG, is related to $\Delta G°$ by

$$\Delta G = \Delta G° + 2.303RT \log Q \qquad (9)$$

where Q is the reaction quotient (Section 16.4). Although activities should be used in the expression for Q, we will continue to use pressure (in atmospheres) for gases, concentrations (in moles/liter) for solutes, and unity for pure substances in their normal physical states (Section 14.10). When a change in the value of Q causes ΔG to become more negative, a reaction will be more favorable; a more positive ΔG will mean the reaction is less favorable under the new conditions.

EXAMPLE 25.10

■ Would changing the conditions of the reaction described in Example 25.8 to $p_{H_2} = 10$ atm and $p_{HCl} = 0.1$ atm increase or decrease the spontaneity of the reaction?

The reaction quotient for the equation is

$$Q = \frac{p_{HCl(l)}}{p_{H_2}} = \frac{0.1 \text{ atm}}{10 \text{ atm}} = 0.01$$

and Equation (9) gives

$$\begin{aligned}
\Delta G &= \Delta G° + 2.303RT \log Q \\
&= -24.79 \text{ kcal} \\
&\quad + (2.303) \left(1.987 \, \frac{\text{cal}}{\text{mole °K}}\right) \left(10^{-3} \frac{\text{kcal}}{\text{cal}}\right) (298°K)(\log 0.01) \\
&= -27.52 \text{ kcal/mole}
\end{aligned}$$

The more negative value of ΔG predicts the reaction to be more favorable under these conditions than under standard state conditions. Again, note the proper use of units and conversion factors to give the correct answer. ■

25.6 Relationship of K to $\Delta G°$

As we mentioned earlier, $\Delta G = 0$ under equilibrium conditions and, of course, $Q = K$. Thus, Equation (9) becomes

$$\Delta G° = -(2.303)RT \log K \qquad (10)$$

making it possible to calculate $\Delta G°$ if the equilibrium constant is known or to calculate the equilibrium constant if $\Delta G°$ is known (Example 25.11).

EXAMPLE 25.11

■ Find the equilibrium constant at 25°C for the reaction described in Example 25.8.

Solving Equation (10) for $\log K$ gives

$$\log K = \frac{-\Delta G^\circ}{2.303RT} = \frac{-(-24.79 \text{ keal/mole})(10^3 \text{ eal/keal})}{(2.303)(1.987 \text{ eal/mole °K})(298°\text{K})}$$

$$= 18.18$$

$$K = 1.5 \times 10^{18}$$

The equilibrium constant is 1.5×10^{18}.   ■

25.7 ΔG° and E°

In Section 19.5 we described electromotive force as the chemical driving force of a cell reaction, almost the same words just used to describe free energy. Indeed, the two quantities are related in a simple way:

$$\Delta G = -nFE \tag{11}$$

where n is the number of equivalents in the balanced overall equation, F is Faraday's constant (96,485 coulombs/faraday, or 23,060 cal/equiv V), and E is the cell potential in volts.

Of course, under standard state conditions,

$$\Delta G^\circ = -nFE^\circ \tag{11a}$$

and if we substitute Equations (11) and (11a) into Equation (9) we get

$$(-nFE) = (-nFE^\circ) + (2.303)RT \log Q$$

$$E = E^\circ - \frac{(2.303)RT}{nF} \log Q \tag{12}$$

which is the Nernst equation (Section 19.8).

EXAMPLE 25.12

■ What is ΔG° for the following reaction, given that $E^\circ = 1.10$ V?

$$\text{Zn}(s) + \text{Cu}^{2+}(1M) \longrightarrow \text{Zn}^{2+}(1M) + \text{Cu}(s)$$

Substituting the voltage into Equation (11a), we get

$$\Delta G^\circ = -nFE^\circ = -(2 \text{ equiv})(23,060 \text{ cal/equiv V})(1.10 \text{ V}) = -50.7 \text{ kcal}$$

The standard state free energy change is -50.7 kcal.   ■

EXAMPLE 25.13

■ Using $E° = 0.7996$ V for the Ag^+/Ag half-cell at 25°C and $K_{sp} = 1.8 \times 10^{-10}$ for AgCl at 25°C, calculate $E°$ for $AgCl + e^- \longrightarrow Ag + Cl^-$.

Converting $E°$ and K_{sp} to $\Delta G°$ using Equations (11a) and (10), and combining the equations and ΔG values gives

$$Ag^+ + e^- \longrightarrow Ag \qquad \Delta G° = -nFE° = -18.44 \text{ kcal}$$
$$\underline{AgCl \longrightarrow Ag + Cl^- \qquad \Delta G° = -2.303RT \log K = 13.3 \text{ kcal}}$$
$$AgCl + e^- \longrightarrow Ag + Cl^- \qquad \Delta G° = -5.1 \text{ kcal}$$

Using Equation (11a) a second time to find $E°$ gives

$$E° = -\frac{\Delta G°}{nF} = -\frac{(-5.1 \text{ kcal})(10^3 \text{ cal/kcal})}{(1 \text{ equiv})(23{,}060 \text{ cal/equiv V})}$$
$$= 0.22 \text{ V}$$

The standard half-cell reduction potential for AgCl is 0.22 V. ■

Temperature dependence of thermodynamic properties

25.8 $\Delta H°$, $\Delta G°$, and K

Suppose the equilibrium constant for a chemical reaction, such as the ionization of water, is measured at several temperatures. Over moderate temperature intervals, it is found by experiment that a plot of $\log K$ versus the reciprocal of the absolute temperature is linear (Figure 25.5). This plot corresponds to the mathematical equation

$$\log K = (\text{slope})\frac{1}{T} + (\text{constant})$$

Analysis of similar graphs for many chemical reactions shows that the slope of the line is equal to the $\Delta H°$ of the reaction divided by $-2.303R$, or

$$\log K = -\left(\frac{\Delta H°}{2.303R}\right)\left(\frac{1}{T}\right) + \text{constant} \tag{13}$$

If the equilibrium constant, K_{T_1}, at one temperature, T_1, and $\Delta H°$ are known for a given reaction, the equilibrium constant, K_{T_2}, at a new temperature T_2 can be calculated using

$$\log\left(\frac{K_{T_2}}{K_{T_1}}\right) = \frac{\Delta H°}{2.303R}\left(\frac{1}{T_1} - \frac{1}{T_2}\right) \tag{14}$$

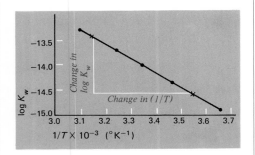

FIGURE 25.5

Plot of log K_w against (1/T) for the ionization constant of water.

provided that $\Delta H°$ does not change significantly over the temperature interval. Equation (14) has the advantage over Equation (13) because the constant need not be evaluated.

If we write Equation (14) as

$$\log K_{T_2} - \log K_{T_1} = \frac{\Delta H°}{2.303R}\left(\frac{1}{T_1} - \frac{1}{T_2}\right)$$

and substitute into it Equation (10), we get an equation for the temperature dependence of $\Delta G°$ for a reaction

$$\frac{\Delta G°_{T_2}}{T_2} - \frac{\Delta G°_{T_1}}{T_1} = \Delta H°\left(\frac{1}{T_2} - \frac{1}{T_1}\right) \tag{15}$$

The use of Equations (13)–(15) is illustrated in Examples 25.14–25.16.

EXAMPLE 25.14

■ Using the data given in Figure 25.5 for the ionization constant of water between 0°C and 50°C, calculate $\Delta H°$ for the ionization of water molecules.

The plot in Figure 25.5 is linear within experimental error. The value of $\Delta H°$ is determined from the slope of the line as calculated using widely spaced points from the line (indicated by ×'s). (See Example 15.6 for an explanation of the graphical method.) The slope of the line is

change in $\log K_w$

$$\text{slope} = \frac{\Delta(\log K_w)}{\Delta(1/T)} = \frac{(-14.58) - (-13.40)}{(3.55 \times 10^{-3}/°K) - (3.15 \times 10^{-3}/°K)}$$

change in $1/T$

coordinates of points marked in Fig. 25.5

$$= -2.95 \times 10^3 \ °K$$

which gives

$$\begin{aligned}
\Delta H° &= -2.303R(\text{slope}) \\
&= (-2.303)(1.987 \text{ cal/mole °K})(-2.95 \times 10^3 \text{ °K})(10^{-3} \text{ kcal/cal}) \\
&= 13.5 \text{ kcal/mole}
\end{aligned}$$

The enthalpy change for the ionization of water is 13.5 kcal/mole. This value compares favorably with the value calculated from enthalpy of formation data, which is 13.35 kcal/mole. The graphical method of finding $\Delta H°$ is preferred to using Equation (14) if enough data are available. By plotting the data and choosing widely spaced points, experimental error is smoothed out and the best possible answer obtained. ■

EXAMPLE 25.15

■ For the following reaction, $K = 3.87 \times 10^{-16}$ at 400°K

$$BeSO_4(s) \longrightarrow BeO(s) + SO_3(g)$$

Calculate K at 600°K assuming that the value of $\Delta H° = 41.9$ kcal for the reaction is constant over this temperature range.

To find K at 600°K, we use Equation (14)

$$\log\left(\frac{K_{T_2}}{K_{T_1}}\right) = \frac{\Delta H°}{2.303R}\left(\frac{1}{T_1} - \frac{1}{T_2}\right)$$

$$\log\left(\frac{K}{3.87 \times 10^{-16}}\right) = \frac{(41.9 \text{ kcal/mole})(10^3 \text{ cal/kcal})}{(2.303)(1.987 \text{ cal/°K mole})}\left(\frac{1}{400°K} - \frac{1}{600°K}\right)$$

$$= 7.63$$

Taking antilogarithms of both sides gives

$$\frac{K}{3.87 \times 10^{-16}} = 4.3 \times 10^7$$

$$K = 1.7 \times 10^{-8}$$

The equilibrium constant is predicted to be 1.7×10^{-8} at 600°K. ■

EXAMPLE 25.16

■ Would the reaction described in Example 25.8 be more or less spontaneous under standard state conditions at 0°C than at 25°C?

To find the answer to this question, we must calculate $\Delta G°_{273}$ and compare it to $\Delta G°_{298}$. Substituting $\Delta G°_{298} = -24.79$ kcal and $\Delta H° = -21.83$ kcal (see Example 25.8) into Equation (15) gives

$$\frac{\Delta G°_{T_2}}{T_2} = \frac{\Delta G°_{T_1}}{T_1} + \Delta H°\left(\frac{1}{T_2} - \frac{1}{T_1}\right)$$

$$\frac{\Delta G°_{273}}{273°K} = \frac{-24.79 \text{ kcal}}{298°K} + (-21.83 \text{ kcal})\left(\frac{1}{273°K} - \frac{1}{298°K}\right)$$

$$= -0.08990 \text{ kcal/°K}$$

$$\Delta G°_{273} = (273°K)(-0.08990 \text{ kcal/°K}) = -24.54 \text{ kcal}$$

Because $\Delta G°_{273}$ is less negative than $\Delta G°_{298}$, the reaction is less spontaneous at 0°C. ■

25.9 Phase equilibria

The chemical equation representing the phase transition between a condensed state of matter—solid or liquid—and the gaseous phase is

$$X(s \text{ or } l) \longrightarrow X(g)$$

For this equation the equilibrium constant is $K = P_x$, where P_x is either the sublimation or vapor pressure, depending on which condensed state was originally present. Substituting $K = P_x$ into Equation (14) gives the equation known as the Clausius–Clapeyron equation

$$\log \frac{P_{T_2}}{P_{T_1}} = \frac{\Delta H°(\text{sublimation or vaporization})}{2.303R} \left(\frac{1}{T_1} - \frac{1}{T_2} \right) \qquad (16)$$

which is a mathematical expression for the temperature dependence of sublimation or vapor pressure.

EXAMPLE 25.17

■ The vapor pressure of tungsten is 1.00 torr at 3990°C and 100.0 torr at 5168°C. Predict the normal boiling point of tungsten.

Using the two sets of data given, Equation (16) can be used to calculate $\Delta H°(\text{vaporization})$ for the tungsten

$$\Delta H° = \frac{2.303R}{(1/T_1 - 1/T_2)} \log \left(\frac{P_{T_2}}{P_{T_1}} \right)$$

$$= \frac{(2.303)(1.987 \text{ cal/mole °K})}{(1/4263°K - 1/5441°K)} \log \left(\frac{100.0 \text{ torr}}{1.00 \text{ torr}} \right)$$

$$= 1.80 \times 10^5 \text{ cal/mole}$$

The normal boiling point corresponds to the temperature at which $P = 760.0$ torr, so rearranging Equation (16) for T_2 gives

$$\frac{1}{T_2} = \frac{1}{T_1} - \frac{2.303R}{\Delta H°(\text{vap})} \log \left(\frac{P_{T_2}}{P_{T_1}} \right)$$

$$= \frac{1}{5441°K} - \frac{(2.303)(1.987 \text{ cal/mole °K})}{(1.80 \times 10^5 \text{ cal/mole})} \log \left(\frac{760.0 \text{ torr}}{100.0 \text{ torr}} \right)$$

$$= 1.614 \times 10^{-4}$$

$$T_2 = 6195°K$$

The predicted normal boiling point is 6195°K or 5922°C. ■

25.10 Reaction rates

The equation describing the temperature dependence of the rate constant, k, takes the same form as Equation (13)

activation energy

$$\log k = - \frac{\Delta E_a}{2.303R} \left(\frac{1}{T} \right) + \text{constant} \qquad (17)$$

A plot of log k versus the reciprocal of the absolute temperature is linear and the slope is equal to $\Delta E_a/2.303R$. Equation (17) is often written as

$$k = (\text{constant})e^{-\Delta E_a/RT}$$

Again, for two temperatures and two rate constants, Equation (17) can be written as

$$\log \frac{k_{T_2}}{k_{T_1}} = \frac{\Delta E_a}{2.303R} \left(\frac{1}{T_1} - \frac{1}{T_2} \right) \tag{18}$$

Equations (17) and (18) are both forms of what is known as the Arrhenius equation.

EXAMPLE 25.18

■ Find ΔE_a for the reaction

$$2C_6H_5CHO(aq) + CN^- \longrightarrow C_6H_5CH(OH)COC_6H_5 + CN^-$$

if $k = 1.40 \times 10^{-4}$ liter^2/mole^2 sec at 25°C and 1.43×10^{-3} liter^2/mole^2 sec at 60°C.

Solving the Arrhenius equation, Equation (18), for the activation energy gives

$$\Delta E_a = \frac{2.303R \log (k_{T_2}/k_{T_1})}{(1/T_1 - 1/T_2)}$$

$$= \frac{(2.303)(1.987 \text{ cal/mole °K}) \left(\dfrac{1 \text{ kcal}}{10^3 \text{ cal}} \right) \log \left(\dfrac{1.43 \times 10^{-3} \text{ liter}^2/\text{mole}^2 \text{ sec}}{1.40 \times 10^{-4} \text{ liter}^2/\text{mole}^2 \text{ sec}} \right)}{\left(\dfrac{1}{298°K} - \dfrac{1}{333°K} \right)}$$

$$= 13.1 \text{ kcal}$$

The energy of activation for this reaction is 13.1 kcal. ■

Application of chemical thermodynamics

25.11 A biochemical example

Usually it is relatively easy to measure the equilibrium constant for a chemical reaction. If this is done at two different temperatures, a great deal of thermodynamic information can be obtained concerning the reaction.

To illustrate how this is done, we will use data for a biochemical reaction, the conversion of L-aspartate ion into fumarate ion and ammonium ion.

This reaction, which is catalyzed by an enzyme, is part of a series of reactions that breaks down larger molecules in human and animal metabolism, so that they can be eliminated as waste products. The overall result of the reaction cycle in which the aspartate–fumarate conversion takes part is the removal of NH_2 groups from amino acids and their conversion to urea, which is eliminated in urine.

The equilibrium constant for the reaction is 7.4×10^{-3} at 29°C and 1.60×10^{-2} at 39°C. From this set of data, we can calculate $\Delta H°$, K, $\Delta G°$, and $\Delta S°$ at 37°C (approximate human body temperature).

From the temperature dependence of K, Equation (14) gives us

$$
\begin{aligned}
\Delta H° &= \frac{2.303R \log (K_{T_2}/K_{T_1})}{(1/T_1) - (1/T_2)} \\
&= \frac{(2.303)(1.987 \text{ cal/mole °K})(1 \text{ kcal}/10^3 \text{ cal}) \log (1.60 \times 10^{-2}/7.4 \times 10^{-3})}{(1/302°K - 1/312°K)} \\
&= 14.4 \text{ kcal}
\end{aligned}
$$

Using this value of $\Delta H°$ and the value of K at 39°C, we can find K at 37°C

$$
\begin{aligned}
\log \left(\frac{K_{T_2}}{K_{T_1}}\right) &= \frac{\Delta H°}{2.303R} \left(\frac{1}{T_1} - \frac{1}{T_2}\right) \\
\log \left(\frac{K_{T_2}}{1.60 \times 10^{-2}}\right) &= \frac{(14.4 \text{ kcal})(10^3 \text{ cal/kcal})}{(2.303)(1.987 \text{ cal/mole °K})} \left(\frac{1}{312°K} - \frac{1}{310°K}\right) \\
&= -6.5 \times 10^{-2} \\
\frac{K_{T_2}}{1.60 \times 10^{-2}} &= 0.86 \\
K &= 1.38 \times 10^{-2}
\end{aligned}
$$

The value of $\Delta G°$ is calculated using Equation (10)

$$
\begin{aligned}
\Delta G° &= -2.303RT \log K \\
&= -(2.303)(1.987 \text{ cal/mole °K}) \left(\frac{1 \text{ kcal}}{10^3 \text{ cal}}\right) (310°K) \log (1.38 \times 10^{-2}) \\
&= 2.6 \text{ kcal/mole}
\end{aligned}
$$

Once $\Delta G°$ and $\Delta H°$ are known, we use Equation (8) to obtain the value of $\Delta S°$

$$
\begin{aligned}
\Delta S° &= \frac{\Delta H° - \Delta G°}{T} = \frac{(14.4 - 2.6) \text{ kcal/mole } (10^3 \text{ cal/kcal})}{310°K} \\
&= 38 \text{ cal/°K mole}
\end{aligned}
$$

It is possible to make a rough common-sense check on the values of some of these numbers. We would expect the entropy content of L-aspartate and fumarate to be about the same because the molecules are quite similar in size and other properties, and so we would expect the ΔS of the reaction to be positive as a result of the formation of a second species (NH_4^+), leading to more randomness in the solution. At 25°C, thermodynamics tables give $S° = 27.1$ cal/mole °K for $NH_4^+(1M)$, which is in good agreement with our approximation of 38 cal/°K for the entropy increase. Likewise, during the reaction we break a C—N bond and a C—H bond, change a C—C bond to a C=C bond, and form an N—H bond. Using bond energies (B) as an estimate (see Table 9.19), we obtain

$$\Delta H = B_{C-N} + B_{C-H} + B_{C-C} - B_{C=C} - B_{N-H}$$
$$= 86 + 99 + 79 - 141 - 86$$
$$= 37 \text{ kcal}$$

which agrees in sign and order of magnitude with our $\Delta H°$ value of $+14.4$ kcal.

THOUGHTS ON
CHEMISTRY

From the preface to a classic book on thermodynamics

THERMODYNAMICS, by
G. N. Lewis and M. Randall

There are ancient cathedrals which, apart from their consecrated purpose, inspire solemnity and awe. Even the curious visitor speaks of serious things, with hushed voice, and as each whisper reverberates through the vaulted nave, the returning echo seems to bear a message of mystery. The labor of generations of architects and artisans has been forgotten, the scaffolding erected for their toil has long since been removed, their mistakes have been erased, or have become hidden by the dust of centuries. Seeing only the perfection of the completed whole, we are impressed as by some superhuman agency. But sometimes we enter such an edifice that is still partly under construction; then the sound of hammers, the reek of tobacco, the trivial jests bandied from workman to workman, enable us to realize that these great structures are but the result of giving to ordinary human effort a direction and a purpose.

Science has its cathedrals, built by the efforts of a few architects and of many workers. In these loftier monuments of scientific thought a tradition has arisen whereby the friendly usages of colloquial speech give way to a certain severity and formality. While this may sometimes promote precise thinking, it more often results in the intimidation of the neophyte. . . .

There are several kinds of audience to which . . . thermodynamics might be addressed. There is the beginner who, in order that he may decide whether the subject will meet his needs or arouse his interest, asks what thermodynamics is and what sorts of problems in physics, chemistry and engineering can be solved by its aid; there is the reader who looks for the philosophical implications of such concepts as energy and entropy; above all there is the investigator who, attacking problems of pure or applied science, seeks the specific thermodynamic methods which are applicable to his problem and the data requisite for its solution. Perhaps we have been over-ambitious in attempting, within the confines of a single volume, to meet all these demands—to lead the beginner through the intricacies of thermodynamic theory and to guide the experienced investigator to the extreme limits now set by existing methods and data.

THERMODYNAMICS, by G.N. Lewis and M. Randall, pp. v–vi. (Copyright © 1923. Used with permission of Mc-Graw Hill Book Company.)

Significant terms defined in this chapter

thermodynamics   free energy
entropy   free energy change
entropy change

Exercises

25.1 The standard enthalpy change at 25°C for dissolving sodium chloride in water is +910 cal/mole

$$NaCl(s) + n\,H_2O \longrightarrow NaCl(aq, 1M)$$

which implies that the process is unfavorable, but yet we know that salt readily dissolves in water (think of the oceans!). (a) For this favorable process, what must be the sign of the numerical value for $\Delta G°$? (b) If the enthalpy change is unfavorable, what is the favorable driving force for this reaction? (c) What must be the sign on the numerical value for $\Delta S°$? (d) Does this sign represent an increase or decrease in randomness? (e) Does the formation from a highly ordered crystal structure and a pure liquid of a mixture in which the sodium and chloride ions are freely moving around in water confirm your answer to (d)?

25.2 Clearly define enthalpy, entropy, and free energy. What are the symbols commonly used for these thermodynamic quantities? Give a suitable set of units for each of these quantities.

25.3 Which statements are true? Rewrite any false statement so that it is correct.
(a) The entropy content of a substance increases as its temperature increases.
(b) Both heat capacity and entropy have the units of (energy/mole °K) and hence are the same thermodynamic property.

(c) Unlike free energy and enthalpy, we can calculate the absolute entropy of a substance.
(d) A chemical reaction in which the entropy decreases will never be spontaneous.
(e) For phase changes, $\Delta G°$ is zero because $\Delta S° = \Delta H°/T$ and thus these changes are examples of equilibrium.

25.4 A flask containing nitrogen and a flask containing oxygen are connected by a small tube with a stopcock. After the stopcock is opened, the gases eventually mix so that a uniform composition is attained even though there is no enthalpy change. What is the driving force of this process?

25.5 Which would you predict to have the higher entropy content—a sample of liquid neon or a sample of liquid hydrogen fluoride (remember those strong hydrogen bonds in HF) at the same temperature and pressure?

25.6 Which set of conditions in each case has the higher entropy (randomness)?
(a) Condition 1—a football team ready to start a play
    Condition 2—the team after the play has run to completion
(b) Condition 1—the individual seeds in a milkweed pod
    Condition 2—the seeds after a strong wind has scattered them
(c) Condition 1—the playing cards in an unopened box
    Condition 2—the cards after being shuffled
(d) Condition 1—the students at a chemistry lecture
    Condition 2—the students studying chemistry in the dorm, in the library, at home, etc.

25.7 Which of the following conditions would predict a process that is (a) always spontaneous, (b) always non-spontaneous, or (c) spontaneous or nonspontaneous depending on the temperature and magnitudes of ΔH and ΔS: (i) $\Delta H = +$, $\Delta S = +$; (ii) $\Delta H = +$, $\Delta S = -$; (iii) $\Delta H = -$, $\Delta S = +$; (iv) $\Delta H = -$, $\Delta S = -$?

25.8 Why can we say that at absolute zero an exothermic reaction will always be spontaneous, but at temperatures above absolute zero we have to consider both heat and entropy before we can predict spontaneity?

25.9 Compare the entropy change for melting a mole of water at 0.00°C ($\Delta H° = 1436.3$ cal/mole for fusion) to that for vaporizing a mole of water at 100.00°C ($\Delta H° = 9717.1$ cal/mole for vaporization). Why is one value so much larger?

25.10 What is the entropy change for heating a mole of water from 0.°C to 100.°C? Assume that $C_p = 18$ cal/mole °K over this temperature range. Is this a large change in randomness?

25.11 Which process results in the larger increase in randomness, expanding a mole of gas from 10.0 liters to 15.0 liters or from 15.0 liters to 20.0 liters?

25.12 What is the ΔS for producing 1 mole of artificial air by mixing 0.80 mole of $N_2(g)$ with 0.20 mole of $O_2(g)$?

25.13 The absolute entropy at 500°K in cal/mole °K of $F_2(g)$ is 52.505, $Br_2(g)$ is 63.171, $BrF(g)$ is 58.941, $BrF_3(g)$ is 78.727, and $BrF_5(g)$ is 91.092. Calculate the entropy change for each of the following reactions:
(a) $Br_2(g) + F_2(g) \longrightarrow 2BrF(g)$
(b) $Br_2(g) + 3F_2(g) \longrightarrow 2BrF_3(g)$
(c) $Br_2(g) + 5F_2(g) \longrightarrow 2BrF_5(g)$

25.14 Calculate the entropy changes for the sublimation of iodine and for forming gaseous iodine atoms from gaseous iodine molecules

$$I_2(s) \longrightarrow I_2(g) \qquad I_2(g) \longrightarrow 2I(g)$$

given $S° = 43.184$ cal/mole °K for $I(g)$, 27.757 cal/mole °K for $I_2(s)$, and 62.28 cal/mole °K for $I_2(g)$. Why are these values similar in sign and magnitude?

25.15 Calculate $\Delta S°$ for the production of ozone from dioxygen

$$3O_2(g) \longrightarrow 2O_3(g)$$

given that $S° = 49.003$ cal/mole °K for $O_2(g)$ and 57.08 cal/mole °K for $O_3(g)$ at 25°C. Is this an increase or decrease in randomness? The heat of formation at 25°C of $O_3(g)$ is 34.1 kcal/mole and of $O_2(g)$ is 0. Calculate $\Delta G°$ for this reaction and $\Delta G°$ of formation for $O_3(g)$. Is this reaction spontaneous under standard state conditions?

25.16 Calculate $\Delta G°$ for the carbon reduction of the oxides of iron and copper at 700°K represented by the reactions

$$2Fe_2O_3(s) + 3C(s) \longrightarrow 4Fe(s) + 3CO_2(g)$$
$$2CuO(s) + C(s) \longrightarrow 2Cu(s) + CO_2(g)$$

given that the standard free energy of $CuO(s)$ is -21.984 kcal/mole, $Fe_2O_3(s)$ is -152.253 kcal/mole, and $CO_2(g)$ is -94.510 kcal/mole. Which oxide can be reduced using carbon in a wood fire (which has a temperature of about 700°K) assuming standard state conditions?

25.17 The standard state free energy of formation at 25°C for $O_2(g)$ is 0, $NO(g)$ is 20.69 kcal/mole, $NO_2(g)$ is 12.26 kcal/mole, $N_2O(g)$ is 24.90 kcal/mole, $N_2O_3(g)$ is 33.32 kcal/mole, and $N_2O_5(g)$ is 27.5 kcal/mole. Calculate $\Delta G°$ for each of the following reactions and predict which one is most favorable under standard state conditions at 25°C:

$$N_2O(g) + \tfrac{1}{2}O_2(g) \longrightarrow 2NO(g)$$
$$N_2O(g) + O_2(g) \longrightarrow N_2O_3(g)$$
$$N_2O(g) + \tfrac{3}{2}O_2(g) \longrightarrow 2NO_2(g)$$
$$N_2O(g) + 2O_2(g) \longrightarrow N_2O_5(g)$$

25.18 At 25°C, $\Delta G° = -22.777$ kcal/mole for the formation of $HCl(g)$:

$$\tfrac{1}{2}H_2(g) + \tfrac{1}{2}Cl_2(g) \longrightarrow HCl(g)$$

What is the value of ΔG for the process if the partial pressure of H_2 is 3.5 atm, of Cl_2 is 1.5 atm, and of HCl is 0.3 atm? Is the process more or less spontaneous under these conditions than under standard state conditions?

25.19 The standard state free energy of formation at 25°C is -32.05 kcal/mole for $H_2O_2(aq)$, 0 for $O_2(g)$, and -56.687 kcal/mole for $H_2O(l)$. Calculate $\Delta G°$ for the reaction

$$2H_2O_2(aq) \longrightarrow 2H_2O(l) + O_2(g)$$

This spontaneous reaction can be stopped by increasing the pressure of the dioxygen. Find the value of this pressure.

25.20 At 25°C, $\Delta G°$ of formation is -56.687 kcal/mole for $H_2O(l)$, 0 for H^+ $(aq, 1M)$, and -37.594 kcal/mole for $OH^-(aq, 1M)$. Find $\Delta G°$ for the reaction

$$H_2O(l) \longrightarrow H^+(aq, 1M) + OH^-(aq, 1M)$$

and calculate the value of equilibrium constant, K_w.

25.21 The standard free energy of formation at 25°C for $NH_4Cl(s)$ is -48.51 kcal/mole, $NH_3(g)$ is -3.94 kcal/mole, and $HCl(g)$ is -22.777 kcal/mole. (a) What is $\Delta G°$ for the reaction

$$NH_4Cl(s) \longrightarrow NH_3(g) + HCl(g)$$

(b) What is the equilibrium constant for this decomposition? (c) Calculate the equilibrium partial pressure of HCl above a sample of NH_4Cl.

25.22 The standard free energy of formation of water at 25°C is -56.687 kcal/mole. What voltage would be expected from a fuel cell consuming hydrogen and oxygen that is operating under standard state conditions?

25.23 The standard reduction half-cell potential for Cu^+ to Cu is 0.522 V and for Cu^{2+} to Cu is 0.3402 V. Predict $E°$ for Cu^{2+}/Cu^+.

25.24 The equilibrium constant for the reaction

$$Cl_2(g) + F_2(g) \longrightarrow 2ClF(g)$$

is 2.10×10^{18} at 298°K and 1.35×10^{11} at 500°K. Find $\Delta H°$ for the reaction.

25.25 For the reaction

$$CO(g) + Cl_2(g) \longrightarrow COCl(g) + Cl(g)$$

$\Delta H° = 40.339$ kcal and $K_p = 9.12 \times 10^{-30}$ at 25°C. What is the value of K_p at 500°K? Has the equilibrium shifted to the reactants or products side by increasing the temperature?

25.26 The standard heat of formation of $O_3(g)$ at 298°K is 34.1 kcal/mole and varies negligibly with temperature up to 6000°K. Calculate $\Delta G°$ of formation of ozone at 1000°K given that $\Delta G°$ of formation is 38.977 kcal/mole at 298°K.

25.27 At the normal boiling point, the heat of vaporization of water (100.00°C) is 9717.1 cal/mole and of heavy water (101.41°C) is 9944 cal/mole. Use these data to calculate the vapor pressure of each liquid at 75°C.

25.28 What is the boiling point of water at a typical laboratory pressure of 745 torr? The heat of vaporization of water at 100.00°C and 760.0 torr is 9717.1 cal/mole.

25.29 The sublimation pressure of NH_3 is 10.0 torr at $-93.2°C$ and 40.0 torr at $-79.2°C$. Calculate the heat of sublimation of NH_3 and use this to find the sublimation pressure at $-77.8°C$, the melting point. The vapor pressure of NH_3 is 100.0 torr at $-66.5°C$ and 400.0 torr at $-43.0°C$. Calculate the heat of vaporization of NH_3 and use this to find the vapor pressure at $-77.8°C$.

25.30 The rate constant for the reaction

$$\begin{matrix} H_2C-C=O \\ | \qquad | \\ H_2C-CH_2 \end{matrix}\ (g) \longrightarrow C_2H_4(g) + H_2C=C=O(g)$$

at 361°C is 4.6×10^{-4} sec^{-1} and at 371°C is 7.2×10^{-4} sec^{-1}. What is the energy of activation for this reaction?

25.31 How much faster would a reaction proceed at 25°C than at 0°C if the activation energy is 15 kcal?

25.32 A laboratory instructor told a student that raising the temperature of a reaction mixture by 10°C will double the rate of reaction. This is true only for reactions having one value of the activation energy. Find the value of ΔE_a for which this statement holds true if the temperature change is between 25°C and 35°C.

25.33 The rate constant for the decomposition of N_2O

$$2N_2O(g) \longrightarrow 2N_2(g) + O_2(g)$$

is 2.6×10^{-11} sec^{-1} at 300.°C and 2.1×10^{-10} sec^{-1} at 330.°C. What is the activation energy for this reaction? Prepare a plot of energy versus progress of reaction using -39.22 kcal as the energy of reaction.

25.34* The absolute entropy of $O_2(\gamma$-solid) at 24°K is 0.04 cal/mole °K. Calculate the entropy change (a) for the solid state transformation

$$O_2(\gamma\text{-solid}) \longrightarrow O_2(\beta\text{-solid})$$

at 24°K, given that $\Delta H° = 22$ cal/mole; (b) for heating $O_2(\beta$-solid) from 24°K to 44°K, given that $C_p = 11.4$ cal/mole °K; (c) for the solid state transformation

$$O_2(\beta\text{-solid}) \longrightarrow O_2(\alpha\text{-solid})$$

at 44°K, given that $\Delta H° = 178$ cal/mole; (d) for heating $O_2(\alpha$-solid) from 44°K to 54°K, given that $C_p = 11.3$ cal/mole °K; (e) for the melting of dioxygen at 54°K, given that the heat of fusion is 106 cal/mole; (f) for heating the liquid from 54°K to 90.2°K, given that $C_p = 13$ cal/mole °K; (g) for evaporating the liquid at 90.2°K given that the heat of vaporization is 1630 cal/mole; and (h) for heating the gas from 90.2°K to 298°K, given that $C_p = 7.016$ cal/mole °K. Add these nine values of entropy changes together to predict the absolute entropy of $O_2(g)$ at 298°K. The accepted value is 49.003 cal/mole °K.

25.35* The average value of $\Delta S°$ for vaporizing a sample of a substance at the normal boiling point is about 21 cal/mole °K (no matter what the substance is) as long as there is no hydrogen bonding in the liquid and dimers are not formed in the gaseous phase. (a) Confirm this value by calculating $\Delta S°$ for the evaporation of the following substances:

(i) $Xe(l) \longrightarrow Xe(g)$ $\Delta H° = 3021$ cal/mole at 165.1°K
(ii) $SO_2(l) \longrightarrow SO_2(g)$ $\Delta H° = 5955$ cal/mole at 263.14°K
(iii) $CCl_4(l) \longrightarrow CCl_4(g)$ $\Delta H° = 7170$ cal/mole at 349.9°K

(b) Would you predict $\Delta S°$ for vaporizing a mole of water to be greater or less than 21 cal/mole °K? Why? Confirm your answer using $\Delta H° = 9171.1$ cal/mole at 373.15°K. (c) Would you predict $\Delta S°$ for vaporizing a mole of methyl alcohol, CH_3OH, to be as large as water since the amount of hydrogen bonding in it is about the same as in water? Confirm your answer using $\Delta H° = 8430$ cal/mole at 337.9°K. (d) Predict the boiling point of lead, given that $\Delta H°$ of vaporization is 43.0 kcal/mole.

25.36* Chlorine and fluorine react to form the interhalogen compound ClF according to the equation

$$Cl_2(g) + F_2(g) \longrightarrow 2ClF(g)$$

(a) What is $\Delta H°$ for this reaction given that the heat of formation of ClF(g) at 25°C is -13.02 kcal/mole? (b) Calculate $\Delta G°$ for this reaction given that the free energy of formation at 25°C of ClF(g) is -13.37 kcal/mole, of $Cl_2(g)$ is 0, and of $F_2(g)$ is 0. (c) Using your values of $\Delta H°$ and $\Delta G°$, calculate $\Delta S°$ for the reaction at 25°C. (d) Compare your answer for (c) to that calculated using absolute entropies of 52.05 cal/mole °K for ClF(g), 53.288 cal/mole °K for $Cl_2(g)$, and 48.44 cal/mole °K for F_2. (e) What would be the value of ΔG if the partial pressure of Cl_2 is 1.52 atm, of F_2 is 1.78 atm, and of ClF is 0.10 atm? (f) Is the reaction more or less spontaneous under these nonstandard state conditions?

25.37* The heat of reaction under standard state conditions at 25°C for the combustion of CO

$$CO(g) + \tfrac{1}{2}O_2(g) \longrightarrow CO_2(g)$$

is -67.635 kcal/mole and $\Delta G°$ for the reaction at 25°C is -61.474 kcal/mole. At what temperature will this reaction no longer be favorable under standard state conditions?

25.38* The standard heat of formation of water at 25°C is -68.315 kcal/mole, of $H^+(aq, 1M)$ is 0, and of $OH^-(aq, 1M)$ is -54.970 kcal/mole. (a) Calculate $\Delta H°$ for the reaction represented by the equation

$$H_2O(l) \longrightarrow H^+(aq, 1M) + OH^-(aq, 1M)$$

The standard free energy of formation of water at 25°C is -56.687 kcal/mole, of $H^+(aq, 1M)$ is 0, and of $OH^-(aq, 1M)$ is -37.594 kcal/mole. (b) Calculate $\Delta G°$ for the above reaction. What is the value of K (c) at 25°C and (d) at 35°C? (e) What is the pH of water at 35°C? (f) Has water become acidic at 35°C?

25.39* The sublimation pressure of W is 9.08×10^{-8} atm at 3000°K and 7.42×10^{-6} atm at 3500°K. Find the heat of sublimation of W. Predict the sublimation pressure of W at 3200°K. How many atoms of W are in a cubic centimeter in the gaseous state near the filament in a light bulb at 3200°K?

25.40* The rate constant for the decomposition of A in solvent X is 4.7×10^{-5} sec^{-1} at 60.0°C and 1.60×10^{-4} sec^{-1} at 70.0°C and in solvent Y is 7.0×10^{-5} sec^{-1} at 65.0°C and 2.30×10^{-4} sec^{-1} at 75.0°C. Calculate the energy of activation for the reaction in each solvent. Are the values of ΔE_a significantly different? Is one solvent acting as a stronger catalyst than the other (as determined by a lower value of ΔE_a)?

25.41** At 25°C the standard state enthalpy of formation of $C_2H_2(g)$ is 54.19 kcal/mole, of $CO_2(g)$ is -94.051 kcal/mole, and of $H_2O(g)$ is -57.796 kcal/mole. (a) Calculate the standard state enthalpy change for the combustion of acetylene

$$C_2H_2(g) + \tfrac{5}{2}O_2(g) \longrightarrow 2CO_2(g) + H_2O(g)$$

The absolute standard state entropy at 25°C of $C_2H_2(g)$ is 48.00 cal/mole °K, of $O_2(g)$ is 49.003 cal/mole °K, of $CO_2(g)$ is 51.06 cal/mole °K, and of $H_2O(g)$ is 45.104 cal/mole °K. (b) Calculate $\Delta S°$ for the equation. (c) Use your answers from (a) and (b) to calculate $\Delta G°$ for the combustion reaction. The standard state free energy of formation at 25°C of $C_2H_2(g)$ is 50.00 kcal/mole, of $CO_2(g)$ is -94.254 kcal/mole, and of $H_2O(g)$ is -54.634 kcal/mole. (d) Calculate $\Delta G°$ for the reaction from these data and compare to your answer for part (c). (e) Is this reaction more or less spontaneous if the partial pressure of the acetylene is 5.2 atm, of O_2 is 52 atm, and of the product gases is 1 atm?

Using your value of $\Delta G°$ found in part (d), (f) calculate the equilibrium constant for this equation. (g) Find the partial pressure of C_2H_2 at equilibrium if the partial pressure of O_2 is 45 atm, of CO_2 is 12 atm, and of H_2O is 6 atm.

Assuming that the heat of reaction does not change significantly with temperature, (h) calculate $\Delta G°$ for the equation at 3000°K, a typical flame temperature. (i) Is this reaction more or less spontaneous at this higher temperature?

26 THE REPRESENTATIVE METALS

The representative metals have in common electronic configurations in which only valence shell s and p electrons are available for bonding. In one sense, zinc, cadmium, and mercury are d transition elements, since in each the final d electron needed to fill the level is added. However, because only the electrons in their outermost shells are involved in chemical reactions, the members of the zinc family are more like representative than transition elements, and we include them in this chapter.

Three of the representative metals—francium ($_{87}$Fr), radium ($_{88}$Ra) and polonium ($_{84}$Po)—are radioactive. Because of their scarcity, their ordinary chemistry is of no interest here. Among the twenty elements considered in this chapter, five are particularly valuable for the physical properties of their metals and alloys. These important metals are magnesium and aluminum, which form very light and corrosion-resistant alloys, and zinc, tin, and lead.

Among the lighter representative metals are four that are essential to animal life. The cations Na^+, K^+, $Ca^{2+,}$ and Mg^{2+} (together with the anions Cl^-, HCO_3^-, PO_4^{3-}, and the anions of organic acids) are responsible for maintaining the electrolyte balance of the body fluids. The Na^+ ion is the principal ion in the fluids that surround cells, and K^+ is the principal ion in the fluids within the cells. A deficiency or overabundance of any of the cations listed above has serious consequences for health. Calcium is also a major component of bone, and magnesium is required for the activity of several enzymes. Zinc is an essential trace element in both plant and animal life, and some recent studies suggest that inadequate zinc in the diet may increase susceptibility to heart disease.

This chapter covers the chemistry of the representative metals in four convenient groupings: (1) the alkali and alkaline earth metals of Representative Groups I and II, (2) the zinc family elements, zinc, cadmium, and mercury, (3) the Representative Group III metals, aluminum, gallium, indium, and thallium, and (4) the three remaining representative group metals, tin, lead, and bismuth. Some of the most interesting aspects of the industrial chemistry and uses of these elements are included.

Representative Groups I and II: The alkali and alkaline earth metals

26.1 Properties of Group I and II metals

The alkali metals—the members of Representative Group I (Table 26.1)—have the highest reactivity of all metals. Atoms of these elements easily give up their single valence electrons to form monopositive ions such as Na^+ and K^+, which have noble gas configurations. We refer to such metals as highly electropositive. The alkali metals form ionic compounds in virtually all of their reactions.

The trends of physical properties expected with increasing atomic weight and atomic size are clearly displayed by Group I metals. The atomic and ionic radii (Table 26.1) increase with increasing atomic weight. Paralleling this trend, the melting points and boiling points decrease as the atoms become larger and heavier, and the forces holding the atoms in the metal lattice decrease. The alkali metals have lower melting and boiling points and also lower densities, than most other metals. In addition, they are soft; all except lithium easily can be cut with a knife. When freshly cut they have a characteristic metallic luster, but on exposure to air they soon tarnish by rapidly reacting with atmospheric gases.

The alkaline earth metals—Representative Group II, also known as the beryllium family elements—also have a high degree of reactivity, second only to that of their Group I neighbors. Beryllium is covalent in most of its compounds because of the large amount of energy needed to form Be^{2+}. Magnesium sometimes forms at least partially covalent bonds, but the remaining alkaline earth elements react almost exclusively to form dipositive ions, for example, Ca^{2+} and Ba^{2+}, and to give ionic compounds. In both groups, the

TABLE 26.1

Properties of the Representative Group I and II metals

Configuration: Group I, [noble gas] ns^1 Group II, [noble gas] ns^2					
	Li Be	Na Mg	K Ca	Rb Sr	Cs Ba
Melting point (°C)	179 1283	97.5 650	63.5 851	39.0 757	28.4 704
Boiling point (°C)	1372 ~1500	892 1107	774 1487	679 1384	690 1640
Density (g/cc)	0.53 1.85	0.70 1.74	0.86 1.54	1.53 2.58	1.87 3.65
Atomic radius (nm)	0.1225 0.0889	0.1572 0.1364	0.2025 0.1736	0.216 0.1914	0.235 0.1981
Ionic radius (nm)	0.071 0.031	0.095 0.066	0.133 0.099	0.147 0.115	0.174 0.136

Li 3 6.941	Be 4 9.01218					B		
Na 11 22.9898	Mg 12 24.305					Al 13 26.9815		
K 19 39.098	Ca 20 40.08	Sc	Cu	Zn 30 65.38	Ga 31 69.72	Ge		
Rb 37 85.4678	Sr 38 87.62	Y	Ag	Cd 48 112.40	In 49 114.82	Sn 50 118.69	Sb	
Cs 55 132.905	Ba 56 137.33	La*	Au	Hg 80 200.59	Tl 81 204.37	Pb 82 207.19	Bi 83 208.980	
Fr 87 (223)	Ra 88 (226)	Ac**						

larger the atom, the more easily it gives up its valence electrons and the greater is its reactivity.

The trends in physical properties are generally the same for the elements of Group II as for those of Group I. Atoms and ions of the Group II elements have smaller radii than their respective alkali metal neighbors, reflecting their greater effective nuclear charge. Group II elements are denser and have higher melting and boiling points, and they are much harder—all properties derived from greater interatomic forces in the solid state.

The ease with which the elements of both groups form ions is reflected both in their ionization energies, which are low, and their standard reduction potentials (Table 26.2), which show that these elements are among the strongest of all reducing agents (compare

The ionization energies can be converted to kJ/mole by multiplying by 4.184 kJ/kcal.

	Li Be	Na Mg	K Ca	Rb Sr	Cs Ba
Ionization energy **(kcal/mole, 0°K)** $M(g) \longrightarrow M^+(g) + e^-$	**124** 215	**118** 176	**100** 141	**96.3** 131	**89.8** 120
Electronegativity	**1.0** 1.5	**0.9** 1.2	**0.8** 1.0	**0.8** 1.0	**0.7** 0.9
Standard reduction potential[a] $M^{n+} + ne^- \longrightarrow M(s)$	**−3.05** −1.70	**−2.71** −2.37	**−2.92** −2.76	**−2.93** −2.89	**−2.92** −2.90

[a] $n = 1$ for alkali metals, $n = 2$ for beryllium family elements.

TABLE 26.2
Electronic properties of elements of the lithium and beryllium families

THE REPRESENTATIVE
METALS [26.1]

TABLE 26.3

Sources of the alkali metals

Element (% in Earth's crust)	Major natural sources
Li (0.002)	Springs and brine lakes Silicate ores, e.g., $Li_2O \cdot Al_2O_3 \cdot SiO_2$
Na (2.36)	Rock salt, NaCl Chile saltpeter, $NaNO_3$ Feldspars and micas Borax, $Na_2B_4O_7 \cdot 10H_2O$ Seawater, brine wells
K (2.6)	Carnallite, $KCl \cdot MgCl_2 \cdot 6H_2O$ Sylvite, KCl K_2SO_4 (with rock salt)
Rb (0.009) and Cs (0.0003)	Often found together. Often impurities in ores of other alkali metals.

TABLE 26.4

Sources of the alkaline earth metals

Element (% in Earth's crust)	Major natural sources
Be (0.00028)	Beryl, $3BeO \cdot Al_2O_3 \cdot 6SiO_2$
Mg (2.1)	Seawater (0.13%) Magnesite, $MgCO_3$ Dolomite, $CaCO_3 \cdot MgCO_3$
Ca (4.15)	$CaCO_3$ chalk, limestone, marble, coral, oyster shells, pearls, egg shells Gypsum, $CaSO_4 \cdot 2H_2O$
Sr (0.0375)	Phosphate rock Celestite, $SrSO_4$ Strontianite, $SrCO_3$
Ba (0.0425)	Barite, $BaSO_4$ Witherite, $BaCO_3$

with values in Table 19.6). The effect of increasing atomic size and mass can be seen in the regular decrease in the ionization energy values for the elements of both groups. This trend is obscured in the standard reduction potential values, to which hydration energies and lattice energies also contribute. The large negative value of the reduction potential of lithium is a result of the much greater hydration energy of the small lithium ion as compared with those of the other ions of the family.

The elements of both groups are excellent <u>conductors</u> of electricity. The loosely held outer electrons of the alkali metals can be set free by light of the correct wavelength, in what is called the photoelectric effect (Section 7.12). Cesium, the least electronegative (most electropositive) of all the elements, releases electrons most readily of all the alkali metals. Cesium also has the lowest first ionization energy of all the elements.

26.2 Sources of Group I and II elements

The alkali metals most abundant in the crust of the Earth are sodium and potassium (Table 26.3), which are the sixth and seventh most abundant of all the elements. The fifth most abundant element is an alkaline earth metal—calcium—while magnesium is eighth in abundance (Table 26.4). All of the elements in these two groups occur in the Earth's crust in combination with other elements; they are too reactive ever to be found free. Sodium and magnesium are also present in relatively large amounts in seawater, brine wells, and a few salt lakes.

26.3 Preparation and uses of Group I and Group II metals

a. Group I. To obtain the free alkali metals from their compounds, the following reduction must be carried out in the absence of water:

$$M^+ + e^- \longrightarrow M(s)$$

Because the alkali metals are the most electropositive of all elements, they are very stable as positive ions and have a great tendency to remain ions. Therefore, the reduction process requires the expenditure of much energy. One convenient way to supply this energy is with electricity. Most sodium is produced by electrolysis of molten sodium chloride by what is called the Downs process (Figure 26.1).

$$2Na^+ + 2e^- \longrightarrow 2Na(l) \quad \text{(cathode)}$$
$$2Cl^- \longrightarrow Cl_2(g) + 2e^- \quad \text{(anode)}$$

In practice, since sodium chloride melts at 808°C, which is close to the boiling point of sodium, calcium chloride is added to lower the melting point to about 600°C. This also permits the electrolysis to be

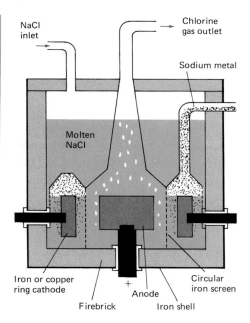

NaCl inlet

Chlorine gas outlet

Sodium metal

Molten NaCl

Iron or copper ring cathode

Firebrick

+ Anode

Iron shell

Circular iron screen

FIGURE 26.1
The Downs cell for the production of sodium.

carried out more economically. Resistance to the flow of electric current is the source of the heat required to keep the salt mixture molten. Lithium is also produced by electrolysis of its chloride.

Recently the largest use of sodium (Table 26.5) has been in the manufacture of tetraethyllead, the gasoline antiknock additive, which is made via a Na–Pb alloy (also the largest use of lead). As we have pointed out, this is a declining market. The major uses of sodium and the other alkali metals are given in Table 26.5.

Potassium is made by reaction of molten potassium chloride with sodium vapor in the absence of air.

$$KCl(l) + Na(g) \xrightarrow{\Delta} K(g) + NaCl(l)$$

The success of the process depends upon the great difference in volatility between sodium and potassium, the former boiling at 892°C and the latter at 774°C. The process is also employed for the preparation of sodium–potassium alloys, which, in liquid form, are useful heat-transfer agents. Rubidium and cesium can also be produced by sodium reduction of their chlorides. There is, however, very little commercial use of these elements.

b. Group II. Like the alkali metals, the metals of Representative Group II are usually prepared either by electrolysis of their molten halides or by reaction of halides or oxides with reducing agents at high temperatures. The most widely used of these metals, magnesium, is obtained from seawater, where it occurs as Mg^{2+} to the extent of 0.13%. The magnesium is first precipitated as the hydroxide by reaction with a slurry of calcium hydroxide, $Ca(OH)_2$.

$$Mg^{2+} + 2OH^- \longrightarrow Mg(OH)_2(s)$$

TABLE 26.5
Major uses of the alkali metals
In addition to the uses of the free metals listed here, each metal is used in the manufacture of other industrial chemicals. See Section 26.4 for use of Li as a scavenger.

Li	Scavenger
	Lithium–magnesium alloys
	Lithium battery
Na	**Tetraethyllead production**
	Coolant
	Na vapor light
	Reducing agent
K	**KO$_2$ (superoxide) production**
	(for gas masks)
	Reducing agent
Rb and Cs	**Photoelectric cells**
	Scavengers

THE REPRESENTATIVE METALS [26.3]

793

The magnesium hydroxide is then dissolved in hydrochloric acid:

$$Mg(OH)_2(s) + 2H^+ + 2Cl^- \longrightarrow Mg^{2+} + 2Cl^- + 2H_2O(l)$$

Concentration of the resulting solution yields magnesium chloride monohydrate, $MgCl_2 \cdot H_2O$. By electrolysis of this chloride in the molten state, magnesium is deposited at the cathode and chlorine is liberated at the anode.

The major use of magnesium is in the production of alloys (Table 26.6). Magnesium is the lightest metal that can be used for structural applications. Its alloys are easily fabricated and welded.

Calcium, barium, and strontium are similar in their properties and uses (Table 26.6). Since calcium is less expensive, more of it is used than either of the others.

26.4 Reactions of Group I and Group II metals

a. With other elements. The alkali metals are so reactive that they combine directly with most nonmetals. A notable exception is that only lithium forms a nitride by direct combination with nitrogen. Among the alkaline earth metals, calcium, strontium, and barium are reactive enough to combine directly with most nonmetals. As would be expected from their ease of oxidation, the metals of both groups combine vigorously with the halogens to form the appropriate halides, MX and MX_2. Most of the binary compounds formed in these reactions are ionic.

Lithium is the only alkali metal that yields the normal oxide (containing O^{2-}), Li_2O, when it is burned with dioxygen. Sodium forms the peroxide, Na_2O_2, and potassium, rubidium, and cesium give superoxides, in which oxygen is present as O_2^-.

$$4Li(s) + O_2(g) \longrightarrow 2Li_2O(s)$$
$$\text{lithium oxide}$$

$$2Na(s) + O_2(g) \longrightarrow Na_2O_2(s)$$
$$\text{sodium peroxide}$$

$$K(s) + O_2(g) \longrightarrow KO_2(s)$$
$$\text{potassium superoxide}$$

Most Group II elements yield the normal oxides, for example, CaO, when heated with oxygen.

The direct combination of some of the elements in these two groups occurs so easily and completely that they are used as **getters,** or **scavengers**—substances that can pull traces of contaminants out of metals, alloys, or gases. For example, lithium and calcium are scavengers for oxygen and nitrogen in molten metals, barium and cesium remove the last traces of oxygen and nitrogen from vacuum tubes.

TABLE 26.6

Major uses of the alkaline earth metals

Strontium is seldom used commercially because calcium and barium serve the same purposes and are less expensive.

Be	Alloys
	(Small amounts as a hardening agent; in nonsparking tools)
	Shielding nuclear reactors (captures neutrons)
	Heat sink for aerospace vehicles
Mg	Structural metal
	aircraft
	portable tools
	industrial machinery
	Deoxidizer and desulfurizer
	Batteries
	Flares, rocket propellants
	Cathodic protection of other metals
Ca	Deoxidizer
	Reducing agent
Ba	Alloys
	Scavenger

b. With water. The products from the reaction of the alkali and alkaline earth metals with water are hydrogen and metal hydroxide.

$$2M(s) + H_2O \longrightarrow 2MOH(aq) + H_2(g) \qquad (M = \text{alkali metal})$$
$$M'(s) + 2H_2O \longrightarrow M'(OH)_2(aq) + H_2(g) \qquad (M' = \text{Group II element})$$

With most alkali metals, the reaction is so exothermic that it may occur with explosive violence as a result of the ignition of the hydrogen. It should be emphasized that these very active metals react even more violently with solutions of acids than they do with water because of the much higher hydrogen ion concentration in the acids.

$$2M(s) + 2H^+ \longrightarrow 2M^+ + H_2(g)$$

Among the metals of Group II, only calcium, strontium, and barium are sufficiently reactive to combine vigorously with water at ordinary temperatures, the rate of reaction increasing from calcium to barium. Magnesium reacts with boiling water, but beryllium is practically inert.

26.5 Cations of Group I and Group II metals

With the exception of the compounds of beryllium, compounds of the alkali and alkaline earth metals are all ionic substances. The alkali halides are the classic examples of ionic compounds.

The cations of Group I and Group II metals possess noble gas configurations and are colorless and diamagnetic. Colored compounds containing any of these cations derive their color from the anion.

Most alkali metal compounds have high water solubility, mostly because, except for Li^+, alkali metal cations are large and of small charge. As a result, they are bound to anions in a crystal by relatively weak electrostatic charges, and the salts have relatively small lattice energies.

In the solubility of its compounds, as in many of its other properties, lithium resembles its diagonal neighbor magnesium to a much greater extent than it does the other alkali metals. For example, the fluorides, carbonates, and phosphates of the other alkali metals are highly soluble in water, but the corresponding lithium salts are relatively insoluble.

The cations formed by the elements of Group II are smaller and more highly charged than the corresponding alkali metal ions. Therefore, a Group II metal cation binds any particular anion in a crystal with greater force than does the corresponding alkali metal cation. For example, sodium fluoride has a lattice energy (at 0°K) of 217 kcal/mole (908 kJ/mole), whereas that for magnesium fluoride is 696 kcal/mole (2912 kJ/mole). The values for potassium chloride and calcium chloride are 168 and 536 kcal/mole (703 and 2240 kJ/mole), respectively.

26.6 Compounds of Group I and Group II metals

a. Oxides and hydroxides. With the exception of beryllium oxide, BeO, which is amphoteric, all of the Group I and II metal oxides are strongly basic. The alkali metal oxides all react vigorously with water to give the hydroxides.

$$M_2O(s) + H_2O(l) \longrightarrow 2MOH(aq) \qquad (M = Li, Na, K, Rb, Cs)$$

Beryllium oxide is essentially unreactive toward water. Magnesium oxide reacts slowly and not very energetically, but the oxides of calcium, strontium, and barium combine with water with progressively greater vigor.

$$MO(s) + H_2O(l) \longrightarrow M(OH)_2(s) \qquad (M = Mg, Ca, Sr, Ba)$$

The alkali metal hydroxides are white, water-soluble solids. All except lithium hydroxide are thermally stable and may be melted and vaporized without decomposition. When heated, lithium hydroxide decomposes before its boiling point is reached to give lithium oxide and water. The alkali metal hydroxides are the most basic metal hydroxides. Their aqueous solutions absorb carbon dioxide from the air to form carbonates.

$$2MOH(aq) + CO_2(g) \longrightarrow M_2(CO_3) + H_2O(l) \qquad (M = Li, Na, K, Rb, Cs)$$

The Group II element hydroxides are much less soluble than those of the alkali metals, the solubility increasing from $Be(OH)_2$ to $Ba(OH)_2$, in line with the decreasing attraction between cation and hydroxide ion in the crystal. The alkalinity of solutions of these hydroxides is limited by their solubility.

Many alkali metal and alkaline earth metal compounds are important because of their commercial applications. Table 26.7 lists some of these compounds. Sodium hydroxide, NaOH, is sixth in the list of industrial chemicals produced in the largest amounts. It is prepared industrially in two ways. In one process, equivalent quantities of sodium carbonate and calcium hydroxide are mixed as a slurry.

$$Na_2CO_3(aq) + Ca(OH)_2(aq) \longrightarrow 2NaOH(aq) + CaCO_3(s)$$

After removal of the calcium carbonate precipitate by filtration, the liquor is evaporated to give solid sodium hydroxide. In the second and more widely used method, a concentrated aqueous solution of sodium chloride (brine) is electrolyzed, the products being sodium hydroxide, chlorine, and hydrogen (Section 24.8).

Potassium hydroxide, KOH, can be prepared by methods similar to those described for the sodium compound. For example, the reaction between barium hydroxide and potassium sulfate solutions yields barium sulfate as a precipitate and potassium hydroxide in solution.

$$Ba(OH)_2(aq) + K_2SO_4(aq) \longrightarrow 2KOH(aq) + BaSO_4(s)$$

Sodium peroxide, Na_2O_2, particularly, and potassium superoxide, KO_2, to a lesser extent, are compounds of industrial significance. They are powerful oxidizing and bleaching agents and form hydrogen peroxide on contact with cold solutions of acids or with an excess of cold water.

$$Na_2O_2(s) + 2H_2O(l) \longrightarrow H_2O_2(aq) + 2NaOH(aq)$$
$$2KO_2(s) + 2H_2O(l) \longrightarrow H_2O_2(aq) + O_2(g) + 2KOH(aq)$$

Potassium superoxide is employed in "breathing" masks as a quick source of oxygen. Such masks are used in rescue work in mines and in other areas where the air is so deficient in oxygen that an artificial atmosphere must be generated. The moisture of the breath reacts with the oxide to liberate oxygen, and at the same time the potassium hydroxide formed removes carbon dioxide as it is exhaled, allowing the atmosphere in the mask to be continuously regenerated.

From the industrial viewpoint, by far the most important among the oxides of Group I and II metals is calcium oxide, CaO, which is commonly known as lime or quicklime. It is the cheapest source of hydroxide ion and among industrial chemicals is second only to sulfuric acid in the amount produced. Lime is produced by heating limestone to drive off CO_2.

$$\underset{\text{limestone}}{CaCO_3(s)} \xrightarrow{\Delta} \underset{\text{lime}}{CaO(s)} + CO_2(g)$$

Some commercial grades of lime are mixtures of calcium oxide and magnesium oxide. Such mixtures are obtained by heating dolomitic limestone, which contains magnesium carbonate, together with calcium carbonate.

Calcium hydroxide, $Ca(OH)_2$, the product of the reaction of water with calcium oxide, is known as *slaked lime*, and a saturated solution of calcium hydroxide (which is slightly soluble) is called *limewater*. The oxide and the hydroxide are used, sometimes interchangeably, in many ways (Table 26.7).

Mortar is a mixture of $Ca(OH)_2$, sand, and water. It sets because the $Ca(OH)_2$ absorbs CO_2 from the air.

$$Ca(OH)_2(aq) + CO_2(g) \longrightarrow CaCO_3(s) + H_2O(g)$$

The $CaCO_3$ crystals that form tend to interlock and bind the mass together. The sand is added to give bulk and to prevent excessive shrinkage.

b. Carbonates and hydrogen carbonates. The alkali metal carbonates, except for lithium carbonate, are quite soluble in water, while the carbonates of the elements of Group II (and most other carbonates) are practically insoluble in water. Aqueous solutions of the soluble carbonates are alkaline due to hydrolysis of the carbonate ion,

$$CO_3^{2-} + H_2O(l) \rightleftharpoons HCO_3^- + OH^-$$

carbonate ion   *hydrogen carbonate ion*

Metal carbonates react with strongly acidic solutions to yield carbon dioxide, for example,

$$MgCO_3(s) + 2H^+ \longrightarrow Mg^{2+} + CO_2(g) + H_2O(l)$$

Insoluble carbonates are brought into solution by water containing dissolved carbon dioxide by the following reaction, which is the reverse of the thermal decomposition of solid metal hydrogen carbonates:

$$MCO_3(s) + H_2O(l) + CO_2(g) \rightleftharpoons M^{2+} + 2HCO_3^-$$

Groundwater always contains carbon dioxide, and limestone and dolomite dissolve in such water,

$$CaCO_3 \cdot MgCO_3(s) + 2H_2O(l) + 2CO_2(aq) \rightleftharpoons Ca^{2+} + Mg^{2+} + 4HCO_3^-$$

dolomite

The calcium and magnesium thus brought into solution make the water "hard" (Section 13.13).

The dissolution of carbonate minerals in groundwater has another interesting effect—if the mineral is below the surface of the Earth and is covered by rocks that do not dissolve, a cave may be formed. Groundwater may then seep into the cave through cracks in the walls or roof. The water contains calcium and magnesium bicarbonates that it has dissolved elsewhere, and when it enters the cave, the pressure on it is lowered appreciably and carbon dioxide escapes from it by a reversal of the reaction shown above. Insoluble calcium and magnesium carbonates are thus formed and deposit around the opening through which the water seeped into the cave, forming an iciclelike deposit called a *stalactite*. Some of the bicarbonate remains in solution until drops of water fall from the tip of the stalactite. The bicarbonate decomposes as the drops fall and forms a deposit on the floor of the cave, building up another "icicle," called a *stalagmite*. After many centuries, the tips of the stalactite and stalagmite meet and the two "icicles" grow into a column that one might think was left there to support the roof of the cave.

Among the Group I and II carbonates and hydrogen carbonates, sodium carbonate and sodium hydrogen carbonate are the most widely used compounds. In 1938, very large deposits of the relatively rare mineral *trona*, $Na_2CO_3 \cdot NaHCO_3 \cdot 2H_2O$, were discovered in Wyoming, and most sodium carbonate is now produced from this mineral. The trona is either purified first by crystallization and then heated to convert the sodium hydrogen carbonate to normal carbonate, or the heating is effected first and then the impure sodium carbonate is purified.

Sodium hydrogen carbonate, also known as sodium bicarbonate or bicarbonate of soda, is made from pure sodium carbonate

$$Na_2CO_3(aq) + H_2O(l) + CO_2(g) \longrightarrow 2NaHCO_3(s)$$

Anhydrous sodium carbonate is known in commerce as *soda ash* (Table 26.7). The carbonate also exists in the form of a number of hydrates, the most common of which is the decahydrate, $Na_2CO_3 \cdot 10H_2O$ (*washing soda*). Solutions of the salt are effective cleaning agents because of the alkalinity resulting from hydrolysis of the carbonate ion. The hydroxide ion is responsible for the cleansing action.

Sodium bicarbonate is used in baking powders as a leavening agent, a substance that produces gas bubbles in dough. This gives rise to its common name, *baking soda*. An acid substance must be present in the dough to liberate carbon dioxide by the following reaction

$$HCO_3^- + H^+ \longrightarrow CO_2(g) + H_2O(l)$$

The liberation of gas causes the dough to rise and gives the product the appropriate lightness and texture. Sour milk is often the source of the acid. Some recipes call for use of a baking powder. These powders are mixtures of sodium hydrogen carbonate and an acidic substance, for example, sodium alum (sodium aluminum sulfate), $NaAl(SO_4)_2 \cdot 12H_2O$; calcium dihydrogen phosphate, $Ca(H_2PO_4)_2$; or potassium hydrogen tartrate, $K(HC_4H_4O_6)$. "Double acting" baking powder is made possible by coated crystals of $Ca(H_2PO_4)_2$; they release about half of their H^+ ion during mixing and half during baking.

The production of metals

26.7 Metallurgy

The word **metallurgy** covers all aspects of the science and technology of metals. We need only learn a little of the terminology of metallurgy—that which applies to beginning with a metal-bearing ore and producing the pure, or relatively pure, metal.

A **mineral** is a naturally occurring substance with a characteristic range of chemical composition. Minerals are frequently the silicates, oxides, sulfides, or carbonates of metals. An **ore** is a mixture of minerals from which a particular metal or several metals can profitably be extracted. For example, aluminum silicates would not usually be referred to as aluminum ores, because no practical method for extracting the aluminum from a silicate has yet been devised.

The three principal steps in winning a metal from an ore are (1) preparing the ore, (2) extracting the metal, and (3) refining or purifying the crude metal.

The ore is prepared by separating from the metal-bearing mineral as much as possible of the unwanted rock, collectively called the **gangue** (pronounced "gang"). Usually the first step is crushing or grinding the ore into smaller pieces so that separating the different components will be easier. A valuable method for such a separation is **flotation**—concentration of the metal-bearing mineral in a froth of bubbles that can be skimmed off.

In flotation, the crushed or pulverized ore is suspended in water containing an appropriate oil or detergent, and the mixture is then beaten or blown into a froth. The oil or detergent preferentially wets the metal mineral particles, which concentrate in the froth; the gangue particles are preferentially wet by the water and sink to the bottom. The froth is skimmed off, permitted to collapse, and dried.

In some cases, a separation by a simple flotation based solely on the difference in density between the gangue and mineral particles is possible. In other cases, minerals with magnetic properties can be separated by magnets.

Before production of the pure metal, other operations may be necessary. A finely divided ore may be **sintered**—heated without melting to cause formation of larger particles—or **calcined**—heated to drive off a gas, usually to convert a carbonate to an oxide, a chemical change. In **roasting**, a term usually applied to sulfide ores, the ore is heated in air or oxygen below its melting point to convert sulfides to oxides.

Production of the pure metal most often requires a chemical reduction. This is accomplished either by heat plus an appropriate reducing agent such as another metal or carbon (*pyrometallurgy*), or by electrolysis (*electrometallurgy*). **Smelting** is melting accompanied by chemical change. The reduction of iron oxide ores with coke to produce molten iron is a smelting process. Some gangue usually remains with an ore, and it is separated during smelting as **slag**—a molten mixture of the gangue minerals, often aided in melting by the addition of a **flux**—a substance that lowers the melting point of the gangue. The molten slag and the molten metal are insoluble in each other. They form two layers, and the slag and metal layers can be separated.

Some metals or their compounds can be separated at various stages of extraction from their ores by **leaching**—washing a soluble compound from insoluble material with an appropriate solvent.

Methods for refining crude metal include further electrolytic purification, oxidation of impurities to be removed, and distillation of low-boiling metals.

Minerals are among what are called **nonrenewable resources,** substances that cannot be manufactured from something else. We have only the copper or iron ore that is here in the crust of the Earth. As high-grade ores with high percentages of metals have been used up, remarkable technological advances have been made in extracting metals from lower-grade ores. It is tempting, but unrealistic, to think that such improvements can continue indefinitely.

The zinc family: Zn, Cd, Hg

26.8 Properties of zinc, cadmium, and mercury

Zinc and cadmium are both silvery, lustrous metals when they are pure. In moist air, zinc becomes covered with a grayish coating of basic carbonate, $Zn_2(OH)_2CO_3$, which protects it from further attack by the atmosphere. Cadmium tarnishes only slightly, also forming a basic carbonate. Zinc is hard and brittle, and becomes malleable and ductile at 150°C, while cadmium is soft enough to be cut with a knife. Mercury is familiar as the silvery, dense liquid of thermometers.

Only the two outer s electrons (Table 26.8) are employed in bonding by atoms of zinc, cadmium, and mercury. Therefore, these elements exhibit no oxidation state higher than +2, and for zinc and cadmium this is the only stable state known. Mercury has a +2 oxidation state in the Hg^{2+} ion and a +1 oxidation state in the Hg_2^{2+} ion,

$$^+Hg:Hg^+$$

in which two Hg^+ ions, each formed by the loss of an s electron, are covalently joined through the remaining s electron of each.

Calcium, strontium, and barium are the representative elements that precede zinc, cadmium, and mercury in the same rows of the periodic table. These two groups of elements differ electronically only in their next to outermost shell, where Ca, Sr, and Ba have eight electrons (s^2p^6) and Zn, Cd, and Hg have eighteen electrons ($s^2p^6d^{10}$). Consider calcium and zinc. Because of the interposition of the transition elements, zinc has a larger atomic number and nuclear charge (Zn, 30; Ca, 20) and a considerably smaller atomic radius (Zn, 0.1249 nm; Ca, 0.1736 nm) than calcium. As a result, the zinc atom holds its valence electrons more tightly than does the calcium atom. For similar reasons, cadmium and mercury lose electrons less readily than strontium and barium. The less negative standard reduction potentials and much larger enthalpies of conversion from the solid state to gaseous ions (Table 26.9) reflect the differences in electronegativities

TABLE 26.8

Properties of zinc, cadmium, and mercury

Configuration: $(n-1)d^{10}ns^2$

	Zn	Cd	Hg
Melting point (°C)	419.5	320.9	−38.87
Boiling point (°C)	907	767	357
Density (g/cc)	7.14	8.64	13.59
Atomic radius (nm)	0.1249	0.1413	0.1440
Ionic radius, M^{2+} (nm)	0.072	0.096	0.110

	$_{20}Ca$ $_{30}Zn$	$_{38}Sr$ $_{48}Cd$	$_{56}Ba$ $_{80}Hg$
Heat of conversion from solid metal to gaseous ion, kcal/mole (kJ/mole) $M(s) \longrightarrow M^{2+}(g) + 2e^-$	463.6 (1940) 665.1 (2783)	427.8 (1790) 627.0 (2623)	395.7 (1656) 690.1 (2887)
Standard reduction potential (V) $M^{2+} + 2e^- \longrightarrow M$	−2.76 −0.763	−2.89 −0.403	−2.90 0.851
Electronegativity	1.0 1.6	1.0 1.7	0.9 1.9

TABLE 26.9

Comparison of electronic properties of alkaline earth and zinc family elements

THE REPRESENTATIVE METALS [26.8]

of these elements. Also, the anhydrous chlorides, bromides, and iodides of zinc, cadmium, and mercury are quite covalent in character, compared to the corresponding ionic halides of calcium, strontium, and barium. For similar reasons, zinc, cadmium, and mercury form oxides and hydroxides that are much less basic than the strongly basic calcium, strontium, and barium oxide and hydroxides.

Both zinc and cadmium have negative standard reduction potentials, while that of mercury is strongly positive. As a result of the interposition between Cd and Hg of two series in which electrons are added to inner shells (La–Lu; Hf–Au), the increase in atomic radius from cadmium to mercury is relatively small (Table 26.8), although there is an increase in nuclear charge caused by the addition of thirty-two protons. Consequently, the valence electrons of mercury are held much more tightly than those of cadmium.

The toxicity of cadmium and mercury is discussed later in this chapter (An Aside: Metals as Poisons); zinc and its compounds are much less harmful.

26.9 Sources, metallurgy, and uses of zinc, cadmium, and mercury

a. Zinc and cadmium. Zinc is not found abundantly in nature (0.007%, twenty-third in abundance), but its occurrence is widespread. The most common ores are *sphalerite* or *zinc blende*, ZnS, and *smithsonite*, $ZnCO_3$. Others of significance are *zincite*, ZnO, and *franklinite* $(Zn,Mn)O \cdot xFe_2O_3$. The expression $(Zn,Mn)O$ indicates that the zinc and manganese are not present in any fixed ratio. Different samples of ore contain the metals in different ratios. The x preceding the formula Fe_2O_3 indicates that the ratio of zinc and manganese to iron is also variable.

Franklinite is of special interest, for upon reduction at a high temperature, it yields zinc, manganese, and iron. The zinc distills from the mixture and is condensed outside of the furnace. The manganese–iron alloy that remains is used directly in the manufacture of alloy steels. Cadmium is present in small amounts (1%) in most zinc ores and is obtained as a by-product in the refining and smelting of zinc. There are no ores in which cadmium is present in major amounts.

Zinc sulfide ores are usually concentrated by flotation, after which the zinc sulfide is converted to the oxide by heating it in air (roasting)

$$2ZnS(s) + 3O_2(g) \longrightarrow 2ZnO(s) + 2SO_2(g)$$

The sulfur dioxide, which is a by-product of the roasting process, is often used in the manufacture of sulfuric acid. Zinc oxide is converted to metal in either of two ways:

1. It is heated to a high temperature with carbon:

$$ZnO(s) + C(s) \longrightarrow Zn(g) + CO(g)$$

TABLE 26.10
Some alloys of zinc and cadmium

The numbers are percentages by weight of the metals in the alloy.

	Zn	Cd	Cu	Bi	Pb	Sn	Ni	In
Brass	18–40	—	60–82	—	—	—	—	—
Lead–tin–yellow brass	24.0	—	72.0	—	1.0	3.0	—	—
Bronze	1–25	—	70–95	—	—	1–18	—	—
Lead–nickel–brass	20.0	—	57.0	—	9.0	2.0	12.0	—
Nickel silver (German silver)	24.0	—	64.0	—	—	—	12.0	
Wood's metal	—	12.5	—	50.0	25.0	12.5	—	—
Lipowitz metal	—	10	—	50.0	26.7	13.3	—	—
Cerrolow	—	5.3	—	44.7	22.6	8.3	—	19.1

At the temperature at which the reaction takes place, zinc is a gas, and it escapes into a cooler part of the apparatus, where it condenses. The zinc usually contains specks of carbon that have been swept out of the furnace by the escaping carbon monoxide gas and zinc vapor. The irregularly shaped pieces of zinc, called "mossy zinc," often used for reactions in the laboratory, are formed by pouring molten zinc into water.

Any cadmium sulfide that was contained in the zinc sulfide ore goes through the metallurgical process with the zinc. Cadmium metal has a lower boiling point than zinc (767°C as compared with 907°C for zinc) and so may be partially separated during the preparation of the metals.

2. The roasted ore is dissolved in dilute sulfuric acid, and the ions of metals with higher reduction potentials are reduced to the metallic state by adding zinc dust. The filtered solution is then electrolyzed to recover metallic zinc:

$$ZnO(s) + 2H^+ \longrightarrow Zn^{2+}(aq) + H_2O(l)$$
$$M^{2+}(aq) + Zn(s) \longrightarrow Zn^{2+}(aq) + M \quad (M = Fe, Cd, etc.)$$
$$Zn^{2+}(aq) + 2e^- \longrightarrow Zn(s)$$

As a result of incomplete roasting, the zinc oxide usually contains some zinc sulfate, which dissolves in the dilute acid and, upon electrolysis, generates sulfuric acid. Thus a plant using this method of making zinc produces sulfuric acid as a by-product. Most of the cadmium that was contained in the ore remains with the zinc until the electrolysis, at which point the cadmium metal plates out at a higher reduction potential than does zinc metal. The separation is not always made, however.

Zinc is one of the four workhorse metals of our modern civilization (the others being iron, copper, and lead). It is used in coating iron to prevent rusting (galvanized iron) and in many important alloys, of which the most widely used are brasses, bronzes, and bearing metals (Table 26.10). Other uses of zinc, together with uses of cadmium and mercury, are listed in Table 26.11.

TABLE 26.11
Some end uses of zinc, cadmium, and mercury

Zn	Electroplating
	Alloys
	Galvanized iron
	Pigments, e.g., ZnO, ZnS (both white)
	Rubber
	Zinc oxide ointment (antiseptic)
	Batteries (dry cells)
Cd	Electroplating
	Alloys
	Ni–Cd batteries
	Pigments, e.g., CdS (yellow)
	Fungicides
Hg	Fungicides
	Pulp and paper industry
	Batteries
	Pigment, e.g., HgS, (red)
	Tanning leather
	Thermometers and scientific instruments
	Hg vapor lights
	Electrical switches

THE REPRESENTATIVE METALS [26.9]

The largest use of metallic cadmium is in alloys (Table 26.10), especially those that melt at low temperatures. Wood's metal and Lipowitz metal, both of which melt at 70°C, are used in automatic fire extinguishers and fire alarms. Cerrolow alloy, which contains indium, melts at 47°C. Cadmium rods are used in nuclear reactors to absorb neutrons, and thus to control the chain reaction.

b. Mercury. Mercury is less abundant in the crust of the Earth than is cadmium ($8 \times 10^{-6}\%$ for mercury and $2 \times 10^{-5}\%$ for cadmium), it is not as widely distributed, and it is not obtained as a by-product. Yet there are important uses for mercury for which there are no adequate substitutes. Hence, mercury is an expensive metal. Most of the mercury used in the United States is imported from Spain, though some is mined in California. The chief ore is the red *cinnabar,* HgS. The ore is concentrated by flotation and then heated to a high temperature in a stream of air, whereupon it is converted to elemental mercury and sulfur dioxide:

$$\text{HgS}(s) + \text{O}_2(g) \longrightarrow \text{Hg}(g) + \text{SO}_2(g)$$

The mercury distills from the furnace (b.p. 356.9°C) and is collected by condensation.

Mercury (m.p. -38.87°C) is liquid at room temperature, and many of its uses in modern civilization depend upon its liquid nature. There are, it has been estimated, more than three thousand uses of mercury (a few are given in Table 26.12). The relatively great change in volume of the metal with changes in temperature makes it useful in thermometers. Its liquidity and high density account for its use in barometers, and its reasonably good electrical conductivity for its use in electric switches. Mercury also conducts electricity in the vapor state, emitting the bright blue light of mercury arc lights. Mercury forms alloys with most of the metals—these are called **amalgams.** One amalgam that is of great importance is the sodium amalgam formed in the manufacture of chlorine (Section 24.8). Another is used in dentistry. When the intermetallic compound Ag$_3$Sn is ground with mercury, it dissolves to form a semisolid amalgam that, on standing, sets to form a hard, solid mixture of the intermetallic compounds Ag$_5$Hg$_8$ and Sn$_7$Hg. During the formation of these compounds, the amalgam expands slightly and fits the walls of a cavity so tightly that bacteria cannot easily get in. These amalgams are not toxic, apparently because the intermetallic compounds are extremely stable.

26.10 Reactions of zinc, cadmium, and mercury

a. With acids and bases. Both zinc and cadmium react with nonoxidizing acids, such as dilute hydrochloric acid and dilute sulfuric acid, to yield hydrogen, for example,

$$\text{Zn}(s) + 2\text{H}^+ \longrightarrow \text{Zn}^{2+} + \text{H}_2(g) \tag{1}$$

Mercury is inert toward such acids.

The reaction of zinc and cadmium with these acids is slow when the metals are pure, but rapid when they contain impurities such as small particles of less reactive metals or of conducting substances such as carbon. This is the result of **couple action**—the formation of a voltaic cell by direct contact between two substances rather than through an external connection. Consider a piece of zinc—when placed in an acidic aqueous medium it tends to liberate electrons

$$Zn \rightleftharpoons Zn^{2+} + 2e^- \tag{2}$$

This reaction proceeds only to a slight extent, for the electrons stay on the metal surface and the Zn^{2+} ions that have gone into solution remain close to the surface preventing further reaction. If some of the electrons are removed from the zinc metal, then the Zn^{2+} ions can wander off into the solution and Reaction (2) will continue to occur.

If a bit of graphite is embedded in the zinc surface some of the electrons generated on the zinc move over to the graphite, which is a conductor. H^+ ions in the solution are attracted by the electrons, and the reaction

$$2H^+ + 2e^- \longrightarrow H_2(g) \tag{3}$$

occurs. The hydrogen seems to come from the carbon particle, but actually it is the result of redox reaction (1). The zinc and carbon have formed an electrochemical couple, with Reaction (2) occurring at the zinc anode and Reaction (3) at the carbon cathode. If a piece of zinc immersed in an acid solution is touched with, say, a piece of platinum wire, the same thing happens. Some of the electrons leave the zinc surface for the other metal, and hydrogen appears to form on the platinum wire. In this case, the platinum wire is the inert cathode.

Zinc, cadmium, and mercury are all attacked by oxidizing acids such as nitric acid and concentrated sulfuric acid. For example,

$$Zn(s) + 2H_2SO_4(conc) \longrightarrow Zn^{2+} + SO_4^{2-} + SO_2(g) + 2H_2O(l)$$

The reaction between mercury and an excess of hot concentrated sulfuric acid yields mercury(II) sulfate, $HgSO_4$, whereas the mercury(I) salt, Hg_2SO_4, is formed if the metal is present in excess.

Zinc, like other elements in the same area of the periodic table, is amphoteric, illustrating the gradual change from metallic to nonmetallic character across the table. Zinc reacts with aqueous solutions of strong bases

$$Zn(s) + 2OH^- + 2H_2O(l) \longrightarrow \underset{\substack{tetrahydroxozincate(II)\\ion}}{[Zn(OH)_4]^{2-}} + H_2(g)$$

while cadmium and mercury are not amphoteric and do not do so.

b. With other elements. When zinc and cadmium are strongly heated in air, they burn to give the corresponding oxides, ZnO and CdO. Mercury reacts slowly at temperatures in the neighborhood of 350°C to give HgO.

All three elements of the family react with the halogens and with sulfur. Zinc and cadmium give products of the expected compositions, for example, $ZnCl_2$, CdS. In the reactions of mercury with the halogens, the mercury(I) halides, Hg_2X_2, are formed when the metal is in excess, and the mercury(II) halides, HgX_2, are formed when the halogens are in excess. Oxidation of mercury by sulfur yields only mercury(II) sulfide, HgS.

26.11 Compounds of zinc, cadmium, and mercury

Mercury(II) oxide, HgO, is either red or yellow, depending upon its method of preparation. The red and yellow forms have identical crystal structures and differ only in particle size, the yellow form being more finely divided than the red.

Mercury(I) oxide is unknown. When a strong base is added to a solution of a mercury(I) salt, a dark precipitate forms, which is a mixture of finely divided elemental mercury and mercury(II) oxide.

$$Hg_2^{2+}(aq) + 2OH^-(aq) \longrightarrow Hg(l) + HgO(s) + H_2O(l)$$

Zinc oxide, ZnO, is white at ordinary temperatures, but turns yellow when heated (Table 26.12). On cooling, it turns white again. Cadmium oxide, CdO, varies in color from green-yellow to black, depending upon the temperature at which it is obtained. Both cadmium and zinc oxides are only slightly soluble in water.

Mercury forms two series of halides, the mercury(I) compounds of the general formula Hg_2X_2 and the mercury(II) compounds of the composition HgX_2. The most widely used of the mercury(I) halides is calomel, Hg_2Cl_2, long known in medicine as a purgative and a drug to kill intestinal worms. The extremely poisonous nature of mercury compounds is masked in this substance by its very small solubility.

The anhydrous mercury(II) halides, HgX_2, are obtained by the reaction of excess halogen with the heated metal. The most important of these compounds, mercury(II) chloride, $HgCl_2$, known as *corrosive sublimate*, can also be prepared by heating a mixture of mercury(II) sulfate and sodium chloride.

$$HgSO_4(s) + 2NaCl(s) \longrightarrow HgCl_2(g) + Na_2SO_4(s)$$

Corrosive sublimate is a violent poison, but in very dilute solution it is used as an antiseptic.

Representative Group III metals: Al, Ga, In, Tl

26.12 Properties of Group III metals

With the exception of boron, a semiconducting element, the Representative Group III elements are all metals. The valence shell configuration in ns^2np^1, and all of the Group III elements form compounds in

TABLE 26.12

Some compounds of zinc, cadmium, and mercury

ZnO	**Zinc oxide** **Chinese white, a pigment**
ZnS	**Zinc sulfide** **in lithopone, a white pigment; TV tube phosphor**
$ZnCl_2$	**Zinc chloride** **deodorant; wood preservative**
CdO	**Cadmium oxide** **storage battery electrodes**
CdS	**Cadmium sulfide** **yellow to red pigment**
HgO	**Mercury(II) oxide** **red and yellow forms**
Hg_2Cl_2	**Mercury(I) chloride** **calomel, a drug**
$HgCl_2$	**Mercury(II) chloride** **corrosive sublimate, antiseptic in dilute solution**

TABLE 26.13
Properties of the Representative Group III metals

Configuration: ns^2np^1				
	Al	Ga	In	Tl
Melting point (°C)	658	29.75	155	303.5
Boiling point (°C)	1800	1700	>1450	1650
Density (g/cc)	2.70	5.9 (solid)	7.30	11.5
Atomic radius (nm)	0.1248	0.1245	0.1497	0.1549
Ionic radius, M^{3+} (nm)	0.05	0.06	0.081	0.095 Tl^+ 0.159
First ionization energy, kcal/mole (kJ/mole) $X(g) \longrightarrow X^+(g) + e^-$	138 (577)	138 (577)	133 (556)	141 (590)
Third ionization energy, kcal/mole (kJ/mole)	656 (2745)	708 (2963)	646 (2704)	680 (2878)
Electronegativity	1.5	1.8	1.5	1.4
Standard reduction potential (V) $M^{3+} + 3e^- \longrightarrow M(s)$	−1.66	−0.560	−0.338	0.719

the +3 oxidation state in which all of the electrons are involved in bonding. Boron forms only covalent compounds, but the metallic members of the group form both covalent and ionic compounds.

Aluminum is a silvery white metal of low density (Table 26.13). When pure, it is rather soft and weak, but its strength can be increased considerably by alloying with other metals, such as copper or magnesium. The pure metal is an excellent conductor of heat and electricity. Aluminum, although it is a reactive metal, is not noticeably corroded in air, nor is it attacked by liquid water, even at high temperatures. In air, the metal becomes covered with a very thin film of adherent oxide, which serves as a protective coating and inhibits further attack.

Gallium, indium, and thallium are also typical silvery white metals similar in appearance to aluminum. These elements all have high boiling points, but note the low melting point of gallium, which gives it an extremely large liquid temperature range. Gallium is also unusual because, like water, it is one of the few substances that expands on freezing.

The ionization energies for removal of three electrons from these elements (Table 26.13) are so large that their compounds are often covalent because the lattice energy cannot compensate for the energy needed for ionization. However, the hydrated Al^{3+}, Ga^{3+}, and In^{3+} ions form readily in aqueous solutions. Despite the large enthalpy of hydration of its +3 ion (−1300.8 kcal/mole, 25°C) thallium has little tendency to lose all of its valence electrons in aqueous solutions, as shown by its positive standard reduction potential. A thallium atom holds onto its electrons more tightly than those of the other Group III metals because it has a high effective nuclear charge—both the d transition elements and the lanthanide elements precede thallium in atomic number and atomic weight. Because of this increase in effec-

TABLE 26.14

End uses of aluminum

In 1934 aluminum was proclaimed to be the "theme metal of the twentieth century."

Structural material
 Buildings
 Airplanes
 Boats

Electrical wire

Kitchen utensils

Foil wrap

Alloys
 Alnico (magnetic)
 Duraluminum (light, but strong)

Welding (thermite reaction)

Pigments

Fireworks, flares, and rocket fuel

Alumina—various forms
 Watch bearings (synthetic ruby)
 Dehydrating agent ("activated")
 Refractory bricks
 Catalyst support
  (catalyst deposited on surface of finely divided alumina)

Aluminum sulfate
 Water purification
 Pulp and paper sizing

tive nuclear charge, the lower oxidation states increase in stability relative to higher oxidation states in going down Representative Groups III–VI. Thallium is stable in aqueous solutions as well as in compounds as the thallium(I) ion, in which only the valence shell p electron has been lost. Gallium and indium form a few compounds in the +1 oxidation state, but they are not stable in aqueous solution. The tendency of Tl^{3+} to revert to the more stable Tl^+ ion is so great, that compounds of tripositive thallium are strong oxidizing agents:

$$Tl^{3+} + 2e^- \longrightarrow Tl^+ \qquad E° = 1.25 \text{ V}$$

26.13 Sources, preparation, and uses of Group III metals

a. Aluminum. Aluminum occurs to the extent of 8.23% in the crust of the Earth and is the third most abundant element, being exceeded in abundance only by oxygen and silicon. In usage, aluminum is the most important light metal (Table 26.14).

The most plentiful aluminum-containing minerals are the aluminum silicates, which include granites and clays. At present there is no economical method for the extraction of aluminum from the silicate minerals, and the source of practically all of the metal is *bauxite*, which is a hydrated oxide, $Al_2O_3 \cdot xH_2O$.

Bauxite is generally found associated with large amounts of silica (SiO_2) and iron(III) oxide (Fe_2O_3), and these substances must be removed before the aluminum can be obtained in metallic form. The purification of bauxite is accomplished by the Bayer process, which takes advantage of the difference in properties of the oxides. Aluminum oxide is amphoteric, whereas ferric oxide is almost entirely basic and silica is a relatively inert acidic oxide. The crude bauxite is digested under pressure with hot sodium hydroxide solution, which dissolves the aluminum oxide as sodium aluminate, $Na[Al(OH)_4]$, and leads to precipitation of silica as a complex sodium aluminum silicate. Iron(III) oxide and other impurities are unaffected by the sodium hydroxide treatment. Insoluble materials are removed by filtration, and the filtrate is then diluted with water and cooled. This treatment hydrolyzes the sodium aluminate, and aluminum hydroxide is precipitated. The precipitate is filtered and converted to the pure anhydrous oxide, Al_2O_3, by heating. The sodium hydroxide solution is concentrated and used again.

Aluminum metal is obtained from the anhydrous oxide by an electrolytic method known in this country as the Hall process. While he was a student at Oberlin College, Charles Martin Hall, inspired by a professor's remark that a fortune awaited the man who could invent a cheap process for producing aluminum, set this as his goal. At the age of 21, shortly after he graduated, he succeeded. Hall did become a rich man, and a memorial to him in the form of an aluminum statue stands on the Oberlin campus. The Hall process is used today with only minor modifications. Before its invention, aluminum was rare and highly prized. Denmark's King Christian X wore an aluminum crown, and Napoleon III set the table with aluminum knives and forks for his most important guests.

In the electrolytic cells (Figure 26.2) for the Hall process, steel boxes lined with graphite are the cathodes. The electrolyte is molten cryolite (Na_3AlF_6, a mineral found only in Greenland) or a molten mixture of sodium, calcium, and aluminum fluorides. The purified aluminum oxide dissolves readily in such electrolytes. Carbon rods, which serve as anodes, are suspended in the molten solution, and electrolysis is carried out at temperatures in the neighborhood of 900–1000°C. During electrolysis, Al^{3+} ions from the oxide migrate toward the cell lining (the cathode) where they are reduced to the liquid metal, which collects at the bottom of the cell.

$$Al^{3+} + 3e^- \longrightarrow Al(l)$$

At the carbon anodes, the main electrolytic reaction is the oxidation of oxide ions (O^{2-}) to molecular oxygen. In addition, elemental fluorine is formed by oxidation of fluoride ions present in the melt. Both the oxygen and the fluorine react with the carbon anodes, which are gradually consumed and must be replaced periodically. The Hall process yields aluminum of a purity between 99.0 and 99.9%.

b. Gallium, indium, and thallium. Gallium is widely distributed in the Earth's crust as the hydrated oxide $Ga_2O_3 \cdot H_2O$. It occurs in association with aluminum-bearing ores. As the thirty-seventh element in abundance (0.0015%), gallium is in much greater supply than indium (1×10^{-5}%) or thallium (4.5×10^{-5}%). Both indium and thallium occur in lead- and zinc-bearing ores, and they are obtained by extraction from flue dust and other by-products of lead and zinc smelting and refining. Gallium is recovered by electrolysis from the concentrated sodium hydroxide solution used to digest the bauxite from which aluminum is obtained.

Gallium and indium are important to the solid-state electronics industry. They are used primarily in compound semiconductors, compounds between Representative Group III and Representative Group V elements that have semiconducting properties, for example, gallium arsenide or indium antimonide. The intense red light in the numerical display in some hand-held calculators is provided by gallium–arsenic–phosphorus light-emitting diodes. Over 80% of all gallium and indium produced is used in the electronics industry.

Thallium compounds are useful as rat and ant poisons, but care must be exercised, because thallium and its compounds are also highly toxic to human beings. Thallium finds other uses in low-melting glasses and, to a much lesser extent than indium and gallium, in semiconductors.

26.14 Reactions of aluminum

At high temperatures, aluminum burns in the air to form the oxide, the reaction being strongly exothermic

$$4Al(s) + 3O_2(g) \longrightarrow 2Al_2O_3(s) \qquad \Delta H^\circ = -801.0 \text{ kcal } (-3350 \text{ kJ})$$
$$\Delta G^\circ = -756.4 \text{ kcal } (-3165 \text{ kJ})$$

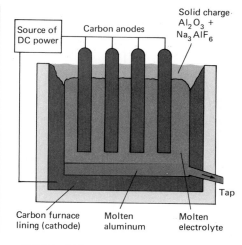

FIGURE 26.2
Electrolytic cell for production of aluminum.

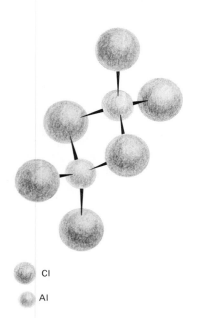

CI

Al

FIGURE 26.3
Aluminum chloride, Al_2Cl_6, molecule, in the vapor state.

The large enthalpy of formation of aluminum oxide makes the finely divided metal an excellent reducing agent toward many metal oxides at elevated temperatures. The reactions of aluminum with metal oxides are called **aluminothermic reactions.** So much heat is liberated in these reactions that the reduced metal is usually obtained in the molten condition. The Thermit process (see Example 5.5) involves the reaction of iron(III) oxide to provide the heat and molten iron needed for welding.

$$Fe_2O_3(s) + 2Al(s) \longrightarrow 2Fe(s) + Al_2O_3(s) \quad \Delta H° = -203.5 \text{ kcal } (-851.4 \text{ kJ})$$

Aluminum stands well above hydrogen in the electromotive series and displaces hydrogen from such strong acids as dilute hydrochloric and sulfuric acids.

$$2Al(s) + 6H^+ \longrightarrow 2Al^{3+} + 3H_2(g)$$

Pure aluminum reacts very slowly with these acids. The presence of small amounts of metal impurities greatly increases the rate of reaction through couple action. Aluminum is passive toward concentrated nitric acid, because this strong oxidizing agent coats the metal with a very thin layer of the metal oxide.

Aluminum reacts with aqueous solutions of strong bases such as NaOH with the formation of the aluminate ion, and the evolution of hydrogen.

$$2Al(s) + 2OH^- + 6H_2O(l) \longrightarrow 2Al(OH)_4^- + 3H_2(g)$$
$$\text{aluminate}$$
$$\text{ion}$$

26.15 Compounds of aluminum

Anhydrous aluminum chloride, $AlCl_3$, is a white, deliquescent substance. It is a catalyst for many organic reactions. In aqueous solution it is used in antiperspirants.

Salts of aluminum with strong acids are obtained as hydrates from aqueous solution under ordinary conditions—for example, $AlCl_3 \cdot 6H_2O$, $Al(NO_3)_3 \cdot 9H_2O$, $Al_2(SO_4)_3 \cdot 18H_2O$—a reflection of the strong tendency of the aluminum ion to combine with water. The salts contain the $[Al(H_2O)_6]^{3+}$ ion. The remaining water molecules are associated with the anions or held in crystal lattices.

In benzene solution and in the vapor state the chloride, bromide, and iodide form Al_2X_6 molecules in which each aluminum atom is surrounded tetrahedrally by four halogen atoms, with two of them common to both tetrahedra (Figure 26.3).

Aluminum oxide, Al_2O_3, occurs in nature in the hydrated form as bauxite and in the anhydrous form as corundum (Table 26.15). Corundum is used as an abrasive and a refractory (a material that is unchanged by high temperatures) because of its hardness and comparative inertness toward chemical attack. Some deposits of corundum are colored by the presence of small amounts of oxides of other metals and are of value as gemstones. In the ruby the impurity is Cr_2O_3; in the sapphire, FeO and TiO_2; in the oriental amethyst, Mn_2O_3; and in the oriental topaz, Fe_2O_3.

Treatment of an aqueous solution of an aluminum salt with a weak base, such as aqueous ammonia, yields a white gelatinous precipitate usually formulated as $Al(OH)_3$ and referred to as aluminum hydroxide. Sometimes the product is written as $Al_2O_3 \cdot xH_2O$. A number of hydrates of aluminum oxide are known, and at least one of them contains OH groups. The hydroxide is converted to the anhydrous oxide at about 800°C.

Corundum (Figure 26.4) and the anhydrous aluminum oxide obtained by dehydrating the hydroxide have the same formula, Al_2O_3, but they have different crystal structures, which impart different properties. As noted above, corundum—the form of the oxide called α-alumina, α-Al_2O_3—is quite inert. However, the other form of the oxide is, like the hydroxide, an amphoteric substance that will dissolve in either acids or bases.

$$Al_2O_3(s) + 6H^+ \longrightarrow 2Al^{3+} + 3H_2O(l)$$
$$Al_2O_3(s) + 3H_2O(l) + 2OH^- \longrightarrow 2Al(OH)_4^-$$

This reactive form of the oxide is known as γ-alumina, or γ-Al_2O_3. Heating γ-alumina above 1000°C converts it to α-alumina. For some uses of the aluminas, see Table 26.14.

Probably the most widely used aluminum salt is aluminum sulfate, $Al_2(SO_4)_3$, which can be obtained directly from bauxite by the action of sulfuric acid.

$$Al_2O_3 \cdot xH_2O(s) + 6H^+ + 3SO_4^{2-} \longrightarrow Al_2(SO_4)_3(aq) + (3 + x)H_2O(l)$$

Aluminum sulfate forms hydrates with eighteen molecules of water, $Al_2(SO_4)_3 \cdot 18H_2O$, or with fewer molecules of water. The major commercial outlet for aluminum sulfate is the pulp and paper industry, where it is used in the sizing process (in which rosin renders paper resistant to water), to adjust the pH of the stock, and to treat waste effluent.

The second most important use of aluminum sulfate is in treatment of municipal water supplies and sewage. The essential step in water treatment is the precipitation of aluminum hydroxide by an alkaline substance, such as lime, naturally present or added with the aluminum sulfate, for example,

$$Al_2(SO_4)_3(aq) + 3Ca(OH)_2(aq) \longrightarrow 2Al(OH)_3(s) + 3CaSO_4(aq)$$
$$Al_2(SO_4)_3(aq) + 3Na_2CO_3(aq) + 3H_2O(l) \longrightarrow$$
$$2Al(OH)_3(s) + 3Na_2SO_4(aq) + 3CO_2(g)$$

The aluminum hydroxide is a gelatinous substance called a flocculant that adsorbs and entangles suspended and colloidal impurities, including bacteria. The "floc" of hydroxide plus impurities is allowed to settle and the clarified water is filtered to remove any remaining particles.

When solutions containing aluminum sulfate and potassium

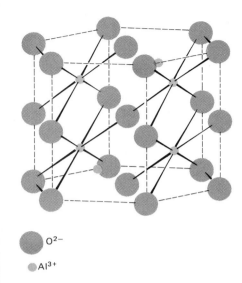

$\bigcirc$ O^{2-}

$\circ$ Al^{3+}

FIGURE 26.4
Crystal structure of corundum, Al_2O_3, or α-alumina.

sulfate in equimolar concentrations are permitted to evaporate, a double salt known as alum, $KAl(SO_4)_2 \cdot 12H_2O$, crystallizes. This substance is one of a general class of compounds, the alums, of the formula $M^+M^{3+}(SO_4)_2 \cdot 12H_2O$, where M^+ can be any one of a large number of singly charged cations and M^{3+} one of several triply charged ions. Alums of Na^+, K^+, NH_4^+ and of Al^{3+}, Cr^{3+}, and Fe^{3+} are the most common.

Aluminum sulfate and alum are extensively used in the dyeing industry, particularly in the dyeing of cotton and linen cloth. These salts serve as sources of Al^{3+} ion, which is precipitated in the fibers of the cloth as the gelatinous hydroxide in the presence of the dye. The hydroxide has great adsorptive power for the dye and fixes the dye firmly to the cloth—an action known as **mordanting.** The aluminum salt is called a *mordant.* Other gelatinous hydroxides are also used as mordants.

Representative Group IV and V metals: Sn, Pb, and Bi

26.16 Properties of tin, lead, and bismuth

Tin and lead are the metals of Representative Group IV and are preceded in the group by the semiconducting element germanium. Their valence shell configurations are ns^2np^2 (Table 26.16), and they form series of compounds with oxidation states +2 and +4 in which, respectively, either the two p electrons or all four valence electrons are involved in bonding. Bismuth is the only metal in Representative

TABLE 26.16

Some properties of tin, lead, and bismuth

	Sn	Pb	Bi
Configuration	$[Kr]4d^{10}5s^25p^2$	$[Xe]4f^{14}5s^25p^65d^{10}6s^26p^2$	$[Xe]4f^{14}5s^25p^65d^{10}6s^26p^3$
Melting point (°C)	231.9	327.4	271
Boiling point (°C)	2270	1620	1560
Density (g/cc)	5.75 (gray) 7.28 (white) 6.52 (brittle)	11.29	9.80
Atomic radius (nm)	0.1508	0.1746	0.152
Ionic radius (nm)	0.071 (+4)	0.084 (+4)	0.074 (+5)
Ionization energy (at 25°C) kcal/mole (kJ/mole) $X(g) \longrightarrow X^+(g) + e^-$	169 (707)	171 (715)	168 (703)
Electronegativity	1.7	1.6	1.9
Standard reduction potential (V) $M^{2+} + 2e^- \longrightarrow M(s)$	−0.136	−0.126	—
$BiO^+ + 2H^+ + 3e^- \longrightarrow$ $Bi(s) + H_2O$	—	—	0.32

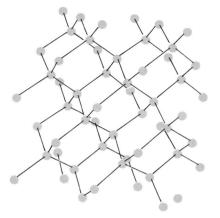

Gray tin (four unit cells)

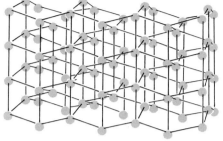

White tin (nine unit cells)

FIGURE 26.5
Crystal structures of gray and white tin.

Group V and is preceded by the semiconducting element antimony. With five valence electrons, bismuth forms compounds with oxidation states of +3 and +5. All three elements also exhibit negative oxidation states in their hydrides.

Tin, lead, and bismuth are the representative metals furthest to the right in the periodic table (except for the rare polonium). They have high electronegativity and low reactivity (Table 26.16). The standard reduction potential for bismuth is actually larger than that of hydrogen, showing that bismuth is a rather poor reducing agent. In their compounds, atoms of all three elements can form both covalent and ionic bonds, depending upon which other atoms are present.

Tin is a soft, low-melting metal that exists in three allotropic forms. The form stable at ordinary temperatures is *white tin* (Figure 26.5), which is distinctly metallic in character. Below 13.2°C, it changes slowly into an amorphous gray powder, *gray tin*, which is a semiconductor and is less dense than the metallic form (Table 26.16). The gradual crumbling away of tin changing to gray tin when exposed to low temperatures is called "tin disease." When the metal is heated to 161°C it changes to *brittle tin*, a material that shatters when it is struck with a hammer. Crystals of white tin are tetragonal, those of gray tin are cubic, and those of brittle tin are rhombic.

Lead, like tin, is soft and easily melted. It is quite dense. When freshly cut, lead has a high luster, but upon exposure to the air, it quickly becomes dull as the result of the formation of a thin coating of oxide or basic carbonate. This coating adheres tightly to the metal and protects it from further corrosion. For this reason, the metal has long been used for roofing, gutters, downspouts, and sewer lines. It should never be used to conduct water intended for human consumption, for lead is slowly dissolved by the water and is toxic (see An Aside: Metals as Poisons, this chapter).

Bismuth is a hard, brittle, reddish-white substance, which for a metal has low electrical and thermal conductivity, values for these properties being less than 2% of the corresponding values for silver. Molten bismuth expands on solidification, and this property is made use of in alloys.

TABLE 26.17

Sources of tin, lead, and bismuth

Sulfides of Cu, As, Bi, or Zn are usually combined with PbS in galena. Note that the unreactive Bi occurs in the free state.

Element (% in Earth's crust)	Major sources
Sn (0.0002)	**Cassiterite, SnO_2**
Pb (0.0013)	**Galena, PbS**
Bi (1.7×10^{-5})	**Free element** **Bi_2S_3 and Bi_2O_3** **Combined with other minerals in Cu and Pb ores**

TABLE 26.18

Major end products of tin, lead, and bismuth

Sn
Tinplate

Alloys, e.g.,
   soft solders (Sn, Pb)
   Babbit metal (Sn, Cu, Pb)
     used in bearings
   bronzes (Cu, Sn)

Pb
Alloys, e.g.,
   in lead storage batteries

Tetraethyllead (antiknock)

Litharge, PbO
   used in rubber, ceramics, etc.

Red lead, Pb_3O_4
   rust-inhibiting pigment

Bi
Low-melting alloys, e.g.,
   Woods metal, m.p. 70–72°C
   (50% Bi, 25% Pb, 12.5% Sn, 12.5% Cd)
   used in, e.g., electrical fuses, automatic
   sprinklers, safety plugs in gas cylinders

26.17 Sources of tin, lead, and bismuth

The ancient Egyptians and Babylonians were familiar with tin and lead, and a record of bismuth goes back to the 1400s, but it may have been known long before that time and confused with lead. The aqueducts that brought water to ancient Rome were lined with lead. Even though they have been known and used for so long, these elements are not abundant in nature (Table 26.17).

26.18 Metallurgy and uses of tin, lead, and bismuth

a. Tin. Tin is obtained by heating cassiterite with charcoal or coke.

$$SnO_2(s) \;+\; 2C(s) \xrightarrow{\Delta} Sn(l) + 2CO(g)$$
cassiterite

The ore is ordinarily prepared for the reduction process by first crushing and washing it to remove lighter rock impurities and then roasting it in air to remove impurities that give volatile oxides (for example, arsenic and sulfur). The crude tin obtained from the reduction is partially purified by placing it on a hot sloping table and permitting the molten metal to flow away from the less easily fusible impurities. It may be further refined by an electrolytic method in which impure tin, serving as anode, is oxidized into solution and electrolytically plated out on a pure tin cathode.

The major uses of tin are in the manufacture of tinplate—low-carbon steel with a thin coating of tin—and alloys (Table 26.18). Tinplate is made by electrolytic deposition of tin on the steel; the coating improves such properties of the metal as its workability and ease of soldering and protects it from corrosion.

b. Lead. In the preparation of lead, the galena is first concentrated by selective flotation processes that remove most of the gangue along with some of the other metal sulfides found in the ore. The concentrated lead sulfide is then roasted in air,

$$2PbS(s) + 3O_2(g) \longrightarrow 2PbO(s) + 2SO_2(g)$$

The roasted material is reduced in a blast furnace by means of coke and scrap iron.

$$PbO(s) + C(s) \xrightarrow{\Delta} Pb(l) + CO(g)$$
$$PbO(s) + CO(g) \xrightarrow{\Delta} Pb(l) + CO_2(g)$$
$$PbS(s) + Fe(s) \xrightarrow{\Delta} Pb(l) + FeS(s)$$

The lead generally contains silver and gold as well as other metal impurities. It is refined not only to give a better grade of lead, but also to recover these precious metals. Purification can be effected by electrolytic means or by extraction of the silver and gold in molten zinc (called the Parkes process).

Almost 4 million tons of lead were used throughout the world in 1970, much of it going into tetraethyllead and lead storage batteries. The uses of lead and its compounds should be under constant scrutiny because of its high toxicity (An Aside: Metals as Poisons).

c. Bismuth. The recovery of bismuth from its native ores (those in which an element occurs in the free state) is a simple process. The ore is heated at least to the melting point of bismuth (271°C) and the metal is permitted to flow away from the impurities. Oxide and sulfide ores are roasted in air and then reduced with carbon. A large amount of bismuth is obtained in the United States as a by-product of copper and lead smelting and refining.

Most of the bismuth that is used commercially appears in low-melting alloys for electrical fuses, automatic sprinkler systems, safety plugs in compressed gas cylinders, and other such devices. Bismuth alloys, for which dimensional changes during casting are predictable, are useful where accurate dimensions in the product are important.

26.19 Reactions of tin, lead, and bismuth

Tin reacts with nonoxidizing acids to give tin(II) compounds

$$Sn(s) + 2HCl(aq) \longrightarrow SnCl_2(aq) + H_2(g)$$

and with oxidizing acids such as nitric acid and sulfuric acid to give tin(IV) compounds. With nitric acid, the tin(IV) oxide rather than the nitrate is formed

$$Sn(s) + 4HNO_3(aq) \longrightarrow SnO_2(s) + 4NO_2(g) + 2H_2O(l)$$

Lead gives only Pb^{2+} compounds with both oxidizing and nonoxidizing acids.

Lead and tin react with alkali metal hydroxides to give the plumbate(II) ion, $Pb(OH)_4^{2-}$, and the stannate(II) ion, $Sn(OH)_4^{2-}$, respectively. For example,

$$Sn(s) + 2OH^- + 2H_2O(l) \longrightarrow Sn(OH)_4^{2-} + H_2(g)$$
stannate(II)
ion

Lead dissolves readily in hot concentrated sulfuric acid,

$$Pb(s) + 3H_2SO_4(conc) \xrightarrow{\Delta} Pb(HSO_4)_2(aq) + SO_2(g) + 2H_2O$$

By contrast, it is attacked only superficially by dilute sulfuric acid, with which it forms lead(II) sulfate, $PbSO_4$. The $PbSO_4$ is insoluble in dilute acid and forms an adherent layer that protects the metal from further action. Because of these solubility relationships, the lead chamber process (Section 22.18b) cannot be used to make concentrated sulfuric acid.

Metals as poisons

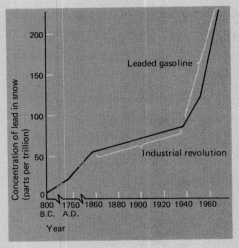

Environmental lead. *Lead levels in isolated Greenland glaciers have been growing since 800 B.C. The actual data are a series of scattered points, and the curve represents the best average line through these points. In addition, there is considerable seasonal variation in lead levels. The trend is significant, however. (Source: G. Tyler Miller, Jr. Living in the Environment, p. 97, Wadsworth, California, 1975)*

Much of the metal that is mined and then used is returned to the atmosphere, hydrosphere, or lithosphere in a finely divided state or as part of synthetic chemical compounds. Along with increasing awareness of environmental pollution has come the recognition that exposure to small concentrations of metals over long periods may threaten human health.

Evaluating the danger from metals in the environment is difficult for several reasons. The concentrations involved may be small and difficult to measure. There are wide variations in amount of exposure and the reaction to exposure of different individuals. Also, the only testing possible is with animals, and much is yet to be learned about how much a human reaction will be like that of, for example, a rat. In addition, as recent experiences with mercury have shown, the natural cycles of metals must be better understood before the fate of metals in the environment can be predicted.

In this section we discuss three representative metals under suspicion as environmental poisons—lead, mercury, and cadmium. Other metals that have been detected in the environment and that may be of concern are beryllium, which is highly poisonous, and nickel, particularly as nickel carbonyl, $Ni(CO)_4$.

a. Lead. Some historians believe that lead poisoning contributed to the fall of the Roman Empire. The Romans stored their wine in pottery vessels glazed with lead compounds, creating acidic conditions that were sure to leach lead into the wine. They also received their water from lead-lined aqueducts. Upper-class Romans, who could afford glazed pottery vessels, apparently suffered from high rates of stillbirth and brain damage, which may have contributed to their downfall. The use of lead glazes is now banned in the United States, but care must be taken with old pottery or pottery made in other countries where such glazes are still in use.

Another current source of lead is the paint and window putty in old buildings put up before the danger of lead was recognized. Children, particularly in old and poor neighborhoods, are often the victims of lead poisoning because they tend to eat chips of paint or putty.

Since soluble lead salts are cumulative poisons, the indiscriminate use of the metal and its compounds represents a serious health hazard. A daily intake of more than 1 mg of the element for a prolonged period apparently can be dangerous. Lead concentrates mainly in the bones; under the right conditions, it is removed along with calcium and brought into the blood. Early stages of the disease are characterized by constipation, anemia, loss of appetite, and pain in the joints. Unfortunately, these symptoms may not immediately be associated with lead poisoning. Later stages of the disease include paralysis of the extremities and mental damage.

Concentrations of airborne lead have risen in recent years mainly because of the use of tetraethyllead in gasoline (Figure A).

b. Mercury. Pure metallic mercury is not as toxic as mercury vapor and soluble mercury compounds, which are highly poisonous. However, mercury metal as well as its compounds must be handled with care.

It has been found in recent years that traces of mercury are present in the air and in lakes, rivers, and other bodies of water. This was first attributed to mercury producers and industries in which mercury is used in large amounts. However, we now know that the natural mercury cycle contributes even more mercury to the air and to the hydrosphere than do the activities of man.

Mercury poisoning, therefore, is most often a local problem where concentrations are high. The mercury reaches man mainly in food. It was thought for years that the discharge of metallic mercury into, say, a lake, was harmless because the mercury would sink and remain at the bottom as part of the sediment. Now it is known that bacteria can convert metallic mercury into methylmercury (CH_3Hg^+), a form in which it is highly poisonous. Mercury is concentrated up the food chain as, for example, bigger fish eat smaller fish that have eaten still smaller fish that contained mercury. A very serious incidence of mercury poisoning occurred in Japan among fishermen and their families who three times a day ate fish from Minamata Bay, which was receiving a mercury-laden effluent from a plastics factory. Other mercury poisonings have occurred when people or their farm animals ate seed grain treated with mercury-containing fungicides.

The symptoms of mercury poisoning, which, like lead poisoning, are hard to diagnose in its early stages, include loss of muscle control and blurred vision, leading ultimately to paralysis and kidney failure. Because the body has a natural mechanism for eliminating mercury, mercury is not a great threat as a low-level cumulative poison.

c. Cadmium. Cadmium compounds are exceedingly poisonous, and several tragic events have resulted from cadmium poisoning. The most notable of these took place in Toyama Prefecture, Japan, where the poisoning went on for several years before the cause was discovered, and where several hundred people died of cadmium poisoning. A smelter handling large amounts of cadmium-rich ore discharged wastes into the Jintsu River, the waters of which were used to irrigate the rice fields downstream. The cadmium in this water was absorbed into the rice, and so was introduced into the diet. The body has no mechanism for eliminating cadmium, which in this instance rapidly accumulated and caused serious kidney trouble and disintegration of the bones.

Recently, fear has been expressed that enough cadmium gets into the atmosphere to cause health problems. For example, as an automobile tire wears, the zinc oxide and the associated cadmium oxide in the rubber are liberated as fine dusts, which may not settle for some time and so may be drawn into the lungs. Tobacco contains a small amount of cadmium, which in smoking is carried into the lungs of both the smoker and those around him. The human body, at birth, does not contain cadmium, but as we grow older, cadmium compounds accumulate from our use of galvanized water pipes and kitchen utensils (the zinc used in galvanizing always contains some cadmium), from automobile tires, and in other ways. Eventually, this may bring on chronic disease. There seems to be a definite correlation between the incidence of hypertension (high blood pressure) and the concentration of cadmium in the body.

In direct combination with any of the halogens, tin yields the tin(IV) halides, SnX_4. Lead yields only the +2 compound with chlorine

$$Pb(s) + Cl_2(g) \longrightarrow PbCl_2(s)$$

and bismuth combines directly with all of the halogens to give the trihalides, BiX_3. Only fluorine, one of the strongest of the oxidizing agents, is capable of oxidizing bismuth to its maximum oxidation state of +5; BiF_5 is produced by fluorination of the fused metal at 600°C.

26.20 Compounds of tin, lead, and bismuth

Tin(II) chloride, a common laboratory chemical, finds some use as a weak reducing agent. Its aqueous solutions are slowly oxidized to tin(IV) by air and often some of the free metal is kept in contact with such solutions to maintain the tin in the +2 state.

In contrast to the tin(II) halides, which are soluble in water, the lead(II) halides are water-insoluble. A common source of Pb^{2+} ion in solution is the soluble nitrate $Pb(NO_3)_2$.

The crystal structures of the tin(II) halides have not yet been completely determined. The solid lead(II) compounds with the more electronegative halogens—PbF_2, $PbCl_2$, $PbBr_2$—have known ionic structures (Figure 26.6).

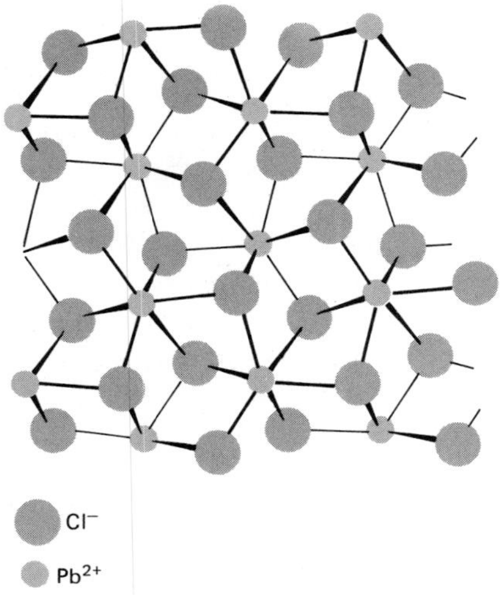

Bismuth, through the use of the three p electrons in the valence shell of its atom, forms bismuth trihalides with all of the halogens. While the solid fluoride, BiF_3, is ionic in character, the bonding in the other trihalides is essentially covalent. However, the structures of the solids are complex and show no discrete BiX_3 molecules.

Tin and lead, like the other members of Representative Group IV (carbon, silicon, and germanium), can form four covalent single bonds through the use of all four (s^2p^2) valence electrons in their atoms. The tin(IV) halides are monomolecular, volatile (except for SnF_4) covalent species with the expected tetrahedral geometry.

Of the possible lead(IV) halides, only two have been prepared—PbF_4 and $PbCl_4$. When heated, these compounds lose halogen atoms to give the dihalides. The nonexistence of the tetrabromide and tetraiodide and the thermal instability of the other two tetrahalides are reflections of the tendency of lead(IV) to act as a strong oxidizing agent.

Bismuth(V) fluoride, BiF_5, the only +5 bismuth halide definitely known, is a potent fluorinating agent, readily supplying fluorine in chemical reactions and being itself converted to the trifluoride.

Tin(II) oxide, SnO, and lead(II) oxide, PbO, are amphoteric substances that dissolve in solutions of strong acids to form hydrated Sn^{2+} and Pb^{2+} ions and in solutions containing high concentrations of hydroxide ion to give stannate(II) and plumbate(II) ions, for example,

$$SnO(s) + H_2O(l) + 2OH^- \longrightarrow Sn(OH)_4^{2-}$$
stannate(II) ion

Cl$^-$
Pb^{2+}

FIGURE 26.6
Crystal structure of lead chloride, $PbCl_2$.

PbO, known as *litharge*, is used in the glaze for ceramic vessels, and the molten oxide is an excellent medium for growing garnet crystals useful in electronic devices and as gems. A paste of litharge and glycerine hardens on standing and yields a cement that is stable toward water and is often employed to seal drain pipes to sinks.

Lead(IV) oxide, PbO_2, does not exist in nature but may be obtained as a dark brown material by the oxidation of lead(II) in alkaline solution with very strong oxidizing agents such as hypochlorite ion, OCl^-. The compound isolated never has the stoichiometric formula PbO_2, the atomic ratio of oxygen to lead ordinarily being about 1.88. This is an example of a compound with a lattice defect. In this case, the crystal has some vacancies at lattice points where there should be oxygen atoms. Because of this defect, lead(IV) oxide is able to conduct electricity and function as an electrode in the lead storage battery (see MnO_2, Section 29.12).

For many years, white lead, a basic carbonate of the approximate composition $Pb_3(OH)_2(CO_3)_2$, was the most important paint pigment, largely because of its excellent covering power and protection of wood. However, it poisons children who eat paint peelings. Moreover, it darkens on exposure to air as a result of the formation of black lead sulfide and therefore has been replaced by less toxic and more stable compounds, such as titanium dioxide, TiO_2. White lead is also used as a component of glazes for ceramic vessels, but not, of course, for use with food (see An Aside: Metals as Poisons, this chapter).

Other lead compounds that see service as pigments include the yellow chromate $PbCrO_4$; the red compound $PbO \cdot PbCrO_4$; and the red oxide Pb_3O_4, known as red lead, a component of paints used to protect structural steel from corrosion. The arsenate, $Pb_3(AsO_4)_2$, is an insecticide.

Some of the compounds of tin, lead, and bismuth are listed in Table 26.19.

TABLE 26.19

Some compounds of tin, lead, and bismuth

$SnCl_2$	Tin(II) chloride tinplating
SnF_2	Tin(II) fluoride toothpaste
$SnCl_4$	Tin(IV) chloride
SnO_2	Tin(IV) oxide white enamel
SnO	Tin(II) oxide
PbO	Lead(II) oxide litharge, pottery glaze
$Pb(C_2H_3O_2)_2 \cdot 3H_2O$	Lead acetate trihydrate
PbO_2	Lead(IV) oxide a nonstoichiometric compound; lead storage battery
Pb_3O_4	Red lead anticorrosion paint for structural steel
$PbCrO_4$	Lead chromate chrome yellow pigment
BiF_3	Bismuth(III) fluoride
BiF_5	Bismuth(V) fluoride
Bi_2O_3	Bismuth(III) oxide yellow enamel
$Bi(NO_3)_3 \cdot 5H_2O$	Bismuth(III) nitrate pentahydrate
$BiOCl$	Bismuth oxychloride

THOUGHTS ON CHEMISTRY

Of things metallic–1556 A.D.

Many persons hold the opinion that the metal industries are fortuitous and that the occupation is one of sordid toil, and altogether a kind of business requiring not so much skill as labour. But as for myself, when I reflect carefully upon its special points one by one, it appears to be far otherwise. For a miner must have the greatest skill in his work, that he may know first of all what mountain or hill, what valley or plain, can be prospected most profitably, or what he should leave alone; moreover, he must understand the veins, stringers and seams in the rocks. Then he must be thoroughly familiar with the many and varied species of earths, juices [soluble salts], gems, stones, marbles, rocks, metals, and compounds. He must also have a complete knowledge of the method of making all underground works. Lastly, there are the various systems of assaying substances and of preparing

DE RE METALLICA, *by Georgius Agricola*

THE REPRESENTATIVE METALS [26.20]

them for smelting; and here again there are many altogether diverse methods. For there is one method for gold and silver, another for copper, another for quicksilver, another for iron, another for lead, and even tin and bismuth are treated differently from lead. Although the evaporation of juices is an art apparently quite distinct from metallurgy, yet they ought not to be considered separately, inasmuch as these juices are also often dug out of the ground solidified, or they are produced from certain kinds of earth and stones which the miners dig up, and some of the juices are not themselves devoid of metals. Again, their treatment is not simple, since there is one method for common salt, another for soda, another for alum, another for vitriol, another for sulphur, and another for bitumen.

Furthermore, there are many arts and sciences of which a miner should not be ignorant. First there is Philosophy, that he may discern the origin, cause, and nature of subterranean things; for then he will be able to dig out the veins easily and advantageously, and to obtain more abundant results from his mining. Secondly, there is Medicine, that he may be able to look after his diggers and other workmen, that they do not meet with those diseases to which they are more liable than workmen in other occupations, or if they do meet with them, that he himself may be able to heal them or may see that the doctors do so. Thirdly follows Astronomy, that he may know the divisions of the heavens and from them judge the direction of the veins. Fourthly, there is the science of Surveying that he may be able to estimate how deep a shaft should be sunk to reach the tunnel which is being driven to it, and to determine the limits and boundaries in these workings, especially in depth. Fifthly, his knowledge of Arithmetical Science should be such that he may calculate the cost to be incurred in the machinery and the working of the mine. Sixthly, his learning must comprise Architecture, that he himself may construct the various machines and timber work required underground, or that he may be able to explain the method of the construction to others. Next, he must have knowledge of Drawing, that he can draw plans of his machinery. Lastly, there is the Law, especially that dealing with metals, that he may claim his own rights, that he may undertake the duty of giving others his opinion on legal matters, that he may not take another man's property and so make trouble for himself, and that he may fulfill his obligations to others according to the law.

DE RE METALLICA, by Georgius Agricola, 1556 (Translated by Herbert C. Hoover and Lou H. Hoover, 1912. Quoted from p. 1ff, Dover, New York, 1950)

Significant terms defined in this chapter

scavengers	sintered	leaching
metallurgy	calcined	nonrenewable resources
mineral	roasting	amalgam
ore	smelting	couple action
gangue	slag	aluminothermic reactions
flotation	flux	mordanting

Exercises

26.1 The general electronic configuration for elements in Representative Group I is ns^1. (a) Write similar configurations for elements in Group II, Group III, the zinc family, Sn and Pb, and Bi. Predict (b) the oxidation states for these families, (c) the formulas of the oxides, and (d) the formulas of the binary halides.

First year students in chemistry laboratories are usually surprised to find that metals in Group I such as Na are stored under a liquid, metals in Group II such as Ca are covered by a white powder, and that Al reacts with H_2O and HCl much more slowly than expected. (e) Explain these observations.

26.2 Distinguish clearly between (a) an ore and a mineral; (b) flotation and leaching; (c) alloys and amalgams; (d) slaked and unslaked lime; and (e) sintering, calcining, roasting, and smelting.

26.3 Which statements are true? Rewrite any false statement so that it is correct.
(a) For representative elements that can form two or more oxidation states, the lower oxidation states are more favorable if the element is located near the bottom of the periodic table.
(b) Many salts having the empirical formula of $M^{2+}X^{2-}$ are much less soluble than those having the empirical formula of M^+X^- because of the large increase in hydration energy.
(c) In many respects, compounds of Li^+ are like those of Group II, and those of Be^{2+} are like those of Group III elements.
(d) A sample of frozen mercury would be expected to be malleable instead of brittle like ice.
(e) A common name for NaOH is "lye."
(f) Calcium oxide is an acidic anhydride.

26.4 In addition to the usual β^- emission of $^{227}_{89}Ac$, a very small amount of α emission also takes place. The product of the α emission, AcK, is also radioactive in that it undergoes β^- emission. Write the nuclear equations for the formation of AcK and its subsequent decay. What is the identity of AcK?

26.5 Write chemical equations describing the reaction of O_2 with each of the alkali metals.

26.6 Write balanced equations and give the experimental conditions necessary for preparing (a) NaOH, (b) Na_2CO_3, (c) $NaHCO_3$, (d) CaO, (e) $Ca(OH)_2$, (f) KOH, (g) K_2CO_3, (h) MgO, and (i) $MgSO_4$.

26.7 Write and discuss the chemical equations describing the formation of limestone caverns and the growth of stalagmites and stalactites in these caverns.

26.8 Mercury reacts with a mixture of concentrated nitric and hydrochloric acids to produce mercury(II) chloride, nitric oxide, and water. Write the oxidation and reduction half reactions describing this reaction and write the overall equation. What is the oxidizing agent?

26.9 Write the electronic configurations of Hg, Hg^+ and Hg^{2+}. Which of these are diamagnetic? One piece of evidence that mercury(I) exists as Hg_2^{2+} is that solutions containing mercury(I) are diamagnetic. Sketch the Lewis diagram for Hg_2^{2+} and show why this ion is diamagnetic.

26.10 Why is there less difference in chemical behavior between Zn and Cd than between Cd and Hg?

26.11 When zinc sulfide ore is roasted, the major products are ZnO and $ZnSO_4$. However, when mercury(II) sulfide, cinnabar, is treated in the same way, metallic mercury is formed. Explain this difference.

26.12 Explain the use of aluminum sulfate and calcium hydroxide in water purification.

26.13 As a student added a solution of NaOH dropwise to a test tube containing a solution of Al^{3+}, he noticed that a gelatinous percipitate was formed that disappeared upon adding more of the alkaline solution. Explain his observations and write the chemical equations for the reactions involved.

821

26.14 Give a brief definition of each of the following terms used in metallurgy: (a) flotation, (b) gangue, (c) sintering, (d) calcining, (e) roasting, (f) smelting, (g) slag, (h) flux, (i) leaching, (j) electrolysis, and (k) reduction.

26.15 An abbreviated activity series for some of the representative metals studied in this chapter is $Mg \mid Zn \mid Sn \mid Pb \mid H_2 \mid Hg$ with the most "active" metal on the left and the least "active" metal on the right. Use this series to predict whether the following reactions will occur or not:
(a) $Sn + HCl \longrightarrow$
(b) $Pb^{2+} + Mg \longrightarrow$
(c) $Hg + HCl \longrightarrow$
(d) $Zn^{2+} + Mg^{2+} \longrightarrow$
(e) $Mg + ZnO \longrightarrow$
(f) $Hg_2O + H_2 \longrightarrow$
For those that you predicted to occur, complete and balance the equation.

26.16* Derive a procedure by which in each of the following groups the ions can be separated from each other: (a) Zn^{2+}, Cd^{2+}, Hg^{2+}, Hg_2^{2+}; (b) Zn^{2+}, Ba^{2+}, Al^{3+}; and (c) Na^+, Be^{2+}, Mg^{2+}.

26.17 Write the equation that describes the electrolysis of a brine solution to form $NaOH$, Cl_2, and H_2. What mass of each substance will be produced in an electrolysis cell for each faraday of electricity passed through the cell?

26.18 Seawater contains 0.13 wt % Mg^{2+}. How much water would have to be processed to yield 1.0 ton of the metal if the recovery process is 75% efficient? Describe the steps required to obtain the metal.

26.19 What is the concentration of Mg^{2+} in solution under equilibrium conditions in a buffer solution containing $0.10M$ NH_3 and $0.10M$ NH_4Cl? $K_b = 1.6 \times 10^{-5}$ for NH_3 and $K_{sp} = 7.1 \times 10^{-12}$ for $Mg(OH)_2$.

26.20 What volume of "dry" hydrogen gas at STP would be generated by the reaction of 1.00 g of strontium hydride with water?

26.21 A sample of impure zinc was to be purified by electroplating. What mass of metal can be transferred from an electrode to another in an hour's time for each ampere of current passed through the system?

26.22 The standard state free energy of formation at 25°C of $Hg(l)$ is 0, of $Hg^{2+}(aq)$ is 39.30 kcal/mole, and of $Hg_2^{2+}(aq)$ is 36.70 kcal/mole. Calculate $\Delta G°$ and the equilibrium constant for the reaction

$$Hg(l) + Hg^{2+} \longrightarrow Hg_2^{2+}$$

Explain why a drop of Hg is added to a solution of Hg_2^{2+} to make it more stable.

26.23 With all the controversy about the United States changing to the metric system of units, one person remarked, "The only time that the two sides involved in the controversy will agree will be when their mercury thermometers read the same temperature." At what temperature will thermometers calibrated in °F and °C have exactly the same reading? Look at Table 26.8 and see if you can spot a flaw in the observer's statement.

26.24 Describe the various steps by which metallic aluminum is obtained from bauxite and write equations describing the chemical reactions. What weight of Al can be obtained from 1.00 ton of ore assuming the ore to contain 70. wt % Al_2O_3?

26.25 What weight of "white lead," $Pb(OH)_2 \cdot 2PbCO_3$, can be made from 10.0 g of Pb?

26.26* A mixture was believed to contain $Ca(HCO_3)_2$, $CaCO_3$, and CaO. Exactly 100 g of this mixture was heated and 2.00 g of H_2O and 15.00 g of CO_2 were collected. Determine the composition of the original mixture. The equations for the reactions are

$$Ca(HCO_3)_2(s) \xrightarrow{\Delta} CaO(s) + H_2O(g) + 2CO_2(g)$$

$$CaCO_3(s) \xrightarrow{\Delta} CaO(s) + CO_2(g)$$

$$CaO(s) \xrightarrow{\Delta} \text{no reaction}$$

Name all the substances involved in the above reactions.

26.27* A sample containing 40 g of Ca and 60 g of Al is thoroughly mixed and heated to 1200°C, resulting in the formation of a liquid solution. The solution is cooled to about 1050°C and crystals start forming that contain 43 wt % Ca and 57 wt % Al. As the mixture is cooled further, a second solid phase appears at 72.5°C, and it contains 32 wt % Ca and 68 wt % Al. Each of these solid phases corresponds to an intermetallic compound formed between aluminum and calcium. Find the empirical formulas of these substances.

26.28* In a certain qualitative analysis scheme, Pb^{2+}, Ag^+, and Hg_2^{2+} are separated from other ions by adding hydrochloric acid and precipitating the cations as insoluble chlorides. However, a large excess of HCl cannot be used, because AgCl and $PbCl_2$ will form soluble complex compounds and a complete separation will not occur. What is the concentration of each cation in solution if the concentration of Cl^- is $0.3M$? $K_{sp} = 2 \times 10^{-5}$ for $PbCl_2$, 1.8×10^{-10} for AgCl and 1.3×10^{-18} for Hg_2Cl_2. The various tests in the scheme for each cation are sensitive to $1 \times 10^{-4}M$. Are any of these concentrations large enough that subsequent tests might be in error?

26.29** A smelter processes 500. tons of concentrated zinc ore (90. wt % ZnS) each day. As the ore is roasted, two reactions take place: 80.% of the ZnS reacts to form ZnO and SO_2, and 20.% of the ZnS reacts to form $ZnSO_4$. (a) Write chemical equations for these processes and calculate the masses of ZnO, $ZnSO_4$, and SO_2 produced each day.

The $ZnSO_4$ is leached from the roasted ore using 250. tons of water each day, and the solution is filtered, purified, and electrolyzed, forming Zn and a dilute solution of H_2SO_4. (b) Write a chemical equation for this process and calculate the mass of Zn and H_2SO_4 produced each day.

The ZnO is reduced with coal, and the metal that is formed distills out of the reaction vessel and is collected. (c) Write the chemical equation for this process and calculate the masses of Zn and 96.4% pure coal involved in this process each day. (d) Assuming the coal impurity to be S, what mass of SO_2 is produced each day from this process?

The SO_2 produced by the smelting plant is prevented from entering the atmosphere by passing the stack gases through a slurry of air, water, and $CaCO_3$. The reactions may be summarized as

$$SO_2(g) + CaCO_3(s) + 2H_2O(l) \xrightarrow{H_2O}$$

$$CaSO_3 \cdot 2H_2O(s) + CO_2(g)$$

$$SO_2(g) + \tfrac{1}{2}O_2(g) + CaCO_3(s) + 2H_2O(l) \xrightarrow{H_2O}$$

$$CaSO_4 \cdot 2H_2O(s) + CO_2(g)$$

Assuming that 95% of the sulfur is converted to $CaSO_3 \cdot 2H_2O$ and the remainder to $CaSO_4 \cdot 2H_2O$, (e) calculate the masses of $CaSO_3 \cdot 2H_2O$, $CaSO_4 \cdot 2H_2O$, and $CaCO_3$ involved in this process each day.

There is no local market for the dilute sulfuric acid solution, and it cannot be dumped into a nearby river any longer because of pollution regulations. So $CaCO_3$ is added to form $CaSO_4 \cdot 2H_2O$, which, in turn, is filtered off as a solid. (f) Write a chemical equation for this process and calculate the masses of $CaCO_3$ and $CaSO_4 \cdot 2H_2O$ involved in the process each day. Some of the $CaSO_4 \cdot 2H_2O$ remains in the solution (4.8 lb per ton of H_2O) that is dumped into the river. (g) Calculate the mass of $CaSO_4 \cdot 2H_2O$ dumped each day. If the flow rate of the river is 500 tons of water each hour, (h) calculate the change in water hardness (parts per million by weight) caused by adding this solution to the river.

There is no market for the $CaSO_4 \cdot 2H_2O$ and $CaSO_3 \cdot 2H_2O$ by-products of this process, and these materials are taken to a landfill for dumping. (i) What mass of solid materials is dumped each day? (j) What mass of $CaCO_3$ is required each day?

27 SEMICONDUCTING ELEMENTS

This chapter presents the properties of the semiconducting elements and their hydrides, oxygen compounds, and halides. Three-center bonding in boron compounds is described, and the nature of bonding and conductivity in semiconductors is explained. Brief descriptions are given of the preparation of ultrapure silicon and germanium for the electronics industry, and of the nature of semiconducting devices. Some of the natural and man-made silicon–oxygen compounds and their uses are discussed.

A M/FM portable radio, solid state chassis," "100% solid state small screen color TV," "100% solid state electronic clock, no moving parts,"—these are all advertisements from a recent mail-order catalog. What does "solid state" really mean? In electronics, it means that the electric signals are generated, or amplified, or controlled by tiny devices made of the semiconducting elements—semiconductor devices. These transistors, rectifiers, diodes, and so on have no moving parts, no heated wires, and need no vacuum. They perform their functions because of properties imparted by their internal electronic structure.

The solid state revolution began in 1947 with the discovery of the transistor by Walter Brattain, John Bardeen, and William Shockley of the Bell Telephone Laboratories in Murray Hill, New Jersey. Time magazine (July 12, 1948) described the transistor as a "little brain cell." Newsweek (September 6, 1948) hailed it as "an innovation which may revolutionize electronics and communication as the original three-element vacuum tube did 35 years ago." Today we have the space ships, the calculators, the computers, and the solid state color TV sets to prove that this statement was correct.

The semiconducting elements would be important and valuable for their role in the electronics industry alone. However, their bonding and structural properties make these elements special in many other ways. Bonding in boron compounds is unique. Naturally occurring silicon–oxygen compounds form much of the crust of the Earth. Furthermore, we reshape silicon–oxygen compounds to give structural materials such as glass and cement, and to make beautiful objects of cut glass, pottery, and fine china.

The seven semiconducting elements

27.1 General properties

The properties of the semiconducting elements, like their positions in the periodic table, are intermediate between those of the metals and the nonmetals. Often called the *metalloids*, these seven elements and their Representative Groups are Group III, boron; IV, silicon and germanium; V, arsenic and antimony; and VI, selenium and tellurium. Some of their properties are given in Table 27.1.

These elements resemble metals in appearance but are more like nonmetals in their chemical behavior. The semiconducting elements conduct electric current, but much less effectively than metals. It is the electrical conductivity of the semiconducting elements that puts them in a class by themselves—their conductivity is raised under conditions that lower conductivity in metals. This distinctive property is explained in terms of bonding (Section 27.6).

The sizes of the atoms of these elements (Table 27.1) are in line with their positions in the periodic table groups. In electronic properties they are intermediate between metals and nonmetals. Their electronegativities all fall between 1.9 and 2.4. Semiconducting elements generally form covalent compounds. Their halides, like those of the nonmetals, are volatile and covalent. The bonds between the semiconducting element and the halogen are strong, and the heats of formation of the compounds are high; for example, $SiCl_4$, $\Delta H_f^\circ = -164.2$ kcal/mole (-667.0 kJ/mole); $AsCl_3$, $\Delta H_f^\circ = -72.9$ kcal/mole (-305 kJ/mole). The oxides are acidic, except for two, As_4O_6 and Sb_4O_6, which are amphoteric. The free elements exist in many polyatomic forms with varying degrees of covalent bonding.

The following sections describe some of the sources, preparation, properties, and uses of the semiconducting elements other than those connected with their use as semiconductors, which are covered in Sections 27.6–27.8.

Semiconducting elements
B, Si, Ge, As, Sb, Se, Te

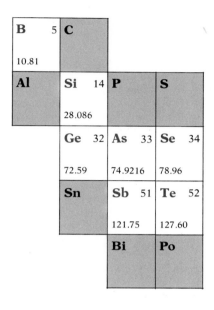

TABLE 27.1
Properties of the semiconducting elements

Element	Configuration	Atomic radius (nm)	Melting point (°C)	Boiling point (°C)
B	[He]$2s^2 2p^1$	0.088	2300	2500
Si	[Ne]$3s^2 3p^2$	0.117	1420	2355
Ge	[Ar]$3d^{10}4s^2 4p^2$	0.122	958	2700 (?)
As	[Ar]$3d^{10}4s^2 4p^3$	0.121	814 (32 atm)	610 (sublimes)
Sb	[Kr]$4d^{10}5s^2 5p^3$	0.141	630.5	1325
Se	[Ar]$3d^{10}4s^2 4p^4$	0.117	217.4 (gray form)	684.8
Te	[Kr]$4d^{10}5s^2 5p^4$	0.137	449.8	1390

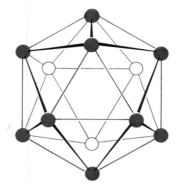

B_{12} icosahedron

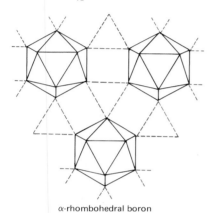

α-rhombohedral boron

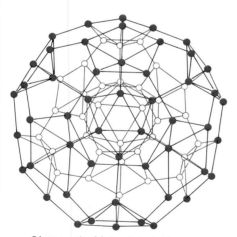

84-atom unit of β-rhombohedral boron

FIGURE 27.1
Elemental boron.

27.2 Boron

Boron occurs only to the extent of about 0.001% in the crust of the Earth, mainly as orthoboric acid, H_3BO_3, and borates. The borates most important as boron ores are *borax* ($Na_2B_4O_7 \cdot 10H_2O$), *kernite* ($Na_2B_4O_7 \cdot 4H_2O$), and *colemanite* ($Ca_2B_6O_{11} \cdot 5H_2O$).

The free element is obtained as an impure microcrystalline brown powder, "amorphous" boron, by the reduction of boric oxide, B_2O_3, with magnesium at high temperatures.

$$B_2O_3(s) + 3Mg(s) \longrightarrow 2B(s) + 3MgO(s)$$

Pure boron, in the form of black lustrous crystals, can be prepared by reduction of boron trichloride with hydrogen above 1000°C.

$$2BCl_3(g) + 3H_2(g) \xrightarrow{\Delta} 2B(s) + 6HCl(g)$$

Elemental boron is an extremely hard, high-melting (2300°C) substance that exists in at least three allotropic forms. All three forms have giant-molecule structures in which the fundamental structural unit is an icosahedral (twenty-sided) arrangement of twelve boron atoms (Figure 27.1) united by strong covalent single bonds. Weaker covalent bonds join the icosohedra to each other (these are three-center bonds, explained in Section 27.9).

The boron atom has a small atomic radius and a high ionization energy. It forms no cationic compounds and usually is in the +3 oxidation state. Boron is an unusual element—the structure and properties of boron and many of its compounds are not predictable on the basis of what is known of other elements and their compounds.

Boron is added in small amounts to aluminum alloys destined for electrical wires and to steel, where it aids in hardening. The element has a high affinity for oxygen and is used as an oxygen scavenger in the production of metals. A newer application of boron is the production of fibers for use in fiber-reinforced materials. Pure boron fibers or boron-coated tungsten fibers embedded in aluminum or magnesium produce a light, very stiff and strong material used in aircraft parts.

At ordinary temperatures, boron is inert to all chemical reagents except very strong oxidizing agents such as fluorine and concentrated nitric acid. (Compare this with the behavior of aluminum, the element that follows boron in Group III, Section 26.14). At high temperatures, boron is much more reactive, combining with oxygen to give B_2O_3, with all the halogens to form the trihalides (BX_3), and with many metals to yield hard high-melting compounds known as borides. The element also reacts with molten alkali or alkaline earth hydroxides with the evolution of hydrogen and the formation of orthoborates.

$$2B(s) + 6NaOH(l) \longrightarrow 2Na_3BO_3(s) + 3H_2(g)$$
sodium
orthoborate

27.3 Silicon and germanium

Silicon is the second most abundant element in the Earth's crust (28.15%; oxygen is first), while germanium is among the less abundant elements ($1.5 \times 10^{-4}\%$). Silicon is not found naturally as the free element, but it is everywhere in the form of **silica**, SiO_2, or **silicates**, compounds containing silicon–oxygen groups and metals. Roughly 85% of the Earth's crust is composed of silica and silicate minerals (described further in Sections 27.13 and 27.14). Elemental silicon of about 98% purity is obtained by reduction of sand, which is largely SiO_2, with coke in an electric furnace

$$SiO_2(s) + 2C(s) \xrightarrow{3000°C} Si(l) + 2CO(g)$$

Silicon made in this way is used in alloys, for example, in spring steel (22% silicon), in corrosion-resistant iron alloy (about 15%), and in aluminum alloys for fine casting (about 17%). It is also the starting material in the manufacture of **silicones**—polymers composed of silicon, carbon, hydrogen, and oxygen (Section 27.16). The preparation of the ultrapure silicon and germanium needed in semiconductor devices is discussed in Section 27.7.

Germanium is found mainly as a sulfide in association with other metal sulfides, for example, those of lead and zinc. It is extracted from the ore with hydrochloric acid, and the resulting germanium tetrachloride, $GeCl_4$, is purified by fractional distillation. Hydrolysis and then heating yields germanium dioxide, GeO_2, which is reduced by carbon or hydrogen to yield the free element. Germanium is also recovered from coal ashes and flue dust.

Most germanium is used in semiconductors. Other specialty applications include infrared-transmitting glass, low-melting gold–germanium alloys (sometimes used in dental work), and red-fluorescing phosphors.

Silicon and germanium have chemical properties intermediate between those of carbon and of tin and lead, but they resemble carbon more than they do the metals. Both silicon and germanium crystals have the diamond structure (see Figure 23.1). At room temperature, both elements are rather inert toward all elemental substances except fluorine, with which they react vigorously to form tetrafluorides. At higher temperatures, they combine with all the halogens to give tetrahalides. Both elements burn in oxygen to give dioxides that, unlike carbon dioxide, are not simple molecular compounds, but have distinctive crystal structures. Like carbon, silicon and germanium do not liberate hydrogen from acids. However, in contrast to carbon but like tin and lead, both elements react with solutions of alkali metal hydroxides to give hydrogen.

$$Si(s) + 4OH^- \longrightarrow SiO_4^{4-} + 2H_2(g)$$

Germanium, but not silicon, reacts readily with concentrated nitric acid to form the dioxide (GeO_2), behavior analogous to that of tin.

Silicon—the element
Silica—SiO_2
Silicates—compounds of Si, O, and metals
Silicones—C, H, O, Si polymers

27.4 Arsenic and antimony

Arsenic and antimony are not abundant elements (As, $1.8 \times 10^{-4}\%$; Sb, $2 \times 10^{-5}\%$). Their principal minerals are sulfides, the common ones being *realgar*, As_4S_4; *orpiment*, As_2S_3; *enargite*, $CuAs_2S_6$; *arseno-pyrite*, FeAsS; and *stibnite*, Sb_2S_3. The elements, like their Group V neighbors bismuth and nitrogen, but unlike phosphorus, are also found in the free state in nature.

Arsenic and antimony can be prepared from sulfide ores by a two-step process. In the first step, the sulfide is converted to the corresponding oxide by roasting. The oxide is then reduced by heating with coke. For example,

$$2Sb_2S_3(s) + 9O_2(g) \xrightarrow{\Delta} Sb_4O_6(s) + 6SO_2(g)$$

$$Sb_4O_6(s) + 6C(s) \xrightarrow{\Delta} 4Sb(l) + 6CO(g)$$

Arsenic (III) oxide, As_4O_6, a compound analogous to P_4O_6, is recovered from the flue dusts of copper and lead smelters.

An alternate method for converting stibnite to elemental antimony utilizes the direct reduction of the mineral by iron at high temperatures.

$$Sb_2S_3(s) + 3Fe(s) \xrightarrow{\Delta} 2Sb(l) + 3FeS(l)$$

The antimony and iron(II) sulfide form two liquid layers that are easily separated.

From arsenopyrite, arsenic is obtained simply by heating the mineral with the exclusion of air. Under these conditions the arsenic sublimes (at 610°C).

$$FeAsS(s) \xrightarrow{\Delta} FeS(s) + As(g)$$

Elemental arsenic and antimony each have two allotropic modifications. The common semiconducting forms, often referred to as the metallic forms, are gray, lustrous, and crystalline, while the much less stable nonmetallic allotropes are yellow. Yellow arsenic, which is obtained on the rapid cooling of arsenic vapor, is very unstable and reverts quickly to the semiconducting form. Like white phosphorus, yellow arsenic is soluble in carbon disulfide. In solution in this solvent it exists as tetrahedral molecules, As_4.

Arsenic and antimony are both of use in hardening lead alloys destined for lead shot, bullets, bearings, battery grids, and cable sheathings. Antimony has been a component of type metal since the fifteenth century. Because it expands on solidifying, antimony helps to produce type that has sharp edges and prints clear images.

Yellow arsenic and arsenic compounds are highly poisonous, leading quickly in high doses to convulsions and death. Chronic arsenic poisoning causes fatigue and a battery of unpleasant effects, including hair loss, visual disturbances and blindness, a garlic odor on the breath, paralysis, and anemia. By taking small doses of arsenic over an extended period, tolerance to arsenic can be built up. "Arsenic eating" was thought to increase the vigor of persons living in high altitudes. The "arsenic" of detective stories is the oxide, As_4O_6, which is soluble, odorless, tasteless, and fatal in doses of 0.1 g or more.

Antimony compounds are much less toxic than arsenic compounds, but more so than bismuth compounds. Antimony compounds played a role in the early history of medicine, sometimes with disastrous results, as arsenic and antimony compounds were often confused with each other.

In their general chemical behavior, the common forms of arsenic and antimony resemble to some extent both phosphorus and bismuth (in reactivity, they are more similar to red phosphorus than to the white form). Arsenic and antimony are inert toward oxygen and most other elements at ordinary temperatures. When heated, they burn in air to give As_4O_6 and Sb_4O_6. With the more active metals they give, respectively, arsenides and stibnides, for example, Na_3As, Na_3Sb. The elements are unaffected by such nonoxidizing acids as hydrochloric acid but are attacked by such oxidizing acids as nitric and concentrated sulfuric. For example, hot dilute nitric acid oxidizes arsenic and antimony to the +3 state, as it does bismuth. However, bismuth forms the hydrated Bi^{3+} ion, while arsenic and antimony yield H_3AsO_3 and Sb_4O_6, respectively. Hot concentrated nitric acid converts arsenic and antimony to the +5 state (to H_3AsO_4 and Sb_2O_5).

27.5 Selenium and tellurium

Selenium and tellurium are among the ten to fifteen rarest elements (Se, $5 \times 10^{-6}\%$; Te, $\sim 2 \times 10^{-7}\%$). No ores are mined specifically for their selenium or tellurium content. These elements generally occur as selenides [e.g., PbSe, $(AgCu)_2Se$, Ag_2Se] and tellurides [e.g., PbTe, Cu_4Te_3, Ag_3AuTe_2] associated with sulfide ores of such heavy metals as copper and lead and are obtained as by-products of the metallurgy of the heavy metals. For example, when a metal sulfide is heated in air, any selenides present are converted to selenium dioxide, SeO_2, which is collected as a dust in the flues leading from the chambers in which the sulfide is roasted. The selenium dioxide is then dissolved in water and reduced by sulfur dioxide to give brick-red elemental selenium. Tellurium is also readily obtained in the free state by reduction of its compounds with sulfur dioxide in acidic aqueous solution. The principal commercial source of both selenium and tellurium is the anode mud formed in electrolytic copper refining.

Selenium, like sulfur, is known in several allotropic forms, including two crystalline forms. One is red selenium (m.p. 170–180°C), which is soluble in carbon disulfide and is an insulator. It is generally obtained when selenium compounds are reduced to the free element in solution. The selenium molecule is octaatomic and has the same puckered ring structure as the S_8 molecule. When red selenium is heated below its melting point, it changes to the more stable gray semiconducting form (m.p. 217.4°C), called metallic se-

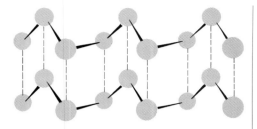

FIGURE 27.2

Elemental tellurium. *Gray selenium has the same structure.*

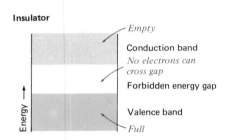

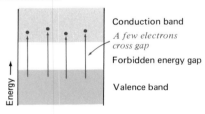

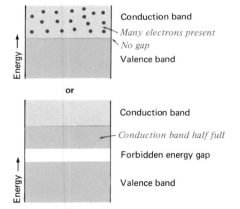

FIGURE 27.3

Energy bands in insulators, semiconductors, and metals.

lenium. This substance consists of spiral chains of selenium atoms held by covalent bonds; the chains are parallel to each other in the crystal (see Figure 27.2). Gray selenium is most unusual in that its electrical conductivity is roughly proportional to the square root of the intensity of the light striking it. This is not a photoelectric effect, but is related to the semiconducting properties of the selenium. At one time, this property had wide application in photoconductive devices for regulating current, but selenium photocells have been replaced by other devices, for example, the silicon solar cell (Section 27.8).

Crystalline tellurium, unlike sulfur and selenium, does not exhibit allotropy. It is a silvery white solid that has the same crystal structure as gray selenium (Figure 27.2). The electrical conductivity of tellurium is little affected by light.

Selenium has a variety of uses in alloys. It is a decolorizer for glass at low concentrations, is a coloring agent for ruby glass at high concentrations, and is a vulcanizing agent for rubber. Tellurium is also a component of many different alloys, including various steels, malleable cast iron, and copper, lead, and tin alloys used in automotive bearings.

Semiconductivity

27.6 Bonding and semiconductivity

For an explanation of the electrical properties of the semiconducting elements we must go back to the band picture of metallic bonding (Section 21.13). Recall that by combination of many atomic orbitals, energy bands are formed. The highest band fully occupied by electrons is the valence band, and the band above the valence band is called the conduction band. In a metal the conduction band is either half-filled, or is empty but right next to the valence band, from which electrons readily enter it. In either case, under the influence of an electrical field, current can flow as electrons move freely in the conduction band.

An energy gap in which no electrons reside (because these energy levels are forbidden) can lie between the valence band and the conduction band. In an **insulator**—a substance that does not conduct electricity—the energy gap is so large that electrons from the valence band cannot cross it (Figure 27.3). And since the valence band is full, no conduction occurs in an insulator, because no net flow of electrons is possible.

In a semiconducting element, an energy gap between the valence band and the conduction band is also present, but it is smaller than in an insulator. Even at room temperature, a few electrons have enough energy to jump the gap and enter the conduction band, where they are free to move. Some energy gaps are given in Table 27.2.

Semiconductors are not as conductive as metals, because fewer electrons are available. Put another way, we say that semiconductors are more resistive to the passage of electrical current. Electrical resistance is measured in ohms. The *resistivity* of a substance is the electrical resistance per centimeter of a conductor of 1 cm² cross-sectional area and has units of ohm-centimeters (ohm-cm). *Conductivity* is the reciprocal of resistivity and has units of ohm⁻¹-cm⁻¹.

Aluminum, a typical metal, has an electrical resistivity of 2.7×10^{-6} ohm-cm at 20°C. Pure silicon has a resistivity of 10^5 ohm-cm, while pure diamond, an insulator, is highly resistive—10^{14} ohm-cm at 15°C. The resistivity range for semiconductors is roughly 10^{-2} to 10^8 ohm-cm.

An increase in temperature causes the ions in a metallic crystal lattice to vibrate more within their lattice positions. This increases the chances for an electron moving through the metal under the influence of an electric field to collide with the ions. For a metal, the net result of an increase in temperature is an increase in resistivity (Figure 27.4).

Now we come to the essential difference between metals and semiconductors. With an increase in temperature, more electrons gain the energy needed to jump out of the valence band and into the conduction band. Therefore, with rising temperature, the resistivity of a semiconductor decreases (Figure 27.4). The amount of decrease is different for each semiconductor. At low enough temperatures, the conductivity of semiconductors is the same as that of insulators, and at high enough temperatures it is like that of metals.

The conductance of a semiconductor is aided both by the free flow of electrons in the conduction band and by what is thought of as the migration of holes in the valence band in a direction opposite to that of the electron flow. Understanding this may take a moment's thought. Look at it this way. An electron jumping out of the valence band leaves behind a hole, in the same way as a person getting up from his seat in a theater leaves a vacant seat. Suppose that the empty seat is on the end of a row. The person in the second seat can move over to the end seat. Then the person in the third seat from the end can decide to move into the second seat. If everyone in the row moves over by one seat the vacant seat—the hole—will have moved across the entire row. Under the influence of an electric field, holes move in this way through the valence band. The net effect of the motion of the holes is that of positive charges moving in the opposite direction from the conduction electrons. Note that the holes and the electrons need not move at the same rate.

In a pure semiconductor at room temperature, the number of electrons in the conduction band and the number of holes in the valence band are equal. An **intrinsic semiconductor** contains equal numbers of current-carrying holes and electrons; the conduction is an intrinsic property of the material. At high enough temperatures, even insulators such as diamond can become intrinsic semiconductors.

In an **extrinsic semiconductor** the number of current-carrying

TABLE 27.2

Energy gaps

The energy gap for diamond is 120 kcal/mole.

	Energy gap (kcal/mole)
Semiconducting elements	
B	76
Si	25
Ge	16
As (gray)	28
β-Sb	2.5
Te	8.8
Some semiconducting compounds	
InP	30
GaAs	33
InSb	4.9

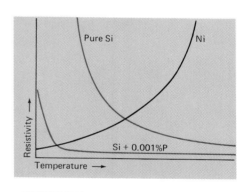

FIGURE 27.4

Variation of resistivity with temperature. *Note the large change in resistivity with the addition of only 0.001% of an impurity.*

n-type semiconductor—
    more electrons (negative)
    than holes
p-type semiconductor—
    more holes (positive)
    than electrons

n-type

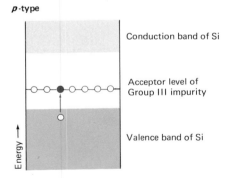

FIGURE 27.5
n- and p-type semiconductors.

holes and electrons is not equal, and conduction depends on extrinsic materials—on the addition of appropriate impurities. The impurities are of two types—donors and acceptors. A donor provides electrons in the following way. In pure silicon and germanium, each atom is joined to four neighboring atoms by covalent bonds. An atom of a Representative Group V element—for example, phosphorus, arsenic, or antimony—enters the germanium or silicon crystal lattice without distorting the lattice very much and bonds to its four germanium atom neighbors. Since the atoms of the Group V element have the ns^2np^3 configuration, one electron is left over for every atom. These extra electrons enter an occupied valence band, called a **donor level,** that usually lies slightly below the conduction band of the host semiconductor (Figure 27.5). Electrons in the donor level are easily promoted to the host semiconductor conduction band. This greatly increases the conductivity.

In a semiconductor, a **donor impurity** contributes electrons to the conduction band and it does so without leaving holes in the valence band.

The addition of controlled amounts of impurities to semiconducting elements is called **doping.** The impurity is usually added at concentrations of 100 to 1000 parts per million. A crystal doped with a donor impurity is called an **n-type semiconductor**—negative electrons are the majority of the current carriers.

An atom of a Representative Group III element—for example, boron, aluminum, or indium—has the ns^2np^1 configuration, and therefore when added to silicon or germanium can bond to only three of its neighboring atoms. This situation contributes an electron deficiency, in other words, a hole. An **acceptor impurity** contributes holes to a vacant **acceptor level** of energy, which appears slightly above the valence band (Figure 27.5). Electrons are easily promoted from the valence band to the acceptor level. In such a semiconductor, called **p-type semiconductor,** positive holes are the majority of the current carriers.

27.7 Purification of silicon and germanium

Silicon and germanium are the elements most extensively used as host semiconductors. As starting materials for semiconductor device manufacture, the elements must contain less than 1 part per *billion* of impurities. This is comparable to one pinch of salt in 10 tons of potato chips. Conversion of the elements as they are first isolated to materials of such purity requires exacting procedures.

The 98% pure metallurgical silicon (Section 27.3) is purified by a series of chemical and physical processes such as the following. The crude silicon is converted to the chloride, $SiCl_4$, by reaction with elemental chlorine. The chloride is purified by repeated fractional distillation and is then reduced to elemental silicon by means of pure magnesium or zinc. In the process, the reducing agent is converted to its chloride ($MgCl_2$ or $ZnCl_2$), which is sublimed from the silicon. For further purification, the iodide, SiI_4, is prepared from the silicon and then decomposed by heat. The silicon is then formed into ingots for the remaining steps in the purification process. Elemental germanium ingots ready for further purification are obtained directly from the preparation of the free metal described in Section 27.3.

Zone refining, or zone melting, of the ingots is the next step. **Zone refining** is a method for purifying solids in which a melted zone carries impurities out of the solid. A rod of the solid is heated at a point near one end until it melts. The melted zone is moved slowly along the rod (either the heater or the rod is moved). Because dissolved impurities lower the melting point of the solid, the melt is always less pure than the solid. Impurities therefore collect in the molten zone and eventually are concentrated at the end of the rod.

Not only must materials for semiconducting devices be ultrapure, but they must also be free of crystalline imperfection. Boundaries between microcrystals would act as barriers to the desired free flow of electrons and holes. A semiconducting device is usually made from a piece of a single crystal. Among the many techniques of growing single crystals, one of the most successful is the "pulling" of a large crystal from a melt of the same composition as the crystal. A small seed crystal is touched to the melt and gradually withdrawn at such a rate that the melt solidifies slowly onto the seed crystal (Figure 27.6). Single crystals of germanium 5 cm in diameter and 25 cm long and even larger silicon crystals are made by this technique.

27.8 Semiconductor devices

Fabrication of semiconductors into transistors and rectifiers is the basis for the solid-state electronics industry. The semiconductors are either doped silicon or germanium, or compound semiconductors, many of them compounds between a Group III and a Group V element (so-called III–V compounds), such as gallium arsenide, aluminum arsenide, or indium antimonide (see Table 27.2). The Group III elements contribute three electrons per atom, and the Group V elements contribute five electrons per atom, giving a four electron per atom average. These compounds form crystals with a diamondlike structure similar to that of silicon and germanium, and can be doped with the same results as for the elemental semiconductors.

Transistors are used to amplify and control electric current, and **rectifiers,** or diodes, to convert alternating to direct current. Each of these solid-state electronic devices makes use of both n- and p-type semiconductors. For example, in a rectifier, an n-type of semiconductor crystal is joined to a p-type semiconductor and the device is said to have an n–p **junction.** Often the n–p junction is created by different doping of adjacent areas of the same crystal. Metal electrodes are attached to the ends of the semiconducting material and an alternating current is passed through. During part of the current cycle, electrons from the n-type (electron-rich) part of the semiconductor are attracted to one electrode and the positive holes of the p-type part are attracted to the other electrode, leaving the junction region bare of electrical carriers and thus effectively stopping current flow.

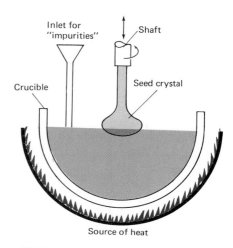

FIGURE 27.6
Apparatus for pulling a single crystal.

FIGURE 27.7

A _p–n_ junction. _Current can flow in only one direction (right to left as drawn here)._

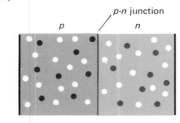

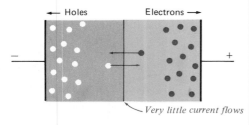

← _Very little current flows_

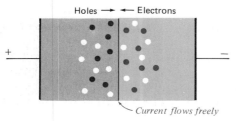

↳ _Current flows freely_

TABLE 27.3

Some compounds of boron

B₂H₆	Diborane (m.p. −165.5°C, b.p. −92.5°C)
H₃BO₃	Orthoboric acid (at 169°C ⟶ HBO₂)
HBO₂	Metaboric acid (m.p. 236°C)
B₂O₃	Boric oxide (m.p. 460°C, b.p. ~1860°C)
BCl₃	Boron trichloride (m.p. −107°C, b.p. 13°C)
B₄C	Boron carbide (m.p. 2350°C, b.p. >3500°C; very hard)
BN	Boron nitride (~3000°C, sublimes; very hard)
Na₂B₄O₇·10H₂O	Sodium tetraborate decahydrate (borax) (m.p. 320°C)

During the other part of the current cycle, when the polarity of the electrodes is reversed, electrons are repelled from the pole to which they were previously attracted and are now pulled to the other electrode, while the positive "holes" are attracted to the electrode from which the electrons are repelled. Now the two processes reinforce each other with respect to flow of current. The rectifier thus prevents flow of current in one direction and offers a low resistance to flow in the other (Figure 27.7).

Transistors, resistors, and capacitors can be created by proper combination of _p_-type and _n_-type semiconductors. The latest technical advance to result from this capability is the **integrated circuit**—several elements of an electrical circuit combined in the same piece of silicon.

To choose just one more example from the array of devices made possible by semiconductors, consider solar cells, which convert sunlight to electricity. In a typical solar cell, a disk or wafer of silicon impregnated with a Group V impurity such as arsenic (_n_-type semiconductor) is coated with a transparent layer of silicon containing a Group III element such as boron as an impurity (_p_-type semiconductor). The wafer is made part of an electrical circuit. When sunlight strikes the silicon wafer, electrons are expelled from the electron-rich _n_-type region across the _n–p_ junction into the holes of the _p_-type conductor and pass into the electrical circuit. Solar cells provide power for space vehicles.

Compounds of the semiconducting elements

27.9 Compounds of boron

With the exception of the boron hydrides, the binary compounds of the semiconducting elements with hydrogen have formulas and properties that can be predicted from periodic relationships. The boron hydrides are not predictable—they are unique compounds.

From the $2s^2 2p^1$ configuration of a boron atom, a hydride of the composition BH_3 would be expected. However, the simplest of the boron hydrides is diborane, B_2H_6 (Table 27.3). Diborane is the first member of a series of **boranes**—boron-hydrogen compounds—among them B_4H_{10}, B_5H_9, B_5H_{11}, $B_{10}H_{14}$, and fifteen or so others that have been prepared so far.

The boron hydrides are generally volatile substances (B_2H_6, b.p. −92.5°C; B_4H_{10}, b.p. 17.6°C; B_5H_{11}, b.p. 65°C) and highly reactive. They decompose when heated in the absence of air to give boron and hydrogen, many of them are spontaneously flammable, and they are decomposed by water to yield hydrogen and boric acid (H_3BO_3).

The nature of the bonding in the boron hydrides has been a

matter of great interest. Although the boranes are covalent compounds, they are **electron-deficient compounds**—they possess too few valence electrons for the atoms to be held together by ordinary covalent bonds. In diborane, for example, there are twelve valence electrons, three from each of the two boron atoms and one from each of the six hydrogen atoms. For each boron atom to be joined covalently to three hydrogen atoms and to the other boron atom by covalent single bonds would require fourteen electrons. Therefore, the configuration

not a possible configuration

which is found in the organic compound ethane, C_2H_6, is impossible in diborane.

The experimentally determined structure of diborane is shown in Figure 27.8. The boron atoms and the four terminal hydrogen atoms lie in one plane, whereas the two remaining hydrogen atoms lie above and below the plane and connect the two boron atoms. The four terminal hydrogen atoms are bonded to the boron atoms by ordinary covalent single bonds. Each "bridging" hydrogen atom is bonded to the two boron atoms by a **three-center bond**—a bond in which a single pair of electrons bonds three atoms covalently.

The three-center bond is considered to be the result of the overlap of three atomic orbitals to form one bonding molecular orbital and several antibonding molecular orbitals. Boron forms both B—H—B and B—B—B three-center bonds (Figure 27.9), and both types are present in the allotropic forms of elemental boron and in the boranes. Pentaborane (Figure 27.10), for example, has four B—H—B bonds and one B—B—B bond in addition to two conventional B—B bonds.

Orthoboric acid, H_3BO_3, more commonly just called boric acid, is a white, crystalline substance usually made by acidification of a cold

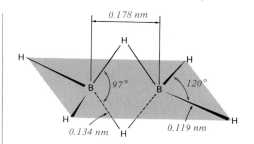

FIGURE 27.8
Structure of diborane.

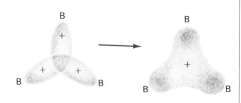

B—B—B three-center bond

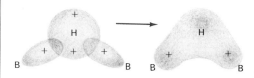

B—H—B three-center bond

FIGURE 27.9
Boron three-center bonds.

$B_{10}H_{14}$ decaborane

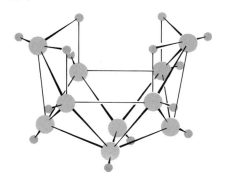

B_5H_9 pentaborane

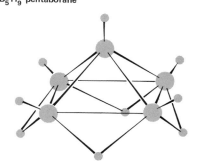

FIGURE 27.10
Structures of a few boranes.

B_4H_{10} tetraborane

aqueous solution of borax. Although it is a triprotic acid, in aqueous solution it behaves as a weak monoprotic acid.

$$H_3BO_3(aq) \xrightleftharpoons{H_2O} H^+ + \quad H_2BO_3^- \qquad K_{a_1} = 6.0 \times 10^{-10}$$
orthoboric acid   *orthoborate ion*

A boric acid solution is a mild antiseptic, often used as an eyewash.

Carefully controlled heating of boric acid yields <u>metaboric acid</u>, HBO_2, an intermediate dehydration product.

$$H_3BO_3(s) \xrightarrow{169°C} HBO_2(s) + H_2O(g)$$
     metaboric acid

Boric acids more complicated than metaboric and boric acid have not been isolated. However, it is evident from the naturally occurring boron compounds that salts more complex in character than the acids are readily formed.

Neutralization of solutions of orthoboric acid yields salts containing the <u>tetraborate ion</u>, $B_4O_7{}^{2-}$,

$$4H_3BO_3(aq) + 2OH^- \longrightarrow \quad B_4O_7{}^{2-} \quad + 7H_2O(l)$$
     tetraborate ion

Many of the boron minerals are tetraborates. In water, tetraborates give alkaline solutions as a result of hydrolysis. This reaction makes borax a useful cleansing agent and water softener. Borax is also necessary in the manufacture of Pyrex glass (Section 27.15a).

Only a few simple salts containing the orthoborate anion, $BO_3{}^{3-}$, are known. Most borates contain planar BO_3 units or tetrahedral BO_4 units in complex polymeric structures.

As would be expected for molecules with three shared pairs and no unshared pairs of electrons, the <u>boron halides</u> are planar.

$$\ddot{\underset{\displaystyle \ddot{B}}{:\ddot{X}:}}$$
$$:\!\ddot{X}\quad\ddot{X}\!:$$

Because the boron halides have only six electrons in the valence shell, they can function as electron-pair acceptors (Lewis acids). Boron trifluoride and trichloride are particularly strong Lewis acids. Reactions of the trifluoride with fluoride ion and with ammonia illustrate its ability to act as an electron-pair acceptor

$$:\!\ddot{F}\!: \atop \ddot{F}\!:\!\ddot{B}\!:\!\ddot{F}: \quad + :\!\ddot{F}\!:^- \longrightarrow \left[:\!\ddot{F}\!:\!\ddot{B}\!:\!\ddot{F}\!: \atop :\!\ddot{F}\!: \right]^-$$

tetrafluoroborate ion

$$:\!\ddot{F}\!: \atop \ddot{F}\!:\!\ddot{B}\!:\!\ddot{F}\!: \quad + \quad \overset{H}{\underset{H}{N}}\!:\!H \longrightarrow :\!\ddot{F}\!:\!\ddot{B}\!:\!\ddot{N}\!:\!H \atop :\!\ddot{F}\!:\!H$$

In each case, the acceptance of an electron pair by the boron atom results in a tetrahedral arrangement of the groups surrounding it [four shared pairs and no unshared pairs of electrons (Section 9.26)].

Boron chloride, bromide, and iodide are irreversibly decomposed by water to orthoboric acid and the hydrohalic acid.

$$BX_3 + 3H_2O(l) \longrightarrow H_3BO_3(s) + 3H^+ + 3X^-$$

The trifluoride hydrolyzes to orthoboric and tetrafluoroboric acids.

$$4BF_3(g) + 3H_2O(l) \longrightarrow H_3BO_3(s) + 3HBF_4(aq)$$

27.10 Compounds of silicon and germanium

Silicon and germanium form both monoxides and dioxides, but only the dioxides are stable. The natural and man-made silicon–oxygen compounds are so numerous and important that they are discussed separately (Sections 27.13 to 27.15). Germanium dioxide, GeO_2, the major commercially available germanium compound, is a solid that exists in two crystalline forms, one more inert and less soluble than the other (Table 27.4).

Silicon and germanium, like carbon, the nonmetallic member of Group IV, can form four covalent single bonds using the four ns^2np^2 valence electrons in each atom. Atoms of these elements do not self-link to the great extent that carbon atoms do, so they form a smaller number of hydrides than the huge number available for carbon—the hydrocarbons. Moreover, silicon and germanium atoms have never been found to self-link through multiple bonds to yield compounds like the unsaturated hydrocarbons.

Silicon–hydrogen compounds are called **silanes.** The parent compound is monosilane, SiH_4, and other members of the series that have been definitely characterized include Si_2H_6, Si_3H_8, and Si_4H_{10}. The general formula for the series is Si_nH_{2n+2}, with compounds known up to $n = 10$. The higher silanes are unstable and decompose to lower silanes and hydrogen. Most silanes are quite flammable and must be handled in the absence of air. For example, silane itself is a gas and is spontaneously flammable in air

$$\underset{silane}{SiH_4(g)} + 2O_2(g) \longrightarrow SiO_2(s) + 2H_2O(l)$$

and violently decomposed by water.

$$SiH_4(g) + (2 + x)H_2O(l) \longrightarrow SiO_2 \cdot xH_2O(aq) + 4H_2(g)$$

The germanium hydrides, Ge_nH_{2n+2}, called **germanes,** are known up to about $n = 10$ also. The germanes are generally less flammable and less easily hydrolyzed than the silanes.

The tetrahalides of silicon and germanium are volatile substances; their boiling points increase with the atomic weight of the halogen in each series. Silicon also forms a number of halides of the

TABLE 27.4

Some compounds of silicon and germanium

SiH_4	**Silane** (m.p. $-185°C$, b.p. $-112°C$)
SiO_2	**Silica (silicon dioxide)** (quartz, m.p. $1610°C$, b.p. $2230°C$)
SiF_4	**Silicon tetrafluoride** (m.p. $-90.2°C$, b.p. $-86°C$)
$SiCl_4$	**Silicon tetrachloride** (m.p. $-70°C$, b.p. $57.6°C$)
Si_2Cl_6	**Disilicon hexachloride** (m.p. $-1°C$, b.p. $145°C$)
GeH_4	**Germane** (m.p. $-165°C$, b.p. $-89°C$)
GeO_2	**Germanium dioxide** (insoluble, m.p. $1087°C$; soluble, m.p. $1115°C$)
$GeCl_4$	**Germanium tetrachloride** (m.p. $-49.5°C$, b.p. $84°C$)
Si_2H_6	**Disilane** (m.p. $-132.5°C$, b.p. $-14.5°C$)
Si_3H_8	**Trisilane** (m.p. $-117.4°C$, b.p. $52.9°C$)

TABLE 27.5

Some compounds of arsenic and antimony

AsH_3	Arsine (m.p. $-116°C$, b.p. $-55°C$)
As_4O_6	Arsenic(III) oxide (sublimes at $193°C$)
As_2O_5	Arsenic(V) oxide (m.p. $315°C$, dec, amorphous)
$AsCl_3$	Arsenic(III) chloride (m.p. $-8.5°C$, b.p. $63°C$)
$AsOCl$	Arsenic(III) oxychloride (dec)
SbH_3	Stibine (m.p. $-88°C$, b.p. $-17°C$)
Sb_4O_6	Antimony(III) oxide (m.p. $656°C$, b.p. $1550°C$)
Sb_2O_5	Antimony(V) oxide (loses O_2 at $930°C$)
$SbCl_5$	Antimony(V) chloride (m.p. $2.8°C$, b.p. $79°C$)
$SbOCl$	Antimony(III) oxychloride (m.p. $170°C$, dec)
$NaSb(OH)_6$	Sodium hydroxo- antimonate(V) (difficultly soluble in water)

[27.10] SEMICONDUCTING
ELEMENTS

composition Si_nX_{2n+2} in which there are silicon–silicon bonds. (Compare with the silicon hydrides.) For example, molecular species such as Si_2Cl_6, Si_2Br_6, Si_2I_6, Si_5Cl_{12}, and $Si_{10}Cl_{22}$ have been characterized. All of the tetrahalides are readily hydrolyzed in reactions quite similar to those observed for the boron trihalides.

$$SiX_4 + 2H_2O(l) \longrightarrow SiO_2(s) \text{ (hydrated)} + 4H^+ + 4X^- \quad (X = Cl, Br, I)$$

27.11 Compounds of arsenic and antimony

Arsine, AsH_3, and stibine, SbH_3, are unstable gases that are easily decomposed into their elements by heat. They are much less thermally stable than ammonia and phosphine, the hydrides of the nonmetallic members of the family. Neither arsine nor stibine appears to be significantly basic, and cations analogous to the ammonium and phosphonium ions are not known. Arsine and stibine are both very poisonous.

Arsenic and antimony form oxides in both the $+3$ and $+5$ oxidation states (Table 27.5). Burning the free elements in air gives the $+3$ oxides. In contrast to phosphorus, arsenic and antimony do not yield $+5$ oxides by direct union with oxygen. The oxides of the $+3$ state are assigned the formulas As_4O_6 and Sb_4O_6, since it has been shown that such molecules are present in the solids. No structural information is available for oxides corresponding to the $+5$ state. These are given the formulas As_2O_5 and Sb_2O_5.

Arsenic(III) oxide, As_4O_6, is amphoteric, with the acidic character predominant. Aqueous solutions of the oxide are definitely acidic and contain arsenous acid, H_3AsO_3 (an arbitrarily assigned formula). However, the acid undergoes both acidic and basic dissociation.

$$H_3AsO_3 \rightleftharpoons H^+ + H_2AsO_3^- \qquad K_{a_1} = 5.08 \times 10^{-10}$$
$$H_3AsO_3 \rightleftharpoons AsO^+ + OH^- + H_2O(l) \qquad K_{b_1} = 5 \times 10^{-15}$$

Antimony(III) oxide, Sb_2O_3, is also amphoteric, with the basic character much more pronounced than in the corresponding arsenic compound. Arsenic(V) oxide, As_2O_5, and antimony(V) oxide, Sb_2O_5, are acidic oxides obtained by oxidation of the lower oxides with concentrated nitric acid. The arsenic compound dissolves in water to give a solution from which orthoarsenic acid, H_3AsO_4, can be obtained. This acid, like orthophosphoric acid, is decomposed by heat to give pyro ($H_2As_2O_7$) and meta ($HAsO_3$) acids. Salts of these acids, the arsenates, are similar to the corresponding phosphates. Antimony(V) oxide is relatively insoluble in water, and no free antimonic acid has been prepared. However, the oxide reacts with alkaline substances to give salts known as antimonates which contain the octahedral $Sb(OH)_6^-$ ion. [Note the difference in formulas of oxyanions corresponding to the $+5$ state: AsO_4^{3-} and $Sb(OH)_6^-$. This difference is a reflection of the difference in size of the arsenic and antimony atoms; the latter, being larger, can accommodate more surrounding oxygen atoms without crowding.]

In acidic solution, arsenic(V) and antimony(V) are strong oxidizing agents, but, as expected not so strong as bismuth(V).

$$H_3AsO_4 + 2H^+ + 2e^- \rightleftharpoons HAsO_2 + 2H_2O \qquad E^\circ_{298} = 0.58 \text{ V}$$
$$HSb(OH)_6 + 3H^+ + 2e^- \rightleftharpoons SbO^+ + 5H_2O \qquad E^\circ_{298} = 0.58 \text{ V}$$

Arsenic and antimony form complete series of trihalides with the use of the three p electrons in the valence shells of their atoms. The trihalides are liquids or low-melting solids, and in the vapor state, at least, exist as simple molecules with trigonal pyramidal structures. They are rapidly hydrolyzed in reactions that may be reversed by addition of the hydrohalic acid

$$AsX_3(s \text{ or } l) + 3H_2O(l) \rightleftharpoons H_3AsO_3(aq) + 3H^+ + 3X^-$$

In contrast, the hydrolysis of the boron trihalides and the tetrahalides of silicon and germanium are complete and irreversible.

Only three pentahalides—AsF_5, SbF_5, and $SbCl_5$—have been prepared and characterized unequivocally. The inability of these elements to form the other possible pentahalides is probably due to their strong oxidizing power in the +5 oxidation state.

27.12 Compounds of selenium and tellurium

The semiconducting elements of Group VI, selenium and tellurium, form the hydrides hydrogen selenide, H_2Se, and hydrogen telluride, H_2Te. In their chemistry, these compounds have a close resemblance to hydrogen sulfide (Section 22.16). The compounds are gaseous, have vile odors, and are poisonous.

Hydrogen selenide and telluride are stronger reducing agents than the sulfides, with the reducing power increasing in the order H_2S, H_2Se, H_2Te, in line with the decreasing thermal stability of the hydrides.

Selenium and tellurium form dioxides and trioxides. In contrast to SO_2, which is a colorless gas at room temperature, SeO_2 and TeO_2 are white solids (Table 27.6). Selenium dioxide has a molecular structure of polymeric chains

Tellurium dioxide is known in several allotropic modifications, none of which is based upon discrete molecules.

Selenium dioxide is very soluble in water, and selenous acid, H_2SeO_3, can be isolated from the aqueous solutions. Tellurium dioxide is only slightly soluble in water, and no acid derived from the oxide is known. Both selenites, such as $NaHSeO_3$, Na_2SeO_3, and tellurites, such as Na_2TeO_3, have been prepared by reaction of the dioxides with solutions of alkali metal hydroxides.

TABLE 27.6

Some compounds of selenium and tellurium

H_2Se	Hydrogen selenide (m.p. −60°C, b.p. −41.5°C)
SeO_2	Selenium(IV) oxide (sublimes at ~345°C)
SeO_3	Selenium(VI) oxide (m.p. 118°C, b.p. 180°C, dec)
H_2SeO_3	Selenous acid (m.p. 70°C, dec)
$SeBr_4$	Selenium tetrabromide (m.p. 75°C, dec)
H_2Te	Hydrogen telluride (m.p. −49°C, b.p. −2.2°C)
TeO_2	Tellurium(IV) oxide (m.p. 733°C, b.p. 1245°C)
TeO_3	Tellurium(VI) oxide (m.p. 395°C, dec)
TeI_4	Tellurium(IV) iodide (m.p. 280°C)
H_2SeO_4	Selenic acid (m.p. 58°C, strong acid)
H_6TeO_6	Telluric acid (m.p. 136°C, weak acid)

Selenic acid, H_2SeO_4, has been obtained by the oxidation of selenous acid with strong oxidizing agents. The corresponding acid of tellurium, telluric acid, H_6TeO_6, is made in a similar manner. This acid may be dehydrated to TeO_3 by heating. Both selenic and telluric acids are diprotic acids, with the former being much stronger than the latter (K_{a_1}: $H_2SeO_4 > 1$; $H_6TeO_6 \sim 10^{-7}$). Both acids are much stronger oxidizing agents than sulfuric acid. Salts corresponding to the replacement of both one and two hydrogen atoms are known, for example, $NaHSeO_4$, Na_2SeO_4, $Na(H_5TeO_6)$, $Na_2H_4TeO_6$. Many such salts are quite toxic; for example, about 5 mg of sodium selenate, Na_2SeO_4, is a fatal dose.

The difference in the formulas of selenic and telluric acids is worth noting. Here, as in the case of the arsenate and antimonate ions (see above), size factors are important. The tellurium atom is large enough to bind six —OH groups in an octahedral fashion; the smaller selenium atom can accommodate only four oxygen atoms, and selenic acid, like sulfuric acid, has a tetrahedral molecular structure.

Silicon–oxygen compounds

27.13 Natural silica

The common crystalline form of silica, SiO_2, is quartz, which is present in most igneous, sedimentary, and metamorphic rocks. Sand, flint, and agate are other familiar natural forms of silica. All of the crystalline modifications of silica are three-dimensional polymeric substances consisting of linked SiO_4 tetrahedra, with each silicon atom joined to four oxygen atoms and each oxygen atom attached to two silicon atoms. Indeed, all silicon-containing substances in the Earth's crust have the SiO_4 tetrahedron as the fundamental building unit.

At high temperatures (about 1600°C for quartz), silica melts to a viscous liquid that has a strong tendency to supercool and form a glass. Quartz glass undergoes very small changes in volume with changes in temperature and is highly transparent to both ultraviolet and visible light. Because of these properties, it is useful for chemical apparatus where large temperature changes occur and for optical instruments where transparency to ultraviolet radiation is necessary. (Ordinary glass absorbs ultraviolet light.)

Silica is a substance of considerable chemical stability [ΔH_f° (quartz) $= -217.72$; $\Delta G_f^\circ = -204.75$ kcal/mole]. It is inert toward all the halogens except fluorine and all acids but hydrofluoric.

$$SiO_2(s) + 4HF(aq \text{ or } g) \longrightarrow SiF_4(g) + 2H_2O(l)$$

At high temperatures, silica is reduced by active metals and also by carbon, with which it gives either elemental silicon or silicon carbide, SiC, a very hard, diamondlike substance known as *carborundum* and used as an abrasive.

$$SiO_2(s) + 3C(s) \longrightarrow \underset{carborundum}{SiC(s)} + 2CO(g)$$
$$\underset{silica}{}$$

Sodium hydroxide solution converts the oxide to <u>water-soluble silicates</u>.

$$x\,SiO_2(s) + 2NaOH(aq) \longrightarrow Na_2O \cdot x\,SiO_2(aq) + H_2O(l)$$

In the mixture of silicates formed, the ratio of sodium to silicon ranges from 0.5 to 4. Concentrated, syrupy solutions of the silicate mixtures are sold under the name of *water glass*. The solutions are useful as adhesive, cleansing, waterproofing, and fireproofing agents. Treatment of a solution of a soluble silicate with dilute acid yields a gelatinous material, $SiO_2 \cdot xH_2O$, called silicic acid. Although silica is an acidic oxide, no protonic acids derived from it have been definitely characterized.

A variety of anions have been detected in the soluble silicates, including SiO_4^{4-}, $Si_2O_7^{6-}$, $Si_3O_{10}^{8-}$, and others. The soluble silicates are relatively simple substances compared to most silicates.

27.14 Natural silicates

The simplest naturally occurring silicates contain SiO_4^{4-} ions and are known as orthosilicates. The gemstone zircon, $ZrSiO_4$, is an example of an orthosilicate. In SiO_4^{4-} ions, the silicon is at the center of a tetrahedron of oxygen atoms, which is often represented just by drawing the tetrahedron (Figure 27.11).

The ions of many different metals occur in natural silicates. The most common ones are Al^{3+}, Fe^{3+}, Ti^{4+}, Mg^{2+}, Ca^{2+}, Fe^{2+}, Li^+, Mn^{2+}, Na^+, and K^+. Few silicate minerals are homogeneous substances, one reason being that metal ions of appropriate sizes can substitute for each other in the crystal lattices. For example, olivine, $(Mg,Fe)_2SiO_4$, a simple orthosilicate like zircon, can be found with magnesium and iron present in all proportions.

In structure, most silicates are polymeric. They are easily classified by how the SiO_4 tetrahedra are to linked to each other. Beginning with the disilicate ion, $Si_2O_7^{6-}$, in which two tetrahedra have a common oxygen corner

and which occurs in the rare scandium mineral thortveitite, $Sc_2Si_2O_7$, they range through rings, chains, and sheets to three-dimensional

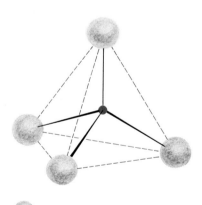

FIGURE 27.11
A silicate anion, SiO_4^{4-}.

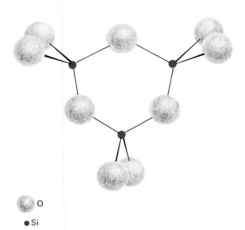

FIGURE 27.12
The $Si_3O_9^{6-}$ ion. *This ion contains three SiO_4 tetrahedra.*

FIGURE 27.13
Silicon–oxygen chain. *View (b) is 90° from view (a).*

networks like that of silica. These variations are simply represented by drawing the SiO_4 tetrahedra as in Table 27.7, which summarizes some of the varieties of silicate minerals.

Benitoite, $BaTiSi_3O_9$, contains a ring structure made up of three SiO_4 tetrahedra (Figure 27.12). Beryl, known as a gemstone, has a larger ring (Table 27.7).

Since the oxygen atoms that surround each silicon atom are arranged tetrahedrally, the SiO_4 chains are not straight, but zigzag (Figure 27.13). Although each silicon atom is attached to four oxygen atoms, the composition of the chain is expressed by the formula $(SiO_3)_n^{2n-}$, for two of the four oxygen atoms are shared with other silicon atoms.

Double silicate chains are also known (see Table 27.7). In these, half of the silicon atoms share three oxygen atoms with other silicon atoms, whereas the other half share only two. The repeating anionic silicate unit is $(Si_4O_{11})^{6-}$. Minerals containing this group, the *amphiboles*, are complex in structure, for most of them contain two or more cations, and all of them have hydroxide groups linked to the cations. The *asbestos* minerals, tremolite, $Ca_2(OH)_2Mg_5(Si_4O_{11})_2$, for example, are typical members of the amphibole family. The internal structure of these minerals is reflected in their highly fibrous nature. The strong Si—O bonds in the chains remain intact, but the weaker bonds to the ions between the chains can be broken. Asbestos is not flammable and the fibers are incorporated in many minerals, ranging from wall board to fireproof fabrics. However, airborne asbestos fiber is a cancer-causing agent and care must be exercised that the asbestos material is not worn down and released to the atmosphere.

There are many silicate minerals in which the anions are in sheets formed by extension of the double silicate chains. Each silicon atom shares three of its four oxygen atoms with adjacent silicon

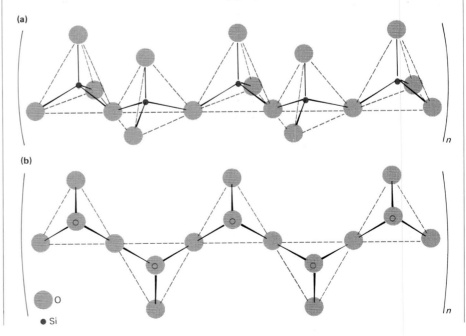

TABLE 27.7

Silicate structures[a]

Structural type	Number of vertices shared	Anion composition	General class	Anion structure	Examples
Discrete anionic groups	0	SiO_4^{4-}	**Orthosilicates**		**Be_2SiO_4** (phenacite) **$(Mg, Fe)_2SiO_4$** (olivine)
	1	$Si_2O_7^{6-}$	**Pyrosilicates**	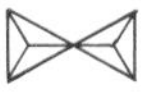	**$Sc_2Si_2O_7$** (thortveitite)
	2	$Si_3O_9^{6-}$ $Si_6O_{18}^{12-}$	**(Rings)**	 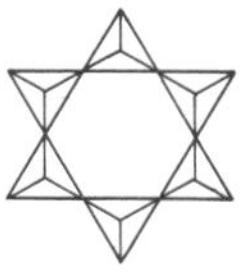	**$BaTiSi_3O_9$** (benitoite) **$Be_3Al_2Si_6O_{18}$** (beryl)
One-dimensional chains	2	$[SiO_3]_n^{2n-}$	**Pyroxenes (linear chains)**	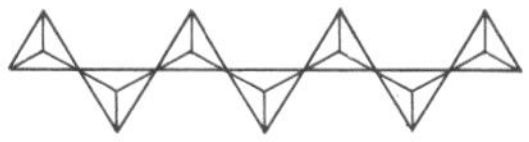	**$CaMgSi_2O_6$** (diopside)
	2	$[(Si_4O_{11})_n]^{6n-}$	**Amphiboles (double chains)**	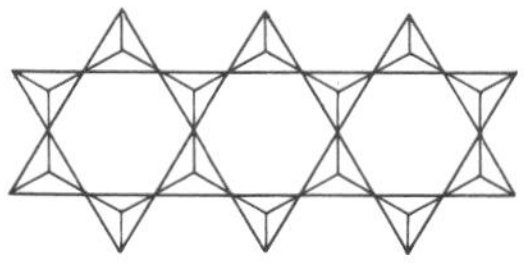	**$Ca_2(OH)_2Mg_5$ $(Si_4O_{11})_2$** (tremolite, an asbestos)
Two-dimensional sheets	3	$[Si_2O_5]^{2-}$	**Mica and talc; clays**	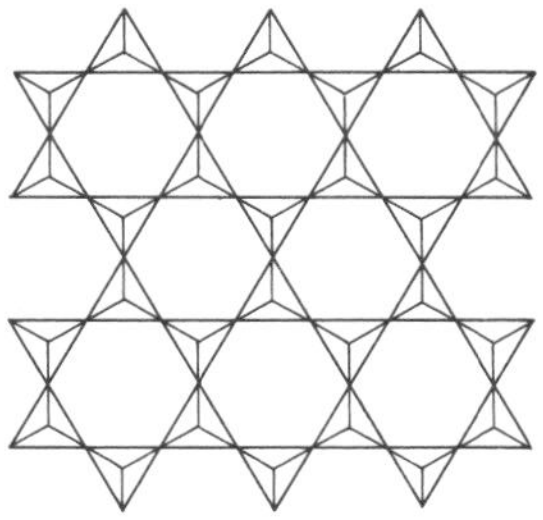	**$Mg_3[Si_4O_{10}](OH)_2$** (talc) **$Al_2[Si_4O_{10}](OH)_2$** (pyrophyllite)
Three-dimensional networks	4	SiO_2	**Silica**	—	**Quartz, tridymite, cristobalite**
	4	$[AlSi_3O_8]^-$ $[Al_2Si_2O_8]^{2-}$	**Feldspars and zeolites**	—	**$NaAlSi_3O_8$** (albite) **$CaAl_2Si_2O_8$** (anorthite)

[a] Adapted from P. Powell and P. L. Timms, *The Chemistry of the Non-Metals*, p. 115. Chapman & Hall, London, 1974.

atoms. Talc, some clay minerals, and some micas contain sheets of this sort, but the structures of these minerals are complicated by the presence of hydroxide groups, which are bonded to the silicate sheets through magnesium or aluminum ions. The bonds between the hydroxide groups and the metal ions are rather weak and are easily broken. This accounts for the ease with which these minerals can be split into thin layers.

The mica phlogopite is an *aluminosilicate,* a silicate in which aluminum atoms replace part of the silicon atoms. Many clays and feldspars are also aluminosilicates. Zeolites form a family of three-dimensional aluminosilicates with much more open structures than silica or the feldspars. Within the aluminum–silicon–oxygen framework, the zeolites contain cations such as Na^+, Ca^{2+}, or K^+ and water. The aluminum-to-silicon ratio, the amount of water, and the number and type of ions vary in different zeolites.

Because of their open structures, the zeolites have two interesting and useful properties. When heated, zeolites become dehydrated, in a change that is reversed when the zeolite is again exposed to water. The dehydrated zeolites can take up other small molecules in the open spaces left by the water molecules, a property for which zeolites are referred to as *molecular sieves.* Such small molecules as ammonia, carbon dioxide, and ethyl and methyl alcohols are reversibly absorbed by zeolites. By choice of a zeolite with the correct size of openings, the process can be made quite selective. This property is used industrially, for example, in drying gases and liquids and in separating saturated and unsaturated hydrocarbons.

The ions in zeolites can also be reversibly replaced by other ions. The best known use of this property is in water softening. A sodium zeolite will release Na^+ ions to the water and take up Ca^{2+} ions instead, thereby converting hard water to soft water. The Ca^{2+}-saturated zeolite can be regenerated for further use by washing it with a concentrated salt solution.

Zeolites that retain the useful properties of the natural substances have been synthesized.

27.15 Man-made silicate materials

a. Glass. Many of the silicate minerals, once melted, do not crystallize readily when cooled, but form hard, noncrystalline, transparent substances called glasses. (Other materials also form glasses, but the term is most often used for the silicate glass of everyday use.) Since glasses are extremely viscous, rearrangement of the structure into the more stable crystalline forms proceeds only very slowly, and may require centuries.

Common window glass is a mixture of sodium and calcium silicates. It is made by melting sodium carbonate, calcium carbonate (or oxide), and silicon dioxide together.

$$x\,Na_2CO_3(l) + x\,SiO_2(s) \longrightarrow Na_{2x}(SiO_3)_x(l) + x\,CO_2(g)$$
$$x\,CaCO_3(s) + x\,SiO_2(s) \longrightarrow Ca_x(SiO_3)_x(l) + x\,CO_2(g)$$

White sand serves as the source of the silica. Even the best grade of sand contains a small proportion of iron(III) compounds, which give it a yellow or brown shade. When this sand is made into glass, the iron is converted into a mixture of light green iron(II) silicates. The undesirable color of these compounds may be offset by the inclusion of manganese dioxide (MnO_2) in the glass mix. The manganese dioxide is reduced to form manganese(II) silicates, the pink color of which balances the green color due to the iron. This makes the glass appear to be colorless.

Countless variations in glass composition are possible. For example, the replacement of part of the sodium by potassium makes the glass harder and raises its melting point. The addition of transition metal compounds to the glass mix forms silicates that are colored—chromium(III) oxide gives a deep green glass, cobalt(II) oxide gives a blue one, and so on. The inclusion of colloidal materials may give a glass a color (Au gives ruby glass) or make it translucent or opaque (SnO_2 gives opaque glass). The addition of lead(II) oxide increases the refractive index of the glass so that it gives a play of colors when exposed to the rays of white light. Lead glass (flint glass) is used for cut-glass dishes, lenses, and artificial gems.

If part of the silicon dioxide is replaced by boric oxide in the form of borax, the resultant glass has a very low coefficient of expansion and can undergo rapid changes in temperature without breaking. Pyrex, the trade-marked glass common in laboratories, is a borosilicate glass.

Ordinary glass is only slightly soluble in water, but it is soluble enough that water that has stood in a vessel of such glass for some time gives an alkaline reaction and, upon evaporation, yields a weighable residue. Alkali and alkaline earth metal hydroxides attack glass markedly, for the polymeric silicate anions are degraded by alkalies. Acids, except hydrofluoric, are almost without action on most glasses. Hydrofluoric acid attacks glass rather rapidly and is used for etching designs on objects made of glass.

A very interesting glass, which is only slightly etched by hydrofluoric acid, contains a large proportion of aluminum phosphate, $AlPO_4$. Aluminum and phosphorus flank silicon in the periodic table, the aluminum atom having one proton and one electron less than the silicon atom and the phosphorus atom having one more of each. Aluminum phosphate is, therefore, isoelectronic with silicon dioxide. Indeed, it has the same crystal structure as quartz, with alternate silicon atoms replaced by aluminum and phosphorus atoms.

b. Ceramics. Glass is noncrystalline and has sand as its major silicate ingredient. In contrast, most of the products that we think of as ceramics—brick, tile, earthenware, pottery, and porcelain—have clay as their major ingredient and are mixtures of crystalline and glass phases. The term "ceramics" is hard to define precisely. One encyclopedia includes as ceramics all solid materials that are not organic or metallic. A common definition is materials made from clay.

Clays are formed by the weathering of aluminosilicate minerals and generally contain the original mineral and sand as impurities.

Most clays also have iron(III) oxide as an impurity, and the reddish color of many clays and ceramic products is due to this oxide. Clays of greatest importance to the ceramics industry contain kaolin group minerals [$Al_2Si_2O_5(OH)_4$, for example] as the main components.

Ceramic objects are made by shaping, drying, and then heating to high temperature plastic mixtures of clay, additives such as other silicate minerals or quartz, and water. The changes that occur on firing are complex and have not been studied in a truly scientific manner.

A vitreous coating or glaze is often put on ceramic pieces after a preliminary period of firing. The glaze material—it may be a metal oxide or a salt, or a metal silicate, or a mixture of these—is applied to the surface, and the object is reheated to a temperature where the additive either melts or reacts with the clay to form a glassy coating. The appearance and properties of ceramic ware may vary widely depending upon the type of clay employed, the nature and quantity of the additives, the nature of the glaze material, the time and temperature of firing, and the presence of an oxidizing or reducing atmosphere during firing.

c. Cement. The main raw materials of cement, a mixture often called Portland cement (because of its similarity to a stone native to the Isle of Portland in England) or hydraulic cement (because it sets even under water), are limestone ($CaCO_3$) and clay. A powdered mixture of these materials is heated in a rotary kiln to 1400–1600°C. Carbon dioxide is evolved from the limestone, and the resulting mixture sinters together into small lumps called "clinker." This material, a mixture of calcium aluminates and silicates with lime (CaO), is mixed with a little gypsum ($CaSO_4 \cdot 2H_2O$) and powdered.

When cement is mixed with water, hydrates are formed. This gives a plastic mass. When the material first hardens, within about 24 hours, it is still reasonably soft and can even be cut with a knife for a day or two. The gypsum extends the hardening time, permitting a longer period for working with the cement. The process of hardening appears to involve hydrolysis of the calcium aluminate to give calcium hydroxide and hydrated aluminum oxide, which react with the calcium silicates to form calcium aluminosilicates. Hardening continues for many years. The compounds first formed when cement sets apparently undergo slow hydration, followed by crystallization of the hydrates.

In quick-hardening cements, aluminum oxide in the form of bauxite is substituted for some of the clay. These cements set more slowly but harden much more rapidly than Portland cement. In use, cement is usually mixed with sand and gravel or crushed rock. This prevents excessive shrinkage and gives a hard, strong material known as concrete.

27.16 Silicones

The usefulness of the mineral silicates and similar manmade substances is largely attributable to their high stability to heat and their relative inertness toward the action of other chemicals—properties related to the great strength of the silicon–oxygen bond (88 kcal/mole at 298°K). In recent years, extremely useful polymeric substances that combine the —Si—O—Si— framework with hydrocarbon groups have been synthesized. Silicones are polymers of the general formula $(R_2SiO)_n$, where R is a hydrocarbon group such as —CH_3 (methyl), —C_2H_5 (ethyl), or —C_6H_5 (phenyl). The simplest silicones are linear

However, they may be cyclic or there may be cross-linking of the linear polymers

By varying the nature of the hydrocarbon groups, the degree of cross-linking, and the length of the chains, a great variety of silicone polymers is obtained.

These polymers have properties that are to a considerable extent a combination of those of hydrocarbons and of the —Si—O—Si— framework. They are resistant to heat and to attack by chemical agents. They are not susceptible to atmospheric oxidation and are not wet by water. Their inertness has made silicones valuable in surgical applications. Those silicones that are liquid are used as brake fluids and as water repellants, those that are greaselike are good lubricants, the plastic ones have special applications in place of rubber, and the cross-linked, rigid ones are employed as insulators.

THOUGHTS ON CHEMISTRY

Buck Rogers: An autobiography

[In 1929 the comic strip, Buck Rogers in the 25th Century, first appeared. We remind you that in 1929 space ships, television, tape recorders, transistor radios, and the science that has made them possible were as yet unknown.]

I entered into the life of the 25th Century with a mighty zest. On every hand were marvels almost unbelievable. Cities of towering pinnacles. Others that had been roofed over with great domes of metalloglass, *a transparent product with a strength greater than steel. And still others that were in reality one single great building, spreading for miles with mazes of thoroughfares, internal corridors and external galleries, along which shot automatically controlled* floating *cars.*

The lift in these cars was furnished by an over-balance of inertron, but the cosmomagnetic *grip of the guide rails embedded in the pavements held them down to within twelve inches of the ground. One had only to enter one of these cars, locate his destination as to avenue, cross-corridor and level on the triple dial and then relax. An amazingly complex system of car and power-house controls guided the vehicle promptly and safely by the shortest available route to the recorded destination.*

But I never ceased to wonder at the amazing number of these marvels whose real beginnings, back in the 20th Century, I could actually recall.

Radio? *It was basically and fundamentally woven into the whole fabric and structure of the 25th Century civilization. But* such *radio! Radio that embraced myriad types and varieties of* electronic, subelectronic, infra-magnetic *and* cosmic *oscillations. Matter could be formed out of force with it. And even as the 20th Century scientists conceived and executed great scientific advances, so the 25th Century scientists to an even greater extent developed new,* synthetic *elements of strange properties, not existing naturally in any part of the universe. Inertron, for instance, was one of these. It had weight; but its weight caused it to fall* up *instead of* down. . . .

The weapons and equipment of the military service were most interesting to me. Men and girls wore close-fitting uniforms of a synthetically fabricated material, *not a woven cloth, that had the consistency of soft leather and yet was most difficult to cut or tear. For service in cold climates, uniform cloth was* electronically *treated to radiate inward a continuous glow of heat, while the outside surface was heat resistant. For warm climates the cloth was given a spongy texture for aeration, and a high ratio of heat conductivity. . . .*

The telev-eye, *used either as a weapon of destruction or for scouting, was an aerial torpedo, its weight eliminated by inertron counterbalancing, radio-controlled, with a great "eye" or lens, behind which was located a television transmitter that relayed back to its operator, who was safely entrenched miles in the rear, the picture which this "eye" picked up. Once the telev-eye picked up a fugitive aircraft, that ship was doomed, for no ship could outmaneuver or outspeed these terrible projectiles of destruction, which were so small that they could seldom be hit by enemy guns. . . .*

For interplanetary travel was an accomplished fact in the 25th *Century. Even back in 1933 aviation engineers constructed a craft to fly the* stratosphere, *that upper section of the Earth's atmosphere in which the air is too rare for breathing, and from which its density declines gradually to the vacuum of interplanetary space.*

The first space ships in which we, Wilma and I, feeling infinitely less than microscopic, dared the immensity of outer void, were rocket propelled. In a vacuum, whirling fan blades are futile for propulsion, for there is nothing for the blade to pull or push against. But the rocket, so to speak, provides "air" against which to push. The blazing gas, roaring out of the rocket tube, piles up against that which was emitted the preceding instant, and has not yet had time to expand to extreme rarefaction. The reaction of this piling up shoves the ship ahead. And since there is no air friction in space to retard the ship a single impulse would give it a momentum that would continue forever, or until it was altered by some such event as entering the gravitational field of some planet, or colliding with a planetoid. Such speed, however, would be very slow, and the enormous distances to be covered in space made it imperative to attain speeds undreamed of in the antiquated days of 1933—five hundred years in the dim past.

BUCK ROGERS: AN AUTOBIOGRAPHY, by Phil Nowlan and Dick Calkins, 1932 (Quoted from *The Collected Works of Buck Rogers in the 25th Century* by permission of Robert C. Dille, Carmel, California)

Significant terms defined in this chapter

silica	doping	$n-p$ junction
silicates	n-type semiconductor	integrated circuit
silicones	acceptor impurity	boranes
insulator	acceptor level	electron-deficient compounds
intrinsic semiconductor	p-type semiconductor	three-center bond
extrinsic semiconductor	zone refining	silanes
donor level	transistor	germanes
donor impurity	rectifier	

Exercises

27.1 A physics laboratory instructor caught a chemistry major daydreaming in an electronics laboratory. The professor wanted to know what the student was thinking about instead of doing the experiment, which called for characterizing the electronic properties of a $p-n-p$ transistor. The student explained that he was thinking about everything that went into the making of the transistor such as material preparation and theory of operation. The instructor insisted that the student share his knowledge with the rest of the class and the student discussed (a) the natural sources of Ge, As, and B; (b) the winning and purification of the elements; (c) the doping processes; and (d) the theory of operation of each type of semiconducting device. Pretend that you are the student giving this short report and make some notes for your talk.

27.2 Which statements are true? Rewrite any false statement so that it is correct.
(a) The semiconducting elements generally look like metals and act like metals in most of their chemical behavior.
(b) At very low temperatures, the semiconducting elements approach the nonmetallic elements in behavior, and at high temperatures, they approach the metallic elements in behavior.
(c) The addition of controlled amounts of impurities to semiconducting elements is known as sintering.
(d) During zone refining, an ultrapure zone of liquid is formed and is then used for making electronic parts.
(e) The resistivity of a semiconducting element decreases with increasing temperature while that of metals increases.

27.3 Distinguish clearly between: (a) insulators, semi-

849

conductors, and metals; (b) intrinsic and extrinsic semi-conductors, (c) n-type and p-type semiconductors, and (d) "ane" and "ate" endings as in borane and borate or silane and silicate.

27.4 What does the term "three-center bond" mean? How is it different from our ordinary concept of covalent bonding? Would we find three-center bonds in diborane or disilane?

27.5 Write chemical equations illustrating how the halides of the semiconducting elements resemble the nonmetallic halides in their behavior toward water much more than they do the metallic halides.

27.6 Write the formulas of the oxides of the seven semiconducting elements discussed in this chapter. Write an equation for the chemical reaction each undergoes with water. Name the substance formed.

27.7 Write balanced equations for the reactions which occur as (a) boric oxide is heated with magnesium metal, (b) elemental boron is added to molten potassium hydroxide, (c) orthoboric acid is heated gently, (d) a solution of orthoboric acid is neutralized by hydroxide ion, and (e) boron trifluoride and hydrogen fluoride are brought together.

27.8 Describe how an n-type semiconductor device operates and how this differs from the operation of a p-type device. How does a rectifier work? How is a transistor different from a rectifier?

27.9 What properties does a glass have that are similar to those of a crystalline solid? What properties are common to a liquid? Describe the processes involved in making ordinary glass from Na_2CO_3, $CaCO_3$ and SiO_2. Write chemical equations for any reactions that take place.

27.10 What structural characteristic do all metal silicates have in common? What types of structural configurations are found among the natural silicates (minerals)?

27.11* In a student-designed qualitative analysis scheme, As^{3+} and Sb^{3+} are separated from other ions in the form of highly insoluble sulfide precipitates by adding either $Na_2S(aq)$ or $H_2S(aq)$ to the solution. After filtering, these two elements are separated from each other by adding excess $HCl(aq)$ to the precipitate. The basis of this separation is that the Sb_2S_3 reacts with HCl to form the soluble $SbCl_4^-$ anion, whereas the As_2S_3 does not react with the HCl and remains as a precipitate. After filtering, the presence of As^{3+} is confirmed by adding $H_2O_2(aq)$ in the presence of OH^- to the As_2S_3, which forms AsO_4^{3-}, and then by adding Ag^+ to form the red precipitate Ag_3AsO_4. The

presence of Sb^{3+} is confirmed by adding NH_3 to remove enough H^+ to allow the orange sulfide Sb_2S_3 to precipitate again. Write chemical equations describing the above reactions.

27.12 Write the electronic structure for (a) an atom of boron and (b) a molecule of B_2. Based on the electronic structure for the atom, (c) predict whether boron in the gaseous atomic state will be paramagnetic or diamagnetic. (d) What common oxidation state would we expect for this element? Based on the electronic structure for the diatomic molecule, (e) predict whether the molecule is paramagnetic or diamagnetic. (f) What type of bonding is present in the molecule? The standard enthalpy of formation at 25°C is 198.5 kcal/mole for $B_2(g)$ and 134.5 kcal/mole for $B(g)$. (g) Calculate the bond dissociation energy for B_2.

27.13 A gaseous sample of a boron hydride had a density of 0.57 g/liter at 25°C and 0.500 atm. What is the molecular weight of the compound?

27.14 A graduate student prepared some diborane using the reaction

$$4BF_3(g) + 3LiAlH_4(s) \xrightarrow{ether} 2B_2H_6(g) + 3LiF(s) + 3AlF_3(s)$$

Assuming 100% yield, what mass of B_2H_6 could be produced from the reaction of 5.0 g of BF_3 with 10.0 g of $LiAlH_4$? The student used ether that had not been carefully dried and lost some of the diborane to the following reaction

$$B_2H_6(g) + 6H_2O(l) \longrightarrow 2H_3BO_3(ether) + 6H_2(g)$$

How much of the diborane would react with 0.01 g of water?

27.15 The equation for the carbon reduction of SiO_2 is

$$SiO_2(l) + 2C(s) \xrightarrow{3300°K} Si(l) + 2CO(g)$$

The standard state heats of formation of 3300°K are zero for the elements, −30.786 kcal/mole for $CO(g)$, and −217.519 kcal/mole for $SiO_2(l)$. (a) Calculate ΔH°_{3300} for this reaction. Based on energy alone, would you expect this reaction to be favorable? The standard state entropies at 3300°K are 66.220 cal/mole °K for $CO(g)$, 26.253 cal/mole °K for $Si(l)$, 12.708 cal/mole °K for $C(s)$, and 51.565 cal/mole °K for $SiO_2(l)$. (b) Calculate ΔS°_{3300} for this reaction. Based on entropy alone, would you expect this reaction to be favorable? Using your values of ΔH°_{3300} and ΔS°_{3300} (c) calculate ΔG°_{3300} for this reaction and discuss your result.

27.16 The standard state heat of formation at 25°C of $SiH_4(g)$ is 8.2 kcal/mole, of $O_2(g)$ is 0, of $SiO_2(g)$ is -217.27 kcal/mole, and of $H_2O(g)$ is -57.796 kcal/mole. What is the enthalpy change for 1.00 g of $SiH_4(g)$ igniting in air?

27.17 Calculate the equilibrium constant for the reaction

$$H_3AsO_4(aq) + 2H^+ + 2I^- \rightleftharpoons HAsO_2(aq) + I_2(s) + 2H_2O(l)$$

given the following half reactions and associated standard reduction potentials:

$$H_3AsO_4 + 2H^+ + 2e^- \rightarrow HAsO_2 + 2H_2O \qquad E° = 0.58 \text{ V}$$
$$I_2 + 2e^- \rightarrow 2I^- \qquad E° = 0.535 \text{ V}$$

27.18 The standard state free energies of formation at 25°C are -6.66 kcal/mole for $H_2S(aq)$, 0 for $H^+(aq)$, 20.5 kcal/mole for $S^{2-}(aq)$, 5.3 kcal/mole for $H_2Se(aq)$ and 30.9 kcal/mole for $Se^{2-}(aq)$. Calculate $\Delta G°$ and K for the reactions

$$H_2S(aq) \longrightarrow 2H^+ + S^{2-}$$
$$H_2Se(aq) \longrightarrow 2H^+ + Se^{2-}$$

Which of these substances is the stronger acid?

27.19* Many metals and semiconducting elements can be purified by subliming them and collecting the condensed metal that forms in a cooler part of the apparatus. The vapor pressure of As is 10.0 torr at 437°C and 40.0 torr at 483°C. Calculate the heat of sublimation of As. If the temperature of the As is raised to 518°C, what will be the pressure above the metal?

27.20* A manufacturer wanted to prepare some semiconductor devices by using radioactive isotopes of Ge that would form the Ga and As impurities upon decay. In one sample of ultrapure Ge (^{70}Ge, ^{72}Ge, ^{73}Ge, ^{74}Ge, and ^{76}Ge are the naturally occurring stable isotopes) a sufficient amount of ^{69}Ge ($t_{1/2} = 39$ hours for β^+ decay) was added so that its concentration was 4 ppm. (a) Write the equation for the nuclear decay process and (b) identify the doping impurity. (c) Will this sample of Ge be used for a p- or an n-type semiconductor? (d) How long will the device have to remain unused until the doping impurity reaches a concentration of 1 ppm? In the second sample of ultrapure Ge, a sufficient amount of ^{77}Ge ($t_{1/2} = 11.3$ hours for β^- decay) was added so that its concentration was 2 ppm. (e) Write the equation for the nuclear decay process and (f) identify the doping impurity. (g) What will be the concentration of this doping impurity at the time the first doping impurity reaches 1 ppm?

27.21* A sample of arsenic was burned in a limited amount of dioxygen to form an oxide containing 76 wt % As. (a) What is the empirical formula of this oxide? The approximate molar weight of this oxide is 400 g/mole. (b) What is the molecular formula of the oxide? This oxide was dissolved in water to form an acid that contained 59.5 wt % As, 38.1 wt % O and 2.4 wt % H. (c) What is the empirical formula of this acid? (d) Write chemical equations for both reactions.

27.22** The unit cell of elemental silicon is similar to that of diamond (face-centered cube with four atoms inside in the shape of a tetrahedron). (a) Draw a sketch of the unit cell. (b) How many atoms are in the unit cell? (c) How many nearest-neighbor atoms are around an atom? (d) Calculate the theoretical density of Si given that $a = 0.54305$ nm. A careful analysis of the unit cell shows that the body diagonal of the cube ($a\sqrt{3}$) is equal to eight times the radius of a bonded atom in the structure. (e) Calculate the single-bond covalent radius of a Si atom. The sublimation energy of Si is 108.9 kcal/mole at 25°C. On the average, there must be two bonds broken for each atom that undergoes sublimation. (f) Calculate the average bond energy for a Si—Si single bond.

As does carbon, silicon forms a diatomic molecule in the gaseous phase. The bonding consists of a σ bond and a π bond. (g) Calculate the bond dissociation energy for a Si—Si double bond given that the standard heat of formation at 25°C of $Si_2(g)$ is 142 kcal/mole and of $Si(g)$ is 108.9 kcal/mole. (h) Which bond energy is larger—that of the single bond or that of the double bond?

Depending on reaction conditions, Si reacts with O_2 to form six allotropes of $SiO_2(s)$ known as α- and β-quartz, α- and β-cristobalite, and α- and β-tridymite. The respective heats of sublimation at 25°C for the α forms are 140.72 kcal/mole, 140.37 kcal/mole and 140.27 kcal/mole. (i) Calculate the heat of formation of each crystalline form given that $\Delta H_f° = -77.00$ kcal/mole for $SiO_2(g)$ and (j) predict which form is stable at 25°C.

As does carbon, silicon forms a dioxide in the gaseous phase. The molecule contains two double bonds. (k) Calculate the average bond energy for a Si—O double bond given that the standard heat of formation at 25°C of $SiO_2(g)$ is -77.00 kcal/mole, of $Si(g)$ is 108.9 kcal/mole and of $O(g)$ is 59.553 kcal/mole. (l) Which bond energy is larger, the Si—O double bond value you calculated or the Si—O single bond value of 88 kcal/mole?

Compare the values for the Si—Si and Si—O single bond energies and the values for Si—Si and Si—O double bond energies. (m) Which are larger and why?

28 THE CHEMISTRY OF COMPLEXES

This chapter presents the nomenclature of complexes and describes their geometry, isomerism, and general properties. Three different approaches to the explanation of bonding in complexes are introduced. Two of them—valence bond theory and molecular orbital theory—we have used before. The third—crystal field theory—has not been mentioned before in this book. The equilibria among the components of complexes are discussed, and a few practical applications of complexes are mentioned.

Although the properties of ionic and covalent substances have been well known for more than one hundred and fifty years, that there are two kinds of binding forces was not recognized until 1893. The chemists of that day were puzzled by the fact that, for example, the valences of the elements in copper(II) chloride and in ammonia seemed to be completely satisfied, yet these two substances combine to form a new substance, the formula of which they wrote $CuCl_2 \cdot 4NH_3$. Alfred Werner suggested that there are two kinds of valence, which he called "principal" and "auxiliary," but which we now call "ionic" and "covalent." In $CuCl_2 \cdot 4NH_3$, the chloride ions are held to the copper ion by ionic valences, and the ammonia molecules are held by covalences. Nowadays, we write the formula of the compound $[Cu(NH_3)_4]Cl_2$ to indicate that the ammonia molecules and the copper ion form a complex; the chloride ions are not part of the complex, but exist independently. Our knowledge of coordination complexes now constitutes a large portion of our understanding of inorganic chemistry.

Inorganic complexes show a wide range of properties—some of them are extremely stable, while others are unstable; many have colors that are quite different from the colors of their constituents; some are readily soluble in water, whereas others are insoluble in that medium but dissolve easily in nonpolar solvents such as benzene. Some are volatile, whereas others are not. By proper "tailoring" of a complex molecule or ion, almost any combination of properties can be built into it.

Because of this wide range of properties, complexes are used in many ways. Some are used as paint pigments, some as catalysts, and some as drugs. The formation of complexes is valuable in water softening and in the separation of metal ions from each other. Some naturally occurring complexes are essential to life. These include hemoglobin, chlorophyll, vitamin B_{12}, and carboxypeptidase, which contain iron, magnesium, cobalt, and zinc, respectively.

Structure, nomenclature, and properties of complexes

28.1 Some definitions

To review, a coordinate covalent bond (Section 9.11) is formed between an atom or an ion that has an unshared pair of electrons—a donor atom—and an atom or ion that has room for an electron pair—an acceptor atom. Both of these can be nonmetals

$$H:\overset{\displaystyle H}{\underset{\displaystyle H}{N}}: + H^+ \longrightarrow \left[H:\overset{\displaystyle H}{\underset{\displaystyle H}{N}}:H \right]^+ \quad \text{or} \quad NH_4^+$$

donor   acceptor
atom   atom

or the acceptor can be a metal

$$2H:\overset{\displaystyle H}{\underset{\displaystyle H}{N}}: + Ag^+ \longrightarrow \left[H:\overset{\displaystyle H}{\underset{\displaystyle H}{N}}:Ag:\overset{\displaystyle H}{\underset{\displaystyle H}{N}}:H \right]^+ \quad \text{or} \quad [Ag(NH_3)_2]^+ \tag{1}$$

A **complex** is formed between a metal atom or ion that accepts one or more electron pairs and ions or neutral molecules that donate electron pairs. The term is usually reserved for metals combined with donors that also can exist independently either in the pure state or as ions in solution. Most metal atoms or ions can accept more than one pair of electrons, for example,

$$4:NH_3 + Ni^{2+} \longrightarrow [Ni(:NH_3)_4]^{2+} \tag{2}$$

$$6:\overset{..}{\underset{..}{Cl}}:^- + Pt^{4+} \longrightarrow \left[Pt(:\overset{..}{\underset{..}{Cl}}:)_6 \right]^{2-} \tag{3}$$

$$4:NO_2^- + Pt^{2+} \longrightarrow [Pt(:NO_2)_4]^{2-} \tag{4}$$

$$4:CO + Ni \longrightarrow [Ni(:CO)_4] \tag{5}$$

The molecule or ion that contains the donor atom is called the **ligand,** or coordinating agent, and the ligand is said to be "coordinated to" the metal atom which is in the center of the new structure. In the silver–ammonia complex shown in Equation (1), the ammonia molecules are the ligands, and they are coordinated to the silver(I) ion. We might point out here that the terminology of complexes derives from coordinate covalent bonding, but, as with all types of compounds, complexes cover the range from covalent to ionic bonding.

The donor atom may be part of a molecule (Equations 1, 2, and 5), it may be an ion (Equation 3), or it may be part of an ion (Equation 4). Complexes can be either positively (Equation 2) or negatively charged (Equations 3 and 4), in which cases they are called **complex ions;** they can also be neutral (Equation 5). The charge on a complex

TABLE 28.1

Common coordination numbers of cations

Ag^+	2	Ag^{2+}	4	Al^{3+}	4, 6	Pd^{4+}	6
Au^+	2, 4	Ca^{2+}	6	Au^{3+}	4	Pt^{4+}	6
Cu^+	2, 4	Co^{2+}	4, 6	Co^{3+}	6		
Li^+	4	Cu^{2+}	4, 6	Cr^{3+}	6		
Na^+	4	Fe^{2+}	6	Fe^{3+}	6		
Tl^+	2	Ni^{2+}	4, 5, 6	Ir^{3+}	6		
		Pb^{2+}	4	Os^{3+}	6		
		Pd^{2+}	4	Sc^{3+}	6		
		Pt^{2+}	4				
		V^{2+}	6				
		Zn^{2+}	4				

TABLE 28.2

Common ligands

The nitrito ligand is attached though the O atom; the nitro ligand through the N atom.

Ligand	Name
F^-	fluoro
Cl^-	chloro
Br^-	bromo
I^-	iodo
CN^-	cyano
NCS^-	thiocyanato
ONO^-	nitrito
NO_2^-	nitro
OH^-	hydroxo
O^{2-}	oxo
CO_3^{2-}	carbonato
$C_2O_4^{2-}$	oxalato
SO_4^{2-}	sulfato
NH_3	ammine
H_2O	aquo
CO	carbonyl
NO	nitrosyl

ion is the sum of the charges of the constituent parts. For example, in $[PtCl_4]^{2-}$ the charge is found by adding $+2$ (for the platinum) and $4 \times (-1)$ for the four chloro groups to obtain the charge of -2 that the complex carries.

The neutral compound formed between a complex ion and other ions or molecules is called a **coordination compound,** for example, $[Ag(NH_3)_2]^+Cl^-$ or $(K^+)_2[Pt(NO_2)_4]^{2-}$. It is common practice to enclose the formula of the complex in square brackets.

The **coordination number** is the number of nonmetal atoms surrounding the central metal atom or ion in a complex (Table 28.1). The most common coordination numbers are two, four, and six. For Ag^+, the coordination number is commonly two, as in $[Ag(CN)_2]^-$, but sometimes it is three, as in $[Ag(CN)_3]^{2-}$. Nickel(II) can have a coordination number of four, five, or six, as in $[Ni(CN)_4]^{2-}$, $[Ni(CN)_5]^{3-}$, and $[Ni(NH_3)_6]^{2+}$; with cobalt(III), it is almost invariably six; and with zirconium(IV) and hafnium(IV), it is often, but not invariably, eight, as in $[ZrF_8]^{4-}$ and $[HfF_8]^{4-}$.

A number of complexes in which the metal has a coordination number of five are known. Other odd coordination numbers, particularly seven and nine, are less common than even ones, although they do occur in such complexes as $[TaF_7]^{2-}$ and $[ReH_9]^{2-}$. Coordination numbers of ten and twelve are also rare.

28.2 Nomenclature

A few simple rules will cover what it is necessary to know about naming complexes:

1. The ligands are named first; the prefixes *di, tri, tetra,* and so on are used to indicate the number of each kind of ligand present. [Sometimes the prefixes *bis* (2), *tris* (3), and *tetrakis* (4) are also used, especially when the ligand name is complicated or already includes *di*, etc.]

2. Negative ligands are named first, with the ending -o, neutral ones second, and positive ones last. Within each group the ligand names

are given in alphabetical order. Some ligands have familiar names also used in naming other types of compounds (e.g., chloro, cyano); others have names special to complexes (e.g., carbonato, CO_3^{2-}; aqua, H_2O). Table 28.2 lists the names of the most common coordinating groups. Note that the ammine group—two m's—is NH_3.

3. The name of the central metal atom followed by its oxidation state in parentheses is given after the ligand names. The metal name is not separated from the ligand names by a space.

4. When a complex ion is negative, the name of the central metal atom is given the ending *ate*. For some of the elements, the ion name is based on the Latin name from which the symbol is derived, for example, ferrate for iron, Fe; plumbate for lead, Pb. When naming only the ion, the word "ion" is always used in the name.

Using these rules, the complexes formed in Reactions (1)–(5) receive the following names

$[Ag(NH_3)_2]^{2+}$	diamminesilver(I) ion
$[Ni(NH_3)_4]^{2+}$	tetraamminenickel(II) ion
$[Pt(Cl_6)]^{2-}$	hexachloroplatinate(IV) ion
$[Ni(CO)_4]$	tetracarbonylnickel(0)
	(commonly called "nickel carbonyl")

In naming a coordination compound, the name of the cation is given first as usual, followed by the name of the anion.

$K^+[Pt(NH_3)Cl_5]^-$	potassium pentachloroammineplatinate(IV)
$[Co(NH_3)_4SO_4]^+NO_3^-$	sulfatotetraamminecobalt(III) nitrate

28.3 Chelation

In many cases, two or more atoms with unshared pairs of electrons are present in the same ion or molecule, and if their spatial properties are favorable (that is, if they are not too close together or too far apart), they may coordinate to the same metal ion to form a ring. The carbonate (CO_3^{2-}) and oxalate ($C_2O_4^{2-}$) ions usually behave in this way

carbonato complex   oxalato complex

but the carbonate ion does not form a very stable complex, for the four-membered ring is highly strained. Five- and six-membered rings are much less strained and are very common. Most of the ligands that form them are organic anions. A few examples are listed in Table 28.3.

The phenomenon of ring formation by a ligand in a complex is called **chelation** and the ring formed is called a **chelate ring** (from the Greek *kela*, meaning "crab's claw," and pronounced "key-late.") Chelation greatly increases the stability of the complex. This is of tremendous value in altering the properties of metal ions by com-

TABLE 28.3
Some common chelating agents
The electron pairs available for donation are shown in color.

Formula	Abbreviation
Form five-membered rings	
$NH_2CH_2CH_2NH_2$	en
ethylenediamine	
$(NH_2CH_2CO)^-$ with O double bond	gly
glycinate ion	
$\begin{bmatrix} :O-C=O \\ :O-C=O \end{bmatrix}^{2-}$	ox
oxalate ion	
$NH_2CH_2CH_2NHCH_2CH_2NH_2$	dien
diethylenetriamine	
Form six-membered rings	
$NH_2CH_2CH_2CH_2NH_2$	tm
trimethylenediamine	
$\begin{bmatrix} CH_3-C \\ || \\ O \end{bmatrix}$ acetylacetonate structure	acac
acetylacetonate ion	

FIGURE 28.1

Ethylenediaminetetraacetic acid and complex formation. *The EDTA can form five chelate rings (b), but one of the rings opens up (c) if the structure is strained.*

(a) Ethylenediamminetetraacetic acid (edta) anion

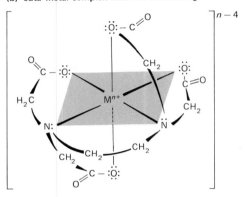

(b) edta–metal complex with five chelate rings

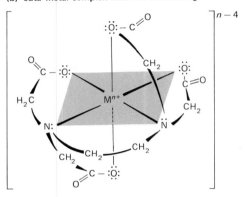

(c) edta–metal complex with four chelate rings

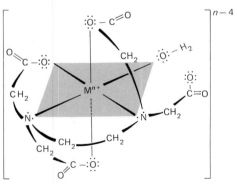

plexation. If a molecule or ion contains more than two potential donor atoms, it may attach itself to the metal ion through all of them, and thus form several "fused" chelate rings.

One of the most powerful chelating ligands known is the anion of ethylenediaminetetraacetic acid (EDTA) (Figure 28.1a). This ion is able to attach itself to a single metal ion through both nitrogen atoms and one oxygen atom of each CH_2COO^- (acetate) group, forming five chelate rings, with a single metal atom common to all of them (Figure 28.1b). The six atoms coordinated to the metal are located at the corners of an octahedron (see Figure 28.2). There is some strain in this structure, and it may overcome the stabilizing effect of the five rings. In such cases, one —CH_2COO— "arm" remains unattached, and a molecule of water or some other ligand takes its place (Figure 28.1c). If this ligand is water or another volatile neutral ligand, it can be driven off by heating, with the formation of the fifth chelate ring. In aqueous solution this process may be reversed.

28.4 Molecular geometry and isomerism

The metal ions and the ligands that make up a complex have a definite spatial relationship with each other. As with the geometry of covalent molecules (Table 9.22), the geometry of complexes is generally governed by maintenance by the ligands of the maximum distance from each other. Complexes containing two unchelated coordinated groups, such as $[ClAgCl]^-$, are linear, and those containing four coordinated groups are either square-planar, as is $[Pt(NH_3)_4]^{2+}$ or tetrahedral, as is $[Be(H_2O)_4]^{2+}$ (Figure 28.2). Most six-coordinate complexes are octahedral.

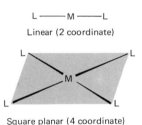

Linear (2 coordinate)

Square planar (4 coordinate)

FIGURE 28.2
Geometry of complexes.

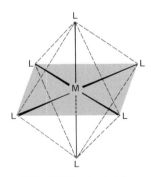

Octahedral (6 coordinate)

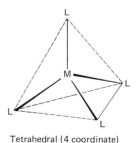

Tetrahedral (4 coordinate)

[28.3] THE CHEMISTRY
OF COMPLEXES

The fixed positions of the ligands in a complex give rise to the possibility of both geometrical and optical isomerism. Square-planar complexes such as [Pt(NH$_3$)$_2$Cl$_2$] can form geometrical isomers in which the two like atoms are on adjacent (cis) or opposite (trans) corners of the square plane (Figure 28.3). Six-coordinate, octahedral complexes such as [Co(NH$_3$)$_4$Cl$_2$]$^+$ also exhibit geometrical isomerism that depends upon the location of the identical ligands on adjacent or opposite corners of the octahedron (Figure 28.3).

We have encountered (Section 23.7) optical isomerism as a property of asymmetrically substituted carbon atoms. Complexes can also be optically active, due to asymmetry, and many of them are.

Asymmetry is most readily achieved in complexes containing two or three chelate rings. In four-coordinate tetrahedral complexes, optical activity is possible if the rings are unsymmetrical, but with octahedral complexes, either two or three chelate rings of any sort suffice to produce optical activity. If there are only two rings, the isomer in which the two nonchelated groups are in cis positions will have optical isomers. The trans geometrical isomer of the same compound will not be optically active—it is superimposable on its mirror image (Figure 28.4).

All six-coordinate, complexes with three chelate rings are opti-

FIGURE 28.4
Isomers of a dichelate octahedral complex.

Dichlorobis(ethylenediammine)cobalt(III) ion

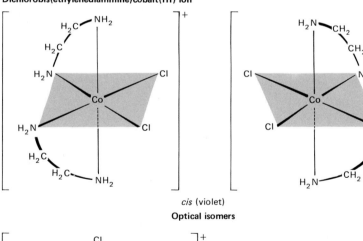

cis (violet)
Optical isomers

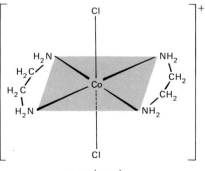

trans (green)

Dichlorodiammineplatinum(II)

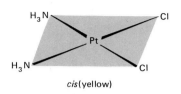

cis (yellow)

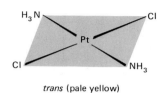

trans (pale yellow)

Dichlorotetraamminecobalt(III) ion

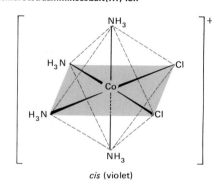

cis (violet)

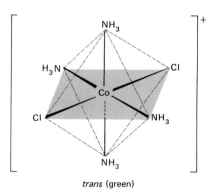

trans (green)

THE CHEMISTRY
OF COMPLEXES [28.4]

FIGURE 28.5
Isomers of a trichelate octahedral complex.

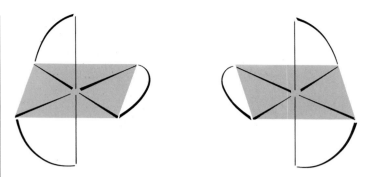

cally active. Figure 28.5 shows a shorthand method frequently used to draw such complexes.

28.5 Formation and properties of complexes

All metal ions have the ability to form coordination compounds. Those with small radius and high charge, and those that have vacant d orbitals—the transition metal ions—do so most readily, but even the sodium ion, with its comparatively large radius and its small charge, forms coordination compounds. All of the sodium complexes, however, are immediately destroyed by water, probably because of the formation of the more stable $[Na(H_2O)_x]^+$. The complexes of the beryllium ion, Be^{2+}, and the aluminum ion, Al^{3+}, are more stable toward dissociation in water than are those of the sodium ion. Ions of the heavier representative elements yield still more stable complexes. For example, the ions $[Zn(CN)_4]^{2-}$, $[Cd(NH_3)_4]^{2+}$, and $[SnS_3]^{2-}$ remain intact in aqueous solution, and their salts can be crystallized from these solutions.

Typical of the very stable coordination ions of the transition metal ions are $[Co(NH_3)_6]^{3+}$, $[Fe(CN)_6]^{3-}$, $[Fe(CN)_6]^{4-}$, $[Pt(NH_3)_4]^{2+}$, and $[Co(NH_3)_3(NO_2)_3]$. None of these, in aqueous solution, gives any indication of the presence of the component parts of the complexes. No change occurs when $[Co(NH_3)_6]Cl_3$ is recrystallized from concentrated hydrochloric acid, and aqueous solutions of $[Co(NH_3)_3(NO_2)_3]$ show no electrical conductivity, which they would if any free NO_2^- ions were present. The nature and the strength of the bonding between the metal and the nonmetal groups in such complexes varies a great deal, depending upon the nature of the metal atom or ion and the ligand.

Complexation greatly changes the properties of the constituents of a complex. For example, anhydrous $NiCl_2$ is light brown, $[Ni(H_2O)_6]Cl_2$ is light green, and $[Ni(NH_3)_6]Cl_2$ is deep blue. Other properties may be changed in equally striking ways. When a metal is part of a complex, the ease with which it is oxidized or reduced is changed, and its catalytic properties toward a given reaction may be either enhanced or diminished. Complex ion salts usually have quite different solubilities from the simple metal salts. And the acid–base character of a metal is greatly modified in a complex, as is shown by the fact that $Co(OH)_3$ is a very weak base, but $[Co(NH_3)_6](OH)_3$ is as strong a base as sodium hydroxide. Some metal complexes are volatile at moderate temperatures. Others, even though they are ionic, are soluble in nonpolar solvents. In short, a coordinated metal ion is part of a new species, and by proper coordination, ions of a great variety of properties can be formed.

As the discussion of bonding in the next three sections makes clear, the presence of unpaired electrons is an important characteristic of many transition metal ions and their complexes. Substances with unpaired electrons are paramagnetic—they are weakly attracted to a magnetic field, but lose their magnetism in the absence of the field. Diamagnetic substances contain no unpaired electrons and are repelled by a magnetic field.

The number of unpaired electrons in a compound can be estimated quite accurately by measuring its degree of paramagnetism, usually called the "magnetic susceptibility" of the compound. Complexes are often colored. In many cases, electronic transitions caused by the absorption of energy by unpaired electrons are the cause of the color (see Section 29.2).

Bonding in complexes

To give even a simple discussion of bonding in complexes, it is necessary to take into account three different approaches to the subject. Two of these ways of describing bonding have already been introduced (Chapter 21)—valence bond theory and molecular orbital theory. The third is crystal field theory, which views ions and groups as point charges and examines their interactions due to electronic attraction and repulsion.

Bonding in complexes, as in other compounds, is rarely strictly ionic or strictly covalent. The valence bond approach to complex bonding emphasizes covalent bonding, the crystal field theory emphasizes ionic bonding, and the molecular orbital theory brings about a compromise between the two.

28.6 Valence bond theory

In valence bond theory, the metal–ligand bonds are looked upon as primarily coordinate covalent bonds. Electron pairs from the donor atom are shared by entering vacant spaces in the valence shell of the metal atom or ion or by forming new shells. Bonding is the overlap of atomic orbitals, and the atomic orbitals on the metal that are filled by the ligand electrons are viewed as hybridized and of equal energy (Section 21.4)

The Be atom and Be^{2+} have the following configurations:

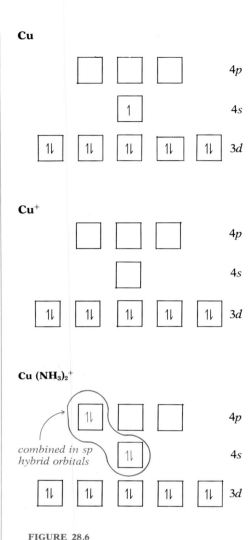

Cu

Cu⁺

Cu (NH₃)₂⁺

combined in sp hybrid orbitals

FIGURE 28.6

Valence bond picture of configuration in a copper(I) complex. *The* [Ar] *configuration is 1s²2p²2p⁶3s²3p⁶. The electrons from the* NH₃ *ligands in the complex are shown in color.*

In the $[Be(H_2O)_4]^{2+}$ ion, the $2s$ and $2p$ orbitals are filled by a total of eight electrons, two from each of the four water molecules.

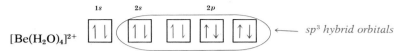

The four pairs of ligand electrons have filled four equal sp^3 hybrid orbitals.

The copper(I) ion (Figure 28.6), like the beryllium ion, has each of its orbitals occupied by two electrons, in this case including completely filled $3d$ orbitals. Here again, the pairs of electrons from the donor groups must enter levels outside the filled ones. In the complex ion $[Cu(NH_3)_2]^+$, the two pairs of electrons donated by the ammonia molecules enter the $4s$ and one of the $4p$ orbitals, which have formed sp hybrids.

The cobalt(III) ion contains six $3d$ electrons (Figure 28.7). Since there are only five orbitals to accommodate them, two electrons must be paired. The four unpaired electrons make the ion highly paramagnetic. This property is maintained in the cobalt(III) complex $[CoF_6]^{3-}$; therefore, it must be concluded that the six pairs of electrons furnished by the fluoride ligands do not interact with the unpaired electrons of cobalt. Instead, they occupy outer orbitals—the $4s$, the three $4p$, and two of the $4d$ orbitals, forming an sp^3d^2 hybrid.

The $[Co(NH_3)_6]^{3+}$ complex, however, is diamagnetic, which indicates that it contains no unpaired electrons. The ammonia complex is much more stable than the fluoride complex, and it is believed that the energy liberated by coordination of ammonia with cobalt(III) is sufficient to cause the unpaired electrons to form pairs. This empties two of the $3d$ orbitals and allows pairs of electrons from two of the ammonia molecules to enter $3d$ orbitals. The pairs from the other four ammonia molecules then occupy the $4s$ and the three $4p$ orbitals, giving a d^2sp^3 hybrid electronic configuration.

Complexes in which $(n-1)d$ orbitals remain only partly filled while ligand electrons enter orbitals in the n, or outermost, shell are called **outer orbital complexes.** Complexes in which ligand electrons fill the $(n-1)d$ orbitals as well as some of the n shell orbitals are called **inner orbital complexes.** Table 28.4 gives additional examples of inner and outer orbital complexes, as well as some in which ligand electrons use only s and p orbitals. (Bonding in the inner and outer orbital complexes $[Co(NH_3)_6]^{3+}$ and $[CoF_6]^{3-}$ is discussed from the crystal field and molecular orbital viewpoints in the following two sections.)

A somewhat different arrangement of electrons is possible in ions such as chromium(III) (Table 28.4). The simple ion contains three $3d$ electrons, so there are two vacant $3d$ orbitals. In the formation of hexacoordinate chromium(III) complexes, two pairs of electrons from the ligands can enter these orbitals, the other four going into the $4s$ and the three $4p$ orbitals. Thus, the $3d$ orbitals are "filled," although three of them contain only one unpaired electron each. The presence of three unpaired electrons in the chromium complex is disclosed by magnetic data.

The dsp hybrids are inner orbital complexes; the spd hybrids are outer orbital complexes.

Ion	Complex	Number of electrons in each energy level[a]						Type of hybrid[b]
		3s	3p	3d	4s	4p	4d	
Cr^{3+}	$[Cr(NH_3)_6]^{3+}$	2	6	3 + 4	2	6		$d^2sp^{3\,c}$
Fe^{2+}	$[Fe(CN)_6]^{4-}$	2	6	6 + 4	2	6		d^2sp^3
Fe^{3+}	$[Fe(CN)_6]^{3-}$	2	6	5 + 4	2	6		$d^2sp^{3\,c}$
Fe^{3+}	$[FeF_6]^{3-}$	2	6	5	2	6	4	sp^3d^2
Co^{3+}	$[Co(NH_3)_6]^{3+}$	2	6	6 + 4	2	6		d^2sp^3
Co^{3+}	$[CoF_6]^{3-}$	2	6	6	2	6	4	sp^3d^2
Ni^{2+}	$[Ni(CN)_4]^{2-}$	2	6	8 + 2	2	4		dsp^2
Cu^+	$[Cu(NH_3)_2]^+$	2	6	10	2	2		sp
Cu^+	$[Cu(CN)_4]^{3-}$	2	6	10	2	6		sp^3
Cu^{2+}	$[Cu(NH_3)_4]^{2+}$	2	6	8 + 2	2	4 + 1		$dsp^{2\,c}$

[a] Electrons supplied by the ligand are shown in color.
[b] Most complexes containing six coordinated groups are octahedral, whether the binding is sp^3d^2 or d^2sp^3. In four coordinate complexes, dsp^2 binding is characteristic of the square-planar configuration, and sp^3 binding, of the tetrahedral configuration. sp binding is linear. These configurations result from hybridization interactions of the electrons in the s, p, and d orbitals.
[c] Since the metal ions in these complexes have an odd number of electrons, the complexes must have an odd number. On that account, they must have unpaired electrons and be slightly paramagnetic, even though they are inner orbital complexes. One electron in the $4p$ level of $[Cu(NH_3)_4]^{2+}$ has been promoted from the $3d$ level.

When several pairs of electrons unite with a metal ion, a great concentration of negative charge is put on that ion, and chemical theorists were long puzzled as to how this could lead to a stable structure. It is now generally believed that in the most stable complexes, some back-bonding from the metal ion takes place. **Back-bonding** is the formation of double bonds between the metal and the ligands in a complex, electrons coming from both the nonmetal and the metal. In $[Fe(CN)_6]^{4-}$, for example, some (perhaps about half) of the cyanide groups are held to the iron by a single pair of electrons furnished by the carbon, Fe:C::N:, but others are held by two pairs, one furnished by the carbon and one by the iron. Since the carbon atom cannot have more than eight electrons in the valence shell, a rearrangement of electrons within the cyanide groups that are doubly bonded to the iron is necessary, as in Fe::C::N:. It is impossible to distinguish between the cyanide groups held to the iron by single bonds and those held by double bonds. Undoubtedly, resonance gives them identical properties. This theory of back-bonding is supported by x-ray measurements that show the iron–carbon distance to be less than would be expected if all of the iron–carbon bonds were single bonds, but greater than if they were all double bonds. Bonding of this type is possible with any donor group that contains a double or triple bond so that rearrangement of the electrons can take place. It is also

TABLE 28.4

Arrangement of electrons in complexes of ions of some transition metals, according to the valence bond theory

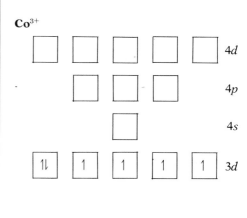

Co^{3+}

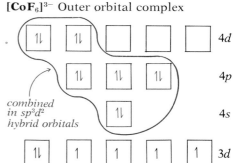

$[CoF_6]^{3-}$ Outer orbital complex

combined in sp^3d^2 hybrid orbitals

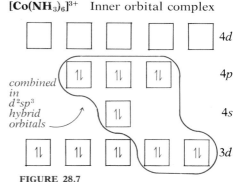

$[Co(NH_3)_6]^{3+}$ Inner orbital complex

combined in d^2sp^3 hybrid orbitals

FIGURE 28.7
Valence bond picture of configuration in cobalt(III) outer and inner complexes. *Ligand electrons are shown in color.*

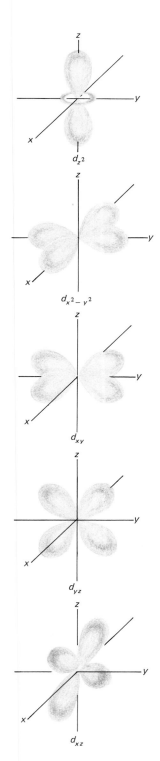

FIGURE 28.8
The five _d_ orbitals.

possible if the donor atom can expand its valence shell. For example, in the [SbS$_4$]$^{3-}$ complex some of the antimony–sulfur bonds probably consist of four electrons, Sb::S:. The ability of sulfur to have an outer shell of more than eight electrons is shown by the great thermal stability of sulfur hexafluoride, SF$_6$.

The formation of a double bond is not possible when ammonia, water, or fluoride ion coordinates with metal ions, for (according to our present theories) there is no mechanism by which these substances can accept additional electrons. In such cases the high concentration of negative charge imposed upon the metallic ion is thought to be distributed over the complex, but how this happens is not clear. The concentration of negative charge on the metal ion is less effective in reducing stability if the metal ion initially had a large positive charge. As a result, the most stable ammines and hydrates are those derived from species of high ionic charge.

28.7 Crystal field theory

Crystal field theory views the ions and groups in complexes in terms of their spatial relationships and the interactions based on their electrostatic attraction and repulsion. It is called "crystal field" theory because it derives from a theory developed for the energy relationships of ions in crystals.

We shall describe the crystal field theory for the formation of an octahedral complex of the Ti^{3+} ion. Application of the theory to complexes of other coordination numbers and geometry would be similar.

We begin with an isolated Ti^{3+} ion, which has one $3d$ electron in addition to its filled shells. The d electron can reside in any of the five d orbitals (Figure 28.8), for they are of equal energy. As six ligands bearing electron pairs approach the metal ion, it becomes more difficult for an electron to occupy the d orbitals, because there is a repulsion between the electrons of the metal and the approaching ligand electrons. In other words, the energy level of the d orbitals is raised. If the approaching ligand electrons had a perfectly symmetrical effect on the metal ion d orbitals, all of the d orbitals would be raised in energy equally.

However, the ligands approach along the x, y, z coordinates as shown in Figure 28.9, producing an octahedral field. Comparison with the arrangement of the d orbitals in Figure 28.8 shows that the two lobes of the d_{z^2} orbital and the four lobes of the $d_{x^2-y^2}$ orbital point directly toward the corners of the octahedron, where the negative charge of the approaching ligands is concentrated. These orbitals are therefore raised in energy relative to their positions in a symmetrical field. The other three orbitals are further from the negative charge of the donor atoms and are at a lower energy than they would be in a symmetrical field. The three orbitals of lower energy are called t_{2g} orbitals (pronounced "t-two-g") and the two of higher energy are called e_g orbitals (pronounced "e-g"). (These names are derived from spectroscopic terms. One of the successes of crystal field theory has been in explaining the spectra of complexes.) The changes in energy levels can be pictured as follows:

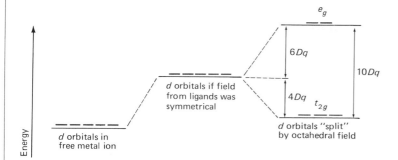

The single d electron in a Ti^{3+} complex will avoid the areas of higher energy and higher electronic repulsion at the corners of the octahedron and enter one of the t_{2g} orbitals. This increases the stability of the system.

The energy difference between the t_{2g} and e_g orbitals is called the *octahedral crystal field splitting* and is represented by the symbol $10Dq$ (or sometimes by Δ). Crystal field splitting is a measure of the "crystal field strength" of the ligand. In splitting into two levels, no energy is gained or lost; the loss of energy by one set of orbitals must be balanced by a gain by the other set. The energy gain by four electrons occupying e_g orbitals must equal the energy lost by six electrons in t_{2g} orbitals. Therefore, the energy of each of the two e_g orbitals is $6Dq$ higher, and that of each of the three t_{2g} orbitals is $4Dq$ lower than if the separation had not taken place.

The amount of stabilization provided by the splitting of the d orbitals into two levels can be calculated in terms of $10Dq$. It is the **crystal field stabilization energy (CFSE)**—the algebraic sum of $-4Dq$ per electron in a t_{2g} level and $+6Dq$ for each electron in an e_g level.

The value of $10Dq$ is measured spectroscopically. It is equivalent to the wavelength of light absorbed when an electron that was in the t_{2g} level is excited up to the e_g level. Therefore, the position of the major absorption band for a complex is a measure of the tightness of the bond between the metal ion and the ligand.

For $[Ti(H_2O)_3]^{3+}$, $10Dq$ is 57.2 kcal/mole, or Dq is 5.72 kcal/mole. Since one electron has entered the t_{2g} level, the CFSE for this complex is

$$CFSE = -4(5.72 \text{ kcal/mole})$$
$$= -22.9 \text{ kcal/mole}$$

The system is thus stabilized due to the crystal field splitting by having given up 22.9 kcal/mole of energy. The amount of increased stability depends upon both the nature of the metal ion and of the ligand. For a ligand that coordinated more strongly than water, the effect would be more pronounced.

The arrangement of the electrons and the CFSE for d^1 through d^4

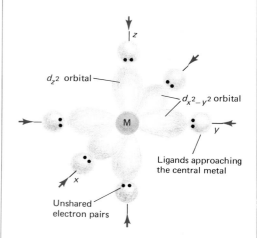

FIGURE 28.9
Six ligands about to form an octahedral complex.

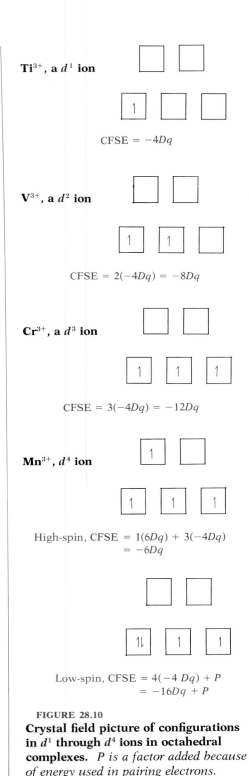

Ti³⁺, a *d*¹ ion

CFSE = −4*Dq*

V³⁺, a *d*² ion

CFSE = 2(−4*Dq*) = −8*Dq*

Cr³⁺, a *d*³ ion

CFSE = 3(−4*Dq*) = −12*Dq*

Mn³⁺, *d*⁴ ion

High-spin, CFSE = 1(6*Dq*) + 3(−4*Dq*)
= −6*Dq*

Low-spin, CFSE = 4(−4 *Dq*) + *P*
= −16*Dq* + *P*

FIGURE 28.10
Crystal field picture of configurations in *d*¹ through *d*⁴ ions in octahedral complexes. *P is a factor added because of energy used in pairing electrons.*

metal ions in octahedral complexes are summarized in Figure 28.10. In metal ions with four *d* electrons there are two possibilities. Three of the electrons, of course, occupy places in the t_{2g} orbitals. The fourth electron may go into an e_g orbital or pair up with an electron in a t_{2g} orbital. If the fourth electron goes in at the higher level, then

$$\text{CFSE} = 1(6Dq) - 3(4Dq) = -6Dq$$

There is stabilization energy, but it is decreased by the occupancy of a higher energy orbital. The $[\text{Mn}(\text{H}_2\text{O})_6]^{3+}$ ion is a typical example of an ion with this configuration. Since the four electrons occupy separate orbitals, their spins are not paired. A complex in which *d* electrons remain unpaired is called a **high-spin complex** (or a spin-free complex).

In $[\text{Mn}(\text{CN})_6]^{3-}$ and in many other d^4 complexes, the fourth electron pairs up with one of the electrons in a t_{2g} orbital. A complex in which *d* electrons are paired is called a **low-spin complex** (or a spin-paired complex). It might be expected that having all of the electrons in the lower energy orbitals would give added stability, but a good deal of energy (represented by *P*, see Figure 28.10) is consumed in pairing the spins of two electrons, so some stability is lost.

Whether a complex of a d^4 ion will be high-spin or low-spin depends upon the nature of the ligand. If the ligand is a strong coordinating agent like CN⁻—a **strong-field ligand**—it will force a pairing of electrons and form the spin-paired system. A less strongly coordinating ligand—a **weak-field ligand**—will form the high-spin system. It is easy to distinguish between the two cases, for the high-spin complexes are more highly paramagnetic than the low-spin complexes. Also, there is a marked difference in the position of the spectroscopic absorption bands.

To examine the crystal field approach for one more case, consider the d^6 cobalt complexes with ammine groups and with fluoride ion that were described from the valence bond viewpoint in the preceding section. In each case, at least two of the six *d* electrons must be paired, for there are only five orbitals. If only two electrons are paired, the complex has a high spin $(t_{2g})^4(e_g)^2$ (where the superscripts indicate the number of electrons in each level). The $[\text{CoF}_6]^{3-}$ ion, which shows the presence of the unpaired electrons by its paramagnetism, is not particularly stable, for its CFSE is low

$$\text{CFSE} = 2(6Dq) + 4(-4Dq) + P = -4Dq + P$$

where *P* is the energy consumed in pairing two electrons.

Most of the well-known Co³⁺ complexes are of the low-spin variety, with all of the electrons paired and occupying t_{2g} orbitals. The two e_g orbitals are entirely empty. These complexes are diamagnetic and the crystal field stabilization energy is high (Figure 28.11). The compound $[\text{Co}(\text{NH}_3)_6]\text{Cl}_3$ does not lose ammonia even at 200°C and can be recrystallized from hot concentrated hydrochloric acid without decomposition. In the valence bond nomenclature it is a d^2sp^3 complex. The two *d* orbitals are, of course, of the e_g type. (Compare Figure 28.11 with Figure 28.7.) The high-spin complexes are equivalent to the valence bond outer orbital complexes, and the low-spin complexes are equivalent to the valence bond inner orbital complexes.

The preceding discussion has covered d^1 through d^4 and d^6 complexes and was illustrated by first transition series elements. A similar approach is used for ions with other numbers of electrons in their d levels, and the crystal field approach can equally well be illustrated by examples from the second and third transition series, for the electron distributions are the same. The differences in the chemistry of the corresponding ions of the three d series depend chiefly upon nuclear charge and ion size, rather than upon electron arrangement.

28.8 Molecular orbital theory

Through molecular orbital theory, a somewhat different explanation of inner and outer orbital complexes and the presence of unpaired electrons is achieved. (You might want to review Sections 21.8–21.11 at this time.) Recall that molecular orbitals are pictured as being formed by the overlap and combination of regular or hybridized orbitals on the species to be bonded. Two atomic orbitals combine to form one bonding and one antibonding orbital.

In the formation of an octahedral complex, molecular orbitals are formed by combination of the atomic orbitals of the metal with the orbitals that contain the available electron pairs on the ligands. When six ligands approach a d transition metal atom or ion, the $4s$ and $4p$ orbitals of the metal are available for overlap with the ligand orbitals. In addition, as we discussed in connection with crystal field theory, two of the five d orbitals are oriented in space (Figure 28.8) so that they point toward the approaching ligands, and these two are also available. The other three d orbitals cannot overlap ligand orbitals.

What happens is pictured this way: the $4s$, the three $4p$, and the two available d orbitals (a total of six orbitals) combine with the orbitals that hold the bonding electron pairs in each of the six ligands to give six bonding and six antibonding molecular orbitals. The remaining three d orbitals from the metal are nonbonding molecular orbitals—they are not properly oriented in space, and electrons residing in these orbitals make no contribution to bonding, nor is there any change in their energy levels.

The energy levels of the orbitals can be calculated from spectroscopic data and can be depicted in the usual energy level diagram. Figure 28.12 shows the diagrams for the two cobalt(III) complexes previously examined from the valence bond and crystal field viewpoints (Sections 28.6 and 28.7)—the low-spin, inner orbital hexaamminecobalt(III) complex and the high-spin, outer orbital hexafluorocobalt(III) complex. For the hexaamminecobalt(III) ion, $[Co(NH_3)_6]^{3+}$, considering only the orbitals that are concerned in the binding of the Co^{3+} ion and the six ammonia molecules, we have, for the cobalt ion, five $3d$, one $4s$, and three $4p$ orbitals. The orbitals of the Co^{3+} ion are

Co^{3+}, a d^6 ion, in $[CoF_6]^{3-}$, $(t_{2g})^4(e_g)^2$

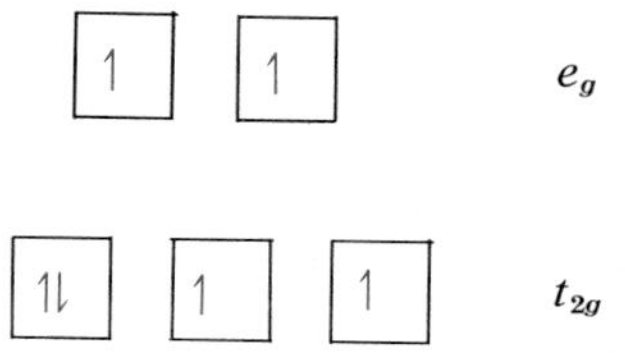

High-spin, CFSE $= 2(6Dq) + 4(-4Dq) + P$
$= -4Dq + P$

Co^{3+}, a d^6 ion, in $[Co(NH_3)_6]^{3+}$, $(t_{2g})^6$

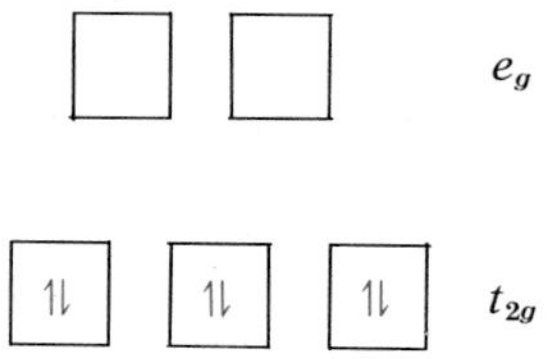

Low-spin, CFSE $= 6(-4Dq) + 3P$
$= -24Dq + 3P$

FIGURE 28.11

Crystal field picture of configurations in d^6 high-spin and low-spin Co^{3+} octahedral complexes.

shown at the left, with the electrons occupying the orbitals of lowest energy, that is, the $3d$ orbitals. The ammonia molecules are identical, so their valence orbitals all have the same energy. These are shown on the right. They are at a lower energy level than the orbitals from the metal ion because the Co^{3+} has transferred some electronic charge to the ammonia molecules. This is where molecular orbital theory can make allowance for the ligand contribution to the bonding. The total number of orbitals for the cobalt ion and the ammonia molecules is fifteen, so there must be fifteen molecular orbitals. These are shown in the center of the diagram, their levels on the page indicating the energy content of each. The electrons that were furnished by the cobalt ion are paired in the complex and are shown in the t_{2g} orbitals at the same level as they were in the $3d$ atomic orbitals. The electrons from the ammonia molecules are now in the molecular orbitals of the lowest energy, all six of which are below the original atomic orbitals and are therefore bonding orbitals. The antibonding orbitals are shown at higher energy levels, but all of them are empty. This reflects the great stability of the complex $[Co(NH_3)_6]^{3+}$ ion.

Note that the molecular orbital approach incorporates the concept of crystal field splitting, the value of $10Dq$ being the energy difference between the nonbonding t_{2g} orbitals and the antibonding e_g orbitals.

For the high-spin, outer orbital $[CoF_6]^{3-}$ ion, spectroscopic evidence indicates that the electrons from the Co^{3+} ion are not all paired, and only four of them occupy the molecular orbital at the original energy level. The other two go into the empty orbital of lowest energy, but this is an antibonding orbital, so the stability of the complex is lowered.

FIGURE 28.12
Molecular orbital energy levels for high-spin and low-spin Co^{3+} octahedral complexes.

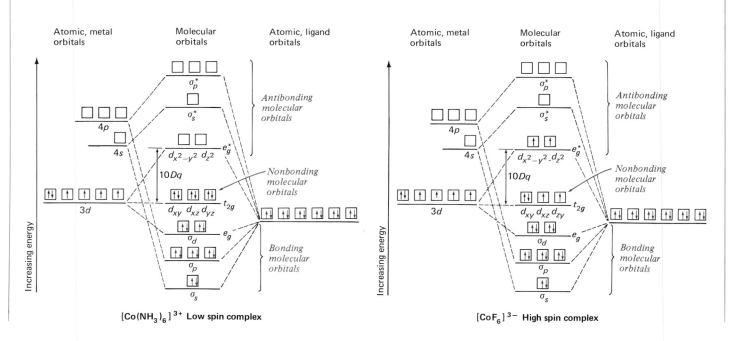

$[Co(NH_3)_6]^{3+}$ Low spin complex

$[CoF_6]^{3-}$ High spin complex

28.9 Dissociation constants, K_d

In solution, all complexes tend to come to equilibrium with their component parts. However, the ligands do not all become attached to the metal at the same time, nor do they dissociate from the metal at the same time—instead, a stepwise process occurs. Consider, for example, the diamminesilver ion, $[Ag(NH_3)_2]^+$. When ammonium hydroxide is added to a solution of a silver salt in water solution, the basic silver hydroxide precipitates first, then it rapidly loses water to give a silver oxide precipitate. Addition of more ammonia takes this precipitate back into solution, forming first $[Ag(NH_3)]^+$ and then $[Ag(NH_3)_2]^+$. The equations for the reactions are

$$2Ag^+ + 2OH^- \rightleftharpoons Ag_2O(s) + H_2O(l)$$
$$Ag_2O(s) + 2NH_3(aq) + H_2O(l) \rightleftharpoons 2[Ag(NH_3)]^+ + 2OH^-$$
$$[Ag(NH_3)]^+ + NH_3(aq) \rightleftharpoons [Ag(NH_3)_2]^+$$

Since these are equilibrium reactions, the same principles that were discussed for acid–base equilibria (Sections 17.8, 17.9) and solubility product constants (Sections 18.12–18.14) also supply to these reactions.

The stability of complex ions is conveniently compared in terms of **dissociation constants, K_d**—equilibrium constants for the dissociation of complex ions into their components.

Writing the equilibrium constants for the stepwise dissociation of the diamminesilver ion

$$[Ag(NH_3)_2]^+ \rightleftharpoons Ag(NH_3)^+ + NH_3$$

$$K_1 = \frac{[NH_3][Ag(NH_3)^+]}{[Ag(NH_3)_2^+]} = 1.45 \times 10^{-4}$$

$$[Ag(NH_3)]^+ \rightleftharpoons Ag^+ + NH_3$$

$$K_2 = \frac{[NH_3][Ag^+]}{[Ag(NH_3)^+]} = 4.3 \times 10^{-4}$$

and combining them

$$[Ag(NH_3)_2]^+ \rightleftharpoons Ag^+ + 2NH_3$$

$$K_d = K_1 \times K_2 = (1.45 \times 10^{-4})(4.3 \times 10^{-4}) = 6.2 \times 10^{-8}$$
$$= \frac{[NH_3][Ag(NH_3)^+]}{[Ag(NH_3)_2^+]} \times \frac{[NH_3][Ag^+]}{[Ag(NH_3)^+]}$$
$$= \frac{[NH_3]^2[Ag^+]}{[Ag(NH_3)_2^+]}$$

gives the overall dissociation constant, usually simply referred to as the dissociation constant of the complex ion.

For a complex containing four ligands, such as $[Cu(NH_3)_4]^{2+}$, there are four stepwise dissociation reactions and four stepwise dissociation constants, K_1, K_2, K_3, and K_4. The values at room temperature are $K_1 = 5 \times 10^{-3}$; $K_2 = 1 \times 10^{-3}$; $K_3 = 2 \times 10^{-4}$; and $K_4 = 1 \times$

10^{-4}. The first is, of course, the largest, for there is a greater probability of an ammonia molecule escaping from an ion containing four such molecules than there is for it to escape from an ion containing three or less. The four equilibrium equations can be combined into a single one

$$[Cu(NH_3)_4]^{2+} \rightleftharpoons Cu^{2+} + 4NH_3$$

for which

$$\frac{[Cu^{2+}][NH_3]^4}{[Cu(NH_3)_4^{2+}]} = K_1K_2K_3K_4 = 1 \times 10^{-13} = K_d$$

What is referred to in these equations as Cu^{2+} is probably the tetraaquocopper(II) ion, $[Cu(H_2O)_4]^{2+}$, and the reaction with ammonia probably occurs via a stepwise displacement of one ligand by the other.

$$[Cu(H_2O)_4]^{2+} + NH_3(aq) \rightleftharpoons [Cu(H_2O)_3(NH_3)]^{2+} + H_2O(l)$$
$$[Cu(H_2O)_3(NH_3)]^{2+} + NH_3(aq) \rightleftharpoons [Cu(H_2O)_2(NH_3)_2]^{2+} + H_2O(l)$$
$$\text{etc.}$$

As with other equilibria, we can omit the water of hydration in writing complex ion equations and disregard the concentration of water in equilibrium expressions.

The overall dissociation constants of some complexes are shown in Table 28.5. The data given in Table 28.5 illustrate several interesting facts about the stability of complexes:

1. Metal ions vary widely in their ability to form complex ions, even with the same ligand and in complexes with the same coordination number (compare 3 and 5; 2 and 3; 7 and 8).

2. The higher the oxidation state of the metal, the less dissociated are the complexes it forms with a given ligand. This is illustrated here by 16 and 17, 25 and 26, 27 and 28.

3. The stability of halide complexes falls in the order $I^- > Br^- > Cl^-$ (compare 7, 10, and 12). This is often, but not always, true.

4. The cyanide ion is like the halide ions in many respects, but far exceeds them in the stability of its complexes (compare 7, 10, and 12 with 14). This is doubtless due to the fact that the cyanide ion can readily accept electrons from metal ions (back-bonding, Section 28.6).

5. Chelate ring formation (Section 28.3) greatly decreases the dissociation constants of complexes. Monoammines are rather poor coordinating agents (35), but diammines, if they can form five-membered rings, form extremely stable complexes (24).

6. The formation of fused chelate rings gives complexes that are more stable than those containing rings that are not fused (compare

TABLE 28.5

1.	$[Ag(NH_3)_2]^+$	6.2×10^{-8}	19.	$[Zn(OH)_4]^{2-}$	5×10^{-21}
2.	$[Cd(NH_3)_4]^{2+}$	1×10^{-7}	20.	$[CaP_2O_7]^{2-}$	$1 \times 10^{-5\ a}$
3.	$[Cu(NH_3)_4]^{2+}$	1×10^{-13}	21.	$[MgP_2O_7]^{2-}$	2×10^{-6}
4.	$[Ni(NH_3)_6]^{2+}$	1×10^{-9}	22.	$[CuP_2O_7]^{2-}$	2.0×10^{-7}
5.	$[Zn(NH_3)_4]^{2+}$	3.46×10^{-10}	23.	$[Ag(en)]^+$	1×10^{-5}
6.	$[AgCl_2]^-$	9×10^{-6}	24.	$[Cd(en)_2]^{2+}$	2.60×10^{-11}
7.	$[CdCl_4]^{2-}$	9.3×10^{-3}	25.	$[Co(en)_3]^{2+}$	1.52×10^{-14}
8.	$[PdCl_4]^{2-}$	6×10^{-14}	26.	$[Co(en)_3]^{3+}$	2.04×10^{-49}
9.	$[AgBr_2]^-$	7.8×10^{-8}	27.	$[Fe(C_2O_4)_3]^{4-}$	$6 \times 10^{-6\ b}$
10.	$[CdBr_4]^{2-}$	2×10^{-4}	28.	$[Fe(C_2O_4)_3]^{3-}$	3×10^{-21}
11.	$[PdBr_4]^{2-}$	8.0×10^{-14}	29.	$[Cu(gly)_2]$	5.6×10^{-16}
12.	$[CdI_4]^{2-}$	8×10^{-7}	30.	$[Zn(gly)_2]$	1.1×10^{-10}
13.	$[Ag(CN)_2]^-$	1×10^{-22}	31.	$[Cu(edta)]^{2-}$	$1.38 \times 10^{-19\ c}$
14.	$[Cd(CN)_4]^{2-}$	8.2×10^{-18}	32.	$[Zn(edta)]^{2-}$	2.63×10^{-17}
15.	$[Au(CN)_2]^-$	5×10^{-39}	33.	$[Ca(nta)_2]^{4-}$	$2.44 \times 10^{-12\ d}$
16.	$[Fe(CN)_6]^{4-}$	1.3×10^{-37}	34.	$[Mg(nta)_2]^{4-}$	6.3×10^{-11}
17.	$[Fe(CN)_6]^{3-}$	1.3×10^{-44}	35.	$[Cd(CH_3NH_2)_4]^{2+}$	2.82×10^{-7}
18.	$[Cu(OH)_4]^{2-}$	7.6×10^{-17}			

TABLE 28.5

Dissociation constants of some common complexes

The abbreviations of chelating agents are given in Table 28.3.

^{a} $P_2O_7^{4-}$ is the anion of pyrophosphoric acid, $(HO)_2POP(OH)_2$.

$$\overset{O\ \ O}{\underset{\parallel\ \ \parallel}{}}$$

^{b} $C_2O_4^{2-}$ is the anion of oxalic acid, HOOC-COOH.

^{c} edta is the anion of ethylenediaminetetraacetic acid (Fig. 28.1a).

^{d} nta is the anion of nitrilotriacetic acid, $N(CH_2COOH)_3$.

29 and 31; 30 and 32). The complexing agents in these cases are amino acids that form five-membered chelate rings, but with glycine, no fused rings are formed; EDTA forms five fused rings if all of its possible donor atoms coordinate.

The equilibria between metal ions and complexing agents in solution and the dissociation constants of complexes are important in analytical chemistry. They are used in finding the conditions under which one metal ion can be separated as a precipitate while another metal ion is left in solution as a complex and in the determination of whether a precipitate will dissolve in a solution of a complexing agent. Dissociation equilibria are also important in biochemistry. The metal ions in body fluids and plant fluids are always in equilibrium with naturally occurring coordinating agents such as sugars and proteins.

28.10 Using K_d

As is usually the case, the principles of equilibria are best understood by applying them to specific situations. Examples 28.2 and 28.3 illustrate how dissociation equilibria are used in the separation of ions in qualitative analysis.

EXAMPLE 28.1

■ What is the concentration of silver ion in a 0.1M solution of Na[Ag(CN)$_2$]? For Ag(CN)$_2^-$, $K_d = 1 \times 10^{-22}$.

This problem can be solved by the usual equilibrium problem techniques.

$$[Ag(CN)_2]^- \rightleftharpoons Ag^+ + 2CN^-$$

$$K_d = \frac{[Ag][CN^-]^2}{[Ag(CN)_2^-]}$$

Let $x = [Ag^+]$

	$[Ag(CN)_2]^-$	$\rightleftharpoons$	Ag^+	$+$	$2CN^-$
initial	0.1		—		—
change	$-x$		$+x$		$+2x$
equilibrium	$0.1-x$		$+x$		$+2x$

$$K_d = \frac{(x)(2x)^2}{0.1 - x}$$

Since only a minute amount of the complex will dissociate, we take $0.1 - x \approx 0.1$. Thus

$$K_d = \frac{(x)(2x)^2}{0.1 - x} \approx \frac{4x^3}{0.1} = 1 \times 10^{-22}$$

Solving for x gives

$$x = \sqrt[3]{\frac{(1 \times 10^{-22})(0.1)}{4}} = 1 \times 10^{-8}\,M$$

The approximation is valid. The concentration of uncomplexed silver ion in an 0.1M solution of Na[Ag(CN)$_2$] is $1 \times 10^{-8}\,M$. ■

EXAMPLE 28.2

■ Bismuth can be separated from solutions also containing copper and cadmium by adding sufficient ammonia to convert the initially formed copper and cadmium hydroxides to soluble complexes, leaving behind the insoluble bismuth hydroxide, which does not form a complex with ammonia.

Assuming typical cation concentrations of 0.01M, what must [NH$_3$] be in order to assure formation of complex ions by Cu^{2+} and Cd^{2+}? For Cd(OH)$_2$, $K_{sp} = 8.1 \times 10^{-15}$; for Cu(OH)$_2$, $K_{sp} = 1.3 \times 10^{-20}$; for Cd(NH$_3$)$_4^{2+}$, $K_d = 1 \times 10^{-7}$; for Cu(NH$_3$)$_4^{2+}$, $K_d = 1 \times 10^{-13}$.

First, we find the overall equilibrium constant for the formation of the ammine complex ion from cadmium hydroxide.

$$Cd(OH)_2(s) \rightleftharpoons Cd^{2+} + 2OH^- \qquad K = K_{sp} = [Cd^{2+}][OH^-]^2$$

$$Cd^{2+} + 4NH_3(aq) \rightleftharpoons Cd(NH_3)_4{}^{2+} \qquad K = \frac{1}{K_d} = \frac{[Cd(NH_3)_4{}^{2+}]}{[Cd^{2+}][NH_3]^4}$$

$$\overline{Cd(OH)_2(s) + 4NH_3(aq) \rightleftharpoons} \qquad K = \frac{K_{sp}}{K_d} = \frac{[Cd(NH_3)_4{}^{2+}][OH^-]^2}{[NH_3]^4}$$

$$Cd(NH_3)_4{}^{2+} + 2OH^-$$

Solving for $[NH_3]$ and substituting $[Cd(NH_3)_4{}^{2+}] = 0.01M$ and $[OH^-] = 0.02M$ (two OH^- ions for each Cd^{2+}), we get

$$[NH_3] = \sqrt[4]{\frac{K_d}{K_{sp}}[Cd(NH_3)_4{}^{2+}][OH^-]^2}$$

$$= \sqrt[4]{\left(\frac{1 \times 10^{-7}}{8.1 \times 10^{-15}}\right)(0.01)(0.02)^2}$$

$$= 2.7M$$

A similar calculation for copper hydroxide yields a smaller value, $2.4M$. Therefore, NH_3 must be added until the concentration is greater than $2.7M$. ■

EXAMPLE 28.3

■ Copper(II) and cadmium(II) ions both form complexes with the cyanide ion, copper being reduced in the process to copper(I). Both ions also form insoluble sulfides.

From the K_{sp} values (Cu_2S, $K_{sp} = 3 \times 10^{-48}$; CdS, $K_{sp} = 2 \times 10^{-28}$) and K_d values ($[Cu(CN)_3]^{2-}$, $K_d = 1.0 \times 10^{-35}$; $[Cd(CN)_4]^{2-}$, $K_d = 8.2 \times 10^{-18}$) we can prove what experiment shows to be true—that in the presence of excess CN^- and S^{2-}, most of the cadmium ion precipitates as cadmium sulfide, but most of the copper remains in solution as the complex ion.

First, we find the overall reaction and equilibrium constants for the precipitation of the sulfides from the complexes in solution.

$$[Cd(CN)_4]^{2-} \rightleftharpoons Cd^{2+} + 4CN^- \qquad K = K_d = \frac{[Cd^{2+}][CN^-]^4}{[Cd(CN)_4{}^{2-}]}$$

$$Cd^{2+} + S^2 \rightleftharpoons CdS(s) \qquad K = \frac{1}{K_{sp}} = \frac{1}{[Cd^{2+}][S^{2-}]}$$

$$(1) \quad \overline{[Cd(CN)_4]^{2-} + S^{2-} \rightleftharpoons CdS(s) + 4CN^-} \qquad K_1 = \frac{K_d}{K_{sp}} = \frac{8.2 \times 10^{-18}}{2 \times 10^{-28}}$$

$$= 4 \times 10^{10}$$

$$2([Cu(CN)_3]^{2-} \rightleftharpoons Cu^+ + 3CN^-) \qquad K = K_d{}^2 = \frac{[Cu^+]^2[CN^-]^6}{[Cu(CN)_3]^{2-}}$$

$$2Cu^+ + S^{2-} \rightleftharpoons Cu_2S(s) \qquad K = \frac{1}{K_{sp}} = \frac{1}{[Cu^+]^2[S^{2-}]}$$

$$(2) \quad 2[Cu(CN)_3]^{2-} + S^{2-} \rightleftharpoons Cu_2S + 6CN^- \qquad K_2 = \frac{K_d{}^2}{K_{sp}} = \frac{(1.0 \times 10^{-35})^2}{3 \times 10^{-48}}$$

$$= 3 \times 10^{-23}$$

Comparison of the equilibrium constants shows that formation of CdS is quite favorable, but that Cu_2S formation is not.

By assuming typical values of $0.1M$ for $[CN^-]$ and $[S^{2-}]$ and solving the overall equilibrium expressions for the complex concentrations, we can find indirectly how much of each ion will form a sulfide.

$$[Cd(CN)_4{}^{2-}] = \frac{[CN^-]^4}{K_1[S^{2-}]} = \frac{(0.1)^4}{(4 \times 10^{10})(0.1)} = 3 \times 10^{-14}\ M$$

$$[Cu(CN)_3{}^{2-}] = \sqrt{\frac{[CN^-]^6}{K_2[S^{2-}]}} = \sqrt{\frac{(0.1)^6}{(3 \times 10^{-23})(0.1)}} = 6 \times 10^8\ M$$

In the case of cadmium, nearly all of the complex reacts to form sulfide. On the other hand, none of the copper forms sulfide— precipitation of sulfide would not begin until the complex concentration reached $6 \times 10^8\ M$, an impossibility if $[Cu^+]$ is about $0.1M$. ∎

Complexes in nature and in practical applications

Many of the uses of complexes are based on their ability to maintain a metal in solution. As we have mentioned, complexing agents are often used in analytical chemistry, where an ion that would otherwise interfere is held in solution as a soluble complex while another ion is removed as a precipitate. Also, soluble complexes make trace elements available to plants under growing conditions in which the element is scarce or unavailable. For reasons that are not well understood, metals such as silver and gold can be electroplated much more smoothly and evenly from solutions of the complexes $[Ag(CN)_2]^-$ and $[Au(CN)_2]^-$, than from solutions of the simple metal ions.

Complexes are also used to snatch metal ions out of solution under circumstances where the metal ions are harmful. The calcium, magnesium, and iron ions can be removed from hard water by complexing agents. The chelating agent British anti-Lewisite (BAL)

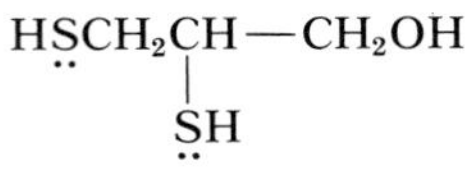

$$HSCH_2CH-CH_2OH$$
$$|$$
$$SH$$

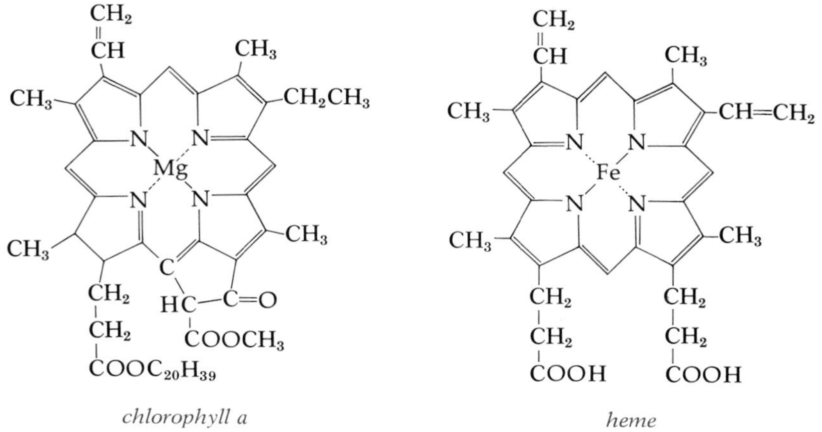

The structures of chlorophyll a and heme are shown in Figure 28.13.

chlorophyll a

heme

FIGURE 28.13
Structures of chlorophyll *a* and heme. *These compounds are both members of a class called porphyrins.*

was developed during wartime as an antidote to an arsenic-containing poisonous gas (called Lewisite). BAL is now used to treat poisoning by arsenic, mercury, gold, bismuth, antimony, thallium, tellurium, and chromium, and it is possibly useful against other metals such as lead and cadmium.

Most of the transition metals that are essential trace elements are complexed with proteins in metalloenzymes essential to various body functions. For example, cobalt is part of glutamate mutase, an enzyme that functions in amino acid metabolism. Zinc, as it is incorporated in a carboxypeptidase, is necessary to the digestion of proteins by aiding in cleavage of the protein carbon–nitrogen bond. The two best known biochemically important metal complexes are heme, part of hemoglobin, the red pigment in blood that transports oxygen, and chlorophyll *a*, vital to the photochemical transfer of energy in green plants. Both of these compounds (Figure 28.13) are members of a biochemically important and structurally similar class of compounds called porphyrins.

Phthalocyanine blue (Figure 28.14), an extremely stable copper(II) complex, is one of many complex compounds used as dyes or pigments. This deep blue paint pigment is made by heating a mixture of a copper(II) salt, phthalic anhydride (the anhydride of an organic acid), a nitrogen compound such as urea, and a trace of catalyst (usually a molybdenum compound). Other metals can be used in place of copper, but the product is not as stable toward acids, bases, or sunlight. The color is essentially the same no matter what metal is used, showing that the color is due to the large organic group rather to the metal. Replacement of all the hydrogen atoms in phthalocyanine blue by chlorine atoms gives a green paint pigment.

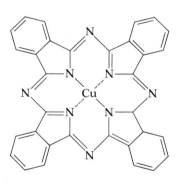

FIGURE 28.14
Copper phthalocyanine blue.

Some coordination compounds in biochemistry

Biochemists have long known that living bodies contain small amounts of metallic elements and have sought to learn the structures and functions of the compounds that contain these metals. In very recent years, this has become an exciting part of inorganic chemistry as well. . . . Chelating agents are widely prevalent in nature and are more powerful than is generally believed. Certain strains of soybeans growing in alkaline soil generate and secrete into the soil chelating agents that solubilize the iron needed by the plant. Similarly, mosses and lichens growing on solid rock extract the metals necessary for their growth through chelating agents which they generate. Microbiologists have often observed that bacteria growing in stainless-steel culture tanks etch the steel in order to get the iron they need for growth. . . . Many soils are deficient in cobalt, copper, zinc, and molybdenum, and the plants that grow in them are likewise deficient. This may not be important in the growth of the plant, but it may become important if the plants are used for human or animal food. For example, sheep raisers in Australia have known for many years that the sheep became ill if they grazed in certain areas. This illness, called sheep-sick, was finally traced to a deficiency of cobalt in the soil, and hence in the grass that grew there. The difficulty was quickly rectified by sprinkling the soil with solutions of cobalt salts, but this had to be done periodically and was inefficient in the use of cobalt. The problem is now avoided by forcing each sheep to swallow a pellet of cobalt metal and a small screw. These objects, being heavy, remain in the rumen indefinitely and give the animal all the cobalt it needs. The screw rubs against the pellet of cobalt and thus prevents it from becoming covered with an insoluble coat. When the sheep dies or is slaughtered, the cobalt pellet is recovered and given to another sheep—it may even become a family heirloom.

SOME COORDINATION COMPOUNDS IN BIOCHEMISTRY, by John C. Bailar, Jr.
(*American Scientist*, Vol. 59, p. 586, 1971)

Significant terms defined in this chapter

complex	inner orbital complexes
ligand	back-bonding
complex ions	crystal field stabilization energy
coordination compound	high-spin complex
coordination number	low-spin complex
chelation	strong-field ligand
chelate ring	weak-field ligand
outer orbital complexes	dissociation constant

Exercises

28.1 Iron in different oxidation states forms three different coordination compounds with CN^-

$Fe_4[Fe(CN)_6]_3$
Iron(III) ferrocyanide
Dark blue
  (Prussian blue)

$Fe[Fe(CN)_6]$
Iron(III) ferricyanide
Green (Berlin green)

$Fe_2[Fe(CN)_6]$
Iron(II) ferrocyanide
Blue-white

(a) Name each of these compounds using the conventional system presented in this chapter. (b) Write the electronic configuration of Fe, Fe(II), and Fe(III), and (c) describe the bonding in $[Fe(CN)_6]^{4-}$ and $[Fe(CN)_6]^{3-}$ using the valence bond theory. Suppose you have available solutions of Fe^{3+}, Fe^{2+}, $K_4[Fe(CN)_6]$ and $K_3[Fe(CN)_6]$. (d) Devise a qualitative analysis scheme that could be used to determine the oxidation state of iron in an unknown solution.

28.2 Distinguish clearly between: (a) donor and acceptor atoms, (b) outer- and inner-orbital complexes, (c) high- and low-spin complexes, (d) strong- and weak-field ligands, and (e) ionic and covalent bonding within a coordination compound.

28.3 Which statements are true? Rewrite any false statement so that it is correct.
(a) Those metal ions with small radius and high charge and those that have vacant d orbitals readily form coordination compounds.
(b) Chelation greatly increases the stability of a complex compared to a similar complex not having chelation.
(c) Valence bond theory describes metal–ligand bonding as coordinate covalent bonding in terms of overlap of atomic orbitals. The atomic orbitals on the metal that are filled by the ligand electrons are hybridized.
(d) Crystal field theory views the ions and groups in complexes in terms of the spatial relationships of the orbitals and the interactions of the ions and groups based on their electrostatic attraction and repulsion.
(e) Molecular orbital theory describes the bonding in terms of the formation of bonding, antibonding, and non-bonding molecular orbitals formed by the combination of orbitals on the metal and ligands.

28.4 How many unpaired electrons would you predict to be in each of the following: (a) $[Fe(CN)_6]^{3-}$, (b) $[Fe(H_2O)_6]^{3+}$, (c) $[Mn(H_2O)_6]^{3+}$, and (d) $[Co(NH_3)_6]^{3+}$? (e) Name these complexes.

28.5 Draw a sketch of the crystal field splitting for octahedral complexes and label each of the energy levels. How many t_{2g} electrons and how many e_g electrons are there in (a) $[Mn(H_2O)_6]^{3+}$, (b) $[CoF_6]^{3-}$, and (c) $[Ti(H_2O)_6]^{3+}$? (d) Name these complexes.

28.6 Why is the dissociation constant less for (a) $[Co(en)_3]^{3+}$ than $[Co(en)_3]^{2+}$, (b) $[Cu(gly)_2]$ than $[Zn(gly)_2]$, and (c) $[Cd(en)_2]^{2+}$ than $[Cd(NH_3)_4]^{2+}$?

28.7 Werner studied the electrical conductance of aqueous solutions containing a series of platinum(IV) complexes having the general formula $Pt(NH_3)_xCl_4$, where x is an integer that varied from 2 to 6. His results can be summarized as

Formula of complex	Number of ions produced upon complete dissociation
$Pt(NH_3)_6Cl_4$	5
$Pt(NH_3)_5Cl_4$	4
$Pt(NH_3)_4Cl_4$	3
$Pt(NH_3)_3Cl_4$	2
$Pt(NH_3)_2Cl_4$	0

Assuming that Pt(IV) forms octahedral complexes, (a) write the formulas for the five compounds based on the dissociation results, (b) draw three-dimensional sketches of the complexes, (c) draw sketches of any isomers that are possible, and (d) name each compound.

28.8* Consider the compound having the formula $[Co(NH_3)_5(H_2O)]^{3+}[Co(NO_2)_6]^{3-}$. (a) Name this compound. In terms of valence bond theory, (b) describe the bonding in each ion. (c) Would you expect this substance to be paramagnetic or diamagnetic?

28.9* At 25°C, $\Delta H° = -11.1$ kcal, $\Delta G° = -10.24$ kcal, and $\Delta S° = -2.8$ cal/°K for the reaction

$$[Cu(H_2O)_4]^{2+} + 2NH_3(aq) \longrightarrow [Cu(NH_3)_2(H_2O)_2]^{2+} + 2H_2O(l)$$

and $\Delta H° = -6.0$ kcal, $\Delta G° = -11.7$ kcal and $\Delta S° = 19$ cal/°K for the reaction

$$[Cu(H_2O)_4]^{2+} + gly^- \longrightarrow [Cu(gly)(H_2O)_2]^+ + 2H_2O(l)$$

In each case, two Cu–water bonds are broken, but in the first reaction two Cu–NH_3 bonds are formed and in the second reaction a Cu–NH_2 and a Cu–O bond are formed. Based on the $\Delta H°$ values, (a) which set of bonds is stronger? In each case, two water molecules are replaced in the hydrated Cu^{2+} ion by the ligands—in the first reaction by two ligands and in the second reaction by one ligand. (b) Which reaction should have the larger increase in randomness (entropy)? (c) Do the values of $\Delta S°$ confirm your answer to (b)? Based on values of $\Delta G°$, (d) which complex ion is more stable? (e) Name all of the complexes.

28.10* People are born with almost no Cd^{2+} in their bodies, but this chemical is accumulated throughout their lives by those living in the United States and other "advanced" countries. It is believed that most of the cadmium ion enters the lungs in the form of dust from the wear of rubber tires on pavement (there is a high percentage of ZnO in rubber and Cd^{2+} is a common contaminant in Zn^{2+}). Certain complexing agents, taken orally, can react with the Cd^{2+} so that it can be eliminated from the body through the urinary system. (In fact, the body can slowly excrete Cd^{2+} in the urine, but not as rapidly as it is absorbed.) What characteristics must these oral complexing agents have?

28.11 Consider the formation of the triiodosilver(I) ion

$$Ag^+ + 3I^- \longrightarrow [AgI_3]^{2-}$$

Would you expect an increase or decrease in the entropy of the system as the complex is formed? The standard state absolute entropies as 25°C are 17.37 cal/mole °K for Ag^+, 26.6 cal/mole °K for I^- and 60.5 cal/mole °K for AgI_3^{2-}. Calculate $\Delta S°$ for the reaction and confirm your prediction.

28.12 A compound synthesized in an inorganic preparation experiment was analyzed and found to contain 17.0 wt % Fe, 21.9 wt % C, 25.5 wt % N, and 35.6 wt % K. Find the empirical formula of the compound. The osmotic pressure exerted by a $0.0010M$ solution at 25°C of this compound was 0.095 atm. How many ions are present in this compound? Write the formula of the compound as a coordination compound. Name this substance.

28.13 Write the electronic configuration for (a) an atom of Co and (b) a Co^{3+} ion. In terms of valence bond theory, show the electronic distribution in (c) $[Co(CN)_6]^{3-}$, a low-spin (spin-paired) complex ion, and in (d) $[CoF_6]^{3-}$, a high-spin (spin-free) complex ion. (e) Express the crystal field stabilization energy for each complex and (f) name these complex ions.

28.14 Molecular iodine reacts with I^- to form a complex ion

$$I_2(aq) + I^- \rightleftharpoons [I_3]^-$$

Calculate the equilibrium constant for this reaction given the following data at 25°C:

$I_2(aq) + 2e^- \rightleftharpoons 2I^-$ $E° = 0.535$ V
$[I_3]^- + 2e^- \rightleftharpoons 3I^-$ $E° = 0.5338$ V

28.15 Find the concentration of the metal cation in equilibrium with (a) $0.010M$ $[Fe(CN)_6]^{3-}$, $K_d = 1.3 \times 10^{-44}$; (b) $0.10M$ $[Zn(CN)_4]^{2-}$, $K_d = 2.4 \times 10^{-20}$; (c) $1.0M$ $[Co(C_2O_4)_3]^{4-}$, $K_d = 2.2 \times 10^{-7}$. (d) Name the complexes.

28.16 Calculate the concentration of Ag^+ in equilibrium with $0.010M$ $[Ag(NH_3)_2]^+$ if $K_d = 6.2 \times 10^{-8}$. Use this value to calculate the ion product for AgCl if the solution is also $0.010M$ Cl^-. If $K_{sp} = 1.8 \times 10^{-10}$ for AgCl, will any AgCl form?

28.17 A solution is $0.010M$ with respect to Cd^{2+} and also with respect to Pd^{2+}. Solid KBr is added to the solution (assume no change in volume) until $[Br^-] = 0.10M$. What will be the concentrations of Cd^{2+} and Pd^{2+} once equilibrium is reached? $K_d = 2 \times 10^{-4}$ for $[CdBr_4]^{2-}$ and 8.0×10^{-14} for $[PdBr_4]^{2-}$. Name the complexes.

28.18 As part of its water purification program, a certain city aerates the water. During this process, any $Ca(HCO_3)_2$ present in the water is changed to $CaCO_3$. To prevent the formation of scale in the treatment plant equipment, $Na_4P_2O_7$ is added to keep the concentration of Ca^{2+} low enough in the form of $[CaP_2O_7]^{2-}$ so that the $CaCO_3$ doesn't precipitate. What weight of $Na_4P_2O_7$ must be added to a thousand liters of the water if the bicarbonate concentration in the water is 50 ppm by weight? $K_{sp} = 3.84 \times 10^{-9}$ for $CaCO_3$ and $K_d = 1 \times 10^{-5}$ for $[CaP_2O_7]^{2-}$.

28.19* If $K = 3.9 \times 10^{16}$ for

$$Zn^{2+} + EDTA^{4-} \rightleftharpoons [Zn(EDTA)]^{2-}$$

where $EDTA^{4-}$ is the ethylenediaminetetraacetate ion and if $K_{sp} = 2 \times 10^{-24}$ for ZnS, will ZnS form in a solution originally containing $0.100M$ $EDTA^{4-}$, $0.01M$ S^{2-}, and $0.010M$ Zn^{2+}?

28.20* It is possible to dissolve a precipitate of AgCl by adding concentrated aqueous NH_3, but not a precipitate of AgI. To prove this, calculate the silver halide ion product for (a) a solution containing $0.010M$ Ag^+, $0.010M$ Cl^-, $5.0M$ NH_3 and (b) a solution containing $0.010M$ Ag^+, $0.010M$ I^-, and $5.0M$ NH_3. (c) In which solution will the ion product exceed the K_{sp}? $K_{sp} = 8.3 \times 10^{-17}$ for AgI, $K_{sp} = 1.8 \times 10^{-10}$ for AgCl, and $K_d = 6.2 \times 10^{-8}$ for $Ag(NH_3)_2^+$.

28.21* The general rate expression for the chemical reaction

$$[Co(NH_3)_5F]^{2+} + H_2O(l) \longrightarrow [Co(NH_3)_5(H_2O)]^{3+} + F^-$$

is given by

$$rate = k[Co(NH_3)_5F^{2+}]^x[H_2O]^y$$

Because the amount of water present is so large, the rate is essentially independent of $[H_2O]$, so we write the pseudo-order expression

$$rate = k'[Co(NH_3)_5F^{2+}]^x$$

Experimentally it is found that doubling the concentration of $[Co(NH_3)_5F]^{2+}$ doubles the rate of reaction. (a) What is the order of reaction with respect to the complex? A second experiment shows that the rate doubles as the temperature is changed from 25°C to 35°C. (b) What is the energy of activation for the process? (c) Name the complexes.

28.22** An inorganic chemist synthesized a compound that contained 65 wt % Pt, 24 wt % Cl, 6 wt % NH_3, and 6 wt % H_2O. (a) Determine the empirical formula of the compound and (b) name the compound. Assuming this to be a complex of platinum with four ligands, he drew a tetrahedral structure and two square-planar structures for the compound. (c) Draw these structures and (d) decide which arrangement allows for isomerism. After a careful analysis of the inorganic compound, two isomers were separated and studied. The standard state free energy of formation at 25°C is −94.7 kcal/mole for the cis isomer and −96.0 kcal/mole for the trans isomer. (e) Which isomer is more stable? (f) Calculate the equilibrium constant for the reaction

cis isomer ⇌ trans isomer

If a 0.0100M solution of the cis isomer were prepared, (g) what would be the concentrations of each isomer once equilibrium conditions are reached?

Describe the bonding in the compound using (h) valence bond theory and (i) crystal field theory. The crystal field splitting in terms of Dq is

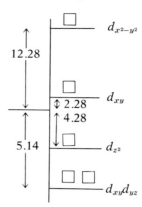

In this chapter we take a periodic perspective look at the chemistry of the *d* and *f* transition metals, continuing what was begun in Chapter 10. The descriptive chemistry of the most important transition metals—chromium, manganese, iron, cobalt, nickel, copper, silver, and gold—is presented in separate sections. And a brief look is taken at some of the most interesting aspects of the chemistry and uses of the other transition metals.

Think of something made of metal. The chances are good that whatever you pictured in your mind—an automobile, a cooking pot, a water pipe, or a piece of jewelry—contains either a transition metal or several transition metals combined in an alloy. The beautiful and precious metals that occasionally arouse violent disputes over their possession—platinum, silver, and gold—are transition metals, as is that most important workhorse metal, iron.

Many transition elements that are close neighbors in the periodic table have very similar chemical and physical properties. As a result, they often occur together in minerals, and it is difficult to separate the pure elements. This made proof of the existence of some of them quite difficult. For example, elements 72 and 75 were hunted for many years. The search was inspired by the work of Moseley on atomic numbers and x-ray spectra, which showed that there were gaps in the periodic table at these spots (and also at atomic numbers 43 and 61).

The search for element 72 was first carried out in the wrong place—among the rare earth minerals. Then, on the basis of quantum

Ca	Sc 21	Ti 22	V 23	Cr 24	Mn 25	Fe 26	Co 27	Ni 28	Cu 29	Zn	Ga
	44.959	47.90	50.9414	51.996	54.9380	55.847	58.9332	58.70	63.546		
Sr	Y 39	Zr 40	Nb 41	Mo 42	Tc 43	Ru 44	Rh 45	Pd 46	Ag 47	Cd	In
	88.9059	91.22	92.9064	95.94	(97)	101.07	102.905	106.4	107.868		
Ba	La* 57	Hf 72	Ta 73	W 74	Re 75	Os 76	Ir 77	Pt 78	Au 79	Hg	Tl
	138.905	178.49	180.948	183.85	186.207	190.2	192.22	195.09	196.967		
Ra	Ac** 89	Rf 104	Ha 105								
	(227)	(257)	(260)								

f-transition elements intervene

theory, Niels Bohr predicted that element 72 should have four, rather than three, valence electrons and therefore should be found in zirconium minerals. This prediction was proven correct by the identification of the x-ray spectrum of element 72 (named hafnium after the Latin name for Copenhagen) in samples of what had been thought to be pure zirconium, as well as in zirconium ores.

Element 75, rhenium (named by its German discoverers for the Rhine River), was also identified by its x-ray spectrum. It was necessary to concentrate a mineral extract 100,000 times to get a sample that contained enough rhenium for identification.

The other two missing elements predicted by Moseley are element 43, technetium, and element 61, promethium. Both of these elements are radioactive and were first identified as the products of nuclear reactions.

Periodic perspective: d transition elements

29.1 Members and configurations

The three d transition series of elements, in which the d energy levels are being filled with electrons, are as follows:

First transition series—$3d$ level being filled

$_{21}$Sc–$_{29}$Cu

Second transition series–$4d$ level being filled

$_{39}$Y–$_{47}$Ag

Third transition series—$5d$ level being filled

$_{57}$La–$_{79}$Au

The outermost configurations of these elements are shown in Table 29.1. Atoms of each d transition metal, with the exception of palla-

TABLE 29.1

Electronic configurations of d transition elements
Deviations from the idealized configuration occur when a different configuration is more stable.

Idealized configuration:	$(n-1)d^1ns^2$	$(n-1)d^2ns^2$	$(n-1)d^3ns^2$	$(n-1)d^4ns^2$	$(n-1)d^5ns^2$	$(n-1)d^6ns^2$	$(n-1)d^7ns^2$	$(n-1)d^8ns^2$	$(n-1)d^{10}ns^1$
$[3s^23p^6]$	$_{21}$Sc $3d^14s^2$	$_{22}$Ti $3d^24s^2$	$_{23}$V $3d^34s^2$	$_{24}$Cr $3d^54s^1$	$_{25}$Mn $3d^54s^2$	$_{26}$Fe $3d^64s^2$	$_{27}$Co $3d^74s^2$	$_{28}$Ni $3d^84s^2$	$_{29}$Cu $3d^{10}4s^1$
$[4s^24p^6]$	$_{39}$Y $4d^15s^2$	$_{40}$Zr $4d^25s^2$	$_{41}$Nb $4d^45s^1$	$_{42}$Mo $4d^55s^1$	$_{43}$Tc $4d^55s^2$	$_{44}$Ru $4d^75s^1$	$_{45}$Rh $4d^85s^1$	$_{46}$Pd $4d^{10}$	$_{47}$Ag $4d^{10}5s^1$
$[5s^25p^6]$	$_{57}$La $5d^16s^2$	$_{72}$Hf $4f^{14}5d^26s^2$	$_{73}$Ta $4f^{14}5d^36s^2$	$_{74}$W $4f^{14}5d^46s^2$	$_{75}$Re $4f^{14}5d^56s^2$	$_{76}$Os $4f^{14}5d^66s^2$	$_{77}$Ir $4f^{14}5d^76s^2$	$_{78}$Pt $4f^{14}5d^96s^1$	$_{79}$Au $4f^{14}5d^{10}6s^1$
$[6s^26p^6]$	$_{89}$Ac $6d^17s^2$	104	105						

dium ($4s^2 4p^6 4d^{10}$), contain one or two *s* electrons, and these elements are all metals. Except for elements near the ends of the series, each transition metal atom also has an incompletely filled $(n - 1)d$ shell.

Lanthanum can be thought of as the first member of the third transition series, but also as the first member of the *f* transition series, called collectively the lanthanides. Following the addition of the $5d$ electron to form the lanthanum atom ($5d^1 6s^2$), the next fourteen electrons enter the $4f$ shell ($_{58}$Ce–$_{71}$Lu), and it is only after that shell is filled that electrons proceed to enter the $5d$ shell again at hafnium ($4d^{14} 5d^2 6s^2$).

The *d* electrons in the transition metals are susceptible to many influences, and generalizations about how they will behave, about configurations, and about trends of properties are harder to make for the *d* transition metals than for the representative elements. The ready availability of *d* electrons is, however, responsible for many of the characteristic properties of the *d* transition elements.

29.2 Characteristic transition metal properties

The *d* transition metal atoms can contain up to five unpaired *d* electrons, and their compounds often also contain unpaired electrons. In contrast to inorganic compounds containing only representative elements, compounds of many of the *d* transition elements are paramagnetic and colored. The unpaired electrons are often responsible for the color in such compounds. Unpaired electrons are so readily elevated from the *d* level to higher energy levels that absorption of visible light provides enough energy to cause this change. White light falls on the compound, either in the solid state or in solution, a portion of the spectrum is absorbed, and we see a color corresponding to the frequency of the unabsorbed light (the absorbed energy is dispersed as vibrational energy of the atoms and molecules, not as re-emitted light). Among the transitions that may occur are promotion of electrons from the *d* level to the higher *s* level or from the lower t_{2g} to the higher e_g energy level in complexes (called crystal-field transitions).

There are some exceptions—for example, titanium(IV) compounds are usually colorless because the electrons are all paired and are not readily raised to higher energy levels. Many colorless ions are also found among the compounds of the second and third transition series, for example, $[Pd(NH_3)_4]^{2+}$ and $[Pt(NH_3)_6]^{4+}$. In such compounds the electrons are so tightly held that they are not readily raised to orbitals of higher energy.

Another general property of *d* transition metals and their compounds is their high catalytic activity, which is usually due to the ease with which electrons are lost and gained or moved from one shell to another. Some typical applications of this property are the use of nickel as a hydrogenation catalyst, the ability of V_2O_5 to serve as a catalyst in the contact process for the manufacture of sulfuric acid, and the use of MnO_2 in the catalytic decomposition of potassium chlorate. The catalytic activity of the transition metals and their compounds gives them great importance in the chemical industry. Chemists are always attempting to modify the rates of reactions by catalytic action. In many cases, the mechanism by which a well-known catalyst operates has not yet been discovered, and there is continuous effort to better understand how catalysts behave. In the meantime, empirical research is carried on in an attempt to improve the catalysts that are available or to find new and better ones. ("Empirical" research is aimed not necessarily at understanding, but at just finding a better way.) These two types of research overlap to a considerable extent, and, not infrequently, discoveries based on empirical research throw light on the more fundamental problems of the "why and how" of catalytic behavior.

A third important property of transition metals is their ready formation of <u>coordination compounds</u> (as discussed in the preceding chapter). Many transition metal ions, both in solution and in their crystalline compounds, prefer to be associated with coordinated water molecules.

29.3 Atomic and ionic radii

Across the periodic table from left to right there is a general, but not regular, decrease in the atomic radii of the *d* transition elements (Figure 29.1). This trend is consistent with the steady increase in effective nuclear charge as electrons are added to an inner *d* energy level, while the configuration in the outermost shell remains essentially fixed. The effective nuclear charge increases because *d* electrons are relatively poor at screening and do not completely neutralize the increased nuclear charge caused by the addition of a proton.

Electrons in the *f* level are also poor screeners and for the same reason there is (with a few exceptions) a general decrease in atomic radii across the *f* transition series from lanthanum to lutetium—the **lanthanide contraction.**

In going down a periodic table family, an increase in size is expected as new shells of electrons are added. This increase is evident

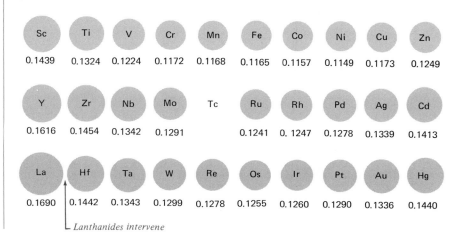

Sc	Ti	V	Cr	Mn	Fe	Co	Ni	Cu	Zn
0.1439	0.1324	0.1224	0.1172	0.1168	0.1165	0.1157	0.1149	0.1173	0.1249

Y	Zr	Nb	Mo	Tc	Ru	Rh	Pd	Ag	Cd
0.1616	0.1454	0.1342	0.1291		0.1241	0. 1247	0.1278	0.1339	0.1413

La	Hf	Ta	W	Re	Os	Ir	Pt	Au	Hg
0.1690	0.1442	0.1343	0.1299	0.1278	0.1255	0.1260	0.1290	0.1336	0.1440

└ *Lanthanides intervene*

FIGURE 29.1

Atomic radii (in nanometers) of the *d* transition elements. *Observe the effect of the interposition of the lanthanides between lanthanum and hafnium on the comparative sizes of the elements of the second and third sequences.*

TRANSITION
ELEMENTS [29.3]

881

between the first and second members of each transition metal family (e.g., compare Ti and Zr, or Ni and Pd). However, the lanthanide contraction just about counteracts the further expected increase, and atoms of the second and third members of the transition metal families are nearly the same in size (compare Zr and Hf or Pd and Pt). Their similarity in both configuration and size causes the second and third members of each transition metal family to resemble each other more closely in chemical behavior and to be more similar to each other than to the first members of the families.

For ions of the same charge, a similar gradual decrease in ionic radius occurs across both the d and f transition series (for example, for M^{3+} in the first transition series: Sc, 0.081; Ti, 0.076; V, 0.074; Cr, 0.069; Mn, 0.066; Fe, 0.064; Co, 0.063; and Ni, 0.062 nm).

29.4 Ionization energies and standard reduction potentials

Along with the decrease in size and increase in effective nuclear charge from left to right in each transition series goes a general, although not regular, increase in electronegativities and a decrease in ease of electron loss. The first ionization energies (Table 29.2) illustrate this irregular decrease in ease of electron loss and also show that no predictable trend exists in going down each transition family. The only firm generalization to be made about the families is that, due to the interposition of the lanthanides, the elements hafnium through gold have higher ionization energies than the elements that precede them in the same families (zirconium through silver). A comparison of the ionization energies for formation of M^{2+} and the standard reduction potentials for the M^{2+} ions demonstrates that the decrease in ease of electron loss across the series holds both in the gaseous state and in solution (Table 29.3).

29.5 Oxidation states and compound formation

All of the d transition metals except scandium exhibit a number of oxidation states. The three valence electrons of scandium, although they fall in two different shells, have such similar energies that it is difficult to bring the $4s$ electrons into reaction without also involving a $3d$ electron. Therefore, scandium always has oxidation state +3.

In most cases, it is difficult to predict from the electronic configuration of a transition element what its most stable, and therefore, most common oxidation state will be. For Sc through Mn in the $3d$ series, the maximum oxidation state (Table 29.4) can be correlated with electronic configuration—it is the sum of the number of $3d$ and $4s$ electrons. Beyond manganese there is no such correlation. At present, the highest known oxidation state of iron is +6 and the highest states known for cobalt and nickel are +4. Apparently, the higher the nuclear charge, the more tightly the electrons are held.

For vanadium, as for titanium, the maximum oxidation state is

TABLE 29.2
Additional properties of d transition elements

The ionization energies may be converted to kJ/mole by multiplying by 4.184 kJ/kcal.

	Density (g/cm³, 20°C)	Electro-negativity	First ionization energy (0°K) (kcal/mole)
First series			
Scandium (Sc)	3.0	1.3	151
Titanium (Ti)	4.5	1.5	157
Vanadium (V)	6.1	1.6	155
Chromium (Cr)	7.2	1.6	156
Manganese (Mn)	7.4	1.5	171
Iron (Fe)	7.8	1.8	182
Cobalt (Co)	8.90	1.8	181
Nickel (Ni)	8.91	1.8	176
Copper (Cu)	8.94	1.9	178
Second series			
Yttrium (Y)	4.3	1.2	147
Zirconium (Zr)	6.5	1.4	158
Niobium (Nb)	8.6	1.6	158
Molybdenum (Mo)	10.2	1.8	164
Technetium (Tc)	11.5	1.9	167
Ruthenium (Ru)	12.45	2.2	170
Rhodium (Rh)	12.41	2.2	172
Palladium (Pd)	12.02	2.2	192
Silver (Ag)	10.5	1.9	174
Third series	*lanthanide elements intervene*		
Lanthanum (La)	6.2	1.1	129
Hafnium (Hf)	13.3	1.3	161
Tantalum (Ta)	16.7	1.5	182
Tungsten (W)	19.3	1.7	184
Rhenium (Re)	21.0	1.9	182
Osmium (Os)	22.61	2.2	201
Iridium (Ir)	22.65	2.2	214
Platinum (Pt)	21.45	2.2	207
Gold (Au)	19.3	2.4	213

TABLE 29.3
Ionization energies and standard reduction potentials for first series transition metals

	Ionization energy (0°K) kcal/mole (kJ/mole) $M(g) \rightleftharpoons M^{2+}(g) + 2e^-$	Standard reduction potential (25°C) (V) $M^{2+} + 2e^- \rightarrow M$
Sc	446 (1870)	—
Ti	470 (1970)	−1.63
V	493 (2060)	−1.2
Cr	536 (2240)	−0.56
Mn	532 (2225)	−1.029
Fe	555 (2320)	−0.41
Co	574 (2400)	−0.28
Ni	595 (2490)	−0.23
Cu	646 (2700)	+0.34

TABLE 29.4
Oxidation states of the 3d transition elements

Element	Electronic configuration (Ar core not shown)	Most common oxidation state	Maximum known oxidation state
Sc	$3d^1 4s^2$	+3	+3
Ti	$3d^2 4s^2$	+4	+4
V	$3d^3 4s^2$	+5	+5
Cr	$3d^5 4s^1$	+3	+6
Mn	$3d^5 4s^2$	+2	+7
Fe	$3d^6 4s^2$	+2, +3	+6
Co	$3d^7 4s^2$	+2	+4
Ni	$3d^8 4s^2$	+2	+4
Cu	$3d^{10} 4s^1$	+2	+3

also the most stable and most common. Beyond vanadium in the $3d$ series (Table 29.4), the most common states are lower than the maximum oxidation states.

The maximum oxidation state encountered for any transition element is +8, and this is attained by ruthenium and osmium, members of the iron family, which have $4d^7 5s^1$ and $5d^6 6s^2$ configurations, respectively. For these two elements, the maximum oxidation state is also the sum of the number of $(n - 1)d$ and ns electrons.

Do the elements of the $3d$ sequence in their most common oxidation states form ionic or covalent species when combined with nonmetals? We know from ionization energies that removal of the first electron is easier than removal of the second, which is easier than removal of the third, and so on. Therefore, ionic compounds are most likely to be formed when the $3d$ transition elements are in their lowest oxidation states. For example, there are many ionic compounds of dipositive manganese, iron, cobalt, and nickel, but in most binary chromium(III) and iron(III) compounds, the bonding is essentially covalent.

A striking fact emerges from examination of the outer configuration of these ions

Mn^{2+}	Fe^{2+}	Co^{2+}	Ni^{2+}	Cu^{2+}
$3d^5$	$3d^6$	$3d^7$	$3d^8$	$3d^9$

Although many representative metal ions have noble-gas configurations, none of the $3d$ dipositive transition metal ions do. In fact, among the transition metals, only scandium and the other members of its family (yttrium, lanthanum, and actinium) form ions with noble gas configurations (M^{3+} ions). Most other transition metal ions have irregular configurations in which the d level is incomplete, although in some cases the d level has its maximum of ten electrons and the ion then has a pseudo-noble-gas configuration (e.g., the Cu^+ ion, of configuration $3d^{10}$, one of the less common oxidation states of copper).

An important point to remember is that in the formation of $3d$ transition metal ions it is the $4s$ electrons that are lost first. Similarly, ns electrons are lost first in other transition-series elements.

29.6 $3d$ series elements

The first-series metals titanium, vanadium, chromium, manganese, iron, cobalt, and nickel have a great deal in common economically and industrially, as well as chemically. All of them are used in alloys in which strength, hardness, and resistance to corrosion are important. And all of them readily form carbides at high temperature, making it impractical to use carbon as the reducing agent in winning them from their ores (except for iron, because iron carbide is readily destroyed by oxidation). Iron, cobalt, nickel, and gadolinium are the only elements that are **ferromagnetic;** that is, they exhibit magnetism in the absence of an external magnetic field. Scandium is more like the lanthanide elements in its chemistry (as are yttrium and lanthanum).

Titanium, chromium, and manganese occur in nature almost exclusively as oxides; vanadium and iron as both oxides and sulfides (and, for iron, to a lesser extent as carbonate); and cobalt and nickel as sulfides.

Titanium is ninth in abundance of the elements in the Earth's crust and was found in unusually high percentage (12%) in rocks brought back from the Moon by Apollo 11. Titanium metal is expensive because its production is difficult—the hot metal combines with oxygen, nitrogen, and moisture from the air. As a result, most titanium is used in military and aerospace applications, where the high cost can be borne. Titanium dioxide is the most important titanium compound; it is a bright, white substance used as a pigment in paint, paper, and many other products. The minerals ilmenite ($FeO \cdot TiO_2$) and rutile (TiO_2) are converted directly to pure TiO_2 for commercial use.

The outstanding use of vanadium, which can be produced more easily than titanium, is in high strength steels. Eighty percent of all vanadium is used in steel in small percentages, where it imparts desirable mechanical properties.

The chemistry and uses of chromium, manganese, iron, cobalt, and nickel are discussed in following sections of this chapter, as is the chemistry of copper, silver, and gold, which together make up the family sometimes called the coinage metals because of their use in coins since ancient times.

29.7 $4d$ and $5d$ transition series elements

The second and third elements in each transition metal family resemble each other more in physical and chemical properties than they resemble the first member of the series. Comparable compounds of second and third series elements also tend to be very similar.

Hafnium, which immediately follows the lanthanide elements, is amazingly similar to zirconium. Hafnium and zirconium always occur together in nature, and their separation has been called the most difficult in chemistry. It can now be accomplished in various ways, including the distillation of similar compounds of the elements that have slightly different boiling points or fractional solvent extraction.

Niobium and tantalum are also very similar to each other, if slightly less so than zirconium and hafnium. In moving to the right across the periodic table, the second and third series elements in each family gradually become more different from each other, with palladium and platinum, and with silver and gold showing the greatest differences. Technetium is a radioactive element. It has been produced artificially in kilogram quantities and has been observed in stars, but it has never been found in the Earth's crust.

Horizontal similarities also exist in these series, as in the first series. The six elements following technetium and rhenium are all quite alike and occur together in various combinations in nature. Ruthenium, osmium, rhodium, iridium, palladium, and platinum are collectively called the **platinum metals.**

TABLE 29.5

Elements in the telephone handset

Source: *Committee on Survey of Materials Science and Engineering, Appendix to COSMAT Report, Vol. 2, Natl. Acad. Sci., Washington, D.C.*

Element	How used
Aluminum	Metal alloy in dial mechanism, transmitter, and receiver
Antimony	Alloy in dial mechanism
Arsenic	Alloy in dial mechanism
Beryllium	Alloy in dial mechanism
Bismuth	Alloy in dial mechanism
Boron	Touch-Tone dial mechanism
Cadmium	Color in yellow plastic housing
Calcium	In lubricant for moving parts
Carbon	Plastic housing, transmitter steel parts
Chlorine	Wire insulation
Chromium	Color in green plastic housing, metal plating, stainless steel parts
Cobalt	Magnetic material in receiver
Copper	Wires, plating, brass piece parts
Fluorine	Plastic piece parts
Germanium	Transistors in Touch-Tone dial mechanism
Gold	Electrical contacts
Hydrogen	Plastic housing, wire insulation
Indium	Touch-Tone dial mechanism
Iron	Steel, magnetic materials
Krypton	Ringer in Touch-Tone set
Lead	Solder in connections
Lithium	In lubricant for moving parts
Magnesium	Die castings in transmitter, ringer
Manganese	Steel in piece parts
Mercury	Color in red plastic housing
Molybdenum	Magnet in receiver
Nickel	Magnet in receiver, stainless steel parts
Nitrogen	Hardened heat-treated piece parts
Oxygen	Plastic housing, wire insulation
Palladium	Electrical contacts
Phosphorus	Steel in piece parts
Platinum	Electrical contacts
Silicon	Touch-Tone dial mechanism
Silver	Plating
Sodium	In lubricant for moving parts
Sulfur	Steel in piece parts
Tantalum	Integrated circuit in Trimline set
Tin	Solder in connections, plating
Titanium	Color in white plastic housing
Tungsten	Lights in Princess and key sets
Vanadium	Receiver
Zinc	Brass, die casting in transmitter, ringer

[29.7] TRANSITION ELEMENTS

Typically metallic in appearance, the second and third series metals range from the bluish gray tantalum, to platinum, which is a beautiful, silvery white metal that takes a high polish, to the familiar silver and gold. Whether it be hardness, corrosion resistance, temperature resistance, activity as a catalyst, or electrical properties, almost every one of these metals possesses some property that makes it desirable and useful in some practical application.

Zirconium and zirconium oxide and alloys of niobium and molybdenum have been developed because of their great resistance to high temperatures and their possible use in space vehicles that must reenter the atmosphere. Niobium and molybdenum are also important in steel-making. Tantalum is very resistant to body fluids and is used as a bone replacement; for example, a tantalum plate can be put into the skull when a piece of bone has been crushed or removed surgically.

All six of the platinum metals are used as catalysts in a wide variety of reactions. Palladium has the incredible ability to absorb 800 to 900 times its own volume of hydrogen, the hydrogen atoms apparently entering interstitial positions in the palladium crystal lattice. Very pure hydrogen gas is produced by diffusion of the hydrogen at high temperatures through a palladium–silver alloy.

The variety of ways in which the desirable properties of all of the elements are put to work is at least partially illustrated by Table 29.5, which is a list of the forty-two elements that are present in a telephone handset as constituents of thirty-five metals or alloys, fourteen types of plastics, twelve different adhesives, and twenty different semiconductor devices.

Chromium

29.8 Sources, metallurgy, and uses of chromium

Chromium occurs chiefly as the mineral chromite, $FeCr_2O_4$, which is a mixed oxide and can be written $FeO \cdot Cr_2O_3$. The term "mixed oxide" does not imply that the iron and chromium atoms are randomly distributed in the crystal. Rather there is a definite arrangement—the oxygen atoms lie between the metal atoms and are shared by them. Each iron atom is surrounded tetrahedrally by four oxygen atoms, and each chromium atom is surrounded octahedrally by six oxygen atoms. Several naturally occurring minerals have this same structure, which is indicative of its great stability. Such minerals are known as spinels—they are all represented by the formula $M^{2+}M^{3+}_2O_4$.

Metallic chromium is a bluish white, hard and brittle metal, which does not tarnish, is very resistant to corrosion, and takes a high polish. Its principal uses are in alloys, particularly steel, and in **chrome plating**—the electrolytic deposition of a layer of chromium on another metal (Table 29.6). Chrome plating produces a shiny,

TABLE 29.6

Uses of chromium and chromium compounds

Chromium
Steel

Other alloys

Chrome plating

Chromium(VI) oxide (CrO_3)
Chrome plating baths

Other metal treatment

Oxidizing agent

Sodium chromate (Na_2CrO_4) and sodium dichromate ($Na_2Cr_2O_7$)
Pigments

Leather tanning

Corrosion inhibitors

Aluminum anodizing

Pigment manufacture

Chromite ($FeCr_2O_4$ ore)
Refractories

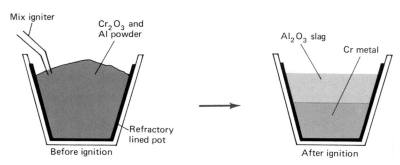

Mix igniter

Cr_2O_3 and
Al powder

Refractory
lined pot

Before ignition

Al_2O_3 slag

Cr metal

After ignition

FIGURE 29.2
Aluminothermic production of chromium.

attractive surface with high resistance to wear and corrosion. The chromium surface is protected by a thin layer of oxide from tarnishing or reaction with corrosive materials.

The method of treating chromite ore depends upon the use for which the chromium is destined. If chromium compounds are to be prepared, the ore is heated with sodium carbonate or calcium carbonate in the presence of air or some other oxidizing agent to give iron(III) oxide and chromate salts. In these reactions both iron and chromium are oxidized.

$$4FeCr_2O_4(s) + 8CaCO_3(s) + 7O_2(g) \longrightarrow$$
$$2Fe_2O_3(s) + 8CaCrO_4(s) + 8CO_2(g)$$
$$6FeCr_2O_4(s) + 12Na_2CO_3(l) + 7NaClO_3(l) \longrightarrow$$
$$3Fe_2O_3(s) + 12Na_2CrO_4(s) + 7NaCl(s) + 12CO_2(g)$$

The insoluble iron(III) oxide is easily separated from the chromates, which are soluble—$CaCrO_4$ in dilute acid and Na_2CrO_4 in water.

For the manufacture of stainless steel, which is chiefly iron, chromium (18%), and nickel (8%), with small amounts of other constituents, separation of the metals in the chromite ore is not necessary. An alloy of iron and chromium called *ferrochrome* is produced from chromite by reduction with carbon or silicon in an electric furnace and is used directly in steelmaking.

Chromium is obtained by reduction of the oxide with aluminum (an aluminothermic reaction; Section 26.14), carbon, or silicon.

$$2Al(s) + Cr_2O_3(s) \longrightarrow 2Cr(l) + Al_2O_3(l) \qquad \Delta H° = -126 \text{ kcal}$$

The chromium melts and is separated from the molten aluminum oxide by the difference in their densities (Figure 29.2).

29.9 Compounds of chromium

In its most commonly encountered compounds (Table 29.7) chromium has the oxidation states of +2, +3, or +6. Chromium(II) compounds are strong reducing agents and chromium(VI) compounds are strong oxidizing agents. In the +6 state, chromium is normally combined with oxygen in the oxide, CrO_3, the chromate ion, CrO_4^{2-}, or the dichromate ion, $Cr_2O_7^{2-}$.

Chromium and its compounds are extremely poisonous. Contact with the skin can cause open sores; inhalation can damage the respiratory system and may cause cancer of the bronchia. And the ingestion of chromium or its compounds can be fatal.

Chromium(III) oxide, Cr_2O_3, is a green, stable compound that is used as a pigment in glass and porcelain, in fabrics, and in printing. Chromium(III) salts can be formed by reaction of the oxide, Cr_2O_3, with acids, or by reduction of chromium(VI) compounds in the presence of the appropriate acid.

The crystalline chromium(III) chloride of commerce, commonly labeled $CrCl_3 \cdot 6H_2O$, is green. It is either $[Cr(H_2O)_5Cl]Cl_2 \cdot H_2O$ or $[Cr(H_2O)_4Cl_2]Cl \cdot 2H_2O$, or a mixture of the two, depending upon the method of manufacture. In each case, six groups (those shown inside the square brackets) have formed coordinate bonds with the chromium ion. The coordinated chloride ion is only slowly precipitated by silver ion, whereas the noncoordinated chloride ion (shown outside the bracket) is precipitated at once. The two green compounds can be distinguished from each other by their different shades of green, and more certainly when solutions of the two are treated with silver ion, one of them immediately precipitating one-third of its chlorine, and the other one, two-thirds of its chlorine. The coordinated water molecules are bound tightly to the chromium, but the others are not bound at all; they evidently fit into holes in the crystal lattice in a stoichiometric way without being attached.

The violet chloride $[Cr(H_2O)_6]Cl_3$ is also well known, but is more difficult to prepare in crystalline form than either of the green chlorides. It is precipitated as fine crystals when an ice-cold, saturated solution of either of the green forms is treated with hydrogen chloride gas. These three chromium(III) chlorides are examples of a type of isomerism known as **hydrate isomerism**—isomerism involving coordinated and noncoordinated water molecules (Table 29.8).

If a solution containing chromium(III) sulfate and an equivalent weight of an alkali metal sulfate, say, K_2SO_4, is allowed to evaporate at room temperature, large, octahedral crystals of the alum, $KCr(SO_4)_2 \cdot 12H_2O$, separate. Their violet color indicates that the chromium is coordinated to six molecules of water, and therefore, the formula might well be written $K[Cr(H_2O)_6](SO_4)_2 \cdot 6H_2O$. Studies of the crystals by x-ray analysis confirm this and also indicate that each of the other six molecules of water is hydrogen-bonded to two sulfate ions.

Chromium(III) hydroxide, usually written as $Cr(OH)_3$, is actually a highly polymeric, hydrated material in which OH groups are coordinated to two chromium atoms (Figure 29.3). Such bridges often, but not always, occur in pairs. Each chromium atom is coordinated to six oxygen atoms. Theoretically, chromium(III) hydroxide is $[Cr(OH)_3(H_2O)_3]$, but part of the water is lost upon exposure to air or heating, making the composition somewhat variable. On that account, the formula is simply written as $Cr(OH)_3$ or as the hydrated

Common oxidation states
Cr, +2, +3, +6

TABLE 29.8

Hydrate isomerism of chromium(III) chloride, $CrCl_3 \cdot 6H_2O$

$[Cr(H_2O)_6]Cl_3$	(violet)
$[Cr(H_2O)_5Cl]Cl_2 \cdot H_2O$	(green)
$[Cr(H_2O)_4Cl_2]Cl \cdot 2H_2O$	(green)

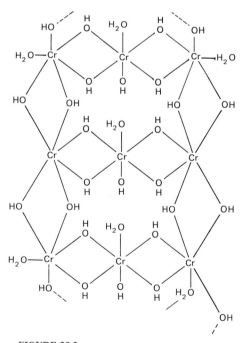

FIGURE 29.3
Structure of chromium(III) hydroxide. *The amount of water incorporated in the polymeric structure is variable.*

oxide, $Cr_2O_3 \cdot xH_2O$. Chromium(III) hydroxide is amphoteric—it is soluble in acids because the addition of protons to the hydroxide groups converts them to water molecules, which are not able to form bridges,

$$Cr(OH)_3(s) + 3H^+ \longrightarrow Cr^{3+} + 3H_2O(l)$$

and it dissolves in solutions of strong bases because removal of protons from the bridges converts the neutral polymer to a negative ion and degrades it to the monomeric form.

$$[Cr(OH)_3](s) + 3OH^- \longrightarrow [Cr(OH)_6]^{3-}$$

Chromium(VI) oxide, CrO_3, or chromium trioxide, is a red crystalline compound that is a strong oxidizing agent and is used in chrome plating. It is made commercially from sodium dichromate

$$Na_2Cr_2O_7(s) + H_2SO_4(aq) \longrightarrow \underset{\substack{\text{chromium(VI)} \\ \text{oxide}}}{2CrO_3(s)} + H_2O(l) + Na_2SO_4(aq)$$
$$\underset{\substack{\text{sodium} \\ \text{dichromate}}}{}$$

Chromium trioxide is the anhydride of chromic acid, H_2CrO_4, which is the parent of a series of polychromic acids and polychromates

$$2H_2CrO_4(aq) \rightleftharpoons H_2Cr_2O_7(aq) + H_2O(l)$$
$$3H_2CrO_4(aq) \rightleftharpoons H_2Cr_3O_{10}(aq) + 2H_2O(l)$$
etc.

The dichromates are the only important members of the polychromate series.

The condensation of oxo anions to polyanions in acidic solution is characteristic of many anions containing metals. Polyvanadates, polymolybdates, and polytungstates are common, the degree of polymerization in some cases being so great that the polyacids precipitate from solution. The existence of these polyanions is reminiscent of the formation of the pyrosulfate ion, $S_2O_7^{2-}$, and the polyphosphate ions. However, there are some differences. The pyrosulfate ion is not stable in aqueous solution, even at very low pH. The polyphosphates cannot be formed in aqueous solution, but only by heating the monophosphates to relatively high temperatures. Unlike the polychromates, they are only slowly destroyed in basic solution.

Chromate ion, CrO_4^{2-}, and dichromate ion, $Cr_2O_7^{2-}$ (Figure 29.4), exist in solution in an equilibrium dependent upon the pH.

$$H^+ + \underset{\substack{\text{chromate} \\ \text{ion}}}{CrO_4^{2-}} \rightleftharpoons \underset{\substack{\text{hydrogen} \\ \text{chromate ion}}}{HCrO_4^-}$$

$$2HCrO_4^- \rightleftharpoons \underset{\substack{\text{dichromate} \\ \text{ion}}}{Cr_2O_7^{2-}} + H_2O(l)$$

The shifting of the chromate ion–dichromate ion equilibrium as the pH of the solution is changed is readily observed, for the chromate ion is yellow and the dichromate ion is a deep orange.

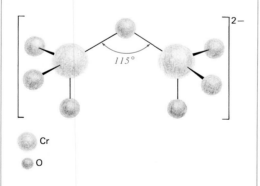

Cr

O

FIGURE 29.4
Dichromate ion.

29.10 Oxidation and reduction of chromium

The dichromate ion is a powerful oxidizing agent, the chromium(VI) being reduced to chromium(III). Dichromates in acidic solution are frequently used as oxidizing agents, for example,

$$Cr_2O_7^{2-} + 6Fe^{2+} + 14H^+ \longrightarrow 2Cr^{3+} + 6Fe^{3+} + 7H_2O(l)$$

$$Cr_2O_7^{2-} + 6Br^- + 14H^+ \longrightarrow 2Cr^{3+} + 3Br_2 + 7H_2O(l)$$

$$Cr_2O_7^{2-} + 3CH_3CHO(aq) + 8H^+ \longrightarrow$$
$$\text{acetaldehyde}$$
$$+ 2Cr^{3+} + 3CH_3COOH(aq) + 4H_2O(l)$$
$$\text{acetic acid}$$

Some interrelationships of the oxidation states of chromium are summarized in Figure 29.5.

Manganese

29.11 Sources, metallurgy, and uses of manganese

Manganese is widely distributed on Earth and is the eleventh most abundant element (0.095%). Its primary ore is the mineral known as pyrolusite, $MnO_2 \cdot xH_2O$. An intriguing possible source of manganese is nodules that have been found on the ocean floor. These nodules, 1–15 cm across, contain a high percentage of manganese, together with compounds of other metals, such as iron, nickel, copper, and cobalt. The nodules were apparently laid down very slowly in concentric rings, perhaps at the rate of 1–100 mm per million years. Microorganisms are thought to play a role in depositing MnO_2 in the nodules. Whether the sea floor can successfully be mined for manganese nodules is not yet known.

Manganese imparts hardness and strength to steel and is also used in other alloys, for example, manganese bronze (copper and manganese) and a nonconducting alloy (with nickel and copper) called *manganin* (Table 29.9). For steelmaking, production of pure manganese is not necessary. Instead, mixed iron and manganese oxides are reduced by carbon to give ferromanganese, an iron–manganese carbide, which is used directly. The free metal can be made from the ore by electrolysis or reduction by, for example, aluminum.

Manganese is a silvery, brittle metal with a slightly pink appearance. It has several allotropic forms that vary in brittleness and ductility. Unlike chromium, manganese corrodes in moist air. Manganese compounds are much less toxic than those of chromium. Manganese is one of the essential trace elements for both plants and animals, and manganese sulfate is added to some fertilizers.

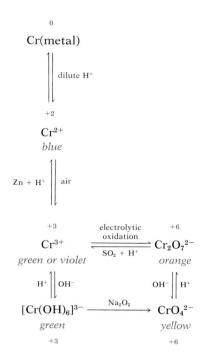

FIGURE 29.5
Oxidation and reduction of chromium.

TABLE 29.9

Uses of manganese and its compounds

Manganese
Steel

Other alloys

Manganese dioxide (MnO_2, black)
Glassmaking

Paint (drying agent)

Dry cell batteries

Potassium permanganate ($KMnO_4$, purple)
Metal treatment

Medicine and pharmaceuticals

Water and waste treatment

Chemical oxidizing agent

TABLE 29.10
Some compounds of manganese

Mn(OH)$_2$	**Manganese(II) hydroxide** (flesh colored)
MnS	**Manganese(II) sulfide** (salmon colored)
MnSO$_4$	**Manganese(II) sulfate** (reddish)
MnCl$_2$	**Manganese(II) chloride** (pink)
MnCl$_2$·4H$_2$O	**Manganese(II) chloride** **tetrahydrate** (rose colored)
MnO$_2$	**Manganese(IV) oxide** (black; nonstoichio- metric)
K$_2$MnO$_4$	**Potassium manga- nate(VI)** (deep green)
KMnO$_4$	**Potassium permanganate** (purple; oxidizing agent)

29.12 Compounds of manganese

Compounds are known containing manganese in oxidation states +2, +3, +4, +5, +6, and +7. (Some compounds are listed in Table 29.9.) The pale pink manganese(II) salts are formed when manganese in any higher oxidation state is reduced in acidic solution, and manganese is most stable in its +2 state. Compounds of manganese in the +3 and +5 states are not common. Manganese in the +6 and +7 oxidation states is best known as the dark green manganate, MnO_4^{2-}, and the deep purple permanganate, MnO_4^- ions. Permanganate ion is a strong oxidizing agent, and the most interesting aspect of manganese chemistry is the behavior of manganese compounds and ions in oxidation–reduction reactions. Manganese dioxide, MnO_2, and potassium permanganate, $KMnO_4$, are the two most frequently used manganese compounds (Table 29.10).

Manganese(IV) oxide, or manganese dioxide, MnO_2, is the most abundant ore from which compounds of the metal are obtained. When prepared in the laboratory, manganese oxide is a dark brown or black powder. It dissolves slowly in solutions of strong bases, or rapidly in fused basic media, to form manganate(IV) salts. These may contain the anion MnO_3^{2-}

$$MnO_2(s) + 2OH^- \longrightarrow MnO_3^{2-} + H_2O(l)$$

Polymanganate(IV) anions such as $Mn_2O_5^{2-}$ and $Mn_5O_{11}^{2-}$ are also known.

Stoichiometric manganese dioxide is extremely rare, if it exists at all. Analysis of the material found in nature as well as of that prepared in the laboratory shows a Mn to O ratio of 1 to 1.85 (approximately), the exact ratio depending on the mode of preparation. (Compare lead dioxide; Section 26.20.) X-ray analysis shows that this material has the crystal structure calculated for pure manganese(IV) dioxide, but that several percent of the oxygen atoms are missing, leaving holes in the crystal lattice. Oxide ions from adjacent sites are able to move into these holes, thus leaving new holes; this process makes the oxide conductive. It is both the oxidizing power and conductivity of the manganese(IV) oxide that make it valuable in dry cell batteries.

Manganese(VI) is of interest chiefly because of the deep green potassium manganate(VI), K_2MnO_4, which is an intermediate in the preparation of potassium permanganate, $KMnO_4$, in which Mn is in the +7 state. Potassium manganate(VI) is prepared by fusing a compound in which manganese is in a lower oxidation state with a basic material (e.g., KOH or K_2CO_3) and an oxidizing agent (e.g., KNO_3, $KClO_3$, or air)

$$MnO_2(s) + 2KOH(l) + KNO_3(l) \xrightarrow{\Delta} K_2MnO_4(s) + KNO_2(s) + H_2O(g)$$

$$4MnO(OH)(s) + 4K_2CO_3(l) + 3O_2(g) \xrightarrow{\Delta}$$
$$4K_2MnO_4(s) + 4CO_2(g) + 2H_2O(g)$$

Manganate(VI) ion is stable only in alkaline medium; if the solution is acidified, self oxidation–reduction (disproportionation) immediately takes place, producing manganese(IV) oxide and permanganate.

$$3MnO_4^{2-}(aq) + 4H^+ \longrightarrow 2MnO_4^-(aq) + MnO_2(s) + 2H_2O(l)$$

Potassium manganate(VII), $KMnO_4$, usually called <u>potassium permanganate</u>, crystallizes in dark purple, almost black, needles, which dissolve in water to give a purple solution. Unlike manganate(VI), manganate(VII) is more stable in acidic solution than in alkaline and is always prepared in an acidic medium.

$$2MnO_4^{2-} + O_3(g) + 2H^+ \longrightarrow 2MnO_4^- + O_2(g) + H_2O(l)$$
$$2MnO_2(s) + 3PbO_2(s) + 4H^+ \longrightarrow 2MnO_4^- + 3Pb^{2+} + 2H_2O(l)$$

Even in acidic solution, however, permanganate ion slowly decomposes, especially when exposed to light, leaving a residue of MnO_2. Potassium permanganate is a strong oxidizing agent, and its use in treating athlete's foot and rattlesnake bite and as an antidote for poisons depend upon that property. It is also widely employed as an oxidant in organic chemistry and in quantitative analysis. Because of its instability, solutions that are to be used in quantitative work must be freshly prepared and standardized by titration with a pure, stable reducing agent such as sodium oxalate, $Na_2C_2O_4$.

29.13 Oxidation and reduction of manganese

When permanganate ion is reduced in very strongly alkaline solution, the dark green manganate(VI) ion is formed, for example,

$$\underset{\substack{permanganate \\ ion}}{2MnO_4^-} + OH^- + \underset{\substack{formate \\ ion}}{HCOO^-} \longrightarrow \underset{\substack{manganate \\ ion}}{2MnO_4^{2-}} + CO_2(g) + H_2O(l)$$

In less alkaline, or in neutral solution, manganese(IV) oxide is the chief product.

$$2MnO_4^- + 3H_2S(g) \longrightarrow 2MnO_2(s) + 3S(s) + 2OH^- + 2H_2O(l)$$

In weakly acidic solutions containing excess fluoride or phosphate, manganese(III) complex ions are formed, for example,

$$MnO_4^- + 2\underset{oxalate}{C_2O_4^{2-}} + 2H_2PO_4^- + 4H^+ \longrightarrow$$
$$[Mn(PO_4)_2]^{3-} + 4CO_2(g) + 4H_2O(l)$$

but from other acidic solutions, manganese(II) is obtained, two examples being

$$2MnO_4^- + 5H_2S(g) + 6H^+ \longrightarrow 2Mn^{2+} + 5S(s) + 8H_2O(l)$$
$$2MnO_4^- + 5C_2O_4^{2-} + 16H^+ \longrightarrow 2Mn^{2+} + 10CO_2(g) + 8H_2O(l)$$

FIGURE 29.6

Oxidation and reduction of manganese.

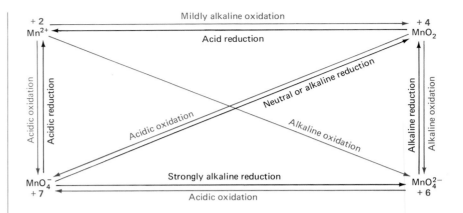

TABLE 29.11

Some manganese standard reduction potentials

Couple	Reaction	$E°(V)$
Mn(7)/Mn(4)	$MnO_4^- + 4H^+ + 3e^- \rightleftharpoons MnO_2 + 2H_2O$	1.68
Mn(7)/Mn(2)	$MnO_4^- + 8H^+ + 5e^- \rightleftharpoons Mn^{2+} + 4H_2O$	1.49
Mn(3)/Mn(2)	$Mn^{3+} + e^- \rightleftharpoons Mn^{2+}$	1.51
Mn(7)/Mn(4)	$MnO_4^- + 2H_2O + 3e^- \rightleftharpoons MnO_2 + 4OH^-$	0.59
Mn(6)/Mn(4)	$MnO_4^{2-} + 2H_2O + 2e^- \rightleftharpoons MnO_2 + 4OH^-$	0.60
Mn(7)/Mn(6)	$MnO_4^- + e^- \rightleftharpoons MnO_4^{2-}$	0.56

The oxidation–reduction reactions of manganese in alkaline and acidic solution are summarized in Figure 29.6. Table 29.11 gives some of the pertinent standard reduction potentials.

Iron

29.14 Sources and uses of iron

In many respects, our industrial civilization is built upon iron. It is the fourth in abundance of all the elements and makes up about 5.5% of the crust of the Earth. The abundant ores are readily reduced to the metal, and by metallurgical working, iron can be made extremely strong. Iron can also be made soft or hard, flexible or stiff, malleable or brittle, depending on the treatment to which it is subjected.

During the long history of the universe, according to present theory, processes of nuclear synthesis have brought about the transformation of less stable elements into more stable ones. Since the greatest nuclear stability is found in iron and the elements adjoining it in the periodic table (Section 8.7), it is not surprising that iron is very abundant throughout the universe. Many meteorites consist of metallic iron or alloys containing large percentages of iron, and it is believed that the core of the Earth is made up largely of iron or an iron–nickel alloy. If this is true, the Earth contains more iron than all of the other elements combined. In the crust of the Earth, the metal is widely distributed as oxides, silicates, and sulfides, and in other forms. The yellow sea sand, the red rocks of the Rocky Mountains, and the red sand and soil of many parts of the United States all owe their color to the presence of iron(III) compounds.

The common ores of iron are oxides. Of these, *hematite*, Fe_2O_3, is the most abundant. *Magnetite*, Fe_3O_4, which is a spinel, is a valuable ore, for it contains a higher percentage of iron than does iron(III) oxide. As the name implies, magnetite is attracted by a magnet; some samples, in fact, act as magnets. These are called *lodestone*.

Iron(II) carbonate, or *siderite*, $FeCO_3$, is a good source of iron, for it is converted by heat to oxides, which can be reduced to metal. Siderite is present in many soils and contributes to hardness in water through its ready conversion to the soluble bicarbonate.

$$FeCO_3(s) + CO_2(g) + H_2O(l) \rightleftharpoons Fe(HCO_3)_2(aq)$$

The Fe(II) is thus free to react, as it was not when trapped in the insoluble carbonate. In air, iron(II) bicarbonate in solution is oxidized to the insoluble iron(III) oxide.

$$4Fe(HCO_3)_2(aq) + O_2(g) \longrightarrow 2Fe_2O_3(s) + 8CO_2(g) + 4H_2O(l)$$

This accounts for the brown stain so often seen under dripping faucets and in other places where hard water is in contact with air.

Taconite ores, which are chiefly iron oxides containing silica, are now increasingly used in the United States as sources of iron. They are extremely hard and difficult to handle, but metallurigical research has overcome most of the problems.

The term **steel** is used for alloys of iron that contain carbon (up to ~1.5%) and often also contain other metals. Without carbon, iron is not strong enough or hard enough for many modern applications. The properties of steel depend upon the percentage of carbon present, the heat treatment of the steel, and the alloying metals present. Low-carbon, or mild, steel contains up to 0.2% carbon. It is malleable and ductile and is used in making wire, pipe, and sheet steel. Medium steel (0.2–0.6% carbon) is used in rails, boiler plate, and structural pieces. High-carbon steel (0.6–1.5% carbon) is hard, but lacks ductility and flexibility. It is used for tools, springs, and cutlery. Wrought iron contains less than 0.2% of carbon. It was once used for decorative objects, chains, and other applications where great ductility is required, but it has now been largely replaced by mild steel.

Almost all metallic iron produced is used in steel or other alloys. Some of the nonsteel alloys of iron are cast iron (>1.5% C), wrought iron, nickel iron, and iron–silicon alloys.

29.15 Corrosion of iron

Iron corrodes only in the presence of both air and moisture, and the process is accelerated by acids or by contact with less active metals, such as tin or copper. Certain salt solutions also hasten corrosion, not only because they are acidic by hydrolysis, but also because of specific catalytic effects or reactions of the anion. Chlorides are particularly bad in this respect.

A piece of iron rusts more in some spots than in others. Often, deep pits are formed. If corrosion were merely the combination of iron with oxygen, rusting would be more uniform than it is. Rusting, or corrosion, is an electrochemical process. Some portions of the surface attract electrons away from others because of the presence of impurities, strains, or other influences that change the reactivity of the metal. If electrons migrate from one spot on the surface to another, the first becomes anodic (electron loss—oxidation) and the second, cathodic (electron gain—reduction). A miniature cell is set up by couple action, with iron corroding to iron(II) ions at the anode and water being reduced to hydroxide ions and dihydrogen at the cathode. The iron(II) hydroxide, once formed, is readily oxidized by the air to hydrated iron(III) oxide, $Fe_2O_3 \cdot xH_2O$, the red compound that is "rust." Rust does not adhere to the metal, and so the iron surface is continually exposed to corrode further.

$$2Fe(s) \longrightarrow 2Fe^{2+}(aq) + 4e^- \qquad \text{(anode)}$$
$$2H_2O + 2e^- \longrightarrow 2OH^-(aq) + H_2(g) \qquad \text{(cathode)}$$
$$2Fe^{2+}(aq) + 4OH^-(aq) \longrightarrow 2Fe(OH)_2(s)$$
$$2Fe(OH)_2(s) + \tfrac{1}{2}O_2(g) + xH_2O(l) \longrightarrow Fe_2O_3 \cdot xH_2O(s) + 2H_2O(l)$$

Methods of preventing rust formation depend upon keeping air and moisture away from the metal (by a protective layer of paint, grease, or some other metal), lowering the reactivity by alloying (as in stainless steel), or making the iron the cathode of a cell. When connected to a piece of a more active metal, the iron becomes the cathode and the other metal the anode. Zinc, magnesium, and aluminum are often used in this way. The active metal is, of course, corroded rapidly, and so it is sometimes referred to as the *sacrificial anode*. The hulls of steel ships, for example, are often protected from corrosion by bolting strips of magnesium to them below the water level, and oil tanks are protected by blocks of magnesium, which can be replaced easily (Figure 29.7).

Galvanized iron, which is iron coated with a thin layer of zinc, resists corrosion even if a portion of the iron is not covered with zinc, because of the formation of an iron–zinc couple in which the iron is the cathode. With tin plate, however, the opposite effect occurs; once a bit of iron is exposed, rusting proceeds rapidly, for a cell is set up in which the iron is the anode and the tin the cathode. Even a tiny pinhole through the tin plate will allow accelerated corrosion.

29.16 Iron making: The blast furnace

The raw materials for iron making are (1) iron ore that has been pretreated by crushing, grinding, washing, sintering, and so on, (2) coke, and (3) limestone, which serves as a flux. Crude iron, called pig iron or cast iron, is produced in a blast furnace—a tower about 100 ft high

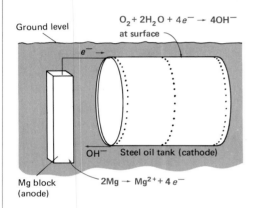

Ground level

$O_2 + 2H_2O + 4e^- \rightarrow 4OH^-$
at surface

$e^- \rightarrow$

OH^- Steel oil tank (cathode)

Mg block
(anode)

$2Mg \rightarrow Mg^{2+} + 4e^-$

FIGURE 29.7

Cathodic protection of an oil tank.
The magnesium is embedded in sand to aid flow of ions away from surface. The electrons used at the surface of the oil tank come through the conductor from the magnesium, and the tank is undisturbed.

and 25 ft in diameter that is lined with special refractory (heat-resistant) brick.

The furnace is charged from the top with a mixture of iron ore, coke, and limestone (Figure 29.8). A strong (about 350 mph) blast of preheated air (or oxygen) is blown in at the bottom, where the coke burns to carbon monoxide, which is the reducing agent. The charge is heated gradually as it descends. First the moisture is driven off. Then the ore is partially reduced by carbon monoxide. In the hotter part of the furnace the reduction of the ore to metallic iron is completed, and the limestone loses CO_2 and reacts with the impurities in the ore (mainly silicon dioxide, but also manganese and phosphorus oxides) to produce the molten slag. The molten iron and slag are immiscible and form separate layers at the bottom of the furnace.

The reduction reactions are reversible, and complete reduction takes place only if the carbon dioxide formed is destroyed. This is effected by reduction to carbon monoxide with an excess of coke.

$$CO_2(g) + C(s) \longrightarrow 2CO(g)$$

The gas that escapes from the top of the furnace consists largely of carbon monoxide and the nitrogen that was introduced in the air blast. This hot nitrogen–carbon monoxide mixture is combined with air, so that the carbon monoxide can burn and the products of this combustion pass through a heat exchanger to help heat the incoming gas.

Relatively pure oxygen is used in place of air in some blast furnaces. This allows for a smaller furnace and somewhat higher temperatures, and the carbon monoxide formed is a much better fuel than a carbon monoxide–nitrogen mixture. The obvious disadvantage is the cost of the oxygen.

29.17 Steel making

The iron withdrawn from the blast furnace contains small amounts of carbon, sulfur, phosphorus, silicon, manganese, and other impurities. These cause the iron to be so brittle that it is useless for most purposes. Although iron making is a reduction process, steel making is an oxidation process. The two objectives in steel making are to burn out the unwanted impurities from pig iron and to add the exact amounts of alloying materials desired.

The manganese, phosphorus, and silicon in molten pig iron are converted by air or oxygen to oxides, which react with appropriate fluxes to give slags. Sulfur enters the slag as a sulfide and carbon is burned to carbon monoxide or carbon dioxide. If the chief impurity is manganese, an acidic flux—the oxide of a nonmetal (an acid anhydride)—must be used. Silicon dioxide is the usual acidic flux.

$$MnO(s) + SiO_2(s) \longrightarrow MnSiO_3(l)$$

If the chief impurity is silicon or phosphorus (the more common

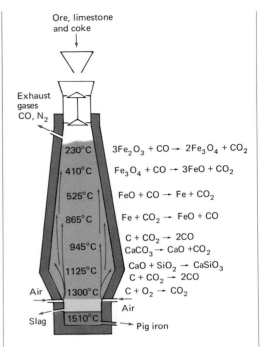

FIGURE 29.8
The blast furnace.

Ore, limestone and coke

Exhaust gases CO, N_2

230°C	$3Fe_2O_3 + CO \rightarrow 2Fe_3O_4 + CO_2$
410°C	$Fe_3O_4 + CO \rightarrow 3FeO + CO_2$
525°C	$FeO + CO \rightarrow Fe + CO_2$
865°C	$Fe + CO_2 \rightarrow FeO + CO$
945°C	$C + CO_2 \rightarrow 2CO$
	$CaCO_3 \rightarrow CaO + CO_2$
1125°C	$CaO + SiO_2 \rightarrow CaSiO_3$
	$C + CO_2 \rightarrow 2CO$
1300°C	$C + O_2 \rightarrow CO_2$

Air

Air

Slag

1510°C

Pig iron

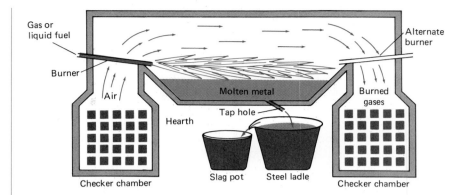

FIGURE 29.9

Open hearth furnace for steel making.
*The checker chambers are arrangements
of hot bricks. Heated air from the
chambers joins the flame, raising its
temperature. Other chambers are
heated by gases from the furnace, and
the flow of air and gases is periodically
reversed.*

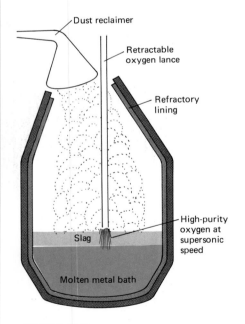

FIGURE 29.10

**Basic oxygen process steel-making fur-
nace.**

case), a basic flux must be used. This is usually magnesium oxide or
calcium oxide.

$$MgO(s) + SiO_2(s) \longrightarrow MgSiO_3(l)$$
$$6MgO(s) + P_4O_{10}(s) \longrightarrow 2Mg_3(PO_4)_2(l)$$

The steel-making furnace is lined with brick made of the fluxing
material, and this lining absorbs part of the oxide to be removed.

Most steel in the United States is made either by the older, open
hearth method, or the newer method that is displacing it, the basic
oxygen method. In 1974, 55% of the steel produced in the world was
made by the basic oxygen method.

The essential feature in the open hearth process furnace (Figure 29.9) is a large, dish-shaped container into which 100 or 200 tons
of molten iron is placed. This dish is underneath a concave roof,
which reflects heat onto the surface of the metal. A blast of oxygen is
passed over the surface of the molten crude iron to burn out the
impurities. The impurities beneath the surface are brought to the top
by convection and diffusion and the process requires several hours.
Some iron is also oxidized, of course, and the iron oxide is recovered
and returned to the blast furnace.

The basic oxygen process of removing the impurities from iron
follows the same chemical principles as the open hearth method, but
is quite different in engineering design (Figure 29.10). The molten pig
iron, along with scrap iron and materials to form a slag, is contained
in a barrel-shaped furnace that might hold up to 300 tons of material.
A blast of high-purity oxygen under a pressure of 150–180 lb/sq inch
is directed against the surface of the liquid, and the barrel can be
tipped and rotated to bring fresh material to the surface. The oxidation of the impurities is very rapid, and the escape of gaseous products so agitates the mass that even the iron at the bottom of the vessel
is brought into reaction. The temperature of the material rises almost
to the boiling point of the iron without the application of any external
heat. At such a temperature, the reactions are extremely rapid, and
the whole process is completed in an hour or less, leaving a product
that is uniform and of high quality.

The molten purified iron is converted to steel by addition of the correct amount of carbon and alloying metals such as vanadium, chromium, titanium, manganese, and nickel. Special steels may contain molybdenum, tungsten, or other metals to give them the desired properties.

The electric arc method of steel making is also being used more and more. An electric arc between carbon electrodes in the roof over a large saucer containing the molten metal provides the heat. Oxygen is added in controlled amounts and also released from impurities and oxides of alloying metals that are added at the beginning of the process. Advantages of this process are that the amount of oxide impurities in the steel can be controlled and also that there is less loss of alloying materials, some of which are quite expensive.

29.18 Heat treatment of steel

At high temperatures, iron and carbon combine to form iron carbide, Fe_3C, called cementite. The reaction of formation is reversible, but unlike most reactions of combination, it is endothermic. As the equation

$$3Fe + C + heat \rightleftharpoons Fe_3C$$

indicates, the stability of cementite increases as the temperature rises, at least in the temperature range involved in the heat treatment of steel. When steel containing cementite is cooled slowly, the equilibrium shifts toward the formation of iron and carbon, and the carbon separates as flakes of graphite. These give the metal a gray color. If, however, the steel is cooled very rapidly, equilibrium is not attained and the carbon remains largely in the form of cementite, which is light in color. At ordinary temperatures the decomposition of cementite is so slow that, for all practical purposes, it does not take place at all. Steel containing cementite is harder and much more brittle than that containing graphite. By "tempering," that is by heating the steel to a suitable temperature for a short time and then cooling it rapidly, the proportion of carbon in each form can be adjusted within rather wide limits. It is also possible to vary the total amount of carbon in different parts of a single piece of steel, and thus modify its properties. Ball bearings, for example, are made of a medium steel to give them toughness and strength, but the surface is *case hardened* by heating the bearings in a bed of carbon to give a thin surface coating containing cementite.

29.19 Compounds of iron

The most common of the soluble iron(II) salts is iron(II) sulfate, which crystallizes from water in the form of large, light green crystals, $FeSO_4 \cdot 7H_2O$ (Table 29.12). It is a by-product of the steel industry, for sheet steel that is to be plated or enameled is often cleaned of corrosion products by treatment with dilute sulfuric acid.

Common oxidation states
Fe, +2, +3

TABLE 29.12
Some compounds of iron

$FeSO_4 \cdot 7H_2O$	Iron(II) sulfate heptahydrate (light green; from "pickle liquor")
FeO	Iron(II) oxide (black; nonstoichiometric)
FeS	Iron(II) sulfide (brownish black)
$FeCl_3$	Iron(III) chloride (anhydrous) (brown black)
Fe_2O_3	Iron(III) oxide (red-brown, nonstoichiometric)
$K_4Fe(CN)_6$	Potassium ferrocyanide (yellow)
$K_3Fe(CN)_6$	Potassium ferricyanide (red)
$Fe(SCN)_3$	Iron(III) thiocyanate (bright red)
FeS_2	Iron pyrite ("fool's gold")

TRANSITION
ELEMENTS [29.19]

899

(Cleaning metal in an acid bath is called pickling; the leftover acid is called pickle liquor.) In this process some of the metal is dissolved, along with the rust. The iron(II) sulfate that forms is recovered in the form of the green hydrate, which is used as the source of other iron compounds, as well as in weed killers and wood preservatives. On exposure to air, its solutions soon become turbid and deposit a precipitate of hydrated iron(III) compounds.

Iron(II) oxide, FeO, and also Fe_2O_3, tend to be nonstoichiometric. In both compounds, the oxygen atoms form a lattice with iron atoms in the interstices.

Iron pyrite, FeS_2, a mineral, is often called "fool's gold" because of its yellow color and bright metallic luster. The iron in FeS_2 is in the +2 state. Upon heating, pyrite decomposes to iron(II) sulfide and sulfur, and it is sometimes used as a source of elemental sulfur.

Iron(III) chloride, $FeCl_3$, in the anhydrous form is prepared by the action of chlorine on hot iron, by heating a mixture of iron oxide and carbon in a stream of chlorine, or by passing the vapor of carbon tetrachloride over hot oxide:

$$2Fe(s) + 3Cl_2(g) \xrightarrow{\Delta} Fe_2Cl_6(g)$$

$$Fe_2O_3(s) + 3C(s) + 3Cl_2(g) \xrightarrow{\Delta} Fe_2Cl_6(g) + 3CO(g)$$

$$Fe_2O_3(s) + 3CCl_4(g) \xrightarrow{\Delta} Fe_2Cl_6(g) + 3COCl_2(g)$$

In the vapor, iron(III) chloride, due to the tendency of iron to increase its coordination number to four, exists as the dimer Fe_2Cl_6. Unlike iron(II) chloride, which is a typical salt, iron(III) chloride is covalent, volatile, and somewhat soluble in nonpolar solvents.

In both the +2 and +3 states, iron forms many complexes. The hydrated ion $[Fe(H_2O)_6]^{3+}$ is, of course, one such complex. The addition of cyanide ion to a solution of an iron(II) salt gives a gray, slimy precipitate of iron(II) cyanide, which dissolves when excess cyanide is added to form a clear yellow solution:

$$Fe^{2+} + 2CN^- \longrightarrow Fe(CN)_2(s)$$

$$Fe(CN)_2(s) + 4CN^- \longrightarrow [Fe(CN)_6]^{4-}$$
$$\textit{hexacyanoferrate(II) ion}$$

Potassium hexacyanoferrate(II), $K_4Fe(CN)_6 \cdot 3H_2O$, is a common laboratory chemical, known as potassium ferrocyanide. Oxidizing agents readily change potassium hexacyanoferrate(II) to potassium hexacyanoferrate(III), $K_3Fe(CN)_6$, or potassium ferricyanide. The ferricyanide ion contains one less electron than the corresponding iron(II) complex ion.

When $K_4Fe(CN)_6$ is treated with an iron(III) salt, a precipitate of $KFe[Fe(CN)_6]$ is formed. This very slightly soluble, deep-blue substance is known as Prussian blue. Once used as a paint pigment and laundry blueing, it has been displaced by phthalocyanine blue, which is somewhat more stable. The absorption of light by Prussian blue, resulting in its intense color, is facilitated by the ease with which electrons can migrate from one iron atom to another. This is a good example of the general rule that a molecule that contains atoms of the same metal in two different oxidation states is deeply colored—a phenomenon known as *interaction absorption*.

The formation of Prussian blue is utilized in making blueprints. Paper is moistened in total darkness with a solution containing iron(III) ion, hexacyanoferrate(III) ion, and ammonium citrate, a mild reducing agent. The paper is then dried and is ready for use. As long as it is kept in darkness, no reaction occurs, but upon exposure to light the citrate reduces part of the iron(III) to iron(II), and when the paper is moistened, the insoluble Prussian blue is formed. Any parts of the paper protected from the light remain unaffected. The soluble compounds that have not reacted are then washed away, leaving only the Prussian blue.

Cobalt

29.20 Sources and uses of cobalt

Cobalt is much less abundant than chromium, manganese, or iron, and there are very few ores that are valuable mainly for their cobalt content. The metal and its salts are obtained chiefly as by-products in the metallurgy of nickel. Nonetheless, cobalt is an important metal and is a component of many valuable alloys. Among these are Alnico (Fe, Al, Ni, Co), which is highly magnetic, Stellite (Cr, Co, W), used in surgical instruments because it is hard and corrosion resistant, and Hastelloy B (chiefly Co and Ni), which is very strong and hard, even at high temperatures. Cobalt is much less reactive than iron and dissolves in acids only slowly. It corrodes very slightly in ordinary air. Cobalt compounds are widely used in industry as catalysts. One of the most important of these is cobalt naphthenate, which catalyzes the drying of paint. It is deep blue, but is used in such small concentration that, even in a white paint, the color is not perceptible. A cobalt-molybdenum-alumina catalyst is used in removing sulfur from crude oil.

29.21 Compounds of cobalt

In its compounds, cobalt is almost always in the +2 or +3 state (Table 29.13). Simple salts contain the pink, hydrated ion $[Co(H_2O)_6]^{2+}$. The water is expelled from most of these salts by heating somewhat above 100°C. In general, tetracoordinated cobalt(II) is blue and hexacoordinated cobalt(II) is pink. The color change of a hydrated cobalt complex is used in devices that are supposed to tell when it is going to rain. At high humidity, a blue complex takes up water and turns pink.

In complexes with cyanide ion, nitrite ion, ammonia, organic amines, and many other ligands, cobalt assumes the +3 oxidation

Common oxidation states
Co, +2, +3

TABLE 29.13
Some compounds of cobalt

$[Co(H_2O)_6]Cl_2$	Hexaaquacobalt(II) chloride (pink)
$[Co(H_2O)_6](NO_3)_2$	Hexaaquacobalt(II) nitrate (red)
CoS	Cobalt(II) sulfide (black)
CoO	Cobalt(II) oxide (green-brown)
Co_2O_3	Cobalt(III) oxide (gray-black)
$K_3[Co(CN)_6]$	Potassium hexacyanocobaltate(III) (yellow)
$[Co(NH_3)_6]Cl_3$	Hexaaminecobalt(III) chloride (yellow)
$K_3[Co(NO_2)_6]$	Potassium hexanitrocobaltate(III) (yellow)

TABLE 29.14

Standard reduction potentials for the Co(+2)/Co(+3) couple

Reaction	$E°(V)$
$[Co(H_2O)_6]^{3+} + e^- \longrightarrow [Co(H_2O)_6]^{2+}$	+1.842
$[Co(NH_3)_6]^{3+} + e^- \longrightarrow [Co(NH_3)_6]^{2+}$	+0.1
$[Co(CN)_6]^{3-} + e^- \longrightarrow [Co(CN)_6]^{4-}$	−0.83

state. The ease with which this happens varies widely, as is shown by the reduction potentials given in Table 29.14.

The best known of the cobalt(II) salts is the pink chloride, $[Co(H_2O)_6]Cl_2$, which dissolves in water to give a pink solution and in alcohol to give a blue solution of $[Co(C_2H_5OH)_4]Cl_2$. It is the source of most other cobalt salts.

When an aqueous solution containing the cobalt(II) ion is treated with potassium cyanide, a slimy gray precipitate of $Co(CN)_2$ is formed, but this dissolves upon addition of more cyanide. In a few moments, the solution grows warm, and bubbles of hydrogen gas escape. When this reaction is ended, yellow crystals of potassium hexacyanocobaltate(III) can be obtained from the solution:

$$Co^{2+}(aq) + 2CN^-(aq) \longrightarrow Co(CN)_2(s)$$
$$2Co(CN)_2(s) + 8KCN(aq) + 2H_2O(l) \longrightarrow$$
$$2K_3[Co(CN)_6](aq) + 2KOH(aq) + H_2(g)$$

If an aqueous solution of cobalt(II) ion, ammonium chloride, and ammonium hydroxide is treated with a mild oxidizing agent (e.g., air, iodine, or hydrogen peroxide) and is then acidified with hydrochloric acid, a mixture of complex cobalt(III) salts is formed. Among these are substances with the following empirical formulas: $CoCl_3 \cdot 6NH_3$ (yellow), $CoCl_3 \cdot 5NH_3$ (purplish red), and $CoCl_3 \cdot 4NH_3$ (either green or violet). In the first, all of the chlorine is ionic; in the second, two-thirds of it is ionic; and in the last, only one-third is ionic. The formulas are, therefore, $[Co(NH_3)_6]Cl_3$, $[Co(NH_3)_5Cl]Cl_2$, and $[Co(NH_3)_4Cl_2]Cl$. These compounds are called, respectively, hexaamminecobalt(III) chloride, chloropentaamminecobalt(III) chloride, and dichlorotetraamminecobalt(III) chloride. In each case, the cobalt has a coordination number of six, which seems to be almost universal in cobalt(III) complexes. If the hexaammine chloride is heated to a high enough temperature to cause a molecule of ammonia to escape (about 225°C), one of the chloride ions simultaneously takes its place in the coordination sphere.

$$[Co(NH_3)_6]Cl_3(s) \longrightarrow [Co(NH_3)_5Cl]Cl_2(s) + NH_3(g)$$

There are two isomeric forms of the dichlorotetraammine ion (see Figure 28.3).

Nickel

29.22 Sources and uses of nickel

The importance of nickel in modern civilization can hardly be overestimated, for this metal is widely used as a protective plate on iron, as a catalyst for hydrogenation and petroleum refining, and as a constituent of many valuable alloys (Table 29.15). The nickel steels are hard, tough, and corrosion resistant and are widely used in making machinery and safes. Stainless steel (18% Cr, 8% Ni) is particularly resistant to corrosion. Nichrome (80% Ni, 20% Cr) is used for electrical heating elements. Monel metal (Ni, Cu, and a little iron and manganese) can be made directly from the Canadian nickel ores, which are largely mixed copper and nickel sulfides. Monel metal is strong and highly resistant to corrosion. The "nickel" coin, which is actually 25% nickel and 75% copper, is bright and hard. It resists corrosion and wear.

In view of its usefulness, it is unfortunate that nickel is not more widely distributed in nature; its abundance in the crust is 0.0075%.

29.23 Compounds of nickel

Nickel has the +2 oxidation state in most of its compounds, often as the light green hydrated ion, $[Ni(H_2O)_6]^{2+}$ (Table 29.16). The halide salts, upon heating above 100°C, are converted to the anhydrous salts $NiCl_2$ (yellow), $NiBr_2$ (yellow-brown), and NiI_2 (black). These dissolve readily in water, giving green solutions. Nickel(II) sulfate, nickel(II) nitrate, and other nickel(II) salts containing oxo anions retain their green color upon dehydration, and therefore, this color must be due to the coordination of oxygen atoms with the nickel(II) ion.

The addition of ammonium hydroxide to a solution containing nickel(II) ion gives an apple green precipitate of nickel(II) hydroxide, which dissolves upon the addition of excess ammonium hydroxide to yield the deep blue hexaamminenickel(II) ion, $[Ni(NH_3)_6]^{2+}$.

$$Ni^{2+} + 2OH^- \longrightarrow Ni(OH)_2(s)$$
$$Ni(OH)_2(s) + 6NH_3(aq) \longrightarrow [Ni(NH_3)_6]^{2+} + 2OH^-$$

Many organic amines yield similar products.

In nickel carbonyl, $Ni(CO)_4$, the metal shows a formal oxidation state of zero. This interesting compound, which long baffled theoretical chemists, is a volatile, highly refractive, *extremely toxic* liquid. It is obtained by the action of carbon monoxide gas on nickel powder at slightly elevated temperatures.

$$Ni(s) + 4CO(g) \longrightarrow Ni(CO)_4(g)$$
$$\text{\textit{nickel}}$$
$$\text{\textit{carbonyl}}$$

At somewhat higher temperatures, it decomposes with the liberation of metallic nickel. This property once was employed in the separation of nickel from other metals, but nickel is now purified mainly by electrolytic refining.

The two outer electron shells of the nickel atom have the configuration $3s^23p^63d^84s^2$. The result of reaction with carbon monoxide, which is a very powerful coordinating agent, is a structure having the configuration $3s^23p^63d^{10}4s^24p^6$. Evidently the electrons in the two $3d$ orbitals that contained one electron each are paired during the reaction, and four pairs of electrons from the four carbon monoxide mole-

TABLE 29.16

Some nickel compounds

NiCl₂	Nickel(II) chloride (yellow)
NiCl₂·6H₂O	Nickel(II) chloride hexahydrate (green)
NiO	Nickel(II) oxide (green black)
NiS	Nickel(II) sulfide (green black)
NiSO₄·6H₂O	Nickel(II) sulfate hexahydrate (green)
Ni(CO)₄	Nickel carbonyl (volatile, flammable liquid, poisonous)
[Ni(NH₃)₆](NO₃)₂	Hexaamminenickel(II) nitrate (bright blue)

CH₃—C—C—CH₃

HON   NOH

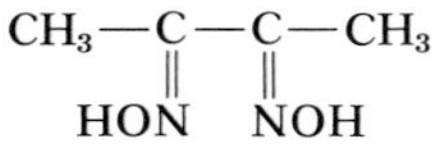

dimethylglyoxime

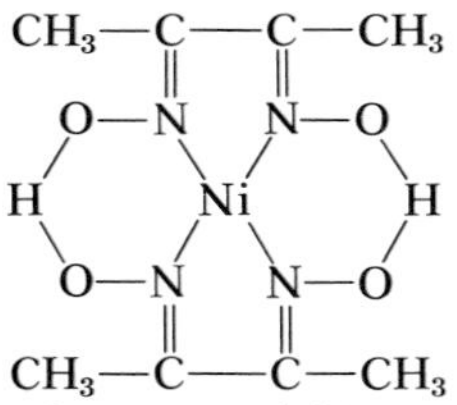

*complex between nickel ion and two
molecules of dimethylglyoxime*

FIGURE 29.11

**Nickel complex with dimethyl-
glyoxime.**

cules are shared with the nickel atom. The complex is stabilized by double bond formation to give

$$Ni(::C::\ddot{O}:)_n(:C:::O:)_{4-n}$$

Several of the transition metals form carbonyls of this sort; all are volatile, highly poisonous, and soluble in nonpolar solvents.

Most nickel complexes are six-coordinate and octahedral. There are, however, many nickel(II) complexes in which the metal is truly four-coordinate. Some of these have the spin-paired $3d4s4p^2$ configuration and are planar; others have the spin-free $4s4p^3$ configuration and are tetrahedral. The planar configuration is more stable and is achieved wherever the ligands are not too large to fit at the corners of a square.

One of the most striking of the nickel complexes is that formed with the organic compound dimethylglyoxime (DMG), which has the structural formula shown in Figure 29.11. When nickel(II) ion is added to a neutral solution of this material, it coordinates with the two nitrogen atoms in each of the two DMG molecules. At the same time one hydrogen ion escapes from each molecule of DMG, and the other forms a hydrogen bond with an oxygen atom of the other DMG molecule. This complex is only very slightly soluble in water and is bright red. Since no other common metal ion gives a precipitate with dimethylglyoxime, the formation of a red precipitate when DMG is added to a solution is almost infallible evidence for the presence of nickel(II) ion in that solution.

Copper, silver, and gold

29.24 Properties and uses of copper, silver, and gold

Copper, silver, and gold atoms each have one electron in the outer s subshell and ten electrons in the underlying d subshell (see Table 29.1). As would be expected, an atom of each of these elements can lose its s electron; under suitable conditions, each can also lose one or two of the d electrons. Thus, all show oxidation states of +1, +2, and +3. However, in most of its important compounds, copper is in the +2 state; silver rarely has any oxidation state but +1, and gold is practically always in the +1 or +3 state.

Copper, silver, and gold all lie below hydrogen in the electromotive series. They do not react with nonoxidizing acids such as hydrochloric acid and they are not readily attacked by oxygen at ordinary temperatures. On heating with air, copper, which is the most reactive of the three, forms copper(II) oxide, CuO; silver slowly forms silver(I) oxide, Ag_2O, which, however, decomposes into its elements on strong heating. Gold does not react with oxygen.

Copper and silver readily react with sulfur and sulfur-containing compounds, forming either Cu_2S or CuS and Ag_2S, respectively. This reaction is particularly evident in the case of silver, which darkens rapidly when left in contact with sulfur-containing substances such as eggs, rubber, or mustard. Both metals tarnish slowly when exposed to the atmosphere, for there are nearly always traces of hydrogen sulfide in the air.

Both copper and silver readily react with oxidizing acids, such as nitric acid, copper going to the +2 state, and silver to the +1 state:

$$3Cu(s) + 8HNO_3(dil) \longrightarrow 3Cu(NO_3)_2(aq) + 2NO(g) + 4H_2O(l)$$
$$3Ag(s) + 4HNO_3(dil) \longrightarrow 3AgNO_3(aq) + NO(g) + 2H_2O(l)$$

Gold, by far the least reactive of the metals in this family, does not tarnish noticeably in the air or when exposed to sulfur or its compounds. It is not attacked by nitric acid, but it is dissolved by the mixture of concentrated nitric acid and concentrated hydrochloric acid known as aqua regia.

The cations of these metals readily form complex ions, especially with groups containing nitrogen and sulfur donor atoms. The Cu^+, Ag^+, and Au^+ ions usually show a coordination number of two, whereas Cu^{2+} and Au^{3+} commonly show a coordination number of four. The formation of complexes is of great importance in the chemistry of these metals; advantage is taken of it in the metallurgy and the electroplating of all three, in analysis, and, in the case of silver, in the developing of photographs.

Metals are, we have learned, distinguished from nonmetals by several physical properties—metals are ductile, malleable, good conductors of heat and electricity, and have metallic luster. In these physical criteria, copper, silver, and gold are the most metallic of all of the elements. If a heavy weight is hung on a fairly thin copper wire, the wire will continue to stretch slowly for several days until it becomes too small in diameter to support the weight. Gold is the most malleable of all metals; it can be rolled into sheets that are so thin as to be transparent (these are green by transmitted light). Gold leaf is used in signs on windows and office doors, and in covering the domes of capitol buildings.

Copper, silver, and gold are all excellent conductors of heat and electricity. Copper is used in cooking utensils because of its heat conductivity and in electrical apparatus because of its high electrical conductivity. Both silver and gold are better conductors than copper, but their higher cost rules them out for most purposes. Even copper is becoming so expensive that substitutes are being sought, although its major use is still in electrical applications. Aluminum is not as good a conductor of electricity as copper volume for volume. However, it is comparable to it, weight for weight, and has been used in many electrical applications. Some question has arisen about the safety of aluminum wiring in private homes—it may eventually deteriorate and cause short circuits.

All three of the metals of this family are soft, and are often al-

Name	Composition (%)
Brass	Cu, 20–97; Zn, 2–80; Sn, 0–14; Pb, 0–12; Mn, 0–25
Bronze	Cu, 50–98; Sn, 0–35; Zn, 0–29; Pb, 0–50; P, 0–3
German silver	Cu, 46–93; Zn, 20–36; Ni, 6–30
Bell metal	Cu, 75–80; Sn, 20–25
Nickel coin	Cu, 75; Ni, 25
Sterling silver	Cu, 7.5; Ag, 92.5
Gold (18 carat)	Cu, 5–14; Au, 75; Ag, 10–20
Gold (14 carat)	Cu, 12–28; Au, 58; Ag, 4–30
Purple gold	Au, 78; Al, 22
White gold	Au with varying amounts of Pd, Ni, or Zn

TABLE 29.18

End uses of copper, silver, and gold

Copper

Electrical applications

Pipes, plumbing, gutters

Industrial machinery

Coinage

Silver

Sterling silver tableware

Photography

Mirror backing

Heat-exchange equipment

Pharmaceuticals

Electronic devices

Jewelry

Gold

Electronic devices

Photography

Electrolyte in electroplating

Jewelry

Reflective coating, e.g., on spaceships, windows

loyed with other metals for hardness, rigidity, and other desirable properties (Table 29.17).

Because of their beautiful luster and color, silver and gold have been used for many centuries for ornaments, jewelry, and tableware. The comparative scarcity of silver has prompted its use in coinage—a use that many governments have now abandoned because the metal has become too scarce and costly. Most silver is now used in photography, in the form of silver halide salts.

The major uses of copper, silver, and gold are summarized in Table 29.18.

29.25 Sources and metallurgy of copper, silver, and gold

Copper, silver, and gold are all found both in the free state and in ores. Copper is most abundant (twenty-fifth among the elements in the Earth's crust), while silver and gold are much less abundant (they are among the fifteen *least* abundant elements).

a. Copper. The production of copper from ores containing the free metal is relatively simple. The free metal is separated from the rock that accompanies it, melted, and cast into ingots. This metal is often nearly pure.

The sulfide ores of copper, *chalcocite* (Cu_2S) and *chalcopyrite* ($CuFeS_2$), which are by far the most common, are concentrated by flotation and then heated to a high temperature in a stream of air, which converts part of the sulfide to oxide:

$$2Cu_2S(l) + 3O_2(g) \longrightarrow 2Cu_2O(l) + 2SO_2(g)$$

The Cu_2O, as it is formed, reacts with the remaining Cu_2S, to give metallic copper

$$Cu_2S(l) + 2Cu_2O(l) \longrightarrow 6Cu(l) + SO_2(g)$$

If the conversion to oxide has gone too far, carbon is added to reduce the excess copper(I) oxide

$$Cu_2O(l) + C(s) \longrightarrow 2Cu(l) + CO(g)$$

Any iron that was present in the ore is also converted to oxide, which immediately forms a molten slag with the silicon dioxide that was present in the ore and that was not removed in the flotation process.

$$2FeS(l) + 3O_2(g) + 2SiO_2(s) \longrightarrow 2FeSiO_3(l) + 2SO_2(g)$$

If the sulfur dioxide produced in the metallurgy of all sulfide ions is allowed to escape into the atmosphere (as it usually is), it not only creates an unpleasant odor, but affects the health of those who inhale it. If it is present in high concentration, it may kill growing plants for several miles downwind from the smelter. Moreover, it is slowly converted in the air to sulfur trioxide, which, in turn, reacts with moisture in the atmosphere to produce sulfuric acid. Smelters in some localities in the United States have been forced by government action to remove the sulfur dioxide from stack gases and convert it to nonpolluting products. This can be done either by catalytically oxidizing the sulfur dioxide in the stack to sulfuric acid (via sulfur trioxide) or by reducing it to elemental sulfur by passing it over red hot carbon

$$SO_2(g) + 2C(s) \longrightarrow S(l) + 2CO(g)$$

The products of these reactions (H_2SO_4 and S) are useful materials, but commonly they cannot be sold for enough to pay for their recovery. However, the pollution problem in the vicinity of smelters is so great that removal of the sulfur dioxide from the stack gases is necessary.

The copper from smelting a sulfide ore is quite impure, the chief impurities being silver, gold, iron, zinc, lead, arsenic, sulfur, copper(I) oxide, and bits of slag. Since even traces of impurities reduce its conductivity greatly, the copper must be purified. By heating the molten metal in a stream of air, arsenic and sulfur are converted to the volatile oxides and escape. The other impurities are removed by an electrolytic process. Bars of the partially purified copper serve as anodes in the electrolysis and plates of pure copper as the cathodes; a mixture of dilute sulfuric acid, sodium chloride, and copper(II) sulfate is the electrolyte (Figure 29.12). By careful control of the voltage across the cell, only copper and the more electropositive metal impurities (e.g., iron, zinc, lead) in the anode are oxidized and dissolved. Metallic impurities less electropositive than copper, such as silver and gold, are unaffected and drop from the anode as it disintegrates. These anode sludges are worked over, and valuable metal by-products of the refining process are recovered. Voltage control is so maintained that only copper, the least electropositive of the various metals that dissolve from the anode, is plated out on the pure copper cathode. The refining process yields electrolytic copper, which has a purity of greater than 99.9%.

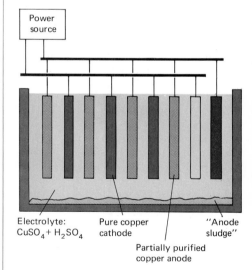

Power source

Electrolyte: $CuSO_4 + H_2SO_4$

Pure copper cathode

Partially purified copper anode

"Anode sludge"

FIGURE 29.12
Electrolytic cell for refining copper.

b. Silver. Since silver ion forms extremely strong complexes with sulfur atoms, it is not surprising that the most common ores of silver contain silver sulfide. Other ores are of little consequence in the economy of silver. The naturally occurring Ag_2S is nearly always mixed in small amounts with the sulfides of copper, lead, and other heavy metals, and most of the silver of commerce is a by-product of the metallurgy of other metals. Much silver is obtained as a by-product in the refining of copper. Often, silver is present in large enough amounts to pay the cost of refining the copper.

c. Gold. Gold nearly always occurs in the metallic state. It is found as fine grains in river sand and in quartz veins or sometimes as nuggets, which vary from the size of a pinhead to those weighing several hundred pounds. The latter, of course, are extremely rare.

The early methods of separating gold from sand or rock depended upon the fact that gold is much heavier than the rock (specific gravity of Au is 19.3; that of sand and most rocks is about 2.5). The methods involved panning, sluicing, and other variants of washing the sand or powdered rock away and leaving the heavier gold behind. Such methods are inefficient and have been used only under primitive conditions.

The best method of separating gold is to render it soluble by "cyaniding" or "chlorinating." In spite of its low reactivity, gold will dissolve in solutions of cyanide ion in the presence of dioxygen and in chlorine solutions, forming stable complexes in both cases:

$$4Au(s) + 8NaCN(aq) + O_2(g) + 2H_2O(l) \longrightarrow$$
$$4Na[Au(CN)_2](aq) + 4NaOH(aq)$$
$$2Au(s) + 2NaCl(aq) + 3Cl_2(g) \longrightarrow 2Na[AuCl_4](aq)$$

Gold dissolves rather slowly, but eventually nearly all of it can be taken into solution. After filtering the solution from the rocky material, the gold is recovered by electrolytic reduction or by reduction with another metal, for example, zinc,

$$[Au(CN)_2]^- + e^- \longrightarrow Au(s) + 2CN^-$$
$$2[Au(CN)_2]^- + Zn(s) \longrightarrow 2Au(s) + [Zn(CN)_4]^{2-}$$

29.26 Compounds of copper

Stable copper(I) compounds include the halides, the sulfide, and the cyanide (Table 29.19). These compounds all occur as anhydrous salts and are more covalent than ionic. Upon addition of I^- or CN^-, copper(II) ion in solution can be reduced to yield the insoluble copper(I) cyanide or copper(I) iodide. Both of these compounds go into solution with excess anion by the formation of complexes.

$$2Cu^{2+} + 4CN^- \longrightarrow 2CuCN(s) + (CN)_2(g)$$
$$2Cu^{2+} + 6CN^- \longrightarrow 2[Cu(CN)_2]^- + (CN)_2(g)$$
$$2Cu^{2+} + 4I^- \longrightarrow 2CuI(s) + I_2(aq)$$
$$CuI(s) + I^- \longrightarrow [CuI_2]^-$$

The [Cu(CN)₂]⁻ complex ion is so stable that hydrogen sulfide does not decompose it, even though copper(I) sulfide is a highly insoluble substance.

The Cu^+ ion is not at all stable in solution, for it is subject to disproportionation

$$2Cu^+ \rightleftharpoons Cu + Cu^{2+} \qquad K = \frac{[Cu^{2+}]}{[Cu^+]^2} = 1.1 \times 10^6$$

Copper(I) salts containing oxo anions are known, but they are also unstable. For example, Cu_2SO_4 can be prepared and kept in an anhydrous medium, but upon exposure to moisture, it reacts vigorously.

$$Cu_2SO_4(s) \longrightarrow Cu(s) + CuSO_4(s)$$

Compounds of copper(II) are much more common than those of copper(I). Copper(II) oxide, CuO, is a black, insoluble substance formed by gentle heating of the metal in air or by the addition of a base to a hot solution of a copper(II) salt. The blue-green copper(II) hydroxide is obtained either as a gelatinous material with variable water content or as a crystalline substance of the composition $Cu(OH)_2$ when a base is added to a copper(II) solution maintained at room temperature. Both the oxide and the hydroxide are predominantly basic in character and react with most acids to give solutions of copper(II) salts. Such solutions are generally blue or blue-green and probably contain the hydrated copper(II) ion, $[Cu(H_2O)_4]^{2+}$.

Soluble simple copper(II) salts crystallize from aqueous solution as hydrates (Table 29.19).

The most frequently encountered insoluble compound of copper(II) is the copper(II) sulfide, CuS, usually obtained by the action of hydrogen sulfide on a solution of a copper(II) salt. Copper(II) sulfide is highly insoluble in water ($K_{sp} = 6 \times 10^{-36}$ at 25°C) but dissolves readily in a solution of sodium sulfide because of the formation of the still less dissociated complex, $[CuS_2]^{2-}$.

$$Na_2S(aq) + CuS(s) \longrightarrow Na_2CuS_2(aq)$$

This reaction is often used in analytical chemistry to separate copper from the ions of other heavy metals that do not form such stable sulfide complexes. Copper(II) ion has a great tendency to form both cationic and anionic complexes. The most familiar cation is the deep blue, stable tetraamminecopper(II) ion, $[Cu(NH_3)_4]^{2+}$.

By far the most common salt of copper(II) is copper(II) sulfate pentahydrate, $CuSO_4 \cdot 5H_2O$, commonly named blue vitriol. Like most other hydrated copper(II) salts, this one is light blue in color, which is characteristic of the square-planar $[Cu(H_2O)_4]^{2+}$ ion. The fifth molecule of water is held to the sulfate ion by hydrogen bonding. When the salt is dehydrated by gentle heating or placing it is a desiccator, the blue crystals crumble to a fine white powder. On dissolution in water, this powder again gives a blue solution.

TABLE 29.19

Some compounds of copper

Cu_2O	Copper(I) oxide (red)
Cu_2S	Copper(I) sulfide (black)
CuCl	Copper(I) chloride (white)
$CuCl_2 \cdot 2H_2O$	Copper(II) chloride dihydrate (green)
$Cu(NO_3)_2 \cdot 6H_2O$	Copper(II) nitrate hexahydrate (blue)
$CuSO_4 \cdot 5H_2O$	Copper(II) sulfate pentahydrate (blue)
CuO	Copper(II) oxide (black)

29.27 Compounds of silver

Silver nitrate, $AgNO_3$ (once referred to as "lunar caustic," because in alchemical lore, silver was related to the moon; "caustic" because in the solid form, it burns flesh) is made by dissolving metallic silver in nitric acid. It can be crystallized from the resulting solution as large, white crystals that are readily soluble in water. It is the usual source of other silver compounds (Table 29.20).

Most of the simple compounds of silver are insoluble in water and can be prepared by mixing solutions of silver nitrate with solutions of the appropriate anion. For example,

$$Ag^+(aq) + Cl^-(aq) \longrightarrow AgCl(s)$$

The formation of the white, curdy precipitate of silver chloride, AgCl, is used as a sensitive test for the presence of either Ag^+ or Cl^- in solution. The pale yellow bromide and yellow iodide are even less soluble than the chloride. Although silver chloride is not very soluble in water ($K_{sp} = 1.8 \times 10^{-10}$), it is dissolved by solutions of ammonia, with which it forms a stable soluble complex ($K_d = 6.2 \times 10^{-8}$), diamminesilver chloride, $[Ag(NH_3)_2]Cl$,

$$AgCl(s) + 2NH_3(aq) \longrightarrow [Ag(NH_3)_2]Cl(aq)$$

Upon addition of acid, this reaction is reversed:

$$[Ag(NH_3)_2]Cl(aq) + 2H^+(aq) \longrightarrow AgCl(s) + 2NH_4^+(aq)$$

The diamminesilver ion is typical of the majority of silver(I) complexes in which silver has a coordination number of two. Silver cyanide, for example, goes into solution as a complex ion, $[Ag(CN)_2]^-$, in the presence of cyanide ions.

29.28 Compounds of gold

As with copper, the simple gold compounds in which the metal is in the +1 state are the halides, the sulfide, and the cyanide. All of these are insoluble and show a tendency to disproportionate into metallic gold and the gold(III) compound. All of them are readily reduced to metallic gold. The most important soluble compound of gold(I) is the cyano complex, $Na[Au(CN)_2]$, which is used in electroplating gold onto metallic objects.

Among the gold(III) compounds, the commonest simple substance is gold(III) chloride, $AuCl_3$. Other salts can be prepared from it. The tetrachloroaurate ion, $[AuCl_4]^-$, is obtained by the action of aqua regia on metallic gold.

In its complexes, gold(I), like copper(I) and silver(I), usually shows a coordination number of two, and forms linear complexes. Gold(III), like copper(II), shows a coordination number of four and forms square-planar complexes.

TABLE 29.20

Some compounds of silver and gold

$AgNO_3$	Silver(I) nitrate (white)
AgCl	Silver(I) chloride (white)
AgCN	Silver(I) cyanide (white)
AuCl	Gold(I) chloride (yellow)
$AuCl_3$	Gold(III) chloride (red)
Au_2S_3	Gold(III) sulfide (brown-black)

The photographic process

When exposed to light of short wavelength, silver halides are "activated," after which they are more easily reduced to the metallic state than unactivated silver halides. This is the basis of the photographic process.

$$AgX(s) + \text{light energy} \longrightarrow AgX^* \text{ (activated silver halide)}$$
$$AgX^* + \text{reducing agent} \longrightarrow Ag(s)$$

The silver halide (usually a mixture of halides) is suspended in gelatin, which is spread in a uniform layer on the film. Exposure activates those parts of the coating on which light falls. The image is developed by placing the film in a solution of a weak organic reducing agent. This reduces the activated silver halide to metallic silver in a finely divided form, which is black. (If the reducing action is too strong or if the film is left in the reducing agent too long, both the activated and the unactivated silver halide are reduced and the entire film becomes black.) Once the film has been developed, the unreduced silver halide must be removed before the film is taken into the light; otherwise, it will slowly turn black. This silver halide is removed by washing the film in a solution of sodium thiosulfate, $Na_2S_2O_3$, which forms a stable, soluble complex with the silver halide. This is the fixing process, and sodium thiosulfate is called *hypo*.

$$AgX(s) + 2S_2O_3^{2-}(aq) \longrightarrow [Ag(S_2O_3)_2]^{3-}(aq) + X^-(aq)$$

After rinsing in water, the negative is dried. It is black where light fell on it and clear where no light struck it. To make a print, the negative is projected onto paper coated with silver halide–gelatin. Where the negative is black, no light falls through to the paper, and vice versa. Once the paper is developed, fixed, and dried, the print is ready.

Portraits can be toned by converting the silver in the picture to some color other than black. This is done by immersing the picture in a solution of a material that will react with silver and leave a suitable deposit on the paper.

$$Ag(s) + [AuCl_2]^-(aq) \longrightarrow AgCl(s) + Cl^- + Au(s) \quad \text{(sepia)}$$
$$2Ag(s) + PtCl_4^{2-}(aq) \longrightarrow 2AgCl(s) + 2Cl^- + Pt(s) \quad \text{(gray)}$$

The silver chloride formed in toning with gold or platinum is removed by treatment with hypo.

The chemistry of color photography is extremely complex, but the initial steps rely on the photosensitivity of the silver halides, just as black and white photography does. There are three layers of the AgX–gelatin suspension. Each layer also contains an organic dye, which reacts with the silver to give a characteristic color. The silver salt formed by reaction with the dye is later washed out; the final picture contains dyes but no silver.

Many reactions are effected by exposure to light, and the photographic companies' researchers are continually looking for a photographic process that does not use silver salts. But any new process will have to compete with the silver halide process, which has been brought to a high degree of efficiency.

Periodic perspective:
f transition elements

29.29 Members and electronic configurations

The elements from lanthanum, $_{57}$La, through lutetium, $_{71}$Lu, are collectively called the **lanthanides,** or the **rare earth elements.** To the ancient Greeks, oxides of metals were known as "earths." Because the first elements in the series were found in rare minerals as oxides, these elements came to be known as the "rare earth elements." Although they are often hard to isolate, many of the "rare earth" metals are not particularly rare. Thirteen of the fifteen rank between twenty-ninth and sixtieth in the abundance of the elements. Cerium is the most abundant rare earth element, and thulium and promethium are the least abundant.

The elements from actinum, $_{89}$Ac, through lawrencium, $_{103}$Lr, are collectively called the **actinides.** The actinides are all radioactive. The elements with atomic numbers greater than 93 (uranium) are called the **transuranium elements.**

In the lanthanides and actinides the *f* energy levels in the second from the outermost shell are being filled to capacity with fourteen electrons (Tables 29.21, 29.22). For the lanthanides, the 4*f* level is being filled; for the actinides, it is the 5*f* level. The outermost shell contains a pair of *s* electrons, and in some cases the second from outermost shell contains a single *d* electron. All energy levels below the $(n - 2)f$ level are filled to capacity and their electrons are unavailable for bonding.

29.30 Properties of *f* transition elements

Since atoms of the lanthanide elements differ from each other to only a slight extent in size and electronic configuration (Table 29.21), they are remarkably similar in chemical properties. Because of increasing effective nuclear charge, there is a gradual decrease in atomic radii across the group. Presumably, there is a similar decrease in atomic radii across the actinide sequence, but few data on the sizes of these

La* 57	Ce 58	Pr 59	Nd 60	Pm 61	Sm 62	Eu 63	Gd 64	Tb 65	Dy 66	Ho 67	Er 68	Tm 69	Yb 70	Lu 71
138.905	140.12	140.907	144.24	(145)	150.35	151.96	157.25	158.925	162.50	164.930	167.26	168.934	173.04	174.97
Ac** 89	Th 90	Pa 91	U 92	Np 93	Pu 94	Am 95	Cm 96	Bk 97	Cf 98	Es 99	Fm 100	Md 101	No 102	Lr 103
(227)	232.038	(231)	238.029	(237)	(244)	(243)	(247)	(247)	(251)	(254)	(257)	(258)	(255)	(260)

⌐ *number in parentheses is atomic mass number of isotope of longest known half-life*

Europium and ytterbium apparently have larger atomic radii because they contribute only two electrons to metallic bonding instead of three (see their configurations).

TABLE 29.21

Properties of the lanthanides

Element	Symbol	Electronic configuration	Atomic radii (nm)	Ionic radii (M^{3+}; nm)
Lanthanum	$_{57}$La	$5d^1\ 6s^2$	—	0.1061
Cerium	$_{58}$Ce	$4f^1\ 5d^1\ 6s^2$	0.1646	0.1034
Praseodymium	$_{59}$Pr	$4f^3\quad 6s^2$	0.1648	0.1013
Neodymium	$_{60}$Nd	$4f^4\quad 6s^2$	0.1642	0.0995
Promethium	$_{61}$Pm	$4f^5\quad 6s^2$	—	0.0979
Samarium	$_{62}$Sm	$4f^6\quad 6s^2$	0.166	0.0964
Europium	$_{63}$Eu	$4f^7\quad 6s^2$	0.185	0.0950
Gadolinium	$_{64}$Gd	$4f^7\ 5d^1\ 6s^2$	0.1614	0.0938
Terbium	$_{65}$Tb	$4f^9\quad 6s^2$	0.1592	0.0923
Dysprosium	$_{66}$Dy	$4f^{10}\quad 6s^2$	0.1589	0.0908
Holmium	$_{67}$Ho	$4f^{11}\quad 6s^2$	0.1580	0.0894
Erbium	$_{68}$Er	$4f^{12}\quad 6s^2$	0.1567	0.0881
Thulium	$_{69}$Tm	$4f^{13}\quad 6s^2$	0.1562	0.0869
Ytterbium	$_{70}$Yb	$4f^{14}\quad 6s^2$	0.1699	0.0858
Lutetium	$_{71}$Lu	$4f^{14}\ 5d^1\ 6s^2$	0.1557	0.0848

TABLE 29.22

Properties of the actinides

Element	Symbol	Electronic configuration	Ionic radii (M^{3+}; nm)
Actinium	$_{89}$Ac	$6d^1\ 7s^2$	0.111
Thorium	$_{90}$Th	$6d^2\ 7s^2$	0.108[a]
Protactinium	$_{91}$Pa	$5f^2\ 6d^1\ 7s^2$	0.105[a]
Uranium	$_{92}$U	$5f^3\ 6d^1\ 7s^2$	0.103
Neptunium	$_{93}$Np	$5f^4\ 6d^1\ 7s^2$	0.101
Plutonium	$_{94}$Pu	$5f^6\quad 7s^2$	0.100
Americium	$_{95}$Am	$5f^7\quad 7s^2$	0.099
Curium	$_{96}$Cm	$5f^7\ 6d^1\ 7s^2$	
Berkelium	$_{97}$Bk	$5f^9\quad 7s^2$	
Californium	$_{98}$Cf	$5f^{10}\quad 7s^2$	
Einsteinium	$_{99}$Es	$5f^{11}\quad 7s^2$	
Fermium	$_{100}$Fm	$5f^{12}\quad 7s^2$	
Mendelevium	$_{101}$Md	$5f^{13}\quad 7s^2$	
Nobelium	$_{102}$No	$5f^{14}\quad 7s^2$	
Lawrencium	$_{103}$Lr	$5f^{14}\ 6d^1\ 7s^2$	

[a] The oxidation state of the ion is unknown.

Superheavy elements

How far can the periodic table be extended? Is it possible that elements beyond 106 will be found in nature or produced by nuclear bombardment reactions? These are not fanciful questions. On the basis of nuclear theory, periodic table trends, and computer-assisted calculations, predictions have been made about the configurations and properties of elements all the way to atomic number 168.

The elements most recently produced, 104, 105, and 106, are very unstable, with half-lives of less than 1 second. Elements 107–109 will probably also be of this type.

However, what has been termed an "island of stability" (Figure A) lies ahead in the region of atomic number 114 and mass number 298. Nuclei of element $^{294}_{110}X$ will contain a magic number of neutrons (184), and nuclei of $^{298}_{114}X$ will be doubly magic, with magic numbers of both neutrons (184) and protons (114). These nuclei will have the closed nuclear shell configurations known to impart stability (Section 8.7). Half-lives of about 10^3 yr are thought possible for elements of the island of stability.

Calculations of possible electronic configurations of the superheavy atoms suggest that after element 106 (probably of outer configuration $5f^{14}6d^47s^2$) will be regular completion of $6d$, $7p$, and $8s$ orbitals. Next will come complete series of transition elements in which the $6f$ and $8g$ energy levels are being filled.

Element 114 would be a member of Representative Group IV and would resemble lead in its properties. Proposed similarities between known elements and the superheavy elements are being used in the examination of potential naturally occurring or artificial sources of these elements. Several flurries of excitement have been caused by tentative identification of superheavy elements in minerals and meteorites. Although conclusive proof of the existence of superheavy, stable elements has not yet been presented, it remains possible that they will eventually be found.

FIGURE A
Regions of stability and instability of nuclei.

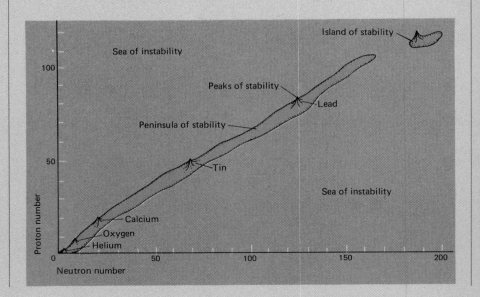

TABLE 29.23

Electronic configurations of some lanthanide element ions

Each ion has a xenon noble gas core.

Some +3 ions		Most common ions other than +3 ions			
Ce³⁺	$4f^1$	Ce⁴⁺	[Xe]	Sm²⁺	$4f^6$
Nd³⁺	$4f^3$	Pr⁴⁺	$4f^1$	Eu²⁺	$4f^7$
Gd³⁺	$4f^7$	Tb⁴⁺	$4f^7$	Yb²⁺	$4f^{14}$
Tb³⁺	$4f^8$				
Lu³⁺	$4f^{14}$				

are currently available. For ions of comparable charge, a regular decrease in size occurs across each series.

The elements of both f transition series form positive ions very readily. The <u>standard reduction potentials</u> of the lanthanide elements for the reaction

$$M^{3+} + 3e^- \rightleftharpoons M$$

range from -2.52 V for lanthanum to -2.25 V for lutetium, with a slight, regular decrease in electropositive character across the series. (The value for cerium is lower than the trend would require, -2.48 V.)

The standard reduction potentials for the actinide elements for various ions in solution also have large negative values, for example,

$$Th^{4+} + 4e^- \rightleftharpoons Th \qquad E° = -1.90 \text{ V}$$
$$U^{3+} + 3e^- \rightleftharpoons U \qquad E° = -1.80 \text{ V}$$
$$Pu^{3+} + 3e^- \rightleftharpoons Pu \qquad E° = -2.03 \text{ V}$$
$$Am^{3+} + 3e^- \rightleftharpoons Am \qquad E° = -2.38 \text{ V}$$

Electrons in the $4f$ and $5f$ energy levels of the lanthanide and actinide elements, respectively, lie deep within the atom and are less available for chemical bonding than are the d electrons of the d transition elements. This is especially true of the lanthanide elements, which do not exhibit the great variety of oxidation states found for d transition elements. The lanthanides, with outer configurations $4f^{1-14}5d^{0-1}6s^2$, characteristically form +3 ions. This occurs through the loss of the two $6s$ electrons and a $5d$ or $4f$ electron. In addition to the common +3 state, a number of the lanthanides give ions of +4 and +2 charge. The ions best known for these oxidation states and their configurations are listed in Table 29.23. Cations of +4 charge are rare and are found for only a few metals of relatively large atomic radius. The scarcity of such ions is consistent with the large energy requirements for the removal of four electrons from an atom.

The actinide elements (outer electronic configurations $5f^{1-14}6d^{0-1}7s^2$) exhibit a greater variety of oxidation states, at least early in the series, than do the lanthanide elements. The atoms of the actinide elements are significantly larger than those of the corresponding lanthanides, and in them the $5f$ electrons are more effectively screened from the attractive force of the nucleus by underlying levels of electrons than are the $4f$ electrons in the lanthanides. As a re-

TABLE 29.24

In the +3 and +4 states, the elements can exist as simple ions.

Element	Electronic configuration (Rn core omitted)		Most common oxidation state	Oxidation states
$_{90}$Th		$6d^2\ 7s^2$	+4	+4
$_{91}$Pa	$5f^2$	$6d^1\ 7s^2$	+5	+4, +5
$_{92}$U	$5f^3$	$6d^1\ 7s^2$	+6	+3 to +6
$_{93}$Np	$5f^4$	$6d^1\ 7s^2$	+5	+3 to +7
$_{94}$Pu	$5f^6$	$7s^2$	+4	+3 to +6
$_{95}$Am	$5f^7$	$7s^2$	+3	+3 to +6
$_{96}$Cm	$5f^7$	$6d^1\ 7s^2$	+3	+3, +4
$_{97}$Bk	$5f^9$	$7s^2$	+3	+3, +4
$_{98}$Cf	$5f^{10}$	$7s^2$	+3	+3

sult, the 5f electrons are more readily utilized for chemical bonding. In Table 29.24 are listed the oxidation states known for the elements thorium (atomic number 90) through californium (atomic number 98). The +3 state becomes increasingly stable across the sequence because, as the 5f level comes closer to being half-filled, electrons in this level are held more tightly and are less readily available for bonding. No simple cationic species of greater than +4 charge are known for the actinide elements.

29.31 The lanthanides and their compounds

The lanthanides are all shiny, silver, reactive metals. Most tarnish readily in air by formation of oxides, although gadolinium and lutetium are quite stable. Some form white oxides and colorless ions in aqueous solution, while others have colored ions and oxides. The pure metals range in density from 6.2 g/cm³ for lanthanum to 9.8 g/cm³ for lutetium, and their melting points all fall between about 800°C and 1600°C.

The largest use for compounds of the lanthanides (12 million pounds in 1969) is in petroleum cracking catalysts. The glass and metallurgical industries also consume lanthanide compounds. In some alloys, rare earths are used to impart desirable properties and in others to react with and remove undesirable impurities. Praseodymium and neodymium are added to the glass in welders goggles to absorb the bright yellow light of the sodium flame. Cerium oxide is effective in polishing camera and eyeglass lenses. And pure neodymium oxide is added to glass to produce a beautiful purple color.

A mixed oxide of europium and yttrium (Y_2O_3, Eu_2O_3) gives a brilliant red phosphor which is used in color television screens. To mention just one more application of a lanthanide, yttrium–aluminum garnets (YAG) are used both in electronic equipment (as microwave filters) and as synthetic gems.

Energy: The key to natural resource abundance

Basically the problem of providing adequate resources for the perpetuation and expansion of civilization is the problem of providing adequate quantities of energy of the right type at the right place at the right time. This is true no matter whether the resource is food, minerals, structural materials, clean air, or energy itself.

It is often said that "it takes money to make money." Similarly it takes energy to make energy. When our energy comes from petroleum, oil fields must be found, wells must be drilled, and the oil must be pumped to the surface and transported to the refinery where it is modified to suit the needs of the users. All of this requires energy, not only to drill, pump, and move the material, but to make the equipment, pipes, rails, trains, and cracking plants. Vast quantities of energy are required to mine and transport coal, to lay pipelines for gas, to construct and move oil tankers, to enrich isotopes and build nuclear reactors. . . .

With respect to minerals we find it necessary to dig ever deeper into the earth and to process ores of lower and lower grade. At one time in the United States we had very large deposits of high-grade iron ore. As time passed our better grades were consumed, so we moved to lower-grade ores such as taconites that require beneficiation and agglomeration to make them usable. Naturally this requires increased expenditures of energy. An alternative is to import high-grade iron ore from other countries. This, too, results in higher energy costs, in this case for transportation. . . .

Is there any limit to how low the grade of ore can be and still provide the metals and other substances necessary for the continued functioning of our society? Here the ultimate answer is that there is no limit, provided the necessary quantities of energy are available for undertaking the physical and chemical operations. *In practice, of course, there are limits to grade, below which we need not concern ourselves. Those limits are represented by the most extensive and yet the lowest-grade ore deposits on the surface of the earth: seawater and ordinary sedimentary and igneous rocks. In principle, a high level of civilization could be nourished for many millions of years by those substances. From them society could obtain all that it needs, including the energy itself. But the energy costs per capita of such a system would obviously be much higher than those of today.*

ENERGY IN OUR FUTURE, by Harrison Brown [Quoted from Vol. 1,
Annual Review of Energy (J. M. Hollander and M. K. Simmons, eds.),
Annual Reviews Inc., Palo Alto, Calif., © 1976]

ENERGY IN OUR FUTURE, *by Harrison Brown*

TRANSITION
ELEMENTS

Significant terms defined in this chapter

lanthanide contraction
ferromagnetic
platinum metals
chrome plating

hydrate isomerism
steel
lanthanides

rare earth elements
actinides
transuranium elements

Exercises

29.1 A sample of copper metal was treated as follows: (i) concentrated nitric acid was poured on it, giving at first a green and then a blue solution with a brown gas coming off, (ii) a dilute solution of ammonia was added forming a very dark blue solution, (iii) zinc was added producing a colorless solution and a slime of finely divided metal, (iv) the metal was separated and dried, and then heated in the air to form a black powder, (v) the black powder was added to a solution of HCl, forming a blue solution, (vi) a solution of NaOH was added, forming a blue gelatinous precipitate, (vii) the blue precipitate changed to a black precipitate when the mixture was boiled, (viii) the black precipitate reacted with a solution of H_2SO_4 to form a blue solution, (ix) the blue solution was slowly evaporated to form blue crystals, and (x) the blue crystals were heated to form very light blue, nearly white crystals. Identify the form of the copper after each process and write chemical equations for each process.

29.2 Distinguish clearly between open hearth and basic oxygen processes for making steel, acidic and basic fluxes, and tempering and case hardening of steel.

29.3 Which statements are true? Rewrite any false statement so that it is correct.
(a) Similarity in electronic configuration and size causes the first and second members of each transition metal family to more closely resemble each other in chemical behavior than to resemble the third member of the family.
(b) Ionic compounds are most likely to form between transition metals and the representative nonmetals when the transition metal is in its higher oxidation state(s).
(c) Copper is an excellent "sacrificial anode" for protecting iron and steel structures because of its high electrical conductivity.
(d) Aqueous solutions of Fe^{3+} are rather unstable because the Fe^{3+} can be air-oxidized to Fe^{2+}.
(e) The compound $[Fe(H_2O)_6]^{3+}[Fe(CN)_6]^{3-}$ provides an example of interaction absorption.

29.4 A common laboratory experiment in general chemistry is to prepare $FeS(s)$ by strongly heating a mixture of Fe and S. The iron is carefully weighed, excess sulfur is thoroughly mixed with the iron, the mixture is heated and cooled, the iron sulfide is carefully weighed, and the weight of combined sulfur is found by difference. From the weight of Fe and of S combined in the compound, the empirical formula of the compound can be determined. A usual result is something like $Fe_{0.83}S$. Some students recall enough high school chemistry to recognize that the answer should be FeS, Fe_2S_3, or FeS_2 and "change" their data to obtain one of these answers. Why was the first answer probably correct?

29.5 Which solution would have a lower pH: $0.1M$ $FeCl_3$ or $0.1M$ $FeCl_2$? Why?

29.6 Describe the operation of a blast furnace using Fe_2O_3, $CaCO_3$, C, SiO_2, and O_2 as raw materials. Write equations representing the reactions that take place.

29.7 The hexaaquacobalt(II) ion is pink and the tetraaquacobalt(II) ion is blue. Describe the bonding in these ions in terms of the valence bond theory. Is there any difference in the magnetic properties of these ions?

29.8 The presence of Fe, Co, and Ni in an alloy can be confirmed by dissolving the alloy in a nonoxidizing acid to form Fe^{2+}, Co^{2+}, and Ni^{2+} and then performing the following series of qualitative analysis separations and tests on the resulting solution: (i) The metal ions are precipitated as FeS, CoS, and NiS by H_2S; (ii) dilute HCl is added to dissolve the more soluble FeS, and the Fe^{2+} is separated from the other sulfides; (iii) the CoS and NiS are dissolved in concentrated HNO_3 forming Co^{2+} and Ni^{2+}; (iv) the Fe^{2+} is oxidized to Fe^{3+} by HNO_3; (v) a red complex, $Fe(CNS)^{2+}$, is formed by adding a few crystals of NH_4CNS to Fe^{3+}; (vi) ammonia is added to the solution containing Co^{2+} and Ni^{2+} to form the hexaammine complexes; (vii) to half of the solution of the hexaammines, dimethylglyoxime is added to form the red $Ni(DMG)_2$; and (viii) to the other half of the solution of the hexaammines, HCl is added to destroy the complex and then a few crystals of NH_4CNS are added to form the blue complex $Co(CNS)_4^{2-}$. Write chemical equations for the reactions involved in these eight steps.

29.9 Which of the following would you expect to be paramagnetic: (a) $[Mn(PO_4)_2]^{3-}$ (b) $[Co(NH_3)_6]^{3+}$, (c) $[Co(NH_3)_6]^{2+}$, (d) $[Fe(CN)_6]^{4-}$, (e) $[Fe(CN)_6]^{3-}$, (f) $[Cr(NH_3)_6]^{3+}$, (g) $[Ni(CN)_4]^{2-}$, (h) $[Ni(H_2O)_6]^{2+}$?

29.10 Describe the chemistry of tarnishing of silver upon exposure to the atmosphere.

29.11 Write chemical equations describing the reaction that occurs as (a) aqueous solutions of silver nitrate and sodium hydroxide are mixed, (b) gold is treated with a mixture of concentrated nitric and hydrochloric acids to form $HAuCl_4$, (c) aqueous solutions of copper(II) ion and cyanide ion are mixed, (d) aqueous ammonia is added to a solution of copper(II) ion, (e) silver chloride is treated with aqueous ammonia, (f) copper(I) sulfate is treated with water, and (g) copper(II) sulfate is treated with water.

29.12 A student placed clean samples of Mn, Fe, Ni, and Cu into test tubes containing hydrochloric acid and observed gas bubbles being formed in the test tubes containing Mn, Fe, and Ni. (a) Write chemical equations for these reactions. A student with somewhat poorer laboratory technique performed the same experiment without cleaning the surfaces of the metals and reported gas bubbles in the test tubes containing Mn, Fe, and Ni and a change of color on the Cu surface. (b) What reaction did the student see in the test tube containing the Cu? Write an equation describing this reaction. A surprising number of students each year insist that Cu reacts with HCl. However, Cu will react with an oxidizing acid such as HNO_3. (c) Write the chemical equation for this reaction and explain the difference in reactivity.

29.13 How would you explain the lanthanide contraction to your younger brother who is taking a high school chemistry course? How would you answer his questions about the influence of this contraction on chemical behavior? How would you answer his question about whether you would predict an "actinide" contraction?

29.14 The total surface area of a car bumper is 1750 sq inches. The bumper is to be chrome plated to a thickness of 0.0003 inch in a Cr(VI) solution in one hour's time. What must the current be to perform this plating assuming 20% efficiency? The density of Cr is 7.20 g/cm³.

29.15 The standard state enthalpy of formation at 25°C is -51.1 kcal/mole for freshly precipitated MnS(s, pink) and -51.2 kcal/mole for MnS(s, green). (a) Calculate the enthalpy change for the reaction

$$MnS(s, pink) \longrightarrow MnS(s, green)$$

(b) Based on the enthalpy change, which form is more stable? The pink solid is amorphous and the green form is crystalline. (c) Based on randomness (entropy), which form is more stable? (d) Discuss these opposite driving forces in the light of the knowledge that the reaction as written is spontaneous.

29.16 Two identical sheets of iron from a mill were chosen for testing. One of the sheets was tin-plated and the other was galvanized. Both were subjected to a series of tests determining resistance to oxidation. What results could *you* have predicted simply using the following data

$$\begin{array}{ll} Sn^{2+} + 2e^- \rightleftharpoons Sn & E° = -0.1364 \text{ V} \\ Fe^{2+} + 2e^- \rightleftharpoons Fe & E° = -0.409 \text{ V} \\ Zn^{2+} + 2e^- \rightleftharpoons Zn & E° = -0.7628 \text{ V} \end{array}$$

and saved the expense of the testing?

29.17 Iron(II) can be oxidized to iron(III) by MnO_4^- in acid solution, the latter being reduced to Mn^{2+}. Write the half reactions for the oxidation and reduction processes and combine them to obtain the overall equation. A 0.302 g sample of iron ore is dissolved, the iron in it is reduced to Fe^{2+}, and the Fe^{2+} is oxidized using $0.0205M$ $KMnO_4$. If 42.79 ml of the $KMnO_4$ is required to oxidize all of the Fe^{2+}, what is the percentage of iron in the sample?

29.18 Air oxidation of solutions of Fe^{2+} and Cu^+ presents a problem in keeping these solutions in the laboratory. If $E° = -0.409$ V for the Fe^{2+}/Fe couple and 0.770 V for Fe^{3+}/Fe^{2+}, show that simply putting a piece of Fe in solutions of Fe^{2+} will stabilize them, whereas putting a piece of Cu in solutions of Cu^+ will not stabilize them. The corresponding data for copper are $E° = 0.522$ V for Cu^+/Cu and 0.158 V for Cu^{2+}/Cu^+.

29.19 A photograph is toned with a solution of $[AuCl_2]^-$. Write the equation for the toning process and state what color the tone will be. If 2.1 mg of silver is oxidized to Ag^+, what weight of gold is reduced to metal?

29.20 In most activity series of metals, silver is listed above platinum. Show that the reaction between Ag and Pt^{2+} is favorable, given that the standard reduction potentials for the Ag^+/Ag and Pt^{2+}/Pt couples are 0.799 V and 1.2 V, respectively. However, a student observed that Ag does not reduce Pt(II) in basic solution. Confirm this results using $E° = 0.16$ V for the $Pt(OH)_2/Pt$ couple and 0.342 V for the Ag_2O/Ag couple.

29.21* The standard state free energy of formation at 25°C of $CrO_4^{2-}(aq)$ is -173.96 kcal/mole, of $Cr_2O_7^{2-}(aq)$ is -311.0 kcal/mole, of $H^+(aq)$ is 0 and of $H_2O(l)$ is -56.687 kcal/mole. What is the equilibrium constant for the reaction

$$2CrO_4^{2-} + 2H^+ \rightleftharpoons Cr_2O_7^{2-} + H_2O(l)$$

To what pH must a $0.100M$ CrO_4^{2-} solution be adjusted so that the molar concentrations of CrO_4^{2-} and $Cr_2O_7^{2-}$ are equal?

29.22* Solid iron is rather interesting in that it undergoes three solid–solid phase transitions before it melts at 1535°C. Up to 760°C, the α form is stable (body-centered cubic, $a = 0.286106$ mm); from 760°C to 907°C, the β form is stable (body-centered cubic, $a = 0.290$ mm); from 907°C to 1400°C, the γ form is stable (face-centered cubic, $a = 0.363$ mm); and from 1400°C to the melting point, the δ form is stable (body-centered cubic, $a = 0.293$ mm). Calculate the radius of an iron atom in each of these structures.

29.23* The shorthand designation for the "alkaline accumulator" (or Edison or ferro–nickel cell) is

$$\text{steel} \mid \text{Fe}(s) \mid \text{Fe(OH)}_2(s) \mid \text{KOH}(aq) \mid \text{Ni(OH)}_2(s) \mid$$
$$\text{NiOOH}(s) \mid \text{steel}$$

Write the half reactions for the oxidation and reduction processes and write the overall equation for the cell. The standard reduction potential for the Fe(OH)_2/Fe couple is -0.87 V and the overall voltage of the cell is 1.4 V. What is the standard reduction potential for the NiOOH/Ni(OH)_2 couple?

29.24** The major source of nickel is the mineral pentlandite, (Ni,Fe)S, in which the nickel-to-iron ratio is about one to one. The sulfides are separated by selective flotation methods and the NiS is roasted to give the oxide. There are two methods commonly used to obtain the metal from the oxide. The first involves a reduction of the oxide with carbon to form the metal and subsequent electrolytic purification. The second process, known as the Mond process, involves passing warm producer gas (H_2 and CO at 50°C) over the oxide to reduce any oxides of impurities to the free metals while reacting selectively with the nickel to form gaseous Ni(CO)_4. The Ni(CO)_4 is separated and heated to 200°C to decompose it into metallic nickel and CO, which can be recycled.

(a) Write chemical equations for all of the processes described. (b) Find the values of the densities of FeS and NiS in a suitable handbook and discuss a possible method used for the separation of the sulfides. (c) Show that FeO is reduced to Fe by CO in the Mond process given the following ΔG_f° values: FeO(s), -58.6 kcal/mole; CO(g), -32.780 kcal/mole; and $\text{CO}_2(g)$, -94.254 kcal/mole. (d) Show that forming Ni(CO)_4 and water is more favorable in the Mond process than forming Ni and CO_2 given that ΔG_f° is -50.6 kcal/mole for NiO, -56.687 kcal/mole for $\text{H}_2\text{O}(l)$ and -140.36 kcal/mole for $\text{Ni(CO)}_4(g)$. (e) What mass of Ni will be obtained from 1.0 kg of mineral assuming an overall efficiency of 82%?

(f) Write the electronic configuration for an atom of Ni. (g) How does this change for a Ni^{2+} ion? (h) Discuss the bonding in Ni(CO)_4 in terms of valence bond theory. (i) Is the formation of Ni(CO)_4 from Ni and CO an oxidation–reduction process? (j) What is the pressure of CO in equilibrium with a sample of Ni(CO)_4 at a pressure of 0.35 atm?

$$\text{Ni(CO)}_4(g) \longrightarrow \text{Ni}(s) + 4\text{CO}(g)$$

30 QUALITATIVE ANALYSIS FOR CATIONS I. CHEMICAL PRINCIPLES REVIEWED

Chemical equilibrium is the basis of qualitative analysis. One guiding principle in making equilibrium reactions work for us appears over and over again —Le Chatelier's principle: If a system in equilibrium is subjected to a stress, a change that will offset the stress will occur in the system.

The man responsible for this principle, Henry Louis Le Chatelier, was born in Paris in 1850. He grew up in a home where the leading chemists of the day frequently came and went, visiting his father, who was Inspector General of Mines and who also had interests in the steel industry. His mother was a disciplinarian and enforced a strict schedule in her home.

Throughout his career, Le Chatelier maintained an equal interest in the practical developments of technology and their intrinsic scientific meaning. Much of his work grew from trying to solve industrial problems. His doctoral thesis was on the setting of cement, a process that includes complex and varied chemical reactions. This work has been called a classic of inorganic chemistry.

Le Chatelier went on to study the combustion of gases that causes mine explosions. From what he learned about acetylene gas came the development of the oxyacetylene torch used in welding. Further studies of reactions at high temperatures in the blast furnace led to his interest in equilibrium reactions. The outcome of that work, in 1884, was the famous principle now bearing Le Chatelier's name.

This chapter opens with a brief introduction to cation analysis and a description of the cation groups and how they are separated. The remainder of the chapter is devoted to a review of equilibrium reactions—for acids and bases, redox reactions, complex formation, and the precipitation and dissolution of solids. In this review we concentrate on how the different types of equilibrium reactions are utilized in the qualitative analysis for cations.

Cations in aqueous solution

30.1 Cations for analysis

The qualitative analysis scheme for cations is based upon characteristic reactions of cations in aqueous solution. It is a highly systematic and experimental approach to problem solving. The cations are separated from each other and then unequivocally identified by characteristic reactions. (If you have not read Section 22.19, we suggest that you do so before going on with this chapter.)

At this point, let's briefly go over some of the language of qualitative analysis. The *unknown* is the material whose composition we want to determine. In a laboratory course in general chemistry, you probably will receive the unknown from the instructor. A *reagent* is any chemical used to bring about a desired chemical reaction. In cation analysis, the *group reagent* simultaneously reacts with all of the cations in a particular group. When we say an excess (*xs*) of a reagent is used, we mean that more than the stoichiometric amount of the reagent is made available.

A *precipitate* is any solid formed by a chemical reaction in solution. At many points in qualitative analysis we deal with the separation of solids and liquids. Precipitates are often driven to the bottom of a glass tube by centrifugation. The *supernatant solution*, which lies above the precipitate after centrifugation, can be removed by *decanting* it, which means carefully pouring the liquid off to separate it from the precipitate. This decanted solution is sometimes called the *centrifugate*. Precipitates can also be collected on a filter. The liquid that passes through the filter is called the *filtrate*.

What is left after part of a solid has dissolved or after a solution has been evaporated to dryness is often called the *residue*. And, in addition to separating solids from liquids, we frequently have to dissolve solids, a process called *dissolution*.

The steps in cation analysis fall into the following general sequence:

1. *Separation of the cations into a series of groups.* The cations of each successive group are precipitated as compounds with anions supplied by the group reagents. The precipitate containing the cations of one group is separated (usually by centrifugation followed by decantation). Then the group reagent for the next group is added to the remaining solution.

2. *Separation of the cations in each group from each other.* A series of reactions is carried out that eventually leads to the separation of each cation in a group from all of the others in that group. The reactions are carefully chosen to take advantage of similarities and differences in chemical properties.

3. *Identification of individual cations.* The presence of a cation is confirmed by one or more reactions characteristic of that cation.

Our cation analysis scheme, which is discussed in detail in the next chapter, deals with the 22 commonly encountered cations listed in Table 30.1. These cations are divided into five groups based on their reactions with the Cl^-, S^{2-}, OH^-, and CO_3^{2-} ions: The group reagents for Groups II to IV precipitate all of the members of each of the preceding groups as well as the members of their own groups.

TABLE 30.1

Cations for analysis

Ion	Symbol	Color in solution
Aluminum(III)	Al^{3+}	Colorless
Ammonium	NH_4^+	Colorless
Antimony(III)	Sb^{3+} or SbO^+	Colorless
Barium(II)	Ba^{2+}	Colorless
Calcium(II)	Ca^{2+}	Colorless
Chromium(III)	Cr^{3+}	Green or violet
Cobalt(II)	Co^{2+}	Pink or red
Copper(II)	Cu^{2+}	Blue
Iron(III)	Fe^{3+}	Yellow
Iron(II)	Fe^{2+}	Pale green
Lead(II)	Pb^{2+}	Colorless
Magnesium(II)	Mg^{2+}	Colorless
Manganese(II)	Mn^{2+}	Pale pink
Mercury(II)	Hg^{2+}	Colorless
Mercury(I)	Hg_2^{2+}	Colorless
Nickel(II)	Ni^{2+}	Green
Potassium(I)	K^+	Colorless
Silver(I)	Ag^+	Colorless
Sodium(I)	Na^+	Colorless
Tin(IV)	Sn^{4+}	Colorless
Tin(II)	Sn^{2+}	Colorless
Zinc(II)	Zn^{2+}	Colorless

Systematic analysis of samples that may contain cations from all of the groups must therefore proceed by removal of each group in numerical order. (A flow chart summarizing the group separations is given in Figure 31.1.)

Cation Group I. Cations precipitated as chlorides from cold dilute acidic solution.

Pb^{2+}, Hg_2^{2+}, Ag^+

Cation Group II. Cations not precipitated as chlorides from cold dilute acidic solution, but precipitated as sulfides from such a solution.

SbO^+, Cu^{2+}, Pb^{2+}, Hg^{2+}, Sn^{4+}, Sn^{2+}

Cation Group III. Cations not precipitated as chlorides or sulfides from dilute acidic solutions, but precipitated as hydroxides or sulfides from alkaline solutions containing ammonia and ammonium ion.

Al^{3+}, Cr^{3+}, Co^{2+}, Fe^{3+}, Fe^{2+}, Mn^{2+}, Ni^{2+}, Zn^{2+}

Cation Group IV. Cations not precipitated as chlorides or sulfides from dilute acidic solutions or as hydroxides or sulfides from alkaline solutions containing ammonia and ammonium ion but precipitated as carbonates from alkaline solutions containing ammonia and ammonium ion.

Ba^{2+}, Ca^{2+}

Cation Group V. Cations not precipitated under any of the conditions described above.

NH_4^+, Mg^{2+}, K^+, Na^+

The cation groups, the group reagents, and the color of the initial precipitate of each cation are summarized in Table 30.2. As we review the chemical principles applied in qualitative analysis we will frequently use cations from this list in our examples, and we will indicate where in the scheme the various principles are applied.

30.2 Net ionic equations

Most of the reactions that are important in inorganic qualitative analysis involve ions. Therefore, it is best to describe these reactions with net ionic equations rather than molecular equations (Section 4.4). A net ionic equation includes only those ions and molecules that are necessary to describe completely the observed chemical reaction. The following conventions are used:

1. All soluble, strong and moderately strong electrolytes are written as ions.

2. All soluble, very weak electrolytes (e.g., H_2O, CH_3COOH) are written as molecules.

TABLE 30.2

The cation groups

Cation group	Group reagent	Group behavior	
		Ion	Product with group reagent
I	**Cold dilute HCl**	Ag^+	$AgCl(s)$, white
		Pb^{2+}	$PbCl_2(s)$, white
		Hg_2^{2+}	$Hg_2Cl_2(s)$, white
II	$CH_3C(S)NH_2{}^a$ in *ca.* 0.3M HCl	Hg^{2+}	$HgS(s)$, black (or red)
		Pb^{2+}	$PbS(s)$, deep brown
		Cu^{2+}	$CuS(s)$, black
		SbO^+ or Sb^{3+}	$Sb_2S_3(s)$, orange
		$Sn^{4+}(Sn^{2+})$	$SnS_2(s)$, yellow
			SnS, brown
III	$CH_3C(S)NH_2{}^a +$ $NH_3(aq) + NH_4^+$	Mn^{2+}	$MnS(s)$, pink
		$Fe^{2+}(Fe^{3+})$	$FeS(s)$ $Fe_2S_3(s)$, black
		Co^{2+}	$CoS(s)$, black
		Ni^{2+}	$NiS(s)$, black
		Zn^{2+}	$ZnS(s)$, white
		Al^{3+}	$Al(OH)_3(s)$, white
		Cr^{3+}	$Cr(OH)_3(s)$, gray-green
IV	$(NH_4)_2CO_3 + NH_3(aq)$ $+ NH_4^+$	Ba^{2+}	$BaCO_3(s)$, white
		Ca^{2+}	$CaCO_3(s)$, white
V	**None**	Mg^{2+}	
		NH_4^+	
		K^+	
		Na^+	

^{a} Thioacetamide, $CH_3C(S)NH_2$, is a source of H_2S and S^{2-} in solution.

3. All solids and gases (e.g., PbS, CO_2) are written as molecules, even though they may be ionic (e.g., $BaSO_4$, $CaCO_3$).

In addition, we have chosen to omit (*aq*) for ions in solution in most cases. You can assume that all ions are in aqueous solution. Remember that *an ionic equation is not correctly balanced unless both atoms and ionic charges balance.*

Some typical reactions from cation qualitative analysis and their corresponding equations illustrate these points:

1. Mercury(II) sulfide precipitates when an acidic mercury(II) salt solution is saturated with hydrogen sulfide.

$$Hg^{2+} + H_2S(g) \xrightarrow{H^+} HgS(s) + 2H^+$$
colorless   *black*

2. Manganese(IV) oxide is oxidized to permanganate ion by sodium bismuthate(V) in nitric acid solution.

$$2MnO_2(s) + 3NaBiO_3(s) + 10H^+ \rightleftharpoons 2MnO_4^- + 3Bi^{3+} + 5H_2O(l) + 3Na^+$$

black   *violet*
                                              or
                                            purple

3. A zinc salt solution when treated with excess sodium hydroxide solution gives an initial precipitate that redissolves.

$$Zn^{2+} + 2OH^- \rightleftharpoons Zn(OH)_2(s)$$

colorless   *white*

$$Zn(OH)_2(s) + 2OH^- \rightleftharpoons [Zn(OH)_4]^{2-} \text{ or } ZnO_2^{2-} + 2H_2O$$

                              colorless   *colorless*

4. Copper(II) sulfide dissolves in hot dilute nitric acid.

$$3CuS(s) + 8H^+ + 2NO_3^- \rightleftharpoons 3Cu^{2+} + 2NO(g) + 4H_2O(l) + 3S(s)$$

black   *blue*
or
brown

EXAMPLE 30.1

■ The separation and confirmation of Ag^+ involve (a) the precipitation of AgCl by adding HCl, (b) the formation of $[Ag(NH_3)_2]^+$ from the AgCl precipitate by adding NH_3, and (c) the reprecipitation of AgCl from the complex by adding HNO_3, which also forms NH_4NO_3. Write the net ionic equations for these reactions.

First we write the balanced equations for these reactions, using the complete formulas for the reactants and the products:

$$Ag^+ + HCl(aq) \longrightarrow AgCl(s) + H^+$$
$$AgCl(s) + 2NH_3(aq) \longrightarrow [Ag(NH_3)_2]^+Cl^-(aq)$$
$$[Ag(NH_3)_2]^+Cl^- + 2HNO_3(aq) \longrightarrow AgCl(s) + 2NH_4NO_3(aq)$$

The formulas of HCl, $[Ag(NH_3)_2]^+Cl^-$, HNO_3, and NH_4NO_3 should be written in ionic form in a net ionic equation because these compounds are strong electrolytes. However, the formulas of AgCl and NH_3 should not be written in the ionic form because AgCl is a solid and aqueous NH_3 is a weak electrolyte. Rewriting the equations and canceling ions common to each side of the equations,

$$Ag^+ + \cancel{H^+} + Cl^- \longrightarrow AgCl(s) + \cancel{H^+}$$
$$AgCl(s) + 2NH_3(aq) \longrightarrow [Ag(NH_3)_2]^+ + Cl^-$$
$$[Ag(NH_3)_2]^+ + Cl^- + 2H^+ + \cancel{2NO_3^-} \longrightarrow AgCl(s) + 2NH_4^+ + \cancel{2NO_3^-}$$

gives the net ionic equations

(a) $Ag^+ + Cl^- \longrightarrow AgCl(s)$
(b) $AgCl(s) + 2NH_3(aq) \longrightarrow [Ag(NH_3)_2]^+ + Cl^-$
(c) $[Ag(NH_3)_2]^+ + Cl^- + 2H^+ \longrightarrow AgCl(s) + 2NH_4^+$

Note that each equation is balanced with respect to number and type of each atom and with respect to ionic charges. ■

30.3 Chemical equilibrium—the basis for qualitative analysis

The equilibrium constant expression is our major aid in understanding and controlling what happens in the reactions of ions in aqueous solution. For the general reaction

$$aA + bB \cdots \rightleftharpoons cC + dD \cdots$$

the concentrations of the reactants and products are related to the equilibrium constant K by the expression

$$K = \frac{[C]^c[D]^d \cdots}{[A]^a[B]^b \cdots}$$

In previous chapters we have studied the adaptations of this expression to water (K_w; Section 17.4), acids and bases (K_a, K_b; Sections 17.5 to 17.9; 18.1 to 18.9), and solids in saturated solutions (K_{sp}; Sections 18.10 to 18.12). We know that chemical reactions occur so that equilibria are achieved. We also know that whatever the concentrations of reactants and products are at equilibrium, they are related to each other so that their substitution into the equilibrium constant expression gives a value equal to K. The equilibrium constant expression can be used to find unknown concentrations from known concentrations at equilibrium. And nonequilibrium concentrations can be used to calculate the reaction quotient, or the ion product, thus allowing the prediction of whether a reaction will take place.

EXAMPLE 30.2

■ What is the concentration of acetate ion in equilibrium with a solution having $[H^+] = 0.01M$ and $[CH_3COOH] = 0.5M$? $K_a = 1.754 \times 10^{-5}$.

The equilibrium expression for the reaction

$$CH_3COOH(aq) \rightleftharpoons CH_3COO^- + H^+$$

is

$$K_a = \frac{[H^+][CH_3COO^-]}{[CH_3COOH]} = 1.754 \times 10^{-5}$$

Solving for $[CH_3COO^-]$ and substituting values of the concentrations gives

$$[CH_3COO^-] = K_a \frac{[CH_3COOH]}{[H^+]} = \frac{(1.754 \times 10^{-5})(0.5)}{(0.01)} = 9 \times 10^{-4}M$$

The acetate ion concentration is $9 \times 10^{-4}M$. ■

EXAMPLE 30.3

■ Will a precipitate of $PbCl_2$ form if $[Pb^{2+}] = 0.01M$ and $[Cl^-] = 6M$? K_{sp} for $PbCl_2$ is 2×10^{-5}.

The reaction quotient (Q), commonly called the ion product for precipitation reactions, takes the same form as the equilibrium constant expression. For

$$PbCl_2(s) \underset{\longleftarrow}{\overset{\longrightarrow}{\rightleftharpoons}} Pb^{2+} + 2Cl^-$$

the reaction quotient is given by

$$Q = [Pb^{2+}][Cl^-]^2$$

Substituting the values of the concentrations gives

$$Q = (0.01)(6)^2 = 0.4$$

Because $Q > K_{sp}$, the precipitation process will occur. ■

Cation analysis is based on the controlled displacement of a series of equilbria in aqueous solution. We begin with an unknown solution that contains many cations and, of course, also anions. These ions in solution are definitely not independent of each other, and Le Chatelier's principle is a powerful tool in controlling chemical reactions in such solutions. As soon as we change the concentration of one ion, other concentrations must change.

By adding more of a product ion, we cause the formation of more reactant as well as a decrease in the concentration of any other products.

$$A + B \underset{\overset{\longleftarrow C}{}}{\rightleftharpoons} C + D$$

For example, at a point in cation analysis in which we want Sb^{3+} to stay in solution, the precipitation of $SbOCl$ is inhibited by the presence of an excess of hydrogen ion.

$$Sb^{3+} + Cl^- + H_2O(l) \underset{\overset{\longleftarrow H^+}{}}{\rightleftharpoons} SbOCl(s) + 2H^+$$

By adding more reactant, we can increase the concentration of the products of a reaction. This technique is used to ensure the complete removal of Ba^{2+} from solution as $BaSO_4$.

$$Ba^{2+} + SO_4^{2-} \underset{\overset{SO_4^{2-} \longrightarrow}{}}{\rightleftharpoons} BaSO_4(s)$$

A single ion may participate simultaneously in several equilibria. In fact, this is quite common. For example, carbonate ion

(CO_3^{2-}) in a solution containing barium ion, lead ion, and hydrogen ion may participate in (a) the precipitation of barium carbonate

$$Ba^{2+} + CO_3^{2-} \rightleftharpoons BaCO_3(s) \tag{1}$$

(b) the precipitation of lead carbonate

$$Pb^{2+} + CO_3^{2-} \rightleftharpoons PbCO_3(s) \tag{2}$$

(c) the formation of the conjugate acid of CO_3^{2-}

$$CO_3^{2-} + H^+ \rightleftharpoons HCO_3^- \tag{3}$$

(d) or, with sufficient acid, in decomposition.

$$CO_3^{2-} + 2H^+ \rightleftharpoons CO_2(g) + H_2O(l) \tag{4}$$

At equilibrium in this solution, the concentration of carbonate ions remaining in solution will govern all of the equilibria in which carbonate ions participate. In other words, the same value of $[CO_3^{2-}]$ would be used to calculate how much Pb^{2+} is in solution [Equation (2)], how much HCO_3^- has been formed [Equation (3)], and so on.

Throughout cation analysis we want to control competing equilibria so that the desired behavior of the ions can be achieved. We do this by adding other ions that lead to (1) formation of precipitates, (2) altered solubilities, (3) formation of complexes, changes in acid or base concentration, or sometimes (4) changes in oxidation numbers. The ultimate goal in most cases is the complete removal (or as nearly complete as possible) of all of one cation or of several cations from solution, while leaving behind other cations yet to be detected.

Competing equilibria are governed by a very important general principle: *An ion (X^+ or X^-) will form the precipitate, complex ion, or other species that is in equilibrium with the smallest concentration of that ion (X^+ or X^-) in solution.* This principle allows us to predict which of two or more products will form in a particular solution in which there are competing equilibria. For example, the K_{sp} value for two precipitates both containing, say, Hg^{2+} ion can be used to find out which precipitate will, at equilibrium, leave the smaller concentration of Hg^{2+} in solution. We do this by starting with comparable initial concentrations of the ions.

Table 30.3 gives sequences of precipitates and complex ions that have been determined in this way, using K_{sp} and K_d values. The table is based on initial ion concentrations normally encountered in qualitative analysis—$0.001M$ for the unknown ions. Within each sequence the substance on the left leaves the largest concentration of the metal cation in solution at equilibrium and the substance on the right leaves the smallest.

What happens if we add a reagent containing Hg^{2+} to a solution that contains both Br^- and I^-? From their relative positions in Table 30.3, we can predict that mercury(II) iodide, HgI_2, would precipitate rather than mercury(II) bromide, $HgBr_2$. A moment's thought will

TABLE 30.3
Equilibrium concentrations of cations

Cation	Equilibrium concentrations of cation
Pb^{2+}	$PbCl_2(s) > PbBr_2(s) > PbI_2(s) > PbSO_4(s) > Pb(OH)_2(s) > PbCO_3(s)$ $> PbCrO_4(s) > PbS(s)$
Ag^+	$Ag_2SO_4(s) > Ag_2CO_3(s) > Ag_2CrO_4(s) > Ag_2O(s) > AgCl(s) > AgNCS(s)$ $> AgBr(s) > [Ag(NH_3)_2]^+ > AgCN(s) > AgI(s) > [Ag(CN)_2]^- > Ag_2S(s)$
Hg_2^{2+}	$Hg_2SO_4(s) > Hg_2Cl_2(s) > Hg_2(NCS)_2(s) > Hg_2Br_2(s) > Hg_2I_2(s)$
Hg^{2+}	$HgBr_2(s) > Hg(NCS)_2(s) > HgO(s) > HgI_2(s) > [HgI_4]^{2-} > HgS(s)$
Cu^{2+}	$CuCO_3(s) > Cu(OH)_2(s) > [Cu(NH_3)_4]^{2+} > CuS(s)$
Sn^{2+}	$SnI_2(s) > Sn_3(PO_4)_2(s) > Sn(OH)_2(s) > [Sn(OH)_4]^{2-} > SnS(s)$
Sn^{4+}	$SnI_4(s) > SnO_2(s) > [Sn(OH)_6]^{2-} > SnS_2(s) > [SnS_3]^{2-}$
Mn^{2+}	$Mn(OH)_2(s) > MnCO_3(s) > MnNH_4PO_4(s) > MnS(s)$
Fe^{2+}	$Fe(OH)_2(s) > FeCO_3(s) > [Fe(CN)_6]^{4-} > KFe[Fe(CN)_6](s) > FeS(s)$
Fe^{3+}	$FePO_4(s) > Fe(OH)_3(s) > [Fe(CN)_6]^{3-} > KFe[Fe(CN)_6](s) > Fe_2S_3(s)$
Co^{2+}	$[Co(NCS)_4]^{2-} > Co(OH)_2(s) > CoCO_3(s) > [Co(NH_3)_6]^{2+} > CoS(s-\alpha)$ $> CoS(s-\beta)^a$
Ni^{2+}	$NiCO_3(s) > Ni(OH)_2(s) > [Ni(NH_3)_6]^{2+} > NiS(s-\alpha) > NiS(s-\beta)^a$
Zn^{2+}	$ZnCO_3(s) > Zn(OH)_2(s) > [Zn(OH)_4]^{2-} > [Zn(NH_3)_4]^{2+} > ZnS(s)$
Al^{3+}	$AlPO_4(s) > Al(OH)_3(s) > [Al(OH)_4]^-$
Cr^{3+}	$CrPO_4(s) > Cr(OH)_3(s) > [Cr(OH)_4]^- > [Cr(NH_3)_5(OH)]^{2+}$

^{a} The more soluble α form which precipitates initially changes on standing to the much less soluble β form.

show why this is reasonable. Suppose that $HgBr_2$ formed instead. The concentration of Hg^{2+} in equilibrium with this solid

$$HgBr_2(s) \rightleftharpoons Hg^{2+} + 2Br^-$$

will be greater than that required for precipitation of HgI_2. (Recall that precipitation begins when the product of the concentrations of the ions just exceeds the K_{sp} value.) Therefore, HgI_2 will begin to precipitate. As the Hg^{2+} concentration decreases, the $HgBr_2$ will dissolve to replenish the equilibrium concentration of Hg^+ ions. Essentially, the reaction will be

$$HgBr_2(s) + 2I^- \rightleftharpoons HgI_2(s) + 2Br^-$$

Therefore, HgI_2, and *not* $HgBr_2$, will be the precipitate formed. (If an *excess* of Hg^{2+} is added, some $HgBr_2$ will also form.)

EXAMPLE 30.4

■ Calculate $[Zn^{2+}]$ in equilibrium with a saturated solution of $Zn(OH)_2$ in which $[OH^-] = 0.1M$. $K_{sp} = 3.3 \times 10^{-13}$ for $Zn(OH)_2$. Likewise, find $[Zn^{2+}]$ in equilibrium with a solution containing $[OH^-] = 0.1M$ and $[Zn(OH)_4]^{2-} = 0.1M$. $K_d = 5 \times 10^{-21}$ for

$[Zn(OH)_4]^{2-}$. In which form, $Zn(OH)_2(s)$ or $[Zn(OH)_4]^{2-}$, would the majority of the zinc(II) ion be found?

The concentration of zinc(II) ion in equilibrium with $Zn(OH)_2(s)$ is

$$Zn(OH)_2(s) \rightleftharpoons Zn^{2+} + 2OH^- \qquad K_{sp} = [Zn^{2+}][OH^-]^2 = 3.3 \times 10^{-13}$$

$$[Zn^{2+}] = \frac{K_{sp}}{[OH^-]^2} = \frac{3.3 \times 10^{-13}}{(0.1)^2} = 3 \times 10^{-11}M$$

and the concentration of zinc(II) in equilibrium with $[Zn(OH)_4]^{2-}$ is

$$[Zn(OH)_4]^{2-} \rightleftharpoons Zn^{2+} + 4OH^- \qquad K_d = \frac{[Zn^{2+}][OH^-]^4}{[Zn(OH)_4^{2-}]} = 5 \times 10^{-21}$$

$$[Zn^{2+}] = \frac{K_d[Zn(OH)_4^{2-}]}{[OH^-]^4} = \frac{(5 \times 10^{-21})(0.1)}{(0.1)^4} = 5 \times 10^{-18}M$$

Because $[Zn^{2+}]$ in solution at equilibrium is less with the hydroxo complex than with the hydroxide precipitate, the formation of $[Zn(OH)_4]^{2-}$ is preferred. ∎

Thermodynamics also supplies a reason for the predominance of the equilibrium with the smallest concentration of the ion in solution. Such a reaction produces the most stable system. Recall that once equilibrium is established, the free energy undergoes no further change (Sections 25.4 to 25.6). We have shown that at equilibrium, the standard free energy change ($\Delta G°$) for a reaction is related to the equilibrium constant as follows:

$$\Delta G° = -2.303RT \log K \tag{5}$$

Let's compare two precipitation reactions with dramatically different values for K. These precipitation reactions are the reverse of the solubility reactions for silver chloride and silver sulfide. Consequently, the K values for these reactions are the reciprocals of the K_{sp} values.

(i) $Ag^+ + Cl^- \rightleftharpoons AgCl(s)$

$$K_{(i)} = \frac{1}{K_{sp}} = \frac{1}{1.8 \times 10^{-10}} = 5.6 \times 10^9$$

(ii) $2Ag^+ + S^{2-} \rightleftharpoons Ag_2S(s)$

$$K_{(ii)} = \frac{1}{K_{sp}} = \frac{1}{7.1 \times 10^{-50}} = 1.4 \times 10^{49}$$

With such a great difference in the K values, we can assume that the second reaction leaves the smaller concentration of Ag^+ in solution (see also Table 30.3). Calculation of $\Delta G°$ using Equation (5) shows that the reaction with the larger K value, the precipitation of silver sulfide, has the much larger negative value for $\Delta G°$

$(\Delta G^{\circ}_{(i)} = -13.30$ kcal/mole; $\Delta G^{\circ}_{(ii)} = -67.05$ kcal/mole). Therefore, the silver sulfide precipitation which leaves behind the smaller cation concentration in solution, is by far the preferable reaction because it leads to the more thermodynamically stable system.

The equilibrium principle that we are discussing is illustrated in the progress of Zn^{2+} ion through the first few steps of the Cation Group III analysis. The colorless Zn^{2+} ion is present in the unknown solution. The group reagent solution contains $NH_3(aq)$, NH_4^+, and OH^- ions. Upon the first addition of the reagent, zinc hydroxide precipitates

$$Zn^{2+} + 2OH^- \rightleftharpoons \underset{\text{white solid}}{Zn(OH)_2(s)}$$

However, further addition of the reagent dissolves the hydroxide to give the zinc(II)–ammonia complex, which is in equilibrium with a smaller $[Zn^{2+}]$ (see Table 30.3).

$$Zn(OH)_2(s) + 4NH_3(aq) \rightleftharpoons \underset{\text{colorless ion}}{Zn(NH_3)_4^{2+}} + 2OH^-$$

Hydrogen sulfide, the next reagent in the sequence, leads to the reappearance of Zn^{2+} in the form of a white precipitate, that of zinc sulfide, one of the least soluble sulfides of the Group III cations.

$$Zn(NH_3)_4^{2+} + S^{2-} \rightleftharpoons \underset{\text{white}}{ZnS(s)} + 4NH_3(aq)$$

Such sequences of reactions in which cation concentration is gradually decreased are frequently used in qualitative analysis. From our knowledge of the principles of equilibrium and the equilibrium constant values, the decrease in concentration can be shown mathematically, as demonstrated in Example 30.5.

EXAMPLE 30.5

■ Consider the following sequences of reactions performed on a $0.01M$ solution of Ni^{2+}:

$$\underset{\text{green}}{Ni^{2+}} \xrightarrow{\text{NaOH}} \underset{\text{pale green}}{Ni(OH)_2(s)} \xrightarrow{\text{H}_2\text{S}} \underset{\text{black}}{NiS(s)}$$

Assuming that stoichiometric amounts of NaOH and H_2S were added, show that the concentration of Ni^{2+} at equilibrium decreases in each case.

The first reaction, the formation of nickel(II) hydroxide, involves the equilibrium

$$Ni(OH)_2(s) \rightleftharpoons Ni^{2+} + 2OH^-$$

for which

$$K_{sp} = [Ni^{2+}][OH^-]^2 = 3 \times 10^{-16}$$

Letting $x = [Ni^{2+}]$ and recognizing that $[OH^-] = 2[Ni^{2+}] = 2x$ gives

$$(x)(2x)^2 = 3 \times 10^{-16}$$
$$x = [Ni^{2+}] = 4 \times 10^{-6}M$$

The concentration of Ni^{2+} has decreased from $0.01M$ to $4 \times 10^{-6}M$.

The second reaction, the conversion of $Ni(OH)_2$ to NiS, involves the equilibrium

$$NiS(s) \rightleftharpoons Ni^{2+} + S^{2-}$$

for which

$$K_{sp} = [Ni^{2+}][S^{2-}] = 3 \times 10^{-19}$$

Letting $x = [Ni^{2+}]$ and recognizing that $[S^{2-}] = [Ni^{2+}] = x$, we write

$$(x)(x) = 3 \times 10^{-19}$$
$$x = [Ni^{2+}] = 5 \times 10^{-10}M$$

The concentration of Ni^{2+} has decreased from $4 \times 10^{-6}M$ to $5 \times 10^{-10}M$. ∎

The concentrations of ions in solution may be decreased by formation of precipitates and complex ions, as we have seen. They may also be decreased by formation of weakly ionized conjugate acids and bases, or by conversion to oxidation or reduction products. These ionic equilibria are of the same types already discussed in earlier chapters: acid-base equilibria (Chapters 17 and 18); heterogeneous equilibria (Sections 18.12 to 18.15); redox equilibria (Section 19.9); and complex ion equilibria (Sections 28.9 and 28.10).

The remainder of this chapter is devoted to a review of chemical equilibria, with special emphasis on applications in the cation analysis scheme.

Acid-base equilibria

30.4 Strengths of acids and bases

As we have pointed out, the Brønsted-Lowry definitions of acids and bases are most useful to us when we deal with aqueous solutions (Section 17.2). Remember that the water molecule can serve both as a proton acceptor, that is, as a *base*

water as a base

$$HCl(aq) + H_2O(l) \xrightleftharpoons{} H_3O^+ + Cl^-$$
$$NH_4^+ + H_2O(l) \xrightleftharpoons{} H_3O^+ + NH_3(aq)$$
$$[Cu(H_2O)_4]^{2+} + H_2O(l) \xrightleftharpoons{} H_3O^+ + [Cu(H_2O)_3(OH)]^+$$

and as a proton donor, that is, as an *acid*

water as a acid

$$CO_3^{2-} + H_2O(l) \xrightleftharpoons{} HCO_3^- + OH^-$$
$$NH_3(g) + H_2O(l) \xrightleftharpoons{} NH_4^+ + OH^-$$

Thus, water is often an important component of equilibrium acid-base reactions.

Although water may be an important reactant, the water molecule is itself only slightly ionized. The ionization of water (Section 17.5)

$$H_2O(l) + H_2O(l) \xrightleftharpoons{} H_3O^+ + OH^-$$

written more simply as

$$H_2O(l) \xrightleftharpoons{} H^+ + OH^-$$

leads to the ion product constant (at 25°C) of

$$K_w = [H^+][OH^-] = 1.00 \times 10^{-14} \qquad (6)$$

These relationships make it clear that (1) in any aqueous solution both hydrogen and hydroxide ions are present, (2) the concentrations of the hydrogen and hydroxide ions are completely interdependent, and (3) in pure water (at 25°C)

$$[H^+] = [OH^-] = 1 \times 10^{-7}$$

To review briefly, we use K_w to establish a relative scale of acidity. By definition (Section 17.6),

$$pH = -\log [H^+] \qquad (7)$$
$$pOH = -\log [OH^-] \qquad (8)$$
$$pK_w = -\log K_w \qquad (9)$$
$$= pH + pOH = 14.000 \quad \text{(at 25°C)} \qquad (10)$$

Pure water is said to be neutral at $pH = pOH = 7$. Acidic solutions have $[H^+] > [OH^-]$ and alkaline solutions have $[OH^-] > [H^+]$ (Table 17.5).

The strong acids (e.g., HCl, HBr, HI, HNO_3, $HClO_4$) are assumed to be completely ionized in $0.1M$, or less concentrated, solution (Table 17.8). Thus, in a $0.1M$ solution of each of these five acids noted, $[H^+] = 0.1$ mole/liter. The weak acids (e.g., HSO_4^-, H_3PO_4, HNO_2, HF, CH_3COOH, H_2CO_3, H_2S, NH_4^+, HS^-) are only partially ionized in terms of equilibria such as

$$HA(aq) + H_2O(l) \rightleftharpoons H_3O^+ + A^- \tag{11}$$

or, neglecting hydration of the proton,

$$HA(aq) \rightleftharpoons H^+ + A^- \tag{12}$$

For weak acids, the equilibrium constant, or acid dissociation constant (Table 17.9), is

$$K_a = \frac{[H^+][A^-]}{[HA]} \tag{13}$$

Therefore, the hydrogen ion concentration for, say, a $0.1M$ solution of a weak acid is less than $0.1M$ (Table 17.8). The relative strengths of weak acids are shown by their hydrogen ion concentrations in solutions of the same molarity.

The hydroxide ion is the strongest base that can exist in aqueous solution. Soluble ionic hydroxides (e.g., those of the alkali metals, Ba^{2+}, Sr^{2+}, and Ca^{2+}) give the hydroxide ion directly by dissolution and are essentially 100% dissociated; for example, $[OH^-] = 0.1$ mole/liter in $0.1M$ NaOH or KOH, and $[OH^-] = 0.02$ mole/liter in $0.01M$ $Ba(OH)_2$. Note, however, that the less ionic hydroxides do not always act as bases (as further discussed in Section 30.7 under amphoterism).

The anions of very weak acids, such as H_2CO_3, H_2S, and HCN, are relatively strong bases. They give alkaline aqueous solutions in their reactions with water as follows:

$$CO_3^{2-} + H_2O(l) \rightleftharpoons HCO_3^- + OH^-$$
$$S^{2-} + H_2O(l) \rightleftharpoons HS^- + OH^-$$
$$CN^- + H_2O(l) \rightleftharpoons HCN(aq) + OH^-$$

Such acid-base equilibria are sometimes referred to as hydrolysis equilibria (Section 18.2). Most other anions, ammonia, and many amines are somewhat weaker bases. For aqueous solutions of weak bases, the general equilibria are

$$B(aq) + H_2O(l) \rightleftharpoons BH^+ + OH^- \tag{14}$$

for example,

$$CH_3NH_2(aq) + H_2O(l) \rightleftharpoons CH_3NH_3^+ + OH^-$$
methylamine

or

$$B^{n-}(aq) + H_2O(l) \rightleftharpoons BH^{(n-1)-} + OH^- \tag{15}$$

for example,

$$F^-(aq) + H_2O(l) \rightleftharpoons HF + OH^-$$

(Some K_b values for weak bases are given in Table 17.12.)

EXAMPLE 30.6

■ Compare the values of $[H^+]$ for $0.10M$ solutions of HCl and of HCN. $K_a = 6.2 \times 10^{-10}$ for HCN.

Because HCl is a strong acid and is virtually completely ionized, $[H^+] = [HCl] = 0.10M$. For HCN, a weak acid, letting $x = [H^+]$, we find $[H^+]$ by our usual method for solving equilibrium problems. (To review this method, see Section 16.12.)

$$HCN(aq) \rightleftharpoons H^+ + CN^-$$

initial	$0.10M$		
change	$-x$	$+x$	$+x$
equilibrium	$0.10 - x$	x	x

$$K_a = \frac{[H^+][CN^-]}{[HCN]} = \frac{(x)(x)}{(0.10 - x)} \approx \frac{x^2}{0.1} = 6.2 \times 10^{-10}$$

Assuming $(0.10 - x) \approx 0.10$, the K_a expression gives $x = 7.9 \times 10^{-6}M$. Since the assumption is valid, $[H^+] = 7.9 \times 10^{-6}M$. Thus the equilibrium concentration of H^+ in $0.10M$ HCl is about 1.3×10^{-4} times greater than that found in $0.10M$ HCN. ■

EXAMPLE 30.7

■ The separation of Fe^{3+} from Co^{2+} and Ni^{2+} is done by adding excess $NH_3(aq)$ so that Fe^{3+} precipitates as $Fe(OH)_3$, while Co^{2+} and Ni^{2+} are complexed as $[Co(NH_3)_6]^{2+}$ and $[Ni(NH_3)_6]^{2+}$. Assume that we have a solution $0.01M$ in each of these ions and that sufficient concentrated ammonia solution has been added to make this solution $6M$ in NH_3. Calculate the OH^- concentration in this solution.

Because K_{sp} for $Fe(OH)_3(s)$ is 3×10^{-39}, we can assume the reaction

$$Fe^{3+} + 3NH_3(aq) + 3H_2O(l) \longrightarrow Fe(OH)_3(s) + 3NH_4^+$$

to be essentially complete. From the stoichiometry we can see that if the initial concentration of Fe^{3+} was $0.01M$, the concentration of NH_4^+ is $0.03M$.

The OH^- concentration is controlled by the equilibrium involving the aqueous ammonia, because OH^- does not enter into the formation of the cobalt and nickel complexes.

$$NH_3(aq) + H_2O(l) \rightleftharpoons NH_4^+ + OH^-$$

$$K_b = \frac{[NH_4^+][OH^-]}{[NH_3]} = 1.6 \times 10^{-5}$$

Letting $x = [OH^-]$

$$NH_3(aq) + H_2O(l) \rightleftharpoons NH_4^+ + OH^-$$

initial		0.03	
change		$+x$	$+x$
equilibrium	6	$0.03 + x$	x

$$K_b = \frac{[NH_4^+][OH^-]}{[NH_3]} = \frac{(0.03 + x)(x)}{6} \approx \frac{0.03x}{6} = 1.6 \times 10^{-5}$$

$$x = 3 \times 10^{-3}M$$

The concentration of OH^- is $3 \times 10^{-3}M$. ■

In cation analysis, acids and bases are most often used in situations in which several simultaneous equilibria are taking place. For example, acids and bases are essential in the buffered solutions used to control pH and thereby to control precipitation; these topics are discussed in Sections 30.6 and 30.10. Acids, in particular, are important in the displacement of equilibria which leads to the dissolution of solids, which is discussed in Section 30.11.

30.5 Effect of an added common ion

The addition of an ion common to a weak acid or a weak base such as ammonia in aqueous solution increases the equilibrium concentration of the undissociated acid or base. This change in concentration, called the common ion effect, occurs by displacement of the dissociation equilibrium of the acid or the base, for example,

$$\overset{\longleftarrow \qquad\qquad A^-}{HA \rightleftharpoons H^+ + A^-}$$

As you can see, this is the principle of Le Chatelier at work again. As $[A^-]$ increases, $[H^+]$ decreases and $[HA]$ increases.

Ionization constant expressions for weak acids or weak bases can be used to find the changes in concentration that occur in the presence of a common ion, as illustrated in Example 30.8. The common ion effect is applied in buffer solutions, which are discussed in the next section. The effect is, of course, also applicable to the dissolution of slightly soluble salts and to any other equilibrium reaction.

EXAMPLE 30.8

■ Find the effect on $[OH^-]$ and $[NH_3]$ of making an $0.010M$ aqueous solution of ammonia also $0.050M$ in ammonium chloride.

$$NH_3(aq) + H_2O(l) \rightleftharpoons NH_4^+ + OH^- \qquad K_b = 1.6 \times 10^{-5}$$

For the solution containing only aqueous ammonia, letting $x = [OH^-]$ gives

$$NH_3(aq) + H_2O(l) \rightleftharpoons NH_4^+ + OH^-$$

initial	0.010		
change	$-x$	$+x$	$+x$
equilibrium	$0.010 - x$	x	x

$$K_b = \frac{[NH_4^+][OH^-]}{[NH_3]}$$

$$= \frac{(x)(x)}{(0.010 - x)} = 1.6 \times 10^{-5}M$$

We will assume that $(0.010 - x)$ is approximately equal to 0.01. Therefore

$$\frac{x^2}{0.010} = 1.6 \times 10^{-5}$$

$$x = 4.0 \times 10^{-4}M = [OH^-]$$

For the solution containing ammonia and additional NH_4^+, letting $x = [OH^-]$ gives

$$NH_3(aq) + H_2O(l) \rightleftharpoons NH_4^+ + OH^-$$

initial	0.010	0.050	
change	$-x$	$+x$	$+x$
equilibrium	$0.010 - x$	$0.050 + x$	x

$$K_b = \frac{(0.050 + x)(x)}{(0.010 - x)} = 1.6 \times 10^{-5}$$

Again assuming $(0.010 - x) \approx 0.010$ and also assuming $(0.050 + x) \approx 0.050$, we find

$$\frac{0.050x}{0.010} = 1.6 \times 10^{-5}$$

$$x = 3.2 \times 10^{-6}M = [OH^-]$$

The $[OH^-]$ has been decreased more than a hundredfold (from 1.6×10^{-5} to 3.2×10^{-6} mole/liter) by the addition of the NH_4^+ ion, which is common to the NH_3 solution. Since in each case $[NH_3] = 0.010 - x$, the ammonia concentration is increased in the NH_4^+-containing solution. ∎

30.6 Buffer solutions

When a salt containing the conjugate base of a weak acid is added to a solution of that acid, a buffer solution is produced. Similarly, the addition of a salt containing the conjugate acid of a weak base such as ammonia to a solution of that base yields a buffer solution. You can see that buffer solutions are thus made by adding a common ion. Buffer solutions serve the very important purpose of controlling the [H$^+$] or the [OH$^-$]. The addition of small amounts of a strong acid or a strong base to a buffer solution causes only minor changes in the pH. (This effect is illustrated in Table 18.3.)

Acetic acid/sodium acetate is a frequently used buffer system. When a strong acid is added to this buffer, it reacts with the basic acetate ion

$$H^+ + CH_3COO^- \rightleftharpoons CH_3COOH(aq)$$

When the strongly basic OH$^-$ ion is added to the same solution, it reacts with the hydrogen ion to give water

$$H^+ + OH^- \rightleftharpoons H_2O(l)$$

In each case the equilibrium between H$^+$ and CH$_3$COO$^-$ is quickly restored, and the pH of the original solution comes back to nearly its original value.

By the appropriate choice of the ion/acid or ion/base buffer pair, pH can be maintained in any desired range. Buffers are most effective in the pH = pK_a (or pK_b) $\pm$ 1 range.

Buffer solutions play an essential role in cation analysis. It is only by controlling the pH with buffers that we are able to, for example, precipitate some sulfides while leaving others in solution. The applications of the common ion effect and buffer solutions to separations in cation analysis usually involve this kind of situation—preventing the precipitation of one cation by a given anion while allowing another cation to precipitate. These applications also involve heterogeneous equilibria and are discussed in Section 30.10.

EXAMPLE 30.9

■ We wish to make a buffer solution using HF and NaF. (a) What will be the most effective pH range of this buffer? (b) Knowing that buffers best resist change when the ratio of the acid concentration to the conjugate base concentration (or the base to the conjugate acid concentration) is between 1/10 and 10/1 (see Section 18.9), what range of concentration can our NaF solution have if we wish to make the buffer solution by mixing equal volumes of the NaF solution and a 0.200M HF solution? (c) What will be the pH of the buffer solution made with equal volumes of 0.200M HF and 0.200M NaF? For HF, $K_a = 6.5 \times 10^{-4}$.

(a) From the relationship

$$pH = pK_a \pm 1$$

we find that the most effective pH range of the HF/F$^-$ buffer is

$$pH = -\log (6.5 \times 10^{-4}) \pm 1$$
$$= 3.19 \pm 1$$

(b) Our buffer solution should have an [HF]/[F$^-$] ratio of between 10/1 and 1/10, or

$$\frac{0.200}{[F^-]} = \frac{10}{1} \qquad \frac{0.200}{[F^-]} = \frac{1}{10}$$
$$[F^-] = 0.020M \qquad [F^-] = 2.00M$$

between $0.020M$ and $2.00M$.

(c) For a $0.200M$ HF/$0.200M$ NaF buffer the [H$^+$] is found in the usual way, letting $x = $ [H$^+$]. Note that we have doubled the volume and therefore divided the molarity in half.

	HF	$\rightleftharpoons$ H$^+$ +	F$^-$
initial	0.100		0.100
change	$-x$	$+x$	$+x$
equilibrium	$0.100 - x$	x	$0.100 + x$

$$K_a = \frac{[H^+][F^-]}{[HF]} = \frac{(x)(0.100 + x)}{(0.100 - x)} \approx \frac{0.100\, x}{0.100} = 6.5 \times 10^{-4}$$
$$x = 6.5 \times 10^{-4} = [H^+]$$
$$pH = -\log (6.5 \times 10^{-4}) = 3.19$$

Whenever the ratio of the concentrations of the buffer pair is 1/1, the pH is equal to the pK_a. ∎

30.7 Amphoterism

Any molecule or ion that can behave either as an acid or a base is said to be amphoteric (Section 17.2). In the Brønsted-Lowry treatment of acids and bases, an amphoteric substance can either donate or accept one or more protons. The water molecule is thus a typical amphoteric molecule. As you might expect, in the Lewis acid-base system, an amphoteric substance can either accept or donate electron pairs. Most commonly we encounter amphoterism in the oxides and hydroxides of representative metals and d transition metals. Using aluminum oxide and hydroxide as examples, the following equations describe the amphoteric behavior of such compounds:

as bases
$$Al_2O_3(s) + 6H^+ \longrightarrow 2Al^{3+} + 3H_2O(l) \qquad (16)$$
$$Al(OH)_3(s) + 3H^+ \longrightarrow Al^{3+} + 3H_2O(l) \qquad (17)$$

as acids

$$Al_2O_3(s) + 2OH^- + 3H_2O(l) \longrightarrow 2[Al(OH)_4]^- \tag{18}$$
$$Al(OH)_3(s) + OH^- \longrightarrow [Al(OH)_4]^- \tag{19}$$

Do not be confused if in various places you see the oxide, the hydroxide, or the hydrate written in the same reaction, such as

$$2Al^{3+} + 6OH^- \rightleftharpoons Al_2O_3(s) + 3H_2O \tag{20}$$
$$Al^{3+} + 3OH^- \rightleftharpoons Al(OH)_3(s) \tag{21}$$
$$2Al^{3+} + 6OH^- \rightleftharpoons Al_2O_3 \cdot 3H_2O \tag{22}$$

The hydroxide $Al(OH)_3(s)$ and the oxide $Al_2O_3(s)$ are chemically interchangeable, because the hydroxide is a hydrate of the oxide.

The thick gelatinous precipitate of aluminum hydroxide that forms when OH^- is added to Al^{3+} in aqueous solution is transformed to aluminum oxide upon drying.

A similar relationship exists between the hydroxo complex ions such as $[Al(OH)_4]^-$ and the oxoanions such as AlO_2^-. Here again, the complex and the oxoanion are equivalent—dehydration of the complex yields the oxoanion. In writing an equation the two are interchangeable.

The relationships among the amphoteric oxides and hydroxides that are part of the cation analysis scheme are summarized in Table 30.4.

The acidic properties of amphoteric metal oxides and hydroxides can be utilized in separating them from nonamphoteric oxides and hydroxides. For example, in cation analysis Al^{3+} is separated from Fe^{3+} in this way. As aqueous sodium or potassium hydroxide solution is added slowly to a solution containing these ions, precipitation occurs first

$$Al^{3+} + 3OH^- \rightleftharpoons Al(OH)_3(s)$$
colorless   *white*

$$Fe^{3+} + 3OH^- \rightleftharpoons Fe(OH)_3(s)$$
yellow   *red-brown*

TABLE 30.4
Amphoteric oxides and hydroxides

Compound	Reaction with H^+ produces	Reaction with OH^- produces[a]
Sb_2O_3 or $Sb(OH)_3$	Sb^{3+} or SbO^+	$[Sb(OH)_4]^-$ or SbO_2^-
SnO or $Sn(OH)_2$	Sn^{2+}	$[Sn(OH)_4]^{2-}$ or SnO_2^{2-}
PbO or $Pb(OH)_2$	Pb^{2+}	$[Pb(OH)_4]^{2-}$ or PbO_2^{2-}
SnO_2 or $Sn(OH)_4$	Sn^{4+}	$[Sn(OH)_6]^{2-}$ or SnO_3^{2-}
Al_2O_3 or $Al(OH)_3$	Al^{3+}	$[Al(OH)_4]^-$ or AlO_2^-
Cr_2O_3 or $Cr(OH)_3$	Cr^{3+}	$[Cr(OH)_4]^-$ or CrO_2^-
ZnO or $Zn(OH)_2$	Zn^{2+}	$[Zn(OH)_4]^{2-}$ or ZnO_2^{2-}

[a] Written as hydroxo complex ions or as their dehydration products.

Then, as more hydroxide ion is added, aluminum(III) hydroxide dissolves

$$Al(OH)_3(s) + OH^- \xrightleftharpoons{\text{excess OH}^-} [Al(OH)_4]^-$$
$$\text{colorless}$$

but the less acidic iron(III) hydroxide does not. This separation can be formulated in a single step as

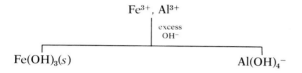

EXAMPLE 30.10

■ Some amphoteric oxides and hydroxides are stronger bases than acids, while others are stronger acids than bases. Find [H^+] assuming $Al(OH)_3$ dissociates as an acid as follows:

$$Al(OH)_3(s) \rightleftharpoons 3H^+ + AlO_3^{3-} \qquad K_a = [H^+]^3[AlO_3^{3-}] = 4 \times 10^{-13}$$

and [OH^-] assuming $Al(OH)_3$ acts as a base

$$Al(OH)_3(s) \rightleftharpoons Al^{3+} + 3OH^- \qquad K_b = K_{sp} = [Al^{3+}][OH^-]^3 = 3.5 \times 10^{-34}$$

Is the reaction of $AlOH_3$ more favorable as an acid or as a base?

For the acid reaction, if we let $x = [AlO_3^{3-}]$, then [H^+] $= 3x$. Substituting these values into the expression for K_a gives

$$K_a = [H^+]^3[AlO_3^{3-}] = (3x)^3(x) = 4 \times 10^{-13}$$
$$x = 3 \times 10^{-4}M$$

Thus [H^+] $= 3x = 9 \times 10^{-4}M$. For the base reaction, if we let $x = [Al^{3+}]$, then [OH^-] $= 3x$ and

$$K_b = [Al^{3+}][OH^-]^3 = (x)(3x)^3 = 3.5 \times 10^{-34}$$
$$x = 1.9 \times 10^{-9}M$$

Thus [OH^-] $= 3x = 5.7 \times 10^{-9}M$. The acidic reaction is more favorable for $Al(OH)_3$ than the reaction as a base—in fact, the concentration of OH^- produced by the $Al(OH)_3$ is less than that formed by the ionization of water. ■

Amphoterism is not restricted to metal oxides and hydroxides. Sulfur, the element just below oxygen in the periodic table, also forms amphoteric compounds. The acid-base reactions of the amphoteric metal sulfides are illustrated by the following reactions of tin(IV) sulfide:

as a base
$$SnS_2(s) + 4H^+ \rightleftharpoons Sn^{4+} + 2H_2S(aq) \qquad (23)$$
yellow   *colorless*

as an acid
$$SnS_2(s) + 2SH^- \rightleftharpoons [SnS_3]^{2-} + H_2S(aq) \qquad (24)$$
sulfostannate(IV)
ion
(yellow)

The elegant word "sulfoamphoteric" can be used for such compounds.

Certain metal cyanides are also amphoteric, for example,

as a base
$$Fe(CN)_2 + 2H^+ \rightleftharpoons Fe^{2+} + 2HCN(aq) \qquad (25)$$
white

as an acid
$$Fe(CN)_2 + 4CN^- \rightleftharpoons [Fe(CN)_6]^{4-} \qquad (26)$$
pale yellow

The equations involving sulfide and cyanide reactions as bases are consistent with Brønsted-Lowry acid-base definitions. The reactions as acids are more readily interpreted by Lewis definitions, for $Fe(CN)_2$ accepts electron pairs from the CN^- ions and SnS_2 accepts an electron pair from S^{2-}, making them Lewis acids. ∎

EXAMPLE 30.11

∎ Write equations showing the sulfoamphoteric behavior of Sb_2S_3 to form Sb^{3+} and $[SbS_2]^-$.

The base reaction is

$$Sb_2S_3(s) + 6H^+ \rightleftharpoons 2Sb^{3+} + 3H_2S(aq)$$

and the acid reaction is

$$Sb_2S_3(s) + 2HS^- \rightleftharpoons 2[SbS_2]^- + H_2S(aq) \qquad ∎$$

Redox equilibria

A redox reaction is an equilibrium reaction that represents the summation of two electron-transfer equilibria—one involving the loss of electrons in the oxidation of a chemical species, the other involving the gain of electrons in the reduction of a chemical species. For example, the equation for the reaction of hydrogen sulfide with dilute nitric acid

$$3H_2S(aq) + 2NO_3^- + 2H^+ \longrightarrow 2NO(g) + 4H_2O(l) + 3S(s) \qquad (27)$$

is the summation of equations for two half-reactions:

reduction
$$NO_3^- + 4H^+ + 3e^- \longrightarrow NO + 2H_2O$$

oxidation
$$H_2S \longrightarrow 2H^+ + S + 2e^-$$

In balancing redox equations, electron gain and loss are equalized.

Quantitatively, the strengths of oxidizing and reducing agents under standard state conditions are measured by standard reduction potentials (Section 19.6). Recall that these potentials are always recorded for reduction reactions. Therefore, the relevant standard reduction potentials for the H_2S–HNO_3 reaction are

for the reduction of NO_3^- to NO
$$NO_3^- + 4H^+ + 3e^- \longrightarrow NO + 2H_2O \qquad E° = 0.96 \text{ V} \qquad (28)$$

for the reduction of S(s) to H_2S
$$S + 2H^+ + 2e^- \longrightarrow H_2S \qquad E° = 0.142 \text{ V} \qquad (29)$$

A larger and more positive $E°$ value indicates a better oxidizing agent, that is, a substance that is more easily reduced. The $E°$ values show that in the H_2S–HNO_3 system, NO_3^- ion will be reduced, while H_2S will be oxidized to sulfur.

EXAMPLE 30.12

■ Would adding Sn^{4+} to an aqueous solution of H_2S cause the H_2S to be oxidized to S? $E° = 0.070$ V for the half-reaction

$$Sn^{4+} + 2e^- \xrightarrow{H^+} Sn^{2+}$$

Combining the half-reactions involving the tin and the sulfur (Equation 29) gives

$Sn^{4+} + 2e^- \longrightarrow Sn^{2+}$	$E° = 0.070$ V
$H_2S \longrightarrow S + 2H^+ + 2e^-$	$E° = -0.142$ V
$Sn^{4+} + H_2S \longrightarrow Sn^{2+} + S + 2H^+$	$E° = -0.072$ V

Because $E°$ is negative, this reaction would not be spontaneous under standard state conditions. ■

The reduction potential values reveal which states are likely to be the more stable oxidation states of ions in solution. An ion that is resistant to both oxidation and reduction in solution will be found more commonly than an ion that is easily oxidized and/or reduced. For example, manganese(II) is more stable in aqueous solution than manganese(III), for manganese(II) is quite resistant to reduction

$$Mn^{2+} + 2e^- \longrightarrow Mn \qquad E° = -1.029 \text{ V}$$

and also quite resistant to oxidation

$$Mn^{3+} + e^- \longrightarrow Mn^{2+} \qquad E° = +1.51 \text{ V}$$

(Remember that oxidation is the reverse of this reaction as it is written and that for the reverse reaction, $E° = -1.51$ V.)

As we have said, redox reactions are equilibrium reactions. Using thermodynamics and our knowledge that at equilibrium $\Delta G = 0$ and $\Delta E = 0$, we have shown (Sections 19.9 and 25.7) that the equilibrium constant and the standard reduction potential are related by the following equation, derived from the Nernst equation

$$\log K = \frac{n}{0.0592} E° \qquad (30)$$

no. of electrons transferred

This relationship can be used to calculate equilibrium constants from the potential data. For example, suppose that we wish to find the equilibrium constant for the reduction of nitric acid by hydrogen sulfide. First, we must find the overall $E°$. Note that the $E°$ values are not multiplied by any factors because the standard state voltage does not depend on the actual quantities of the substances involved.

$$2X(NO_3^- + 4H^+ + 3e^- \longrightarrow NO + 2H_2O) \qquad E° = 0.96 \text{ V}$$
$$\underline{3X(H_2S \longrightarrow S + 2H^+ + 2e^-) \qquad E° = -0.142 \text{ V}}$$
$$3H_2S + 2NO_3^- + 2H^+ + 6e^- \longrightarrow 2NO + 4H_2O + 3S + 6e^- \qquad E° = 0.82 \text{ V}$$

The positive $E°$ value shows that the reaction will be spontaneous in the direction written.

From Equation (30)

$$\log K = \left(\frac{6}{0.0592}\right)(0.82) = 83$$

$$K = \frac{[NO]^2}{[H_2S]^3[NO_3^-]^2[H^+]^2} = 1 \times 10^{83} \qquad (31)$$

The very large value of K shows that equilibrium is established when the reaction is almost completely displaced toward nitrogen(II) oxide and sulfur. The next example makes use of this K value.

EXAMPLE 30.13

■ Show that HNO_3 will dissolve $CuS(s)$, but not $HgS(s)$.

The overall equations for the dissolution process are

$$3HgS(s) + 2NO_3^- + 8H^+ \rightleftharpoons 3Hg^{2+} + 3S(s) + 2NO(g) + 4H_2O(l) \quad \text{(a)}$$
$$3CuS(s) + 2NO_3^- + 8H^+ \rightleftharpoons 3Cu^{2+} + 3S(s) + 2NO(g) + 4H_2O(l) \quad \text{(b)}$$

and the respective overall equilibrium constants are

$$K_{(a)} = \frac{[Hg^{2+}]^3[NO]^2}{[NO_3^-]^2[H^+]^8}$$

$$K_{(b)} = \frac{[Cu^{2+}]^3[NO]^2}{[NO_3^-]^2[H^+]^8}$$

Each of the overall reactions can be considered to occur by the series of reactions described below. The overall equilibrium constants are the products of the respective equilibrium constants (see Section 16.5).

The first reactions in the series to consider are those involving the solubility products of the sulfide precipitates:

$$HgS(s) \rightleftharpoons Hg^{2+} + S^{2-} \qquad K_{(c)} = K_{sp} = [Hg^{2+}][S^{2-}] = 4 \times 10^{-53} \quad \text{(c)}$$
$$CuS(s) \rightleftharpoons Cu^{2+} + S^{2-} \qquad K_{(d)} = K_{sp} = [Cu^{2+}][S^{2-}] = 6 \times 10^{-36} \quad \text{(d)}$$

The second reaction in the series is the dissociation of H_2S

$$H_2S(aq) \rightleftharpoons 2H^+ + S^{2-} \qquad K_{(e)} = K_a = \frac{[H^+]^2[S^{2-}]}{[H_2S]} = 3 \times 10^{-20} \quad \text{(e)}$$

The third reaction in the series involves the oxidation of H_2S by NO_3^- for which K was found in Equation (31).

$$3H_2S(aq) + 2NO_3^- + 2H^+ \rightleftharpoons 2NO(g) + 4H_2O(l) + 3S(s)$$
$$K_{(f)} = 1 \times 10^{83} \quad \text{(f)}$$

The overall Equation (a) can be obtained by combining Equation (c) multiplied by 3 with Equation (e) inverted and multiplied by 3 and Equation (f)

$$
\begin{array}{ll}
3HgS(s) \rightleftharpoons 3Hg^{2+} + 3S^{2-} & K = K_{(c)}^3 \\
6H^+ + 3S^{2-} \rightleftharpoons 3H_2S(aq) & K = 1/K_{(e)}^3 \\
3H_2S(aq) + 2NO_3^- + 2H^+ \rightleftharpoons 2NO(g) + 4H_2O(l) + 3S(s) & K = K_{(f)} \\
\hline
3HgS(s) + 2NO_3^- + 8H^+ \rightleftharpoons 3Hg^{2+} + 3S(s) + 2NO(g) + 4H_2O(l) & \\
& K_{(a)} = K_{(c)}^3 K_{(f)} / K_{(e)}^3
\end{array}
$$

Thus

$$K_{(a)} = \frac{K_{(c)}^3 K_{(f)}}{K_{(e)}^3} = \frac{(4 \times 10^{-53})^3 (1 \times 10^{83})}{(3 \times 10^{-20})^3} = 2 \times 10^{-16}$$

This extremely low value of $K_{(d)}$ implies that greater concentrations of the reactants will be present at equilibrium, and for all practical purposes the reaction will not occur. For the overall Equation (b), we find

$$K_{(b)} = \frac{K_{(d)}^3 K_{(f)}}{K_{(e)}^3} = \frac{(6 \times 10^{-36})^3(1 \times 10^{83})}{(3 \times 10^{-20})} = 8 \times 10^{35}$$

and because $K_{(b)}$ is very large, this reaction will occur readily. ■

In separating and identifying cations, we take advantage of selective oxidation and reduction. For example, in Cation Group II, $2M$ nitric acid oxidizes lead(II) and copper(II) sulfides, but not mercury(II) sulfide. This difference in oxidizability arises because of differences in solubility among these compounds. The lead(II) and copper(II) compounds are soluble enough to provide high enough concentrations of sulfide ion in solution to allow for its oxidation. The mercury(II) sulfide is so slightly soluble that oxidation in this manner is not possible. Analysis of Cation Group III provides another example of selective oxidation—$[Cr(OH)_4]^-$ is oxidized (to CrO_4^{2-}) by hydrogen peroxide, but $(Al(OH)_4]^-$ is not.

Complexation Equilibria

In aqueous solution, each complex ion is in equilibrium with its components, for example,

$$[Ag(NH_3)_2]^+ \rightleftharpoons Ag^+ + 2NH_3(aq)$$
$$[SnS_3]^{2-} \rightleftharpoons Sn^{4+} + 3S^{2-}$$
$$[Cr(OH)_4]^- \rightleftharpoons Cr^{3+} + 4OH^-$$
$$[Fe(CN)_6]^{4-} \rightleftharpoons Fe^{2+} + 6CN^-$$

Each such reaction is described by an equilibrium constant which we call a dissociation constant (K_d) because the equilibria are written as dissociation processes. (Typical dissociation constants are given in Table 28.5.)

The smaller the numerical magnitude of K_d, the smaller is the equilibrium concentration of the metal ion in question. In addition, a smaller K_d indicates that the complex is more stable and forms more readily. Both separations and identifications of cations may be effected via complexation.

EXAMPLE 30.14

■ As aqueous NH_3 is added to the mixture of Fe^{3+}, Co^{2+}, and Ni^{2+} described in Example 30.7, $Co(OH)_2$ and $Ni(OH)_2$ also form; but because the cation concentrations in equilibrium with the ammonia complexes are less than those in equilibrium with the hydroxide precipitates, these cations eventually form the ammonia complexes. Calculate the $[Co^{2+}]$ in equilibrium with $Co(OH)_2(s)$ and $0.01M$ $[Co(NH_3)_6]^{2+}$ in $6M$ NH_3. For $Co(OH)_2$, $K_{sp} = 2 \times 10^{-16}$; for $[Co(NH_3)_6^{2+}]$, $K_d = 9 \times 10^{-6}$.

For the hydroxide precipitate that forms first

$$Co(OH)_2(s) \rightleftharpoons Co^{2+} + 2OH^-$$
$$K_{sp} = [Co^{2+}][OH^-]^2 = 2 \times 10^{-16}$$

In Example 30.7 we found that $[OH^-] = 3 \times 10^{-3}$ in this solution. Therefore,

$$[Co^{2+}] = \frac{K_{sp}}{[OH^-]^2} = \frac{2 \times 10^{-16}}{(3 \times 10^{-3})^2} = 2 \times 10^{-11} M$$

For the dissociation of the ammonia complex

$$[Co(NH_3)_6]^{2+} \rightleftharpoons Co^{2+} + 6NH_3(aq)$$
$$K_d = \frac{[Co^{2+}][NH_3]^6}{[Co(NH_3)_6^{2+}]} = 9 \times 10^{-6}$$

which upon substituting $[NH_3] = 6M$ and $[Co(NH_3)_6^{2+}] = 0.01M$,

$$[Co^{2+}] = \frac{K_d[Co(NH_3)_6^{2+}]}{[NH_3]^6}$$
$$= \frac{(9 \times 10^{-6})(0.01)}{(6)^6} = 2 \times 10^{-12} M$$

The concentration of CO^{2+} in equilibrium with the complex is about one-tenth that of CO^{2+} in equilibrium with $Co(OH)_2$. ∎

A separation can be achieved if one cation forms a complex under a particular set of conditions while another cation forms a precipitate. For example, cobalt(II), nickel(II), and zinc(II) ions form soluble ammine complexes, whereas under the same conditions, iron(II or III), aluminum, and chromium(III) ions are precipitated as hydroxides. Also, tin(II or IV), lead(II), zinc, and aluminum ions form soluble hydroxo complexes with excess hydroxide ion, whereas iron(II or III), nickel(II), and cobalt(II) ions form insoluble hydroxides.

Certain complex species are highly and distinctively colored and are thus useful for identifying the cations they contain. For example, in Cation Group III, each of the d transition metals iron, cobalt, and nickel is positively identified by the characteristic colors of complexes. The Fe^{3+} ion is identified by the formation of both Prussian blue, $KFe[Fe(CN)_6]$—the salt between Fe^{3+} and K^+—and the hexacyanoferrateII complex ion, and by the thiocyanatoferrateIII ion, $[Fe(NCS)]^{2+}$, which gives a blood red solution. Cobalt is also identi-

TABLE 30.5
Colors of some complex species

Species[a]	Color	Species[a]	Color
$[Cu(H_2O)_4]^{2+}$	Pale blue	$[Co(H_2O)_2(NCS)_4]^{2-}$	Blue to blue-green
$[Cu(NH_3)_4]^{2+}$	Dark blue	$[Co(NH_3)_6]^{3+}$	Yellow-brown
$[Cu(py)_2(NCS)_2]$	Light green	$[Co(NO_2)_6]^{3-}$	Yellow
$[Fe(H_2O)_6]^{2+}$	Pale green	$[Ni(H_2O)_6]^{2+}$	Green
$[Fe(H_2O)_6]^{3+}$	Yellow	$[Ni(NH_3)_6]^{2+}$	Deep blue
$[Fe(CN)_6]^{4-}$	Yellow	$[Ni(DMG)_2]$	Bright red
$[Fe(CN)_6]^{3-}$	Red	$[Mn(H_2O)_6]^{2+}$	Pale pink
$KFe[Fe(CN)_6]$	Dark blue	$[Cr(OH)_4]^-$	Green
$[Co(H_2O)_6]^{2+}$	Pink	$[Cr(NH_3)_6]^{3+}$	Yellow
$[Co(NH_3)_6]^{2+}$	Tan	$[Cr(NH_3)_5(OH)]^{2+}$	Violet

[a] Idealized formulas based on common coordination numbers.
py = pyridine; DMG = dimethylglyoxime

fied by the color of a thiocyanato complex ion in solution— $Co(NCS)_4^{2-}$, which is blue. The presence of Ni^{2+} is confirmed by the formation of the highly distinctive bright red precipitate of the complex between nickel and dimethylglyoxime (DMG). (The structures of DMG and the complex are shown in Figure 29.11.) The colors of some characteristic complex ions and compounds are listed in Table 30.5.

Heterogeneous Equilibria

30.8 The solubility product constant

Equilibria between substances in different phases are called **heterogeneous equilibria.** To maintain such equilibria molecules or ions pass from one phase to another. In qualitative analysis we repeatedly produce precipitates or dissolve solids. These processes depend upon establishing equilibria in which ions leave and rejoin the surface of a solid in a solution.

When the ions of a slightly soluble salt such as silver(I) chloride are brought together in aqueous solution in sufficient concentration, a precipitate forms

$$Ag^+ + Cl^- \rightleftharpoons AgCl(s)$$

Once equilibrium is established and as long as the temperature remains unchanged, the mass of the precipitate does not change. But evidence for the movement of ions from the solution to the solid and back again can be seen in the gradual change from tiny crystals to larger ones.

The equilibrium between a slightly soluble solid and its ions in solution is described by the solubility product constant, K_{sp}, which we have already used several times in this chapter. For silver(I) chloride

$$AgCl \rightleftharpoons Ag^+ + Cl^-$$
$$K_{sp} = [Ag^+][Cl^-] = 1.8 \times 10^{-10}$$

As for all equilibrium constants, the concentration (activity) of the solid is taken as 1 and does not appear in the equilibrium expression.

For the general reaction

$$M_xA_y(s) \rightleftharpoons xM^{m+} + yA^{n-}$$

we have

$$K_{sp} = [M^{m+}]^x[A^{n-}]^y \tag{32}$$

where $[M^{m+}]$ and $[A^{n-}]$ are the equilibrium concentrations of the ions in a saturated solution.

Some solubility product constants that are particularly significant to the separation and detection of cations are listed in Table 30.6. Three kinds of information particularly useful for cation analysis can be found from K_{sp} values: (1) the concentrations of ions necessary for precipitation; (2) the means of controlling a precipitation by a second reaction; and (3) the conditions under which a solid will dissolve.

30.9 Concentrations of ions necessary for precipitation

For a given cation to be precipitated by a particular anion, the ion product of their molar concentrations, which takes the same form as the equilibrium constant expression, must exceed the solubility product constant. When solutions of the reacting cation and anion are mixed, precipitation begins when the product of the concentrations of the ions just exceeds the appropriate K_{sp} value. Precipitation continues as more ions are added but stops when the final ion product again equals the K_{sp} value.

To illustrate this, let's consider the addition, with adequate stirring, of $0.1M$ aqueous ammonia solution to 25 ml of $0.10M$ aluminum nitrate solution. The ultimate equilibrium that will be established is

$$Al(OH)_3(s) \rightleftharpoons Al^{3+} + 3OH^-$$

for which

$$K_{sp} = [Al^{3+}][OH^-]^3 = 3.5 \times 10^{-34}$$

A permanent precipitate of $Al(OH)_3$, that is, one that does not go back into solution, will result only when the product of the ion concentrations just exceeds the K_{sp} value. The $[OH^-]$ at which this occurs for the

TABLE 30.6

Solubility product constants at 25°C

Anion	Equilibrium equation	K_{sp}	pK_{sp}
Carbonate	$MgCO_3(s) \rightleftharpoons Mg^{2+} + CO_3^{2-}$	1×10^{-5}	5.0
	$NiCO_3(s) \rightleftharpoons Ni^{2+} + CO_3^{2-}$	1.3×10^{-7}	6.89
	$CaCO_3(s) \rightleftharpoons Ca^{2+} + CO_3^{2-}$	3.84×10^{-9}	8.416
	$BaCO_3(s) \rightleftharpoons Ba^{2+} + CO_3^{2-}$	2×10^{-9}	8.70
	$SrCO_3(s) \rightleftharpoons Sr^{2+} + CO_3^{2-}$	5.2×10^{-10}	9.28
	$MnCO_3(s) \rightleftharpoons Mn^{2+} + CO_3^{2-}$	5×10^{-10}	9.30
	$CuCO_3(s) \rightleftharpoons Cu^{2+} + CO_3^{2-}$	2.3×10^{-10}	9.64
	$CoCO_3(s) \rightleftharpoons Co^{2+} + CO_3^{2-}$	1×10^{-10}	10.0
	$FeCO_3(s) \rightleftharpoons Fe^{2+} + CO_3^{2-}$	2.1×10^{-11}	10.68
	$ZnCO_3(s) \rightleftharpoons Zn^{2+} + CO_3^{2-}$	1.7×10^{-11}	10.77
	$Ag_2CO_3(s) \rightleftharpoons 2Ag^+ + CO_3^{2-}$	8.1×10^{-12}	11.09
	$PbCO_3(s) \rightleftharpoons Pb^{2+} + CO_3^{2-}$	7.4×10^{-14}	13.13
Chloride	$PbCl_2(s) \rightleftharpoons Pb^{2+} + 2Cl^-$	2×10^{-5}	4.7
	$AgCl(s) \rightleftharpoons Ag^+ + Cl^-$	1.8×10^{-10}	9.74
	$Hg_2Cl_2(s) \rightleftharpoons Hg_2^{2+} + 2Cl^-$	1.3×10^{-18}	17.88
Chromate	$CaCrO_4(s) \rightleftharpoons Ca^{2+} + CrO_4^{2-}$	6×10^{-4}	3.2
	$BaCrO_4(s) \rightleftharpoons Ba^{2+} + CrO_4^{2-}$	1.2×10^{-10}	9.92
	$Ag_2CrO_4(s) \rightleftharpoons 2Ag^+ + CrO_4^{2-}$	2.5×10^{-12}	11.60
	$PbCrO_4(s) \rightleftharpoons Pb^{2+} + CrO_4^{2-}$	2.8×10^{-13}	12.55
Hydroxide	$Ag_2O(s) + H_2O \rightleftharpoons 2Ag^+ + 2OH^-$	2×10^{-8}	7.7
	$Mg(OH)_2(s) \rightleftharpoons Mg^{2+} + 2OH^-$	7.1×10^{-12}	11.15
	$Zn(OH)_2(s) \rightleftharpoons Zn^{2+} + 2OH^-$	3.3×10^{-13}	12.48
	$Mn(OH)_2(s) \rightleftharpoons Mn^{2+} + 2OH^-$	2×10^{-13}	12.7
	$Pb(OH)_2(s) \rightleftharpoons Pb^{2+} + 2OH^-$	1.2×10^{-15}	14.92
	$Fe(OH)_2(s) \rightleftharpoons Fe^{2+} + 2OH^-$	8×10^{-16}	15.1
	$Ni(OH)_2(s) \rightleftharpoons Ni^{2+} + 2OH^-$	3×10^{-16}	15.5
	$Co(OH)_2(s) \rightleftharpoons Co^{2+} + 2OH^-$	2×10^{-16}	15.7
	$SbOOH(s) \rightleftharpoons SbO^+ + OH^-$	1×10^{-17}	17.0
	$Cu(OH)_2(s) \rightleftharpoons Cu^{2+} + 2OH^-$	1.3×10^{-20}	19.89
	$Hg(OH)_2(s) \rightleftharpoons Hg^{2+} + 2OH^-$	4×10^{-26}	25.4
	$Sn(OH)_2(s) \rightleftharpoons Sn^{2+} + 2OH^-$	6×10^{-27}	26.2
	$Cr(OH)_3(s) \rightleftharpoons Cr^{3+} + 3OH^-$	6×10^{-31}	30.2
	$Al(OH)_3(s) \rightleftharpoons Al^{3+} + 3OH^-$	3.5×10^{-34}	33.46
	$Fe(OH)_3(s) \rightleftharpoons Fe^{3+} + 3OH^-$	3×10^{-39}	38.5
	$Sn(OH)_4(s) \rightleftharpoons Sn^{4+} + 4OH^-$	$ca.\ 10^{-57}$	$ca.\ 57$

TABLE 30.6 (*Continued*)

Anion	Equilibrium equation	K_{sp}	pK_{sp}
Sulfate	$CaSO_4(s) \rightleftharpoons Ca^{2+} + SO_4^{2-}$	2.5×10^{-5}	4.60
	$AgSO_4(s) \rightleftharpoons 2Ag^+ + SO_4^{2-}$	1.5×10^{-5}	4.82
	$Hg_2SO_4(s) \rightleftharpoons Hg_2^{2+} + SO_4^{2-}$	6.8×10^{-7}	6.17
	$PbSO_4(s) \rightleftharpoons Pb^{2+} + SO_4^{2-}$	2.2×10^{-8}	7.66
	$BaSO_4(s) \rightleftharpoons Ba^{2+} + SO_4^{2-}$	1.7×10^{-10}	9.77
Sulfide	$MnS(s) \rightleftharpoons Mn^{2+} + S^{2-}$	2.3×10^{-13}	12.64
	$FeS(s) \rightleftharpoons Fe^{2+} + S^{2-}$	4.2×10^{-17}	16.38
	$NiS(s-\alpha) \rightleftharpoons Ni^{2+} + S^{2-}$	3×10^{-19}	18.5
	$ZnS(s) \rightleftharpoons Zn^{2+} + S^{2-}$	2×10^{-24}	23.7
	$CoS(s-\alpha) \rightleftharpoons Co^{2+} + S^{2-}$	2×10^{-25}	24.7
	$SnS(s) \rightleftharpoons Sn^{2+} + S^{2-}$	3×10^{-27}	26.5
	$PbS(s) \rightleftharpoons Pb^{2+} + S^{2-}$	1×10^{-28}	28.0
	$CuS(s) \rightleftharpoons Cu^{2+} + S^{2-}$	6×10^{-36}	35.2
	$Ag_2S(s) \rightleftharpoons 2Ag^+ + S^{2-}$	7.1×10^{-50}	49.15
	$HgS(s) \rightleftharpoons Hg^{2+} + S^{2-}$	4×10^{-53}	52.4
	$Fe_2S_3(s) \rightleftharpoons 2Fe^{3+} + 3S^{2-}$	*ca.* 1×10^{-88}	*ca.* 88.0

aluminum hydroxide solution is found as follows:

$$K_{sp} = [Al^{3+}][OH^-]^3$$
$$3.5 \times 10^{-34} = (0.10)[OH^-]^3$$
$$[OH^-]^3 = \frac{3.5 \times 10^{-34}}{(0.10)}$$
$$[OH^-] = 1.5 \times 10^{-11} M$$

As aqueous ammonia is added beyond this point the $[OH^-]$ increases and the $[Al^{3+}]$ is reduced accordingly until it is ultimately reduced to a small value. However, at all times the conditions of equilibrium defined by the K_{sp} expression must be satisfied.

EXAMPLE 30.15

■ Is the $[OH^-]$ concentration in a $6M$ $NH_3(aq)$ solution great enough to produce precipitation of $Fe(OH)_3$ if the original solution is $0.01M$ in Fe^{3+}?

In Example 30.7 we calculated $[OH^-] = 3 \times 10^{-3} M$ for the $6M$ aqueous ammonia solution. For the equilibrium

$$Fe(OH)_3(s) \rightleftharpoons Fe^{3+} + 3OH^-$$
$$K_{sp} = [Fe^{3+}][OH^-]^3 = 3 \times 10^{-39}$$

the ion product under these conditions is

$$[Fe^{3+}][OH^-]^3 = (0.01)(3 \times 10^{-3})^3 = 3 \times 10^{-10}$$

Because the ion product exceeds the value of K_{sp}, precipitation of $Fe(OH)_3$ will occur. ∎

30.10 Controlled precipitation

A particularly valuable use of precipitation reactions is in *controlled precipitation*. The precipitation of cations can be controlled by controlling the concentrations of anions. Suppose two cations that form compounds of different solubilities with the same anion are present in a solution. Controlling the concentration of the anion permits the precipitation of one cation while leaving the other unprecipitated. The separation of the cations is achieved because the K_{sp} of one of the possible compounds is never exceeded. Although a controlled precipitation can sometimes be carried out by merely limiting the quantity of the added precipitating reagent, this procedure is not generally useful. Localized high concentrations result where the reagent enters the solution, and subsequent equilibrations are slow and not necessarily complete.

A much more effective approach is to utilize a second chemical reaction to limit the concentration of the anion in solution. The second reaction maintains the concentration of the anion at the predetermined level that allows the precipitation of the less soluble compound while preventing that of the more soluble compound. In qualitative analysis cations are frequently separated by the controlled precipitations of hydroxides, carbonates, and sulfides. In the following sections we examine how these precipitations can be controlled by adding appropriate amounts of additional ions.

Hydroxide precipitations. Aqueous ammonia is the reagent of choice for the precipitation of hydroxides. The hydroxide ion is provided by the acid-base reaction between ammonia and water.

$$NH_3(aq) + H_2O(l) \rightleftharpoons NH_4^+ + OH^-$$
$$K_b = \frac{[NH_4^+][OH^-]}{[NH_3]} = 1.6 \times 10^{-5}$$

The concentration of the OH^- ion, and thereby the precipitation of hydroxides, is controlled by adding NH_4^+ to the solution. The added compound must be a salt between NH_4^+ and an anion that does not react with water and therefore does not affect the pH of the solution.

When we add NH_4^+ in this way we are adding an ion common to the reaction of NH_3 as a base with water and we are creating an NH_3/NH_4^+ buffer pair in the solution. This buffer pair maintains the $[OH^-]$ required for precipitation of the desired compounds while preventing the precipitation of other hydroxides. The "second reaction" being used to control the precipitation is the reaction between the added NH_4^+ and OH^- in the aqueous ammonia solution. Let us calculate, for example, the concentration of ammonium ion necessary to prevent the precipitation of magnesium hydroxide when a 0.0010M solution of Mg^{2+} is made 0.1M in NH_3. The solubility product equilibrium is

$$Mg(OH)_2(s) \rightleftharpoons Mg^{2+} + 2OH^-$$

for which (Table 30.6),

$$K_{sp} = [Mg^{2+}][OH^-]^2 = 7.1 \times 10^{-12}$$

The equilibrium $[OH^-]$, which must be exceeded for precipitation to occur, is then

$$[OH^-] = \sqrt{\frac{7.1 \times 10^{-12}}{[Mg^{2+}]}} = \sqrt{\frac{7.1 \times 10^{-12}}{1.0 \times 10^{-3}}} = 8.4 \times 10^{-5} \text{ mole/liter}$$

From the K_b for ammonia, the $[NH_4^+]$ needed to limit $[OH^-]$ to this value is then

$$[NH_4^+] = (1.6 \times 10^{-5}) \times \frac{[NH_3]}{[OH^-]} = (1.6 \times 10^{-5}) \times \frac{1 \times 10^{-1}}{8.4 \times 10^{-5}}$$

$$= 0.02 \text{ mole/liter}$$

The precipitation of magnesium(II) hydroxide is prevented by first adding to the 0.0010M Mg^{2+} solution sufficient ammonium salt to give $[NH_4^+]$ = 0.02 mole/liter or more, and then making the solution 0.1M in ammonia.

Similar calculations for a number of metal hydroxides give the data summarized in Table 30.7. Precipitation of the more soluble metal hydroxides can be prevented by ammonium ion concentrations readily attainable in the laboratory, but not precipitation of the less

TABLE 30.7

Effect of ammonium ion upon precipitation of hydroxides

Hydroxide	K_{sp}	$[OH^-]$ for precipitation of $M(OH)_n$ ($[M^{n+}]$ = 0.001M) (mole/liter)	$[NH_4^+]$ to prevent precipitation ($[NH_3]$ = 0.10M) (mole/liter)
$Mg(OH)_2$	7.1×10^{-12}	8×10^{-5}	$>2 \times 10^{-2}$
$Zn(OH)_2$	3.3×10^{-13}	1.8×10^{-5}	>0.09
$Mn(OH)_2$	2×10^{-13}	1×10^{-5}	$>1 \times 10^{-1}$
$Pb(OH)_2$	1.2×10^{-15}	1.1×10^{-6}	>1.5
$Fe(OH)_2$	8×10^{-16}	9×10^{-7}	>2
$Ni(OH)_2$	3×10^{-16}	5×10^{-7}	>3
$Co(OH)_2$	2×10^{-16}	4×10^{-7}	>4
$Cu(OH)_2$	1.3×10^{-20}	3.6×10^{-9}	$>4.4 \times 10^2$
$Sn(OH)_2$	6×10^{-27}	2×10^{-12}	$>7 \times 10^5$
$Cr(OH)_3$	6×10^{-31}	8×10^{-10}	$>2 \times 10^3$
$Al(OH)_3$	3.5×10^{-34}	7.0×10^{-11}	$>2.3 \times 10^4$
$Fe(OH)_3$	3×10^{-39}	1×10^{-12}	$>1 \times 10^6$

soluble ones. This technique is used to prevent the precipitation of $Mg(OH)_2$ in Cation Group III while allowing $Al(OH)_3$ and $Cr(OH)_3$ to precipitate.

Carbonate precipitations. The precipitation of carbonates can be controlled by controlling the carbonate ion concentration with ammonium ion by use of the equilibrium

$$NH_4^+ + CO_3^{2-} \rightleftharpoons NH_3(aq) + HCO_3^-$$

for which

$$K = \frac{[NH_3][HCO_3^-]}{[NH_4^+][CO_3^{2-}]} = 11.7$$

We can use the precipitation of magnesium carbonate as an example

$$MgCO_3(s) \rightleftharpoons Mg^{2+} + CO_3^{2-}$$
$$K_{sp} = [Mg^{2+}][CO_3^{2-}] = 1 \times 10^{-5}$$

For a $0.001M$ magnesium salt solution, precipitation of the carbonate should occur if

$$[CO_3^{2-}] > \frac{K_{sp}}{[Mg^{2+}]} = \frac{1 \times 10^{-5}}{1 \times 10^{-3}} = 1 \times 10^{-2} \text{ mole/liter}$$

From K for the ammonium ion-carbonate ion reaction used to control $[CO_3^{2-}]$, the quantity of ammonium ion required to reduce $[CO_3^{2-}]$ to this value for a solution with $[NH_3] = 0.1M$ and $[HCO_3^-] = 0.1M$ is

$$[NH_4^+] > \frac{[NH_3][HCO_3^-]}{[CO_3^{2-}]K} = \frac{(1 \times 10^{-1})(1 \times 10^{-1})}{(1 \times 10^{-2})(11.7)} = 0.09 \text{ mole/liter}$$

Under these conditions, addition of sufficient ammonium salt to give $[NH_4^+] > 0.09$ mole/liter will prevent precipitation of magnesium carbonate from a solution $0.001M$ in Mg^{2+}.

Similar calculations give the additional data summarized in Table 30.8.

An ammonium ion-controlled carbonate precipitation is used in Cation Group IV to prevent the precipitation of $MgCO_3$ while allowing $CaCO_3$ and $BaCO_3$ to precipitate.

TABLE 30.8

Effect of ammonium ion upon precipitation of carbonates

Carbonate	K_{sp}	$[CO_3^{2-}]$ for precipitation of MCO_3 ($[M^{2+}] = 0.001M$) (mole/liter)	$[NH_4^+]$ to prevent precipitation of MCO_3 ($[NH_3] = [HCO_3^-] = 0.1M$) (mole/liter)
MgCO$_3$	1×10^{-5}	1×10^{-2}	$>9 \times 10^{-2}$
CaCO$_3$	3.84×10^{-9}	3.84×10^{-6}	$>2.23 \times 10^2$
BaCO$_3$	2.0×10^{-9}	2.0×10^{-6}	$>4.3 \times 10^2$
SrCO$_3$	5.2×10^{-10}	5.2×10^{-7}	$>1.6 \times 10^3$
PbCO$_3$	7.4×10^{-14}	7.4×10^{-11}	$>1.2 \times 10^7$

Sulfide precipitations. Hydrogen sulfide in aqueous solution dissociates in two steps (these equilibria are discussed in detail in Section 18.7). Writing them in simplified form, these equilibria are

$$H_2S(aq) \rightleftharpoons HS^- + H^+ \qquad K_{a_1} = 1.0 \times 10^{-7}$$
$$HS^-(aq) \rightleftharpoons S^{2-} + H^+ \qquad K_{a_2} = 3 \times 10^{-13}$$

The precipitation of sulfides can be controlled by controlling the hydrogen ion concentration. Let's start with the overall equation for H_2S ionization

$$H_2S(aq) \rightleftharpoons 2H^+ + S^{2-} \tag{33}$$

for which the equilibrium constant is the product of K_{a_1} and K_{a_2}.

$$K = \frac{[H^+]^2[S^{2-}]}{[H_2S]} = \frac{[H^+][HS^-]}{[H_2S]} \times \frac{[H^+][S^{2-}]}{[HS^-]} = K_{a_1}K_{a_2} \tag{34}$$
$$= (1.0 \times 10^{-7})(3 \times 10^{-13})$$
$$= 3 \times 10^{20}$$

At atmospheric pressure an aqueous solution saturated with gaseous hydrogen sulfide has an H_2S concentration of 0.1 mole/liter. Incorporating this into the preceding equation gives us an expression, K_{H_2S}, that shows the relationship of $[H^+]$ to $[S^{2-}]$

$$\frac{[H^+]^2[S^{2-}]}{0.1} = 3 \times 10^{-20}$$
$$K_{H_2S} = [H^+]^2[S^{2-}] = 3 \times 10^{-21} \tag{35}$$

It is apparent from this equation that the sulfide ion concentration is strictly controlled by the hydrogen ion concentration. The smaller the $[H^+]$, the greater is the $[S^{2-}]$ and vice versa.

Table 30.9 summarizes the concentrations of hydrogen ion that must be exceeded to prevent precipitation of various sulfides from so-

TABLE 30.9

Effect of hydrogen ion upon precipitation of sulfides

Sulfide	K_{sp}	$[S^{2-}]$ for precipitation of M_xS_y ([M^{m+}] = 0.0010M) (mole/liter)	$[H^+]$ to prevent precipitation of M_xS_y (mole/liter)
MnS	2.3×10^{-13}	2.3×10^{-10}	$>4 \times 10^{-6}$
FeS	4.2×10^{-17}	4.2×10^{-14}	$>3 \times 10^{-4}$
NiS(α)	3×10^{-19}	3×10^{-16}	$>3 \times 10^{-3}$
ZnS	2×10^{-24}	2×10^{-21}	>1
CoS(α)	2×10^{-25}	2×10^{-22}	>4
SnS	3×10^{-27}	3×10^{-24}	$>3 \times 10^{1}$
PbS	1×10^{-28}	1×10^{-25}	$>2 \times 10^{2}$
CuS	6×10^{-36}	6×10^{-33}	$>7 \times 10^{5}$
Ag$_2$S	7.1×10^{-50}	7.1×10^{-44}	$>2 \times 10^{11}$
HgS	4×10^{-53}	4×10^{-50}	$>3 \times 10^{14}$

QUALITATIVE
ANALYSIS I. [30.10]

lutions $0.0010M$ in metal ions (the average concentration in solutions used for cation analysis). Almost half of the cations in our analysis scheme are first precipitated as sulfides. The Group II sulfides, those of Hg^+, Cu^{2+}, Pb^{2+}, Sb^{3+}, Sn^{2+}, and Sn^{4+}, are precipitated from an acidic solution that does not provide a large enough $[S^{2-}]$ for the precipitation of the other cations present. The sulfides of the Group III cations, those of Mn^{2+}, Fe^{2+} or Fe^{3+}, Ni^{2+}, Zn^{2+}, and Co^{2+} are precipitated from an alkaline solution of hydrogen sulfide.

EXAMPLE 30.16

■ What concentration of $[H^+]$ will keep Mn^{2+} (initially $0.1M$) in solution at a concentration of $0.1M$ while Pb^{2+} (initially $0.1M$) precipitates as PbS?

Both precipitation reactions can be represented by the following general expressions:

$$MS(s) \rightleftharpoons M^{2+} + S^{2-}$$
$$K_{sp} = [M^{2+}][S^{2-}]$$

As described above, for the H_2S–S^{2-} equilibrium in a solution saturated with H_2S at atmospheric pressure

$$K_{H_2S} = [H^+]^2[S^{2-}] = 3 \times 10^{-21}$$

which upon solving for $[S^{2-}]$ and substituting into the expression for K_{sp} gives

$$K_{sp} = [M^{2+}]\frac{K_{H_2S}}{[H^+]^2}$$

Solving for $[H^+]$ gives

$$[H^+] = \sqrt{\frac{K_{H_2S}[M^{2+}]}{K_{sp}}}$$

For MnS, $K_{sp} = 2.3 \times 10^{-13}$; therefore

$$[H^+] = \sqrt{\frac{(3 \times 10^{-21})(0.1)}{2.3 \times 10^{-13}}} = 4 \times 10^{-5}M$$

For PbS, $K_{sp} = 1 \times 10^{-28}$, giving $[H^+] = 2 \times 10^3M$ by a similar calculation. We can see that by keeping $[H^+]$ at a reasonable value between 2×10^3M and $4 \times 10^{-5}M$, for example, $1M$, the MnS will not precipitate and the Pb^{2+} will precipitate as PbS. ■

Hydrogen sulfide and thioacetamide. At one time inorganic qualitative analysis laboratories were recognizable from great distances by the strong odor of rotten eggs in the air. The aroma is caused by hydrogen sulfide. In the laboratories hydrogen sulfide was often produced in Kipp generators—large pieces of glassware each containing an unsavory-looking mass of hydrochloric acid and ferrous sulfide. The gas was generated as it was needed and bubbled directly into the solutions.

However, awareness grew that hydrogen sulfide is a highly toxic substance. Exposure to 600 ppm of hydrogen sulfide in the air for 30 minutes can be fatal. The OSHA (Occupational Safety and Health Administration) standard for peak concentrations in the air is 20 ppm. Hydrogen sulfide should always be treated with respect; brief exposure to it deadens the sense of smell, making it possible to breathe large amounts of H_2S without realizing that it is still in the air.

We now use the organic compound thioacetamide as the precipitating reagent for the sulfides. Heating an acidic thioacetamide solution yields directly an aqueous solution of hydrogen sulfide.

$$\underset{\text{thioacetamide}}{CH_3\overset{\displaystyle \overset{S}{\|}}{C}NH_2} + H^+ + 2H_2O \rightleftharpoons \underset{\text{acetic acid}}{CH_3\overset{\displaystyle \overset{O}{\|}}{C}OH} + NH_4^+ + H_2S(aq) \qquad (36)$$

The hydrogen sulfide thus generated goes on to dissociate as discussed above. In alkaline solution, thioacetamide reacts to yield sulfide ion directly.

$$CH_3\overset{\displaystyle \overset{S}{\|}}{C}NH_2 + 3OH^- \rightleftharpoons CH_3\overset{\displaystyle \overset{O}{\|}}{C}O^- + NH_3(aq) + H_2O(l) + S^{2-} \qquad (37)$$

The conditions for controlled precipitation of the sulfides are affected very little by the different sources of sulfide ion.

(A government agency has cited thioacetamide as a compound that may possibly cause cancer. There is no conclusive proof that this is correct, nor is there any information yet available about the conditions of exposure necessary. However, this situation serves as a reminder that *all* chemicals should be handled with care.)

30.11 Dissolution of solids

A moment's thought about the general equilibrium

$$M_xA_y(s) \rightleftharpoons xM^{m+} + yA^{n-}$$

will reveal that the solid M_xA_y can be dissolved if the equilibrium concentration of either the cation or the anion, or both, can be decreased sufficiently. Once again Le Chatelier's principle applies. As the product concentrations decrease, the equilibrium is shifted away from the solid. The concentration of the cation can be decreased by complexation or by oxidation or reduction. The concentration of the anion can be decreased by formation of a weak acid, by complexation, or by oxidation or reduction. The principle is the same whether the cation or the anion concentration is decreased. It is generally more practical to decrease the equilibrium concentration of the anion.

Dissolution of a solid by conversion of the anion into a weak acid depends upon a favorable relationship between K_{sp} and K_a and is best achieved, of course, when the anion is strongly basic, that is, when K_a is small. Dissolution of oxides, hydroxides, carbonates, and sulfites by adding strong acids is particularly effective for this reason.

EXAMPLE 30.17

■ To what pH must a solution containing a precipitate of $Cr(OH)_3$ be adjusted so that all of the precipitate dissolves, forming a solution such that $[Cr^{3+}] = 0.1M$?

The dissolution process is represented by the equilibrium

$$Cr(OH)_3(s) \rightleftharpoons Cr^{3+} + 3OH^-$$
$$K_{sp} = [Cr^{3+}][OH^-]^3 = 6 \times 10^{-31}$$

Solving for $[OH^-]$ and substituting $[Cr^{3+}] = 0.1M$ gives

$$[OH^-] = \sqrt[3]{\frac{K_{sp}}{[Cr^{3+}]}} = \sqrt[3]{\frac{6 \times 10^{-31}}{0.1}} = 2 \times 10^{-10}M$$

The corresponding $[H^+]$ is

$$[H^+] = \frac{K_w}{[OH^-]} = \frac{1.00 \times 10^{-14}}{2 \times 10^{-10}} = 5 \times 10^{-5}M$$

which gives

$$pH = -\log [H^+] = -\log (5 \times 10^{-5}) = 4.3$$

The precipitate of $Cr(OH)_3$ will be dissolved by adding H^+ until the pH ≤ 4.3. ■

If either an anion or a cation can be oxidized or reduced and the resulting product does not itself give a product of low solubility with any other ion that is present, then dissolution will occur. An example of dissolution by a redox equilibrium in cation analysis is the oxidation of sulfur in the dissolution of CuS

$$3CuS(s) + 8H^+ + 2NO_3^- \rightleftharpoons 3Cu^{2+} + 2NO(g) + 4H_2O(l) + 3S(s)$$

Dissolution by complexation is usually restricted to decreasing the equilibrium concentration of the cation and is dependent upon a favorable relationship between K_{sp} and K_d, the dissociation constant for the complex ion. Cations derived from the d transition metals are most commonly involved. At various points in cation analysis, dissolution by complexation is carried out as follows:

$$AgCl(s) + 2NH_3(aq) \rightleftharpoons [Ag(NH_3)_2]^+ + Cl^-$$
$$Al(OH)_3(s) + OH^- \rightleftharpoons [Al(OH)_4]^-$$
$$SnS_2(s) + S^{2-} \rightleftharpoons [SnS_3]^{2-}$$

EXAMPLE 30.18

■ Compare the solubilities of silver(I) chloride and silver(I) iodide in $1.0M$ NH_3. For AgCl, $K_{sp} = 1.8 \times 10^{-10}$; for AgI, $K_{sp} = 8.3 \times 10^{-17}$; for $[Ag(NH_3)_2]^+$, $K_d = 6.2 \times 10^{-8}$.

Using **X** for the halide ions, the dissolution of these silver halides in aqueous ammonia, which occurs by complex formation, is

$$AgX(s) + 2NH_3(aq) \rightleftharpoons [Ag(NH_3)_2]^+ + X^-$$

$$K = \frac{[Ag(NH_3)_2^+][X^-]}{[NH_3]^2}$$

We can show that, for this reaction, K is equal to the solubility product constant for the silver halide

$$AgX(s) \rightleftharpoons Ag^+ + X^-$$
$$K_{sp} = [Ag^+][X^-]$$

divided by the dissociation constant of the complex

$$[Ag(NH_3)_2]^+ \rightleftharpoons Ag^+ + 2NH_3(aq)$$

$$K_d = \frac{[Ag^+][NH_3]^2}{[Ag(NH_3)_2^+]}$$

We have

$$K_{sp} \times \frac{1}{K_d} = [Ag^+][X^-] \times \frac{[Ag(NH_3)_2^+]}{[Ag^+][NH_3]^2}$$

$$= \frac{[Ag(NH_3)_2^+][X^-]}{[NH_3]^2} = K$$

Letting s represent the solubility of AgX, we can see that $s = [Ag(NH_3)_2]^+ = [X^-]$ and $[NH_3] = 1.0 - 2s \approx 1.0M$. Thus

$$\frac{K_{sp}}{K_d} = \frac{(s)(s)}{(1.0)^2}$$

$$s^2 = (1.0)^2 \frac{K_{sp}}{K_d}$$

$$s_{AgCl} = \sqrt{(1.0)^2 \frac{(1.8 \times 10^{-10})}{(6.2 \times 10^{-8})}} = 5.4 \times 10^{-2}M$$

$$s_{AgI} = \sqrt{(1.0)^2 \frac{(8.3 \times 10^{-17})}{(6.2 \times 10^{-8})}} = 3.6 \times 10^{-5}M$$

Solid AgCl is 1500 times more soluble in NH_3 than is AgI. ■

Le Chatelier's Principle

The Original Statement in 1884

Every system in stable chemical equilibrium submitted to the influence of an exterior force which tends to cause variation either in its temperature or its condensation (pressure, concentration, number of molecules in the unit of volume) in its totality or only in some one of its parts can undergo only those interior modifications which, if they occur alone, would produce a change of temperature, or of condensation, of a sign contrary to that resulting from the exterior force.

A Simplified Statement Made in 1888

Every change in one of the factors of an equilibrium occasions a rearrangement of the system in such a direction that the factor in question experiences a change in the sense opposite to the original change.

DICTIONARY OF SCIENTIFIC BIOGRAPHY, C. C. Gillispie, ed. (Quoted from Vol. 8, p. 116, Charles Scribner's Sons, New York, 1973.)

Exercises

30.1 A "chemical magician" performed the following series of "tricks" on a solution of $AgNO_3$: forming a black precipitate by adding a base, dissolving the precipitate by adding $S_2O_3^{2-}$, forming a cream-colored precipitate by adding Br^-, dissolving the precipitate by adding concentrated $NH_3(aq)$, and forming a yellow precipitate by adding I^-. Explain the theory behind this series of reactions and write equations describing each step.

30.2 Write the net ionic equations describing (a) S^{2-} reacting with $SnS_2(s)$ to produce $[SnS_3]^{2-}$; (b) HCl reacting with $SnS_2(s)$ to produce $[SnCl_6]^{2-}$; (c) Pb^{2+} reacting with H_2SO_4 to give $PbSO_4(s)$; (d) Cu^{2+} reacting with $NH_3(aq)$ to produce the copper-ammonia complex, which, in turn, is converted to Cu^{2+} by acetic acid, followed by the Cu^{2+} reacting with $K_4[Fe(CN)_6]$ to form $Cu_2[Fe(CN)_6](s)$; (e) Fe^{2+} reacting with $NH_3(aq)$ to give $Fe(OH)_2(s)$; (f) $NiS(s)$ reacting with HNO_3 to produce Ni^{2+}, $NO(g)$, and $S(s)$; (g) Co^{2+} reacting with NaOH to give $Co(OH)_2(s)$.

30.3 Write chemical equations illustrating the amphoteric behavior of HS^-.

30.4 What is the effect of adding each of the following substances to a saturated solution of calcium oxalate, CaC_2O_4: (a) $Na_2C_2O_4$, (b) $H_2C_2O_4$, (c) NaCl, (d) $CaCl_2$, (e) additional CaC_2O_4, (f) HCl? $K_{sp} = 6 \times 10^{-4}$ for CaC_2O_4 and $K_{a_1} = 5.60 \times 10^{-2}$ and $K_{a_2} = 6.2 \times 10^{-5}$ for $H_2C_2O_4$.

30.5 Write the expression for the equilibrium constant for each of the following reactions:

(a) $Hg_2^{2+} + 2Cl^- \rightleftharpoons Hg_2Cl_2(s)$
(b) $Hg_2Cl_2(s) + 2NH_3(aq) \rightleftharpoons Hg(l) + Hg(NH_2)Cl(s) + NH_4^+ + Cl^-$
(c) $AgCl(s) + 2NH_3(aq) \rightleftharpoons [Ag(NH_3)_2]^+ + Cl^-$
(d) $[Ag(NH_3)_2]^+ + Cl^- + 2H^+ \rightleftharpoons AgCl(s) + 2NH_4^+$
(e) $3HgS(s) + 8H^+ + 12Cl^- + 2NO_3^- \rightleftharpoons 3[HgCl_4]^{2-} + 2NO(g) + 4H_2O(l) + 3S(s)$
(f) $[HgCl_4]^{2-} + Sn^{2+} + 2Cl^- \rightleftharpoons Hg(l) + [SnCl_6]^{2-}$
(g) $[Pb(OH)_4]^{2-} + 4H^+ \rightleftharpoons Pb^{2+} + 4H_2O(l)$

30.6 Nitrous acid, which is used in the analysis of Mn^{2+}, is unstable and must be freshly prepared by mixing $NaNO_2$ and HNO_3. Does the presence of Na^+ or NO_3^- affect the equilibrium

$$HNO_2(aq) \rightleftharpoons H^+ + NO_2^-$$

30.7 Explain why $Al(OH)_3$ is formed when $(NH_4)_2S$ is added to a solution containing Al^{3+}.

30.8 Compare the concentrations of Pb^{2+} and Hg^{2+} in a solution containing $[S^{2-}] = 0.01$ M. $K_{sp} = 1 \times 10^{-28}$ for PbS and 4×10^{-53} for HgS.

30.9 Calculate the solubility in grams per liter of AgCl in water. $K_{sp} = 1.8 \times 10^{-10}$ for AgCl.

30.10 Will a precipitate of Ag_2SO_4 form if $[Ag^+] = 0.01M$ and $[SO_4^{2-}] = 0.1M$? $K_{sp} = 1.5 \times 10^{-5}$ for Ag_2SO_4.

30.11 A solution is $0.1M$ in Ba^{2+} and $0.1M$ in Ag^+. As CrO_4^{2-} is added, which cation will begin to precipitate first? $K_{sp} = 2.5 \times 10^{-12}$ for Ag_2CrO_4 and 1.2×10^{-10} for $BaCrO_4$.

30.12 A solution is $0.01M$ in CrO_4^{2-} and $0.01M$ in Cl^-. Describe what will happen as a solution of Ag^+ is added dropwise. $K_{sp} = 1.8 \times 10^{-10}$ for AgCl and 2.5×10^{-12} for Ag_2CrO_4.

30.13 What is the concentration of HCO_3^- in equilibrium with $[H_2CO_3] = 0.01M$ and $[H^+] = 0.01M$? What is the concentration of CO_3^{2-} in this system? $K_{a_1} = 4.5 \times 10^{-7}$ and $K_{a_2} = 4.8 \times 10^{-11}$ for H_2CO_3.

30.14 Is the concentration of $C_2O_4^{2-}$ in a $0.1M$ $H_2C_2O_4$ solution large enough to produce a precipitate of CaC_2O_4 if the original solution was $0.01M$ in Ca^{2+}? $K_{sp} = 1 \times 10^{-9}$ for CaC_2O_4 and $K_{a_1} = 5.60 \times 10^{-2}$ and $K_{a_2} = 6.2 \times 10^{-5}$ for $H_2C_2O_4$.

30.15 Calculate $[Ni^{2+}]$ in equilibrium with a saturated solution of $Ni(OH)_2$ such that $[NH_3] = 6M$. Likewise, find $[Ni^{2+}]$ in equilibrium with a solution in which $[Ni(NH_3)_6^{2+}] = 0.1M$ and $[NH_3] = 6M$. In which form, $Ni(OH)_2$ or $[Ni(NH_3)_6]^{2+}$, would the majority of the nickel be found? $K_b = 1.6 \times 10^{-5}$ for $NH_3(aq)$, $K_{sp} = 3 \times 10^{-16}$ for $Ni(OH)_2$, and $K_d = 1 \times 10^{-9}$ for $[Ni(NH_3)_6]^{2+}$.

30.16 What is the pH of a buffer solution prepared by adding 10.0 ml of $0.100M$ NH_4NO_3 to 15.0 ml of $0.100M$ NH_3? $K_b = 1.6 \times 10^{-5}$ for NH_3.

30.17 What mole ratio of CH_3COOH to $Na(CH_3COO)$ would be needed to produce a buffer solution of pH = 5.0? What would be the ratio for pH = 5.5? $K_a = 1.754 \times 10^{-5}$ for acetic acid.

30.18 Assuming $Pb(OH)_2(s)$ to act as an acid, find $[H^+]$

$$Pb(OH)_2(s) \rightleftharpoons 2H^+ + PbO_2^{2-}$$
$$K_a = [H^+]^2[PbO_2^{2-}] = 4.6 \times 10^{-16}$$

and assuming $Pb(OH)_2(s)$ to act as a base, find $[OH^-]$

$$Pb(OH)_2(s) \rightleftharpoons Pb^{2+} + 2OH^-$$
$$K_b = K_{sp} = [Pb^{2+}][OH^-]^2 = 1.2 \times 10^{-15}$$

What is the pH of the solution?

30.19 Would adding Fe^{3+} to an acidic aqueous solution of H_2S cause the H_2S to be oxidized to S? $E° = 0.770$ V for

$$Fe^{3+} + e^- \xrightarrow{H^+} Fe^{2+}$$

and 0.142 V for

$$S + 2H^+ + 2e^- \longrightarrow H_2S(aq)$$

30.20 Calculate the solubility of AgBr in $1.0M$ NH_3. $K_{sp} = 4.9 \times 10^{-13}$ for AgBr and $K_d = 6.2 \times 10^{-8}$ for $[Ag(NH_3)_2]^+$.

30.21 To what pH must a solution containing a precipitate of $Ni(OH)_2$ be adjusted so that 0.1 mole of the precipitate dissolves in 1 liter? $K_{sp} = 3 \times 10^{-16}$ for $Ni(OH)_2$ and $K_w = 1.00 \times 10^{-14}$.

30.22 To what final concentration of $NH_3(aq)$ must a solution be adjusted to just dissolve (a) 1 g of AgCl in 1 liter of solution, (b) 1 g of AgBr in 1 liter of solution, and (c) 1 g of AgI in 1 liter of solution? Which of these silver halides can be separated from the others using $6M$ $NH_3(aq)$? $K_{sp} = 1.8 \times 10^{-10}$ for AgCl, 4.9×10^{-13} for AgBr, and 8.3×10^{-17} for AgI, and $K_d = 6.2 \times 10^{-8}$ for $[Ag(NH_3)_2]^+$.

30.23 Confirm that $[NH_4^+]$ must be larger than $0.1M$ to prevent precipitation of $Mn(OH)_2$ in a solution in which $[NH_3] = 0.1M$ and $[Mn^{2+}] = 0.0010M$. $K_{sp} = 2 \times 10^{-13}$ for $Mn(OH)_2$ and $K_b = 1.6 \times 10^{-5}$ for $NH_3(aq)$.

30.24 What concentration of H^+ is needed to keep Ni^{2+} in solution at a concentration of $0.1M$ while Zn^{2+} precipitates as ZnS? $K_{sp} = 3 \times 10^{-19}$ for NiS, $K_{sp} = 2 \times 10^{-24}$ for ZnS, and $[H^+]^2[S^{2-}] = 3 \times 10^{-21}$.

30.25 Show that concentrated HNO_3 will dissolve $PbS(s)$ to form a dilute solution. $K_{sp} = 1 \times 10^{-28}$ for PbS, $K_a = 3 \times 10^{-20}$ for H_2S, and $K = 1 \times 10^{83}$ for the reaction

$$3H_2S(aq) + 2NO_3^- + 2H^+ \rightleftharpoons 2NO(g) + 4H_2O(l) + 3S(s)$$

30.26 Calculate the equilibrium constant for the reaction

$$HgS(s) + 2H^+ + 4Cl^- \rightleftharpoons [HgCl_4]^{2-} + H_2S(aq)$$

using the following equations and respective equilibrium constants:

$$HgS(s) \rightleftharpoons Hg^{2+} + S^{2-} \quad K_{sp} = [Hg^{2+}][S^{2-}] = 4 \times 10^{-53}$$
$$H_2S(aq) \rightleftharpoons 2H^+ + S^{2-}$$
$$K_a = [H^+]^2[S^{2-}]/[H_2S] = 3 \times 10^{-20}$$

$$[HgCl_4]^{2-} \rightleftharpoons Hg^{2+} + 4Cl^-$$

$$K_d = [Hg^{2+}][Cl^-]^4/[HgCl_4^{2-}] = 2 \times 10^{-16}$$

Will HgS dissolve in HCl?

30.27** Although Cd^{2+} is not considered among the cations in the scheme used in this book, it is often included in more nearly complete schemes.

(a) Calculate K_{sp} for $CdCl_2$, given that the solubility of $CdCl_2$ is 1400 g/liter. Will $CdCl_2$ precipitate from a $0.01M$ Cd^{2+} solution using $6M$ HCl?

(b) The solubility of the sulfide of cadmium is quite low, and CdS is usually removed (along with HgS, PbS, CuS, Sb_2S_3, and SnS_2) from a cation mixture by sulfide precipitation. What is the concentration of S^{2-} needed to start the precipitation of CdS in a $0.01M$ solution of Cd^{2+}? $K_{sp} = 2 \times 10^{-28}$ for CdS.

(c) Cadmium sulfide is not sulfoamphoteric, and so it is separated (along with HgS, PbS, and CuS) from $[SbS_2]^{2-}$ and $[SnS_3]^{2-}$ in the presence of excess S^{2-}. However, CdS (along with PbS and CuS) dissolves in concentrated HNO_3

and so it can be separated from HgS. Write an equation describing this reaction and show that the reaction is favorable by calculating the equilibrium constant for the reaction.

(d) Upon addition of H_2SO_4, $PbSO_4$ can be removed leaving only Cu^{2+} and Cd^{2+}. Both of these cations form tetraammine complexes with $NH_3(aq)$. Find $[Cd^{2+}]$ in equilibrium with $[Cd(NH_3)_4^{2+}] = 0.01M$ and $[NH_3] = 1M$. $K_d = 1 \times 10^{-7}$ for $[Cd(NH_3)_4]^{2+}$.

(e) The tetraammine complexes are treated with CN^- to produce $[Cu(CN)_3]^{2-}$ and $[Cd(CN)_4]^{2-}$. Write the net ionic equations for both reactions (assume CN^- is oxidized to cyanogen, $(CN)_2$, in the copper reaction).

(f) The final separation between Cd^{2+} and Cu^{2+} is made by a sulfide precipitation with the cyano complexes. Assuming $[S^{2-}] = 0.10M$, $[Cu(CN)_3^{2-}] = 0.01M$ and $[Cd(CN)_4^{2-}] = 0.01M$, show that CdS is formed while the copper(I) remains as the cyano complex. $K_d = 8.2 \times 10^{-18}$ for $[Cd(CN)_4]^{2-}$, $K_{sp} = 3 \times 10^{-48}$ for Cu_2S, and $K_d = 1 \times 10^{-35}$ for $[Cu(CN)_3]^{2-}$.

31 QUALITATIVE ANALYSIS FOR CATIONS II. THE CATIONS AND THE SCHEME

Most inorganic qualitative analysis is done today by methods based on the classic work of Fresenius in the 1800s (see Chapter 22). The cation analysis methods described in this chapter are also derived from that source. From time to time, however, entirely different schemes have been developed to meet the demands of specific situations. One rather interesting scheme arose out of a wartime need.

During World War II the U.S. government felt that we should be prepared against the possibility of chemical warfare. If any toxic agents were used in the field, identifying them rapidly would be vitally important. A battlefield was one place where analytical instruments such as spectrometers were not likely to be available. The California Institute of Technology was commissioned to develop a qualitative analysis scheme that could detect on the spot any of the elements and also any organic compounds that might be present.

The comprehensive analysis plan that they devised was based on the periodic table. The elements were divided into three groups: the basic elements, the amphoteric elements, and the acidic elements. The first step of the analysis was a fusion with sucrose and sodium peroxide. Procedures were included for identifying organic functional groups. Fortunately, this ambitious analysis scheme was never needed in a battlefield. But after the war it was used in identifying chemicals abandoned by departing armies.

This chapter introduces the analysis scheme and flow charts for the identification of 22 cations. The general properties of the cations in each group are briefly discussed, as well as the properties of the individual ions. In conjunction with the flow charts, chemical and practical comments are given about each step in the analysis.

Flow charts and the group separations

A systematic method for presenting the steps in cation analysis and for keeping track of what is happening is essential. We use flow charts for this purpose. A flow chart is a schematic outline that begins with the ions in question, indicates the reagents and conditions for each step, and gives the formulas for the chemical products that result. Flow charts can be written in various ways. In this book we adopt the following style:

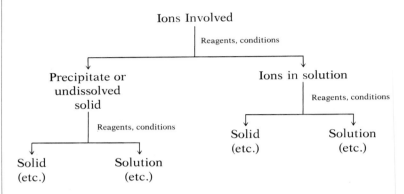

Figure 31.1 is the flow chart for the separation of the groups from each other. Individual flow charts for the groups are given with the discussion of the ions in each group. In these group flow charts, each step is identified by a procedure number. Paragraphs in the text give further information about the chemistry and practical aspects of these steps. (The specific step-by-step directions for the laboratory work are given in the separate laboratory manual.)

Cation Group I (Hg_2^{2+}, Pb^{2+}, Ag^+)

Cation Group I brings together three metal ions that form chlorides that are insoluble in acidic solution. The group reagent is hydrochloric acid, and this group is sometimes known as the hydrochloric acid group, the chloride group, or the silver group. The chlorides of all the other cations in our cation analysis scheme are soluble in acidic solution.

The use of a moderate excess of hydrochloric acid to precipitate the group serves two purposes: (1) the excess chloride ion encourages the precipitation of the chlorides (by Le Chatelier's principle) and (2) the hydrogen ion prevents the interfering precipitation of the bismuth and the antimony oxochlorides, which would form if these ions were present.

$$Bi^{3+} + Cl^- + H_2O \rightleftharpoons BiOCl(s) + 2H^+ \qquad (1)$$
$$Sb^{3+} + Cl^- + H_2O \rightleftharpoons SbOCl(s) + 2H^+ \qquad (2)$$

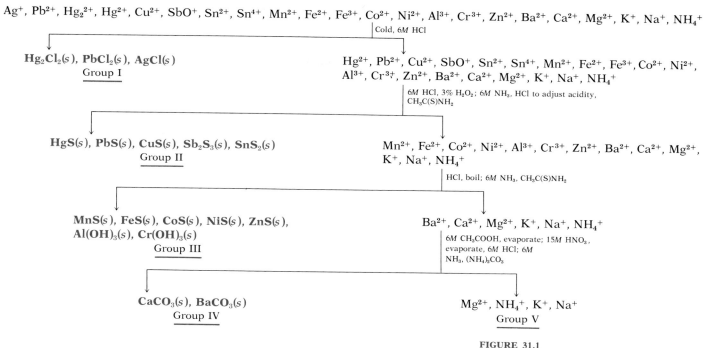

Ag^+, Pb^{2+}, Hg_2^{2+}, Hg^{2+}, Cu^{2+}, SbO^+, Sn^{2+}, Sn^{4+}, Mn^{2+}, Fe^{2+}, Fe^{3+}, Co^{2+}, Ni^{2+}, Al^{3+}, Cr^{3+}, Zn^{2+}, Ba^{2+}, Ca^{2+}, Mg^{2+}, K^+, Na^+, NH_4^+

Cold, 6M HCl

$Hg_2Cl_2(s)$, $PbCl_2(s)$, $AgCl(s)$
Group I

Hg^{2+}, Pb^{2+}, Cu^{2+}, SbO^+, Sn^{2+}, Sn^{4+}, Mn^{2+}, Fe^{2+}, Fe^{3+}, Co^{2+}, Ni^{2+}, Al^{3+}, Cr^{3+}, Zn^{2+}, Ba^{2+}, Ca^{2+}, Mg^{2+}, K^+, Na^+, NH_4^+

6M HCl, 3% H_2O_2; 6M NH_3, HCl to adjust acidity,
$CH_3C(S)NH_2$

$HgS(s)$, $PbS(s)$, $CuS(s)$, $Sb_2S_3(s)$, $SnS_2(s)$
Group II

Mn^{2+}, Fe^{2+}, Co^{2+}, Ni^{2+}, Al^{3+}, Cr^{3+}, Zn^{2+}, Ba^{2+}, Ca^{2+}, Mg^{2+}, K^+, Na^+, NH_4^+

HCl, boil; 6M NH_3, $CH_3C(S)NH_2$

$MnS(s)$, $FeS(s)$, $CoS(s)$, $NiS(s)$, $ZnS(s)$, $Al(OH)_3(s)$, $Cr(OH)_3(s)$
Group III

Ba^{2+}, Ca^{2+}, Mg^{2+}, K^+, Na^+, NH_4^+

6M CH_3COOH, evaporate; 15M HNO_3,
evaporate, 6M HCl; 6M
NH_3, $(NH_4)_2CO_3$

$CaCO_3(s)$, $BaCO_3(s)$
Group IV

Mg^{2+}, NH_4^+, K^+, Na^+
Group V

FIGURE 31.1
Flow chart for separation of cations into groups. *Details of the group separation procedures are discussed in the sections on the individual group analyses.*

Too large an excess of the acid, however, dissolves the silver and lead chlorides by complex formation. Mercury(I) chloride does not dissolve because the chloro complexes of mercury(I) are very unstable.

Lead chloride ($PbCl_2$) is by several thousandfold the most soluble of the three chlorides in this group. The use of a cool solution and the presence of excess chloride ion help to precipitate the maximum possible amount of $PbCl_2$ at this stage. However, lead chloride is soluble enough that it is impossible to avoid carrying some lead(II) over into the filtrate and into Cation Group II. Therefore, provision is also made for the detection and removal of lead in the next analytical group.

(The metals and ions in Cation Group I have been discussed in previous chapters: Hg and Hg_2^{2+}, Chapter 26; Pb and Pb^{2+}, Chapter 26; Ag and Ag^+, Chapter 29.)

31.1 Mercury(I) ion

The colorless Hg_2^{2+} ion is unique among the cations in our qualitative scheme because it contains a covalent metal–metal bond (Section 26.8). The Hg_2^{2+} ion is found in a number of solid compounds. In aqueous solutions, its existence is limited to the pH region of about 3–4. At higher pH values, a disproportionation reaction with water or hydroxide ion takes place, that is, the Hg_2^{2+} is simultaneously oxidized and reduced

$$Hg_2^{2+} + 2OH^- \rightleftharpoons Hg(l) + HgO(s) + H_2O(l) \qquad (3)$$

This reaction provides an example of the displacement of an equilibrium as the concentration of one ion is decreased by a competing reaction. The general equilibrium reaction is

$$Hg(l) + Hg^{2+} \rightleftharpoons Hg_2^{2+} \tag{4}$$

for which

$$K = \frac{[Hg_2^{2+}]}{[Hg^{2+}]} = 166 \tag{5}$$

At equilibrium, the concentration of Hg_2^{2+} is only 166 times that of Hg^{2+}. As $[Hg^{2+}]$ is decreased by the formation of the slightly soluble HgO, $[Hg_2^{2+}]$ is also decreased by further disproportionation. Disproportionation rather than mercury(I) compound formation occurs in several other reactions in which the concentration of Hg^{2+} in solution is decreased by the formation of slightly soluble compounds or of stable complexes:

$$Hg_2^{2+} + CO_3^{2-} \rightleftharpoons Hg(l) + HgO(s) + CO_2(g) \tag{6}$$
$$Hg_2^{2+} + H_2S(aq) \rightleftharpoons Hg(l) + HgS(s) + 2H^+ \tag{7}$$
$$Hg_2^{2+} + 4CN^- \rightleftharpoons Hg(l) + [Hg(CN)_4]^{2-} \tag{8}$$
$$Hg_2Cl_2(s) + 2NH_3(aq) \rightleftharpoons Hg(l) + HgNH_2Cl(s) + NH_4^+ + Cl^- \tag{9}$$

The elemental mercury formed in these reactions is so finely divided that it is black.

Of the common mercury(I) compounds, only the nitrate and perchlorate are soluble in water. Aqueous solutions of these compounds can be stabilized against disproportionation by adding elemental mercury, which converts any Hg^{2+} formed back to Hg_2^{2+}. Standard electrode potential data

$$2Hg^{2+} + 2e^- \longrightarrow Hg_2^{2+} \qquad E° = 0.905 \text{ V} \tag{10}$$
$$Hg^{2+} + 2e^- \longrightarrow Hg \qquad E° = 0.851 \text{ V} \tag{11}$$
$$Hg_2^{2+} + 2e^- \longrightarrow 2Hg \qquad E° = 0.7961 \text{ V} \tag{12}$$

indicate that reduction of Hg_2^{2+} to elemental mercury occurs more easily than oxidation of Hg_2^{2+} to Hg^{2+}. The standard potentials of the Hg(II)–Hg(I) and Hg(II)–Hg(0) couples are so close to that of the Hg(I)–Hg(0) couple that oxidation of mercury invariably gives mercury(II) unless an excess of the metal is present.

Reactions Important in the Separation and Identification of Hg_2^{2+}

Group precipitation

$$Hg_2^{2+} + 2Cl^- \rightleftharpoons Hg_2Cl_2(s) \tag{13}$$
$$\text{white}$$

Confirmatory test

$$Hg_2Cl_2(s) + 2NH_3(aq) \rightleftharpoons Hg(l) + Hg(NH_2)Cl(s) + NH_4^+ + Cl^- \tag{14}$$
$$\qquad\qquad\qquad\qquad\text{black} \qquad \text{white}$$

31.2 Lead(II) ion

Both lead(II) and lead(IV) compounds are known in the solid state. However, only lead(II) compounds are found in aqueous solution, for the lead(IV) compounds are very strong oxidizing agents and are easily reduced. These relationships are indicated by the standard potential data

$$PbSO_4 + 2e^- \longrightarrow Pb + SO_4^{2-} \qquad E° = -0.356 \text{ V} \qquad (15)$$
$$Pb^{2+} + 2e^- \longrightarrow Pb \qquad E° = -0.1263 \text{ V} \qquad (16)$$
$$PbO_2 + 4H^+ + 2e^- \longrightarrow Pb^{2+} + 2H_2O \quad E° = +1.46 \text{ V} \qquad (17)$$
$$PbO_2 + 4H^+ + SO_4^{2-} + 2e^- \longrightarrow$$
$$PbSO_4 + 2H_2O \qquad E° = +1.685 \text{ V} \qquad (18)$$

The $+2$ oxidation state results even when elemental lead is treated with a relatively strong oxidizing agent such as dilute nitric acid

$$3Pb(s) + 8H^+ + 2NO_3^- \longrightarrow 3Pb^{2+} + 2NO(g) + 4H_2O(l) \qquad (19)$$

Of the common lead(II) compounds, only the nitrate, acetate, and perchlorate are soluble in water. Lead(II) acetate is a weak electrolyte in solution because of the formation of acetato complexes such as $[Pb(CH_3COO)_4]^{2-}$. Lead(II) sulfate, an insoluble compound, dissolves in the presence of an excess of acetate ion by formation of such acetate complexes. Similarly, lead(II) hydroxide dissolves in strongly alkaline solutions by formation of the hydroxo complex ion, $[Pb(OH)_4]^{2-}$. In the presence of an excess of chloride ions, lead(II) chloride also forms a complex ion

$$PbCl_2(s) + 2Cl^-(xs) \rightleftharpoons [PbCl_4]^{2-} \qquad (20)$$

Reactions Important in the Separation and Identification of Pb^{2+} in Cation Group I

Group precipitation

$$Pb^{2+} + 2Cl^- \xrightarrow{\text{cold}} PbCl_2(s) \qquad (21)$$
$$\text{white}$$

Confirmatory tests

$$Pb^{2+} + SO_4^{2-} \rightleftharpoons PbSO_4(s) \qquad (22)$$
$$\text{white}$$
$$Pb^{2+} + CrO_4^{2-} \rightleftharpoons PbCrO_4(s) \qquad (23)$$
$$\text{yellow}$$

31.3 Silver(I) ion

Silver(II) and silver(III) compounds exist, but they are such strong oxidizing agents that they are difficult to prepare and are readily reduced. Even the silver(I) ion is readily reduced in acidic solution,

$$Ag^+ + e^- \longrightarrow Ag \qquad E° = 0.7996 \text{ V} \tag{24}$$

In solution, the Ag^+ ion is colorless, but many solid silver(I) compounds that form with colorless anions are colored (e.g., AgBr, cream; AgI, Ag_3PO_4, yellow; Ag_2S, Ag_2O, black; Ag_3AsO_4, reddish brown). The color of simple metal compounds deepens as the bonding becomes less ionic and more covalent. Apparently this is due to changes in the distribution of electron density in the anions.

The only common easily water-soluble silver(I) compounds are the nitrate, fluoride, and perchlorate; the nitrite, acetate, and sulfate are moderately soluble. Silver(I) nitrate solution is the usual source of Ag^+ in the laboratory. As a d transition metal ion, silver(I) forms a number of complex ions, of which the most commonly encountered are $[Ag(NH_3)_2]^+$ and $[Ag(CN)_2]^-$.

Reactions Important in the Separation and Identification of Ag$^+$

Group precipitation

$$Ag^+ + Cl^- \rightleftharpoons \underset{white}{AgCl(s)} \tag{25}$$

Dissolution by complex formation

$$AgCl(s) + 2NH_3(aq) \rightleftharpoons [Ag(NH_3)_2]^+ + Cl^- \tag{26}$$

Confirmatory test

$$[Ag(NH_3)_2]^+ + Cl^- + 2H^+ \rightleftharpoons AgCl(s) + 2NH_4^+ \tag{27}$$

31.4 Cation Group I analysis

The flow chart for Cation Group I is given in Figure 31.2. The procedures identified in the flow chart are discussed in the following paragraphs.

Procedure 1. Solubility product constants (Table 30.6) indicate that lead(II) chloride is much more soluble than the chlorides of mercury(I) or silver(I). Because the solubility in water of $PbCl_2$ increases from about $0.036M$ at 20°C to about $0.12M$ at 100°C, cold solutions are used in this procedure to maximize precipitation of the lead(II) ion. Addition of excess hydrochloric acid initially reduces the solubility of each chloride by the common ion effect, but a large excess then increases the solubilities of AgCl and $PbCl_2$ as a consequence of the formation of chloro complex ions. In any event, a sufficiently large concentration of Pb^{2+} ion remains in solution to form PbS(s) in Cation Group II.

Procedure 2. Boiling the precipitate with water increases the rate of dissolution of lead(II) chloride and maximizes its removal. Mere washing with hot water is seldom completely effective.

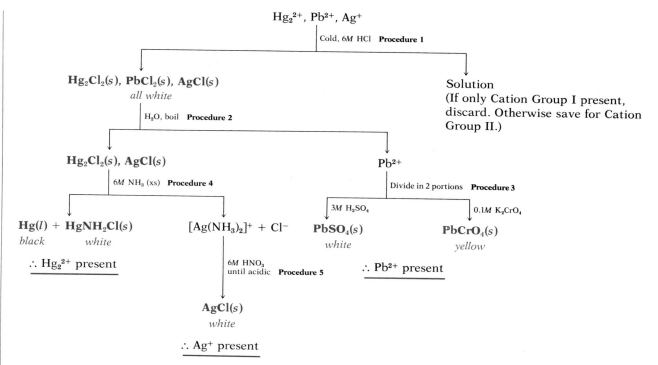

$$Hg_2^{2+}, Pb^{2+}, Ag^+$$

Cold, 6*M* HCl **Procedure 1**

Hg₂Cl₂(s), PbCl₂(s), AgCl(s)
all white

Solution
(If only Cation Group I present, discard. Otherwise save for Cation Group II.)

H₂O, boil **Procedure 2**

Hg₂Cl₂(s), AgCl(s)

Pb²⁺

6*M* NH₃ (xs) **Procedure 4**

Divide in 2 portions **Procedure 3**

3*M* H₂SO₄

0.1*M* K₂CrO₄

Hg(*l*) + HgNH₂Cl(s)
black *white*

∴ Hg₂²⁺ present

[Ag(NH₃)₂]⁺ + Cl⁻

PbSO₄(s)
white

PbCrO₄(s)
yellow

6*M* HNO₃ until acidic **Procedure 5**

∴ Pb²⁺ present

AgCl(s)
white

∴ Ag⁺ present

FIGURE 31.2
Flow chart for analysis of Cation Group I. *The notation "xs" means "excess"; the symbol ∴ reads "therefore."*

Procedure 3. If crystals of lead(II) chloride separate as the solution cools, it is necessary to reheat until they dissolve so that enough Pb²⁺ is in solution to give a clear test. The tests are done on the hot solution. Lead(II) chromate is both a more distinctive precipitate and less soluble than lead(II) sulfate. Chromate precipitation is thus the better confirmatory test for the lead(II) ion.

Procedure 4. Any lead(II) chloride not dissolved in Procedure 2 is converted by aqueous ammonia to a finely divided white basic salt [possibly Pb(OH)Cl] which gives a turbid suspension. Since this material dissolves in nitric acid, it causes no interference with the confirmatory test for the Ag⁺ ion.

Formation of a gray to black residue is the confirmatory test for the Hg₂²⁺ ion. The easily complexed Ag⁺ ion remains in solution. Therefore, aqueous ammonia serves to confirm both the presence or the absence of the Hg₂²⁺ ion and to separate silver(I) from mercury(I). Note that this solution should not be allowed to stand (see Procedure 5).

Procedure 5. Ammoniacal silver(I) solutions may deposit explosive residues. (Azides, containing N₃⁻, may be formed. These can explode even when the solution is being poured.) Therefore, this procedure should be carried out without delay. Slow addition of nitric acid initially causes localized destruction of the complex ion and precipitation of white silver (I) chloride, which then redissolves upon agitation because of the presence of excess ammonia. Consequently, nitric acid must be added until the solution is definitely acidic before the presence or absence of the Ag^+ ion can be confirmed.

EXAMPLE 31.1

■ One of each of the following pairs of substances is known to be in a sample. What single reagent will show which species is present in each case?

(a) $PbCl_2$ and Hg_2Cl_2; (b) AgCl and $PbCl_2$; (c) $[Ag(NH_3)_2]^+Cl^-$ and $Pb^{2+}(NO_3^-)_2$.

(a) Both compounds are water insoluble. Treatment of the sample with aqueous ammonia will give a black material if Hg_2Cl_2 is present and no apparent change if the substance is $PbCl_2$.

$$Hg_2Cl_2(s) + 2NH_3(aq) \rightleftharpoons \underset{black}{Hg(l)} + \underset{white}{HgNH_2Cl(s)} + NH_4^+ + Cl^- \qquad (28)$$

(b) AgCl will dissolve in aqueous ammonia; $PbCl_2$ will not. (Also, $PbCl_2$ will dissolve in hot water, but AgCl will not.)

$$AgCl(s) + 2NH_3(aq) \longrightarrow [Ag(NH_3)_2]^+ + Cl^- \qquad (29)$$

(c) Both $[Ag(NH_3)_2]^+Cl^-$ and $Pb^{2+}(NO_3^-)_2$ are water soluble. However, acidification of an aqueous solution of the sample with dilute HNO_3 will precipitate white AgCl if $[Ag(NH_3)_2]^+Cl^-$ is present. No effect is observed if HNO_3 is added to a solution of $Pb^{2+}(NO_3^-)_2$

$$Ag(NH_3)_2^+ + Cl^- + 2H^+ \longrightarrow AgCl(s) + 2NH_4^+ \qquad (30)$$

■

EXAMPLE 31.2

■ A solution contains both Hg_2^{2+} and Pb^{2+} ions. A student suggests that these ions can be separated by adding a large excess of sodium hydroxide to the solution and thoroughly shaking the mixture. Was the student correct? Justify your answer.

The student was correct. The addition of a small amount of sodium hydroxide will cause both ions to precipitate.

$$Hg_2^{2+} + 2OH^- \longrightarrow$$
$$Hg(l) + HgO(s) + H_2O \qquad \text{(disproportionation)} \qquad (31)$$
$$Pb^{2+} + 2OH^- \longrightarrow Pb(OH)_2(s) \qquad (32)$$

The $Pb(OH)_2$ is amphoteric and will dissolve with complex formation when an excess of the base is added.

$$Pb(OH)_2(s) + 2OH^- \longrightarrow [Pb(OH)_4]^{2-} \qquad (33)$$

■

Cation group II
(Hg^{2+}, Pb^{2+}, Cu^{2+}, Sn^{2+} or Sn^{4+}, Sb^{3+} or SbO$^+$)

The sulfides of the cations in this group are precipitated by hydrogen sulfide in the presence of dilute (0.25 to 0.3M) acid. This group is also sometimes called the copper–tin group or the acidic hydrogen sulfide group. Although we have not included them in our analysis scheme, cadmium, arsenic, and bismuth (all very toxic) would, if present, be part of this group.

The sulfides in this group are those with the smallest K_{sp} values (see Table 30.6). Therefore, they precipitate in the presence of a sulfide ion concentration that is kept low enough to avoid precipitation of the more soluble sulfides. Tin(IV) sulfide is the most soluble compound in this group. The sulfide ion concentration must exceed that required for precipitation of SnS$_2$, while remaining low enough to avoid the precipitation of ZnS, the least soluble sulfide in Cation Group III. As we have seen (Section 30.10), the sulfide ion concentration is controlled by controlling [H$^+$].

This large group is further separated by differences in the solubilities of the sulfides. The amphoteric sulfides of antimony and tin dissolve in alkaline solution in the presence of S^{2-}, but the sulfides of mercury, lead, and copper remain undissolved in alkaline solution. Lead and copper sulfides dissolve when the concentration of sulfide ion is decreased by its oxidation to elemental sulfur by nitric acid, while mercury sulfide is too slightly soluble to be affected by this procedure.

(The metals and metal ions of this group have been discussed in previous chapters: Hg, Hg^{2+}, Chapter 26; Sn, Sn^{2+}, Sn^{4+}, Pb, Pb^{2+}, Chapter 26; Sb, Sb^{3+}, SbO$^+$, Chapter 27; Cu, Cu^{2+}, Chapter 29.)

31.5 Mercury(II) ion

Most mercury(II) compounds are more covalent than ionic. Mercury(II) ion probably exists in only a few compounds, such as the perchlorate or nitrate or in the aqueous solutions of these two compounds. In aqueous solution, mercury(II) either reacts extensively with water or is strongly complexed. For example, in the presence of Cl$^-$, a series of complexes is formed: [HgCl]$^+$, [HgCl$_2$], [HgCl$_3$]$^-$, [HgCl$_4$]$^{2-}$. Although the mercury(II) ion is colorless, solid compounds with colorless anions are often intensely colored as a consequence of covalency (e.g., HgO, red or yellow; HgS, red or black; HgI$_2$, red).

In aqueous solutions containing complex ions, the equilibrium concentration of Hg^{2+} *decreases* in the order

$$[HgCl_4]^{2-} > [Hg(NCS)_4]^{2-} > [HgBr_4]^{2-} > [HgI_4]^{2-}$$
$$> [Hg(CN)_4]^{2-} > [HgS_2]^{2-}$$

The halo complex ions are sufficiently stable that even mercury(II) oxide can be dissolved in alkali metal chloride, bromide, or iodide solutions

$$HgO(s) + 4X^- + H_2O(l) \longrightarrow [HgX_4]^{2-} + 2OH^- \tag{34}$$

Mercury(II) sulfide is the least soluble sulfide ($K_{sp} = 4 \times 10^{-53}$). It is not soluble in either dilute nitric acid or hydrochloric acid, but dissolves in hot aqua regia (Section 22.8c) with the formation of a chloro complex

Reactions Important in the Separation and Identification of Hg^{2+}

Group precipitation

$$Hg^{2+} + H_2S(aq) \underset{\longleftarrow}{\overset{H^+}{\longrightarrow}} \underset{black}{HgS(s)} + 2H^+ \tag{35}$$

Dissolution by oxidation of S^{2-} and complex formation

$$3HgS(s) + 8H^+ + 12Cl^- + 2NO_3{}^- \longrightarrow$$
$$3[HgCl_4]^{2-} + 2NO(g) + 4H_2O(l) + 3S(s) \tag{36}$$

Confirmatory tests

$$[HgCl_4]^{2-} + Sn^{2+}(xs) + 2Cl^- \longrightarrow \underset{black}{Hg(l)} + [SnCl_6]^{2-} \tag{37}$$

$$[HgCl_4]^{2-} + 4NCS^- + Co^{2+} \longrightarrow \underset{deep\ blue}{Co[Hg(NCS)_4](s)} + 4Cl^- \tag{38}$$

31.6 Lead(II) ion

The chemistry of lead(II) ion presented in the discussion of Cation Group I need only be supplemented by equations for additional reactions that take place in Cation Group II.

Reactions Important in the Separation and Identification of Pb^{2+} in Cation Group II

Group precipitation

$$Pb^{2+} + H_2S(aq) \underset{\longleftarrow}{\overset{H^+}{\longrightarrow}} \underset{dark\ brown}{PbS(s)} + 2H^+ \tag{39}$$

Dissolution by oxidation of S^{2-}

$$3PbS(s) + 8H^+ + 2NO_3{}^- \longrightarrow 3Pb^{2+} + 2NO(g) + 4H_2O(l) + 3S(s) \tag{40}$$

Precipitation to separate from Cu^{2+}

$$Pb^{2+} + SO_4^{2-} \rightleftharpoons PbSO_4(s) \qquad (41)$$
$$\text{white}$$

Dissolution as complex

$$PbSO_4(s) + 4OH^- \rightleftharpoons [Pb(OH)_4]^{2-} + SO_4^{2-} \qquad (42)$$
$$\text{colorless}$$

Confirmatory test

$$[Pb(OH)_4]^{2-} + 4H^+ \rightleftharpoons Pb^{2+} + 4H_2O(l) \qquad (43)$$
$$Pb^{2+} + CrO_4^{2-} \rightleftharpoons PbCrO_4(s) \qquad (44)$$
$$\text{yellow}$$

31.7 Copper(II) ion

Copper(I), copper(II), and copper(III) are all known in solid compounds. However, copper(III) compounds are relatively rare, and copper(III) is so strongly oxidizing that it is reduced by water. Copper(II) is the only common species in aqueous solution. Copper(I) and copper(II) are related in terms of standard potentials by the following equilibria:

$$
\begin{array}{lll}
Cu_2O + H_2O + 2e^- \longrightarrow 2Cu + 2OH^- & E^\circ = -0.361 \text{ V} & (45) \\
2Cu(OH)_2 + 2e^- \longrightarrow Cu_2O + 2OH^- + H_2O & E^\circ = -0.09 \text{ V} & (46) \\
Cu^{2+} + e^- \longrightarrow Cu^+ & E^\circ = +0.158 \text{ V} & (47) \\
Cu^{2+} + 2e^- \longrightarrow Cu & E^\circ = +0.3402 \text{ V} & (48) \\
Cu^+ + e^- \longrightarrow Cu & E^\circ = +0.522 \text{ V} & (49)
\end{array}
$$

In acidic solution, Cu^+ and Cu^{2+} are related by a redox disproportionation equilibrium like that of mercury

$$2Cu^+ \rightleftharpoons Cu(s) + Cu^{2+} \qquad (50)$$

for which

$$K = \frac{[Cu^{2+}]}{[Cu^+]^2} = 1.4 \times 10^6 \qquad (51)$$

At equilibrium, $[Cu^{2+}]$ is always $(1.4 \times 10^6)[Cu^+]$, and Cu^+ can exist in solution only if its concentration is extremely small (e.g., in equilibrium with complex ion $[Cu(CN)_3]^{2-}$, the $K_d = 1 \times 10^{-35}$).
   Water-soluble copper(II) compounds include the acetate, bromide, chloride, chromate, nitrate, perchlorate, and sulfate. Most other anions either form insoluble compounds with Cu^{2+} or reduce Cu^{2+} to Cu^+ or Cu^0. Equilibrium $[Cu^{2+}]$ decreases in copper complex solutions as follows:

$$[Cu(H_2O)_4]^{2+} > [CuCl_4]^{2-} > [Cu(NH_3)_4]^{2+}$$

pale blue   *yellow*   *deep blue*

The volatile compounds of some elements give characteristic colors when the compound or its solution is exposed to a flame. This is usually done by dipping a clean platinum wire into the compound or its solution and holding the wire in the oxidizing part of a Bunsen burner flame (the blue flame). (Note that if the wire is held in the reducing flame, a brittle carbide is formed and the wire is ruined.) Copper(II) nitrate gives a bright blue flame color and copper(II) chloride a green flame color.

Reactions Important in the Separation and Detection of Cu²⁺

Group precipitation

$$Cu^{2+} + H_2S(aq) \rightleftharpoons CuS(s) + 2H^+ \tag{52}$$
black

Dissolution by oxidation of S²⁻

$$3CuS(s) + 8H^+ + 2NO_3^- \longrightarrow$$
$$3Cu^{2+} + 2NO(g) + 4H_2O(l) + 3S(s) \tag{53}$$

Confirmatory tests

$$Cu^{2+} + 4NH_3(aq) \rightleftharpoons [Cu(NH_3)_4]^{2+} \tag{54}$$
dark blue-purple

$$[Cu(NH_3)_4]^{2+} + 4H^+ \rightleftharpoons Cu^{2+} + 4NH_4^+ \tag{55}$$
$$2Cu^{2+} + [Fe(CN)_6]^{4-} \rightleftharpoons Cu_2[Fe(CN)_6](s) \tag{56}$$
reddish

$$Cu^{2+} + 2NCS^- + 2C_5H_5N(aq) \rightleftharpoons [Cu(NCS)_2(NC_5H_5)_2](s) \tag{57}$$
pyridine   *green*

31.8 Tin(II) and tin(IV) ions

Both tin(II) and tin(IV) are common, and both are encountered in aqueous solutions. These two oxidation states are related to each other and to elemental tin in terms of standard potentials as follows:

for acidic solutions
$$Sn^{2+} + 2e^- \longrightarrow Sn \qquad E° = -0.1364 \text{ V} \tag{58}$$
$$Sn^{4+} + 2e^- \longrightarrow Sn^{2+} \qquad E° = +0.15 \text{ V} \tag{59}$$

for alkaline solutions
$$[Sn(OH)_4]^{2-} + 2e^- \longrightarrow Sn + 4OH^- \qquad E° = -0.909 \text{ V} \tag{60}$$
$$[Sn(OH)_6]^{2-} + 2e^- \longrightarrow [Sn(OH)_4]^{2-} + 2OH^- \qquad E° = -0.96 \text{ V} \tag{61}$$

Thus, tin(II) is a moderately strong reducing agent in acidic or alkaline solution, and tin(IV) is a weak oxidizing agent in acidic solution.

Aqueous acidic tin(II) solutions are protected from atmospheric oxidation by adding elemental tin. Tin in both oxidation states reacts extensively with water. Both Sn^{2+} and Sn^{4+} oxides and hydroxides are amphoteric. For example, tin(II) hydroxide reacts with both acids and bases as follows:

$$Sn(OH)_2(s) + 2H^+ \rightleftharpoons Sn^{2+} + 2H_2O(l) \qquad (62)$$
$$Sn(OH)_2(s) + 2OH^- \rightleftharpoons Sn(OH)_4{}^{2-} \qquad (63)$$

Tin(IV) compounds are both more covalent and more acidic than the corresponding tin(II) compounds. Covalency in the solid compounds of tin is responsible for color (e.g., SnS, brown; SnS_2, yellow; SnI_4, red).

Reactions Important in the Separation and Identification of Sn^{2+} and Sn^{4+}

Group precipitation

$$Sn^{2+} + H_2O_2(aq) + 2H^+ \rightleftharpoons Sn^{4+} + 2H_2O(l) \qquad (64)$$
$$Sn^{4+} + 2H_2S(aq) \xrightarrow{H^+} \underset{yellow}{SnS_2(s)} + 4H^+ \qquad (65)$$

Dissolution by sulfoamphoterism

$$SnS_2(s) + S^{2-} \rightleftharpoons \underset{yellow}{[SnS_3]^{2-}} \qquad (66)$$

Reprecipitation as sulfide

$$[SnS_3]^{2-} + 2H^+ (dil.) \rightleftharpoons SnS_2(s) + H_2S(g) \qquad (67)$$

Dissolution as complex

$$SnS_2(s) + 4H^+ + 6Cl^- \rightleftharpoons [SnCl_6]^{2-} + 2H_2S(g) \qquad (68)$$

Confirmatory test

$$3[SnCl_6]^{2-} + 4Al \xrightarrow{HCl} 3Sn(s) + 4Al^{3+} + 18Cl^- \qquad (69)$$
$$Sn(s) + 2H^+ \longrightarrow Sn^{2+} + H_2(g) \qquad (70)$$
$$Sn^{2+} + 2Hg^{2+}(xs) + 2Cl^- \longrightarrow \underset{white}{Hg_2Cl_2(s)} + Sn^{4+} \qquad (71)$$

31.9 Antimony(III) ion

Both antimony(III) and antimony(V) are known; however, antimony(V) is too strongly oxidizing to be stable in aqueous solution unless complexed (e.g., as $[Sb(OH)_6]^-$ or $[SbS_4]^{3-}$). Although we commonly describe antimony(III) in acidic solutions as Sb^{3+}, reaction with water to give oxoantimony(III), SbO^+, is pronounced

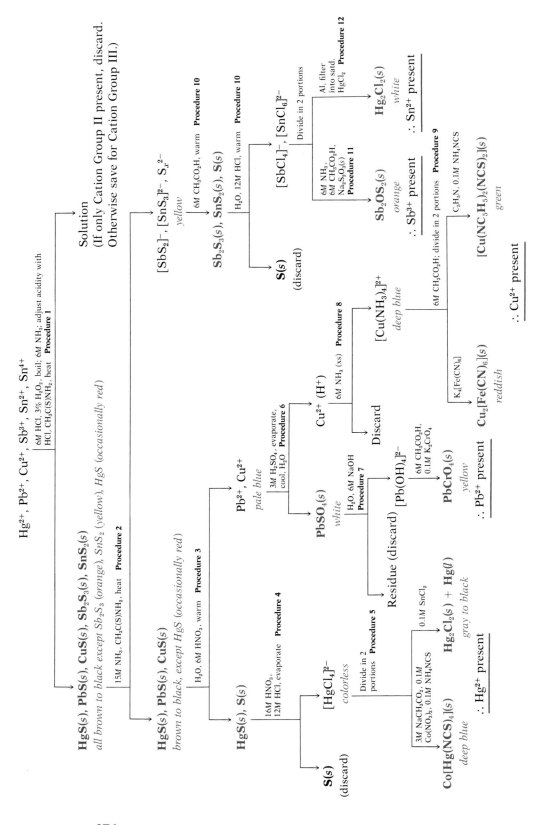

976

FIGURE 31.3
Flow chart for analysis of Cation Group II. *The initial solution of ions from separation of Cation Group I is acidic.*

$$Sb^{3+} + H_2O(l) \rightleftharpoons SbO^+ + 2H^+ \tag{72}$$

and upon dilution such solutions precipitate oxoantimony(III) compounds, also known as antimonyl compounds. For example,

$$Sb^{3+} + Cl^- + H_2O(l) \rightleftharpoons \underset{white}{SbOCl(s)} + 2H^+ \tag{73}$$

The oxide and anhydrous halides of antimony are known, but salts with oxoanions are not known.

Reactions Important in the Separation and Identification of Sb³⁺

Group precipitation

$$2Sb^{3+} + 3H_2S(g) \underset{\longrightarrow}{\rightleftharpoons} \underset{orange}{Sb_2S_3(s)} + 6H^+ \tag{74}$$

Dissolution by sulfoamphoterism

$$Sb_2S_3(s) + S^{2-} \rightleftharpoons 2[SbS_2]^- \tag{75}$$

Reprecipitation as sulfide

$$2[SbS_2]^- + 2H^+ \text{ (dil.)} \rightleftharpoons Sb_2S_3(s) + H_2S(g) \tag{76}$$

Dissolution as complex

$$Sb_2S_3(s) + 6H^+ + 8Cl^- \rightleftharpoons 2[SbCl_4]^- + 3H_2S(g) \tag{77}$$

Confirmatory test

$$2[SbCl_4^-]^- + 2S^{2-} + H_2O(l) \rightleftharpoons \underset{orange}{Sb_2OS_2} + 2H^+ + 8Cl^- \tag{78}$$

31.10 Cation Group II analysis

The flow chart for Cation Group II is given in Figure 31.3. The procedures identified in the flow chart are discussed in the following paragraphs.

Procedure 1. Boiling with hydrogen peroxide oxidizes tin(II) to tin(IV), which forms a much less soluble sulfide than does tin(II). Only small quantities of tin(II) could coexist with mercury(II) without reaction to give Hg_2^{2+} or $Hg(l)$ and tin(IV). Adjustment of $[H^+]$ gives the approximately 0.3 mole/liter concentration required to precipitate Group II sulfides without precipitating those of Fe^{2+}, Mn^{2+}, Co^{2+}, Ni^{2+}, or Zn^{2+}. For an unknown sample, the color of the precipitate can indicate the cations that are present or absent. If Group II cations are present, the precipitate will be yellow to brown or black. A light-colored precipitate is sulfur from the oxidation of thioacetamide.

Procedure 2. The separation is based on the acidic properties of Sb_2S_3 and SnS_2, which are amphoteric, as opposed to the very slightly acidic properties of HgS, PbS, and CuS, which are only slightly amphoteric. In the reaction between $SnS_2(s)$ and S^{2-}, for example, $SnS_2(s)$ is a Lewis acid and S^{2-} is a Lewis base. Polysulfide ion, S_x^{2-}, forms by oxidation of sulfide ion under alkaline conditions.

Procedure 3. Both PbS and CuS dissolve as the S^{2-} ions in solution are oxidized to S(s). However, the solubility of HgS is so much less than the solubilities of PbS and CuS (Table 30.6) that there is not enough sulfide ion from HgS present to be oxidized by the warm (not boiling) approximately $2M$ HNO_3 solution that remains at the end of the PbS and CuS dissolution.

Procedure 4. The solubility of HgS is increased by both the more concentrated acids and the removal of mercury(II) ions as the stable complex $[HgCl_4]^{2-}$.

Procedure 5. Either test or both tests may be used to confirm the presence of mercury(II). Reduction by tin(II) is progressive, that is,

$$[HgCl_4]^{2-} \xrightarrow[\text{stoichiometric}]{Sn^{2+}} \underset{white}{Hg_2Cl_2(s)} \xrightarrow[\text{excess}]{Sn^{2+}} \underset{black}{Hg(l)} \tag{79}$$

Thus confirmation is positive if a white, a gray, or a black insoluble material results.

Procedure 6. Evaporation with H_2SO_4 removes nitric acid in which $PbSO_4$ is soluble. Complete removal of nitric acid is assured if evaporation is continued until a dense white cloud of SO_3 is produced, showing that nitric acid and water have evaporated, the temperature has risen, and sulfuric acid is decomposing.

$$H_2SO_4(l) \xrightarrow{\Delta} SO_3(g) + H_2O(g) \tag{80}$$

(Note that SO_3 and HNO_3 vapors are highly corrosive, and this evaporation must be done under the hood.) Pouring the residue into water displaces the equilibrium

$$Pb^{2+} + HSO_4^- + H_2O(l) \rightleftharpoons \underset{white}{PbSO_4(s)} + H_3O^+ \tag{81}$$

Procedure 7. The amphoteric nature of lead(II) is responsible for its ready solubility in a high concentration of hydroxide ion. The reaction of $[Pb(OH)_4]^{2-}$ with acetic acid yields Pb^{2+} in solution. Lead(II) chromate cannot be precipitated from strongly alkaline solution because the $[Pb^{2+}]$ in equilibrium with the hydroxo complex is lower than the $[Pb^{2+}]$ that would be in equilibrium with the chromate.

Procedure 8. This procedure serves a double purpose: (1) small quantities of lead(II) ion or mercury(II) ion that remain are removed as hydroxides and discarded and (2) the presence of copper(II) ion is indicated by the color of the $[Cu(NH_3)_4]^{2+}$ complex ion. If a deep blue color results, it is unnecessary to carry out Procedure 9. However, small quantities of copper(II) ion cannot be positively detected in this way.

Procedure 9. Neither of the two confirmatory tests given is effective unless the $[Cu(NH_3)_4]^{2+}$ ion is first converted by acid to the Cu^{2+} ion. Either or both tests can then be used. The complex $[Cu(NC_5H_5)_2NCS)_2]$ can be extracted into chloroform ($CHCl_3$) to give a green solution.

Procedure 10. Acetic acid is strong enough to destroy the $[SbS_2]^-$ and $[SnS_3]^{2-}$ ions, but too weak to dissolve the resulting sulfides. Polysulfide ion reacts to give H_2S and elemental sulfur.

$$S_x^{2-} + 2H^+ \longrightarrow H_2S(g) + (x-1)S(s) \tag{82}$$

Hydrochloric acid redissolves the sulfides of antimony and tin but does not dissolve precipitated sulfur. Heating removes liberated hydrogen sulfide and prevents reprecipitation of the sulfides.

Procedure 11. The thiosulfate ion serves as a source of sulfide ion

$$S_2O_3^{2-} + H_2O(l) \underset{}{\overset{H^+}{\rightleftharpoons}} SO_4^{2-} + H_2S(aq) \tag{83}$$

If solid $Na_2S_2O_3$ is added to the solution in a small test tube, two layers develop, with orange $Sb_2OS_2(s)$ forming at the interface where the ions come into contact. Tin(IV) does not interfere. A white to yellow precipitate of sulfur always forms, but is hidden if antimony(III) is present.

Procedure 12. Elemental aluminum reduces tin(IV) to elemental tin, which is then oxidized by hydrogen ion to tin(II). The oxidation of $Sn(s)$ by hydrogen ion occurs *only after all of the aluminum has dissolved*, because the $Al(s)$ is oxidized first, taking preference over $Sn(s)$ oxidation. The tin(II) ion reacts as noted in Procedure 5, except that mercury(II) is always in excess and only white $Hg_2Cl_2(s)$ forms. Antimony(III) is reduced by aluminum to elemental antimony, but the latter is not oxidized by hydrogen ion.

EXAMPLE 31.3

■ A colorless solution containing ions of Cation Group II in dilute HCl was treated with 3% H_2O_2 and the solution was then boiled. After adjustment of the acidity to about $0.3M$, thioacetamide was added and the solution heated. A black precipitate formed. Addition of concentrated aqueous ammonia and thioacetamide, followed by heating, left a precipitate that was still black (A) and a yellow solution (B). Precipitate A dissolved in warm dilute HNO_3, leaving behind a small amount of a whitish-yellow material that proved to be sulfur. The resulting solution gave a white precipitate with dilute H_2SO_4. When solution B was made acidic with dilute acetic acid, a yellow precipitate formed. This precipitate was treated with concentrated HCl to give a solution (C), and the small residue of sulfur was discarded. Addition of powdered aluminum to solution C and filtration into a saturated

$HgCl_2$ solution gave a white precipitate. Identify the ions present in the original solution.

The ions present were Pb^{2+} and Sn^{2+} or Sn^{4+} (or both of the latter two ions).

The fact that the original solution was colorless shows that Cu^{2+} ion was absent. The solubility of the black precipitate A in warm dilute HNO_3 eliminated Hg^{2+}. The formation of the white precipitate with dilute H_2SO_4 showed Pb^{2+} to be present. The yellow precipitate produced when solution B was treated with acetic acid was probably SnS_2. If Sb^{3+} had been present, the precipitate would have had an orange (Sb_2S_3) color. The white precipitate (Hg_2Cl_2) produced when solution C was reduced by aluminum and added to $HgCl_2$ solution confirmed the presence of tin

$$Sn^{2+} + 2Hg^{2+} + 2Cl^- \longrightarrow Hg_2Cl_2(s) + Sn^{4+}$$

■

EXAMPLE 31.4

■ Name a chemical reagent which will
  (a) dissolve SnS_2, but not PbS;
  (b) oxidize Sn^{2+} to Sn^{4+}, but not Sb^{3+} to Sb^{5+};
  (c) distinguish between Hg_2^{2+} and Hg^{2+} ions;
  (d) precipitate Pb^{2+}, but not Cu^{2+}.

(a) Sulfide ion, S^{2-}, will convert SnS_2 to the $[SnS_3]^{2-}$ complex, but will not react with PbS.

(b) In an acidic solution, 3% H_2O_2 will oxidize Sn^{2+} to Sn^{4+}, but not Sb^{3+} to Sb^{5+}. The Sb^{5+} ion is a strong oxidizing agent in aqueous solution and is readily reduced to Sb^{3+}.

(c) The Hg_2^{2+} ion is a member of Cation Group I and is precipitated as Hg_2Cl_2 by Cl^- ion. The Hg^{2+} ion forms water-soluble $HgCl_2$ with Cl^-.

(d) Dilute sulfuric acid will precipitate Pb^{2+} as white $PbSO_4$ but will leave Cu^{2+} in solution. Aqueous ammonia in excess also precipitates Pb^{2+}, but forms a complex ion with Cu^{2+}.

$$Pb^{2+} + 2NH_3 + 2H_2O \longrightarrow Pb(OH)_2(s) + 2NH_4^+ \qquad (84)$$
$$\text{white}$$
$$Cu^{2+} + 4NH_3(aq) \longrightarrow [Cu(NH_3)_4]^{2+} \qquad (85)$$
$$\textit{deep blue}$$

■

Cation Group III (Zn^{2+}, Mn^{2+}, Fe^{2+} or Fe^{3+}, Co^{2+}, Ni^{2+}, Al^{3+}, Cr^{3+})

The ions of Cation Group III are all precipitated by an ammonia/ammonium chloride-buffered hydrogen sulfide solution. This group has been called the basic hydrogen sulfide group, also the aluminum-iron group. The sulfides that did not precipitate in Group II appear here. These sulfides have larger K_{sp} values than the Group II cation sulfides and therefore require for precipitation the higher concentration of S^{2-} available in an alkaline hydrogen sulfide solution (pH about 9). None of the cations that remain form slightly soluble sulfides.

Two of the ions in this group, aluminum(III) and chromium(III), form very slightly soluble hydroxides. The $[OH^-]$ available in the group reagent ammonia solution is more than enough to almost completely precipitate these hydroxides. Why do the hydroxides form and not the sulfides? This is another case where the species is formed that is in equilibrium with the smallest cation concentration in solution (Section 30.3). The only remaining cation in our scheme that forms a precipitable hydroxide is magnesium. However, its hydroxide is much more soluble than those of aluminum and chromium, and for precipitation would require a larger $[OH]^-$ than is available in the presence of NH_4^+ ion.

With the exception of aluminum, which is a representative metal, and zinc, which has a filled d electron level, the elements in this group are transition metals and their atoms have incompletely filled d electron levels. This leads to a variety of oxidation states. Some of the commonly encountered ions and compounds of elements of Cation Group III are listed in Table 31.1.

The presence of unpaired electrons leads to a delightful array of colors for ions of the elements in this group (Table 31.2). Most of these elements have a great tendency to form complex ions. Careful observation of the color of the unknown solution and of colors produced along the way can provide useful clues in the analysis of this group. But remember, as we pointed out earlier (Section 22.20), colors can be misleading.

(The metals and ions of Cation Group III have been discussed in previous chapters: Al and Al^{3+}, Zn and Zn^{2+} in Chapter 26; Cr and Cr^{3+}, Mn and Mn^{2+}, Fe and Fe^{2+} or Fe^{3+}, Co and Co^{2+}, Ni and Ni^{2+} in Chapter 29.)

31.11 Zinc(II) ion

Zinc has the $+2$ oxidation state in all of its compounds; oxidation beyond $+2$ is not possible. Elemental zinc is a moderately strong reducing agent. Zinc compounds are substantially covalent when anhydrous, but are saltlike if hydrated. They yield the colorless, hydrated Zn^{2+} ion. Zinc hydroxide precipitates from a slightly alkaline solution, but dissolves as the $[OH^-]$ increases due to the formation of hydroxo complexes. The hydroxide also dissolves in aqueous ammonia by formation of the complex ion, $[Zn(NH_3)_4]^{2+}$. Zinc hydroxide is amphoteric, and in acidic solution yields the Zn^{2+} ion

$$Zn(OH)_2(s) + H^+ \rightleftharpoons Zn^{2+} + H_2O \tag{86}$$

The Zn^{2+} ion forms complexes with numerous ligands, but the re-

TABLE 31.1

Common ions, oxides, and hydroxides of the Cation Group III elements[a]

			Oxidation state		
	2+	3+	4+	6+	7+
Al		Al^{3+} Al_2O_3 $Al(OH)_3$ $[Al(OH)_4]^-$			
Cr		Cr^{3+} Cr_2O_3 $[Cr(OH)_4]^-$		CrO_4^{2-} $Cr_2O_7^{2-}$ CrO_3	
Mn	Mn^{2+} $Mn(OH)_2$	$MnO(OH)$	MnO_2	MnO_4^{2-}	MnO_4^-
Fe	Fe^{2+} $Fe(OH)_2$ $[Fe(CN)_6]^{4-}$	Fe^{3+} $FeO(OH)$ Fe_2O_3 $[Fe(CN)_6]^{3-}$			
Co	Co^{2+} $Co(OH)_2$ $[Co(NH_3)_6]^{2+}$	$Co[OH_3]$ $[Co(NH_3)]^{3+}$ $[Co(NO_2)_6]^{3-}$ $[Co(CN)_6]^{3-}$			
Ni	Ni^{2+} $Ni(OH)_2$ $[Ni(NH_3)_6]^{2+}$	$NiO(OH)$ Ni_2O_3	NiO_2		
Zn	Zn^{2+} $Zn(OH)_2$ $[Zn(NH_3)_4]^{2+}$ $[Zn(OH)_4]^{2-}$				

[a] Adapted from E. J. King, *Ionic Reactions and Separations*, Harcourt Brace Jovanovich, New York, 1973, p. 165, Table 9.2.

TABLE 31.2

Colors of ions of Cation Group III elements[a]

$[Al(H_2O)_6]^{3+}$	Colorless	$Fe(H_2O)_6^{3+}$	Pale violet
$[Cr(H_2O)_6]^{3+}$	Blue-violet	$FeOH(H_2O)_5^{2+}$	Amber
$[CrCl(H_2O)_5]^{2+}$	Green	$FeCl(H_2O)_5$	Yellow
CrO_4^{2-}	Yellow	$FeSCN^{2+}$	Red (blood)
$Cr_2O_7^{2-}$	Orange	$Mn(H_2O)_6^{2+}$	Very pale pink
$[Co(H_2O)_6]^{2+}$	Rose red	MnO_4^{2-}	Deep green
$[Co(NH_3)_6]^{2+}$	Tan	MnO_4^-	Purple
$[Co(NCS)_4]^{2-}$	Blue-green	$Ni(H_2O)_6^{2+}$	Pale green
$[Co(DMG)_3]^{-b}$	Brown	$Ni(NH_3)_6^{2+}$	Violet blue
$[Co(NH_3)_5(H_2O)]^{3+}$	Red	$Zn(H_2O)_4^{2+}$	Colorless
$[Fe(H_2O)_6]^{2+}$	Pale green	$Zn(NH_3)_4^{2+}$	Colorless

[a] From Hogness, Johnson, and Armstrong, *Qualitative Analysis and Chemical Equilibrium*, Holt, Rinehart and Winston, New York, 1966, 5th ed., p. 418, Table 19.2.
[b] DMG = dimethylglyoxime.

[31.11] QUALITATIVE ANALYSIS II.

sulting species have much larger dissociation constants than the corresponding mercury(II) complexes. Association with halide ions is very weak. The species $[Zn(NH_3)_4]^{2+}$ $(K_d = 3.4 \times 10^{-10})$ and $[Zn(CN)_4]^{2-}$ $(K_d = 2.4 \times 10^{-20})$ are more stable than the halide complexes.

Water-soluble zinc salts include the acetate, bromide, chloride, iodide, nitrate, sulfate, and thiocyanate. Common water-insoluble compounds are the carbonate, hydroxide, double potassium hexacyanoferrate(II) $(K_2 Zn_3[Fe(CN)_6]_2)$ and sulfide (Table 30.3).

Reactions Important in the Separation and Identification of Zn^{2+}

Group precipitation

$$Zn^{2+} + 4NH_3(aq) \xrightarrow{\text{excess } NH_3} \underset{colorless}{[Zn(NH_3)_4]^{2+}} \qquad (87)$$

$$[Zn(NH_3)_4]^{2+} + S^{2-} \rightleftharpoons \underset{white}{ZnS(s)} + 4NH_3(aq) \qquad (88)$$

Dissolution in acid

$$ZnS(s) + 2H^+ \rightleftharpoons Zn^{2+} + H_2S(g) \qquad (89)$$

Complex formation

$$Zn^{2+} + 4OH^- \xrightarrow{\text{excess } OH^-} [Zn(OH)_4]^{2-} \qquad (90)$$

Confirmatory test

Repetition of the group precipitation reactions to give $ZnS(s)$

31.12 Aluminum(III) ion

Only aluminum(III) species are stable at ordinary temperatures and in aqueous solution. Anhydrous aluminum(III) compounds are generally covalent, but when hydrated or dissolved in water, they yield the colorless, hydrated aluminum ion. Because of its large charge and comparatively small size, this ion reacts extensively with water. However, only in the presence of such strongly basic anions as CO_3^{2-}, CN^-, or S^{2-} does hydrolysis result in the precipitation of the hydroxide. Aluminum hydroxide is amphoteric and dissolves in alkaline solutions above a pH of about 10 with the formation of $Al(OH)_4^-$. When first precipitated, aluminum hydroxide is a gelatinous substance that is hydrated. Upon standing in an open container, it gradually loses its water of hydration. The oxide, Al_2O_3, is obtained by heating the hydroxide.

Only a few complexes of aluminum are important in qualitative analysis, such as, $[Al(OH)_4]^-$, $[AlF_6]^{3-}$, and $[Al(H_2O)_6]^{3+}$. Aluminon, the ammonium salt of aurintricarboxylic acid, gives an insoluble red complex with the Al^{3+} ion. This red complex is one of a type of colored complexes called "lakes." The term originated in the dye industry as the name for the colored substance formed by adding a metal hydroxide to an animal or vegetable dye. A lake is a colored precipitate, usually produced by making a solution of a dye and a metal ion alkaline. The ratio of dye to metal ion in the lake is variable. The aluminon lake with Al^{3+} can be represented as follows to show that two bonds to aluminum are associated with each dye molecule.

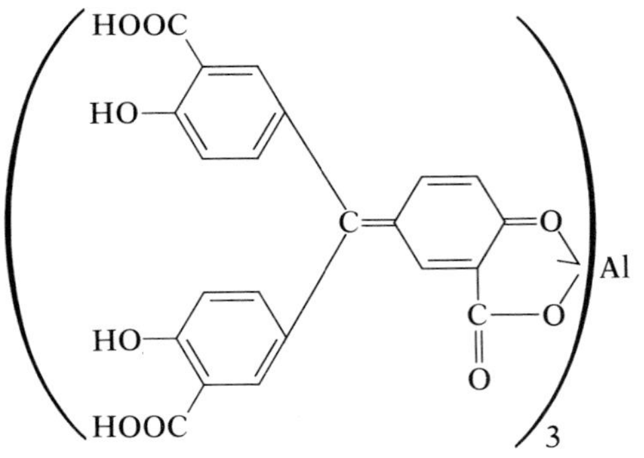

Ammine complex species of aluminum are unknown, and aqueous ammonia precipitates the hydroxide. The aluminum(III) ion is so weak an oxidizing agent that it gives no significant redox reactions in aqueous solutions.

Water-soluble aluminum(III) compounds include the acetate, bromide, chloride, iodide, nitrate, perchlorate, sulfate, and thiocyanate. Water-insoluble compounds include the hydroxide and phosphate.

Reactions Important in the Separation and Identification of Al^{3+}

Group precipitation

$$Al^{3+} + 3OH^- \rightleftharpoons Al(OH)_3(s) \tag{91}$$
$$\text{white}$$

Dissolution in acid

$$Al(OH)_3(s) + 3H^+ \rightleftharpoons Al^{3+} + 3H_2O(l) \tag{92}$$

Complex formation and dissociation

$$Al^{3+} + 4OH^-(xs) \rightleftharpoons [Al(OH_4)]^- \tag{93}$$
$$\text{alkaline}$$
$$[Al(OH)_4]^- + 4H^+(xs) \xrightarrow{\text{soln.}} Al^{3+} + 4H_2O(l) \tag{94}$$

Confirmatory test

$$Al^{3+} + \text{aluminon} \longrightarrow \text{red precipitate}$$

31.13 Chromium(III) ion

Both chromium(II) and chromium(III) can exist as cations. However, standard potential data

$$Cr^{3+} + e^- \longrightarrow Cr^{2+} \qquad\qquad E° = -0.41 \text{ V} \qquad (95)$$
$$Cr(OH)_3(s) + e^- \longrightarrow Cr(OH)_2(s) + OH^- \qquad E° = -1.1 \text{ V} \qquad (96)$$

indicate clearly that chromium(II) is readily oxidized. Exposure of an aqueous solution containing chromium(II) to air is sufficient to oxidize the chromium(II) ions.

Anhydrous chromium(III) compounds are covalent, but the hydrated compounds are saltlike, and aqueous solutions contain hydrated cations. The ion $[Cr(H_2O)_6]^{3+}$ is violet, but in the presence of chloride ion, the green complex ions $[Cr(H_2O)_5Cl]^{2+}$ and $[Cr(H_2O)_4Cl_2]^+$ form. Chromium(III) probably forms more stable complex species than any other common cation except cobalt(III). Many of these complexes are distinctively colored and have very small dissociation constants. Common examples are $[Cr(NH_3)_6]^{3+}$, yellow; $[Cr(NH_3)_5Cl]^{2+}$, red; $[Cr(NH_3)_5(OH)]^{2+}$, pink; and $[Cr(NH_3)_2(SCN)_4]^-$, red.

Water-soluble chromium(III) salts include the acetate, bromide, chloride, iodide, nitrate, perchlorate, and sulfate. Difficulty soluble compounds include the hydroxide and the phosphate. The hydroxide is amphoteric and is readily soluble in strongly alkaline solutions and in acidic solutions. Its solubility is small enough ($K_{sp} = 6 \times 10^{-31}$) that strongly basic anions such as S^{2-} and CO_3^{2-} yield sufficient hydroxide ion concentration by reaction with water to precipitate it.

$$2Cr^{3+} + 3S^{2-} + 6H_2O(l) \longrightarrow 2Cr(OH)_3(s) + 3H_2S(g) \qquad (97)$$
$$2Cr^{3+} + 3CO_3^{2-} + 3H_2O(l) \longrightarrow 2Cr(OH)_3(s) + 3CO_2(g) \qquad (98)$$

Oxidation to chromium(VI) is much easier to carry out in alkaline media than in acidic media, as indicated by standard potential data

$$Cr_2O_7^{2-} + 14H^+ + 6e^- \longrightarrow 2Cr^{3+} + 7H_2O \qquad E° = +1.33 \text{ V} \qquad (99)$$
$$CrO_4^{2-} + 4H_2O + 3e^- \longrightarrow Cr(OH)_3 + 5OH^- \qquad E° = -0.12 \text{ V} \qquad (100)$$

We have discussed the dichromate and chromate anions in Section 22.21.

Reactions Important in the Separation and Identification of Cr^{3+}

Group precipitation

$$Cr^{3+} + 3OH^- \rightleftharpoons Cr(OH)_3(s) \qquad (101)$$
$$\text{gray-green}$$

Dissolution in acid

$$Cr(OH)_3(s) + 3H^+ \longrightarrow Cr^{3+} + 3H_2O(l) \qquad (102)$$

Complex formation and oxidation

$$Cr^{3+} + 4OH^-(xs) \underset{}{\overset{}{\rightleftharpoons}} [Cr(OH)_4]^- \qquad (103)$$
$$\text{dark green}$$

$$2[Cr(OH)_4]^- + 2OH^- + 3H_2O_2(aq) \xrightarrow{OH^-} 2CrO_4^{2-} + 8H_2O(l) \qquad (104)$$
$$\text{yellow}$$

$$2CrO_4^{2-} + 2H^+ \rightleftharpoons Cr_2O_7^{2-} + H_2O(l) \qquad (105)$$
$$\text{orange}$$

Confirmatory tests

$$Ba^{2+} + CrO_4^{2-} \rightleftharpoons BaCrO_4(s) \qquad (106)$$
$$\text{yellow}$$
$$2BaCrO_4(s) + 2H^+ \rightleftharpoons 2Ba^{2+} + Cr_2O_7^{2-} + H_2O(l) \qquad (107)$$
$$Cr_2O_7^{2-} + 4H_2O_2(aq) + 2H^+ \rightleftharpoons 2CrO_5 \ (ether) + 5H_2O(l) \qquad (108)$$
$$\text{blue}$$

31.14 Manganese(II) ion

Manganese(II) and manganese(III) both exist as cations. From the standard potential data, however, it is clear that manganese(II) is resistant to both oxidation and reduction and is therefore the stable species in aqueous solution.

$$Mn^{2+} + 2e^- \longrightarrow Mn \qquad\qquad E° = -1.029\ V \qquad (109)$$
$$Mn(OH)_2(s) + 2e^- \longrightarrow Mn + 2OH^- \qquad E° = -1.47\ V \qquad (110)$$
$$Mn^{3+} + e^- \longrightarrow Mn^{2+} \qquad\qquad E° = +1.51\ V \qquad (111)$$
$$MnO_2(s) + 4H^+ + 2e^- \longrightarrow Mn^{2+} + 2H_2O \qquad E° = +1.208\ V \qquad (112)$$
$$MnO_4^- + 8H^+ + 5e^- \longrightarrow Mn^{2+} + 4H_2O \qquad E° = +1.491\ V \qquad (113)$$

Manganese(II) ion does not react extensively with water. Its hydroxide is among the more soluble and more basic of the precipitable hydroxides, and in these respects it closely resembles magnesium hydroxide. Manganese(II) ion is pale pink in hydrated salts and in aqueous solution, but the color is so delicate as to be undetectable in dilute solutions. The ion forms few stable complexes. Water-soluble compounds include the acetate, bromide, chloride, iodide, nitrate, sulfate, and thiocyanate. The hydroxide, carbonate, and sulfide are insoluble in water, but the sulfide is soluble in acidic solutions. Manganese(II) compounds resemble magnesium compounds very closely. Like magnesium hydroxide, manganese(II) hydroxide fails to precipitate in the presence of moderate concentrations of ammonium ion.

Oxidation is effected only by powerful oxidants. Conversion to purple MnO_4^- ion by oxidants like sodium bismuthate(V) or lead(IV) oxide is an excellent confirmatory reaction. Oxidation to black, insoluble MnO_2 by chlorate ion in nitric acid solution is specific to Mn^{2+} ion and is an excellent means of separating manganese from other species in this cation group. The manganese(II) ion does not produce a flame color.

Reactions Important in the Separation and Identification of Mn^{2+}

Group precipitation

$$Mn^{2+} + NH_3(aq) + NH_4^+ \longrightarrow \text{no observed reaction} \qquad (114)$$
$$Mn^{2+} + S^{2-} \rightleftharpoons MnS(s) \qquad (115)$$
$$\text{pink}$$

Dissolution in acid

$$MnS(s) + 2H^+ \longrightarrow Mn^{2+} + H_2S(g)$$

Oxidation and reduction of Mn (Procedures 3 and 4)

$$Mn(OH)_2(s) + H_2O_2(aq) \longrightarrow MnO_2(s) + 2H_2O \qquad (116)$$
$$\text{white} \qquad\qquad\qquad \text{black}$$
$$MnO_2(s) + H_2O_2(aq) + 2H^+ \longrightarrow Mn^{2+} + 2H_2O(l) + O_2(g) \qquad (117)$$

Oxidation and reduction of Mn (Procedures 5 and 6)

$$3Mn^{2+} + 3H_2O(l) + ClO_3^- \xrightarrow[\Delta]{HNO_3} 3MnO_2(s) + 6H^+ + Cl^- \qquad (118)$$
$$\text{black}$$
$$MnO_2(s) + HNO_2 + H^+ \longrightarrow Mn^{2+} + H_2O(l) + NO_3^- \qquad (119)$$

Confirmatory test

$$2Mn^{2+} + 14H^+ + 5NaBiO_3(s) \xrightarrow[\Delta]{HNO_3}$$
$$2MnO_4^- + 5Bi^{3+} + 7H_2O(l) + 5Na^+ \qquad (120)$$
$$\text{purple to} \atop \text{violet}$$

31.15 Iron(II) and iron(III) ions

Both Fe^{2+} and Fe^{3+} cations are encountered in aqueous solution, where they are hydrated, and in a variety of compounds. Pertinent standard potential data are

$$Fe^{2+} + 2e^- \longrightarrow Fe \qquad\qquad E° = -0.409 \text{ V} \qquad (121)$$
$$Fe(OH)_2 + 2e^- \longrightarrow Fe + 2OH^- \qquad E° = -0.877 \text{ V} \qquad (122)$$
$$Fe^{3+} + e^- \longrightarrow Fe^{2+} \qquad\qquad E° = +0.770 \text{ V} \qquad (123)$$
$$Fe(OH)_3 + e^- \longrightarrow Fe(OH)_2 + OH^- \quad E° = -0.56 \text{ V} \qquad (124)$$

Oxidation of elemental iron to iron(II) and of iron(II) to iron(III) are both easier to carry out under alkaline conditions than under acidic conditions. Elemental iron reacts with H^+ ion to form Fe^{2+} ion, but conversion of Fe^{2+} to Fe^{3+} ion under acidic conditions requires a much stronger oxidizing agent. Because of the reduction of iron(III) by elemental iron,

$$Fe(s) + 2Fe^{3+} \rightleftharpoons 3Fe^{2+} \qquad K = 7.3 \times 10^{39} \qquad (125)$$

addition of elemental iron to an iron(II) salt solution prevents the oxidation of the solution by reducing any Fe^{3+} that is formed.

Iron(II) compounds closely resemble those of manganese(II), cobalt(II), and nickel(II). Iron(III) compounds most closely resemble those of aluminum(III) and chromium(III). In aqueous solution, the hydrated iron(II) ion is green, but the color is apparent only in concentrated solution. The hydrated iron(III) ion probably has a faint violet color, but in solution sufficient colloidal iron(III) oxide is formed to impart a yellow or even reddish color. Neither hydroxide dissolves in excess OH^- ion, but the iron(III) compound is more acidic than the iron(II) compound. Both cations form a variety of complexes. Characteristic ones are the anions $[Fe(CN)_6]^{4-}$, hexacyanoferrate(II), which is yellow; $[Fe(CN)_6]^{3-}$, hexacyanoferrate(III), which is red; and $Fe(NCS)^{2+}$, which is red. Ammine complexes do not form in aqueous solution.

Water-soluble iron(II) and iron(III) compounds include the acetates, chlorides, bromides, nitrates, perchlorates, and sulfates. Water-insoluble iron(II) and iron(III) compounds include the hydroxides, phosphates, and sulfides. Iron(II) carbonate is insoluble, but, as with the Al^{3+} and Cr^{3+} ions, reaction of CO_3^{2-} ion with Fe^{3+} ion yields the hydroxide as a consequence of hydrolysis. Under acidic conditions, reaction with hydrogen sulfide or thioacetamide reduces iron(III) to iron(II) but does not precipitate iron(II) sulfide. Under alkaline conditions, however, both FeS and Fe_2S_3 are precipitated.

Either the Fe^{2+} or the Fe^{3+} ion or both ions can be present in a sample received for analysis. Any Fe^{2+} is ultimately oxidized in the Cation Group III analysis to Fe^{3+}, which is separated and identified.

Reactions Important in the Separation and Identification of Fe^{2+} and Fe^{3+}

Group precipitation

$$Fe^{2+} + 2OH^-(xs) \rightleftharpoons Fe(OH)_2(s) \qquad (126)$$
$$\text{\textit{green} } \longrightarrow \text{\textit{black} } \longrightarrow \text{\textit{reddish brown on}}$$
$$\text{\textit{exposure to air}}$$

$$Fe^{3+} + 3OH^-(xs) \rightleftharpoons Fe(OH)_3(s) \qquad (127)$$
$$\text{\textit{reddish brown}}$$

$$Fe(OH)_2(s) + S^{2-} \xrightarrow{OH^-} FeS(s) + 2OH^- \qquad (128)$$
$$\text{\textit{black}}$$

$$2Fe(OH)_3(s) + 3S^{2-} \xrightarrow{OH^-} Fe_2S_3(s) + 6OH^- \qquad (129)$$
$$\text{\textit{black}}$$

Dissolution in acid

$$FeS(s) + 2H^+ \longrightarrow Fe^{2+} + H_2S(g) \qquad (130)$$
$$Fe_2S_3(s) + 4H^+ \longrightarrow 2Fe^{2+} + 2H_2S(g) + S(s) \qquad (131)$$

Oxidation

$$3Fe^{2+} + NO_3^- + 4H^+ \longrightarrow 3Fe^{3+} + NO(g) + 2H_2O(l) \qquad (132)$$

Confirmatory tests

$$Fe^{3+} + K^+ + [Fe(CN)_6]^{4-} \underset{\longleftarrow}{\longrightarrow} \underset{\text{dark blue}}{KFe[Fe(CN)_6](s)} \qquad (133)$$

$$Fe^{3+} + NCS^- \underset{\longleftarrow}{\longrightarrow} \underset{\text{blood red}}{[Fe(NCS)]^{2+}} \qquad (134)$$

31.16 Cobalt(II) ion

Although both cobalt(II) and cobalt(III) are known in many compounds, only cobalt(II) is stable as the simple hydrated ion in aqueous solution. In the presence of many complexing ligands, however, cobalt(II) is readily oxidized to cobalt(III), and the cobalt(III) complexes are among the most stable and most numerous of all known complexes. These relationships are indicated by standard potential data such as

$$
\begin{array}{lll}
Co^{3+} + e^- \longrightarrow Co^{2+} & E° = +1.842\ V & (135) \\
Co(OH)_3(s) + e^- \longrightarrow Co(OH)_2(s) + OH^- & E° = +0.17\ V & (136) \\
[Co(NH_3)_6]^{3+} + e^- \longrightarrow [Co(NH_3)_6]^{2+} & E° = +0.11\ V & (137) \\
[Co(CN)_6]^{3-} + e^- \longrightarrow [Co(CN)_6]^{4-} & E° = -0.83\ V & (138)
\end{array}
$$

Thus, we are concerned primarily with reactions of the Co^{2+} ion, but oxidation in the presence of OH^- ion or complexing groups is readily effected, even by atmospheric oxygen.

In many of its reactions, the rose-colored Co^{2+} ion so closely resembles the pale green Fe^{2+} and bright green Ni^{2+} ions that separations are not easy to carry out. Water-soluble cobalt(II) compounds include the acetate, bromide, chloride, iodide, nitrate, sulfate, and thiocyanate. Water-insoluble compounds include the hydroxide, carbonate, and sulfide. Two forms of the sulfide exist—initial precipitation gives black α–CoS, which is readily soluble in $6M$ HCl, but on standing this form converts spontaneously to black β-CoS, which is only very slightly soluble in $6M$ HCl. Cobalt(II) ion forms numerous complex ions, for example, $[Co(NH_3)_6]^{2+}$, tan; $[CoCl_4]^{2-}$, blue; and $[Co(CN)_6]^{4-}$, brown. In aqueous solution, these species are generally unstable with respect to either conversion to the hydrated species, for example,

$$\underset{\text{blue}}{[CoCl_4]^{2-}} + 6H_2O(l) \underset{\longleftarrow}{\longrightarrow} \underset{\text{rose}}{[Co(H_2O)_6]^{2+}} + 4Cl^- \qquad (139)$$

or oxidation, for example,

$$\underset{\text{tan}}{4[Co(NH_3)_6]^{2+}} + 2H_2O(l) + O_2(g) \underset{\longleftarrow}{\longrightarrow}$$

$$\underset{\text{yellow}}{4[Co(NH_3)_6]^{3+}} + 4OH^- \qquad (140)$$

Cobalt(II) ion is not extensively hydrolyzed, and its hydroxide is not amphoteric.

Reactions Important in the Separation and Identification of Co²⁺

Group precipitation

$$Co^{2+} + 2OH^-(xs) \rightleftharpoons Co(OH)_2(s) \tag{141}$$

rose   *blue* → *pink*

$$Co(OH)_2(s) + 6NH_3(aq) \rightleftharpoons [Co(NH_3)_6]^{2+} + 2OH^- \tag{142}$$

tan

$$[Co(NH_3)_6]^{2+} + S^{2-} \rightleftharpoons CoS(s) + 6NH_3(aq) \tag{143}$$

black

Dissolution by oxidation of S^{2-}

$$3CoS(s) + 8H^+ + 2NO_3^- \longrightarrow$$
$$3Co^{2+} + 2NO(g) + 4H_2O(l) + 3S(s) \tag{144}$$

Oxidation

$$4Co(OH)_2(s) + 2H_2O_2(aq) \xrightarrow[\text{neutral soln.}]{\substack{\text{alkaline} \\ \text{or}}} 4Co(OH)_3(s) \tag{145}$$

blue or pink   *black*

Reduction

$$2Co(OH)_3 + 4H^+ + H_2O_2(aq) \xrightarrow[\text{soln.}]{\text{acidic}} 2Co^{2+} + 6H_2O(l) + O_2(g) \tag{146}$$

Confirmatory test

$$Co^{2+} + 4NCS^- \rightleftharpoons [Co(NCS)_4]^{2-} \tag{147}$$

blue to blue-green

31.17 Nickel(II) ion

Only the $+2$ oxidation state of nickel is known in solution. One example of a higher oxidation state is found in the black solid formulated variously as Ni_2O_3, $NiO \cdot NiO_2$, or NiO_2,

$$NiO_2 + H_2O + 2e^- \longrightarrow Ni(OH)_2 + 2OH^- \qquad E^\circ = +0.490 \text{ V} \tag{148}$$

However, in solution in acids NiO_2 gives the Ni^{2+} ion. Except in resistance to oxidation and in color, nickel(II) compounds resemble those of cobalt(II), and most of what was said about cobalt in the preceding section also applies to nickel. A precipitate of nickel(II) sulfide, unlike cobalt(II) sulfide, is readily converted into a brown colloidal sol by sulfide ion in alkaline solutions. A sol of this type is flocculated by adding an ammonium salt and then heating. Nickel(II) complexes closely resemble those of cobalt(II), but do not undergo atmospheric oxidation. A complex characteristic of the nickel(II) ion is the bright red dimethylglyoxime (HDMG) derivative, with two DMG$^-$ ions for each Ni^{2+}

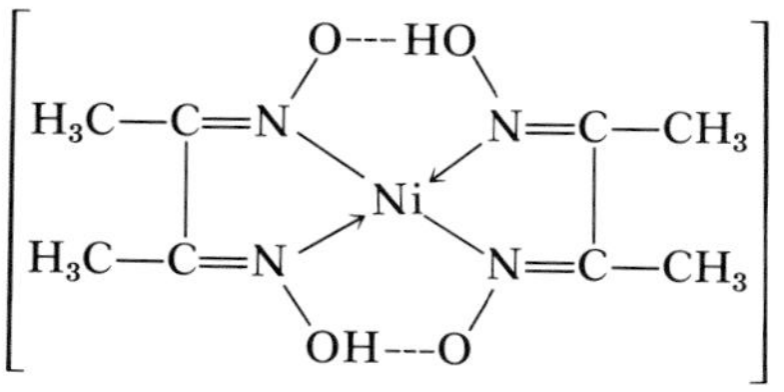

Precipitation of this compound from a buffered acetate solution can detect 5×10^{-5} mole Ni^{2+}/liter without interference from other common cations.

Reactions Important in the Separation and Identification of Ni^{2+}

Group precipitation

$$Ni^{2+} + 2OH^-(xs) \rightleftharpoons Ni(OH)_2(s) \tag{149}$$
green   green
$$Ni(OH)_2(s) + 6NH_3 \rightleftharpoons [Ni(NH_3)_6]^{2+} + 2OH^- \tag{150}$$
                                    deep blue
$$[Ni(NH_3)_6]^{2+} + S^{2-} \rightleftharpoons NiS(s) + 6NH_3(aq) \tag{151}$$
                            black

Dissolution by oxidation of S^{2-}

$$3NiS(s) + 8H^+ + 2NO_3^- \longrightarrow 3Ni^{2+} + 2NO(g) + 4H_2O(l) + 3S(s) \tag{152}$$

Dissolution in acid

$$Ni(OH)_2(s) + 2H^+ \longrightarrow Ni^{2+} + 2H_2O(l) \tag{153}$$

Confirmatory test

$$Ni^{2+} + 2HDMG(aq) + 2CH_3CO_2^- \rightleftharpoons$$
$$Ni(DMG)_2(aq) + 2CH_3CO_2H(aq) \tag{154}$$

31.18 Cation Group III analysis

The flow chart for Cation Group III is given in Figure 31.4. The procedures identified in Figure 31.4 are discussed in the following paragraphs.

Procedure 1. The centrifugate from Group II is first boiled to remove sulfide ion as hydrogen sulfide so that the rather distinctive reactions with aqueous ammonia can be observed before being obscured by the formation of sulfide precipitates. These reactions are

$$Fe^{2+} \xrightarrow{NH_3} Fe(OH)_2(s) \xrightarrow{NH_3(xs)} \text{no change} \tag{155}$$
                    dark green $\xrightarrow{O_2}$ red brown

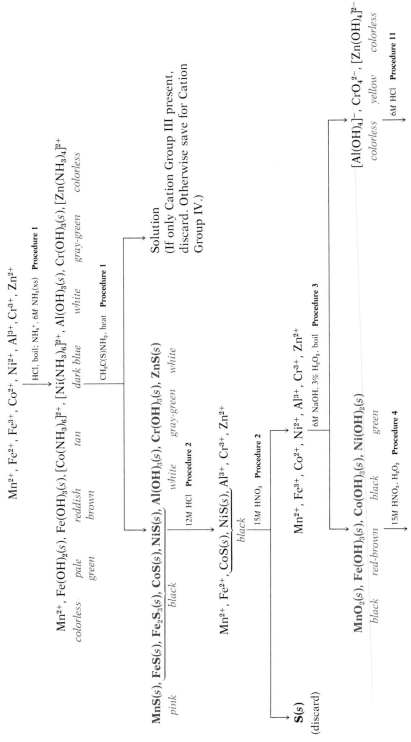

FIGURE 31.4
Flow chart for analysis of Cation Group III. *The initial solution of ions from the separation of Cation Group II is acidic.*

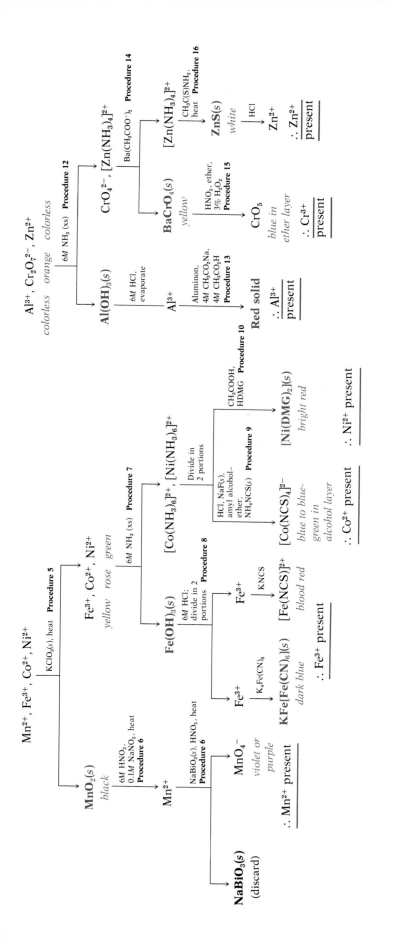

$$Co^{2+} \xrightarrow{NH_3} Co(OH)_2(s) \xrightarrow{NH_3(xs)} [Co(NH_3)_6]^{2+} \qquad (156)$$

blue $\longrightarrow$ pink $\xrightarrow[\text{slowly}]{O_2}$ brown $\qquad$ tan

$$Ni^{2+} \xrightarrow{NH_3} Ni(OH)_2(s) \xrightarrow{NH_3(xs)} [Ni(NH_3)_6]^{2+} \qquad (157)$$
$\qquad\qquad\qquad$ pale green $\qquad\qquad$ deep blue

$$Zn^{2+} \xrightarrow{NH_3} Zn(OH)_2(s) \xrightarrow{NH_3(xs)} [Zn(NH_3)_4]^{2+} \qquad (158)$$
$\qquad\qquad\qquad$ white $\qquad\qquad$ colorless

$$Al^{3+} \xrightarrow{NH_3} Al(OH)_3(s) \xrightarrow{NH_3(xs)} \text{no change} \qquad (159)$$
$\qquad\qquad\qquad$ white

$$Cr^{3+} \xrightarrow{NH_3} Cr(OH)_3(s) \xrightarrow{NH_3(xs)} \text{no change} \qquad (160)$$
$\qquad\qquad\qquad$ gray-green

$Mn(OH)_2$ does not precipitate because the $[OH^-]$ is controlled by the NH_4^+ ion present (see Table 30.7). The precipitation of magnesium hydroxide in this group is also prevented by the presence of ammonium ion (see Table 30.7).

Procedure 2. The β forms of CoS and NiS are insoluble in $12M$ hydrochloric acid, but dissolve when nitric acid is added because of the oxidation of sulfide to elemental sulfur. A black residue remaining after addition of hydrochloric acid suggests the presence of Co^{2+} or Ni^{2+} ion or both (unless you have failed to dissolve all of the FeS or Fe_2S_3).

Procedure 3. This separation is based on the amphoteric properties of aluminum, chromium, and zinc hydroxides as compared to the nonamphoteric properties of the other hydroxides. Slow addition of the sodium hydroxide solution precipitates all of the hydroxides first and then causes dissolution of the amphoteric ones. Careful observations at this point may give useful information when an unknown sample is being analyzed. The appreciable solubility of $Mn(OH)_2$ and the tendency of $[Cr(OH)_4]^-$ ion to yield $Cr(OH)_3(s)$ decrease the effectiveness of a separation involving only sodium hydroxide. Oxidation with hydrogen peroxide yields insoluble, black $MnO_2(s)$ and the stable, distinctively colored CrO_4^{2-} ion. Simultaneously, distinctive red-brown $Fe(OH)_3(s)$ and black $Co(OH)_3(s)$ are produced by oxidation.

Procedure 4. Neither $MnO_2(s)$ nor $Co(OH)_3(s)$ is appreciably soluble in nitric acid. Reduction of each with hydrogen peroxide enhances solubility.

Procedure 5. Oxidation of Mn^{2+} to $MnO_2(s)$ by ClO_3^- ion in acidic solution is a unique reaction and provides adequate proof of the presence of manganese in a sample.

Procedure 6. Manganese(IV) oxide again dissolves in acidic solution only as a consequence of reduction—this time by nitrous acid. Oxidation to intensely colored MnO_4^- ion is also a unique reaction of manganese.

Procedure 7. The separation utilizes the equilibrium

$$NH_3(aq) + H_2O(l) \rightleftharpoons NH_4^+ + OH^- \qquad (161)$$

The Fe^{3+} ions react with the OH^- ions, and the Co^{2+} and Ni^{2+} ions react with the NH_3 molecules. When the ammonia solution is first added, the hydroxides all precipitate. Excess ammonia takes cobalt(II) and nickel(II) back into solution as complexes.

Procedure 8. Reactions of Fe^{3+} ions with $[Fe(CN)_6]^{4-}$ and NCS^- ions are specific and extremely sensitive. The Fe^{2+} ion gives a bluish white precipitate with the $[Fe(CN)_6]^{4-}$ ion and gives no apparent reaction with the NCS^- ion.

Procedure 9. The Ni^{2+} ion does not interfere with the formation of blue to blue-green $[Co(NCS)_4]^{2-}$. Fluoride ion removes any Fe^{3+} ion as the $[FeF_6]^{3-}$ complex, and thus cuts out interference by the red $[Fe(NCS)]^{2+}$ ion.

Procedure 10. Dimethylglyoxime gives only a brown color with the Co^{2+} ion, so that the Ni^{2+} ion can be detected in the presence of the Co^{2+} ion.

Procedure 11. The addition of HCl neutralizes the hydroxo complexes of Al^{3+} and Zn^{2+}, and converts CrO_4^{2-} to $Cr_2O_7^{2-}$.

Procedure 12. The same principle noted for Procedure 7 applies here—the Al^{3+} ions react with the OH^- ions and the Zn^{2+} ions react with the NH_3 molecules. Making the solution alkaline forms the CrO_4^{2-} ion from $Cr_2O_7^{2-}$.

$$Cr_2O_7^{2-} + 2OH^- \underset{}{\rightleftharpoons} 2CrO_4^{2-} + H_2O(l) \qquad (162)$$
orange   yellow

Procedure 13. The formation of the red lake is described in Section 31.12.

Procedure 14. Barium chromate is not as intensely colored as lead(II) chromate (Cation Groups I and II). A yellow acid-soluble precipitate at this point can only be $BaCrO_4$.

Procedure 15. Formation of the blue peroxochromate is inhibited by excess acid, excess peroxide, or heat. Assignment of the formula CrO_5 to this compound is not certain. Even a fleeting blue color confirms the presence of chromium in a sample.

Procedure 16. Zinc sulfide is the only common insoluble white sulfide. It dissolves readily in hydrochloric acid. A faint white cloudiness that is unchanged upon the addition of acid is colloidal sulfur formed by oxidation of sulfide ion by chromate ion.

EXAMPLE 31.5

■ A solution was known to contain the following Group III cations: Mn^{2+}, Fe^{3+}, and Al^{3+}. Show by a flow chart how you would separate these ions.

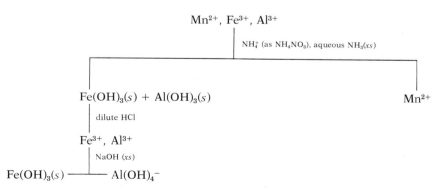

Manganese(II) hydroxide is insoluble, but in the presence of an excess of NH_4^+, insufficient OH^- ion is supplied to precipitate the hydroxide. Separation of Fe^{3+} and Al^{3+} is effected by taking advantage of the amphoteric nature of $Al(OH)_3$. ■

EXAMPLE 31.6

■ Show the chemical reagent or reagents that will separate each of the following pairs of species:
  (a) Pb^{2+} and Zn^{2+}
  (b) Hg^{2+} and Fe^{2+}
  (c) Mn^{2+} and Co^{2+}
  (d) Al^{3+} and $Cr_2O_7^{2-}$

(a) An excess of aqueous ammonia will precipitate Pb^{2+} as $Pb(OH)_2$ and leave Zn^{2+} in solution as the $[Zn(NH_3)_4]^{2+}$ complex.
(b) In a solution $0.3M$ in HCl, Hg^{2+} can be precipitated as HgS, while Fe^{2+} remains in solution. (Hg^{2+} is a member of Cation Group II, and Fe^{2+} is a member of Group III.)
(c) In concentrated nitric acid solution, Mn^{2+} is oxidized to black, insoluble MnO_2 by $KClO_3$, and Co^{2+} is unaffected.
(d) Aqueous ammonia will precipitate $Al(OH)_3$ and convert the $Cr_2O_7^{2-}$ (dichromate) ion to CrO_4^{2-} (chromate) ion.

$$Cr_2O_7^{2-} + 2OH^- \rightleftharpoons 2CrO_4^{2-} + H_2O(l) \qquad (163)$$

■

Cation Group IV (Ca^{2+}, Ba^{2+})

31.19 Calcium(II) and barium(II) ions

Calcium and barium are members of the same periodic family. They have very similar chemical properties and, as a result, are difficult to separate. Because there are only two ions in this cation group, and because as individuals they have few distinctive characteristics, we have combined the discussion of the group and the properties of the individual ions.

The chlorides, sulfides, and hydroxides of barium and calcium are so soluble that these ions are in no danger of being separated with the earlier groups. We precipitate these ions as carbonates with an ammonium chloride/ammonia-buffered ammonium carbonate solution. Sometimes this group of cations is known as the ammonium carbonate group or the alkaline earth group.

Strontium, which we have chosen not to include in our analysis scheme, would also, if present, precipitate as a carbonate in this group. Magnesium is the only ion remaining in our scheme that might interfere here. However, precipitation of $MgCO_3$ is suppressed by the presence of ammonium ion (see Table 30.8).

(Calcium and barium, and their ions, have been discussed in Chapter 26.)

Only the +2 oxidation states of calcium and barium are known. Almost without exception, calcium and barium compounds are ionic. Both cations are colorless and give white or colorless salts unless the anion is colored. Water-soluble salts include the acetates, bromides, chlorides, iodides, perchlorates, and nitrates. Slightly soluble compounds include the carbonates, fluorides, oxalates, and sulfates. The equilibrium concentrations of calcium and barium ions in saturated aqueous solutions *decrease* in the following series:

$$CaCrO_4(s) > Ca(OH)_2(s) > CaSO_4(s) > CaF_2(s) > CaCO_3(s) > CaC_2O_4(s)$$
$$Ba(OH)_2(s) > BaF_2(s) > BaC_2O_4(s) > BaCO_3(s) > BaSO_4(s) > BaCrO_4(s)$$

An analytically useful difference in solubility lies between the two chromates—barium chromate can be precipitated while calcium ion is left in solution. Almost no complex species of calcium and barium are known.

These ions give distinctively different flame colors. The Ca^{2+} flame is brick red, and the Ba^{2+} flame is yellow-green.

Reactions Important in the Separation and Identification of Ba^{2+} and Ca^{2+}

Group precipitation (M = Ba^{2+}, Ca^{2+})

$$M^{2+} + CO_3^{2-} \rightleftharpoons MCO_3(s) \tag{164}$$

Dissolution by acid (M = Ba^{2+}, Ca^{2+})

$$MCO_3(s) + 2H^+ \rightleftharpoons M^{2+} + CO_2(g) + H_2O(l) \tag{165}$$

Confirmatory tests

$$Ba^{2+} + CrO_4^{2-} \rightleftharpoons BaCrO_4(s) \tag{166}$$
$$Ca^{2+} + C_2O_4^{2-} \rightleftharpoons CaC_2O_4(s) \tag{167}$$

31.20 Cation Group IV analysis

The flow chart for Cation Group IV is given in Figure 31.5. The procedures that are identified on the flow chart are discussed in the following paragraphs.

Procedure 1. This procedure is designed to remove ammonium ion primarily by the oxidation-reduction (Section 22.4d)

$$NH_4^+ + NO_3^- \xrightarrow{\Delta} N_2O(g) + 2H_2O(g) \tag{168}$$

It may be omitted if ammonium salts are known to be absent. If the solution used is a centrifugate from Cation Group III and is brown due to colloidal NiS, prior neutralization with acetic acid and evaporation to flocculate and remove this material may be necessary.

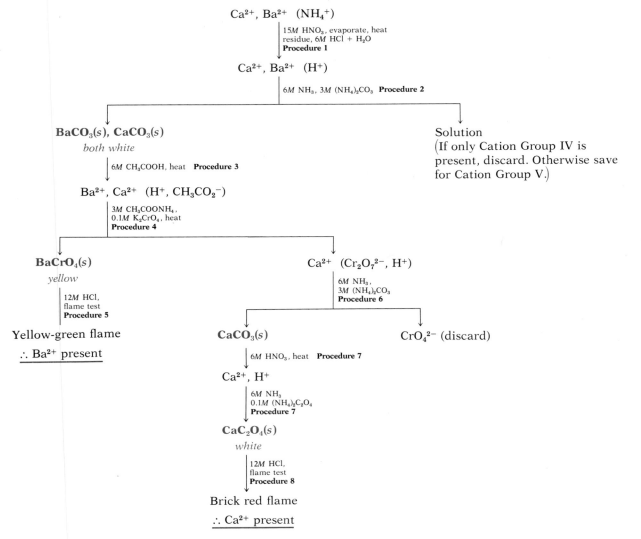

Ca²⁺, Ba²⁺ (NH₄⁺)

15M HNO₃, evaporate, heat
residue, 6M HCl + H₂O
Procedure 1

Ca²⁺, Ba²⁺ (H⁺)

6M NH₃, 3M (NH₄)₂CO₃ **Procedure 2**

BaCO₃(s), CaCO₃(s)
both white

6M CH₃COOH, heat **Procedure 3**

Ba²⁺, Ca²⁺ (H⁺, CH₃CO₂⁻)

3M CH₃COONH₄,
0.1M K₂CrO₄, heat
Procedure 4

BaCrO₄(s)
yellow

12M HCl,
flame test
Procedure 5

Yellow-green flame

∴ Ba²⁺ present

Ca²⁺ (Cr₂O₇²⁻, H⁺)

6M NH₃,
3M (NH₄)₂CO₃
Procedure 6

CaCO₃(s)

6M HNO₃, heat **Procedure 7**

Ca²⁺, H⁺

6M NH₃
0.1M (NH₄)₂C₂O₄
Procedure 7

CaC₂O₄(s)
white

12M HCl,
flame test
Procedure 8

Brick red flame

∴ Ca²⁺ present

CrO₄²⁻ (discard)

Solution
(If only Cation Group IV is
present, discard. Otherwise save
for Cation Group V.)

FIGURE 31.5

**Flow chart for analysis of
Cation Group IV.**

Procedure 2. The concentration of the NH₄⁺ ion is adjusted to prevent precipitation of magnesium carbonate, but allows precipitation of the less soluble carbonates of calcium and barium.

Procedure 3. Acetic acid, rather than a strong acid, is used to permit buffering to an optimum pH in the next step.

Procedure 4. Because of the solubility of BaCrO₄ at low pH, an acetic acid/acetate ion buffer is used to ensure complete precipitation of the Ba²⁺ ion.

Procedure 5. Although the formation of a yellow precipitate in the preceding step is usually adequate proof of the presence of the Ba²⁺ ion, a flame test is more sensitive and provides confirmation. Flame tests are usually made on chlorides because of their larger volatilities. The only other cation in the series studied here that gives a green flame color is the Cu²⁺ ion. That flame is bright green to blue.

Procedure 6. This step is necessary only to separate Ca^{2+} from the yellow CrO_4^{2-} ion that was introduced in Procedure 4.

Procedure 7. Calcium oxalate, CaC_2O_4, is best precipitated from neutral or slightly alkaline solution.

Procedure 8. Although the formation of a precipitate in the previous step proves the presence of the Ca^{2+} ion, the flame test gives added evidence. This flame color is not particularly distinctive and should be compared with that obtained with pure calcium chloride.

EXAMPLE 31.7

■ The Mg^{2+}, Ca^{2+}, and Ba^{2+} ions exhibit similar behavior in the solubilities of their salts; for example, all three form insoluble carbonates. Explain why Ca^{2+} and Ba^{2+} ions can be separated from Mg^{2+} by means of an appropriate NH_4Cl–$(NH_4)_2CO_3$–NH_3 aqueous mixture.

The compounds $CaCO_3$ and $BaCO_3$ (K_{sp}, 4.7×10^{-9} and 1.6×10^{-9}, respectively; Table 30.6) are much less soluble than $MgCO_3$ (K_{sp}, 4.0×10^{-5}). The CO_3^{2-} concentration in solution is controlled with NH_4^+ ion:

$$NH_4^+ + CO_3^{2-} \rightleftharpoons NH_3 + HCO_3^- \tag{169}$$

The additional NH_4^+ ion from the NH_4Cl reduces the CO_3^{2-} ion concentration (common ion effect) to the point where the ion product $[Mg^{2+}][CO_3^{2-}]$ does not exceed the solubility product and Mg^{2+} remains in solution. On the other hand, the CO_3^{2-} concentration remains sufficiently high that the K_{sp} values of $CaCO_3$ and $BaCO_3$ are exceeded by the corresponding ion products. ■

Cation Group V (Mg^{2+}, Na^+, K^+, NH_4^+)

The collection of these somewhat different ions into a single qualitative analysis group is based solely on the sizable solubilities of many of their compounds. Not surprisingly, this is sometimes referred to as the soluble group. Because of the high degree of solubility of most compounds of these cations with simple anions, no single group reagent is available. Ammonium ion is identified in the original unknown sample by production of gaseous ammonia. For the other ions, we resort to unusual reagents for the formation of individually distinctive precipitates. Additional confirmation of the presence of sodium and potassium can be made by their colors in flame tests.

(The metals and ions of this group have been discussed in previous chapters: Mg and Mg^{2+}, Na and Na^+, K and K^+ in Chapter 26; NH_4^+ salts in Chapter 22 and in other places cited in the index.)

31.21 Magnesium(II) ion

Magnesium is a member of Representative Group II, suggesting that the properties of the colorless Mg^{2+} ion closely resemble those of the Ca^{2+} and Ba^{2+} ions. The smaller Mg^{2+} ion is more acidic and more readily complexed than the Ca^{2+} and Ba^{2+} ions. Magnesium sulfate and chromate are very soluble in water, but the hydroxide is slightly soluble except in the presence of the ammonium ion (Section 30.10). The same is true of the carbonate (Section 30.10). The double ammonium phosphate, $MgNH_4PO_4 \cdot 6H_2O$, has limited solubility in water. In saturated aqueous solutions, the equilibrium concentration of Mg^{2+} *decreases* in the order of

$$MgC_2O_4(s) > MgCO_3(s) > MgF_2(s) > MgHPO_4(s)$$
$$> Mg(OH)_2(s) > MgNH_4PO_4 \cdot 6H_2O(s)$$

A distinctive reaction of the Mg^{2+} ion is the formation of an intensely blue lake of $Mg(OH)_2$ with *p*-nitrobenzeneazoresorcinol, known as Magneson I or as S and O reagent (for Suitzu and Okuma, who studied it). The formula for this organic dye is

31.22 Sodium and potassium ions

The Na^+ and K^+ cations are derived from closely related Representative Group I metals and have very similar properties. Both are large, colorless ions that cannot be reduced chemically to the free metals in aqueous solution. They are very weak acids and thus do not react appreciably with water. They do not form complexes in aqueous systems. With few exceptions, their compounds dissolve extensively in water. A few compounds have reduced solubilities, as noted in Table 31.3, but precipitation reactions are not particularly sensitive. However, selective precipitations of the two cations can be used for identification. Flame tests (bulky yellow for Na^+, fleeting violet for K^+) are much more sensitive, but contamination by Na^+ ion is so common that its identification in this manner presents problems. The sodium flame color masks that of potassium. The potassium color can best be seen under this circumstance by viewing through a didymium glass filter, which absorbs yellow light.

31.23 Ammonium ion

The NH_4^+ ion has roughly the same radius as the K^+ ion (0.143 nm compared to 0.133 nm for K^+) and thus forms many compounds of

TABLE 31.3

Formula	Solubility (mole/liter)	Formula	Solubility (mole/liter)
KClO$_4$	0.15	NH$_4$ClO$_4$	2.0
K[BF$_4$]	0.036	(NH$_4$)$_2$[PtCl$_6$]	ca. 0.005
KIO$_4$	0.03	(NH$_4$)$_2$Na[Co(NO$_2$)$_6$]	0.001
K$_2$[PtCl$_6$]	0.016	NH$_4$[B(C$_6$H$_5$)$_4$]	Very small
K$_2$[SiF$_6$]	0.006		
K$_2$Na[Co(NO$_2$)$_6$]	ca. 0.001	Na$_2$SiF$_6$	0.03
		NaZn(UO$_2$)$_3$·(CH$_3$CO$_2$)$_9$·9H$_2$O	0.02
K[B(C$_6$H$_5$)$_4$]	ca. 10^{-6}	Na[Sb(OH)$_6$]	0.002

Solubilities of more difficultly soluble Na$^+$, K$^+$, and NH$_4^+$ compounds in water at 20 to 25°C

comparable crystal structures and solubilities (Table 31.3). For this reason, removal of the NH$_4^+$ before the K$^+$ ion can be identified by precipitation is essential. Similarities to the Na$^+$ ion are much less striking.

All solid ammonium compounds undergo thermal decomposition. If the anion present is nonoxidizing, ammonia is a product, for example,

$$NH_4Cl(s) \xrightarrow{\Delta} NH_3(g) + HCl(g) \tag{170}$$
$$(NH_4)_2CO_3(s) \xrightarrow{\Delta} 2NH_3(g) + H_2O(g) + CO_2(g) \tag{171}$$

If the anion present is oxidizing, an oxidation product of the NH$_4^+$ ion results, for example,

$$NH_4NO_3(s) \xrightarrow{\Delta} N_2O(g) + 2H_2O(g) \tag{172}$$
$$3(NH_4)_2SO_4(s) \xrightarrow{\Delta} N_2(g) + 4NH_3(g) + 3SO_2(g) + 6H_2O(g) \tag{173}$$
$$(NH_4)_2Cr_2O_7(s) \xrightarrow{\Delta} N_2(g) + 4H_2O(g) + Cr_2O_3(s) \tag{174}$$

A distinctive reaction of the NH$_4^+$ ion is that with OH$^-$ ion, which liberates ammonia, particularly when heated

$$NH_4^+ + OH^- \xrightarrow{\Delta} H_2O(l \text{ or } g) + NH_3(g)$$

Liberated ammonia is identified by odor (with CARE, for ammonia in high concentrations is toxic) by the change in moist red litmus to blue, or by its reaction with an alkaline solution containing the [HgI$_4$]$^{2-}$ ion (Nessler's reagent)

$$4NH_3 + 2[HgI_4]^{2-} \underset{\longleftarrow}{\overset{OH^-}{\longrightarrow}} \underset{\substack{\text{yellow to orange} \\ \text{to brown}}}{Hg_2NI(s)} + 7I^- + 3NH_4^+ \tag{175}$$

Reactions of Importance in the Separation and Detection of Mg²⁺, Na⁺, K⁺, and NH₄⁺

Confirmatory tests

$$NH_4^+ + OH^- \xrightarrow{\Delta} NH_3(g) + H_2O(g) \qquad (176)$$

$$Mg^{2+} + NH_4^+ + PO_4^{3-} + 6H_2O(l) \rightleftharpoons$$
$$MgNH_4PO_4 \cdot 6H_2O(s) \qquad (177)$$
magnesium ammonium phosphate hexahydrate

$$MgNH_4PO_4 \cdot 6H_2O(s) + 2H^+ \rightleftharpoons$$
$$Mg^{2+} + NH_4^+ + H_2PO_4^- + 6H_2O(l) \qquad (178)$$

$$Mg^{2+} + \text{Magneson I} \longrightarrow \text{blue lake} \qquad (179)$$

$$K^+ + [B(C_6H_5)_4]^- \longrightarrow K[B(C_6H_5)_4](s) \qquad (180)$$
potassium tetraphenylborate

$$Na^+ + Zn^{2+} + 3UO_2^{2+} + 9CH_3COO^- + 6H_2O \rightleftharpoons$$
$$NaZn(UO_2)_3(CH_3COO)_9 \cdot 6H_2O(s) \qquad (181)$$
sodium zinc uranyl acetate hexahydrate

31.24 Cation Group V analysis

The flow chart for the separation and identification of the cations of this group is given in Figure 31.6. The following procedures are identified in Figure 31.6.

FIGURE 31.6
Flow chart for analysis of Cation Group V.

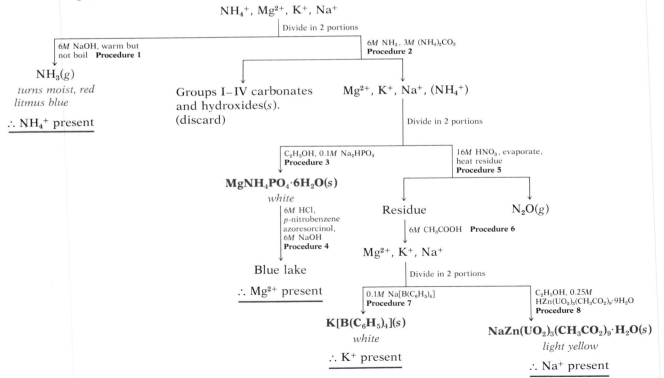

Procedure 1. Only a sample to which neither aqueous ammonia nor an ammonium salt has been added can be used for this procedure. It is recommended that a portion of the original sample be used directly for this test. Boiling must be avoided because the alkaline solution can spatter on the litmus paper and thus give misleading results. Proper procedure requires that the moist red litmus contact only the vapors released from the warm solution or suspension.

Procedure 2. This procedure is designed only to remove any remaining traces of cations from Groups I to IV which would precipitate in the next step. It may be omitted for samples containing only Group V cations.

Procedure 3. White crystalline $MgNH_4PO_4 \cdot 6H_2O$ is less soluble in ethanolic solutions than in aqueous systems. Supersaturation sometimes occurs and can be relieved by scratching the inner walls of the reaction vessel with a glass rod.

Procedure 4. A blue solid, a lake, identifies the Mg^{2+} ion. The reagent itself may give a bluish solution when sodium hydroxide is added.

Procedure 5. Complete removal of any NH_4^+ ion present is necessary because, like the K^+ ion, it forms an insoluble tetraphenylborate.

Procedure 6. A weak acid is used to avoid subsequent problems with precipitations. Flame tests for the K^+ and Na^+ ions can be run on the residue before dissolution in acetic acid.

Procedure 7. No other common ion except the NH_4^+ ion will form a precipitate at this point. A flame test for the K^+ ion can be made for confirmation.

Procedure 8. This reaction is specific for the Na^+ ion. Adding ethanol improves its sensitivity. Again a flame test can be made.

THOUGHTS ON CHEMISTRY

The epigrams of Karl Remigius Fresenius

[Karl Remigius Fresenius loved to take walks. He built a long covered gallery beside his house and walked there when the weather was bad. Thinking as he walked, he occasionally stopped to write an epigram—a distillation of his philosophy of life—on the wall. Some of these epigrams are given here in both the original German and in a free translation.]

Forsche gründlich,
Rede wahr,
Schreibe bündig,
Lehre klar.

Research thoroughly,
speak truthfully,
write concisely,
teach clearly.

Nach goldner Frucht vom Baum des Lebens
Greift der Müssiggang vergebens.

The golden fruit from the tree of life
will not be reached by laziness.

Musst Du Dich in der Jugend
   müh'n,
So wolle nicht darüber klagen,
Nur Bäumme, die im Lenze blühn,
Können im Herbste Früchte tragen.

Do not complain if you have to
struggle in your youth;
only trees that bloom in spring
will bear fruit in fall.

Trittst Du frisch an's Werk heran,
Ist die Arbeit halb getan.

With enthusiasm start your tasks
and you will find them half
   completed.

Wünschest Du ein froh Geschick,
Nütze auch den Augenblick.

To lead a happy life
take advantage of each moment.

Glauben was man hofft,
Führt zu Täuschung oft.

To believe what one expects
often leads to disappointment.

Soll das Leben Dir gelingen,
Halte Mass in allen Dingen.

To be successful in your life
keep everything in moderation.

Das Ganze ist in guter Hut,
Wenn jedes Glied das seine tut.

The whole is certainly protected,
if every member does its part.

Freundschaft treu, fest und rein,
Im Lebensring der Edelstein.

Friendship, true, loyal, and pure,
is the diamond in the ring of life.

Legst Du Abends Dich zur Ruh',
Mazh' das Sorgenkästchen zu.
Drückst Du noch so fest darauf,
Morgens springt von selbst es auf.

When you go to bed at night,
close your little box of sorrow,
but even if you lock it tight,
'twill open up again tomorrow.

THE EPIGRAMS OF REMIGIUS FRESENIUS. The German quoted from an article by Ralph E. Oesper, in the Journal of Chemical Education, Vol. 14, p. 313, 1937. English translation by Winfried J. Fremuth.

Exercises

31.1 Why does Pb^{2+} appear in both Cation Groups I and II? Is this the same reason that mercury appears in both of these groups? Explain.

31.2 Write chemical formulas for the following substances: (a) mercury(I) chloride, (b) diamminesilver(I) ion, (c) antimony(III) sulfide, (d) tetrachlorohydrargyrate(II) ion, or tetrachloromercurate(II) ion, (e) manganese(IV) oxide, (f) chromate ion, (g) sodium nitrite, (h) potassium hexacyanoferrate(II), (i) ammonium carbonate, (j) thioacetamide, (k) hydrosulfuric acid.

31.3 Write names for the following substances: (a) $PbCl_2$, (b) SnS_2, (c) $[SnS_3]^{2-}$, (d) $[Cu(NH_3)_4]^{2+}$, (e) $PbCrO_4$, (f) NH_4NCS, (g) $Cr_2O_7^{2-}$, (h) $[Zn(OH)_4]^{2-}$, (i) $KFe[Fe(CN)_6]$, (j) $NaBiO_3$, (k) $(NH_4)_2C_2O_4$, (l) $MgNH_4PO_4 \cdot 6H_2O$.

31.4 Why must a large excess of Cl^- be avoided during the precipitation of the cations in Group I? Write equations illustrating (a) the precipitation and (b) what happens if the Cl^- concentration is too high.

31.5 If the concentration of Ag^+ in equilibrium with $AgCl(s)$ is larger than $[Ag^+]$ in equilibrium with $[Ag(NH_3)_2]^+$, why does AgCl reprecipitate upon the addition of HNO_3 to a solution containing the ammine complex?

31.6 If a light-colored sulfide precipitate is obtained for Cation Group II, what cations are probably not present in the sample?

31.7 Why must any NH_4^+ in an unknown be removed before performing the confirmation test for K^+? How is this done? Why should the test for NH_4^+ be performed on the original sample?

31.8 Why do certain metallic ions give flame tests, whereas others do not?

31.9 What is meant by a "lake"? Cite two cases where lakes are used in the confirmatory tests for the scheme presented in this book.

31.10 Write equations illustrating the amphoteric behavior of $Sn(OH)_2$ with HCl and NaOH.

31.11 Why will CuS dissolve in HNO_3, but not in HCl?

31.12 Why does aluminum hydroxide readily dissolve in concentrated NaOH? Why is it insoluble in concentrated $NH_3(aq)$?

31.13 Write the net ionic equations describing (a) Al^{3+} reacting with excess NaOH to give $[Al(OH)_4]^-$; (b) Ni^{2+} reacting with NH_3 to produce the nickel-ammonia complex, which, in turn, is converted to Ni^{2+} by acetic acid and then the Ni^{2+} reacting with HDMG to form the bright red precipitate; (c) ZnS(s) reacting with HCl; (d) Ca^{2+} reacting with $(NH_4)_2CO_3$ to give $CaCO_3(s)$; (e) K^+ reacting with $Na[B(C_6H_5)_4]$ to give a white precipitate; (f) $Hg_2Cl_2(s)$ reacting with excess $NH_3(aq)$; (g) $Sb_2S_3(s)$ reacting with S^{2-} to give $[SbS_2]^-$, which, in turn, reacts with acetic acid to give $Sb_2S_3(s)$.

31.14 Using the flow chart for Cation Group III, identify those steps in which oxidation-reduction reactions occur and write the overall net ionic equation for each case.

31.15 What reagent is specific only for Na^+ among the Group V cations?

31.16 Which reagent, Na_2CO_3, K_2CrO_4, $NH_3(aq)$, NaOH, or $(NH_4)_2C_2O_4$, can be used to distinguish Ca^{2+} from Ba^{2+}?

31.17 Which reagent, NaOH, $NH_3(aq)$, or $(NH_4)_2S$, would you use to distinguish between (a) Co^{2+} and Ni^{2+}, (b) Zn^{2+} and Fe^{2+}, (c) Fe^{2+} and Ni^{2+}, (d) Mn^{2+} and Fe^{3+}?

31.18 A student decided to determine the presence of NH_4^+ in an acidic solution by adding Cu^{2+} and observing whether or not the dark blue copper-ammonia complex formed. The results were negative. Did this experiment prove the absence of the ammonium ion? Explain.

31.19 A student felt that his separation of Hg^{2+} from Pb^{2+} and Cu^{2+} by treatment of the sulfides with warm HNO_3 might have been in error and he wanted to run a one-step test on the filtrate as a check. He added excess $NH_3(aq)$ and found that a white precipitate and blue solution formed. Were Pb^{2+} and Cu^{2+} in the filtrate?

31.20 An unknown was found to contain either Sn^{2+} or Sn^{4+}. A student decided to precipitate the sulfide using $Na_2S(aq)$ and determine the oxidation state by the color of the sulfide. How would this test have distinguished between the two ions?

31.21* A student analyzed a solid sample and reported that it contained Hg_2^{2+} and S^{2-}. Are her results reasonable? Explain.

31.22 An unknown solution was known to contain only Pb^{2+}, Ag^+, or Hg_2^{2+}. A student decided to use only one confirmatory test to identify the ion present—namely the addition of K_2CrO_4 to the unknown solution. Look up the colors of the chromate precipitates of Pb^{2+}, Ag^+, and Hg_2^{2+} and comment on this procedure. A second student decided to use an excess of $NH_3(aq)$. Explain her expected observations. A third student decided to use an excess of NaOH. Explain his expected observations.

31.23 You are given a sample of a solder which could contain one or more of the elements Pb, Sn, Cu, and Ag. Devise a flow chart to determine which metals are present. The first step will be the dissolving of the metal in aqua regia.

31.24 Which cations in Group II would you report for an unknown for which you obtained the following results:
(a) Part of the sulfide precipitate dissolved in ammoniacal thioacetamide, leaving a dark precipitate A and a supernatant B. (An ammoniacal solution is one that contains NH_3.)
(b) The supernatant B was treated with acetic acid and concentrated HCl and then divided into two portions.
(c) One portion gave an orange precipitate with NH_3, acetic acid, and $Na_2S_2O_3$. The other portion did not form a precipitate with Al and $HgCl_2$.
(d) Precipitate A completely dissolved in HNO_3, forming solution C.
(e) Solution C did not form a precipitate upon addition of H_2SO_4.
(f) Solution C formed a deep blue complex with NH_3. After acidification with acetic acid, the solution was divided into two portions. A red precipitate formed in one portion when treated with $K_4[Fe(CN)_6]$ and a green precipitate formed in the other portion when treated with C_5H_5N and NH_4NCS.

31.25 A Group III unknown formed a dark precipitate A upon the addition of NaOH and H_2O_2. The colorless supernatant B was treated with HCl and then NH_3, with no effect. When the supernatant B was heated with thioacetamide, no precipitate formed. Precipitate A dissolved in HNO_3 and H_2O_2 to form a colored solution C. No precipitate formed when $KClO_3$ was added to C and the mixture heated. Solution C was treated with NH_3, without appar-

1005

ent effect. What ion(s) are possibly present? What additional tests would you like to perform on this sample?

31.26 A sample of German silver was dissolved in acid and subjected to analysis. The following results were obtained:
(a) No precipitate was formed in the presence of Cl^-.
(b) A dark precipitate A was formed with acidic thioacetamide, leaving supernatant B.
(c) A dark precipitate C was formed from B in ammoniacal thioacetamide.
(d) Subsequent treatment of the supernatant from (c) indicated no other ions present.
(e) Treatment of A with ammoniacal thioacetamide produced no change.
(f) Precipitate A dissolved in HNO_3 to form solution D.
(g) Treatment of D with H_2SO_4 had no effect, but addition of NH_3 produced a bluish solution E.
(h) Solution E was separated into two portions. Treatment of one portion with H^+ and $K_4[Fe(CN)_6]$ gave a reddish solid and treatment of the other portion with C_5H_5N and NH_4NCS gave a green precipitate.
(i) Treatment of C with HCl caused part of the precipitate to dissolve and the remainder dissolved in HNO_3, forming solution F.
(j) Addition of NaOH and H_2O_2 to F gave a green precipitate G and a colorless solution H.
(k) Precipitate G dissolved in HNO_3 and H_2O_2. The resulting solution did not form a precipitate with $KClO_3$ nor with NH_3. Division of the solution into two portions and treatment of one portion with HDMG gave a red color. No color was formed in the other portion upon addition of HCl, NaF, amyl alcohol, ether, and NH_4NCS.
(l) Solution H was acidified and NH_3 was added without effect. A white precipitate was formed upon heating the solution with thioacetamide. The precipitate dissolved when HCl was added. Identify the metals that make up this alloy.

31.27 Identify the metals found in Wood's metal (which is used in automatic sprinkler systems, electric fuses, and safety valves for steam boilers) from the following information:
(a) A sample of the metal was dissolved in acid to produce solution A.
(b) A white precipitate B was formed upon addition of HCl to A, leaving supernatant C.
(c) A dark precipitate D was formed upon treatment of C with HCl and a source of S^{2-}, leaving solution E.
(d) Treatment of E with the NH_4Cl–NH_3 buffer and a source of S^{2-} gave no precipitate.
(e) No precipitate formed upon the addition of $(NH_4)_2CO_3$ to the buffered solution prepared in (d).
(f) Flame tests and other evidence indicated that ions from Group V were absent.
(g) Precipitate B dissolved in hot water. Half of the resulting solution gave a yellow precipitate upon addition of K_2CrO_4 and the other half gave a white precipitate upon addition of H_2SO_4.
(h) Precipitate D was treated with additional S^{2-}, and a new precipitate F was formed in contact with solution G.
(i) Addition of HCl to G gave a precipitate. The precipitate dissolved in excess HCl.
(j) The solution of G found in (i) gave a grey precipitate after reduction with aluminum and addition of $HgCl_2$.
(k) Precipitate F dissolved in HNO_3. A white precipitate H was separated from the solution I upon addition of H_2SO_4.
(l) Further analysis of the precipitate H showed it to be soluble in NaOH. A yellow precipitate formed upon acidification of the solution with acetic acid and addition of K_2CrO_4.
(m) A white precipitate J formed upon addition of NH_3 to solution I, indicating the presence of Bi^{3+}.
(n) A colorless solution K remained after removal of J.
(o) A yellow precipitate formed upon addition of CN^- and S^{2-} to solution K, indicating the presence of Cd^{2+}.

31.28 How would you prepare 15 ml of a solution that is $0.1M$ in Ag^+, $0.1M$ in Pb^{2+}, and $0.1M$ in Hg_2^{2+} from separate $0.5M$ stock solutions of each ion?

31.29 Approximately how many drops (20 drops = 1 ml) of $1M$ $NH_3(aq)$ are needed to neutralize the acid in 5 ml of a $0.1M$ HCl solution?

31.30 A solution is $0.01M$ in Ag^+, $0.01M$ in Pb^{2+}, and $0.01M$ in Hg_2^{2+}. As HCl is added dropwise, which cation precipitates first and at what Cl^- concentration? $K_{sp} = 2 \times 10^{-5}$ for $PbCl_2$, 1.8×10^{-10} for AgCl, and 1.3×10^{-18} for Hg_2Cl_2.

31.31 At large concentrations of Cl^-, Ag^+ forms the complex $[AgCl_4]^{3-}$. Calculate the concentration of Ag^+ in equilibrium with $6M$ HCl. Assume the complex is at a concentration of $0.01M$. $K_d = 5 \times 10^{-6}$ for $[AgCl_4]^{3-}$.

31.32 What mass of thioacetamide is needed to quantitatively precipitate all of the Pb^{2+} and Sn^{4+} in a 5 ml sample originally containing these ions each at a concentration of $0.1M$?

31.33 Calculate the concentration of Pb^{2+} remaining in a $2M$ HCl solution. Is this concentration large enough to precipitate PbS in a $0.001M$ S^{2-} solution? $K_{sp} = 2 \times 10^{-5}$ for $PbCl_2$ and 1×10^{-28} for PbS.

31.34 The solubility of CO_2 in water is about $0.034M$. Calculate $[CO_3^{2-}]$ and predict whether an appreciable amount of $PbCO_3$ will form in a $0.1M$ Pb^{2+} solution. $K_{a_1} = 4.5 \times 10^{-7}$ and $K_{a_2} = 4.8 \times 10^{-11}$ for H_2CO_3 and $K_{sp} = 7.4 \times 10^{-14}$ for $PbCO_3$.

31.35 Using values of $E°$, show that the oxidation of $Sn(s)$ by H^+ occurs only after any excess Al has reacted.

$Sn^{2+} + 2e^- \longrightarrow Sn \qquad E° = -0.1364 \text{ V}$
$2H^+ + 2e^- \longrightarrow H_2 \qquad E° = 0.0000 \text{ V}$
$Al^{3+} + 3e^- \longrightarrow Al \qquad E° = -1.67 \text{ V}$

31.36 An unknown was found to contain I^- and Fe. Using the following half reactions and potentials, predict the formula of the compound.

$I_2 + 2e^- \longrightarrow 2I^- \qquad E° = 0.535 \text{ V}$
$Fe^{3+} + e^- \longrightarrow Fe^{2+} \qquad E° = 0.770 \text{ V}$

31.37** In an extended scheme of cation analysis, additional ions could be separated in the groups considered in the scheme presented in this book. Look up solubility data for the compounds (such as chlorides and sulfides) of the following elements and predict which group they would belong to: As, Au, Be, Cs, Li, Mo, Pt, Rb, Sb, Se, Sr, Te, Ti, Tl, U, V, Zr, and lanthanides.

32 ORGANIC CHEMISTRY: FUNCTIONAL GROUPS AND THE MOLECULES OF BIOCHEMISTRY

In Chapter 23 we had an introduction to organic chemistry in our study of the hydrocarbons. As we pointed out there, all organic compounds can be thought of as derived from hydrocarbons. Very small differences in structure can have remarkably great effects on the properties of organic compounds. One type of simple structural difference might be in the addition or subtraction of a few carbon atoms of the hydrocarbon skeleton of a compound. Another might be the difference between two isomers of the same compound. A third difference might be in the location or type of groups attached to the hydrocarbon skeleton.

In this chapter we discuss some of the functional groups of atoms that, when attached to hydrocarbons, impart specific properties to the molecules. A primary purpose of this chapter is to give you some familiarity with the most common of these groups, which include atoms of elements other than carbon, most often halogen, oxygen, nitrogen, or sulfur atoms.

To convey how varied organic compounds are, we include with the discusson of each group a few examples of interesting or practical compounds that contain these groups. We have no intention that you should memorize the structures of such compounds.

Never underestimate the power of a functional group. Urea, the first synthetic organic chemical, a component of urine, and a common fertilizer, differs by only one functional group, C=O versus C=S, from thiourea, a compound suspected of being a cancer-causing agent.

$$H_2N-\overset{\overset{\textstyle O}{\|}}{C}-NH_2 \qquad H_2N-\overset{\overset{\textstyle S}{\|}}{C}-NH_2$$

urea   *thiourea*

In this chapter the most common functional groups of the two main types—groups with only covalent single bonds and groups that include covalent double bonds—are discussed. The nomenclature of compounds containing these groups is explained. Examples are given of some simple compounds and some more complex compounds of specific interest containing each of the groups. A few of the most common preparations and reactions of the various types of compounds are included. The chapter (and the book) ends with a brief introduction to proteins, carbohydrates, lipids, and nucleic acids—the molecules of biochemistry.

The arrangement of the functional groups —NO_2 or —NH_2 makes a vast difference in the taste of the following three compounds.

| | | |
| tasteless | 4100 times sweeter than sugar | bitter |

Chemistry has come a long way in predicting the effects on chemical and physical properties of variations in the groups attached to organic compounds. However, we do not yet know the explanation for many of the subtle relationships between structure and biochemical activity. The study of such relationships is at the frontier of scientific research today.

32.1 Functional groups

If one of the hydrogen atoms of the water molecule is replaced by one of the various saturated alkyl groups, the series that begins with CH_3OH results.

$$HOH$$
$$CH_3OH$$
$$CH_3CH_2OH$$
$$CH_3CH_2CH_2OH$$
$$CH_3CH_2CH_2CH_2OH$$
etc.

These compounds can all be represented by the general expression, ROH, where, except for water, R is some collection of carbon and hydrogen atoms.

The chemistry of these compounds is dominated by the OH group. All compounds of the formula ROH exhibit similar chemical behavior, no matter how long the continuous chain of carbon atoms may be. Saturated hydrocarbons, you will recall (Section 23.4), are relatively unreactive, so we should not expect the hydrocarbon part to overshadow the waterlike part. The compounds of the family represented by the formula ROH are collectively called alcohols, and the OH group is referred to as a functional group. A **functional group** is an atom or group that bonds to one or more carbon atoms in an organic molecule and contributes a characteristic chemical behavior to the molecule.

Table 32.1 lists the most common functional groups. Large and even not so large molecules can contain many functional groups of different types.

TABLE 32.1

Some common functional groups

Compound structure (functional group[a] in color)	Compound name
R—X	Alkyl halide
R—OH	Alcohol
Ar—OH	Phenol
ROR	Ether
R—NH₂	Amine

R—C=O
   |
   H
	Aldehyde

R₂C=O	Ketone

R—C=O
   |
  OH
	Carboxylic acid

R—C=O
   |
  NH₂
	Amide

R—C=O
   |
  X
	Acyl halide

R—C—O—C—R
   ‖   ‖
   O   O
	Acid anhydride

R—C=O
   |
  OR
	Ester

R—NO₂	Nitroalkane
R—SO₃H	Sulfonic acid
R—CN	Nitrile

[a] Ar is the abbreviation used for an aromatic group, such as the phenyl group (C_6H_5). X represents any halogen atom.

Functional groups with covalent single bonds

32.2 Halogen derivatives of hydrocarbons

Fluorine, chlorine, bromine, and iodine can all be found in organic molecules. The halides of hydrocarbons may be **alkyl halides, RX,** where R is a saturated group; **aryl halides, ArX,** where Ar is an aromatic group, or they may contain unsaturated R groups; X usually represents any halogen. Some common halogen derivatives of hydrocarbons are listed in Table 32.2 to illustrate their nomenclature (compare with the saturated hydrocarbons, Table 23.2) and the effect of molecular structure on boiling point.

Halogen atoms are more electronegative than the carbon atom, so the halogen atom draws the electron pair in a halogen–carbon bond toward itself. The result is a polar covalent bond, as, for example, in methyl bromide

$$\overset{\delta+}{CH_3}\!-\!\overset{\delta-}{Br}$$

The effect of this polarization is shown in the reactions of compounds containing such bonds. The halogen atom is rather easily removed as a halide ion. Furthermore, the carbon atom is readily attacked by anions or by molecules that possess an unshared pair of electrons (nucleophilic reagents, Section 23.8). The following equations illustrate a few reactions in which the halogen is displaced by a nucleophilic group.

$$Na^+OH^- + CH_3\!-\!Br \xrightarrow{H_2O} \underset{an\ alcohol}{CH_3OH} + NaBr \qquad (1)$$

$$Na^+CN^- + CH_3\!-\!Br \xrightarrow{H_2O} \underset{a\ nitrile}{CH_3CN} + NaBr \qquad (2)$$

$$:NH_3 + CH_3\!-\!Br \xrightarrow{H_2O} \underset{an\ amine}{CH_3NH_2} + HBr \qquad (3)$$

Halogen atoms on aromatic rings are inert to such displacement reactions.

Alkyl halides can also be induced to lose HX (where X is Cl, Br, or I) and form alkenes, for example,

$$\underset{propyl\ bromide}{CH_3CH_2CH_2Br} \xrightarrow[\text{alcohol, heat}]{NaOH} \underset{propylene}{CH_3CH=\!CH_2} + NaBr + H_2O$$

Many pesticides are highly chlorinated compounds. There can be no doubt as to the value of these compounds in improving the yield of good crops by preventing insects from destroying the crops. How-

---

Alkyl halides, RX
Aryl halides, ArX

TABLE 32.2

Halogen derivatives of hydrocarbons
The common name is given first; the systematic name is given in parentheses. The general molecular structure of these hydrocarbon halides is R—X

	Boiling point (°C)
CH_3Br *methyl bromide* *(bromomethane)*	**5**
CH_3I *methyl iodide* *(iodomethane)*	**42**
CH_3CH_2Br *ethyl bromide* *(bromoethane)*	**38**
$CH_3CH_2CH_2Br$ *propyl bromide* *(1-bromopropane)*	**71**
$\underset{\quad\ \ Br\ \ Br}{CH_3CHCH_2}$ *propylene dibromide* *(1,2-dibromopropane)*	**141**
$CH_2\!=\!CHCl$ *vinyl chloride* *(chloroethene)*	**14**
Cl *chlorobenzene*	**132**

Insecticides

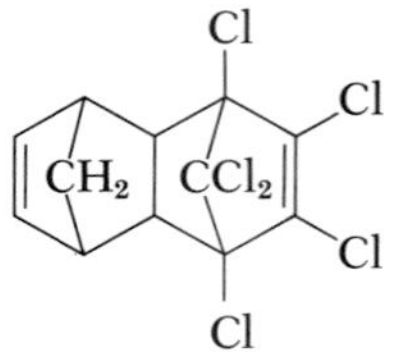

CH_2 CCl_2

Cl
Cl
Cl
Cl

Aldrin

H
Cl
C
Cl
Cl—C—Cl
Cl

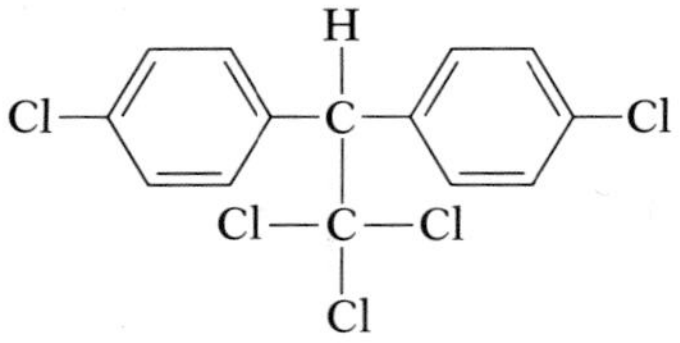

DDT
[1,1,1-trichloro-2,2-bis
(p-chlorophenyl)ethane]

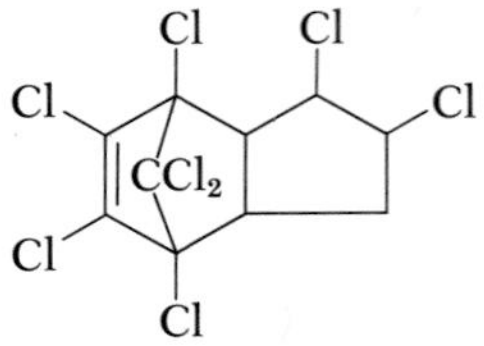

Cl Cl
Cl
CCl_2
Cl
Cl
Cl

Chlordane

Herbicides

Cl
Cl
—OCH₂C
O
OH

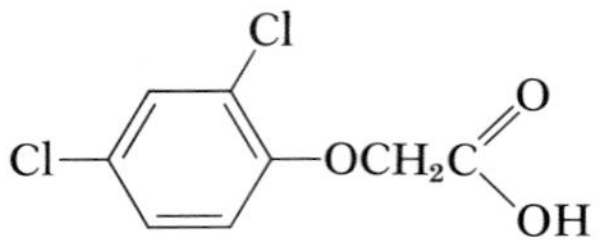

2,4-D
[(2,4-dichlorophenoxy)acetic acid]

Cl
Cl
—OCH₂C
O
OH
Cl

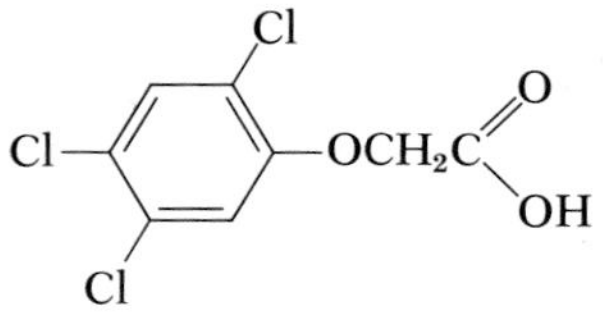

2,4,5-T
[(2,4,5-trichlorophenoxy)acetic acid]

Fungicide

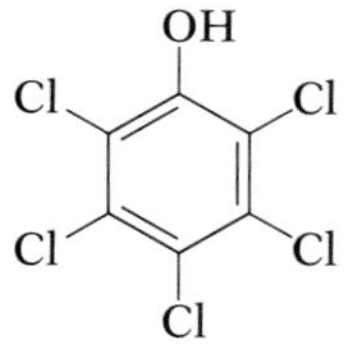

OH
Cl
Cl
Cl
Cl
Cl

PCP
(pentachlorophenol)

TABLE 32.3
Chlorinated hydrocarbon pesticides

ever, some of the compounds listed in Table 32.3 have come under criticism for (1) their long persistence in the environment and/or (2) their toxicity to both man and animals. Aldrin and a structurally similar compound, dieldrin, have now been banned, and the use of DDT and chlordane is carefully restricted. 2,4,5-T, an herbicide used as a defoliant during the Vietnam war, has been suspected of causing genetic defects. However, these effects seem to be due to the presence of small amounts of a very highly toxic impurity, dioxin.

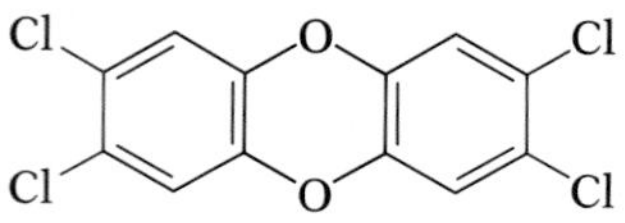

Cl
O
Cl
Cl
O
Cl

dioxin
a very toxic substance

32.3 Alcohols

In organic compounds, the OH group is usually covalently bonded and is called the **hydroxyl group.** The ending *ol* indicates the presence of a functional —OH in an organic molecule. **Alcohols** have the general formula ROH. The generic name for alcohols is *alkanol* when R is saturated, *alkenol* when R is olefinic, and *alkynol* when R is acetylenic. (When R is aromatic, the compound is a phenol; next section.)

In the common system of nomenclature, the R group is named and the word *alcohol* follows, as in methyl alcohol (CH_3OH) and cyclohexyl alcohol ($C_6H_{11}OH$). Ethyl alcohol (C_2H_5OH) is the common beverage alcohol. In the IUPAC system of nomenclature, the name is derived from the longest hydrocarbon chain that includes the OH group by dropping the final *e* and adding *ol*, as in methanol, ethanol, and cyclohexanol. When necessary, a number is used to show the position of the OH group. Numbering starts at the end of the chain nearest to the OH group. Table 32.4 provides a few examples.

Alcohols are classified as primary, secondary, or tertiary, depending on whether the carbon atom to which the hydroxyl group is attached is primary, secondary, or tertiary. Like water the alcohols exhibit hydrogen bonding, which causes their boiling points to be relatively high.

Methanol, once known as wood alcohol because of its formation during wood distillation, is now made synthetically from carbon monoxide and hydrogen in quantities exceeding those of most other synthetic organic chemicals (5000 million pounds in 1975).

$$CO(g) + 2H_2(g) \xrightarrow{\Delta} CH_3OH(g)$$

Most of this methanol is converted to formaldehyde. Methanol is toxic and causes blindness or death when taken internally.

Ethanol, in addition to being a component of many beverages, is a valuable solvent and a frequently used reagent in organic synthesis. Ethylene glycol is used as a permanent antifreeze in automobiles because of its relatively low volatility. Isopropyl alcohol is rubbing alcohol, and large quantities are converted into acetone (Section 32.7). Glycerol has a strong attraction for water and, because of this property, it is used in lotions and other applications where moisture retention is desired. Glycerol is also used in the manufacture of paints, varnishes, explosives, and a variety of other products.

The formation of an alcohol from the reaction of an alkyl halide with an aqueous base was shown in Equation (1). The reaction of one class of organic compound often becomes a method of preparation for another class of compound. Alcohols can also be prepared by the treatment of olefins with sulfuric acid, followed by hydrolysis.

$$CH_2{=}CH_2 + H_2SO_4 \longrightarrow CH_3CH_2OSO_3H \xrightarrow{H_2O} CH_3CH_2OH + H_2SO_4$$

Fermentation of various grains, fruits, and other natural products that contain starch or sugars produces ethyl alcohol. Enzymes, which are the catalysts in biochemical systems, first convert starch to sugars, and the sugars are then converted to ethyl alcohol and carbon dioxide by other enzymes.

At one time all ethyl alcohol was produced commercially by fermentation. It is now more economical to make ethyl alcohol by the

Alcohols, ROH

TABLE 32.4
Alcohols
The general molecular formula for an alcohol is ROH

	Boiling point (°C)
CH_3OH *methyl alcohol* *(methanol)*	65
CH_3CH_2OH *ethyl alcohol* *(ethanol)*	78
$CH_3CH_2CH_2OH$ *propyl alcohol* *(1-propanol)*	97
$CH_3\underset{\underset{\textstyle OH}{\mid}}{C}HCH_3$ *isopropyl alcohol* *(2-propanol)*	82
$HOCH_2CH_2OH$ *ethylene glycol* *(1,2-ethanediol)*	197
$\underset{\underset{\textstyle OH}{\mid}}{C}H_2\underset{\underset{\textstyle OH}{\mid}}{C}H\underset{\underset{\textstyle OH}{\mid}}{C}H_2$ *glycerol (glycerin)* *(1,2,3-propanetriol)*	290 (dec)
⬡—CH_2CH_2OH *β-phenylethyl alcohol* *(2-phenylethanol)*	219

reaction of water with ethylene, which is readily available from petroleum refineries.

$$CH_2{=}CH_2 + H_2O \xrightarrow[\text{pressure}]{\text{heat, catalysts,}} CH_3CH_2OH$$

Fermentation, however, is still the preferred method of obtaining ethyl alcohol for beverages, many of which retain flavors characteristic of the material that was fermented (Table 32.5). After fermentation, a solution contains up to 14% ethyl alcohol. Except for wines and beer, this solution is distilled to remove water and increase the alcohol content. Up to 95% ethyl alcohol by weight, which is the composition of a constant-boiling mixture of alcohol and water, can be obtained by distillation. The usual 80-proof beverage is a 40% solution of ethyl alcohol.

Absolute ethyl alcohol, 100% or 200 proof, is made by removing the remaining water from 95% ethyl alcohol by distillation with benzene or by chemical means. (An azeotropic mixture of benzene, alcohol, and water distills at 65°C and removes the last few percent of water.) Ethyl alcohol for beverages is subject to a federal tax. In order to render industrial ethyl alcohol unfit for drinking, various toxic or objectionable materials (e.g., methanol) that are difficult to remove from ethyl alcohol are added. This *denatured alcohol* is tax-free, hence much cheaper for industrial users.

Like water, alcohols can act as either acids or bases. As acids, they are even weaker than water, however. In the presence of relatively strong bases, alcohols donate protons. For example, reaction of alcohols with active metals such as Li, Na, K, Mg, or Al gives a class of compounds known as **alkoxides,** RO^-M^+. The reaction of ethyl alcohol with sodium

$$2CH_3CH_2OH + 2Na \longrightarrow 2CH_3CH_2O^-Na^+ + H_2$$
$$\textit{sodium ethoxide}$$

is sufficiently moderate to permit the disposal of waste sodium by converting it to sodium ethoxide, an alkaline, ionic compound comparable to sodium hydroxide.

In the presence of concentrated acids, alcohols act as proton acceptors in overall reactions such as

$$CH_3OH + HBr \longrightarrow CH_3Br + H_2O$$

Similar results are obtained with concentrated HCl, HI, H_2SO_4, and HNO_3. The reaction of glycerol with nitric acid yields nitroglycerin, an explosive and a component of dynamite

$$
\begin{array}{l}
CH_2OH \\
| \\
CHOH \\
| \\
CH_2OH \\
\textit{glycerol}
\end{array}
+ 3HNO_3 \longrightarrow
\begin{array}{l}
CH_2ONO_2 \\
| \\
CHONO_2 \\
| \\
CH_2ONO_2 \\
\textit{glyceryl trinitrate} \\
\textit{(nitroglycerin)}
\end{array}
\xrightarrow[\text{shock}]{\text{heat or}} \tfrac{3}{2}N_2 + 3CO_2 + \tfrac{5}{2}H_2O + \tfrac{1}{4}O_2
$$

TABLE 32.5
Alcoholic beverages

Beverage	Starting material
Whiskey	Grains (rye, corn, wheat, oats, barley)
Rum	Molasses
Brandy	Grapes and other fruits
Gin	Grains (distilled and flavored with juniper berry)
Vodka	Potatoes and corn
Wines	Grapes
Sake	Rice
Mead	Honey
Beer	Barley and hops
Tequila	Cactus

Nitroglycerin is also useful in the treatment of heart disease because of its action in dilating blood vessels.

32.4 Phenols

The attachment of a hydroxyl group to an aromatic ring gives a class of organic compounds known as **phenols,** ArOH. The simplest compound in the group is phenol itself, C_6H_5OH (Table 32.6). The structural similarity of phenols to alcohols is apparent. However, the aromatic ring modifies the chemical behavior of the hydroxyl group. Phenols are weak acids; most of them have pK_a values of about 10, which makes them much stronger acids than alcohols. The acidity of phenols arises in part because the aromatic ring delocalizes the negative charge on the anion; for example,

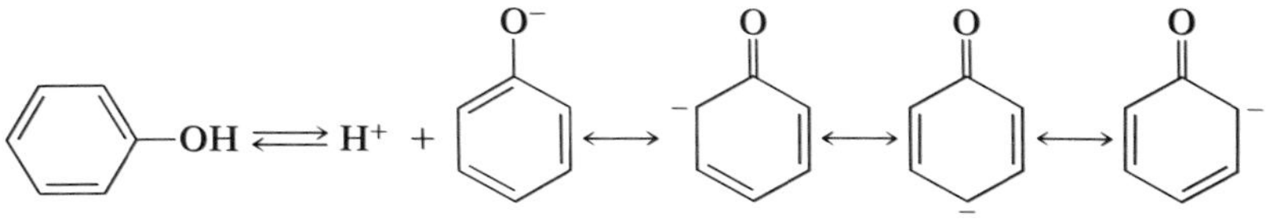

phenol   resonance structures of the
                          phenoxide ion

Furthermore, the hydroxyl group causes the aromatic ring to be much more reactive toward substitution and oxidation. The Br, NO_2, and SO_3H groups are readily substituted on ring carbon atoms in phenols under conditions that do not cause any reaction to occur with benzene.

Phenol itself is poisonous and will cause blisters on the skin. The biggest use for phenol is in the manufacture of phenol–formaldehyde resins (among them the product known as Bakelite). Another important use of phenols is in bactericidal products. "Carbolic acid" is a name given to phenol and its aqueous solutions. Joseph Lister, applying Pasteur's theories about bacteria and infection in the late 1800s, did surgery under a spray of carbolic acid. Some of the more common antiseptic phenols are hexylresorcinol (Table 32.7) and the cresols

TABLE 32.7

Some additional phenols

Urushiol is a mixture of phenols that vary only in the amount of unsaturation in the side chain.

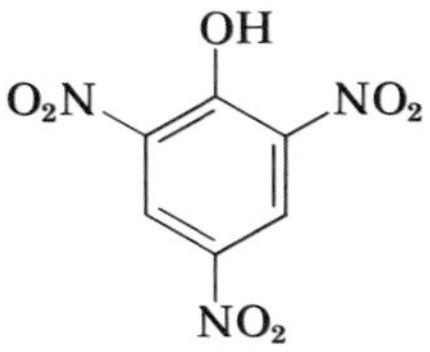

picric acid 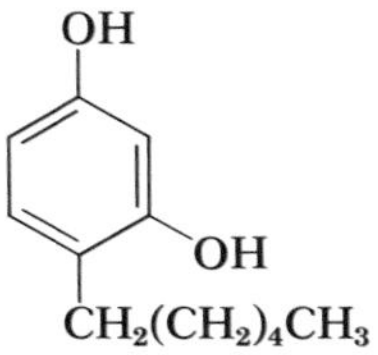   hexylresorcinol 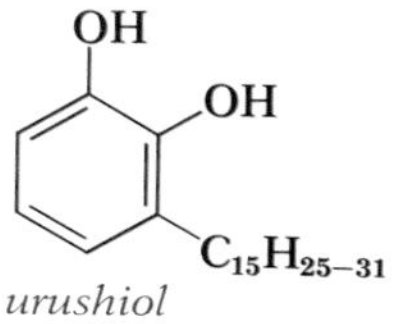   urushiol

TABLE 32.6

Phenols

The general molecular formula of a phenol is ArOH

	Melting point (°C)
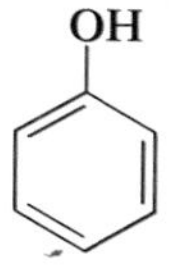 phenol	42
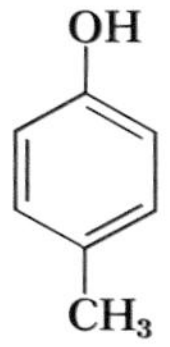 p-cresol	36
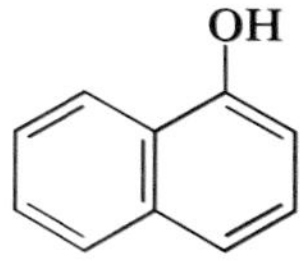 α-naphthol (1-naphthol)	94
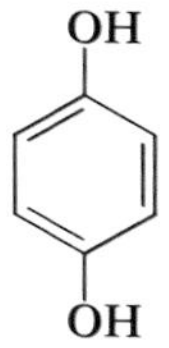 hydroquinone	169
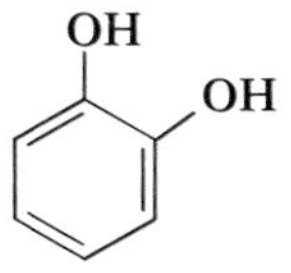 catechol	104

TABLE 32.8

Ethers

The general molecular formula for an ether is ROR

	Boiling point (°C)
CH_3OCH_3 *diethyl ether (ethoxyethane)*	−24
$CH_3CH_2OCH_2CH_3$ *diethyl ether (ethoxyethane)*	35
$CH_3OCH_2CH_2OCH_3$ *ethylene glycol dimethyl ether ("glyme") (1,2-dimethoxyethane)*	85
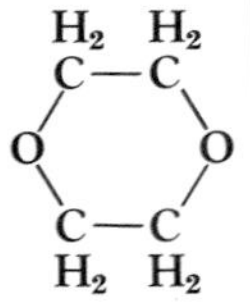 *dioxane (1,4-dioxin)*	101
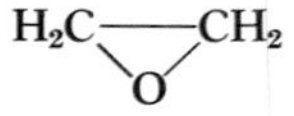 *ethylene oxide (oxirane)*	11
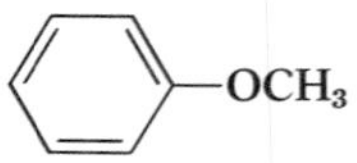 *anisole (methoxybenzene)*	154

(methylphenols) (Table 32.6). Phenols are also used in the manufacture of dyes, drugs, photographic developers, adhesives, and a large variety of other products. Picric acid (Table 32.7) is a strong acid and a high explosive. Many phenols occur naturally, for example, the active component in poison ivy (urushiol, Table 32.7) and several essential oils (oil of cloves, oil of aniseed).

One method for the preparation of phenol itself is illustrated by the following equations.

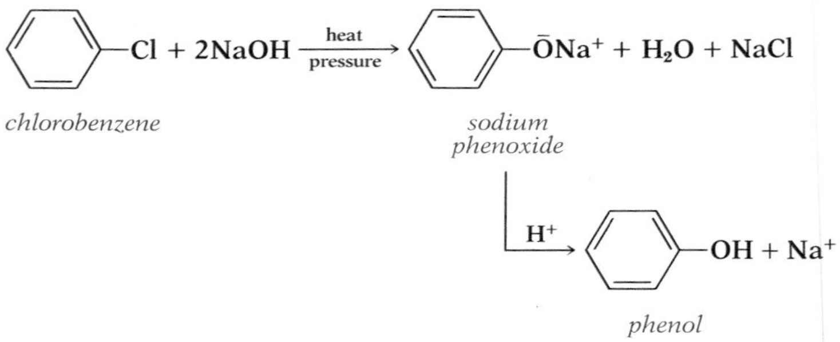

chlorobenzene   *sodium phenoxide*

phenol

Millions of pounds of phenol are made annually by this and other processes (1700 million pounds in 1975).

32.5 Ethers

An ether can be viewed as derived from water by replacing both hydrogen atoms with hydrocarbon groups. An **ether** has the general formula ROR. If the two hydrocarbon groups are alike, the ether is a *simple* or *symmetrical* ether; if they are different, the ether is a *mixed* or *unsymmetrical* ether.

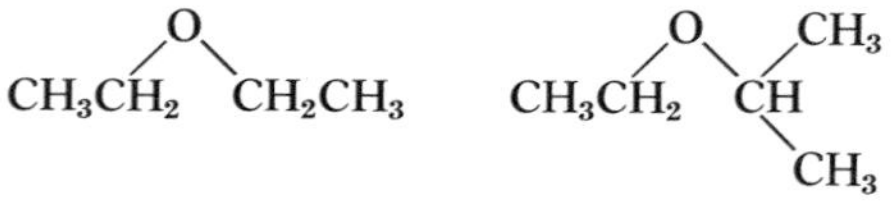

diethyl ether a simple ether   *ethyl isopropyl ether a mixed ether*

The common system of nomenclature names the groups attached to the oxygen atom followed by the word *ether,* as just shown. Sometimes the *di* is omitted when the groups are identical. Diethyl ether is then simply named ethyl ether. RO— is an **alkoxy group,** and in the IUPAC system of nomenclature, the ether is named as an alkoxy derivative of the longest-chain hydrocarbon to which the alkoxy group is attached. The position of attachment is given by a number, starting at the end of the chain nearest the alkoxy group. Thus, ethyl isopropyl ether is 2-ethoxypropane. Table 32.8 gives several examples. Note that cyclic ethers are also known.

An ether cannot form hydrogen bonds with itself. As a consequence, the boiling points of ethers are near those of hydrocarbons having the same molecular weight. The effect of hydrogen bonding is

Note that the two isomers differ in structure only in the position of one double bond, but differ considerably in activity.

dramatically exposed by a comparison of the boiling points of ethylene glycol ($HOCH_2CH_2OH$, b.p. 197°C) and 1,2-dimethoxyethane ($CH_3OCH_2CH_2OCH_3$, b.p. 85°C).

The ethers are rather unreactive and are often used as solvents in carrying out reactions. However, ethers do have an unfortunate tendency to form peroxides (compounds containing —O—O— bonds) when exposed to the oxygen in air. These peroxides are susceptible to explosive decomposition and for this reason special precautions must always be used when working with ethers.

Diethyl ether, a volatile and flammable liquid, is often used as a general anesthetic and is an excellent solvent. The ether linkage is found in many naturally occurring compounds, among them tetrahydrocannabinol, which in its two isomeric forms (Table 32.9) is the active component in marijuana and hashish from the hemp plant.

Ethers can be obtained from alcohols by intermolecular dehydration, a method that is used industrially.

$$2CH_3CH_2OH \xrightarrow[\text{heat}]{\text{conc. } H_2SO_4} CH_3CH_2OCH_2CH_3 + H_2O$$
ethyl alcohol   *diethyl ether*

32.6 Amines

Replacement of the hydrogen atoms of an ammonia molecule by hydrocarbon groups gives amines (not to be confused with ammines, Chapter 28). **Amines** are classified (Table 32.10) as primary (RNH_2), secondary (R_2NH), or tertiary (R_3N), according to the number of hydrogen atoms of the ammonia molecule that have been replaced by bonds to carbon atoms. Addition of a fourth hydrocarbon group yields a quaternary ammonium ion. The common system of nomenclature for amines is illustrated in Table 32.10. Aniline, pyridine, and morpholine (Table 32.11) are nonsystematic names.

Aromatic amines are those with the amine group attached to an aromatic ring, as in aniline. Such amines are usually prepared by

Amines, RNH$_2$, R$_2$NH, R$_3$N

TABLE 32.12

Some physiologically active amines

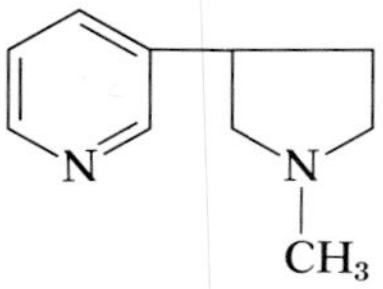

nicotine

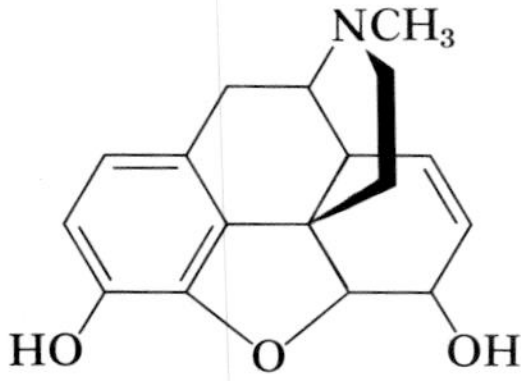

morphine

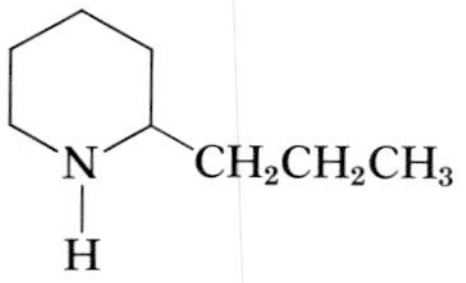

coniine

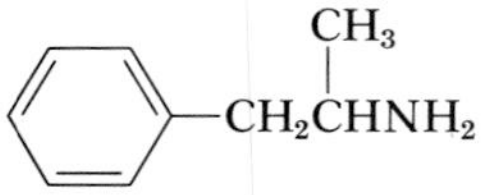

amphetamine
(benzedrine)

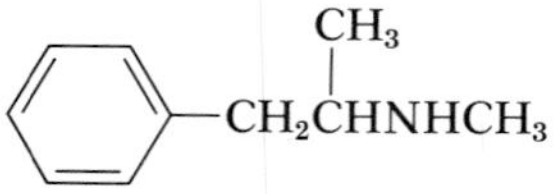

methedrine
("speed")

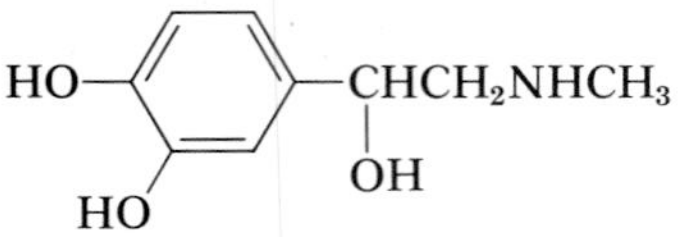

epinephrine
(adrenalin)

the nitration of an aromatic compound and subsequent reduction of the nitro group, NO$_2$, with, for example, tin and concentrated hydrochloric acid.

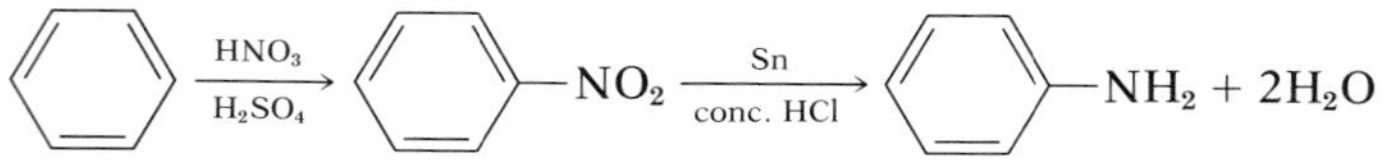

benzene   nitrobenzene   aniline

The reaction of ammonia with alkyl halides also produces amines.

$$CH_3CH_2Cl + NH_3 \longrightarrow CH_3CH_2\overset{+}{N}H_3Cl^- \xrightarrow{NH_3} CH_3CH_2NH_2 + NH_4Cl$$

ethyl
chloride

ethylammonium
chloride

ethylamine

In this reaction the ethylamine can react further with ethyl chloride to form diethylamine, which can then react to form triethylamine. The mixture from the reaction of an alkyl halide with ammonia normally contains all of the possible amines (RNH$_2$, R$_2$NH, R$_3$N), as well as the quarternary ammonium salt, R$_4$NX. Amines with unlike groups attached to the nitrogen atom are attainable by an extension of this method.

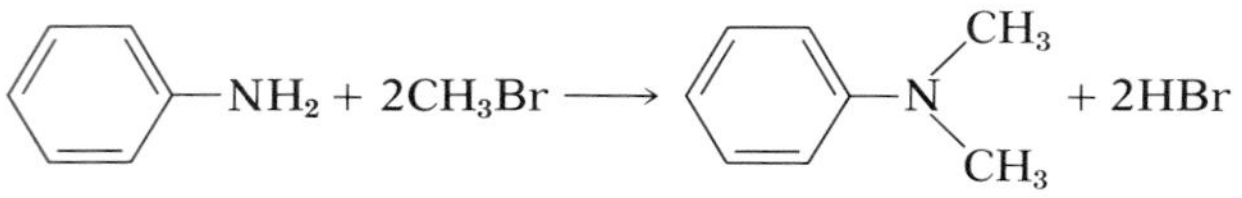

aniline

N,N-dimethylaniline
b.p. 193°C

Amines are basic. Like ammonia, they form salts with acids.

$$CH_3NH_2 + HCl \longrightarrow CH_3NH_3^+Cl^-$$

methylamine

methylammonium
chloride

The aliphatic amines are comparable to ammonia in basic strength. Aromatic amines are considerably weaker bases, because, in theory, the unshared electron pair on the nitrogen can be delocalized around the ring.

Most amines have fairly strong, unpleasant odors; the salts are odorless, water-soluble solids. Amine salts are easily converted to the amine by treating with a base.

$$CH_3\overset{+}{N}H_3\overset{-}{C}l + NaOH \longrightarrow CH_3NH_2 + NaCl + H_2O$$

Physiologically active amines (Table 32.12) found in plants are known as *alkaloids*. Nicotine is present in tobacco and is used as an insecticide. Morphine, a sedative and analgesic (pain killer), is a major component of opium. Codeine, another popular analgesic, is the methyl ether (at the phenolic hydroxyl group) of morphine. Heroin is also a derivative of morphine. Coniine, a relatively simple alkaloid, is the toxic substance in the hemlock that poisoned Socrates. The "uppers," or stimulating drugs, amphetamine and methedrine are also amines. Note the structural similarity of these stimulants to epinephrine, also known as adrenalin, a naturally occurring stimulant that is released by the adrenal glands under conditions of fear or stress.

Aliphatic primary amines react quantitatively with nitrous acid and liberate nitrogen.

$$\underset{\substack{methylamine}}{CH_3NH_2} + \underset{(HCl + NaNO_2)}{HNO_2} \longrightarrow CH_3OH + N_2 + H_2O$$

(Nitrous acid, HNO_2, is not stable, so to carry out this reaction, sodium nitrite is added to an aqueous solution of the amine salt in dilute hydrochloric acid. Nitrous acid is formed in solution and reacts with the amine salt.) Aromatic primary amines react with a chilled nitrous acid solution to form water-soluble and fairly unstable **diazonium salts,** $ArN_2^+X^-$

aniline   benzenediazonium chloride

Much of the early incentive to study organic compounds arose from the desire to make synthetic dyes. An important class of dyes, the azo dyes, is formed by the coupling of diazonium salts with phenols or aromatic amines.

p-nitrobenzene-
diazonium chloride   2-naphthol

para red
an azo dye

The *azo group,* —N=N—, is a **chromophoric group,** that is, a color-producing group. Any organic molecule that possesses a chromophoric group is colored. However, not all colored compounds are good dyes. For example, a satisfactory fabric dye must adhere to the fabric and be stable to light, heat, and soap. Para red has these qualities.

A dye that is suitable as an indicator of the pH of a solution must change color within a narrow pH range (Section 17.7). Methyl orange is yellow in a solution having a pH above 4.4 and becomes red when the solution is more acidic than a pH of 3.1. Obviously, some change in molecular structure must accompany the change in color. This change is expressed in the following equation.

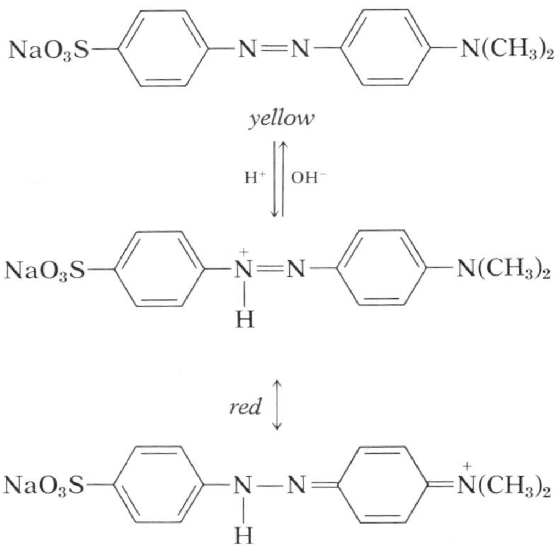

yellow

H⁺ ‖ OH⁻

H

red ↓

H

At a pH of 3.1 a nitrogen atom of the azo group adds a proton, and the resulting positive charge is delocalized as shown. The azo group is not present in either of the resonance structures, but the structural feature

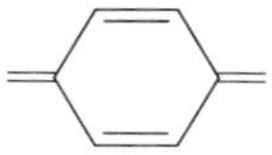

is also a chromophoric group, and the contribution of the resonance structure containing it accounts for the red color in the more acidic solution. This new chromophoric group is referred to as a *quinonoid* structure, and all compounds possessing this particular arrangement of atoms and bonds are colored.

Functional groups with covalent double bonds

32.7 Aldehydes and ketones

The **carbonyl group** consists of a carbon atom and an oxygen atom joined by a covalent double bond.

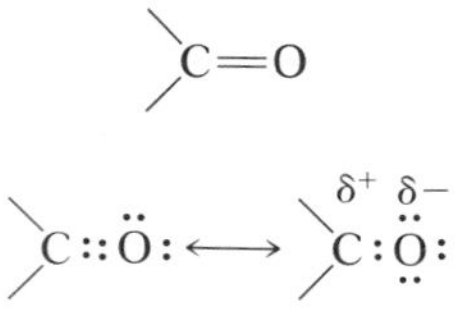

a carbonyl group

Like the covalent double bond between two carbon atoms, the covalent double bond in a carbonyl group is a combination of a σ bond and a π bond. The carbonyl group and its two attached groups lie

---

TABLE 32.13

Aldehydes

The general molecular formula for an aldehyde, often written as RCHO, *is*

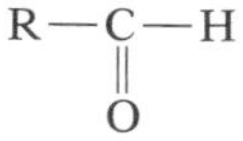

R—C—H
‖
O

	Boiling point (°C)
HCHO	−21
formaldehyde	
(methanal)	
CH₃CHO	21
acetaldehyde	
(ethanal)	
CH₃(CH₂)₄CHO	128
caproaldehyde	
(hexanal)ᵃ	
CH₂=CHCHO	52
acrolein	
(propenal)	
—CHO	179
benzaldehyde	

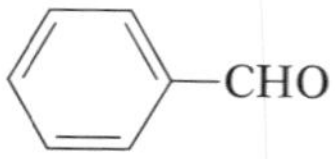

ᵃ Note that the aldehyde group must be at the end of a chain, so it is unnecessary to write 1-hexanal.

in the same plane and include an sp^2-type carbon atom with valence angles near 120°. The actual molecular structure of the carbonyl group is a hybrid of the two principal resonance structures shown above. In effect, the carbon atom is slightly positive and the oxygen atom is slightly negative. The greater electronegativity of the oxygen atom plays a role in establishing this bond polarity.

The carbonyl group is present in aldehydes and ketones, and in the rest of the functional groups discussed in this chapter.

An **aldehyde**, RCHO, is a compound in which a hydrogen atom and a hydrocarbon group are bonded to a carbonyl group (Table 32.13). In the simplest possible case, R is a hydrogen atom also. The common names of the aldehydes are derived from the names of the carboxylic acids (next section) containing the same carbon skeleton by replacing the final *ic* by *aldehyde.* Thus, formaldehyde, HCHO, is the aldehyde corresponding to formic acid, HCOOH. The IUPAC system of nomenclature for aldehydes requires only the replacement of the final *e* of the name of the longest-chain hydrocarbon including the aldehyde group by *al.* Substitutions or branches along this main chain are designated by numbers (the carbonyl carbon atom is numbered 1) and names by prefixes. Table 32.13 gives several examples.

Formaldehyde is used mainly in the manufacture of phenol–formaldehyde resins. It also finds some use as a disinfectant and a preservative, often as *formalin,* a 37% aqueous solution of formaldehyde. Trichloroacetaldehyde, Cl_3CCHO, forms a hydrate, chloral hydrate $Cl_3CCH(OH)_2$, that is a sleep-inducing drug.

Several aldehydes have been isolated from fruits, to which they contribute characteristic odor and taste. Citral, a terpene aldehyde, is found in the oil from citrus fruits and in oil of lemon grass. It has a strong lemonlike odor. Vanillin is used as a vanilla flavoring agent.

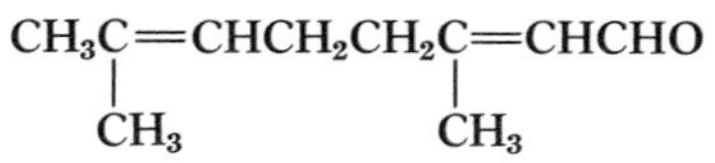

$$CH_3C=CHCH_2CH_2C=CHCHO$$

citral
(3,7-dimethyl-2,6-octadienal)

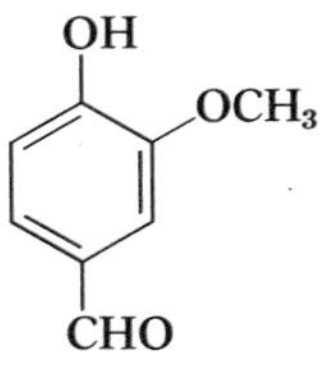

vanillin

A **ketone**, $R_2C=O$, contains two hydrocarbon groups attached to the carbon atom of the carbonyl group (Table 32.14). In the common system of nomenclature, the two hydrocarbon groups are named, followed by the word *ketone,* as in methyl ethyl ketone, $CH_3COCH_2CH_3$. Dimethyl ketone is almost always referred to by its trivial name, *acetone.* In the IUPAC system of nomenclature, a ketone is designated by the suffix *one,* which replaces the final *e* of the name of the longest-chain hydrocarbon containing the carbonyl group. A number is normally used to show the position of the carbonyl group, starting at the end of the chain nearest the carbonyl group. Thus, methyl ethyl ketone is 2-butanone in the IUPAC system. Substitutions and branches on the main chain are treated as before (Section 23.4). Table 32.14 gives several examples.

Ketones are used extensively as solvents, especially in lacquers

---

Aldehydes, RCHO

Ketones, $R_2C=O$

TABLE 32.14

Ketones

The general molecular formula for a ketone, often written as R_2CO or $RCOR'$, and in which R and R' may be the same or different, is

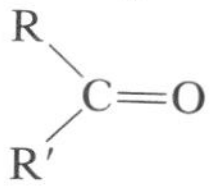

$$R{\diagdown} \atop R'{\diagup} C=O$$

	Boiling point (°C)
CH_3COCH_3 *dimethyl ketone (acetone)* *(2-propanone)*	56
$CH_3CH_2COCH_2CH_3$ *diethyl ketone* *(3-pentanone)*	102
$CH_3COCH_2CH_2CH_3$ *methyl n-propyl ketone* *(2-pentanone)*	102
$CH_3COCOCH_3$ *biacetyl[a] (dimethylglyoxal)* *(2,3-butanedione)*	88
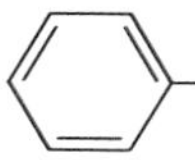 ⬡—$COCH_3$ *methyl phenyl ketone* *(acetophenone)*	200

---

[a] This is one of the few colored (yellowish green) organic liquids. Practically all *pure* organic liquids are colorless; many pure solids are colored, however.

Infrared and ultraviolet spectroscopy

$\mathbf{A}$bsorption spectra arise when energy from incoming radiation is absorbed, causing atoms, molecules, or ions to move into a higher energy state. Among the valuable tools of the organic chemist are infrared (IR) and ultraviolet (UV) absorption spectroscopy.

The infrared region of the electromagnetic spectrum extends from 2.5 μm to 15 μm (1 μm = 1×10^{-4} cm = 1000 nm). Infared spectra originate in an increase in energy level of the motions of atoms in a molecule relative to each other. The atoms in a molecule are never motionless. They may, for example, bend or stretch (Figure A) or undergo other motions known as wagging, rocking, deformation, and so on. When the motion involves a change in dipole moment, and when the frequency of incoming radiation matches the frequency of the motion, absorption occurs and a "peak" appears in the infrared spectrum. Figure B shows the spectra of ethylene glycol and ethylenediamine with the origins of many of the peaks identified (str = stretching; def = deformation; wag = wagging). Customarily in infrared spectra the percent of the radiation *transmitted* (% transmittance) is plotted versus the wavelength (in μm, for example) or wave numbers (in cm^{-1}).

The more complicated a molecule is, the more possibilities there are for relative motions of the atoms. Therefore, infrared spectra can be quite complex, and the larger the molecule, the more peaks will appear in the spectrum. In spectra with many peaks, identification of the origin of all the peaks is difficult.

Infrared spectra are most useful in two ways. One depends upon the fact that the same functional groups cause absorption at the same or close to the same wavelength in the spectra of all the molecules in which they occur. From the study of the spectra of thousands of compounds "correlation charts" have been assembled. These charts show the location of peaks that result from the presence of specific groups of atoms. A small portion of a correlation chart is shown in Figure C. This segment includes the location of the peaks of functional groups that contain a carbon-nitrogen single bond.

In the determination of the structure of an unknown compound, the infrared spectrum is used to indicate which functional groups are present by comparison of the spectrum with such charts. This is often not as simple and straightforward as it sounds, but with experience, much useful information about structure can be gleaned from infrared spectra. For example, an experienced observer would quickly know from the two spectra in Figure B that one compound contained OH groups and the other, NH_2 groups.

Infrared spectra, like mass spectra, are also used as fingerprints. It has been said that the infrared spectrum of a compound may be its

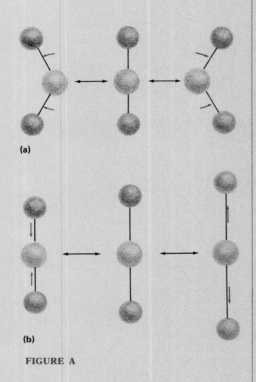

(a)

(b)

FIGURE A

FIGURE B
**Infrared spectra of ethylene glycol
(1,2-ethanediol) and ethylenediamine
(1,2-diaminoethane).**

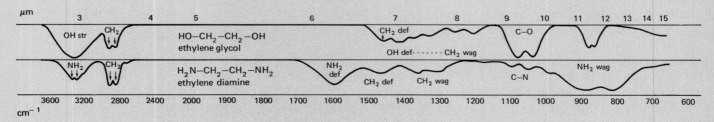

C—N ——M—— ——M——NH₂ Primary amine NH₂— ——M—— aryl-NH₂— ——S CH₂—NH₂— ——M—— NH₂— ——S—— *(broad)*
   ——W—— —NH Secondary amine   aryl-NH— ——S   ——M—— —CH₂—NH—CH₂ CH₂—NH—CH₂— ——M——
—N(CH₂)₂, —N(CH₃)₂ — — Tertiary amine   aryl-N(CH₃)₂— ——S
       ——S——NH₄⁺
          ——S——NH₃⁺   Ammonium ion
              —NH₂⁺   Primary amine salt   —NH₃⁺ ——M—— NH₄⁺— ——S
       ——S——   —NH⁺ Secondary amine salt NH₂⁺— ——M—— ——M——
                        Tertiary amine salt

| 3600 | 3200 | 2800 | 2400 | 2000 | 1900 | 1800 | 1700 | 1600 | 1500 | 1400 | 1300 | 1200 | 1100 | 1000 | 900 | 800 | 700 | 600 |

cm⁻¹

most characteristic physical property. By comparison of the spectrum of a compound, say a reaction product that has not been identified, with the spectra of known compounds that might be products, identification can often be made. For example, if a compound was known to be either ethylenediamine or ethylene glycol, comparing its spectrum with the two known spectra in Figure B would quickly show which compound had been produced.

The near-ultraviolet region of the electromagnetic spectrum extends from 200 to 400 nm. (Far-ultraviolet spectra, at longer wavelengths, are difficult to measure experimentally and are of less practical value.) Ultraviolet radiation is of higher energy than infared radiation. At such energy, absorption occurs by transitions of outer π or nonbonding electrons into higher energy levels, usually into unoccupied antibonding, σ^* or π^* orbitals.

A simple saturated hydrocarbon does not absorb in the near-ultraviolet region. The presence of carbon-carbon multiple bonds or of functional groups containing multiple bonds, that is the presence of π electrons, and π orbitals, is required for UV absorption. Ultraviolet spectra are much more simple than infrared spectra. They show a single peak or a few peaks for each isolated functional group or carbon-carbon multiple bond present in the molecule. Ultraviolet spectra are described in terms of λ_{max}, the wavelength of maximum absorption. Usually measured in solution, ultraviolet spectra are presented as plots of absorption versus wavelength or, as in the pyridine spectrum shown in Figure D, as the logarithm of the molar absorptivity, log ϵ, versus wavelength. Molar absorptivity, ϵ, is a function of the amount of radiation transmitted, the concentration of the substance in solution, and the length of the path of the light through the sample. Ultraviolet absorption peaks are shifted to longer wavelengths when a molecule contains conjugated systems, for example, —CH=CH—CH=CH— or —CH=CH—CH=O.

Ultraviolet spectra can be used for structure analysis and compound identification in the same way as infrared spectra are used. However, their application in these ways is limited because there are fewer peaks in ultraviolet spectra and the peaks are broader. Spectra in the near-ultraviolet region are more often used in quantitative analysis, as are spectra in the visible region. The spectra are measured in solution and the intensity of the absorption varies with the concentration of the absorbing species. The relationship between the ultraviolet absorption and concentration is first determined experimentally. Then an unknown concentration can be found by comparison with the data for known concentrations. Titrations can be followed by spectral changes, and frequently, the concentration of a substance in a complex mixture can be measured by its characteristic absorption.

FIGURE C

Portion of an infrared correlation chart. *This segment shows the location of peaks for groups that contain carbon-nitrogen single bonds. The letters S, M, and W indicate strong, medium, and weak bands, respectively.*

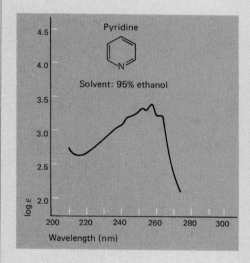

FIGURE D

Ultraviolet spectrum of pyridine. (*From R. M. Silverstein et al.,* Spectrometric Identification of Organic Compounds, *3rd ed., John Wiley & Sons, Inc., New York, 1974.*)

and other coatings. They frequently serve as starting materials for organic synthesis. Phenacyl chloride (Table 32.15), a potent lachrymator (tear inducer), is used in tear gas (Mace and CN). Camphor, from the camphor tree, is a terpene ketone that has some use in medicine and plastics. Methadone is an addictive analgesic and is used in a controlled program to combat heroin addiction.

Alcohols are often the starting materials for the preparation of aldehydes and, especially, ketones. Sodium dichromate easily oxidizes a primary alcohol to an aldehyde and a secondary alcohol to a ketone.

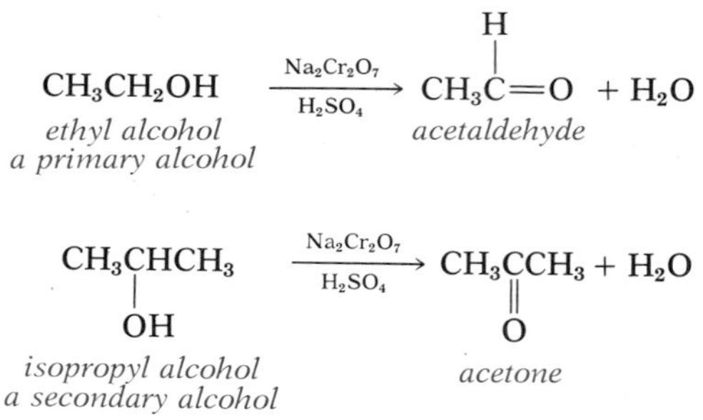

TABLE 32.15

Some additional ketones

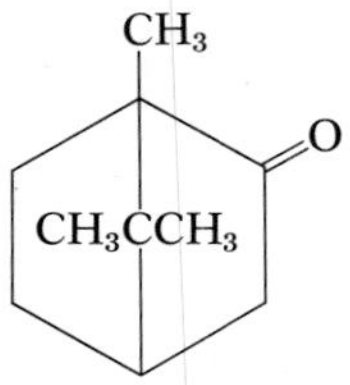

camphor

phenacyl chloride

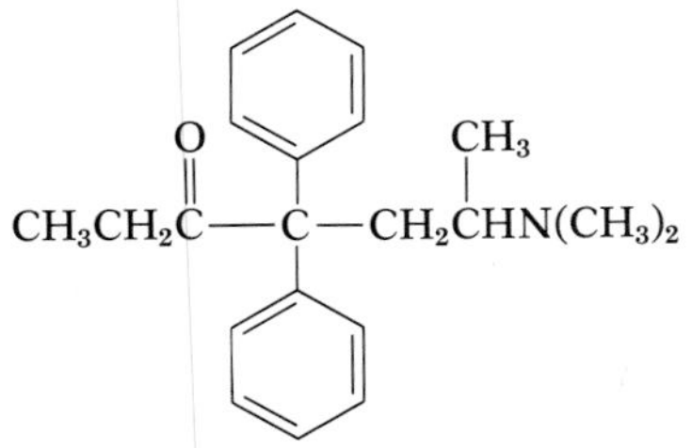

methadone

$$CH_3CH_2OH \xrightarrow[H_2SO_4]{Na_2Cr_2O_7} CH_3\overset{\displaystyle H}{\underset{}{C}}{=}O + H_2O$$

ethyl alcohol
a primary alcohol
acetaldehyde

$$CH_3\underset{\underset{OH}{|}}{C}HCH_3 \xrightarrow[H_2SO_4]{Na_2Cr_2O_7} CH_3\underset{\underset{O}{\|}}{C}CH_3 + H_2O$$

isopropyl alcohol
a secondary alcohol
acetone

A ketone is resistant to further oxidation under these conditions, but an aldehyde is readily oxidized to the corresponding carboxylic acid (next section). This method for the preparation of aldehydes requires the immediate removal of the aldehyde from the reaction medium, a circumstance that limits the method to quite volatile aldehydes.

Commercially, the conversion of an alcohol to an aldehyde or ketone is often effected by passing the alcohol vapor and air over a heated copper catalyst in a **dehydrogenation reaction**—a reaction in which H_2 is removed.

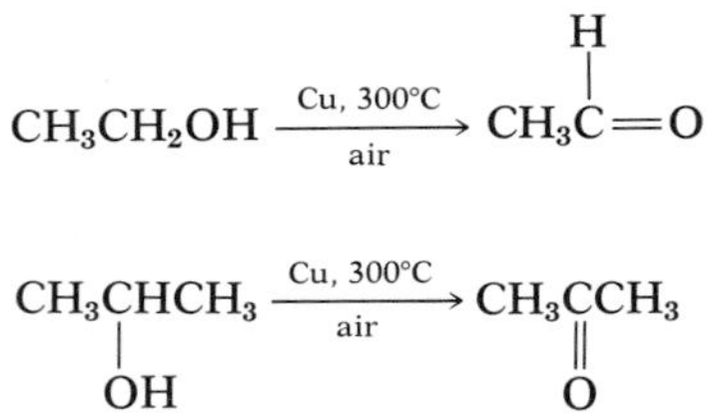

$$CH_3CH_2OH \xrightarrow[air]{Cu,\ 300°C} CH_3\overset{\displaystyle H}{\underset{}{C}}{=}O$$

$$CH_3\underset{\underset{OH}{|}}{C}HCH_3 \xrightarrow[air]{Cu,\ 300°C} CH_3\underset{\underset{O}{\|}}{C}CH_3$$

Most acetone is made in this manner.

Aldehydes, but not ketones, are easily oxidized.

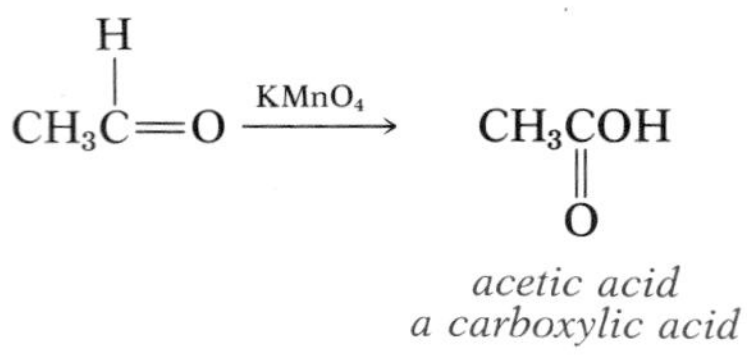

$$CH_3\overset{\displaystyle H}{\underset{}{C}}{=}O \xrightarrow{KMnO_4} CH_3\underset{\underset{O}{\|}}{C}OH$$

acetic acid
a carboxylic acid

Even atmospheric oxygen will bring about the conversion of many aldehydes to the corresponding acids. A very mild oxidizing agent,

[Ag(NH₃)₂]OH, called Tollens' reagent, provides a diagnostic test for aldehydes.

$$RCHO + 2[Ag(NH_3)_2]OH \longrightarrow 2Ag + RCOONH_4 + H_2O + 3NH_3$$

When this reaction is carried out in a clean test tube, the elemental silver that forms deposits as a silver mirror on the walls of the test tube. A process for the silvering of mirrors utilizes this reaction.

32.8 Carboxylic acids

A **carboxylic acid**, RCOOH, has a functional group that is a combination of a carbonyl group and a hydroxyl group

$$\overset{\displaystyle O}{\underset{}{\overset{\|}{R C O H}}}$$

a carboxylic acid

The more familiar carboxylic acids carry common names that are associated with the sources from which they were first isolated. In the IUPAC system of nomenclature, the ending *oic* replaces the final *e* of the hydrocarbon name that is formed from the longest continuous chain of carbon atoms including the carboxyl group. The word *acid* follows this derived name. When this chain of carbon atoms is numbered, the carboxylic carbon atom is numbered 1. Table 32.16 shows the structures and names of some of the more common carboxylic acids.

Simple, unsubstituted monocarboxylic acids have pK_a values in the range 4–5; they are weaker as acids than sulfonic acids (RSO_3H) but stronger than phenols (Section 32.4). The ionization of a carboxylic acid, as shown, produces the **carboxylate ion**, $RCOO^-$.

$$RC{=}O \rightleftharpoons H^+ + RC\overset{O}{\underset{O^-}{\big<}} \longleftrightarrow RC\overset{O^-}{\underset{O}{\big<}}$$

a carboxylate ion

This ion is stabilized by the delocalization of the negative charge over the two oxygen atoms.

All of the carboxylic acids form salts when they are treated with bases (e.g., hydroxides, carbonates, or bicarbonates).

$$\underset{\substack{\text{acetic acid} \\ \text{(ethanoic acid)}}}{CH_3COOH} + NaOH \longrightarrow \underset{\substack{\text{sodium acetate} \\ \text{(sodium ethanoate)}}}{CH_3COO^-Na^+} + H_2O$$

The name of the salt is obtained from the name of the cation, followed by the name of the carboxylic acid with the final *ic* replaced by *ate*.

Carboxylic acids, RCOOH

TABLE 32.16

Carboxylic acids

The general molecular formula for a carboxylic acid, often written RCOOH or RCO_2H, and in which R can be a hydrogen atom, is

$$\overset{\displaystyle}{\underset{O}{\overset{\|}{RC{-}OH}}}$$

	Boiling point (°C)
HCOOH *formic acid* *(methanoic acid)*	101
CH₃COOH *acetic acid* *(ethanoic acid)*	118
CH₃CH₂COOH *propionic acid* *(propanoic acid)*	140
CH₃CH₂CH₂COOH *butyric acid* *(butanoic acid)*	163
CH₂=CHCOOH *acrylic acid* *(propenoic acid)*	140

	Melting point (°C)
⬡—COOH *benzoic acid*	121
⬡ with COOH, COOH *phthalic acid*	230

TABLE 32.17

Some additional carboxylic acids

$$H-\underset{\underset{CH_3}{|}}{\overset{\overset{COOH}{|}}{C}}-OH$$

lactic acid

$$H-\underset{\underset{HO-\underset{\underset{COOH}{|}}{C}-H}{|}}{\overset{\overset{COOH}{|}}{C}}-OH$$

tartaric acid

$$HO-\underset{\underset{CH_2COOH}{|}}{\overset{\overset{CH_2COOH}{|}}{C}}-COOH$$

citric acid

prostaglandin $F_{1\alpha}$
(one of several prostaglandins)

TABLE 32.18

Esters of carboxylic acids

The general molecular formula for a carboxylic acid ester, often written as RCOOR' *or* RCO₂R', *and in which* R *and* R' *may be alike or different, is*

$$R-\underset{\underset{O}{||}}{C}-O-R'$$

	Boiling point (°C)
HCOOCH₃ *methyl formate* *(methyl methanoate)*	32
CH₃COOCH₂CH₃ *ethyl acetate* *(ethyl ethanoate)*	77
CH₂=CHCOOCH₃ *methyl acrylate* *(methyl propenoate)*	85
⬡—COOCH₃ *methyl benzoate*	198
⬡ COOC₂H₅ / COOC₂H₅ *diethyl phthalate*	298

[32.8] ORGANIC CHEMISTRY

Acetic acid is produced in large quantities (2000 million pounds in 1975) for use in making plastics and solvents. Phthalic acid is also widely used in the production of varnishes, paints, and plastics (4000 million pounds in 1975).

Many carboxylic acids occur in nature, either free or combined in some form. The combination of hydroxyl groups and carboxylic acid groups in the same molecule to give hydroxy acids is very common in nature. Lactic acid (Table 32.17) is found in sour milk (racemic form) and in the muscles of man and animals (dextrorotatory isomer). Tartaric acid (dextrorotatory isomer) occurs in many fruits. Citric acid occurs in both plants and animals; lemon juice contains 5–8% citric acid.

Currently there is considerable interest in some physiologically active substances called prostaglandins (Table 32.17). These substances are found in many body tissues and fluids, but they seem to occur in greatest amounts in human and sheep seminal plasma. Because of their broad physiological activity, high hopes are held for their use in medicine, particularly with respect to induced abortion, prevention of conception, and the regulation of menstruation and fertility.

32.9 Esters

A compound in which the acidic hydrogen atom of a carboxylic acid molecule is replaced by a hydrocarbon group is an **ester**, RCOOR', for example

$$CH_3\overset{\overset{O}{||}}{C}OCH_3$$

methyl acetate

In the two-word name for an ester, the first word is the name of the R' group in RCOOR'. This word could be methyl (CH_3), ethyl (C_2H_5), phenyl (C_6H_5), or the like. The second word of the name is the name of the carboxylic acid with the final *ic* replaced by *ate*, identical with the name of the carboxylate anion. Table 32.18 illustrates the common and IUPAC system names.

Esters of the lower carboxylic acids and lower alcohols, for example, ethyl acetate and butyl acetate, are used extensively as solvents. Most esters of carboxylic acids have quite pleasant odors, even when derived from rather foul-smelling acids. Many fruits and flowers owe their flavor and fragrance to the esters present (Table 32.19); esters are often used as artificial flavors.

Aspirin, the popular fever reducer and pain killer, is an ester formed between acetic acid, CH_3COOH, and salicylic acid. When the aspirin reaches the alkaline environment of the intestines, the salicylic acid is liberated. Salol (phenyl salicylate) is used to coat a pill when it is desired that the pill pass through the stomach unchanged and release its contents in the intestines. (Such a coating is called an enteric coating.) The salol coating is inert to the acidic conditions in

TABLE 32.19

Characteristic odors of some esters

The natural flavoring and odor of a food arise from a combination of a large group of substances and not from just a single compound. For example, the total number of flavoring substances in the strawberry is over 250. These substances are esters, alcohols, carboxylic acids, aldehydes, ketones, and several other classes of organic compounds.

Structure and common name	Characteristic odor
$CH_3COOCH_2CH_2CH(CH_3)_2$ *isoamyl acetate*	**Banana**
$CH_3CH_2CH_2COO(CH_2)_4CH_3$ *amyl butyrate*	**Apricot**
$(CH_3)_2CHCH_2COOCH_2CH_2CH(CH_3)_2$ *isoamyl isovalerate*	**Apple**
$CH_3CH_2CH_2COOC_2H_5$ *ethyl butyrate*	**Pineapple**
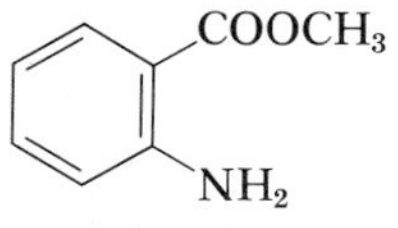 ⬡—$COOCH_3$ / NH_2 *methyl anthranilate*	**Grape**
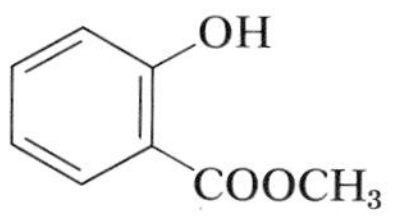 ⬡—OH / $COOCH_3$ *methyl salicylate*	**Wintergreen**

the stomach, but it is hydrolyzed and dissolves in the alkaline intestines.

salicylic acid | aspirin (acetylsalicylic acid) | salol (phenyl salicylate)

Inorganic acids also form esters, such as glyceryl trinitrate and dimethyl sulfate, $(CH_3)_2SO_4$, as well as sulfites, phosphites, phosphates, nitrites, borates, and others. Malathion, an important insecticide, is an ester of both an inorganic acid ($H_3PO_2S_2$) and a carboxylic acid.

$$(C_2H_5O)_2\underset{\underset{S}{\parallel}}{P}\underset{\underset{CH_2COOC_2H_5}{|}}{S}CHCOOC_2H_5$$

malathion

One advantage in the use of malathion as an insecticide is that it is not persistent and is degraded in a few weeks after application. This behavior is in contrast to that of DDT (Section 13.11), which remains unchanged in the environment for years. On the other hand, malathion is an enzyme poison and is highly toxic to animals, whereas DDT is less hazardous in this respect.

A carboxylic acid and an alcohol form an ester in an **esterification reaction** when they are heated together in the presence of an acid catalyst.

$$\underset{\underset{OH}{|}}{CH_3C}=O + CH_3CH_2OH \underset{H_2SO_4}{\rightleftharpoons} CH_3\overset{\overset{O}{\parallel}}{C}-OCH_2CH_3 + H_2O$$

acetic acid   ethyl alcohol   ethyl acetate

The hydrogen ion from the sulfuric acid attaches to the oxygen atom of the carbonyl group, thereby enhancing the positive character and the reactivity of the carbonyl carbon atom.

The hydrolysis of an ester is the reversal of the reaction leading to its formation.

$$CH_3\overset{\overset{O}{\parallel}}{C}-OCH_2CH_3 + H_2O \underset{H^+}{\rightleftharpoons} CH_3COOH + CH_3CH_2OH$$

ethyl acetate   acetic acid   ethyl alcohol

Hot aqueous sodium hydroxide will also hydrolyze esters; the reaction is referred to as **saponification,** a term that is associated with the process for making soap (Section 13.12).

$$CH_3COOCH_2CH_3 + NaOH \xrightarrow{H_2O} CH_3COONa + CH_3CH_2OH$$

ethyl acetate   *sodium acetate*   *ethyl alcohol*

The saponification reaction is initiated by the attachment of the hydroxide ion, a relatively good nucleophile, to the carbonyl carbon atom.

32.10 Acyl halides and carboxylic acid anhydrides

The replacement of the hydroxyl group of a carboxylic acid by a halogen atom gives an **acyl halide,** RCOX, also called an acid halide (Table 32.20). An **acyl group**

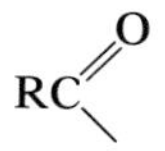

an acyl group

is named by dropping the final *ic* from the name of the corresponding carboxylic acid, and adding *yl*. The complete name of the acyl halide is then the name of the acyl group, followed by the anionic name for the halogen atom.

An **acid anhydride,** (RCO)$_2$O—the anhydride of a carboxylic acid—results from the loss of water between two carboxyl groups.

The anhydrides are given the name of the corresponding acid, followed by the word *anhydride*. Mixed anhydrides, from two different carboxylic acids, and cyclic anhydrides, are named similarly, for instance, acetic butyric anhydride, (CH$_3$CO)O(COC$_3$H$_7$), or phthalic anhydride (Table 32.20).

Acyl halides, RCOX
Acid anhydrides, (RCO)$_2$O

TABLE 32.20

Acyl halides and carboxylic acid anhydrides

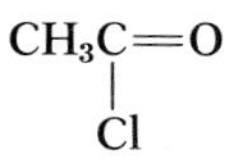

acetyl chloride
(ethanoyl chloride)
b.p. 55°C

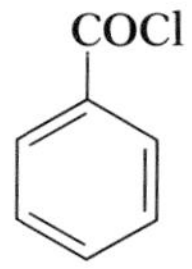

benzoyl chloride
b.p. 197°C

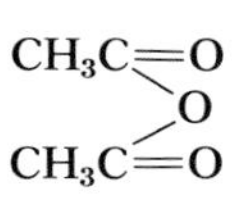

acetic anhydride
(ethanoic anhydride)
b.p. 138°C

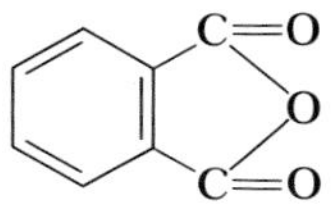

phthalic anhydride
m.p. 131°C

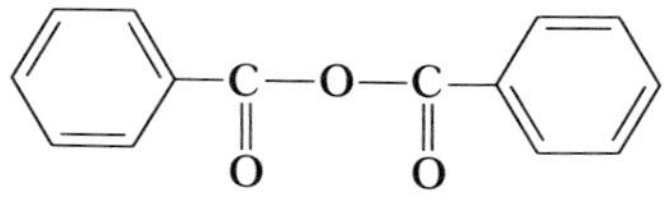

benzoic anhydride
m.p. 42°C

Amides, RCONH$_2$

TABLE 32.21

Amides

The general molecular formula for an amide, often written RCONH$_2$, *is*

$$\underset{\underset{NH_2}{|}}{RC\!\!=\!\!O}$$

$$\underset{\underset{NH_2}{|}}{CH_3C\!\!=\!\!O}$$

acetamide
(ethanamide)
m.p. 82°C

$$\underset{\underset{N(CH_3)_2}{|}}{HC\!\!=\!\!O}$$

N,N-dimethyl-
formamide
b.p. 153°C

benzamide
m.p. 128°C

N-phenylacetamide
(acetanilide)
m.p. 114°C

TABLE 32.22

Some additional amides

The acyl halides, usually the chlorides, are used almost exclusively for the synthesis of other compounds. Acid anhydrides, particularly acetic anhydride and phthalic anhydride, are valuable in the synthesis of esters, for example,

$$\underset{\substack{acetic\\anhydride}}{CH_3\overset{\overset{O}{||}}{C}O\overset{\overset{O}{||}}{C}CH_3} + \underset{\substack{ethyl\\alcohol}}{CH_3CH_2OH} \longrightarrow \underset{\substack{ethyl\\acetate}}{CH_3COOCH_2CH_3} + \underset{\substack{acetic\\acid}}{CH_3COOH}$$

32.11 Amides

The replacement of a hydrogen atom in an ammonia molecule by an alkyl group gives an amine (RNH$_2$). Replacement of a hydrogen atom in an ammonia or an amine molecule by an acyl group gives an **amide,** RCONH$_2$. The common name for an amide is derived by replacing the *yl* of the acyl name by *amide*. Thus, acetamide is the name for CH$_3$CONH$_2$. Since the acyl group is named by replacing the *ic* of the carboxylic acid name by *yl*, one can derive the amide name from the corresponding acid name by dropping the *ic* (or *oic* of the IUPAC name) and adding *amide* (Table 32.21).

If one or more of the hydrogen atoms on the amide nitrogen atom are replaced by hydrocarbon groups, the structure is named as an *N*-substituted amide.

Amide molecules form intermolecular hydrogen bonds with themselves, or with other appropriate molecules such as water. Practically all of the amides are solids. *N,N*-Dimethylformamide (DMF), one of the few liquid amides, is an excellent solvent. Urea, H$_2$NCONH$_2$, is the diamide of carbonic acid.

An amide of considerable current interest is the hallucinogenic drug LSD (lysergic acid diethylamide) (Table 32.22), also known as "acid" and by several other names. Lysergic acid, from which the diethylamide is derived, is present in ergot, a fungus found on rye.

p-Aminobenzenesulfonamide (sulfanilamide) (Table 32.22) and

lysergic acid
diethylamide
(LSD)

sulfanilamide

sulfadiazine

penicillin G (R is C$_6$H$_5$CH$_2$)

Aureomycin

some of its N-substituted derivatives are known as the *sulfa drugs*, an important group of drugs that combat the growth of bacteria.

The *penicillins*, which are amides, were the first of the antibiotics to be discovered; they have to a great extent displaced sulfa drugs in the treatment of infectious diseases. Many penicillins are known; they differ only in the R group of the structure shown. *Aureomycin* is a representative of a group of tetracycline antibiotics.

Compounds with two acyl groups bonded to the nitrogen atom of an ammonia molecule are known as imides. *Barbiturates* are cyclic imides that are depressants. They are the "downers" that have a sedative or soporific action, typified by Nembutal, Luminal, and Seconal.

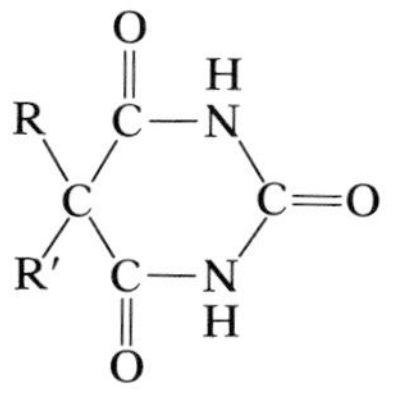

Nembutal (R, C_2H_5; R', $CH_3CH_2CH_2CHCH_3$)
Luminal (*phenobarbital*) (R, C_2H_5; R', C_6H_5)
Seconal (R, $CH_2=CHCH_2$; R', $CH_3CH_2CH_2CHCH_3$)

The molecules of biochemistry

It may be going too far to introduce a subject as broad as biochemistry in the last section of a book as long as this one. And yet, everyone should know at least a little about the marvelous molecular machinery that keeps us alive. So, particularly for those of you who will not have the opportunity to study the subject at another time, we give here a brief introduction to some of the molecules of biochemistry and what their functions are.

The four major classes of organic compounds in living cells are proteins, carbohydrates, lipids, and nucleic acids. Table 32.23 shows the amounts of these substances that, together with water, inorganic ions and smaller molecules, are present in the cells of a microorganism and an animal. The ratios vary from species to species and among different types of cells. But the proteins, carbohydrates, lipids, and nucleic acids perform similar functions in all plants and animals.

Many of the large molecules of biochemistry are polymers, in which many simple structural units are held together by similar types of bonds. However, while polystyrene, for example, is made up of thousands of identical repeating units

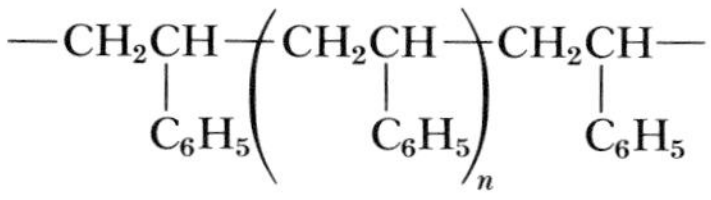

polystyrene

TABLE 32.23
Composition of cells

Component	Bacterium, E. coli (%)	Rat liver (%)
Water	70	69
Protein	15	21
Amino acids	0.4	—
Nucleic acids	7	1.2
Nucleotides	0.4	—
Carbohydrates	3	3.8
Lipids	2	6
Other small molecules	0.2	—
Inorganic ions	1	0.4

ORGANIC CHEMISTRY [32.11]

polymeric biochemical molecules are often made up of repeating units that are similar but different. The subtle and highly specific functions of many biochemical molecules are made possible by the combination of different small molecules in an exact sequence.

32.12 Proteins

Amino acids are the building blocks of proteins. As the name implies, an *amino acid* contains both an amine group and a carboxylic acid group. In solution and in the crystalline state, amino acids exist in an ionic, or inner salt, form.

$$R-\underset{\underset{NH_2}{|}}{\overset{\overset{H}{|}}{C}}-\overset{\overset{O}{\|}}{C}-OH \qquad R-\underset{\underset{NH_3^+}{|}}{\overset{\overset{H}{|}}{C}}-\overset{\overset{O}{\|}}{C}-O^-$$

an α-amino acid   *ionic form of an α-amino acid*

The α indicates that the amino group is attached to the carbon atom next to the carboxyl group. As the general formula shows, α-amino acids contain a carbon atom with four different groups attached. Therefore, these amino acids form optically active isomers (except when R = H), which are designated as either D or L forms. All amino acids found in proteins are α-L-amino acids. (For historical reasons, D and L in amino acid nomenclature refer to absolute configurations, not to the direction of rotation of polarized light.)

Amino acids combine with each other by the formation of a *peptide bond*, which can be pictured as the loss of water between a carboxyl group and an amine group. Because each amino acid has both functional groups, two amino acids can combine to form dipeptides in two different ways:

glycine   *alanine*

$$\underset{glycylalanine}{H_2NCH_2\overset{\overset{O}{\|}}{C}NH\underset{\underset{CH_3}{|}}{C}HCHCOOH} \quad \text{or} \quad \underset{alanylglycine}{H_2NCH\overset{\overset{O}{\|}}{C}NHCH_2COOH}$$

peptide linkage   *peptide linkage*

dipeptides

Twenty α-amino acids are the major constituents of all proteins. This is quite remarkable when we realize that proteins serve functions as diverse as supporting the body and regulating metabolism and growth (Table 32.24). The shapes of protein molecules vary with

TABLE 32.24
Some functions of proteins

Enzymes	Catalyze biochemical reactions
Hormones	Regulate growth and metabolism
Storage proteins	Release amino acids when they are needed
Transport proteins	Bind and transport other molecules in bloodstream
Structural proteins	Form bone, connective tissue, cartilage
Contractile proteins	Form muscle tissue
Protective proteins	Act as antibodies, blood-clotting agents

their size and function. Figure 32.1 shows some of the possible forms of a protein composed of 300 amino acids.

32.13 Carbohydrates

Carbohydrate molecules are either simple sugars, called *monosaccharides;* or polymers of two to ten sugars, called *oligosaccharides;* or polymers of more than ten sugars, called *polysaccharides.* The monosaccharides are either polyhydroxy aldehydes or polyhydroxy ketones (Table 32.25). Most monosaccharides contain several asymmetrically substituted carbon atoms, and like amino acids, monosaccharides and their derivatives form optically active isomers.

Sugars with five or six carbon atoms are more stable in a ring structure than in the open structures of Table 32.25. The ring is formed by the reaction of a carbonyl group with a hydroxyl group to give what is called a *hemiacetal.* The general reaction is

$$\underset{R'}{\overset{H}{{\diagdown}}}C{=}O + ROH \longrightarrow \underset{R' \quad OR}{\overset{H \quad OH}{{\diagdown}\ C\ {\diagup}}}$$

a hemiacetal

For D-glucose, writing the structures as though the ring is perpendicular to the plane of the paper, the ring forms as follows

β-D-glucose ⇌ D-glucose aldehyde form ⇌ α-D-glucose

The α and β indicate the two different configurations possible at carbon atom 1.

TABLE 32.25
Some monosaccharides

Aldoses (aldehydes)				Ketoses (ketones)	

D-glyceraldehyde   D-ribose   D-glucose   D-ribulose   D-fructose

FIGURE 32.1

Some shapes a protein molecule of 300 amino acids might assume. (*From D. E. Metzler,* Biochemistry: The Chemical Reactions of Living Cells, *p. 76, Academic Press, New York, 1977.*)

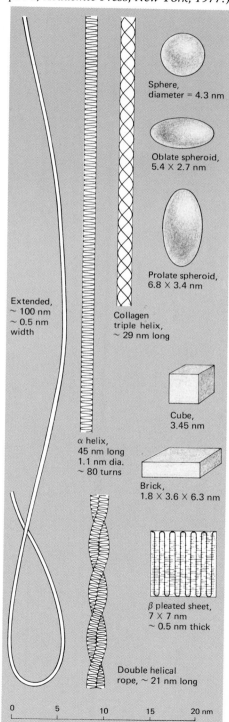

Extended,
~ 100 nm
~ 0.5 nm
width

α helix,
45 nm long
1.1 nm dia.
~ 80 turns

Sphere,
diameter = 4.3 nm

Oblate spheroid,
5.4 X 2.7 nm

Prolate spheroid,
6.8 X 3.4 nm

Collagen
triple helix,
~ 29 nm long

Cube,
3.45 nm

Brick,
1.8 X 3.6 X 6.3 nm

β pleated sheet,
7 X 7 nm
~ 0.5 nm thick

Double helical
rope, ~ 21 nm long

0   5   10   15   20 nm

1033

Derivatives of monosaccharides	Metabolism intermediates
	Plant pigments
	Blood anticoagulant (heparin)
	Vitamin C
Structural polysaccharides	Plant cell walls (cellulose)
	Animal cell coating
	Skin (keratin)
Storage polysaccharides	Release monosaccharides (starch in plants; glycogen in animals)
	Release energy when oxidized

Monosaccharides combine with each other by the loss of water between two OH groups to form a glycosidic linkage. For example, sucrose, or common table sugar, is a disaccharide of glucose and fructose.

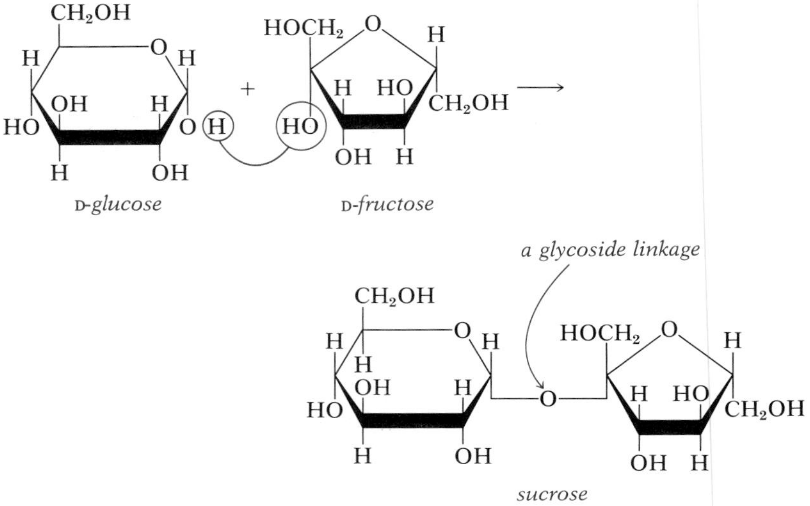

D-*glucose*   D-*fructose*

a glycoside linkage

sucrose

Cellulose and starch are both polymers of D-glucose. In cellulose, which may contain from hundreds to thousands of glucose units, linear molecules are organized into bundles. Cotton is 98% cellulose. Starch, which is the major source of energy in plants, contains two different polymers of D-glucose—amylose and amylopectin. Amylose is a linear polymer and amylopectin is a highly branched, treelike polymer. The functions of carbohydrates are summarized in Table 32.26.

32.14 Lipids

There are several types of *lipids*—they are all nonpolar compounds that contain large hydrocarbon segments and are insoluble in water, but soluble in organic solvents. In contrast to proteins and carbohydrates, lipids are not polymeric.

We have already encountered fatty acids—long, straight-chain carboxylic acids—in soaps (Section 13.12). The naturally occurring esters of fatty acids form one class of lipids. The most common of these are esters of glycerol (1,2,3-propanetriol) called *glycerides*

$$
\begin{array}{l}
\quad\quad\ \ \overset{\displaystyle O}{\overset{\displaystyle \|}{}} \\
CH_2OCR \\
\quad\ \ |\quad\ \ \overset{\displaystyle O}{\overset{\displaystyle \|}{}} \\
CHOCR \\
\quad\ |\quad\ \ \overset{\displaystyle O}{\overset{\displaystyle \|}{}} \\
CH_2OCR
\end{array}
$$
a glyceride

Animal fats, such as lard, tallow, and butter, and vegetable oils, such as corn oil, soybean oil, linseed oil, olive oil, and peanut oil are all glycerides. In simple glycerides the R groups are all alike, and in mixed glycerides the R groups are not all alike. There are monoglycerides, in which only one OH group of glycerol has been esterified, and diglycerides and triglycerides. Most fats and oils are mixtures of mixed glycerides. In fats the hydrocarbon chains of the fatty acids are largely saturated, and in oils they are predominantly unsaturated. The fatty acids found in glycerides nearly always have an even number of carbon atoms.

The major function of glycerides is to store energy. In animals fats consumed in the diet are partially broken down to mono- and diglycerides in the intestines and then pass through the intestine walls. Triglycerides are reformed and carried to the liver and other tissues. At these sites the triglycerides are either chemically transformed to release energy or deposited as fat droplets for future conversion to energy. Sugars can be converted to fats in the liver and then are either used to produce energy or are stored in the body.

Several other types of compounds are included in the lipid class. *Phospholipids* are generally derived from glycerol by esterification of two OH groups by fatty acids and of one OH group by a phosphoric acid derivative.

$$
\begin{array}{l}
\quad\quad\ \ \overset{\displaystyle O}{\overset{\displaystyle \|}{}} \\
CH_2OCR \\
\quad\ \ |\quad\ \ \overset{\displaystyle O}{\overset{\displaystyle \|}{}} \\
CHOCR \\
\quad\ | \\
CH_2OPO_3R'
\end{array}
$$
a phospholipid

Phospholipids have important functions in cell membranes. *Waxes*, which form protective coatings on, for example, leaves or feathers, are lipids that are esters of long-chain fatty acids and long-chain monohydroxy alcohols. *Steroids* all contain a four-ring hydrocarbon unit. Cholesterol, which has been implicated in "hardening" of the arteries and heart disease, is one example of a steroid.

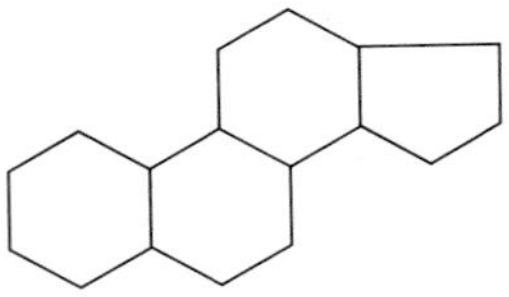

cyclopentanoperhydrophenanthrene,
basis for all steroids

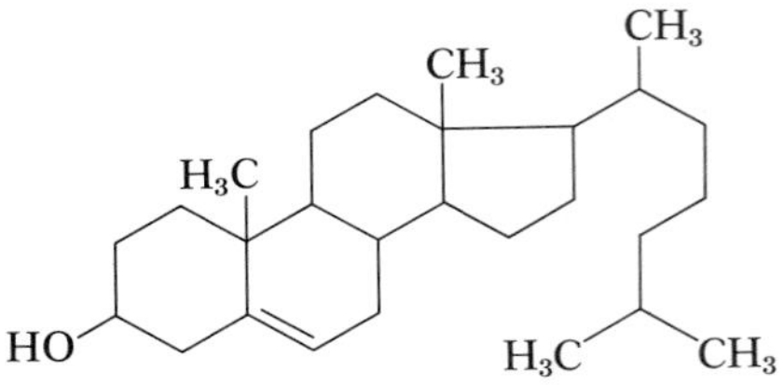

cholesterol, a steroid

Terpense, such as the unsaturated hydrocarbons that occur in plants (Sections 23.5), are also lipids.

32.15 Nucleotides and nucleic acids

Nucleic acids are polymers of great structural complexity. The repeating units in a nucleic acid are *nucleotides*. Uncombined nucleotides are also present in many cells and play major biochemical roles. Each nucleotide has three parts: (1) a heterocyclic base that contains either a purine or a pyrimidine ring

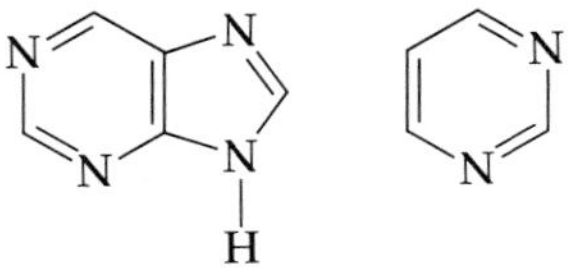

purine   *pyrimidine*

(2) a sugar—either D-ribose (see Table 32.25) or 2-deoxy-D-ribose, which is ribose in which the CHOH group at the second carbon atom

FIGURE 32.2

Examples of a nucleoside, a nucleotide, and the linkage of nucleotides in a nucleic acid. *In each structure, the monosaccharide is shown in color.*

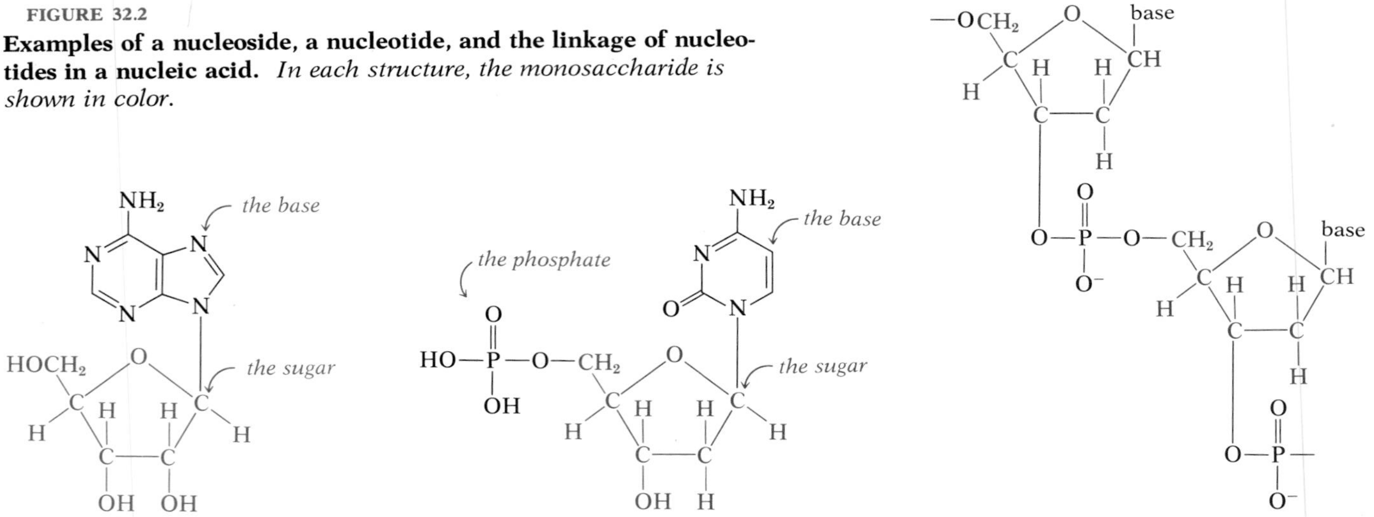

a nucleoside
(adenosine)

a nucleotide
(deoxycytidylic acid)

two nucleotide units
in a DNA molecule

has been transformed to a CH_2 group; and (3) one or more phosphate groups.

A *nucleic acid* consists of several thousand nucleotides joined together by ester linkages between the third carbon atom of the sugar of one nucleotide and the phosphate group of another. *Nucleosides,* which do not play a biochemical role as free molecules but are intermediates in metabolism, are nucleotides without the phosphate unit. A nucleoside, nucleotide, and two units of a nucleic acid are shown in Figure 32.2.

The nucleotide adenosine triphosphate, commonly referred to as ATP, is the primary source of energy within the cell. The transfer of one phosphate group from ATP to another molecule releases energy for use as heat or to support some chemical process that requires energy.

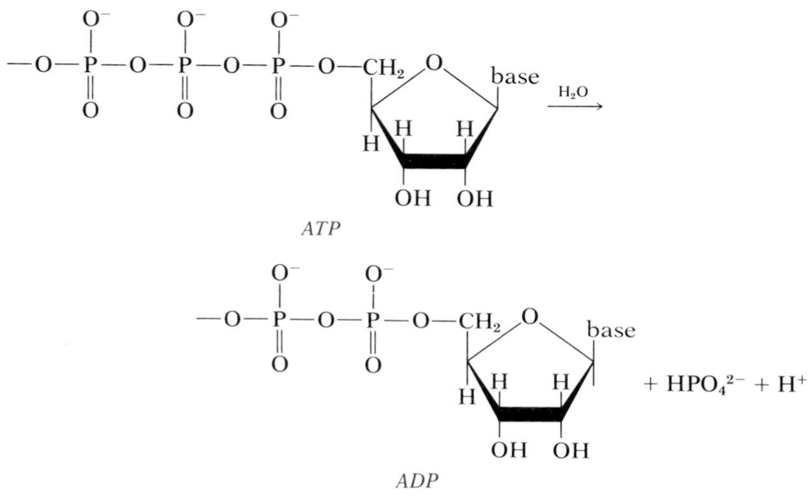

ATP

$+ HPO_4^{2-} + H^+$

ADP

$$\Delta G° = -7.3 \text{ kcal } (-31.\text{kJ})$$

When excess energy is available to be stored, the preceding reaction is reversed. The three phosphate groups in ATP can store or release a total of about 16 kcal/mole of energy.

The role of nucleic acids is to store genetic information in code form and to regulate protein synthesis. The code resides in the sequence of bases in the deoxyribonucleic acids (DNAs) and ribonucleic acids (RNAs) of all the different species of living things, and the three-dimensional molecular structure of DNA is all important in allowing the transfer of genetic information. In 1953 James Watson and Francis Crick proposed that DNA has the structure of a double helix and explained how the sequence of nucleotides is duplicated in the synthesis of new DNA molecules and passed along by RNA to the sites of protein synthesis. This was one of the most exciting and fundamental discoveries of biochemistry.

In DNA the phosphate–sugar portions of the nucleotides form two intertwined helical chains that are connected by successive pairs

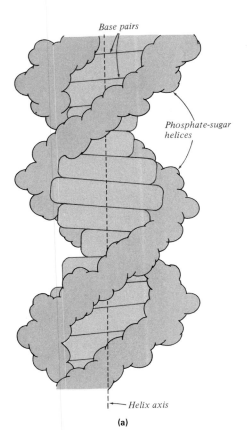

Base pairs

Phosphate-sugar helices

Helix axis

(a)

Helix axis

(b)

FIGURE 32.3
The double helix of DNA. (*a*) *Flattened view of the structure of DNA,* (*b*) *the intertwined helices.*

of bases, one from each of the adjacent nucleotides. The bases are attracted to each other by hydrogen bonds. A flattened view of the structure is given in Figure 32.3, together with a simple sketch of the double helix connected by the base pairs.

**THOUGHTS ON
CHEMISTRY**

*Kekulé describes the origins
of his theories*

KEKULÉ'S DREAM, *by August Kekulé*

It is also said that genius thinks in leaps. Gentlemen, the growing intellect does not think in leaps. That is not given to it.

Perhaps it will interest you, if I let you know through highly indiscreet disclosures from my inner life, how I arrived at some of my ideas.

During my stay in London I resided for a considerable time in Clapham Road in the neighborhood of Clapham Common. I frequently, however, spent my evenings with my friend Hugo Müller at Islington at the opposite end of the metropolis. We talked of many things but most often of our beloved chemistry. One fine summer evening I was returning by the last bus, "outside," as usual, through the deserted streets of the city, which are at other times so full of life. I fell into a reverie (Traümerei), and lo, the atoms were gamboling before my eyes! Whenever, hitherto, these diminutive beings had appeared to me, they had always been in motion; but up to that time I had never been able to discern the nature of their motion. Now, however, I saw how, frequently, two smaller atoms united to form a pair; how a larger one embraced the two smaller ones; how still larger ones kept hold of three or even four of the smaller; whilst the whole kept whirling in a giddy dance. I saw how the larger ones formed a chain, dragging the smaller ones after them but only at the ends of the chain. I saw what our past master, Kopp, my highly honored teacher and friend, has depicted with such charm in his "Molekular-welt"; but I saw it long before him. The cry of the conductor: "Clapham Road," awakened me from my dreaming; but I spent a part of the night in putting on paper at least sketches of these dream forms. This was the origin of the "Structural Theory."

Something similar happened with the benzene theory. During my stay in Ghent I resided in elegant bachelor quarters in the main thoroughfare. My study, however, faced a narrow side-alley and no daylight penetrated it. For the chemist who spends his day in the laboratory this mattered little. I was sitting writing at my textbook but the work did not progress; my thoughts were elsewhere. I turned my chair to the fire and dozed. Again the atoms were gamboling before my eyes. This time the smaller groups kept modestly in the background. My mental eye, rendered more acute by repeated visions of the kind, could now distinguish larger structures of manifold conformation: long rows, sometimes more closely fitted together all twining and twisting in snake-like motion. But look! What was that? One of the snakes had seized hold of its own tail, and the form whirled mockingly before my eyes. As if by a flash of lightning I awoke; and this time also I spent the rest of the night in working out the consequences of the hypothesis.

Let us learn to dream, gentlemen, then perhaps we shall find the truth.

KEKULE'S DREAM, by August Kekulé, 1890 (*Berichte der Deutschen Chemischen Gesellschaft*. Translated by P. Theodor Benfey; from *Journal of Chemical Education*, Vol. 35, p. 21, 1958)

Significant terms defined in this chapter

functional group
alkyl halides
aryl halides
hydroxyl group
alcohols
alkoxide
phenols
ether
alkoxy group

amines
diazonium salts
chromophoric group
carbonyl group
aldehyde
ketone
dehydrogenation reaction
carboxylic acid

carboxylate ion
ester
esterification reaction
saponification
acyl halide
acyl group
acid anhydride
amide

ORGANIC CHEMISTRY

Exercises

32.1 A student was given three bottles labeled A, B, and C. One of these contained acetic acid, one contained acetaldehyde, and one contained ethyl alcohol. (a) Write a structural formula for each of these compounds and (b) name them using the IUPAC nomenclature. The student observed that substance A reacted with substance B to form an ester under certain conditions and that substance B formed an acidic solution when dissolved in water. (c) Identify which compound is in each bottle and (d) write balanced chemical equations for the reaction involving the formation of the ester and for the ionization of the acid. To confirm the identification, the student treated the compounds with a strong oxidizing agent and found that compound A required roughly twice as much oxidizing agent as compound C. (e) Write chemical equations for the reactions of compounds A and C with an acidic solution of MnO_4^-.

32.2 Distinguish clearly among: (a) alkyl, aryl, and acyl halides; (b) alkanols, alkenols, alkynols, and phenols; (c) simple and mixed ethers; (d) primary, secondary, and tertiary amines; and (e) amines, amides, and imides.

32.3 Which statements are true? Rewrite any false statement so that it is correct.
(a) The covalent bond between the carbon atom and the halogen atom in an alkyl halide is polar because of the larger electronegativity of the carbon atom.
(b) The boiling point of ethyl alcohol is significantly greater than that of dimethyl ether because there is considerable hydrogen bonding in the alcohol.
(c) Denatured alcohol contains various toxic or objectionable materials that are difficult to remove so that it cannot be used for internal consumption.
(d) Phenols act as weak acids in many reactions, alcohols as weak acids and bases, and amines as weak bases.
(e) All alkanols can be oxidized easily to carboxylic acids.

32.4 Alcohols can be used as starting reagents for producing many other types of compounds: (a) alkoxides, (b) ethers, (c) aldehydes and ketones, (d) acids, (e) esters, and (f) alkyl halides. Write a chemical equation showing each of these reactions.

32.5 Ethanol, like water, can act as a weak acid or a weak base. Write chemical equations showing CH_3CH_2OH acting (a) as a base with HCl and (b) as an acid with Na(s). (c) Name the compounds formed.

32.6 Draw the molecular structure for each of the following compounds: (a) p-bromotoluene, (b) cyclohexanol, (c) o-cresol, (d) 2-methoxy-3-methylbutane, (e) diethylamine, (f) o-chlorophenol, (g) 1,4-butanediol, (h) propionalde-hyde, (i) cyclobutanone, (j) 2,2-dimethylpropanoic acid, (k) ammonium benzoate, (l) acetic propionic anhydride, and (m) isopropyl formate.

32.7 Give a name (common or IUPAC) for each of the following compounds:

(a) $CH_3CH_2CH_2CH_2OH$

(b) $CH_3-CH-CH_3$
       $|$
       NH_2

(c) Br—⟨benzene ring⟩—Br

(d) ⟨benzene ring⟩—O—⟨benzene ring⟩

(e) ⟨cyclopentane ring⟩—OH

(f) $CH_3-C=CH_2$
          $|$
          Cl

(g) $(CH_3CH_2)_3N$

(h) Br—⟨benzene ring with NH_2 top, Br positions⟩—Br, Br
     structure with NH_2 at top, Br at 2,6 positions and Br at para

(i) CH_3CH_2CHO

(j) ⟨benzene ring⟩—$\overset{O}{\overset{||}{C}}CH_2CH_3$

(k) $CH_3-CH-COOH$
          $|$
          CH_3

(l) ⟨benzene ring⟩—$O-\overset{O}{\overset{||}{C}}CH_3$

(m) $CH_3CH_2\overset{O}{\underset{||}{C}}NH_2$

(n) ⟨benzene ring⟩—$\overset{O}{\overset{||}{C}}-NHCH_3$

32.8 A laboratory procedure called for oxidizing 2-propanol to acetone using an acidic solution of $K_2Cr_2O_7$. However, an insufficient amount of $K_2Cr_2O_7$ was on hand, so the laboratory instructor decided to use an acidic solution of $KMnO_4$ instead. How many grams of $KMnO_4$ in place of 1.0 g of $K_2Cr_2O_7$ were required to carry out the same oxidation process?

32.9 The equilibrium constant for the equation

$$CH_3CH_2OH + CH_3COOH \rightleftharpoons CH_3CH_2OOCCH_3 + H_2O$$

is

$$K = \frac{[CH_3CH_2OOCCH_3][H_2O]}{[CH_3CH_2OH][CH_3COOH]} = 4$$

(Note that in this case water is not the solvent and is not in large excess. Its concentration appears in the equilibrium expression because the activity is not unity.) What fraction of the alcohol reacts if the original reaction mixture contained equal concentrations of alcohol and acid?

32.10 Complete the following equations:

(a) $CH_3CH_2I + NaOH \xrightarrow{H_2O}$

(b) 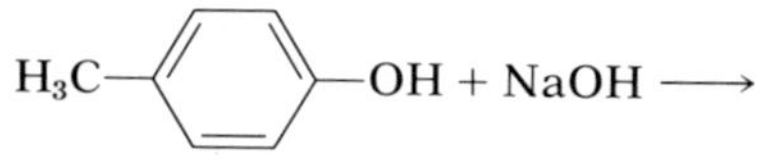 $H_3C-\text{⟨⟩}-OH + NaOH \longrightarrow$

(c) 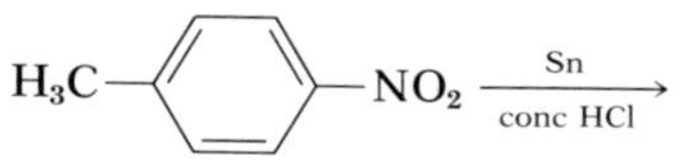 $H_3C-\text{⟨⟩}-NO_2 \xrightarrow[\text{conc HCl}]{Sn}$

(d) 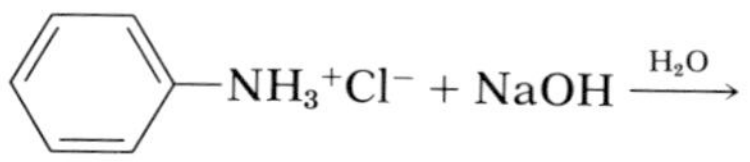 $\text{⟨⟩}-NH_3^+Cl^- + NaOH \xrightarrow{H_2O}$

(e)

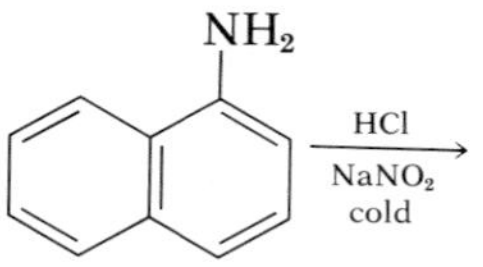

$\text{(naphthalene with } NH_2) \xrightarrow[\substack{NaNO_2 \\ cold}]{HCl}$

(f) 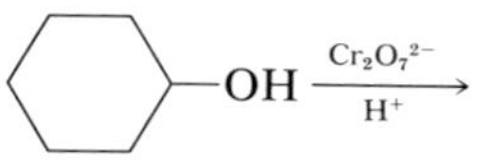 $\text{(cyclohexyl)}-OH \xrightarrow[H^+]{Cr_2O_7^{2-}}$

(g) $2CH_3COOH + Ba(OH)_2 \longrightarrow$

(h) $CH_3CH_2CHO \xrightarrow[H^+]{MnO_4^-}$

(i)

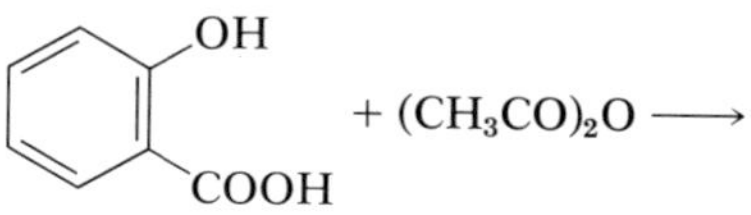

$\text{(benzene with OH and COOH)} + (CH_3CO)_2O \longrightarrow$

(j)

$\text{(benzene with } CH_2CH_2OH) + \text{(phthalic anhydride)} \longrightarrow$

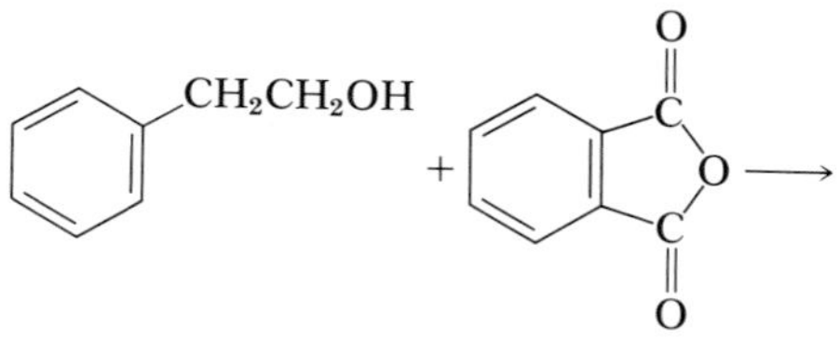

(k) $CH_3CH_2COOCH_3 \xrightarrow[\Delta]{conc\ HCl}$

(l)

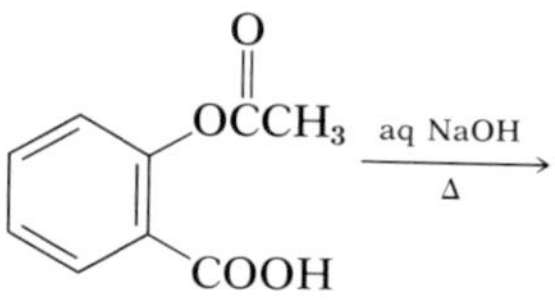

$\text{(benzene with } OCCH_3 \text{ and COOH)} \xrightarrow[\Delta]{aq\ NaOH}$

32.11 Show by equations how you might be able to carry out the following conversions. You may use any other organic or inorganic compounds you need. More than one step may be necessary.

(a) $CH_3-\underset{\underset{OH}{|}}{CH}-CH_3$ to $CH_3-\underset{\underset{Br}{|}}{CH}-CH_3$

(b) 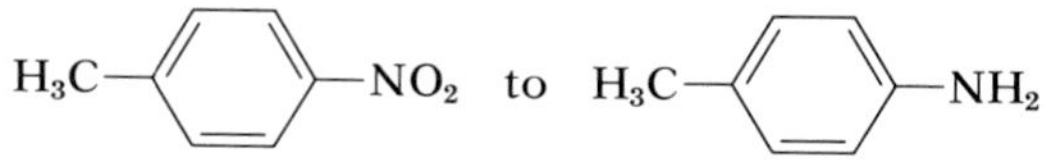 $H_3C-\text{⟨⟩}-NO_2$ to $H_3C-\text{⟨⟩}-NH_2$

(c) $CH_3CH_2CH_2Br$ to $CH_3\underset{\underset{Br}{|}}{\overset{\overset{Br}{|}}{CH}}CH_2$

(d) 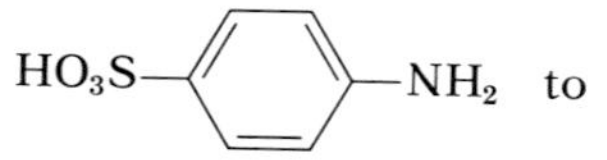 $HO_3S-\text{⟨⟩}-NH_2$ to

$HO_3S-\text{⟨⟩}-N=N-\text{(naphthol with HO)}$

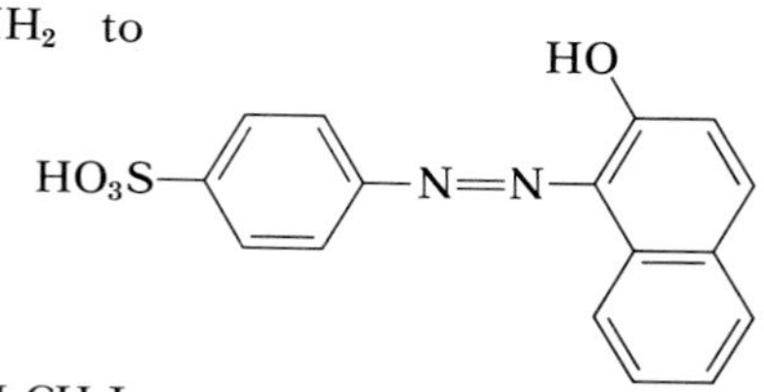

(e) CH_3CH_2OH to CH_3CH_2I
(f) CH_3CH_2OH to $(CH_3CH_2)_4N^+Br^-$

32.12* Many of the classes of compounds discussed in this chapter can be "derived" from the water molecule

$$H_1-O-H_2$$

by changing one or both of the hydrogens for various organic groups. Name the class of compound formed when (a) H_1 is replaced by an alkyl group, (b) H_1 is replaced by an aromatic group, (c) H_1 and H_2 are replaced by alkyl groups, (d) H_1 is replaced by $R-C=O$ group, (e) H_1 is replaced by a $R-C=O$ group and H_2 is replaced by an alkyl group, and (f) H_1 and H_2 are replaced by $R-C=O$ groups.

32.13 What is the approximate pH of a $0.10M$ solution of sodium benzoate? $K_a = 6.6 \times 10^{-5}$ for benzoic acid, C_6H_5COOH. Would this solution be more or less acidic than a $0.10M$ solution of sodium acetate? $K_a = 1.754 \times 10^{-5}$ for acetic acid, CH_3COOH.

32.14* A piece of glass 12 inches by 18 inches is to be silvered using the following reaction

$$CH_3CHO + 2[Ag(NH_3)_2]OH \longrightarrow$$
$$2Ag + CH_3COONH_4 + H_2O + 3NH_3$$

If the thickness of the silver is to be 0.0003 inch, what mass of acetaldehyde is needed for the reaction? The density of Ag is 10.5 g/cm³.

32.15* In aqueous solution, acetic acid exists mainly in the molecular form ($K_a = 1.745 \times 10^{-5}$). (a) Calculate the freezing point depression for a 0.10 molal aqueous solution of acetic acid neglecting any ionization of the acid. $K_{fp} = 1.86$ for water. In nonpolar solvents such as benzene, acetic acid exists mainly as dimers

as a result of hydrogen bonding. (b) Calculate the freezing point depression for a 0.10 molal solution of acetic acid in benzene. $K_{fp} = 4.90$ for benzene.

32.16* A "chemist" decided to make some nitroglycerin using a home chemistry set. Without taking any precautions, he heated a mixture of 25 g of glycerol with 25 ml of 12M HNO$_3$ in a 100 ml flask. (a) Write the equation for the reaction and (b) calculate the theoretical yield of nitroglycerin. However, the nitroglycerin underwent detona-

tion and the room was filled with flying pieces of glass as a result of the explosion. (c) Write the equation for the explosion (note that no additional reactants such as O$_2$ from the air are required). Calculate the number of moles of gaseous products formed (d) for each mole of nitroglycerin that reacted and (e) for the sample that exploded.

An explosion takes place so fast because the energy released cannot be dissipated to the surroundings and therefore the products are formed at very high temperatures and rapidly expand. We can find the approximate temperature by dividing the heat of reaction, -342 kcal/mole of nitroglycerin, by the heat capacity of the product gases and adding 25°C to this value. (f) Calculate the heat capacity of the product gases by adding together the number of moles of each gas formed multiplied by the heat capacity of that gas using the following data for heat capacity: 8.87 cal/mole °C for CO$_2$, 6.961 cal/mole °C for N$_2$, 8.025 cal/mole °C for H$_2$O, and 7.016 cal/mole °C for O$_2$. (g) Calculate the approximate temperature of the product gases. (h) Assuming the ideal gas law to be valid, what pressure did the product gases exert on the inside of the flask before the explosion?

APPENDIX A
Chemical arithmetic

Chemistry, like most sciences, requires the use of mathematics in its study, development, and application. Even though you may use an electronic calculator for many calculations, it is important to be able to (1) recognize which numbers in an answer are "significant," (2) use scientific notation for the very large and very small numbers encountered in chemistry, and (3) find and use logarithms when necessary.

A.1 Significant figures

A student once wrote the following answer on an examination:

$$(1.0 \text{ lb})\left(\frac{454 \text{ g}}{\text{lb}}\right)\left(\frac{1 \text{ mole C}}{12.01 \text{ g}}\right) = 37.80183180682 \text{ moles C}$$

To get that many digits in his final answer required a lot of work that was not needed. In fact, based on the rules of significant figures that we will discuss, the correct answer to the problem is simply 38 moles C.

As an illustration of how many significant figures a number should have, consider the following measurements of the mass of a piece of candy made by the members of a class:

23.1 g 23.3 g 23.0 g 23.3 g 23.2 g 23.2 g
23.3 g 23.4 g 23.2 g 23.1 g 23.3 g 23.2 g

As you can see, there is some uncertainty in the value of the mass.

This uncertainty is in the third figure from the left—the tenths of a gram place. The uncertainty is the result of the reproducibility of the balance or having to estimate the digit between 1-g markings on the balance scale.

The average of the weights measured by the twelve students is 23.2 g. There is no doubt that the piece of candy weighs 23 g plus something more, but here is where the uncertainty is. To reflect the uncertainty in the mass of the candy, we could report the mass as 23.2 ± 0.2 g, which includes all of the class data. (The highest value is 23.4 and the lowest is 23.0.)

Another method of reporting the uncertainty is by giving the *relative error* that the uncertainty makes in the number. Relative error is calculated by dividing the uncertainty in a number by the number itself. For example, the relative error in the mass of the candy is the uncertainty divided by the average mass.

$$\frac{0.2 \text{ g}}{23.2 \text{ g}} = 0.01$$

The relative error is usually expressed in percent (%), parts per thousand (ppt), or parts per million (ppm), so the above answer would be given as 1%, or 10 ppt, or 10,000 ppm.

The number of significant figures in any given number includes all of the digits that are of definite value plus the first digit to the right with an uncertain value. Scientific data should not be reported in numbers that go beyond the number of figures that are significant.

Many times the actual uncertainty in a number is not given. In these cases we assume that all figures cited are significant and that the uncertainty is, for example, roughly ± 2 in the place represented by the right-hand digit. In this case, if we had simply stated that the mass of the candy was 23.2 g, we would assume that all three numbers are significant and that the uncertainty is ± 0.2 g. Similarly, the uncertainty in 5285 is ± 2, or in 0.042 is ± 0.002, etc. *The number of significant figures in a given number is found by counting the number of figures from the left to right in the number beginning with the first non-zero digit and continuing until reaching the digit that contains the uncertainty.* Each of the following numbers has three significant figures:

1 2 3	1 2 3	1 2 3	1 2 3	1 2 3
454	0.296	7.31	0.00846	10.7

Sometimes zeros at the end of a number present a problem. These zeros can be significant figures or they can be space-fillers used to locate the decimal point. The terminal zero in the number 1520 simply locates the decimal point and is not significant—the uncertainty is ± 20, and we say that there are three significant figures. However, if the terminal zero in the number 1520 is significant the number is written 1520. to show that the uncertainty is ± 2 and the number of significant figures is four. Likewise, in the number 1.520 there are four significant figures and the uncertainty is ± 0.002. Check to see that the following numbers contain the indicated number of significant figures.

$$\overset{1\,2}{340} \qquad \overset{1\,2\,3}{700.} \qquad \overset{1\,2\,3\,4\,5}{10.000} \qquad \overset{1}{0.000002} \qquad \overset{1\,2\,3}{0.0402} \qquad \overset{1\,2\,3\,4\ \ 5}{1062.0}$$

The number of significant figures in conversion factors such as 365 days/yr and 60 min/hr also depends on whether or not the number contains an uncertainty. In the first conversion factor, we are interested in knowing how many days are in *exactly* 1 yr—there is no uncertainty here. However, the number 365 contains only three significant figures because there is some uncertainty in the third figure—think of leap year. A more precise value, one having more significant figures, is 365.24, which has five significant figures. The second conversion factor can be treated as having an infinite number of significant figures, because there are *exactly* 60 min in *exactly* 1 hr. This means that the number of significant figures in the answer to a calculation that includes 60 min/hr is not limited by the number of significant figures in this factor—it can be treated as having as many significant figures as are needed. Think for a moment about why the following conversion factors contain the indicated numbers of significant figures.

$$\overset{1\ \ 2\,3}{2.54}\,cm/inch \qquad \overset{\infty}{12}\ inches/ft \qquad \overset{1\,2\,3}{238}\ g\ of\ uranium/mole$$

Conversion factors that appear in equations such as

$$diameter = \overset{\infty}{2} \times radius$$

are usually exact, although there are exceptions such as

$$k = (\overset{1\,2\,3}{0.693})t_{1/2}$$

If we add the numbers

$$
\begin{array}{r}
23.2\vert \quad g \\
6.0\vert52 \\
\underline{139.4\vert} \\
\end{array}
$$

we get 168.652 g as an answer. However, this is not the correct answer because of the uncertainty in each of the numbers: ±0.2 in 23.2, ±0.002 in 6.052, and ±0.2 in 139.4. The old adage about a chain being only as strong as its weakest link applies here. We know that our answer cannot be more exact than the least exact of these numbers, which have uncertainties of ±0.2. Thus, we round the answer to 168.7 g. *In addition or subtraction, we perform the mathematical operation and round the answer to the digit that has the greatest amount of uncertainty.* We can apply this rule simply by drawing a vertical line after the least exact number, performing the addition or subtraction, and rounding our answer so that it "fits" in the space before the vertical line. Confirm for yourself that the following examples are done correctly.

						0.29|6	atm
14.|	cal	0.0097| V	0.|62°	4.41|	
−0.|072		0.0563|	273.|	8.92|73	
13.|928		0.0660|	273.|62	13.63|33	
Answer:	14|	cal	0.0660| V	274°|	13.63|	atm

In this book we use the following rule for rounding off an answer: *If the numbers following the desired place are between 000. . . and 499. . ., we drop the numbers and do nothing to the number in the desired place; if the numbers following are between 500. . . and 999. . ., we drop the numbers and increase the number in the desired place by one.*

If we multiply two numbers such as

$$(23.2 \text{ g})(0.1257 \text{ cm}^3/\text{g})$$

we get 2.91624 cm³ as an answer. However, this is not the correct answer because of the uncertainty in each of the numbers: ±0.2 in 23.2 (1% relative error) and ±0.0002 in 0.1257 (0.2% relative error). Again, our answer cannot be more exact than the least exact of these numbers. During multiplication and division the decimal point changes place. Therefore, we cannot tell by looking at the decimal points in the factors in the problem where the decimal point should be in the answer. *In multiplication or division, we perform the mathematical operation and round the answer to the number of significant figures so that the answer has the same relative error as does the least exact factor.* For the problem above, the answer becomes 2.92 cm³, which has roughly a 1% relative error.

Usually, the rule above can be applied simply by performing the multiplication or division, counting the number of significant figures in each factor, and rounding the answer to the same number of significant figures as are in the factor with the smallest number of significant figures. Some additional examples follow:

$$(1.\overset{1}{6}\overset{2}{2}\overset{3}{1}\text{ moles})(\overset{1}{1}\overset{2}{6}\text{ g/mole}) = \overset{1}{2}\overset{2}{6}\text{ g} \qquad \frac{\overset{1}{2}\overset{2}{6}.\overset{3}{9}\text{ g}}{\underset{1\ 2\ 3}{2.69\text{ g/cm}^3}} = \overset{1}{1}\overset{2}{0}.\overset{3}{0}\text{ cm}^3$$

$$(0.\overset{1}{0}\overset{2}{7}\overset{3}{2}9\text{ m})(\overset{\infty}{1}00\text{ cm/m}) = \overset{1}{7}.\overset{2}{2}\overset{3}{9}\text{ cm}$$

For calculations involving several steps, we apply the above rules to each step as it is performed.

$$V = \frac{(1.08 \text{ moles})(27° + 273.15°\text{K})(0.0821 \text{ liter-atm/mole °K})}{1.649 \text{ atm}}$$

$$= \frac{(1.08)(300.)(0.0821 \text{ liters})}{1.649}$$

$$= 16.1 \text{ liters}$$

$$X = \frac{45821}{(2.303)(1.987)}\left(\frac{1}{25 + 273.15} - \frac{1}{35 + 273.15}\right)$$

$$= \frac{45821}{(2.303)(1.987)}\left(\frac{1}{298} - \frac{1}{308}\right)$$

$$= \frac{45821}{(2.303)(1.987)}(0.00336 - 0.00325)$$

$$= \frac{45821}{(2.303)(1.987)}(0.00011)$$

$$= 1.1$$

When supplying conversion factors or values of physical constants in a calculation, always make sure that the values you have supplied contain enough significant figures so that the uncertainty in the answer is the same as the uncertainty in the original number. You do not want to decrease the number of significant figures just by choosing a conversion factor with too few significant figures. Confirm for yourself that the following examples illustrate this point.

```
  23°          14.3  °         (16 mole)(16 g/mole) = 260 g
 273          273.15
 296°         287.45          (15.74 mole)(15.999 g/mole) = 251.8 g
              287.5  °
```

Exercises

1 Express the uncertainty for each of the following numbers assuming the uncertainty in the last significant figure is ± 2:

(a) 273   (d) 632.2   (g) 6000
(b) 0.5649   (e) 710.   (h) 0.0003
(c) 470   (f) 12.529   (i) 8.0

2 Determine the relative error that the uncertainty makes in each of the numbers in Exercise 1.

3 How many significant figures are in each of the numbers in Exercise 1?

4 Perform the following calculations and express the answers in the proper number of significant figures:

(a) 423.1
    0.256
    100

(b) 52.987
    9.3545
    6.12

(c) 14.3920
    -4

(d) (5183)(2)

(e) $\dfrac{14.000}{6.1}$

(f) $(6.11)(\pi)$

(g) (14.3 hr)(60 min/hr)

(h) $\dfrac{1020 \text{ inches}}{12 \text{ inches/ft}}$

(i) $V = \dfrac{(3.2 \text{ lb})(454 \text{ g/lb})}{(8.6214 \text{ g/cm}^3)}$

(j) $V = (4/3)\pi(2.16 \text{ cm})^3$

(k) m = (6.0 liter + 9.57 liter + 0.61 liter)(1.113 g/liter)

(l) (2.93)(14.7) + (1203)(0.0296) + (9.38)(5.2)

Answers

1 (a) ± 2, (b) ± 0.0002, (c) ± 20, (d) ± 0.2, (e) ± 2, (f) ± 0.002, (g) ± 2000, (h) ± 0.0002, (i) ± 0.2 **2** (a) 0.007, (b) 0.0004, (c) 0.04, (d) 0.0003, (e) 0.003, (f) 0.0002, (g) 0.3, (h) 0.7, (i) 0.03 **3** (a) 3, (b) 4, (c) 2, (d) 4, (e) 3, (f) 5, (g) 1, (h) 1, (i) 2 **4** (a) 500, (b) 68.46, (c) 10, (d) 10000, (e) 2.3, (f) 19.2, (g) 858 min, (h) 85.0 ft, (i) 170 cm³, (j) 42.2 cm³, (k) 18.0 g, (l) 127

A.2 Scientific, or exponential, notation

If we attempt to enter 2,000,000,000 or 0.0000000004 on most electronic calculators, we run into difficulty, because there are usually not enough digits available. In chemistry we find that we are often using very large and very small numbers, and for convenience we must be able to represent and use these numbers in some different fashion. *In standard scientific notation the significant figures of a number are retained in a factor between 1.00. . . and 10.00. . . and the decimal point is given by a power of ten.*

For numbers greater than unity, we move the decimal point to the left until we reach a number that is between 1 and 10. The power of ten is equal to the number of digits that the decimal point was moved. For example, we would rewrite 2,000,000 and 120 as follows:

$$2{,}000{,}000 \text{ becomes } 2 \times 10^6 \qquad 120 \text{ becomes } 1.2 \times 10^2$$

For numbers less than unity, we move the decimal point to the right in the number until we reach a number that is between 1 and 10. The power of ten is equal to the number of digits that the decimal point was moved and a negative sign appears in the exponent. For example, we would rewrite 0.0006 and 0.0263 as follows:

$$0.0006 \text{ becomes } 6 \times 10^{-4} \qquad 0.0263 \text{ becomes } 2.63 \times 10^{-2}$$

Confirm for yourself that the following examples follow these rules.

$$6100 = 6.1 \times 10^3 \qquad 4{,}921{,}000 = 4.921 \times 10^6 \qquad 10.2 = 1.02 \times 10^1$$
$$0.392 = 3.92 \times 10^{-1} \qquad 0.000870 = 8.70 \times 10^{-4} \qquad 6.3 = 6.3 \times 10^0$$

Usually we do not write a number in scientific notation if the result is more cumbersome than the original number. For example, the numbers 1.32, 0.9624, and 42 are really simpler in their original form, and we would probably not use scientific notation for them.

$$1.32 = 1.32 \times 10^0 \qquad 0.9624 = 9.624 \times 10^{-1} \qquad 42 = 4.2 \times 10^1$$

To add or subtract numbers expressed in scientific notation, we rewrite all numbers so that they have the same power of ten. Some of the numbers may no longer be in standard scientific notation. Next we add or subtract the factors obeying the rules of significant figures. The power of ten in the answer is the same as that in the problem. For example, if we were to add 6.35×10^{14} to 7.29×10^{12}, we would write

$$6.35 \quad \times 10^{14}$$
$$\underline{0.0729 \times 10^{14}}$$
$$6.4229 \times 10^{14}$$
$$6.42 \quad \times 10^{14}$$

To multiply or divide numbers expressed in scientific notation, we multiply or divide the factors obeying the rules of significant figures.

We find the power of ten for the answer by adding or subtracting the powers of ten on the various numbers. For example,

$$(6.022 \times 10^{23})(4.2 \times 10^3) = 25 \times 10^{26} = 2.5 \times 10^{27}$$

Think about how these rules apply in the following examples:

$$(5.3 \times 10^8)(9.62 \times 10^{-4}) = 51 \times 10^4 = 5.1 \times 10^5$$

$$(3.14 \times 10^{-4})(5 \times 10^{-5}) = 20 \times 10^{-9} = 2 \times 10^{-8}$$

$$\frac{7.11 \times 10^{-3}}{4.26 \times 10^4} = 1.67 \times 10^{-7} \qquad \frac{2.996 \times 10^6}{4.32 \times 10^{14}} = 0.694 \times 10^{-8} = 6.94 \times 10^{-9}$$

$$\frac{6.934 \times 10^{48}}{3.496 \times 10^{47}} = 1.983 \times 10^1 \qquad \frac{4.58 \times 10^4}{5.44 \times 10^{-14}} = 0.842 \times 10^{18} = 8.42 \times 10^{17}$$

When numbers with units frequently appear to the same power of 10, prefixes can be used to give the decimal point information. The symbol for a prefix is placed before the unit on the number instead of writing ten to a power. The commonly used prefixes are

decimal location		prefix	prefix symbol
1,000,000,000,000	$= 10^{12}$	tera	T
1,000,000,000	$= 10^9$	giga	G
1,000,000	$= 10^6$	mega	M
1,000	$= 10^3$	kilo	k
100	$= 10^2$	hecto	h
10	$= 10^1$	deka	da
0.1	$= 10^{-1}$	deci	d
0.01	$= 10^{-2}$	centi	c
0.001	$= 10^{-3}$	milli	m
0.000 001	$= 10^{-6}$	micro	μ
0.000 000 001	$= 10^{-9}$	nano	n
0.000 000 000 001	$= 10^{-12}$	pico	p
0.000 000 000 000 001	$= 10^{-15}$	femto	f
0.000 000 000 000 000 001	$= 10^{-18}$	atto	a

Confirm for yourself that the following examples illustrate this technique:

$$5200 \, \text{cal} = 5.2 \times 10^3 \, \text{cal} = 5.2 \, \text{kcal}$$

$$0.072 \, \text{liter} = 72 \times 10^{-3} \, \text{liter} = 72 \, \text{ml}$$

$$6278 \, \text{kJ} = 6.278 \times 10^3 \, \text{kJ} = 6.278 \times 10^6 \, \text{J} = 6.278 \, \text{MJ}$$

$$0.000 \, 000 \, 000 \, 0512 \, \text{m} = 51.2 \times 10^{-12} \, \text{m} = 51.2 \, \text{pm}$$

$$\$1,200,000 = 1.2 \, \text{``megabucks''}$$

Exercises

1 Express the following numbers in standard scientific notation:

(a) 6500
(b) 0.0041
(c) 999
(d) 14,000,000,000
(e) 0.03050
(f) 810.
(g) 3.1416
(h) 0.00000030
(i) 102.096
(j) 6.3 kcal
(k) 14.9 cm
(l) 0.1 ps

2 Perform the following calculations and express the answers in standard scientific notation:
(a) $(6.057 \times 10^3) + 9.35$
(b) $(4.51 \times 10^{-3})(8.78 \times 10^4)$
(c) $(2.35 \times 10^{-14}) - (7.1 \times 10^{-15})$
(d) $(1.4 \times 10^5)/(1.1 \times 10^5)$
(e) $(1812)(1492)/(1979)$
(f) $(4.3 \times 10^3) + (5.2 \times 10^2) + (6.1 \times 10^1)$

(g) $(7.33 \times 10^{-3} + 4.29 \times 10^1)/(5.88 \times 10^{-3} + 4.29 \times 10^1)$
(h) $(52.6 \times 10^3)(3.86 \times 10^{-4})/(4 \times 10^4 + 5 \times 10^5)$
(i) $(10.3 \text{ cm}^2)(10^8 \text{Å/cm})^2$ (j) $(5.3 \text{ km})/\pi$

(k) $V = (\frac{4}{3})\pi(1.962 \times 10^{-10} \text{ m})^3$
(l) $(2.5 \times 10^6 \text{ gal})/(0.75 \text{ million people})$

Answers

1 (a) 6.5×10^3, (b) 4.1×10^{-3}, (c) 9.99×10^2, (d) 1.4×10^{10}, (e) 3.050×10^{-2}, (f) 8.10×10^2, (g) 3.1416×10^0, (h) 3.0×10^{-7}, (i) 1.02096×10^2, (j)

6.3×10^3 cal, (k) 1.49×10^{-1}m, (l) 1×10^{-13}s **2** (a) 6.066×10^3, (b) 3.96×10^2, (c) 1.64×10^{-14}, (d) 1.3×10^0, (e) 1.366×10^3, (f) 4.9×10^3, (g) 1.00×10^0, (h) 4×10^5, (i) 1.03×10^{16} Å^2, (j) 1.7×10^3m, (k) 3.164×10^{-29}m^3, (l) 3.3×10^0 gal/person

A.3 Logarithms

Logarithms and antilogarithms are used in chemistry when we work with pH, with some of the equations in kinetics like radioactive decay, and with some of the more complicated equations in thermodynamics. Even though your electronic calculator may have logarithm–antilogarithm keys, it is important that you know how to work with logarithms and antilogarithms.

In this book we have used logarithms to the base 10, called "common logarithms," for all calculations. "Common logarithms" may be abbreviated "log." There are different base numbers that are also used such as, 2, 12, or e, where $e = 2.718281828459$, but common logarithms meet our needs quite well. Sometimes the mathematics by which an equation is derived gives it in the form of a logarithm to be the base e, usually written $\ln x$. In such equations, since

$$\ln x = 2.302585092994 \log x$$

we replace $\ln x$ by, as it is usually rounded off, $2.303 \log x$.

Logarithms can be used to multiply and divide numbers, to raise a number to some power, and to find the nth root of a number. In each case, the desired calculation is written in logarithmic form using the following equations:

desired calculation	logarithmic form
$Z = a \times b$	$\log Z = \log(a \times b) = \log a + \log b$
$Z = a/b$	$\log Z = \log(a/b) = \log a - \log b$
$Z = a^n$	$\log Z = \log(a^n) = n \log a$
$Z = \sqrt[n]{a}$	$\log Z = \log(\sqrt[n]{a}) = \dfrac{\log a}{n}$

The logarithm is calculated, and the answer is found by taking the antilogarithm of the logarithm. (A very common mistake is to use $\log a/\log b$ when the desired calculation is $Z = a/b$; don't make this mistake.)

We have laid the groundwork for logarithms in the previous section on scientific notation. *The logarithm of a number is the power to which 10 must be raised to get that number.* In other words, instead of (factor) $\times 10^n$, a number becomes 10^m when m is the logarithm. The logarithm of any number has two parts called the *characteristic* and the *mantissa*. In a logarithm, the characteristic is the number on the left of the decimal point and the mantissa is the number on the right of the decimal point:

$$\overset{\textit{characteristic}}{\log 5.000 = 0.} \overset{\textit{mantissa}}{6990}$$

read "*the logarithm of 5 equals 0.6990*"

which means that

$$10^{0.6990} = 5.000$$

The characteristic identifies the location of the decimal point in the original number, and the mantissa is given to the same number of significant figures as the original number. Logarithms of negative numbers are undefined because there is no power to which 10 can be raised that will generate a negative number.

For a number that is equal to or greater than 1 but less than 10, the logarithm is a number equal to or greater than 0 but less than 1. In other words, the characteristic is 0 and the logarithm is the same as the mantissa. If the number contains four significant figures or less, a table of four-place common logarithms (Table A.1) can be used to find the mantissa as follows:

1. Move downward in the column labeled "number" until you reach the row containing the first two significant figures of the original number.

2. Move across the row until you reach the column containing the third significant figure of the original number. Write down the four-digit number that appears.

3. Move further across the row into the proportional parts section until you reach the column containing the fourth significant figure of the original number. Add this number to the four-digit number you found in step 2.

4. Place the decimal point before the final four-digit number.

If the original number does not contain four significant figures, the procedure is terminated at the appropriate place and the mantissa is rounded off to the correct number of significant figures. The following examples illustrate this technique:

$x = \log 4.683$ 1. Move down the number column to 46.

                              2. Move across the row to 8 and write down 6702.

                              3. Move further across the row to 3 in the proportional parts section and add 3 to get 6705.

                              4. $x = 0.6705$.

$y = \log 7.2$ 1. Move down the number column to 72 and write down 8573.

                              2. Rounding off gives $y = 0.86$.

To find the mantissa of a number having more than four significant figures, we must use "five-place log tables" or calculators.

Four-place common logarithms

Number	0	1	2	3	4	5	6	7	8	9	1	2	3	4	5	6	7	8	9
10	0000	0043	0086	0128	0170	0212	0253	0294	0334	0374	4	8	12	17	21	25	29	33	37
11	0414	0453	0492	0531	0569	0607	0645	0682	0719	0755	4	8	11	15	19	23	26	30	34
12	0792	0828	0864	0899	0934	0969	1004	1038	1072	1106	3	7	10	14	17	21	24	28	31
13	1139	1173	1206	1239	1271	1303	1335	1367	1399	1430	3	6	10	13	16	19	23	26	29
14	1461	1492	1523	1553	1584	1614	1644	1673	1703	1732	3	6	9	12	15	18	21	24	27
15	1761	1790	1818	1847	1875	1903	1931	1959	1987	2014	3	6	8	11	14	17	20	22	25
16	2041	2068	2095	2122	2148	2175	2201	2227	2253	2279	3	5	8	11	13	16	18	21	24
17	2304	2330	2355	2380	2405	2430	2455	2480	2504	2529	2	5	7	10	12	15	17	20	22
18	2553	2577	2601	2625	2648	2672	2695	2718	2742	2765	2	5	7	9	12	14	16	19	21
19	2788	2810	2833	2856	2878	2900	2923	2945	2967	2989	2	4	7	9	11	13	16	18	20
20	3010	3032	3054	3075	3096	3118	3139	3160	3181	3201	2	4	6	8	11	13	15	17	19
21	3222	3243	3263	3284	3304	3324	3345	3365	3385	3404	2	4	6	8	10	12	14	16	18
22	3424	3444	3464	3483	3502	3522	3541	3560	3579	3598	2	4	6	8	10	12	14	15	17
23	3617	3636	3655	3674	3692	3711	3729	3747	3766	3784	2	4	6	7	9	11	13	15	17
24	3802	3820	3838	3856	3874	3892	3909	3927	3945	3962	2	4	5	7	9	11	12	14	16
25	3979	3997	4014	4031	4048	4065	4082	4099	4116	4133	2	3	5	7	9	10	12	14	15
26	4150	4166	4183	4200	4216	4232	4249	4265	4281	4298	2	3	5	7	8	10	11	13	15
27	4314	4330	4346	4362	4378	4393	4409	4425	4440	4456	2	3	5	6	8	9	11	13	14
28	4472	4487	4502	4518	4533	4548	4564	4579	4594	4609	2	3	5	6	8	9	11	12	14
29	4624	4639	4654	4669	4683	4698	4713	4728	4742	4757	1	3	4	6	7	9	10	12	13
30	4771	4786	4800	4814	4829	4843	4857	4871	4886	4900	1	3	4	6	7	9	10	11	13
31	4914	4928	4942	4955	4969	4983	4997	5011	5024	5038	1	3	4	6	7	8	10	11	12
32	5051	5065	5079	5092	5105	5119	5132	5145	5159	5172	1	3	4	5	7	8	9	11	12
33	5185	5198	5211	5224	5237	5250	5263	5276	5289	5302	1	3	4	5	6	8	9	10	12
34	5315	5328	5340	5353	5366	5378	5391	5403	5416	5428	1	3	4	5	6	8	9	10	11
35	5441	5453	5465	5478	5490	5502	5514	5527	5539	5551	1	2	4	5	6	7	9	10	11
36	5563	5575	5587	5599	5611	5623	5635	5647	5658	5670	1	2	4	5	6	7	8	10	11
37	5682	5694	5705	5717	5729	5740	5752	5763	5775	5786	1	2	3	5	6	7	8	9	10
38	5798	5809	5821	5832	5843	5855	5866	5877	5888	5899	1	2	3	5	6	7	8	9	10
39	5911	5922	5933	5944	5955	5966	5977	5988	5999	6010	1	2	3	4	5	7	8	9	10
40	6021	6031	6042	6053	6064	6075	6085	6096	6107	6117	1	2	3	4	5	6	8	9	10
41	6128	6138	6149	6160	6170	6180	6191	6201	6212	6222	1	2	3	4	5	6	7	8	9
42	6232	6243	6253	6263	6274	6284	6294	6304	6314	6325	1	2	3	4	5	6	7	8	9
43	6335	6345	6355	6365	6375	6385	6395	6405	6415	6425	1	2	3	4	5	6	7	8	9
44	6435	6444	6454	6464	6474	6484	6493	6503	6513	6522	1	2	3	4	5	6	7	8	9
45	6532	6542	6551	6561	6571	6580	6590	6599	6609	6618	1	2	3	4	5	6	7	8	9
46	6628	6637	6646	6656	6665	6675	6684	6693	6702	6712	1	2	3	4	5	6	7	7	8
47	6721	6730	6739	6749	6758	6767	6776	6785	6794	6803	1	2	3	4	5	5	6	7	8
48	6812	6821	6830	6839	6848	6857	6866	6875	6884	6893	1	2	3	4	4	5	6	7	8
49	6902	6911	6920	6928	6937	6946	6955	6964	6972	6981	1	2	3	4	4	5	6	7	8
50	6990	6998	7007	7016	7024	7033	7042	7050	7059	7067	1	2	3	3	4	5	6	7	8
51	7076	7084	7093	7101	7110	7118	7126	7135	7143	7152	1	2	3	3	4	5	6	7	8
52	7160	7168	7177	7185	7193	7202	7210	7218	7226	7235	1	2	2	3	4	5	6	7	7
53	7243	7251	7259	7267	7275	7284	7292	7300	7308	7316	1	2	2	3	4	5	6	6	7
54	7324	7332	7340	7348	7356	7364	7372	7380	7388	7396	1	2	2	3	4	5	6	6	7

Proportional Parts

Number	0	1	2	3	4	5	6	7	8	9	Proportional Parts								
											1	2	3	4	5	6	7	8	9
55	7404	7412	7419	7427	7435	7443	7451	7459	7466	7474	1	2	2	3	4	5	5	6	7
56	7482	7490	7497	7505	7513	7520	7528	7536	7543	7551	1	2	2	3	4	5	5	6	7
57	7559	7566	7574	7582	7589	7597	7604	7612	7619	7627	1	2	2	3	4	5	5	6	7
58	7634	7642	7649	7657	7664	7672	7679	7686	7694	7701	1	1	2	3	4	4	5	6	7
59	7709	7716	7723	7731	7738	7745	7752	7760	7767	7774	1	1	2	3	4	4	5	6	7
60	7782	7789	7796	7803	7810	7818	7825	7832	7839	7846	1	1	2	3	4	4	5	6	6
61	7853	7860	7868	7875	7882	7889	7896	7903	7910	7917	1	1	2	3	4	4	5	6	6
62	7924	7931	7938	7945	7952	7959	7966	7973	7980	7987	1	1	2	3	3	4	5	6	6
63	7993	8000	8007	8014	8021	8028	8035	8041	8048	8055	1	1	2	3	3	4	5	5	6
64	8062	8069	8075	8082	8089	8096	8102	8109	8116	8122	1	1	2	3	3	4	5	5	6
65	8129	8136	8142	8149	8156	8162	8169	8176	8182	8189	1	1	2	3	3	4	5	5	6
66	8195	8202	8209	8215	8222	8228	8235	8241	8248	8254	1	1	2	3	3	4	5	5	6
67	8261	8267	8274	8280	8287	8293	8299	8306	8312	8319	1	1	2	3	3	4	5	5	6
68	8325	8331	8338	8344	8351	8357	8363	8370	8376	8382	1	1	2	3	3	4	4	5	6
69	8388	8395	8401	8407	8414	8420	8426	8432	8439	8445	1	1	2	2	3	4	4	5	6
70	8451	8457	8463	8470	8476	8482	8488	8494	8500	8506	1	1	2	2	3	4	4	5	5
71	8513	8519	8525	8531	8537	8543	8549	8555	8561	8567	1	1	2	2	3	4	4	5	5
72	8573	8579	8585	8591	8597	8603	8609	8615	8621	8627	1	1	2	2	3	4	4	5	5
73	8633	8639	8645	8651	8657	8663	8669	8675	8681	8686	1	1	2	2	3	4	4	5	5
74	8692	8698	8704	8710	8716	8722	8727	8733	8739	8745	1	1	2	2	3	4	4	5	5
75	8751	8756	8762	8768	8774	8779	8785	8791	8797	8802	1	1	2	2	3	3	4	5	5
76	8808	8814	8820	8825	8831	8837	8842	8848	8854	8859	1	1	2	2	3	3	4	5	5
77	8865	8871	8876	8882	8887	8893	8899	8904	8910	8915	1	1	2	2	3	3	4	4	5
78	8921	8927	8932	8938	8943	8949	8954	8960	8965	8971	1	1	2	2	3	3	4	4	5
79	8976	8982	8987	8993	8998	9004	9009	9015	9020	9025	1	1	2	2	3	3	4	4	5
80	9031	9036	9042	9047	9053	9058	9063	9069	9074	9079	1	1	2	2	3	3	4	4	5
81	9085	9090	9096	9101	9106	9112	9117	9122	9128	9133	1	1	2	2	3	3	4	4	5
82	9138	9143	9149	9154	9159	9165	9170	9175	9180	9186	1	1	2	2	3	3	4	4	5
83	9191	9196	9201	9206	9212	9217	9222	9227	9232	9238	1	1	2	2	3	3	4	4	5
84	9243	9248	9253	9258	9263	9269	9274	9279	9284	9289	1	1	2	2	3	3	4	4	5
85	9294	9299	9304	9309	9315	9320	9325	9330	9335	9340	1	1	2	2	3	3	4	4	5
86	9345	9350	9355	9360	9365	9370	9375	9380	9385	9390	1	1	2	2	3	3	4	4	5
87	9395	9400	9405	9410	9415	9420	9425	9430	9435	9440	0	1	1	2	2	3	3	4	4
88	9445	9450	9455	9460	9465	9469	9474	9479	9484	9489	0	1	1	2	2	3	3	4	4
89	9494	9499	9504	9509	9513	9518	9523	9528	9533	9538	0	1	1	2	2	3	3	4	4
90	9542	9547	9552	9557	9562	9566	9571	9576	9581	9586	0	1	1	2	2	3	3	4	4
91	9590	9595	9600	9605	9609	9614	9619	9624	9628	9633	0	1	1	2	2	3	3	4	4
92	9638	9643	9647	9652	9657	9661	9666	9671	9675	9680	0	1	1	2	2	3	3	4	4
93	9685	9689	9694	9699	9703	9708	9713	9717	9722	9727	0	1	1	2	2	3	3	4	4
94	9731	9736	9741	9745	9750	9754	9759	9763	9768	9773	0	1	1	2	2	3	3	4	4
95	9777	9782	9786	9791	9795	9800	9805	9809	9814	9818	0	1	1	2	2	3	3	4	4
96	9823	9827	9832	9836	9841	9845	9850	9854	9859	9863	0	1	1	2	2	3	3	4	4
97	9868	9872	9877	9881	9886	9890	9894	9899	9903	9908	0	1	1	2	2	3	3	4	4
98	9912	9917	9921	9926	9930	9934	9939	9943	9948	9952	0	1	1	2	2	3	3	4	4
99	9956	9961	9965	9969	9974	9978	9983	9987	9991	9996	0	1	1	2	2	3	3	3	4

CHEMICAL ARITHMETIC

We can use the same table to find the value of the *number to which the logarithm corresponds—the antilogarithm*—by essentially reversing the above steps. The following examples illustrate this technique.

$\log x = 0.6952$
$x = $ antilog 0.6952

1. Look for 6952 in the table and write down 49 as the first two significant figures, because 6952 would be in this row if it appeared in the table.

2. 6952 is between 6946 and 6955, so, taking the number at the top of the column, write down 5 as the third significant figure.

3. The difference $6952 - 6946 = 6$ appears in the proportional parts section, giving the fourth significant figure as 7.

4. $x = 4.957$.

$\log y = 0.438$
$y = $ antilog 0.438

1. Look for 4380 in the table and write down 27 as the first two significant figures.

2. 4380 is between 4378 and 4393, so write down 4 as the third significant figure.

3. The difference $4380 - 4378 = 2$ appears in the proportional parts section giving the fourth significant figure as 1.

4. Rounding off 2.741 gives $y = 2.74$.

Remember, the antilogarithm is a number equal to or greater than 1 but less than 10 if the logarithm is a number equal to or greater than 0 but less than 1.

For a number equal to or greater than 10, the logarithm is a number equal to or greater than 1. To see this, we rewrite the number in standard scientific notation and apply the rule concerning the logarithm of a product,

$$\log (a \times 10^n) = \log a + \log 10^n = \log a + n$$

Next find the logarithm of the factor using the technique described above for a number equal to or greater than 1 but less than 10, and add n to the result. The number n is exact. The desired logarithm consists of a mantissa given by $\log a$ and a characteristic given by n. The following examples illustrate the technique:

$$x = \log 107 = \log (1.07 \times 10^2)$$
$$= \log 1.07 + \log 10^2$$
$$= 0.029 + 2 = 2.029$$

$$y = \log (6.022 \times 10^{23})$$
$$= \log 6.022 + \log 10^{23}$$
$$= 0.7797 + 23 = 23.7797$$

Antilogarithms are found essentially by reversing the above procedure, as shown by the following examples:

$$\begin{aligned}
\log x &= 6.9372 \\
x &= \text{antilog } 6.9372 \\
&= \text{antilog } (0.9372 + 6) \\
&= (\text{antilog } 0.9372)(\text{antilog } 6) \\
&= 8.654 \times 10^6
\end{aligned}
\qquad
\begin{aligned}
\log y &= 173.26 \\
y &= \text{antilog } 173.26 \\
&= \text{antilog } (0.26 + 173) \\
&= (\text{antilog } 0.26)(\text{antilog } 173) \\
&= 1.8 \times 10^{173}
\end{aligned}$$

Remember, the antilogarithm is a number equal to or greater than 10 if the logarithm is a number equal to or greater than 1.

For a number less than 1 but greater than 0, the logarithm is a negative number. To see this, rewrite the number in standard scientific notation and apply the rule concerning the logarithm of a product.

$$\log(a \times 10^{-n}) = \log a + \log 10^{-n} = \log a + (-n)$$

Next find the logarithm of the factor, and add a negative n to the result, which gives a negative number as the final answer. The following examples illustrate the technique:

$$\begin{aligned}
x = \log 0.362 &= \log (3.62 \times 10^{-1}) \\
&= \log 3.62 + \log 10^{-1} \\
&= 0.559 + (-1) \\
&= 0.559 - 1 = -0.441
\end{aligned}
\qquad
\begin{aligned}
y &= \log (4.7 \times 10^{-16}) \\
&= \log 4.7 + \log 10^{-16} \\
&= 0.67 + (-16) \\
&= 0.67 - 16 = -15.33
\end{aligned}$$

(Some of you may have learned other ways to represent logarithms, such as $\overline{1}.559$ and $\overline{16}.67$, or $9.559 - 10$ and $4.67 - 20$. These are very difficult to use in the types of calculations performed in chemistry.)

Antilogarithms are again found essentially by reversing the procedure. Remember, the antilogarithm is a number less than 1 but greater than 0 if the logarithm is a negative number. The following examples illustrate the finding of an antilogarithm:

$$\begin{aligned}
\log x &= -1.36 \\
x &= \text{antilog } (-1.36) \\
&= \text{antilog } (0.64 - 2) \\
&= \text{antilog } [0.64 + (-2)] \\
&= (\text{antilog } 0.64)[\text{antilog } (-2)] \\
&= 4.4 \times 10^{-2}
\end{aligned}
\qquad
\begin{aligned}
\log y &= -17.936 \\
y &= \text{antilog } (-17.936) \\
&= \text{antilog } (0.064 - 18) \\
&= \text{antilog } [0.064 + (-18)] \\
&= (\text{antilog } 0.064)[\text{antilog } (-18)] \\
&= 1.16 \times 10^{-18}
\end{aligned}$$

The first step in the above technique is usually hardest to understand. It is like knowing the punch line to a joke but having no idea how the story goes. What we are doing is finding a positive number that is less than 1 from which we subtract an integer to obtain the original negative number. The equation

$$-X.Y = (1 - 0.Y) - (X + 1)$$

shows how this is done in the following examples:

$$\begin{aligned}
-1.36 &= (1 - 0.36) - (1 + 1) = 0.64 - 2 \\
-17.936 &= (1 - 0.936) - (17 + 1) = 0.064 - 18
\end{aligned}$$

Exercises

1 Find the logarithm of
(a) 7.42
(c) 0.276
(e) 5×10^{26}
(b) 5.288
(d) 4.99×10^{-132}
(f) 6.022×10^{317}

2 Find the antilogarithm of
(a) 0.2635
(c) −5.2955
(e) 5.3153
(b) 0.9988
(d) −104.7
(f) 421.2

3 Perform the following calculations using logarithms only:
(a) (7.290)(18.26)
(c) $(9.35)^{3.5}$
(b) (9.435)/(0.8888)
(d) $\sqrt[\pi]{6.293}$

Answers

1 (a) 0.870, (b) 0.7233, (c) −0.559, (d) −131.302, (e) 26.7, (f) 317.7797 **2** (a) 1.834, (b) 9.972, (c) 5.064×10^{-6}, (d) 2×10^{-105}, (e) 2.067×10^5, (f) 2×10^{421} **3** (a) 133.1, (b) 10.62, (c) 2.50×10^3, (d) 1.796

APPENDIX B
Units and constants

Most of the published research work and tables of data contain measurements expressed in a modified cgs system that has been used in chemistry for many years. Although many of the current textbooks retain this system, there is a strong trend to adopt the International System of Units (SI), or at least a modified form of it. Values of the fundamental constants in both systems, as well as conversion factors to change non-SI units to SI units, are given in the three tables that follow.

TABLE B.1
Basic units

Physical quantity	Modified cgs system		Modified SI system		Conversion factor (to change from cgs to SI, multiply by:)[a]
Length	Centimeter	cm	Meter	m	10^{-2}*
	Angstrom	Å			10^{-10}*
Mass	Gram	g	Kilogram	kg	10^{-3}*
	Atomic mass unit	u, amu			$1.6605655(86) \times 10^{-27}$
Time	Second	sec	Second	s	1*
Electrical current	Ampere	amp, a, A	Ampere	A	1*
Temperature	Degree Kelvin, absolute	°K, °A	Degree Kelvin	K	1*
	Degree centigrade, Celsius	°C			$T(K) = T(°C) + 273.15$
Amount of substance	Mole	mole	Mole	mol	1*
			Kilomole	kmol	10^{-3}*
Luminous intensity	—	—	Candela	cd	—

[a]The numbers in parentheses represent the standard deviation in the last two significant figures cited. Factors marked with an asterisk are exact. Values based on *J. Phys. Chem. Ref. Data* **2**, 663–734 (1973).

TABLE B.2
Derived units

Physical quantity	Modified cgs system		Modified SI system		Conversion factor (to change from cgs to SI, multiply by:)[a]
Area	Square centimeter	cm^2	Square meter	m^2	10^{-4*}
	Square angstrom	$Å^2$			10^{-20*}
Density		$g\,cm^{-3} = g\,ml^{-1}$		$kg\,m^{-3}$	10^{3*}
Dipole moment	Debye	$D = 10^{-18}\,esu\,cm$		$C\,m$	$3.335641(14) \times 10^{-30}$
Electrical resistance	Ohm	$\Omega = ohm = V\,A^{-1}$	Ohm	$\Omega = V\,A^{-1} = kg\,m^2\,A^{-2}\,s^{-3}$	1^*
Electricity, quantity	Coulomb	$C = A\,sec$	Coulomb	$C = A\,s$	1^*
	Electrostatic unit	$esu = cm^{3/2}\,g^{1/2}\,sec^{-1}$			$3.335641(14) \times 10^{-10}$
Electromotive force	Volt	V, v	Volt	$V = kg\,m^2\,A^{-1}\,s^{-3}$	1^*
Energy	Erg	$erg = g\,cm^2\,sec^{-2}$	Joule	$J = kg\,m^2\,s^{-2}$	10^{-7*}
	Calorie	cal			4.184^*
	Electron volt	eV			$1.602192(46) \times 10^{-19}$
	Liter atmosphere	$1\,atm = 24.217256\,cal$			$1.01325 \times 10^{2*}$
	Wave number	cm^{-1}			$1.986477(10) \times 10^{-23}$
	Atomic mass unit	amu			$1.492442(6) \times 10^{-10}$
Entropy	Entropy unit	$eu = cal\,{}^{\circ}K^{-1}$	Entropy unit	$EU = J\,K^{-1}$	4.184^*
Force	Dyne	$dyne = g\,cm\,sec^{-2}$	Newton	$N = kg\,m\,s^{-2}$	10^{-5*}
Frequency	Cycle per second	$cps = sec^{-1}$	Hertz	$Hz = s^{-1}$	1^*
Heat (see Energy)					
Power	Horsepower	hp	Watt	$W = kg\,m^2\,s^{-3}$	7.46×10^2
Pressure	Atmosphere	atm	Pascal	$Pa = N\,m^{-2} = kg\,m^{-1}\,s^{-2}$	$1.01325 \times 10^{5*}$
	Bar	bar			10^{5*}
	Pound per square inch	psi			$6.89475729\overline{3}167 \times 10^3$
	Torr, millimeter of mercury	$torr = mm\,Hg$			$1.33322368421 \times 10^2$
Radiation	Rad	rad		$J\,kg^{-1}$	10^{-2*}
	Roentgen	r		$C\,kg^{-1}$	$2.57976 \times 10^{-4*}$
Viscosity	Poise	$poise = g\,cm^{-1}\,sec^{-1}$	Poiseuille	$Pl = kg\,m^{-1}\,s^{-1}$	10^{-1*}
Volume	Cubic centimeter	cm^3, cc		m^3	10^{-6*}
	Liter	l			10^{-3*}
Work (see Energy)					

[a]The numbers in parentheses represent the standard deviation in the last two significant figures cited. Factors marked with an asterisk are exact.
Values based on *J. Phys. Chem. Ref. Data* **2**, 663–734 (1973).

TABLE B.3

Fundamental constants

Symbol	Quantity	Value[a]	Modified cgs system	Modified SI system
a_0	Bohr radius	5.2917706(44)	10^{-1} Å	10^{-11} m
c	Velocity of light	2.99792458(12)	10^{10} cm sec^{-1}	10^8 m s^{-1}
e	Electronic charge	1.6021892(46)		10^{-19} C
		4.803242(14)	10^{-10} esu	
F	Faraday constant	9.648456(27)	10^4 C mole^{-1}	10^4 C mol^{-1}
		2.306036(6)	10^4 cal mole^{-1}	
g	Gravitational acceleration	9.80665	10^2 cm sec^{-2}	m s^{-2}
h	Planck's constant	6.626176(36)	10^{-27} erg sec	10^{-34} J s
k	Boltzmann's constant	1.380662(44)	10^{-16} erg °K^{-1}	10^{-23} J °K^{-1}
L	Avogadro's number	6.022045(31)	10^{23} mole^{-1}	10^{23} mol^{-1}
m_e	Electron rest mass	9.109534(47)	10^{-28} g	10^{-31} kg
m_n	Neutron rest mass	1.6749543(86)	10^{-24} g	10^{-27} kg
m_p	Proton rest mass	1.6726485(86)	10^{-24} g	10^{-27} kg
R	Gas constant	8.31441(26)	10^7 erg °K^{-1} mole^{-1}	J mol^{-1} K^{-1}
		1.98719(6)	cal mole^{-1} °K^{-1}	
		8.20568(26)	10^{-2} l atm mole^{-1} °K^{-1}	10^{-5} m^3 atm mol^{-1} K^{-1}
$\mathcal{R}$	Rydberg constant	1.097373177(83)	10^5 cm^{-1}	10^7 m^{-1}
		2.179907(12)	10^{-11} erg	10^{-18} J
V	Ideal gas molar volume	22.41383(70)	l mole^{-1}	10^{-3} m^3 mol^{-1}

[a] The numbers in parentheses represent the standard deviation in the last two significant figures cited. Values based on *J. Phys. Chem. Ref. Data* **2**, 663–734 (1973).

TABLE C.1

Standard Reduction Potentials

$F_2 + 2e^- \rightleftharpoons 2F^-$	2.87
$O_3 + 2H^+ + 2e^- \rightleftharpoons O_2 + H_2O$	2.07
$H_2O_2 + 2H^+ + 2e^- \rightleftharpoons 2H_2O$	1.776
$Au^+ + e^- \rightleftharpoons Au$	1.68
$MnO_4^- + 4H^+ + 3e^- \rightleftharpoons MnO_2 + 2H_2O$	1.679
$Mn^{3+} + e^- \rightleftharpoons Mn^{2+}$	1.51
$MnO_4^- + 8H^+ + 5e^- \rightleftharpoons Mn^{2+} + 4H_2O$	1.491
$ClO_3^- + 6H^+ + 5e^- \rightleftharpoons \frac{1}{2}Cl_2 + 3H_2O$	1.47
$ClO_3^- + 6H^+ + 6e^- \rightleftharpoons Cl^- + 3H_2O$	1.45
$ClO_4^- + 8H^+ + 8e^- \rightleftharpoons Cl^- + 4H_2O$	1.37
$Cl_2(g) + 2e^- \rightleftharpoons 2Cl^-$	1.3583
$Cr_2O_7^{2-} + 14H^+ + 6e^- \rightleftharpoons 2Cr^{3+} + 7H_2O$	1.33
$O_2 + 4H^+ + 4e^- \rightleftharpoons 2H_2O$	1.229
$MnO_2 + 4H^+ + 2e^- \rightleftharpoons Mn^{2+} + 2H_2O$	1.208
$Br_2(aq) + 2e^- \rightleftharpoons 2Br^-$	1.087
$Br_2(l) + 2e^- \rightleftharpoons 2Br^-$	1.065
$NO_3^- + 4H^+ + 3e^- \rightleftharpoons NO + 2H_2O$	0.96
$NO_3^- + 3H^+ + 2e^- \rightleftharpoons HNO_2 + H_2O$	0.94
$2Hg^{2+} + 2e^- \rightleftharpoons Hg_2^{2+}$	0.905
$ClO^- + H_2O + 2e^- \rightleftharpoons Cl^- + 2OH^-$	0.90
$Hg^{2+} + 2e^- \rightleftharpoons Hg$	0.851
$\frac{1}{2}O_2 + 2H^+(10^{-7}M) + 2e^- \rightleftharpoons H_2O$	0.815
$2NO_3^- + 4H^+ + 2e^- \rightleftharpoons N_2O_4 + 2H_2O$	0.81
$Ag^+ + e^- \rightleftharpoons Ag$	0.7996
$Hg_2^{2+} + 2e^- \rightleftharpoons 2Hg$	0.7961
$Fe^{3+} + e^- \rightleftharpoons Fe^{2+}$	0.770
$Fe(CN)_6^{3-} + e^- \rightleftharpoons Fe(CN)_6^{4-}(1M\ H_2SO_4)$	0.69
$O_2 + 2H^+ + 2e^- \rightleftharpoons H_2O_2$	0.682
$MnO_4^- + 2H_2O + 3e^- \rightleftharpoons MnO_2 + 4OH^-$	0.58
$IO_3^- + 2H_2O + 4e^- \rightleftharpoons IO^- + 4OH^-$	0.56
$I_2 + 2e^- \rightleftharpoons 2I^-$	0.535
$Cu^+ + e^- \rightleftharpoons Cu$	0.522
$Fe(CN)_6^{3-} + e^- \rightleftharpoons Fe(CN)_6^{4-}(0.01M\ NaOH)$	0.46

$O_2 + 2H_2O + 4e^- \rightleftharpoons 4OH^-$ 0.401

$ClO_3^- + H_2O + 2e^- \rightleftharpoons ClO_2^- + 2OH^-$ 0.35

$Ag_2O + H_2O + 2e^- \rightleftharpoons 2Ag + 2OH^-$ 0.342

$Cu^{2+} + 2e^- \rightleftharpoons Cu$ 0.3402

$AgCl + e^- \rightleftharpoons Ag + Cl^-$ 0.2223

$SO_4^{2-} + 4H^+ + 2e^- \rightleftharpoons H_2SO_3 + H_2O$ 0.20

$ClO_4^- + H_2O + 2e^- \rightleftharpoons ClO_3^- + 2OH^-$ 0.17

$Cu^{2+} + e^- \rightleftharpoons Cu^+$ 0.158

$Sn^{4+} + 2e^- \rightleftharpoons Sn^{2+}$ 0.15

$AgBr + e^- \rightleftharpoons Ag + Br^-$ 0.0713

$2H^+ + 2e^- \rightleftharpoons H_2$ 0.0000

$AgCN + e^- \rightleftharpoons Ag + CN^-$ -0.02

$Fe^{3+} + 3e^- \rightleftharpoons Fe$ -0.036

$Pb^{2+} + 2e^- \rightleftharpoons Pb$ -0.1263

$Sn^{2+} + 2e^- \rightleftharpoons Sn$ -0.1364

$2SO_4^{2-} + 4H^+ + 2e^- \rightleftharpoons S_2O_6^{2-} + 2H_2O$ -0.2

$Ni^{2+} + 2e^- \rightleftharpoons Ni$ -0.23

$Fe^{2+} + 2e^- \rightleftharpoons Fe$ -0.409

$Cr^{3+} + e^- \rightleftharpoons Cr^{2+}$ -0.41

$NO_2^- + H_2O + e^- \rightleftharpoons NO + 2OH^-$ -0.46

$2CO_2 + 2H^+ + 2e^- \rightleftharpoons H_2C_2O_4$ -0.49

$Fe(OH)_3 + e^- \rightleftharpoons Fe(OH)_2 + OH^-$ -0.56

$Ni(OH)_2 + 2e^- \rightleftharpoons Ni + 2OH^-$ -0.66

$Cr^{3+} + 3e^- \rightleftharpoons Cr$ -0.74

$Zn^{2+} + 2e^- \rightleftharpoons Zn$ -0.7628

$2H_2O + 2e^- \rightleftharpoons H_2 + 2OH^-$ -0.8277

$Sn(OH)_6^{2-} + 2e^- \rightleftharpoons HSnO_2^- + 3OH^- + H_2O$ -0.96

$Mn^{2+} + 2e^- \rightleftharpoons Mn$ -1.029

$ZnO_2^- + 2H_2O + 2e^- \rightleftharpoons Zn + 4OH^-$ -1.216

$Mn(OH)_2 + 2e^- \rightleftharpoons Mn + 2OH^-$ -1.47

$Be^{2+} + 2e^- \rightleftharpoons Be$ -1.70

$Al^{3+} + 3e^- \rightleftharpoons Al(0.1M\ NaOH)$ -1.706

$\frac{1}{2}H_2 + e^- \rightleftharpoons H^-$ -2.23

$Mg^{2+} + 2e^- \rightleftharpoons Mg$ -2.375

$Na^+ + e^- \rightleftharpoons Na$ -2.7109

$Ca^{2+} + 2e^- \rightleftharpoons Ca$ -2.76

$Sr^{2+} + 2e^- \rightleftharpoons Sr$ -2.89

$Ba^{2+} + 2e^- \rightleftharpoons Ba$ -2.90

$Cs^+ + e^- \rightleftharpoons Cs$ -2.923

$K^+ + e^- \rightleftharpoons K$ -2.924

$Rb^+ + e^- \rightleftharpoons Rb$ -2.925

$Ca(OH)_2 + 2e^- \rightleftharpoons Ca + 2OH^-$ -3.02

$Li^+ + e^- \rightleftharpoons Li$ -3.045

TABLE C.2

Ionization Constants of Acids, K_a, at 25°C [Data from *Stability Constants of Metal-Ion Complexes*, Special Publications 17 (1964) and 25 (1971), The Chemical Society, London. Some values are theoretical.]

Formula	Name	K_a	Formula	Name	K_a
Inorganic acids			H_2SO_3	Sulfurous	
H_3AsO_4	Arsenic		H_2SO_3		$K_{a1} = 1.43 \times 10^{-2}$
			HSO_3^-		$K_{a2} = 5.0 \times 10^{-8}$
H_3AsO_4		$K_{a1} = 6.5 \times 10^{-3}$	$H_2S_2O_3$	Thiosulfuric	
$H_2AsO_4^-$		$K_{a2} = 1.1 \times 10^{-7}$	$H_2S_2O_3$		$K_{a1} = 2.0 \times 10^{-2}$
$HAsO_4^{2-}$		$K_{a3} = 3 \times 10^{-12}$	$HS_2O_3^-$		$K_{a2} = 3.2 \times 10^{-3}$
H_3BO_3	Boric	$K_{a1} = 6.0 \times 10^{-10}$			
H_2CO_3	Carbonic		*Organic acids*		
H_2CO_3		$K_{a1} = 4.5 \times 10^{-7}$	CH_3COOH	Acetic	1.754×10^{-5}
HCO_3^-		$K_{a2} = 4.8 \times 10^{-11}$	C_6H_5COOH	Benzoic	6.6×10^{-5}
$HClO_3$	Chloric	5×10^2	$HCOOH$	Formic	1.772×10^{-4}
$HClO_2$	Chlorous	1.1×10^{-2}	HCN	Hydrocyanic	6.2×10^{-10}
H_2CrO_4	Chromic		$H_2C_2O_4$	Oxalic	
H_2CrO_4		$K_{a1} = 1.8 \times 10^{-1}$	$H_2C_2O_4$		$K_{a1} = 5.60 \times 10^{-2}$
$HCrO_4^-$		$K_{a2} = 3.2 \times 10^{-7}$	$HC_2O_4^-$		$K_{a2} = 6.2 \times 10^{-5}$
HBr	Hydrobromic	10^9	CH_3CH_2COOH	Propionic	1.3×10^{-5}
HCl	Hydrochloric	3×10^8	$HNCS$	Thiocyanic	6.9×10^1
HF	Hydrofluoric	6.5×10^{-4}			
H_2O_2	Hydrogen peroxide	$K_{a1} = 2.2 \times 10^{-12}$	*Amphoteric hydroxides*		
HI	Hydroiodic	3×10^9	$Al(OH)_3$	Aluminum hydroxide	4×10^{-13}
H_2S	Hydrosulfuric		$SbO(OH)$	Antimony(III) hydroxide	1×10^{-11}
H_2S		$K_{a1} = 1.0 \times 10^{-7}$	$Cr(OH)_3$	Chromium(III) hydroxide	9×10^{-17}
HS^-		$K_{a2} = 3 \times 10^{-13}$	$Cu(OH)_2$	Copper(II) hydroxide	
$HBrO$	Hypobromous	2.2×10^{-9}	$Cu(OH)_2$		$K_{a1} = 1 \times 10^{-19}$
$HClO$	Hypochlorous	2.90×10^{-8}	$HCuO_2^-$		$K_{a2} = 7.0 \times 10^{-14}$
HIO	Hypoiodous	2.3×10^{-11}	$Pb(OH)_2$	Lead(II) hydroxide	4.6×10^{-16}
HIO_3	Iodic	1.6×10^{-1}	$Sn(OH)_4$	Tin(IV) hydroxide	10^{-32}
HNO_3	Nitric	2×10^2	$Sn(OH)_2$	Tin(II) hydroxide	3.8×10^{-15}
HNO_2	Nitrous	7.2×10^{-4}	$Zn(OH)_2$	Zinc hydroxide	1.0×10^{-29}
$HClO_4$	Perchloric	3.5×10^1			
HIO_4	Periodic	5.6×10^{-9}	*Metal cations*		
$HMnO_4$	Permanganic	2.0×10^3	Al^{3+}	Aluminum ion	1.4×10^{-5}
H_3PO_4	Phosphoric		NH_4^+	Ammonium ion	6.3×10^{-10}
H_3PO_4		$K_{a1} = 7.5 \times 10^{-3}$	Bi^{3+}	Bismuth(III) ion	1×10^{-2}
$H_2PO_4^-$		$K_{a2} = 6.6 \times 10^{-8}$	Cr^{3+}	Chromium(III) ion	1×10^{-4}
HPO_4^{2-}		$K_{a3} = 1 \times 10^{-12}$	Cu^{2+}	Copper(II) ion	1×10^{-8}
H_2SiO_3	Metasilicic		Fe^{3+}	Iron(III) ion	4.0×10^{-3}
H_2SiO_3		$K_{a1} = 3.2 \times 10^{-10}$	Fe^{2+}	Iron(II) ion	1.2×10^{-6}
$HSiO_3^-$		$K_{a2} = 1.5 \times 10^{-12}$	Mg^{2+}	Magnesium ion	2×10^{-12}
H_2SO_4	Sulfuric	Large	Hg^{2+}	Mercury(II) ion	2×10^{-3}
H_2SO_4		Large	Zn^{2+}	Zinc ion	2.5×10^{-10}
HSO_4^-		$K_{a2} = 1.0 \times 10^{-2}$			

TABLE C.3

Ionization constants of bases, K_b, at 25°C

Formula	Name	K_b		Formula	Name	K_b	
CH_3COO^-	Acetate ion	5.701	$\times 10^{-10}$	$C_2O_4^{2-}$	Oxalate ion	1.6	$\times 10^{-10}$
NH_3	Ammonia	1.6	$\times 10^{-5}$	$HC_2O_4^{2-}$		1.79	$\times 10^{-13}$
$C_6H_5NH_2$	Aniline	4.2	$\times 10^{-10}$	MnO_4^-	Permanga-nate ion	5.0	$\times 10^{-17}$
AsO_4^{3-}	Arsenate ion	3.3	$\times 10^{-12}$	PO_4^{3-}	Phosphate ion	1	$\times 10^{-2}$
$HAsO_4^{2-}$		9.1	$\times 10^{-8}$	HPO_4^{2-}		1.5	$\times 10^{-7}$
$H_2AsO_4^-$		1.5	$\times 10^{-12}$	$H_2PO_4^-$		1.3	$\times 10^{-12}$
$H_2BO_3^-$	Borate ion	1.6	$\times 10^{-5}$	SiO_3^{2-}	Metasilicate ion	6.7	$\times 10^{-3}$
$B_4O_7^{2-}$			10^{-3}	$HSiO_3^-$		3.1	$\times 10^{-5}$
Br^-	Bromide ion		10^{-23}	SO_4^{2-}	Sulfate ion	1.0	$\times 10^{-12}$
CO_3^{2-}	Carbonate ion	2.1	$\times 10^{-4}$	SO_3^{2-}	Sulfite ion	2.0	$\times 10^{-7}$
HCO_3^-		2.2	$\times 10^{-8}$	HSO_3^-		6.99	$\times 10^{-13}$
Cl^-	Chloride ion	3	$\times 10^{-23}$	S^{2-}	Sulfide ion	3	$\times 10^{-2}$
CrO_4^{2-}	Chromate ion	3.1	$\times 10^{-8}$	HS^-		1.0	$\times 10^{-7}$
CN^-	Cyanide ion	1.6	$\times 10^{-5}$	NCS^-	Thiocyanate ion	1.4	$\times 10^{-11}$
$(C_2H_5)_2NH$	Diethylamine	9.5	$\times 10^{-4}$	$S_2O_3^{2-}$	Thiosulfate ion	3.1	$\times 10^{-12}$
$(CH_3)_2NH$	Dimethyl-amine	5.9	$\times 10^{-4}$	$(C_2H_5)_3N$	Triethyl-amine	5.2	$\times 10^{-4}$
$C_2H_5NH_2$	Ethylamine	4.7	$\times 10^{-4}$	$(CH_3)_3N$	Trimethyl-amine	6.3	$\times 10^{-5}$
F^-	Fluoride ion	1.5	$\times 10^{-11}$				
$HCOO^-$	Formate ion	5.643	$\times 10^{-11}$				
I^-	Iodide ion	3	$\times 10^{-24}$				
CH_3NH_2	Methylamine	3.9	$\times 10^{-4}$				
NO_3^-	Nitrate ion	5	$\times 10^{-17}$				
NO_2^-	Nitrite ion	1.4	$\times 10^{-11}$				

Acetates			BaCrO$_4$	1.2	$\times 10^{-10}$
Ag(CH$_3$COO)	4.4	$\times 10^{-3}$	Ag$_2$CrO$_4$	2.5	$\times 10^{-12}$
Hg$_2$(CH$_3$COO)$_2$	4	$\times 10^{-10}$	PbCrO$_4$	2.8	$\times 10^{-13}$
Arsenates			Cyanides		
Ag$_3$AsO$_4$	1	$\times 10^{-22}$	AgCN	2.3	$\times 10^{-16}$
Bromides			Fluorides		
PbBr$_2$	3.9	$\times 10^{-5}$	BaF$_2$	1.0	$\times 10^{-6}$
CuBr	5.2	$\times 10^{-9}$	MgF$_2$	6.8	$\times 10^{-9}$
AgBr	4.9	$\times 10^{-13}$	SrF$_2$	2.5	$\times 10^{-9}$
Hg$_2$Br$_2$	5.8	$\times 10^{-23}$	CaF$_2$	2.7	$\times 10^{-11}$
Carbonates			ThF$_4$	4	$\times 10^{-28}$
MgCO$_3$	1	$\times 10^{-5}$	Ferrocyanides		
NiCO$_3$	1.3	$\times 10^{-7}$	KFe[Fe(CN)$_6$]	3	$\times 10^{-41}$
CaCO$_3$	3.84	$\times 10^{-9}$	Ag$_4$[Fe(CN)$_6$]	2	$\times 10^{-41}$
BaCO$_3$	2.0	$\times 10^{-9}$	K$_2$Zn$_3$[Fe(CN)]$_2$	1	$\times 10^{-95}$
SrCO$_3$	5.2	$\times 10^{-10}$	Hydroxides ($= K_b$)		
MnCO$_3$	5.0	$\times 10^{-10}$	Ba(OH)$_2$	1.3	$\times 10^{-2}$
CuCO$_3$	2.3	$\times 10^{-10}$	Sr(OH)$_2$	6.4	$\times 10^{-3}$
CoCO$_3$	1.0	$\times 10^{-10}$	Ca(OH)$_2$	4.0	$\times 10^{-5}$
FeCO$_3$	2.1	$\times 10^{-11}$	Ag$_2$O	2	$\times 10^{-8}$
ZnCO$_3$	1.7	$\times 10^{-11}$	Mg(OH)$_2$	7.1	$\times 10^{-12}$
Ag$_2$CO$_3$	8.1	$\times 10^{-12}$	BiO(OH)	1	$\times 10^{-12}$
CdCO$_3$	1.0	$\times 10^{-12}$	Be(OH)$_2$	4	$\times 10^{-13}$
PbCO$_3$	7.4	$\times 10^{-14}$	Zn(OH)$_2$	3.3	$\times 10^{-13}$
Chlorides			Mn(OH)$_2$	2	$\times 10^{-13}$
PbCl$_2$	2	$\times 10^{-5}$	Cd(OH)$_2$	8.1	$\times 10^{-15}$
CuCl	1.2	$\times 10^{-6}$	Pb(OH)$_2$	1.2	$\times 10^{-15}$
AgCl	1.8	$\times 10^{-10}$	Fe(OH)$_2$	8	$\times 10^{-16}$
Hg$_2$Cl$_2$	1.3	$\times 10^{-18}$	Ni(OH)$_2$	3	$\times 10^{-16}$
Chromates			Co(OH)$_2$	2	$\times 10^{-16}$
CaCrO$_4$	6	$\times 10^{-4}$	SbO(OH)	1	$\times 10^{-17}$
SrCrO$_4$	2.2	$\times 10^{-5}$	Cu(OH)$_2$	1.3	$\times 10^{-20}$
Hg$_2$CrO$_4$	2.0	$\times 10^{-9}$			

Hg(OH)$_2$	4 $\times 10^{-26}$		*Phosphates*		
Sn(OH)$_2$	6 $\times 10^{-27}$		Li$_3$PO$_4$	3 $\times 10^{-13}$	
Cr(OH)$_3$	6 $\times 10^{-31}$		Mg(NH$_4$)PO$_4$	3 $\times 10^{-13}$	
Al(OH)$_3$	3.5 $\times 10^{-34}$		Ag$_3$PO$_4$	1.4 $\times 10^{-16}$	
Fe(OH)$_3$	3 $\times 10^{-39}$		AlPO$_4$	5.8 $\times 10^{-19}$	
Sn(OH)$_4$	10^{-57}		Mn$_3$(PO$_4$)$_2$	1 $\times 10^{-22}$	

Iodides

PbI$_2$	7.1 $\times 10^{-9}$		Ba$_3$(PO$_4$)$_2$	3 $\times 10^{-23}$	
CuI	1.1 $\times 10^{-12}$		BiPO$_4$	1.3 $\times 10^{-23}$	
AgI	8.3 $\times 10^{-17}$		Ca$_3$(PO$_4$)$_2$	10^{-26}	
HgI$_2$	3 $\times 10^{-26}$		Sr$_3$(PO$_4$)$_2$	4 $\times 10^{-28}$	
Hg$_2$I$_2$	4.5 $\times 10^{-29}$		Mg$_3$(PO$_4$)$_2$	10^{-32}	
			Pb$_3$(PO$_4$)$_2$	7.9 $\times 10^{-43}$	

Nitrates

BiO(NO$_3$)	2.8 $\times 10^{-3}$		*Sulfates*		
			CaSO$_4$	2.5 $\times 10^{-5}$	

Nitrites

Ag(NO$_2$)	6.0 $\times 10^{-4}$		Ag$_2$SO$_4$	1.5 $\times 10^{-5}$	
			Hg$_2$SO$_4$	6.8 $\times 10^{-7}$	

Oxalates

MgC$_2$O$_4$	8 $\times 10^{-5}$		SrSO$_4$	3.5 $\times 10^{-7}$	
CoC$_2$O$_4$	4 $\times 10^{-6}$		PbSO$_4$	2.2 $\times 10^{-8}$	
FeC$_2$O$_4$	2 $\times 10^{-7}$		BaSO$_4$	1.7 $\times 10^{-10}$	
NiC$_2$O$_4$	1 $\times 10^{-7}$				
SrC$_2$O$_4$	5 $\times 10^{-8}$		*Sulfides*		
CuC$_2$O$_4$	3 $\times 10^{-8}$		MnS	2.3 $\times 10^{-13}$	
BaC$_2$O$_4$	2 $\times 10^{-8}$		FeS	4.2 $\times 10^{-17}$	
CdC$_2$O$_4$	2 $\times 10^{-8}$		NiS	3 $\times 10^{-19}$	
ZnC$_2$O$_4$	2 $\times 10^{-9}$		ZnS	2 $\times 10^{-24}$	
CaC$_2$O$_4$	1 $\times 10^{-9}$		CoS	2 $\times 10^{-25}$	
Ag$_2$C$_2$O$_4$	3.5 $\times 10^{-11}$		SnS	3 $\times 10^{-27}$	
PbC$_2$O$_4$	4.8 $\times 10^{-12}$		CdS	2 $\times 10^{-28}$	
Hg$_2$C$_2$O$_4$	2 $\times 10^{-13}$		PbS	1 $\times 10^{-28}$	
MnC$_2$O$_4$	1 $\times 10^{-15}$		CuS	6 $\times 10^{-36}$	
La$_2$(C$_2$O$_4$)$_3$	2 $\times 10^{-28}$		Cu$_2$S	3 $\times 10^{-48}$	
			Ag$_2$S	7.1 $\times 10^{-50}$	
			HgS	4 $\times 10^{-53}$	
			Fe$_2$S$_3$	1 $\times 10^{-88}$	

TABLE C.5

Dissociation constants for complexes, K_d, at 25°C.[a]

Complex	K_d	Complex	K_d	Complex	K_d
$[Ag(NH_3)_2]^+$	6.2×10^{-8}	$[Cd(en)_4]^{2+}$	2.60×10^{-11}	$[Fe(C_2O_4)_3]^{3-}$	3×10^{-21}
$[AgBr_2]^-$	7.8×10^{-8}	$[CdI_4]^{2-}$	8×10^{-7}	$[Fe(C_2O_4)_3]^{4-}$	6×10^{-6}
$[AgCl_2]^-$	9×10^{-6}	$[Cd(SCN)_4]^{2-}$	1×10^{-3}	$[Fe(SCN)_3]$	5×10^{-7}
$[AgCl_4]^{3-}$	5×10^{-6}	$[Co(NH_3)_6]^{2+}$	9×10^{-6}	$[HgCl_4]^{2-}$	2×10^{-16}
$[Ag(CN)_2]^-$	1×10^{-22}	$[Co(en)_3]^{2+}$	1.52×10^{-14}	$[Hg(SCN)_4]^{2-}$	2.0×10^{-22}
$[Ag(en)]^+$	1×10^{-5}	$[Co(en)_3]^{3+}$	2.04×10^{-49}	$[Mg(nta)_2]^{4-}$	6.3×10^{-11}
$[Ag(OH)_3]^{2-}$	1.7×10^{-5}	$[Co(C_2O_4)_3]^{4-}$	2.2×10^{-7}	$[Mg(P_2O_7)]^{2-}$	2×10^{-6}
$[Ag(SCN)_4]^{3-}$	2.1×10^{-10}	$[Cu(NH_3)_4]^{2+}$	1×10^{-13}	$[Ni(NH_3)_6]^{2+}$	1×10^{-9}
$[AlF_6]^{3-}$	3×10^{-20}	$[CuCl_2]^-$	1.15×10^{-5}	$[Pb(SCN)_2]$	3×10^{-3}
$[Au(CN)_2]^-$	5×10^{-39}	$[Cu(edta)]^{2-}$	1.38×10^{-19}	$[PdBr_4]^{2-}$	8.0×10^{-14}
$[Ca(nta)_2]^{4-}$	2.44×10^{-12}	$[Cu(gly)_2]$	5.6×10^{-16}	$[PdCl_4]^{2-}$	6×10^{-14}
$[Ca(P_2O_7)]^{2-}$	1×10^{-5}	$[Cu(OH)_4]^{2-}$	7.6×10^{-17}	$[Zn(NH_3)_4]^{2+}$	3.46×10^{-10}
$[Cd(CH_3NH_2)_4]^{2+}$	2.82×10^{-7}	$[Cu(C_2O_4)_2]^{2-}$	6×10^{-11}	$[Zn(CN)_4]^{2-}$	2.4×10^{-20}
$[Cd(NH_3)_4]^{2+}$	1×10^{-7}	$[Cu(P_2O_7)]^{2-}$	2.0×10^{-7}	$[Zn(edta)]^{2-}$	2.63×10^{-17}
$[Cd(NH_3)_6]^{2+}$	1×10^{-5}	$[Cu(SCN)_2]$	1.8×10^{-4}	$[Zn(gly)_2]$	1.1×10^{-10}
$[CdBr_4]^{2-}$	2×10^{-4}	$[Fe(CN)_6]^{4-}$	1.3×10^{-37}	$[Zn(OH)_4]^{2-}$	5×10^{-21}
$[CdCl_4]^{2-}$	9.3×10^{-3}	$[Fe(CN)_6]^{3-}$	1.3×10^{-44}		
$[Cd(CN)_4]^{2-}$	8.2×10^{-18}	$[Fe(C_2O_4)]^{2-}$	2×10^{-8}		

[a] (en) ethylenediamine, $H_2NCH_2CH_2NH_2$ (nta) nitrilotriacetate ion, $N(CH_2COO^-)_2$ (gly) glycine ion, $(H_2NCH_2COO^-)$ (edta) ethylenediaminetetraacetate ion, $(-OOCCH_2)_2NCH_2CH_2N(CH_2COO^-)_2$

TABLE C.6

Additional tables of data in this book

Standard enthalpies of formation ($\Delta H°$)	**5.1**
Electronic configurations of the elements	**7.10 (also inside cover)**
Periodic table	**7.11 (also inside cover)**
Average bond energies	**9.19**
Some bond lengths	**9.20**
Some typical atomic and ionic radii	**9.21**
First ionization energies of the representative elements	**10.4**
Ionization energies of Na–Ar period elements	**10.5**
Electron affinities of the representative elements	**10.6**
Electronegativities of the representative elements	**10.7**
Atomic radii of the representative elements	**Figure 10.1**
Molal boiling point elevation (K_b) and molal freezing point lowering (K_f) constants	**14.6**
Standard reduction potentials	**19.6 (also Appendix C.1)**
Dissociation constants of some common complexes	**28.5 (also Appendix C.5)**
Properties of d transition elements	**29.2, 29.3**
Atomic radii of d transition elements	**Figure 29.1**

APPENDIX D
Inorganic nomenclature

D.1 Symbols for the elements

Each element has a distinctive name and symbol; these names and symbols are listed in alphabetical order of the names inside the front cover of this book and in the order of atomic numbers inside the back cover. Most symbols consist of two letters: the first letter is always a capital and the second one is always lower case. The symbols for most of the elements are based on the modern names of the elements, for example,

Ac	Al	Am	Ar	As	At
actinium	*aluminum*	*americium*	*argon*	*arsenic*	*astatine*

Some elements have symbols that do not appear to be related to their names. Most of these are elements that have been known for a long time, and the symbols are derived from their Greek or Latin names. Two ancient elements, sulfur and carbon, have single-letter symbols.

Ag	Au	Sb	S	C
silver	*gold*	*antimony*	*sulfur*	*carbon*
(argentum)	*(aurum)*	*(stibium)*	*(sulfurium)*	*(carbo)*

All the elements that have symbols based on older names are listed in Table D.1.

Some of the elements occur naturally as molecules, not as atoms (Section 2.4). When referring to the naturally occurring form of these elements, the name of the element can be modified by adding an appropriate prefix, called a multiplying prefix, that shows the number of atoms combined in the molecule, for example,

N_2 dinitrogen O_2 dioxygen P_4 tetraphosphorus

TABLE D.1

Elements with symbols based on older names

All of these elements except sodium (discovered in 1807) and tungsten (discovered in 1783) have been known since antiquity.

Modern name	Symbol	Old name
Antimony	Sb	*stibium*
Copper	Cu	*cuprum*
Gold	Au	*aurum*
Iron	Fe	*ferrum*
Lead	Pb	*plumbum*
Mercury	Hg	*hydrargyrum*
Potassium	K	*kalium*
Silver	Ag	*argentum*
Sodium	Na	*natrium*
Tin	Sn	*stannum*
Tungsten	W	*wolfram*

TABLE D.2
Multiplying prefixes

Number indicated	Prefix
1	mono
2	di
3	tri
4	tetra
5	penta
6	hexa
7	hepta
8	octa
9	nona
10	deca

The first ten multiplying prefixes used in chemical nomenclature are listed in Table D.2. The distinction between atomic and molecular forms of an element is also made as follows:

N atomic nitrogen N_2 molecular nitrogen

D.2 Isotopes

During the pioneer days of nuclear research, before anyone knew how many isotopes would eventually be discovered, each new isotope was given a unique name. These names were sometimes based on the parent element from which the new isotope was generated. For example, isotopes derived from thorium were called radiothorium, thorium A, thorium B, thorium X, and mesothorium. Although you may occasionally encounter such names, we now prefer to name isotopes by the name of the element (based on the atomic number of the isotope), followed by the mass number, such as, carbon-14 and uranium-235, or ^{14}C and ^{235}U (Section 7.9). The isotopes of hydrogen are an exception, because they are frequently known by distinctive names.

H hydrogen ^2H or D deuterium
^1H protium ^3H or T tritium

Exercises

1 Look up the names of the following elements:
(a) Ta (e) Tl (i) Ti (m) N_2
(b) Tc (f) Th (j) W (n) I_2
(c) Te (g) Tm (k) ^{14}C
(d) Tb (h) Sn (l) ^{37}Cl

2 Look up the symbols of the following elements:
(a) cadmium (f) cesium (k) curium
(b) calcium (g) chlorine (l) fluorine-18
(c) californium (h) chromium (m) dioxygen
(d) carbon (i) cobalt (o) molecular sulfur
(e) cerium (j) copper

D.3 Binary acids

A binary acid is an aqueous solution of a compound containing hydrogen and another element. The formula for a binary acid consists of the symbols for hydrogen and of the second element, for example,

$HF(aq)$ $H_2S(aq)$ $HI(aq)$
hydrofluoric acid   *hydrosulfuric acid*   *hydroiodic acid*

Note that each name consists of three parts: (1) the prefix "hydro," which refers to the hydrogen atom; (2) the name of the second element modified with the suffix "ic"; (3) the word "acid,"

hydro . . . ic acid

Exercises

1 Name the following binary acids: (a) HF, (b) HBr, (c) H_2S, and (d) H_2Se.

2 Write the formulas for the following binary acids: (a) hydrochloric acid; (b) hydroiodic acid; and (c) hydrotelluric acid.

D.4 Simple cations

A cation is a positively charged ion. Simple cations are formed by the removal of one or more electrons from an atom (Section 2.4). The International Union of Pure and Applied Chemistry (IUPAC) sets standards for chemical nomenclature. For simple cations, the Stock system of nomenclature is recommended by IUPAC. The name in this system consists of (1) the elemental name and (2) the charge on the ion given inside parentheses as a Roman numeral, followed by (3) the word "ion," for example,

$$Li^+ \quad lithium(I)\ ion \qquad Fe^{2+} \quad iron(II)\ ion$$
$$Cr^{3+} \quad chromium(III)\ ion \qquad Fe^{3+} \quad iron(III)\ ion$$

Note that, although we say an ion has, for instance, a $+3$ charge, the number always precedes the $+$ or $-$ sign in the symbol for the ion. For elements that form only one cation, the numerical part of the name is commonly omitted, for example,

$$H^+ \quad hydrogen\ ion \qquad Na^+ \quad sodium\ ion \qquad Al^{3+} \quad aluminum\ ion$$

An older system of naming cations uses word endings instead of numerals. The cation with the lower charge has the suffix "ous" and the cation with the higher charge has the suffix "ic" following the name of the element.

Mn^{2+}	manganous ion	manganese(II) ion
Mn^{3+}	manganic ion	manganese(III) ion
Cu^+	cuprous ion	copper(I) ion
Cu^{2+}	cupric ion	copper(II) ion

Frequently, the ancient name of the metal is used as the root in this system, as, for example, in cuprous and cupric, which are based on *cuprum*. Although you may encounter such names, their use should be avoided whenever possible.

The simple cations of mercury present an unusual case. The mercury(I) ion does not exist as Hg^+ but instead occurs as a dimer, Hg_2^{2+}. This ion is simply called the mercury(I) ion, and the two known ions of mercury are thus the mercury(I), Hg_2^{2+}, and mercury(II), Hg^{2+} ions.

Exercises

1 Name the following cations:
(a) Cr^{3+}
(b) Sc^{2+}
(c) Sc^{3+}
(d) Cs^+
(e) Sn^{4+}
(f) Pb^{2+}
(g) Au^+
(h) Au^{3+}
(i) Ba^{2+}

2 Write the formulas for the following cations:
(a) sodium(I) ion
(b) zinc(II) ion
(c) lead(IV) ion
(d) silver(I) ion
(e) mercury(II) ion
(f) iron(III) ion
(g) mercury(I) ion
(h) calcium ion
(i) magnesium(II) ion

3 Write the formulas for the following cations:
(a) stannic ion, stannous ion
(b) mercurous ion, mercuric ion

4 Name the following cations, using the "ous" and "ic" endings:
(a) Fe^{2+}, Fe^{3+}
(b) Cu^+, Cu^{2+}

D.5 Simple anions

An anion is a negatively charged ion. Simple anions are formed by the addition of one or more electrons to an atom (Section 2.4). Such anions are named by writing (1) the name of the element modified by (2) the ending "ide," followed by (3) the word "ion." Here, as for simple cations, in the symbol the number is followed by the charge. Most elements form only one anion, and it is not necessary to include the charge in the name. Some examples of simple anions are

Cl^- chloride ion O^{2-} oxide ion N^{3-} nitride ion

A few ions have unique names, such as

N_3^- azide ion O_2^{2-} peroxide ion

Exercises

1 Name the following anions:
(a) N^{3-}
(b) O^{2-}
(c) Se^{2-}
(d) F^-
(e) Br^-

2 Write the formulas for the following anions:
(a) phosphide ion
(b) sulfide ion
(c) telluride ion
(d) chloride ion
(e) iodide ion

D.6 Binary ionic compounds

A binary ionic compound is named by giving the cation name first, followed by the anion name. The word "ion" does not appear in the name of the compound. For example,

$CuCl$	copper(I) chloride (cuprous chloride)
$CuCl_2$	copper(II) chloride (cupric chloride)
NaF	sodium fluoride
Al_2O_3	aluminum oxide
Na_3P	sodium phosphide
NaN_3	sodium azide
TlI	thallium(I) iodide

Exercises

1 Name the following compounds:
(a) Li_2S
(b) SnO_2
(c) RbI
(d) Li_2O_2
(e) UO_2
(f) Hg_2S
(g) Li_2O
(h) $MnCl_2$
(i) $ZrBr_4$

2 Write the formulas for the following compounds:
(a) sodium fluoride
(b) zinc oxide
(c) barium peroxide
(d) magnesium bromide
(e) hydrogen iodide
(f) calcium phosphide
(g) iron(II) oxide
(h) nickel(II) chloride
(i) silver fluoride

D.7 Binary covalent compounds

A binary covalent compound contains atoms of two different elements, held together by covalent bonding (Section 9.9). Such compounds can be named by the Stock system of nomenclature, exactly as it was described for binary ionic compounds. The Roman numeral in this case represents the positive oxidation number (Section 4.11) of the element with the higher electronegativity (Section 10.7), and in order to name the compound it is not essential to know whether it is ionic or covalent. An alternate and classical system for naming binary covalent compounds is based on the prefixes listed in Table D.2. Note that the prefix "mono" is quite often omitted.

	Classical system	Stock system
N_2O	Dinitrogen monoxide	Nitrogen(I) oxide
N_2O_3	Dinitrogen trioxide	Nitrogen(III) oxide
N_2O_5	Dinitrogen pentoxide	Nitrogen(V) oxide
ICl	Iodine monochloride	Iodine(I) chloride
ICl_3	Iodine trichloride	Iodine(III) chloride

Sometimes the Stock system name is not complete enough; for example, both NO_2 and N_2O_4 would be "nitrogen(IV) oxide." To give each substance a distinctive name, the word "dimer" (and, occasionally, "monomer") is used as part of the name.

NO_2	nitrogen dioxide	nitrogen(IV) oxide
N_2O_4	dinitrogen tetroxide	nitrogen(IV) oxide dimer

Of course, a few common names persist, such as water for H_2O and ammonia for NH_3.

Exercises

1 Name the following compounds:
(a) CO
(b) CO_2
(c) SF_6
(d) $SiCl_4$
(e) IF
(f) IF_3
(g) IF_7
(h) P_4O_{10}
(i) SO_3

2 Write the formulas for the following compounds:
(a) boron(II) oxide
(b) silicon dioxide
(c) phosphorus trichloride
(d) sulfur(IV) chloride
(e) bromine trifluoride
(f) phosphorus(III) oxide
(g) iodine(V) oxide
(h) xenon tetrafluoride
(i) hydrogen sulfide

D.8 Oxoacids

Oxoacids contain hydrogen, oxygen, and a third, central element. The common oxoacids are listed in Table D.3. Some of the acids listed, such as carbonic acid, have never been isolated, but their salts are known. The names of the oxoacids are based on the use of prefixes ("per" and "hypo") and suffixes ("ic" and "ous"). The basis for choosing the names is the number of oxygen atoms surrounding the central atom, or the oxidation number of the central atom. In the following table, n ordinarily represents the number of oxygen atoms in the most common oxoacid of a particular element. This acid is given the ending "ic," and the names of the other acids of the same element are derived from this name as shown. The chlorine acids (see Table D.3) are an exception, because perchloric acid is the most common. The oxidation number of the central atom in the most common acid of an element is represented here by m.

No. of oxygen atoms	Name	Oxidation number of central atom
$(n + 1)$	per . . . ic acid	$+(m + 2)$
n	. . . ic acid	$+m$
$(n - 1)$	. . . ous acid	$+(m - 2)$
$(n - 2)$	hypo . . . ous acid	$+(m - 4)$

Note that chlorine (Table D.3) forms all four types of acids.

A slightly different naming system distinguishes oxoacids on the basis of the number of hypothetical water molecules that they contain. This system begins with the prefix "ortho" for the acid that contains the largest number of hydroxo (OH) units. For instance, the structure of H_3PO_4

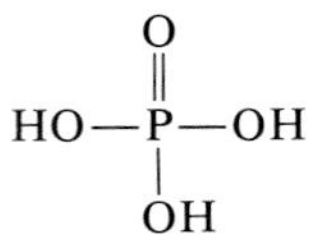

$$
\begin{array}{c}
\quad\quad O \\
\quad\quad \| \\
HO-P-OH \\
\quad\quad | \\
\quad\quad OH
\end{array}
$$

shows that it contains as many OH units as possible; thus the prefix is added before the usual name, and it is commonly called "orthophosphoric acid." Certain ortho acids can be changed to "meta" acids by the "removal" of a water molecule; for example,

$$H_3BO_3 \quad - H_2O = \quad HBO_2$$
orthoboric acid $\quad\quad\quad$ metaboric acid

$$H_4SiO_4 \quad - H_2O = \quad H_2SiO_3$$
orthosilicic acid $\quad\quad\quad$ metasilicic acid

or they can be changed to "pyro" acids by removing a water molecule from two "ortho" molecules, such as

$$2H_4SiO_4 \quad - H_2O = \quad (H_6SO_2O_7)$$
orthosilicic acid $\quad\quad\quad$ pyrosilicic acid

$$2H_3PO_4 \quad - H_2O = \quad H_4P_2O_7$$
orthophosphoric acid $\quad\quad\quad$ pyrophosphoric acid

$$2H_2SO_4 \quad - H_2O = \quad H_2S_2O_7$$
orthosulfuric acid $\quad\quad\quad$ pyrosulfuric acid

The substitution of a sulfur atom for an oxygen atom in an oxoacid gives rise to "thio" acids, for example,

TABLE D.3

Oxoacids of representative elements (*see Periodic Table inside front cover of book*)

Acids given in parentheses have never been isolated.

H_3BO_3 boric acid	(H_2CO_3) (carbonic acid)	HNO_3 nitric acid HNO_2 nitrous acid $H_2N_2O_2$ hyponitrous acid		
(H_3AlO_3) (aluminic acid)	(H_4SiO_4) (silicic acid)	H_3PO_4 phosphoric acid H_3PO_3 phosphorous acid H_3PO_2 hypophosphorous acid	H_2SO_4 sulfuric acid (H_2SO_3) (sulfurous acid)	$HClO_4$ perchloric acid $HClO_3$ chloric acid $HClO_2$ chlorous acid $HClO$ hypochlorous acid
(H_3GaO_3) (gallic acid)	(H_4GeO_4) (germanic acid) H_2GeO_3 (germanous acid)	H_3AsO_4 arsenic acid (H_3AsO_3) arsenous acid	H_2SeO_4 selenic acid H_2SeO_3 selenous acid	$HBrO_3$ bromic acid $HBrO_2$ bromous acid $HBrO$ hypobromous acid
	(H_4SnO_4) (stannic acid)	H_3SbO_4 antimonic acid (H_3SbO_3) (antimonous acid)	H_6TeO_6 telluric acid H_2TeO_3 tellurous acid	HIO_4 periodic acid HIO_3 iodic acid HIO hypoiodous acid
	(H_4PbO_4) (plumbic acid)	H_3BiO_3 bismuthic acid		

H_2SO_4 sulfuric acid
$H_2S_2O_3$ thiosulfuric acid

And the presence of an oxygen–oxygen bond gives rise to "peroxy" acids, for example,

H_2SO_4 sulfuric acid

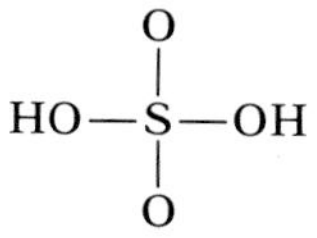

$$HO-\overset{\displaystyle O}{\underset{\displaystyle O}{S}}-OH$$

INORGANIC NOMENCLATURE

1073

$$H_2SO_5 \qquad \text{peroxysulfuric acid} \qquad \overset{\overset{\displaystyle O}{|}}{\underset{\underset{\displaystyle O}{|}}{HO-S-OOH}}$$

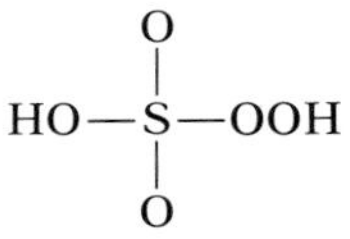

To identify these special cases of the oxoacids, we usually need to know something about the structure of the compound.

Exercises

1 Name the following oxoacids:

(a) HNO_2 (d) $HClO_4$ (g) $H_2S_2O_3$

(b) H_2SO_3 (e) HIO_3 (h) H_2SO_3

(c) $HClO$ (f) H_2SeO_4 (i) $H_4P_2O_7$

2 Write the formulas for the following oxoacids:

(a) antimonic acid   (f) gallic acid

(b) tellurous acid   (g) phosphorous acid

(c) iodic acid   (h) pyrosulfuric acid

(d) bromic acid   (i) metasilicic acid

(e) hyponitrous acid

D.9 Oxoanions and salts of oxoacids

The oxoanions formed by the removal of all acidic hydrogen atoms from the oxoacids are named as follows:

per . . . ic acid	gives	*per . . . ate ion*
. . . ic acid	gives	*. . . ate ion*
. . . ous acid	gives	*. . . ite ion*
hypo . . . ous acid	gives	*hypo . . . ite ion*

For instance, some of the ions generated from the oxoacids in Table D.3 would be

$HBrO_3$	gives	BrO_3^-	bromate ion
$HBrO_2$	gives	BrO_2^-	bromite ion
$HBrO$	gives	BrO^-	hypobromite ion
H_3PO_4	gives	PO_4^{3-}	phosphate ion
H_3PO_3	gives	HPO_3^{2-}	phosphite ion (one H is not ionizable)

To name salts of oxoacids we simply give the name of the cation first (Section D.4), followed by the name of the oxoanion. For example,

Na_2SO_4	sodium sulfate
$NbPO_4$	niobium(III) phosphate
$K_4P_2O_7$	potassium pyrophosphate
$Ba(BrO)_2$	barium hypobromite

A few oxoanions of transition metals (from acids not included in Table D.3) are often encountered and have common names:

CrO_4^{2-}	chromate ion
$Cr_2O_7^{2-}$	dichromate ion
MnO_4^-	permanganate ion
MnO_4^{2-}	manganate ion

Exercises

1 Name the following oxoanions and salts of oxoacids:
(a) $Ga_2(SeO_4)_3$ (d) $TlNO_3$ (g) $K_2S_2O_7$
(b) Co_2SrO_4 (e) $ThSiO_4$ (h) $KAsO_2$
(c) $Zn(BrO_3)_2$ (f) $Ag_2S_2O_3$ (i) $PbCrO_4$

2 Write formulas for the following substances:
(a) lead metarsenate   (f) mercury(I) chromate
(b) cobalt(II) sulfite   (g) manganese(III) phosphate
(c) cadmium iodate   (h) nickel(II) carbonate
(d) cesium perchlorate   (i) silver thiosulfate
(e) copper(I) sulfite

D.10 Special cases

Common names are used for certain ions:

NH_4^+	ammonium ion	OH^-	hydroxide ion
H_3O^+	hydronium ion	CN^-	cyanide ion
OCN^-	cyanate ion	SCN^-	thiocyanate ion
CH_3COO^-	acetate ion	$HCOO^-$	formate ion
$C_2O_4^{2-}$	oxalate ion		

Polyprotic acids can form acid oxoanions by incomplete ionization (see Section 18.7), and "hydrogen" is included in their names:

HSO_4^-	hydrogen sulfate ion (bisulfate ion)
$H_2AsO_4^-$	dihydrogen arsenate ion
$NaHS_2O_3$	sodium hydrogen thiosulfate

A few compounds exist that contain the same cation in two oxidation states. For such compounds the Roman numerals for both oxidation states are included in the name:

$Fe_3O_4 = FeO \cdot Fe_2O_3$ iron(II, III) oxide

Hydrated salts (Section 13.5) can be of two types. If the water molecules are known to be bonded to the ion, the ion is named as a complex ion (Section 28.2), but if the water molecules are held by lesser forces, the salt is named in the usual fashion, followed by the term "hydrate," with a prefix or just a number indicating the number of water molecules present. For example,

$BaCl_2 \cdot 2H_2O$	barium chloride dihydrate (barium chloride 2-hydrate)
$Na_2SO_3 \cdot 7H_2O$	sodium sulfite heptahydrate (sodium sulfite 7-hydrate)
$Mo_3Cl_4(OH)_2 \cdot 2H_2O$	molybdenum dihydroxytetrachloride dihydrate (molybdenum dihydroxytetrachloride 2-hydrate)
$FeF_2 \cdot 8H_2O$	iron(II) fluoride octahydrate (iron(II) fluoride 8-hydrate)

Exercises

1 Name the following compounds:

(a) $Cr(CH_3COO)_3 \cdot H_2O$ (c) $(NH_4)_2C_2O_4 \cdot H_2O$ (e) $KSCN$

(b) $Cu(BrO_3)_2 \cdot 6H_2O$ (d) $LiHC_2O_4 \cdot H_2O$ (f) $NOBr$

2 Write formulas for the following substances:

(a) sodium acetate trihydrate

(b) zinc dihydrogen phosphate

(c) ammonium acetate

(d) calcium thiocyanate trihydrate

(e) manganese(II) pyrophosphate trihydrate

D.11 Complexes and coordination compounds

If a complex is an ion, the rules given in this section (and in Section 28.2) are followed when naming the complex. This name is followed by the word "ion." If the complex is a neutral compound, the same rules are followed, but no word "ion" appears. If the compound consists of a cation and an anion, these rules are used in naming both ions separately and the compound is named by giving the cation name first, followed by the anion name (without the word "ion").

When naming a complex, (1) the negative ligands are named first, usually modified with the suffix "o," (2) followed by the name of neutral ligands, and then (3) the names of positive ligands, (4) followed by the name of the metal, and (5) the oxidation state given in parentheses. The number of ligands is indicated by the usual prefix and, if there is more than one of the same class of ligands, they are named in alphabetical order. If the ion is an anion, the metal name is modified with the suffix "ate." For example,

$[Cr(NH_3)_6]Cl_3 \cdot H_2O$	hexaamminechromium(III) chloride monohydrate
$[Cr(NH_3)_5Cl]Cl_2$	chloropentaaminechromium(III) chloride
$[Co(NH_3)_5 \cdot H_2O]^{3+}$	pentaamineaquocobalt(III) ion
$[Co(NH_3)_2(NO_2)_4]^-$	tetranitrodiamminecobaltate(III) ion
$K_4[Fe(CN)_6]$	potassium hexacyanoferrate(II)
$Ni(CO)_4$	tetracarbonylnickel(0)

Exercises

1 Write formulas for the following substances:

(a) diamminezinc(II) chloride

(b) tin(IV) hexacyanoferrate(II)

(c) tetracyanoplatinate(II) ion

(d) potassium hexacyanochromate(III)

(e) tetraammineplatinum(II) ion

(f) tetrachlorodiammineplatinum(IV)

(g) hexaamminenickel(II) iodide

(h) magnesium tetracyanoplatinate(II) heptahydrate

(i) lead tetrafluoroborate(III)

(j) iron(III) hexacyanoferrate(II)

2 Name the following substances:

(a) $K_2[Fe(C_2O_4)_2] \cdot 2H_2O$ (f) $K_3[Mn(CN)_6]$

(b) $Na_2[Pt(CN)_4] \cdot 3H_2O$ (g) $[Ag(CN)_2]^-$

(c) $Na[Au(CN)_2]$ (h) $[Pd(NH_3)_2](OH)_2$

(d) $[Ag(NH_3)_2]^+$ (i) $[Ni(NH_3)_4(H_2O)_2](NO_3)_2$

(e) $[Co(NO_2)_6]^{3-}$

CHAPTER 1

1.1 (a) demonstration is trick; (b) visual observations, taste, smell, feel, chemical tests, biological action; (c) each liquid is different with different properties to give desired magic results; (d) liquid state, color, clarity, odor; (e) biological action, chemical test results, formation of colors upon mixing; (f) all; (g) all; (h) all four original, first and third final; (i) second final; (j) none; (k) none; (l) none

1.3 (a) F, (b) T, (c) F, (d) F, (e) T

1.4 (a) ix, xvi, xvii; (b) i, vi, xiii; (c) ii, v, vii, xi, xiv; (d) iii, viii, xii, xv; (e) iv, x

1.10 (a) 1.0 inch, (b) 35°C, (c) 85°F, (d) 1.0 cg, (e) 1.0 m, (f) 20.0 mi, (g) 325 kcal, (h) 0.5 m, (i) 1.0 gal, (j) 55 mi/hr, (k) 10 kg, (l) 1.0 kg, (m) same

1.11 (a) i, iii, v, vi; (b) i, iii, iv; (c) i, iii, iv; (d) i, iii, iv; (e) i, iii, v, vi; (f) i, iii, v, vii; (g) i, iii, iv; (h) i, iii, v, vi; (i) i, ii

1.14 10.3 sec

1.15 (a) 8.314×10^7 erg/mole °K, (b) 1.987 cal/mole °K, (c) 0.0821 liter-atm/mole °K

1.17 8.3 lb

1.18 (a) 37.0 °C, (b) −12 °C, (c) 26 °C, (d) 120 °C

1.21 1.89 oz, 28¢

1.23 3.9×10^{-22}

1.26 mass of metal is 21.073 g, volume of metal is 7.451 ml, 2.828 g/ml

1.27 (a) −40. °F, (b) 320. °F, (c) +11.4 °F(−11.4 °C)

1.30 (a) 2.7 g/cm³, (b) 2.70 g/cm³, (c) essentially the same, (d) physical, (e) 168 lb/ft³, (f) maybe samples are not solid, (g) density measurements, (h) samples were solid

CHAPTER 2

2.1 (a) C_2H_5Cl, $Pb(C_2H_5)_4$, NaCl, Na_4Pb, C_2H_5OH, HCl, H_2O; (b) NaCl, C_2H_5Cl, $Pb(C_2H_5)_4$, $Na_{31}Pb_8$, C_2H_5OH, HCl, H_2O; (c) Na^+, H^+; (d) Cl^-; (e) water; (f) HCl, NaCl; (g) Na_{31}, Pb_8, HCl, H_2O, NaCl

2.2 (a) T, (b) T, (c) F, (d) F, (e) F, (f) F, (g) F

2.5 (a) 2 atoms of H compared to 1 molecule containing 2 atoms of H; (b) a molecule containing 2 C atoms and 2 H atoms com- pared to a molecule containing 6 C atoms and 6 H atoms; (c) a molecule containing 2 Hg atoms and 2 Cl atoms compared to a molecule containing 1 Hg atom and 2 Cl atoms; (d) an O atom, an O atom which has gained two negative charges (electrons), a molecule containing 2 O atoms, a molecule containing 2 O atoms which has gained two electrons, a molecule containing 3 O atoms

2.6 (a) 1 barium atom, 10 oxygen atoms, 18 hydrogen atoms; (b) 4 copper atoms, 2 bromine atoms, 6 oxygen atoms, 6 hydrogen atoms; (c) 3 magnesium atoms, 2 nitrogen atoms

2.7 (a) 12.011 amu, (b) 24.31 amu, (c) 32.06 amu

2.8 (a) 98.08 amu, (b) 74.10 amu, (c) 310.18 amu, (d) 101.96 amu, (e) 132.14 amu, (f) 70.906 amu, (g) 55.847 amu, (h) 342.30 amu, (i) 122.55 amu, (j) 46.07 amu, (k) 249.68 amu

2.11 (a) 3.01×10^{23}, (b) 1.5×10^{23}, (c) 1.97×10^{24}, (d) 2.7×10^{24}, (e) 2.69×10^{24}

2.13 (a) 153.823 amu, (b) 153.823 g/mole, (c) 2.55433×10^{-22} g, (d) 0.1166 mole, (e) 7.022×10^{22} molecules, (f) 7.022×10^{22} C atoms, (g) 2.8088×10^{23} Cl atoms

2.16 (a) 19.97%, (b) 26.46%, (c) 25.26%, (d) 43.64%, (e) 26.73%

2.18 77.73%, 69.94%, 72.36%, FeO

2.22 (a) $Mg_3Si_4O_{10}(OH)_2$

2.23 (a) C_2H_6O, (b) $C_4H_{12}O_2$, (c) 92.138 amu

2.26 4.05 wt%

2.27 (a) 19 g, (b) 22 g, (c) 70 g, (d) 32 g, (e) 4.3 g

2.31 0.6916 of 62.9298 amu and 0.3084 of 64.9278 amu

2.34 55.31%, 79.86%, 34.63%, 66.47%, 88.83%, 57.48%, cuprite

2.36 C_4H_8O

2.40 1.54M

2.41 (a) $(NH_4)_2Cr_2O_7$, (b) Cr_2O_3, (c) 0.0397 mole, (d) 4.54×10^{23} atoms, (e) 4.65 ml, (f) N_2, (g) 1.1 g

CHAPTER 3

3.1 (a) 2 ml; (b) 1 ml; (c) Dalton's law; (d) F_2 volume decreases, some XeF_4 becomes solid; (e) F_2 volume increases, some XeF_4 becomes gas; (f) XeF_4, Xe, F_2; (g) XeF_4

3.2 b, d, e, f

3.5 (a) F, (b) F, (c) F, (d) F, (e) T

3.7 502°F would really be twice as hot as 21°F

3.9 Graham's law; d, molar wt, mass, etc., based on ideal gas law; reaction volume based on Gay-Lussac's law

3.10 (a) 831 torr, (b) 53 torr, (c) 749 torr

3.13 − 23°C

3.15 (a) 480. torr, (b) 124 °C, (c) 100. °C

3.17 6 psig

3.19 8

3.22 48.01 amu, O_3

3.24 7.55 g/liter

3.26 27°K

3.29 2.0×10^8 liters

3.31 0.00290 atm ft³/mole °K

3.32 4960 torr

3.35 (a) p = 2.78 atm for He, 1.78 atm for N_2; (b) 4.56 atm; (c) 0.610

3.38 $rate_{H_2}/rate_{HD}/rate_{D_2}$ = 1.4136/1.2244/1.0000

3.40 (a) 3×10^{-23} cm³, (b) 20 cm³, (c) 9×10^{-2}%, (d) 99.91% of gas is empty space

3.43 $(P_{atm} + h_{Hg})(h_{air})$ = 18,880 in each case

3.45 140°C

3.46 10. gal

3.47 Ne, He

3.49 46.6 g/mole

3.51 (a) 0.958 atm, (b) 0.940 atm, (c) 1.9%, (d) small error

3.52 Intercept is 2.858114 g/liter-atm, 64.0612 g/mole

3.53 (a) 28.962 amu, (b) 28.96 amu, (c) same, (d) air is mixture and components would effuse at different rates, (e) 1.16 g/liter, (f) 1.15 g/liter, (g) some of the heavier N_2, O_2, CO_2, and Ar molecules are replaced by lighter H_2O molecules, (h) 3×10^{-4} atm, (i) 6 g, (j) 10.1 atm, (k) 9.11 atm, (l) about 10% difference

CHAPTER 4

4.1 (a) CO_2, H_2O; (b) $C_{12}H_{22}O_{11}$, O_2; (c) chlorophyll; (d) light; (e) $12CO_2 + 11H_2O \rightarrow C_{12}H_{22}O_{11} + 12O_2$; (f) 12; (g) 11; (h) 12; (i) 11; (j) 12; (k) 22; (l) 11; (m) yes; (n) no

4.2 (a) T, (b) F, (c) T, (d) F, (e) F, (f) T, (g) F, (h) F, (i) F, (j) F

4.5 cannot change formula, should be $2H_2O_2$ reacting

4.7 (a) 0.1; (b) at 210°C, water is a gas, so 4 liters; (c) 6×10^{23}; (d) none; (e) 3.6×10^{24}; (f) H_3PO_4

4.8 (a) 1.88, (b) 1.50, (c) 2.25

4.10 (a) $2HBr(aq) + H_2SO_4(aq) \rightarrow 2H_2O(l) + SO_2(g) + Br_2(g)$; (b) (i) 0.500, (ii) 6.02×10^{23}, (iii) 3.01×10^{23}

4.12 (a) +6, (b) +6, (c) 0, (d) +3, (e) +2, (f) +3, (g) +3, (h) +2, (i) +4, (j) +3, (k) +6, (l) +2, (m) +3

4.14 (i) (a) O_2, (b) Al, (c) 0 to + 3, 0 to − 2; (ii) (a) $Cr_2O_7^{2-}$, (b) SO_3^{2-}, (c) +6 to + 3, +4 to + 6

4.16 (a) H_2O; (b) add $2H_2O$ on right side; (c) 1.2×10^{24}; (d) heat reactants at 190°–200°C for 10 hours in the presence of anhydrous (water-free) $ZnCl_2$ which serves as a catalyst; (e) 148 g + 2(110 g) = 332 g + 2(18 g)

4.20 48.5 g

4.21 3.87 g

4.23 67.8%

4.24 52.4 g

4.27 (a) $CCl_3CHO + 2C_6H_5Cl \rightarrow (ClC_6H_4)_2CHCCl_3 + H_2O$; (b) 157.4 g; (c) 0.444; (d) nothing

4.28 750. g

4.30 CH

4.34 5.00 liters

4.36 11.2 liters, 12.6 liters

4.37 28.0 liters

4.39 13.9 liters, 3.48 liters, 6.95 liters

4.41 3.0×10^6 liters

4.43 12.6 g

4.47 0.6340N

4.49 0.23 equiv, 0.23 equiv, 0.046N

4.50 0.1096M

4.53 1.63 g

4.55 Zn is 1.82 times more expensive than Al.

4.58 8.50 g HNF_2, 1.65 g $ClNF_2$, 6.97 g HCl, 0.504 g NH_4Cl, 0.377 g HF

4.60 6.4 g, heating in presence of MnO_2 catalyst

4.63 (a) $4HNO_3(aq) + Cu(s) \rightarrow Cu(NO_3)_2(aq) + 2NO_2 (g) + 2H_2O(l)$; (b) 0.87 mole Cu, 0.212 mole HNO_3; (c) 6.38×10^{22}; (d) 2.56 liters; (e) nothing

4.64 (a) a mole of water as a product in second reaction; (b) 0.5 mole of each; (c) 0.36 mole diacetone alcohol, 0.23 mole mesityl oxide; (d) 46%

4.65 Note that O_2 is a reactant in the first two reactions, thus 1.00 mole O_2 will give 0.571 mole NO_2 which gives 0.381 mole HNO_3; recycle NO into the second reaction.

4.67 (a) Ca is + 2, O is − 2 except in O_2 where it is 0, H is + 1, C is + 4 in $CaCO_3$ and CO_2, C is 0 in C, C is − 1 in CaC_2 and C_2H_2, C is + 2 in CO; (b) ii and iv; (c) ii needs 3C, iii needs $2H_2O$, iv needs $2.5O_2$ and $2CO_2$; (d) CaO is limiting reagent, 17.8 mole CaO = 17.8 mole CaC_2 = 17.8 mole C_2H_2, 444 liters; (e) 1 vol C_2H_2 requires 2.5 vol O_2; (f) 1 vol C_2H_2 requires 12.5 vol air

CHAPTER 5

5.1 (a) i, ii, iv, v, vi, viii, ix, x: (b) i, ii, iv, v, vi, viii, ix, x; (c) i, ii; (d) i, iv, vi; (e) i, ii, v, vi, viii, ix, x; (f) i, ii, iv, v, vi, viii, ix, x; (g) increase in temperature; (h) boiling occurs in shorter length of time

5.2 (a) T, (b) F, (c) T, (d) T, (e) F

5.7 ii, heat of vaporization is released

5.9 required energy to add negative charge to a negative ion

5.11 (a) 1.1×10^3 cal (b) 539.8 cal, (c) − 5 cal, (d) 1070 cal

5.14 31.2 °C

5.15 16 cal/°C

5.19 − 80.45 kcal

5.22 − 12.5 MJ

5.23 − 88.9 kcal/mole Fe, − 101.8 kcal/mole Fe

5.28 16.4 kcal/mole, solid

5.30 (a) W, (b) Mo

5.32 15 mi

5.34 + 25 kJ, cold beaker

5.36 (a) − 1514 cal, (b) − 22,062 cal, (c) 1410 cal; − 22.17 kcal/mole, more exothermic

5.37 (a) 2.514 kcal/mole, (b) 74.119 kcal/mole, (c) 76.633 kcal/mole, (d) heat of sublimation = heat of fusion + heat of vaporization

5.38 (a) 7.37 cal/ °C; (b) 7.3 cal/ °C; (c) − 13.4 kcal/mole, − 88.7 kcal/mole; (d) + 4.7 kcal/mole, − 93.4 kcal/mole

CHAPTER 6

6.1 (a) N_2, Ar; (b) CO_2, CO, SO_2, NO_2, (c) CO, SO_2, NO_2, fly ash; (d) CO, sulfur oxides, nitrogen oxides, particulates; (e) SO_2; (f) CO

6.5 (a) F, (b) F, (c) T, (d) F, (e) F

6.8 18 torr

6.10 29.08 g/mole

6.14 1.56 g

6.15 − 154.15 kcal

6.18 3.20 g Mg_3N_2, 0.685 g MgO, 0.0097 liter of gas left

6.19 1200 g, flammable, heavier than air so will not lift

6.21 1×10^{-4} g

6.24 (a) − 40 °C (b) 0.25 atm, (c) 0.38 g/liter (d) 0.38 g/liter (e) 0.38 g/liter (f) 3.3×10^{-6} m, (g) 1.0×10^{-6} m, (h) 3.3

CHAPTER 7

7.3 (a) T, (b) T, (c) T, (d) T, (e) T, (f) T, (g) T, (h) F

7.5 (a) i, iii, v, vi, vii, ix; (b) ii, iv, vi, viii, x, xi; (c) ii and iii, iv and v, vii and viii, ix and x

7.9 b, c, f

7.12 (a) 0, ±1, ±2; (b) 0; (c) 0, ±1

7.13 e, f, k, p, r, and t are different resulting from atomic orbitals that are very close to each other in energy

7.14 (a) +1, +3, +1, +2, +2, +3, +2, +2 or +3, +2 and +4; (b) −1, −4, −3, −1

7.15 (a) $1s^2 2s^2 2p^6 3s^2 3p^6$, diamagnetic; (b) $1s^2 2s^2 2p^6 3s^2 3p^6 3d^6$, paramagnetic; (c) $1s^2 2s^2 2p^6 3s^2 3p^6 3d^5$, paramagnetic; (d) $1s^2 2s^2 2p^6 3s^2 3p^6 4s^2 3d^{10} 4p^6 5s^2 4d^{10} 5p^6 6s^2 4f^{14} 5d^{10}$, diamagnetic; (e) $1s^2 2s^2 2p^6 3s^2 3p^6 4s^2 3d^{10} 4p^6 4d^{10}$, diamagnetic

7.17 $1s^2 2s^2 2p^6 3s^2 3p^6 4s^2 3d^9$; $1s^2 2s^2 2p^6 3s^2 3p^6 4s^1 3d^{10}$; 0, +1, +2; each has 1 unpaired electron—so paramagnetic

7.20 6.941 amu, 24.31 amu, 87.61 amu

7.21 97,491 cm^{-1}, 1.0257×10^{-7} m, no, ultraviolet

7.23 1.7×10^{-14}

7.27 75.77% ^{35}Cl and 24.23% ^{37}Cl

7.28 (a) 87; (b) 87; (c) 136; (d) no change in electrons or protons, neutrons change from 117 to 137; (e) $1s^2 2s^2 2p^6 3s^2 3p^6 4s^2 3d^{10} 4p^6 5s^2 4d^{10} 5p^6 6s^2 4f^{14} 5d^{10} 6p^6 7s^1$; (f) 0, +1; (g) paramagnetic; (h) . . . $5d^{10} 6p^6$, diamagnetic; (i) $7p \rightarrow 7s$ at 9500 Å, $8p \rightarrow 7s$ at 5000 Å (j) yes, blue-green; (k) MCl; (l) M_2O; (m) MH; (n) FrCl, Fr_2O, FrH; (o) alkali metals; (p) 25°C

CHAPTER 8

8.1 (a) $^{96}_{42}Mo + ^2_1H \rightarrow ^{97}_{43}Tc + ^1_0n$, (b) n, (c) 43, (d) 54, (e) 97, (f) $^{98}_{43}Tc$, (g) $^{97}_{43}Tc + ^0_{-1}\beta^- \rightarrow ^{97}_{42}Mo$, (h) $^{97}_{42}Mo$, (i) artificial, (j) p changes to n, (k) 54/43 = 1.26 changing to 55/42 = 1.31, (l) no, (m) Mo, (n) Ru

8.3 (a) T, (b) F, (c) F, (d) T, (e) F, (f) T, (g) F

8.5 β^- to lower the ratio

8.7 (a) $^{228}_{90}Th \rightarrow ^4_2He + ^{224}_{88}Ra$, (d) $^{127}_{53}I + ^1_1H \rightarrow 7^1_0n + ^{121}_{54}Xe$, (g) $^{95}_{42}Mo + ^1_1H \rightarrow ^1_0n + ^{95}_{43}Tc$

8.10 $^{235}_{92}U \xrightarrow{-\alpha} ^{231}_{90}Th \xrightarrow{-\beta^-} ^{231}_{91}Pa \xrightarrow{-\alpha} ^{227}_{89}Ac \xrightarrow{-\beta^-} ^{227}_{90}Th$

8.12 (a) $n/p = 11/8 = 1.375$, too large, so β^- decay; (b) $^{19}_8O \rightarrow ^0_{-1}\beta + ^{19}_9F$;
(c)

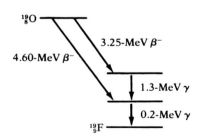

8.13 (a) 104.66 MeV, 7.476 MeV/nucleon; (e) 492.3 MeV, 8.791 MeV/nucleon, largest; (j) 1095.5 MeV, 8.427 MeV/nucleon

8.15 (a) $^7_3Li + ^1_1H \rightarrow 2^4_2He$, (b) 17.35 MeV

8.17 (a) 7.4680 MeV/nucleon; (b) 7.9769 MeV/nucleon, most stable; (e) 7.570 MeV/nucleon

8.19 1.42 cm, 9.42 cm

8.20 (a) $^{206}_{82}Pb + ^1_0n \rightarrow ^4_2He + ^{203}_{80}Hg$, $^{203}_{80}Hg \rightarrow ^0_{-1}\beta^- + ^{203}_{81}Tl$, $^{203}_{81}Tl + ^1_0n \rightarrow ^4_2He + ^{200}_{79}Au$, $^{200}_{79}Au \rightarrow ^0_{-1}\beta^- + ^{200}_{80}Hg$, $^{200}_{80}Hg + ^1_0n \rightarrow ^4_2He + ^{197}_{78}Pt$, $^{197}_{78}Pt \rightarrow ^0_{-1}\beta^- + ^{197}_{79}Au$; (b) 22.8 MeV; (c) on the left the n/p ratio is too low so we expect EC, β^+ or α decay, on the right the n/p ratio is too high so we expect β^- decay

CHAPTER 9

9.1 (a) metallic; (b) covalent; (c) ionic; (d) covalent; (e) polar covalent; (f) :Ö=Ö: ; (g) Cu^{2+} :Ö: $^{2-}$; (h) H—H; (i) H—Ö—H; (j) linear; (k) bent; (l) Cu, O_2; (m) H_2O; (n) London forces; (o) London forces; (p) hydrogen bonding, dipole–dipole interactions, London forces

9.3 (a) T, (b) T, (c) F, (d) F, (e) T, (f) T, (g) F, (h) T, (i) T

9.8 e, f

9.9 (a) i, (b) ii, (c) i, (d) iii, (e) ii, (f) ii, (g) ii, (h) i

9.10 (a) H—Ö—Ö—H, (b) [:Ö—H]⁻, (c) H—S̈—H,

(d)
$$\left[\begin{array}{c} :\ddot{O}: \\ | \\ :\ddot{O}—I—\ddot{O}: \\ | \\ :\ddot{O}: \end{array}\right]^-,$$
(e)
a structure showing C=C with H and H on left carbon, H and :Cl: on right carbon

9.11 (f) linear, (g) square pyramidal, (h) bent, (i) bent, (j) linear

9.13 (m) dipole, (p) dipole, (q) H bonding, (r) dipole, (s) London forces

9.16 (a) decreasing dipole moment and increasing molecular weight, (b) about 4 kcal/mole, (c) additional effect of hydrogen bonding in H_2O

9.19 (a) dipole and London, (b) dipole and London, (c) London

9.21 Yes, average bond energy

9.23

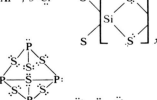

(b) dipole–dipole interactions for CH_3Cl, CH_2Cl_2, and $CHCl_3$ and London forces for CH_4 and CCl_4; (c) CH_4, CH_3Cl, CH_2Cl_2, $CHCl_3$, CCl_4

9.24 A water molecule can participate in two hydrogen bonds, HF only one.

9.26 (a) 91%, (c) 20%, (e) 0%

9.28 78 kcal/mole

9.30 2.4, 1.8

9.32 6600°C, significant error in assuming specific heat to be constant

9.34 (a) Na· Mg: Äl· ·S̈i· ·P̈· ·S̈: :C̈l· :Är:

(b) Na—Na

structures: P_4 tetrahedron with :P—P: bonds; S_8 ring S—S; :C̈l—C̈l:

(c) metallic; (d) network covalent; (e) London forces; (f) London forces; (g) v–xii; (h) xiii–xvii; (i) none; (j) Na_3P, NaCl, Mg_2Si, Mg_3P_2, MgS, $MgCl_2$, Al_2S_3, $AlCl_3$, SiS_2, $SiCl_4$, P_4S_6, PCl_3, SCl_2;
(k)

3 Na^+, :P̈: $^{3-}$ Na^+, :C̈l: $^-$ 2 Mg^{2+}, :S̈i: $^{4-}$

3 Mg^{2+}, 2 :P̈: $^{3-}$ Mg^{2+}, :S̈: $^{2-}$ Mg^{2+}, 2 :C̈l: $^-$

2 Al^{3+}, 3 :S̈: $^{2-}$ Al^{3+}, 3 :C̈l: $^-$

structures for $SiCl_4$, SiS_2 chain, P_4S_6 cage, SCl_2, PCl_3

(l) tetrahedral, tetrahedral, P are in tetrahedron with S's between and on outside, trigonal pyramid, bent; (m) xvi, xvii, (n) filled octet; (o) no; (p) none

CHAPTER 10

10.1 (a) ⌐; (b) E, Ǝ; (c) ⌐, □, ▽, △; (d) ⊔; (e) ⊏, ∧, ∨, E, Ǝ, ⊔; (f) Ǝ; (g) ⊏; (h) C

10.2 (a) F, (b) T, (c) T, (d) F, (e) T, (f) F, (g) F

10.4 (a) ii, (b) ii, (c) i, (d) i, (e) i, (f) i, (g) ii; (a) i, (b) i, (c) ii, (d) ii, (e) ii, (f) iii, (g) iii

10.6 Be^{2+}, Mg^{2+}, Na^+, F^-, Cl^-, S^{2-}

10.10 (a) i, (b) ii, (c) v, (d) ii, (e) iv, (f) iv, (g) v, (h) iii

10.13 (a) 11, $1s^2\,2s^2\,2p^6\,3s^1$; (b) 20, $1s^2\,2s^2\,2p^6\,3s^2\,3p^6\,4s^2$; (c) 35, $1s^2\,2s^2\,2p^6\,3s^2\,3p^6\,4s^2\,3d^{10}\,4p^5$; (d) 18, $1s^2\,2s^2\,2p^6\,3s^2\,3p^6$

10.15 (a) about 30 amu, (b) 15, (c) 6, (d) nonmetal, (e) Na_3E, (f) ECl_5 or ECl_3, (g) 0, ±3, +5

10.17 (a) $HClO_3$, (b) $HClO_4$, (c) H_2S

10.19 Cl_2, H_2, Ne, N_2, Rn

10.22 −898 kcal/mole, −202.5 kcal/mole, ionic charge is greater in MgO than in NaF

10.23 −94 kcal/mole, −126 kcal/mole, RbF

10.24 (a) $1s^2\,2s^2\,2p^6\,3s^2\,3p^6\,4s^2\,3d^{10}\,4p^6\,5s^2\,4d^{10}\,5p^4$; (b) 5th period, Group VI; (c) semiconducting elements; (d) no; (e) larger; (f) larger; (g) yes; (h) about 125 amu; (i) 0, +1, +4, +6, −2; (j) smaller; (k) larger; (l) EO_2 and EO_3; (m) EX_4 and EX_6; (n) H_2E; (o) 46 kcal/mole; (p) 208 kcal/mole; (q) 875.5 kJ/mole or 209 kcal/mole, about 2% error

CHAPTER 11

11.1 (a) 0 in H_2, −1 in NaH and $NaBH_4$, +1 in rest; (b) 0 in O_2, −2 in rest; (c) i, iii, iv, v; (d) i; (e) iv; (f) NaH, $NaBH_4$, $NaOCH_3$, Na_2SO_4; (g) H_2, $B(OCH_3)_3$, H_2SO_4, B_2H_6, O_2, H_2O, B_2O_3; (h) saltlike; (i) covalent; (j) covalent; (k) yes; (l) none; (m) iv, v

11.2 (a) F, (b) T, (c) T, (d) F, (e) F, (f) T, (g) F

11.7 (a) yes, (b) yes, (c) yes, (d) no, (e) no, (f) no, (g) no

11.9 MO, the lower oxides of M are more basic

11.10 dioxygen, ozone; :Ö=Ö: , :Ö=Ö—Ö: ↔ :Ö—Ö=Ö: ; ozone; linear, bent; ozone; ozone

11.12 no; NO_3^- and SO_4^{2-} are the oxidizing agents, not H^+

11.14 (a) no difference; (b) 0–2 neutrons, 1 proton; (c) 0.015%

11.16 0.021 g

11.20 (a) −41.5 kcal/mole, (b) 20.71 g

11.23 9, $(22/18)^{1/2} = 1.1$

11.25 (a) +1; (b) −1 in H_2O_2, −2 in H_2O, 0 in O_2; (c) H_2O_2; (d) H_2O_2; (e) H bonding and London; (f) H bonding, dipole and London; (g) H bonding, dipole and London; (h) London; (i) 14 atm; (j) −189.32 kJ; (k) 463.42 kJ; (l) 143.57 kJ

CHAPTER 12

12.1 (a) 2, 1, 1, 1; (b) vaporization, fusion; (c) ice, sugar, coffee; (d) ice, sugar; (e) water, unsweetened coffee, sweetened coffee

12.2 (a) T, (b) F, (c) F, (d) T, (e) T, (f) T, (g) F, (h) T

12.5 Temperature will remain at boiling point, well below kindling temperature of the paper, until all liquid has evaporated.

12.9 some water condenses (rain) until pressure is 14.7 torr

12.10 in both cases solid expands slightly, sublimes, gas expands

12.11 CO_2 remains solid, H_2O melts—the reason you can skate on ice, but not on dry ice

12.18 4.4 torr, 45.9 torr

12.19 36 g

12.22 Amount of vapor remains the same because of the fixed volume, 1.25 g of ice melts.

12.23 0.524

12.26 1.00 g/cm³

12.27 NaCl structures for both

12.29 Cu Zn, Cu_5Zn_8, $Cu Zn_3$

12.30 50. wt %

12.33 33.6 torr

12.35 7.0 times as long

12.37 (a) 0.1602 nm, (b) 2, (c) 8.074×10^{-23} g, (d) 1.749 g/cm³, (e) 3.444×10^{-23} cm³, (f) 0.7460

12.40 (a) graphite; (b) sublime at 3600°C, cool slightly; (c) 4.6×10^{289} liters; (d) very little; (e) 0.077070 nm; (f) 3.537 g/cm³, 2.28 g/cm³

CHAPTER 13

13.3 (a) F, (b) F, (c) T, (d) T, (e) T, (f) F, (g) T, (h) F, (i) F

13.5 (a) $H_2O(g) \rightarrow H(g) + OH(g)$, (b) $2H_2O(l) \rightleftharpoons H_3O^+ + OH^-$, (c) $AlCl_3(s) + 3H_2O(l) \rightarrow Al(OH)_3(s) + 3HCl(aq)$, (d) $2Na(s) + 2H_2O(l) \rightarrow 2NaOH(aq) + H_2(g)$, (e) $Cu(s) + H_2O(l) \rightarrow$ no reaction, (f) $P_4O_{10}(s) + 6H_2O(l) \rightarrow 4H_3PO_4(aq)$, (g) $CaO(s) + H_2O(l) \rightarrow Ca(OH)_2(aq)$, (h) $CaCl_2 \cdot H_2O(s) + 5H_2O(l) \rightarrow CaCl_2 \cdot 6H_2O(s)$

13.10 (a) scum formation; (b) scaling; (c) Ca^{2+}, Mg^{2+}, HCO_3^-, CO_3^{2-}; (d) Ca^{2+}, Mg^{2+}, Fe^{2+}, SO_4^{2-}; (e) $Ca^{2+} + 2HCO_3^- + Ca(OH)_2(s) \rightarrow 2CaCO_3(s) + 2H_2O(l)$, $Mg^{2+} + Ca(OH)_2(s) + Na_2CO_3(s) \rightarrow Mg(OH)_2(s) + CaCO_3(s) + 2Na^+$

13.12 12 g

13.13 (a) 20.92 wt %, (b) 51.17 wt %

13.15 0.052 g

13.17 0.033 g Na_2CO_3, 0.086 g $Ca(OH)_2$

13.19 100°C

13.21 (a) 10.851 kcal/mole; (b) 10.519 kcal/mole; (c) yes; (d) D_2O, same H bonding, but heavier; (e) 111.78°C; (f) 112.538 kcal/mole; (g) 110.770 kcal/mole; (h) D–O by about 2%; (i) 1.05

CHAPTER 14

14.1 (a) impure air, nitrogen (solvent), chlorine and oxygen (solute); (b) CCl_4 solution containing Cl_2, CCl_4(solvent), Cl_2 (solute); (c) charcoal containing Cl_2, C (solvent), Cl_2 (solute); (d) impure CCl_4, CCl_4 (solvent), chloroform (solute); (e) impure CCl_4, CCl_4 (solvent), naphthalene (solute); (f) stainless steel, iron (solvent), chromium (solute)

14.3 (a) F, (b) T, (c) F, (d) F, (e) F, (f) T

14.4 (a) high, strong; (b) low, nonelectrolyte; (c) high, nonelectrolyte; (d) high, nonelectrolyte; (e) high, nonelectrolyte; (f) high, weak

14.8 (a) decreases, (b) increases, (c) no, (d) the relative strengths of the intermolecular forces between like and unlike molecules, (e) solubility increases with increasing temperature

14.9 molarity, normality, vol %

14.14 (a) hydration energy is much larger than lattice energy in $MgBr_2$, about the same in $MgBr_2 \cdot 6H_2O$; (b) decrease; (c) $MgBr_2$

14.15 (a) 1, (b) 1, (c) concentration usually in molarity, (d) partial pressure in atm, (e) mole fraction

14.16 1.40×10^{-3} g N_2/100 g H_2O, 0.79×10^{-3} g O_2/100 g H_2O; ratio in water is 0.56 and in air is 0.286, greater in water

14.19 $CaCl_2$

14.22 207 torr acetone, 118 torr chloroform, 325 torr total

14.24 1044°C

14.25 methyl alcohol by a factor of nearly 2, both are same

14.27 0.053 M

14.29 $i = 3.01$ giving 4 ions, $3K^+$ and $Fe(CN)_6^{3-}$

14.32 193 proof

14.34 84.4 g $KClO_3$ will precipitate

14.38 drain 2 qt of original solution, add 1.7 qt pure antifreeze, add some original solution back, will protect slightly below −40°F (no wonder service stations have charts for these calculations)

14.40 0.5545m Cl$^-$, 0.4758m Na$^+$, 0.02859m SO$_4{}^{2-}$, 0.05419m Mg^{2+}, 0.0103m Ca^{2+}, 0.0101m K$^+$, 0.00241m HCO$_3{}^-$, 0.00084m Br$^-$; $-2.11°$C, 100.58°C; 27.8 atm

14.42 59 g/mole, 112 g/mole, dimer formation in nonpolar solvent

14.43 (a) 0.837M, 0.876m, 0.0155; (b) 0.01; (c) 20.5 atm; (d) 23.4 torr H$_2$O, 0.24 torr CH$_3$COOH, 23.6 torr total; (e) -1.0 kJ/mole

CHAPTER 15

15.1 (a) ii; (b) i, iii; (c) i; (d) ii, iii; (e) ii, iii; (f) i; (g) i, ii

15.3 (a) F, (b) F, (c) T, (d) T, (e) F, (f) T, (g) T

15.5 HI

15.9 (a) 2B + F → 2E; (b) A; (c) C, D; (d) rate = k[B][C]; (e) rate = k'[A][B]; (f) second

15.12 (a) second, (b) first, (c) third

15.14 second

15.16 18 hr

15.18 0.88

15.21 (a) similar to Figure 15.8a; (b) 0.105M; (e) 1.1 × 10^{-4} mole/liter sec; (d) 1.3 × 10^{-4} and 8.5 × 10^{-5} mole/liter sec, yes; (e) 6.4 × 10^{-4} and 3.7 × 10^{-3}, 5.9 × 10^{-4}, and 2.6 × 10^{-3}, 6.6 × 10^{-4} and 4.6 × 10^{-3}, first order

15.23 6.0 × 10^9 yr

15.24 (b) first; (c) 0.023 sec^{-1}; (d) ii and iv

15.26 (a) $x = 2$, $y = -1$, inhibitor; (b) rate = $kp_{O_2}^2$, rate = kp_{O_3}, rate = $kp_{O_2}^2/p_{O_2}$, rate = $kp_{O_3}^2$, rate = $kp_{O_3}^2/p_{O_2}$, 3 and 5; (c) first, 1.98 × 10^{-4} sec^{-1}, 3.0 × 10^{-4} sec^{-1}, yes; (d) 1.7 × 10^{-6}, 0.72 ≈ 1

CHAPTER 16

16.1 (a) i, (b) ii, (c) iii, (d) iii, (e) i, (f) iii

16.2 (a) T, (b) F, (c) F, (d) T, (e) F, (f) F, (g) F

16.3 (a) $K_c = 1/[Ag^+][Cl^-]$, (b) $K_p = p_{SO_2}p_{H_2O}/p_{SO_3}p_{H_2}$, (d) $K_p = p_{CO_2}$, (f) $K_c = [NO]^4[H_2O]^6/[NH_3]^4[O_2]^5$

16.4 more AgCl forms, [Cl$^-$] < [Ag$^+$] but [Ag$^+$][Cl$^-$] remains constant

16.6 (a) ii, (b) i, (c) iii

16.8 SO$_2$ and SO$_3$ eventually contain radioactive oxygen because of dynamic equilibrium

16.9 0.13

16.11 1.3

16.13 1.48 × 10^{-39}, no

16.14 0.51

16.17 (a) 2 × 10^9; (b) $Q > K$, no

16.19 4.90 × 10^{-15}

16.22 3 × 10^{-21}M

16.24 (a) 0.507%, (b) 0.235%, yes higher pressure results in less NOCl dissociating

16.27 (b) $X_{NO_2} = 0.69$, $X_{N_2O_4} = 0.31$, $K_p = 0.65$

16.29 $K_X = X_{SO_3}^2/X_{SO_2}^2X_{O_2}$, 3.41

16.30 [Fe^{2+}] = 7M, [Fe^{3+}] = 9 × 10^{-15}M, 7M

16.32 (a) 0.172 mole; (b) 0.064 mole BrF$_5$, 0.018 mole Br$_2$, 0.090 mole F$_2$; (c) 0.79 atm BrF$_5$, 0.22 atm Br$_2$, 1.11 atm F$_2$; (d) $K_c = 2.59 \times 10^{-9}$, $K_p = 0.59$

16.34 (a) increase, increase, decrease; decrease, decrease, increase; increase, decrease, increase; decrease, decrease, increase; no change, no change, no change; (b) 1.3 × 10^7 atm, 0.030 atm, 0.41 atm; (c) 1.90 × 10^4; (d) decrease; (e) 6.02 × 10^5

CHAPTER 17

17.1 (a) Physically remove much of the irritant, cool it and dilute it; (b) neutralize acid, HCO$_3{}^-$ + H$^+$ = H$_2$O + CO$_2$; (c) neutralizes base, CH$_3$COOH + OH$^-$ = H$_2$O + CH$_3$COO$^-$; (d) remove neutralization products and excess weak acid or base; (e) strong acids or bases would cause more damage; (f) common sense

17.2 (a) α, iv; (b) β, ii, (c) γ, i; (d) γ, iii; (e) β, i; (f) α, iii; (g) α, i and ii; (h) γ, iv; (i) β, iii; (j) γ, ii; (k) β, iv

17.3 (a) T, (b) F, (c) F, (d) T, (e) T

17.5 (a) ii, (b) i, (c) i, (d) i, (e) i, (f) iii, (g) ii, (h) iii, (i) i

17.7 (a) i, ii, iv, v, vi, vii; (b) i, iii, v, vi; (c) iv, v, vi, vii

17.10 (a) i, ii, iv, v, vi, vii; (b) i, iii, v, vi; (c) iv, v, vi, vii

17.12 (a) H$_3$O$^+$, H$_2$O; (b) neutralize H$_3$O$^+$ and shift equilibrium to left; (c) Brønsted-Lowry, Lewis and solvent system theories

17.13 acids are NH$_4$I, NH$_3$; NONO$_3$, N$_2$O$_3$; H$_2$O, HOCN; Zn(H$_2$O)$_4{}^{2+}$, H$_3$O$^+$; CO$_2$; Al^{3+}

17.15 (a) NH$_3$, (b) HI, (c) HIO$_3$

17.16 (a) H$_3$PO$_4$, H$_2$SO$_4$, HClO$_4$; (b) HIO, HIO$_2$, HIO$_3$, HIO$_4$; (c) H$_2$TeO$_3$, H$_2$SeO$_3$, H$_2$SO$_3$

17.18 (a) HCO$_3{}^-$ + OH$^-$ ⇌ CO$_3{}^{2-}$ + H$_2$O, HCO$_3{}^-$ + H$_3$O$^+$ ⇌ H$_2$CO$_3$ + H$_2$O; (b) acids are HCO$_3{}^-$ and H$_2$O and bases are OH$^-$ and CO$_3{}^{2-}$, acids are H$_3$O$^+$ and H$_2$CO$_3$ and bases are HCO$_3{}^-$ and H$_2$O

17.19 HCl + H$_2$O ⇌ H$_3$O$^+$ + Cl$^-$, HNO$_3$ + H$_2$O ⇌ H$_3$O$^+$ + NO$_3{}^-$

17.22 (a) red, (b) yellow, (c) green

17.24 (a) 5.02, (b) 3.8321

17.25 (a) 10.54, (b) 2.04

17.26 (a) 2.1 × 10^{-5}, (b) 8.9 × 10^{-10}

17.27 (a) 0.93, (b) 1.3 × 10^{-14}

17.29 [H$^+$] = 6M, [OH$^-$] = 2 × 10^{-15}M, pH = -0.8, pOH = 14.8 for HCl

17.32 (a) 1.32%, (b) 4.19%

17.35 1.1 × 10^{-4}

17.37 3.100 × 10^{-7}M; 6.5085; [H$_3$O$^+$] = [OH$^-$] so water is neutral

17.39 (a) 2HCl + Na$_2$CO$_3$ = 2NaCl + H$_2$O + CO$_2$; (b) 2.0506 × 10^{-2} mole Na$_2$CO$_3$, 1.731M

17.41 $i = 1.12$, $\alpha = 12\%$

17.43 (a) HA(aq) + H$_2$O = H$_3$O$^+$ + A$^-$; (b) (i) HA, (ii) H$_2$O, (iii) H$_3$O$^+$, (iv) A$^-$; (c) 5.2

17.45 (a) 0.1577M, not very close to 0.25M; (b) 0.1147 g

17.48 (a) 1.35 × 10^{-15}; (b) [D$_3$O$^+$] = [OD$^-$] = 3.68 × 10^{-8}M; (c) 7.435; (d) pD + pOD = 14.869; (e) if pD < 7.435, acid solution; (f) CH$_3$COOH, HD$_2$O$^+$; (g) none; (h) CH$_3$COOH, HD$_2$O$^+$; (i) D$_2$O, CH$_3$COO$^-$; (j) D$_2$O, NH$_3$D$^+$; (k) D$_2$O; (l) 27.0M

CHAPTER 18

18.1 (a) Ca(HCO$_3$)$_2$, CaCO$_3$, CaC$_2$O$_4$, Na$_2$CO$_3$; (b) Ca(HCO$_3$)$_2$; (c) iii; (d) iii; (e) NaOH in vii

18.4 (a) T, (b) F, (c) T, (d) T, (e) T

18.5 (a) i, iv, viii; (b) iii, vi, vii, ix; (c) ii, v

18.8 (a) 8.08, (b) 5.10, (c) 6.20

18.10 pH = 2.82 for C$_6$H$_5$NH$_3{}^+$ solution, pH = 5.80 for CH$_3$NH$_3{}^+$ solution

18.13 [H$^+$] = [HSO$_3{}^-$] = 0.0313M, [H$_2$SO$_3$] = 0.069M, [SO$_3{}^{2-}$] = 5.0 × 10^{-8}

18.15 1 × 10^{-10}M

18.17 (a) 8.92, hydrolysis of CN$^-$; (b) 1.00

18.18 0.0953M

18.20 4.12 wt%, yes

18.21 (a) neutral, (b) 7.92, (c) 11.68, pH approaches 7 as K_a increases

18.22 (a) 1.000 liter, (b) nearly 0.500M, (c) 0.250M, (d) nearly 0.500M, (e) 1.75 × 10^{-5}, (f) 4.757

18.24 (a) 4.754, (b) 4.758, (c) 4.756

18.26 7.85, 8.43

18.28 (a) 2.4 × 10^{-12}, (b) 2.3 × 10^{-8}

18.29 (a) 5 × 10^{-4} g/liter

18.32 $2 \times 10^{-35} M$

18.34 less than $3 \times 10^{-6} M$, greater than $4.8 \times 10^{-10} M$

18.37 pH < 5.0

18.39 PbS will dissolve

18.40 4.25

18.42 (a) 6.9×10^{-12}, (b) 7.0×10^{-4} g/100 g H_2O

18.44 (a) 2.0, (b) $1 \times 10^{-12} M$

18.46 $[Ca^{2+}] = 5 \times 10^{-4} M$, $[CO_3{}^{2-}] = 8 \times 10^{-6} M$

18.47 (a) 2.5×10^{-4}, (b) 5.0×10^{-4}, (c) 5.0×10^{-3}, (d) 1.3×10^{-4}, (e) 1.3×10^{-4}, (f) 3.7×10^{-4}, (g) 3.7×10^{-4}, (h) 4.6×10^{-3}, (i) $1.3 \times 10^{-3} M$, (j) $4.6 \times 10^{-2} M$, (k) 6.0×10^{-5}, (l) ion product greater than K_{sp}

CHAPTER 19

19.1 (a) 0, (b) +4, (c) +6, (d) +6, (e) +6, (f) 0, (g) −2, (h) −2, (i) −2, (j) −2, (k) i, ii, (l) O_2 in both, (m) 1

19.3 (a) H^+ and H^- to H_2; (c) ClO^- to Cl^- and $ClO_3{}^-$; (d) Cu(1) to Cu^{2+}, S(−2) to S and $NO_3{}^-$ to NO

19.5 (a) 0, (b) +2, (c) −2, (d) −2, (e) 0, (f) two oxidations

19.7 (a) T, (b) T, (c) F, (d) F, (e) T

19.10 (a) $NO_2{}^-$, I^-, I^-, $NO_2{}^-$; (b) $SO_4{}^{2-}$, Al, Al, $SO_4{}^{2-}$; (c) $NO_3{}^-$, Zn, Zn, $NO_3{}^-$; (d) I_2, I_2, I_2, I_2; (e) $Cr_2O_7{}^{2-}$, H_2S, H_2S, $Cr_2O_7{}^{2-}$

19.11 (a) $p_{NO}^2 /[H^+]^4[NO_2{}^-]^2[I^-]^2$, 2; (b) $[Al^{3+}]^2 p_{SO_2}^3/[H^+]^4[SO_4{}^{2-}]^3$, 6; (c) $p_{NH_3}[Zn(OH)_4{}^{2-}]^4/[OH^-]^7[NO_3{}^-]$, 8; (d) $[I^-]^{10}[IO_3{}^-]^2/[OH^-]^{12}$, 10; (e) $[Cr^{3+}]^{16}/p_{H_2S}^{24}[H^+]^{64}[Cr_2O_7{}^{2-}]^8$, 48

19.14 (a) $K^+ < Na^+ < Cu^{2+} < Cu^+ < F_2$; (b) $Cu^+ \rightarrow Cu^{2+} + e^-$; (c) Cu, Fe, Na, K

19.17 solvation of the ion

19.18 Numbers given are coefficients of reactants and products in order: (a) $6,1,10OH^- = 6,2,5H_2O$; (b) $5,1,2OH^- = 5,2,H_2O$; (c) $1,1 = 1,2H_2O$; (d) $1,2,2OH^- = 1,2,H_2O$; (e) $1,2 = 1,2,4H_2O$; (f) $4,1,10H^+ = 4,1,3H_2O$; (g) $3,2,8H^+ = 2,3,4H_2O$; (h) $3,20,8H_2O,20H^+ = 12,20$; (i) $3,2,2H^+ = 3,2,4H_2O$; (j) $10,6, 36H^+ = 10,3,18H_2O$.

19.19 (a) $Br^- + 3H_2O \rightarrow HBrO_3 + 5H^+ + 6e^-$, (b) $2Co(OH)_2 + 2OH^- \rightarrow Co_2O_3 + 3H_2O + 2e^-$, (c) $Cl^- + 6OH^- \rightarrow ClO_3{}^- + 3H_2O + 6e^-$, (d) $Pb^{2+} + 2H_2O \rightarrow PbO_2 + 4H^+ + 2e^-$, (e) $Pb(OH)_2 + 2OH^- \rightarrow PbO_2 + 2H_2O + 2e^-$

19.20 0.40 V

19.22 −0.470 V

19.23 (a) $E° = 0.859$ V, favorable; (b) $E° = 0.036$ V, favorable; (c) $E° = -0.03$ V, unfavorable; (d) $E° = 0.76$ V, favorable

19.24 (a) $E = 0.095$ V, more spontaneous; (c) $E = 0.11$ V, more spontaneous; (d) $E = 0.72$ V, less spontaneous

19.26 (a) −1.1030 V, 5×10^{-38}; (b) 2.096 V, 6×10^{70}; (c) 0.271 V, 1.4×10^9

19.28 163 g

19.29 69.8%

19.32 (a) $Ag + Cl^- \rightarrow AgCl + e^-$, $6e^- + 14H^+ + Cr_2O_7{}^{2-} \rightarrow 2Cr^{3+} + 7H_2O$; (b) $6Ag + 6Cl^- + 14H^+ + Cr_2O_7{}^{2-} \rightarrow 6AgCl + 2Cr^{3+} + 7H_2O$; (c) $E° = 1.11$ V; (d) 1.10 V; (e) 10^{113}

19.34 2.057 V for the basic hydrogen electrode connected to the acidic oxygen electrode

19.35 (a) $Fe_2O_3 + 6HCl \rightarrow 2Fe^{3+} + 6Cl^- + 3H_2O$, $Sn^{2+} + 2Fe^{3+} \rightarrow Sn^{4+} + 2Fe^{2+}$, $2HgCl_2 + Sn^{2+} \rightarrow Hg_2Cl_2 + Sn^{4+} + 2Cl^-$, $MnO_4{}^- + 8H^+ + 5Fe^{2+} \rightarrow 5Fe^{3+} + Mn^{2+} + 4H_2O$; (b) the dissolving of hematite by HCl; (c) Fe^{3+}, $HgCl_2$, $MnO_4{}^-$; (d) 0.721 V; (e) 10^{21}; (f) 0.654 V; (g) 31 wt %; (h) 45 wt %; (i) $MnO_4{}^-$ would oxidize Sn^{2+}

CHAPTER 20

20.1 (a) calcium hypochlorite chloride, calcium chloride, calcium hypochlorite; (b) 0 in Cl_2, −1 in $CaCl_2$, +1 in Ca $(OCl)_2$, −1 and +1 in $Ca(OCl)Cl$; (c) Cl_2; (d) Cl_2; (e) Cl; (f) Cl; (g) increases

20.3 (a) F, (b) F, (c) T, (d) F, (e) T

20.5 (a) $2NaI + Cl_2 \rightarrow 2NaCl + I_2$, (b) no reaction, (c) $2NaI + Br_2 \rightarrow 2NaBr + I_2$, (d) $2NaBr + Cl_2 \rightarrow 2NaCl + Br_2$, (e) no reaction

20.7 of the halogens, only fluorine will form strong hydrogen bonds

20.10 (a) bond energy, (b) electronegativity, (c) induced polarity

20.12 keep $[H_2]$ and $[Br_2]$ high and [HBr] low

20.14 2.49 liters, 7.09 g

20.15 1.75 g

20.17 $CaCl_2$

20.19 1.396, β^-, $^{131}_{53}I \rightarrow {}^{0}_{-1}\beta + {}^{131}_{54}Xe$

20.20 $6e^- + Cr_2O_7{}^{2-} + 14H^+ \rightarrow 2Cr^{3+} + 7H_2O$, $2I^- \rightarrow I_2 + 2e^-$; 0.79 V $= E°$; 0.60 V, decrease

20.22 (a) $^{209}_{83}Bi + {}^{4}_{2}He \rightarrow 2{}^{1}_{0}n + {}^{211}_{85}At$; (b) $^{211}_{85}At \rightarrow {}^{4}_{2}He + {}^{207}_{83}Bi$, $^{211}_{85}At + {}^{0}_{-1}\beta \rightarrow {}^{211}_{84}Po$; (c) 5×10^{-15} g; (d) solid, (e) 300°C; (f) 0.25 Å; (g) 25 kcal/mole; (h) 0, −1, +1, +5, +7; (i) most; (j) AtI, AtBr, AtCl; (k) HAt; (l) AgAt; (m) HAt; (n) AtO^-, $AtO_3{}^-$; (o) astatine iodide, astatine bromide, astatine chloride, hydrogen astatide, silver astatide, hypoastatite ion, astatate ion; (p) $HAt(aq) + AgNO_3(aq) \rightarrow AgAt(s) + H^+ + NO_3{}^-$; (q) $At_2 + KBr(aq) \rightarrow NR$; (r) $F_2 + 2At^- \rightarrow At_2 + 2F^-$

CHAPTER 21

21.1 (a) 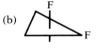 sp^3d, sp^3d^2, sp^3d^3

(b)

linear, (c) square planar

21.3 (a) F, (b) T, (c) T, (d) F, (e) F, (f) F

21.5 (a) linear with two unhybridized p orbitals, (b) 120° with an unhybridized p orbital, (c) tetrahedral, (d) trigonal bipyramid, (e) octahedron

21.7 $sp^3, sp^3d, sp^3d^2, sp^3d^2, sp^3d^3$

21.9 σ, none, π, 2π, σ and 2π, σ and π, σ, none; Al_2, S_2; 1, 0, 1, 2, 3, 2, 1, 0

21.11 $CN^- < CN < CN^+$; 3, 2.5, 2

21.14 (a) sp, (b) sp, (c) sp^2; CO and CO_2; $CO_3{}^{2-}$

21.16 the band is half-filled

21.18 (a) sp^2, (b) sp^2, (c) sp, (d) sp^2, (e) sp^2, (f) sp^2, (g) sp^2, (h) sp, (i) sp^2, (j) sp^2, (k) sp^2

21.21 78.24 kcal/mole, 192.224 kcal/mole, 257.260 kcal/mole; C—H bond is 99.39 kcal/mole and O—H bond is 110.770 kcal/mole; yes

21.23 (a) $1s^2 1s^2 \sigma_{2s}^2 \sigma_{2s}^{*2} \sigma_{2p}^2 \pi_{2p}^2$, (b) $1s^2 1s^2 \sigma_{2s}^2 \sigma_{2s}^{*2} \pi_{2p}^4$, (c) b, (d) no, (e) paramagnetic, (f) 140.912 kcal/mole, (g) 142.358 kcal/mole, (h) yes

CHAPTER 22

22.1 $1s^2 2s^2 2p^3$, −3, +3, +5; $1s^2 2s^2 2p^6 3s^2 3p^3$, −3, +3, +5; $1s^2 2s^2 2p^6 3s^2 3p^4$, −2, +4, +6

22.3 (a) $3Mg + N_2 \rightarrow Mg_3N_2$, $3Mg + 2P \rightarrow Mg_3P_2$, $Mg + S \rightarrow MgS$; (b) $2Al + N_2 \rightarrow 2AlN$, $Al + P \rightarrow AlP$, $2Al + 3S \rightarrow Al_2S_3$; (c) $N_2 + 2O_2 \rightarrow 2NO_2$, $4P + 3O_2 \rightarrow P_4O_6$, $S + O_2 \rightarrow SO_2$

22.6 (a) T, (b) T, (c) F, (d) T, (e) T, (f) T

22.8 (a) $N_2O(g) + 2H_2O(g)$, $NO_3{}^-$; (b) $N_2(g) + 2H_2O(g)$, O_2; (h) $S_2O_3{}^{2-}$, $SO_3{}^{2-}$

22.12 4 P atoms in a tetrahedron, P atoms in center of tetrahedron of O atoms

22.13

$sp^3, sp^3, sp^3d, sp^3d^2;$

trigonal pyramid, tetrahedron, trigonal bipyramid, octahedron

22.15 (a) (i) SO_2, (ii) H_2SO_3, (iii) FeS, (iv) H_2S, (v) S

22.18 $E° = 0.96$ V for $SO_3^{2-} + 2S^{2-} + 6H^+ \rightarrow 3S + 3H_2O$

22.20 for NO_3^- $6I^- + 2NO_3^- + 8H^+ \rightarrow 3I_2 + 2NO + 4H_2O$

22.21 S^{2-}, CO_3^{2-}, PO_4^{3-}

22.24 $E° = -0.10$ V for Br^- reacting with HNO_2, but $E° = 0.5$ V for I^- reacting with HNO_2; only I_2 will be formed if present

22.26 Br^-, I^-, CO_3^{2-}

22.29 0.36%

22.31 (a) 1.3, (b) 11.6, (c) 2.1

22.33 170 g

22.35 rhombic by 80. cal/mole

22.37 4.0

22.39 (a) $6.5 \times 10^{-9}M$, (b) $4.2 \times 10^{-16}M$, (c) $1 \times 10^{-3}M$

22.40 $[SO_4^{2-}]/[CO_3^{2-}] = 0.085$, $0.13M$

22.42 $7 \times 10^{-5}M$

22.43 0.01 g

22.45 12.49, 6.50, 4.09

22.47 (a) -1, $+3$, $-\frac{1}{3}$; (b) HNO_2; (c) 3.3, (d) 79 kcal/mole, 41 kcal/mole; (e) 9 kcal/mole; (f) 0.1449 nm to 0.123 nm, 0.123 nm to 0.108758 nm;

(g)

CHAPTER 23

23.1 15 octanes and 9 heptanes

23.3 $:N\equiv N:$, $[:C\equiv N:]^-$, $:C\equiv O:$, $[:C\equiv C:]^{2-}$; 3; $1s^2 1s^2 \sigma_{2s}^2$, $\sigma_{2s}^{*2}\pi_{2p}^4$, σ_{2p}^2; sp

23.5 (a) F, (b) T, (c) T, (d) F, (e) F, (f) T

23.7 (a) $CH_3CH_2CH_2CH_2CH_3$ (b) $CH_2-CH-CH_3$, CH_2-CH_2

(c) $CH_3-\overset{CH_3}{\underset{CH_3}{\overset{|}{C}}}-CH_3$, (d) $CH_3-CH=CH-CH=CH-CH_3$

23.9 (a) ethyl, (b) isopropyl, (c) phenyl, (d) *tert*-butyl

23.10 (a), (b), (e) [ring structures]

23.11 (a) *n*-propane, $CH_3CH_2CH_2Cl$ 1-chloropropane, $CH_3CHClCH_3$ 2-chloropropane; (b) 2,2-dimethylpropane, $CH_3-C(CH_3)_2-CH_2Cl$ 1-chloro-2,2-dimethylpropane; (c) *n*-pentane, $CH_3CH_2CH_2CH_2CH_2Cl$ 1-chloropentane, $CH_3CH_2CH_2CHClCH_3$ 2-chloropentane, $CH_3CH_2CHClCH_2CH_3$ 3-chloropentane

23.12

(a) $CH_3CH_2CH_2CH_2CH_3$, $CH_3CH_2CHCH_3$, $CH_3-\overset{CH_3}{\underset{CH_3}{\overset{|}{C}}}-CH_3$;

(b) CH_3-O-CH_3, CH_3CH_2OH; (c) $CH_2=CH-CH_3$, H_2C-CH_2 CH_2

23.13 (a) sp^3, sp^2, sp^2; (b) sp^2, sp^2, sp, sp; (c) sp^3, sp^2

23.14

(a) [cis and trans isomers]; (b) none; (c) none;

(d) [cyclopropane structures]

23.16 -453 cal/mole, graphite

23.17 Na_2CO_3, 5.3 g

23.20 99.39 kcal/mole, 79.05 kcal/mole, 140.91 kcal/mole, 193.80 kcal/mole

23.22 (a) -4 in Al_4C_3 and CH_4, -1 in CaC_2, C_2H_2, and C_6H_6, 0 in C, $+2$ in CO, $+4$ in CO_2; (b) CH_4, -17.6 Mcal; (c) CH_4, 13.2 kcal/g fuel

CHAPTER 24

24.1 (a) 0, (b) $+2$, (c) $+4$, (d) $+2$, (e) NiO_2, (f) Cd, (g) Cd, (h) NiO_2, (i) $NiO_2,H_2O/Ni(OH)_2,OH^-$, $Cd(OH)/Cd,OH^-$, (j) $Cd | Cd(OH)_2 | H_2O, OH^- | Ni(OH)_2 | NiO_2$, inert, (k) equilibrium, (l) $Cd(OH)_2 + Ni(OH)_2 \rightarrow Cd + NiO_2 + 2H_2O$, (m) yes

24.2 (a) T, (b) F, (c) T, (d) T, (e) T, (f) T

24.7 $2NiO\cdot OH + 2H_2O + 2e^- \rightarrow 2Ni(OH)_2 + 2OH^-$, $Fe + 2OH^- \rightarrow Fe(OH)_2 + 2e^-$, $2NiO\cdot OH + Fe + 2H_2O \rightarrow 2Ni(OH)_2 + Fe(OH)_2$

24.8 1.179 V, 1.456 V

24.10 0.892 V

24.11 (a) at the anode $2H_2O \rightarrow O_2 + 4H^+ + 4e^-$ and at the cathode $Ag^+ + e^- \rightarrow Ag$, (c) at the anode $2H_2O \rightarrow O_2 + 4H^+ + 4e^-$ and at the cathode $2H^+ + 2e^- \rightarrow H_2$

24.12 40 g Na, 134 g Br_2

24.14 110 amperes

24.15 3.4 liters

24.17 (a) $2Hg \rightarrow Hg_2^{2+} + 2e^-$, $Fe^{3+} + e^- \rightarrow Fe^{2+}$, $2Hg + 2Fe^{3+} \rightarrow 2Fe^{2+} + Hg_2^{2+}$ (b) $Hg|Hg_2^{2+}\|Fe^{3+},Fe^{2+}|$ inert electrode, (c) -0.026 V, no, (d) 0.122 V, more

24.18 1.08 g Ag, 0.383 g In, 0.294 g Ni

24.19 (a) 2.7 amperes, (b) 11 amperes, (c) 0.055 amperes

24.22 (a) 65.7 g/equiv, (b) $+3$, (c) 8.69 min

24.23 (a) $+4$, (b) $+4$, (c) 0, (d) UF_4, (e) UF_4, (f) 0.88 V, (g) 20.5 ampere, (h) 18.7 liter (i) 92.6 g required, yes

CHAPTER 25

25.1 (a) $-$, (b) $\Delta S°$, (c) $+$, (d) increase, (e) yes

25.3 (a) T, (b) F, (c) T, (d) F, (e) T

25.5 Ne

25.7 (a) iii; (b) ii; (c) i, iv

25.9 $\Delta S° = 5.2583$ cal/mole °K for fusion, 26.041 cal/mole °K for vaporization; larger change in randomness for liquid changing to gas than for solid changing to liquid

25.11 10 to 15 liters, 0.806 cal/mole °K

25.13 (a) 2.206 cal/mole °K, (b) -63.232 cal/mole °K, (c) -143.512 cal/mole °K

25.15 $\Delta S° = -32.85$ cal/°K, decrease; $\Delta H° = 68.2$ kcal; $\Delta G° = 78.0$ kcal, 39.0 kcal/mole; nonspontaneous

25.17 16.48 kcal, 8.42 kcal, -0.38 kcal, 2.6 kcal; formation of NO_2

25.19 -49.27 kcal, 1.3×10^{36} atm

25.21 (a) 21.79 kcal, (b) 1.1×10^{-16}, (c) 1.0×10^{-8} atm

25.23 0.158 V

25.25 8.1×10^{-18}, products

25.27 295 torr and 274 torr

25.29 $\Delta H°(sub) = 6870$ cal/mole, $\Delta H°(vap) = 5580$ cal/mole, 45.5 torr

25.31 10. times as fast $= k_2/k_1$

25.33 48 kcal, see Figure 15.5

25.35 (a) 18.30, 22.63, 20.5, average $= 20.47$ cal/mole °K; (b) greater, hydrogen bonding in the liquid, 24.578 cal/mole °K; (c) about the same, 24.9 cal/mole °K; (d) 2050°K

25.37 3270°K

25.39 184 kcal/mole, 6.2×10^{-7} atm, 1.4×10^{12} atoms

25.41 (a) -300.09 kcal; (b) -23.28 cal/mole °K; (c) -293.15 kcal; (d) -293.14 kcal, 0.003% difference; (e) -299.97 kcal, more; (f) 7.5×10^{214}; (g) 8×10^{-217} atm; (h) -230.12 kcal; (i) less

CHAPTER 26

26.1 (a) ns^2, $ns^2(n-1)d^{10}np^1$, $ns^2(n-1)d^{10}$, $ns^2(n-1)d^{10}np^2$, $ns^2(n-1)d^{10}np^3$; (b) $+1, +2, +3, +2, +4$ and $+2, +3$ and $+5$; (c) M_2O, MO, M_2O_3, MO, MO_2 and MO, Bi_2O_3 and Bi_2O_5; (d) MX, MX_2, MX_3, MX_2, MX_4 and MX_2, BiX_3 and BiX_5; (e) metals form oxides in air

26.3 (a) T, (b) F, (c) T, (d) T, (e) T, (f) F

26.4 $^{227}_{85}At \rightarrow ^4_2\alpha + ^{223}_{87}Fr$, $^{223}_{87}Fr \rightarrow ^0_{-1}\beta + ^{223}_{88}Ra$, Fr-223

26.8 $Hg + 2Cl^- \rightarrow HgCl_2 + 2e^-$, $3e^- + NO_3^- + 4H^+ \rightarrow NO + 2H_2O$, $3Hg + 6Cl^- + 8H^+ + 2NO_3^- \rightarrow 3HgCl_2 + 2NO + 4H_2O$

26.11 HgS is converted to HgO which decomposes to Hg

26.13 $Al^{3+} + 3OH^- \rightarrow Al(OH)_3(s)$, $Al(OH)_3 + OH^- \rightarrow Al(OH)_4^-$

26.15 (a) $Sn + 2HCl \rightarrow H_2 + SnCl_2$, (b) $Mg + Pb^{2+} \rightarrow Mg^{2+} + Pb$, (c) no reaction, (d) no reaction, (e) $Mg + ZnO \rightarrow MgO + Zn$, (f) $Hg_2O + H_2 \rightarrow H_2O + 2Hg$

26.17 $2NaCl + 2H_2O \rightarrow H_2 + 2NaOH + Cl_2$; 35.5 g Cl_2, 1.0 g H_2, 40.0 g NaOH

26.19 $2.8 \times 10^{-2} M$

26.21 1.22 g

26.23 $-40°$, Hg freezes

26.25 12.5 g

26.27 $CaAl_2$, $CaAl_3$

26.29 (a) $ZnS + \frac{3}{2}O_2 \rightarrow ZnO + SO_2$, $ZnS + 2O_2 \rightarrow ZnSO_4$, 300 tons ZnO, 240 tons SO_2, 150 tons $ZnSO_4$; (b) $Zn^{2+} + H_2O \rightarrow 2H^+ + \frac{1}{2}O_2 + Zn$, $ZnSO_4 + H_2O \rightarrow H_2SO_4 + \frac{1}{2}O_2 + Zn$, 91 tons H_2SO_4, 61 tons Zn; (c) $ZnO + C \rightarrow Zn + CO$, 240 tons C, 46 tons C; (d) 3.3 tons SO_2 (e) 560 tons $CaSO_3 \cdot 2H_2O$, 32 tons $CaSO_4 \cdot 2H_2O$, 370 tons $CaCO_3$; (f) $H_2SO_4 + CaCO_3 + H_2O \rightarrow CaSO_4 \cdot 2H_2O + CO_2$, 93 tons $CaCO_3$, 160 tons $CaSO_4 \cdot 2H_2O$; (g) 0.60 tons $CaSO_2 \cdot 2H_2O$; (h) 49 ppm; (i) 750 T; (j) 460 T

CHAPTER 27

27.2 (a) F, (b) T, (c) F, (d) F, (e) T

27.4 see Section 27.9

27.7 (a) $B_2O_3 + 3Mg \rightarrow 2B + 3MgO$, (b) $2B + 6KOH \rightarrow 2K_3BO_3 + 3H_2$, (c) $H_3BO_3 \rightarrow HBO_2 + H_2O$, (d) $4H_3BO_3 + 2NaOH \rightarrow Na_2B_4O_7 + 7H_2O$, (e) $BF_3 + HF \rightarrow HBF_4$

27.10 SiO_4 tetrahedron either single or joined at corners; chains that are linear, planar, or three dimensional

27.12 (a) $1s^2 2s^2 2p^1$, (b) $1s^2 1s^2 \sigma^2_{2s} \sigma^{*2}_{2s} \pi^2_{2p}$, (c) paramagnetic, (d) $+3$, (e) paramagnetic, (f) π bonding, (g) 70.5 kcal/mole

27.14 1.0 g; 0.003 g

27.16 -10.6 kcal

27.18 $\Delta G° = 27.2$ kcal, $\Delta G° = 25.6$, H_2Se

27.20 (a) $^{69}_{32}Ge \rightarrow ^0_{+1}\beta + ^{69}_{31}Ga$, (b) ^{69}Ga, (c) p-type, (d) 16 hr, (e) $^{77}_{32}Ge \rightarrow ^0_{-1}\beta + ^{77}_{33}As$, (f) ^{77}As, (g) 1.25 ppm

27.22 (a) see figure 12.17; (b) 8; (c) 4; (d) 2.3298 g/cm³; (e) 0.11757 nm; (f) 54.5 kcal/mole; (g) 76 kcal; (h) double bond; (i) -217.72, -217.37, -217.27 kcal/mole; (j) α-quartz; (k) 152.2 kcal/mole; (l) double bond, (m) SiO bonds are larger, O is more electronegative

CHAPTER 28

28.1 (a) iron(III) hexacyanoferrate(II), iron(II) hexacyanoferrate(II), iron(III) hexacyanoferrate(III); (b) $1s^2 2s^2 2p^6 3s^2 3p^6 4s^2 3d^6$, $1s^2 2s^2 2p^6 3s^2 3p^6 3d^6$, $1s^2 2s^2 2p^6 3s^2 3p^6 3d^5$; (c) all are d^2sp^3; (d) to known samples of Fe^{2+} add $K_4[Fe(CN)_6]$ and $K_3[Fe(CN)_6]$ and repeat for known sample of Fe^{3+} to see what positive tests look like. To samples of the unknown add each complex and compare.

28.3 (a) T, (b) T, (c) T, (d) T, (e) T

28.4 (a) 1; (b) 5; (c) 4; (d) 0; (e) hexacyanoferrate(III) ion, hexaaquoiron(III) ion, hexaaquomanganese(III) ion, hexaamminecobalt(III) ion

28.5 see Fig 28.10; (a) 3, 1; (b) 4, 2; (c) 1, 0; (d) hexaaquomanganese(III) ion, hexafluorocobaltate(III) ion, hexaaquotitanium(III) ion

28.8 (a) pentaammineaquocobalt(III) hexanitrocobaltate(III), (b) both are d^2sp^3, (c) diamagnetic

28.11 decrease, -36.7 cal/°K

28.13 (a) $1s^2 2s^2 2p^6 3s^2 3p^6 4s^2 3d^7$; (b) $1s^2 2s^2 2p^6 3s^2 3p^6 3d^6$; (c) d^2sp^3; (d) sp^3d^2; (e) $-24Dq + 3P$, $-4Dq + P$; (f) hexacyanocobaltate(III) ion, hexafluorocobaltate(III) ion

28.15 (a) $6.0 \times 10^{-8} M$; (b) $2.5 \times 10^{-5} M$; (c) $9.5 \times 10^{-3} M$; (d) hexacyanoferrate(III) ion, tetracyanozincate(II) ion, trioxalatocobaltate(II) ion

28.17 $7 \times 10^{-3} M$, $8.0 \times 10^{-12} M$, tetrabromocadmate(II) ion, tetrabromopalladate(II) ion

28.19 yes

28.22 (a) $PtCl_2(NH_3)(H_2O)$; (b) dichloroammineaquoplatinum(II); (c)

(d) square planar; (e) trans; (f) 8.9; (g) [cis] $0.0010M$, [trans] $0.0090M$; (h) dsp^2; (i) $-24.56Dq +4P$

CHAPTER 29

29.1 (i) $[Cu(H_2O)_4]^{2+}$, $Cu(s) + 2HNO_3(aq) \rightarrow Cu^+ + NO_2(g) + H_2O(l) + NO_3^-$, $Cu^+ + 2HNO_3(aq) \rightarrow Cu^{2+} + NO_2(g) + H_2O(l) + NO_3^-$; (ii) $[Cu(NH_3)_4]^{2+}$, $[Cu(H_2O)_4]^{2+} + 4NH_3(aq) \rightarrow [Cu(NH_3)_4]^{2+} + 4H_2O(l)$; (iii) $Cu(s)$, $[Cu(NH_3)_4]^{2+} + Zn(s) \rightarrow Cu(s) + [Zn(NH_3)_4]^{2+}$; (iv) $CuO(s)$, $Cu(s) + \frac{1}{2}O_2(g) \rightarrow CuO(s)$; (v) $[Cu(H_2O)_4]^{2+}$, $CuO(s) + 2HCl(aq) \rightarrow H_2O(l) + Cu^{2+} + 2Cl^-$; (vii) $Cu(OH)_2(s)$, $Cu^{2+} + 2OH^- \rightarrow Cu(OH)_2(s)$; (vii) $CuO(s)$, $Cu(OH)_2(s) \underset{\Delta}{\rightarrow} CuO(s) + H_2O(l)$; (viii) $[Cu(H_2O)_4]^{2+}$, $CuO(s) + H_2SO_4(aq) \rightarrow Cu^{2+} + H_2O(l) + SO_4^{2-}$; (ix) $CuSO_4 \cdot 5H_2O(s)$; (x) $CuSO_4(s)$, $CuSO_4 \cdot 5H_2O(s) \underset{\Delta}{\rightarrow} 5H_2O(g) + CuSO_4(s)$

29.3 (a) F, (b) F, (c) F, (d) F, (e) F

29.5 $FeCl_3$, loss of one or more H^+ from coordinated water molecules by hydrolysis

29.7 sp^3d^2, sp^3, no

29.9 a, c, e, f, h

29.11 (a) $2Ag^+ + 2OH^- \rightarrow Ag_2O(s) + H_2O(l)$; (b) $Au(s) + 3HNO_3 + 4HCl(aq) \rightarrow HAuCl_4(aq) + 3NO_2(g) + 3H_2O(l)$; (c) $2Cu^{2+} + 4CN^- \rightarrow 2CuCN(s) + (CN)_2(g)$, $2Cu^{2+} + 6CN^- \rightarrow 2[Cu(CN)_2]^- + (CN)_2(g)$; (d) $[Cu(H_2O)_4]^{2+} + 4NH_3(aq) \rightarrow [Cu(NH_3)_4]^{2+} + 4H_2O(l)$; (e) $AgCl(s) + 2NH_3(aq) \rightarrow [Ag(NH_3)_2]^+ + Cl^-$; (f) $Cu_2SO_4(s) \xrightarrow{H_2O} Cu(s) + CuSO_4(s)$; (g) $CuSO_4(s) + 5H_2O(l) \rightarrow CuSO_4 \cdot 5H_2O(s)$

29.14 900 amperes

29.16 0.354 V for Zn being sacrificed, -0.273 V for Sn being sacrificed

29.18 1.179 V for iron reaction, -0.364 V for copper reaction

29.20 0.4 V, -0.18 V

29.22 0.123888 nm, 0.126 nm, 0.128 nm, 0.127 nm

29.24 (a) $NiS(s) + \frac{3}{2}O_2(g) \xrightarrow{\Delta} NiO(s) + SO_2(g)$, $NiO(s) + C(s) \rightarrow Ni(s) + CO(g)$, $MO(s) + CO(g) \rightarrow M(s) + CO_2(g)$, $NiO(s) + 4CO(g) + H_2(g) \rightarrow Ni(CO)_4(g) + H_2O(l)$, $Ni(CO)_4(g) \xrightarrow{\Delta} Ni(s) + 4CO(g)$; (b) 4.74 g/cm³, 5.5 g/cm³; (c) -2.9 kcal; (d) -15.3 kcal, -10.9 kcal; (e) 270 g; (f) $1s^2 2s^2 2p^6 3s^2 3p^6 4s^2 3d^8$; (g) $4s^2$ removed; (h) $4s^2$ becomes ninth and tenth in $3d^{10}$, sp^3 for CO electrons, some double bond character; (i) no; (j) $K = 5.6 \times 10^{-7}$, 0.021 atm

CHAPTER 3

30.1 $[Ag^+]$ decreases in order of $AgNO_3(aq) > Ag_2O(s) > [Ag(S_2O_3)_2]^{3-} > AgBr(s) > [Ag(NH_3)_2]^+ > AgI(s)$; $2Ag^+ + 2OH^- \rightleftharpoons Ag_2O(s) + H_2O(l)$; $Ag_2O(s) + 4S_2O_3^{2-} + 2H^+ \rightleftharpoons 2[Ag(S_2O_3)_2]^{3-} + H_2O(l)$; $[Ag(S_2O_3)_2]^{3-} + Br^- \rightleftharpoons AgBr(s) + 2S_2O_3^{2-}$; $AgBr(s) + 2NH_3(aq) \rightleftharpoons [Ag(NH_3)_2]^+ + Br^-$; $[Ag(NH_3)_2]^{2+} + I^- \rightleftharpoons AgI(s) + 2NH_3(aq)$.

30.4 More CdC_2O_4 formed in (a) and (d); more CdC_2O_4 dissolves in (b) and (f); no change in (c) and (e).

30.7 $S^{2-} + 2H_2O(l) \rightleftharpoons H_2S(aq) + 2OH^-$, $Al^{3+} + 3OH^- \rightleftharpoons Al(OH)_3(s)$.

30.9 1.9×10^{-3} g/liter

30.11 Ag_2CrO_4

30.14 $[C_2O_4^{2-}] = 4 \times 10^{-3}M$, $Q = 4 \times 10^{-5}$, $Q > K_{sp}$, yes

30.16 9.38

30.18 $[H^+] = 9.8 \times 10^{-6}M$, $[OH^-] = 1.3 \times 10^{-5}M$, 9.11

30.20 $2.8 \times 10^{-3}M$

30.21 6.7

30.22 $[NH_3] = 0.1M$ for AgCl, $2M$ for AgBr, $100M$ for AgI; AgI can be separated from the others using $6M$ NH_3.

30.25 $K = 4 \times 10^{57}$, favorable

30.27 (a) $K_{sp} = 1.8 \times 10^3$, $[Cl^-] = 400M$, no; (b) $[S^{2-}] = 2 \times 10^{-26}M$; (c) $3CdS(s) + 2NO_3^- + 8H^+ \rightleftharpoons 3Cd^{2+} + 3S(s) + 2NO(g) + 4H_2O(l)$, $K = 3 \times 10^{58}$, favorable; (d) $[Cd^{2+}] = 1 \times 10^{-9}M$; (e) $[Cd(NH_3)_4]^{2+} + 4CN^- \rightleftharpoons [Cd(CN)_4]^{2-} + 4NH_3(aq)$, $2[Cu(NH_3)_4]^{2+} + 8CN^- \rightleftharpoons 2[Cu(CN)_3]^{2-} + (CN)_2 + 8NH_3(aq)$; (f) $[Cd(CN)_4^{2-}] = 7 \times 10^{-16}M$ at equilibrium, so negligible amount of complex remains and $[Cu(CN)_3^{2-}] = 8 \times 10^{-16}M$ reacts, so negligible amount of precipitate forms.

CHAPTER 31

31.1 $PbCl_2$ is soluble enough that an appreciable amount is carried over into Group II (see Exercise 31.33); mercury has different oxidation states in the two groups

31.2 (a) Hg_2Cl_2, (b) $[Ag(NH_3)_2]^+$, (c) Sb_2S_3, (d) $[HgCl_4]^{2-}$

31.3 (a) lead chloride, (b) tin(IV) sulfide, (c) trisuldifostannate(IV) ion, (d) tetraamminecopper(II) ion, (e) lead chromate

31.6 Hg^{2+}, Pb^{2+}, Cu^{2+}

31.8 electronic configuration and spacing of energy levels

31.11 HNO_3 is an oxidizing agent.

31.15 $HZn(UO_2)_3(C_2H_3O_2)_9 \cdot 9H_2O$ in ethanol

31.16 K_2CrO_4

31.17 (a) NaOH gives pink $Co(OH)_2$ and green $Ni(OH)_2$ or NH_3 gives tan $[Co(NH_3)_6]^{2+}$ and blue $[Ni(NH_3)_6]^{2+}$

31.18 no, NH_4^+ does not form $[Cu(NH_3)_4]^{2+}$ because NH_3 is bonded more strongly by H^+.

31.20 SnS is brown, SnS_2 is yellow

31.22 yellow $PbCrO_4$, red-brown Ag_2CrO_4 and red Hg_2CrO_4; white $Pb(OH)_2$, colorless $[Ag(NH_3)_2]^+$, and black-white mixture of Hg and $HgNH_2NO_3$; colorless PbO_2^{2-}, brown Ag_2O, and black HgO

31.24 Sb^{3+}, Cu^{2+}

31.28 3 ml of each stock solution, dilute to 15 ml

31.30 Hg_2^{2+} at $[Cl^-] = 1 \times 10^{-8}M$

31.32 0.11 g

31.36 FeI_2, Fe^{3+} oxidizes I^- to I_2

31.37 T1 in Group I; As, Au, Mo, Pt, Sb, Se, Te in Group II; Be, Ti, U, V, Zr, and rare earths in Group III; Sr in Group IV; Cs, Li, Rb in Group V

CHAPTER 32

32.1 (a)

H—C—C—O—H, H—C—C—H, H—C—C—O—H,

(b) ethanoic acid, ethanal, ethanol; (c) A is ethanol, B is acetic acid, C is acetaldehyde; (d) $CH_3CH_2OH + HOOCCH_3 \rightarrow H_2O + CH_3CH_2OOCCH_3$, $CH_3COOH \rightarrow CH_3COO^- + H^+$; (e) $5CH_3CH_2OH + 4MnO_4^- + 12H^+ \rightarrow 5CH_3CHO + 4Mn^{2+} + 11H_2O$, $5CH_3CHO + 6H^+ + 2MnO_4^- \rightarrow 2Mn^{2+} + 3H_2O + 5CH_3COOH$

32.3 (a) F, (b) T, (c) T, (d) T, (e) F

32.5 (a) $CH_3CH_2OH + HCl \rightarrow CH_3CH_2Cl + H_2O$, (b) $2CH_3CH_2OH + 2Na \rightarrow H_2 + 2CH_3CH_2ONa$, (c) ethyl chloride, sodium ethoxide, water, dihydrogen

32.6 (a) Br—⬡—CH₃,

(b) ⬡—OH, (g) HO—CH₂—CH₂—CH₂—CH₂—OH,

(h) CH_3CH_2CHO, (l) $CH_3CH_2COCCH_3$, (m) CH_3—CH—O—CH | CH_3

32.8 0.64 g

32.10 (a) $CH_3CH_2OH + NaI$,

(d) ⬡—NH₂ + H₂O + NaCl, (e) [naphthalene]—$N_2^+Cl^-$ + 2H₂O,

(k) $CH_3CH_2COOH + CH_3OH$, (l) [benzene ring with O⁻Na⁺ and COO⁻Na⁺] + $CH_3CO_2^-Na^+ + H_2O$

32.11 (a) HBr; (b) Sn, conc HCl; (c) NaOH, alcohol, heat remove H and Br forming double bond, add Br_2; (e) HI

32.13 8.59, more acidic (pH = 8.88)

32.15 (a) 0.19°C, (b) 0.25°C

32.16 (a) CH_2OH | $CHOH$ + $3HNO_3 \rightarrow 3H_2O$ + CH_2ONO_2 | $CHONO_2$ | CH_2OH | CH_2ONO_2

(b) 23g, (c) CH_2ONO_2 | $CHONO_2 \rightarrow 3/2 N_2 + 3CO_2 + \frac{5}{2}H_2O + \frac{1}{4}O_2$ | CH_2ONO_2

(d) 7.25 mole product, (e) 0.73 mole, (f) 5.9 cal/°C, (g) 5800°C, (h) 3700 atm

absolute zero $-273.15°C$; the lowest possible temperature

absorption spectrum spectrum of radiation remaining after a continuum of radiation has passed through a substance

acceptor atom atom that accepts an electron pair in a coordinate covalent bond

acceptor impurity contributes holes to the acceptor level in a semiconductor

acceptor level energy level slightly above the valence band that appears when an acceptor impurity is added to a host semiconducting element

acetylenes hydrocarbons with covalent triple bonds; alkynes

acid a substance that in aqueous solution increases the hydrogen ion concentration

acid anhydride $(RC{=}O)_2O$, where R is an alkyl or an aryl group

acid anhydride (acidic anhydride) a nonmetal oxide that combines with water to give an acid

acid–base reaction, Brønsted–Lowry transfer of a proton from a proton donor to a proton acceptor

acid–base reaction, Lewis electron-pair donation; formation of a coordinate covalent bond

acid–base reaction, solvent system reaction of the solvent anion with the solvent cation to give the solvent molecule

acid, Brønsted–Lowry a molecule or ion that can act as a proton donor

acidic aqueous solution contains a greater concentration of H^+ than OH^- ions; $pH < 7$

acid, Lewis a molecule or ion that can act as an electron-pair acceptor

acid salts salts that contain hydrogen

acid, solvent system any substance that gives the cation of the solvent

actinides actinium through lawrencium

activated complex combination of reacting molecules intermediate between reactants and products; also called transition state

activation energy the energy that reactants must have for reaction to occur

activity the effective concentration of a component in a solution

activity coefficient the ratio between the activity and the actual concentration of a component in a solution

acyl halide RCOX, where R is an alkyl or an aryl group

addition reactions addition of atoms or groups to the carbon atoms in a covalent double or triple bond

adsorbent the solid phase in adsorption chromatography

adsorption adherence of atoms, molecules, or ions to a surface

adsorption chromatography chromatography in which the stationary phase is a solid

aerobic decomposition decomposition of organic matter by bacteria in the presence of oxygen

aerosol a colloidal suspension of solid or liquid particles in air

alcohol ROH, where R is an alkyl group

aldehyde RCHO, where R is an alkyl or an aryl group

aliphatic hydrocarbons hydrocarbons that contain no aromatic rings

alkaline aqueous solution contains a greater concentration of OH^- ions than H^+ ions; pH > 7

alkanes saturated hydrocarbons with the general molecular formula C_nH_{2n+2}

alkenes hydrocarbons with covalent double bonds; olefins

alkoxide RO^-M^+, where R is an alkyl group

alkoxy group RO^-, where R is an alkyl group

alkyl groups groups containing one less hydrogen than an alkane

alkyl halides RX, where R is an alkyl group and X is a halogen atom

alkynes hydrocarbons with triple covalent bonds; acetylenes

allotropes different forms of the same element in the same state

alloy intimate mixture of two or more metals or metals plus nonmetals in a substance that has metallic properties

α-decay (alpha-decay) emission of an α-particle by a radioactive nuclide

α-particle (alpha-particle) a dipositive helium ion, He^{2+}

aluminothermic reactions reactions of aluminum with metal oxides

amalgam alloy of mercury and another metal

amide $RCONH_2$, where R is an alkyl or an aryl group

amines RNH_2, R_2NH, R_3N, where R is an alkyl or an aryl group

ammine group the NH_3 group in complexes

amorphous solid substance in which the atoms, molecules, or ions have a random and nonrepetitive three-dimensional arrangement

amphiprotism the ability to either gain or lose a proton

amphoterism the ability to act as either an acid or a base

anaerobic decomposition decomposition of organic matter by bacteria in the absence of oxygen

anhydrous free of water

anion negatively charged ion

anode electrode at which oxidation occurs

antibonding molecular orbital molecular orbital in which most electron density is located away from the internuclear axis

antineutrino massless, chargeless particle emitted with an electron

aqua regia 3 to 1 by volume mixture of HCl and HNO_3

aqueous solution a solution of any substance in water

aromatic character stability to oxidation plus reaction by substitution instead of addition

aromatic hydrocarbons unsaturated hydrocarbons with resonance-stabilized ring systems

artificial isotope isotope not found in nature

artificial radioactivity decay of man-made radioactive isotopes

aryl halides ArX, where Ar is an aryl group and X is a halogen atom

asymmetric atom an atom bonded to four different groups

atmosphere all of the gases that surround the Earth

atmosphere (unit) 760 torr

atom smallest particle of an element that can participate in a chemical reaction

atomic mass unit $\frac{1}{12}$ of the weight of one carbon-12 atom; the unit of atomic and molecular weights

atomic number the number of protons in the nucleus of each atom of an element

atomic orbital the space in which an electron with a specific energy is most likely to be found

atomic radius radius of an atom in a single covalent bond

atomic weight the average weight of the atoms of the naturally occurring element relative to $\frac{1}{12}$ the weight of an atom of carbon-12

Avogadro's law equal volumes of gases, measured at the same T and P, contain equal numbers of molecules

Avogadro's number the number of atoms in exactly 12 grams of carbon-12; 6.022×10^{23}

azeotropes constant-boiling mixtures that distill without change in composition

back-bonding formation of double bonds between the metal and ligands in a complex, with electrons coming from both the nonmetal and the metal

band spectrum radiation at certain wavelength ranges

barometer an instrument that measures the pressure of the atmosphere

base a substance that in aqueous solution increases the hydroxide ion concentration

base, Brønsted–Lowry a molecule or ion that can act as a proton acceptor

base, Lewis a molecule or ion that can act as an electron-pair donor

base, solvent system any substance that gives the anion of the solvent

basic anhydride metal oxide that yields a base with water

basic salts salts that contain oxygen or hydroxy groups in addition to the usual anions

battery two or more voltaic cells combined to provide electrical energy for a practical purpose

β-decay (beta-decay) electron emission, positron emission, or electron capture by a radioactive nuclide

binary compounds compounds containing only two elements

binding energy per nucleon nuclear binding energy of a nucleus divided by number of nucleons in that nucleus

biochemical oxygen demand total amount of oxygen consumed by microorganisms in decomposing waste

biodegradable can be decomposed by natural bacterial decay

biological sciences study mainly of things that are alive

boiling point temperature at which the vapor pressure of a liquid equals the vapor pressure of the gas above the liquid, and bubbles of vapor form throughout the liquid

bombardment striking a nucleus with a moving particle

bond angle angle between two atoms that are both bonded to the same third atom

bond dissociation energy enthalpy per mole for breaking exactly one bond of the same type per molecule

bond energy average enthalpy per mole for breaking one bond of the same type per molecule

bonding molecular orbital molecular orbital in which most of the electron density is located between the nuclei of the bonded atoms

bond length average distance from nucleus to nucleus in a stable compound

bond order number of electron pairs shared between two atoms

boranes boron–hydrogen compounds

Boyle's law at constant T, V is inversely proportional to P

breeder reactor a nuclear reactor that produces more fissionable atoms than it consumes

buffer solution a solution that resists changes in pH when small amounts of acid or base are added to it

calcined heated to drive off a gas; a chemical change

calorimeter a device for measuring the heat given off or absorbed in a chemical reaction

calorimeter constant the total heat capacity of a calorimeter

carbides binary compounds of carbon

carbonyl group $C{=}O$

carboxylate ion $RCOO^-$, where R is an alkyl or an aryl group

carboxylic acid RCOOH, where R is an alkyl or an aryl group

carcinogens cancer-causing agents

catalyst a substance that increases the rate of a reaction, but may be recovered from the reaction unchanged

catenation covalent bonding between atoms of the same element

cathode electrode at which reduction occurs

cathode rays streams of electrons flowing from the cathode to the anode in a gas discharge tube

cation positively charged ion

chain carriers species that propagate a chain reaction

chain reaction a reaction or series of reaction steps that initiates repetition of itself

changes of state interconversions between the solid, liquid, and gaseous states

Charles' law at constant P, V is directly proportional to T (absolute)

chelate ring the ring formed by a chelated ligand in a complex

chelation ring formation by a ligand in a complex

chemical adsorption adsorption in which forces between surface and adsorbate are of the magnitude of chemical bond forces

chemical bond a force that acts strongly enough between two atoms or groups of atoms to hold them together in a different, stable species of measurable properties

chemical equation symbols and formulas representing the total chemical change that occurs in a chemical reaction

chemical equilibrium an equilibrium in which the rates of simultaneous forward and reverse chemical reactions are the same, and the amounts of species present do not change with time

chemical formula the symbols for the elements combined in a compound, with subscripts indicating how many atoms of each element are included

chemical kinetics the study of reaction rates and reaction mechanisms

chemical nomenclature the rules and regulations that govern naming chemical compounds

chemical properties properties that can only be observed in chemical reactions

chemical reaction process in which at least one substance is changed in composition and identity

chemical shift variation in the nuclear magnetic resonance absorption of protons in different chemical environments

chemistry the branch of science that deals with matter, with the changes that matter can undergo, and with the laws that describe these changes

chirality the property of having nonsuperimposable mirror images

chromatogram a plot of the response of a detector in chromatography versus time

chromatography   distribution of a solute between a stationary phase and a mobile phase

chrome plating   the electrolytic deposition of a layer of chromium on another metal

chromophoric group   a color-producing group

cis isomer   has identical groups on the same side

cis–trans isomerism   the existence of compounds with identical groups arranged in different ways on either side of a covalent double bond or some other rigid bond, as in a ring

colligative property   any property of a solvent that is affected proportionately to the concentration of solute particles, but which is independent of the nature of those particles

collision theory   reactions occur when atoms, ions, or molecules collide; only a very small portion of collisions result in reaction

colloid   a substance made up of particles larger than most molecules, but too small to be seen in an optical microscope; *roughly* 1 nm to 200 nm in diameter

colloidal suspension   a mixture in which particles remain suspended and cannot be removed by filtration

combined gas law $\dfrac{P_1 V_1}{T_1} = \dfrac{P_2 V_2}{T_2}$

combustion   any chemical change in which both heat and light are given off

common ion effect   displacement of an ionic equilibrium by the addition of more of one of the ions involved

complex   compound formed between a metal atom or ion that accepts one or more electron pairs, and ions or neutral molecules that donate electron pairs

complex ions   positively or negatively charged complexes

compound   a substance of definite composition in which two or more elements are chemically combined

condensation   movement of molecules from the gaseous phase to the liquid phase

conduction band   energy band in which electrons are free to move

conjugated double bonds   covalent double bonds that alternate with covalent single bonds

continuous spectrum   radiation at all wavelengths

coordinate covalent bond   a single covalent bond in which both electrons in the shared pair come from the same atom

coordination compound   the neutral compound formed between a complex ion and other ions or molecules

coordination number   number of nonmetal atoms surrounding the central metal atom or ion in a complex

Coulomb force   force of electrical attraction or repulsion between charged particles

couple action   the formation of a voltaic cell by direct contact between two substances rather than through an external connection

covalent bonding   bonding based upon electron-pair sharing; the attraction between two atoms that share electrons

critical mass   smallest mass that will support a self-sustaining nuclear chain reaction under a given set of conditions

critical point   the temperature above which no amount of pressure is great enough to cause liquefaction of a gas

critical pressure   pressure that will cause liquefaction of a gas at the critical temperature

critical temperature   temperature above which a substance cannot exist as a liquid

cryogenics   the study of phenomena at low temperatures

crystal   a solid that has a shape bounded by plane surfaces intersecting at fixed angles

crystal coordination number   number of nearest neighbors of an atom, molecule, or ion in a particular crystal structure

crystal field stabilization energy   the algebraic sum of $-4Dq$ per electron in a t_{2g} level and $+6Dq$ per electron in an e_g level; amount of stabilization by the splitting of d orbitals into two energy levels

crystalline solid   substance in which the atoms, molecules, or ions have a characteristic, regular, and repetitive three-dimensional arrangement

crystal structure   complete geometrical arrangement of the particles that occupy the space lattice of a crystal

cubic closest packing   closest-packed layers of atoms arranged in an ABCABCABC . . . sequence

cycloalkanes   cyclic saturated hydrocarbons with the general formula, C_2H_{2n}

Dalton's law of partial pressures   in a mixture of gases, the total pressure exerted is the sum of the pressures that each gas would exert if it were present alone under the same conditions

dehydrogenation reaction   a reaction in which H_2 is removed

deionized water   water with almost all ions removed

deliquescent   takes up enough water from the air to dissolve itself

delocalization   spreading of an electron cloud into a larger space

delocalized orbitals   orbitals spread over an entire molecule

denitrification   conversion of nitrites and nitrates to dinitrogen

density   mass per unit volume

desalination   removal of ions from water

descriptive chemistry description of the elements and compounds, their physical states, and how they behave

deuterium hydrogen-2; heavy hydrogen

diamagnetism the property of repulsion by a magnetic field, showing the absence of unpaired electrons

diatomic molecule a molecule made of two atoms

diazonium salts $ArN_2^+X^-$, where Ar is an aryl group

diffraction bending of waves around an obstacle in their path

diffusion mixing of molecules of different gases by random motion and collisions until the mixture becomes homogeneous

dihydrogen H_2

dioxygen O_2

dipole a pair of opposite charges of equal magnitude at a specific distance from each other

dipole–dipole interaction electrostatic attraction between molecules with permanent dipoles

dipole moment measure of the polarity of a chemical bond or molecule

direct combination formation of a compound from its elements

dissociation constant equilibrium constant for the dissociation of a complex ion into its components

dissociation of an ionic compound transformation of a neutral ionic compound into positive and negative ions in solution

distillate the product of distillation

distillation heating a liquid or solution to boiling, and collecting and condensing the vapors

distribution coefficient the equilibrium constant for the distribution of a solute between two solvents

donor atom atom that provides both electrons to a coordinate covalent bond

donor impurity an impurity that contributes electrons to the conduction band of a semiconductor without leaving holes in the valence band

donor level energy level usually slightly below the conduction band of a host semiconductor

doping addition of impurities to semiconducting elements

double covalent bond two electron pairs shared between the same two atoms; one σ bond plus one π bond

dynamic equilibrium a state of balance between exactly opposite changes occurring at the same rate

ecosphere the portions of the air, water, and soil spheres that support life

effective nuclear charge portion of the nuclear charge that acts on an electron; influenced by screening and penetration effects

efflorescence loss of water of hydration

effusion the escape of molecules in the gaseous state one by one, without collisions, through a hole of molecular dimensions

electrochemical cell any device in which an electrochemical reaction takes place

electrochemistry study of oxidation–reduction reactions that either produce or utilize electrical energy

electrode a conductor through which electrical current enters or leaves a conducting medium

electrode potential the emf of a half-cell reaction, usually written for a reduction reaction

electrolyte any substance that gives a solution that conducts electricity

electrolytic cell a cell that uses electrical energy from outside the cell to cause a redox reaction to occur

electromotive force (emf) chemical driving force of a cell reaction

electromotive series a list of elements, usually metals, in the order of decreasing ease of electron loss

electron fundamental, subatomic, negatively charged particle

electron affinity energy liberated in addition of an electron to an atom in the gaseous state; reported in tables as a positive number

electron capture capture by the nucleus of one of its own inner shell electrons

electron-deficient compounds compounds that possess too few valence electrons for the atoms to be held by ordinary covalent bonds

electronegative atom an atom that tends to have a partial negative charge in a covalent bond, or to form a negative ion by gaining electrons

electronegativity the ability of an atom to attract electrons to itself; given numerical values on various scales

electronic configuration the designation of the orbitals occupied by all of the electrons in an atom

electrophilic electron-poor

electropositive atom an atom that tends to have a partial positive charge in a covalent bond, or to form a positive ion by losing electrons

element a substance composed of only one kind of atom

elementary reaction a simple reaction that occurs in a single step

emission spectrum the spectrum of radiation emitted by a substance

empirical formula an experimentally determined simplest formula

empirical relationship a relationship based solely on experimental facts and derived without the use of any theory or explanation of the facts

emulsion a colloidal suspension of one liquid in another

enantiomers compounds that exhibit optical isomerism; also called optical isomers

endothermic process a process that absorbs heat

end point the point at which chemically equivalent amounts of reactants in a titration have been combined; also called equivalence point

enthalpy the heat content of a substance

enthalpy change difference between heat content of substances in the final state and the initial state, often called just "enthalpy"

entropy a measure of the randomness or disorder of a system

entropy change the change in disorder accompanying any change in a system

equilibrium state of balance between equal and opposing forces

equilibrium constant product of the concentrations of the reaction products, each raised to the power equal to its coefficient, divided by the product of the concentrations of the reactants, each raised to the power equal to its coefficient; $K = \dfrac{[R]^r \, [S]^s}{[A]^a \, [B]^b}$

equivalence point the point at which chemically equivalent amounts of reactants in a titration have been combined; also called end point

equivalent weight of a base weight of a base that will give one mole of OH^- ions

equivalent weight of an acid weight of an acid that will give one mole of H^+ ions

equivalent weight, redox molar weight of the compound divided by the algebraic change in oxidation number of the atom that is oxidized or reduced

ester RCOOR, where R's are alkyl or aryl groups

esterification reaction reaction between a carboxylic acid and an alcohol to form an ester

eutrophication natural process in which a lake grows rich in nutrients and fills with organic sediment and aquatic plants

evaporation escape of molecules from a liquid in an open container to the gaseous phase

excited state state of energy higher than the ground state reached by electrons when the atom or molecule has absorbed extra energy

exothermic process a process that releases heat

extrinsic semiconductor number of current-carrying holes and electrons is not equal; conduction depends on addition of impurities

family the elements in a single vertical column in the periodic table; also called a group

faraday the amount of electricity represented by one mole of electrons; 96,500 coulombs

ferromagnetic exhibits magnetism in the absence of an external magnetic field

first ionization energy energy necessary to remove the least tightly bound electron from a gaseous atom

flotation concentration of metal-bearing mineral in a froth of bubbles that can be skimmed off

fluorescent reemits radiation when exposed to light, but stops after taken out of light

flux a substance that lowers the melting point, in metallurgy, the melting point of the gangue

foam a colloidal suspension of gas bubbles in a liquid

forbidden energy gap area of forbidden electron energies

formula unit the simplest unit indicated by the formula of a nonmolecular compound

fossil fuels natural gas, coal, and petroleum

fractional distillation a process that separates liquid mixtures into fractions that differ in boiling points

free energy the thermodynamic quantity that interrelates enthalpy and entropy

free energy change the overall chemical driving force in a chemical reaction

free radicals reactive species that contain unpaired electrons

Freons trade name for various chlorofluorocarbons

frequency the number of wavelengths passing a given point in unit time

fuel cell cell that produces electrical energy directly from the oxidation of a fuel continuously supplied to the cell

functional group an atom or group that bonds to one or more carbon atoms in an organic molecule and contributes a characteristic chemical behavior to the molecule

fundamental particle a particle present in all matter

fusion conversion of a solid to a liquid

gangue unwanted rock from an ore

gas chromatography chromatography in which a gas is the mobile phase

gas-discharge tube a glass tube that can be evacuated and into which are sealed two electrodes

gas–liquid chromatography chromatography which involves a stationary liquid phase and a mobile gas phase

Gay-Lussac's law of combining volumes when gases react or gaseous products are formed, the ratios of the volumes of the gases involved, measured at the same temperature and pressure, are small, whole numbers

gel   a rigid sol

germanes   compounds of germanium and hydrogen

Graham's law of diffusion   the rates of diffusion of two gases at constant pressure and temperature are inversely proportional to the square roots of their densities

Graham's law of effusion   the rates of effusion of two gases at constant pressure and temperature are inversely proportional to the square roots of their densities

greenhouse effect   warming by absorption and reemission of radiation

ground state   lowest energy state of an electron in an atom

half-cell   one electrode and its surrounding solution in an electrochemical cell

half-life   the time it takes for one-half of the nuclei in a sample of a radioactive isotope to decay

half-reaction   a reaction that is either oxidation only or reduction only

halides   binary compounds of halogen atoms with other elements

halogenation   introduction of a halogen atom or atoms into a covalent compound

hard water   contains dipositive ions that form precipitates with soap or upon boiling

heat capacity   the amount of heat required to raise the temperature of a given amount of material by 1 Celsius degree

heavy water   deuterium oxide, D_2O

Heisenberg uncertainty principle   it is impossible to know simultaneously both the exact momentum and the exact position of an electron

Henry's law   at constant temperature the solubility of a gas is directly proportional to the pressure of that gas over the solution

Hess' law   the total enthalpy change in a chemical reaction is independent of the number and nature of intermediate steps that may be involved in the reaction

heterogeneous catalyst   a catalyst present in a different phase than the reactants

heterogeneous mixture   a mixture in which the individual components remain physically separate

heterogeneous reaction   a reaction between substances in different phases

heteronuclear   referring to atoms of different elements

hexagonal closest packing   closest-packed layers of atoms arranged in an ABABAB . . . sequence

high-spin complex   a complex in which d electrons remain unpaired; also called a spin-free complex

homogeneous catalyst   a catalyst present in the same phase as the reactants

homogeneous mixture   a mixture of uniform composition and appearance

homogeneous reaction   a reaction between substances in the same gaseous or liquid phase

homologous series   a series of compounds that can be represented by a general formula

homonuclear   referring to atoms of the same element

hybridization   mixing of the atomic orbitals on a single atom to give a new set of orbitals on that atom; sp, two hybrid orbitals from one s and one p atomic orbital; sp^2, three hybrid orbitals from one s and two p atomic orbitals; sp^3, four hybrid orbitals from one s and three p atomic orbitals

hydrate isomerism   isomerism involving coordinated and noncoordinated water molecules

hydrates   substances that have combined with a definite proportion of water molecules

hydration   combination of solute with water molecules; interaction of water with other substances without splitting of the water molecule

hydrocarbons   compounds of the two elements carbon and hydrogen

hydrogenation   addition of the two atoms of dihydrogen to a molecule

hydrogen bond   attraction of a hydrogen atom covalently bonded to an electronegative atom for a second electronegative atom

hydrohalic acid   a binary halogen acid, HX

hydrolysis   any reaction in which the water molecule is split and H or O atoms or OH groups from water are added to the products

hydrolysis constant   equilibrium constant for the reaction of an ion with water

hydrolysis of an ion   reaction of an ion with water to give either H^+ or OH^- ions, plus the hydrolysis product of the ion

hydrosphere   all the water on the surface of the Earth

hydroxyl group   —OH

hygroscopic   takes up water from the air

hypohalous acid   a halogen oxo acid, HOX

hypothesis   a tentative explanation for a set of observations or phenomena

ideal gas   a gas that perfectly obeys the gas laws

ideal gas law $PV = nRT$

ideal solution a solution in which the forces between all particles of both solvent and solute are identical

immiscible mutually insoluble, or very nearly so

indicator an organic acid or base that has in its structure a group that reacts with hydrogen or hydroxide ion so that the color of the compound changes

infinitely miscible completely soluble in each other

inhibitors substances that slow down a catalyzed reaction

initiation production of the first reactive intermediate in a chain reaction

inner orbital complexes complexes in which ligand electrons fill $(n-1)d$ orbitals as well as some of the n shell orbitals

inorganic chemistry the chemistry of *all* the elements and their compounds, with the exception of hydrocarbons and hydrocarbon derivatives

insulator a substance that does not conduct electricity

integrated circuit several elements of an electrical circuit combined in the same piece of silicon

interhalogens covalent compounds between two different halogens

intermetallic compounds phases of more or less fixed composition in alloys

intermolecular forces the various forces of attraction or repulsion between individual molecules

internuclear axis the line between two nuclei

interstitial solid solution alloy in which small atoms occupy holes in the crystal lattice of a metal

intramolecular forces the forces that hold the atoms together in a molecule

intrinsic semiconductor contains equal numbers of current-carrying holes and electrons; conduction is an intrinsic property of the material

ion species formed from an atom by the loss or gain of one or more electrons

ion–electron equation equation that includes only the species directly involved in either the oxidation or reduction of a single type of atom, molecule, or ion

ion exchange replacement of one ion by another

ionic bonding bonding based on electron transfer; the attraction between positive and negative ions

ionic radius radius of an anion or a cation

ionization formation of ions from a nonionic substance

ionization energy energy for removal of one electron from an atom or ion in the gaseous state

ion product product of the concentration of ions in a nonsaturated, or nonequilibrium, solution

ion product constant for water $K_w = [H^+][OH^-]$

isoelectronic having identical electronic configurations

isomeric transition delayed γ-ray emission by an excited nuclide

isomers compounds that differ in structure but have the same molecular formula

isomorphous having the same crystal structure

isotopes atoms with different mass numbers but the same atomic number

ketone $R_2C{=}O$

lanthanide contraction general decrease in radii across the $4f$ transition series

lanthanides lanthanum through lutetium; the rare earth metals

lattice energy energy liberated when gaseous ions combine to give a crystalline, ionic substance

law statement of a relation between phenomena that, so far as is known, is always the same under the same conditions

law of conservation of mass in chemical reactions, matter is neither created nor destroyed

law of definite proportions in a pure compound, two or more elements are combined in definite proportions by weight

law of multiple proportions if two elements combine to form more than one compound, a fixed weight of element A will combine with two or more different weights of element B so that the different weights of B are in the ratio of small whole numbers

leaching washing a soluble compound from insoluble material

Le Chatelier's principle if a system in equilibrium is subjected to a stress, a change that will offset the stress will occur in the system

ligand the molecule or ion that contains the donor atom in a complex

limiting reactant a single reactant that determines the maximum amount of product because it is present in the smallest stoichiometric amount

line spectrum radiation at only certain wavelengths

liquefaction condensation of a gas to a liquid

lithosphere the solid crust of the Earth

London force attraction between fluctuating dipoles in atoms and molecules that are very close together.

London smog initiated by sulfur oxides, particulates, and high humidity

lone pairs pairs of valence shell electrons not involved in bonding

low-spin complex a complex in which d electrons are paired; also called a spin-paired complex

magic numbers numbers of neutrons or protons that impart great nuclear stability

manometer an instrument that measures gas pressure other than that of the atmosphere

mass an intrinsic property; the quantity of matter in a body

mass number the sum of the number of protons and the number of neutrons in a nucleus; the whole number closest to the atomic weight of an element

mass spectrometer instrument that spreads a beam of positive ions into a spectrum based on charge-to-mass ratio of the ions

melting point temperature at which the solid and liquid phases of a substance are in equilibrium

metallic bonding the attraction between positive metal ions and surrounding freely mobile electrons

metallic radius radius of a metal ion in the pure metal

metallurgy all aspects of the science and technology of metals

mineral a naturally occurring substance with a characteristic range of chemical composition

mixture any combination of two or more substances in which the substances combined retain their identity

molality the number of moles of solute per kilogram of solvent

molar heat of fusion amount of heat needed to convert one mole of a solid to the liquid

molar heat of vaporization amount of heat needed to convert one mole of a solid or liquid to the vapor

molarity moles of solute per liter of solution

molar weight weight in grams of one mole of a substance; numerically equal to molecular weight in atomic mass units

mole an Avogadro's number of anything (see Section 2.12 for formal SI definition)

molecularity the number of individual atoms or molecules that must simultaneously react in an elementary reaction

molecular orbital space in which an electron with a specific energy is likely to be found in the vicinity of two or more nuclei that are bonded together

molecular orbital theory explains bond formation as the occupation by electrons of orbitals characteristic of the whole molecule

molecular weight sum of the atomic weights of the total number of atoms in the formula of a chemical compound

molecule smallest particle of a pure substance that has the composition and chemical properties of that substance and is capable of independent existence

mole fraction number of moles of one component in a solution divided by the total number of moles of all components in the solution

momentum mass times velocity

monatomic made of uncombined atoms

mordanting fixing a dye to the cloth

multiple covalent bond more than one pair of electrons shared between the same two atoms

mutagens genetic-malformation-causing agents

natural isotope stable or radioactive isotope found in nature

natural radioactivity decay of radioactive isotopes found in nature

net ionic equation an equation that shows only the species involved in the chemical change; spectator ions not included

network covalent substance three-dimensional array of covalently bonded atoms

neutral aqueous solution contains equal numbers of H^+ and OH^- ions

neutralization the reaction of an acid with a base

neutrino massless, chargeless particle emitted with a positron

neutron fundamental, subatomic particle that has a mass almost the same as the mass of the proton and has no charge

neutron number the number of neutrons in the nucleus; difference between mass number and atomic number

nitrification conversion of ammonia to nitrite and nitrate

nitrogen fixation combination of molecular nitrogen with other atoms

noble gas core two or eight electrons in the $n-1$ shell

nonbonding electron pairs pairs of valence shell electrons not involved in bonding; also called lone pairs

nonelectrolyte any substance that gives a solution that does not conduct electricity

nonpolar covalent bond a bond in which electrons are shared equally between two atoms

nonrenewable resources substances that cannot be manufactured from something else

normal boiling point boiling point at 1 atm pressure

normal freezing point freezing point at 1 atm pressure

normal hydrocarbon chain an unbranched chain

normality concentration of a solution expressed in equivalents per liter

normal salts salts that are the products of the complete reaction between a base and an acid

n–p junction combination of an n-type and a p-type semiconductor

n-type semiconductor negative electrons are the majority of the current carriers; contains a donor impurity

nuclear binding energy energy that would be released in the combination of nucleons to form a nucleus

nuclear fission splitting of a heavy nucleus into two lighter nuclei of intermediate mass number

nuclear force force of attraction between nucleons

nuclear fusion combination of two light nuclei to give a heavier nucleus of intermediate mass number

nuclear magnetic resonance study of the absorption of radio frequency radiation by nuclei

nuclear reactions reactions that result in changes in atomic number, mass number, or energy state of nuclei

nuclear reactor equipment in which nuclear fission is carried out at a controlled rate

nucleons protons and neutrons

nucleophilic electron-rich

nucleus a central region in the atom, very small by comparison with the total size of the atom, in which all of the mass and positive charge of the atom are concentrated

nuclide any isotope of any element

octet rule non-noble-gas atoms tend to combine by gain, loss, or sharing of electrons so that the outermost energy level of each atom holds or shares four pairs of electrons in an $ns^2\,np^6$ configuration

olefins hydrocarbons with covalent double bonds; alkenes

optical isomerism occurrence of pairs of molecules of the same molecular formula that rotate plane-polarized light equally in opposite directions

optical isomers compounds that exhibit optical isomerism; also called enantiomers

optically active rotates the plane of vibration of plane-polarized light

ore a mixture of minerals from which a particular metal or several metals can profitably be extracted

organic chemistry the chemistry of compounds of carbon and of their derivatives

osmosis the passage of solvent molecules through a semipermeable membrane from a more dilute solution into a more concentrated solution

osmotic pressure the external pressure exactly sufficient to oppose osmosis and stop it

outer orbital complexes complexes in which $(n-1)d$ orbitals remain only partly filled while ligand electrons enter orbitals in the n shell orbitals

overvoltage the voltage in addition to the calculated voltage used by an electrode reaction in a cell

oxidation any process in which oxidation number increases algebraically

oxidation number a number that represents the positive or negative character of atoms in compounds; found by a set of rules

oxidation–reduction reactions reactions in which oxidation and reduction occur together; also called redox reactions

oxidation state same as oxidation number

oxides binary compounds of oxygen

oxidizing agent an atom, molecule, or ion that can cause another substance to undergo an increase in oxidation number

paraffin hydrocarbons alkanes

paramagnetism the property of attraction to a magnetic field shown by substances containing unpaired electrons

partially miscible having limited solubility in each other

partial pressure pressure of a single gas in a mixture

particulates airborne solid particles and liquid droplets

partition chromatography chromatography in which the stationary phase is a liquid

penetration effect increase in the nuclear charge acting on electrons that approach the nucleus more closely

percentage composition percent by weight of each element in a compound

percent yield weight of product obtained divided by theoretical yield

period a horizontal row of elements in the periodic table

periodic law the physical and chemical properties of the elements are periodic functions of their atomic numbers

permanent hardness caused by the presence in hard water of sulfate ions or other ions that do not precipitate upon boiling; also called noncarbonate hardness

petroleum alkylation combination of lower molecular weight alkanes and alkenes to form molecules in the gasoline range

petroleum cracking breaking large molecules into small molecules, usually in the gasoline range

petroleum isomerization conversion of straight-chain alkanes into branched-chain alkanes

petroleum reforming conversion of noncyclic hydrocarbons to aromatic compounds

pH $-\log[H^+]$

phase homogeneous part of a system in contact with but separate from other parts of the system

phenol ArOH, where Ar is an aryl group

phenyl group C_6H_5; the benzene ring less one H atom

phosphate rock mineral deposits containing calcium

phosphate and silica; source of phosphate fertilizers and phosphorus

phosphorescent reemits radiation when exposed to light and continues to emit radiation after taken out of light

photochemical reactions reactions that are initiated by radiation

photochemical smog initiated by sunlight, nitrogen oxides, and hydrocarbons; see London smog

photoelectric effect the giving off of electrons by certain metals when light shines on them

photon a single quantum of radiant energy (1 hν)

photosynthesis reaction of carbon dioxide and water in green plants to give carbohydrates

physical adsorption adsorption in which forces between surface and adsorbate are van der Waals forces

physical properties properties that can be exhibited, measured, or observed without resulting in a change in the composition and identity of a substance

physical sciences study mainly of things that are not alive

π-bonds (pi bonds) bonds that do not completely surround the internuclear axis

plane-polarized light light with vibrations moving in only one plane

platinum metals ruthenium, osmium, rhodium, iridium, palladium, and platinum

pOH $-\log [OH^-]$

polar covalent bond a bond in which electrons are shared unequally

pollutant an undesirable substance added to the environment, usually by the activities of Earth's human inhabitants; see secondary pollutant

polyatomic ions ions that incorporate more than one atom; also called radicals

polyatomic molecule molecule containing more than two atoms

polymer a large molecule made of many units of the same structure linked together

polymorphous able to crystallize in more than one crystal system

positron particle identical to an electron in all properties except charge, which is $+1$

precipitate solid that forms during a reaction in solution

precipitation appearance of a solid when a reaction occurs in solution

pressure force per unit area

primary carbon atom joined to only one other carbon atom

primary cell electrochemical cell in which the reactants are used up irreversibly

primitive cell in crystal structure, a unit cell in which only the corners are occupied

principles of chemistry explanations of chemical facts

products new substances produced in a chemical reaction

promoters substances that make a catalyst more effective

propagation production of the species that initiate further steps in a chain reaction

proton fundamental, subatomic particle with a positive charge equal in magnitude to the negative charge of the electron

protonic acid any substance that can lose a proton

pseudo–noble–gas core s, p, and d levels completely filled with a total of 18 electrons in the $n - 1$ shell

p-type semiconductor a semiconductor in which positive holes are the majority of the current carriers; contains an acceptor impurity

pure substance a form of matter that has identical physical and chemical properties regardless of its source

qualitative analysis the identification of the substances in a mixture

quantitative analysis the determination of the amounts of substances in a mixture

quantized restricted to amounts that are multiples of the basic unit, or quantum, for a particular system

quantum numbers whole-number (or half-whole-number) multipliers that specify the amounts of energy

racemic mixture mixture of equal amounts of optical isomers; shows no rotation of plane-polarized light

radiation energy traveling through space

radicals ions that incorporate more than one atom; also called polyatomic ions

radioactive isotopes isotopes that decay spontaneously

radioactivity spontaneous emission by unstable nuclei of particles, or of electromagnetic radiation, or of both

Raoult's law the vapor pressure of a liquid in a solution is directly proportional to the mole fraction of that liquid in the solution

rare earth elements lanthanum through lutetium; the lanthanides

rate constant the proportionality constant between the rate and the reactant concentrations

rate-determining step the slowest step in a reaction mechanism

rate equation gives the relationship between the reaction rate and the concentration of the reactants

reactants substances that are changed in a chemical reaction

reaction mechanism the exact pathway from reactants to products

reaction order sum of the exponents of the concentrations in the rate equation

reaction quotient an expression of the same form as the equilibrium expression, but for a reaction not at equilibrium; not a constant

reaction rate speed with which products are produced and reactants consumed in a particular reaction

reagent any chemical used to bring about a desired chemical reaction

rectifier converts alternating to direct current

redox couple interchangeable oxidized and reduced forms of the same species

redox reactions oxidation–reduction reactions

reducing agent an atom, molecule, or ion that can cause another substance to undergo a decrease in oxidation number

reduction any process in which oxidation number decreases algebraically

refluxing the process of vapor moving up a distillation column, condensing, and trickling down the column

refractive index ratio of the velocity of light in a vacuum or in air to the velocity of light in a given substance

resolution separation of a racemic mixture

resonance the intermediate electronic state of a molecule for which several reasonable electronic arrangements are possible

resonance hybrid actual molecular structure of molecule for which resonance structures can be written

reversible reaction a chemical reaction that can proceed in either direction

roasting heating in air or oxygen below the melting point, usually to convert sulfides to oxides

salts neutral ionic compounds composed of the cations of bases and the anions of acids

saponification hydrolysis of an ester by a base

saturated hydrocarbons hydrocarbons containing only single covalent bonds

saturated solution a solution in which the concentration of dissolved solute is equal to that which would be in equilibrium with undissolved solute under the given conditions

scavengers substances that can pull traces of contaminants out of metals and alloys, or gases; also called getters

screening effect a decrease in the nuclear charge acting on an electron due to the effect of other electrons in inner shells

secondary carbon atom joined to two other carbon atoms

secondary pollutant harmful material formed by chemical reaction in the atmosphere or hydrosphere

second ionization energy energy necessary to remove the least tightly bound electron from a gaseous +1 ion

semipermeable membrane a membrane that allows diffusion of solvent and some small solute molecules, but not larger solute molecules

σ-bonds (sigma bonds) bonds that surround the internuclear axis

silanes compounds of silicon and hydrogen

silica SiO_2

silicates compounds containing silicon–oxygen groups and metals

silicones polymers composed of silicon, carbon, hydrogen, and oxygen; general formula, $(R_2SiO)_n$

simplest formula formula giving the simplest whole-number ratio of atoms in a compound

single covalent bond a bond in which two atoms are held together by sharing of two electrons

sintered heated without melting, to cause formation of larger particles

slag molten mixture of gangue minerals

smelting melting accompanied by chemical change

sol a colloidal suspension of solid particles in a liquid

solubility the amount of solute that can dissolve in a given amount of solvent

solubility product equilibrium constant for a solid electrolyte in equilibrium with its ions in solution

solute the component of a solution usually present in the smaller amount

solution homogeneous mixture of the molecules, atoms, or ions of two or more substances; a single-phase mixture

solvation combination of solute with solvent molecules

solvent the component of a solution usually present in the larger amount

space lattice system of points representing sites in a crystal with identical environments

specific heat the amount of heat required to raise the temperature of 1 g of a substance by 1 Celsius degree

spectator ions ions that are present during a reaction but are unchanged

spectrometer an instrument used to produce a spectrum

spectrum an array of waves or particles spread out according to the increasing or decreasing magnitude of some physical property

stable isotopes isotopes that do not decay spontaneously

standard electrode potential the potential of 0.0000 V for the H_2/H^+ couple and all other potentials established relative to it under standard state conditions; also called standard reduction potential

standard enthalpies enthalpies expressed for changes of substances in their standard states

standard enthalpy of formation heat of formation of one mole of a compound by direct combination of the elements in their standard states at a specified temperature

standardized solution a solution whose exact concentration has been found

standard molar volume volume occupied by one mole of a substance at STP

standard reduction potentials the potential of 0.0000 V for the H_2/H^+ couple and all other potentials established relative to it under standard state conditions; also called standard electrode potentials

standard solution a solution of known concentration

standard state the physical state in which a substance is most stable at 1 atm and a specified temperature

standard state conditions all substances in solution at $1M$ concentration; all gases at 1 atm pressure; all solids and liquids pure and in their most stable or most common state at 1 atm pressure

standard temperature and pressure (STP) 0°C (273°K) and 760 torr

states of matter the gaseous state, the liquid state, and the solid state

static equilibrium a state of balanced forces with no motion occurring

stationary states orbits in the Bohr hydrogen atom

steel alloys of iron that contain carbon (up to 1.5%) and usually other metals as well

steric hindrance interference between large groups in the same molecule

stoichiometric amount exact amount of a substance required according to a chemical equation

stoichiometry the derivation of quantitative information from symbols, formulas, and equations

storage cell an electrochemical cell that can be recharged by electrical energy from an external source

strong acid virtually 100% ionized in dilute aqueous solution

strong base completely dissociated in aqueous solution

strong electrolytes substances 100% present as ions in solution

strong-field ligand a strongly coordinating ligand that forces electron pairing and forms spin-paired complexes

structural isomerism the existence of compounds with the same molecular formula but with the atoms joined in a different order

subatomic particle a particle smaller than the smallest atom

sublimation vaporization of a solid; also, vaporization of a solid followed by condensation of the vapor back to the solid state

substitutional solid solution alloy in which atoms of one metal replace atoms in the crystal lattice of another metal

substitution reactions replacement of one or more hydrogen atoms in an aromatic ring

supercooled cooled to a temperature below the freezing point without the occurrence of freezing

superheated heated to a temperature above the boiling point, without the occurrence of boiling

supersaturated solution a solution that holds more dissolved solute than would be in equilibrium with undissolved solute

support the inert solid that supports a liquid stationary phase in chromatography

surface tension property of a surface that imparts membranelike behavior to the surface

synthetic detergents synthetic substitutes for soap

temporary hardness caused by the presence of HCO_3^- in hard water and can be removed by boiling the water; also called carbonate hardness

termination ending of a chain reaction by recombination of the reactive intermediates

tertiary carbon atom joined to three other carbon atoms

theoretical yield maximum amount of product that can, *according to the chemical equation*, be obtained from a known amount of reactants

theory unifying principle or group of principles that explains a body of facts or phenomena and the laws that are based on them

thermochemistry the study of the thermal energy changes associated with physical or chemical changes of pure substances

thermodynamics the study of the transformation of energy from one form to another

thermonuclear reactions nuclear reactions at very high temperatures, roughly $> 10^6$ °C

three-center bond a bond in which a single pair of electrons bonds three atoms covalently

threshold frequency minimum frequency below which the photoelectric effect does not occur

titration measurement of the volume of a solution of one reactant that is required to react completely with a measured amount of another reactant

torr pressure exerted by a column of mercury 1 mm high

trans isomer has identical groups on opposite sides

transistor semiconducting device used to amplify and control electric current

transition state combination of reacting molecules intermediate between reactants and products; also called an activated complex

transuranium elements elements with atomic numbers greater than 92 (uranium)

triple covalent bond three electron pairs between the same two atoms; one σ-bond plus two π-bonds

triple point point at which three phases are in equilibrium; where three lines intersect in a phase diagram

tritium hydrogen-3

true formula formula giving the actual number of atoms in each molecule of a compound

true mass of an atom mass of an atom in mass units, such as grams

unit cell part of a space lattice that, if repeated in three dimensions, will generate the entire lattice

unsaturated hydrocarbons hydrocarbons with double or triple covalent bonds between carbon atoms

unsaturated solution a solution that can still dissolve more solute under the given conditions

valence band highest completely filled energy band in a metal

valence bond theory explains bond formation as the overlap of atomic orbitals

valence electrons electrons that take part in chemical bonding of any type

valence shell the highest, or n level, electron shell

van der Waals forces short-range intermolecular forces: dipole–dipole forces, hydrogen bonding, and London forces

vaporization escape of molecules from liquid or solid phase to the gas phase

vapor pressure pressure exerted by the vapor over a liquid once evaporation and condensation have come to equilibrium

viscosity resistance of liquid to flow

voltaic cell a cell that generates electrical energy from a spontaneous redox reaction

volume percent (for solutions)

$$\frac{\text{volume of solute}}{\text{volume of solution}} \times 100\%$$

water softening removal of the ions that cause hardness in water

wavelength the distance between any two points in the same relative location on adjacent waves

wave number number of wavelengths per unit length; reciprocal of the wavelength

weak electrolytes substances only partially ionized in solution

weak-field ligand a weakly coordinating ligand that forms high-spin, or spin-free, complexes

weight the force a body exerts because of the pull of gravity on the body's mass

weight percent (for solutions)

$$\frac{\text{weight of solute}}{\text{weight of solution}} \times 100\%$$

zone refining method for purifying solids in which a melted zone carries impurities out of the solid

INDEX

I   3

Heat capacity, specific heat and, 120–121
Heat of vaporization, for water, 359
Heavy hydrogen, 302
Heavy water, 303
Heisenberg, Werner, 158, 160, 178
Heisenberg uncertainty principle, 160, 179
Helium
  in atmosphere, 152
  chemical formula for, 23
  escape of, 300
  "inertness" of, 236
  liquid, 331
  molecular orbitals in, 644–646
  as monatomic molecule, 23
Helmont, Johann van, 52
Hematite, 895
Heme, structure of, 873
Hemiacetal, 1033
Hemlock, 1018
Henderson-Hasselbach equation, 551, 561–562
Henry's law, 384
Herbicides, 620
  chlorinated hydrocarbon, 1012
Heroin, 1018
Hess' law, 128, 140, 772
Heterogeneous catalysis, 427
Heterogeneous equilibria, 451–452, 565–576,
    948–959
  see also Equilibrium (-ia)
  anions and, 697
Heterogeneous mixture, defined, 4
Heterogeneous reactions, 426
Heterogeneous solid-state reactions, 348
Heteronuclear diatomic molecules, 652
Hexaamminenickel(II) ion, 903
Hexaaquacobalt(II) chloride, 901
Hexachloroplatinate(IV) ion, 855
Hexacyanoferrate complex ion, 947
Hexacyanoferrate(III) ion, 900–901
Hexagonal closest packing, in crystalline
    solids, 337
Hexagonal crystal system, 339
Hexane, 725
  normal or n-hexane, 726
Hexylresorcinol, 1015
High-spin complex, 864
H + ion, 301–302
Holes, migration of in semiconductors, 833
Holmium, 913
Homogeneous equilibrium, 452
Homogeneous gas-phase reactions, 449
Homogeneous mixture, defined, 4
Homogeneous reaction, 425
Homogeneous solid-state reactions, 347
Homogenized milk, as emulsion, 415
Homonuclear gaseous diatomic molecules,
    649
Hormones, function of, 1032
Hund principle, in electronic configuration,
    185
Hybridization
  of atomic orbitals, 639
  sp, sp^2, and sp^3, 641
Hybrid orbitals, 639
  formation of, 641
Hydrargum (mercury), 27
Hydrated iron(III) oxide, 896
Hydrate isomerism, 889
Hydrates, 360
Hydration, defined, 43, 360
Hydrazine, 414
  anhydrous, 665
Hydrides, 304
  covalent, 324
  interstitial, 345
  saltlike, 324
Hydriodic acid, 486–487

Hydrobromic acid, 486–487, 620
  as strong acid, 503
Hydrocarbons, 722–739
  additive reactions for, 736
  in air pollution, 146–147
  aliphatic, 731
  aromatic, 731–732
  combustion of, 735
  with covalent double bonds, 726
  defined, 7
  as fuel, 738
  halogen derivatives of, 1011–1012
  organic chemistry and, 921
  paraffin, 724
  polycyclic aromatic, 732
  properties of, 735–737
  reactions of, 735–737
  saturated, 722–726
  sources of, 737–739
  steam reforming of, 305
  substitution reactions in, 737
  terpene, 731
  unsaturated, 726–727, 736–737
  uses of, 738–739
Hydrochloric acid, 486
  as Brønsted-Lowry acid, 484
  cation Group I and, 964
  oxidizing of, 620
  pH and, 499, 548
  sodium hydroxide and, 565
  solution of, 391
  as strong acid, 503
Hydrocyanic acid, 486, 516
Hydrofluoric acid, 486, 504
  glass and, 845
  ionization constant for, 515
  as weak acid, 626
Hydrogen, 301–309
  abundance of, 300
  atomic weight of, 28
  binary compounds of, 308–309
  bond dissociation energy of, 301
  chlorine and, 627
  coke process for, 306
  compound formation by, 290
  in conjugate acid-base process, 486–487
  diffusion through palladium-silver alloy,
    887
  displacement of from hydroxides, 306
  for electrolysis, 306–307
  electromotive series and, 305
  escape of, 300–301
  free, 300
  as fuel, 307–308
  in fuel cell, 759
  Group I–VII compounds and, 308
  hydride ion and, 486
  isotopes of, 302
  molecular orbitals in, 644–646
  origin of, 300–301
  oxidation state of, 302
  in periodic system, 290
  as plant nutrient, 678
  preparation of, 320–322
  properties of, 301–302
  reactions of, 303, 305, 627
  reactive metals and, 304
  as reducing agent, 302
  solubility of, 383
  storage of, 307
  thermochemical cycle for production of, 308
  uses of, 306–307
  as "water-former," 27
Hydrogenation, 304–305
Hydrogen atom
  Bohr model of, 177–178
  noble-gas-type anion of, 241

orbitals of, 181
Hydrogen bonding, 256–257
  acid strength and, 491
Hydrogen bromide, 303, 628
Hydrogen carbonate ion, 738
  in water, 363
Hydrogen carbonates, 719, 798
Hydrogen chloride, 431
  collision with ammonia and tripropylamine,
    421
  properties of, 256
  reaction with water, 484
  solubility of, 383
Hydrogen cyanide, 486, 721
Hydrogen economy, 307
Hydrogen fluoride
  as acid, 309
  properties of, 256
Hydrogen halides
  aqueous solutions of, 625–629
  preparation of, 626–627
  properties of, 625–627
Hydrogen iodide, 628
Hydrogen ion, in sulfide precipitation, 955
Hydrogen ion concentration, 497
  in blood serum, 498
  pH and, 496–497
Hydrogen-metal compounds, 323
Hydrogen molecular orbitals, 644–646
Hydrogen-oxygen fuel cell, 759
Hydrogen peroxide, 316–318
  concentration of, 317
  decomposition of, 317
  as oxidizing agent, 317
  preparation and uses of, 317–318
  properties and reactions of, 316–318
  as reducing agent, 317
  as weak acid, 504
Hydrogen selenide, 839
Hydrogen sulfate ion, 486–487, 504, 685
Hydrogen sulfide, 414, 504, 537, 931
  odor of, 956–957
  properties of, 256, 682
Hydrogen sulfide ion, 486, 504
Hydrogen telluride, 839
Hydrohalic acid, 613, 616, 626, 837
Hydroiodic acid, 486
  as strong acid, 503
Hydrolysis
  defined, 359–360
  equations for, 359
  equilibria in, 934
  of ions, 526
  in soapmaking, 369
Hydronium ion, 486
Hydroquinone, 1015
Hydrosphere, 152, 361–365
Hydrosulfuric acid, 486
  ionization of, 539
Hydroxide ion, 513
  as base in aqueous solution, 934
  from calcium oxide, 797
  ionization constant for, 513
Hydroxide precipitations, common ion and,
    952–953
Hydroxides
  dissociation of, 522
  precipitation of, 952–953
  solubility of, 566
Hydroxy acids, 1026
Hydroxyapatite, 622
Hydroxyl group, 1012
  on aromatic ring, 1015
Hygroscopic compounds, 361
"Hypo," developer in photography, 688, 911
Hypochlorite ion, 628, 631
Hypochlorous acid, 491

A
B
C
D
E
F
G
H
I
J
0
1
2
3
4
5

Table of atomic weights and electronic configurations listed by atomic number

Scaled to the relative atomic mass $^{12}C = 12$ exactly. A number in parentheses is the atomic mass number of the isotope of longest known half-life.

Atomic number	Element	Symbol	Atomic weight	Electronic configuration
1	Hydrogen	H	1.0079	$1s^1$
2	Helium	He	4.0026	$1s^2$
3	Lithium	Li	6.941	$1s^2\ 2s^1$
4	Beryllium	Be	9.01218	$1s^2\ 2s^2$
5	Boron	B	10.81	$1s^2\ 2s^2\ 2p^1$
6	Carbon	C	12.011	$1s^2\ 2s^2\ 2p^2$
7	Nitrogen	N	14.0067	$1s^2\ 2s^2\ 2p^3$
8	Oxygen	O	15.9994	$1s^2\ 2s^2\ 2p^4$
9	Fluorine	F	18.9984	$1s^2\ 2s^2\ 2p^5$
10	Neon	Ne	20.179	$1s^2\ 2s^2\ 2p^6$
11	Sodium	Na	22.9898	$1s^2\ 2s^2\ 2p^6\ 3s^1$
12	Magnesium	Mg	24.305	$1s^2\ 2s^2\ 2p^6\ 3s^2$
13	Aluminum	Al	26.9815	$1s^2\ 2s^2\ 2p^6\ 3s^2\ 3p^1$
14	Silicon	Si	28.086	$1s^2\ 2s^2\ 2p^6\ 3s^2\ 3p^2$
15	Phosphorus	P	30.9738	$1s^2\ 2s^2\ 2p^6\ 3s^2\ 3p^3$
16	Sulfur	S	32.06	$1s^2\ 2s^2\ 2p^6\ 3s^2\ 3p^4$
17	Chlorine	Cl	35.453	$1s^2\ 2s^2\ 2p^6\ 3s^2\ 3p^5$
18	Argon	Ar	39.948	$1s^2\ 2s^2\ 2p^6\ 3s^2\ 3p^6$
19	Potassium	K	39.098	$1s^2\ 2s^2\ 2p^6\ 3s^2\ 3p^6\ 4s^1$
20	Calcium	Ca	40.08	$1s^2\ 2s^2\ 2p^6\ 3s^2\ 3p^6\ 4s^2$
21	Scandium	Sc	44.959	$1s^2\ 2s^2\ 2p^6\ 3s^2\ 3p^6\ 3d^1\ 4s^2$
22	Titanium	Ti	47.90	$1s^2\ 2s^2\ 2p^6\ 3s^2\ 3p^6\ 3d^2\ 4s^2$
23	Vanadium	V	50.9414	$1s^2\ 2s^2\ 2p^6\ 3s^2\ 3p^6\ 3d^3\ 4s^2$
24	Chromium	Cr	51.996	$1s^2\ 2s^2\ 2p^6\ 3s^2\ 3p^6\ 3d^5\ 4s^1$
25	Manganese	Mn	54.938	$1s^2\ 2s^2\ 2p^6\ 3s^2\ 3p^6\ 3d^5\ 4s^2$
26	Iron	Fe	55.847	$1s^2\ 2s^2\ 2p^6\ 3s^2\ 3p^6\ 3d^6\ 4s^2$
27	Cobalt	Co	58.9332	$1s^2\ 2s^2\ 2p^6\ 3s^2\ 3p^6\ 3d^7\ 4s^2$
28	Nickel	Ni	58.70	$1s^2\ 2s^2\ 2p^6\ 3s^2\ 3p^6\ 3d^8\ 4s^2$
29	Copper	Cu	63.546	$1s^2\ 2s^2\ 2p^6\ 3s^2\ 3p^6\ 3d^{10}\ 4s^1$
30	Zinc	Zn	65.38	$1s^2\ 2s^2\ 2p^6\ 3s^2\ 3p^6\ 3d^{10}\ 4s^2$
31	Gallium	Ga	69.72	$1s^2\ 2s^2\ 2p^6\ 3s^2\ 3p^6\ 3d^{10}\ 4s^2\ 4p^1$
32	Germanium	Ge	72.59	$1s^2\ 2s^2\ 2p^6\ 3s^2\ 3p^6\ 3d^{10}\ 4s^2\ 4p^2$
33	Arsenic	As	74.9216	$1s^2\ 2s^2\ 2p^6\ 3s^2\ 3p^6\ 3d^{10}\ 4s^2\ 4p^3$
34	Selenium	Se	78.96	$1s^2\ 2s^2\ 2p^6\ 3s^2\ 3p^6\ 3d^{10}\ 4s^2\ 4p^4$
35	Bromine	Br	79.904	$1s^2\ 2s^2\ 2p^6\ 3s^2\ 3p^6\ 3d^{10}\ 4s^2\ 4p^5$
36	Krypton	Kr	83.80	$1s^2\ 2s^2\ 2p^6\ 3s^2\ 3p^6\ 3d^{10}\ 4s^2\ 4p^6$
37	Rubidium	Rb	85.4678	[Krypton core] $5s^1$
38	Strontium	Sr	87.62	[Krypton core] $5s^2$
39	Yttrium	Y	88.9059	[Krypton core] $4d^1\ 5s^2$
40	Zirconium	Zr	91.22	[Krypton core] $4d^2\ 5s^2$
41	Niobium	Nb	92.9064	[Krypton core] $4d^4\ 5s^1$
42	Molybdenum	Mo	95.94	[Krypton core] $4d^5\ 5s^1$
43	Technetium	Tc	(97)	[Krypton core] $4d^5\ 5s^2$
44	Ruthenium	Ru	101.07	[Krypton core] $4d^7\ 5s^1$
45	Rhodium	Rh	102.905	[Krypton core] $4d^8\ 5s^1$
46	Palladium	Pd	106.4	[Krypton core] $4d^{10}$
47	Silver	Ag	107.868	[Krypton core] $4d^{10}\ 5s^1$
48	Cadmium	Cd	112.40	[Krypton core] $4d^{10}\ 5s^2$
49	Indium	In	114.82	[Krypton core] $4d^{10}\ 5s^2\ 5p^1$
50	Tin	Sn	118.69	[Krypton core] $4d^{10}\ 5s^2\ 5p^2$
51	Antimony	Sb	121.75	[Krypton core] $4d^{10}\ 5s^2\ 5p^3$
52	Tellurium	Te	127.60	[Krypton core] $4d^{10}\ 5s^2\ 5p^4$
53	Iodine	I	126.904	[Krypton core] $4d^{10}\ 5s^2\ 5p^5$
54	Xenon	Xe	131.30	[Krypton core] $5s^2\ 5p^6$